MECHANICAL DESIGN OF MACHINE ELEMENTS AND MACHINES

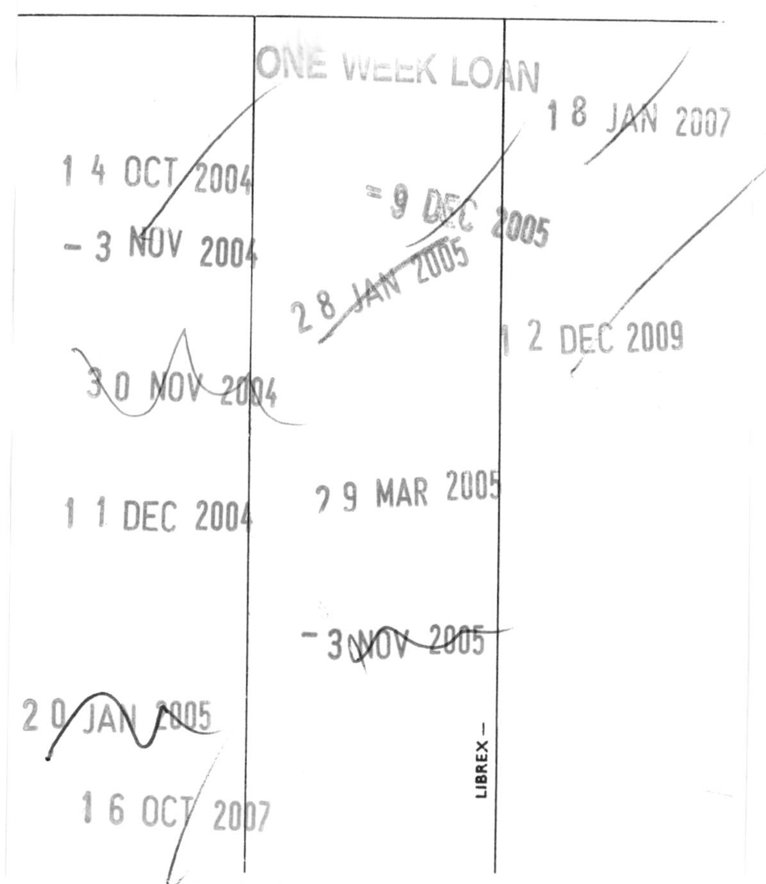

MECHANICAL DESIGN OF MACHINE ELEMENTS AND MACHINES

A Failure Prevention Perspective

Jack A. Collins

The Ohio State University

John Wiley & Sons

ACQUISITIONS EDITOR Joseph Hayton
MARKETING MANAGER Katherine Hepburn
SENIOR PRODUCTION EDITOR Norine M. Pigliucci
SENIOR DESIGNER Karin Kincheloe
PHOTO EDITOR Sara Wight
PRODUCTION MANAGEMENT SERVICES Suzanne Ingrao
EDITORIAL ASSISTANT Mary Moran

This book was set in Times Roman by Automated Composition Services, Inc. and printed and bound by Donnelley Willard. The cover was printed by Phoenix Color.

This book is printed on acid-free paper.

ISBN 0-471-03307-3
WIE ISBN: 0-471-42890-6

Printed in the United States of America

10 9 8 7 6 5 4 3 2 1

To my wife, Jo Ann,
my children, Mike, (Julie), Jennifer, (Larry), Joan, Greg, (Heather),
and my grandchildren, Michael, Christen, David, Erin, Caden

Preface

This new undergraduate book, written primarily to support a Junior-Senior level sequence of courses in Mechanical Engineering Design, takes the viewpoint that *failure prevention* is the cornerstone concept underlying all mechanical design activity. The text is presented in two parts, *Part I—Engineering Principles*, containing 7 chapters, and *Part II—Design Applications*, containing 13 Chapters. Because of the way the book is organized, it also may be conveniently used as the basis for continuing education courses or short-courses directed toward graduate engineers, as well as a reference book for mechanical designers engaged in professional practice.

Organization

Part I introduces the design viewpoint and provides analytical support for the mechanical engineering design task. *Analysis* is characterized by known material, known shape, known dimensions and known loading. The results of analyses usually include the calculation of stresses, strains or *existing* safety factors. Techniques are presented for failure mode assessment, material selection, and safety factor selection. A unique chapter on geometry determination provides basic principles and guidelines for creating efficient shapes and sizes. A case is made for integration of manufacturing, maintenance, and critical point inspection requirements at the design stage, *before* the machine is built.

Part II expands on the design viewpoint introduced in Part I. *Design* is a task characterized by known specifications, and nothing more. The results of design usually include picking a material, picking a *design* safety factor, conceiving a shape, and determining dimensions that will safely satisfy the design specifications in the "best" possible way.

Key Text Features

1. Comprehensive coverage of failure modes (Chapter 2). Basic tools are introduced for recognizing potential failure modes that may govern in any specific design scenario.[1] At a minimum, the topics of elastic deformation, yielding, brittle fracture, fatigue, buckling, and impact, should be considered by the instructor.

1. Chapter 2 presents a condensed and simplified version of my earlier book, *Failure of Materials in Mechanical Design; Analysis Prediction, Prevention, 2nd ed., Wiley, 1993.*

2. Modern coverage of materials selection (Chapter 3). The materials selection concepts presented introduce some new ideas and are a virtual necessity for any competent design engineer.
3. Safety factor and reliability concepts (Chapter 5). Some new notions are presented about the distinction between *design* safety factor and *existing* safety factor, and their relationship to reliability.
4. Guidelines for creating efficient shapes and sizes for components and machines (Chapter 6). This important chapter, covering material rarely discussed in other design text books, is a "must" for any modern course covering the design of machine elements.
5. Concurrent engineering and "Design-for-X" ideas (Chapter 7). These are important in modern manufacturing practice and should be introduced in a well-rounded course in mechanical engineering design.
6. Conceptual introductions to machine elements (Chapters 8 through 19). Organized and designed to be especially helpful to students who may have had little or no exposure to machines, structures, or industrial practice, each chapter in Part II follows a consistent introductory pattern:
   - "Uses and Characteristics"—What does it look like? What does it do? What variations are available?
   - "Probable failure modes"—based on practical experience.
   - "Typical materials used for the application"—based on common design practice.

   These introductory sections are followed in each chapter by detailed discussions about analyzing, selecting, or designing the component under consideration.
7. Inclusion of latest available revisions of applicable codes and standards for well-standardized elements such as gears, rolling-element bearings, V-belts, precision roller-chain, and others. Selected up-to-date supporting data have been included for many commercially available components, such as rolling-element bearings, V-belts, wire rope, and flexible shafts. Many manufacturers' catalogs have been included in the reference lists.
8. Clear sketches and detailed tables to support virtually all of the important design and selection issues discussed.
9. Illuminating footnotes, anecdotes, experience-based observations, and contemporary-event illustrations, to demonstrate the importance of good design decision-making.

Worked Examples and Homework Problems

Nearly 100 *worked examples* have been integrated with the text. Of these worked examples, about half are presented from a *design* viewpoint, including about ¼ of the examples given in Part I, and about ¾ of the examples given in Part II. The remainder are presented from the more traditional *analysis* viewpoint.

End-of-chapter problems have been distilled, in great measure, from real design projects encountered by the author in consulting, research, and short-course interaction with engineers in industry, then filtered through more than three decades of student homework assignments and design-course examinations. It is the author's hope that students (and instructors) will find the problems interesting, realistic, instructional, challenging, and solvable.

To supplement the worked examples, a companion web site at **www.wiley.com/college/collins** has been developed to provide more than 100 additional variations and extensions of the examples worked in the text. Many of the website variations and extensions require solution techniques based on standard computer codes such as MATLAB® or *Mathcad*®.

Additional instructor and student resources, such as errata listings, also are posted at the website.

Suggestions for Course Coverage

Although it is presumed that the user has had basic courses in *Physics, Materials Engineering, Statics* and *Dynamics*, and *Strength of Materials*, most concepts from these courses that are needed for basic mechanical engineering design activity have been summarized and included in Part I, primarily in Chapters 2, 3, and 4. Accordingly, an instructor has great flexibility in selecting material to be covered, depending upon the preparation of students coming to the course. For example, if students are well prepared in strength-of-materials concepts, only the last half of Chapter 4 needs to be covered. Sections 4.1 through 4.6 may readily be skipped, yet the material is available for reference. Sections 4.7 through 4.14 contain important design related material not ordinarily covered in standard strength-of-materials courses.

The three-part introduction to each "elements" chapter makes it possible to offer a (superficial) descriptive survey course on machine elements by covering only the first few sections of each chapter in Part II. Although such an approach would not, by itself, be especially appropriate in educating a competent designer, it would provide the potential for remarkable flexibility in tailoring a course sequence that could *introduce* the student to *all* machine elements of importance (by assigning the first few sections of each chapter of Part II), then covering in depth the chapters selected by the supervisory design-faculty-group, or the instructor, to fit into the designated curricular time frame.

With few exceptions, the machine element chapters (8 through 19) have been written as stand-alone units, independent of each other, each resting upon pertinent principles discussed in Part I. This presentation philosophy affords an instructor great flexibility in formulating a sequence of machine-element topics, in any order, that is compatible with his or her priorities, philosophy, and experience.

Supplements

An instructor's solution manual is available, providing comprehensive solutions for all end-of-chapter problems. Please contact your local Wiley representative for details.

Acknowledgments

As the years have passed, I have found it impossible to distinguish my own original thoughts from the thoughts gathered through reading and discussing the works of others. For the many contributions of others who find their essence in these pages without specific reference, I wish to express my appreciation. I would like to express particular appreciation to two of my own professors during my years as a student: Professor Walter L. Starkey and the late Professor S. M. Marco. No doubt much of their philosophy has been adopted as my own. Professor Starkey's fertile mind has created many of the innovative concepts presented in this book, especially in the material presented in Chapter 3, 5, 6 and 7. I hold

Professor Starkey in high esteem as an outstanding engineer, innovative designer, inspirational teacher, gentleman, mentor, and friend.

I also recognize, with sincere thanks, two of my colleagues at Ohio State, Professor H. R. Busby and Professor G. H. Staab, who, together, have created the companion web site and authored all of the more than one hundred variations and extensions (to worked examples in the text) found there. I am also grateful for their help in winnowing out manuscript errors, both typographical and conceptual. Other colleagues at Ohio State who have reviewed and contributed to various parts of the manuscript include Professor E. O. Doebelin, Professor D. R. Houser, and Professor R. Parker. I express my appreciation to them as well.

Reviewers always play an important role in the development of any textbook. I would like to express my appreciation to all those who reviewed this text at its various stages of development, including Professor G. K. Ananthasuresh, University of Pennsylvania; Professor R. Dippery, Kettering University; Professor W. B. Hall, University of Illinois; Professor K. R. Halliday, Ohio University; Professor F. Kelso, University of Minnesota; Professor T. Lee, McGill University; Professor G. R. Pennock, Purdue University; Professor P. Sheth, University of Virginia; Professor J. S. Strenkowski, North Carolina State University; and Professor Charles Yang, Wichita State University.

Thanks are also due to Joseph P. Hayton, Editor, and many other individuals in the John Wiley & Sons, Inc., organization, who have contributed their talents and energy to the production of this book.

Finally, but not least, I would like to express special thanks to my wife of more than four decades, Jo Ann, for her understanding and support during the nearly ten years that the preparation of this book significantly interfered with important family and social activities. Not only was she unwavering in her support, but she conquered unfamiliar Greek symbols and the vagaries of mathematical language to transform more than two-thousand of my barely-legible hand-written pages into the manuscript that has become this book. I am grateful for her support, her patience, and her skill.

Jack A. Collins
Columbus, Ohio
April, 2002

Contents

PART ONE ENGINEERING PRINCIPLES

APPENDIX

Worked Examples

Chapter 1

Chapter 2

Chapter 3

Chapter 4

Chapter 5

Chapter 6

Chapter 7

Chapter 8

Chapter 16

Chapter 17

Chapter 18

Chapter 19

Chapter 20

MECHANICAL DESIGN OF MACHINE ELEMENTS AND MACHINES

PART ONE

ENGINEERING PRINCIPLES

Chapter 1

Keystones of Design: Materials Selection and Geometry Determination

Do not go where the path may lead,
go instead where there is no path and leave a trail.

—Ralph Waldo Emerson

1.1 Some Background Philosophy

The first objective of any engineering design project is the fulfillment of some human need or desire. Broadly, *engineering* may be described as a judicious blend of science and art in which natural resources, including energy sources, are transformed into useful products, structures, or machines that benefit humankind. *Science* may be defined as any organized body of knowledge. *Art* may be thought of as a skill or set of skills acquired through a combination of study, observation, practice, and experience, or by intuitive capability or creative insight. Thus engineers utilize or apply scientific knowledge together with artistic capability and experience to produce products or plans for products.

A *team approach* is nearly always used in modern industrial practice, enabling engineers from many disciplines, together with marketing specialists, industrial designers, and manufacturing specialists, to integrate their special credentials in a cooperative cross-functional *product design team* effort.[1] *Mechanical engineers* are almost always included in these teams, since mechanical engineers have broad training in principles and concepts relating to products, machines, and systems that perform mechanical work or convert energy into mechanical work.

One of the most important professional functions of mechanical engineers is *mechanical design*, that is, creating new devices or improving existing devices in an attempt to provide the "best," or "optimum" design consistent with the constraints of time, money, and safety, as dictated by the application and the marketplace. Newcomers to mechanical

[1]See 1.2.

design activity, even those with well-developed analytical skills, are often at first frustrated to find that most design problems do not have unique solutions; design tasks typically have many possible approaches from which an "optimum" must be chosen. Experienced designers, on the other hand, find challenge and excitement in the art of extracting a "best" choice from among the many potential solutions to a design problem. Transformation of the frustrations of a newcomer into the excitement experienced by a successful seasoned designer depends upon the adoption of a broadly based design methodology and practice in using it. It is the objective of this text to suggest a broadly based design methodology and demonstrate its application by adapting it to many different important engineering design scenarios. Practice in using it must be supplied by the reader.

1.2 The Product Design Team

Before any of the engineering design methods, concepts, or practices described in this textbook can be put to productive use, it is necessary to first translate customer needs or desires, often vague or subjective, into quantitative, objective *engineering specifications*. After clear specifications have been written, the methods presented in this text provide solid guidelines for selecting materials, establishing geometries, and integrating parts and subassemblies into a whole machine configuration that will safely and reliably meet both engineering and marketing goals. The task of translating marketing ideas into well-defined engineering specifications typically involves interaction, communication, and understanding among marketing specialists, industrial designers, financial specialists, engineering designers, and customers,[2] cooperatively participating in a cross-functional *product design team*.[3] For smaller companies, or smaller projects, the *team functions* just listed may be vested in *fewer* team members by assigning *multiple-function* responsibility to one or more participants.

The first steps in translating customer needs or marketplace opportunities into engineering design specifications are usually managed by marketing specialists and industrial designers. *Marketing specialists* on the product design team typically work directly with customers to bring a sharper focus to perceived needs, to establish marketing goals, to supply supportive research and business decision-making data, and to develop customer confidence that their needs can be efficiently met on schedule.

Industrial designers on the team are responsible for creating an initial broad-based functional description of a proposed product design, together with the essentials of a *vi-*

[2]It has become common practice to include *customers* in product design teams. The argument for doing so is the belief that products should be designed to reflect customers' desires and tastes, so it is efficient to interactively incorporate customer perceptions from the beginning (see ref. 1). On the other hand, an argument has been made that customers do not lead companies to *innovation*, but rather into *refining* existing products. Since technical innovation often wins the marketplace in today's business world, companies that concentrate *solely* on following customer perceptions and desires, rather than *leading* customers to innovative new ideas, are at risk.

[3]An interesting side issue related to the formation of a product design team lies in the task of choosing a team leader without generating interpersonal conflicts among the team members. It has been argued that choosing a team leader is the most important decision that management will make when setting up a product design team (see ref. 1, p. 50). Others have observed that *good followership* is as important to team success as *good leadership* (see ref. 2). The qualities that typically characterize good leaders are, in great measure, the same qualities found in effective followers: intelligence, initiative, self-control, commitment, talent, honesty, credibility, and courage. Followership is not a person but a role. Recognition that leaders and followers are equally important in the activities of an effective cross-functional product design team avoids many of the counterproductive conflicts that arise in teams of diverse participants.

sual concept that embodies appealing external form, size, shape, color, and texture.[4] Artistic renderings and physical models[5] are nearly always developed as a part of this process. In developing an initial product design proposal, industrial designers must consider not only broad functional requirements and marketing goals, but also aesthetics, style, ergonomics,[6] company image, and corporate identity. The result of this effort is usually termed a *product marketing concept*.

A good product marketing concept contains all pertinent information about the proposed product that is essential to its marketing, but as little information as possible about details of engineering design and manufacturing, so as not to artificially constrain the ensuing engineering decision-making processes. This policy, sometimes called the *policy of least commitment*, is recommended for application throughout the engineering design and manufacturing stages as well, to allow as much freedom as possible for making downstream decisions without imposing unnecessary constraints.

Engineering designers on the product design team have the responsibility of identifying the *engineering characteristics* that are directly related to customer perceptions and desires. Describing the potential influences of engineering characteristics on the marketing goals, and evaluating the product design proposal in *measurable* terms, is also an engineering design function. Ultimately, engineering specifications for designing a practical, manufacturable product that is safe, cost-effective, and reliable are primarily the responsibility of the engineering designer on the team.

To implement the work of a cross-functional product design team, it is usually necessary to establish a set of planning and communication routines that focus and coordinate skills and experience within the company. These routines are formulated to stimulate design, manufacturing, and marketing departments to propose products that customers want to purchase, and will continue to purchase. One matrix-based model for interfunctional planning, communication, and evaluation is called the *house of quality*.[7] The principles underlying the house of quality paradigm apply to any effort toward establishing clear relationships between manufacturing functions and customer satisfaction that are not easy to visualize directly. Figure 1.1 illustrates a fraction of one subchart[8] that embodies many of the house of quality concepts, and provides a sequence of steps for answering the following questions:

1. What do customers want?
2. Are all customer preferences equally important?
3. Will delivering perceived needs yield a competitive advantage?
4. How can the product be effectively changed?
5. How much do engineering proposals influence customer-perceived needs?
6. How does an engineering change affect other characteristics?

Building a house of quality matrix to answer these questions begins with customer perceptions, called *customer attributes (CAs)*. Customer attributes are a collection of customer *phrases* describing product characteristics deemed to be important. For the car door example of Figure 1.1, the CAs shown at the left boundary include "easy to close," "stays

[4]See ref. 1, p. 8.

[5]At this conceptual stage, models are usually crude and nonfunctional, although some may have a few moving parts.

[6]*Ergonomics* is the study of how tools and machines can best be fitted to human capabilities and limitations. The terms *human factors engineering* and *human-machine systems* have also been used in this context.

[7]See ref. 1. The *house of quality* concepts presented here are extensively paraphrased or quoted from ref. 3.

[8]Extracted from ref. 3.

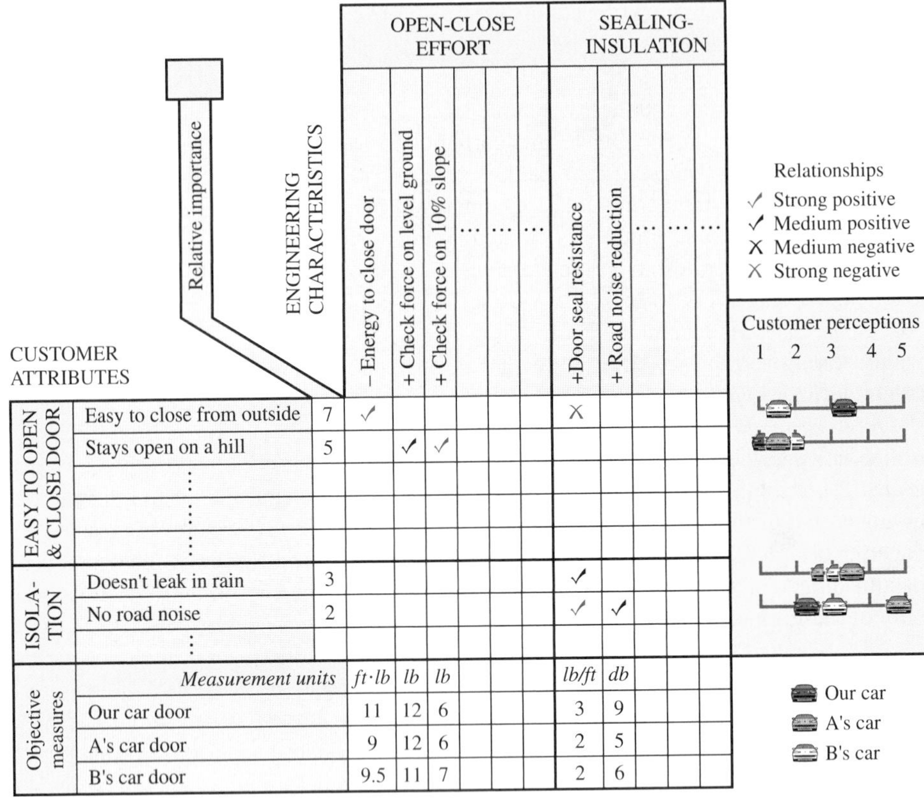

Figure 1.1
Example of a *house of quality* matrix related to the redesign of an automotive door. (Reprinted by permission of *Harvard Business Review*. Exhibit from ref. 3. Copyright © 1998 by the Harvard Business School Publishing Corporation; all rights reserved.)

open on a hill," "doesn't leak in rain," and "allows no road noise." Typical product applications would define 30 to 100 CAs. The *relative importance* of each attribute, as evaluated by the customer, is also included, as shown in Figure 1.1. The *importance-weighting numbers*, shown next to each attribute, are usually expressed as percentages, where the complete list of all attributes totals 100 percent.

Customer evaluations of how the proposed product (car door) compares with *competitive products* are listed at the right side of the matrix. These evaluations, ideally based on scientific surveys of customers, identify opportunities for improvement and ways to gain competitive advantage.

To integrate pertinent engineering characteristics (ECs) into the house of quality, the product design team lists across the top side of the matrix the ECs that are thought likely to affect one or more of the CAs. Engineering characteristics should describe the product in calculable or measurable terms, and should be related directly to one or more customer perceptions.

The cross-functional design team next fills in the *body* of the house (the relationship matrix), reaching a consensus about how much each engineering characteristic affects each of the customer attributes. Semiquantitative symbols or numerical values are inserted into the matrix to establish the *strengths* of the relationships. In Figure 1.1 the semiquantitative symbols represent the relationships as "strong positive," "medium positive," "medium negative," or "strong negative."

Once the product design team has established the relationship strengths linking engineering characteristics to customer attributes, governing variables and objective measures are listed, and *target values* are established. Compromises in target values are commonplace because *all* target values cannot usually be reached at the same time in any real machine.

Finally, the team consensus on *quantitative* target values is summarized and compiled into *initial engineering specifications.* As noted throughout this textbook, engineering specifications provide the basis for in-depth engineering design tasks required to produce a practical, manufacturable product that is safe, cost-effective, reliable, and responsive to customer needs and desires.

1.3 Function and Form; Aesthetics and Ergonomics

Traditionally, the connection between function and form has been direct; the *form* of a product need only suit its *function.* Historically, standardized simple geometry, without ornamentation, was nearly always chosen to accommodate the engineering design and production of reliable, durable, cost-effective products that would meet the engineering specifications. More recently, however, it has been recognized that the *demand* for a new or revised product depends heavily upon *customer perceptions* and *marketplace acceptance*, as well as technical functionality. This recognition has led many contemporary companies to organize cross-functional product design teams[9] that include marketing specialists and industrial designers as well as design and manufacturing engineers, to bring the marketing aspects more to the foreground. This approach seems to result in enhanced customer appeal engendered by integrating aesthetic appearance, perspective, proportion, and style at an early design stage; the attractive shell of a product often plays an important marketing role. To implement decisions on appearance and style, three-dimensional-graphics computer programs now make it possible to simulate a proposed product's appearance on the screen and rapidly make desired changes with vivid clarity.

In addition to assuring that technical performance specifications are met, and that the product has customer appeal, it is also necessary for a designer to make sure that the proposed machine configuration and control features are well matched to human operator performance capabilities.

The activity of designing user-friendly machines for safe, easy, productive use is called *ergonomics* or *human factors engineering*. A key concept in ergonomic design is that *human operators* exhibit a *wide variation* in stature, weight, physical strength, visual acuity, hearing ability, intelligence, education, judgment, endurance, and other human attributes. It becomes necessary, therefore, to provide *machine system features* that *match* potential *user attributes*, and protect operators against injury resulting from operator error or machine malfunction. Because most products and systems are designed for use by an array of people, rather than for use by one specific individual, it becomes necessary to accommodate the whole range of strengths and weaknesses of the potential user population. To accomplish this objective, a designer must be well informed about *anthropometrics*,[10] about the psychology of human behavior,[11] and about how to integrate these factors with technical requirements in order to achieve a safe, productive machine.

Anthropometric constraints upon the configuration of products or systems are widely discussed in the literature.[12] Typically, to properly design a machine for efficient human

[9]See 1.2.

[10]The study, definition, and measurement of human body dimensions, motions, and limitations.

[11]Few engineers are trained in the concepts of industrial psychology. Designers are well advised to consult industrial psychology specialists to help with this task.

[12]See, for example, ref. 1 or ref. 4.

interaction, anthropometric data on human body size, posture, reach, mobility, force, power, foot strength, hand strength, whole-body strength, response speed, and/or response accuracy may be required. Quantitative information on most of these human attributes is available. In some cases, computer simulation models have been developed[13] to help evaluate the physical demands placed upon the operator by a proposed design scenario, and to supply the necessary anthropometric data to evaluate the proposed design (and possible redesigns).

Anticipating potential operator errors, and designing a machine or system to accommodate them without serious consequences, is also an important part of effective ergonomic design. Guidelines for avoiding serious consequences resulting from operator errors include:

1. Survey the machine system to identify potential hazards, then *design the hazards out* of the product. Be vigilant in prototype testing in order to uncover and correct any overlooked hazards.
2. Design equipment so that it is *easier to use safely* than unsafely.
3. Make design decisions that are compatible with stereotypical human expectations. For example,
   a. Clockwise rotation of rotary control knobs should correspond to increased output.
   b. Moving a control lever forward, upward, or to the right should correspond to increased output.
4. Locate and orient controls in such a way that the operator is unlikely to accidentally strike them, or inadvertently move them, in a normal operational sequence.
5. Where needed, recess or shield controls, or provide physical barriers to avoid inadvertent actuation.
6. Provide extra resistance when a control reaches a hazardous range of operation, so that an unusual human effort is required for further actuation.
7. Provide interlocks between or among controls so that prior operation of a related control is required before the critical control can be activated.
8. When consequences of inadvertent actuation are potentially grave, provide covers, guards, pins, or locks that must be removed or broken before the control can be operated.[14]

1.4 Concepts and Definitions of Mechanical Design

Mechanical design may be defined as an iterative decision-making process that has as its objective the *creation* and *optimization* of a *new* or *improved* mechanical engineering system or device for the *fulfillment of a human need or desire*, with due regard for *conservation of resources* and *environmental impact*. The definition just given includes several key ideas that characterize all mechanical design activity. The essence of engineering, especially mechanical design, is the fulfillment of human (customer) needs and desires. Whether a design team is creating a new device or improving an existing design, the ob-

[13]See, for example, ref. 4.

[14]For example, provide a padlock feature on an electrical power switchbox so a maintenance person may install a *personal lock* to assure that it cannot be changed from the *off* position by someone else. Also, integral warning tags should advise that the personal lock must be removed by the *same* maintenance person who installed it before power is restored.

jective is always to provide the "best," or optimum, combination of materials and geometry. Unfortunately, an absolute optimum design can rarely be realized because the criteria of performance, life, weight, cost, safety, and so on place counter-opposing requirements upon the materials and geometry proposed by the designer. Yet competition often demands that performance be enhanced, life be extended, weight be reduced, cost be lowered, or safety be improved. Not only must a design team compete in the marketplace by optimizing the design with respect to the criteria just noted, but it must respond responsibly to the clear and growing obligation of the global technical community to conserve resources and preserve the earth's environment.

Iteration pervades design methodology. The *keystone objectives* of all mechanical design activity are (1) *selection of the best possible material* and (2) *determination of the best possible geometry* for each part. During the *first* iteration, engineering designers concentrate on meeting functional performance specifications[15] by selecting potential materials and geometric arrangements that will provide strength and life adequate for the loads, environment, and potential failure modes governing the application. A reasonable *design safety factor* is typically chosen at this stage to account for uncertainties (see 1.5). Preliminary considerations of manufacturing methods are also included in the first iteration. The *second* iteration usually establishes all nominal dimensions and detailed material specifications to safely satisfy performance, strength, and life requirements. The *third* iteration audits the second-iteration design from the perspectives of fabrication, assembly, inspection, maintenance, and cost. The *fourth* iteration includes careful establishment of fits and tolerances, modifications resulting from the third-iteration audits, and a final check on the safety factor to assure that strength and life are suitable for the application, but that materials and resources are not being wasted.

1.5 Design Safety Factor

Uncertainties and variabilities always exist in design predictions. Loads are often variable and inaccurately known, strengths are variable and sometimes inaccurately known for certain failure modes or certain states of stress, calculation models embody assumptions that may introduce inaccuracies in determining dimensions, and other uncertainties may result from variations in quality of manufacture, operating conditions, inspection procedures, and maintenance practices. To provide safe, reliable operation in the face of these variations and uncertainties, it is common practice to utilize a *design safety factor* to assure that the minimum strength or capacity safely exceeds the maximum stress or load for all foreseeable operating conditions.[16] Design safety factors, always greater than 1, are usually chosen to have values that lie in the range from about 1.15 to about 4 or 5, depending on particular details of the application, as discussed in Chapter 5.

1.6 Stages of Design

Mechanical design activity in an industrial setting embodies a continuum effort from initial concept to development and field service. For discussion, the continuum of design activity may be subdivided into four stages, arbitrarily designated here as (1) *preliminary*

[15]Translating perceived customer needs or desires into quantitative engineering performance specifications is a responsibility of the *product design team*. See 1.1 and 1.2.

[16]As discussed in Chapter 5, statistical methods (reliability methods) may be used in some cases to achieve the same goal.

design, (2) *intermediate design*, (3) *detail design*, and (4) *development and field service*. Although some might argue that stage (4), development and field service, goes beyond design activity, it is clear that in the total life cycle of a product, development and field service data play important roles in product improvement, and therefore become an important part of the iterative design procedure.

Preliminary design, or *conceptual design*, is primarily concerned with synthesis, evaluation, and comparison of proposed machines or system concepts. A "black-box" approach is often used, in which reasonable experience-based performance characteristics are assigned to components or elements of the machine or system, followed by an investigation of integrated system behavior, without much regard for the details within the "black boxes." Gross simplifying assumptions and sound experience-based engineering judgments are usually necessary to complete preliminary design analyses in an acceptably short period of time. Overall system analyses, including force analysis, deflection analysis, thermodynamic analysis, fluid mechanic analysis, heat transfer analysis, electromechanic analysis, or control system analysis may be required at the preliminary design stage. Configurational drawings, or perhaps just free-hand sketches, are usually sufficient to communicate preliminary design concepts. Proprietary software has been developed by many organizations to implement the preliminary design and proposal presentation stage, especially for cases in which *existing product lines* need only be modified to meet new specifications. *The result of the preliminary design stage is the proposal of a likely-successful concept to be designed in depth to meet specified criteria* of performance, life, weight, cost, safety, or others.

Intermediate design, or *embodiment design*, embraces the spectrum of in-depth engineering design of individual components and subsystems for the already preselected machine or system. Intermediate design is vitally concerned with the *internal* workings of the black boxes, and must make them work as well or better than assumed in the preliminary design proposal. Material selection, geometry determination, and component arrangement are important elements of the intermediate design effort, and appropriate consideration must be given to fabrication, assembly, inspection, maintenance, safety, and cost factors as well. Gross simplifying assumptions cannot be tolerated at this stage. Good engineering assumptions are required to produce a good design and careful attention must be paid to performance, reliability, and life requirements, utilizing basic principles of heat transfer, dynamics, stress and deflection analysis, and failure prevention analysis. Either a carefully chosen safety factor must be incorporated into the design at this stage or, if data are available for doing so, properly established reliability specifications may be quantitatively reflected in the selection of materials and dimensions. Engineering drawings made to scale are an integral part of intermediate design. They may be made with instruments or by utilizing a computer-aided drafting system. Computer codes are widely used to implement all aspects of intermediate design activity. *The result of the intermediate design stage is establishment of all critical specifications* relating to function, manufacturing, inspection, maintenance, and safety.

Detail design is concerned mainly with configuration, arrangement, form, dimensional compatibility and completeness, fits and tolerances, standardization, meeting specifications, joints, attachment and retention details, fabrication methods, assemblability, producibility, inspectability, maintainability, safety, and establishing bills of material and purchased parts. The activities of detail design usually *support* the critical intermediate design decisions, but detail design does not usually involve making critical simplifying assumptions or selecting materials or dimensions that are critical in terms of strength, deflection, or life of a component. Although detail design is done largely by nonengineers, it is important that the engineering designer remain informed and vigilant throughout the detail design phase. *The result of the detail design stage is a complete set of drawings and specifications*, including detail drawings of all parts, or an electronic CAD file, *approved* by engineering design, production, marketing, and any other interacting departments, *ready for production* of a prototype machine or system.

Development and field service activities follow in sequence after the production of a prototype machine or system. *Development* of the prototype from a first model to an approved production article may involve many iterations to achieve a product suitable for marketing. The product design team should remain fully engaged with all design modifications required during the development phase, to achieve an optimum production article. *Field service information*, especially warranty service data on failure modes, failure rates, maintenance problems, safety problems, or other user-experience performance data, should be channeled back to the product design team for future use in product improvement and enhancement of life cycle performance. The *lessons-learned* strategy discussed in 1.10 should be made an integral part of the life cycle product improvement effort.

1.7 Steps in the Design Process

Another perspective on design methodology may be gained by examining the steps an engineering designer might take in designing a machine, a machine part, or a mechanical system. Although the sequence of steps presented will be found suitable for many design scenarios, the order may change, depending upon the details of the design task. The real usefulness of the list of basic design steps presented in Table 1.1 lies in the suggestion of a generalized methodology that may be used to implement the design process. In following through the list of steps in Table 1.1, it becomes clear that step VII has special significance, since it must be completely repeated for each and every part of a machine. Step VII, therefore, is outlined in greater detail in Table 1.2.

1.8 Fail Safe and Safe Life Design Concepts

Catastrophic failures of machines or systems that result in loss of life, destruction of property, or serious environmental degradation are simply unacceptable to the human community, and, in particular, unacceptable to the designers of such failed machines or systems. Yet it is evident from studying the probability distributions of material strengths corresponding to all failure modes, of loading spectra in all real applications, of environmental interactions, and of many other possible uncertain influences that *a designer can never provide a design of 100 percent reliability*, that is, she or he *can never provide a design absolutely guaranteed not to fail.* There is always a *finite* probability of failure. To address this frustrating paradox the design community has developed two important design concepts, both of which depend heavily upon *regular inspection of critical points* in a machine or structure. These design concepts are called *fail safe design* and *safe life design.*

The *fail safe design* technique provides redundant load paths in the structure so that if failure of a primary structural member occurs, a secondary member is capable of carrying the load on an emergency basis until failure of the primary structure is detected and a repair can be made.

The *safe life design* technique is to carefully select a large enough safety factor and establish inspection intervals to assure that the stress levels, the potential flaw sizes, and the governing failure strength levels of the material combine to give such a slow crack growth rate that the growing crack will be detected before reaching a critical size for failure.

Both fail safe and safe life design depend upon *inspectability*, the ability to inspect critical points in a machine after it is fully assembled and placed in service. It is impera-

TABLE 1.1 Fundamental Steps in the Design of a Machine

I. *Determine* precisely the *function* to be performed by the machine, and, in turn, by each subassembly and part.

II. *Select the energy source* best suited to driving the machine, giving special attention to availability and cost.

III. *Invent or select suitable mechanisms and control systems* capable of providing the functions defined, utilizing the selected energy source.

IV. *Perform* pertinent *supporting engineering analyses*, as required, including thermodynamics, heat transfer, fluid mechanics, electromechanics, control systems, and others.

V. *Undertake kinematic and dynamic analyses* to determine the important displacements, velocities, and accelerations throughout the machine and all of its parts.

VI. *Conduct a global force analysis* to determine or estimate all forces acting on the machine, so that subsequent local force analyses may be undertaken, as needed, in the design of the component parts.

VII. Carry through the *design of each of the individual parts* required to make up the complete machine. Remember that the *iterative* nature of the design process implies that, for each part, several tries and changes are usually necessary before determining final specifications for the best material and geometry. The important aspects of designing each part are shown in Table 1.2.

VIII. *Prepare layout drawings* of the entire machine by incorporating all parts as designed and sketched in step VII. This task requires attention not only to function and form, but careful attention as well to potential fabrication, assembly, maintenance, and inspection problems; also details of bases, mountings, isolation, shielding, interlocking, and other safety considerations.

IX. *Complete the detailed drawings* to be used as working drawings, for each individual part in the machine. These detail drawings are developed from the sketches of step VII by incorporating the changes generated during preparation of the layout drawings. Specifications for all fits, tolerances, finishes, environmental protection, heat treatment, special processing, imposition of company standards or industry standards, and code requirements are also incorporated.

X. *Prepare assembly drawings* of the entire machine by updating the layout drawings to include final-version detail drawing information from step IX. Subassembly drawings, casting drawings, forging drawings, or other special-purpose drawings are prepared as necessary to be included in the assembly drawing package.

XI. *Conduct a comprehensive design review* in which the product design team and all supporting departments carefully scrutinize the proposed design as depicted by the assembly drawings and detail drawings. Participation by engineering, production, foundry, industrial design, marketing, sales, and maintenance departments is usual. Modify the drawings as required.

XII. *Carefully follow prototype construction and development* to eliminate the problems that appear in experimental testing and evaluation of the machine. *Redesign is typically necessary* to develop the prototype machine into an acceptable product suitable for production and delivery.

XIII. *Monitor field service and maintenance records, failure rate and failure mode data, warranty maintenance and field inspection data*, and *customer service complaints* to identify significant design problems, and if necessary, design modification or retrofit packages to solve serious problems or eliminate design defects.

XIV. *Communicate* all significant field data on failure modes, failure rates, design defects, or other pertinent design factors back to engineering management and, in particular, the preliminary design department. The lessons-learned strategy discussed in 1.10 should be integrated into this communication process.

tive that designers consider inspectability at all stages of design, starting with machine component design, carrying through subassembly design, and design of the whole machine.

1.9 The Virtues of Simplicity

Beginning the design of a machine, a subassembly, or an individual part requires a clear understanding of the intended *function* of the device to be designed. Typically, the function of an individual part is *not identical* to the function of the machine as a whole; individual parts, with their inherent special functions, *combine* to produce the desired overall function of their assembly or machine. Each part in a machine is important to the whole, but each part also has a life (functionality) of its own.

Before determining the numerical dimensions of a part, its *configuration* must be established *qualitatively*. The configuration of a part is usually visualized by making a

TABLE 1.2 Steps in the Design of Each Individual Part

1. *Conceive a tentative* geometrical *shape* for the part. (See Chapter 6.)
2. *Determine the local forces and moments* on the part, based on global force analysis results from step VI. (See Chapter 4.)
3. *Identify probable governing failure modes* based on the function of the part, forces and moments on the part, shape of the part, and operational environment. (See Chapter 2.)
4. *Select a tentative material* for the part that seems to be best suited to the application. (See Chapter 3.)
5. *Select a tentative manufacturing process* that seems to be best suited to the part and its material. (See Chapter 7.)
6. *Select potential critical sections and critical points* for detailed analysis. Critical points are those points in the part that have a high probability of failure because of high stresses or strains, low strength, or a critical combination of these. (See Chapter 6.)
7. *Select appropriate equations of mechanics* that properly relate forces or moments to stresses or deflections, and calculate the stresses or deflections at each critical point considered. The *selection of a particular force-stress or force-deflection relationship* will be greatly influenced by the *shape of the part, the orientation of forces and moments* on the part, and the *choice of pertinent simplifying assumptions*. In later design iterations, more powerful analyses may be involved (such as finite element analyses) if the precision is needed and the cost warranted. (See Chapter 4.)
8. *Determine the dimensions of the part* at each critical point by assuring that the operating stress is always safely below the failure strength at each of these points. The safety margin between operating stress levels and failure strength levels may be established either by determining an appropriate *design safety factor* or by giving a proper *reliability specification*. (See Chapter 5.)
9. *Review the material selection, the shape, and the dimensions* of the designed part from the standpoints of *manufacturing processes* required, potential *assembly* problems, potential *maintenance* problems, and access of critical points to scheduled *inspections* intended to detect and eliminate incipient failures before they occur.
10. *Generate a sketch or drawing* of the designed part, *embodying all* of the *results* from the nine design aspects just listed, supplying the numerous minor decisions about size and shape required to complete a coherent drawing of the part. Such sketches or drawings may be generated either by neat free-hand sketching, by manual drafting using instruments, by using a computer-aided drafting system, or by some combination of these techniques.

sketch,[17] approximately to scale, that embodies the proposed geometric features,[18] and suggests location and retention means within its host assembly.

At this early conceptual stage, a guiding principle should be to *keep it simple*.[19] *Unnecessary* complexity usually leads to increased effort and time, more difficult and more costly manufacturing, slower and more costly assembly, and more difficult and more costly maintenance of the product. Limiting the functions of a part (or a machine) to those *actually required by the specifications* is a good first step in keeping a configuration simple. There is often a built-in desire on the part of the designer, especially an inexperienced designer, to keep adding seemingly desirable functions beyond those specified. Each of these add-on functions generates the need for a "small" increase in size, strength, or complexity of the part under consideration. Unfortunately, such noble efforts usually translate into longer times-to-market, cost overruns, increased difficulty in manufacturing and maintenance, and, in some cases, loss of market share to a competitor that delivers a reliable product to the marketplace earlier, even though it "only" meets the product specifications. The virtues of simplicity, therefore, potentially include on-time, on-budget delivery of a product to the marketplace, improved manufacturability, easier maintenance, gain in market share, and enhanced company reputation.

Design simplicity usually implies simple no-frills geometry, minimum number of individual parts, use of standard parts and components, and ease-of-assembly alignment fea-

[17]This may be accomplished either by hand sketching on paper or using a CAD system.

[18]See Chapter 6.

[19]In training sessions sponsored by the Boy Scouts of America, the *KISS* method is often promoted as a tool for preparing demonstrations and learning experiences. *KISS* is an acronym for *Keep It Simple, Stupid.* Designers could benefit from the same strategy.

tures that allow assembly maneuvers from a single direction.[20] Finally, *fits*, *tolerances*, and *finishes* should be no more restrictive than necessary for properly meeting specification requirements.[21]

1.10 Lessons-Learned Strategy

Most designers would agree that "reinventing the wheel" is a waste of time, yet failure to capitalize on experience is a pervasive problem. In the past decade the U.S. Army has formulated a "lessons-learned system" for improving combat effectiveness by implementing an organized effort to observe in-action problems, analyze them in after-action reviews, distill the reviews into *lessons learned*, and disseminate the lessons learned so the same mistakes are not repeated.[22] The system has proved to be an efficient process for correcting mistakes and sustaining successes through application of the lessons learned.

While the concept of "learning from experience" is not new, *organized* efforts in this direction are rare in most companies. Effective assessment of service failures, an important part of any design-oriented lesson-learned strategy, usually requires the intense interactive scrutiny of a team of specialists, including at least a mechanical designer and a materials engineer, both trained in failure analysis techniques, and often a manufacturing engineer and a field service engineer as well. The mission of the failure-response team is to discover the initiating cause of failure, identify the best solution, and redesign the product to prevent future failures. As undesirable as service failures may be, the results of a well-executed failure analysis may be transformed directly into improved product quality by designers who capitalize on service data and failure analysis results. The ultimate challenge is to assure that the lessons learned are applied. The lessons-learned strategy cannot succeed unless the information generated by the failure-response team is compiled and disseminated. No project is complete until systematically reviewed and its lessons communicated, especially to the preliminary design department.

1.11 Machine Elements, Subassemblies, and the Whole Machine

A well-designed machine is much more than an interconnected group of individual *machine elements*. Not only must the individual parts be carefully designed to function efficiently and safely for the specified design lifetime without failure, but parts must be effectively clustered into *subassemblies*. Each subassembly must function without internal interference, should permit easy disassembly for maintenance and repair, should allow easy critical point inspection without extensive downtime or hazard to inspectors, and should interface effectively with other subassemblies to provide the best possible integrated system configuration to fulfill the function of the *whole machine*. Completing the assembly of the whole machine always requires a frame or supporting structure into or upon which all subassemblies and support systems are mounted. Although design of the machine frame may be based upon either strength requirements or deflection requirements, the need for rigidity to prevent unacceptable changes in dimensions between one subassembly and another is a more usual design criterion for a machine frame. As in the

[20]See also 7.2 through 7.6.

[21]See 6.7.

[22]See ref. 5.

case for proper subassembly design, frames and structures must be designed to allow easy access for critical point inspection, maintenance, and repair procedures, as well as shielding and interlocks for safety of personnel. The basic principles for designing machine frames or structures are no different from the principles for designing any other machine part, and the methodology of Table 1.2 is valid.

Although the emphasis in this text is upon the design of machine elements (the traditional approach by most engineering design textbooks), recognition is given to the growing need for integration of manufacturing, assembly, and inspection requirements into the design process at an early stage, a philosophy widely referred to as "simultaneous engineering."

1.12 The Role of Codes and Standards in the Design Process

No matter how astute a designer may be, and no matter how much experience she or he may have, familiarity with the *codes* and *standards* pertinent to a particular design project is essential. Adherence to applicable codes and standards can provide experience-based guidance for the designer as to what constitutes *good practice* in that field, and assures that the product conforms to applicable legal requirements.

Standards are consensus-based documents, formulated through a cooperative effort among industrial organizations and other interested parties, that define *good practice* in a particular field. The basic objective in developing a standard is to assure interchangeability, compatibility, and acceptable performance within a company (company standard), within a country (national standard), or among many cooperating countries (international standard). Standards usually represent a minimum level of acceptance by the formulating group, and are usually regarded as *recommendations* to the user for *how to do* the task covered by the standard. Standards are prepared, compiled, and distributed by *ANSI*,[23] *ISO*,[24] and other similar organizations.

Codes are usually legally binding documents, compiled by a governmental agency, that are aimed at protecting the general welfare of its constituents and preventing property damage, injury, or loss of life. The objectives of a code are accomplished by requiring the application of accumulated knowledge and experience to the task of avoiding, eliminating, or reducing definable hazards. Codes are usually regarded as *mandatory requirements* that tell the user *what to do* and *when to do it*. Codes often incorporate one or more standards, giving them the force of law.

A designer's responsibility includes seeking out all applicable codes and standards relating to her or his particular design project. Failure of a designer to acquire a complete and comprehensive collection of applicable documents is extremely risky in today's litigious environment. Since customers, and the general public, expect that all marketed products will be safe for intended use (as well as unintended use, or even misuse), a designer, and his or her company, who does not follow code requirements, may be accused of professional malpractice,[25] and may be subject to litigation.

1.13 Ethics in Engineering Design

Like all professionals, engineers have a profound obligation to protect the public welfare by bringing the highest standards of honesty and integrity to their practice. That is, engi-

[23]The American National Standards Institute (see ref. 6).

[24]The International Organization for Standardization (see ref. 7).

[25]See also 1.13.

neers must be bound by adherence to the highest principles of *ethical* or *moral* conduct. Ethics and morality are formulations of what we *ought* to do and how we *ought* to behave as we practice engineering. Engineering designers have a special responsibility for ethical behavior because the health and welfare of the public often hang on the quality, reliability, and safety of their designs.

In the broadest sense, *ethics* are concerned with belief systems about good and bad, right and wrong, or appropriate and inappropriate behavior.[26] As simple as these concepts may seem, *ethical dilemmas* often arise because moral reasons can be offered to support two or more opposing courses of action. It is sometimes a difficult task to decide which competing moral viewpoint is the most compelling or most correct.[27]

To address ethical issues in the workplace, *ethics committees* are often formed to study and resolve ethical dilemmas within a company. Ethics committee consensus opinions and recommendations are usually tendered only after formulating the dilemma, collecting all relevant facts, and then examining the competing moral considerations. Such committee opinions usually disclose the level of consensus within the committee.

To help engineers practice their profession ethically, principles and rules of ethical behavior have been formulated and distributed by most engineering professional societies. The *Model Guide for Professional Conduct*[28] and the *Code of Ethics for Engineers*[29] are two good examples.

The code developed by *NSPE* includes a *Preamble*, six *Fundamental Canons*, five lengthy *Rules of Practice*, and nine *Professional Obligations*. The Preamble and the Fundamental Canons are shown in Figure 1.2.[30] In the end, however, ethical behavior translates into a combination of common sense and responsible engineering practice.

1.14 Units

In engineering design, numerical calculations must be made carefully, and any given set of calculations must employ a consistent *system of units*.[31] The systems of units commonly used in the United States are the *inch-pound-second (ips) system, foot-pound-second (fps) system*, and the *Système International d'Unités* or the *International System (SI)*.[32] All systems of units derive from Newton's second law

$$F = \frac{mL}{t^2} \tag{1-1}$$

[26]See ref. 8.

[27]For engineers (and others) engaged in competing in the *international marketplace*, the task of adhering to ethical behavior may be especially troublesome. This is true because certain practices that are legal and considered proper in some countries are considered to be unacceptable and illegal in the United States. An example is the locally acceptable business practice of giving "gifts" (bribes) to secure contracts in some countries. Without the "gift," no contract is awarded. Such practices are considered unethical and illegal in the United States.

[28]Developed by the American Association of Engineering Societies (see ref. 9).

[29]Developed by the National Society of Professional Engineers (see ref. 10).

[30]The more extensive details of the *Rules of Practice* and *Professional Obligations* are available from NSPE, and are reproduced in the appendix of this textbook.

[31]Any doubts about this should have been erased by the loss of NASA's $125 million Mars Climate Orbiter in September 1999. Two separate engineering teams, each involved in determining the spacecraft's course, failed to communicate that one team was using U.S. units while the other team was using metric units. The result was, apparently, that thrust calculations made using U.S. units were substituted into metric-based thrust equations without converting units, and the error was embedded in the orbiter's software. As a consequence, the spacecraft veered too close to the Martian surface, where it either landed hard, broke up, or burned (see ref. 11).

[32]See ref. 12.

NSPE Code of Ethics for Engineers

Preamble

Engineering is an important and learned profession. As members of this profession, engineers are expected to exhibit the highest standards of honesty and integrity. Engineering has a direct and vital impact on the quality of life for all people. Accordingly, the services provided by engineers require honesty, impartiality, fairness and equity, and must be dedicated to the protection of the public health, safety, and welfare. Engineers must perform under a standard of professional behavior that requires adherence to the highest principles of ethical conduct.

I. Fundamental Canons

Engineers, in the fulfillment of their professional duties, shall:

1. Hold paramount the safety, health and welfare of the public.
2. Perform services only in areas of their competence.
3. Issue public statements only in an objective and truthful manner.
4. Act for each employer or client as faithful agents or trustees.
5. Avoid deceptive acts.
6. Conduct themselves honorably, responsibly, ethically, and lawfully so as to enhance the honor, reputation, and usefulness of the profession.

Figure 1.2 **Preamble and Fundamental Canons of the NSPE *Code of Ethics for Engineers* (reproduced with permission of the National Society of Professional Engineers).**

in which any three of the four quantities F (force), m (mass), L (length), and t (time) may be chosen as *base units*, determining the fourth, called, therefore, a *derived unit*. When force, length, and time are chosen as the base units, making mass the derived unit, the system is called a *gravitational system*, because the magnitude of the mass depends on the local gravitational acceleration, g. Both the ips and fps systems are gravitational systems. When m, L, and t are chosen as base units, making force F the derived unit, the system is called an *absolute system*, because the mass, a base unit, is not dependent upon local gravity. In the ips gravitational system the base units are force in pounds (more properly pounds-force, but in this text lb $\equiv$ lbf), length in inches, and time in seconds, making the derived mass unit, which is given no special name, lb-sec^2/in since (1-1) yields

$$m = \frac{Ft^2}{L} \frac{\text{lb} - \text{sec}^2}{\text{in}} \tag{1-2}$$

Similarly, for the fps gravitational system

$$m = \frac{Ft^2}{L} \frac{\text{lb} - \text{sec}^2}{\text{ft}} \tag{1-3}$$

For the fps system, the mass unit is given the special name *slug*, where

$$\frac{\text{lb} - \text{sec}^2}{\text{ft}} \equiv \text{slug} \tag{1-4}$$

For the SI absolute system, the base units are mass in kilograms, length in meters, and time in seconds, making the derived force unit, from (1-1),

$$F = \frac{mL}{t^2} \frac{\text{kg} - \text{m}}{\text{sec}^2} \tag{1-5}$$

By definition, the force unit is given the special name *newton* (N), where

$$\frac{\text{kg} - \text{m}}{\text{sec}^2} \equiv \text{N} \tag{1-6}$$

The weight W of an object is defined as the force exerted on it by gravity. Thus,

$$W = mg \tag{1-7}$$

or

$$m = \frac{W}{g} \tag{1-8}$$

where g is the acceleration due to gravity. On earth at sea level, the value of g is approximately 386 in/sec^2 in the ips system, 32.17 ft/sec^2 in the fps system, and 9.81 m/sec^2 in the SI system.

Thus, when using Newton's second law to determine acceleration forces in a dynamic system, the equation may be expressed as

$$F = ma = \frac{W}{g} a \tag{1-9}$$

If using the ips system, F = force in lb, m = mass in lb-sec^2/in, a = acceleration in in/sec^2, g = 386 in/sec^2, and W = weight in lb; if using the fps system, F = force in lb, m = mass in slugs, a = acceleration in ft/sec^2, g = 32.17 ft/sec^2, and W = weight in lb; if using the SI system, F = force in newtons, m = mass in kg, a = acceleration in m/sec^2, g = 9.81 m/sec^2, and W = weight in newtons.

When using the SI system, several rules and recommendations of the international standardizing agency[33] should be followed to eliminate confusion among differing customs used in various countries of the world. These include:

1. Numbers having four or more digits should be placed in groups of three, counting from the decimal marker toward the left and the right, separated by spaces rather than commas. (The space may be omitted in four-digit numbers.)

2. A *period* should be used as a decimal point. (Centered periods and commas should not be used.)

3. The decimal point should be preceded by a zero for numbers less than unity.

4. Unit prefixes designating multiples or submultiples in steps of 1000 are recommended; for example, one millimeter equals 10^{-3} meter, or one kilometer equals 10^3 meters. Prefixes should not be used in the denominators of *derived* units. For example, N/mm^2

[33]International Bureau of Weights and Measures.

TABLE 1.3 A Truncated List of Standard SI Prefixes[1]

Name	Symbol	Factor
giga	G	10^9
mega	M	10^6
kilo	k	10^3
centi	c	10^{-2}
mili	m	10^{-3}
micro	μ	10^{-6}
nano	n	10^{-9}

[1]Other standard prefixes exist, but this list covers most engineering design cases. See ref. 2 for complete list.

should not be used; N/m^2, Pa (pascals), or MPa should be used instead. Prefixes should be chosen to make numerical values manageable. For example, using MPa (megapascals) for stress or GPa (gigapascals) for modulus of elasticity, rather than using Pa, gives more compact numerical results. A limited list of prefixes is given in Table 1.3. Table 1.4 lists the variables commonly used in engineering design practice, showing their units in the ips, fps, and SI systems. Table 1.5 gives a short list of conversion factors among the three systems of units. These various systems of units are used in this text as the need arises.

TABLE 1.4 Commonly Used Engineering Design Variables and Their Units (*Base Units* are shown in boldface)

Variable	Symbol	ips Units	fps Units	SI Units
Force	F	**lb** (pounds)	**lb**	N (newtons)
Length	l	**in** (inches)	**ft** (feet)	**m** (meters)
Time	t	**sec** (seconds)	**sec**	**sec**
Mass	m	lb-sec^2/in	slugs (lb-sec^2/ft)	**kg** (kilograms)
Weight	W	lb	lb	N
Pressure	p	psi	psf	Pa (pascals)
Stress	σ, τ	psi	—	Mpa
Velocity	v	in/sec	ft/sec	m/sec
Acceleration	a	in/sec^2	ft/sec^2	m/sec^2
Angle	θ	rad (radians)	rad	rad
Angular velocity	ω	rad/sec	rad/sec	rad/sec
Angular acceleration	α	rad/sec^2	rad/sec^2	rad/sec^2
Moment or torque	M, T	in-lb	ft-lb	N-m
Area	A	in^2	ft^2	m^2
Volume	V	in^3	ft^3	m^3
Area moment of inertia	I	in^4	—	m^4
Mass moment of inertia	I	in-lb-sec^2	ft-lb-sec^2	kg = m^2
Specific weight	w	lb/in^3	lb/ft^3	N/m^3
Energy	E	in-lb	ft-lb	N-m = J (joule)
Power	P	in-lb/sec	ft-lb/sec	N-m/sec (watt)
Spring rate	k	lb/in	lb/ft	N/m
Stress intensity	K	ksi $\sqrt{\text{in}}$	—	MPa $\sqrt{\text{m}}$

TABLE 1.5 Selected Conversion Relationships

Quantity	Conversion
Force	1 lb = 4.448 N
Length	1 in = 25.4 mm
Area	1 in^2 = 645.16 mm^2
Volume	1 in^3 = 16 387.2 mm^3
Mass	1 slug = 32.17 lb
	1 kg = 2.21 lb
	1 kg = 9.81 N
Pressure	1 psi = 6895 Pa
	1 Pa = 1 N/m^2
Stress	1 psi = 6.895×10^{-3} MPa
	1 ksi = 6.895 MPa
Modulus of elasticity	10^6 psi = 6.895 GPa
Spring rate	1 lb/in = 175.126 N/m
Velocity	1 in/sec = 0.0254 m/sec
Acceleration	1 in/sec^2 = 0.0254 m/sec^2
Work, energy	1 in-lb = 0.1138 N-m
Power	1 hp = 745.7 W (watts)
Moment, torque	1 in-lb = 0.1138 N-m
Stress intensity	1 ksi $\sqrt{\text{in}}$ = 1.10 MPa$\sqrt{\text{m}}$
Area moment of inertia	1 in^4 = $4.162 \times 10^{-7} \text{m}^4$
Mass moment of inertia	1 in-lb-sec^2 = 0.1138 N-m-sec^2

Example 1.1 Hitch Pin Bending: ips Units

A clevis-to-cable connection embodies a one-inch diameter hitch pin to be used for towing a large log from the backyard to the street. As illustrated in Figure E1.1, the pin may be modeled as a simply support beam of circular cross section, loaded by a concentrated mid-span load of 10,000 lb. Calculate the maximum bending stress in the pin if the maximum mid-span load is estimated to be 10,000 lb, and the pin is 2.0 inches long end-to-end with simple supports 0.25 inch from each end, as shown in Figure E1.1.

Solution

Referring to (4-5) of Chapter 4, the maximum bending stress is given by

$$\sigma_{max} = \frac{M_{max}c}{I} \tag{1}$$

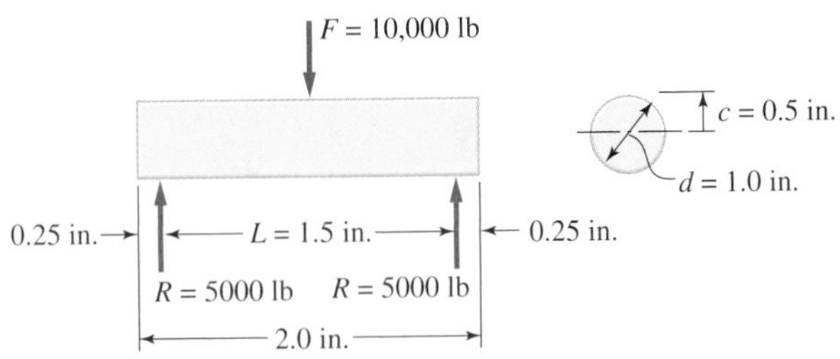

Figure E1.1
Clevis pin modeled as a simply supported beam in bending (refer to 4.4).

where M is the maximum bending moment, $c = d/2$ is the distance from the central axis (neutral axis of bending) to the top fiber, and I is the area moment inertia of the cross section. Utilizing Tables 4.1 and 4.2,

$$M_{max} = \frac{FL}{4} \tag{2}$$

and

$$I = \frac{\pi d^4}{64} \tag{3}$$

Hence the maximum stress may be calculated as

$$\sigma_{max} = \frac{\left(\frac{FL}{4}\right)\frac{d}{2}}{\left(\frac{(\pi d^4)}{64}\right)} = \frac{8FL}{\pi d^3} \tag{4}$$

Since the data are supplied in terms of inches and pounds, the calculation may be conveniently made using the ips system of units. Hence

$$\sigma_{max} = \frac{8(10{,}000 \text{ lb})(1.5 \text{ in})}{\pi(1.0 \text{ in})^3} = 38{,}200 \text{ psi} \tag{5}$$

Example 1.2 Hitch Pin Bending: SI Units

With the same general scenario as given in Example 1.1, the data are as follows:

$F = 44\,480$ N
$L = 38.1$ mm
$d = 25.4$ mm

Again, calculate the maximum bending stress.

Solution

The expression for maximum bending stress given in (4) of Example 1.1 remains valid. Since the data are supplied in terms of millimeters and newtons, the calculation may be conveniently made using the SI system of units. Hence

$$\sigma_{max} = \frac{8(44\,480 \text{ N})(38.1 \times 10^{-3} \text{ m})}{\pi(25.4 \times 10^{-3} \text{ m})^3} = 2.63 \times 10^8 \frac{\text{N}}{\text{m}^2}$$

$$= 2.63 \times 10^8 \text{ Pa} = 263 \text{ MPa} \tag{1}$$

Example 1.3 Units Conversion

It is now suggested that Examples 1.1 and 1.2 may be the same problem framed in two different systems of units. Check to find out if this is the case.

Solution

Using Table 1.5, check the equivalency of the data and the result.

Example 1.3 Continues

The data from Example 1.1, as given in Figure E1.1, are

$$\begin{aligned} F &= 10{,}000 \text{ lb} \\ L &= 1.5 \text{ in} \\ d &= 1.0 \text{ in} \end{aligned} \tag{1}$$

Using conversion factors from Table 1.5, these convert to

$$\begin{aligned} F &= 44\ 480 \text{ N} \\ L &= 38.1 \text{ mm} \\ d &= 25.4 \text{ mm} \end{aligned} \tag{2}$$

Comparing with data given in Example 1.2, they are found to be identical.

Converting the resulting stress from Example 1.1,

$$\sigma_{max} = 38{,}200(6.895 \times 10^{-3}) = 263 \text{ MPa} \tag{3}$$

The results are in agreement, as they should be.

Problems

1-1. Define *engineering design* and elaborate on each important concept in the definition.

1-2. List several factors that might be used to judge how well a proposed design meets its specified objectives.

1-3. Define the term *optimum design*, and briefly explain why it is difficult to achieve an optimum solution to a practical design problem.

1-4. When to stop calculating and start building is an engineering judgment of critical importance. Write about 250 words discussing your views on what factors are important in making such a judgment.

1-5. The stages of design activity have been proposed in 1.6 to include *preliminary design*, *intermediate design*, *detail design*, and *development and field service*. Write a two- or three-sentence descriptive summary of the essence of each of these four stages of design.

1-6. What conditions must be met to guarantee a *reliability* of 100 percent?

1-7. Distinguish between *fail safe design* and *safe life design*, and explain the concept of *inspectability*, upon which they both depend.

1-8. *Iteration* often plays a very important role in determining the material, shape, and size of a proposed machine part. Briefly explain the concept of *iteration*, and give an example of a design scenario that may require an iterative process to find a solution.

1-9. Write a short paragraph defining the term "simultaneous engineering" or "concurrent engineering."

1-10. Briefly describe the nature of codes and standards, and summarize the circumstances under which their use should be considered by a designer.

1-11. Define what is meant by *ethics* in the field of engineering.

1-12. Explain what is meant by an *ethical dilemma*.

1-13.[34] A young engineer, having worked in a multinational engineering company for about five years, has been assigned the task of negotiating a large construction contract with a country where it is generally accepted business practice, and totally legal under the country's laws, to give substantial gifts to government officials in order to obtain contracts. In fact, without such a gift, contracts are rarely awarded. This presents an ethical dilemma for the young engineer because the practice is illegal in the United States, and clearly violates the *NSPE Code of Ethics for Engineers* [see Code Section 5(b) documented in the appendix]. The dilemma is that while the gift-giving practice is unacceptable and illegal in the United States, it is totally proper and legal in the country seeking the services. A friend, who works for a different firm doing business in the same country, suggests that the dilemma may be solved by subcontracting with a local firm based in the country, and letting the local firm handle gift giving. He reasoned that he and his company were not party to the practice of gift giving, and therefore were not acting unethically. The local firm was acting ethically as well, since they were abiding by the practices and laws of that country. Is this a way out of the dilemma?

1-14.[35] Two young engineering graduate students received their Ph.D. degrees from a major university at about the same

[34]The substance of this question and the answer to it have been extracted from ref. 8, with permission.

[35]The substance of this question and the answer to it have been extracted from ref. 8, with permission.

time. Both sought faculty positions elsewhere, and they were successful in receiving faculty appointments at two different major universities. Both knew that to receive tenure they would be required to author articles for publication in scholarly and technical journals.

Engineer A, while a graduate student, had developed a research paper that was never published, but he believed that it would form a sound basis for an excellent journal article. He discussed his idea with his friend, Engineer B, and they agreed to collaborate in developing the article. Engineer A, the principal author, rewrote the earlier paper, bringing it up to date. Engineer B's contributions were minimal. Engineer A agreed to include Engineer B's name as co-author of the article as a favor in order to enhance Engineer B's chances of obtaining tenure. The article was ultimately accepted and published in a refereed journal.

a. Was it ethical for Engineer B to accept credit for development of the article?

b. Was it ethical for Engineer A to include Engineer B as co-author of the article?

1-15. If you were given the responsibility for calculating the stresses in a newly proposed "Mars Lander," what system of units would you probably choose? Explain.

1-16. Explain how the *lessons-learned strategy* might be applied to the NASA mission failure experienced while attempting to land the Mars Climate Orbiter on the Martian surface in September 1999. The failure event is briefly described in footnote 31 to the first paragraph of 1.14.

1-17. A special payload package is to be delivered to the surface of the moon. A prototype of the package, developed, constructed, and tested near Boston, has been determined to have a mass of 23.4 kg.

a. Estimate the weight of the package in newtons, as measured near Boston.

b. Estimate the weight of the package in newtons on the surface of the moon, if $g_{moon} = 17.0$ m/sec^2 at the landing site.

c. Reexpress the weights in pounds.

1-18. Laboratory crash tests of automobiles occupied by instrumented anthropomorphic dummies are routinely conducted by the automotive industry. If you were assigned the task of estimating the force in newtons at the mass center of the dummy, assuming it to be a rigid body, what would be your force prediction if a head-on crash deceleration pulse of 60 *g*'s (*g*'s are multiples of the standard acceleration of gravity) is to be applied to the dummy? The nominal weight of the dummy is 150 pounds.

1-19. Convert a shaft diameter of 2.25 inches to mm.

1-20. Convert a gear-reducer input torque of 20,000 in-lb to N-m.

1-21. Convert a tensile bending stress of 876 MPa to psi.

1-22. It is being proposed to use a standard W10 × 45 (wide-flange) section for each of four column supports for an elevated holding tank. (See Appendix Table A.3 for symbol interpretation and section properties.) What would be the cross-sectional area in mm^2 of such a column cross section?

1-23. What is the *smallest* standard equal-leg angle-section that would have a cross-sectional area at least as large as the W10 × 45 section of problem 1-22? (From Table A.3, the W10 × 45 section has a cross-sectional area of 13.3 in^2.)

Chapter 2

The Failure Prevention Perspective[1]

2.1 Role of Failure Prevention Analysis in Mechanical Design

A primary responsibility of any mechanical designer is to ensure that the proposed design will function as intended, safely and reliably, for the prescribed design lifetime and, at the same time, compete successfully in the marketplace. Success in designing competitive products while averting premature mechanical failures can be consistently achieved only by recognizing and evaluating all potential modes of failure that might govern the design of a machine and each individual part within the machine. If a designer is to be prepared to recognize potential failure modes, he or she must at least be acquainted with the array of failure modes actually observed in the field and with the conditions leading to those failures. For a designer to be effective in averting failure, he or she must have a good working knowledge of analytical and/or empirical techniques for predicting potential failures at the design stage, before the machine is built. These predictions must then be transformed into selection of a material, determination of a shape, and establishment of the dimensions for each part to ensure safe, reliable operation throughout the design lifetime. It is clear that failure analysis, prediction, and prevention perspectives form the basis for successful design of any machine element or machine.

2.2 Failure Criteria

Any change in the size, shape, or material properties of a machine or machine part that renders it incapable of performing its intended function must be regarded as a mechanical failure. It should be carefully noted that the key concept here is that *improper functioning* of a machine or machine part constitutes failure. Thus, a shear pin that does *not* separate into two or more pieces upon the application of a preselected overload must be regarded as having failed as surely as a drive shaft has failed if it *does* separate into two pieces under normal expected operating loads.

Failure of a machine or machine part to function properly might be brought about by any one or a combination of many different responses to loads and environments while in service. For example, too much or too little elastic deformation might produce failure. A load-carrying member that fractures or a shear pin that does not shear under overload conditions each would constitute failure. Progression of a crack due to fluctuating loads or an aggressive environment might lead to failure after a period of time if resulting excessive

[1]Chapter 2 is a condensed version of ref. 1, Copyright © 1993, by permission of John Wiley & Sons, Inc.

deflection or fracture of the part interferes with proper function. A list of potential mechanical failure modes that have been observed in various machine parts and machines is presented in the next section, followed by a brief description of each one.

2.3 Modes of Mechanical Failure

Failure modes are the physical processes that take place or combine their effects to produce failure, as just discussed. The following list[2] includes the failure modes most commonly observed in machines and machine parts.

1. Force- and/or temperature-induced elastic deformation
2. Yielding
3. Brinnelling
4. Ductile rupture
5. Brittle fracture
6. Fatigue:
   a. High-cycle fatigue
   b. Low-cycle fatigue
   c. Thermal fatigue
   d. Surface fatigue
   e. Impact fatigue
   f. Corrosion fatigue
   g. Fretting fatigue
7. Corrosion:
   a. Direct chemical attack
   b. Galvanic corrosion
   c. Pitting corrosion
   d. Intergranular corrosion
   e. Selective leaching
   f. Erosion corrosion
   g. Cavitation corrosion
   h. Hydrogen damage
   i. Biological corrosion
   j. Stress corrosion
8. Wear:
   a. Adhesive wear
   b. Abrasive wear
   c. Corrosive wear
   d. Surface fatigue wear
   e. Deformation wear
   f. Impact wear
   g. Fretting wear
9. Impact:
   a. Impact fracture
   b. Impact deformation
   c. Impact wear
   d. Impact fretting
   e. Impact fatigue
10. Fretting:
   a. Fretting fatigue
   b. Fretting wear
   c. Fretting corrosion
11. Creep
12. Thermal relaxation
13. Stress rupture
14. Thermal shock
15. Galling and seizure
16. Spalling
17. Radiation damage
18. Buckling
19. Creep buckling
20. Stress corrosion
21. Corrosion wear
22. Corrosion fatigue
23. Combined creep and fatigue

[2]Extracted from ref. 1, with permission.

As these terms are used in this text, and as commonly used in engineering practice, the failure modes just listed may be defined and described briefly as follows. It should be emphasized that these potential failure modes only produce failure when they generate a set of circumstances that interferes with the proper functioning of a machine or device.

Force- and/or temperature-induced elastic deformation failure occurs whenever the elastic (recoverable) deformation in a machine member, brought about by the imposed operational loads or temperatures, becomes great enough to interfere with the ability of the machine to satisfactorily perform its intended function.

Yielding failure occurs when the plastic (unrecoverable) deformation in a ductile machine member, brought about by the imposed operational loads or motions, becomes great enough to interfere with the ability of the machine to satisfactorily perform its intended function.

Brinnelling failure occurs when the static forces between two curved surfaces in contact result in local yielding of one or both mating members to produce a permanent surface discontinuity of significant size. For example, if a ball bearing is statically loaded so that a ball is forced to permanently indent the race through local plastic flow, the race is brinnelled. Subsequent operation of the bearing might result in intolerably increased vibration, noise, and heating, and, therefore, failure would have occurred.

Ductile rupture failure occurs when the plastic deformation, in a machine part that exhibits ductile behavior, is carried to the extreme so that the member separates into two pieces. Initiation and coalescence of internal voids slowly propagate to failure, leaving a dull, fibrous rupture surface.

Brittle fracture failure occurs when the elastic deformation, in a machine part which exhibits brittle behavior, is carried to the extreme so that the primary interatomic bonds are broken and the member separates into two or more pieces. Preexisting flaws or growing cracks provide initiation sites for very rapid crack propagation to catastrophic failure, leaving a granular, multifaceted fracture surface.

Fatigue failure is a general term given to the sudden and catastrophic separation of a machine part into two or more pieces as a result of the application of fluctuating loads or deformations over a period of time. Failure takes place by the initiation and propagation of a crack until it becomes unstable and propagates suddenly to failure. The loads and deformations that cause failure by fatigue are typically far below the static failure levels. When loads or deformations are of such magnitude that more than about 50,000 cycles are required to produce failure, the phenomenon is usually termed *high-cycle fatigue*. When loads or deformations are of such magnitude that less than about 10,000 cycles are required to produce failure, the phenomenon is usually termed *low-cycle fatigue*. When load or strain cycling is produced by a fluctuating temperature field in the machine part, the process is usually termed *thermal fatigue. Surface fatigue* failure, usually associated with rolling surfaces in contact (but sometimes associated with sliding contact), manifests itself as pitting, cracking, and spalling of the contacting surfaces as a result of the cyclic Hertz contact stresses that result in maximum values of cyclic shear stresses slightly below the surface. The cyclic subsurface shear stresses generate cracks that propagate to the contacting surface, dislodging particles in the process, to produce surface pitting. This phenomenon is often viewed as a type of wear. Impact fatigue, corrosion fatigue, and fretting fatigue are described later.

Corrosion failure, a very broad term, implies that a machine part is rendered incapable of performing its intended function because of the undesired deterioration of the material as a result of chemical or electrochemical interaction with the environment. Corrosion often interacts with other failure modes such as wear or fatigue. The many forms of corrosion include the following: *Direct chemical attack*, perhaps the most common type of corrosion, involves corrosive attack of the surface of the machine part exposed to the corrosive medium, more or less uniformly over the entire exposed surface. *Galvanic corrosion* is an accelerated electrochemical corrosion that occurs when two dissimilar metals in electrical contact are made part of a circuit completed by a connecting pool or film of electrolyte or corrosive medium, leading to current flow and ensuing corrosion. *Crevice corrosion* is the accelerated corrosion process highly localized within crevices, cracks, or joints where small volume regions of stagnant solution are trapped in contact with the corroding meal. *Pitting corrosion* is a very localized attack that leads to the development of an array of holes or pits that penetrate the metal. *Intergranular corrosion* is the localized attack occurring at grain boundaries of certain copper, chromium, nickel, aluminum, magnesium, and zinc alloys when they are improperly heat treated or welded. Formation of local galvanic cells that precipitate corrosion products at the grain boundaries seriously degrades the material strength because of the intergranular corrosive process.

Selective leaching is a corrosion process in which one element of a solid alloy is removed, such as in dezincification of brass alloys or graphitization of gray cast irons. *Erosion corrosion* is the accelerated chemical attack that results when an abrasive or viscid material flows past a containing surface continuously baring fresh, unprotected material to the corrosive medium. *Cavitation corrosion* is the accelerated chemical corrosion that results when, because of differences in vapor pressure, certain bubbles and cavities within a fluid collapse adjacent to the pressure vessel walls, causing particles of the surface to be expelled, baring fresh, unprotected surface to the corrosive medium. *Hydrogen damage*, while not considered to be a form of direct corrosion, is induced by corrosion. Hydrogen damage includes hydrogen blistering, hydrogen embrittlement, hydrogen attack, and decarburization. *Biological corrosion* is a corrosion process that results from the activity of living organisms, usually by virtue of their processes of food ingestion and waste elimination, in which the waste products are corrosive acids or hydroxides. *Stress corrosion*, an extremely important type of corrosion, is described separately later.

Wear is the undesired cumulative change in dimensions brought about by the gradual removal of discrete particles from contacting surfaces in motion, usually sliding, predominantly as a result of mechanical action. Wear is not a single process, but a number of different processes that can take place independently or in combination, resulting in material removal from contacting surfaces through a complex combination of local shearing, plowing, gouging, welding, tearing, and others. *Adhesive wear* takes place because of high local pressure and welding at asperity contact sites, followed by motion-induced plastic deformation and rupture of asperity junctions, with resulting metal removal or transfer. *Abrasive wear* takes place when the wear particles are removed from the surface by plowing, gouging, and cutting action of the asperities of a harder mating surface or by hard particles entrapped between the mating surfaces. When the conditions for either adhesive wear or abrasive wear coexist with conditions that lead to corrosion, the processes interact synergistically to produce *corrosive wear*. As described earlier, *surface fatigue wear* is a wear phenomenon associated with curved surfaces in rolling or sliding contact, in which subsurface cyclic shear stresses initiate microcracks that propagate to the surface to spall out macroscopic particles and form wear pits. *Deformation wear* arises as a result of repeated *elastic* deformation at the wearing surface that produces a matrix of cracks that grow in accordance with the surface fatigue description just given. Fretting wear is described later.

Impact failure results when a machine member is subjected to nonstatic loads that produce in the part stresses or deformations of such magnitude that the member no longer is capable of performing its function. The failure is brought about by the interaction of stress or strain waves generated by dynamic or suddenly applied loads, which may induce local stresses and strains many times greater than would be induced by static application of the same loads. If the magnitudes of the stresses and strains are sufficiently high to cause separation into two or more parts, the failure is called *impact fracture*. If the impact produces intolerable elastic or plastic deformation, the resulting failure is called *impact deformation*. If repeated impacts induce cyclic elastic strains that lead to initiation of a matrix of fatigue cracks, which grow to failure by the surface fatigue phenomenon described eariler, the process is called *impact wear*. If fretting action, as described in the next paragraph, is induced by the small lateral relative displacements between two surfaces as they impact together, where the small displacements are caused by Poisson strains or small tangential "glancing" velocity components, the phenomenon is called *impact fretting*. *Impact fatigue* failure occurs when impact loading is repetitively applied to a machine member until failure occurs by the nucleation and propagation of a fatigue crack.

Fretting action may occur at the interface between any two solid bodies whenever they are pressed together by a normal force and subjected to small-amplitude cyclic relative motion with respect to each other. Fretting usually takes place in joints that are not intended to move but, because of vibrational loads or deformations, experience minute cyclic relative motions. Typically, debris produced by fretting action is trapped between the surfaces because of the small motions involved. *Fretting fatigue* failure is the premature fatigue fracture of a machine member subjected to fluctuating loads or strains together with conditions that simultaneously produce fretting action. The surface discontinuities and microcracks generated by the fretting action act as fatigue crack nuclei that propagate to failure under conditions of fatigue loading that would otherwise be acceptable. Fretting fatigue failure is an insidious failure mode because the fretting action is usually hidden within a joint where it cannot be seen, leading to premature, and often unexpected, fatigue failure of a sudden and catastrophic nature. *Fretting wear* failure results when the changes in dimensions of the mating parts, because of the presence of fretting action, become large enough to interfere with proper design function or large enough to produce geometrical stress concentration of such magnitude that failure ensues as a result of excessive local stress levels. *Fretting corrosion* failure occurs when a machine part is rendered incapable of performing its intended function because of the surface degradation of the material from which the part is made, as a result of fretting action.

Creep failure results whenever the plastic deformation in a machine member accrues over a period of time under the influence of stress and temperature until the accumulated dimensional changes interfere with the ability of the machine part to satisfactorily perform its intended function. Three stages of creep are often observed: (1) transient or primary creep, during which time the rate of strain decreases, (2) steady-state or secondary creep, during which time the rate of strain is virtually constant, and (3) tertiary creep, during which time the creep strain rate increases, often rapidly, until rupture occurs. This terminal rupture is often called creep rupture and may or may not occur, depending on the stress-time-temperature conditions.

Thermal relaxation or *stress relaxation* failure occurs when the dimensional changes due to the creep process result in the relaxation of a prestrained or prestressed member until it no longer is able to perform its intended function. For example, if the prestressed flange bolts of a high-temperature pressure vessel relax over a period of time because of creep in the bolts, so that finally the peak pressure surges exceed the bolt preload to violate the flange seal, the bolts will have failed because of thermal relaxation.

Stress rupture failure is intimately related to the creep process except that the combination of stress, time, and temperature is such that rupture into two parts is assured. In stress rupture failures the combination of stress and temperature is often such that the period of steady-state creep is short or nonexistent.

Thermal shock failure occurs when the thermal gradients generated in a machine part are so pronounced that differential thermal strains exceed the ability of the material to sustain them without yielding or fracture.

Galling failure occurs when two sliding surfaces are subjected to such a combination of loads, sliding velocities, temperatures, environments, and lubricants that massive surface destruction is caused by welding and tearing, plowing, gouging, significant plastic deformation of surface asperities, and metal transfer between the two surfaces. Galling may be thought of as a severe extension of the adhesive wear process. When such action results in significant impairment to intended surface sliding, or in seizure, the joint is said to have failed by galling. *Seizure* is an extension of the galling process to such a level of severity that the two parts are virtually welded together, and relative motion is no longer possible.

Spalling failure occurs whenever a particle is spontaneously dislodged from the surface of a machine part so as to prevent the proper function of the member. Armor plate fails by spalling, for example, when a striking missile on the exposed side of an armor shield generates a stress wave that propagates across the plate in such a way as to dislodge or *spall* a secondary missile of lethal potential on the protected side. Other examples of spalling failure are manifested in rolling contact bearings and gear teeth because of the action of surface fatigue as described earlier.

Radiation damage failure occurs when the changes in material properties induced by exposure to a nuclear radiation field are of such a type and magnitude that the machine part is no longer able to perform its intended function, usually as a result of the triggering of some other failure mode, and often related to loss in ductility associated with radiation exposure. Elastomers and polymers are typically more susceptible to radiation damage than are metals whose strength properties are sometimes enhanced rather than damaged by exposure to a radiation field, though ductility is usually decreased.

Buckling failure occurs when, because of a critical combination of magnitude and/or point-of-load application, together with the geometrical configuration of a machine member, the deflection of the member suddenly increases greatly with only a slight increase in load. This nonlinear response results in a buckling failure if the buckled member is no longer capable of performing its design function.

Creep buckling failure occurs when, after a period of time, the creep process results in an unstable combination of the loading and geometry of a machine part so that the critical buckling limit is exceeded and failure ensures.

Stress corrosion failure occurs when the applied stresses on a machine part in a corrosive environment generate a field of localized surface cracks, usually along grain boundaries, that render the part incapable of performing its function, often because of triggering some other failure mode. Stress corrosion is a very important type of corrosion failure mode because so many different metals are susceptible to it. For example, a variety of iron, steel, stainless steel, copper, and aluminum alloys are subject to stress corrosion cracking if placed in certain adverse corrosive media.

Corrosion wear failure is a combination failure mode in which corrosion and wear combine their deleterious effects to incapacitate a machine part. The corrosion process often produces a hard, abrasive corrosion product that accelerates the wear, while the wear process constantly removes the protective corrosion layer from the surface, baring fresh metal to the corrosive medium and thus accelerating the corrosion. The two modes combine to make the result more serious than the sum of the modes would have been otherwise.

Corrosion fatigue is a combination failure mode in which corrosion and fatigue combine their deleterious effects to cause failure of a machine part. The corrosion process often forms pits and surface discontinuities that act as stress raisers that in turn accelerate fatigue failure. Further, cracks in the usually brittle corrosion layer also act as fatigue crack nuclei that propagate into the base material. On the other hand, the cyclic loads or strains cause cracking and flaking of the corrosion layer, which bares fresh metal to the corrosive medium. Thus, each process accelerates the other, often making the result disproportionately serious.

Combined creep and fatigue failure is a combination failure mode in which all of the conditions for both creep failure and fatigue failure exist simultaneously, each process influencing the other to produce accelerated failure. The interaction of creep and fatigue is probably synergistic but is not well understood.

Identification of the most probable governing failure mode (in many cases there may be more than one candidate) by a designer is an essential step that should be underaken early in the design of any machine part. Selection of material and establishment of the shape and size of the part must then be tailored to provide safe, reliable, cost-effective operation throughout the design lifetime. The following sections provide some basic concepts and equations useful in designing to avoid failure by the more commonly encountered failure modes.

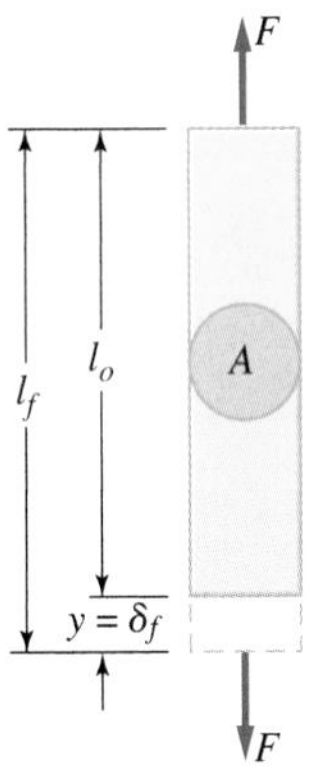

F, l_f, l_o, A, $y = \delta_f$, F

Figure 2.1
Straight cylindrical bar loaded by direct axial force.

2.4 Elastic Deformation, Yielding, and Ductile Rupture

In the seventeenth century it was experimentally established that if a direct axial external force F is applied to a machine element, whether it is a traditional spring (see Chapter 14) or a straight cylindrical bar such as the one shown in Figure 2.1, changes in the length of the machine element are produced. Further, for a broad class of materials, a *linear* relationship exists between the applied force, F, and the induced change in length, y, as long as the material is not stressed beyond its elastic range.[3] The elastic deflection, y, and the

[3]See Hooke's Law discussion of 4.5.

corresponding elastic strain, ε, produced by the aggregation of small changes in interatomic spacing within the material, are fully recoverable[4] as long the applied forces do not produce stresses that exceed the *yield strength* of the material.

For any linear spring, the relationship between force and deflection may be plotted as shown in Figure 2.2, and expressed as

$$F = ky \tag{2-1}$$

where k is called the *spring constant* or *spring rate*.

The change in length induced by the applied force is

$$y = \delta_f = l_f - l_o \tag{2-2}$$

where l_o is the original bar length with no force applied, l_f is the bar length after external force F has been applied, and δ_f is *force-induced elastic deformation*, so defined to distinguish it from temperature-induced elastic deformation, discussed later.

If, in a machine element, δ_f exceeds the *design allowable* axial deformation, failure will occur. For example, if the blade-axial deformation of an aircraft gas turbine blade, caused by the centrifugal force field, exceeds the tip clearance gap, failure will occur because of force-induced elastic deformation.

Based on (2-1), the spring rate, k_{ax}, for the uniform bar shown in Figure 2.1 may be written as

$$k_{ax} = \frac{F}{y} = \frac{F}{\delta_f} \tag{2-3}$$

where k_{ax} is the *spring rate*, or *stiffness*, of the axially loaded bar. It should again be emphasized that all *real* machine parts and structural elements behave as springs because they all have finite stiffnesses. Thus the concept of spring rate is important not only when discussing "traditional" springs, as in Chapter 14, but also when considering potential failure by force-induced elastic deformation, or when examining the consequences of load sharing, preloading, and/or residual stresses (see Chapter 4).

The uniaxial force-deflection plot of Figure 2.2 may be *normalized* by dividing the force coordinate by the original cross-sectional area A_o and the deflection coordinate by original length l_o. The resulting plot, shown in Figure 2.3, is the familiar *engineering*

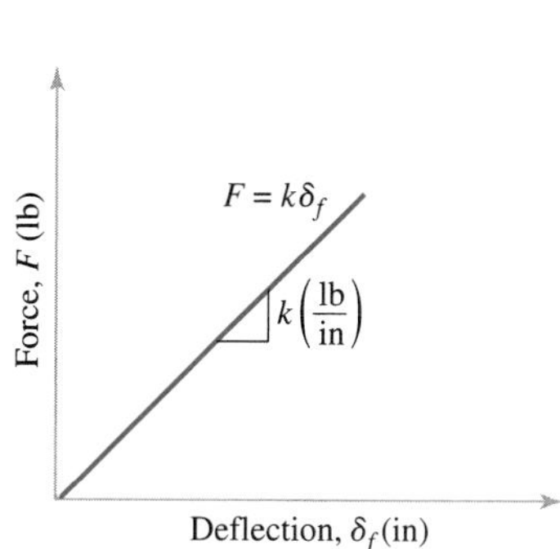

Figure 2.2
Force-deflection curve for linear elastic behavior.

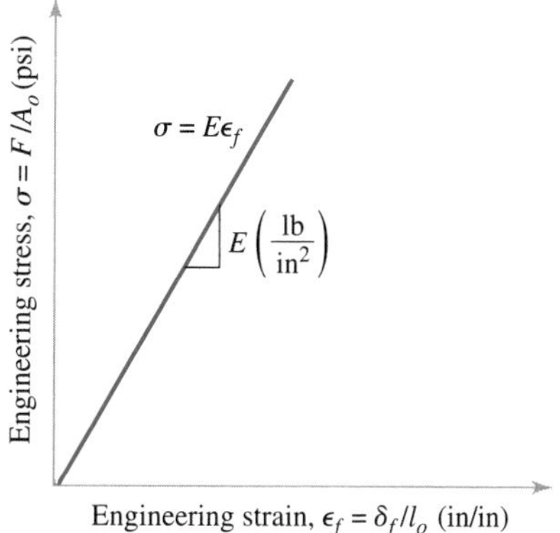

Figure 2.3
Engineering stress-strain diagram for linear elastic behavior.

[4]See ref. 2.

stress-strain diagram,[5] with slope equal to Young's modulus of elasticity, E. For the uniform bar shown in Figure 2.2, the engineering stress, σ, is

$$\sigma = \frac{F}{A_o} \tag{2-4}$$

where A_o is the original cross-sectional area of the bar.

The engineering strain induced by the applied force is

$$\varepsilon_f = \frac{\delta_f}{l_o} \tag{2-5}$$

From the engineering stress-strain curve of Figure 2.3,

$$\sigma = E\varepsilon_f \tag{2-6}$$

where E is Young's modulus of elasticity and ε_f is force-induced elastic strain.

Combining (2-3), (2-4), (2-5), and (2-6),

$$k_{ax} = \frac{A_o E}{l_o} \tag{2-7}$$

where k_{ax} is the axial spring rate, a function of material and geometry. Thus if it were of interest, the force-induced elastic deformation δ_f could be easily calculated for the axial loading case, as

$$\delta_f = \frac{F}{k_{ax}} \tag{2-8}$$

If the limiting or allowable design value of deflection, δ_{allow}, in a particular design situation were exceeded by δ_f, the design would be unacceptable; hence elastic deformation *failure is predicted to occur if (FIPTOI)*

$$\delta_{f-max} \geq \delta_{allow} \tag{2-9}$$

and redesign would be necessary if failure were predicted.

Example 2.1 Force-Induced Elastic Deformation

An axially loaded straight bar with a rectangular cross secction will fail to perform its design function if its length increases 0.0060 inch or more. It is desired to operate with the stress level in the bar never greater than one-half the yield strength, S_{yp}. The bar, made of 1040 HR steel (S_{yp} = 42,000 psi), is 12.0000 inches long.

a. What is the probable governing failure mode?

b. Is failure predicted to occur?

Solution

a. The probable governing failure mode is force-induced elastic deformation.

b. From (2-9), FIPTOI (failure is predicted to occur if)

$$\delta_{f-max} \geq \delta_{allow} \tag{1}$$

[5]*Engineering stress and strain* are based on original values of area and length, in contrast with *true stress* and *true strain*, which are based on instantaneous values of area and length.

Now,

$$\delta_{f-max} = \frac{F_{max}}{k_{ax}} = \frac{\sigma_{max} A_o}{\left(\frac{A_o E}{l_o}\right)} = \frac{S_{yp} l_o}{2E} = \frac{4200(12.0000)}{2(30)10^6} = 0.0084 \text{ inch} \tag{2}$$

Hence, FIPTOI

$$0.0084 \geq 0.0060 \tag{3}$$

Therefore, failure *is* predicted to occur.

Temperature fluctuations may also produce dimensional changes and associated recoverable elastic strains in a machine part. If temperature changes are slow enough so that significant gradients are not generated, and if no external constraints are imposed, the machine part remains stress-free as temperature-induced elastic deformations are produced. The change in length, δ_t, induced by the temperature difference $\Delta\Theta$, may be calculated as

$$\delta_t = l_o \alpha \Delta\Theta \tag{2-10}$$

where l_o is the initial length of the part and a is the linear coefficient of thermal expansion for the material. From (2-10), then, the temperature-induced elastic strain (thermal strain) is

$$\varepsilon_t = \frac{\delta_t}{l_o} = \alpha \Delta\Theta \tag{2-11}$$

Temperature-induced strains are recoverable (elastic), normal (as opposed to shear), and usually close to linear functions of temperature over a wide temperature range (often a few hundred °F). The temperature-induced elastic strain in a given direction, i, may be found from (2-11) as

$$\varepsilon_{ti} = \alpha_i \Delta\Theta \tag{2-12}$$

where ε_{ti} is the thermal or temperature-induced elastic strain in the ith direction due to temperature change $\Delta\Theta$, and α_i is the thermal expansion coefficient in the ith direction.

Example 2.2 Temperature-Induced Elastic Deformation

The straight rectangular 1040 HR steel bar of Example 2.1 is to be evaluated for an entirely different application in which there are no applied forces at all but the temperature is to be increased by 300°F. Again, in this new application, the bar will cease to properly perform its design function if its length increases 0.0060 inch or more. The linear coefficient of thermal expansion for steel is approximately 6.3×10^{-6} in/in/°F.

a. What is the governing failure mode?
b. Is failure predicted to occur?

Solution

a. The governing failure mode is temperature-induced elastic deformation.
b. Following the logic leading to (2-9), FIPTOI (failure is predicted to occur if)

$$\delta_{t-max} \geq \delta_{allow} \tag{1}$$

From (2-10),

$$\delta_{t-max} = l_o \alpha \Delta\Theta = (12.0000)(6.3 \times 10^{-6})(300) = 0.0227 \text{ inch} \tag{2}$$

Example 2.2 Continues

Hence, FIPTOI

$$0.0227 \geq 0.0060 \tag{3}$$

Therefore failure *is* predicted to occur.

It is worth noting that the change in length of this 12-inch steel bar produced by a 300° temperature change is nearly three times the length change produced by an axial force corresponding to a stress equal to half the yield strength of the steel. (Temperature-induced elastic strains are often large enough to be important design criteria.)

The *principle of superposition* may be utilized when governing equations are linear. Since force-induced elastic strain and temperature-induced elastic strain are both linear functions, the principle of superposition may be used to give the total strain ε_i in any given ith direction as

$$\varepsilon_i = \varepsilon_{fi} + \varepsilon_{ti} \tag{2-13}$$

where ε_{fi} and ε_{ti} are force- and temperature-induced components of strain in the ith direction, respectively. This expression may be used to assess the consequences of constraint to thermal expansion. For example, if a prismatic steel bar of length l_o were axially constrained between rigid walls so that its total change in length were forced to remain zero, and the temperature increased by $\Delta\Theta$, equation (2-13) would require

$$0 = \varepsilon_{fi} + \varepsilon_{ti} \tag{2-14}$$

or

$$\varepsilon_{fi} = -\varepsilon_{ti} \tag{2-15}$$

Utilizing the uniaxial Hooke's Law and (2-12), the temperature-induced stress (thermal stress), σ, in this fully constrained bar, would be

$$\sigma = -E\alpha_i\Delta\Theta \tag{2-16}$$

a compressive stress induced in the steel bar because it attempts to increase in length as temperature increases. If the walls are not fully rigid (a more realistic case), the calculations become more complicated, as discussed later in 4.11.

When the state of stress is more complicated than the uniaxial cases just discussed, it becomes necessary to calculate the elastic strains induced by the multiaxial state of stress in three mutually perpendicular directions, say x, y, z, through the use of the generalized Hooke's Law equations given in 4.5. The resulting elastic strains, ε_{fx}, ε_{fy}, and ε_{fz}, may be used to calculate the total force-induced elastic deformation of a member in any of the coordinate directions by integrating the strain over the member's length in that direction. Temperature-induced strains may be included as shown in (2-13). If the change in length of the member in any direction exceeds the design-allowable deformation in that direction, elastic deformation failure will occur.

If the externally applied forces produce stresses that exceed the *yield strength* of a ductile material, the induced strains are not fully recovered upon release of the loads and permanent or *plastic strains* remain. Such permanent deformations usually (but not always) render a machine part incapable of performing its intended function, whereupon failure by yielding is said to occur. For example, if the axially loaded bar of Figure 2.1 were to reach a stress σ, calculated from (2-1), that exceeds the material's yield strength, S_{yp}, the design would be unacceptable; hence yielding failure would be predicted to occur if (FIPTOI)

$$\sigma \geq S_{yp} \tag{2-17}$$

and redesign would be necessary.

Example 2.3 Yielding

An axially loaded straight bar of circular cross section will fail to perform its design function if applied axial loads produce permanent changes in length when the load is removed. The bar has a diameter of 0.500 inch, a length of 7.00 inches, and is made of 1020 HR steel (S_u = 55,000 psi; S_{yp} = 30,000 psi; e = 25 percent in 2 inches). The applied axial load required for this particular case is 5800 lb.

a. What is the governing failure mode?

b. Is failure predicted to occur?

Solution

a. Since elongation in 2 inches is 25 percent, the material is ductile and the governing failure mode is yielding.

b. From (2-17), FIPTOI

$$\sigma \geq S_{yp} \tag{1}$$

Now,

$$\sigma = \frac{F}{A_o} = \frac{F}{\left(\frac{\pi d^2}{4}\right)} = \frac{5800}{\pi\left[\frac{(0.500)^2}{4}\right]} = \frac{5800}{0.196} = 29{,}590 \text{ psi} \tag{2}$$

Hence, FIPTOI

$$29{,}590 \geq 30{,}000 \tag{3}$$

Therefore failure is *not* predicted to occur. It must be noted, however, that failure *is imminent*, and as a practical matter a designer would nearly always redesign to lower the stress. Typically, a *safety factor* would be utilized to determine the properly redesigned dimensions, as discussed in Chapter 5.

For the case of uniaxial loadng, the onset of yielding may be accurately predicted to occur when the uniaxial maximum normal stress reaches a value equal to the yield strength of the material, as shown in (2-17) and Example 2.3. If the loading is more complicated, and a multiaxial state of stress is produced by the loads, the onset of yielding may no longer be predicted by comparing any one of the normal stress components with uniaxial material yield strength, not even the maximum principal normal stress. Onset of yielding for multiaxially stressed critical points in a machine or structure is more accurately predicted through the use of a *combined stress theory of failure* that has been experimentally validated for the prediction of yielding. The two most widely accepted theories for predicting the onset of yielding under multiaxial states of stress are the *distortion energy theory* (also known as the *octahedral shear stress theory*, or the *Huber–von–Mises–Hencky theory*), and the *maximum shearing stress theory*. The use of these theories is discussed in 4.6.

If the externally applied forces are so large that a ductile material not only experiences plastic deformation but proceeds to separate into two pieces, the process is called *ductile rupture*. In most cases, such a separation renders a machine part incapable of performing its intended function, and failure by ductile rupture is said to occur. For example, if the axially loaded bar of Fig. 2.1 reaches a stress σ, calculated from (2-1), that attempts to exceed the material's ultimate strength, S_u, the design would be unacceptable; hence failure by ductile rupture would be predicted to occur if (FIPTOI)

$$\sigma \geq S_u \tag{2-18}$$

and redesign would be necessary.

Example 2.4 Ductile Rupture

The axially loaded straight cylindrical bar of Example 2.3 is to be used in a different application for which permanent deformations are acceptable but separation of the bar into two pieces destroys the ability of the device to perform its function. The required axial load for this application is to be 11,000 lb.

a. What is the governing failure mode?

b. Is failure predicted to occur?

Solution

a. From material specifications in Example 2.3, the 1020 HR material is ductile (e = 25% in 2 inches); hence the governing failure mode is ductile rupture.

b. From (2-18), FIPTOI

$$\sigma \geq S_u \tag{1}$$

Now,

$$\sigma = \frac{F}{A_o} = \frac{11{,}000}{\pi\left[\dfrac{(0.500)^2}{4}\right]} = \frac{11{,}000}{0.196} = 56{,}120 \text{ psi} \tag{2}$$

Also, from Example 2.3, S_u = 55,000 psi, hence FIPTOI

$$56{,}120 \geq 55{,}000 \tag{3}$$

Therefore failure *is* predicted to occur by ductile rupture, and redesign would be necessary.

Again, if the loading is more complicated, and a multiaxial state of stress is produced by the loads, ductile rupture may no longer be accurately predicted by comparing any of the normal stress components with the uniaxial ultimate strength, not even the maximum principal normal stress. A combined stress theory of failure is required for prediction of ductile rupture under multiaxial states of stress. The distortion energy theory is usually chosen for such cases, as discussed in 4.6.

2.5 Brittle Fracture and Crack Propagation; Linear Elastic Fracture Mechanics

When the material behavior of a machine part is brittle rather than ductile, the mechanics of the failure process are much different. As described in 2.3, in brittle fracture the part separates into two or more pieces due to the breaking of primary interatomic bonds, with little or no plastic flow. High-velocity crack propagation from preexisting flaws results in sudden and catastrophic failure. If material behavior is clearly brittle, and geometry and loading are simple, as for the axially loaded bar of Figure 2.1, failure by brittle fracture may be predicted when the stress σ, from (2-1), exceeds the material's ultimate strength

S_u. Hence failure is predicted to occur if (FIPTOI)

$$\sigma \geq S_u \tag{2-19}$$

Thus for a uniaxial state of stress the failure prediction expression (2-19) for brittle fracture is formally the same as (2-18) for ductile rupture. If the loading is more complicated, and a multiaxial state of stress is produced by the loads, fracture may be predicted with reasonable accuracy through the use of the maximum normal stress theory of failure, as discussed in 4.6.

On the other hand, it has now been well established that *nominally ductile* materials may also fail by a brittle fracture response in the presence of cracks or flaws if the combination of crack size, geometry of the part, temperature, and/or loading rate lies within certain critical ranges. The prediction of brittle fracture in these circumstances has been based on the assumptions that the stress at a crack tip, where failure is initiated, may be calculated as if the material behavior is linear elastic and the state of stress is two-dimensional; thus the procedure is often referred to as *linear elastic fracture mechanics (LEFM).*

Three basic types of stress fields have been defined for crack-tip stress analysis, each one associated with a distinct mode of crack deformation, as illustrated in Figure 2.4. The crack opening mode, Mode I, is associated with local displacement in which the crack surfaces move directly apart. Modes II and III are forward sliding and tearing displacements respectively. Mathematical expressions have been developed for the *intensity* and *distribution* of stress near the crack tips for each of the three modes shown in Figure 2.4.[6] For failure prediction purposes, the crack-tip *stress-intensity factors*, developed from these mathematical expressions, provide a good measure of the seriousness of loading and geometry in any particular case. In general, the expressions for stress intensity factor, K, are of the form

$$K = C\sigma\sqrt{\pi a} \tag{2-20}$$

where σ is gross-section nominal stress, a is a crack-length parameter, and C is dependent upon the type of loading and the geometry away from the crack. Many values of C have been published,[7] and several typical charts for selecting proper C values for through-the-thickness cracks are given in Figures 2.5 through 2.9. Figure 2.10 gives a chart for *surface flaw shape parameter* Q to be used in finding stress-intensity factors for part-through thumbnail-shaped surface cracks. The stress-intensity factor, K, calculated from (2-20), is a single-parameter measure of the seriousness of the stress field around the crack tip. The magnitude of K associated with the onset of rapid crack extension (initiation of brittle fracture) has been designated as *critical stress intensity*, K_c. Thus failure by brittle fracture may be predicted to occur for through-the-thickness cracks if (FIPTOI)

$$K = C\sigma\sqrt{\pi a} \geq K_c \tag{2-21}$$

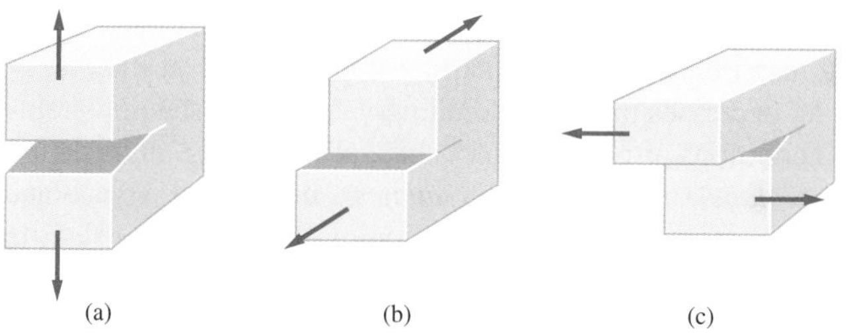

Figure 2.4
Basic modes of crack displacement. (a) Mode I, (b) Mode II, (c) Mode III. (To be read "Mode-one," "Mode-two," and "Mode-three.")

[6]See, for example, the Westergaard equations, ref. 1, pp. 54–55.

[7]See, for example, ref. 3.

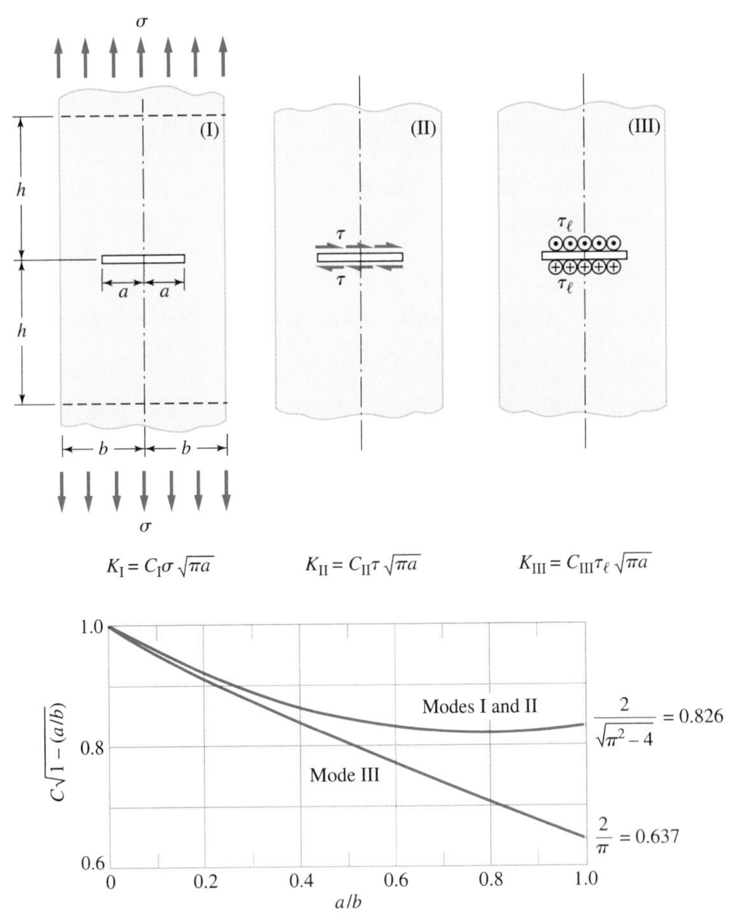

Figure 2.5
Stress-intensity factors K_I, K_{II} and K_{III}, for center-cracked test specimen. (Source: ref. 3, Del Research Corp.)

or, for thumbnail-shaped surface cracks if

$$K = \frac{1.12}{\sqrt{Q}}\sigma\sqrt{\pi a} \geq K_c \tag{2-22}$$

For a given cracked plate, for example the case shown in Figure 2.7, the stress-intensity factor K increases proportionally with gross nominal stress σ, and also is a function of instantaneous crack length a. For the infinite plate shown in Figure 2.11, the crack is oriented so that its plane is perpendicular to the direction of applied uniaxial stress, σ, hence pure Mode 1 loading exists. The crack front extends uniformly all the way through the plate, so the crack is a through-the-thickness crack. The usual coordinate system for defining state of stress at the crack tip is shown. In studying material behavior it has been found that for a given material, depending upon the state of stress at the crack tip, the critical stress intensity K_c decreases to a lower limiting value as the state of strain at the crack tip approaches the conditions of plane strain.[8] This lower limiting value defines a basic material property, the *plane strain fracture toughness*, designated[9] K_{Ic}. Standard test methods have been established for the determination of K_{Ic} values.[10] A few data are given in

[8]For thick plates the surrounding material constrains the crack-tip zone to near-zero strain in the thickness direction, resulting in plane (biaxial) strain. See also 4.4 and 4.5.

[9]To be read "K-one-c."

[10]See ref. 4.

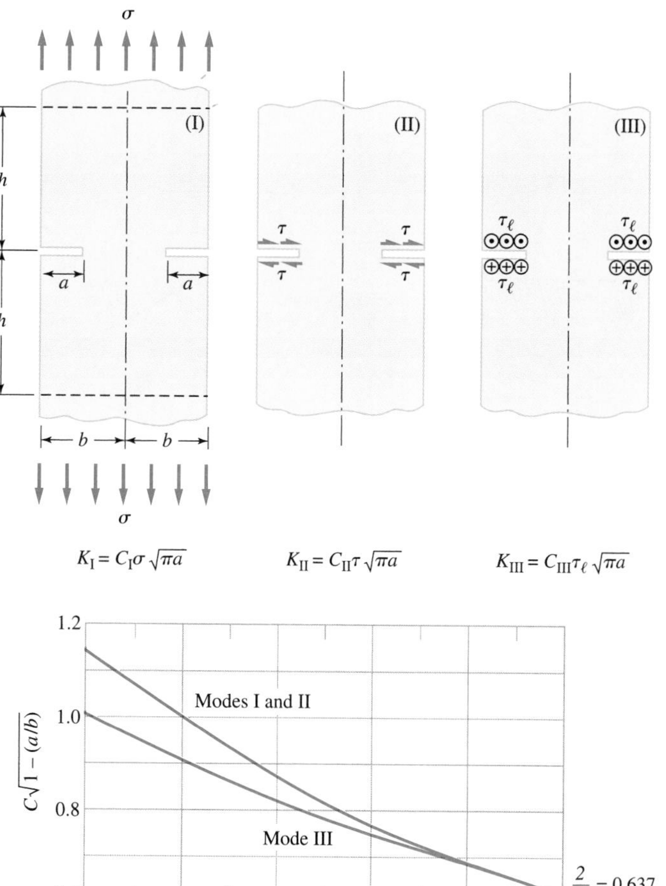

Figure 2.6
Stress-intensity factors K_I, K_{II} and K_{III}, for double-edge notch test specimen. (Source: ref. 3, Del Research Corp.)

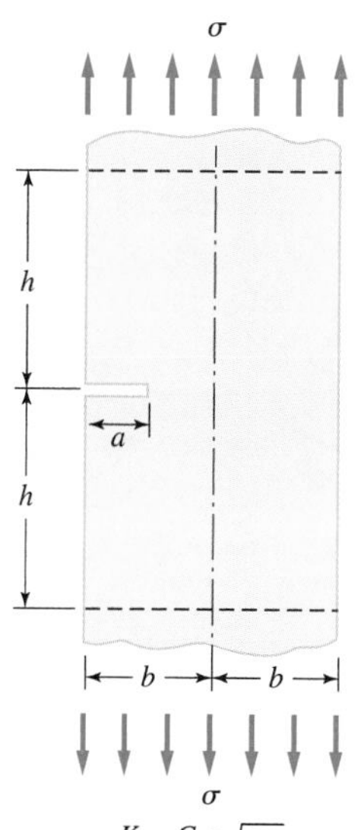

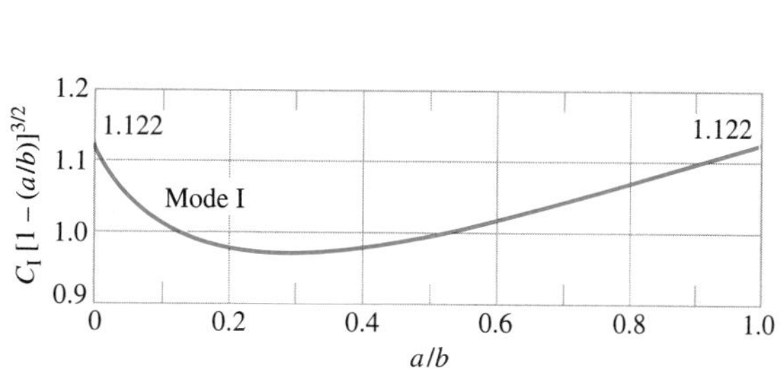

Figure 2.7
Stress-intensity factors K_I, for single-edge notch test specimen. (Source: ref. 3, Del Research Corp.)

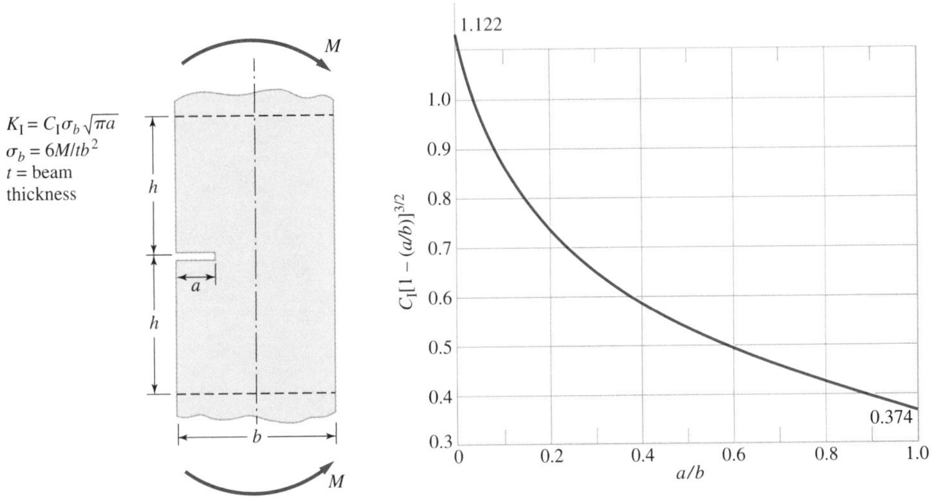

Figure 2.8
Stress-intensity factors K_I, for single through-the-thickness edge crack under pure bending moment. (Source: ref. 3, Del Research Corp.)

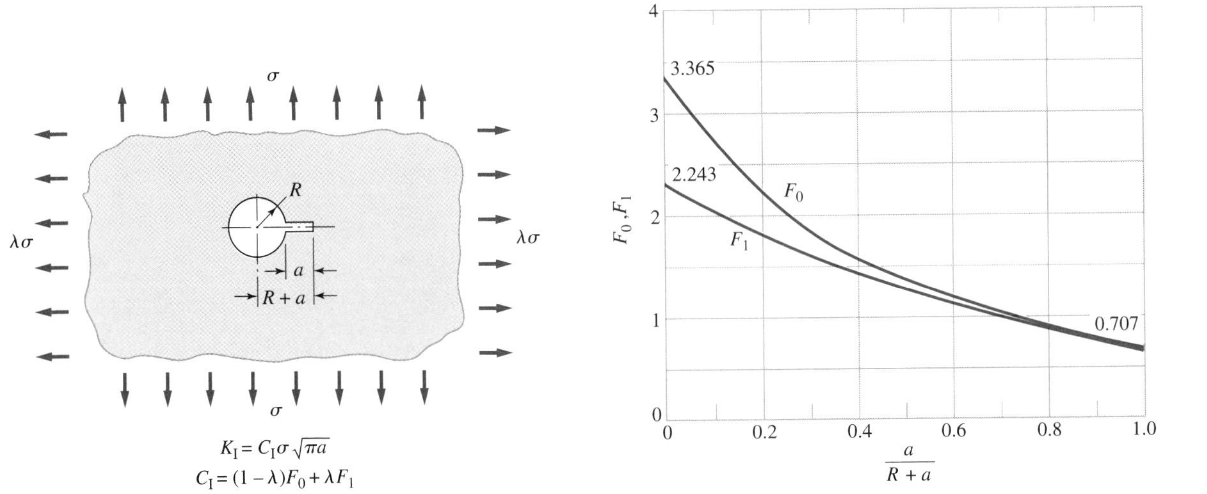

Figure 2.9
Stress-intensity factors K_I, for a through-the-thickness crack emanating from a circular hole in an infinite plate under biaxial tension. (Source: ref. 3, Del Research Corp.)

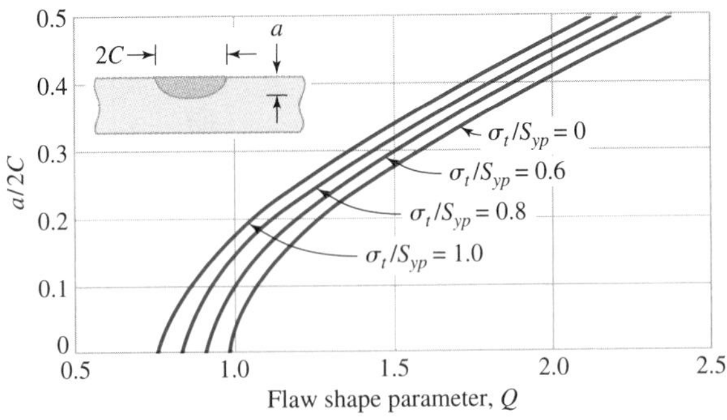

Figure 2.10
Surface flaw shape parameter. (From ref. 27; adapted by permission of Pearson Education, Inc., Upper Saddle River, N.J.)

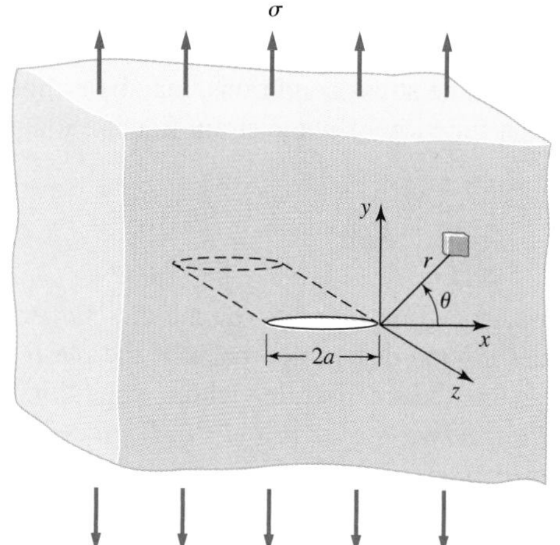

Figure 2.11
Coordinate system for infinite plate containing a through-the-thickness crack of length 2*a*.

Table 2.1. For the plane strain fracture toughness K_{Ic} to be a valid failure prediction criterion for a machine part, *plane strain* conditions[11] must exist at the crack tip; that is, the material must be *thick* enough to ensure plane strain conditions. It has been estimated empirically that for plane strain conditions the minimum material thickness B must be

$$B \geq 2.5\left(\frac{K_{Ic}}{S_{yp}}\right)^2 \tag{2-23}$$

TABLE 2.1 Yield Strength and Plane Strain Fracture Toughness Data for Selected Engineering Alloys[1]

Alloy	Form	Test Temperature °F	Test Temperature °C	S_{yp} ksi	S_{yp} MPa	K_{Ic} ksi$\sqrt{\text{in}}$	K_{Ic} MPa$\sqrt{\text{m}}$
AISI 1045 steel	Plate	25	−4	39	269	46	50
AISI 1045 steel	Plate	0	−18	40	276	46	50
4340 steel (500°F temper)	Plate	70	21	217–238	1495–1640	45–57	50–63
4340 steel (800°F temper)	Forged	70	21	197–211	1360–1455	72–83	79–91
D6AC steel (1000°F temper)	Plate	70	21	217	1495	93	102
D6AC steel (1000°F temper)	Plate	−65	−54	228	1570	56	62
18 Ni maraging steel (300)	Plate	600	316	236	1627	80	87
18 Ni maraging steel (300)	Plate	70	21	280	1931	68	74
18 Ni maraging steel (300)	Plate	−100	−73	305	2103	42	46
A 538 steel	—	—	—	250	1722	100	111
2014-T6 aluminum	Forged	75	24	64	440	28	31
2024-T351 aluminum	Plate	80	27	54–56	370–385	28–40	31–44
6061-T651 aluminum	Plate	70	21	43	296	26	28
6061-T651 aluminum	Plate	−112	−80	45	310	30	33
7075-T6 aluminum	—	—	—	75	517	26	28
7075-T651 aluminum	Plate	70	21	75–81	515–560	25–28	27–31
7075-T7351 aluminum	Plate	70	21	58–66	400–455	28–32	31–35
Ti-6Al-4V titanium	Plate	74	23	119	820	96	106

[1]From refs. 5–7.

[11]*Plane strain (biaxial state of strain)* occurs when nonzero strain components exist in only two coordinate directions. See also 4.4 and 4.5.

If the material is *not* thick enough to meet the criterion of (2-23), *plane stress* better characterizes the state of stress at the crack tip, and K_c, the critical stress-intensity factor for failure prediction under plane stress conditions, may be estimated using a semiempirical relationship for K_c as a function of plane strain fracture toughness K_{Ic} and thickness B.[12] This relationship is

$$K_c = K_{Ic}\left[1 + \frac{1.4}{B^2}\left(\frac{K_{Ic}}{S_{yp}}\right)^4\right]^{1/2} \tag{2-24}$$

As long as the crack-tip plastic zone is in the regime of *small-scale yielding*,[13] this estimation procedure provides a good design approach. If the plastic zone size ahead of the crack tip becomes so large that the small-scale yielding condition is no longer satisfied, an appropriate *elastic-plastic fracture mechanics (EPFM)* procedure would give better results. For example, a *failure assessment diagram*[14] might be utilized; however such EPFM procedures are beyond the scope of this text.

To utilize (2-22) as a design or failure prediction tool, the stress-intensity factor K must be determined for the particular loading and geometry of the part or structure under consideration. The critical stress intensity is set equal to K_{Ic} if the minimum thickness criterion (2-23) is met, otherwise K_c is estimated from (2-24). If failure is predicted to occur using (2-22), redesign becomes necessary.

Example 2.5 Brittle Fracture

Two "identical" support straps of forged 2014-T6 aluminum, shown in Figure E2.5, have been inspected and found to contain through-the-thickness cracks. While the total crack length of 0.20 inch is exactly the same for both straps, one strap (case A) involves two edge cracks, each with 0.10-inch length, opposite each other, while the other strap (case B) involves a single crack of 0.20-inch length at the center. The straps are of rectangular cross section 2.00 inches wide and 0.50 inch thick, and the straps are 26 inches in length. The straps are loaded by a direct tensile force P in the 26-inch direction, as shown. The required axial load for this application is $P = 50{,}000$ lb.

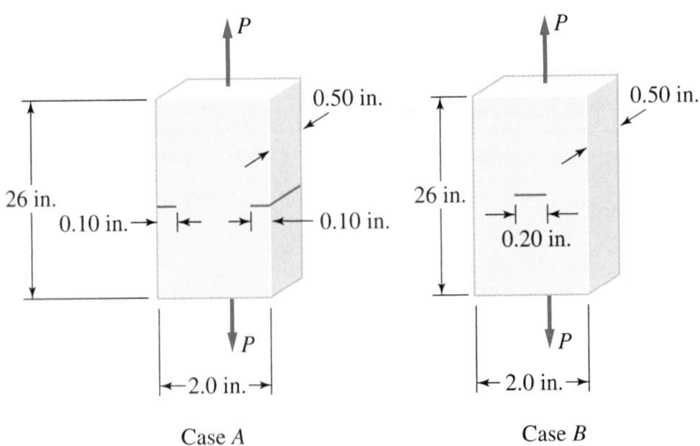

Figure E2.5
Cracked support straps.

[12]See refs. 8 and 9.

[13]*Small-scale yielding* means that the crack-tip plastic zone size is small compared to the dimensions of the crack.

[14]See ref. 1, pp. 70–76.

a. For each case, what is the governing failure mode?

b. For each case, is failure predicted to occur?

c. Is center cracking or edge cracking the more serious configuration?

d. How would the anlaysis change if the strap thickness for each case had been only 0.44 inch instead of 0.50 inch?

Solution

a. Both yielding and brittle fracture should be checked for each case. Property values from Table 2.1 are

$$S_{yp} = 64{,}000 \text{ psi}$$
$$K_{Ic} = 28 \text{ ksi}\sqrt{\text{in}}$$

b. For each case, failure by both yielding and brittle fracture must be checked.

For case A (edge cracks), to check for yielding failure, from (2-17) FIPTOI

$$\sigma \geq S_{yp} \tag{1}$$

Now

$$\sigma = \frac{P}{A_{net}} = \frac{50{,}000}{(2 - 0.2)(0.50)} = \frac{50{,}000}{0.90} = 55{,}560 \text{ psi} \tag{2}$$

Hence, FIPTOI

$$55{,}560 \geq 64{,}000 \tag{3}$$

Therefore, failure by yielding is *not* predicted for case A.

For case B (center crack), the calculation for yielding failure is identical to the case A calculation, since net cross-sectional area is the same in both cases. Therefore, failure by yielding is *not* predicted for case B either.

For case A (edge cracks), to check for brittle fracture, from (2-22) FIPTOI

$$C_A \sigma \sqrt{\pi a} \geq K_c \tag{4}$$

Reading from Figure 2.6, using the Mode I curve, with $a = 0.1$, $b = 1.0$, and $a/b = 0.1$,

$$C_A \sqrt{1 - (0.1)} = 1.07 \tag{5}$$

hence

$$C_A = 1.13 \tag{6}$$

The gross-section stress is

$$\sigma = \frac{P}{A_g} = \frac{50{,}000}{(2.0)(0.5)} = \frac{50{,}000}{1.0} = 50{,}000 \text{ psi} \tag{7}$$

Checking for plane strain, from (2-23) plane strain exists if

$$B \geq 2.5\left(\frac{K_{Ic}}{S_{yp}}\right)^2 = 2.5\left(\frac{28}{64}\right)^2 \tag{8}$$

or if

$$B \geq 0.48 \tag{9}$$

Since $B = 0.5$, (9) is satisfied and plane strain conditions prevail. Therefore

$$K_c = K_{Ic} = 28 \text{ ksi}\sqrt{\text{in}} \tag{10}$$

Example 2.5 Continues

and (4) may be evaluated as FIPTOI

$$(1.13)(50{,}000)\sqrt{\pi(0.1)} \geq 28{,}000 \tag{11}$$

or if

$$31{,}668 \geq 28{,}000 \tag{12}$$

Hence, failure by brittle fracture *is* predicted for case A.

For case B (center crack), to check for brittle fracture, again using (2-22) FIPTOI

$$C_B \sigma \sqrt{\pi a} \geq K_c \tag{13}$$

Reading from Figure 2.5, using the Mode I curve, with $a = 0.1$, $b = 1.0$, and $a/b = 0.1$,

$$C_B \sqrt{1 - (0.1)} = 0.96 \tag{14}$$

hence

$$C_B = 1.01 \tag{15}$$

The gross-section stress is the same as for case A, thus

$$\sigma = 50{,}000 \text{ psi} \tag{16}$$

Since material properties are the same as for case A, plane strain conditions prevail, and from (10)

$$K_c = K_{Ic} = 28 \text{ ksi}\sqrt{\text{in}} \tag{17}$$

Now (15) may be evaluated as FIPTOI

$$(1.01)(50{,}000)\sqrt{\pi(0.1)} \geq 28{,}000 \tag{18}$$

or if

$$28{,}300 \geq 28{,}000 \tag{19}$$

Hence failure by brittle fracture *is* also predicted for case B, but only by a slight margin.

c. Since the stress-intensity factor K for case A ($31{,}668 \text{ psi}\sqrt{\text{in}}$) is larger than for case B ($28{,}300 \text{ psi}\sqrt{\text{in}}$), the edge-crack case (case A) is more serious.

d. If strap thickness is reduced from 0.50 to 0.44 inch, nominal stress levels will increase slightly, and from (9), the plane strain criterion is no longer met. Therefore new calculations should be made utilizing (2-24).

To check yielding for both case A and case B, using (2)

$$\sigma = \frac{50{,}000}{(2 - 0.2)(0.44)} = 63{,}130 \text{ psi} \tag{20}$$

and using (3)

$$63{,}130 \geq 64{,}000 \tag{21}$$

Therefore failure by yielding *is not* predicted. However, it is close.

To check brittle fracture, since only the thickness has changed, the values of $C_A = 1.13$ and $C_B = 1.01$ remain unchanged.

The gross-section stress for both cases becomes, using (7),

$$\sigma = \frac{50{,}000}{2.0(0.44)} = 56{,}820 \text{ psi} \tag{22}$$

Since plane strain conditions are not satisfied, from (2-24)

$$K_c = K_{Ic}\left[1 + \frac{1.4}{B^2}\left(\frac{K_{Ic}}{S_{yp}}\right)^4\right]^{1/2} = 28.0\left[1 + \frac{1.4}{(0.44)^2}\left(\frac{28}{64}\right)^4\right]^{1/2} = 31.49 \text{ ksi}\sqrt{\text{in}} \qquad (23)$$

Then for case A, using (4), FIPTOI

$$(1.13)(56{,}280)\sqrt{\pi(0.1)} \geq 31{,}490 \qquad (24)$$

or if

$$35{,}680 \geq 31{,}490 \qquad (25)$$

and failure by brittle fracture *is* again predicted for case A.

Similarly, for case B, FIPTOI

$$(1.01)(56{,}820)\sqrt{\pi(0.1)} \geq 31{,}490 \qquad (26)$$

or if

$$32{,}166 \geq 31{,}490 \qquad (27)$$

So brittle fracture *is* again predicted for case B.

If loading conditions are more complicated, fracture mechanics procedures are available but much more involved. For example, in cases where the plane of a through-the-thickness crack is oriented at some other angle than perpendicular to the direction of σ, the applied stress field induces a combination of Mode I and Mode II loading on the crack. Mixed-mode fracture behavior presents an analytical challenge, and a designer would usually consult with a fracture mechanics specialist to evaluate such a problem. Likewise, a specialist would often be involved in addressing problems associated with high-rate loading, cyclic loading, or cases where LEFM conditions are not valid.

2.6 Fluctuating Loads, Cumulative Damage, and Fatigue Life

In modern engineering practice, repeated loads, fluctuating loads, and rapidly applied loads are far more common than static or quasistatic loads. By far, the majority of engineering design environments involve machine parts subjected to fluctuating or cyclic loads. Such loads induce fluctuating or cyclic stresses that often result in failure by *fatigue*. Fatigue is a progressive failure process that involves the *initiation* and *propagation* of cracks until one reaches an unstable size, triggering a sudden catastrophic separation of the affected part into two or more pieces. It is difficult to detect the progressive changes in material properties that occur during fatigue stressing, and fatigue failure may therefore occur with little or no warning. Periods of rest, with the fatigue stress removed, do not lead to any significant healing or recovery from the effects of the prior cyclic stressing. Hence, the damage done during the fatigue process is *cumulative*. Fatigue failures have been recognized for over 150 years, but only with the advent of high-speed high-performance machinery, and the development of the aerospace industry, has widespread attention been directed toward trying to understand the fatigue process.

In recent years it has been recognized that the fatigue failure process involves three phases. A *crack initiation* phase occurs first, followed by a *crack propagation* phase; finally, when the crack reaches a critical size, the terminal phase of *unstable rapid crack growth to fracture* completes the failure process. Traditionally, the models for analysis and prediction of fatigue failure have lumped all three phases together in the *stress-life (S-N) approach.*

Numerous analytical/empirical procedures and a large database have been developed to support the *S-N* approach. More recently, the separate modeling of each phase has been under intense development, and prediction models have now been set forth for each phase separately. This methodology may be referred to as the *fracture-mechanics (F-M) approach.*

Additionally, two domains of cyclic loading have been identified on the basis of whether the induced *cyclic strains* are predominantly *elastic* or predominantly *plastic*. When cyclic loads are relatively low, strain cycles are confined largely to the elastic range, and long lives or high numbers of cycles to failure are exhibited, the domain is called *high-cycle fatigue*. When cyclic loads are relatively high, significant levels of plastic strain are induced during each cycle, and short lives or low numbers of cycles to failure are exhibited, the domain is called *low-cycle fatigue*, or *cyclic strain-controlled fatigue*. Occasionally, a machine part may be subjected to intermixed loadings from both domains. High-cycle fatigue predominates in most design environments; therefore high-cycle fatigue analysis will be emphasized in this text. Low-cycle fatigue analysis is widely discussed in the literature.[15]

Fluctuating Loads and Stresses

Fluctuating loads and *loading spectra*, producing associated *stress spectra* in a machine part, reflect the design configuration and operational use of the machine. Perhaps the simplest fatigue stress spectrum to which a machine element may be subjected is a zero-mean sinusoidal stress-time pattern of constant amplitude and fixed frequency, applied for a specified number of cycles. Such a stress-time pattern, often referred to as a *completely reversed* or *zero-mean* cyclic stress, is illustrated in Figure 2.12(a). Using the sketches of Figure 2.12, we may define several useful terms and symbols; these include

$$\sigma_{max} = \text{maximum stress in the cycle}$$

$$\sigma_{min} = \text{minimum stress in the cycle}$$

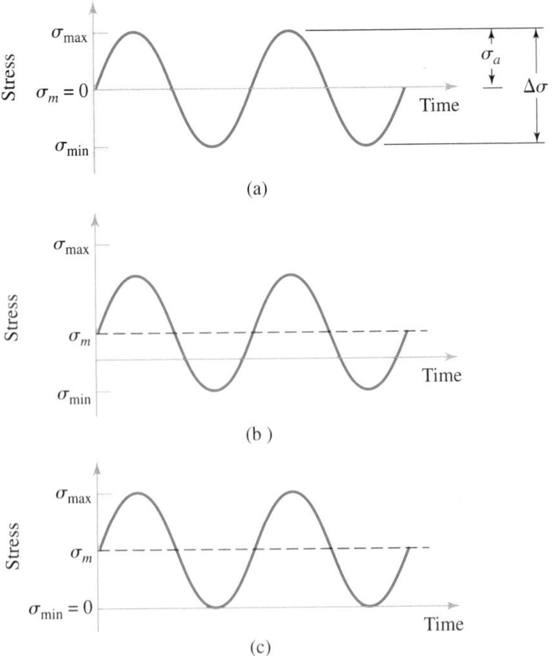

Figure 2.12
Several constant-amplitude stress-time patterns of interest. (a) Completely reversed; *R* = –1. (b) Nonzero-mean stress. (c) Released tension; *R* = 0.

[15]For example, see refs. 1 or 10.

$$\sigma_m = \text{mean cyclic stress} = \frac{\sigma_{max} + \sigma_{min}}{2}$$

$$\sigma_a = \text{alternating stress amplitude} = \frac{\sigma_{max} - \sigma_{min}}{2}$$

$$\Delta\sigma = \text{range of stress} = \sigma_{max} - \sigma_{min}$$

$$R = \text{stress ratio} = \frac{\sigma_{min}}{\sigma_{max}}$$

$$A = \text{amplitude ratio} = \frac{\sigma_a}{\sigma_m}$$

Any two of the quantities just defined, except the combination σ_a and $\Delta\sigma$ or the combination A and R, are sufficient to completely describe the stress-time pattern.

A second type of stress-time pattern often encountered is the *nonzero-mean* spectrum shown in Figure 2.12(b). This pattern is very similar to the completely reversed case except that the mean stress is either tensile or compressive, in any event different from zero. The nonzero-mean case may be thought of as a static stress equal in magnitude to the mean σ_m with a superposed completely reversed cyclic stress of amplitude σ_a.

A special case of nonzero-mean stress, illustrated in Figure 2.12(c), is often encountered in practice. In this special case the minimum stress ranges from zero up to some tensile maximum and then back to zero. This type of stressing is often called *released tension*. For released tension it may be noted that $\sigma_m = \sigma_{max}/2$. A similar but less frequently encountered stress-time pattern is called released compression, where $\sigma_{max} = 0$ and $\sigma_m = \sigma_{min}/2$.

More complicated stress-time patterns are illustrated in Figure 2.13. In Figure 2.13(a) the mean stress is zero but there are two (or more) different stress amplitudes mixed together. In Figure 2.13(b), not only does the stress amplitude vary, but also the magnitude of the mean stress periodically changes, approaching a more realistic condition. Figure 2.14 illustrates a realistic stress-time pattern, as might be observed, for example, in an airframe

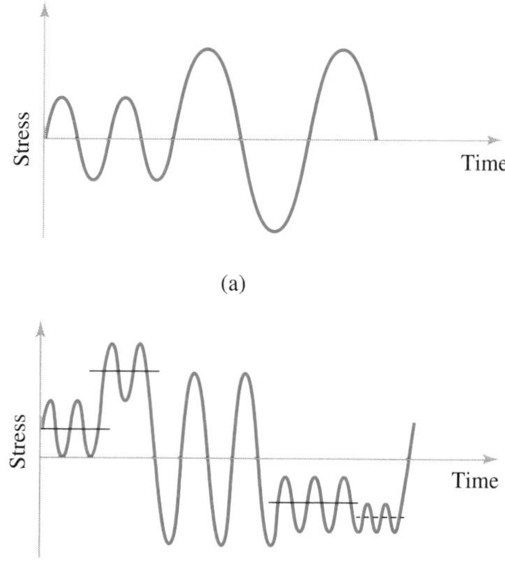

Figure 2.13
Stress-time patterns in which the amplitude changes or both mean and amplitude change to produce a more complicated stress spectrum. (a) Zero mean, changing amplitude. (b) Changing mean and amplitude.

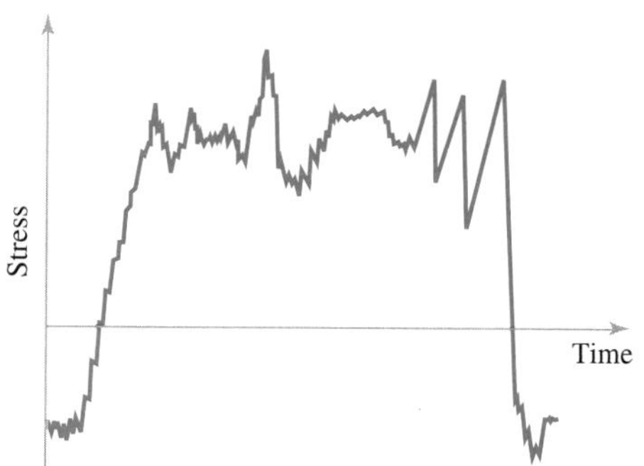

Figure 2.14
A quasi-random stress-time pattern that might be typical of an operational aircraft during any given mission.

structural member during a typical mission including refueling, taxi, takeoff, gusts, maneuvers, and landing. To effectively predict and prevent fatigue failures, procedures for assessing damage due to all of these various stress-time spectra are required.

Fatigue Strength and Fatigue Limit

Designing machine parts or structures that are subjected to fatigue loading is usually based on the results of laboratory fatigue tests using small polished specimens of the material of interest. To be successful in using small-specimen data, a designer must be aware of the merits and limitations of such data, including influences of size, surface finish, geometry, environment, speed, and many other factors. Basic fatigue data in the high-cycle life range are usually displayed on a plot of cyclic stress versus life. These plots, called *S-N curves*, constitute design information of fundamental importance for machine parts subjected to cyclic or repeated loadng.

Figure 2.15 illustrates the characteristic appearance of an *S-N* curve for a ferrous material. The curve is constructed by fitting it to data collected from laboratory tests of a large number of specimens at various cyclic stress amplitudes and zero-mean stress. Because of the *scatter* of fatigue life data at any given stress level, the construction of the most appropriate curve through the data becomes a design issue of importance. A statistical description of fatigue failure data is usually used to facilitate construction of the most appropriate *S-N* curve. Using this approach, one would construct for each test stress level a *histogram*, such as the one shown in Figure 2.16, which shows the *distribution* of failures as a function of the logarithms of lives for the sample tested. Computation of the sample *mean* and *standard deviation*[16] permits estimation of the *probability of failure (P)* at each stress level. Points of equal probability of failure may then be connected to obtain curves of constant probability of failure on the *S-N* plot. A family of such *S-N-P* curves is

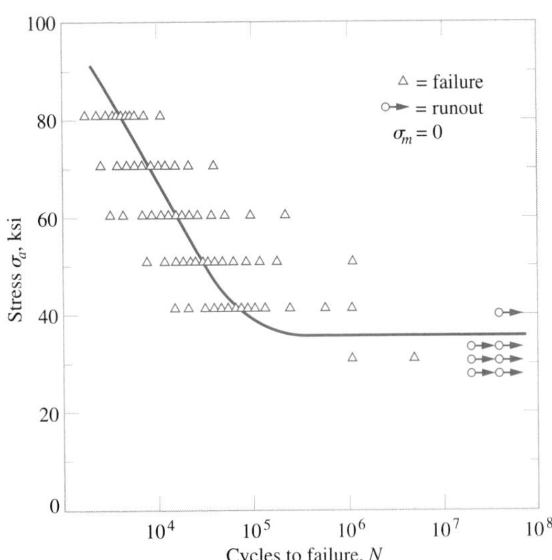

Figure 2.15
Plot of stress-life (*S-N*) data as they might be collected by laboratory fatigue testing of a new alloy.

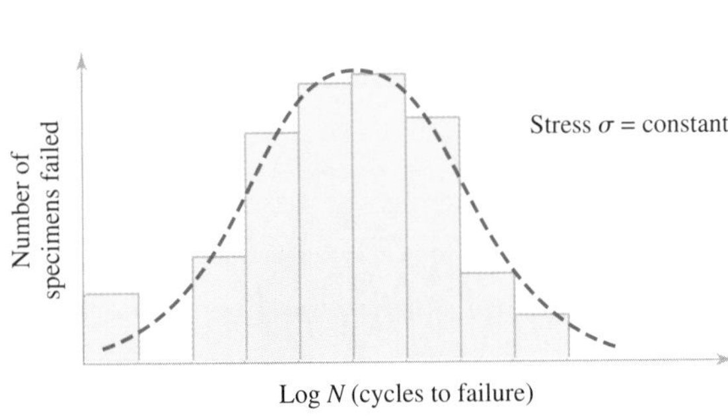

Figure 2.16
Distribution of fatigue specimen failures at a constant stress level as a function of logarithm of life.

[16] See, for example, ref. 1, Ch. 9.

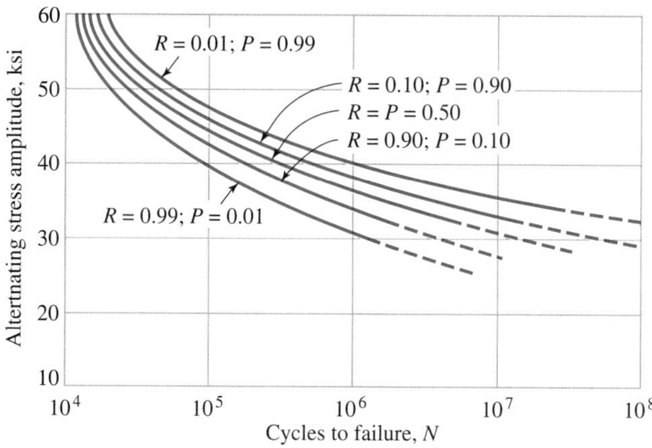

Figure 2.17
Family of *S-N-P* curves, or *R-S-N* curves for 7075-T6 aluminum alloy. Note: *P* = probability of failure; *R* = reliability = 1 – *P*. (Adapted from ref. 11, p. 117, Copyright © 1969, by permission from John Wiley & Sons, Inc.)

shown in Figure 2.17 for 7075-T6 aluminum alloy. It is also of interest that the *reliability (R)* is defined to be 1 minus the probability of failure; hence $R = (1 - P)$. Thus in Figure 2.17, the 10 percent probability of failure curve ($P = 0.10$) may alternatively be designated as the 90 percent reliability curve ($R = 0.90$), and some references may be found in the literature to these so-called *R-S-N* curves.

Usually, references in the literature to "the" *S-N* curve refer to the *mean* or 50 percent probability of failure curve unless otherwise specified. The mean *S-N* curves of Figure 2.18 distinguish two commonly observed types of response to cyclic loading. The ferrous alloys and titanium exhibit a steep branch in the shorter life range, leveling off to approach a stress asymptote at longer lives. This stress asymptote is called the *fatigue limit*, S_f or S'_f (formerly called endurance limit) and is the stress level below which an infinite number of cycles could theoretically be sustained without failure. The symbol S_f is commonly used to denote fatigue limit of an *actual machine part*, while S'_f is fatigue limit of the material based on *small polished specimens*. The nonferrous alloys do not exhibit an asymptote, and the curve of stress versus life continues to drop off indefinitely. For such alloys there is *no fatigue limit*, and failure as a result of cyclic stress is only a matter of applying enough cycles. To characterize the failure response of most nonferrous materials, and of ferrous or titanium alloys in the finite life range, the term *fatigue strength at a specified life*, S_N or S'_N, is used; S_N denotes finite-life fatigue strength of a *component*, while S'_N is finite-life fatigue strength of the material based on *small polished specimens*. The specification of *fatigue strength* without specifying the corresponding life is meaningless. The

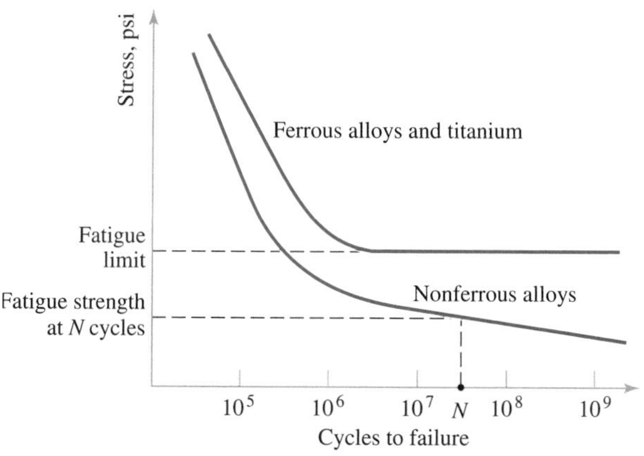

Figure 2.18
Two types of material response to cyclic loading.

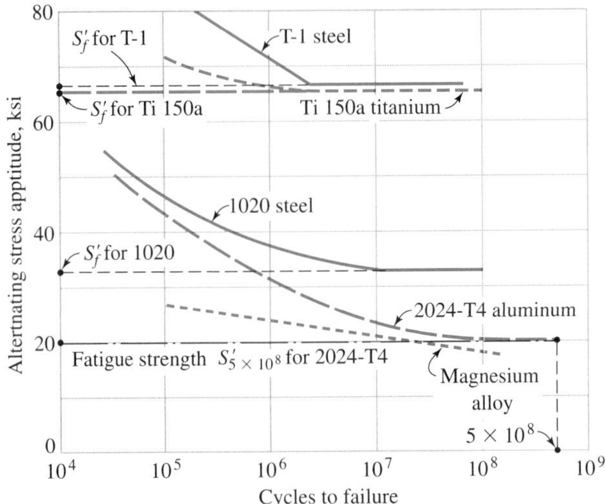

Figure 2.19
Effect of material composition on the *S-N* curve. Note that ferrous and titanium alloys exhibit well-defined fatigue limits whereas other alloy compositions do not. (Data from refs. 12 and 13.)

specification of a *fatigue limit* always implies infinite life. Use of these terms is illustrated in Figure 2.19, which depicts *S-N* curves for several alloys.

Estimating *S-N* Curves

When designing a machine part to be subjected to fluctuating loads it is essential to obtain an *S-N* curve for the candidate material for which heat treatment, operating temperature, and other operating conditions match the application. If such an *S-N* curve can be found in the literature or in an available database, the curve may be used directly to design the part. If not, it may be possible to find an *S-N* curve for small polished specimens of the candidate material and modify it by using a series of appropriate factors as discussed later in connection with Table 2.2. If *S-N* data cannot be found for the candidate material it becomes necessary either to perform laboratory fatigue tests on small polished specimens (a lengthy and expensive process) or to attempt to estimate the basic *S-N* curve for the material. Although there is no direct connection between static strength and fatigue strength, purely empirical relationships *have* been documented by examining large bodies of existing data. These empirical observations allow a designer to estimate *S-N* curves from static ultimate tensile strength (S_{ut}) values. These estimated *S-N* curves are often accurate enough for the design task at hand, especially at the preliminary design stage. For wrought-ferrous alloys, the *mean* *S-N* curve may be estimated for polished specimens by plotting on a semilog plot of strength versus log life the following:

1. Plot $S'_N = S_{ut}$ at $N = 1$ cycle.
2. Plot $S'_f = 0.5S_{ut}$ at $N = 10^6$ cycles if $S_{ut} \leq 200$ ksi

   or,

   Plot $S'_f = 100$ ksi at $N = 10^6$ cycles if $S_{ut} > 200$ ksi
3. Connect (1) with (2) by a straight line.
4. Construct a horizontal straight line from (2) toward very long lives.

This estimated mean (50 percent reliability) *S-N* curve may be supplemented if desired by using the standard normal variable *X* of Table 2.3 to calculate and construct estimated *S-N* curves with higher reliabilities.

TABLE 2.2 Strength-Influencing Factors That May Affect *S-N* Curves

Influencing Factor	Symbol	Approximate Range	"Typical" Value[1]
Material composition	—	—	Specific *S-N* data required
Heat treatment	—	—	Specific *S-N* data required
Operating temperature	—	—	Specific *S-N* data required
Grain size and direction	k_{gr}	0.4–1.0	1.0
Welding	k_{we}	0.3–0.9	0.8
Geometrical discontinuity	k_f	0.2–1.0	Reciprocal of K_f; see 4.8
Surface condition	k_{sr}	0.2–0.9	0.7
Size effect	k_{sz}	0.5–1.0	0.9
Residual surface stress	k_{rs}	0.5–2.5	Specific data required; see 4.11
Fretting	k_{fr}	0.1–0.9	0.35 if fretting exists, 1.0 if no fretting; also see 2.11
Corrosion	k_{cr}	0.1–1.0	Specific data required
Operating speed	k_{sp}	0.9–1.2	1.0
Strength reliability required	k_r	0.7–1.0	0.9; also see Table 2.3
Configuration of stress-time pattern	—	—	See later section titled *Cumulative Damage Concepts and Cycle Counting*
Nonzero-mean stress	—	—	See later section titled *Nonzero-Mean Stress*
Damage accumulation	—	—	See later section titled *Cumulative Damage Concepts and Cycle Counting*

[1]These "typical" values may be used for solving problems in this text or for making preliminary design estimates when actual conditions are poorly known. However, any *critical* design situation would require a literature/database search or supporting laboratory experiments to establish more accurate values.

For cast irons and cast steels a similar procedure may be used except that

$$S'_f = 0.4S_{ut} \quad \text{at} \quad N = 10^6 \text{ cycles} \qquad \text{if} \quad S_{ut} \leq 88 \text{ ksi}$$

or

$$S'_f = 40 \text{ ksi} \qquad \text{if} \quad S_{ut} > 88 \text{ ksi}$$

Guidelines for other alloys have been published as follows:[17]

$$\text{Titanium alloys: } S'_f = 0.45S_{ut} \text{ to } 0.65S_{ut} \quad \text{at} \quad N = 10^6 \text{ cycles}$$

$$\text{Aluminum alloys: } S'_N = 0.4S_{ut} \quad \text{at} \quad N = 5 \times 10^8 \text{ cycles}$$

TABLE 2.3 Strength Reliability Factors as a Function of Reliability Level

Reliability R (percent)	Corresponding Standard Normal Variable X (see Table 5.1)	Strength Reliability Factor k_r
90	1.282	0.90
95	1.645	0.87
99	2.326	0.81
99.9	3.090	0.75
99.995	3.891	0.69

[17]See ref. 15.

$$\text{Magnesium alloys: } S'_N = 0.35S_{ut} \qquad \text{at} \quad N = 10^8 \text{ cycles}$$
$$\text{Copper alloys: } S'_N = 0.2S_{ut} \quad \text{to} \quad 0.5S_{ut} \qquad \text{at} \quad N = 10^8 \text{ cycles}$$
$$\text{Nickel alloys: } S'_N = 0.3S_{ut} \quad \text{to} \quad 0.5S_{ut} \qquad \text{at} \quad N = 10^8 \text{ cycles}$$

It is very important to emphasize that these guidelines provide only *estimates*, to be used when pertinent *S-N* data are unavailable, and must be used with caution.

Stress-Life (*S-N*) Approach to Fatigue

For uniaxial states of cyclic stress the traditional *S-N* approach to fatigue design is straightforward in concept. Data are required for the material of interest, including effects of size, surface finish, geometry, and environment, and information is required on the cyclic loading to be applied to the part. From these, the size and shape of the part to provide the desired cyclic life may be determined. To implement the *S-N* approach, material data are best presented as *S-N* curves, such as those shown in Figure 2.18 or 2.19, or Figure E2.6 of Example 2.6. For example, if a part to be made of 2024-T4 aluminum has a design life requirement of 5×10^8 completely reversed stress cycles, its fatigue strength S'_N at a life $N = 5 \times 10^8$ cycles may be read from Figure 2.19 as 20,000 psi. This assumes that the part operates under conditions matching the conditions used to obtain the data for Figure 2.19. Also, since $S_{N=5\times10^8} = 20{,}000$ psi is a fatigue *failure* strength, an appropriate safety factor would typically be imposed for safe operation, as discussed in Chapter 5.

Factors That May Affect *S-N* Curves

As already mentioned, published *S-N* curves, unless otherwise labeled, are typically *mean value* curves for *small polished specimens*. Actual machine parts subjected to cyclic or fluctuating stress levels exhibit *S-N* responses different from small, polished-specimen curves, depending upon differences in composition, processing, environment, and operational factors. Thus the fatigue strength or fatigue limit of an actual machine part (S_N or S_f) is nearly always different (usually lower) from the fatigue strength or fatigue limit (S'_N or S'_f) read from a published *S-N* curve for small polished specimens of the same material. Furthermore, published *S-N* curves for small polished specimens are somewhat dependent upon the test method used to obtain the underlying data. The test method of choice in modern practice is to use computer-controlled closed-loop axial push-pull fatigue testing machines that apply uniformly distributed cyclic direct stresses to small polished specimens. Most *S-N* data presented in this textbook may be assumed to have been produced by uniform uniaxial cyclic stresses on small polished specimens, unless otherwise noted. Older small specimen *S-N* data found in the literature may have been obtained from cyclic *bending* tests, using either rotating or reciprocating bending machines. If such bending data are used as the basis for obtaining fatigue strength or fatigue limit estimates for an actual machine part, as in (2-25) or (2-26), the fatigue limit values from the cycle bending data should be reduced by about 10 percent (because of the bending-stress gradients; see Figure 4.3) to be compatible with the uniformly stressed push-pull small specimen data used in this textbook.

The factors that may cause differences between the fatigue response of actual machine parts and small polished specimens are shown in Table 2.2 and briefly discussed in the following paragraphs. More detailed discussions may be found in the fatigue literature.[18]

To obtain an estimate for fatigue strength or fatigue limit of an actual machine part from small-polished-specimen *S-N* data and a knowledge of operational and environmental requirements for the part, the usual procedure is to utilize an expression of the form

[18]See for example, ref. 1, 12, 14, 15, 16, 17, 18, 19.

$$S_f = k_\infty S'_f \tag{2-25}$$

or

$$S_N = k_N S'_N \tag{2-26}$$

depending upon whether fatigue limit, or fatigue strength corresponding to a design life of N cycles, is required. From Table 2.2 the expressions for k_∞ and k_N may be written as the products of pertinent influencing factors. Thus

$$k_\infty = (k_{gr}k_{we}k_f k_{sr}k_{sz}k_{rs}k_{fr}k_{cr}k_{sp}k_r)_\infty \tag{2-27}$$

and

$$k_N = (k_{gr}k_{we}k_f k_{sr}k_{sz}k_{rs}k_{fr}k_{cr}k_{sp}k_r)_N \tag{2-28}$$

It should be recognized that the various factors may have different values for finite-life design requirements as compared to infinite-life design specifications. Also, it should be noted that these procedures assume that no interactions occur between or among the various strength-influencing factors, an assumption not always justified. For example, the surface condition effect and corrosion effect may interact, or other effects may interact to produce unknown influences on the fatigue strength or fatigue limit. For this reason it is always advisable in practice to subject the final design to full-scale prototype testing under simulated service conditions.

Approximate ranges and "typical" values for many of the strength-influencing factors are shown in Table 2.2. While these may be helpful guidelines in approximating the fatigue strength or fatigue limit of a machine part, more specific values are usually required for any critical design situation, and a search of fatigue databases or fatigue literature is common practice. The following brief remarks relative to the entries of Table 2.2 are offered as additional guidance.

Material composition is the most basic factor in determining fatigue strength. As noted in the discussion of Figure 2.18, materials divide themselves into two broad groups with respect to *S-N* failure response. The ferrous alloys and titanium exhibit a rather well-defined fatigue limit, which is well established by the time 10^7 cycles of stress have been applied. The other nonferrous alloys do not exhibit a fatigue limit at all, and their *S-N* curves continue to fall off at lives of 10^8, 10^9, and larger numbers of cycles.

Heat treatment is also a strong influencing factor on the fatigue strength, just as it is in the case of static strength. Small-polished-specimen *S-N* data for which the specimen heat treatment corresponds to that proposed for the actual part under consideration is very useful in determining fatigue response of the part.

Operating temperature may have a significant influence on fatigue strength. Generally, the fatigue strength is somewhat enhanced at temperatures below room temperature and diminished at temperatures above room temperature. In the range of temperature from around zero on the Fahrenheit scale to about one-half the absolute melting temperature, the effects of temperature are slight in most cases. At higher temperatures the fatigue strength diminishes significantly. Further, alloys that exhibit a fatigue limit at room temperature tend to lose this characteristic at elevated temperatures, making infinite-life design at elevated temperatures impossible.

Grain size and *grain direction* may play a significant role in fatigue strength. Fine-grained materials tend to exhibit fatigue properties that are superior to coarse-grained materials of the same composition. This superiority becomes less significant at elevated temperatures when the characteristic room temperature transgranular cracking gives way to intergranular cracking paths. When the cyclic loading direction is across the grain (transverse direction) of an anisotropic specimen or machine part, fatigue strength properties are inferior to fatigue strength when the loading direction is along the grain (longitudinal direction).

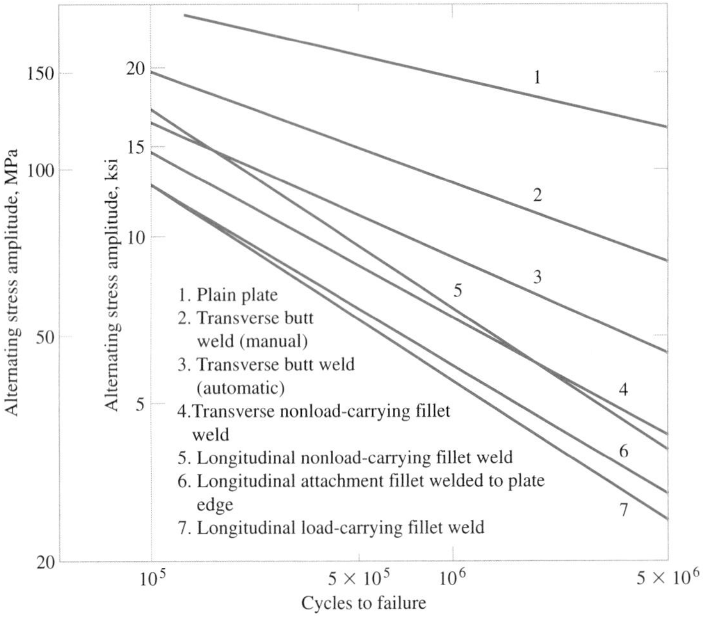

Figure 2.20
Effects of welding detail on the *S-N* curve of structural steels, with yield strengths in the range 30,000–52,000 psi. Tests were released tension ($\sigma_{min} = 0$) (Data from ref. 19.)

Welding produces a metallurgically nonhomogeneous region ranging from unheated parent metal through the heat-affected zoned (HAZ), to the weld metal zone. In some cases the entire welded joint may be post-heat-treated, in which case the structure of weld metal, HAZ, and parent metal may become nearly identical; in some critical cases postweld machining operations may be used to restore geometrical uniformity. Even with such care (and expense), welded joints (as well as bolted, riveted, or bonded joints) tend to have a fatigue strength inferior to that of a monolithic part of the same material. Factors contributing to fatigue strength reduction in welded joints, in addition to the gradient in homogeneity across the weld zone, include cracking in weld metal or base metal due to postcooling shrinkage stresses, incomplete penetration, lack of fusion between weld metal and parent metal from prior welding passes, undercut at the edge of weld metal deposit, overlap of weld metal flowing beyond the fusion zone, slag inclusions, porosity, misshapen welds, or welds with surface defects. Some of the effects of welding on fatigue strength properties are illustrated in Figure 2.20.

Stress concentration effects due to *geometrical discontinuities*, such as changes in shape or joint connections, may seriously diminish the fatigue strength of a machine part, even if the part is made of a ductile material. The severity of notches, holes, fillets, joints, and other stress raisers depends upon the relative dimensions, type of loading, and notch sensitivity of the material. A detailed discussion of stress concentration is presented in 4.8.

Surface condition is a very important factor in fatigue strength of a machine part since a very high proportion of all fatigue failures nucleate at the surface. Irregular surfaces and rough surfaces generally exhibit inferior fatigue properties as compared to smooth or polished surfaces, as illustrated in Figure 2.21. Cladding, plating, or coating may reduce the fatigue strength of the plated member because fatigue nuclei initiated in the plating continue to propagate into the base metal. This may cause a reduction in fatigue strength, as illustrated in Figure 2.22. Usually, however, the corrosion protection afforded by the plating more than offsets the strength loss due to plating.

Larger specimens and machine parts exhibit a *size effect*, having lower fatigue strengths than smaller specimens of the same material. For example, the fatigue strength of a 6-inch-diameter machine part might be as much 15 or 20 percent lower than a one-half-inch-diameter specimen of the same material.

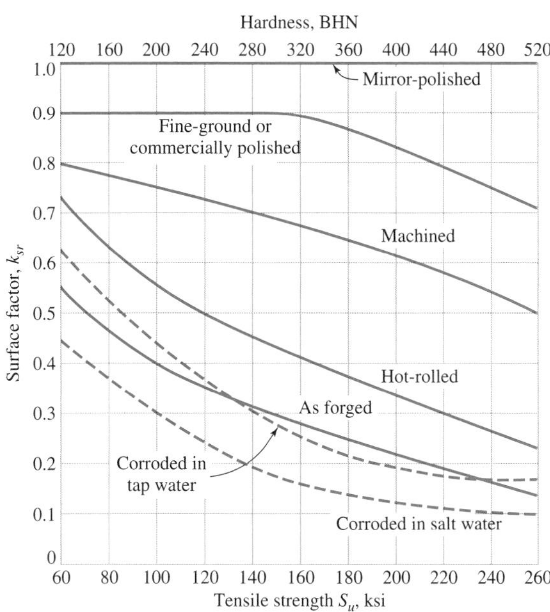

Figure 2.21
Reduction of fatigue strength due to surface finish (steel parts). (From ref. 15, reproduced with permission of The McGraw-Hill Companies.)

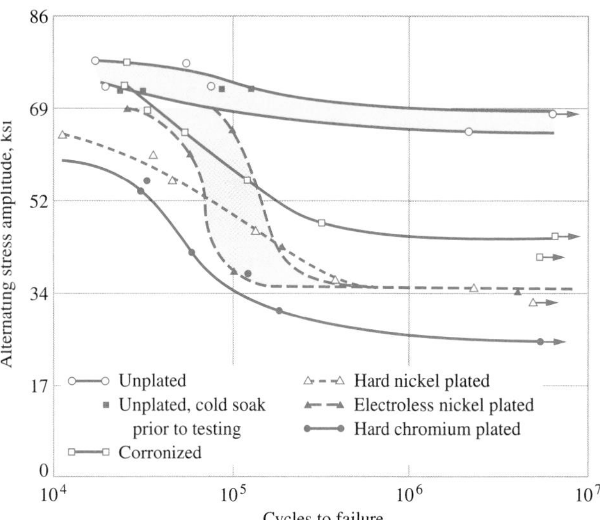

Figure 2.22
Effects of several electrodeposited coatings on the *S-N* curve of low-alloy steel at room temperature under axial tension-tension loading with $R = 0.02$. *Static ultimate strengths:* unplated, 172,300 psi; corronized, 177,700 psi; hard nickel plated, 176,100 psi; electroless nickel plated, 182,100 psi; hard aluminum plated, 162,400 psi. (Data from ref. 11, Copyright © 1969, adapted by permission from John Wiley & Sons, Inc.)

Residual stresses in the surface layer, whether induced intentionally or accidentally, may play an extremely important role in the overall fatigue response of a specimen or machine part. If the induced residual surface stresses are tensile, the fatigue strength is diminished. If the residual surface stresses are compressive, the fatigue strength is improved. Three common methods of inducing compressive residual surface stresses are shot-peening, cold-rolling, and presetting.[19] Figures 2.23 and 2.24 illustrate these effects. It is also notable that surface treatments such as shot-peening and cold-rolling not only improve the mean fatigue strength, but reduce the scatter (standard deviation) as well.

Fretting at the contacting surfaces of joints or connections may lead to a very significant reduction in fatigue strength of a machine part. As discussed in 2.11, the prediction of the fretting effect is difficult because many factors are involved. Experimental testing of the assembly under actual service conditions and loading should be undertaken for critical parts.

Corrosion tends to lower the fatigue strength, often by a large amount. Corrosive effects are specific to the combination of material composition and operating environment. Experimental testing of the machine part under service environment and loading should be employed for critical parts.

Operating speeds in the range from about 200 cycles per minute (cpm) to about 7,000 cpm appear to have little effect on fatigue strength. Below 200 cpm there is often a small decrease in fatigue strength, and in the range 7,000 cpm to around 60,000 cpm many ma-

[19]Shot-peening involves uniform bombardment of the surface with a high-velocity stream of small steel spheres. Cold-rolling is accomplished by pressing a hard contoured-roller against the surface to be treated and uniformly rolling and translating the roller to cover the area to be cold worked. Presetting involves the application of a static overload on the part, in the direction of operational loads, to induce local yielding at stress concentrations and subsequent compressive residual stresses at these sites upon release of the static overload. Also, see 4.13.

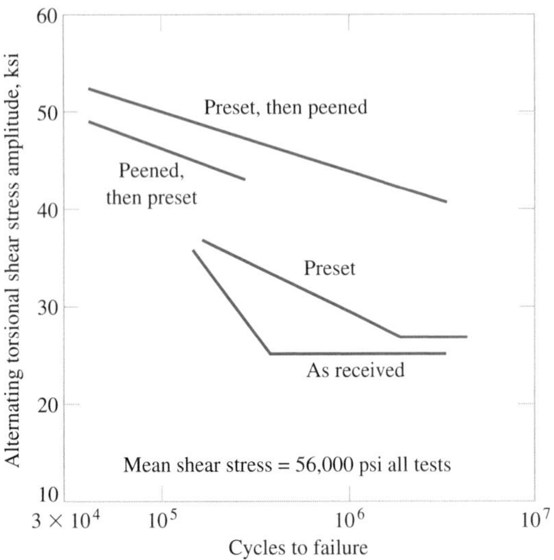

Figure 2.23
Effects of shot-peening and/or presetting on the *S-N* curve of hot wound helical coil springs made of 0.9 percent carbon steel. Spring details: hardness = Vickers DPH 550; wire diameter = $^1/_2$ inch; mean coil diameter = $2^5/_8$ inches; number of turns = 6; free length = $5^1/_{16}$ or 6 inches. (Data from ref. 21.)

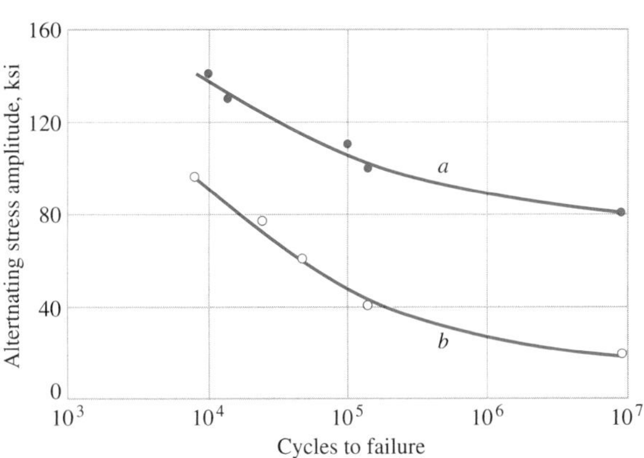

Figure 2.24
Effects of cold-rolling threads before and after heat treatment on the *S-N* curve for 220,000 psi ultimate strength bolts. (a) Rolled after heat treatment. (b) Rolled before heat treatment. (Data from ref. 11, Copyright © 1969, by permission from John Wiley & Sons, Inc.)

terials exhibit a significant increase in fatigue strength. Rest periods have no effect on fatigue strength.

Strength reliability, or reliability of fatigue strength, is usually based on the stresswise distribution of *S-N-P* data for small polished specimens of a particular material. Assuming the stresswise distribution of fatigue strengths to be *normal* is a reasonable and convenient procedure for estimating the strength reliability factor.[20] From any table for the cumulative distribution function[21] for the standard normal variable X, where

$$X = \left| \frac{S'_f(R) - S'_f}{\hat{\sigma}} \right| \tag{2-29}$$

and using the empirical estimate[22] for standard deviation of fatigue strength to be $\hat{\sigma} = 0.08 S'_f$, the strength reliability factor may be expressed as

$$k_r = 1 - 0.08X \tag{2-30}$$

Based on this relationship, values for k_r are shown in Table 2.3 for a range of strength reliability values, R. If statistical data were available for the distributions of other factors listed in Table 2.2, a similar reliability approach could be taken for these influencing factors, but such data are not generally available.

In practice, the *configuration of the stress-time pattern* may take many forms, including sinusoidal, reversed ramp, superposed ripples, or distorted peaks as shown in Figure 2.25. Generally, the fatigue failure response seems to be relatively insensitive to changes in wave shape as long as peak value and period are the same. When practical, however, it

[20]Although the Weibull distribution is widely agreed to be more accurate, its use is more cumbersome. See ref. 1, Ch. 9, for example.

[21]See, for example, Chapter 5, Table 5.1. [22]See ref. 21.

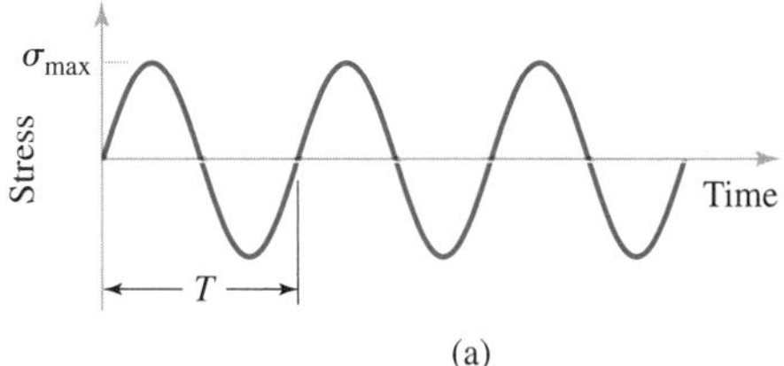

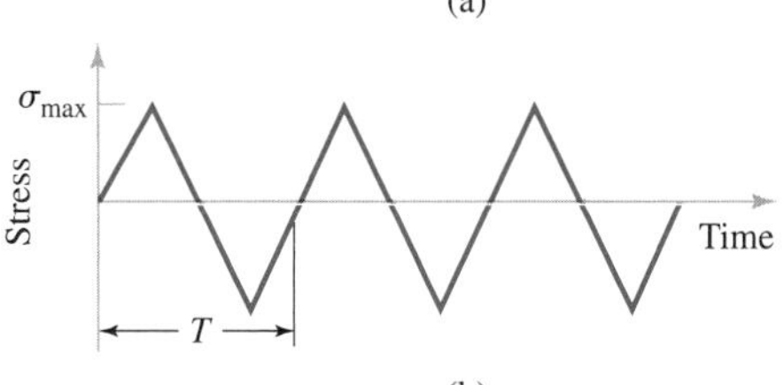

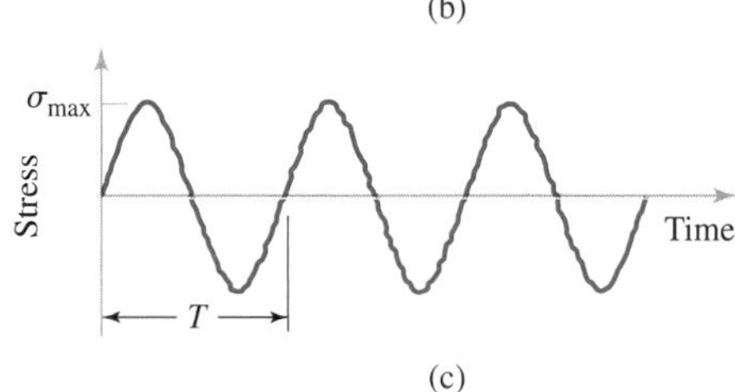

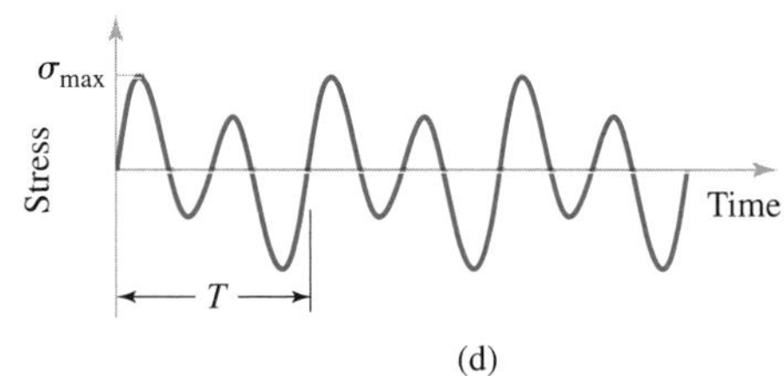

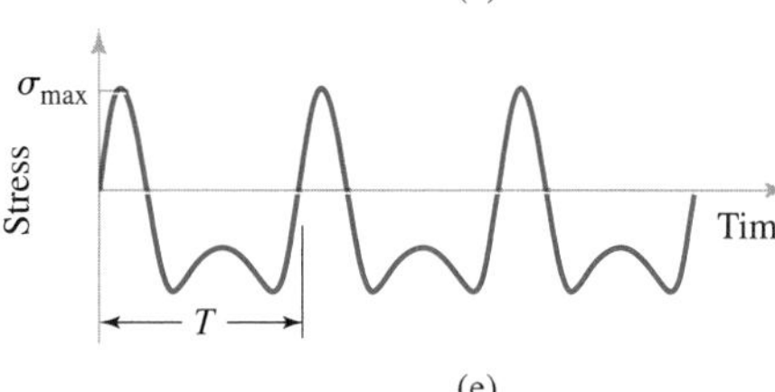

Figure 2.25
A variety of stress-time patterns used in evaluating fatigue behavior. (a) Completely reversed sinusoid. (b) Completely reversed ramp. (c) Superposed ripples. (d) Secondary peaks. (e) Distorted peaks.

is better to account for each and every stress reversal by using a good *cycle counting method* such as the rain flow method discussed in the section "Cumulative Damage Concepts and Cycle Counting" below.

The magnitude of any *nonzero-mean stress* has an important influence on fatigue response and is discussed in detail in the next section. Likewise, the accumulation of fatigue damage caused by cyclic loading, or *cumulative damage*, is fully discussed in a later section.

Example 2.6 Estimating Fatigue Properties

A wrought carbon-steel alloy is known to have the static properties S_u = 76,000 psi, S_{yp} = 42,000 psi, and e (2 inches) = 18 percent, but fatigue properties cannot be located for the material. It is necessary to quickly estimate the fatigue properties for the preliminary design of a machine part for which the fluctuating loads will induce a stress spectrum with cyclic amplitudes in both the finite-life range and the infinite-life range.

a. How could the basic small-polished-specimen "*mean S-N* curve" be estimated for this material?

b. How could the R = 99.9 percent reliability *S-N* curve be estimated for the material?

Solution

a. Using the estimation guidelines for wrought-ferrous alloys with $S_{ut} < 200$ ksi (since $S_{ut} = S_u = 76{,}000$ psi),

$$S'_f = 0.5(S_u) = 0.5(76{,}000) = 38{,}000 \text{ psi} \tag{1}$$

$$S'_{N=1} = S_u = 76{,}000 \text{ psi} \tag{2}$$

Plotting these values on semilog coordinates and connecting them in accordance with the guideline procedures gives the mean (R = 50 percent) *S-N* curve shown in Figure E2.6.

b. To obtain an estimate of the R = 99.9 percent reliability *S-N* curve, the expression for standard normal variable X given in (2-29) may be utilized. Since the R = 99.9 per-

Example 2.6 Continues

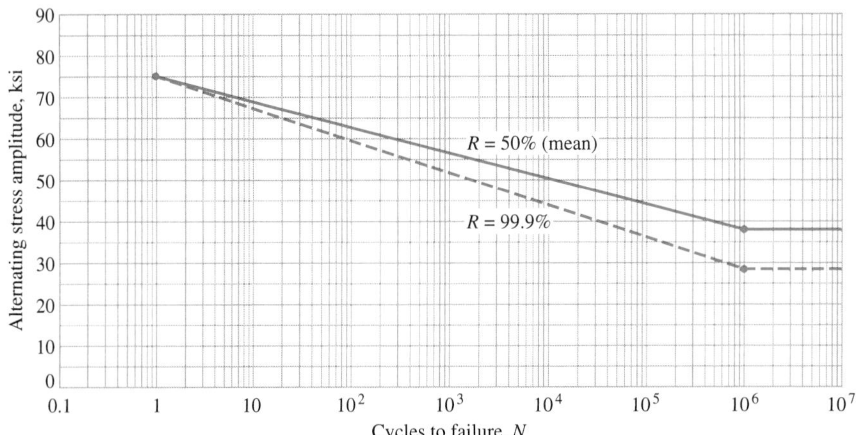

Figure E2.6
Estimated *S-N* curves for fatigue strength reliability levels of 50 percent (mean) and 99.9 percent.

cent curve must lie below the mean curve, S'_f ($R = 99.9$) is less than the mean value S'_f. Hence, for this case, from (2-29)

$$X = \frac{S'_f - S'_f(R = 99.9)}{\hat{\sigma}} \tag{3}$$

Now, from (1)

$$S'_f = 38{,}000 \text{ psi} \tag{4}$$

and using the estimated standard deviation $\hat{\sigma}$ discussed just after (2-29),

$$\hat{\sigma} = 0.08S'_f = 0.08(38{,}000) = 3040 \text{ psi} \tag{5}$$

with the Table 2.3 value for X corresponding to $R = 99.9$ percent reliability

$$X(R = 99.9) = 3.09 \tag{6}$$

equation (3) may be solved for $S'_f(R = 99.9)$ to give

$$S'_f(R = 99.9) = S'_f - X\hat{\sigma} \tag{7}$$

or

$$S'_f(R = 99.9) = 38{,}000 - 3.09(3040) = 28{,}600 \text{ psi} \tag{8}$$

Plotting $S'_f(R = 99.9) = 25{,}600$ psi on the *S-N* coordinates of Figure E2.6 at 10^6 cycles, and connecting to the point S_{ut} at $N = 1$ cycle, results in the estimated 99.9 percent reliability *S-N* curve, as shown.

Example 2.7 Failure Mode Assessment

An axially loaded straight cylindrical bar of diameter $d = 0.500$ inch is to be made of 2024-T4 aluminum with ultimate strength $S_u = 68{,}000$ psi, yield strength $S_{yp} = 48{,}000$ psi, and fatigue properties shown in Figure 2.19. The bar is to be subjected to a completely reversed axial force of 6000 lb, and must last for at least 10^7 cycles.

a. What is the governing failure mode?

b. Is failure predicted to occur?

Solution

a. The two most probable candidates for governing failure mode are yielding and fatigue. Both should be calculated to determine which one governs.

b. For *yielding*, from (2-17), FIPTOI

$$\sigma \geq S_{yp} \tag{1}$$

Now,

$$\sigma = \frac{F}{A_o} = \frac{F}{\left(\frac{\pi d^2}{4}\right)} = \frac{6000}{\pi\left[\frac{(0.500)^2}{4}\right]} = \frac{6000}{0.196} = 30{,}612 \text{ psi} \tag{2}$$

Hence, for yielding, FIPTOI

$$30{,}612 \geq 48{,}000 \tag{3}$$

Therefore, failure by yielding is *not* predicted to occur.
For fatigue, at a design life requirement of $N_d = 10^7$ cycles, FIPTOI

$$\sigma_{\max} \geq S_{N=10^7} \tag{4}$$

From Figure 2.19, using the curve for 2024-T4 aluminum for 10^7 cycles, the fatigue strength at $N = 10^7$ cycles may be read as

$$S_{N=10^7} = 23{,}000 \text{ psi} \tag{5}$$

Hence for fatigue at a design life requirement of $N_d = 10^7$ cycles FIPTOI

$$30{,}612 \geq 23{,}000 \tag{6}$$

Therefore, failure by fatigue *is* predicted. That is, fatigue failure would be expected before the design life of 10^7 cycles is achieved.

Example 2.8 Design for Infinite Fatigue Life

An axially loaded straight cylindrical bar is to be made of 1020 steel, with fatigue properties as shown in Figure 2.19. The bar is to be subjected to a completely reversed axial force of 7000 lb maximum. Fatigue is the governing failure mode.

If infinite life is desired for this part, what is the minimum diameter that the bar should be made?

Solution

Since fatigue is the governing failure mode and infinite life is desired, the stress in the bar must be just below the fatigue limit, read from Figure 2.19 for 1020 steel as

$$S'_f = 33{,}000 \text{ psi} \tag{1}$$

The maximum stress in the bar is

$$\sigma_{max} = \frac{F_{max}}{A_o} = \frac{F_{max}}{\left(\frac{\pi d^2}{4}\right)} \tag{2}$$

or

$$d = \sqrt{\frac{4F_{max}}{\pi \sigma_{max}}} \tag{3}$$

Example 2.8 Continues

Setting $\sigma_{max} = S'_f = 33{,}000$ psi (incipient failure),

$$d = \sqrt{\frac{4(7000)}{\pi(33{,}000)}} = 0.52 \text{ inch} \tag{4}$$

Thus the minimum diameter for infinite life would be 0.52 inch. Again, it must be cautioned that this assumes that the *S-N* curve of Figure 2.19 properly reflects the conditions of the axially loaded bar in terms of size, surface finish, geometry, environment, and other influence factors. Further, statistical scatter or reliability level should also be considered in establishing the bar diameter, using techniques such as the ones described in Example 2.6(b). As a practical matter, the small-polished-specimen *S-N* curve in Figure 2.19 would typically be modified by using appropriate factors from Table 2.2 to reflect the operating conditions, and a safety factor would be used to protect against remaining uncertainties.[23]

Example 2.9 Estimating Fatigue Properties of a Part

The wrought-carbon steel alloy of Example 2.6, with $S_u = 76{,}000$ psi, $S_{yp} = 42{,}000$ psi, and e (2 inches) = 18 percent, is to be considered a candidate material for a proposed machine part to be used under the following operating conditions:

a. The part is to be lathe-turned from a bar of the wrought-steel alloy.

b. The part is uniform in shape at the critical point.

c. Operating speed is 3600 rev/min.

d. A very long life is desired.

e. A strength reliability level of 99.9 percent is desired.

It is desired to make a preliminary design estimate of the pertinent fatigue properties of the candidate material so the approximate size of the part can be established.

Solution

Since a very long life is desired, the fatigue property of primary interest will be the fatigue limit of the part, S_f.

From (2-25),

$$S_f = k_\infty S'_f \tag{1}$$

From (2-27),

$$k_\infty = (k_{gr} k_{we} k_f k_{sr} k_{sz} k_{rs} k_{fr} k_{cr} k_{sp} k_r) \tag{2}$$

From (1) of Example 2.6,

$$S'_f = 38{,}000 \text{ psi} \tag{3}$$

To evaluate k_∞, each of the influencing factors of (2) must be evaluated. Referring to Table 2.2 and the discussion following the table:

$$k_{gr} = 1.0 \text{ (from Table 2.2)}$$

$$k_{we} = 1.0 \text{ (no welding anticipated)}$$

$$k_f = 1.0 \text{ (uniform shape specified)}$$

[23]See Chapter 5.

$k_{sr} = 0.65$ (see Figure 2.21)

$k_{sz} = 0.9$ (size unknown; use Table 2.2)

$k_{rs} = 1.0$ (no information available; later review essential)

$k_{fr} = 1.0$ (no fretting anticipated)

$k_{cr} = 1.0$ (no information available; later review essential)

$k_{sp} = 1.0$ (moderate; use Table 2.2)

$k_r = 0.75$ (from Table 2.3 for $R = 99.9$)

Now (2) may be evaluated as

$$k_\infty = (1.0)(1.0)(1.0)(0.65)(0.9)(1.0)(1.0)(1.0)(1.0)(0.75) \tag{4}$$

or

$$k_\infty = 0.44 \tag{5}$$

Using (3) and (5), (1) becomes

$$S_f = (0.44)(38{,}000) = 16{,}670 \text{ psi} \tag{6}$$

Thus the estimated fatigue limit for the part is 16,670 psi. Using an appropriate safety factor with S_f would allow calculation of the approximate size of the part to produce infinite life.

Nonzero-Mean Stress

Fatigue failure data collected in the laboratory for small polished specimens are most often obtained for completely reversed, or *zero-mean* alternating stresses. Many, if not most, service applications involve *nonzero-mean* cyclic stresses. It is very important, therefore, to be able to predict the influence of mean stress on fatigue behavior so that completely reversed small-polished-specimen *S-N* data may be utilized to design machine parts subjected to nonzero-mean cyclic stresses.

High-cycle fatigue data collected from a series of experiments devised to investigate combinations of alternating stress amplitude σ_a and mean stress σ_m may be characterized by a plot of σ_a versus σ_m for any specified failure life of *N* cycles, such as Figure 2.26. As shown, the failure data points typically tend to cluster about a curve that passes through the point $\sigma_a = S_N$ at $\sigma_m = 0$ and the point $\sigma_m = S_u$ at $\sigma_a = 0$. As shown, failure is very sensitive to the magnitude of mean stress in the tensile mean-stress region but rather in-

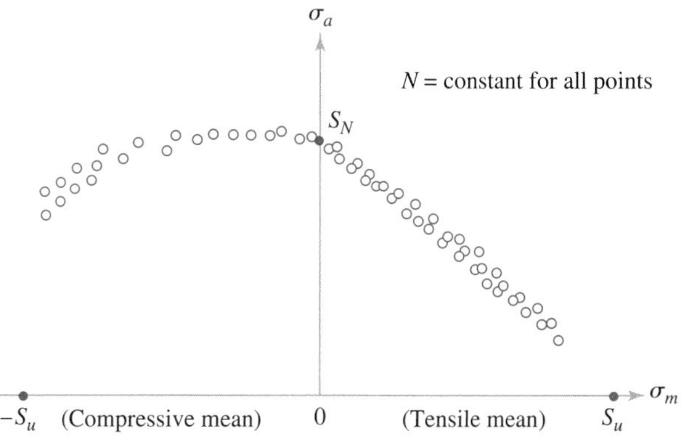

Figure 2.26
Simulated high-cycle fatigue failure data showing the influence of mean stress.

sensitive to the magnitude of mean stress in the compressive mean-stress region. Nonzero-mean failure data are available in the literature for some materials, usually presented as *constant lifetime diagrams*, or *master diagrams*, such as those shown in Figure 2.27. If a designer is fortunate enough to find such data for a proposed material that match the operating conditions for the application of interest, of course these data should be used. If pertinent data are not available, an estimate of the influence of nonzero-mean stress may be made by utilizing a mathematical model that approximates data of the type shown in Figure 2.26. Numerous successful models have been developed for this purpose.[24] Of these, the *modified Goodman relationships* provide a simple linear estimate with reasonably accurate results. Modified Goodman lines used to approximate the data are constructed on a plot of σ_a versus σ_m by connecting the point S_N on the σ_a axis to S_u on the σ_m axis in the tensile mean-stress region and passing a second line horizontally through S_N in the compressive mean-stress region. These lines, shown in Figure 2.28, approximate the data illustrated in Figure 2.26 for a specified constant failure lifetime, N. Other failure lives would give similar curves but slopes would change since the magnitude of S_N depends on failure life. Also it must be recognized that if $|\sigma_{max}|$ or $|\sigma_{min}|$ exceed yield strength values of the material, failure by yielding will occur. Thus yielding failure lines may be plotted from $\sigma_a = S_{yp}$ on the σ_a axis to $\sigma_m = \pm S_{yp}$ on the σ_m axis, as shown in Figure 2.28 by dashed lines.

Any combination of σ_a and σ_m that plots above either the solid or dashed lines in Figure 2.28 represents failure in a life shorter than N cycles, either by fatigue or by yielding. Writing an equation for line S_N-S_u in the point-intercept form for the tensile mean-stress region,

$$y = mx + b \tag{2-31}$$

which becomes

$$\sigma_a = -\frac{S_N}{S_u}\sigma_m + S_N \tag{2-32}$$

Recalling that

$$\sigma_a = \sigma_{max} - \sigma_m \tag{2-33}$$

and substituting into (2-32),

$$\sigma_{max} - \sigma_m + \frac{S_N}{S_u}\sigma_m = S_N \tag{2-34}$$

or

$$\sigma_{max} - \sigma_m\left[\frac{S_u - S_N}{S_u}\right] = S_N \tag{2-35}$$

Defining the terms

$$m_t \equiv \frac{S_u - S_N}{S_u} \tag{2-36}$$

and

$$R_t \equiv \frac{\sigma_m}{\sigma_{max}} \tag{2-37}$$

equation (2-35) may be rewritten as

$$\sigma_{max} = \frac{S_N}{1 - m_t R_t} \tag{2-38}$$

[24]See, for example, ref. 1, p. 232.

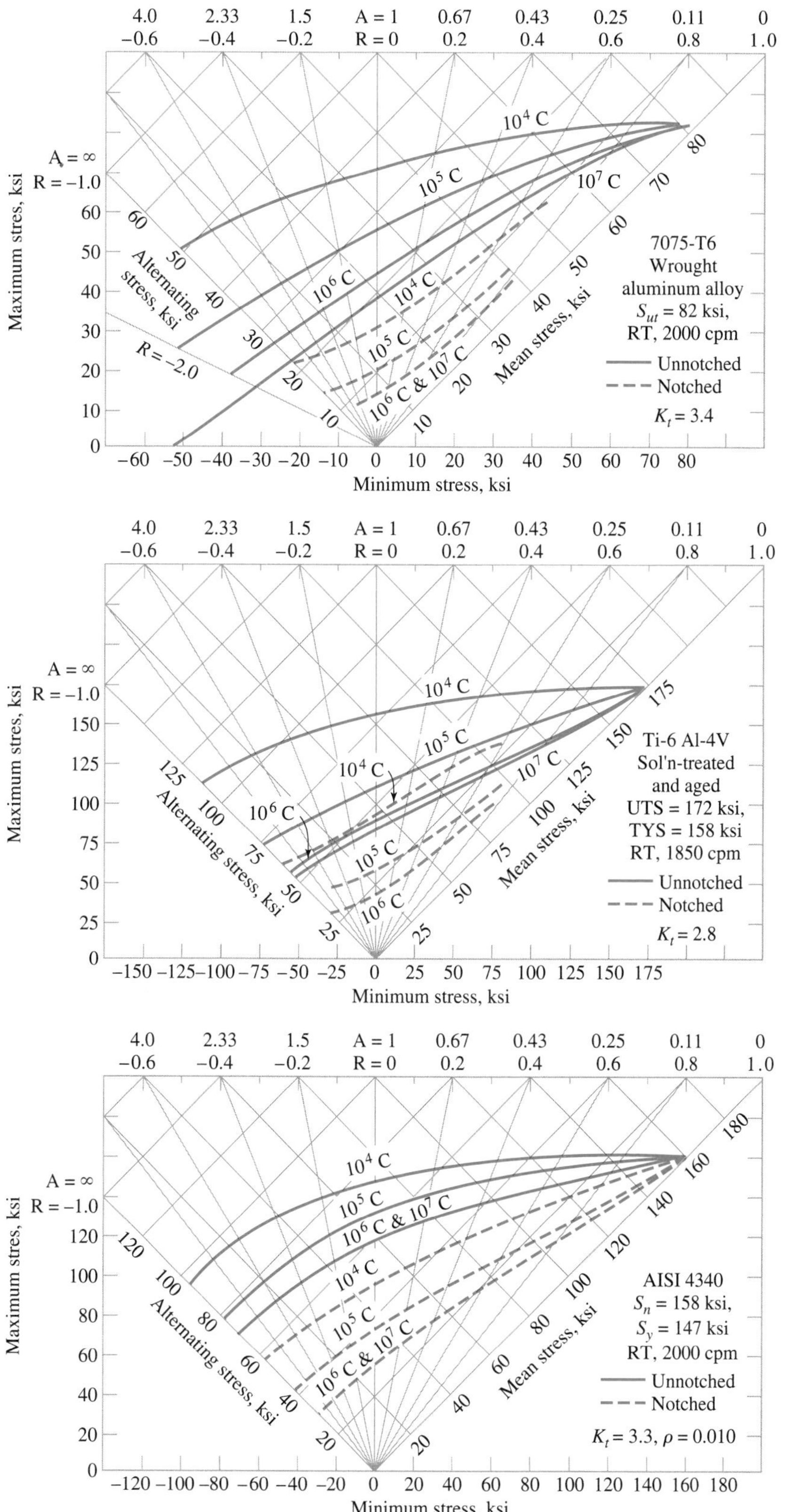

Figure 2.27
Master diagrams for alloys of steel, aluminum, and titanium. (From ref. 22, pp. 317, 322).

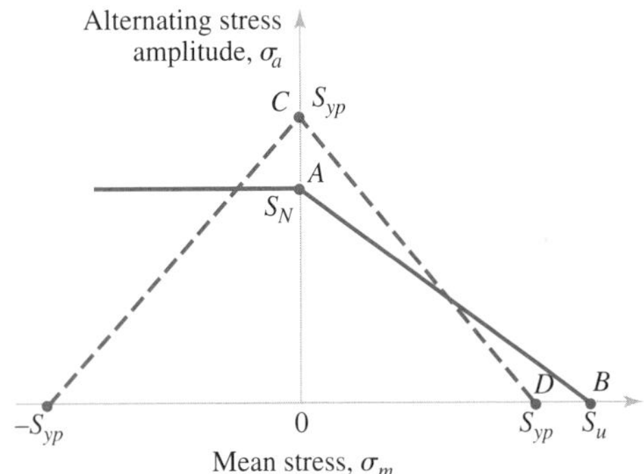

Figure 2.28
Modified Goodman diagram representing fatigue failure at a life of N cycles.

Note 1: Point A may be thought of as a completely reversed cyclic stress amplitude that produces damage equivalent to any combination of $\sigma_m + \sigma_a$ that lies on AB (for $\sigma_m + \sigma_a \leq S_{yp}$)

To clarify, σ_{max} in this equation represents the cyclic maximum stress that, with the presence of mean stress σ_m, will cause failure in the same number of cycles as S_N at zero-mean stress. To recognize that this is therefore a "strength parameter," we set

$$\sigma_{max} \equiv S_{max-N} \tag{2-39}$$

Further, σ_{max} may not exceed S_{yp} either, or yielding will occur. Summarizing then, for the tensile mean-stress region,

$$S_{max-N} = \frac{S_N}{1 - m_t R_t} \quad \text{for} \quad \sigma_m \geq 0 \quad \text{and} \quad S_{max-N} \leq S_{yp}$$

or

$$S_{max-N} = S_{yp} \quad \text{for} \quad \sigma_m \geq 0 \quad \text{and} \quad S_{max-N} \geq S_{yp} \tag{2-40}$$

where

$$m_t = \frac{S_u - S_N}{S_u}$$

and

$$R_t = \frac{\sigma_m}{\sigma_{max}}$$

By a similar reasoning process for the compressive mean-stress region, σ_{min} represents the cyclic minimum stress that, with the presence of mean stress σ_m, will cause failure in the same number of cycles as S_N at zero-mean stress. Again, $|\sigma_{min}|$ may not exceed $|S_{yp}|$ or yielding will occur. Thus for the compressive mean-stress region

$$S_{min-N} = \frac{-S_N}{1 - m_c R_c} \quad \text{for} \quad \sigma_m \leq 0 \quad \text{and} \quad S_{min-N} \geq -S_{yp}$$

or

$$S_{min-N} = -S_{yp} \quad \text{for} \quad \sigma_m \leq 0 \quad \text{and} \quad S_{min-N} \leq -S_{yp} \tag{2-41}$$

where

$$m_c = 1$$

and

$$R_c = \frac{\sigma_m}{\sigma_{min}}$$

Referring to Note 1 of Figure 2.28, and solving (2-32) from a different perspective, modified Goodman expressions may be developed for an *equivalent completely reversed cyclic stress*, σ_{eq-CR}. This stress σ_{eq-CR} is a *calculated* completely reversed cyclic stress that would produce failure in exactly the same number of cycles N as would the actual nonzero-mean cyclic stress that has amplitude σ_a and mean σ_m. Thus any nonzero-mean cyclic stress may be transformed into an *equivalent* completely reversed cyclic stress of amplitude σ_{eq-CR} that induces fatigue damage at the same rate cycle-by-cycle. Using this approach for the tensile mean-stress region

$$\sigma_{eq-CR} = \frac{\sigma_a}{1 - \dfrac{\sigma_m}{S_u}} \quad \text{for} \quad \sigma_m \geq 0 \quad \text{and} \quad \sigma_{max} \leq S_{yp}$$

or

$$\sigma_{eq-CR} = S_{yp} \quad \text{for} \quad \sigma_m \geq 0 \quad \text{and} \quad \sigma_{max} \geq S_{yp} \tag{2-42}$$

and for the compressive mean-stress region

$$\sigma_{eq-CR} = \sigma_a \quad \text{for} \quad \sigma_m \leq 0 \quad \text{and} \quad |\sigma_{min}| \leq |-S_{yp}|$$

or

$$\sigma_{eq-CR} = |-S_{yp}| \quad \text{for} \quad \sigma_m \leq 0 \quad \text{and} \quad |\sigma_{min}| \geq |-S_{yp}| \tag{2-43}$$

In all equations from (2-40) through (2-43) it should be noted that if the design objective is infinite life, S_f should be substituted for S_N everywhere.

Example 2.10 Design to Account for Nonzero-Mean Stress

A low-alloy steel link is to be made from a solid cylindrical bar subjected to an axial cyclic force that ranges from a maximum of 60,000-lb tension to a minimum of 40,000-lb compression. The static material properties are S_u = 100,000 psi, S_{yp} = 76,000 psi, and elongation in 2 inches of 25 percent. Calculate the diameter the link should have to provide infinite life.

Solution

Since the link is cyclically loaded and infinite life is desired, the fatigue limit S_f is the material property of primary interest; yielding should also be checked. Using the estimation guidelines preceding Example 2.6, since $S_u <$ 200 ksi,

$$S'_f = 0.5S_u = 0.5(100{,}000) = 50{,}000 \text{ psi} \tag{1}$$

Following the procedures of Example 2.9, and utilizing Table 2.2, the fatigue limit for the part may be estimated as

$$S_f = k_\infty S'_f = k_\infty(50{,}000) \tag{2}$$

Since little information has been given, the influencing factors of Table 2.2 may be estimated as:

$$\begin{aligned}
k_{gr} &= 1.0 \text{ (from Table 2.2)} \\
k_{we} &= 1.0 \text{ (no welding anticipated)} \\
k_f &= 1.0 \text{ (uniform shape)} \\
k_{sr} &= 0.65 \text{ (lathe-turned)} \\
k_{sz} &= 0.9 \text{ (from Table 2.2)} \\
k_{rs} &= 1.0 \text{ (no information available; later review essential)} \\
k_{fr} &= 1.0 \text{ (no fretting anticipated)} \\
k_{cr} &= 1.0 \text{ (no information available; later review essential)} \\
k_{sp} &= 1.0 \text{ (from Table 2.2)} \\
k_r &= 0.9 \text{ (from Table 2.3 for } R = 90)
\end{aligned}$$

Example 2.10 Continues

hence

$$k_{\infty} = (1.0)(1.0)(1.0)(0.65)(0.9)(1.0)(1.0)(1.0)(1.0)(0.9) = 0.53 \tag{3}$$

and from (2)

$$S_f = (0.53)(50{,}000) = 26{,}300 \text{ psi} \tag{4}$$

From the loading specification it may be noted that we have a case of nonzero-mean cyclic loading. Since stress is proportional to loading,

$$\sigma_m = \frac{P_m}{A} = \frac{\left(\dfrac{P_{max} + P_{min}}{2}\right)}{A} \tag{5}$$

or

$$\sigma_m = \frac{\left[\dfrac{60{,}000 + (-40{,}000)}{2}\right]}{A} = \frac{+\,10{,}000}{A} \tag{6}$$

where A is the unknown cross-sectional area of the link.

Since, from (6), the mean stress is tensile, (2-40) may be used with

$$S_{N=\infty} = S_f = 26{,}300 \text{ psi} \tag{7}$$

$$m_t = \frac{100{,}000 - 26{,}300}{100{,}000} = 0.74 \tag{8}$$

$$R_t = \frac{\sigma_m}{\sigma_{max}} = \frac{\dfrac{P_m}{A}}{\dfrac{P_{max}}{A}} = \frac{P_m}{P_{max}} = \frac{10{,}000}{60{,}000} = 0.167 \tag{9}$$

so that (2-40) gives

$$S_{max-\infty} = \frac{S_N}{1 - m_t R_t} = \frac{26{,}300}{1 - (0.74)(0.167)} = 30{,}000 \text{ psi} \tag{10}$$

Since

$$S_{max-\infty} = 30{,}000 < S_{yp} = 76{,}000 \tag{11}$$

yielding does *not* occur and (10) is valid.

To obtain the required diameter we set the maximum applied stress σ_{max} equal to the infinite-life strength for the specified nonzero-mean loading, to give

$$\sigma_{max} = S_{max-\infty} \tag{12}$$

or

$$\frac{P_{max}}{A} = \frac{P_{max}}{\left(\dfrac{\pi d^2}{4}\right)} = 30{,}000 \tag{13}$$

Solving for link diameter, d,

$$d = \sqrt{\frac{4P_{max}}{\pi(30{,}000)}} = \sqrt{\frac{4(60{,}000)}{\pi(30{,}000)}} = 1.60 \text{ inches} \tag{14}$$

Thus the required link diameter is 1.60 inches to provide infinite life. It should be noted, however, that in practice a *safety factor* would also be imposed, and the required "safe" diameter would be somewhat larger than the 1.60 inches just calculated.

Example 2.11 Design for Selected Reliability

A one-inch-square bar is to be subjected to an axial cyclic force that ranges from a maximum of 36,000 lb (tension) to a minimum of −22,000 lb (compression). The material is the wrought-steel alloy of Example 2.6 and the operating conditions are the same as those specified in Example 2.6. A fatigue strength reliability level of 99.9 percent is desired. How many cycles of operation would be expected before failure occurs?

Solution

Since the material and the operating conditions are exactly the same as for Example 2.6, and a fatigue strength reliability level of 99.9 percent is specified, the $R = 99.9$ percent curve of Figure E2.6 in Example 2.6 represents the appropriate estimate of the *S-N* curve for the one-inch-square bar.

From the known loading and geometry,

$$\sigma_{max} = \frac{P_{max}}{A} = \frac{36{,}000}{(1)^2} = 36{,}000 \text{ psi} \tag{1}$$

$$\sigma_{min} = \frac{P_{min}}{A} = \frac{-22{,}000}{(1)^2} = -22{,}000 \text{ psi} \tag{2}$$

$$\sigma_m = \frac{\sigma_{max} + \sigma_{min}}{2} = \frac{36{,}000 + (-22{,}000)}{2} = 7000 \text{ psi} \tag{3}$$

$$\sigma_a = \frac{\sigma_{max} - \sigma_{min}}{2} = \frac{36{,}000 - (-22{,}000)}{2} = 29{,}000 \text{ psi} \tag{4}$$

Since the mean stress is tensile, from (2-42)

$$\sigma_{eq-CR} = \frac{\sigma_a}{1 - \dfrac{\sigma_m}{S_u}} = \frac{29{,}000}{1 - \left(\dfrac{7000}{76{,}000}\right)} = 31{,}940 \text{ psi} \tag{5}$$

Also,

$$\sigma_{max} = 36{,}000 < S_{yp} = 42{,}000 \tag{6}$$

so yielding *does not* occur and (5) is valid.

Referring to the *S-N* curve of Figure E2.6 for $R = 99.9$ percent, repeated below as Figure E2.11, and reading into the curve from the stress axis at

$$S = \sigma_{eq-CR} = 31{,}940 \text{ psi} \tag{7}$$

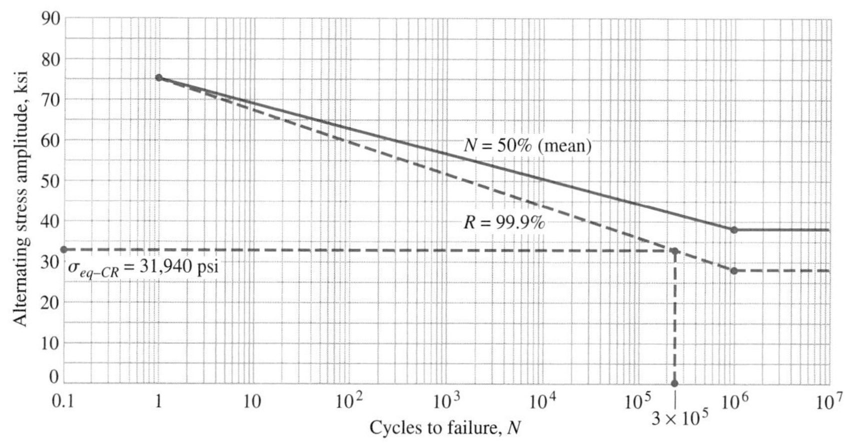

Figure E2.11
Estimated *S-N* curve for fatigue strength reliability levels of 50 percent (mean) and 99.9 percent.

Example 2.11 Continues

then across to intersect the *S-N* curve, the intersection occurs at

$$N = 300{,}000 \text{ cycles} \tag{8}$$

Thus at the fatigue strength reliability level of 99.9 percent for this material operating at the specified conditions, it would be predicted that the bar would operate for 300,000 cycles before fatigue failure occurs.

Cumulative Damage Concepts and Cycle Counting

Nearly all machines operate over a *spectrum* of speeds or loads, giving rise to a *spectrum* of alternating stress amplitudes and mean stresses, as illustrated for example in Figures 2.13 and 2.14. Such variations in stress amplitudes and mean stresses make the direct use of standard *S-N* curves inapplicable because these curves are developed and presented for *constant* stress amplitude operation (and usually for zero-mean stress). Therefore it becomes important to a designer to have available a theory or hypothesis, verified by experimental observations, that will permit good design estimates to be made for operation under conditions of spectrum loading using standard constant-amplitude *S-N* curves.

The basic postulate adopted by all fatigue investigators is that operation at any given cyclic stress amplitude will produce *fatigue damage*, the seriousness of which will be related to the number of cycles of operation at that stress amplitude and also related to the total number of cycles that would be required to produce failure of an undamaged specimen or part at that stress amplitude. It is further postulated that the damage incurred is permanent, and operation at several different stress amplitudes in sequence will result in an accumulation of total damage equal to the sum of damage increments accrued at all the different stress levels. When the total accumulated damage reaches a critical value, fatigue failure occurs.

Of the many damage theories proposed,[25] the most widely used is a linear theory first proposed by Palmgren in 1924 and later developed by Miner in 1945. This linear theory is referred to as the *Palmgren–Miner hypothesis* or the *linear damage rule*. The theory, based on the postulate that there is a linear relationship between fatigue damage and *cycle ratio* n/N, may be described using the *S-N* plot shown in Figure 2.29.

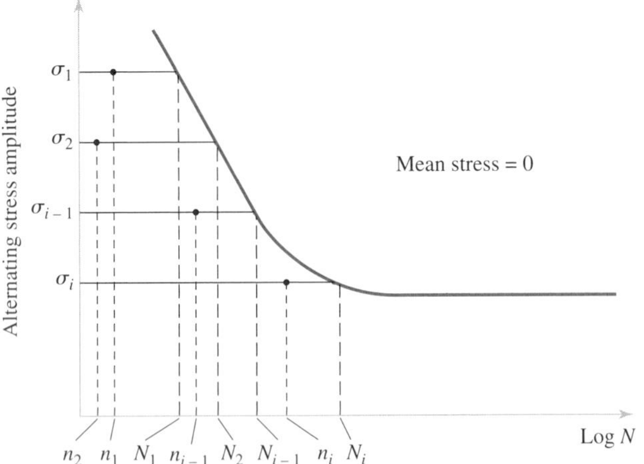

Figure 2.29
Illustration of spectrum loading where n_i cycles of operation are accrued at each of the different corresponding stress levels σ_i, and the N_i are cycles to failure at each σ_i.

[25]See, for example, ref. 1, Ch. 8.

By definition of the *S-N* curve, operation at a constant stress amplitude σ_1 will produce complete damage, or failure, in N_1 cycles. Operation at stress amplitude σ_1 for a number of cycles n_1, smaller than N_1, will produce a smaller fraction of damage, say D_1. D_1 is usually termed the *damage fraction*. Operation over a spectrum of i different stress levels results in a damage fraction D_i for each of the different stress levels σ_i in the spectrum. When these damage fractions sum to unity, failure is predicted; that is, failure is predicted to occur if (FIPTOI)

$$D_1 + D_2 + \cdots + D_i \geq 1 \tag{2-44}$$

The Palmgren–Miner hypothesis asserts that the damage fraction D_i at any stress level σ_i is linearly proportional to the ratio of number of cycles of operation n_i to the total number of cycles N_i that would produce failure at that stress level; that is,

$$D_i = \frac{n_i}{N_i} \tag{2-45}$$

By the Palmgren–Miner hypothesis then, (2-44) becomes FIPTOI

$$\frac{n_1}{N_1} + \frac{n_2}{N_2} + \cdots + \frac{n_i}{N_i} \geq 1 \tag{2-46}$$

or FIPTOI

$$\sum_{j=1}^{n} \frac{n_j}{N_j} \geq 1 \tag{2-47}$$

This is a complete statement of the Palmgren–Miner hypothesis, or the linear damage rule. It has one sterling virtue, namely, *simplicity;* and for this reason it is widely used. Further, other much more complicated cumulative damage theories typically do not yield a significant improvement in failure prediction reliability. Perhaps the most significant shortcomings of the linear damage rule are that no influence of the *order* of application of various stress levels is recognized, and damage is assumed to accumulate at the same rate at a given stress level *without regard to past history*. Nevertheless, the Palmgren–Miner hypothesis may be used with good success most of the time.

Example 2.12 Zero-Mean Spectrum Loading

An alloy steel strut for an experimental aircraft application is fabricated from a supply of material that has an ultimate strength of 135,000 psi, yield strength of 120,000 psi, elongation of 20 percent in 2 inches, and fatigue properties under test conditions that match the actual operating conditions, as shown in the table of experimental results in Table E2.12.

The cross-sectional area of the strut is 0.10 in^2 and buckling has been found not to be a problem due to the selected cross-sectional shape. In service, the strut is to be subjected to the following spectrum of completely reversed axial loads during each duty cycle:

P_a = 11,000 lb for 1000 cycles

P_b = 8300 lb for 4000 cycles

P_c = 6500 lb for 500,000 cycles

This duty cycle is to be repeated three times during the life of the strut.

Assuming fatigue to be the governing failure mode, would this strut be expected to survive all three duty cycles, or would it fail prematurely?

TABLE E2.12 Fatigue Test Data for Strut Material

Stress Amplitude (psi)	Cycles to Failure, N
110,000	6,600
105,000	9,500
100,000	13,500
95,000	19,200
90,000	27,500
85,000	39,000
80,000	55,000
75,000	87,000
73,000	116,000
71,000	170,000
70,000	220,000
69,000	315,000
68,500	400,000
68,000	∞

Example 2.12 Continues

Solution

Based on the data from Table E2.12, an S-N curve may be plotted as shown in Figure E2.12.

Since the completely reversed axial loads are given, and the cross-sectional area of the strut is given as 0.10 in^2, the completely reversed stress spectrum for each duty cycle may be calculated as

$$\sigma_a = \frac{P_a}{A} = \frac{11{,}000}{0.10} = 110{,}000 \text{ psi}; \quad n_a = 1000 \text{ cycles} \tag{1}$$

$$\sigma_b = \frac{P_b}{A} = \frac{8300}{0.10} = 83{,}000 \text{ psi}; \quad n_b = 4000 \text{ cycles} \tag{2}$$

$$\sigma_c = \frac{P_c}{A} = \frac{6500}{0.10} = 65{,}000 \text{ psi}; \quad n_c = 500{,}000 \text{ cycles} \tag{3}$$

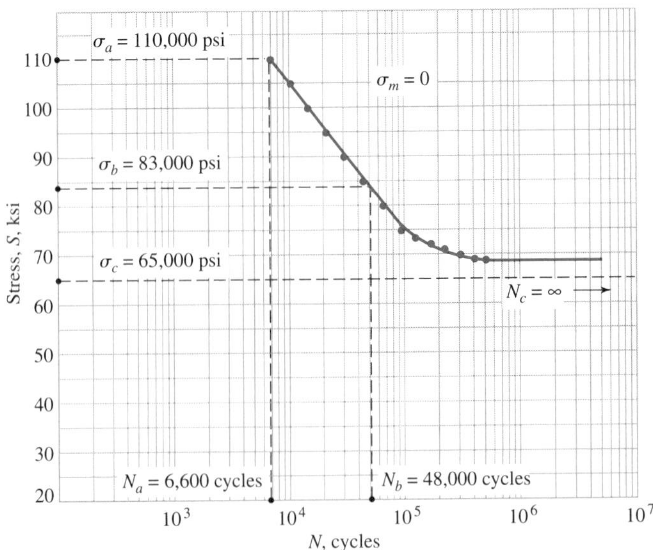

Figure E2.12A
S-N curve for alloy steel data from Table E2.12A.

Reading from the *S-N* curve of Figure E2.12 for completely reversed stresses σ_a, σ_b, and σ_c, respectively,

$$N_a = 6600 \text{ cycles to failure} \tag{4}$$

$$N_b = 48{,}000 \text{ cycles to failure} \tag{5}$$

$$N_c = \infty \text{ (since } \sigma_c < S_f = 68{,}000) \tag{6}$$

Furthermore, three such duty cycles must be sustained without failure.

To determine whether failure would be predicted, the Palmgren–Miner linear damage rule formulated in (2-47) may be used. Hence FIPTOI

$$\sum_{j=1}^{n} \frac{n_j}{N_j} \geq 1 \tag{7}$$

or, FIPTOI

$$3\left[\frac{n_a}{N_a} + \frac{n_b}{N_b} + \frac{n_c}{N_c}\right] \geq 1 \tag{8}$$

Using results of (1) through (6), FIPTOI

$$3\left[\frac{1000}{6600} + \frac{4000}{48{,}000} + \frac{500{,}000}{\infty}\right] \geq 1 \tag{9}$$

or FIPTOI

$$3[0.152 + 0.083 + 0] = 0.706 \geq 1 \tag{10}$$

Since this failure expression is not satisfied, the strut would be predicted to *survive* the three duty cycles.

For the simple case of a stress spectrum composed of a sequence of uniaxial completely reversed stresses of various amplitudes, the estimation of cumulative damage and prediction of failure are relatively straightforward, as illustrated by Example 2.12. If the stress spectrum is more complicated, as shown, for example, in Figures 2.13 or 2.14, the task of evaluating cumulative damage and even the task of *counting the cycles* in the spectrum become much more difficult. Although numerous cycle counting methods have been devised,[26] the *rain flow cycle counting method* is probably more widely used than any other method and will be the only method presented in this text.

To use the rain flow method, the stress-time history (stress-time spectrum) is plotted to scale on a stress-time coordinate system so that the time axis is vertically downward, and the lines connecting the stress peaks are imagined to be a series of sloping roofs, as shown, for example, in Figure 2.30. Several rules are imposed on "raindrops" flowing down these sloping roofs so that the rain flow may be used to define cycles and half-cycles of fluctuating stress in the spectrum.

Rain flow is initiated by placing raindrops successively at the inside of each peak (maximum) or valley (minimum).

The rules are as follows:

1. *The rain is allowed to flow on the roof and drip down to the next slope except that, if it initiates at a valley, it must be terminated when it comes opposite a valley more negative than the valley from which it initiated.* For example, in Figure 2.30 the flow begins

[26]See, for example, ref. 1, p. 289.

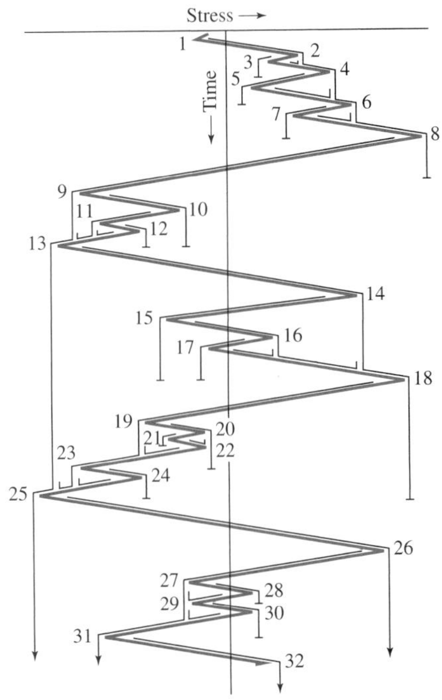

Figure 2.30
Example of rain flow cycle counting method. (After ref. 23.)

at valley 1 and stops opposite valley 9, valley 9 being more negative than valley 1. A half-cycle is thus counted between valleys 1 and peak 8.

2. *Similarly, if the rain flow initiates at a peak, it must be terminated when it comes opposite a peak more positive than the peak from which it initiated.* For example, in Figure 2.30 the flow begins at peak 2 and stops opposite peak 4, thus counting a half-cycle between peak 2 and valley 3.

3. *A rain flow must also stop if it meets the rain from a roof above.* For example, in Figure 2.30, the flow beginning at valley 3 ends beneath peak 2. Using these rules, every part of the stress-time history is counted once and only once.

4. *If cycles are to be counted over the duration of a duty cycle or a "mission profile" block that is to be repeated block after block, the cycle counting should be started by initiating the first raindrop either at the most negative valley or the most positive peak, and continuing until all cycles in a complete block have been counted in sequence.* This procedure assures that a complete stress cycle will be counted between the most positive peak and most negative valley in the block.

The nonzero-mean stress cycles may be converted to equivalent completely reversed cycles by utilizing (2-42) or (2-43).

To determine the fatigue damage in each cycle associated with the equivalent completely reversed spectrum, an *S-N* curve for the material must be available or estimated using the techniques presented in Example 2.6 and the preceding discussion.

Finally, damage is summed by utilizing the Palmgren–Miner linear damage rule (2-47) as illustrated in Example 2.12.

It should be clear that for most real-life design cases the prediction techniques just described are practical only with the help of a digital computer program designed to carry forth the tedious cycle-by-cycle analyses involved. Many such programs are available.[27]

[27]See, for example, ref. 10, App. 5A.

Example 2.13 Cycle Counting, Nonzero-Mean Loading, and Fatigue Life

The stress-time pattern shown in Figure E2.13A is to be repeated in blocks. Using the rain flow cycle counting method and the *S-N* curve of Figure E2.13B, estimate the time in hours of testing required to produce failure.

Solution

The stress-time pattern of Figure E2.13A is first rotated so that the time axis is vertically downward, as shown in Figure E2.13C. Following the rules for rain flow cycle counting, the count is started at a minimum valley, as shown for raindrop (1) of the time-shifted block in Figure E2.13C.

Data for each raindrop, read from Figure E2.13C, are recorded in Table E2.13. Values for σ_{eq-CR} are calculated using (2-42) and (2-43). Values of N corresponding to each rain-

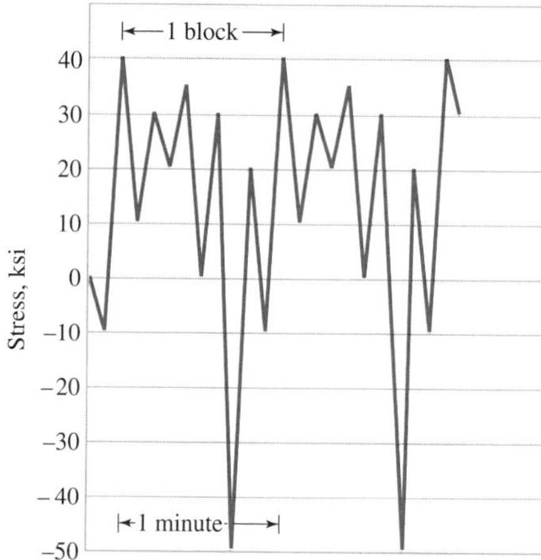

Figure E2.13A
Stress-time pattern applied.

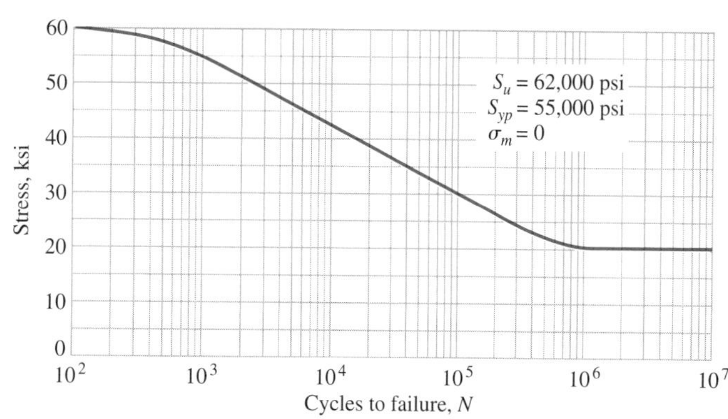

Figure E2.13B
***S-N* curve.**

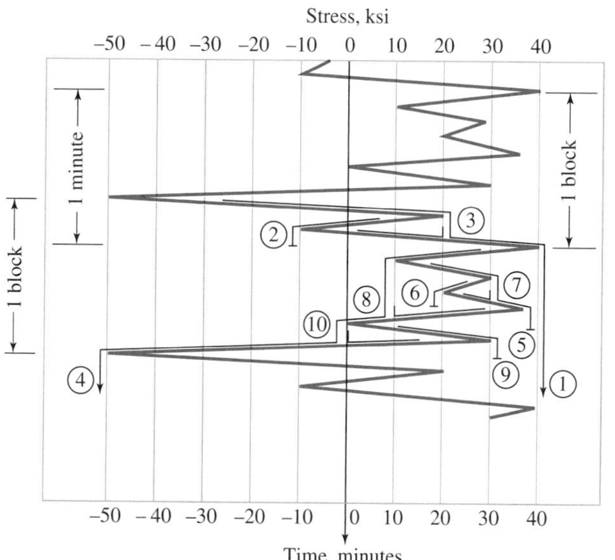

Figure E2.13C
Stress-time pattern oriented for rain flow cycle counting.

Example 2.13 Continues

TABLE E2.13 Rain Flow Data for One "Block" of the Stress-Time Pattern

Raindrop Number	n, cycles	σ_{max}, ksi	σ_{min}, ksi	σ_m, ksi	σ_a, ksi	σ_{eq-CR}, ksi	N, cycles
1,4 @ $\frac{1}{2}$ cycle each	1	40	−50	−5	45	45	6×10^3
2,3 @ $\frac{1}{2}$ cycle each	1	20	−10	5	15	16.3	∞
5,8 @ $\frac{1}{2}$ cycle each	1	35	10	22.5	12.5	19.6	∞
6,7 @ $\frac{1}{2}$ cycle each	1	30	20	25	5	8.4	∞
9,10 @ $\frac{1}{2}$ cycle each	1	30	0	15	15	19.8	∞

drop's σ_{eq-CR} value are read from the completely reversed *S-N* curve of Figure E2.13B. Based on the Palmgren–Miner linear damage rule of (2-47), FIPTOI

$$B_f \sum_{j=1}^{5} \frac{n_j}{N_j} \geq 1 \tag{1}$$

using B_f to denote the number of block to failure.

From Table E2.13 it may be noted that only raindrops 1 and 4 will produce a nonzero cycle ratio.

Thus (1) reduces to FIPTOI

$$B_f\left(\frac{n_{1,4}}{N_{1,4}}\right) \geq 1 \tag{2}$$

Using the equal sign, which corresponds to incipient failure, and reading numerical values from Table E2.13,

$$B_f\left(\frac{1}{6 \times 10^3}\right) = 1 \tag{3}$$

or

$$B_f = 6 \times 10^3 \text{ blocks to failure} \tag{4}$$

At 1 block/minute, as indicated in Figures E2.13A and E2.13C, the predicted time to failure in hours, H_f, would be

$$H_f = \left(\frac{6 \times 10^3 \text{ blocks}}{\text{min}}\right)\left(\frac{1 \text{ min}}{\text{block}}\right)\left(\frac{1 \text{ hr}}{60 \text{ min}}\right) = 100 \text{ hr to failure} \tag{5}$$

Multiaxial Cyclic Stresses

Uniaxial cyclic stressing has been implicit in all of the discussion thus far in 2.6. Most real design situations, including rotating shafts, pressure vessels, power screws, springs, gears, flywheels, and many other machine elements, may involve *multiaxial* states of cyclic stress. Because of the complexities and costs in producing multiaxial fatigue failure data, only a limited body of such data exists. Consequently, no consensus has yet been reached on the best approach to prediction of failure under multiaxial fatigue stresses; however, various proposals have been discussed in recent books on fatigue analysis.[28] The approach adopted in this text involves the concept of *equivalent stress* to define a uniaxial equiva-

[28]See, for example, refs. 1, 14, and 17.

lent to the actual multiaxial state of cyclic stresses, including both the alternating stress amplitudes and mean stresses. Since this method hinges upon the adaptation of *static* combined stress theories of failure, its presentation is deferred to 4.7 following the discussion of multiaxial states of stress and combined stress failure theories.

Fracture Mechanics (*F-M*) Approach to Fatigue

In the introductory paragraphs of section 2.6 it was noted that fatigue is a progressive failure process that involves three phases: the initiation of crack nuclei, propagation of cracks until one reaches an unstable size, and finally, a sudden catastrophic separation of the affected part into two or more pieces. In the discussion of the *S-N* approach to fatigue design and analysis just completed, the three phases are lumped together in an *S-N* curve that represents a final failure locus for a part or specimen subjected to known fluctuating loads. The transitions from initiation to propagation to final fracture cannot be identified from an *S-N* curve, and the *S-N* approach does not require such information. Physically, however, the three phases have been well documented and the separate modeling of each phase in sequence has been under intense development. The results of these efforts have led to an approach to fatigue called the *fracture mechanics (F-M)* approach. An extensive database has now been established to support the *F-M* method and it has in recent years been very successfully used in the analysis and design of machine parts subjected to cyclic loads. A brief discussion of the *F-M* approach is included here, but much more complete discussions may be found in the recent literature on fatigue.[29]

Crack Initiation Phase

The most widely accepted approach to the prediction of crack initiation life is the *local stress-strain* approach. While the details of the local stress-strain approach are complicated, the concepts are not. The basic premise is that the local fatigue response of the small critical zone of material at the crack initiation site, usually at the root of a geometrical discontinuity, is analogous to the fatigue response of a small smooth laboratory specimen subjected to the same cyclic strains and stresses as the critical zone. This concept is illustrated in Figure 2.31. The number of cycles required to *initiate a crack* in the critical zone, N_i, is postulated to be equal to the number of cycles to *produce failure* of the small smooth specimen in a laboratory test under the same cyclic strains and stresses. Digital computer simulation of the smooth specimen failure process then allows the prediction of N_i if appropriate cyclic material response data are available.

To proceed with the analysis, the prediction model must contain the ability to

1. Compute local stresses and strains, including means and ranges, from the applied loads and geometry of the structure or machine part.
2. Count cycles and associate mean and range values of stress and strain with each cycle.
3. Convert nonzero-mean cycles to equivalent completely reversed cycles.
4. Compute fatigue damage during each cycle from stress and/or strain amplitudes and cyclic materials properties.
5. Compute damage cycle-by-cycle and sum the damage to give the desired prediction of N_i.

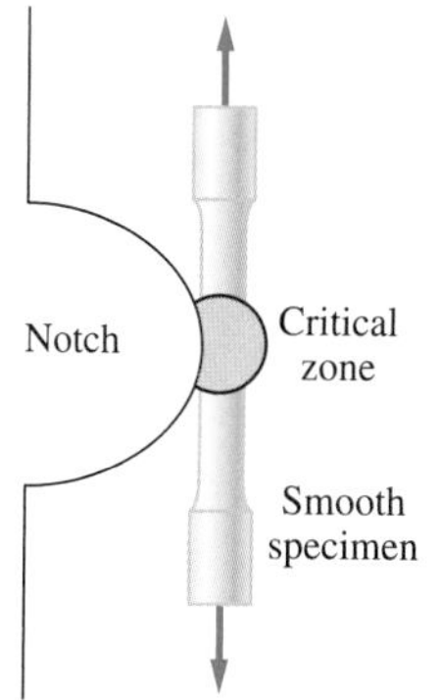

Figure 2.31 Smooth specimen analog of material at critical point in the structure. (See ref. 25.)

Many of these steps may be accomplished using methods already developed earlier in 2.6. Rain flow cycle counting (Example 2.12) may be used for item (2), modified Good-

[29]See, for example, refs. 1, 14, and 17.

man relationships (Example 2.11) for item (3), and Palmgren–Miner linear damage rule (Example 2.12) for item (5). Items (1) and (4) require further discussion.

To compute local stresses and strains from the external loading and geometry, a modified version of the Neuber Rule[30] for estimating theoretical stress concentration factors leads to the expression

$$\Delta\sigma\Delta\varepsilon = \frac{(K_f \Delta S)^2}{E} \tag{2-48}$$

where

$\Delta\sigma$ = local stress *range*

$\Delta\varepsilon$ = local strain *range*

ΔS = nominal[31] stress *range*

K_f = fatigue stress concentration factor[32]

E = Young's modulus of elasticity

From other work,[33] an empirical expression for the cyclic stress-strain curve[34] satisfactory for most engineering metals has been developed as

$$\frac{\Delta\varepsilon}{2} = \frac{\Delta\sigma}{2E} + \left[\frac{\Delta\sigma}{2k'}\right]^{1/n'} \tag{2-49}$$

where k' and n' are material constants, the *cyclic* strength coefficient and *cyclic* strain-hardening exponent, respectively. Values for k' and n' may be found from the intercept and slope of an experimentally determined log-log plot of cyclic stress amplitude $\Delta\sigma/2$ versus cyclic strain amplitude $\Delta\varepsilon/2$. Values for many materials may be found in the literature.[35] With known values of material properties E, k', and n', the fatigue stress concentration factor K_f, and readily calculable value of nominal stress range ΔS, equations (2-48) and (2-49) may be solved simultaneously to find values for local stress range $\Delta\sigma$ and local strain range $\Delta\varepsilon$.

To compute the local fatigue damage associated with equivalent completely reversed stress and strain ranges for each cycle, it is necessary to have experimental failure data for strain amplitude versus cycles to failure (crack initiation), N_i, often plotted as strain amplitude versus *reversals* to failure (crack initiation), $2N_i$, as illustrated in Figure 2.32.

Noting from Figure 2.32 that total strain amplitude may be expressed as the sum of elastic strain amplitude plus plastic strain amplitude, each linear with cycles or reversals to failure (crack initiation) on a log-log plot, an empirical expression for total strain amplitude versus cycles to crack initiation, N_i, has been developed as[36]

$$\frac{\Delta\varepsilon}{2} = \frac{\sigma'_f}{E}(2N_i)^b + \varepsilon'_f(2N_i)^c \tag{2-50}$$

[30]See, for example, ref. 1, p. 283.

[31]Stress calculated assuming no stress concentration effect; note that the use here of S for *stress* deviates from the usual policy in this text of reserving S to denote strength.

[32]See 4.8.

[33]See ref. 25, p. 7.

[34]Usually significantly different from the static stress-strain curve. See, for example, ref. 1, pp. 284–285.

[35]See, for example, ref. 26.

[36]See ref. 23.

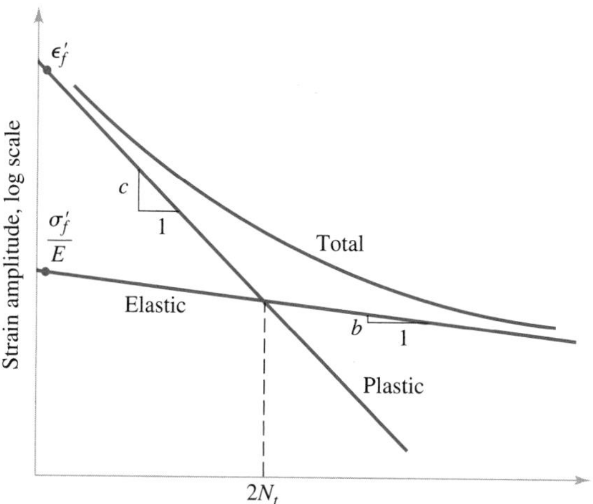

Figure 2.32
Schematic representation of elastic, plastic, and total strain amplitude versus fatigue life. (After ref. 23.)

The constants b and σ'_f/E are the slope and one-reversal intercept of the elastic line in Figure 2.32, and the constants c and ε'_f are the slope and one-reversal intercept of the plastic line in Figure 2.32. The values of the cyclic material properties σ'_f (cyclic true fracture strength) and ε'_f (cyclic true fracture ductility) must be experimentally determined. With known cyclic material properties equation (2-50) may be used to calculate the number of cycles to crack initiation, N_i, for any given total strain amplitude, $\Delta\varepsilon/2$.

All five steps of the local stress-strain approach may therefore be completed if the cyclic material properties data are available and the applied nominal stress spectrum is known.

Example 2.14 Fatigue Crack Initiation Life

Experimentally determined values for the properties of a martensitic steel have been found to be $S_u = 215{,}000$ psi, $S_{yp} = 200{,}000$ psi, $K_{Ic} = 74{,}000 \text{ psi}\sqrt{\text{in}}$, e (2 inches) = 20 percent, $k' = 155{,}000$ psi, $n' = 0.15$, $\varepsilon'_f = 0.48$, $\sigma'_f = 290{,}000$ psi, $b = -0.091$, and $c = -0.60$. A direct tension member made of this alloy has a single semicircular edge notch that results in a fatigue stress concentration factor[37] of 1.6. The rectangular net cross section of the member at the root of the notch is 0.35 inch thick by 1.43 inches wide. A completely reversed axial force of 16,000 lbs amplitude is applied to the tension member.

a. How many cycles would you estimate that it would take to initiate a fatigue crack at the notch root?

b. What length would you estimate this crack to be at the time it is initiated by the calculation of part (a)?

Solution

a. The nominal stress amplitude S_a may be calculated as

$$S_a = \frac{F_a}{A} = \frac{16{,}000}{(0.35)(1.43)} = 31{,}970 \text{ psi} \tag{1}$$

[37]See 4.8.

Example 2.14 Continues

hence the nominal stress range ΔS is given by

$$\Delta S = 2S_a = 2(31{,}970) = 63{,}940 \text{ psi} \tag{2}$$

Using (2-48)

$$\Delta\sigma\Delta\varepsilon = \frac{[1.6(63{,}940)]^2}{30 \times 10^6} = 348.8 \text{ psi} \tag{3}$$

Next, from (2-49), using results from (3),

$$\frac{\Delta\varepsilon}{2} = \frac{\Delta\sigma}{2(30 \times 10^6)}\left(\frac{\Delta\varepsilon}{\Delta\varepsilon}\right) + \left[\frac{\Delta\sigma}{2(155{,}000)}\left(\frac{\Delta\varepsilon}{\Delta\varepsilon}\right)\right]^{1/0.15} \tag{4}$$

or

$$\frac{\Delta\varepsilon}{2} = \frac{348.8}{6.0 \times 10^7 \Delta\varepsilon} + \left[\frac{348.8}{3.1 \times 10^5 \Delta\varepsilon}\right]^{1/0.15} \tag{5}$$

or

$$\frac{(\Delta\varepsilon)^2}{2} = 5.81 \times 10^{-6} + 2.26 \times 10^{-20} \Delta\varepsilon^{(1 - 1/0.15)} \tag{6}$$

or

$$\Delta\varepsilon = \sqrt{1.16 \times 10^{-5} + 4.52 \times 10^{-20} \Delta\varepsilon^{-5.67}} \tag{7}$$

which may be iterated to the solution

$$\Delta\varepsilon = 3.75 \times 10^{-3} \text{ in/in} \tag{8}$$

Then, from (2-50)

$$\frac{3.75 \times 10^{-3}}{2} = \frac{2.9 \times 10^5}{30 \times 10^6}(2N_i)^{-0.091} + 0.48(2N_i)^{-0.60} \tag{9}$$

or

$$1.88 \times 10^{-3} = 9.08 \times 10^{-3} N_i^{-0.091} + 0.73 N_i^{-0.60} \tag{10}$$

or

$$N_i = [0.21 - 80.4 N_i^{-0.60}]^{-1/0.091} \tag{11}$$

which may be iterated to

$$N_i = 3.2 \times 10^7 \text{ cycles to initiation} \tag{12}$$

b. There is no known method for *calculating* the length of a newly initiated fatigue crack. The length must either be *measured* from an experimental test or *estimated* from experience. Often, if no other information is available, a newly initiated crack is *assumed* to be about $a_i = 0.050$ inch.

Crack Propagation and Final Fracture Phases

As discussed in 2.5, the concepts of linear elastic fracture mechanics may be employed to predict the size of a crack in a given structure or machine part that will, under specified loadings, propagate spontaneously to final fracture. This *critical crack size*, a_{cr}, may be determined, for example, by solving (2-22) for crack size, which then (by definition) becomes a_{cr} when the applied stress σ is given its maximum value. Thus, from (2-22)

$$a_{cr} = \frac{1}{\pi}\left[\frac{K_c}{C\sigma_{max}}\right]^2 \tag{2-51}$$

Further, *if plane strain conditions are met*, as defined by (2-23), K_c becomes equal to the plane strain fracture toughness K_{Ic}, and (2-51) becomes

$$a_{cr} = \frac{1}{\pi}\left[\frac{K_{Ic}}{C\sigma_{max}}\right]^2 \tag{2-52}$$

A fatigue crack that has been initiated by cyclic loading, or any other preexisting flaw in the material, may be expected to grow under sustained cyclic loading until it reaches the critical size, a_{cr}, from which it will propagate spontaneously to catastrophic failure in accordance with the laws of fracture mechanics. Typically, the time for a fatigue-initiated crack to grow to critical size is a significant portion of the life of the machine part. Thus, not only is it necessary to understand the crack initiation phase and the definition of critical crack size but an understanding of the growth of a crack from initial size a_i to critical size a_{cr} is also essential.

The crack growth rate, *da/dN*, has been found to be related to the range of stress-intensity factor, ΔK, where

$$\Delta K = C\Delta\sigma\sqrt{\pi a} \tag{2-53}$$

and the range of stress is given by

$$\Delta\sigma = \sigma_{max} - \sigma_{min} \tag{2-54}$$

Most crack propagation data are presented as log-log plots of *da/dN* versus ΔK. Figure 2.33 shows such a plot for various types of steels. The study of such data has led to many different empirical models[38] for the prediction of crack growth rate *da/dN* as a function of ΔK. One such model, developed by Paris and Erdogan,[39] has become widely known as the *Paris Law*,

$$\frac{da}{dN} = C_{PE}(\Delta K)^n \tag{2-55}$$

where *n* is the slope of the log-log plot of *da/dN* versus ΔK, as shown in Figure 2.33, and C_{PE} is an empirical parameter that depends upon material properties, cyclic frequency, mean stress, and perhaps other secondary variables.

If the parameters C_{PE} and *n* are known for a particular application, the crack length, a_N, resulting from the application of *N* cycles of loading after the crack is initiated, may be computed from the expression

$$a_N = a_i + \sum_{j=1}^{N_p} C_{PE}(\Delta K^n) \tag{2-56}$$

or

$$a_N = a_i + \int_1^{N_p} C_{PE}(\Delta K)^n\, dN \tag{2-57}$$

where a_i is the length of a newly initiated crack and N_p the number of loading cycles in the propagation phase following crack initiation. For complicated spectrum loading histories, these computations require block-by-block or cycle-by-cycle analyses, making the

[38]See, for example, ref. 28.

[39]See ref. 29.

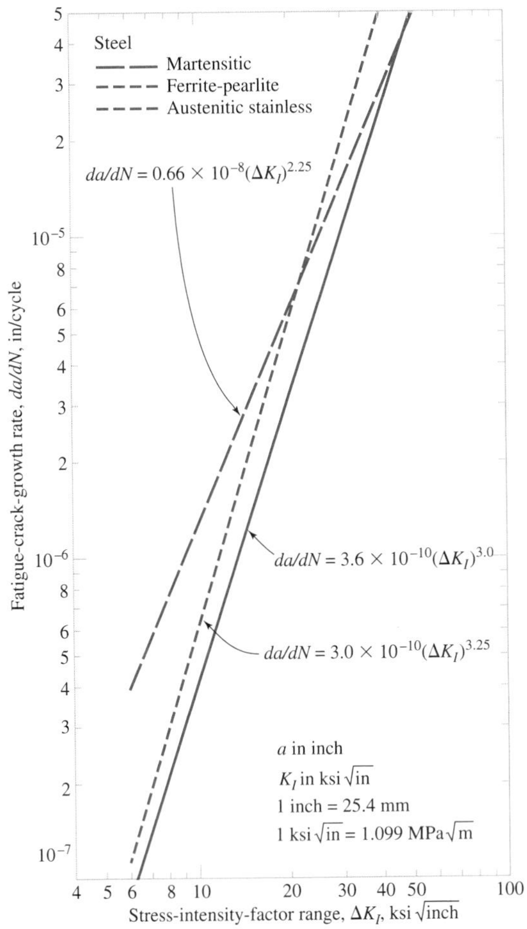

Figure 2.33
Fatigue crack propagation behavior of various steels. (Reprinted from ref. 27 by permission of Pearson Education, Inc., Upper Saddle River, N.J.)

use of modern digital computer systems mandatory for the solution of most practical design problems. It should also be recognized that various other phenomena may take place in the zone of material around the tip of a growing crack to influence the crack growth prediction accuracy.[40] These may include crack growth retardation or acceleration due to plastic zones produced by preceding cyclic load histories, the influence of nonzero-mean stresses, and crack size effects. Typically, a designer would utilize (2-56) or (2-57) to obtain preliminary estimates of crack propagation life, but would enlist the help of a fracture mechanics specialist to improve prediction accuracy by including modifications to account for such factors as loading sequence, environment, frequency, multiaxial states of stress, and determination of applicable ΔK values in view of plasticity at the crack tip.

Example 2.15 Fatigue Crack Propagation Life

The martensitic alloy steel tension member of Example 2.14 is to be subjected to a continuation of the completely reversed cyclic axial force of 16,000-lb amplitude, following the initiation of a fatigue crack. It may be assumed that the length of a newly initiated fatigue crack is 0.050 inch. Further, it may be assumed that the stress concentration effects

[40]See, for example, ref. 1, pp. 293–304.

are negligible for the propagating crack (since the initiated crack tip probably extends through most of the zone of stress concentration). How many cycles of continued loading would you estimate could be applied before catastrophic fracture would occur?

Solution

Since the applied axial force is completely reversed with amplitude P_a = 16,000 lb, the maximum and minimum stresses may be calculated as

$$\sigma_{max} = \frac{P_{max}}{A_g} = \frac{16{,}000}{(0.35)(1.43)} = 31{,}970 \text{ psi} \tag{1}$$

and

$$\sigma_{min} = \frac{P_{min}}{A_g} = \frac{-16{,}000}{(0.35)(1.43)} = -31{,}970 \text{ psi} \tag{2}$$

Using (2-23) to check the plane strain conditions,

$$B = 0.35 \overset{?}{\geq} 2.5\left(\frac{K_{Ic}}{S_{yp}}\right)^2 = 2.5\left(\frac{74{,}000}{200{,}000}\right)^2 \tag{3}$$

or

$$0.35 \overset{?}{\geq} 0.34 \tag{4}$$

Since the expression is satisfied, *plane strain conditions exist* and the critical crack size may be determined from (2-52). Thus

$$a_{cr} = \frac{1}{\pi}\left[\frac{K_{Ic}}{C\sigma_{max}}\right]^2 = \frac{1}{\pi}\left[\frac{74{,}000}{C(31{,}970)}\right]^2 \tag{5}$$

The parameter C may be evaluated from Figure 2.7 using the methods of Example 2.5, but must be iterated together with (5) above since the crack length a_{cr} must be known to find C, and C must be known to find a_{cr}.

To initiate the iteration, assume, for example, a_{cr} = 0.60 inch. Then from Figure 2.7,

$$\frac{a}{b} = \frac{0.60}{1.43} = 0.42 \tag{6}$$

and

$$C(1 - 0.42)^{3/2} = 0.97 \tag{7}$$

or

$$C = 2.20 \tag{8}$$

and from (5)

$$a_{cr} = \frac{1}{\pi}\left[\frac{74{,}000}{(2.20)(31{,}970}\right]^2 = 0.35 \tag{9}$$

Iterating, try a_{cr} = 0.50, giving, from Figure 2.7,

$$\frac{a}{b} = \frac{0.50}{1.43} = 0.35 \tag{10}$$

and

$$C = [1 - 0.35]^{3/2} = 0.98 \tag{11}$$

or

$$C = 1.87 \tag{12}$$

Example 2.15 Continues

and, from (5)

$$a_{cr} = \frac{1}{\pi}\left[\frac{74{,}000}{(1.87)(31{,}970)}\right]^2 = 0.49 \tag{13}$$

This is close enough to the assumed value of a_{cr} = 0.50 inch. Therefore

$$a_{cr} = 0.50 \text{ inch} \tag{14}$$

and correspondingly

$$C_{cr} = 1.87 \tag{15}$$

Also, for the newly initiated crack of length a_i = 0.050 inch, the corresponding C_i may be found from Figure 2.7, with a_i/b = 0.035, as

$$C[1 - 0.035]^{3/2} = 1.07 \tag{16}$$

or

$$C_i = 1.13 \tag{17}$$

Since this is a martensitic steel, from Figure 2.33 the applicable empirical crack growth model is

$$\frac{da}{dN} = 0.66\times10^{-8}(\Delta K)^{2.25} \tag{18}$$

where ΔK is in ksi$\sqrt{\text{in}}$ and *da/dN* is in inch per cycle.

Next, substituting (1) and (2) above into (2-54),

$$\Delta\sigma = 31{,}970 - (-31{,}970) = 63{,}940 \text{ psi} \tag{19}$$

or

$$\Delta\sigma = 63.9 \text{ ksi} \tag{20}$$

Substituting (2-53) into (18) above,

$$\frac{da}{dN} = 0.66 \times 10^{-8}(C\Delta\sigma\sqrt{\pi a})^{2.25} \tag{21}$$

or

$$\frac{da}{a^{1.125}} = 0.66 \times 10^{-8}(C\Delta\sigma\sqrt{\pi})^{2.25}\, dN \tag{21}$$

Integrating both sides,

$$\int_{a_i=0.050}^{a_{cr}=0.50} \frac{da}{a^{1.125}} = \int_0^{N_p} 0.66 \times 10^{-8}\left[\left(\frac{1.13 + 1.87}{2}\right)(63.9)\sqrt{\pi}\right]^{2.25} dN \tag{23}$$

(Note that C is taken as the average of C_i and C_{cr}. A more accurate solution could be found by partitioning the crack growth into smaller increments and separately integrating each increment, summing the results to obtain N_p.)

Evaluating (23) then

$$\left.\frac{a^{-0.125}}{-0.125}\right|_{0.050}^{0.50} = 7.03 \times 10^{-4} N_p \tag{24}$$

or

$$\frac{0.50^{-0.125}}{-0.125} - \frac{0.05^{-0.125}}{-0.125} = 6.88 \times 10^{-4} N_p \tag{25}$$

or

$$N_p = 2.91\times10^4 \text{ propagation cycles to final fracture} \tag{26}$$

Design Issues in Fatigue Life Prediction

Using the *F-M* approach to estimate the total fatigue life of a proposed design configuration is simple in concept. The total fatigue life to failure N_f is the sum of the initiation life N_i plus the propagation life N_p. That is,

$$N_f = N_i + N_p \tag{2-58}$$

The initiation life may be calculated using the local stress-strain approach, as illustrated in Example 2.14. The propagation life may be calculated using an appropriate crack growth model together with LEFM estimates of critical crack size, as illustrated in Example 2.15. As a practical matter, however, several important additional issues must be addressed when attempting to use (2-58) as a design tool. These include the following: (1) determining or specifying the size of an initiated crack corresponding to N_i cycles; (2) accounting for geometric stress concentration effects and stress gradients; (3) accounting for strength gradients, especially as associated with metallurgical or mechanical surface treatments; (4) accounting for residual stress fields; (5) accounting for multiaxial states of stress, three-dimensional effects, and others. Most of these issues remain research topics, requiring designers to consult with fracture mechanics specialists and/or make appropriate simplifying assumptions when utilizing (2-58). In the final analysis, it is essential to conduct full-scale fatigue tests to provide acceptable reliability.

Example 2.16 Estimating Total Fatigue Life

Referring to the solutions of Examples 2.14 and 2.15, estimate the total fatigue life of the martensitic steel tension member subjected to a completely reversed cyclic axial force of 16,000-lb amplitude.

Solution

Total fatigue life may be estimated using (2-58). With results from Examples 2.14 and 2.15,

$$N_f = N_i + N_p = 2.87 \times 10^8 + 2.91 \times 10^4 = 2.87 \times 10^8 \text{ cycles} \tag{1}$$

For this member, the fatigue life N_f is dominated by the crack initiation phase.

2.7 Elastic Instability and Buckling

When steady compressive forces are applied to short "fat" members, such as the one sketched in Figure 2.34(a), they tend to fail by compressive yielding or fracture, as discussed in 2.4 and 2.5. However, when steady compressive forces are applied to long "thin" members, such as the one sketched in Figure 2.34(b), they sometimes induce sudden major changes in geometry, such as bowing, wrinkling, twisting, bending, or *buckling*. For the case of short, fat members, failure is predicted to occur when the compressive stress σ_c exceeds the compressive strength, as formulated in (2-17) and (2-18). However, for the case of long, thin members, even when the compressive stress levels generated by applied compressive forces are well within acceptable strength levels, the large deflections, δ, that suddenly occur may destroy the equilibrium of the structure and produce an unstable configuration that leads to collapse. This type of failure is generally called failure by *elastic instability* or *buckling*.

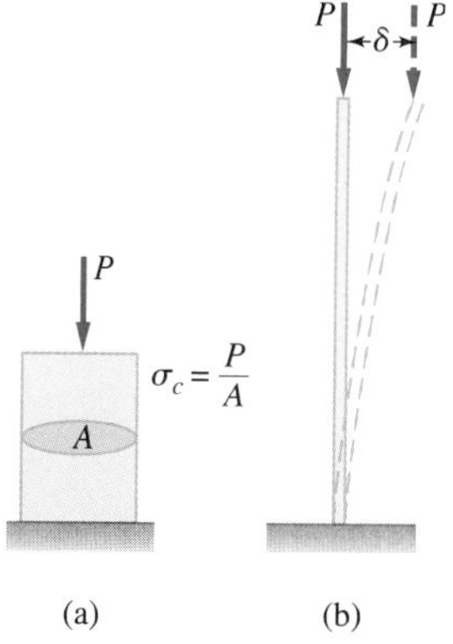

Figure 2.34
Compressive loads on (a) a short "fat" member, and (b) a long "thin" member.

Prediction of the onset of buckling failure is an important task. It is essential to note at the outset that *buckling failure does not depend at all upon the strength of the material, but only on the dimensions of the structure and the modulus of elasticity of the material.* Thus a *high-strength* steel member of a given length and cross-sectional size and shape is *no more able* to withstand a given buckling load *than a low-strength* steel member of the same dimensions.

Buckling of a Simple Pin-Jointed Mechanism

Sidewise buckling of an axially loaded compression member, or *column*, is a case of great practical importance. The basic buckling phenomenon may be understood by considering the perfectly aligned four-bar linkage shown in Figure 2.35, to which auxiliary springs are attached at joint B. When this "column" is perfectly aligned, the springs exert zero force. However, if a lateral displacement of joint B develops for any reason, a lateral resisting force is developed by the springs at B. A lateral deflection might develop because of a slight lateral disturbance, or manufacturing errors might lead to an effective lateral displacement. In any event, viewing the column as shown in Figure 2.35, if moments are taken about point C, it may be observed that the axial force, P_a, leads to an *upsetting* moment, M_u, whereas the spring force, P_s, leads to a *resisting* moment, M_r. As long as the resisting moment is capable of being equal to or greater than the upsetting moment, the linkage is stable. If the upsetting moment exceeds the resisting moment capability, however, the linkage becomes unstable and collapses or buckles. At the point where the maximum available resisting moment exactly equals the upsetting moment, the system is on the verge of buckling and is said to be *critical*. The axial load that produces this condition is called the *critical buckling load*. Thus, the critical buckling load is the value of P_a that satisfies the condition

$$M_r = M_u \tag{2-59}$$

The upsetting moment M_u for the linkage of Figure 2.35 is

$$M_u = \left(\frac{2\delta P_a}{L\cos\alpha}\right)\frac{L}{2}\cos\alpha + P_a\delta = 2P_a\delta \tag{2-60}$$

and the resisting moment M_r is

$$M_r = (k\delta)\frac{L}{2}\cos\alpha \tag{2-61}$$

where k is the spring rate of the lateral spring system.

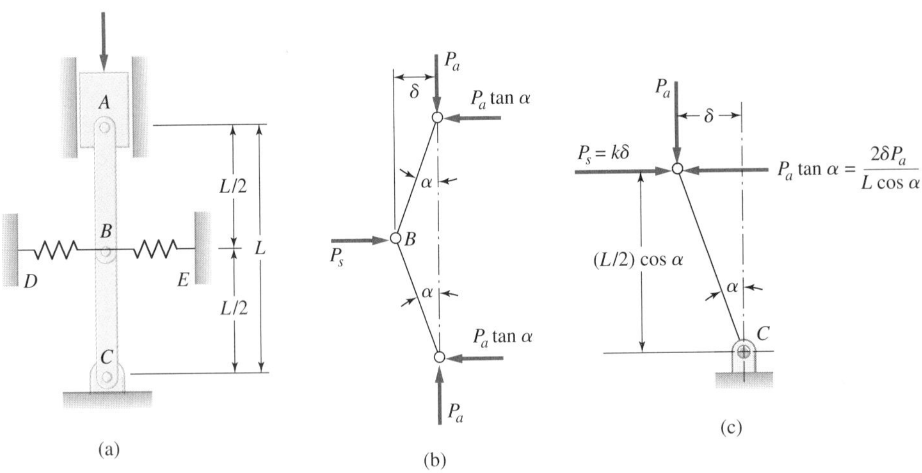

Figure 2.35
Buckling model for a simple pin-jointed mechanism.

Thus, the critical value of axial load P_a is given from (2-59) as

$$\frac{k\delta L\cos\alpha}{2} = 2\delta(P_a)_{cr} \qquad \text{(2-62)}$$

or

$$P_{cr} = \frac{kL}{4}\cos\alpha \approx \frac{kL}{4} \qquad \text{(2-63)}$$

Any axial load that exceeds $kL/4$ in magnitude will cause the mechanism to collapse since the spring is not stiff enough to provide a resisting moment large enough to balance the upsetting moment caused by applied axial force P_a.

Buckling of a Pinned-End Column

The behavior of perfectly straight, ideal elastic columns is quite analogous to the model of Figure 2.35 except that in the case of a column the resisting moment must be provided by the beam itself. Consequently, in the case of column buckling the bending spring rate of the column will become important and the length, cross-sectional dimensions, and modulus of elasticity will all influence the buckling resistance.

A sketch of an axisymmetric ideal pinned-end column loaded axially is shown in Figure 2.36. As long as the axial load P is less than the critical buckling load, the column is stable and any small lateral disturbing force, say at the midspan of the column, will cause a small deflection at midspan that will disappear upon removal of the disturbing force. However, if the axial load P exceeds the critical buckling load, the application of a small lateral disturbing force at midspan leads to large deflections and buckling of the column because the *maximum available resisting moment*, arising from column stiffness, is *not large enough to balance the upsetting moment* generated by axial load P acting at a moment arm equal to the midspan column bending deflection. As can be seen in Figure 2.36, if moments are taken about point m, the upsetting moment M_u is

$$M_u = Pv \qquad \text{(2-64)}$$

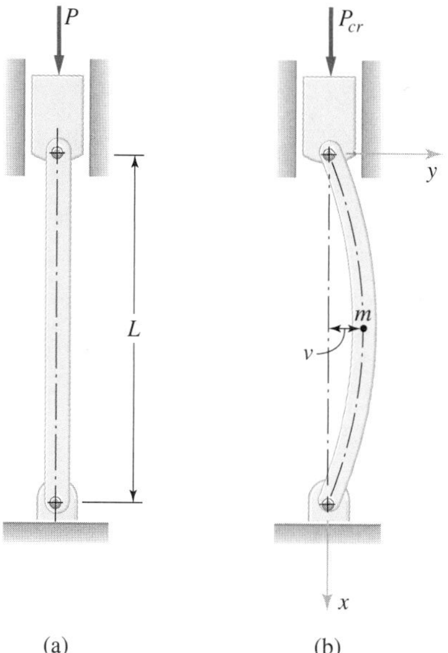

Figure 2.36
Ideal axisymmetric pinned-end column subjected to axial buckling load.

where v is the lateral midspan deflection of the column. Thus, the force P produces an upsetting moment that tends to bend the column even further, which in turn produces a larger eccentricity, v, and therefore a still larger upsetting moment. The elastic forces produced in the column by the bending action tend to resist further bending. When the maximum available resisting moment exactly equals the upsetting moment, the column is at the point of incipient buckling and the axial load at this time is called the *critical buckling load*, P_{cr}, for the column.

The critical buckling load for a column can be calculated by using the basic differential equation[41] for the deflection curve of a beam subjected to a bending moment, which is

$$EI \frac{d^2v}{dx^2} = -M \tag{2-65}$$

where $v(x)$ is the lateral column deflection at any location along the axis, I is the moment of inertia of the cross section about the axis around which bending takes place, and E is the modulus of elasticity for the material. Referring to Figure 2.36(b) and (2-64), at the time when the critical buckling load is applied, the upsetting moment may be calculated as

$$(M_u)_{cr} = P_{cr}v \tag{2-66}$$

whereupon the governing differential equation of (2-65) becomes

$$EI \frac{d^2v}{dx^2} = -P_{cr}v \tag{2-67}$$

or, defining k^2 as

$$k^2 = \frac{P_{cr}}{EI} \tag{2-68}$$

$$\frac{d^2v}{dx^2} + k^2v = 0 \tag{2-69}$$

The general solution of (2-69) is

$$v = A \cos kx + B \sin kx \tag{2-70}$$

where A and B are constants of integration that may be determined from the boundary conditions, which are

$$\begin{aligned} v = 0 \quad &\text{at} \quad x = 0 \\ v = 0 \quad &\text{at} \quad x = L \end{aligned} \tag{2-71}$$

Evaluating constants and eliminating trivial cases, (2-70) yields

$$\sin kL = 0 \tag{2-72}$$

The smallest nonzero value of k that satisfies (2-72) is that which makes the argument equal to π; that is

$$kL = \pi \tag{2-73}$$

or, utilizing (2-68),

$$\sqrt{\frac{P_{cr}}{EI}}\, L = \pi \tag{2-74}$$

[41]See, for example, ref. 30, 31, or 32.

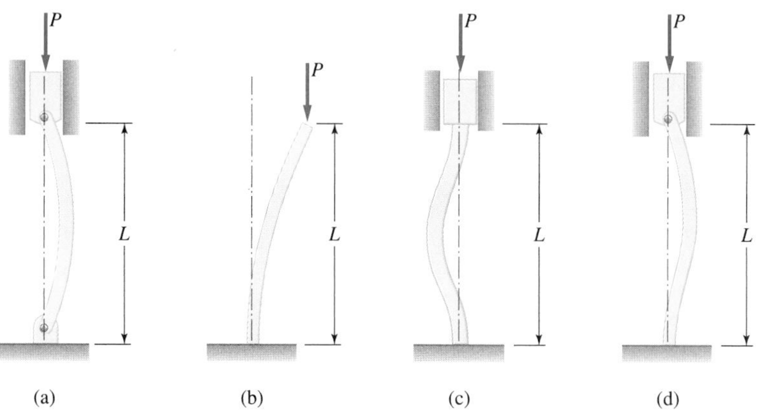

Figure 2.37

Frequently encountered end constraints for columns. (a) Pinned-pinned $L_e = L$. (b) Fixed-free $L_e = 2L$. (c) Fixed-fixed $L_e = L/2$. (d) Fixed-pinned $L_e \approx 0.7L$.

and, solving for the critical buckling load P_{cr},

$$P_{cr} = \frac{\pi^2 EI}{L^2} \tag{2-75}$$

This expression for the smallest critical load that will produce buckling in a pinned-end column is called Euler's equation for buckling of a pinned-end column, and P_{cr} is called *Euler's critical load* for a pinned-end column.

Columns with Other End Constraints

If the column ends are not pinned, the boundary conditions differ from (2-71) and the resulting critical load is different from (2-75). One convenient way to obtain expressions for critical buckling loads for columns with various types of end constraint, such as those shown in Figure 2.37, is to introduce the concept of *effective column length* into the Euler buckling equation (2-75). The *effective length L_e of any column is defined as the length of a pinned-pinned column that would buckle at the same critical load as the actual column.* The direction of the applied column load must remain parallel to the *original* column axis. Using the concept of equivalent length, we can write the critical buckling load expression for any type of column-end constraint by putting the proper value of equivalent length into the Euler equation (2-75). Thus, for any elastic column, buckling load is given by

$$P_{cr} = \frac{\pi^2 EI}{L_e^2} \tag{2-76}$$

where the values of L_e for several types of end constraint are given in Table 2.4.

TABLE 2.4 Effective Lengths for Several Types of Column End Constraint

End Constraint	Effective Length L_e for Actual Column Length L
Both ends pinned	$L_e = L$
One end pinned, one end fixed	$L_e \approx 0.7L$
One end fixed, one end free	$L_e = 2L$
Both ends fixed	$L_e = 0.5L$

For real columns it will often be found that the ends may be partially but not completely fixed, in which case the equivalent length will lie between the pinned-pinned case of $L_e = L$ and the fixed-fixed case of $L_e = 0.5L$, depending on the effective rotational spring rate at the end points.[42]

Inelastic Behavior and Initially Crooked Columns

To account for inelastic behavior of a column if the induced stress levels exceed the elastic range, Engesser suggested in 1889 that the Euler expression for critical buckling load be modified by using the *tangent modulus* E_t, rather than Young's modulus, E. The tangent modulus is defined to be the local slope of the engineering stress-strain curve for the material, or

$$E_t = \frac{d\sigma}{d\varepsilon} \tag{2-77}$$

The tangent modulus may be conveniently determined graphically from a scale plot of the engineering stress-strain diagram or by digital simulation.[43] The critical buckling load equation, often called the *tangent modulus equation* or the *Euler–Engesser equation*, may be written from (2-76) as

$$P_{cr} = \frac{\pi^2 E_t I}{L_e^2} \tag{2-78}$$

or, alternatively,

$$\frac{P_{cr}}{A} = \frac{\pi^2 E_t}{(L_e/k)^2} \tag{2-79}$$

where $k = \sqrt{I/A}$ is the minimum radius of gyration for the cross section of the column, I is the corresponding area moment of inertia for the cross section, A is the cross-sectional area, and the ratio L_e/k is often called the *effective slenderness ratio* for the column. Clearly, the critical buckling load is greatly enhanced by designing a column with a low slenderness ratio.

The Euler–Engesser equation (2-79), a relatively simple expression for determiation of the critical load for a column, gives good agreement with experimental results in both the elastic and inelastic range.[44]

Aside from the Euler–Engesser relationship, a number of empirical column design relationships have been developed to account for the effects of inelastic behavior. Of these, the secant formula is of special interest since it allows direct consideration of initial eccentricity or column crookedness. The secant formula may be expressed as

$$\frac{P_{cr}}{A} = \frac{S_{yp}}{1 + \dfrac{ec}{k^2}\sec\left[\dfrac{L_e}{2k}\sqrt{\dfrac{P_{cr}}{AE_t}}\right]} \tag{2-80}$$

where S_{yp} is the yield strength of the material, e is the eccentricity of the axial load with respect to the centroidal axis of the column cross section, c is the distance from the centroidal axis to the outer fiber, k is the appropriate radius of gyration, L_e is the appropriate equivalent column length, E_t is the tangent modulus, A is the cross-sectional area of the column, and the ratio P_{cr}/A is defined as the *critical unit load* for the column. Although

[42]See, for example, ref. 33, p. 256

[43]Another means of determining the tangent modulus is to differentiate the *Ramberg–Osgood* equation, an empirical stress-strain equation that is valid well into the plastic range. See, for example, ref. 33, p. 20, for the Ramberg–Osgood equation and associated material constants for various steel, aluminum, and magnesium alloys.

[44]See, for example, ref. 34, p. 585.

the critical unit load has the dimensions of stress, *it should not be treated as a stress* because of its nonlinear character.

Because the derivation of (2-80) is based on the premise that equal couples exist at the locations of the column end constraints, and that the maximum lateral deflection occurs at midspan, the secant formula is valid only for columns that meet these conditions. For example, the secant formula is valid for columns of the types shown in Figures 2.37(a) and (c), but not valid for columns of the types shown in Figures 2.37(b) or (d).

The secant formula is not very convenient for calculation purposes since the critical unit load cannot be explicitly isolated, but with the aid of a computer, or by appropriate graphical techniques, it can be employed satisfactorily. It should be observed also that (2-80) is undefined for zero eccentricity, but for very small eccentricities it approaches the Euler buckling curve as a limit for long columns and approaches the simple compressive yielding curve as a limit for short columns. This is shown in Figure 2.38.

Column Failure Prediction and Design Considerations

A designer must be aware that either of two potential failure modes may govern column failure, namely, (1) compressive yielding or (2) buckling. This may be expressed as FIPTOI (failure is predicted to occur if)

$$\frac{P_{cr}}{A} \geq S_{ypc} \tag{2-81}$$

or if

$$\frac{P_{cr}}{A} \geq \frac{\pi^2 E}{\left(\dfrac{L_e}{k}\right)^2} \tag{2-82}$$

Whichever of these expressions gives the lower value of allowable column load P will govern. If (2-81) governs, the column is said to be "short," and compressive yielding is the governing failure mode. If (2-82) governs, the column is said to be "long," and column buckling is the governing failure mode. This may be depicted graphically by plotting (2-81) and (2-82) as unit load versus slenderness ratio, as shown in Figure 2.39. The intersection of the yielding curve with the buckling curve at B represents the slenderness ratio transition from short to long columns, as indicated. As a practical matter, however, the tran-

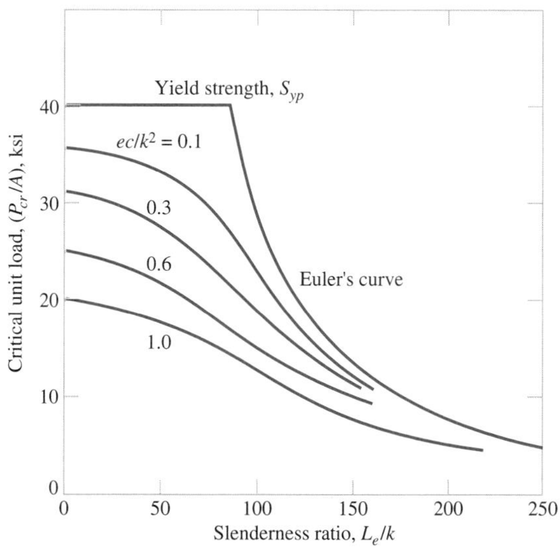

Figure 2.38
Secant buckling equation plotted for various eccentricities. Euler's curve and the compressive yielding curve, as shown, are asymptotes as the column eccentricity approaches zero. Modulus of elasticity $E = 30 \times 10^6$ psi.

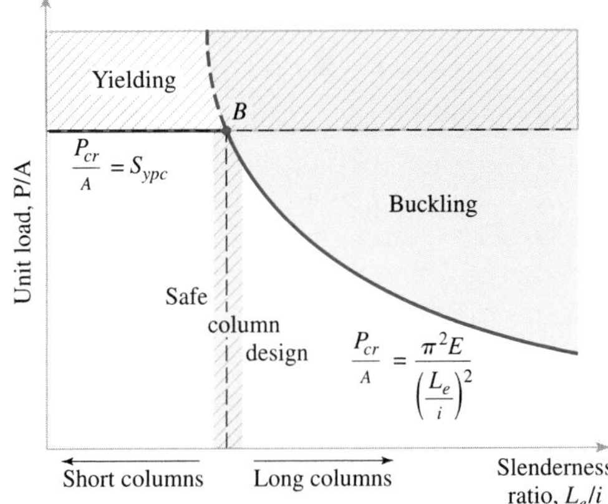

Figure 2.39
Regions of column behavior.

sition from short to long columns is blurred by unknown factors, such as load eccentricity, initial crookedness, local yielding or others, requiring designer judgment when using (2-81) and (2-82) in the vicinity of B. Column-end constraints are sometimes different in two different planes, as for example in a connecting rod end that is essentially pinned in one principal plane and fixed in the other principal plane. It is important to recognize that selection of the appropriate radius of gyration and appropriate effective column length depends on which plane of buckling is under consideration. Two separate column analyses might be required in such a case to define the more critical buckling condition. Finally, it should be noted that all the foregoing developments are for *primary buckling* of the column as a whole, where the shape of the cross section does not change significantly. In certain cases, usually involving thin-walled sections such as tubes or rolled shapes, *local buckling* may occur, in which a significant local change in the cross section takes place. Local buckling must be considered separately, and the final design of the column must be based on its ability to resist both primary buckling and local buckling under the applied loads.

Example 2.17 Critical Column-Buckling Load

The compressive stress-strain curve for 7075-T7351 aluminum alloy is given in Figure E2.17. A hollow cylindrical bar of this material has a 4-inch outside diameter with a $^1/_8$-inch wall thickness. If a 9-ft long column is constructed with one end fixed and one end free [as in Figure 2.37(b)], calculate the critical buckling load according to

a. Euler's equation
b. The Euler–Engesser equation
c. The secant formula, assuming zero eccentricity
d. The secant formula, assuming a load eccentricity of $^1/_8$ inch off the axial centerline of the column
e. The secant formula, assuming a load eccentricity of 1 inch off the axial centerline of the column

Solution

a. Using Euler's equation (2-76), and effective length $L_e = 2L$ from Table 2.3,

$$P_{cr} = \frac{\pi^2 EI}{L_e^2} = \frac{\pi^2(10.5 \times 10^6)\left[\frac{\pi}{64}(4.0^4 - 3.75^4)\right]}{[2(9 \times 12)]^2} = 6350\text{-lb critical load} \quad (1)$$

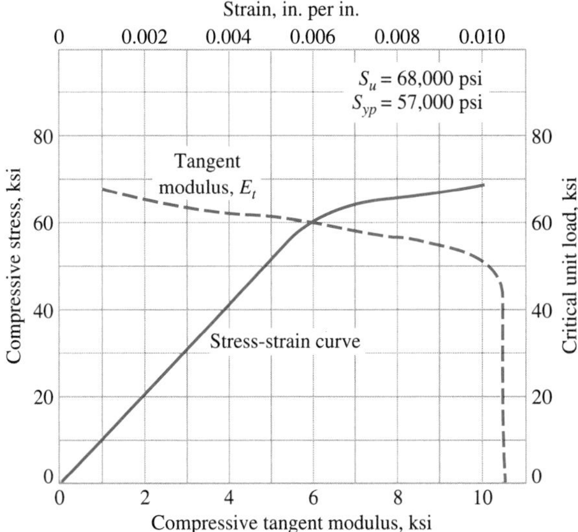

Figure E2.17
Stress-strain and compressive tangent modulus curves for 7075-T7351 aluminum alloy. See ref. 35, p. 3-232.

b. Using the Euler–Engesser equation (2-79)

$$\frac{P_{cr}}{A} = \frac{\pi^2 E_t}{(L_e/k)^2} \tag{2}$$

where

$$k = \sqrt{\frac{I}{A}} = \sqrt{\frac{\frac{\pi}{64}[4.0^4 - 3.75^4]}{\frac{\pi}{4}[4.0^2 - 3.75^2]}} = 1.37 \tag{3}$$

so (2) becomes

$$\frac{P_{cr}}{\frac{\pi}{4}[4.0^2 - 3.75^2]} = \frac{\pi^2 E_t}{\left[\frac{2(9 \times 12)}{1.37}\right]^2} \tag{4}$$

or

$$0.657 P_{cr} = 3.97 \times 10^{-4} E_t \tag{5}$$

Solving (5) simultaneously with the E_t curve of Figure E2.17 requires an iterative process. Using $P_{cr} = 6350$ lb from (1) as a starting point, (5) gives

$$E_t = \frac{0.657(6350)}{3.97 \times 10^{-4}} = 10.5 \times 10^6 \text{ psi} \tag{6}$$

Next, the critical unit load for $P_{cr} = 6350$ lb may be calculated as

$$\frac{P_{cr}}{A} = 0.657(6350) = 4170 \text{ psi} \tag{7}$$

Reading into Figure E2.17 to the E_t curve with a critical unit load of 4170 psi gives

$$E_t = 10.5 \times 10^6 \text{ psi} \tag{8}$$

Since (6) and (8) agree, no further iteration is required, and the compatible solution is

$$E_t = 10.5 \times 10^6 \text{ psi} \tag{9}$$

Example 2.17 Continues

and

$$P_{cr} = 6350\text{-lb critical load} \tag{10}$$

Thus for this case, the Euler solution and Euler–Engesser solution give the same result.

c. For zero eccentricity ($e = 0$), the secant formula (2-80) is undefined and furthermore (2-80) is not valid for fixed-free end condtions.

d. The secant formula (2-80) is not valid for fixed-free end conditions.

e. The secant formula (2-80) is not valid for fixed-free end conditions.

Buckling of Elements Other Than Columns

Although column buckling is an important case of elastic instability, other machine elements may also buckle under certain conditions. For example, elastic instability may be induced when a torsional moment is applied to a long thin rod, when a bending moment is applied to a thin deep beam, or when a thin-walled tube is subjected to external pressure. The long thin rod buckles by trying to coil, the thin deep beam buckles by twisting laterally, and the externally pressurized thin-walled tube buckles by folding or wrinkling.

The equations for the onset of buckling for these three cases may be written as shown here, but many other cases are presented in the literature.[45]

Torsion of a Long Thin Rod

The governing equation for the onset of buckling for the case shown in Figure 2.40(a) is

$$(M_t)_{cr} = \frac{2\pi EI}{L} \tag{2-83}$$

where $(M_t)_{cr}$ = critical torsional buckling moment, in-lb
E = Young's modulus, psi
I = area moment of inertia, in^4
L = length, in

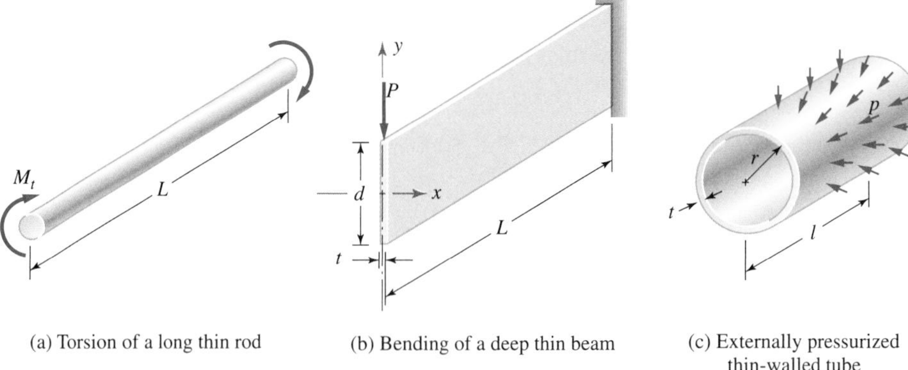

Figure 2.40
Some potential elastic instability cases other than columns.

[45]See, for example, refs. 32, 34, 36, and 37.

TABLE 2.5 Buckling Constant K for Various Beam Loading Conditions

Type of Beam	Type of Loading	K
Cantilever	P (lb) concentrated at free end	4.013
Cantilever	q (lb/in) distributed over length L	12.85
Simply supported	P (lb) concentrated at center	16.93
Simply supported	q (lb/in) distributed over length L	28.3

Bending of a Thin Deep Beam

The governing equation for onset of buckling for the case shown in Figure 2.40(b) is

$$P_{cr} = \frac{K\sqrt{GJ_e EI_y}}{L^2} \tag{2-84}$$

where
P_{cr} = critical buckling load, lb
E = Young's modulus, psi
I_y = area moment of inertia about y-axis, in^4
G = shear modulus, psi
J_e = property of cross section relating torque M_t to angle of twist θ in $\theta = M_t L/GJ_e$; for a thin rectangle $J_e = dt^3/3$
L = beam length, in
d = beam depth, in
t = beam thickness (width), in
K = constant depending on type of loading and type of support (see Table 2.5)

Externally Pressurized Thin-Walled Tube

The governing equation for the onset of buckling for the case shown in Figure 2.40(c) is

$$p_{cr} = \frac{0.25Et^3}{(1 - v^2)r^3} \tag{2-85}$$

valid only for a long tube with free ends, that is, for

$$L > 4.90r\sqrt{\frac{r}{t}} \tag{2-86}$$

where
p_{cr} = critical external buckling pressure, psi
l = length of tube, in
E = Young's modulus, psi
t = wall thickness, in
r = tube radius, in
v = Poisson's ratio

Example 2.18 Buckling of a Beam in Bending

A sheet-steel cantilevered bracket of rectangular cross section 0.125 inch by 4.0 inches is fixed at one end with the 4.0-inch dimension vertical. The bracket, which is 14.0 inches long, must support a vertical load P at the free end. The properties of the steel material are S_u = 82,000 psi, S_{yp} = 45,000 psi, and elongation in 2 inches = 16 percent.

Example 2.18 Continues

a. What is the maximum load P that can be placed vertically at the free end without inducing failure?

b. Identify the governing failure mode.

Solution

a. Both yielding and buckling are potential failure modes. Both should be checked.

For buckling failure, the critical buckling load is given by (2-84) as

$$P_{cr} = \frac{K\sqrt{GJ_e EI_y}}{L^2} \tag{1}$$

where $K = 4.013$ (from Table 2.4)

$G = 11.5 \times 10^6$ psi

$J_e = \frac{dt^3}{3} = \frac{4(0.125)^3}{3} = 2.60 \times 10^{-3}\ \text{in}^4$ (for thin rectangle)

$E = 30 \times 10^6$ psi (Young's modulus for steel)

$I_y = \frac{dt^3}{12} = \frac{4(0.125)^3}{12} = 6.51 \times 10^{-4}\ \text{in}^4$ (about the vertical y-axis)

$L = 14.0$ inches

Hence (1) becomes

$$P_{cr} = \frac{4.013\sqrt{(11.5\times10^6)(2.60\times10^{-3})(30\times10^6)(6.51\times10^{-4})}}{(14)^2} = 495\ \text{lb} \tag{2}$$

For yielding failure, assuming no stress concentration effects, the critical point is at the outer fiber at the fixed end, where bending stress is

$$\sigma = \frac{Mc}{I} = \frac{PLc}{I} = \frac{PL\left(\frac{d}{2}\right)}{\left(\frac{td^3}{12}\right)} = \frac{6PL}{td^2} \tag{3}$$

Setting $\sigma = S_{yp}$, the corresponding critical load P_{yp-cr} becomes

$$P_{yp-cr} = \frac{S_{yp}td^2}{6L} = \frac{(45{,}000)(0.125)(4.0)^2}{6(14.0)} = 1070\ \text{lb} \tag{4}$$

Since the buckling critical load P_{cr} is lower than the yielding critical load P_{yp-cr}, buckling governs and the critical failure load P_f is

$$P_f = P_{cr} = 495\ \text{lb} \tag{5}$$

b. From the calculations just completed, buckling is the governing failure mode.

2.8 Shock and Impact

The rapid application of forces or displacements to a structure or a machine part often produces stress levels and deformations very much larger than would be generated by the same forces and displacements applied gradually. Such rapidly applied loads or displacements are usually called *shock* or *impact loads*. Whether the loading on a structure should be considered *quasistatic* or as impact loading is often judged by comparing the time of application of the load, or *rise time*, with the longest natural period of the structure. If the rise time is more than about three times the longest natural period, the loading may usually be considered as quasistatic. If the rise time is less than about one-half the longest natural period, it is usually necessary to consider the loading as impact or shock loading. For quasistatic loading a designer is usually interested only in the maximum value of the load. For shock loading, not only is the peak load of importance, but also the rise time and the *impulse* (area under the force versus time curve).

Impact loading may be generated in machines or structures in various ways. For example, a *rapidly moving load*, such as a train moving across a bridge, generates impact loading. *Suddenly applied loads*, such as produced during combustion in the power stroke of an internal combustion engine, and *direct impact loads*, such as the drop of a forging hammer, produce shock loads. *Inertial loads* produced by large accelerations, such as during the crashing of an aircraft or automobile, in most cases generate conditions of shock or impact. When impact loading is repetitive, conditions of *impact fatigue, impact wear*, or *impact fretting* may be induced.[46] It is also important to note that material properties, such as ultimate strength, yield strength, and ductility, may be significantly influenced by impact loading conditions.[47]

Stress Wave Propagation Under Impact Loading Conditions

When a force is rapidly applied to a region of the boundary of an elastic body, the particles in a thin layer of material directly under the region of application are set into motion; the remainder of the body, remote from the loading, remains undisturbed for some finite length of time. As time passes the thin region of moving particles expands and propagates into the body in the form of an *elastic deformation wave*. Behind the *wave front* the body is deformed and the particles are in motion. Ahead of the wave front the body remains undeformed and at rest. If the geometry of the body is simple and uniform, and if the applied force is well defined and uniformly applied, classical equations for wave propagation in elastic media may be utilized to calculate stresses and deformations in the body.[48] Complicating factors such as boundary support, material damping, local yielding, irregular geometry, and nonuniform load application may be accounted for by more advanced methods. Using the *wave propagation model*, and considering the influencing factors just listed, it has been found that propagating waves reflect internally from boundaries of the body and interact with each other, sometimes canceling and sometimes reinforcing to produce local regions of high stress or strain due to the impact loading. Although the wave propagation model provides a good concept of the actual physical behavior of a body under impact conditions, calculation of stresses and deformations using this method is complicated, and beyond the scope of this text. A simpler method based on conservatiion of energy will be utilized here to approximate maximum stresses and deflections under impact loading.

Energy Method of Approximating Stress and Deflection Under Impact Loading Conditions

For simple machine members the maximum stress or deflection under impact conditions may be approximated by utilizing the concept of conservation of energy, wherein the ex-

[46]See 2.3. [47]See ref. 1, pp. 562–569. [48]See, for example, ref. 1, pp. 533–561.

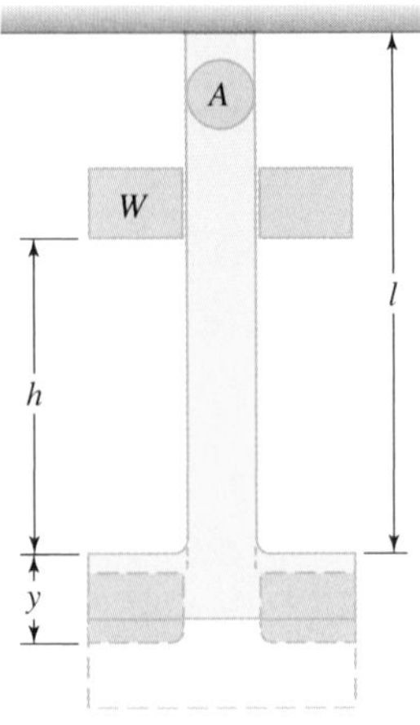

Figure 2.41
Simple tension member subjected to impact loading caused by a falling weight.

ternal work done on a structure must be equal to the potential energy of strain stored in the structure (if losses are assumed to be negligible). To utilize this energy method, a designer proceeds by equating the external work to the stored strain energy, putting the energy expressions in terms of stress or deflection, and solving for the stress or deflection.

For example, in the case of the simple tension member shown in Figure 2.41, the mass of weight W is allowed to fall through height h before it contacts the pan at the end of the tension bar. The resistance offered by the bar brings the weight W to a halt, and in the process it stretches the bar a distance y as shown, storing strain energy in the bar. In using the energy method to estimate the maximum stress in the bar due to this impact loading, it is assumed that:

1. The inertial resistance of tension bar and weight pan is negligible; that is, the mass of the bar and pan is much smaller than the striking mass.
2. The deflection of the bar is directly proportional to the applied force and is not a function of time.
3. The material obeys Hooke's law, that is, remains in the linear elastic range.
4. No energy is lost in the impact.

Under these assumptions, the maximum stress may be estimated by the energy method. The external energy, EE, may be calculated as the change in poential energy of the striking mass during its fall, whence

$$EE = W(h + y_{max}) \tag{2-87}$$

Since the bar is assumed to remain in the linear elastic range, Hooke's Law states that

$$\sigma_{max} = E\varepsilon = \frac{Ey_{max}}{l} \tag{2-88}$$

from which the end deflection y may be expressed in terms of stress σ, length l, and modulus of elasticity E as

$$y_{max} = \frac{\sigma_{max} l}{E} \tag{2-89}$$

Utilizing this expression for the deflection, (2-87) may be written as

$$EE = W\left(h + \frac{\sigma_{max} l}{E}\right) \tag{2-90}$$

The strain energy, SE, stored in the bar at the time of maximum deflection y may be expressed as the product of the average force applied times the deflection, or

$$SE = F_{ave} y_{max} = \left(\frac{0 + F_{max}}{2}\right) y_{max} \tag{2-91}$$

However, for this simple tension bar of cross-sectional area A

$$F_{max} = \sigma_{max} A \tag{2-92}$$

which may be substituted into (2-91), together with (2-89), to give

$$SE = \left(\frac{\sigma_{max} A}{2}\right)\left(\frac{\sigma_{max} l}{E}\right) = \frac{\sigma_{max}^2}{2E}(Al) \tag{2-93}$$

Equating the external energy expression of (2-90) to the strain energy expression of (2-93) then gives

$$W\left(h + \frac{\sigma_{max} l}{E}\right) = \frac{\sigma_{max}^2}{2E}(Al) \tag{2-94}$$

or

$$\sigma_{max}^2\left(\frac{Al}{2E}\right) - \sigma_{max}\left(\frac{Wl}{E}\right) - Wh = 0 \tag{2-95}$$

Dividing equation (2-95) by $(Al/2E)$ gives

$$\sigma_{max}^2 - \left(\frac{2W}{A}\right)\sigma_{max} - \frac{2WhE}{Al} = 0 \tag{2-96}$$

which may be solved by the quadratic formula to give a maximum value of

$$\sigma_{max} = \frac{W}{A}\left[1 + \sqrt{1 + \frac{2hEA}{Wl}}\right] \tag{2-97}$$

This is an energy-method estimate of the maximum stress that will be developed in the tension bar due to the impact of weight W falling from rest through height h. A similar expression for the maximum end deflection y_{max} may be written by combining (2-97) with (2-89) to give

$$y_{max} = \frac{Wl}{AE}\left[1 + \sqrt{1 + \frac{2hEA}{Wl}}\right] \tag{2-98}$$

It is interesting to note in (2-97) that under impact loading the maximum stress may be reduced not only by increasing the cross-sectional area A but also by decreasing the modulus of elasticity E or increasing the length l of the bar. Thus, it is apparent that the impact situation is quite different from the case of static loading in which the stress in the bar is independent of the modulus of elasticity and the length of the bar. The bracketed expression of (2-97) and (2-98) is called the *impact factor*.

It is also interesting to consider the *limiting case* of (2-97) for which the drop height h is zero. This limiting case, in which the striking mass is held just in contact with the weight pan and then released from zero height, is called a *suddenly applied load*. From (2-97), with $h = 0$, the stress developed in a bar subjected to a suddenly applied load is

$$(\sigma_{max})_{\substack{suddenly\\applied}} = 2\frac{W}{A} = 2(\sigma_{max})_{static} \tag{2-99}$$

Thus, the maximum stress developed by suddenly applying load W to the end of the bar is twice as large as the maximum stress developed in the bar by slowly applying the same load W. Similarly, from (2-98) the maximum deflection generated by a suddenly applied load is twice the deflection produced under static loading, whence

$$(y_{max})_{\substack{suddenly\\applied}} = 2(y_{max})_{static} \tag{2-100}$$

If the weight of the bar shown in Figure 2.41 is *not* negligibly small, as was assumed, the expressions of (2-97) and (2-98) are somewhat modified to account for the fact that some of the energy of the falling weight is used to accelerate the mass of the bar. If the bar is assumed to have a weight of q per unit length of the bar, then (2-97) and (2-98) are modified to

$$\sigma_{max} = \frac{W}{A}\left[1 + \sqrt{1 + \frac{2hEA}{Wl}\left(\frac{1}{1 + \frac{ql}{3W}}\right)}\right] \tag{2-101}$$

and

$$y_{max} = \frac{Wl}{AE}\left[1 + \sqrt{1 + \frac{2hEA}{Wl}\left(\frac{1}{1 + \frac{ql}{3W}}\right)}\right] \tag{2-102}$$

If the stress level generated by the impact exceeds the yield strength of the material, Hooke's Law no longer holds and the equations just developed are not valid. However, it is still possible to estimate stress and deflection under such conditions.[49]

To maximize the resistance of machine or structural members to impact loading, it is important to distribute the maximum stress as uniformly as possible. To illustrate, the two specimens shown in Figure 2.42 may be compared under static and impact loading. Note that both specimens have the same length, have the same minimum cross-sectional area, and are made of the same material. The effects of stress concentration will be neglected for this comparison.

Under conditions of static loading, it may be observed that the load P_{fy} required to produce first yielding is, for *both* bars in Figure 2.42,

$$P_{fy} = A_1 S_{yp} \tag{2-103}$$

where S_{yp} is the yield strength of the material.

Next, the total strain energy stored in each of the bars may be calculated at the time of first yielding. This total strain energy stored is, of course, also equal to the external energy required to initiate yielding in each case. Utilizing (2-93), we may write the total strain energy U_a stored in the bar of Figure 2.42(a) as

$$U_a = \frac{S_{yp}^2}{2E}(A_1 l) \tag{2-104}$$

In calculating the total strain energy U_b stored in the bar of Figure 2.42(b), it must be recognized that the stress level in the two end segments will be lower than the stress level in the central segment by the ratio A_1/A_2. Thus, again utilizing (2-93) in piecewise fashion,

$$U_b = \frac{S_{yp}^2}{2E}A_1\left(\frac{l}{3}\right) + 2\left\{\frac{\left[S_{yp}\left(\frac{A_1}{A_2}\right)\right]^2}{2E}A_2\left(\frac{l}{3}\right)\right\} \tag{2-105}$$

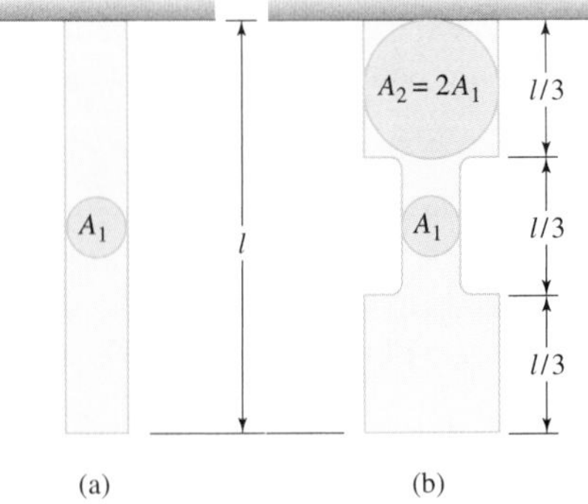

Figure 2.42
Configuration of two specimens used in comparing the effects of geometry under static and impact loading. Configuration (a) is superior because stress is uniformly distributed throughout.

[49]See, for example, ref. 1, p. 530 ff.

Since the area A_2 is given in Figure 2.42 to be twice A_1, (2-105) becomes

$$U_b = \frac{S_{yp}^2}{2E}\left(\frac{A_1 l}{3}\right) + \frac{S_{yp}^2}{8E}\left(\frac{4A_1 l}{3}\right) \tag{2-106}$$

or

$$U_b = \frac{2}{3}\left[\frac{S_{yp}^2}{2E}(A_1 l)\right] \tag{2-107}$$

Thus,

$$U_b = \frac{2}{3}U_a \tag{2-108}$$

From this result it is evident that the energy stored in the specimen of Figure 2.42(b) when yielding begins is only two-thirds of the energy stored in the specimen of Figure 2.42(a) when yielding begins, in spite of the larger total volume in the specimen of Figure 2.42(b). In other words, the impact resistance of design (a), with its smaller total volume, is significantly superior to the impact resistance of design (b) in Figure 2.42. Furthermore, if stress concentration effects had been considered, design (b) would have been even worse by comparison.

The key to success in designing members for maximum resistance to impact loading is to assure that the maximum stress is uniformly distributed throughout the largest possible volume of material. Grooved and notched members are very poor in impact resistance because for these configurations there always exist small, highly stressed volumes of material. A relatively minor impact load may in such cases produce fracture. Thus, members that contain holes, fillets, notches, or grooves may be subject to abrupt fracture under impact or shock loading.

Example 2.19 Impact Loading on a Beam

A simply supported steel beam is 60 inches long and has a rectangular cross section 1.0 inch wide by 3.0 inches deep. A weight of 78 lb is dropped from a height h at midspan.

a. If the yield strength of the material is 40,000 psi, and if beam mass is neglected, what drop height would be required to produce first evidence of yielding in the beam?

b. What is the impact factor under these circumstances?

Solution

a. For a simply supported beam the mid-span deflection under quasistatic loading is (see Table 4.1, case 1)

$$y_{st} = \frac{WL^3}{48EI} \tag{1}$$

and the corresponding maximum bending stress is, at mid-span,

$$\sigma_{max} = \frac{Mc}{I} = \frac{\left(\frac{W}{2}\right)\left(\frac{L}{2}\right)c}{I} = \frac{WLc}{4I} \tag{2}$$

Thus, from (1)

$$y_{max} = \frac{WLcL^2}{(4)(12)\,cEI} = \frac{\sigma_{max}L^2}{12Ec} \tag{3}$$

Example 2.19 Continues

From (2-91)

$$SE = F_{ave}y_{max} = \left(\frac{W}{2}\right) y_{max} \tag{4}$$

or, substituting from (2) and (3)

$$SE = \left(\frac{2I\sigma_{max}}{Lc}\right)\left(\frac{\sigma_{max}L^2}{12Ec}\right) = \frac{\sigma_{max}^2 IL}{6Ec^2} \tag{5}$$

Using (2-87)

$$EE = W(h + y_{max}) = W\left(h + \frac{\sigma_{max}L^2}{12Ec}\right) \tag{6}$$

Equating (5) to (6) and solving for σ_{max}

$$\sigma_{max} = \frac{WLc}{4I}\left[1 + \sqrt{1 + \frac{(2)48EIh}{WL^3}}\right] \tag{7}$$

or, using (1)

$$\sigma_{max} = \frac{WLc}{4I}\left[1 + \sqrt{1 + \frac{2h}{y_{st}}}\right] \tag{8}$$

Using known numerical values, (8) becomes

$$\sigma_{max} = \frac{(78)(60)(3.0/2)}{4[(1)((3.0)^3/12)]}\left[1 + \sqrt{1 + \frac{2h}{\left[\dfrac{78(60)^3}{48(30 \times 10^6)(1.0)(3.0)^3(12)}\right]}}\right] \tag{9}$$

or

$$\sigma_{max} = 780\left[1 + \sqrt{1 + \frac{2h}{0.0052}}\right] \tag{10}$$

Equating (10) to the yield strength and solving for h,

$$h = \frac{0.0052}{2}\left[\left(\frac{40{,}000}{780} - 1\right)^2 - 1\right] = 6.57 \text{ inches} \tag{11}$$

Thus the 78-lb weight must be dropped from a height of at least 6.57 inches to produce initial yielding of the mid-span outer fibers of the beam.

b. The impact factor for this case is, from (10)

$$IF = \left[1 + \sqrt{1 + \frac{2(6.57)}{0.0052}}\right] = 51.3 \tag{12}$$

2.9 Creep and Stress Rupture

Creep in its simplest form is the progressive accumulation of plastic strain in a specimen or machine part under stress at elevated temperature over a period of time. Creep failure occurs when the accumulated creep strain results in a deformation of the machine part that exceeds the design limits. *Creep rupture* is an extension of the creep process to the limit-

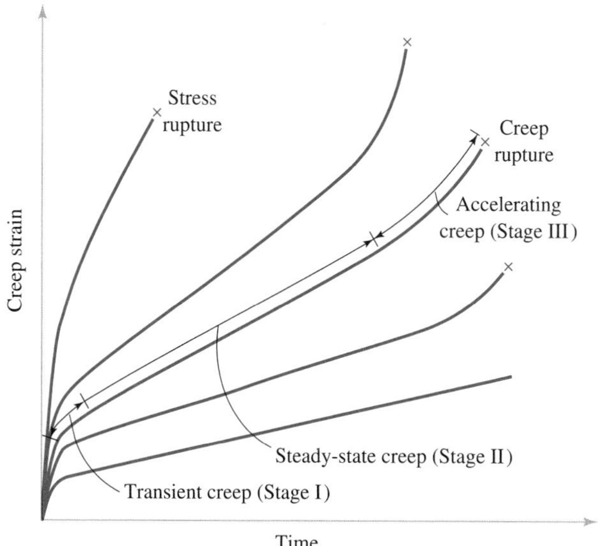

Figure 2.43
Illustration of creep and stress rupture.

ing condition where the stressed member actually separates into two parts. *Stress rupture* is a term used interchangeably by many with *creep rupture*; however, others reserve the term *stress rupture* for the rupture termination of a creep process in which steady-state creep is never reached, and use the term *creep rupture* for the rupture termination of a creep process in which a period of steady-state creep has persisted. Figure 2.43 illustrates these differences.

Creep strains of engineering significance are not usually encountered until the operating temperatures reach a range of approximately 35 to 70 percent of the melting point on a scale of absolute temperature. The approximate melting temperatures for several substances are shown in Table 2.6.

Early creep studies were reported by a French engineer who was motivated to study time-dependent elongations of wire ropes, used for bridge suspension, that exceeded

TABLE 2.6 Melting Temperatures[1]

Material	°F	°C
Hafnium carbide	7030	3887
Graphite (sublimes)	6330	3500
Tungsten	6100	3370
Tungsten carbide	5190	2867
Titanium	3260	1795
Platinum	3180	1750
Chromium	3000	1650
Iron	2800	1540
Stainless steels	2640	1450
Steel	2550	1400
Aluminum alloys	1220	660
Magnesium alloys	1200	650
Lead alloys	605	320

[1]From ref. 15.

elastic predictions. Not until after World War I, however, did creep failure become important as a failure mode. Since that time many applications have been identified in which creep failure may govern the design. Load-carrying members operating in the temperature range from 1000°F to 1600°F are found in power plants, refineries, and chemical processing plants. Furnace parts are routinely exposed to temperatures of 1600°F to 2200°F. Gas turbine rotor blades are subjected to temperatures of 1200°F to 2200°F, together with high centrifugal stresses. Rocket nozzles and spacecraft nose cones are subjected to even higher temperatures for brief periods of time. Skin temperatures of Mach 7 aircraft have been estimated at about 5000°F, with aerodynamic and structural consequences of creep deformation, creep buckling, and stress rupture becoming critical design considerations.

Creep deformation and rupture are initiated in the grain boundaries and proceed by sliding and separation. Thus, creep rupture failures are intercrystalline, in contrast, for example, to the transcrystalline failure topography exhibited by room-temperature fatigue failures. Although creep is a plastic flow phenomenon, the intercrystalline failure path gives a rupture surface that has the appearance of brittle fracture. Creep rupture typically occurs without necking and without warning. Current state-of-the-art knowledge does not permit a reliable prediction of creep or stress rupture properties on a theoretical basis. Further, there seems to be little or no correlation between the creep properties of a material and its room-temperature mechanical properties. Therefore, test data and empirical methods of extending these data are relied on heavily for prediction of creep behavior under anticipated service conditions.

Predictions of Long-Term Creep Behavior

Much time and effort have been expended in attempting to devise good short-time creep tests for accurate and reliable prediction of long-term creep and stress rupture behavior. It appears, however, that really reliable creep data can be obtained only be conducting long-term creep tests that duplicate actual service loading and temperature conditions as nearly as possible. Unfortunately, designers are unable to wait for years to obtain design data needed in creep failure analysis. Therefore, certain useful techniques have been developed for approximating long-term creep behavior based on a series of short-term tests.

Data from creep testing may be cross plotted in a variety of different ways. The basic variables involved are stress, strain, time, temperature, and perhaps strain rate. Any two of these basic variables may be selected as plotting coordinates, with the remaining variables to be treated as parametric constants for a given set of curves. Three commonly used methods for extrapolating short-time creep data to long-term applications are the *abridged method*, the *mechanical acceleration method*, and the *thermal acceleration method*. In the abridged method of creep testing the tests are conducted at several different stress levels and at the contemplated operating temperatures. The data are plotted as creep strain versus time for a family of stress levels, all run at constant temperature. The curves are plotted out to the laboratory test duration and then extrapolated to the required design life.

In the mechanical acceleration method of creep testing, the stress levels used in the laboratory tests are significantly higher than the contemplated design stress levels, so the limiting design strains are reached in a much shorter time than in actual service. The data taken in the mechanical acceleration method are plotted as stress level versus time for a family of constant strain curves, all at a constant temperature. The stress rupture curve may also be plotted by this method. The constant strain curves are plotted out to the laboratory test duration and then extrapolated to the design life.

The thermal acceleration method involves laboratory testing at temperatures much higher than the actual service temperature expected. The data are plotted as stress versus time for a family of constant temperatures where the creep strain produced is constant for the whole plot. Stress rupture data may also be plotted in this way. The data are plotted out to the laboratory test duration and then extrapolated to the design life.

Creep testing guidelines usually dictate that test periods of less than 1 percent of the expected life should not be deemed to give significant results. Tests extending to at least 10 percent of the expected life are preferred where feasible.

In addition to these graphical data-extrapolation methods for predicting long-term creep behavior, several prediction *theories* have been proposed to correlate short-time elevated temperature tests with long-term service performance at more moderate temperatures. The most widely used of these theories is the *Larson–Miller theory*,[50] which postulates that for each combination of material and stress level there exists a unique value of a parameter P that is related to temperature and time by the equation

$$P = (\Theta + 460)(C + \log_{10} t) \tag{2-109}$$

where P = Larson–Miller parameter, constant for a given material and stress level

Θ = temperature, °F

C = constant, usually assumed to be 20

t = time in hours to rupture or to reach a specified value of creep strain

Larson and Miller tested this equation by predicting actual creep and rupture results for some 28 different materials with good success. By using (2-109) it is a simple matter to find a short-term combination of temperature and time that is equivalent to any desired long-term service requirement. For example, for any given material at a specified stress level, the test conditions listed in Table 2.7 should be equivalent to the corresponding operating conditions shown.

Good agreement between theory and experiment has been confirmed by other investigators using the Larson–Miller parameter for a wide variety of materials, including several plastics, in predicting long-term creep behavior and stress rupture performance.

Creep under Uniaxial State of Stress

Uniaxial creep and stress rupture tests of 100 hours (4 days), 1000 hours (42 days), and 10,000 hours (420 days) duration are common, with longer-duration testing of 100,000 hours (11.5 years) being performed in a few cases. Certain recent high-performance applications have given rise to short-term creep testing that measures duration in minutes rather than hours or years. For example, in some cases creep test durations of 1000 minutes, 100 minutes, 10 minutes, and 1 minute have been used.

It is of interest to note that with an increase in temperature the static ultimate strength,

TABLE 2.7 Equivalent Conditions Based on Larson–Miller Parameter

Operating Condition	Equivalent Test Condition
10,000 hours at 1000°F	13 hours at 1200°F
1000 hours at 1200°F	12 hours at 1350°F
1000 hours at 1350°F	12 hours at 1500°F
1000 hours at 300°F	2.2 hours at 400°F

[50]See ref. 39.

yield strength, and modulus of elasticity all tend to decrease, whereas the elongation and reduction in area tend to increase. The stress concentration effect due to a geometrical notch is also reduced at elevated temperatures.

Many relationships have been proposed to relate stress, strain, time, and temperature in the creep process. If one investigates experimental creep strain versus time data, it will be observed that the data are close to linear for a wide variety of materials when plotted on log-strain versus log-time coordinates. Such a plot is shown in Figure 2.44 for three different materials. An equation describing this type of behavior is

$$\delta = At^a \tag{2-110}$$

where δ = true creep strain
t = time
A, a = empirical constants

Differentiating with respect to time and setting $aA = b$ and $(1 - a) = n$, gives

$$\dot{\delta} = bt^{-n} \tag{2-111}$$

This equation represents a variety of different types of creep strain versus time curves, depending on the magnitude of the exponent n. If n is zero, the behavior is termed *constant creep rate*. This type of creep behavior is most commonly found at high temperatures. If

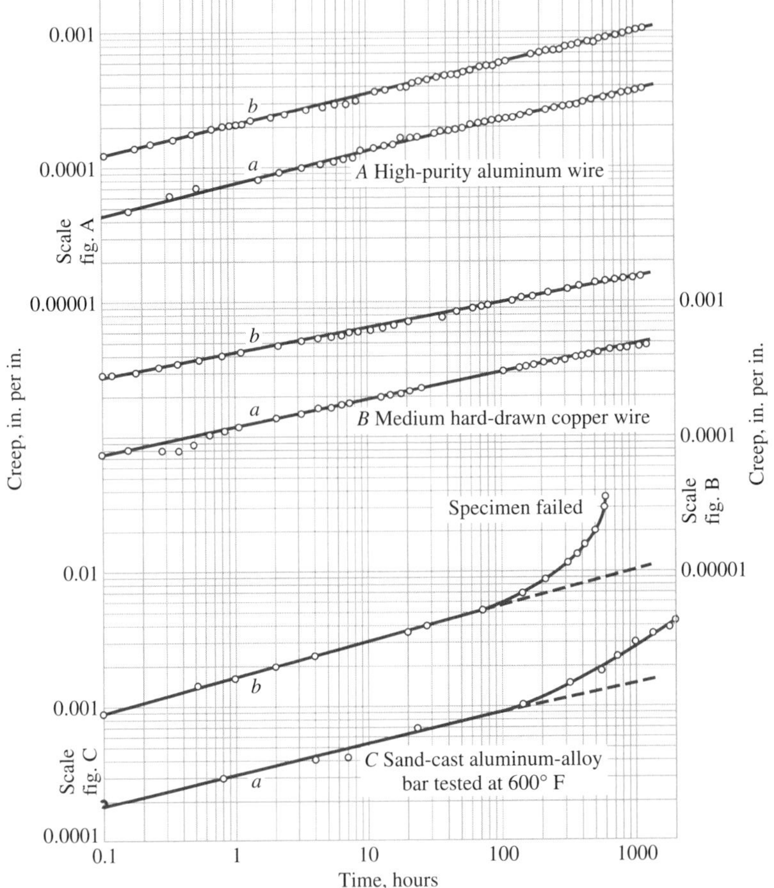

Figure 2.44
Creep curves for three materials plotted on log-log coordinates. (From ref. 39.)

the exponent n is unity, the behavior is termed *logarithmic creep*. This type of creep behavior is displayed by rubber, glass, and certain types of concrete, as well as by metals at lower temperatures. If the exponent n lies between zero and 1, the behavior is termed *parabolic creep*. This type of creep behavior occurs at intermediate and high temperatures.

The influence of stress level σ on creep rate can often be represented by the empirical expression

$$\dot{\delta} = B\sigma^N \tag{2-112}$$

Assuming the stress σ to be independent of time, (2-112) may be integrated to yield the true creep strain

$$\delta = Bt\sigma^N + C' \tag{2-113}$$

If the constant C' is small compared with $Bt\sigma^N$, as it often is, the result is called the *log-log stress-time* creep law, given as

$$\delta = Bt\sigma^N \tag{2-114}$$

As long as the instantaneous deformation upon load application and the Stage I transient creep are small compared to Stage II steady-state creep, (2-114) is useful as a design tool. With this expression a designer may calculate the stress required at a specified temperature to hold creep deformation within specified limits. In Table 2.8 the constants B and N are evaluated for five materials and temperatures, where time t is in days.

Methods are also available for calculating true creep strain under multiaxial states of stress [51] but are beyond the scope of this text.

TABLE 2.8 Constants for Log-Log Stress-Time Creep Law[1]

Material	Temperature	B, $(\text{in}^2/\text{lb})^N$ per day	N
1030 steel	750°F	48×10^{-38}	6.9
1040 steel	750°F	16×10^{-46}	8.6
2Ni-0.8Cr-0.4Mo steel	850°F	10×10^{-20}	3.0
12Cr steel	850°F	10×10^{-27}	4.4
12Cr-3W-0.4Mn steel	1020°F	15×10^{-16}	1.9

[1]See ref. 49, p. 82.

Example 2.20 Design to Prevent Excessive Creep

It is desired to design a solid cylindrical tension member 5 feet long, made of 1030 steel, to support a 10,000-lb load for 10 years at 750°F without exceeding 0.1-inch creep deformation. What diameter should the bar be made to prevent failure prior to the end of the 10-year design life?

Solution

From (2-114), the failure stress, σ_f, may be calculated as

$$\sigma_f = \left[\frac{\delta}{Bt}\right]^{1/N} \tag{1}$$

or

[51]See, for example, ref. 1, p. 471 ff.

Example 2.20 Continues

$$\sigma_f = \left[\frac{0.1/(5 \times 12)}{(48 \times 10^{-38})(10 \times 365)} \right]^{1/6.9} = 23{,}200 \text{ psi} \tag{2}$$

where the constants, B and N, are read from Table 2.8.

The diameter of the bar corresponding to incipient failure in 10 years, d_f, may then be calculated from

$$\sigma_f = \frac{F}{A} = \frac{4F}{\pi d_f^2} \tag{3}$$

whence

$$d_f = \sqrt{\frac{4F}{\pi \sigma_f}} = \sqrt{\frac{4(10{,}000)}{\pi(23{,}200)}} = 0.74 \text{ inch} \tag{4}$$

It should be noted that in actual practice a safety factor would typically be incorporated into the calculation, resulting in a somewhat larger design diameter.

Cumulative Creep Prediction

There is at the present time no universally accepted method for estimating the creep strain accumulated as a result of exposure for various periods of time at different temperatures and stress levels. However, several different techniques for making such estimates have been proposed. The simplest of these is a linear hypothesis suggested by E. L. Robinson. A generalized version of the Robinson hypothesis may be written as follows: If a design limit of creep strain δ_D is specified, it is predicted that the creep strain δ_D will be reached when

$$\sum_{i=1}^{k} \frac{t_i}{L_i} = 1 \tag{2-115}$$

where t_i = time of exposure at the ith combination of stress level and temperature

L_i = time required to produce creep strain δ_D if entire exposure were held constant at the ith combination of stress level and temperature

Stress rupture may also be predicted by (2-115) if the L_i values correspond to stress rupture. This prediction technique gives relatively accurate results if the creep deformation is dominated by Stage II steady-state creep behavior.

Example 2.21 Cumulative Creep

A sand-cast aluminum alloy creep tested at 600°F results in the data shown in Figure 2.44(C), curve b, when subjected to a stress level of 10,000 psi. A solid square support bracket of this material is 1 inch × 1 inch × 4 inches long and is subjected to a direct tensile load of 10,000 lb in the 4-inch direction. The ambient temperature cycle is 380°F for 1200 hours, then 600°F for 2 hours, and then 810°F for 15 seconds. How many of these cycles would you predict could be survived by the bracket if the design criterion is that the bracket must not elongate more than 0.10 inch?

Solution

As given, one ambient temperature cycle is

$$t_A = 1200 \text{ hours at } 380°\text{F} \tag{1}$$

$$t_B = 2 \text{ hours at } 600°\text{F} \tag{2}$$
$$t_C = 15 \text{ seconds at } 810°\text{F} \tag{3}$$

The limiting design elongation is given as

$$\Delta L = 0.10 \text{ inch} \tag{4}$$

hence the limiting design strain is

$$\delta_D = \frac{\Delta L}{L} = \frac{0.1}{4.0} = 0.025 \frac{\text{in}}{\text{in}} \tag{5}$$

Using the Larson–Miller parameter (2-109) to convert (1) and (3) to 600°F equivalent times of exposure,

$$P_A = (380 + 460)(20 + \log_{10} 1200) = 19{,}387 \tag{6}$$

whence, for 600°F operation, the equivalent time, t_{A-eq}, may be calculated from (2-109) as

$$19{,}387 = (600 + 460)(20 + \log_{10} t_{A-eq}) \tag{7}$$

or

$$t_{A-eq} = 0.0195 \text{ hour at } 600°\text{F} \tag{8}$$

Likewise,

$$P_C = (810 + 460)(20 + \log_{10} 0.0042) = 22{,}382 \tag{9}$$

so,

$$22{,}382 = (600 + 460)(20 + \log_{10} t_{C-eq}) \tag{10}$$

or

$$t_{C-eq} = 13.02 \text{ hours at } 600°\text{F} \tag{11}$$

Reading from Figure 2.44(C), curve b, using $\delta_D = 0.025$ from (5), and noting that $\sigma = 10{,}000$ psi for both the test data and the proposed support bracket,

$$L_{0.025} = 530 \text{ hours} \tag{12}$$

Next, using the Robinson hypothesis (2-115) with (8), (2), (11), and (12), and setting N equal to the number of repeated temperature cycles required to produce the limiting design strain δ_D,

$$N\left[\frac{0.0195}{530} + \frac{2}{530} + \frac{13.02}{530}\right] = 1 \tag{13}$$

or

$$N = 35 \text{ cycles} \tag{14}$$

to produce the limiting elongation.

2.10 Wear and Corrosion

Wear and corrosion probably account for a majority of all mechanical failures in the field, and an extensive research literature has built up in both areas. In spite of this, however, widely accepted quantitative life prediction techniques have not yet been developed. During the past three decades, substantial progress has been made in quantitative wear-prediction methods. Some of these prediction models are briefly discussed in the following paragraphs. Quantitative corrosion prediction remains a highly specialized area for which mechanical designers usually seek the help of corrosion engineers.

Wear

Wear may be defined as the undesired cumulative change in dimensions brought about by the gradual removal of discrete particles from contacting surfaces in motion, due predominantly to mechanical action. Corrosion often interacts with the wear process to change the character of the surfaces of wear particles through reaction with the environment.

Wear is, in fact, not a single process but a number of different processes that may take place independently or in combination. It is generally accepted that there are at least five major subcategories of wear,[52] including adhesive wear, abrasive wear, corrosive wear, surface fatigue wear, and deformation wear.

The complexity of the wear process may be better appreciated by recognizing that many variables are involved, including the hardness, toughness, ductility, modulus of elasticity, yield strength, fatigue properties, and structure and composition of the mating surfaces, as well as geometry, contact pressure, temperature, state of stress, stress distribution, coefficient of friction, sliding distance, relative velocity, surface finish, lubricants, contaminants, and ambient atmosphere at the wearing interface. Clearance versus contact-time history of the wearing surfaces may also be an important factor in some cases.

Adhesive wear is often characterized as the most basic or fundamental subcategory of wear since it occurs to some degree whenever two solid surfaces are in rubbing contact, and remains active even when all other modes of wear have been eliminated. The phenomenon of adhesive wear may be best understood by recalling that all real surfaces, no matter how carefully prepared and polished, exhibit a general waviness upon which is superposed a distribution of local protuberances or *asperities*. As two surfaces are brought into contact, therefore, only a relatively few asperities actually touch. Thus, even under very small applied loads the local pressures at the contact sites become high enough to exceed the yield strength of one or both surfaces, and local plastic flow ensues. If the contacting surfaces are clean and uncorroded, the very intimate contact produced by this local plastic flow brings the atoms of the two contacting surfaces close enough together to cause *cold welding*. Then if the surfaces are subjected to relative sliding motion, the cold-welded junctions must be broken. Whether they break at the original interface or elsewhere within the asperity depends upon surface conditions, temperature distribution, strain-hardening characteristics, local geometry, and stress distribution. If the junction is broken away from the original interface, a particle of one surface is transferred to the other surface, marking one event in the adhesive wear process. Later sliding interactions may dislodge the transferred particle as a loose wear particle, or it may remain attached. If properly controlled, the adhesive wear rate may be low and self-limiting, often being exploited in the "wear-in" process to improve mating surfaces such as bearings or gears so that full film lubrication may be effectively used.

One quantitative estimate of the amount of adhesive wear may be made as follows:[53]

If d_{adh} is the average wear depth, A_a the apparent contact area, L_s the total sliding distance, and W the normal force pressing the surfaces together,

$$d_{adh} = \left(\frac{k}{9S_{yp}}\right)\left(\frac{W}{A_a}\right)L_s \tag{2-116}$$

[52]See p. 120 of ref. 41; see also ref. 42.

[53]See ref. 41 and Chs. 2 and 6 of ref. 43.

or

$$d_{adh} = k_{adh}\, p_m L_s \tag{2-117}$$

where $p_m = W/A_a$ is the mean nominal contact pressure between bearing surfaces and $k_{adh} = k/9S_{yp}$ is a wear coefficient that depends on the probability of formation of a transferred fragment and the yield strength (or hardness) of the softer material. Typical values of the wear constant k (Archard constant) for several material pairs are shown in Table 2.9, and the influence of lubrication on the wear constant k is indicated in Table 2.10.

Noting from (2-117) that

$$k_{adh} = \frac{d_{adh}}{p_m L_s} \tag{2-118}$$

it may be observed that if the ratio $d_{adh}/(p_m L_s)$ is experimentally found to be constant, (2-117) should be valid. Experimental evidence has been accumulated[54] to confirm that for a given material pair this ratio is constant up to mean nominal contact pressures approximately equal to the uniaxial yield strength. Above this level the adhesive wear coefficient increases rapidly, with attendant severe galling and seizure.

Thus the average wear depth for adhesive wear conditions may be estimated as

$$d_{adh} = k_{adh} p_m L_s \quad \text{for} \quad p_m < S_{yp} \tag{2-119}$$

$$\text{unstable galling and seizure for } p_m \geq S_{yp}$$

Three major wear-control methods have been defined, as follows[55]: *principle of protective layers*, including protection by lubricant, surface film, paint, plating, phosphate, chemical, flame-sprayed, or other types of interfacial layers; *principle of conversion*, in which wear is converted from destructive to permissible levels through better choice of metal pairs, hardness, surface finish, or contact pressure; and *principle of diversion*, in which the wear is diverted to an economical replaceable wear element that is periodically discarded and replaced as "wearout" occurs. These general wear-control methods pertain not only to adhesive wear but to abrasive wear as well.

In the case of abrasive wear, the wear particles are removed from the surface by plowing and gouging action of the asperities of a harder mating surface or by hard particles en-

TABLE 2.9 Archard Adhesive Wear Constant *k* for Various Unlubricated Material Pairs in Sliding Contact[1]

Material Pair	Wear Constant k
Zinc on zinc	160×10^{-3}
Low-carbon steel on low-carbon steel	45×10^{-3}
Copper on copper	32×10^{-3}
Stainless steel on stainless steel	21×10^{-3}
Copper (on low-carbon steel)	1.5×10^{-3}
Low-carbon steel (on copper)	0.5×10^{-3}
Bakelite on bakelite	0.02×10^{-3}

[1]From Ch. 6 of ref. 43, Copyright © 1966, reprinted by permission of John Wiley & Sons, Inc.

[54]See pp. 124–125 of ref. 41.

[55]See p. 36 of ref. 44.

TABLE 2.10 Order of Magnitude Values for Adhesive Wear Constant *k* Under Various Conditions of Lubrication[1]

Lubrication Condition	Metal (on metal) Like	Metal (on metal) Unlike	Nonmetal (on metal)
Unlubricated	5×10^{-3}	2×10^{-4}	5×10^{-6}
Poor lubrication	2×10^{-4}	2×10^{-4}	5×10^{-6}
Average lubrication	2×10^{-5}	2×10^{-5}	5×10^{-6}
Excellent lubrication	2×10^{-6} to 10^{-7}	2×10^{-6} to 10^{-7}	2×10^{-6}

[1]From Ch. 6 of ref. 43, Copyright © 1966, reprinted by permission of John Wiley & Sons, Inc.

trapped between the rubbing surfaces. This type of wear is manifested by a system of surface grooves and scratches, often called *scoring*. The abrasive wear condition in which the hard asperities of one surface wear away the mating surface is commonly called *two-body wear*, and the condition in which hard abrasive particles between the two surfaces cause the wear is called *three-body wear*.

If d_{abr} is the average wear depth, A_a the apparent contact area, L_s the total sliding distance, and W the normal force pressing the surfaces together, the average abrasive wear depth may be estimated as

$$d_{abr} = \left(\frac{k_1}{9S_{yp}}\right)\left(\frac{W}{A_a}\right) L_s \tag{2-120}$$

or

$$d_{abr} = k_{abr} p_m L_s \tag{2-121}$$

where $p_m = W/A_a$ is mean nominal contact pressure between bearing surfaces, L_s is total sliding distance, and $k_{abr} = k_1/9S_{yp}$ is an abrasive wear coefficient that depends on the roughness characteristics of the surface and the yield strength (or hardness) of the softer material. Values of k_{abr} must be experimentally determined for each material combination and surface condition of interest, although useful data for approximating k_{abr} have been generated for several cases, some of which are shown in Table 2.11.

In selecting materials for abrasive wear resistance, it has been established that both hardness and modulus of elasticity are key properties. Increasing wear resistance is asso-

TABLE 2.11 Abrasive Wear Constant k_1 for Various Materials in Sliding Contact as Reported by Different Investigators[1]

Materials	Wear Type	Particle Size (μ)	Wear Constant k_1
Many	Two-body	—	180×10^{-3}
Many	Two-body	110	150×10^{-3}
Many	Two-body	40–150	120×10^{-3}
Steel	Two-body	260	80×10^{-3}
Many	Two-body	80	24×10^{-3}
Brass	Two-body	70	16×10^{-3}
Steel	Three-body	150	6×10^{-3}
Steel	Three-body	80	4.5×10^{-3}
Many	Three-body	40	2×10^{-3}

[1]See p. 169 of ref. 43, Copyright © 1966, reprinted by permission of John Wiley & Sons, Inc.

ciated with higher hardness and lower modulus of elasticity since both the amount of elastic deformation and the amount of elastic energy that can be stored at the surface are increased by higher hardness and lower modulus of elasticity.

When two surfaces operate in rolling contact, the wear phenomenon is quite different from the wear of sliding surfaces just described. Rolling surfaces in contact result in Hertz contact stresses that produce maximum values of shear stress slightly below the surface. As the rolling contact zone moves past a given location on the surface, the *subsurface peak shear stress* cycles from zero to a maximum value and back to zero, thus producing a cyclic stress field. From 2.6 it is clear that such conditions may lead to fatigue failure by the initiation of a subsurface crack that propagates under repeated cyclic loading, and that the crack may ultimately propagate to the surface to spall out a macroscopic surface particle to form a wear pit. This action, called *surface fatigue wear*, is a common failure mode in rolling element bearings, gears, cams, and all machine parts that involve rolling surfaces in contact. Empirical tests by bearing manufacturers have shown the life N in cycles to be approximately

$$N = \left(\frac{C}{P}\right)^{3.33} \tag{2-122}$$

where P is the bearing load and C is a constant for a given bearing. It may be noted that the life is inversely proportional to approximately the cube of the load. The constant C has been defined by the American Bearing Manufacturer's Association (ABMA) as the *basic load rating*,[56] which is the radial load C that a group of identical bearings can sustain for a rating life of 1 million revolutions of the inner ring, with 90 percent reliability. It must be noted that since surface fatigue wear is basically a fatigue phenomenon, all the influencing factors of 2.6 must be taken into account.

Surface durability and determination of gear-tooth wear-load criteria must also consider the surface fatigue wear phenomenon. In some types of gearing, such as helical gears and hypoid gears, a combination of rolling and sliding exists; thus adhesive wear, abrasive wear, and surface fatigue wear are all potential failure modes. Only through proper design, good manufacturing practice, and use of a proper lubricant can the desired design life be attained. Experimental life-testing is essential in complex applications such as these if proper field operation is to be achieved.

Example 2.22 Design to Prevent Excessive Wear

An experiment was performed using a cylindrical 1045 steel slider heat treated to a hardness of Rockwell C-45 (S_{yp} = 128,000 psi) pressed endwise against a 52100 steel disk with no lubricant. It was found that for a relative sliding velocity of 0.67 feet per second the 0.031-inch diameter slider, loaded by a 40-lb axial force, produced a slider wear volume of 5.8×10^{-8} cubic inches during a test of 40 minutes' duration.

If the same material combination is to be used in a slider bearing application at a sliding velocity of 3.0 feet per second under a bearing load of P = 100 lb, and if the slider is to be square, what side dimension, s, should it have to assure a lifetime of 1000 hours, if a maximum wear depth of 0.050 inch can be tolerated?

Solution

To determine the slider dimensions either (2-119) or (2-121) may be rearranged to give, for the slider bearing application,

[56]See 11.5.

Example 2.22 Continues

$$p_m = \frac{P}{A_a} = \frac{P}{s^2} = \frac{d_w}{k_w L_s} \tag{1}$$

or

$$s^2 = \frac{P k_w L_s}{d_w} \tag{2}$$

The load $P = 100$ lb and wear depth $d_w = 0.050$ inch are known design requirements, and sliding distance L_s may be calculated as

$$L_s = (1000 \text{ hours})\left(3\,\frac{\text{feet}}{\text{second}}\right)\left(3600\,\frac{\text{seconds}}{\text{hour}}\right)\left(12\,\frac{\text{inches}}{\text{foot}}\right) \tag{3}$$

or

$$L_s = 1.296 \times 10^8 \text{ inches} \tag{4}$$

Then

$$s = \sqrt{\frac{(100)(1.296 \times 10^8)\,k_w}{0.050}} = 5.09 \times 10^5 \sqrt{k_w} \tag{5}$$

The value of the wear constant k_w may be determined from an equation of the form of (2-118), together with the experimental data, as

$$k_w = \frac{d_w}{p_m L_s} \tag{6}$$

where

$$d_w = \frac{V}{A_a} = \frac{5.8 \times 10^{-8}}{\left[\pi \dfrac{(0.031)^2}{4}\right]} = 7.68 \times 10^{-5} \text{ inch} \tag{7}$$

$$p_m = \frac{P}{A_a} = \frac{40}{\left[\pi \dfrac{(0.031)^2}{4}\right]} = 5.3 \times 10^4 \text{ psi} \tag{8}$$

$$L_s = \left(0.67\,\frac{\text{feet}}{\text{second}}\right)\left(60\,\frac{\text{seconds}}{\text{minute}}\right)(40 \text{ minutes})\left(12\,\frac{\text{inches}}{\text{foot}}\right) \tag{9}$$

or

$$L_s = 1.93 \times 10^4 \text{ inches} \tag{10}$$

whence

$$k_w = \frac{7.68 \times 10^{-5}}{(5.3 \times 10^4)(1.93 \times 10^4)} = 0.75 \times 10^{-13}\,\frac{\text{inch}^2}{\text{pound}} \tag{11}$$

and from (5), then,

$$s = 5.09 \times 10^5 \sqrt{0.75 \times 10^{-13}} = 0.14 \text{ inch} \tag{12}$$

so the slider tentatively would be made $\frac{9}{64}$ inch on a side.

It still remains to verify the limiting conditions in the use of (1). For the equation to be valid it must be true that

$$p_m \leq S_{yp} \tag{13}$$

or

$$\frac{P}{A_a} = \frac{100}{(0.14)^2} = 5100 \leq 128{,}000 \tag{14}$$

Since (13) is satisfied, the design is valid, and the $\frac{9}{64}$-inch slider is adopted as the final design.

Corrosion

Corrosion may be defined as the undesired deterioration of a material through chemical or electrochemical interaction with the environment, or destruction of materials by means other than purely mechanical action.

Failure by corrosion occurs when the corrosive action renders the corroded device incapable of performing its design function. Corrosion often interacts synergistically with another failure mode, such as wear or fatigue, to produce the even more serious combined failure modes, such as corrosion-wear or corrosion-fatigue. Failure by corrosion and protection against failure by corrosion have been estimated to cost in excess of $30 billion annually in the United States alone.[57] Although much progress has been made in recent years in understanding and controlling this important failure mode, much remains to be learned. It is important for the mechanical engineering designer to become acquainted with the various types of corrosion so that corrosion-related failures can be avoided.

The complexity of the corrosion process may be better appreciated by recognizing that many variables are involved, including environmental, electrochemical, and metallurgical aspects. For example, anodic reactions and rate of oxidation, cathodic reactions and rate of reduction, corrosion inhibition, polarization or retardation, passivity phenomena, effects of oxidizers, effects of velocity, temperature, corrosive concentration, galvanic coupling, and metallurgical structure all influence the type and rate of the corrosion process.

Corrosion processes have been categorized in many different ways. One convenient classification divides corrosion phenomena into the following types:[58] direct chemical attack, galvanic corrosion, crevice corrosion, pitting corrosion, intergranular corrosion, selective leaching, erosion corrosion, cavitation corrosion, hydrogen damage, biological corrosion, and stress corrosion cracking.

Direct chemical attack is probably the most common type of corrosion. Under this type of corrosive attack the surface of the machine part exposed to the corrosive medium is attacked more or less uniformly over its entire surface, resulting in a progressive deterioration and dimensional reduction of sound load-carrying net cross section. The rate of corrosion due to direct attack can usually be estimated from relatively simple laboratory tests in which small specimens of the selected material are exposed to a well-simulated actual environment, with frequent weight change and dimensional measurements carefully taken.

Direct chemical attack may be prevented or reduced in severity by any one or a combination of several means, including selecting proper materials to suit the environment; using plating, flame spraying, cladding, hot dipping, vapor deposition, conversion coatings, organic coatings, or paint to protect the base material; changing the environment by using lower temperature or lower velocity, removing oxygen, changing corrosive concentration, or adding corrosion inhibitors; using cathodic protection in which electrons are supplied

[57]See p. 1 of ref. 46.

[58]See p. 28 of ref. 47.

to the metal surface to be protected either by galvanic coupling to a sacrificial anode or by an external power supply; or adopting other suitable design modifications.

Galvanic corrosion is an accelerated electrochemical corrosion that occurs when two dissimilar metals in electrical contact are made part of a circuit completed by a connecting pool or film of electrolyte or corrosive medium. Under these circumstances, the potential difference between the dissimilar metals produces a current flow through the connecting electrolyte, which leads to corrosion, concentrated primarily in the more anodic (less noble) metal of the pair. This type of action is completely analogous to a simple battery cell. Current must flow to produce galvanic corrosion, and, in general, more current flow means more serious corrosion.

The accelerated galvanic corrosion is usually most severe near the junction between the two metals, decreasing in severity at locations farther from the junction. The ratio of cathodic area to anodic area exposed to the electrolyte has a significant effect on corrosion rate. It is *desirable* to have a *small* ratio of cathode area to anode area. For this reason, if only *one* of two dissimilar metals in electrical contact is to be coated for corrosion protection, the *more* noble or more corrosion-resistant metal should be coated. Although this at first may seem the wrong metal to coat, the area effect, which produces anodic corrosion rates of 10^2 to 10^3 times cathodic corrosion rates for equal areas, provides the logic for this assertion.

Galvanic corrosion may be reduced in severity or prevented by one or a combination of several steps, including the selection of material pairs as close together as possible in the galvanic series (see Table 3.14), electrical insulation of one dissimilar metal from the other as completely as possible; maintaining as small a ratio of cathode area to anode area as possible; proper use and maintenance of coatings; the use of inhibitors to decrease the aggressiveness of the corroding medium; and the use of cathodic protection in which a third metal element anodic to both members of the operating pair is used as a sacrificial anode that may require periodic replacement.

Crevice corrosion is an accelerated corrosion process highly localized within crevices, cracks, and other small-volume regions of stagnant solution in contact with the corroding metal. For example, crevice corrosion may be expected in gasketed joints; clamped interfaces; lap joints; rolled joints; under bolt and rivet heads; and under foreign deposits of dirt, sand, scale, or corrosion product.

To reduce the severity of crevice corrosion, or prevent it, it is necessary to eliminate the cracks and crevices. This may involve caulking or seal-welding existing lap joints; redesign to replace riveted or bolted joints by sound, welded joints; filtering foreign material from the working fluid; inspection and removal of corrosion deposits; or using nonabsorbent gasket materials.

Pitting corrosion is a very localized attack that leads to the development of an array of holes or pits that penetrate the metal. The pits, which typically are about as deep as they are across, may be widely scattered or so heavily concentrated that they simply appear as a rough surface. The mechanism of pit growth is virtually identical to that of crevice corrosion, except that an existing crevice is not required to initiate pitting corrosion. The pit is probably initiated by a momentary attack due to a random variation in fluid concentration or a tiny surface scratch or defect.

Measurement and assessment of pitting corrosion damage is difficult because of its highly local nature. Pit depth varies widely and, as in the case of fatigue damage, a statistical approach must be taken in which the probability of generating a pit of specified depth is established in laboratory testing. Unfortunately, a significant size effect influences depth of pitting, and this must be taken into account when predicting service life of a machine part based on laboratory pitting corrosion data.

The control or prevention of pitting corrosion consists primarily of the wise selection of material to resist pitting. However, since pitting is usually the result of stagnant conditions, imparting velocity to the fluid may also decrease pitting corrosion attack.

Because of the atomic mismatch at the grain boundaries of polycrystalline metals, the stored strain energy is higher in the grain boundary regions than in the grains themselves. These high-energy grain boundaries are more chemically reactive than the grains. Under certain conditions, depletion or enrichment of an alloying element or impurity concentration at the grain boundaries may locally change the composition of a corrosion-resistant metal, making it susceptible to corrosive attack. Localized attack of this vulnerable region near the grain boundaries is called *intergranular corrosion*.

Stress corrosion cracking is an extremely important failure mode because it occurs in a wide variety of different alloys. This type of failure results from a field of cracks produced in a metal alloy under the combined influence of tensile stress and a corrosive environment. The metal alloy is not attacked over most of its surface, but a system of intergranular or transgranular cracks propagates through the matrix over a period of time.

Stress levels that produce stress corrosion cracking may be well below the yield strength of the material, and residual stresses as well as applied stresses may produce failure. The lower the stress level, the longer is the time required to produce cracking, and there appears to be a threshold stress level below which stress corrosion cracking does not occur.[59]

Prevention of stress corrosion cracking may be attempted by lowering the stress below the critical threshold level, choice of a better alloy for the environment, changing the environment to eliminate the critical corrosive element, use of corrosion inhibitors, or use of cathodic protection. *Before cathodic protection is implemented*, care must be taken to ensure that the phenomenon is indeed stress corrosion cracking because *hydrogen embrittlement is accelerated* by cathodic protection techniques.

The remaining corrosion processes are less common and require specific techniques to guard against them.[60] A designer would typically seek consulting advice on corrosion prevention from a qualified corrosion engineer.

2.11 Fretting, Fretting Fatigue, and Fretting Wear

Service failure of mechanical components due to *fretting fatigue* has come to be recognized as a failure mode of major importance, in terms of both frequency of occurrence and seriousness of the failure consequences. *Fretting wear* has also presented major problems in certain applications. Both fretting fatigue and fretting wear, as well as fretting corrosion, are directly attributable to *fretting action*. Basically, fretting action may be defined as a combined mechanical and chemical action in which the contacting surfaces of two solid bodies are pressed together by a normal force and are caused to execute oscillatory sliding relative motion, wherein the magnitude of normal force is great enough and the amplitude of the oscillatory sliding motion is small enough to significantly restrict the flow of fretting debris away from the originating site.[61]

Damage to machine parts due to fretting action may be manifested as corrosive surface damage due to fretting corrosion, loss of proper fit or change in dimensions due to fretting wear, or accelerated fatigue failure due to fretting fatigue. Typical sites of fretting damage include interference fits; bolted, keyed, splined, and riveted joints; points of con-

[59]See p. 96 of ref. 47.

[60]See refs. 46 and 47, for example.

[61]See ref. 1.

tact between wires in wire ropes and flexible shafts; friction clamps; small amplitude-of-oscillation bearings of all kinds; contacting surfaces between the leaves of leaf springs; and all other places where the conditions of fretting persist. Thus, the efficiency and reliability of the design and operation of a wide range of mechanical systems are related to the fretting phenomenon.

Although fretting fatigue, fretting wear, and fretting corrosion phenomena are potential failure modes in a wide variety of mechanical systems, there are very few quantitative design data available, and no generally applicable design procedure has been established for predicting failure under fretting conditions. However, significant progress has been made in establishing an understanding of fretting and the variables of importance in the fretting process.

It has been suggested that there may be more than 50 variables that play some role in the fretting process.[62] Of these, however, there are probably only eight that are of major importance; they are:

1. The magnitude of relative motion between the fretting surfaces
2. The magnitude and distribution of pressure between the surfaces at the fretting interface
3. The state of stress, including magnitude, direction, and variation with respect to time in the region of the fretting surfaces
4. The number of fretting cycles accumulated
5. The material from which each of the fretting members is fabricated, including surface condition
6. Cyclic frequency of relative motion between the two members being fretted
7. Temperature in the region of the two surfaces being fretted
8. Atmospheric environment surrounding the surfaces being fretted

These variables interact so that a quantitative prediction of the influence of any given variable may be dependent upon one or more of the other variables in any specific application. Also, the combinations of variables that produce a very serious consequence in terms of fretting fatigue damage may be quite different from the combinations of variables that produce serious fretting wear damage.

Fretting Fatigue

Fretting fatigue is fatigue damage directly attributable to fretting action. Premature fatigue nuclei may be generated by fretting through either abrasive pit-digging action, asperity-contact microcrack initiation, friction-generated cyclic stresses that lead to the formation of microcracks, or subsurface cyclic shear stresses that lead to surface delamination in the fretting zone.[63] Under *abrasive pit-digging action*, tiny grooves or elongated pits are produced at the fretting interface by the asperities and abrasive debris particles moving under the influence of oscillatory relative motion. A pattern of tiny grooves is produced in the fretted region with their longitudinal axes approximately parallel and in the direction of fretting motion.

The *asperity-contact microcrack initiation mechanism* proceeds by virtue of the contact force between the tip of an asperity on one surface and another asperity on the mating

[62]See ref. 48.

[63]See ref. 1, Ch. 14.

surface as they move back and forth. If the initial contact does not shear one or the other asperity from its base, the repeated contacts at the tips of the asperities give rise to cyclic or fatigue stresses in the region at the base of each asperity. It has been estimated that under such conditions the region at the base of each asperity is subjected to large local stresses that probably lead to the nucleation of fatigue microcracks at these sites. Such microcracks have longitudinal axes generally perpendicular to the direction of fretting motion.

Friction-generated cyclic-stress fretting action is based on the observation that when one member is pressed against the other and caused to undergo fretting motion, the tractive friction force induces a compressive tangential stress component in a volume of material that lies ahead of the fretting motion, and a tensile tangential stress component in a volume of material that lies behind the fretting motion. When the fretting direction is reversed, the tensile and compressive regions change places. Thus, these regions of material adjacent to the contact zone are subjected to cyclic stresses that generate fields of microcracks whose axes are generally perpendicular to the direction of fretting motion.

In the *delamination theory of fretting*, the combination of normal and tangential tractive forces transmitted through the asperity contact sites at the fretting interface produces a complex multiaxial state of stress, accompanied by a cycling deformation field, which produces subsurface peak shearing stresses and subsurface crack nucleation sites. With further cycling, the cracks propagate below the surface and approximately parallel to the surface, finally branching to the surface to produce a thin wear sheet, which "delaminates" to become a particle of debris.

Supporting evidence has been generated to indicate that under various circumstances each of the four mechanisms is active and significant in producing fretting damage.

The influence of the state of stress in the member during the fretting process is shown for several different cases in Figure 2.45, including the superposition of static tensile or static compressive mean stresses during fretting. Local compressive stresses are beneficial in minimizing fretting fatigue damage.

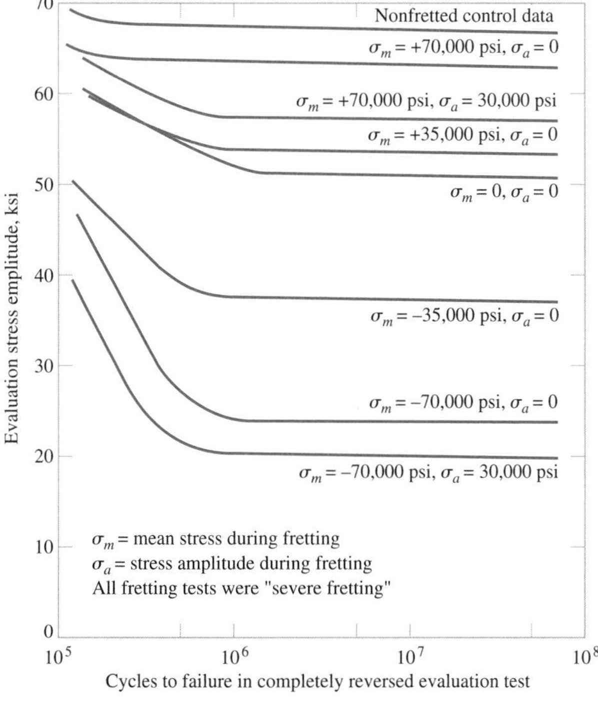

Figure 2.45
Residual fatigue properties subsequent to fretting under various states of stress.

Usually, it is necessary to evaluate the seriousness of fretting fatigue damage in any specific design by running simulated service tests on specimens or components. Within the current state-of-the-art knowledge in the area of fretting fatigue, there is no other safe course of action open to the designer.

Fretting Wear

Fretting wear is a change in dimensions because of wear directly attributable to the fretting process. It is thought that the abrasive pit-digging mechanism, the asperity-contact microcrack initiation mechanism, and the wear-sheet delamination mechanism may all be active in most fretting wear environments. As in the case of fretting fatigue, there has been no good model developed to describe the fretting wear phenomenon is a way useful for design.

Some investigators have suggested that estimates of fretting wear depth may be based on the classical adhesive or abrasive wear equations, in which wear depth is proportional to load and total distance slid, where the total distance slid is calculated by multiplying relative motion per cycle times number of cycles. Although there are some supporting data for such a procedure,[64] more investigation is required before using these estimates as a general approach.

Prediction of wear depth in an actual design application must in general be based on simulated service testing.

Minimizing or Preventing Fretting Damage

The minimization or prevention of fretting damage must be carefully considered as a separate problem in each individual design application because a palliative in one application may significantly accelerate fretting damage in a different application. For example, in a joint that is designed to have no relative motion, it is sometimes possible to reduce or prevent fretting by increasing the normal pressure until all relative motion is arrested. However, if the increase in normal pressure does not *completely* arrest the relative motion, the result may be a nonhomogeneous contact zone in which some regions slip and other regions do not slip. If *partial slip* behavior characterizes the contact zone, it may result in a significant *increase* in fretting damage instead of preventing it. Recent research efforts[65] have established a fretting test methodology based on *fretting maps*. Fretting maps are plots of normal force versus relative displacement amplitude in which either the *running condition* or the *material response* is partitioned into regions of probable fretting behavior.

For the *running condition fretting map (RCFM)*, the partitioned regions include the *partial slip regime (PSR)*, the *gross slip regime (GSR)*, and a transition region between them called the *mixed fretting regime (MFR)*. A fretting test usually starts in the gross slip regime, then transforms to the partial slip regime as the contacting surfaces change character during the fretting process.

For the *material response fretting map (MRFM)*, the partitioned regions include the *crack initiation regime*, the *fretting wear regime*, and an indeterminate transition regime between cracking and wear. Semiquantitative analyses have shown that crack initiation is more probable in the mixed slip regime, but the mapping technique is *configuration specific*, and, so far, cannot be generalized for use as a quantitative design tool.
Nevertheless, there are several basic principles that are generally effective in minimizing or preventing fretting. These include:

1. Complete separation of the contacting surfaces.
2. Elimination of all relative motion between the contacting surfaces.

[64]See ref. 52.

[65]See ref. 50.

3. If relative motion cannot be eliminated, it is sometimes effective to superpose a large unidirectional relative motion that allows effective lubrication. For example, the practice of driving the inner or outer race of an oscillatory pivot bearing may be effective in eliminating fretting.
4. Providing compressive residual stresses at the fretting surface; this may be accomplished by shot-peening, cold-rolling, or interference fit techniques.
5. Judicious selection of material pairs.
6. Use of interposed low shear modulus shim material or plating, such as lead, rubber, or silver.
7. Use of surface treatments or coatings as solid lubricants.
8. Use of surface grooving or roughening to provide debris escape routes and differential strain matching through elastic action.

Of all these techniques, only the first two are completely effective in preventing fretting. The remaining concepts, however, may often be used to minimize fretting damage and may yield an acceptable design.

2.12 Failure Data and the Design Task

The design task is clear: Create new devices or improve existing devices to provide the "best" design configuration consistent with the constraints of time, money, and safety, as dictated by the proposed application and demands of the marketplace. Successful accomplishment of the design task depends heavily upon the designer's ability to recognize and assess the likelihood that any potential failure mode might jeopardize the proposed device's ability to function properly and safely.

Analytical approaches for evaluating the potential seriousness of most failure modes have been presented in the first 11 sections of this chapter. Success in using these approaches hinges directly upon the availability of reliable failure strength data specifically developed for the pertinent failure mode, operating environment, and candidate material.

Failure strength data are readily available for some failure modes and virtually nonexistent for others. If reliable data are available, the design task may be accomplished with confidence. If data are sparse or unavailable, extrapolations or estimates must be made, or pertinent experiments completed, before the design task can proceed. *Data or no data*, the design task must go forward if the competition is to be met. Thus the ability to extrapolate or estimate failure strength data is an important design function, as is the ability of a designer to find access to comprehensive failure strength databases. Professional societies and trade magazines are excellent resources for failure strength data through handbooks, annual compilations, on-line databases, or CD-ROM packages. Sources of failure strength data and methods for selecting materials are discussed in Chapter 3.

2.13 Failure Assessment and Retrospective Design

In spite of all efforts to design and manufacture machines and structures to function properly without failure, *failures do occur*. Whether the failure consequences are simply an annoying inconvenience such as a "binding" support roller on the sliding patio screen door, or a catastrophic loss of life and property as in the crash of a jumbo jet, it is the responsibility of the designer to glean all of the information possible from the failure event so that repeat failures can be avoided in the future. Effective assessment of service failures usually requires the intense interactive scrutiny of a team of specialists, including at least a mechanical designer and a materials engineer trained in failure analysis techniques. The team often includes a manufacturing engineer and a field service engineer as well. The mission of the failure analysis team is to discover the initiating cause of failure, identify the best solution, and redesign the product to prevent future failures.

Techniques utilized in the failure analysis effort include the inspection and documentation of the failure event through direct examination (without touching or altering any failure surface or other crucial evidence); taking photographs; compilation of eyewitness reports; preservation of all parts, especially failed parts; and performing pertinent calculations, analyses, and examinations that may help establish and validate the cause of failure. The materials engineer may utilize macroscopic examination, low-power magnification, microscopic examination, transmission or scanning electron microscopic techniques, energy-dispersive X-ray techniques, hardness tests, spectrographic analysis, metallographic examination, or other methods for determining the type of failure, failure location, any abnormalities in the material, or any other attributes of the failure scenario that may relate to the potential cause of failure. The designer may perform stress and/or deflection analyses, examine geometry, assess service loading, evaluate potential environmental influences, reexamine the kinematics and dynamics of the application, and attempt to reconstruct the sequence of events leading to failure. Other team members may examine the quality of manufacture, the quality of maintenance, the possibility of unusual or unconventional usage by the operator, or other factors that may have played a role in the service failure. Piecing all of this information together, it is the objective of the failure analysis team to identify as accurately as possible the probable cause of failure.

As undesirable as service failures may be, the results of a well-executed failure analysis may be transformed directly into improved product reliability by a design team that capitalizes on service-failure data and failure analysis results. These techniques of retrospective design have become important working tools of the profession.

Problems

2-1. In the context of *machine design*, explain what is meant by the terms *failure* and *failure mode*.

2-2. Distinguish the difference between *high-cycle fatigue* and *low-cycle fatigue*, giving the characteristics of each.

2-3. Describe the usual consequences of *surface fatigue*.

2-4. Compare and contrast *ductile rupture* and *brittle fracture*.

2-5. Carefully define the terms *creep*, *creep rupture*, and *stress rupture*, citing the similarities that relate these three failure modes and the differences that distinguish them from one another.

2-6. Give a definition for *fretting*, and distinguish among the related failure phenomena of *fretting fatigue*, *fretting wear*, and *fretting corrosion*.

2-7. Give a definition of *wear failure* and list the major subcategories of wear.

2-8. Give a definition of *corrosion failure*, and list the major subcategories of corrosion.

2-9. Describe what is meant by a *synergistic* failure mode, give three examples, and for each example describe how the synergistic interaction proceeds.

2-10. Taking a passenger automobile as an example of an engineering system, list all failure modes that you think might be significant, and indicate where in the auto you think each failure mode might be active.

2-11. For each of the following applications, list three of the more likely failure modes, describing why each might be expected: (a) high-performance automotive racing engine, (b) pressure vessel for commercial power plant, (c) domestic washing machine, (d) rotary lawn mower, (e) manure spreader, (f) 15-inch oscillating fan.

2-12. In a tension test of a steel specimen having a 6-mm-by-25-mm rectangular net cross section, a gage length of 20 cm was used. Test data include the following observations: (1) load at the onset of yielding was 37.8 kN, (2) ultimate load was 65.4 kN, (3) rupture load was 52 kN, (4) total deformation in the gage length at 18 kN load was 112 μm. Determine the following:

a. Nominal yield strength

b. Nominal ultimate strength

c. Modulus of elasticity

2-13. A tension test on a 0.505-inch diameter specimen of circular cross section was performed, and the following data were recorded for the test:

TABLE P2.13 Tension Test Data

Load, lb	Elongation,in
1000	0.0003
2000	0.0007
3000	0.0009
4000	0.0012
5000	0.0014
6000	0.002
7000	0.004
8000	0.085
9000	0.150
10,000	0.250
11,000 (maximum load)	0.520

a. Plot an engineering stress-strain curve for this material.

b. Determine the nominal yield strength.

c. Determine the nominal ultimate strength.

d. Determine the approximate modulus of elasticity.

e. Using the available data and the stress-strain curve, make your best guess as to what type of material the specimen was manufactured from.

f. Estimate the axially applied tensile load that would correspond to yielding of a 2-inch diameter bar of the same material.

g. Estimate the axially applied tensile load that would be required to produce ductile rupture of the 2-inch bar.

h. Estimate the axial spring rate of the 2-inch bar if it is 2 feet long.

2-14. An axially loaded straight bar of circular cross section will fail to perform its design function if the applied static axial load produces permanent changes in length after the load is removed. The bar is 12.5 mm in diameter, has a length of 180 cm, and is made of Inconel 601.[66] The axial load required for this application is 25 kN. The operating environment is room-temperature air.

a. What is the probable governing failure mode?

b. Would you predict that failure does take place? Explain your logic.

2-15. A 1.25-inch diameter round bar of material was found in the stock room, but it was not clear whether the material was aluminum, magnesium, or titanium. When a 10-inch length of this bar was tensile-tested in the laboratory, the force-deflection curve obtained was as shown in Figure P2.15. It is being proposed that a vertical deflection-critical tensile support rod made of this material, having a 1.128-inch diameter and 7-foot length, be used to support a static axial tensile load of 8000 pounds. A total deflection of no more than 0.040 inch can be tolerated.

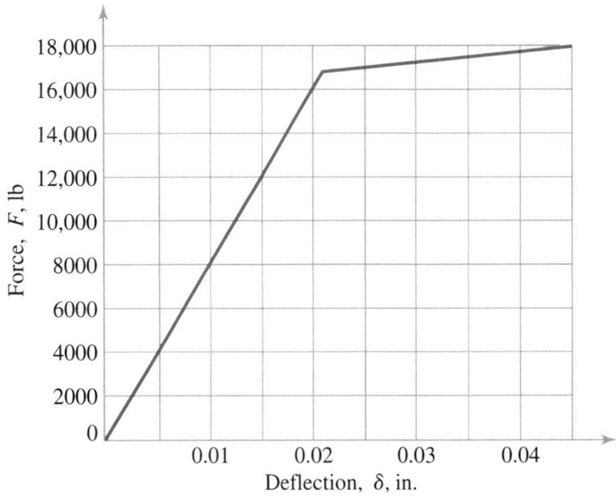

Figure P2.15
Force-deflection curve for unknown material.

a. Using your best engineering judgment, and recording your supporting calculations, what type of material do you believe this to be?

b. Would you approve the use of this material for the proposed application? Clearly show your analysis supporting your answer.

2-16. A 304 stainless-steel alloy, annealed, is to be used in a deflection-critical application to make the support rod for a test package that must be suspended near the bottom of a deep cylindrical cavity. The solid stainless-steel support rod is to have a diameter of 0.750 inch and a precisely manufactured length of 16.000 feet. It is to be vertically oriented and fixed at the top end. The 6000-lb test package is to be attached at the bottom, placing the vertical bar in axial tension. During the test, the bar will experience a temperature increase of 175°F. If the total deflection at the end of the bar must be limited to a maximum of 0.300 inch, would you approve the design?

2-17. A cylindrical 6061-T6 aluminum bar, having a diameter of 1 inch and length of 10 inches, is vertically oriented with a static axial load of 20,000 pounds attached to the bottom end.

a. Write the equation for the maximum tensile stress in the bar, identifying where it occurs, and calculate its numerical magnitude. (Neglect stress concentration.)

b. Write the equation for the elongation of the bar when the 20,000-lb load is applied, and calculate its numerical magnitude.

c. If the bar temperature is nominally 70°F when the axial load is applied, what temperature change would be necessary to bring the bar back to its initial unloaded length of 10.000 inches?

2-18. A portion of a tracking radar unit to be used in an antimissile missile defense system is sketched in Figure P2.18. The radar dish that receives the signals is labeled D and is at-

[66]See Chapter 3 for material properties data.

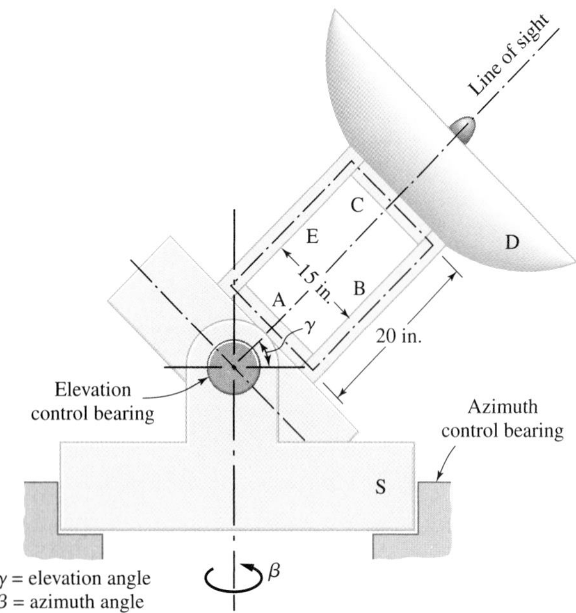

Figure P2.18
Sketch of radar tracking unit.

tached by frame members A, B, C, and E to the tracking structure S. Tracking structure S may be moved angularly in two planes of motion (azimuthal and elevational) so that the dish D can be aimed at an intruder missile and locked on the target to follow its trajectory.

Due to the presence of electronic equipment inside the box formed by frame members A, B, C, and E, the approximate temperature of member E may sometimes reach 200°F while the temperature of member B is about 150°F. At other times, members B and E will be about the same temperature. If the temperature difference between members B and E is 50°F, and joint resistance to bending is negligible, by how many feet would the line of sight of the radar tracking unit miss the intruder missile if it is 40,000 feet away, and

a. the members are made of steel?

b. the members are made of aluminum?

c. the members are made of magnesium?

2-19. Referring to Figure P2.19, it is absolutely essential that the assembly slab be precisely level before use. At room temperature, the free unloaded length of the aluminum support bar is 80 inches, the free unloaded length of the nickel-steel support bar is 40 inches, and a line through A-B is absolutely level before attaching slab W. If slab W is then attached, and the temperature of the entire system is slowly and uniformly increased to 150°F above room temperature, determine the magnitude and direction of the vertical adjustment of support "C" that would be required to return slab surface A-B to a level position. (For material properties, see Chapter 3.)

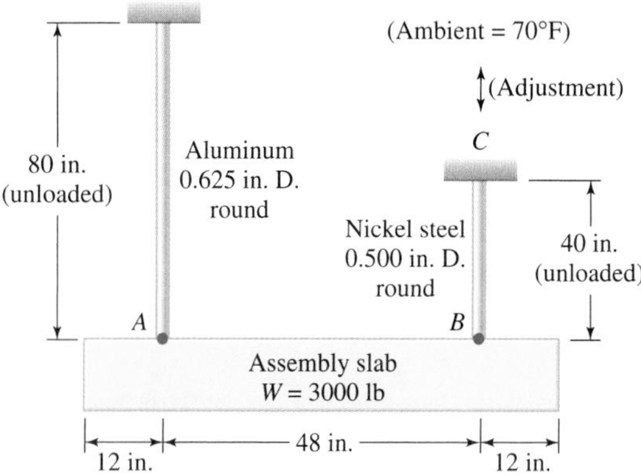

Figure P2.19
Assembly slab configuration.

2-20. Compare and contrast the basic philosophy of failure prediction for *yielding* failure with failure by *rapid crack extension*. As a part of your discussion, carefully define the terms *stress-intensity factor*, *critical stress intensity*, and *fracture toughness*.

2-21. Describe the three basic crack-displacement modes, using appropriate sketches.

2-22. Interpret the following equation, and carefully define each symbol used. *Failure is predicted to occur if:*

$$C\sigma\sqrt{\pi a} \geq K_{Ic}$$

2-23. A very wide sheet of 7075-T651 aluminum plate, $\frac{5}{16}$ inch thick, is found to have a single-edge through-the-thickness crack 1.0 inch long. The loading produces a gross nominal tensile stress of 10,000 psi perpendicular to the plane of the crack.

a. Calculate the stress intensity factor at the crack tip.

b. Determine the critical stress-intensity factor.

c. Estimate the safety factor (see 5.2) that exists in this case, based on rapid crack propagation (brittle fracture) as a failure mode.

2-24. Discuss all parts of problem 2-23 under conditions that are identical to those stated, except that the sheet thickness is $\frac{1}{8}$ inch.

2-25. A steam generator in a remote power station is supported by two tension straps, each one 7.5 cm wide by 11 mm thick by 66 cm long. The straps are made of A538 steel. When in operation, the fully loaded steam generator weights 1300 kN, equally distributed to the two support straps. The load may be regarded as static. Ultrasonic inspection has detected a through-the-thickness center crack 12.7 mm long, oriented perpendicular to the 66-cm dimension (i.e., perpendicular to the tensile load). Would you allow the plant to be put back into operation? Support your answer with clear, complete engineering calculations.

2-26. A pinned-end structural member in a high-performance tanker is made of a 0.375-inch-thick-by-5-inches-wide, rectan-

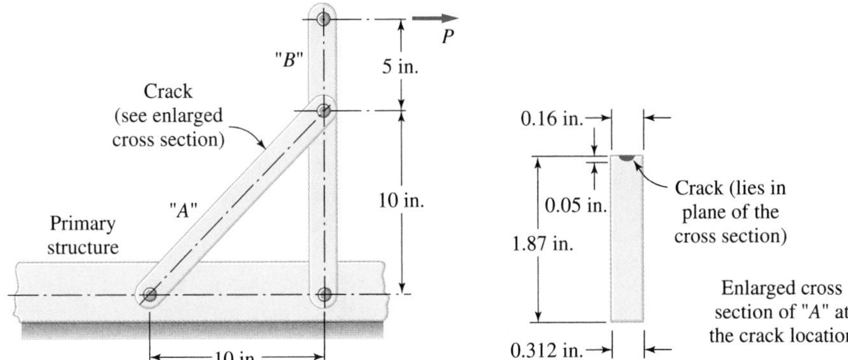

Figure P2.27
Sketch of cracked engine mount.

gular cross-section, titanium 6Al-4V bar, 48 inches long. The member is normally subjected to a pure static tensile load of 154,400 lb. Inspection of the member has indicated a central through-the-thickness crack of 0.50-inch length, oriented perpendicular to the applied load. If a safety factor (see 5.2) of $n = 1.7$ is required, what reduced load limit for the member would you recommend for safe operation (i.e., to give $n = 1.7$)?

2-27. An engine mount on an experimental high-speed shuttle has been inspected, and a thumbnail surface crack 0.05 inch deep and 0.16 inch long at the surface has been found in member A, as shown in Figure P2.27. The structure is pin-connected at all joints. Member A is 0.312 inch thick and 1.87 inches wide, of rectangular cross section, and made of 7075-T6 aluminum alloy. If full power produces a thrust load P of 18,000 lb at the end of member B, as shown in Figure P2-27, what percentage of full-power thrust load would you set as a limit until part A can be replaced, if a minimum safety factor (see 5.2) of 1.2 must be maintained?

2-28. A 90-cm-long structural member of 7075-T6 aluminum has a rectangular cross section 8 mm thick by 4.75 cm wide. The member must support a load of 133 kN static tension. A thumbnail surface crack 2.25 mm deep and 7 mm long at the surface has been found during an inspection.

a. Predict whether failure would be expected.

b. Estimate the existing safety factor (see 5.2) under these conditions.

2-29. A transducer support to be used in a high-flow-rate combustion chamber is to be made of hot-pressed silicon carbide with tensile strength of 110,000 psi, compressive strength of 500,000 psi, fracture toughness of 3100 psi$\sqrt{\text{in}}$, and *nil* ductility. The dimensions of the silicon-carbide support, which has a rectangular cross section, are 1.25 inches wide by 0.094 inch thick by 7.0 inches long. Careful inspection of many such pieces has revealed through-the-thickness edge cracks up to 0.060 inch long, but none longer. If this part is loaded in pure uniform tension parallel to the 7-inch dimension, approximately what maximum tensile load would you predict the part could withstand before failing?

2-30. A newly installed cantilever beam of D6AC steel (1000°F temper) has just been put into use as a support bracket for a large outdoor tank used in processing synthetic crude oil near Ft. McMurray, Alberta, Canada, near the Arctic Circle. As shown in Figure P2.30, the cantilever beam is 25 cm long and has a rectangular cross section 5.0 cm deep by 1.3 cm thick. A large fillet at the fixed end will allow you to neglect stress concentration there. A shallow through-the-thickness crack has been found near the fixed end, as shown, and the crack depth has been measured as 0.75 mm. The load P is static and will never exceed 22 kN. Can we get through the winter without replacing the defective beam, or should we replace it now?

2-31. Identify several problems a designer must recognize when dealing with fatigue loading as compared to static loading.

2-32. Distinguish the differences between high-cycle fatigue and low-cycle fatigue.

2-33. Carefully sketch a typical *S-N* curve, use it to define and distinguish between the terms *fatigue strength* and *fatigue endurance limit*, and briefly indicate how a designer might use such a curve in practice.

2-34. Make a list of factors that might influence the *S-N* curve, and indicate briefly what the influence might be in each case.

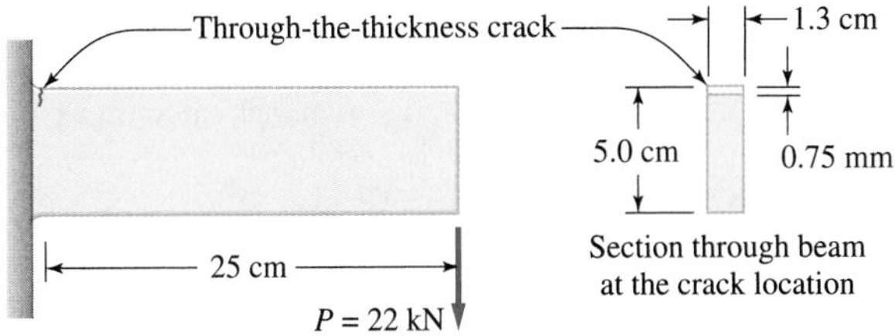

Figure P2.30
Cracked cantilever beam.

2-35. Sketch a family of *S-N-P* curves, explain the meaning and utility of these curves, and explain in detail how such a family of curves would be produced in the laboratory.

2-36. a. Estimate and plot the *S-N* curve for AISI 1020 cold-drawn steel, using the static properties given in Table 3.3.

b. Using the estimated *S-N* curve, determine the fatigue strength at 10^6 cycles.

c. Using Figure 2.19, determine the fatigue strength (of 1020 steel) at 10^6 cycles, and compare with the estimate of part (b).

2-37. a. Estimate and plot the *S-N* curve for 2024-T3 aluminum alloy, using the static properties given in Table 3.3.

b. What is the estimated magnitude of the fatigue endurance limit for this material?

2-38. a. Estimate and plot the *S-N* curve for ASTM A-48 (class 50) gray cast iron, using the static properties given in Table 3.3.

b. On average, based on the estimated *S-N* curve, what life would you predict for parts made from this cast iron material if they are subjected to completely reversed uniaxial cyclic stresses of 30 ksi amplitude?

2-39. It has been suggested that AISI 1060 hot-rolled steel (see Table 3.3) be used for a power plant application in which a cylindrical member is subjected to an axial load that cycles from 78,000 pounds tension to 78,000 pounds compression, repeatedly. The following manufacturing and operating conditions are expected:

a. The part is to be lathe-turned.

b. The cycle rate is 200 cycles per minute.

c. A very long life is desired.

d. A strength reliability of 99 percent is desired.

Ignoring the issues of *stress concentration* (see 4.8) and *safety factor* (see Chapter 5), what diameter would be required for this cylindrical cast iron bar?

2-40. A solid square link for a spacecraft application is to be made of Ti-6Al-4V titanium alloy (see Table 3.3). The link must transmit a cyclic axial load that ranges from 50,000 pounds tension to 50,000 pounds compression, repeatedly. Welding is to be used to attach the link to the supporting structure. The link surfaces are to be finished by using a horizontal milling machine. A design life of 10^5 cycles is required.

a. Estimate the fatigue strength of the *part* in this application.

b. Estimate the required cross-sectional dimensions of this square bar, ignoring the issues of *stress concentration* (see 4.8) and *safety factor* (see Chapter 5).

2-41. An old "standard" design for the cantilevered support shaft for a bicycle pedal has a history of fatigue failure of about one pedal shaft for every 100 pedals installed. If management desires to reduce the incidence of failure to about one pedal shaft for every 1000 pedals installed, by what factor must the operating stress at the critical point be reduced, assuming that all other factors remain constant?

2-42. An axially loaded actuator bar has a solid rectangular cross section 6.0 mm by 18.0 mm, and is made of 2024-T4 aluminum alloy. The loading on the bar may be well approximated as constant-amplitude axial cyclic loading that cycles between a maximum load of 20 kN tension and a minimum load of 2 kN compression. The static properties of 2024-T4 are S_u = 68,000 psi, S_{yp} = 47,000 psi, and e (2 inches) = 20 percent. Fatigue properties are shown in Figure 2.19. Estimate the total number of cycles to failure for this bar. Neglect stress concentration effects. Assume that buckling is not a problem.

2-43. A tie-bar is to be used to connect a reciprocating power source to a remote shaking sieve in an open-pit mine. It is desired to use a solid cylindrical cross section of 2024-T4 aluminum alloy for the tie-bar. See problem 2-42 for material properties. The applied axial load fluctuates cyclically from a maximum of 45,000 pounds tension to a minimum of 15,000 pounds compression. If the tie-bar is to be designed for a life of 10^7 cycles, what diameter should the bar be made? Ignore the issue of *safety factor* (see Chapter 5).

2-44. A 36-inch-long, simply supported horizontal beam is to be loaded at midspan by a vertical cyclic load P that ranges between 20,000 pounds down to 60,000 pounds down. The proposed beam cross section is to be rectangular, 2.0 inches wide by 4.0 inches deep. The material is to be Ti-6Al-4V titanium alloy.

a. What is (are) the governing failure mode(s), and why?

b. Where is (are) the critical point(s) located? How do you come to this conclusion?

c. How many cycles would you predict that the beam could sustain before it fails?

2-45. Explain how a designer might use a *master diagram*, such as the ones shown in Figure 2.27.

2-46. a. An aluminum bar of solid cylindrical cross section is to be subjected to a cyclic axial load that ranges from 5000 pounds tension to 10,000 pounds tension. The material has an ultimate tensile strength of 100,000 psi, a yield strength of 80,000 psi, a mean fatigue strength at 10^5 cycles of 40,000 psi, and an elongation of 8 percent in 2 inches. Calculate the bar diameter that should be used to just produce failure in 10^5 cycles, on the average.

b. If, instead of the loading specified in part (a), the cyclic axial loading ranged from 15,000 pounds tension to 20,000 pounds tension, calculate the bar diameter that should be used to just produce failure in 10^5 cycles, on the average.

c. Compare the results of parts (a) and (b), making any observations you think appropriate.

2-47. The *S-N* data from a series of completely reversed fatigue tests are shown in the chart on page 123.

The ultimate strength is 218,000 psi, and the yield strength is 200,000 psi. Determine and plot the estimated *S-N* curve for

S (psi)	N (cycles)
17×10^4	2×10^4
15.1×10^4	5×10^4
14.1×10^4	1×10^5
12.7×10^4	2×10^5
12.5×10^4	5×10^5
12.3×10^4	1×10^6
12.1×10^4	$2 \times 10^6 \rightarrow \infty$

the material if a mean stress of 40,000 psi tension well characterizes the application.

2-48. The $\sigma_{max} - N$ data for direct stress fatigue tests, in which the mean stress was 25,000 psi *tension* for all tests, are shown in the table.

σ_{max} (psi)	N (cycles)
150,000	2×10^4
131,000	5×10^4
121,000	1×10^5
107,000	2×10^5
105,000	5×10^5
103,000	1×10^6
102,000	2×10^6

The ultimate strength is 240,000 psi, and the yield strength is 225,000 psi.

a. Determine and plot the $\sigma_{max} - N$ curve for this material for a mean stress of 50,000 psi *tension.*

b. Determine and plot, on the same graph sheet, the $\sigma_{max} - N$ curve for this material for a mean stress of 50,000 psi *compression.*

2-49. Discuss the basic assumptions made in using a *linear damage rule* to assess fatigue damage accumulation, and note the major "pitfalls" one might experience in using such a theory. Why, then, is a linear damage theory so often used?

2-50. The critical point in the main rotor shaft of a new VSTOL aircraft of the ducted-fan type has been instrumented, and during a "typical" mission the equivalent completely reversed stress spectrum has been found to be 50,000 psi for 15 cycles, 30,000 psi for 100 cycles, 60,000 psi for 3 cycles, and 10,000 psi for 10,000 cycles.

Ten missions of this spectrum have been "flown." It is desired to overload the shaft to 1.10 times the "typical" loading spectrum. Estimate the number of additional "overloaded" missions that can be flown without failure, if the stress spectrum is linearly proportional to the loading spectrum. An *S-N* curve for the shaft material is shown in Figure P2.50.

2-51. A hollow square tube with outside dimensions of 1.25 inches and wall thickness of 0.125 inch is to be made of 2024-T4 aluminum, with fatigue properties as shown in Figure 2.19. This hollow square tube is to be subjected to the following sequence of completely reversed axial force amplitudes: First, 20,000 lb for 52,000 cycles; next, 10,500 lb for 948,000 cycles; then, 24,900 lb for 11,100 cycles.

After this loading sequence has been imposed, it is desired to change the force amplitude to 19,000 lb, still in the axial direction. How many remaining cycles of life would you predict for the tube at this final level of loading?

2-52. A solid cylindrical bar of 2024-T4 aluminum alloy (see Figure 2.19) is to be subjected to a duty cycle that consists of the following spectrum of completely reversed axial tensile loads: First, 50 kN for 1200 cycles; next, 31 kN for 37,000 cycles; then 40 kN for 4300 cycles. Approximate static properties of 2024-T4 aluminum alloy are S_u = 470 MPa and S_{yp} = 330 MPa.

What bar diameter would be required to just survive 50 duty cycles before fatigue failure takes place?

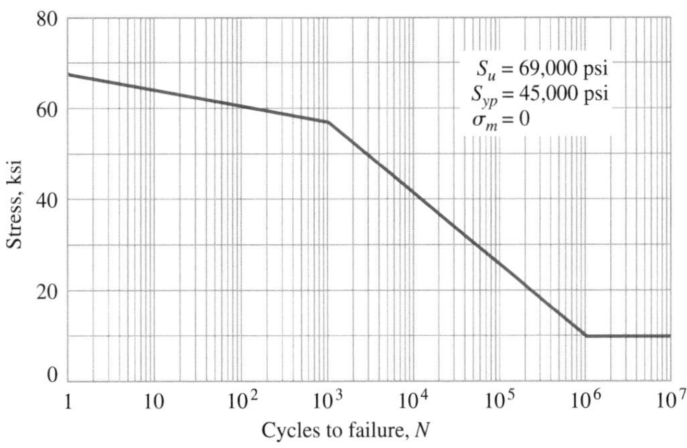

Figure P2.50
S-N curve for rotor shaft material.

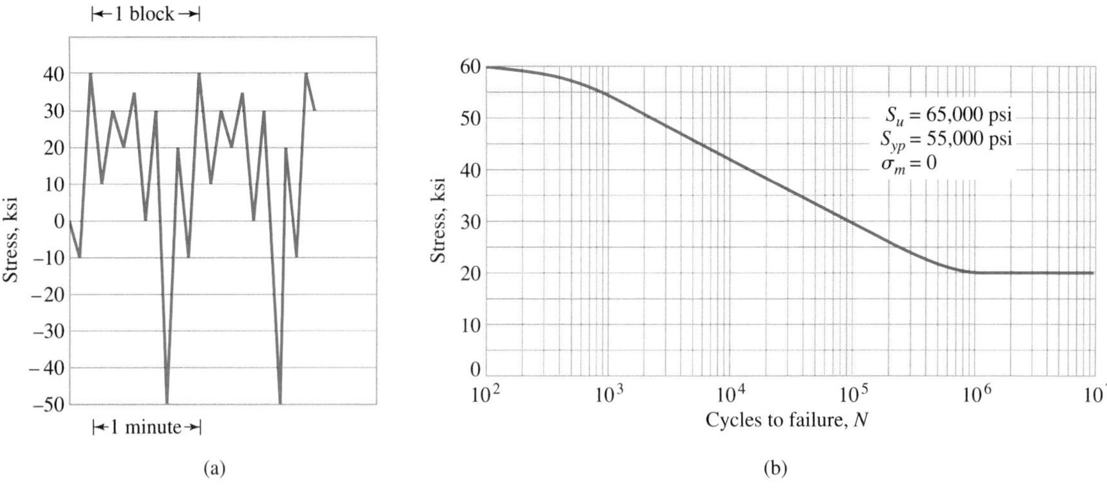

Figure P2.53
Stress-time spectrum and *S-N* curve for test component.

2-53. The stress-time pattern shown in Figure P2.53(a) is to be repeated in blocks until failure of a test component occurs. Using the rain flow cycle counting method, and the *S-N* curve given in Figure P2.53(b), estimate the hours of life until failure of this test component occurs.

2-54. The stress-time spectrum shown in Figure P2.54 is to be repeated in blocks until failure of the component occurs on a laboratory test stand. Using the rain flow cycle counting method, and the *S-N* curve shown in Figure P2.53(b), estimate the time in hours of testing that would be required to produce failure.

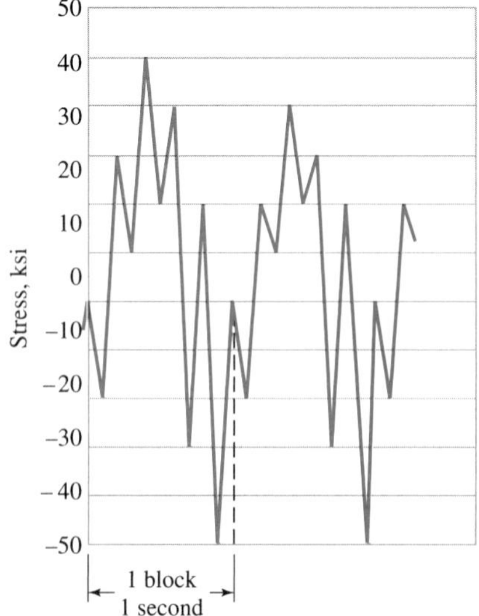

Figure P2.54
Stress-time pattern.

2-55. In "modern" fatigue anlaysis, three separate *phases* of fatigue are defined. List the three phases, and briefly describe how each one is currently modeled and analyzed.

2-56. For the equation $da/dN = C\Delta K^n$, define each term, describe the physical phenomenon being modeled, and tell what the limiting conditions are on the magnitude of ΔK. What are the consequences of exceeding the limits of validity?

2-57. Experimental values for the properties of an alloy steel have been found to be S_u = 1480 MPA, S_{yp} = 1370 MPa, K_{Ic} = 81.4 MPa$\sqrt{\text{m}}$, e = 20 percent in 50 mm, k' = 1070 MPa, n' = 0.15, ε'_f = 0.48, σ'_f = 2000 MPa, b = −0.091, and c = −0.060. A direct tension member made of this alloy has a single semicircular edge notch that results in a fatigue stress concentration factor of 1.6. The net cross section of the member at the root of the notch is 9 mm thick by 36 mm wide. A completely reversed cyclic axial force of 72 kN amplitude is applied to the tension member.

a. How many cycles would you estimate that it would take to initiate a fatigue crack at the notch root?

b. What length would you estimate this crack to be at the time it is "initiated" according to the calculation of part (a)?

2-58. Testing an aluminum alloy has resulted in the following data: S_u = 70 ksi, S_{yp} = 50 ksi, K_{Ic} = 26 ksi$\sqrt{\text{in}}$, e (2 inches) = 22 percent, k' = 95 ksi, n' = 0.065, ε'_f = 0.22, σ'_f = 160 ksi, b = −0.12, c = −0.60, and $E = 10.3 \times 10^6$ psi. A direct tension member made of this alloy is to be 2.0 inches wide, $\frac{3}{8}$ inch thick, and have a $\frac{1}{2}$-inch diameter hole, through the thickness, at the center of the tension member. The hole will produce a fatigue stress concentration factor of K_f = 2.2 (see Figures 4.22 and 4.27 of 4.8). A completely reversed axial force of 6000 lb amplitude is to be applied to the member.

How many cycles would you estimate that it would take to *initiate* a fatigue crack at the edge of the hole?

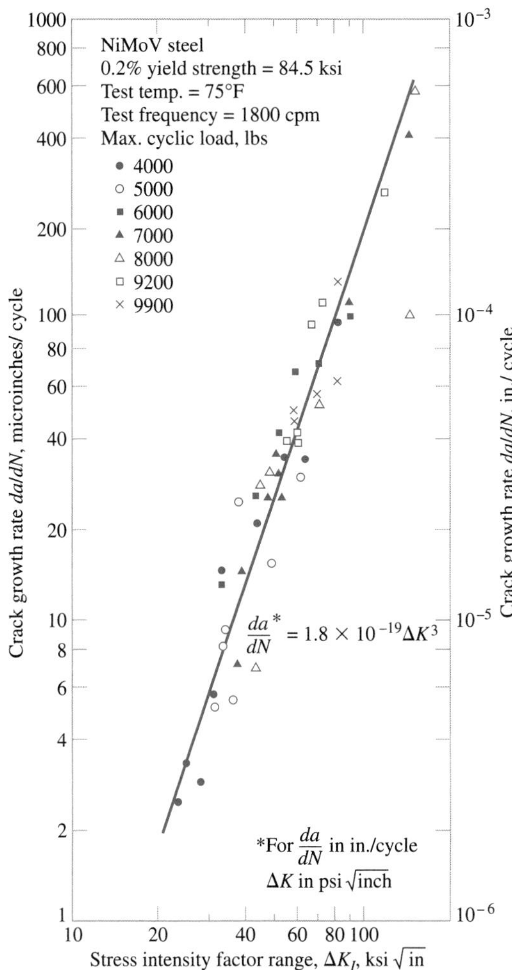

Figure P2.59

Crack-growth data for Ni-Mo-V alloy steel. (From W. G. Clark, Jr., "Fracture Mechanics in Fatigue," *Experimental Mechanics*, Sept. 1971. (Reprinted from "Experimental Mechanics" with the permission of the Society for Experimental Mechanics, Inc., 7 School St., Bethel, CT 06801, (203) 790-6373, www.sem.org.)

2-59. A Ni-Mo-V steel plate with yield strength of 84,500 psi, plane strain fracture toughness of 33,800 psi$\sqrt{\text{in}}$, and crack-growth behavior shown in Figure P2.59, is 0.50 inch thick, 10.0 inches wide, and 30.0 inches long. The plate is to be subjected to a released tensile load fluctuating from 0 to 160,000 lb, applied in the longitudinal direction (parallel to 30-inch dimension). A through-the-thickness crack of length 0.075 inch has been detected at one edge. How many more cycles of this released tensile loading would you predict could be applied before catastrophic fracture would occur?

2-60. A helicopter-transmission support leg (one of three such members) consists of a flat plate of rectangular cross section. The plate is 0.50 inch thick, 6.0 inches wide, and 8.0 inches long. Strain gage data indicate that the load is cycling from 100 lb to 25,000 lb tension each cycle at a frequency of about 5 times per second. The load is applied parallel to the 8.0-inch dimension and is distributed uniformly across the 6.0-inch width. The material is Ni-Mo-V alloy steel with an ultimate strength of 110,000 psi, yield strength of 84,500 psi, plane strain fracture toughness of 33,800 psi$\sqrt{\text{in}}$, and crack-growth behavior as shown in Figure P2.59.

If a through-the-thickness crack at one edge, with crack length of 0.010 inch, is *missed* during an inspection *today*, what length would you estimate that crack might have by the next inspection one month from today? Does the *missed crack* pose a significant threat to the safety of the aircraft? The helicopter operates about 12 hours per day and about 22 days per month.

2-61. Referring to the pinned mechanism with a lateral spring at point B, shown in Figure 2.35, do the following:

a. Repeat the derivation leading to (2-63), using the concepts of *upsetting moment* and *resisting moment*, to find an expression for critical load.

b. Use an energy method to again find an expression for critical load in the mechanism of Figure 2.35, by equating *change in the potential energy* of vertical force P_a to *strain energy stored* in the spring. (*Hint:* Use the first two terms of the series expansion for cos α to approximate cos α.)

c. Compare results of part (a) with results of part (b).

2-62. Verify the value of effective length $L_e = 2L$ for a column fixed at one end and free at the other [see Figure 2.37(b)] by writing and solving the proper differential equation for this case, then comparing the result with text equation (2-75).

2-63. A solid cylindrical steel bar is 2.0 inches in diameter and 12 feet long. If both ends are *pinned*, estimate the axial load required to cause the bar to buckle.

2-64. If the same amount of material used in the steel bar of problem 2-63 had been formed into a hollow cylindrical bar of the same length and supported at the ends in the same way (i.e., same cross-sectional area), what would the critical buckling load be if the tube wall thickness were made (a) $\frac{1}{4}$ inch, (b) $\frac{1}{8}$ inch, and (c) $\frac{1}{16}$ inch. What conclusions would you draw from these results?

2-65. If the solid bar of problem 2-63 were *fixed* at the ends, to what value would the critical buckling load be changed?

2-66. A steel pipe 4 inches in outside diameter, and having 0.226-inch wall thickness, is used to support a tank of water weighing 10,000 pounds when full. The pipe is set vertically in a heavy, rigid concrete base, as shown in Figure P2.66. The pipe material is AISI 1060 cold-drawn steel with S_u = 90,000 psi and S_{yp} = 70,000 psi. A safety factor of 2 *on load* is desired.

a. Derive a design equation for the maximum safe height H above the ground level that should be used for this application. (Use the approximation $I \approx \pi D^3 t/8$.)

b. Compute a numerical value for $(H_{max})_{pipe}$.

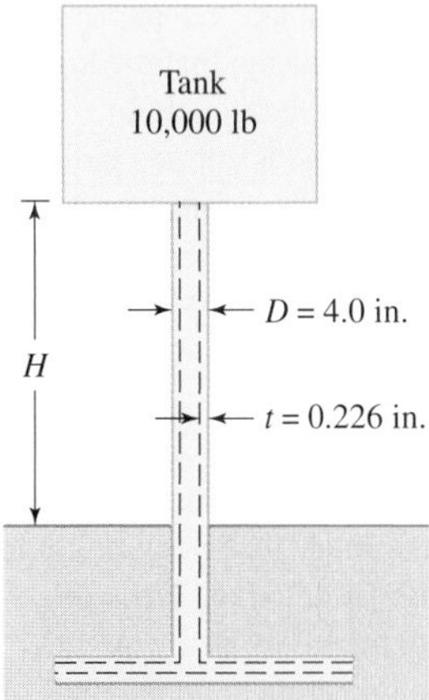

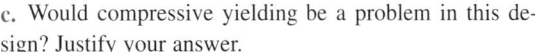

Figure P2.66
Water tank supported by a steel pipe.

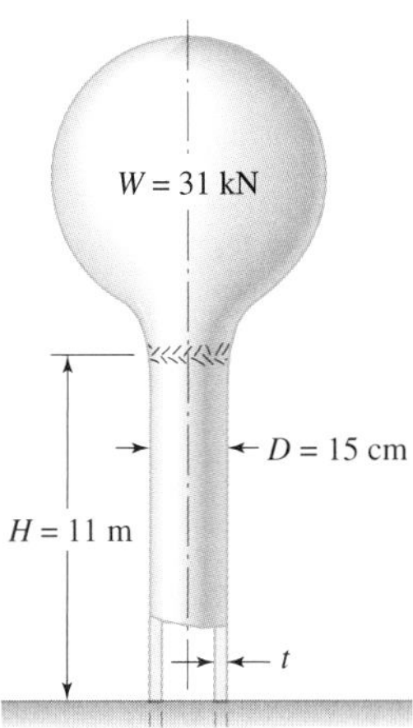

Figure P2.69
Liquid fertilizer tank and support.

c. Would compressive yielding be a problem in this design? Justify your answer.

2-67. Instead of using a steel pipe for supporting the tank of problem 2-66, it is being proposed to use a $W6 \times 25$ wide-flange beam for the support, and a plastic line to carry the water. (See Appendix Table A.3 for beam properties.) Compute the maximum safe height $(H_{max})_{beam}$ above ground level that this beam could support and compare the result with the height $(H_{max})_{pipe} = 145$ inches, as determined in problem 2-66(b).

2-68. A steel pipe is to be used to support a water tank using a configuration similar to the one shown in Figure P2.66. It is being proposed that the height H be chosen so that failure of the supporting pipe by *yielding* and by *buckling* would be equally likely. Derive an equation for calculating the height, H_{eq}, that would satisfy the suggested proposal.

2-69. A steel pipe made of AISI 1020 cold-drawn material (see Table 3.3) is to have an outside diameter of $D = 15$ cm, and is to support a tank of liquid fertilizer weighing 31 kN when full, at a height of 11 meters above ground level, as shown in Figure P2.69. The pipe is set vertically in a heavy, rigid concrete base. A safety factor of $n = 2.5$ *on load* is desired.

a. Using the approximation that $I \approx (\pi D^3 t)/8$, derive a design equation, using symbols only, for the minimum pipe wall thickness that should be used for this application. Write the equation explicitly for t as a function of H, W, n, and D, defining all symbols used.

b. Compute a numerical value for thickness t.

c. Would compressive yielding be a problem in this design? Justify your answer.

2-70. A connecting link for the cutter head of a rotary mining machine is shown in Figure P2.70. The material is to be AISI

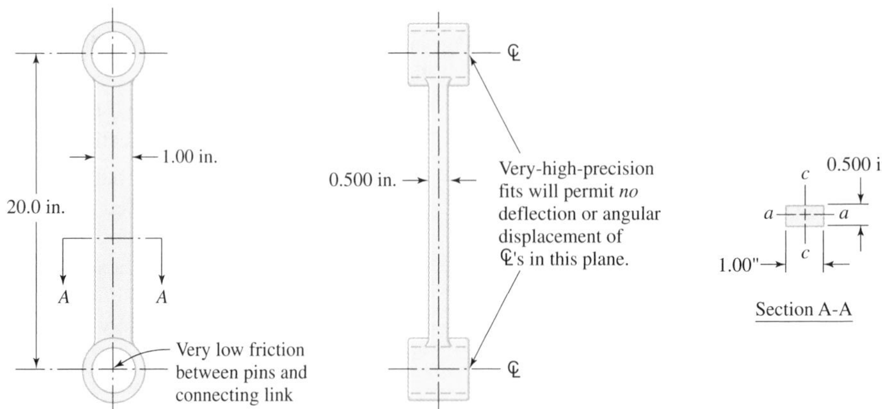

Figure P2.70
Connecting link for rotary mining machine. (Not to scale.)

1020 steel, annealed. The maximum axial load that will be applied in service is P_{max} = 10,000 pounds (compression) along the centerline, as indicated in Figure P2.70. If a safety factor of at least 1.8 is desired, determine whether the link would be acceptable as shown.

2-71. A steel wire of 0.1-inch diameter is subjected to torsion. If the material has a tensile yield strength of 100,000 psi, and the wire is 10 feet long, find the torque at which it will fail, and identify the failure mode.

2-72. A sheet-steel cantilevered bracket of rectangular cross section 0.125 inch by 4.0 inches is fixed at one end with the 4.0-inch dimension vertical. The bracket, which is 14 inches long, must support a vertical load, P, at the free end.

a. What is the maximum load that should be placed on the bracket if a safety factor of 2 is desired? The steel has a yield strength of 45,000 psi.

b. Identify the governing failure mode.

2-73. A hollow tube is to be subjected to torsion. Derive an equation that gives the length of this tube for which failure is equally likely by yielding or by elastic instability.

2-74. A steel cantilever beam 60 inches long with a rectangular cross section 1 inch wide by 3 inches deep is made of steel material that has a yield strength of 40,000 psi. Neglecting the weight of the beam, from what height h would a 13-lb weight have to be dropped on the free end of the beam to produce yielding? Neglect stress concentration effects.

2-75. A utility cart used to transport hardware from a warehouse to a loading dock travels along smooth, level rails. At the end of the line the cart runs into a cylindrical steel bumper bar of 3.0-inch diameter and 10-inch length, as shown in Figure P2.75. Assuming a perfectly "square" contact, frictionless wheels, and negligibly small bar mass, do the following:

a. Use the energy method to derive an expression for maximum stress in the bar.

b. Calculate the numerical value of the compressive stress induced in the bar if the weight of the loaded cart is 1100 lb and it strikes the bumper bar at a velocity of 5 miles per hour.

2-76. If the impact factor, the bracketed expression in (2-97) and (2-98), is generalized, it may be deduced that for any elastic structure the impact factor is given by $[1 + \sqrt{1 + (2h/y_{max\text{-}static})}]$. Using this concept, estimate the reduction in stress level that would be experienced by the beam of Example 2.19 if it were supported by a spring with k = 390 lb\in at each of the simple supports, instead of being rigidly supported.

2-77. A tow truck weighing 5000 lb is equipped with a 1-inch nominal diameter wire tow rope that has a metallic cross-sectional area of 0.404 in^2, a rope modulus of elasticity of 12×10^6 psi, and an ultimate strength of 200,000 psi. The 20-foot-long tow rope is attached to a wrecked vehicle and the driver tries to jerk the wrecked vehicle out of the ditch. If the tow truck is traveling at 5 miles per hour when the slack in the rope is taken up, and the wrecked car does not move, would you expect the rope to break?

2-78. An automobile that weighs 14.3 kN is traveling toward a large tree in such a way that the bumper contacts the tree at the bumper's midspan between supports that are 1.25 m apart. If the bumper is made of steel with a rectangular cross section 1.3 cm thick by 13.0 cm deep, and it may be regarded as simply supported, how fast would the automobile need to be traveling to just reach the 1725 MPa yield strength of the bumper material?

2-79. a. If there is zero clearance between the bearing and the journal (at point B in Figure P2.79), find the maximum stress in the steel connecting rod A–B, due to impact, when the 200-psi pressure is suddenly applied.

b. Find the stress in the same connecting rod due to impact if the bearing at B has a 0.005-inch clearance space between bearing and journal and the 200-psi pressure is suddenly applied. Compare the results with part (a) and draw conclusions.

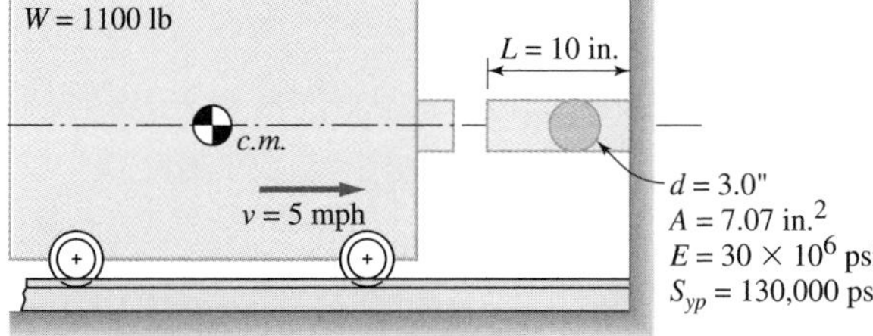

Figure P2.75
Utility cart and bumper bar.

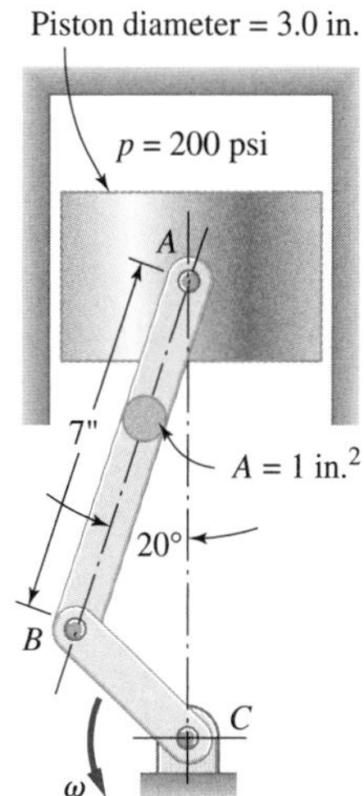

Figure P2.79
Sketch of a connecting rod in an internal combustion engine.

2-80. Carefully define the terms *creep, creep rupture*, and *stress rupture*, citing the similarities that relate these three failure modes and the differences that distinguish them from one another.

2-81. List and describe several methods that have been used for extrapolating short-time creep data to long-term applications. What are the potential pitfalls in using these methods?

2-82. A new high-temperature alloy is to be used for a 0.125-inch diameter tensile support member for an impact-sensitive instrument that weighs 200 lb and costs $300,000. The instrument and its support are to be enclosed in a test vessel for 3000 hours at 1600°F. A laboratory test on the new alloy used a 0.125-inch diameter specimen loaded in tension by a mass weighing 200 lb, and was found to fail by stress rupture after 100 hours at 1800°F. Based on the test results, determine whether the tensile support is adequate for this application.

2-83. A sand-cast aluminum alloy, creep tested at 600°F, results in the data shown in Figure 2.44(C), curve *b*, when subjected to a stress level of 10,000 psi. A solid square support bracket of this material is 1 inch × 1 inch × 4 inches long and is subjected to a direct tensile load of 10,000 lb. The ambient temperature cycle is 380°F for 1200 hours, then 600°F for 2 hours, and then 810°F for 15 seconds. How many cycles would you predict could be survived by the bracket if the design criterion is that the bracket must not elongate more than 0.10 inch?

2-84. From the data plotted in Figure P2.84, evaluate the constants *B* and *N* of (2-112) for the material tested.

2-85. Give a definition of *wear failure* and list the major subcategories of wear.

2-86. One part of the mechanism is a new metering device for a seed-packaging machine is shown in Figure P2.86. Both the slider shoe and the rotating wheel are to be made of stainless steel, with a yield strength of 275 MPa. The contact area of the shoe is 2.5 cm long by 1.3 cm wide. The rotating wheel is 25 cm in diameter and rotates at 30 rpm. The spring is set to exert a constant normal force at the wearing interface of 70 N.

a. If no more than 1.5-mm wear of the shoe surface can be tolerated, and no lubricant may be used, estimate the maintenance interval in operating hours between shoe replacements. (Assume that *adhesive wear* predominates.)

b. Would this be an acceptable maintenance interval?

c. If it were possible to use a lubrication system that would provide "excellent" lubrication to the contact interface, estimate the potential improvement in maintenance interval, and comment on its acceptability.

d. Suggest other ways to improve the design from the standpoint of reducing wear rate.

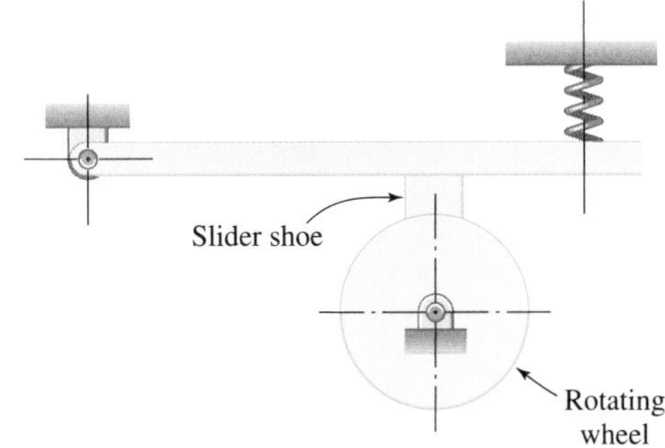

Figure P2.86
Seed-packaging mechanism.

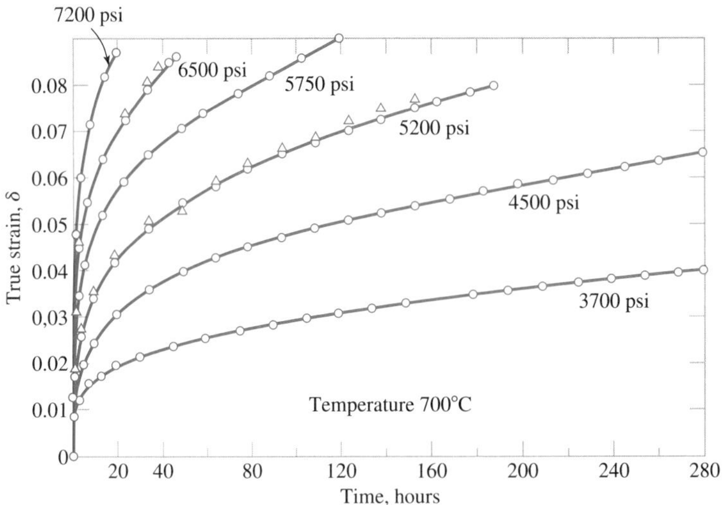

Figure P2.84
Creep data. (From ref. 51.)

2-87. In a cinder block manufacturing plant the blocks are transported from the casting machine on rail carts supported by ball-bearing-equipped wheels. The carts are currently being stacked six blocks high, and bearings must be replaced on a 1-year maintenance schedule because of ball-bearing failures. To increase production, a second casting machine is to be installed, but it is desired to use the same rail cart transport system with the same number of carts, merely stacking the blocks 12 high. What bearing-replacement interval would you predict might be necessary under this new procedure?

2-88. Give a definition of *corrosion failure* and list the major subcategories of corrosion.

2-89. It is planned to thread a bronze valve body into a cast-iron pump housing to provide a bleed port.

a. From the corrosion standpoint, would it be better to make the bronze valve body as large as possible or as small as possible?

b. Would it be more effective to put an anticorrosion coating on the bronze valve or on the cast-iron housing?

c. What other steps might be taken to minimize corrosion of the unit?

2-90. Give a definition for *fretting* and distinguish among the failure phenomena of *fretting fatigue*, *fretting wear*, and *fretting corrosion*.

2-91. List the variables thought to be of primary importance in fretting-related failure phenomena.

2-92. Fretting corrosion has proved to be a problem in aircraft splines of steel on steel. Suggest one or more measures that might be taken to improve the resistance of the splined joint to fretting corrosion.

2-93. List several basic principles that are generally effective in minimizing or preventing fretting.

Chapter 3

Materials Selection

3.1 Steps in Materials Selection

Keystone objectives of mechanical design activity were featured in Chapter 1 as (1) selection of the best possible material and (2) determination of the best possible geometry for each part. In contrast with the materials engineer's task of *developing new and better* materials, a mechanical designer must be effective in *selecting* the best *available* material for each application, considering *all* important design criteria. Although materials engineers are often key members of a design team, the mechanical designer should also be well grounded in the types and properties of materials available to meet specific design needs.

Materials selection is typically carried out as a part of the intermediate design stage, but in some cases must be considered earlier, during the preliminary design stage. The basic steps in selection of candidate materials for any given application are:

1. Analyzing the material-specific requirements of the application.
2. Assembling a list of requirement-responsive materials, with pertinent performance evaluation data, rank-ordered so the "best" material is at the top of the table[1] for each important application requirement.
3. Matching the lists of materials responsive to the pertinent application requirements in order to select the "best" candidate materials for the proposed design.

3.2 Analyzing Requirements of the Application

Material-specific requirements for most applications may be identified by examining the list of possibilities given in Table 3.1. This might be accomplished by first developing a *specification statement* for the machine part under consideration, based upon the anticipated use of the device. For example, the specification statement for the crankshaft to be used in a proposed new design for a one-cylinder belt-driven air compressor might be written as:

The crankshaft should be short, compact, relatively rigid, fatigue resistant, wear resistant at the bearing sites, and capable of low-cost production.

Clearly, such a statement embodies design insights relating not only to operational and functional requirements of the device, but to potential failure modes and market-driven factors as well.

To translate the specification statement into material-specific requirements, for each entry in Table 3.1 the question may be posed, *Is there a special need to consider this requirement in this application?* Considered responses of *yes, no*, and *perhaps* may be en-

[1]An equivalent "graphical" presentation of evaluation data, developed by *Ashby*, is discussed in 3.5.

TABLE 3.1 Potential Material-Specific Application Requirements

Potential Application Requirement	Special Need?
1. Strength/volume ratio	________
2. Strength/weight ratio	________
3. Strength at elevated temperature	________
4. Long-term dimensional stability at elevated temperature	________
5. Dimensional stability under temperature fluctuation	________
6. Stiffness	________
7. Ductility	________
8. Ability to store energy elastically	________
9. Ability to dissipate energy plastically	________
10. Wear resistance	________
11. Resistance to chemically reactive environment	________
12. Resistance to nuclear radiation environment	________
13. Desire to use specific manufacturing process	________
14. Cost constraints	________
15. Procurement time constraints	________

tered in the special need column of Table 3.1. The *yes* and *perhaps* responses then identify the material-specific requirements that should be addressed in the process of selecting good candidate materials for the particular part being designed.

3.3 Assembling Lists of Responsive Materials

To assist in assembling *lists of materials* responsive to each of the specific needs identified from Table 3.1, *performance evaluation indices* must be available by which candidate materials may be ranked in order of their ability to respond to any designated application requirement listed in Table 3.1. Such a list of *performance evaluation indices* is given in Table 3.2, together with the corresponding material-specific requirements from Table 3.1.

TABLE 3.2 Requirement-Responsive Material Characteristics and Corresponding Performance Evaluation Indices

Requirement-Responsive Material Characteristic	Performance Evaluation Index
1. Strength/volume ratio	Ultimate or yield strength
2. Strength/weight ratio	Ultimate or yield strength/density
3. Resistance to thermal weakening	Strength loss/degree temperature
4. Creep resistance	Creep rate at operating temperature
5. Thermal expansion	Strain/degree temperature change
6. Stiffness	Modulus of elasticity
7. Ductility	Percent elongation in 2 inches
8. Resilience	Energy/unit volume at yield
9. Toughness	Energy/unit volume at rupture
10. Wear resistance	Dimensional loss at operating condition; also hardness
11. Corrosion resistance	Dimensional loss in operating environment
12. Susceptibility to radiation damage	Strength or ductility change in operating environment
13. Manufacturability	Suitability for specific process
14. Cost	Cost/unit weight; also machinability
15. Availability	Procurement time and effort

For items from Table 3.1 that have been identified as special needs in any given application, Table 3.2 then provides performance evaluation indices that form the basis for comparison and selection of candidate materials. The procedure for accomplishing this task is discussed further in 3.4.

To use the performance evaluation indices for comparing candidate materials, it is necessary to find quantitative materials data for each of the key parameters that comprise the pertinent evaluation indices. There are many sources for such data.[2] To illustrate the procedure suggested here, a short compilation of selected materials, and their rank-ordered properties relating to the performance evaluation indices of Table 3.2, are given in Tables 3.3 through 3.20. Each rank-ordered table is *arranged with the best material for a given*

TABLE 3.3 Strength Properties of Selected Materials[1]

Material	Alloy	Ultimate Tensile Strength S_u (psi)	Yield Strength S_{yp} (psi)
Ultra-high-strength steel	AISI 4340	287,000	270,000
Stainless steel (age hardenable)	AM 350	206,000	173,000
High-carbon steel	AISI 1095[2]	200,000	138,000
Graphite-epoxy composite	—	200,000	—
Titanium	Ti-6Al-4V	150,000	128,000
Ceramic	Titanium carbide (bonded)	134,000	—
Nickel-based alloy	Inconel 601	102,000	35,000
Medium-carbon steel	AISI 1060 (HR)[3]	98,000	54,000
	AISI 1060 (CD)[4]	90,000	70,000
Low-carbon, low-alloy steel	AISI 4620 (HR)	87,000	63,000
	AISI 4620 (CD)	101,000	85,000
Stainless steel (austenitic)	AISI 304 (annealed)	85,000	35,000
Yellow brass	C 26800 (hard)	74,000	60,000
Commercial bronze	C 22000 (hard)	61,000	54,000
Low-carbon (mild) steel	AISI 1020 (CD)	61,000	51,000
	AISI 1020 (annealed)	57,000	43,000
	AISI 1020 (HR)	55,000	30,000
Phosphor bronze	C 52100 (annealed)	55,000	24,000
Gray cast iron	ASTM A-48 (class 50)	50,000[5]	—
Gray cast iron	ASTM A-48 (class 40)	40,000	—
Aluminum (wrought)	2024-T3 (heat treated)	70,000	50,000
Aluminum (wrought)	2024 (annealed)	27,000	11,000
Aluminum (perm. mold cast)	356.0 (sol'n. treated; aged)	38,000	27,000
Magnesium (extruded)	ASTM AZ80A-T5	50,000	35,000
Magnesium (cast)	ASTM AZ63A	29,000	14,000
Thermosetting polymer	Epoxy (glass reinforced)	—	10,000
Thermoplastic polymer	Acrylic (cast)	—	7000

[1]See, for example, ref. 1–10.
[2]Quenched and drawn to Rockwell C-42.
[3]Hot-rolled.
[4]Cold-drawn.
[5]Ultimate *compressive* strength is 170,000 psi.

[2]See, for example, refs. 1–10 and 16–19.

TABLE 3.4 Strength/Weight Ratios for Selected Materials

Material	Weight Density, w (lb/in^3)	Approx. Ultimate Strength/Wt Ratio, $\frac{S_u}{w}$ (inches $\times 10^3$)	Approx. Yield Strength/Wt Ratio, $\frac{S_{yp}}{w}$ (inches $\times 10^3$)
Graphite-epoxy composite	0.057	3500	—
Ultra-high-strength steel	0.283	1000	950
Titanium	0.160	950	800
Stainless steel (age hardenable)	0.282	750	600
Aluminum (wrought)	0.100	700	500
Titanium carbide	0.260	500	—
Aluminum (perm. mold cast)	0.097	400	300
Medium-carbon steel	0.283	350	200
Nickel-based alloy	0.291	350	100
Stainless steel (austenitic)	0.290	290	120
Yellow brass	0.306	250	200
Low-carbon steel	0.283	200	150
Commercial bronze	0.318	200	150
Gray cast iron (class 50)	0.270	200	—
Epoxy (glass reinf.)	0.042	—	250
Acrylic (cast)	0.043	—	150

TABLE 3.5 Strength at Elevated Temperature for Selected Materials

Material	Temperature Θ (°F)	Ultimate Tensile Strength $(S_u)_\Theta$ (psi)	Yield Strength $(S_{yp})_\Theta$ (psi)
Ultra-high-strength steel (4340)	−200	313,000	302,000
	RT[1]	287,000	270,000
	400	273,000	235,000
	800	221,000	186,000
	1200	103,000	62,000
Stainless steel (AM 350)	RT	206,000	173,000
	400	185,000	144,000
	800	179,000	119,000
	1000	119,000	83,000
Titanium (Ti-6Al-4V)	−200	187,000	155,000
	RT	150,000	128,000
	400	126,000	101,000
	800	90,000	75,000
	1000	81,000	59,000
Titanium carbide	RT	134,000	—
	1500	94,000	—
	1800	72,000	—
Inconel (601)	RT	102,000	35,000
	400	94,000	31,000
	800	84,000	28,000
	1200	66,000	23,000
	1600	20,000	12,000

TABLE 3.5 (*Continued*)

Material	Temperature Θ (°F)	Ultimate Tensile Strength $(S_u)_\Theta$ (psi)	Yield Strength $(S_{yp})_\Theta$ (psi)
Low-carbon steel (1020)	−200	97,000	83,000
	RT	61,000	51,000
	400	61,000	51,000
	800	45,000	38,000
	900	29,000	24,000
Aluminum (2024-T3)	−200	74,000	54,000
	RT	70,000	50,000
	400	52,000	39,000
	800	4,000	4,000
Magnesium (AZ80A-T5)	−200	63,000	53,000
	RT	50,000	35,000
	200	43,000	19,000
	400	22,000	11,000

[1]Room temperature.

index at the top, followed in order of decreasing desirability by several other materials. Although suitable for solving problems in this textbook, real-world design tasks often require more comprehensive data gleaned from the literature and/or other data banks if the design is to be competitive.

TABLE 3.6 Stress Rupture Strength Levels (psi) Corresponding to Various Rupture Times and Temperatures for Selected Materials

Material	Alloy	Temp., Θ (°F)	Rupture Time, t, hours: 10	100	600	1000	10,000
Stainless steel	AM 350	800	—	184,000	—	182,000	—
Iron-base superalloy	A-286	1000	120,000	100,000	—	80,000	76,000
		1200	78,000	68,000	—	50,000	34,000
		1350	50,000	35,000	—	21,000	14,000
		1500	21,000	11,000	—	—	—
Cobalt base superalloy	X-40	1500	61,000	56,000	—	51,000	—
Inconel	601	1200	—	—	—	28,000	—
		1400	—	—	—	9,100	—
		1600	—	—	—	4,200	—
		1800	—	—	—	2,000	—
Carbon steel	1050	750	—	52,500	49,000	—	—
		930	—	22,400	18,000	—	—
Aluminum	Duralumin	300	—	38,000	32,500	—	—
		480	—	11,200	8,300	—	—
		660	—	3,100	2,700	—	—
Brass	60/40	300	—	47,000	42,500	—	—
		480	—	15,700	9,000	—	—

TABLE 3.7 Creep Limited Maximum Stresses (psi) Corresponding to Various Strain Rates and Temperatures for Selected Materials

Material	Alloy	Temp., Θ (°F)	Strain Rate, $\dot{\delta}$, in/in/hr				
			4×10^{-7}	1×10^{-6}	4×10^{-6}	1×10^{-5}	4×10^{-5}
Strainless steel	AM 350	800	—	91,000	—	—	—
Chromium steel (Q&T)	13% Cr	840	23,500	—	33,600	—	41,500
Manganese steel	1.7% Mn	840	23,500	—	27,000	—	36,000
Carbon steel (forged)	1030	930	13,000	—	16,300	—	19,000
Stainless steel	304	1000	—	—	—	10,000	—
		1300	—	—	—	8,000	—
		1500	—	—	—	5,000	—
Phosphor bronze		440	10,000	—	15,700	—	21,300
Magnesium	HZ32A-T5	400	—	—	—	—	10,000
		500	—	—	—	—	8,000
		600	—	—	—	—	5,000
Aluminum	Duralumin	440	5,600	—	7,400	—	9,700
Brass	60/40	440	1,020	—	2,700	—	5,600

TABLE 3.8 Coefficients of Thermal Expansion for Selected Materials

Material	Alloy	Coefficient of Thermal Expansion, α (10^{-6} in/in/°F)	Temperature Range of Validity (°F)
Ceramic	Titanium carbide (bonded)	4.3–7.5	68–1200
Titanium	Ti-6Al-4V	5.3	68–1000
Gray cast iron	ASTM A-48 (class 50)	6.0	32–212
Steel	Most	6.3	0–200
Stainless steel	AM 350	6.3	—
Nickel-base alloy	Inconel 601	7.6	80–200
Nickel-base alloy	Inconel 600	9.3	80–1500
Cobalt base superalloy	X-40	9.2	70–1800
Stainless steel	304	9.6	32–212
Graphite-epoxy composite	—	10	—
Commercial bronze	C 22000	10.2	68–572
Iron-base superalloy	A-286	10.3	70–1000
Yellow brass	C 26800	11.3	68–572
Aluminum (cast)	356.0	11.9	68–212
Aluminum (wrought)	2024-T3	12.9	68–212
Aluminum (wrought)	2024-T3	13.7	68–572
Magnesium	Most	14.0	68
Magnesium	Most	16.0	68–750
Thermosetting polymer	Epoxy (glass reinf.)	10–20	—
Thermoplastic polymer	Acrylic	45	—

TABLE 3.9 Stiffness Properties of Selected Materials

Material	Young's Modulus of Elasticity, *E* (10^6 psi)	Shear Modulus of Elasticity, *G* (10^6 psi)	Poisson's Ratio, *ν*
Tungsten carbide	95	—	0.20
Titanium carbide	42–65 (77°F)	—	0.19
Titanium carbide	33–48 (1600–1800°F)	—	—
Molybdenum	47 (RT)[1]	—	0.29
Molybdenum	33 (1600°F)	—	—
Molybdenum	20 (2400°F)	—	—
Steel (most)	30	11.5	0.30
Stainless steel	28	10.6	0.31
Iron-base superalloy (A-286)	29.1 (RT)	—	0.31
	23.5 (1000°F)	—	—
	22.2 (1200°F)	—	—
	19.8 (1500°F)	—	—
Cobalt-base superalloy	29	—	—
Inconel	31	11.0	—
Cast iron	13–24	5.2–8.5	0.21–0.27
Commercial bronze (C 22000)	17	6.3	0.35
Titanium	16	6.2	0.31
Phosphor bronze	16	6.0	0.35
Aluminum	10.3	3.9	0.33
Magnesium	6.5	—	0.29
Graphite-epoxy composite	6.0	—	—
Acrylic thermoplastic	0.4	—	0.4

[1]Room temperature.

TABLE 3.10 Ductility of Selected Materials

Material	Alloy	Elongation in 2-inch Gage Length, *e* (2 in) (percent)
Phosphor bronze C	C 52100	70
Inconel	601	50 (RT)
Inconel	601	50 (1000°F)
Inconel	601	75 (1400°F)
Stainless steel	AISI 304	60
Copper	Oxygen-free	50
Silver		48
Gold		45
Aluminum (annealed)	1060	43
Low-carbon low-alloy steel	AISI 4620 (HR)[1]	28
	AISI 4620 (CD)[2]	22
Low-carbon steel	AISI 1020 (HR)	25
Low-carbon steel	AISI 1020 (CD)	15
Aluminum (wrought)	2024-T3	22
Stainless steel	AM 350	13
Medium-carbon steel	AISI 1060 (HR)	12
Medium-carbon steel	AISI 1060 (CD)	10

TABLE 3.10 (*Continued*)

Material	Alloy	Elongation in 2-inch Gage Length, e (2 in) (percent)
Ultra-high-strength steel	AISI 4340	11
Titanium	Ti-6Al-4V	10 (RT)
Titanium	Ti-6Al-4V	18 (800°F)
Cobalt-base superalloy	X-40	9 (RT)
Cobalt-base superalloy	X-40	12 (1200°F)
Cobalt-base superalloy	X-40	22 (1700°F)
Magnesium (forged)	AZ80A-T5	6
Aluminum (perm. mold cast)	356.0 (sol'n treated; aged)	5
Commercial bronze	C 22000 (hard)	5
Gray cast iron	All	nil

[1]Hot-rolled.
[2]Cold-drawn.

TABLE 3.11 Modulus of Resilience *R* for Selected Materials Under Tensile Loading

Material	Alloy	$R = \frac{S_{yp}^2}{2E}$ (in-lb/in^3)
Ultra-high-strength steel	AISI 4340	1220
Stainless steel	AM 350	530
Titanium	Ti-6Al-4V	510
Aluminum (wrought)	2024-T3 (heat treated)	120
Magnesium (extruded)	AZ80A-T5	90
Medium-carbon steel	AISI 1060	80
Low-carbon steel	AISI 1020	40
Stainless steel	AISI 304	21
Nickel-base alloy	Inconel 601	20
Phosphor bronze	C 52100 (annealed)	20

TABLE 3.12 Toughness Merit Number *T* for Selected Materials Under Tensile Loading (also see Table 2.1 for plane strain fracture toughness data)

Material	Alloy	$T = S_u[e(2\text{ in})/100]$ (in-lb/in^3)
Nickel-base alloy	Inconel 601	51,000
Stainless steel	AISI 304	51,000
Phosphor bronze	C 52100 (annealed)	38,500
Ultra-high-strength steel	AISI 4340	31,600
Stainless steel	AM 350	26,800
Aluminum (wrought)	2024-T3 (heat treated)	15,400
Low-carbon steel	AISI 1020	15,300
Titanium	Ti-6Al-4V	15,000
Medium-carbon steel	AISI 1060	9,000
Magnesium (extruded)	AZ80A-T5	3,000

TABLE 3.13 Hardness of Selected Materials

Material	Hardness Scale[1]						
	BHN	R_C	R_A	R_B	R_M	V	Mohs
Diamond	8500 (approx.)[2]	—	—	—	—	—	10
Sapphire	—	—	—	—	—	—	9
Tungsten carbide	1850 (approx.)[2]	—	93	—	—	—	8–9
Titanium carbide	1850 (approx.)[2]	—	93	—	—	—	8–9
Case-hardened low-carbon steel	650	62	82.5	—	—	—	—
Ultra-high-strength steel	560	56	79	—	—	—	—
Titanium	315	34	67.5	—	—	—	—
Gray cast iron	262	26	—	—	—	—	—
Low-carbon low-alloy steel	207	15	—	—	—	—	—
Medium-carbon steel (CD)[3]	183	(9)[2]	—	89.5	—	—	—
Low-carbon steel (CD)	121	—	—	68	—	127	—
Aluminum (wrought)	120	—	—	67.5	—	126	—
Nickel-base alloy	114	—	—	64	—	120	—
Magnesium (extruded)	82	—	—	49	—	—	—
Commercial bronze	70	—	—	34	—	—	—
Gold (annealed)	—	—	—	—	—	25	—
Epoxy (glass reinforced)	—	—	—	—	105	—	—
Acrylic (cast)	—	—	—	—	85	—	—

[1]BHN = Brinell hardness number
R_C = Rockwell C scale
R_A = Rockwell A scale
R_B = Rockwell B scale
R_M = Rockwell M scale
V = Vickers hardness number
Mohs = Mohs hardness number

[2]Out of normal range—information only.

[3]Cold drawn.

TABLE 3.14 Galvanic Corrosion Resistance in Sea Water for Selected Materials[1]

↑
Noble or cathodic (protected end)

Platinum
Gold
Graphite
Titanium
Silver

[Chlorimet 3 (62 Ni, 18 Cr, 18 Mo)
Hastelloy C (62 Ni, 17 Cr, 15 Mo)]

[18-8 Mo stainless steel (passive)
18-8 stainless steel (passive)
Chromium stainless steel 11–30% Cr (passive)]

[Inconel (passive) (80 Ni, 13 Cr, 7 Fe)
Nickel (passive)]

Silver solder

[Monel (70 Ni, 30 Cu)
Cupronickels (60–90 Cu, 40–10 Ni)
Bronzes (Cu-Sn)
Copper
Brasses (Cu-Zn)]

TABLE 3.14 (*Continued*)

	[Chlorimet 2 (66 Ni, 32 Mo, 1 Fe) Hastelloy B (60 Ni, 30 Mo, 6 Fe, 1 Mn)]
	[Inconel (active) Nickel (active)]
	Tin
	Lead
	Lead-tin solders
	[18-8 Mo stainless steel (active) 18-8 stainless steel (active)]
	Ni-Resist (high Ni cast iron)
	Chromium stainless steel, 13% Cr (active)
	[Cast iron Steel or iron]
	2024 aluminum (4.5 Cu, 1.5 Mg, 0.6 Mn)
	Cadmium
Active or anodic	Commercially pure aluminum (1100)
(corroded end)	Zinc
↓	Magnesium and magnesium alloys

[1]See p. 32 of ref. 12. (Reprinted with permission of the McGraw-Hill Companies.)

TABLE 3.15 Corrosion-Fatigue Strength of Selected Materials[1]

Material	Ultimate Tensile Strength (psi)	Fatigue Strength in Air (psi)	Corrosion-Fatigue Strength in Salt Spray (psi)	Cycles to Failure
Beryllium bronze[2]	94,000	36,500	38,800	5×10^7
18 Cr 8 Ni steel	148,000	53,500	35,500	5×10^7
17 Cr 1 Ni steel	122,000	73,500	27,500	5×10^7
Phosphor bronze[3]	62,000	22,000	26,000	5×10^7
Aluminum bronze	80,000	32,000	22,000	5×10^7
15 Cr steel	97,000	55,000	20,500	5×10^7
carbon steel	142,000	56,000	8,750	5×10^7
Duralumin	63,000	20,500	7,600	5×10^7
Mild steel	76,000	38,000	2,500	10×10^7

[1]See ref. 13.
[2]The apparent anomoly, that fatigue resistance with salt spray is higher than fatigue resistance in air, is acknowledged by the authors of ref. 13, but they stand by the values shown, and offer a supporting explanation.
[3]Ibid.

TABLE 3.16 Nuclear Radiation Exposure to Produce Significant (over 10%) Changes in Properties of Selected Materials[1]

Material	Amount of Radiation (integrated fast neutron flux) (neutrons/cm^2)	Property Changes
Zirconium alloys	10^{21}	Little change
Stainless steels		Reduced but not greatly impaired ductility
Aluminum alloys	10^{20}	Reduced but not greatly impaired ductility
Stainless steels		Yield strength tripled
Carbon steels		Increased fracture-transition temperature; severe loss of ductility; yield strength doubled

TABLE 3.16 (Continued)

Material	Amount of Radiation (integrated fast neutron flux) (neutrons/cm^2)	Property Changes
All plastics	10^{19}	Unusable as structural materials
Ceramics		Reduced thermal conductivity, density, and crystallinity
Polystyrene		Loss of tensile strength
Carbon steel	10^{18}	Reduction of notch-impact strength
Metals		Most show significant increase in yield strength
Natural rubber		Large change; hardening
Mineral-filled phenolic	10^{17}	Loss of tensile strength
Polyethylene		Loss of tensile strength
Butyl rubber		Large change; softening
Natural and butyl rubber	10^{16}	Loss of elasticity
Polymethyl methacrylate and cellulosics	10^{15}	Loss of tensile strength
Polytetrafluroethylene		Loss of tensile strength

[1]See ref. 14.

TABLE 3.17 Suitability of Selected Materials for Specific Manufacturing Processes

Material	Alloy	Available Forms[1]	Fabrication Properties
Ultra-high-strength steel	AISI 4340	B,b,f,p,S,s,w	Readily machinable (annealed); readily weldable (post-heat required)
Stainless steel	AM350	b,F,S,s,w,t	Readily machinable; readily weldable
Graphite-epoxy composite	—	Injection, compression, and transfer molding	—
Titanium	Ti-6A1-4V	B,b,P,S,s,w,e	Machinable (annealed), formable, weldable
Nickel-base alloy	Inconel 601	b,P,r,S,s,Sh,t	—
Medium-carbon steel	AISI 1060	b,r,f	Readily machinable; welding not recommended
Stainless steel	AISI 304	b,P,f,S,s,t,w	Readily machinable; readily weldable
Commercial bronze	C 22000 (hard)	P,r,S,s,t,w	Machinable; readily weldable
Low-carbon steel	AISI 1020	b,r,f,S,Sh	Readily machinable; readily weldable
Phosphor bronze	C 52100	r,s,w	Machinable; readily weldable
Gray cast iron	ASTM A-48 (class 50)	—	Machinable; weldable
Aluminum (wrought)	2024-T3	b,P,r,S,Sh,t,w	Easily machinable; weldable
Magnesium (extruded)	AZ80A-T5	b,r,f,Sh	Easily machinable (except fire hazard); weldable
Thermosetting polymer	Epoxy (glass reinforced)	Injection, compression, and transfer molding	—
Thermoplastic polymer	Acrylic (cast)	—	Machinable

[1]B = billets
b = bars
e = extrusions
F = foil
f = forgings
P = plates
r = rods
S = sheets
s = strip
Sh = shapes
t = tubing
w = wire

TABLE 3.18 Approximate Material Cost for Selected Materials[1]

Material	Approximate cost (dollars/lb)
Gray cast iron	0.30
Low-carbon steel (HR)[2]	0.50
Low-carbon steel (CD)[3]	0.60
Ultra-high-strength steel (HR)	0.65
Zinc alloy	1.50
Acrylic	2.00
Commercial bronze	2.25
Stainless steel	2.75
Epoxy (glass reinforced)	3.00
Aluminum alloy	3.50
Magnesium alloy	5.50
Titanium alloy	9.50

[1]Material cost varies widely by year of purchase and by quantity required. Designers should always obtain specific price quotations.
[2]Hot-rolled.
[3]Cold-drawn.

TABLE 3.19 Relative Machinability[1] of Selected Materials

Material	Alloy	Estimated Machinability Index[2]
Magnesium alloy	—	400
Aluminum alloy	—	300
Free-machining steel	B1112	100
Low-carbon steel	AISI 1020	65
Medium-carbon steel	AISI 1060 (annealed)	60
Ultra-high-strength steel	AISI 4340 (annealed)	50
Stainless steel alloy	(annealed)	50
Gray cast iron	—	40
Commercial bronze	—	30
Titanium alloy	(annealed)	20

[1]Machinability index is a less-than-exact evaluation of volume of material removal per hour, produced at maximum efficiency, balanced against a minimum rejection rate for reasons of surface finish or tolerance.
[2]Based on rating of 100 for B1112 resulfurized free-machining steel.

TABLE 3.20 Thermal Conductivity Ranges for Selected Materials

Material	Thermal Conductivity k [Btu/hr/ft/°F (W/m/°C)]
Silver	242 (419)
Copper	112 (194)–226 (391)
Pyrolytic graphite	108 (186.9)–215 (372.1)
Beryllium copper	62 (107)–150 (259)
Brass[1]	15 (26)–135 (234)

TABLE 3.20 (*Continued*)

Material	Thermal Conductivity k [Btu/hr/ft/°F (W/m/°C)]
Aluminum alloys[1]	93 (161)–125 (216)
Bronze[1]	20 (35)–120 (207)
Phosphor bronze[1]	29 (50)–120 (207)
Premium graphite	65 (112)–95 (164)
Carbon graphite	18 (31)–66 (114)
Aluminum bronze[1]	39 (68)
Cast iron	25 (43)–30 (52)
Carbon steel	27 (46.7)
Silicon carbide	9 (15)–25 (43)
Lead	16 (28)–20 (35)
Stainless steel	15 (26)
Titanium	4 (7)–12 (21)
Glass	1 (1.7)–2 (3.5)
Wood composition board (tempered hardboard)	1 (1.7)–1.5 (2.6)
Silicon plastics	0.075 (0.13)–0.5 (0.87)
Phenolics	0.116 (0.201)–0.309 (0.535)
Epoxies	0.1 (0.17)–0.3 (0.52)
Teflon	0.14 (0.24)
Nylon	0.1 (0.17)–0.14 (0.24)
Plastic foam	0.009 (0.016)–0.077 (0.133)

[1]For porous metal sintered parts, thermal conductivity values are 35–65% of values shown.

Another effective way of presenting rank-ordered materials data involves the use of a two-parameter graphical format. *Ashby materials selection charts*, recently published,[3] are two-parameter (two-dimensional) log-log graphs on which important performance evaluation indices (strength, stiffness, density, etc.) are cross plotted as shown, for example, in Figures 3.1 through 3.6.[4] A procedure for utilizing Ashby charts for material selection is discussed in 3.5.

3.4 Matching Responsive Materials to Application Requirements: Rank-Ordered-Data Table Method

One procedure for identifying good candidate materials for any specific application may be summarized as follows:

1. Using Table 3.1 as a guide, together with known requirements imposed by operational or functional constraints, postulated failure modes, market-driven factors, and/or management directives, establish a concise *specification statement* as discussed in 3.2. If information about the application is so sketchy that a specification statement cannot be written, and the remaining steps cannot be executed, it is suggested that *1020*

[3]See ref. 3.

[4]Many more Ashby charts are presented in ref. 3, as well as a set of charts that are helpful in selecting a suitable manufacturing process.

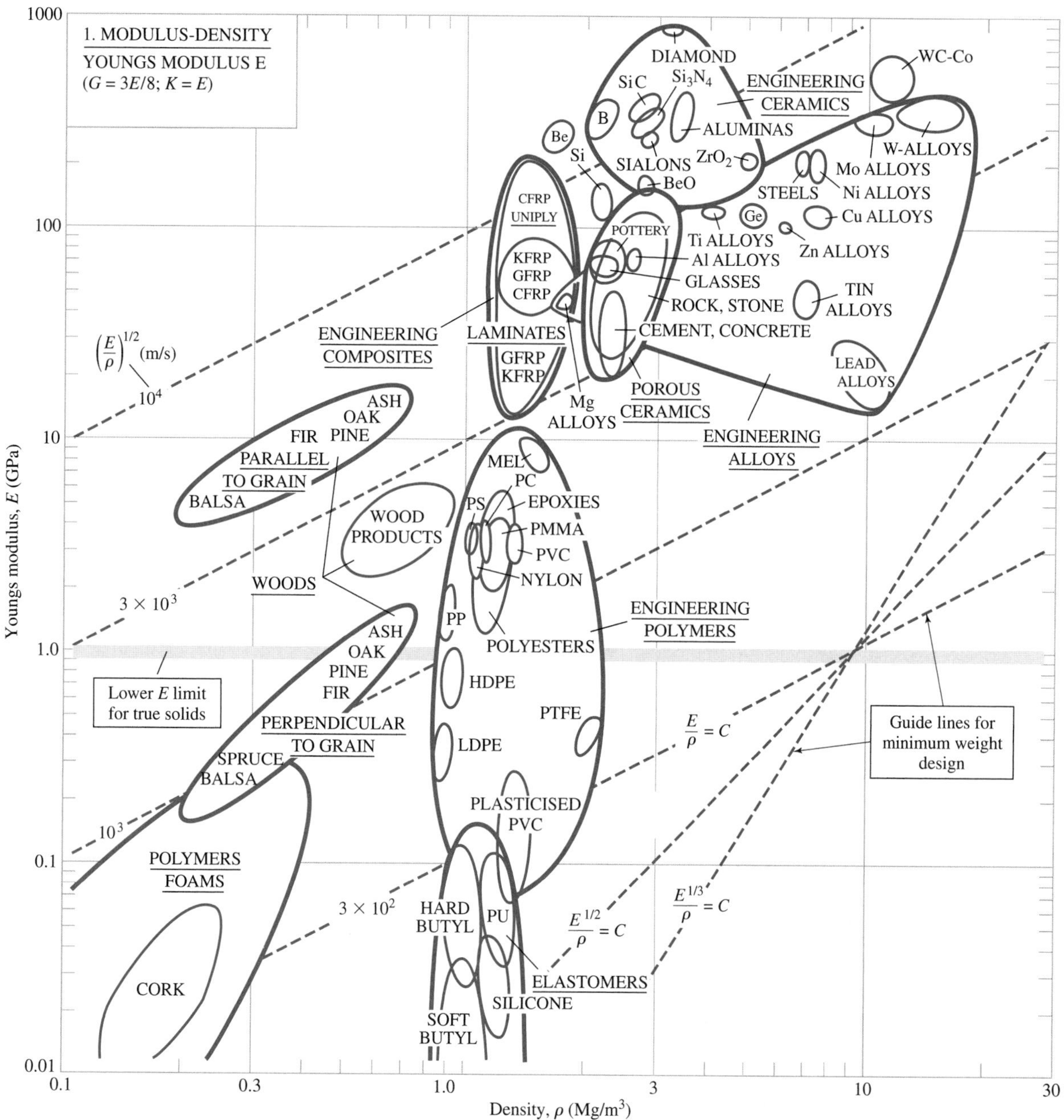

Figure 3.1
Two-parameter Ashby chart for Young's modulus of elasticity, *E*, plotted versus density, *ρ*. The content of the chart roughly corresponds to the data included in Tables 3.4 and 3.9. (The chart is taken from ref. 3, courtesy of M. F. Ashby.)

steel be tentatively selected as the "best" material because of its excellent combination of strength, stiffness, ductility, toughness, availability, cost, and machinability.

2. Based on the information from step 1, and the specification statement, identify all *special needs* for the application, as discussed in 3.2, by writing a response of *yes, no*, or *perhaps* in the blank following each item in Table 3.1.

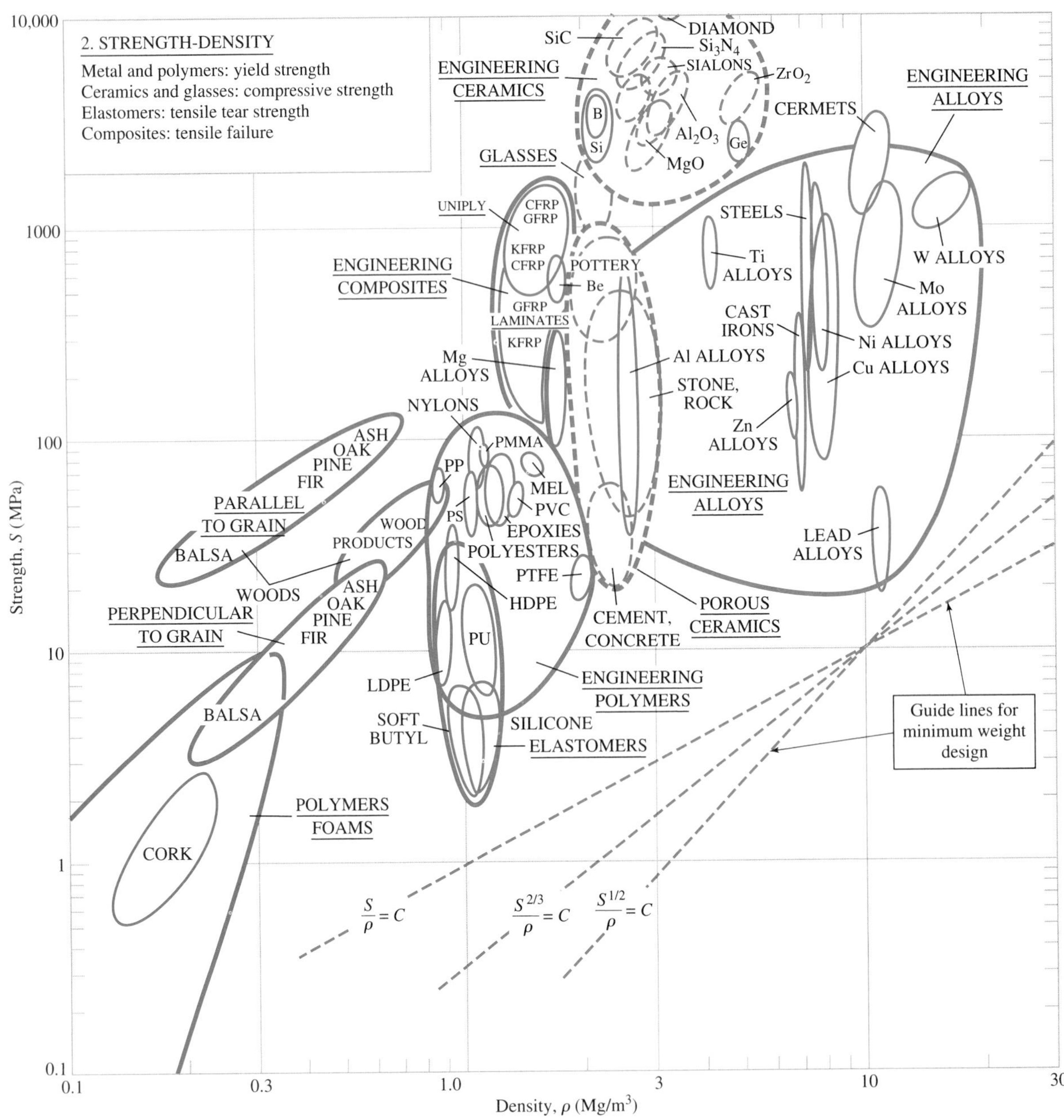

Figure 3.2

Two-parameter Ashby chart for failure strength, S, plotted versus density, ρ. For *metals*, S is yield strength, S_{yp}; for *ceramics and glass*, S is compressive crushing strength; for *composites*, S is tensile strength; for *elastomers*, S is tearing strength. The content of the chart roughly corresponds to the data included in Tables 3.3 and 3.4. (The chart is taken from ref. 3, courtesy of M. F. Ashby.)

3. For each item receiving a *yes* or *perhaps* response, go to Table 3.2 to identify the corresponding *performance evaluation index*, and consult rank-ordered Tables 3.3 through 3.20 for potential material candidates (or similar information from other sources for specific materials data). Using these data sources, write a short list of highly qualified candidate materials corresponding to each identified *special need*.

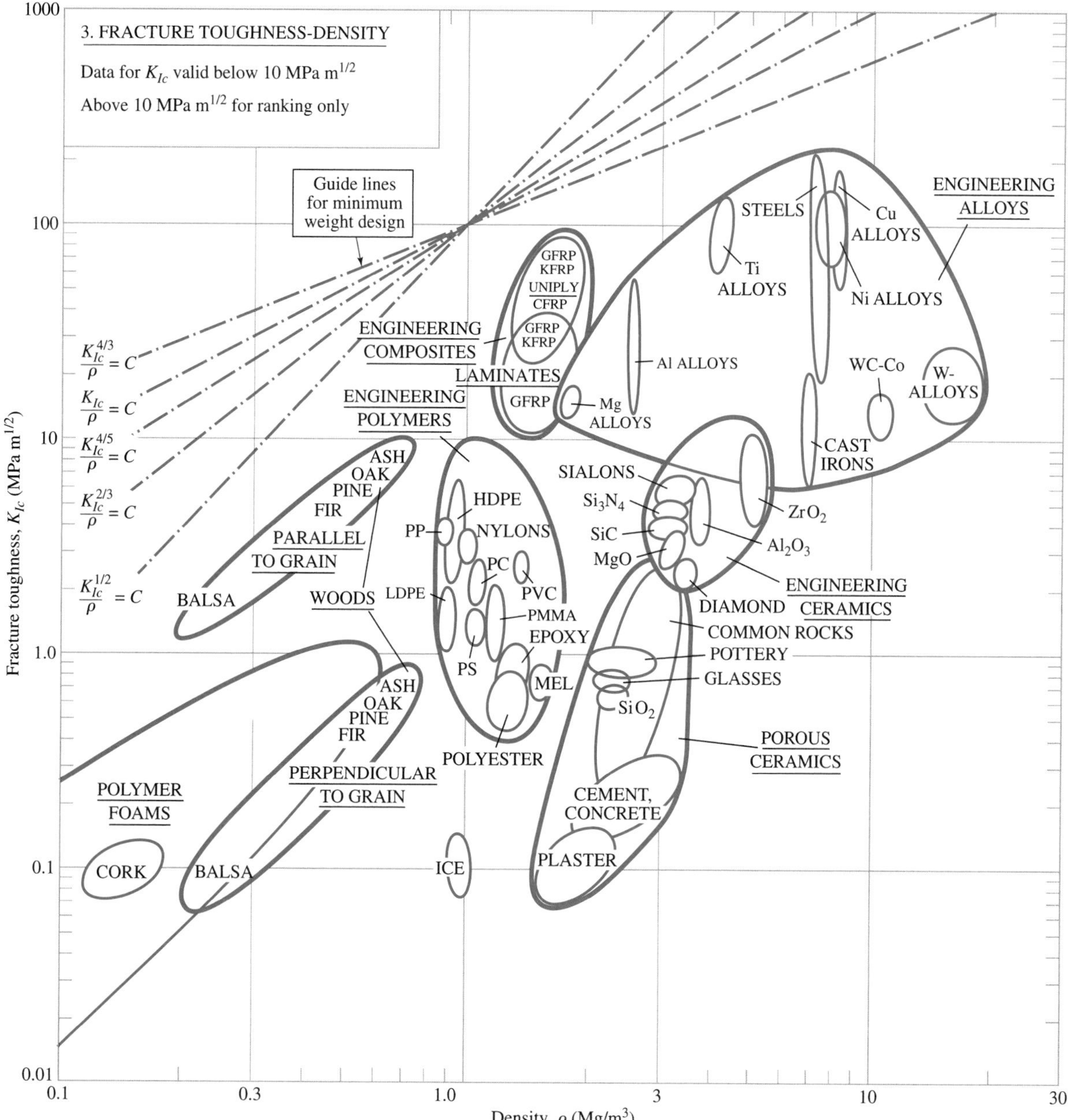

Figure 3.3

Two-parameter Ashby chart for plane strain fracture toughness, K_{Ic}, plotted versus density, ρ. The content of the chart roughly corresponds to the data included in Tables 2.1 and 3.4. (The chart is taken from ref. 3, courtesy of M. F. Ashby.)

4. Comparing all rank-ordered lists written in step 3, establish the two or three better candidate materials by finding those near the tops of all the lists. If a single candidate were to appear at the top of all lists, it would be the clear choice. As a practical matter, compromises are nearly always necessary to identify the two or three better candidates.
5. From the two or three better candidate materials, make a tentative selection for the material to be used. This may require additional data, materials selection software pack-

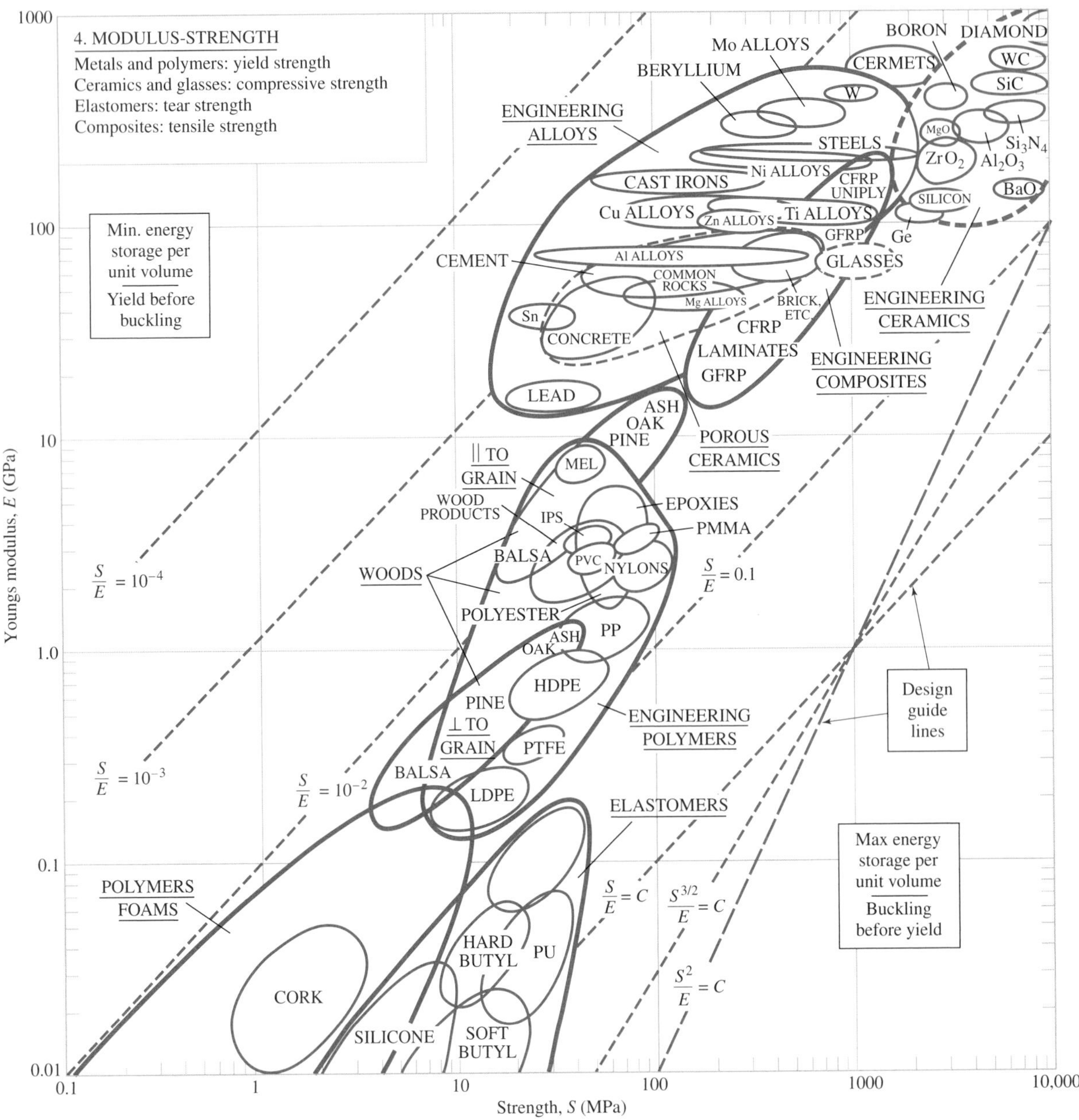

Figure 3.4

Two-parameter Ashby chart for Young's modulus of elasticity, *E*, versus density, *ρ*. The content of the chart roughly corresponds to the data included in Tables 3.9, 3.3, and 3.11. (The chart is taken from ref. 3, courtesy of M. F. Ashby.)

ages, optimization techniques that are more quantitative, discussions with materials specialists, or additional design calculations to confirm the suitability of the selection.[5]

In some cases, mathematical optimization procedures may be available to help establish which of the better candidates should be selected. Such procedures involve writing a

[5] See, for example refs. 16, 17, 18.

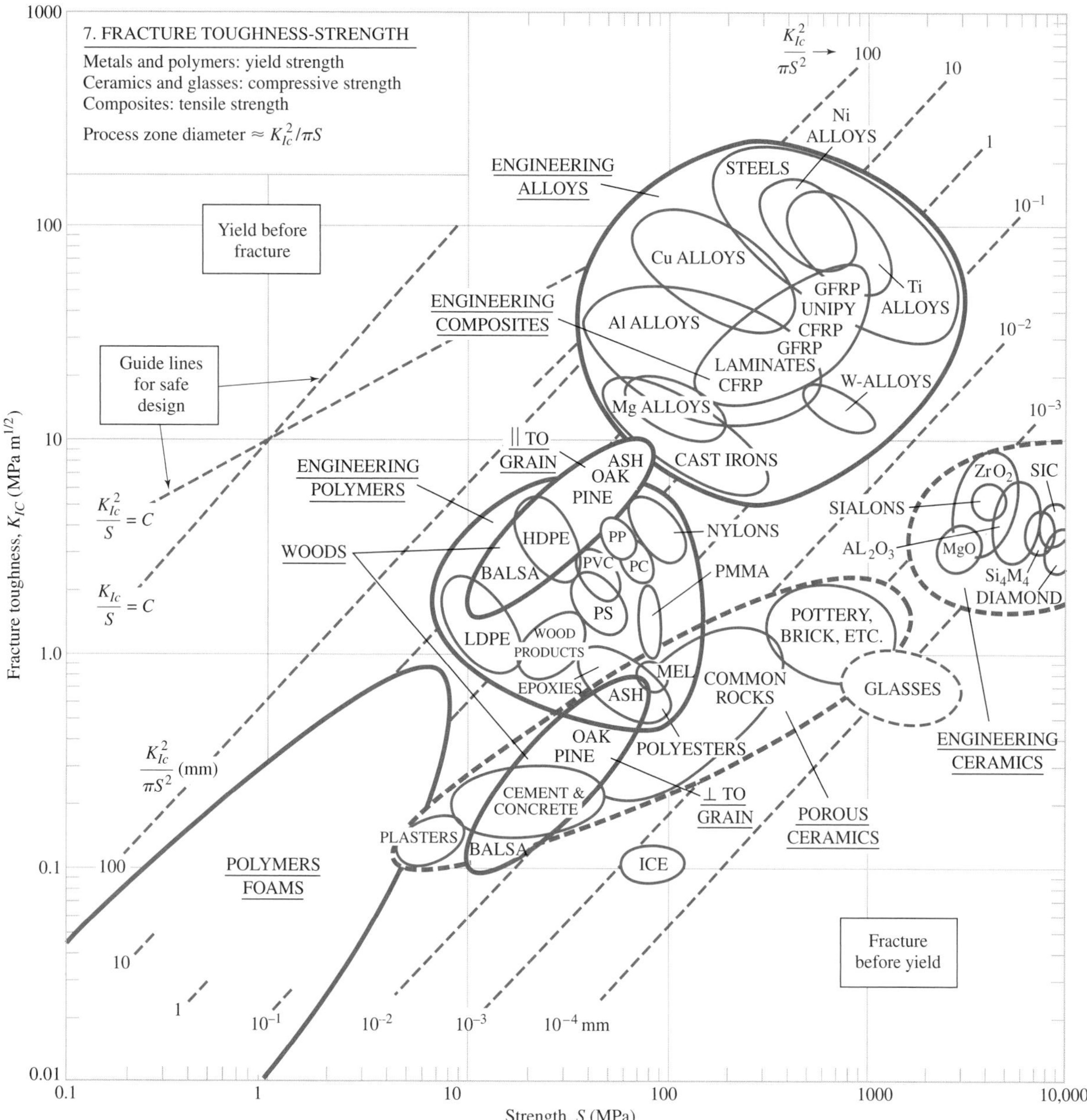

Figure 3.5
Two-parameter Ashby chart for plane strain fracture toughness, K_{Ic}, plotted versus failure strength, S (see legend of Figure 3.2 for definitions of S for various materials classes). The content of the chart roughly corresponds to the data included in Tables 2.1 and 3.4. (The chart is taken from ref. 3, courtesy of M. F. Ashby.)

merit function, defining *performance parameters*, documenting *application constraints*, and partially differentiating the merit function (within the constraints) to calculate a *figure of merit* for each candidate material. The best figure of merit then establishes the best material choice. These optimization procedures are discussed in the literature[6] but are beyond the scope of this text.

[6]See, for example refs. 15 and 3.

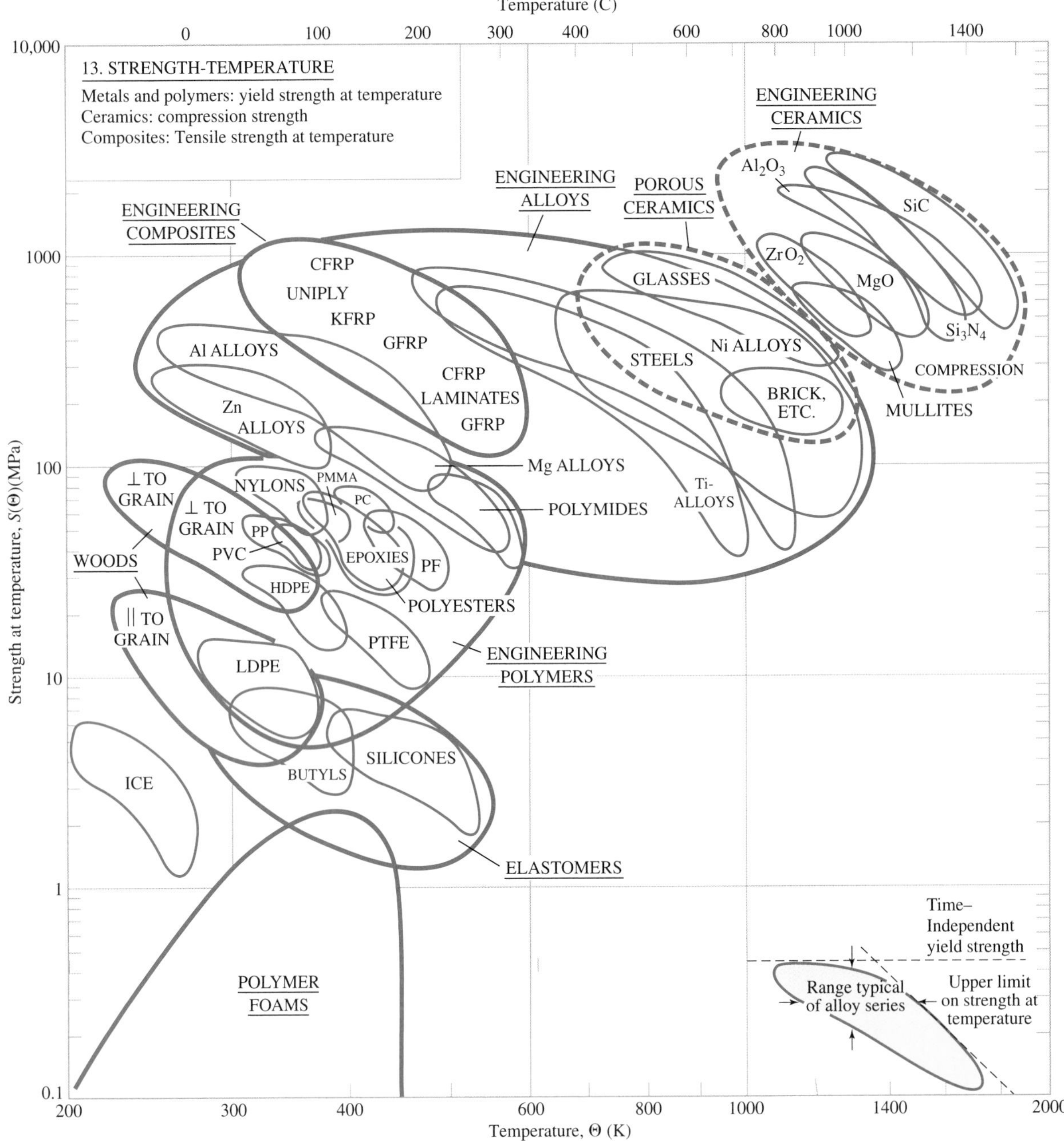

Figure 3.6

Two-parameter Ashby chart for failure strength, S (see legend of Figure 3.2 for definitions of S for various materials classes), plotted versus ambient temperature, Θ. The content of this chart roughly corresponds to the data included in Table 3.5. (The chart is taken from ref. 3, courtesy of M. F. Ashby.)

Computer-aided materials selection systems (CAMSS) are also emerging rapidly as potentially powerful tools for materials selection. Such *expert systems*, capable of interfacing with design teams, consist of three integrated parts connected by search and logic deduction algorithms: *databases, knowledge bases*, and *modeling or analysis capabilities*. Proprietary in-house *databases* exist in many companies, and some are commercially available.[7] *Computerized knowledge bases* are less well developed, requiring a wide range of formulas, design rules, "if-then" rules, manufacturability information, or company-specific "lessons-learned" files.[8]

Example 3.1 Materials Selection: Rank-Ordered-Data Table Method

It is desired to select a material for a proposed design for the crankshaft to be used in a new, compact, one-cylinder air compressor. The crankshaft is to be supported on two main bearings that straddle the connecting rod bearing. A preliminary analysis has indicated that the most probable failure modes of concern are *fatigue, wear*, and *yielding*. Projected production rates are high enough so that cost is an important consideration. Select a tentative material for this application.

Solution

Following the five-step process of 3.4, a specification statement is first formulated as follows:

The crankshaft for this application should be short, compact, relatively rigid, fatigue resistant, wear resistant at the bearing sites, and capable of low cost production.

Using this specification statement as a basis, the "special needs" column of Table 3.1 may be filled in as shown in Table E3.1A.

Surveying these results, special needs have been identified for items 1, 6, 10, and 14. For these special needs, Table 3.2 provides the corresponding performance evaluation indices shown in Table E3.1B.

Materials data for these particular performance indices are given in Table 3.3, 3.9, 3.13, 3.18, and 3.19. Making a short list of candidate materials from each of these tables results in the following array:

TABLE E3.1A Table 3.1 Adapted to Crankshaft Application

Crankshaft Application Requirement	Special Need?
1. Strength/volume ratio	Yes
2. Strength/weight ratio	No
3. Strength at elevated temperature	No
4. Long-term dimensional stability at elevated temperature	No
5. Dimensional stability under temperature fluctuation	No
6. Stiffness	Yes
7. Ductility	No
8. Ability to store energy elastically	No
9. Ability to dissipate energy plastically	No
10. Wear resistance	Yes
11. Resistance to chemically reactive environment	No
12. Resistance to nuclear radiation environment	No
13. Desire to use specific manufacturing process	No
14. Cost constraints	Yes
15. Procurement time constraints	No

[7]For example, *CMS* (see ref. 3) and *PERITUS* (see ref. 4). CMS implements the Ashby chart selection procedure discussed in 3.5, allowing successive application of up to six selection stages. *PERITUS* supports the rank-ordered-data table method discussed in 3.4, with selection based on requesting "high," "medium," or "low" values for pertinent properties.

[8]See 1.10.

Example 3.1 Continues

TABLE E3.1B Performance Evaluation Indices for Special Needs

Special Need	Performance Evaluation Index
1. Strength/volume ratio	Ultimate or yield strength
6. Stiffness	Modulus of elasticity
10. Wear resistance	Hardness
14. Cost constraints	Cost/unit weight; machinability

For high-strength/volume (from Table 3.3):

Ultra-high-strength steel
Stainless steel (age hardenable)
High-carbon steel
Graphite-epoxy composite
Titanium
Ceramic
Nickel-based alloy
Medium-carbon steel
Stainless steel (austenitic)
Yellow brass
Commercial bronze
Low-carbon steel
Phosphor bronze
Gray cast iron

For high stiffness (from Table 3.9):

Tungsten carbide
Titanium carbide
Molybelenum
Steel
Stainless steel
Cast iron

For high hardness (from Table 3.13):

Diamond
Sapphire
Tungsten carbide
Titanium carbide
Case-hardened low-carbon steel
Ultra-high-strength steel
Titanium
Gray cast iron

For low material cost (from Table 3.18):

Gray cast iron
Low-carbon steel
Ultra-high-strength steel
Zinc alloy
Acrylic
Commercial bronze
Stainless steel

For good machinability (from Table 3.19):

Magnesium alloy
Aluminum alloy
Free-machining steel
Low-carbon steel
Medium-carbon steel
Ultra-high-strength steel
Stainless-steel alloy
Gray cast iron

Surveying these five lists, the materials common to all the lists are:

Ultra-high-strength steel
Low-carbon steel (case hardened)
Gray cast iron

For these three candidate materials the specific data from Tables 3.3, 3.9, 3.13, 3,18 and 3.19 are summarized in Table E3.1C.

TABLE E3.1C Evaluation Data for Candidate Materials

Evaluation Index	Candidate Material		
	Ultra-High-Strength Steel	Low-Carbon Steel (case hardened)	Gray Cast Iron
Ultimate strength, S_u, psi	287,000	61,000	50,000
Yield strength, S_{yp}, psi	270,000	51,000	—
Modulus of elasticity, E, psi	30×10^6	30×10^6	$13-24 \times 10^6$
Hardness, BHN	560	650	262
Cost, dollars/lb	0.65	0.50	0.30
Machinability index	50	65	40

Because the specification statement emphasizes short and compact design, the significantly stronger ultra-high-strength steel is probably the best candidate material; however, the case-hardened low-carbon steel is probably worth a more detailed investigation since it has a higher surface hardness (better wear resistance), is cheaper, and is more easily machined prior to heat treatment. If compact design were not an issue, cast iron would probably be the best choice.

3.5 Matching Responsive Materials to Application Requirements: Ashby Chart Method

The Ashby chart method presented here is a simplified version of the complete procedure discussed in reference 3.[9] It is based on using pertinent two-parameter charts, such as those shown in Figures 3.1 through 3.6, to determine which materials make good candidates. In these charts, materials within a given subset (metals, polymers, etc.) tend to cluster together, making it possible to construct a "circumscribing envelope" to delineate each class of material. The dashed families of parallel "design guidelines" shown on many of the *Ashby charts* are presented to help optimize the performance of the component. The *slope* of each family of guidelines is tied to the *degree* of the pertinent performance parameter, as illustrated in Example 3.2. All materials that lie on a given dashed guideline will perform equally well based on the *particular* performance parameter represented by that line. Those materials that lie *above* the line are *better*; those materials that lie *below* the line are *worse*. The "short-name" identifiers used in the Ashby charts are defined in Table 3.21 together with the members of each material class. Similar to the method of 3.4, the two or three better material candidates are then selected from an overall evaluation of the results from all the charts used.

The Ashby procedure for identifying good candidate materials for any specific application may be summarized as follows:

1. Using Table 3.1 as a guide, together with known requirements imposed by operational or functional constraints, postulated failure modes, market-driven factors,

[9]The complete procedure, as presented by Ashby, is beyond the scope of this text. It involves writing performance indices as a function of performance parameters, geometric parameters, and materials properties; writing a merit function (objective function); and optimizing the merit function to determine a *figure of merit*. Guidance is given in ref. 3 for determining appropriate mathematical expressions for many useful performance parameters and merit functions, but since a design nearly always involves optimization with respect to *many* design goals (often contradictory), judgment is frequently required to *rank the goals* before selecting the better materials candidates.

TABLE 3.21 Material Classes and Members of Each Class[1]

Class	Members	Short Name
Engineering Alloys (the metals and alloys of engineering)	Aluminum alloys	Al alloys
	Beryllium alloys	Be alloys
	Copper alloys	Cu alloys
	Lead alloys	Lead alloys
	Magnesium alloys	Mg alloys
	Molybdenum alloys	Mo alloys
	Nickel alloys	Ni alloys
	Steels	Steels
	Tin alloys	Tin alloys
	Titanium alloys	Ti alloys
	Tungsten alloys	W alloys
	Zinc alloys	Zn alloys
Engineering Polymers (the thermoplastics and thermosets of engineering)	Epoxies	EP
	Melamines	MEL
	Polycarbonate	PC
	Polyesters	PEST
	Polyethylene, high density	HDPE
	Polyethylene, low density	LDPE
	Polyformaldehyde	PF
	Polymethylmethacrylate	PMMA
	Polypropylene	PP
	Polytetrafluroethylene	PTFE
	Polyvinylchloride	PVC
Engineering Ceramics (fine ceramics capable of load-bearing application)	Alumina	Al_2O_3
	Diamond	C
	Sialons	Sialons
	Silicon carbide	SiC
	Silicon nitride	Si_3N_4
	Zirconia	ZrO_2
Engineering Composites (the composites of engineering practice) A distinction is drawn between the properties of a ply (UNIPLY) and properties of a laminate (LAMINATES)	Carbon-fiber-reinforced polymer	CFRP
	Glass-fiber-reinforced polymer	GFRP
	Kevlar-reinforced polymer	KFRP
Porous Ceramics (traditional ceramics, cermets, rocks, and minerals)	Brick	Brick
	Cement	Cement
	Common rocks	Rocks
	Concrete	Concrete
	Porcelain	Pcln
	Pottery	Pot
Glasses (ordinary silicate glass)	Borosilicate glass	B-glass
	Soda glass	Na-glass
	Silica	SiO_2
Woods (separate envelopes describe properties parallel to the grain and normal to it, and wood products)	Ash	Ash
	Balsa	Balsa
	Fir	Fir
	Oak	Oak
	Pine	Pine
	Wood products (ply, etc.)	Wood products

TABLE 3.21 (*Continued*)

Class	Members	Short Name
Elastomers	Natural rubber	Rubber
(natural and artificial	Hard butyl rubber	Hard butyl
rubbers)	Polyurethane	PU
	Silicone rubber	Silicone
	Soft butyl rubber	Soft butyl
Polymer foams	*These include:*	
(foamed polymers of	Cork	Cork
engineering)	Polyester	PEST
	Polystyrene	PS
	Polyurethane	PU

[1]From ref. 3, courtesy of M. F. Ashby.

and/or management directives, establish a concise *specification statement* as discussed in 3.2. If information about the application is so sketchy that a specification statement cannot be written, and the remaining steps cannot be executed, it is suggested that *1020 steel* be tentatively selected as the "best" material because of its excellent combination of strength, stiffness, ductility, toughness, availability, cost, and machinability.

2. Based on the information from step 1, and the specification statement, identify all *special needs* for the application, as discussed in 3.2, by writing a response of *yes, no*, or *perhaps* in the blank following each item in Table 3.1.
3. For each item receiving a *yes* or *perhaps* response, go to Table 3.2 to identify the corresponding *performance evaluation index*, and consult the pertinent Ashby charts shown in Figures 3.1 through 3.6. Using these Ashby charts, identify a short list of highly qualified candidate materials corresponding to each selected pair of performance parameters or application constraints.
4. Comparing the results from all charts used, establish the two or three better candidate materials.
5. From the two or three better candidate materials, make a tentative selection for the material to be used. This may require additional data, materials selection software packages, optimization techniques that are more quantitative, discussions with materials specialists, or additional design calculations in order to confirm the suitability of the selection.
6. If a more quantitative optimization technique is desired, consult reference 3.

Example 3.2 Materials Selection: Ashby Chart Method[10]

A preliminary design is being formulated for a solid cylindrical tension rod of diameter d and fixed length L. The rod is to be used in a spacecraft application where weight, strength, and stiffness are all important design considerations. It is to be subjected to a static axial force, F. A safety factor of n_d is desired. The preliminary design analysis has indicated that the most probable failure modes are force-induced elastic deformation and yielding. Further, engineering management has directed that ductile materials be used in this application. Using the Ashby charts shown in Figures 3.1 through 3.6, select a tentative material for this application.

[10]This example adapted from ref. 3, courtesy of M. F. Ashby.

Example 3.2 Continues

TABLE E3.2A Table 3.1 Adapted to Spacecraft Tension Rod Application

Tension Rod Application Requirement	Special Need?
1. Strength/volume ratio	Maybe
2. Strength/weight ratio	Yes
3. Strength at elevated temperature	No
4. Long-term dimensional stability at elevated temperature	No
5. Dimensional stability under temperature fluctuation	No
6. Stiffness	Yes
7. Ductility	No
8. Ability to store energy elastically	No
9. Ability to dissipate energy plastically	No
10. Wear resistance	No
11. Resistance to chemically reactive environment	No
12. Resistance to nuclear radiation environment	No
13. Desire to use specific manufacturing process	No
14. Cost constraints	No
15. Procurement time constraints	No

Solution

Following the step-by-step procedure of 3.5, a specification statement may be formulated as follows:

The tension rod for this application should be light, stiff, and strong.

Using this specification statement as a basis, the special needs column of Table 3.1 may be filled in as shown in Table E3.2A.

Surveying these results, special needs have been identified for items 1, 2, and 6. For these special needs, Table 3.2 provides the corresponding performance evaluation indices as shown in Table E3.2B.

The performance requirements of the tension rod may be described functionally by an equation of the form[11]

$$p = \left[\begin{pmatrix} \text{application} \\ \text{requirements,} \\ A \end{pmatrix}, \begin{pmatrix} \text{geometrical} \\ \text{requirements,} \\ G \end{pmatrix}, \begin{pmatrix} \text{material} \\ \text{properties} \\ M \end{pmatrix} \right] \tag{1}$$

For the tension rod under consideration, it will be assumed that the three groups of parameters in (1) are *separable*[12] and therefore (1) may be reexpressed as

TABLE E3.2B Performance Evaluation Indices for Special Needs

Special Need	Performance Evaluation Index
1. Strength/volume ratio	Ultimate or yield strength
2. Strength/weight ratio	Ultimate or yield strength/density
6. Stiffness	Modulus of elasticity

[11]See ref. 3, p. 58 ff. for a more complete description of this procedure.

[12]Experience has shown that each of these parameter groups is usually independent of the others and therefore mathematically "separable."

$$p = f_1(A) \cdot f_2(G) \cdot f_3(M) \tag{2}$$

Based on (2), the optimum subset of materials for the tension rod can be identified without solving the *entire* design problem.

Using the *specification statement* formulated above, the material selected should be strong, stiff, and light. An expression for the mass of the rod may be written as

$$m = \left(\frac{\pi d^2}{4}\right) L\rho \tag{3}$$

where d is rod diameter, L is length of the rod, and ρ is mass density of the material.

The diameter of the cross section must be large enough to carry the load, F, without yielding, and provide a design safety factor of n_d. Thus

$$\frac{F}{\left(\frac{\pi d^2}{4}\right)} = \frac{S_{yp}}{n_d} \tag{4}$$

Combining (3) and (4)

$$m = (n_d F)(L)\left(\frac{\rho}{S_{yp}}\right) \tag{5}$$

It is of interest to note that (5) has the "separable function" format outlined in (2) and that the materials-based performance index for this case is, therefore,

$$f_3(M) = \frac{S_{yp}}{\rho} \tag{6}$$

A similar expression may be developed based on the need for the rod to be stiff enough to safely carry the load without excessive elastic deformation and to provide a safety factor of n_d. Since the diameter of the rod must be large enough to carry the load, F, without exceeding the critical elastic deformation, $(\Delta L)_{crit}$, and provide a design safety factor of n_d,

$$\frac{F}{\left(\frac{\pi d^2}{4}\right)} = E\varepsilon = E\left(\frac{(\Delta L)_{crit}}{n_d L}\right) \tag{7}$$

where E is Young's modulus of elasticity and ε is axial strain. Combining (3) and (7),

$$m = (n_d F)(L^2)\left(\frac{\rho}{E}\right) \tag{8}$$

Like (5), the expression of (8) has the functional format of (2), so the materials-based performance index for this case is

$$f'_3(M) = \frac{E}{\rho} \tag{9}$$

Materials data for the two performance parameters given in (6) and (9) correspond to the Ashby charts of Figures 3.1 and 3.2. These charts are reproduced again in Figures E3.2A and E3.2B, where each chart has been marked up to isolate a small region that contains materials having a good combination of properties for meeting the pertinent performance-parameter requirements.

Since the performance parameters of (6) and (9) are both of degree 1, a line constructed parallel to the dashed lines having a slope of 1 will be used as a basis for narrowing the charts to relatively small numbers of candidates.

Example 3.2 Continues

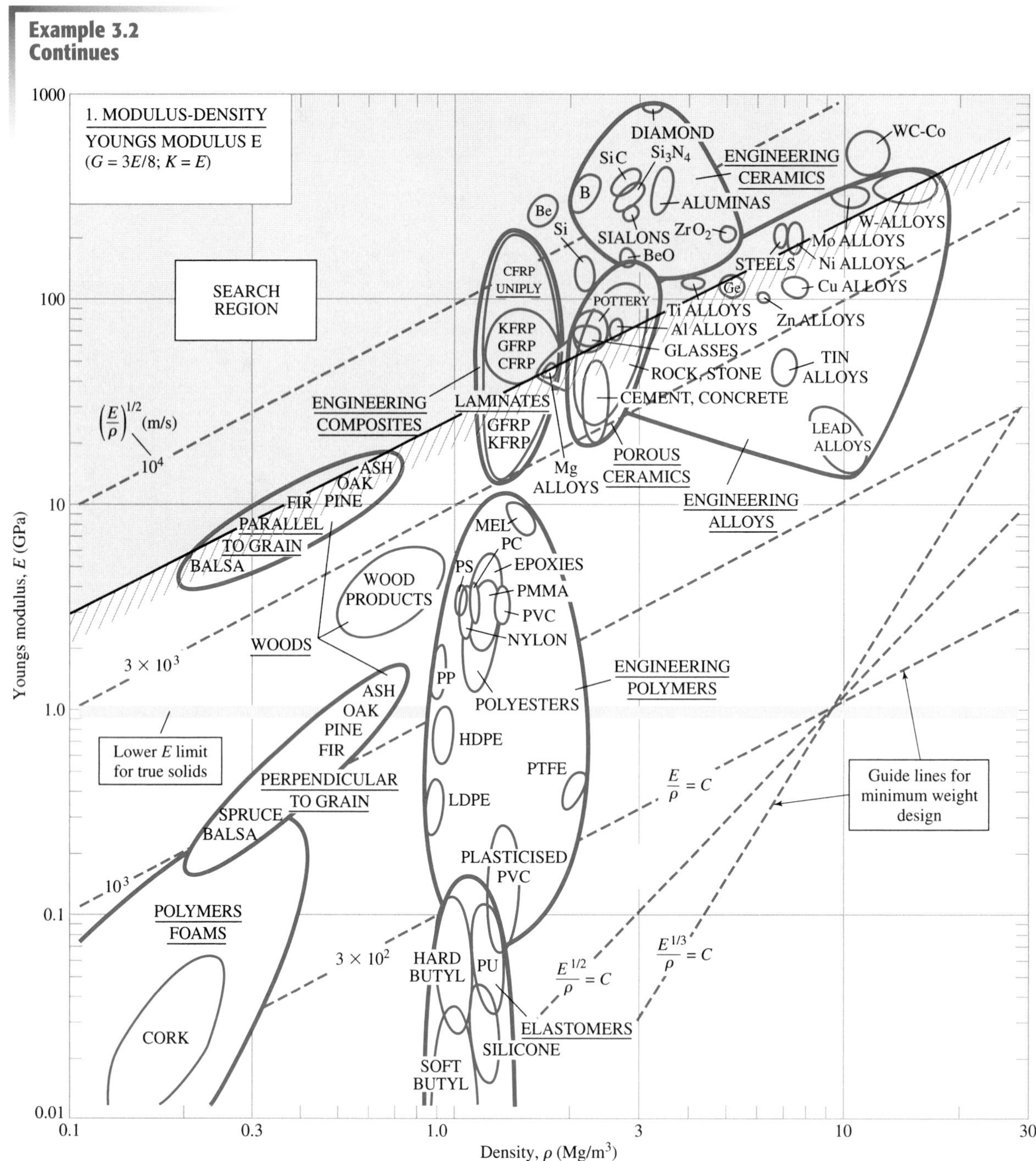

Figure E3.2A
Reproduction of Figure 3.1 showing an acceptable subset of engineering alloys for the tension rod application.

Surveying Figure E3.2A, the group of better material candidates includes:

Steel
Titanium
Molybdenum
Tungsten

Aluminum
Ceramics
Composites

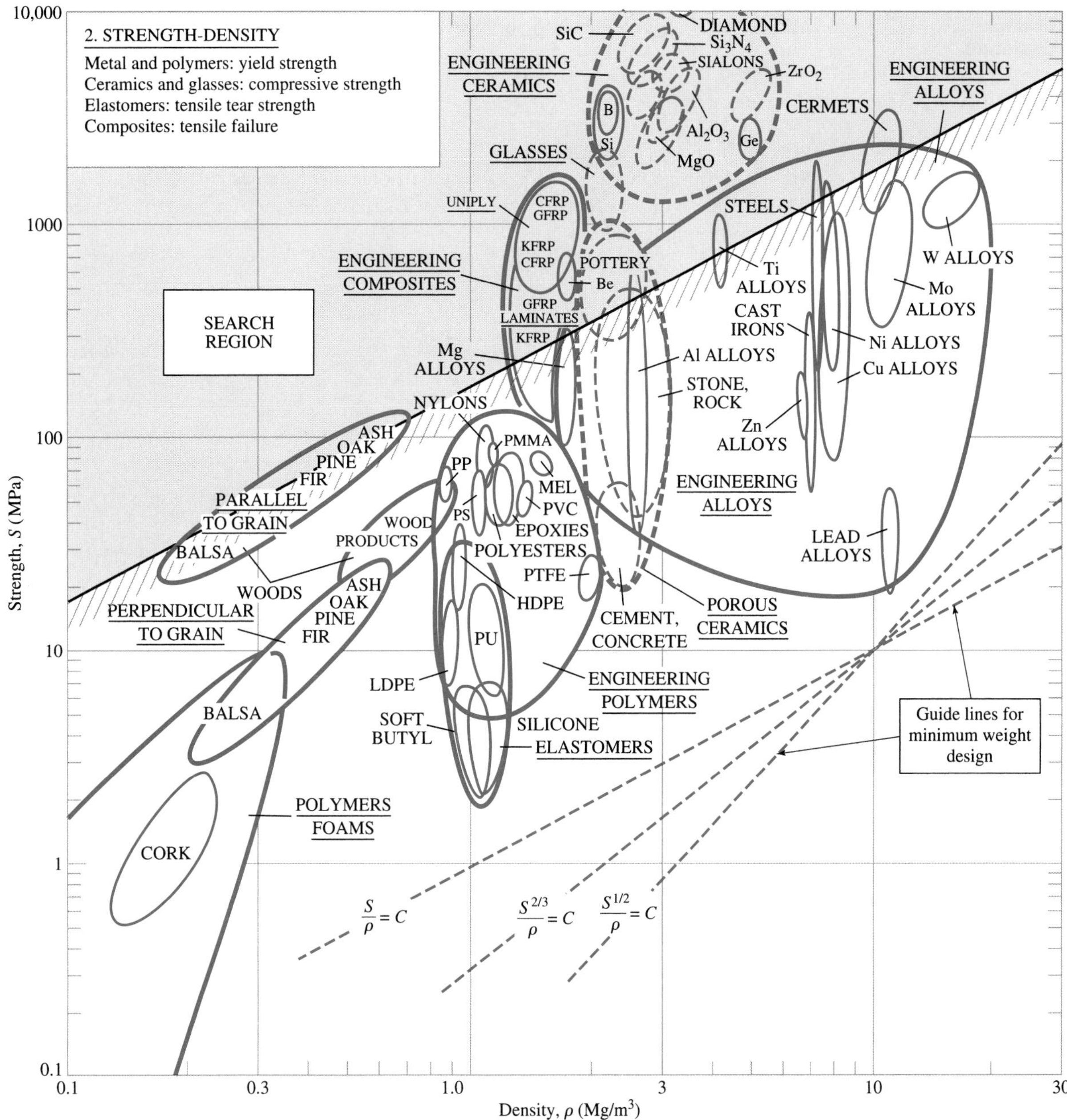

Figure E3.2B

Reproduction of Figure 3.2 showing an acceptable subset of engineering alloys for the tension rod application.

Similarly, from Figure E3.2B, the group of better material candidates includes:

Cermets	Aluminum
Steel	Ceramics
Nickel	Composites
Titanium	

Example 3.2 Continues

Material candidates common to both lists include:

Steel	Ceramics
Titanium	Composites
Aluminum	

Since engineering management has directed that ductile materials be used in this application, ceramics will be dropped from the list. Therefore, the materials worthy of more detailed evaluation include:

Steel	Aluminum
Titanium	Composites

Problems

3-1. A newly graduated mechanical engineer has been hired to work on a weight-reduction project to redesign the clevis connection (see Figure 4.1A) used in the rudder-control linkage of a low-cost high-production surveillance drone. This "new hire" has recommended the use of *titanium* as a candidate material for this application. As her supervisor, would you accept the recommendation or suggest that she pursue other possibilities?

3-2. It is desired to select a material for a back-packable truss-type bridge to be carried in small segments by a party of three when hiking over glacial ice fields. The purpose of the bridge is to allow the hikers to cross over crevasses of up to 12 feet wide. Write a specification statement for such a bridge.

3-3. A very fine tensile support wire is to be used to suspend a 10-lb sensor package from the "roof" of an experimenal combustion chamber operating at a temperature of 850°F. The support wire has a diameter of 0.020 inch. Creep of the support wire is acceptable as long as the creep rate does not exceed 4×10^{-5} in/in/hr. Further, stress rupture must not occur before at least 1000 hours of operation have elapsed. Propose one or two candidate materials for the support wire.

3-4. For an application in which *ultimate strength-to-weight ratio* is by far the dominant consideration, a colleague is proposing to use aluminum. Do you concur with his selection, or can you propose a better candidate material for this application?

3-5. You have been assigned the task of making a preliminary recommendation for the material to be used in the bumper of a new ultra-safe crash-resistant automobile. It is very important that the bumper be able to survive the energy levels associated with low-velocity crashes, without damage to the bumper or the automobile. Even more imporant, for higher energy levels associated with severe crashes, the bumper should be capable of deforming plastically over large displacements without rupture, thereby dissipating crash pulse energy to protect the vehicle occupants. It is anticipated that these new vehicles will be used throughout North America, and during all seasons of the year. A 10-year design life is desired. Cost is also a very important factor. Propose one or a few candidate materials suitable for this application. (Specific alloys need not be designated.)

3-6. A rotor disk to support the turbine blades in a newly designed aircraft gas turbine engine is to operate in a flow of 1000°F mixture of air and combustion product. The turbine is to rotate at a speed of 40,000 rpm. Clearances between rotating and stationary parts must be kept as small as possible and must not change very much when the temperature changes. Disk vibration cannot be tolerated either. Propose one or a few candidate materials for this application. (Specific alloys need not be designated.)

3-7. A material is to be selected for the main landing-gear support for a carrier-based navy airplane. Both weight and size of the support are important considerations, as well as minimal deflection under normal landing conditions. The support must also be able to handle impact loading, both under normal landing conditions and under extreme emergency controlled-crash-landing conditions. Under crash-landing conditions permanent deformations are acceptable, but separation into pieces is not acceptable. What candidate materials would you suggest for this application?

3-8. A job shop manager desires to have a rack built for storing random lengths of pipe, angle iron, and other structural sections. No special considerations have been identified, but the rack should be safe and the cost should be low. What material would you suggest?

3-9. The preliminary specification statement for a new-concept automotive spring application has been written as follows:

The spring should be stiff and light.

Using this specification statement as a basis, special needs have been identified from Table 3.1 as items 2 and 6. From Table 3.2, the corresponding performance evaluation indices have been determined to be *low density* and *high stiffness*.

With these two indices identified, the project manager has requested a report on materials exhibiting values of Young's modulus, E, of more than about 200 GPa, and values of density, ρ, less than about 2 Mg/m^3. Using Figure 3.1, establish a list of candidate materials that meet these criteria.

3-10. By examining Figure 3.3, determine whether the plane strain fracture toughness, K_{Ic}, of common engineering polymers such as PMMA (Plexiglas) is higher or lower than for engineering ceramics such as silicon carbide (SiC).

3-11. It is desired to design a pressure vessel that will *leak before it breaks*[13]. The reason for this is that the leak can be easily detected before the onset of rapid crack propagation[14] that might cause an explosion of the pressure vessel due to brittle behavior. To accomplish the *leak-before-break* goal, the vessel should be designed so that it can tolerate a crack having a length, a, at least equal to the wall thickness, t, of the pressure vessel without failure by rapid crack propagation. A specification statement for design of this thin-walled pressure vessel has been written as follows:

The pressure vessel should experience slow through-the-thickness crack propagation to cause a leak before the onset of gross yielding of the pressure vessel wall.

From evaluation of this specification statement using Tables 3.1 and 3.2, the important evaluation indices have been deduced to be *high fracture toughness* and *high yield strength.*

By combining (2-21) and (9-5), keeping in mind the "separable" quality of the materials parameter $f_3(M)$ discussed in Example 3.2, the materials-based performance index for this case has been found to be

$$f_3(M) = \frac{K_c}{S_{yp}}$$

It is also desired to keep the vessel wall as thin as possible (corresponds to selecting materials with yield strength as high as possible).

a. Using the Ashby charts shown in Figures 3.1 through 3.6, select tentative material candidates for this application.

b. Using the rank-ordered-data tables of Table 2.1 and Tables 3.3 through 3.20, select tentative material candidates for this application.

c. Compare results of parts (a) and (b).

[13]In this application, "break" will be interpreted to mean onset of nominal gross-section yielding of the pressure vessel wall.

[14]See 2.5.

Chapter **4**

Response of Machine Elements to Loads and Environments; Stress, Strain, and Energy Parameters

4.1 Loads and Geometry

During the first design iteration for any machine element, attention must be focused on satisfying functional performance specifications by selecting the "best" material and devising the "best" geometry to provide adequate strength and life. Typically, this effort is based on the loading, environment, and potential failure modes governing the application. A review of the design steps in Table 1.1 will show that a detailed plan for designing each load-carrying machine part is given in step VII. Fundamental to carrying forth the plan is an accurate knowledge of the loads on each machine part, obtained by conducting a careful global force analysis. The importance of an accurate force analysis cannot be overemphasized; highly sophisticated stress or deflection analyses to determine machine element geometry, if based on inaccurate loads, are of little value.

Sometimes the operational loads are well known, or can be accurately estimated, especially if the design task is to improve an existing machine. If a new machine is being designed, or if operating conditions are not well known, however, the accurate determination of the operating loads may be the most difficult task in the design process. Operating loads, as well as reaction forces on the machine, may be encountered at the surfaces (surface forces), in either concentrated or distributed configurations, or generated within the masses of the machine parts (body forces) by gravitational, inertial, or magnetic fields. All significant surface and body forces and moments must be included in the gobal force analysis if the design effort is to be successful.

To proceed with the design of each machine part after having completed the global force analysis, it is usually necessary to determine the local forces and moments on each of the parts. Successfully finding the local forces and moments depends on using the basic concepts and equations of equilibrium.

4.2 Equilibrium Concepts and Free-Body Diagrams

A rigid body is said to be in *equilibrium* when the system of external forces and moments acting on it sum to zero. If the body is being accelerated, inertia forces must be included in the summation. Thus, for a nonaccelerating body the equations of static equilibrium may be written as

$$\begin{aligned} \sum F_x &= 0 \qquad \sum M_x = 0 \\ \sum F_y &= 0 \qquad \sum M_y = 0 \\ \sum F_z &= 0 \qquad \sum M_z = 0 \end{aligned} \tag{4-1}$$

where x, y, and z are three mutually perpendicular coordinate axes with arbitrarily chosen origin and orientation. F_j denotes a force in the jth direction, and M_j denotes a moment about the jth axis. If the body is accelerating, inertia forces must be accounted for in applying the principles of *dynamic analysis*. For a body in *static equilibrium*, no matter how complicated the geometry or how many different external forces and moments are being applied, the force system can always be resolved into three forces along the axes and three moments about the axes of an arbitrarily chosen x-y-z coordinate system, as described by the equations of (4-1).

When a local force analysis is to be undertaken, it is often advantageous to consider only a selected portion of the body or structure of interest. This may be implemented by passing an imaginary *cutting plane* through the body at a desired location to conceptually isolate, or *free*, the selected portion from the rest of the body. (This concept is illustrated for two simple cases in Figures 4.2 and 4.3.) The *internal forces* (*stresses*) that were acting at the cutting plane location before it was cut must be represented by an equivalent system of external forces on the cut face, properly placed and distributed to maintain equilibrium of the selected *free body*. The isolated portion, together with all forces and moments acting *on it*, is called a *free-body diagram*. The construction and use of accurate free-body diagrams is absolutely essential in performing good stress and deflection analyses in the design of a machine part.

4.3 Force Analysis

To perform a good force analysis as the basis for designing any machine element, a designer must pay attention to how forces and moments are applied, how they are transmitted through the body or the structure, and how they are reacted. Furthermore, it is important to identify how applied force systems vary as a function of time, and what types of stress patterns they may induce in the machine part being considered. All of these factors play a role in determining which failure modes must be considered and in identifying locations within the machine elements where failure is more likely.

As a practical matter, in real machines there are no concentrated loads, no pure moments, no simple supports, and no fixed supports. These analytical simplifications are, nevertheless, essential in modeling force systems so that a designer can calculate the stresses and deflections induced in machine parts with sufficient precision but reasonable effort. As the design progresses, the consequences of such simplifications must be reexamined, especially in the local regions of contact where forces, moments, and reactions are applied. Sometimes, more sophisticated analyses may be necessary to properly evaluate these local regions.

To implement the construction of valid free-body diagrams it is often useful to follow the paths along which forces are transmitted through and between parts of a structure. One way to accomplish this is to visualize continuous *lines of force* which *flow* through the machine from applied loads to supporting structure. Following the lines of force as they flow through the structure helps identify stress or deflection patterns and locations where critical analyses should be performed to establish suitable geometry and select material properties to resist failure.

Example 4.1 Force Flow, Free-Body Diagrams, and Failure Mode Assessment

A typical pin-and-clevis joint is sketched in Figure E4.1A. If the joint is subjected to a static tensile force P, do the following:

a. By inspection, visualize how the force is transmitted through the joint, and sketch the "lines of force flow" from left to right.

Example 4.1 Continues

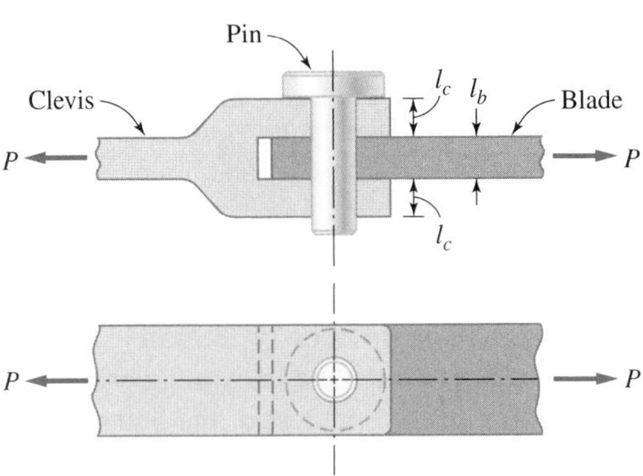

Figure E4.1A
Pin-and-clevis joint under static tensile loading.

b. Isolate and sketch the pin in a free-body diagram, showing all forces on the free body.

c. Identify potential failure modes that should be considered in designing the pin (i.e., in selecting a material and calculating dimensions).

Solution

a. The lines of force passing through the joint structure shown in Figure E4.1A may be sketched as shown in Figure E4.1B as dashed lines.

b. As shown in Figure E4.1C, a free-body diagram of the pin may be sketched by referring to the force-flow diagram of Figure E4.1B. Note that the force P is not concentrated, but *distributed* along bearing contact region C on the left and along D and E on the right. Likewise, it is distributed around the contacting half-circumference in each of these locations. The nature of the distribution depends upon material, geometry, and loading level.

c. Since the load is static, only static failure modes need to considered. Further, it may initially be assumed that environment and temperature are not factors in failure. Depending on specifics of the design requirements then, at least the following failure modes and locations should be considered:

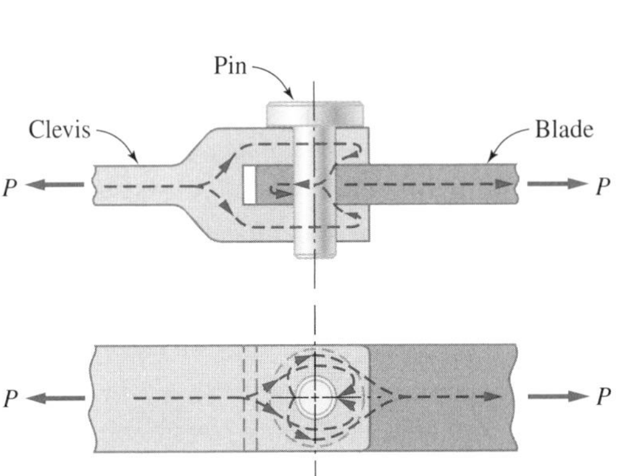

Figure E4.1B
Lines of force flow through the pin-and-clevis joint.

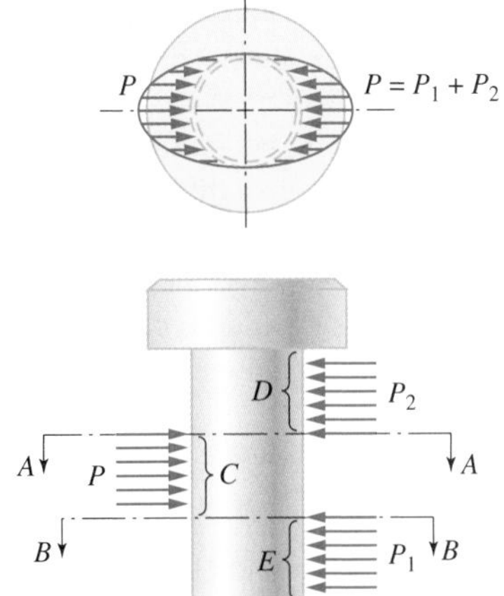

Figure E4.1C
Free-body diagram of pin.

1. *Elastic deformation*, if axial displacement between clevis and blade[1] is critical.
2. *Yielding and/or ductile rupture*, along bearing contact regions *C*, *D*, and *E*, and across shear planes, *A-A* and *B-B*, presuming the pin to be made of a ductile material. Bending of the pin could also contribute.
3. *Brittle fracture*, across planes *A-A* and *B-B*, due to bending and shear.

If other design information is discovered as the design progresses (e.g., vibrational excitations, aggressive environment, etc.), other potential failure modes such as fatigue, corrosion, fretting, and so on, might need to be added to the list.

4.4 Stress and Deflection Analysis; Common Stress Patterns

When designing a machine part to function without failure, a designer must, at an early stage, identify the probable failure mode(s), select a suitable parameter by which severity of loading and environment may be analytically represented, propose a material and geometry for the part, and obtain pertinent critical strength properties related to the probable failure mode. Next, the magnitude of the *loading severity parameter* must be calculated under applicable loading and environmental conditions, and compared with the pertinent *critical strength property*. Failure may be averted by assuring that the loading severity parameter is safely less than the corresponding critical strength property, for each potential failure mode.

States of Stress; Common Types of Loading

The more useful loading severity parameters are stress, strain, and strain energy per unit volume. Of these, *stress* is usually selected for calculation purposes. Strain and strain energy are often expressed as functions of stress. *Stress* is the term used to define the intensity and direction of the internal forces acting across any chosen cutting plane passed through the solid body of interest. To completely define the *state of stress* at any selected point within the solid body, it is necessary to describe the magnitudes and directions of stress vectors on all possible planes that could be passed through the point. One way of defining the state of stress at a point is to determine all of the components of stress that can occur on the faces of an infinitesimal cube of material placed at the origin of an arbitrarily selected right-handed cartesian coordinate system of known orientation. Each of these components of stress may be classified as either a *normal stress*, σ, *normal* to a face of the cube, or a *shear stress*, τ, *parallel* to a face of the cube. Figure 4.1 depicts all of the possible stress components acting on an infinitesimal cubic volume element of dimensions dx-dy-dz. The conventional subscript notations shown are defined as follows:

1. For normal stresses a single subscript is used, corresponding to the direction of the outward drawn normal to the plane on which it acts.
2. For shearing stresses two subscripts are used, the first of which indicates the direction of the plane on which it acts and the second of which indicates the direction of the shear stress in the plane.
3. Normal stresses are called positive (+) when they produce tension and negative (–) when they produce compression.

[1]See Figure E4.1B.

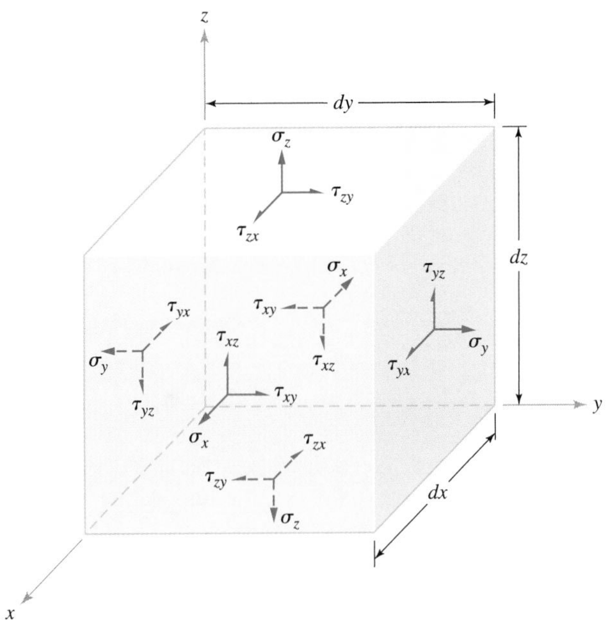

Figure 4.1
Complete definition of the state of stress at a point.

4. Shear stresses are called positive (+) if they are in the direction of an axis whose sign is the same as the sign of the axis in the direction of the outward drawn normal to the plane on which the shear stresses act.

Surveying the sketch of Figure 4.1, it may be noted that in general three normal stresses are active, namely, σ_x, σ_y, and σ_z. It may be further noted that six shear stresses are active, namely, τ_{xy}, τ_{yx}, τ_{yz}, τ_{zy}, τ_{zx}, and τ_{xz}. Thus a total of nine components of stress are apparently necessary to completely define the triaxial state of stress at a point: three normal stresses and six shear stresses. However, it can be shown, for isotropic materials,[2] on the basis of moment equilibrium, that τ_{xy} and τ_{yx} are identical, τ_{yz} and τ_{zy} are identical, and τ_{zx} and τ_{xz} are identical. Consequently, the complete definition of the most general state of stress at a point requires the specification of only six components of stress, three *normal* stresses, σ_x, σ_y, and σ_z, and three *shearing* stresses, τ_{xy}, τ_{yz}, and τ_{zx}. If these six components of stress are known for a particular elemental cube at a given point of interest, it is possible to compute the stresses on *any and all* planes passing through the point, using simple equilibrium concepts. This fact is utilized in determining *principal stresses* as discussed in 4.5 and in developing *combined stress theories of failure*, discussed in 4.6.

The elemental volume shown in Figure 4.1, with components of stress in *all three* coordinate directions, defines the *general triaxial state of stress* at a point. If the forces on the body result in an elemental volume at the point of investigation with stress components in *only two* coordinate directions, it is referred to as a *biaxial state of stress* (*plane stress*) at the point. If there are stress components in *only one* coordinate direction, it is called a *uniaxial state of stress* at the point. Thus simple tension or compression produces a uniaxial state of stress. For reasons of moment equilibrium, shearing stresses are always biaxial.

Before discussing more complicated states of stress, it may be useful to review several common stress patterns encountered in design practice. These include

[2]Properties are not sensitive to orientation.

1. Direct axial stress (tension or compression)
2. Bending stress
3. Direct shear and transverse shear stress
4. Torsional shear stress
5. Surface contact stress

The following brief discussions review the most pertinent equations and present sketches that characterize the loading, geometry, and stress distributions associated with the five common stress patterns just listed.

Direct Axial Stress

As discussed in 2.4, if a direct axial force F is applied to a simple member, as shown in Figure 4.2, the stress σ is uniformly distributed over the area A of the cut surface as shown in Figure 4.2(b). The magnitude of the uniform stress is

$$\sigma = \frac{F}{A} \tag{4-2}$$

If the force is tensile, σ is a uniformly distributed tensile stress. If the force is compressive, and the member does not buckle (see 2.7 for criteria), σ is a uniformly distributed compressive stress. In practice, uniformly distributed stresses may be assumed only if the bar is straight and homogeneous, the line of action of the force passes through the centroid of the cut cross section, and the cutting plane is remote from the ends and from any significant geometric discontinuity in the bar.

Bending; Load, Shear, and Moment Diagrams

Load, shear, and moment diagrams for various types of beam configurations are very useful to a designer because they provide a graphical summary of internal beam forces, which in turn allows a quick visual appraisal of stress distributions and critical points. The first step in any bending analysis should be the careful construction of shear and bending moment diagrams appropriate to the loading and support configuration for the beam of interest.

It is useful to note[3] that for all portions of a beam *between* loads, the shearing force V is equal to the rate of change of bending moment M with respect to x, or

$$\frac{dM}{dx} = V \tag{4-3}$$

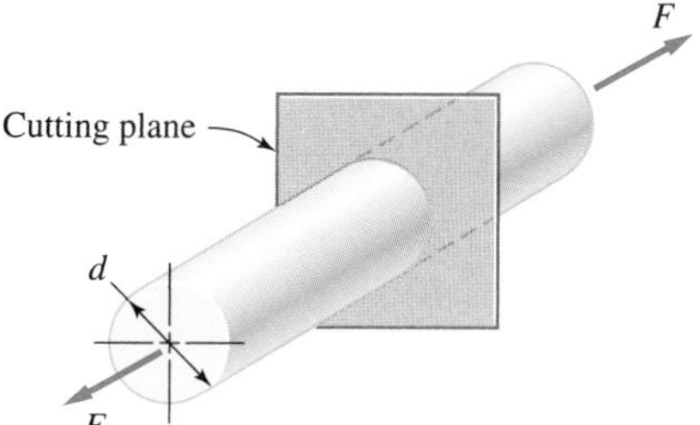

(*a*) Tensile force on solid cylindrical member.

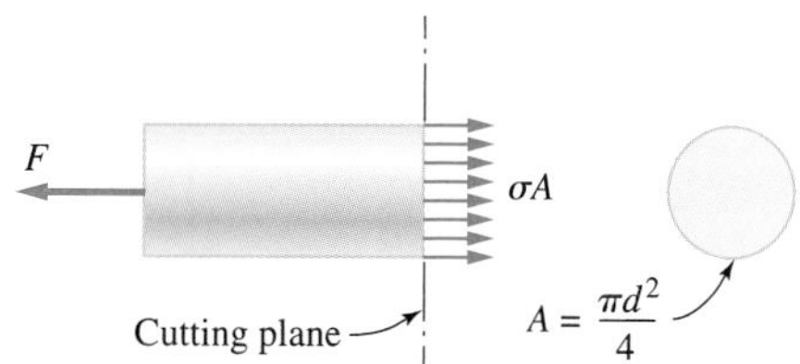

(*b*) Side view of front free body showing uniform normal stress distribution over cut surface of cylinder, producing the force σA.

Figure 4.2
Axially loaded uniform cylinder.

[3]From equilibrium equations.

The details of constructing shear and moment diagrams by writing force and moment equilibrium equations as a function of position x along the beam are presented in most basic strength-of-materials textbooks.[4] Table 4.1 reviews a few common cases, and additional diagrams may be found in the literature.[5]

TABLE 4.1 Loading (*P*), Shear (*V*), and Moment (*M*) Diagrams for Selected Beam Configurations. Note that *y* is transverse deflection and θ is slope.

Case 1. Simple Beam; Concentrated Load P at Center

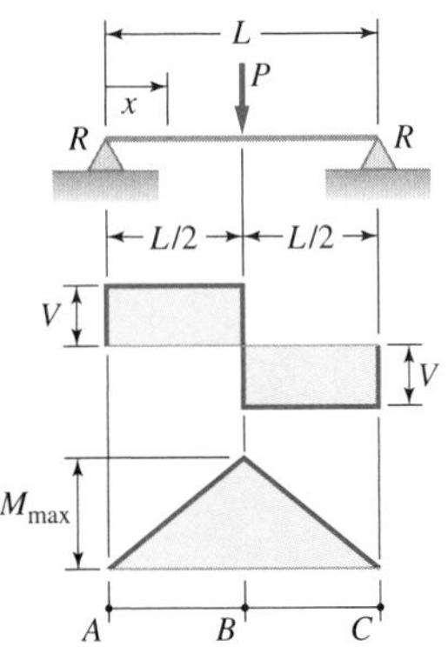

$$R = V = \frac{P}{2}$$

$$M_{max} \text{ (at point of load)} = \frac{PL}{4}$$

$$M_x \left(\text{when } x < \frac{L}{2}\right) = \frac{Px}{2}$$

$$y_{max} \text{ (at point of load)} = \frac{PL^3}{48EI}$$

$$y_x \left(\text{when } x < \frac{L}{2}\right) = \frac{Px}{48EI}(3L^2 - 4x^2)$$

$$\theta_A = -\theta_C = \frac{PL^2}{16EI}$$

Case 2. Simple Beam; Concentrated Load P at Any Point

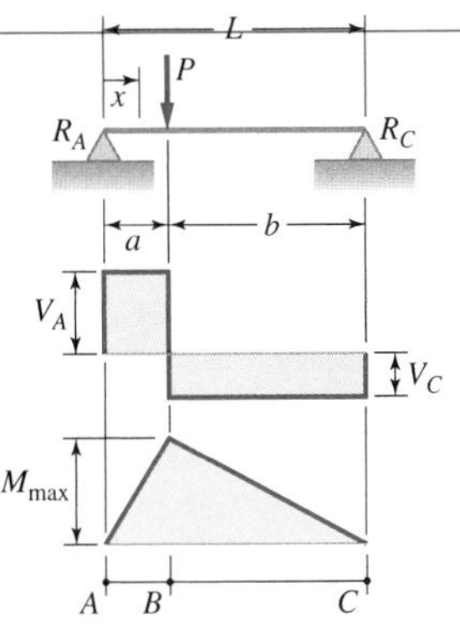

$$R_A = V_A \text{ (max when } a < b) = \frac{Pb}{L}$$

$$R_C = V_C \text{ (max when } a > b) = \frac{Pa}{L}$$

$$M_{max} \text{ (at point of load)} = \frac{Pab}{L}$$

$$M_x \text{ (when } x < a) = \frac{Pbx}{L}$$

$$y_{max} \left(\text{at } x = \sqrt{\frac{a(a + 2b)}{3}} \text{ when } a > b\right) = \frac{Pab(a + 2b)\sqrt{3a(a + 2b)}}{27EIL}$$

$$y_a \text{ (at point of load)} = \frac{Pa^2b^2}{3EIL}$$

$$y_x \text{ (when } x < a) = \frac{Pbx}{6EIL}(L^2 - b^2 - x^2)$$

$$\theta_A = \frac{P}{6EI}\left(bL - \frac{b^3}{L}\right)$$

$$\theta_C = \frac{P}{6EI}\left(2bL + \frac{b^3}{L} - 3b^2\right)$$

[4] See, for example, ref. 1.

[5] See, for example, ref. 2, pp. 2–111 ff.

TABLE 4.1 (*Continued*)

Case 3. Simple Beam; Uniformly Distributed Load w per Unit Length

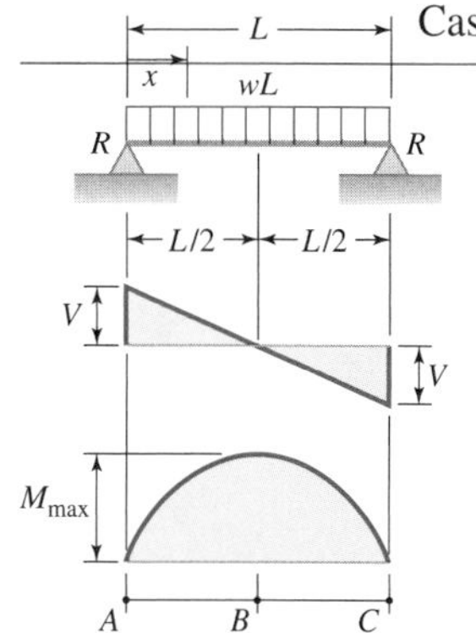

$$R = V = \frac{wL}{2}$$
$$V_x = w\left(\frac{L}{2} - x\right)$$
$$M_{max} \text{ (at center)} = \frac{wL^2}{8}$$
$$M_x = \frac{wx}{2}(L - x)$$
$$y_{max} \text{ (at center)} = \frac{5wL^4}{384EI}$$
$$y_x = \frac{wx}{24EI}(L^3 - 2Lx^2 + x^3)$$
$$\theta_A = -\theta_C = \frac{wL^3}{24EI}$$

Case 4. Simple Beam; End Couple M_0

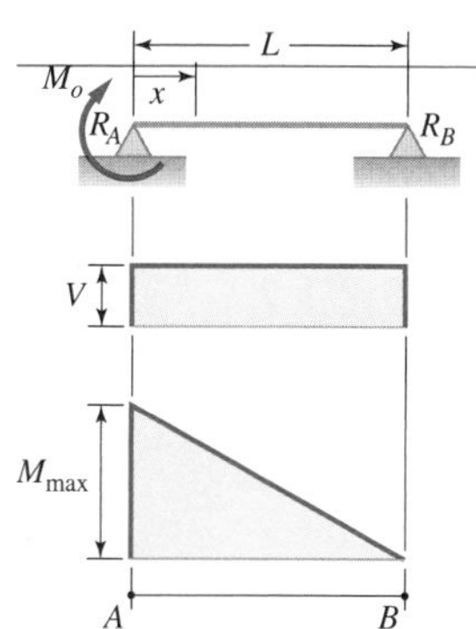

$$R_A = V = -R_B = -\frac{M_0}{L}$$
$$V_x = R_A$$
$$M_{max} \text{ (at A)} = M_0$$
$$M_x = M_0 + R_A x$$
$$y_{max} \text{ (at } x = 0.422L\text{)} = 0.0642\frac{M_0 L^2}{EI}$$
$$y_x = \frac{M_0}{6EI}\left(\frac{x^3}{L} + 2Lx - 3x^2\right)$$
$$\theta_A = \frac{M_0 L}{3EI}$$
$$\theta_B = \frac{M_0 L}{6EI}$$
$$\theta_{midspan} \text{ (at } x = L/2\text{)} = -\frac{M_0 L}{24EI}$$

Case 5. Simple Beam; Intermediate Couple M_0

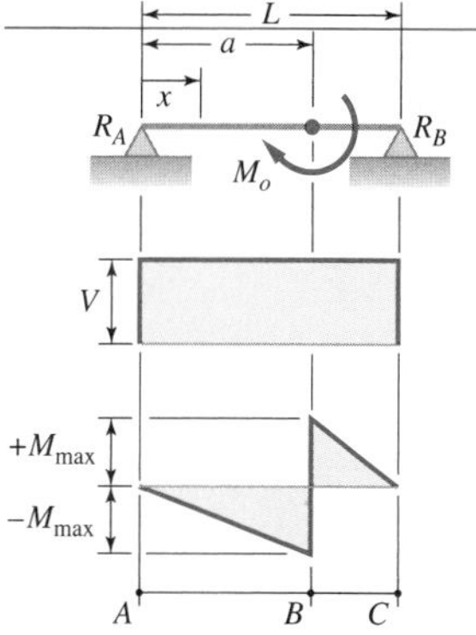

$$R_A = V = -R_C = -\frac{M_0}{L}$$
$$V_x = -R_A$$
$$-M_{max} \text{ (just left of } B\text{)} = R_A a$$
$$+M_{max} \text{ (just right of } B\text{)} = M_0 + R_A a$$
$$M_x\,(0 \le x \le a) = R_A x$$
$$M_x\,(a \le x \le L) = R_A x + M_0$$
$$y_x\,(0 \le x \le a) = \frac{M_0}{6EI}\left[\left(6a - 3\frac{a^2}{L} - 2L\right)x - \frac{x^3}{L}\right]$$

TABLE 4.1 (*Continued*)

Case 5. (*Continued*)

$$y_x\,(a \le x \le L) = \frac{M_0}{6EI}\left[3a^2 + 3x^2 - \frac{x^3}{L} - \left(2L + 3\frac{a^2}{L}\right)x\right]$$

$$\theta_A = +\frac{M_0}{6EI}\left(2L - 6a + 3\frac{a^2}{L}\right)$$

$$\theta_B = -\frac{M_0}{EI}\left(a - \frac{a^2}{L} - \frac{L}{3}\right)$$

$$\theta_C = -\frac{M_0}{6EI}\left(L - 3\frac{a^2}{L}\right)$$

Case 6. Beam Fixed at Both Ends; Concentrated Load P at Center

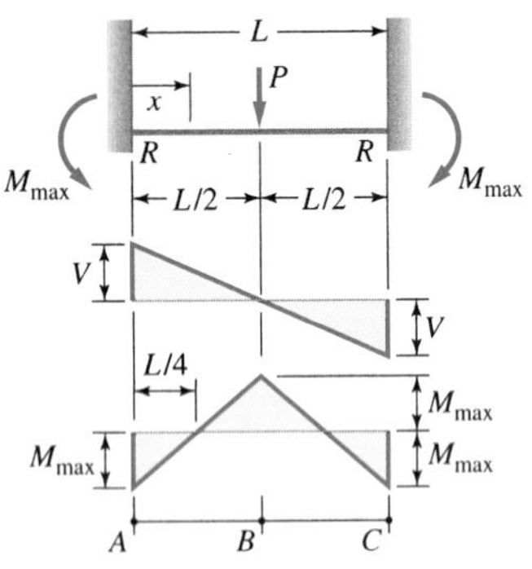

$$R = V = \frac{P}{2}$$

$$M_{max}\text{ (at center and ends)} = \frac{PL}{8}$$

$$M_x\left(\text{when } x < \frac{L}{2}\right) = \frac{P}{8}(4x - L)$$

$$y_{max}\text{ (at center)} = \frac{PL^3}{192EI}$$

$$y_x\left(x < \frac{L}{2}\right) = \frac{Px^2}{48EI}(3L - 4x)$$

$$\theta_A = \theta_C = 0$$

Case 7. Beam Fixed at Both Ends; Uniformly Distributed Load w per Unit Length

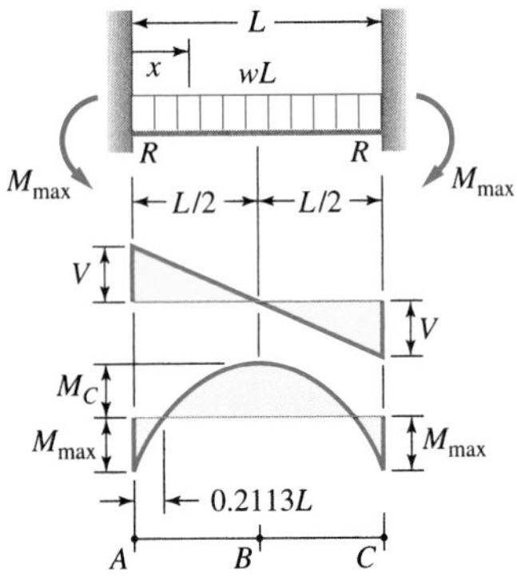

$$R = V = \frac{wL}{2}$$

$$V_x = w\left(\frac{L}{2} - x\right)$$

$$M_{max}\text{ (at ends)} = \frac{wL^2}{12}$$

$$M_B\text{ (at center)} = \frac{wL^2}{24}$$

$$M_x = \frac{w}{12}(6Lx - L^2 - 6x^2)$$

$$y_{max}\text{ (at center)} = \frac{wL^4}{384EI}$$

$$y_x = \frac{wx^2}{24EI}(L - x)^2$$

$$\theta_A = \theta_C = 0$$

TABLE 4.1 (*Continued*)

Case 8. Cantilever Beam; Concentrated Load P at Free End

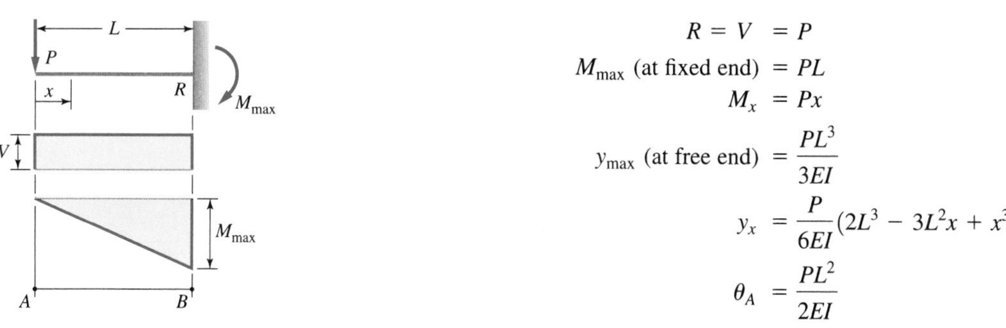

$$R = V = P$$

$$M_{max} \text{ (at fixed end)} = PL$$

$$M_x = Px$$

$$y_{max} \text{ (at free end)} = \frac{PL^3}{3EI}$$

$$y_x = \frac{P}{6EI}(2L^3 - 3L^2x + x^3)$$

$$\theta_A = \frac{PL^2}{2EI}$$

Case 9. Cantilever Beam; Uniformly Distributed Load w per Unit Length

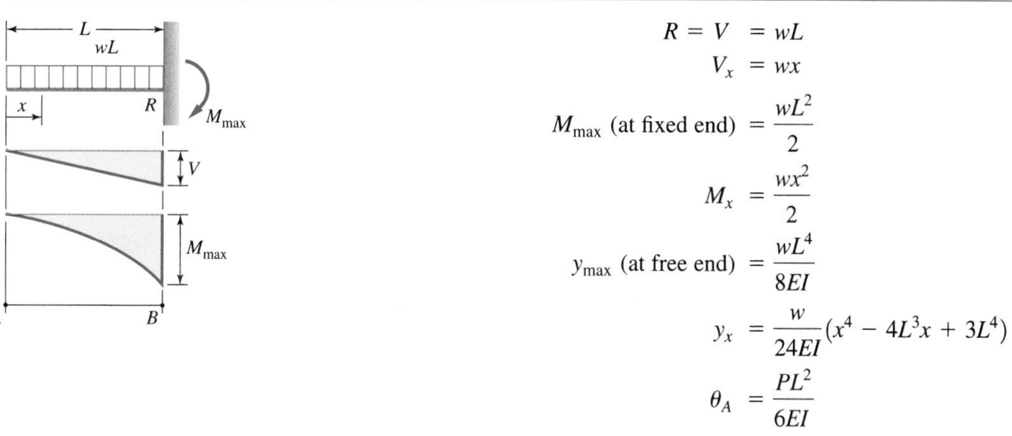

$$R = V = wL$$

$$V_x = wx$$

$$M_{max} \text{ (at fixed end)} = \frac{wL^2}{2}$$

$$M_x = \frac{wx^2}{2}$$

$$y_{max} \text{ (at free end)} = \frac{wL^4}{8EI}$$

$$y_x = \frac{w}{24EI}(x^4 - 4L^3x + 3L^4)$$

$$\theta_A = \frac{PL^2}{6EI}$$

Case 10. Cantilever Beam; End Couple M_o

$$R = 0$$

$$V = 0$$

$$M_x = M_0$$

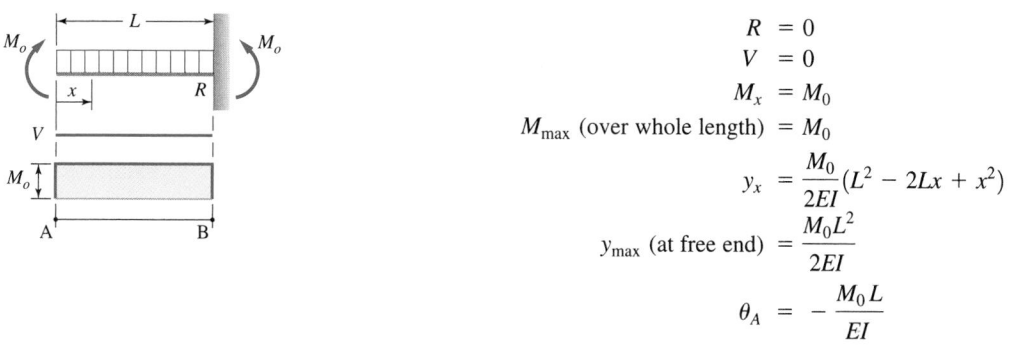

$$M_{max} \text{ (over whole length)} = M_0$$

$$y_x = \frac{M_0}{2EI}(L^2 - 2Lx + x^2)$$

$$y_{max} \text{ (at free end)} = \frac{M_0L^2}{2EI}$$

$$\theta_A = -\frac{M_0 L}{EI}$$

Bending; Straight Beam with Pure Moment

The simplest case of bending involves the application of a pure moment to a symmetrical beam, with the moment applied in a plane containing an axis of symmetry. Under these conditions no shear forces are induced. This case, called *pure bending*, is illustrated in Figure 4.3, where the y-axis is the axis of symmetry contained in the plane of the bending moment M. As shown in Figure 4.3, the bending stress distribution is linear, varying in mag-

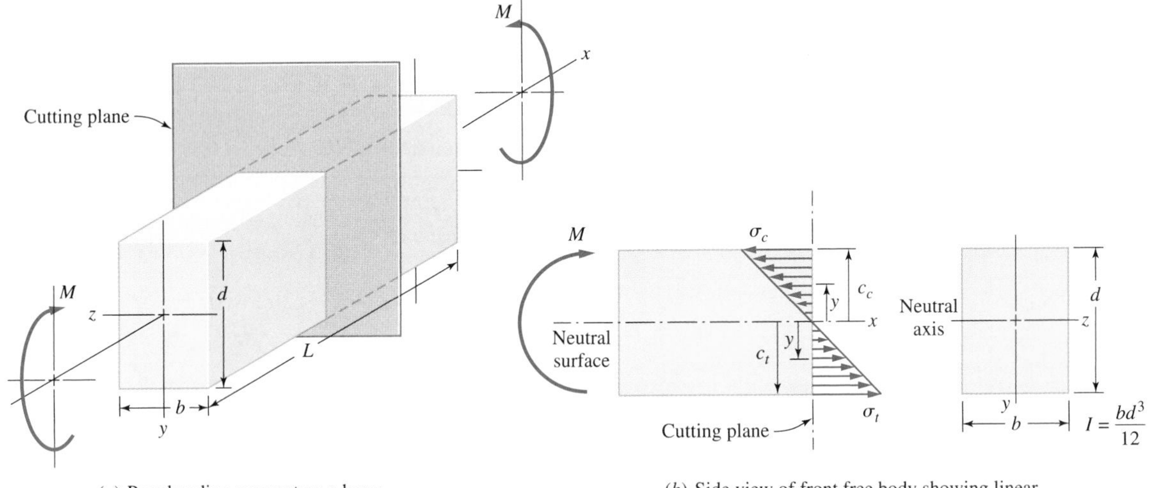

(*a*) Pure bending moment on a beam of rectangular cross section.

(*b*) Side view of front free body showing linear normal stress distribution across cut surface.

Figure 4.3
Straight beam subjected to pure bending moment.

nitude from maximum tensile stress σ_t at one outer fiber, through zero at the neutral axis of bending, to maximum compressive stress σ_c at the other outer fiber. For straight beams the neutral axis passes through the centroid of the beam cross sections. The expression for bending stress at any distance y from the neutral axis within the beam cross section is given by

$$\sigma_y = \frac{My}{I} \tag{4-4}$$

where I is the area moment of inertia of the cross section with respect to the neutral axis. *If the neutral axis is also a symmetry axis* for the cross section, the maximum bending stresses occur at both the tensile and compressive outer fibers where $y = c$. Hence, for this case, the magnitudes of the maximum bending stresses are given by

$$|\sigma_{max}| = \frac{Mc}{I} = \frac{M}{Z} \tag{4-5}$$

Expressions for area moment of inertia I and section modulus Z are shown for selected cross-sectional shapes in Table 4.2. The section modulus is defined as

$$Z = \frac{I}{c} \tag{4-6}$$

If the cross section does not exhibit symmetry with respect to the neutral axis, the distances c_t and c_c from the neutral axis to tensile and compressive outer fibers, respectively, will be unequal, and tensile and compressive outer fiber stresses will differ in magnitude. Depending upon the circumstances, both may need to be calculated. For such cases the outer fiber bending stresses are given by

$$\sigma_t = \frac{Mc_t}{I} \tag{4-7}$$

and

$$\sigma_c = \frac{Mc_c}{I} \tag{4-8}$$

TABLE 4.2 Properties of Plane Cross Sections

Shape	Area, A	Distances c_1 and c_2 to Outer Fibers	Moment of Inertia I About Centroidal Axis 1-1	Section Modulus $Z = I/c$ About Axis 1-1	Radius of Gyration, $\rho = \sqrt{I/A}$
1. Rectangle	bd	$c_1 = c_2 = \frac{d}{2}$	$\frac{bd^3}{12}$	$\frac{bd^2}{6}$	$\frac{d}{\sqrt{12}}$
2. Trapezoid	$\frac{(B+b)d}{2}$	$c_1 = \frac{b+2B}{3(b+B)}d$ $c_2 = \frac{2b+B}{3(b+B)}d$	$\frac{(B^2+4bB+b^2)d^3}{36(b+B)}$	—	$\frac{d}{6(b+B)}\sqrt{2(B^2+4bB+b^2)}$
3. Triangle	$\frac{bd}{2}$	$c_1 = \frac{2d}{3}$ $c_2 = \frac{d}{3}$	$\frac{bd^3}{36}$	$Z_1 = \frac{bd^2}{24}$ $Z_2 = \frac{bd^2}{12}$	$\frac{d}{\sqrt{18}}$
4. Solid circle	$\frac{\pi D^2}{4}$	$c = \frac{D}{2}$	$\frac{\pi D^4}{64}$	$\frac{\pi D^3}{32}$	$\frac{D}{4}$
5. Hollow circle	$\frac{\pi(D_o^2 - D_i^2)}{4}$	$c = \frac{D_o}{2}$	$\frac{\pi(D_o^4 - D_i^4)}{64}$	$\frac{\pi(D_o^4 - D_i^4)}{32D_o}$	$\frac{\sqrt{D_o^2 + D_i^2}}{4}$

In practice, the above discussion, and the associated bending stress equations given, may be assumed valid only if the bar is initially straight, loaded in a plane of symmetry, made of homogeneous material, and stressed only to levels within the elastic range. Also, calculations are valid only for cutting planes remote from sites where loads and reactions are applied, and remote from significant geometric discontinuities. The application of pure moments to any real beam is virtually impossible, but calculations made under the conditions just listed produce reasonably accurate results.

If the beam is initially curved, the neutral axis does not coincide with the centroidal axis, as it does for straight beams. Consequently, the bending stress distribution is nonlinear and calculation of bending stresses becomes more complicated. Initially curved beams are discussed in 4.9.

Bending; Straight Beam with Transverse Forces

If a beam is subjected to transverse loads (perpendicular to the beam axis) acting in a plane of symmetry, the calculation of stresses at any cut cross section involves both *bending stresses*, σ, produced by the bending moment, M, and *transverse shear stresses*, τ, produced by the transverse shear force, V. To calculate the bending stresses it is assumed that the stress distribution is the same as for pure bending (just discussed), and therefore equations (4-4), (4-5), (4-7), and (4-8) remain valid for calculating bending stresses. Calculation of the transverse shearing stresses is discussed in the next section. If the transverse loads are not applied in a plane of symmetry, or if the beam does not have a plane of symmetry, *torsional shear stresses* may also be induced in the beam.[6]

Direct Shear Stress and Transverse Shear Stress

When structures such as bolted, riveted, or pin-and-clevis joints are subjected to the type of nearly colinear loading depicted in Figure 4.4, the bolt or rivet or pin is subjected to *direct shear* on planes such as A-A and B-B. The *average* shear stress on such a shear plane (e.g., A-A), may be calculated as

$$\tau_{ave\,A\text{-}A} = \frac{P_{A\text{-}A}}{A_{A\text{-}A}} \tag{4-9}$$

where $\tau_{ave\,A\text{-}A}$ is average shearing stress, $A_{A\text{-}A}$ is the shear area of the pin at section A-A, and $P_{A\text{-}A}$ is the portion of the force passing through section A-A.

Although *average shearing stress* calculations are often used to roughly estimate bolt or pin dimensions, the *maximum shearing stresses are always higher than the average values* because the actual shearing stress distribution is *not uniform* over the shear plane. This more complicated nonlinear shearing stress distribution results because component *stiffnesses* and *fits between mating members* always plays a role, and some bending is inevitably induced in any real structure. As discussed in the preceding section, bending stresses may be calculated using (4-4), (4-5), (4-7), and (4-8). The distributions and magnitudes of *transverse shearing stresses* require further discussion.

To visualize the transverse shearing stresses in a beam subjected to transverse load-

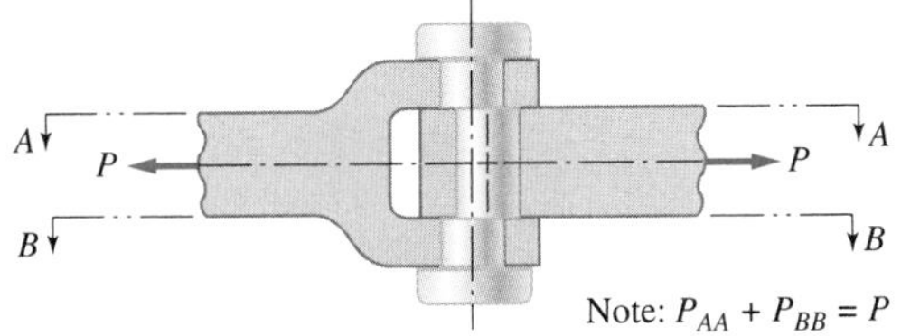

Figure 4.4
Direct shear loading.

[6] See, for example, ref. 1, p. 235 ff.

(*a*) Unloaded beam made of a stack of individual laminae free to slide over each other.

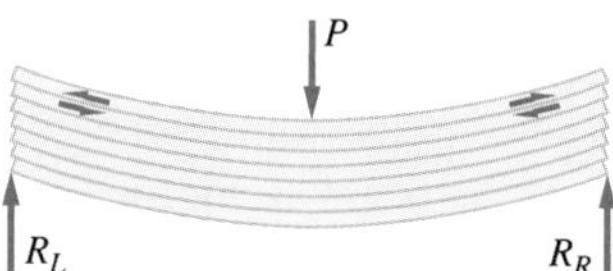

(*b*) Loaded beam showing interfacial sliding and consequent deflection.

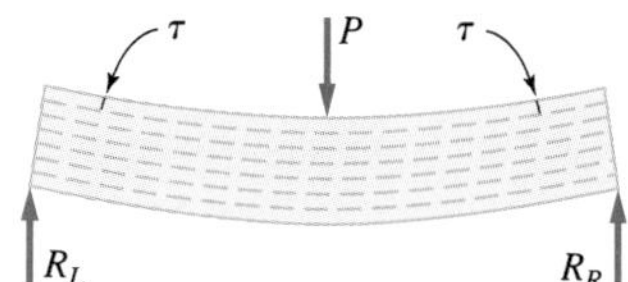

(*c*) Shearing stresses develop in glue layers to resist interfacial sliding in laminated beam to reduce beam deflection.

Figure 4.5
Illustration of transverse loading on a laminated beam before and after interfacial gluing.

ing, the sketches of Figure 4.5 may be helpful. The unloaded simply-supported beam shown in Figure 4.5(a) is made up of a stack of thin laminae free to slide over each other. Figure 4.5(b) shows the deflected configuration of the stacked laminae when a transverse (bending) load P is applied. Note that sliding is induced along each interface between lamina. Figure 4.5(c) schematically indicates the configuration of the stacked laminae under bending load P if a thin glue layer is applied at each laminar interface and cured prior to loading. Intuitively, the cured glue layers resist sliding when the transverse load P is applied, generating resistive shearing stresses within the glue layers. These shearing stresses are called *transverse shearing stresses*.

To assist in developing equations for transverse shearing stresses, Figure 4.6 depicts a cantilever beam of arbitrary cross section lengthwise-symmetrical about a vertical central plane that contains a concentrated transverse end-load P. To investigate the distribution of transverse shearing stresses over any selected plane cross section of the beam, an elemental parallelopiped *A-B-C-D-E-F-G-H* may be defined as shown in Figure 4.6(a). A view of face *A-B-C-D* of the parallelopiped is given in more detail in Figure 4.6(b). Within the parallelopiped, an elemental slab of thickness dy, length dz, and width x_y is defined.

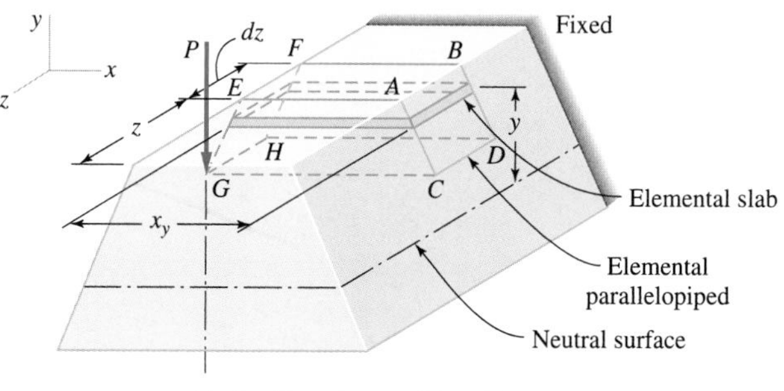

(*a*) Cantilever beam with end load.

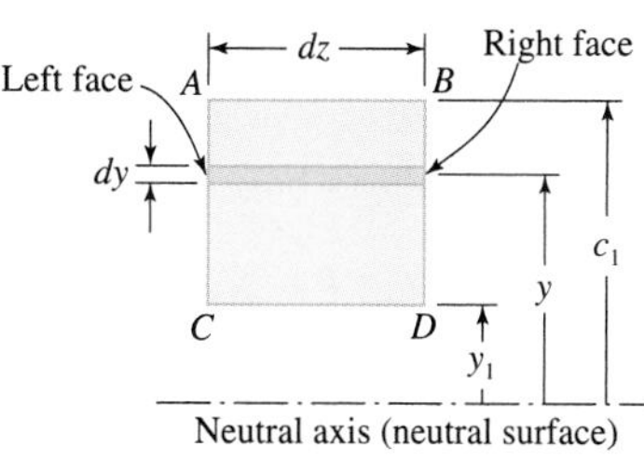

(*b*) Elemental parallelopiped and elemental slab viewed from right side of beam.

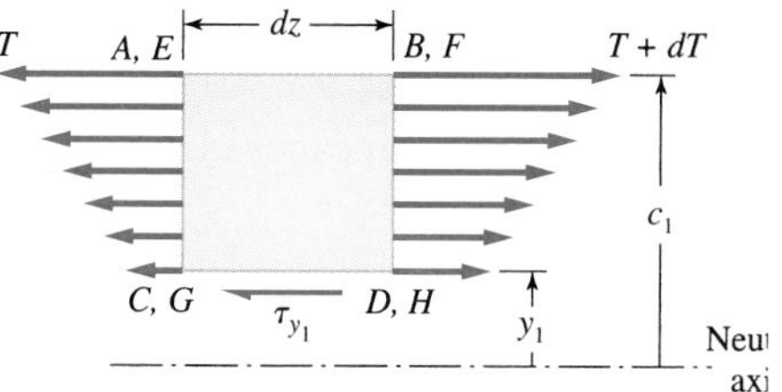

(*c*) Stress distribution on cut faces of elemental parallelopiped.

Figure 4.6
Depiction of transverse shearing stress τ_{y_1} in a cantilever beam with an end-load.

The slab is located a distance y up from the neutral surface. The C-D boundary of the parallelopiped is a distance y_1 up from the neutral surface, and the A-B boundary at the outer fibers of the beam is a distance c_1 up from the neutral surface.

For any cross section of the beam along its length, the bending moment is a function of end-load P and moment arm to the cross section, and the tensile bending stress, in turn, is a function of the bending moment arm. Thus T is the maximum tensile stress along AE (due to bending) at moment arm z from the end-load, and $T + dT$ is the maximum tensile stress along BF at moment arm $z + dz$ from the end-load. Since areas $AEGC$ and $BFHD$ are equal, a net force to the right in Figure 4.6(c) must be balanced by a shearing force to the left acting on shear plane $CDHG$, in accordance with equilibrium requirements.

To formulate these observations mathematically, the elemental slab shown in Figures 4.6(a) and 4.6(b) may be used to find the differential forces $(dF)_l$ on the left face of the slab and $(dF)_r$ on the right face of the slab as

$$(dF)_l = \sigma_{yl} A_{yl} \tag{4-10}$$

and

$$(dF)_r = \sigma_{yr} A_{yr} \tag{4-11}$$

From (4-4)

$$\sigma_y = \frac{My}{I} \tag{4-12}$$

and from Figure 4.6,

$$A_{yl} = A_{yr} = x_y\, dy \tag{4-13}$$

Thus (4-10) becomes

$$(dF)_l = \frac{My}{I} x_y\, dy \tag{4-14}$$

and (4-11) becomes

$$(dF)_r = \frac{(M + dM)}{I} x_y\, dy \tag{4-15}$$

Integrating (4-14) over the left face of the parallelopiped and (4-15) over the right face gives forces F_l to the left and F_r to the right as

$$F_l = \int_{y_1}^{c_1} \frac{M}{I} x_y y\, dy \tag{4-16}$$

and

$$F_r = \int_{y_1}^{c_1} \frac{(M + dM)}{I} x_y y\, dy \tag{4-17}$$

Considering the parallelopiped shown in Figure 4.6(c) as a free body,

$$F_r - F_l - F_{y_1} = 0 \tag{4-18}$$

where F_{y_1} is the shear force to the left on plane $CDHG$. This shear force may be expressed as

$$F_{y_1} = \tau_{y_1} A_{y_1} = \tau_{y_1} x_{y_1}\, dz \tag{4-19}$$

Substituting (4-16), (4-17), and (4-19) into (4-18), noting that moments are constant with respect to integration on y and I is constant,

$$\frac{(M + dM)}{I} \int_{y_1}^{c_1} x_y y\, dy - \frac{M}{I} \int_{y_1}^{c_1} x_y y\, dy = \tau_{y_1} x_{y_1}\, dz \tag{4-20}$$

or

$$\tau_{y_1} = \frac{dM}{Ix_{y_1}\,dz}\int_{y_1}^{c_1} x_y y\,dy \tag{4-21}$$

From (4-3

$$\frac{dM}{dz} = V \tag{4-22}$$

hence

$$\tau_{y_1} = \frac{V}{Ix_{y_1}}\int_{y_1}^{c_1} x_y y\ dy \tag{4-23}$$

This general expression allows the calculation of transverse shearing stress at any distance y_1 above the neutral axis of bending, and, in turn, the transverse shearing stress distribution. This procedure is valid for any transversely loaded beam of any cross-sectional shape as long as the beam is loaded in a vertical plane of symmetry. From (4-23) it may be observed that:

1. At the outer fiber, where $y_1 = c_1$, the transverse shearing stress is zero.
2. At the neutral axis of bending, where $y_1 = 0$, the transverse shearing stress reaches its maximum as long as the net width there is as small as the width anywhere else; if the section is narrower elsewhere the maximum transverse shearing stress may not be at the neutral axis.
3. The integral $\int_{y_1}^{c_1} yx_y\,dy$ represents the moment of the area of *ACGE* (see Figure 4.6) about the neutral axis. This area moment may be expressed in alternative form as $\bar{y}A_{ACGE}$, where $\bar{y}$ is the *y*-distance from the neutral axis to the centroid of area *ACEG*. This alternate expression provides the basis for the *area moment* method of calculating transverse shearing stresses.

In view of (3) above, the transverse shearing stress equation (4-23) may be written in alternate form as

$$\tau_{y_1} = \frac{V}{Ix_{y_1}}\bar{y}A \tag{4-24}$$

and further, if the cross section is irregular but can be divided into several regular parts, each with its own $\bar{y}_i$ and A_i, (4-24) may be reexpressed as

$$\tau_{y_1} = \frac{V}{Ix_{y_1}}\sum_i \bar{y}_i A_i \tag{4-25}$$

Example 4.2 Transverse Shearing Stress Critical Points

It is desired to calculate the maximum transverse shearing stress in the hollow rectangular cross section shown in Figure E4.2A for a cantilever beam loaded at the end by force *F*.

a. Where would the maximum transverse shearing stress τ_{max} be located?
b. What is the magnitude of τ_{max} as a function of average shearing stress?

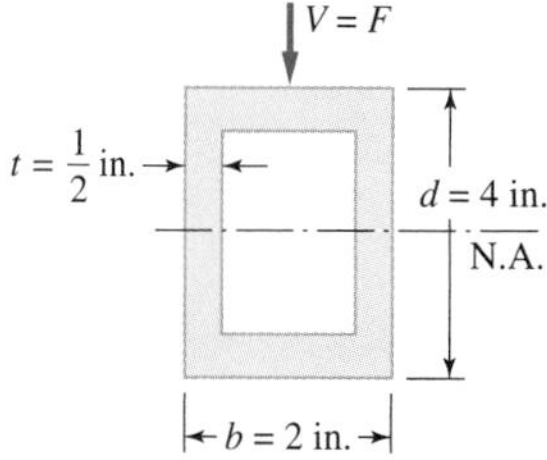

Figure E4.2A
Hollow rectangular beam cross section.

Solution

a. From (4-23) it is clear that τ_{y_1} will reach its maximum when $y_1 = 0$; hence the transverse shearing stress is maximum at the neutral axis of bending.
b. The magnitude of τ_{max} may be calculated directly by integrating (4-23) with $y_1 = 0$, or by using the area moment method of (4-25) with $y_1 = 0$.

For both methods the area moment of inertia of the whole cross section about the neutral axis is

Example 4.2 Continues

$$I = \frac{(2)(4)^3}{12} - \frac{(1)(3)^3}{12} = 8.42 \text{ in}^4 \tag{1}$$

From (4-23) with $y_1 = 0$,

$$\tau_{max} = \frac{V}{(8.42)(1)}\left[\int_0^{1.5}(1)\,y\,dy + \int_{1.5}^{2.0}(2)\,y\,dy\right] \tag{2}$$

or

$$\tau_{max} = \frac{V}{8.42}\left[\left(\frac{y^2}{2}\right)_0^{1.5} + 2\left(\frac{y^2}{2}\right)_{1.5}^{2.0}\right] = 0.34V \tag{3}$$

and

$$\tau_{ave} = \frac{V}{A} = \frac{V}{(2)(4) - (1)(3)} = 0.2V \tag{4}$$

hence

$$\tau_{max} = 1.7\tau_{ave} = 1.7\frac{F}{A} \tag{5}$$

Alternatively, using (4-25),

$$\tau_{max} = \frac{V}{(8.42)(1)}[(1)\{(0.5)2\} + (1)\{(0.5)2\} + (1.75)\{(1)0.5\}] = 0.34\text{V} \tag{6}$$

and again using (4),

$$\tau_{max} = 1.7\tau_{ave} = 1.7\frac{F}{A} \tag{7}$$

c. To sketch the transverse shearing stress distribution for this cross section, equation (4-23) may be utilized. It may be noted from (4-23) that as y_1 varies from zero at the neutral axis to c_1 at the outer fibers, the value of τ_{y_1} varies parabolically from a maximum at the neutral axis to zero at the outer fibers. A discontinuity in the distribution occurs, however, at $y_1 = 1.5$ because width x_{y_1} jumps from 1.0 to 2.0 at that point, giving values from (4-23) of

$$\tau_{y_1=1.5(-)} = \frac{V}{8.42(1.0)}\int_{1.5}^{2.0}(2.0)\,y\,dy = \frac{2.0V}{8.42}\left(\frac{y^2}{2}\right)_{1.5}^{2.0} = 0.208V \tag{8}$$

and

$$\tau_{y_1=1.5(+)} = \frac{V}{8.42(2.0)}\int_{1.5}^{2.0}(2.0)y\,dy = \frac{V}{8.42}\left(\frac{y^2}{2}\right)_{1.5}^{2.0} = 0.104V \tag{9}$$

Using these results and the result from (6), the transverse shearing stress distribution may be plotted as shown in Figure E4.2B

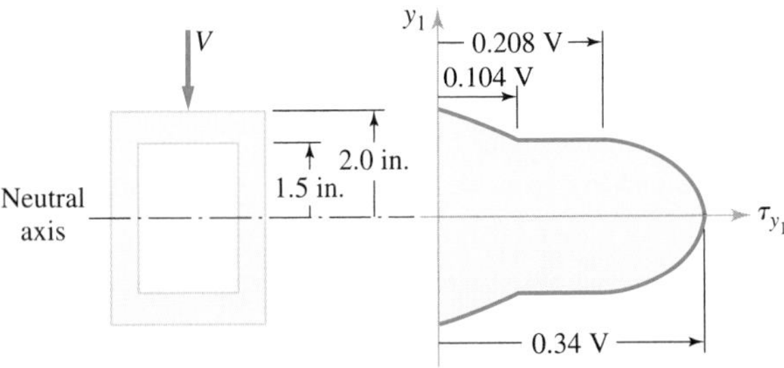

Figure E4.2B
Transverse shearing stress distribution.

TABLE 4.3 Magnitude and Location of Maximum Transverse Shearing Stress for Several Cross-Sectional Shapes

Section Shape	Maximum Transverse Shearing Stress, τ_{max}	Location Where Maximum Occurs
1. Solid rectangle	$\frac{3}{2}\frac{F}{A}$	Neutral axis
2. Solid circle	$\frac{4}{3}\frac{F}{A}$	Neutral axis
3. Hollow circle	$2\frac{F}{A}$	Neutral axis
4. Solid triangle	$\frac{3}{2}\frac{F}{A}$	Half-way between top and bottom
5. Solid diamond	$\frac{9}{8}\frac{F}{A}$	At points $d/8$ above and below the neutral axis

The *maximum* transverse shearing stress can always be expressed as a constant times the *average* transverse shearing stress, where the constant is a function of the cross-sectional shape. The constants for several cross-sectional shapes are given in Table 4.3.

Further, it should be noted that transverse shearing stress typically becomes important, compared to bending stress, only in very short beams. The following rules of thumb are sometimes used. Transverse shearing stress becomes important for

1. Wood beams having a span/depth ratio less than 24
2. Metal beams with thin webs having a span/depth ratio less than 15
3. Metal beams with compact cross sections having a span/depth ratio less than 8

Example 4.3 Design for Transverse Shear Loading

The force analysis of a pin-and-clevis joint (see Example 4.1) has resulted in the free-body diagram for the pin shown in Figure E4.3A. Using this free-body diagram as the basis, complete the following tasks if $P = 10{,}000$ lb, $C = 1.0$ in, $D = E = 0.5$ in, and the pin diameter is d.

a. Construct shear and bending moment diagrams for the pin loaded as shown in Figure E4.3A.

b. Find the location and magnitude of the maximum bending stress in the pin, and estimate the required minimum pin diameter based on bending stress, if the maximum allowable tensile stress for the pin material is 35,000 psi.

c. Using the average shearing stress on pin cross sections *A-A* and *B-B* as the basis, estimate the required minimum pin diameters if the maximum allowable shearing stress for the pin material is 20,200 psi.

Example 4.3 Continues

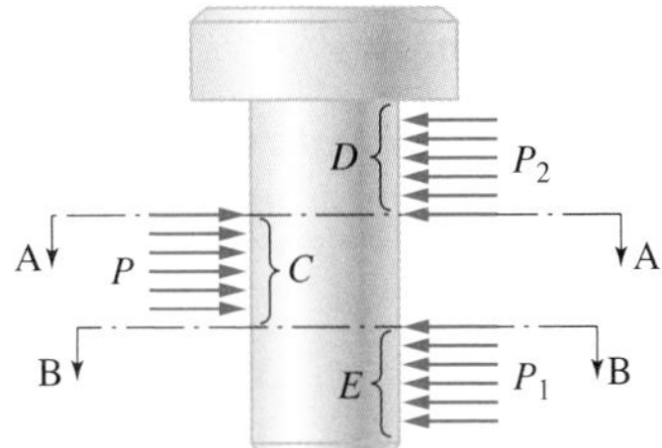

Figure E4.3A
Loading shown on free-body diagram of clevis pin.

d. Calculate the location and magnitude of the maximum transverse shearing stress in the pin, and estimate the required minimum pin diameter on this basis if the maximum allowable shearing stress for the pin material is 20,200 psi.

e. What would be an appropriate first-iteration design recommendation for the pin diameter under these conditions?

Solution

a. Shear and bending moment diagrams for the pin may be plotted directly from governing equations, or constructed using cases 3 and 9 from Table 4.1 as guidelines. The resulting diagrams, based on pin loading shown in Figure E4.3A, are shown in Figure E4.3B.

b. Examination of the bending moment diagram of Figure E4.3B indicates that the maximum bending moment, M_{max} = 2500 in-lb, occurs at midspan of the pin. Since the pin diameter is constant, the maximum bending stress also occurs at midspan of the pin. From (4-5), the maximum bending stress at the midspan outer fibers is

$$\sigma_{max} = \frac{M_{max}c}{I} = \frac{M_{max}}{Z} = \frac{32M_{max}}{\pi d^3} \tag{1}$$

Hence the minimum pin diameter based on bending, with $\sigma_{max} = \sigma_{allowable}$ = 35,000 psi, is

$$d = \sqrt[3]{\frac{32(2500)}{\pi(35{,}000)}} = 0.90 \text{ inch} \tag{2}$$

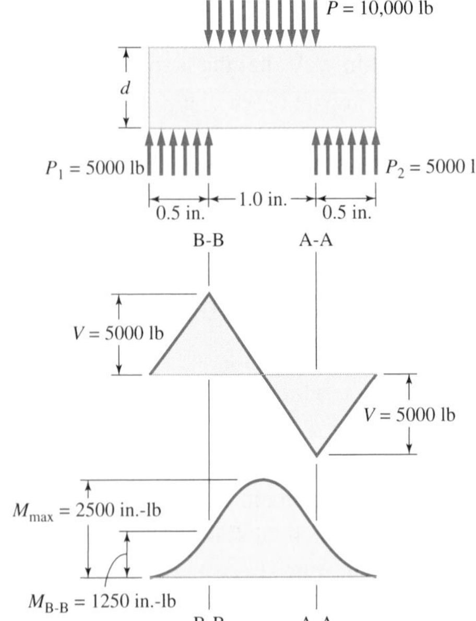

(*a*) Loading diagram (w = 10,000 lb/in. everywhere).

(*b*) Shear diagram.

(*c*) Moment diagram.

Figure E4.3B
Loading, shear, and bending moment diagrams for pin loading of Figure E4.3A.

c. Based on average shearing stress at section *A-A* or *B-B*, equation (4-9) gives

$$\tau_{ave\,A\text{-}A} = \frac{P_{A\text{-}A}}{A_{A\text{-}A}} = \frac{4V_{A\text{-}A}}{\pi d^2} \tag{3}$$

and the minimum pin diameter at *A-A* or *B-B* on the shear diagram of Figure E4.3B with $\tau_{ave} = \tau_{allowable} = 20{,}200$ psi, is

$$d = \sqrt{\frac{4(5000)}{\pi(20{,}200)}} = 0.56 \text{ inch} \tag{4}$$

d. Examination of the shear diagram of Figure E4.3B indicates that the maximum transverse shear force, $V_{max} = 5000$ pounds, occurs at sections *A-A* and *B-B*. Since the pin diameter is constant, the maximum transverse shearing stress also occurs at these sections. From case 2 of Table 4.3, the maximum transverse shearing stress occurs at the neutral axis of bending at *A-A* and *B-B*, and is

$$\tau_{max} = \frac{4}{3}\frac{V_{A\text{-}A}}{A} = \frac{16V_{A\text{-}A}}{3\pi d^2} \tag{5}$$

The minimum pin diameter based on the shear diagram of Figure E4.3B with $\tau_{max} = \tau_{allowable} = 20{,}000$ psi, is

$$d = \sqrt{\frac{16(5000)}{3\pi(20{,}200)}} = 0.65 \text{ inch} \tag{6}$$

e. Examining the results of (b), (c), and (d), the bending-governed diameter is largest and since the transverse shearing stress is zero at the outer-fiber location of maximum bending stress, no combined stress interaction need be considered. The appropriate first-iteration design recommendation for pin diameter, based on bending, would be

$$d_{design} = 0.90 \text{ inch} \tag{7}$$

Torsional Shear; Circular Cross Section

The sketch of Figure 4.7(a) illustrates a torsional moment applied to a straight cylindrical bar. The resulting distribution of shearing stress τ on the cut surface is shown in Figure 4.7(b). It may be noted that the shearing stress appears to present a "swirl" pattern; further, the shearing stress varies linearly from zero at the center to a maximum value τ_{max} at the outer fibers. The magnitude of the shearing stress τ at any radius r is given by

$$\tau = \frac{Tr}{J} \tag{4-26}$$

where

$$J = \int r^2\, dA \tag{4-27}$$

is the polar moment of inertia of the cross section and T is the torsional moment. For a solid circular cross section,

$$J = \frac{\pi D^4}{32} \tag{4-28}$$

and for a hollow circular cross section,

$$J = \frac{\pi(D_o^4 - D_i^4)}{32} \tag{4-29}$$

The maximum shearing stress τ_{max} at the outer fibers is, from (4-26),

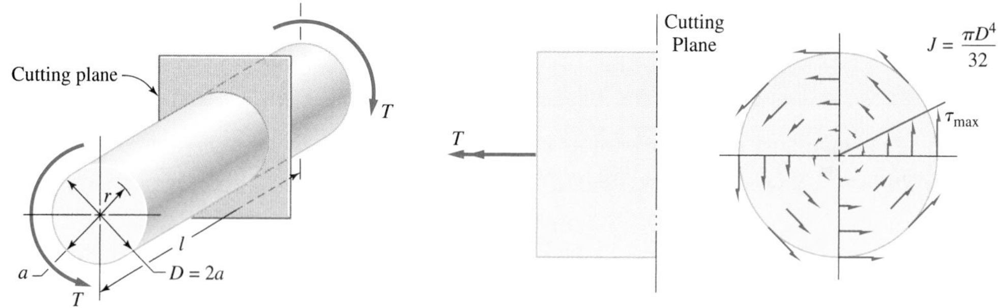

(*a*) Torsional moment on solid cylindrical member.

(*b*) Side view of front free body showing linear shear stress distribution over cut surface of cylinder.

Figure 4.7
Uniform cylinder subjected to torsional moment.

$$\tau_{max} = \frac{Ta}{J} \tag{4-30}$$

where a is the radius to the outer fibers of the cylinder. Equations (4-26) and (4-30) may be assumed valid only if the bar is straight and cylindrical (either solid or hollow), made of homogeneous material, and torque is applied about the longitudinal axis, stresses are within the elastic range, and calculations are made for cutting planes remote from points where loads are applied and remote from significant geometric discontinuities. Under these conditions cross-sectional planes of cylindrical bars remain plane and parallel after twisting.

Because the torque on a shaft is frequently associated with the transmission of power from an electric motor, or other power source, to a machine under load, it is convenient to review the relationship among power, speed, and torque of a rotating shaft. Power is defined as the time-rate of doing work. Horsepower and kilowatts are common measures of power. One horsepower (hp) equals 33,000 ft-lb/min; one kilowatt (kw) equals 60,000 N-m/minute. If a constant torque T is applied to a shaft rotating at constant angular velocity ω rad/min, the work performed each minute is $T\omega$, and the power transmitted is

$$hp = \frac{(T/12)\omega}{33{,}000} = \frac{T(2\pi n)}{(12)(33{,}000)} \tag{4-31}$$

or

$$hp = \frac{Tn}{63{,}025} \tag{4-32}$$

where T = torque, in-lb
n = shaft speed, rev/min
hp = horsepower

or

$$kw = \frac{T\omega}{60{,}000} = \frac{T(2\pi n)}{60{,}000} \tag{4-33}$$

so

$$kw = \frac{Tn}{9549} \tag{4-34}$$

where T = torque, Newton-meters
n = shaft speed, rev/min
kw = kilowatts

Torsional Shear; Noncircular Cross Section

For cases where torsional moments are applied to bars of *noncircular* cross section, *the equations just given do not yield correct results and should not be used.* This is primarily because cross-sectional planes distort significantly when bars of noncircular cross sections are twisted, as illustrated, for example, in Figure 4.8. Although the development of equations for torsional shearing stress in noncircular cross sections is complicated, one means of analyzing such cases is by utilizing the *membrane analogy*.

Analogies are often useful in engineering practice when the *governing equations* for different physical phenomena *are formally the same*. For example, electrical analogies to mechanical components, such as springs, masses, and dampers, permit the use of electric analog computers to solve mechanical vibration problems. In the case of the membrane analogy, it turns out that the governing differential equation and boundary conditions for a bar of any selected cross-sectional shape, subjected to a torsional moment, are formally identical to the differential equation and boundary conditions defining the spatial contour of a thin pressurized membrane protruding above a base datum-plane through a hole having the same shape as the selected bar cross section.[7] The analogy therefore provides a mathematical relationship between the contoured deflection surface and the distribution of torsional shearing stresses in a twisted bar. Consequently, the behavior of a bar in torsion may be deduced by studying the characteristics of the corresponding pressurized-membrane contour map. Such a contour map is illustrated for a circular cross section in Figure 4.9. Three important observations have been established for interpreting results from the membrane analogy.

1. The tangent to a contour line (line of constant elevation) at any point on the deflected membrane gives the *direction* of the torsional shearing stress vector at the corresponding point in the cross section of the twisted bar.
2. The *maximum* slope of the membrane at any point on the deflected membrane is proportional to the *magnitude* of the shearing stress at the corresponding point in the cross section of the twisted bar.
3. The volume enclosed between the datum base-plane and the contoured surface of the deflected membrane is proportional to the torque on the twisted bar.

All of these observations may be readily verified analytically for a twisted bar of circular cross section. Experimental agreement with the observations for noncircular shapes is excellent. For a designer, the importance of the membrane analogy does not lie in its use

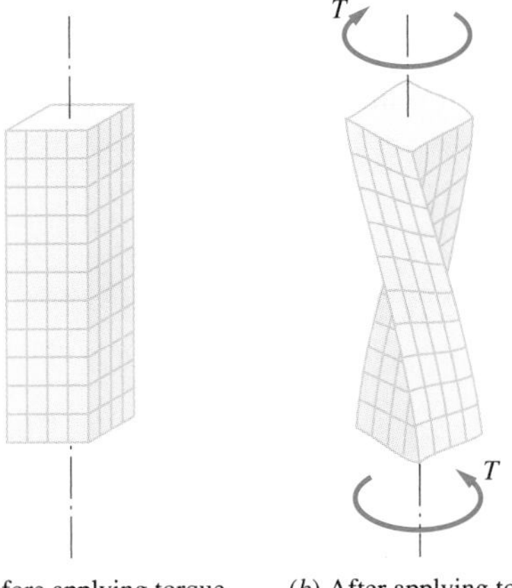

(*a*) Before applying torque. (*b*) After applying torque.

Figure 4.8
Exaggerated depiction of distortions produced by applying torsional moment to a noncircular cross section.

[7]See, for example, ref. 3, pp. 261, 269.

Figure 4.9
Membrane analogy illustrated for torsion applied to a bar of circular cross section.

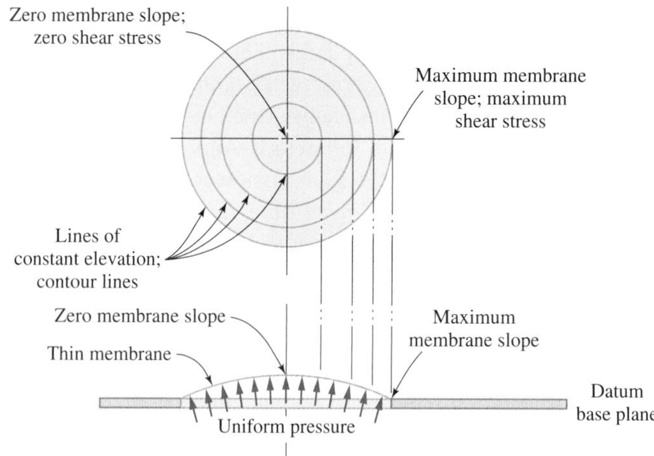

as a calculational tool, but rather, in its use as a quick, *qualitative*, stress analysis tool. By simply *visualizing* a thin pressurized membrane protruding through a hole having the shape of a noncircular bar subjected to torsion, a mental picture of the stress distribution may be formulated, and the points of maximum shearing stress identified. These *critical points* occur where the membrane slope is greatest; that is, where the contour lines are closest together. On the other hand, at points where the membrane slope is near zero, the shearing stress is near zero.

To illustrate, the membrane contour maps corresponding to a rectangular shaft and a circular shaft with a keyway are sketched in Figure 4.10. For the torsionally loaded rectangular bar shown in Figure 4.10(a), the contour lines are closest together (slope of the membrane is greatest) at the midpoint of the longer side; hence *the torsional shearing stress is greatest there*. Also, since the membrane slope is zero at the corners, the *torsional shearing stress is zero at the corners*. It is interesting to note the potential error that might be made if equation (4-30) for a *circular* cross section were applied to this *rectangular* shaft, since (4-30) would seem to predict *maximum shearing stress* at the corners (because the "radius" is largest at the corners), rather than the actual case of *zero shearing stress*. For the circular shaft *with a keyway* depicted in Figure 4.10(b), the contour line pattern is similar to that for the regular circular shaft shown in Figure 4.9, except for the local disruptions around the keyway. The contour lines are crowded together more closely as they flow around the keyway corners, indicating a concentration of stress there. Thus the maximum shearing stress is greatest at the corners of the keyway. Methods of calculating the actual magnitudes of stress at such stress concentration sites are discussed in 4.8.

Typically it is possible to formulate an expression for maximum shearing stress in a

Figure 4.10
Contour lines visualized from the membrane analogy for a rectangular bar and for a circular shaft with a keyway.

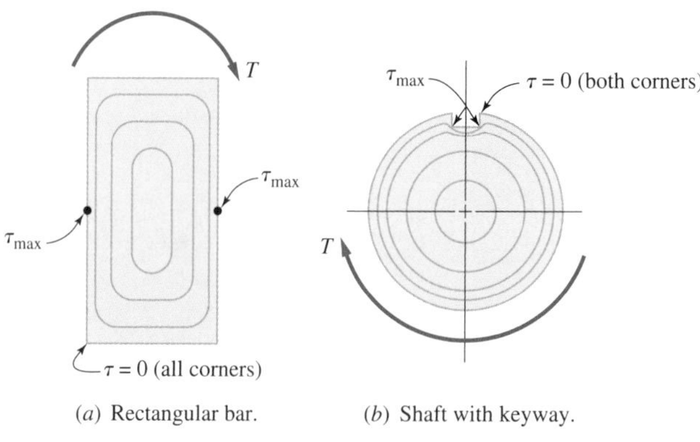

TABLE 4.4 Parameters for Determining Shearing Stress and Angular Deformation for Bars of Various Cross-Sectional Shape Subjected to Torsional Moments

Shape and Dimensions of Cross Section		Q, in^3	K, in^4	Location of Maximum Shearing Stress
1. Solid circular	$2r$	$\dfrac{\pi r^3}{2}$	$\dfrac{\pi r^4}{2}$	On entire boundary
2. Solid elliptical	$2a$, $2b$	$\dfrac{\pi ab^2}{2}$	$\dfrac{\pi a^3 b^3}{a^2+b^2}$	At ends of minor axis
3. Solid rectangular	$2a$, $2b$	$\dfrac{8a^2b^2}{3a+1.8b}$ (for $a \geq b$)	$ab^3\left[\dfrac{16}{3} - 3.36\dfrac{b}{a}\left(1 - \dfrac{b^4}{12a^4}\right)\right]$ (for $a \geq b$)	At midpoint of each longer side
4. Solid equilateral triangle	a	$\dfrac{a^3}{20}$	$\dfrac{\sqrt{3}\,a^4}{80}$	At midpoint of each side
5. Any thin tube of uniform thickness t; U = length of median boundary. A = area enclosed by median boundary.	t, Median boundary	$2At$	$\dfrac{4A^2t}{U}$	Nearly uniform throughout

noncircular bar subjected to torque T as

$$\tau_{max} = \frac{T}{Q} \tag{4-35}$$

where

$$Q = f(\text{cross sectional geometry}) \tag{4-36}$$

Likewise, the angle of twist θ, in radians, may be formulated for noncircular cross sections subjected to torque T as

$$\theta = \frac{TL}{KG} \tag{4-37}$$

where

$$K = g(\text{cross sectional geometry}) \tag{4-38}$$

L is bar length, and G is shear modulus of elasticity. Expressions for Q and K are provided for several cross-sectional shapes in Table 4.4, and for numerous other shapes in the literature.[8]

Example 4.4 Design of Torsionally Loaded Round and Square Shafts

Experimental power measurements made on a new-style rotary garden tiller indicate that under full load conditions the internal combustion engine must supply 4.3 horsepower, steadily, to the mechanical drive train. Power is transmitted through a solid 0.50-inch-

[8]See, for example, refs. 4 and 5.

Example 4.4 Continues

diameter round steel shaft rotating at 1800 rpm. It is being proposed to replace the round steel shaft with a solid square shaft of the same material. Evaluate the proposal by determining the following information:

a. What is the steady full-load torque being transmitted by the round steel shaft?

b. What is the maximum stress in the round shaft, what type of stress is it, and where does it occur?

c. Assuming that the measurements of power transmitted by the round shaft were made under full load, as specified, what *design-allowable stress* was probably used for the shaft?

d. What *design-allowable stress* should be used in estimating the size required for the proposed square shaft?

e. What size should the proposed square shaft be made to be "equivalent" to the existing round shaft in resisting failure?

Solution

a. From equation (4-32),

$$T = \frac{63{,}025\,(hp)}{n} = \frac{63{,}025(4.3)}{1800} = 150.6 \text{ in-lb} \tag{1}$$

b. The torque on the round shaft produces torsional *shearing* stress that reaches a maximum value all around the outer surface. The magnitude of the maximum shearing stress is, from (4-30) and (4-28),

$$\tau_{max} = \frac{Ta}{J} = \frac{T\left(\dfrac{D}{2}\right)}{\left(\dfrac{\pi D^4}{32}\right)} = \frac{16T}{\pi D^3} = \frac{16(150.6)}{\pi(0.50)^3} = 6136 \text{ psi} \tag{2}$$

c. Since the design objective is to size the part so that the maximum stress under "design conditions" is equal to the design-allowable stress τ_{all},

$$\tau_{all} = \tau_{max} = 6136 \text{ psi} \tag{3}$$

d. Since the material for the square shaft is the same as for the round shaft, the design-allowable shearing stress should be the same. Hence, for the square shaft

$$\tau_{all} = 6136 \text{ psi} \tag{4}$$

e. The proposed square shaft is in the category of a noncircular bar subjected to torsion; hence, from (4-35),

$$Q = \frac{T}{\tau_{all}} = \frac{150.6}{6136} = 0.025 \tag{5}$$

For the square shaft, case 3 of Table 4.4 may be utilized by setting $a = b$, giving

$$Q = \frac{8a^4}{4.8a} = 1.67a^3 \tag{6}$$

Equating (5) and (6) then gives

$$a = \sqrt[3]{\frac{0.025}{1.67}} = 0.25 \text{ inch} \tag{7}$$

Hence the dimensions of each side of the square should be

$$s = 2a = 0.50 \text{ inch} \tag{8}$$

Torsional Shear; Shear Center in Bending

In the development of equations (4-4) for bending stress and (4-23) for transverse shearing stress, it was stipulated that the plane of the bending moment produced by a coplanar transverse force couple must coincide with a plane of symmetry for the beam; otherwise the equations are not fully valid. In the case of beam bending produced by a coplanar force couple, where the plane of the transverse force couple is *not* a plane of symmetry for the beam, *bending of the beam is usually accompanied by torsion on the beam.* To eliminate the torsion in such cases, the plane of the transverse force couple must be displaced parallel to itself until it passes through the *shear center*, or *center of twist*, for the cross section. The location of the shear center sometimes may be established by inspection, noting the following:

1. If a cross section has *one axis of symmetry*, the shear center must lie on that axis, but usually *will not* be at the centroid.
2. If a cross section has *two axes of symmetry*, the shear center *will* be at their point of intersection (i.e., at the centroid).
3. If a cross section has a *point of symmetry*, the shear center *will* be at the centroid.

Table 4.5 shows the shear center location for a few beam cross sections, and others are given in the literature.[9]

TABLE 4.5 Location of Shear Center for a Few Beam Cross Sections[1]

<table>
<tr><th>Shape and Dimensions of Cross Section</th><th></th><th>Position of Shear Center Q (for vertical loading)</th></tr>
<tr><td>1. Channel</td><td>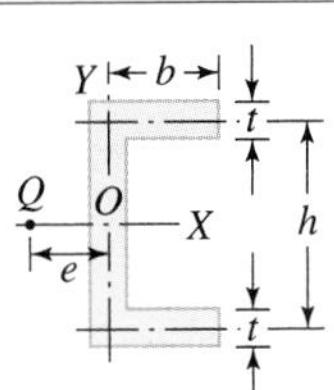
</td><td>$e = h\frac{H_{xy}}{I_x}$

where H_{xy} = product of inertia of the half-section (above X) with respect to axes X and Y, and I_x = moment of inertia of whole section about axis X

If t is uniform, $e = \frac{b^2h^2t}{4I_x}$</td></tr>
<tr><td>2. Angle</td><td>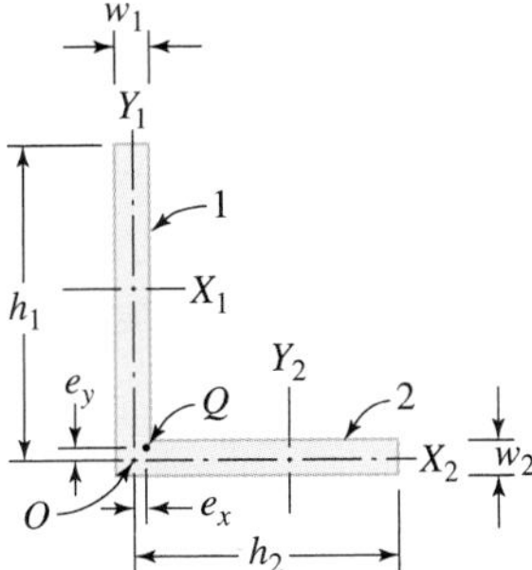
</td><td>$e_x = \frac{1}{2}h_2\left(\frac{I_1}{I_1 + I_2}\right)$

$e_y = \frac{1}{2}h_1\left(\frac{I_1}{I_1 + I_2}\right)$

where Leg 1 = rectangle w_1h_1; Leg 2 = rectangle w_2h_2
I_1 = moment of inertia of Leg 1 about Y_1
I_2 = moment of inertia of Leg 2 about Y_2
If w_1 and w_2 are small. $e_x \approx e_y \approx 0$ and Q is at O</td></tr>
<tr><td>3. Tee</td><td>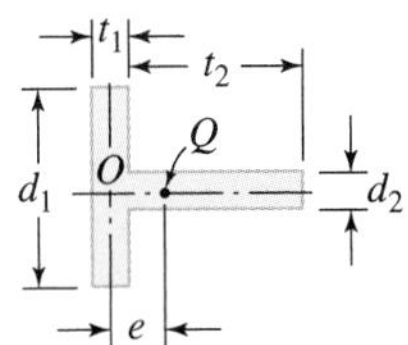
</td><td>$e = \frac{1}{2}(t_1 + t_2)\left[\frac{1}{1 + \frac{d_1^3t_1}{d_2^3t_2}}\right]$

For a T-beam of usual proportions, Q may be assumed to be at O</td></tr>
</table>

[9]See, for example, refs. 4 and 5.

TABLE 4.5 (*Continued*)

Shape and Dimensions of Cross Section		Position of Shear Center Q (for vertical loading)
4. I with unequal flanges and thin web.	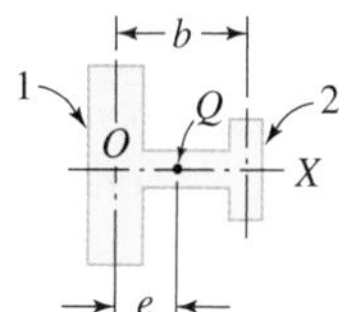	$e = b\left(\frac{I_2}{I_1 + I_2}\right)$ where I_1 and I_2 are moments of inertia of flange 1 and flange 2, respectively, about the X axis

[1]Selected from ref. 4.

Example 4.5 Stresses in a Web-Vertical, End-Loaded, Channel-Section Cantilever Beam

A channel-shaped cantilever bracket, oriented as shown in Figure E4.5A, is loaded at the free end by a force $P = 8000$ lb applied at the middle of the flange. It is desired to review this design configuration by determining the following information:

a. What types of stress patterns are developed, and where the maximum stresses would occur.

b. What are the maximum stresses of each type that should be considered in the failure prevention analysis of the bracket.

c. What measures might be taken to reduce the stresses determined in (b) without changing loading magnitude or cross-sectional geometry.

Solution

a. For this type of loading on a channel section oriented with the web vertical, as shown in Figure E4.5A, three types of stress patterns should be examined. They are:

   1. *Bending stresses*, which reach maximum values at the extreme upper and lower fibers at the wall.

   2. *Transverse shearing stresses*, which reach maximum values at the neutral axis of bending (normally the x-axis in Figure E4.5A, all along the length of the bracket).

   3. *Torsional shearing stresses*, because the load P does not pass through the shear center (see case 1 of Table 4.5). These reach maximum values in the upper and lower flanges, all along the length of the bracket.

b. Maximum stresses of each type may be determined as follows:

   1. Assuming that the bracket does not twist significantly, the maximum bending stress at the outer fibers and at the wall may be calculated from (4-5) as

$$\sigma_{max} = \frac{Mc}{I_x} \tag{1}$$

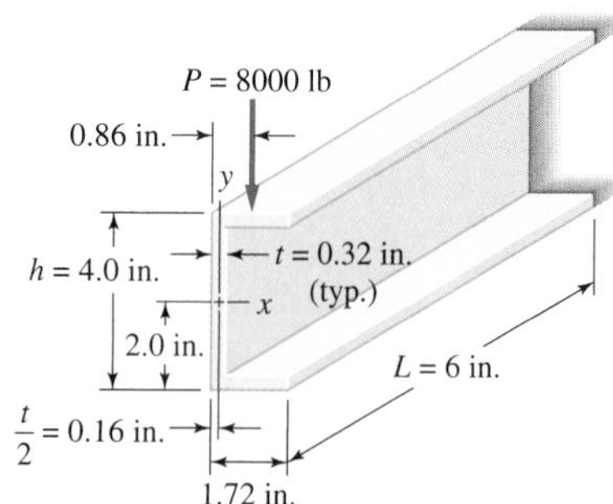

Figure E4.5A
Channel-shaped cantilever bracket subjected to transverse end-load.

where

$$M = PL = (8000)(6) = 48{,}000 \text{ in-lb} \tag{2}$$

$$c = \frac{h}{2} = \frac{4.0}{2} = 2.0 \text{ inches} \tag{3}$$

and[10]

$$I_x = \frac{th^3}{12} + 2A_f d^2 \tag{4}$$

or

$$I_x = \frac{0.32(4)^3}{12} + 2[(1.72 - 0.32)(0.32)](2.0 - 0.16)^2$$
$$= 1.71 + 3.03 = 4.74 \text{ in}^4 \tag{5}$$

hence

$$\sigma_{max} = \frac{48{,}000(2.0)}{4.74} = 20{,}250 \text{ psi} \tag{6}$$

2. Transverse shearing stress τ_{ts} is maximum at the neutral axis (x-axis in Figure E4.5A), and may be calculated using (4-25) as

$$\tau_{ts\text{-}max} = \frac{V}{I_x t}\sum_{i=1}^{2} \bar{y}_i A_i = \frac{8000}{4.74(0.32)}[(1.0)(2.0)(0.32) + (1.84)(1.4)(0.32)]$$
$$= 7720 \text{ psi} \tag{7}$$

3. From case 1 of Table 4.5, the shear center for the channel section is located on the x-axis a distance e to the left of the x-y origin. For the channel of Figure E4.5A,

$$e = \frac{b^2h^2t}{4I_x} = \frac{(1.72 - 0.16)^2(4.0 - 0.16)^2(0.32)}{4(4.74)} = 0.61 \text{ inch} \tag{8}$$

Hence the shear center is on the x-axis a distance $e = 0.61$ inch to the left of the origin. In Figure E4.5B this produces a torsional moment T on the bracket where

$$T = Pa \tag{9}$$

and moment arm a is

$$a = 0.61 + 0.86 - 0.16 = 1.31 \text{ inches} \tag{10}$$

hence

$$T = 8000(1.31) = 10{,}480 \text{ in-lb} \tag{11}$$

This torsional moment is resisted primarily by the opposing horizontal forces R, one in each flange, that form a resisting couple Rd. Based on moment equilibrium

$$T = Rd \tag{12}$$

hence

$$R = \frac{T}{d} = \frac{10{,}480}{3.68} = 2848 \text{ lb} \tag{13}$$

or

$$\tau_{tor\text{-}max} = \frac{R}{A_f} = \frac{2848}{(1.72)(0.32)} = 5170 \text{ psi} \tag{14}$$

Thus, the maximum torsional shearing stress τ_{tor}, in the flanges, is 5170 psi.

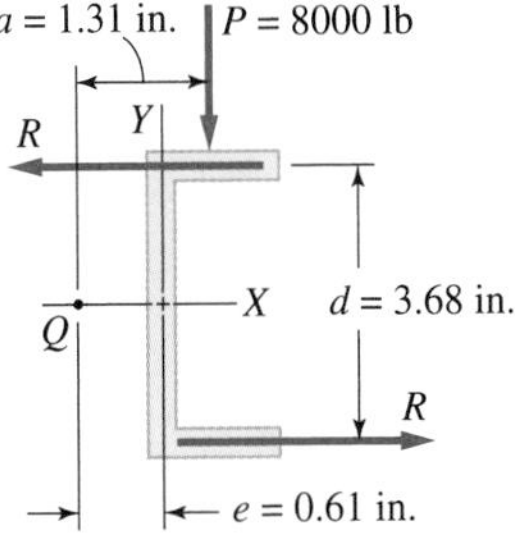

Figure E4.5B
Torsional moment Pa and resisting force couple Rd for the loaded bracket shown in Figure E4.5A.

[10]Equations for area moments of inertia and transfer of axes are available in numerous references (e.g., ref. 4).

Example 4.5 Continues

c. The *torsional* shearing stresses could be eliminated by translating the applied load P to the left a distance $a = 1.31$ inches so that the line of action of P would pass through the shear center. This would probably require welding a small block on the channel web to support the load.

Surface Contact Stress

When loads are transmitted from one machine part to another through pins, fasteners, bearings, gear teeth, or other types of joints, *surface contact pressures* are developed in the areas of contact between mating parts. Since the theoretical contact footprint between *curved* surfaces is a point or a line, for such contacts the operating load must initially be supported on near-zero contact areas, producing very high contact pressures. The high contact pressures, in turn, produce elastic deformations at the contact site that broaden the footprint and enlarge the contact area. The pressure distributions over such contact areas are nonuniform, and the states of stress generated at and below the surface usually are triaxial because of the elastic constraint of surrounding material. The complicated anlaysis of the stress distributions in the regions surrounding the contact site between two curved elastic bodies was originated in 1881 by H. Hertz; therefore such stresses are usually called *Hertz contact stresses*. Even cases of contact between planar bodies may produce nonuniform pressure distributions in and around the contact region. For example, a planar contact between a *small* rigid block pressed against a *large* elastic body produces nonuniform stress distributions near the edges of the block. Since contact stresses are generally multiaxial, further discussion will be deferred until multiaxial states of stress have been discussed[11] in 4.5.

Deflection; Common Types of Loading

When forces are applied to structures or machine parts, not only are stresses generated but strains and deflections are also produced. This concept has already been illustrated for the case of a straight uniform elastic bar loaded by a direct axial force in Figures 2.1 and 2.2. The force-deflection equation (2-1) for this simple axial loading case is repeated here for convenience, as

$$F = ky = k\delta_f \tag{4-39}$$

where

$$k = \frac{AE}{L} \tag{4-40}$$

is axial spring rate of the loaded bar, F is applied force, and δ_f is force-induced elastic deformation.

Torsional moment loading also produces torsional shearing strain and consequent angular deflection of the torsionally loaded member. The angular deflection for torsional loading of elastic members, given by (4-37), is repeated here, as

$$\theta = \frac{TL}{KG} \tag{4-41}$$

where θ is angular deflection in radians, T is applied torque, L is bar length, G is shear modulus of elasticity, and parameter K is a function of the cross-sectional shape, tabulated for several cases in Table 4.4. If the member is cylindrical, K is the polar moment of inertia, J.

Bending loads produce transverse beam deflections that may be found (assuming elastic behavior) by twice integrating the differential equation of the deflection curve and us-

[11]See 4.10 for a more complete discussion of multiaxial Hertz contact stress distributions.

ing the particular boundary conditions of interest. The governing differential equation is [12]

$$\frac{d^2y}{dx^2} = -\frac{M}{EI} \tag{4-42}$$

where y is transverse beam deflection at any location x along the length of the beam, M is applied bending moment, E is modulus of elasticity, and I is area moment of inertia about the neutral axis of bending. The deflection equations for several common cases are given in Table 4.1.

In all of these cases, when an elastic member is loaded slowly by a force, a torsional moment, or a bending moment, to produce corresponding axial, torsional, or bending deflections or displacements, the external forces or moments *do work* on the body (average force or moment times corresponding displacement), which is stored in the strained body as *potential energy of strain*, usually just called *strain energy*. If the strain does not exceed the elastic limit, the stored strain energy can be recovered during a gradual unloading of the body.

Stored Strain Energy

If a material follows Hooke's Law, and if deformations are small, the displacements of an elastic structure are linear functions of the external loads. Thus, if the loads increase in proportion to each other, the displacements increase in like proportion. To illustrate, if a machine part or a structure is simultaneously loaded by external forces $P_1, P_2, \cdots, P_i$, the corresponding displacements $\delta_1, \delta_2, \cdots, \delta_i$ may be found from linear force-displacement diagrams of the type shown in Figure 2.2, and the work done on the machine part, which is equal to the stored strain energy, is

$$U = \frac{P_1\delta_1}{2} + \frac{P_2\delta_2}{2} + \cdots + \frac{P_i\delta_i}{2} \tag{4-43}$$

Forces and displacements in this context are generalized terms which include moments and corresponding angular displacements.

For example, if the cubic element of material shown in Figure 4.11 is subjected to three mutually perpendicular tensile forces P_x, P_y, and P_z, the stored strain energy is

$$U = \frac{P_x\delta_x}{2} + \frac{P_y\delta_y}{2} + \frac{P_z\delta_z}{2} \tag{4-44}$$

or, since

$$\sigma_x = \frac{P_x}{dydz} \tag{4-45}$$

$$\sigma_y = \frac{P_y}{dxdz} \tag{4-46}$$

$$\sigma_z = \frac{P_z}{dxdy} \tag{4-47}$$

and

$$\varepsilon_x = \frac{\delta_x}{dx} \tag{4-48}$$

$$\varepsilon_y = \frac{\delta_y}{dy} \tag{4-49}$$

$$\varepsilon_z = \frac{\delta_z}{dz} \tag{4-50}$$

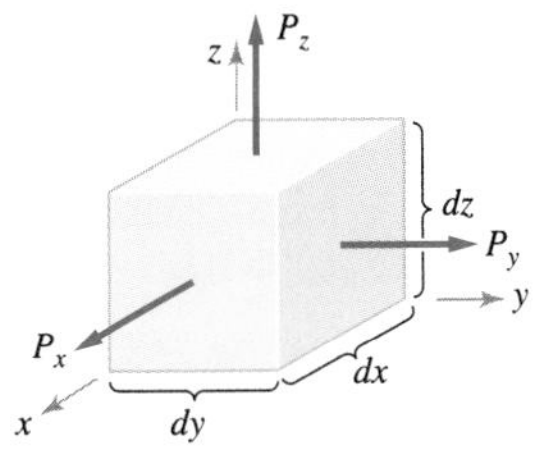

Figure 4.11
Unit cube of material subjected to mutually perpendicular forces P_x, P_y, and P_z.

[12]See, for example, ref. 1, p. 139.

the strain energy expression (4-44) may be rewritten as

$$U = \left[\frac{\sigma_x \varepsilon_x}{2} + \frac{\sigma_y \varepsilon_y}{2} + \frac{\sigma_z \varepsilon_z}{2}\right] dxdydz \tag{4-51}$$

Noting that $dxdydz$ is the volume of the cube, the stored strain energy *per unit volume*, u, may be written as

$$u = \frac{U}{dxdydz} = \frac{\sigma_x \varepsilon_x}{2} + \frac{\sigma_y \varepsilon_y}{2} + \frac{\sigma_z \varepsilon_z}{2} \tag{4-52}$$

Based on the concept of (4-43) and the details of (4-39), (4-41), and (4-42), strain energy expressions for various common loading conditions may be written. For example, tension or compression loading gives, for the case of constant loading and constant geometry[13]

$$U_{tens} = \frac{F}{2}\left(\frac{FL}{AE}\right) = \frac{F^2 L}{2AE} \tag{4-53}$$

For torsion

$$U_{tor} = \frac{T}{2}\left(\frac{TL}{KG}\right) = \frac{T^2 L}{2KG} \tag{4-54}$$

For pure bending

$$U_{bend} = \frac{M}{2}\left(\frac{ML}{EI}\right) = \frac{M^2 L}{2EI} \tag{4-55}$$

and for direct shear

$$U_{dir\text{-}sh} = \frac{P}{2}\left(\frac{PL}{AG}\right) = \frac{P^2 L}{2AG} \tag{4-56}$$

For the case of transverse shear in a beam, expressions for strain energy due to the transverse shear can be developed but are complicated functions of the beam cross section. Except for very short beams, stored strain energy due to the transverse shear is negligible compared to the stored strain energy due to the bending, given in (4-55).

Example 4.6 Stored Strain Energy in a Loaded Beam

A beam is simply supported at each end and loaded by a vertical midspan load P and a pure moment M applied at the left support. Calculate the strain energy stored in the beam if it has an area moment of inertia for the cross section equal to I and modulus of elasticity E.

Solution

For this case, the loading is constant and the cross section all along the beam is constant. Therefore, from Table 4.1, superposing cases 1 and 4, the midspan deflection at $x = L/2$ (under load P) is

$$y = \frac{PL^3}{48EI} + \frac{M}{6EI}\left(\frac{L^3}{8L} + \frac{2L^2}{2} - \frac{3L^2}{4}\right) = \frac{PL^3}{48EI} + \frac{ML^2}{16EI} \tag{1}$$

The end slope at the left end (where the moment is applied) is

$$\theta_A = \frac{PL^2}{16EI} + \frac{ML}{3EI} \tag{2}$$

Based on (4-43), the stored strain energy in the beam with both P and M acting is

$$U = \frac{Py}{2} + \frac{M\theta_A}{2} \tag{3}$$

[13]If load, cross-sectional shape or size, or modulus of elasticity vary along the length of the member, the appropriate integral expression from Table 4.6 must be written instead.

or, substituting (1) and (2) into (3),

$$U = \frac{P}{2}\left(\frac{PL^3}{48EI} + \frac{ML^2}{16EI}\right) + \frac{M}{2}\left(\frac{PL^2}{16EI} + \frac{ML}{3EI}\right) = \frac{1}{EI}\left(\frac{P^2L^3}{96} + \frac{M^2L}{6} + \frac{PML^2}{16}\right) \quad (4)$$

where U is the total stored strain energy in the beam.

Castigliano's Theorem

By writing an expression for stored strain energy in any loaded elastic structure or machine part, a simple *energy method* for calculating the displacements of points in the elastic body may be utilized. This energy method, called *Castigliano's theorem*, may be stated as follows:

> When any combination of external forces and/or moments act on an elastic member, the small displacements induced at any point and in any direction may be determined by computing the partial derivative of total stored strain energy in the loaded member with respect to any selected force or moment to obtain the displacement at the corresponding point and in the corresponding direction of the selected force or moment.

For example, in the case of simple tension in a uniform prismatic member, the stored strain energy is given by (4-53) as

$$U_{tens} = \frac{F^2L}{2AE} \quad (4\text{-}57)$$

Differentiating partially with respect to force F,

$$\frac{\partial U}{\partial F} = \frac{FL}{AE} = \delta_f \quad (4\text{-}58)$$

This expression for force-induced elastic deformation δ_f in the direction of F is confirmed by (2-6).

Castigliano's theorem may also be used to find the deflection at a point in a loaded member even if no force or moment acts there. This is done by placing a *dummy* force or *dummy* moment, Q_i, at the point and in the direction of the deflection desired, writing the expression for total stored strain energy including the energy due to Q_i, and partially differentiating the energy expression with respect to Q_i. This procedure yields the deflection δ_i at the desired point. Since Q_i is a dummy force (or moment), it is set equal to zero in the δ_i expression to obtain the final result. Table 4.6 provides strain energy expressions and deflection equations for common types of loading.

Redundant reactions in statically indeterminate structures or machine parts may also be found by applying Castigliano's theorem.[14] This is accomplished by partially differentiating the total strain energy stored in the loaded member with respect to the chosen redundant reaction force or moment, setting the partial derivative equal to zero (since immovable reactions do no work and store no energy), and solving for the redundant reaction.

Example 4.7 Beam Deflection and Slope Using Castigliano's Theorem

The simply supported beam shown in Figure E4.7(a) is subjected to a load P at midspan and a moment M at the left support. Using Castigliano's theorem, find the following deflections:

a. Midspan beam deflection y_B
b. Angular deflection θ_A of the beam at the left support
c. Angular deflection θ_B of the beam at midspan

[14] See, for example, ref. 1, pp. 340 ff.

TABLE 4.6 Summary of Strain Energy Equations and Deflection Equations for Use with Castigliano's Method Under Several Common Loading Conditions

Type of Load	Strain Energy Equation	Deflection Equation
Axial	$U = \int_0^L \frac{P^2}{2EA} dx$	$\delta = \int_0^L \frac{P(\partial P/\partial Q)}{EA} dx$
Bending	$U = \int_0^L \frac{M^2}{2EI} dx$	$\delta = \int_0^L \frac{M(\partial M/\partial Q)}{EI} dx$
Torsion	$U = \int_0^L \frac{T^2}{2KG} dx$	$\delta = \int_0^L \frac{T(\partial T/\partial Q)}{KG} dx$
Direct shear[1]	$U = \int_0^L \frac{P^2}{2AG} dx$	$\delta = \int_0^L \frac{P(\partial P/\partial Q)}{AG} dx$

[1]Transverse shear associated with bending gives a similar but more complicated function of the particular cross-sectional shape (see Figure 4.6 and associated discussion) and is usually negligible compared to the strain energy of bending.

Example 4.7 Continues

Solution

a. To find the midpsan deflection y_B, where P is acting, the procedure will be to write an expression for total strain energy in the beam under load, U, then differentiate U partially with respect to P. From Table 4.6, the deflection equation for the case of bending may be seen to be

$$\delta = \int_0^L \frac{M(\partial M/\partial Q)}{EI} dx \tag{1}$$

Since the beam of Figure E4.7(a) is not symmetrically loaded, the integration must be separately executed for beam section AB and beam section BC , then added.

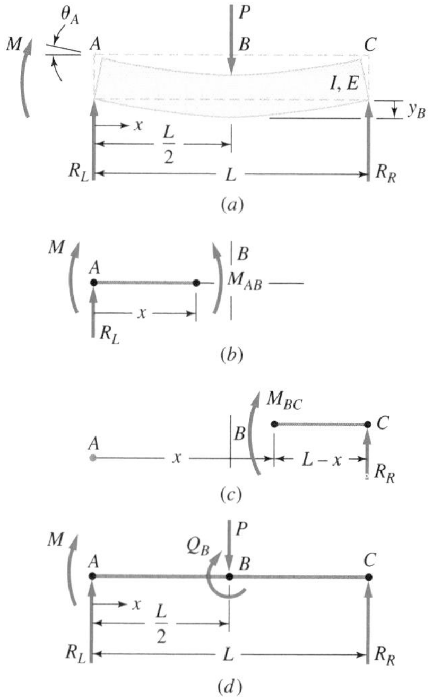

Figure E4.7
Simply supported beam with midspan load P and moment M at the left support.

That is,

$$\delta = \frac{1}{EI}\left[\int_0^{L/2} M_{AB}\frac{\partial M_{AB}}{\partial Q}\,dx + \int_{L/2}^{L} M_{BC}\frac{\partial M_{BC}}{\partial Q}\,dx\right] \tag{2}$$

Using the sketch of Figure E4.7(a), and taking counterclockwise moments as positive, the reactions R_L and R_R may be found from static equilibrium requirements as

$$R_L = \frac{P}{2} - \frac{M}{L} \tag{3}$$

and

$$R_R = \frac{P}{2} + \frac{M}{L} \tag{4}$$

Next, based on the free-body diagrams of Figures E4.7(b) and (c), the moment equations for sections AB and BC of the beam may be found to be

$$M_{AB} = -\frac{Px}{2} + \frac{Mx}{L} - M \qquad 0 \leq x \leq \frac{L}{2} \tag{5}$$

and

$$M_{BC} = \frac{Px}{2} + \frac{Mx}{L} - \frac{PL}{2} - M \qquad \frac{L}{2} \leq x \leq L \tag{6}$$

Also, (5) and (6) may be partially differentiated with respect to midspan load P to give

$$\frac{\partial M_{AB}}{\partial P} = -\frac{x}{2} \tag{7}$$

and

$$\frac{\partial M_{BC}}{\partial P} = \frac{x}{2} - \frac{L}{2} \tag{8}$$

Substituting (5), (6), (7), and (8) into (2) gives

$$y_B = \frac{1}{EI}\left[\int_0^{L/2}\left(-\frac{Px}{2} + \frac{Mx}{L} - M\right)\left(-\frac{x}{2}\right)dx + \int_{L/2}^{L}\left(\frac{Px}{2} + \frac{Mx}{L} - \frac{PL}{2} - M\right)\left(\frac{x}{2} - \frac{L}{2}\right)dx\right] \tag{9}$$

which yields

$$y_B = \frac{PL^3}{48EI} + \frac{ML^2}{16EI} \tag{10}$$

b. To find the angular deflection θ_A at the left support, the approach is similar, except that (5) and (6) must be differentiated with respect to the moment M applied at the left support, to give

$$\frac{\partial M_{AB}}{\partial M} = \frac{x}{L} - 1 \tag{11}$$

and

$$\frac{\partial M_{BC}}{\partial M} = \frac{x}{L} - 1 \tag{12}$$

Substituting (5), (6), (11), and (12) into (2) gives

$$\theta_A = \frac{1}{EI}\left[\int_0^{L/2}\left(-\frac{Px}{2} + \frac{Mx}{L} - M\right)\left(\frac{x}{L} - 1\right)dx + \int_{L/2}^{L}\left(\frac{Px}{2} + \frac{Mx}{L} - \frac{PL}{2} - M\right)\left(\frac{x}{L} - 1\right)dx\right] \tag{13}$$

Example 4.7 Continues

which yields

$$\theta_A = \frac{PL^2}{16EI} + \frac{ML}{3} \tag{14}$$

c. To find θ_B, a dummy moment, Q_B, must be applied at midspan, in addition to the actual loads on the beam, as shown in Figure E4.7(d). Referring to Figure E4-7(d), the reactions R_L and R_R may be found on the basis of static equilibrium to be

$$R_L = \frac{P}{2} - \frac{M}{L} - \frac{Q_B}{L} \tag{15}$$

and

$$R_R = \frac{P}{2} + \frac{M}{L} + \frac{Q_B}{L} \tag{16}$$

With these reactions, the free-body diagrams of Figure E4.7(b) and (c) may again be used to write the moment equations for sections AB and BC of the beam as

$$M_{AB} = -\frac{Px}{2} + \frac{Mx}{L} + \frac{Q_B x}{L} - M \qquad 0 \le x \le \frac{L}{2} \tag{17}$$

and

$$M_{BC} = \frac{Px}{2} + \frac{Mx}{L} + \frac{Q_B x}{L} - \frac{PL}{2} - M - Q_B \qquad \frac{L}{2} \le x \le L \tag{18}$$

From Table 4.6, the total stored strain energy expression, U, for the loaded beam may be written as

$$U = \frac{1}{2EI}\left[\int_0^{L/2} M_{AB}^2 dx + \int_{L/2}^{L} M_{BC}^2 dx\right] \tag{19}$$

Substituting (17) and (18) into (19) gives

$$U = \frac{1}{2EI}\left[\int_0^{L/2}\left(-\frac{Px}{2} + \frac{Mx}{L} + \frac{Q_B x}{L} - M\right)^2 dx + \int_{L/2}^{L}\left(\frac{Px}{2} + \frac{Mx}{L} + \frac{Q_B x}{L} - \frac{PL}{2} - M - Q_B\right)^2 dx\right] \tag{20}$$

which yields

$$U = \frac{P^2L^3}{96EI} + \frac{PML^2}{16EI} + \frac{M^2L}{6EI} - \frac{MQ_BL}{24EI} + \frac{Q_B^2L}{24EI} \tag{21}$$

Differentiating partially with respect to Q_B,

$$\theta_B = \frac{\partial U}{\partial Q_B} = -\frac{ML}{24EI} + \frac{Q_BL}{12EI} \tag{22}$$

Finally, setting dummy load Q_B equal to zero, (22) becomes

$$\theta_B = -\frac{ML}{24EI} \tag{23}$$

This may be verified by superposing cases 1 and 4 of Table 4.1.

4.5 Multiaxial States of Stress and Strain

Under the most complicated loading conditions, a machine part may be subjected to forces and moments produced by various combinations of direct axial loads, bending loads, direct or transverse shearing loads, torsional loads, and/or surface contact loads. As noted in 4.2, no matter how complicated the geometry of the part or how many different external forces and moments are being applied, the force system can always be resolved into just three resultant forces and three resultant moments, each defined with respect to an arbitrarily chosen x-y-z coordinate system. The origin of the selected x-y-z coordinate system may be placed at any desired critical point within the part. The most complicated state of stress that can be produced in a small elemental volume of material at the critical point is the *triaxial state of stress*[15]; *any* triaxial state of stress may be fully specified relative to the chosen x-y-z coordinate system by specifying the three normal stress components $\sigma_x, \sigma_y, \sigma_z$, and the three shearing stress components τ_{xy}, τ_{yz}, and τ_{zx}.

Principal Stresses

It can be shown[16] that if the selected x-y-z coordinate axes, together with the elemental volume of material, are rotated in three-dimensional space, about the fixed origin, a *unique* new orientation may be found for which the shearing stress components vanish on all faces of the newly oriented elemental volume. This unique orientation, for which the shearing stress components on all faces of the elemental cube are zero, is called the *principal orientation*. For the principal orientation, the three mutually perpendicular planes of zero shear are called *principal planes*, and the normal stresses on these principal planes (planes of zero shear) are called *principal normal stresses*, or just *principal stresses*.

The importance of principal stresses lies in the fact that among them will be the *largest normal stress* that can occur on *any* plane passing through the point, for the given loading. The principal stresses are usually designated σ_1, σ_2, and σ_3. There are always *three* principal stresses, but some of them may be zero.

There is a *second* special orientation of coordinate axes that may be found, for which the shearing stresses on the faces of the rotated volume element reach *extreme values*. Of the three extreme values, one is a maximum, one is a minimum, and the third is a minimax value. This orientation is called the *principal shearing orientation*, and the shearing stresses on these three mutually perpendicular *principal shearing planes* are called *principal shearing stresses*. The principal shearing planes may have nonzero normal stresses acting on them, depending upon the type of loading, but *normal* stresses on *planes of principal shear* are *not* principal stresses.

The importance of principal shearing stresses lies in the fact that among them will be the largest or *maximum shearing stress* that can occur on any plane passing through the point, for the given loading. The principal shearing stresses are usually designated τ_1, τ_2, and τ_3.

Stress Cubic Equation

Depending upon whether its material behaves in a *brittle* or a *ductile* manner, failure at the governing critical point of a machine part is dependent upon the principle normal stresses, the principal shearing stresses, or some combination of these. In any event, it is important for a designer to be able to calculate principal normal stresses and principal shearing stresses for any combination of applied loads. To do this, the *general stress cubic equation*[17] may be employed to find the principal stresses σ_1, σ_2, and σ_3 as a function of the readily calculable components of stress $\sigma_x, \sigma_y, \sigma_z, \tau_{xy}, \tau_{yz}$, and τ_{zx} relative to any selected x-y-z coordinate system. The general stress cubic equation, developed from equilibrium

[15]See Figure 4.1. [16]For example, see ref. 6, Ch. 4. [17]For example, see ref. 6, p. 97.

concepts, is

$$\sigma^3 - \sigma^2(\sigma_x + \sigma_y + \sigma_z) + \sigma(\sigma_x\sigma_y + \sigma_y\sigma_z + \sigma_z\sigma_x - \tau_{xy}^2 - \tau_{yz}^2 - \tau_{zx}^2) - (\sigma_x\sigma_y\sigma_z + 2\tau_{xy}\tau_{yz}\tau_{zx} - \sigma_x\tau_{yz}^2 - \sigma_y\tau_{zx}^2 - \sigma_z\tau_{xy}^2) = 0 \quad (4\text{-}59)$$

Since all normal and shearing stress components are real numbers, all three roots of (4-59) are real; these three roots are the principal normal stresses σ_1, σ_2, and σ_3.

It is also possible to find the *directions* of principal stress vectors (and principal shearing stress vectors) if necessary.[18]

Furthermore, it can be shown that the magnitudes of the principal shearing stresses may be calculated from

$$|\tau_1| = \left|\frac{\sigma_2 - \sigma_3}{2}\right| \quad (4\text{-}60)$$

$$|\tau_2| = \left|\frac{\sigma_3 - \sigma_1}{2}\right| \quad (4\text{-}61)$$

$$|\tau_3| = \left|\frac{\sigma_1 - \sigma_2}{2}\right| \quad (4\text{-}62)$$

To summarize, if loads and geometry are known for a machine part, a designer may identify the critical point, arbitrarily select a convenient *x-y-z* coordinate system, and calculate the resultant six stress components σ_x, σ_y, σ_z, τ_{xy}, τ_{yz}, and τ_{zx}. Equations (4-59) through (4-62) may then be solved to find the principal normal stresses and the principal shearing stresses. This procedure works for all cases; if the case is biaxial, one of the principal stresses will be zero, and if it is uniaxial, two principal stresses will be zero.

Example 4.8 Multiaxial States of Stress; Principal Stresses

A hollow cylindrical 4340 steel member has a 1.0-inch outside diameter, wall thickness of 0.25 inch, and length of 30.0 inches. As shown in Figure E4.8A, the member is simply supported at the ends, and symmetrically loaded at the one-third points by 1000-lb loads. The tubular bar is simultaneously subjected to an axial force of 5000-lb tension and a torsional moment of 3000 in-lb. For the critical point at midspan, determine (a) the principal stresses and (b) the principal shearing stresses.

Solution

a. Since the vertical transverse loads and end supports are symmetric, the vertical components (*z*-components) produce pure bending, with the maximum bending moment extending over the entire middle span. Therefore bending stress (tensile) will be maximum at the lower extreme fiber all along the center span between the 1000-lb transverse loads. The axial tensile force produces uniform axial (*x*-component) tensile stress throughout the wall. The torsional shearing stress produced by torque T will be maximum at the outer fibers throughout the tube. Based on these observations, the midspan critical point A is depicted in Figure E4.8B. The resultant state of stress at critical point A due to bending, tension, and torsional shear is shown in Figure E4.8C.

The stress components σ_x and τ_{xy} may be evaluated as follows:

$$\sigma_x = \frac{Mc}{I} + \frac{F_a}{A} \quad (1)$$

[18] See, for example, ref. 6, Ch. 4.

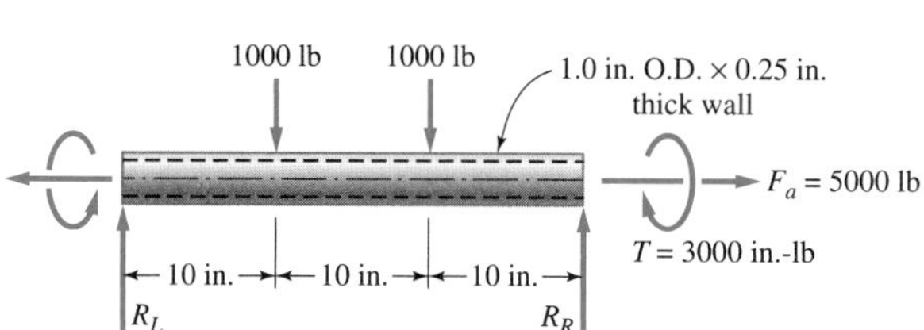
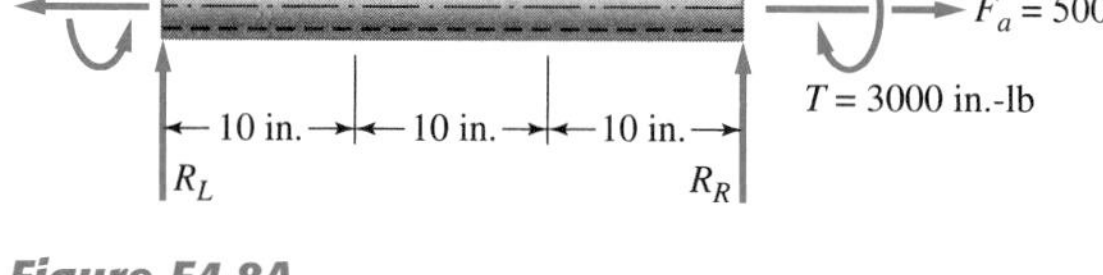

Figure E4.8A
Hollow bar loaded by forces and torques.

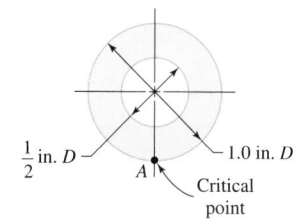

Figure E4.8B
Critical point location.

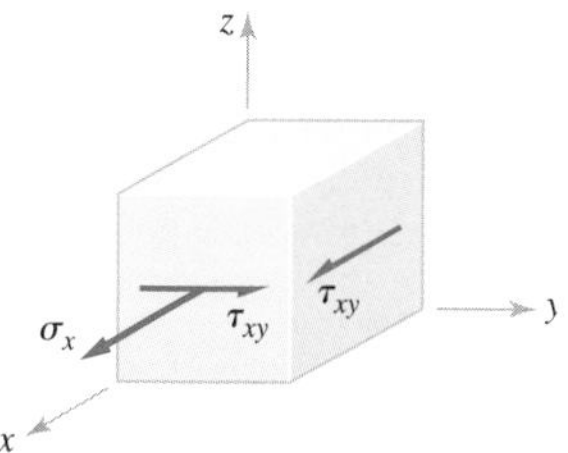

Figure E4.8C
State of stress at critical point A.

and

$$\tau_{xy} = \frac{Ta}{J} \tag{2}$$

From the dimensions of the tube

$$I = \frac{\pi(d_o^4 - d_i^4)}{64} = \frac{\pi(1.0^4 - 0.5^4)}{64} = 0.046 \text{ in}^4 \tag{3}$$

and

$$J = 21 = 0.092 \text{ in}^4 \tag{4}$$

Therefore, from (1)

$$\sigma_x = \frac{[(1000)(10)]\left(\frac{1.0}{2}\right)}{0.046} + \frac{5000}{0.589} = 117{,}200 \text{ psi} \tag{5}$$

and from (2)

$$\tau_{xy} = \frac{(3000)\left(\frac{1.0}{2}\right)}{0.092} = 16{,}300 \text{ psi} \tag{6}$$

For the state of stress at critical point A depicted in Figure E4.8C, the stress cubic (4-59) reduces to

$$\sigma^3 - \sigma^2\sigma_x + \sigma(-\tau_{xy}^2) = 0 \tag{7}$$

giving three roots

$$\sigma = \frac{\sigma_x}{2} \pm \sqrt{\left(\frac{\sigma_x}{2}\right)^2 + \tau_{xy}^2} \tag{8}$$

and

$$\sigma = 0 \tag{9}$$

These may be rewritten as

$$\sigma_1 = \frac{\sigma_x}{2} + \sqrt{\left(\frac{\sigma_x}{2}\right)^2 + \tau_{xy}^2} \tag{10}$$

$$\sigma_2 = \frac{\sigma_x}{2} - \sqrt{\left(\frac{\sigma_x}{2}\right)^2 + \tau_{xy}^2} \tag{11}$$

$$\sigma_3 = 0 \tag{12}$$

Substituting from (5) and (6), the principal stresses become

Example 4.8 Continues

$$\sigma_1 = \frac{117{,}200}{2} + \sqrt{\left(\frac{117{,}200}{2}\right)^2 + (16{,}300)^2} \tag{13}$$
$$= 58{,}600 + 60{,}825 = 119{,}425 \text{ psi}$$

$$\sigma_2 = 58{,}600 - 60{,}825 = -2225 \text{ psi} \tag{14}$$

$$\sigma_3 = 0 \tag{15}$$

b. From (4-60), (4-61), and (4-62), using results from (13), (14), and (15), the principal shearing stresses become

$$|\tau_1| = \left|\frac{-2225 - 0}{2}\right| = 1113 \text{ psi} \tag{16}$$

$$|\tau_2| = \left|\frac{0 - 119{,}425}{2}\right| = 59{,}713 \text{ psi} \tag{17}$$

$$|\tau_3| = \left|\frac{119{,}425 - (-2225)}{2}\right| = 60{,}825 \text{ psi} \tag{18}$$

Mohr's Circle Analogy for Stress

For biaxial states of stress the stress cubic equation (4-59) degenerates to the form

$$\sigma^3 - \sigma^2(\sigma_x + \sigma_y) + \sigma(\sigma_x \sigma_y - \tau_{xy}^2) = 0 \tag{4-63}$$

if stress components lie only in the xy plane. If stress components lie only in the yz plane or the zx plane, the stress cubic yields the same formal expression with solely yz subscripts or solely zx subscripts, respectively. The principal stress solutions for (4-63) are (see Example 4.8)

$$\sigma_1 = \frac{\sigma_x + \sigma_y}{2} + \sqrt{\left(\frac{\sigma_x - \sigma_y}{2}\right)^2 + \tau_{xy}^2} \tag{4-64}$$

$$\sigma_2 = \frac{\sigma_x + \sigma_y}{2} - \sqrt{\left(\frac{\sigma_x - \sigma_y}{2}\right)^2 + \tau_{xy}^2} \tag{4-65}$$

$$\sigma_3 = 0 \tag{4-66}$$

Examination of (4-64) and (4-65) suggests that these equations are formally the same as the equation of a circle plotted on the $\sigma-\tau$ plane, since the equation of such a circle on an xy plane has the form

$$(x - h)^2 + (y - k)^2 = R^2 \tag{4-67}$$

where h and k are the x-y coordinates of the center of the circle, and R is the radius. An *analogy* (refer to the discussion of the "membrane analogy" in 4.4) was developed by *Mohr*[19] in 1882 in which he successfully postulated that a circle plotted on the $\sigma-\tau$ plane could be used to represent *any* biaxial state of stress because the governing equations are formally the same for a circle as for a biaxial state of stress. Such plots have come to be known as *Mohr's circles* for stress. For example, a biaxial state of stress in the xy plane (see principal stress solutions in (4-64), (4-65), and (4-66)) may be represented by a circle[20] with center on the σ-axis of a $\sigma-\tau$ plot if in (4-67) x and y are replaced by σ and τ respectively, and

[19]See ref. 7.

[20]For example, see ref. 1, Ch. 2, or ref. 8, Ch. 10.

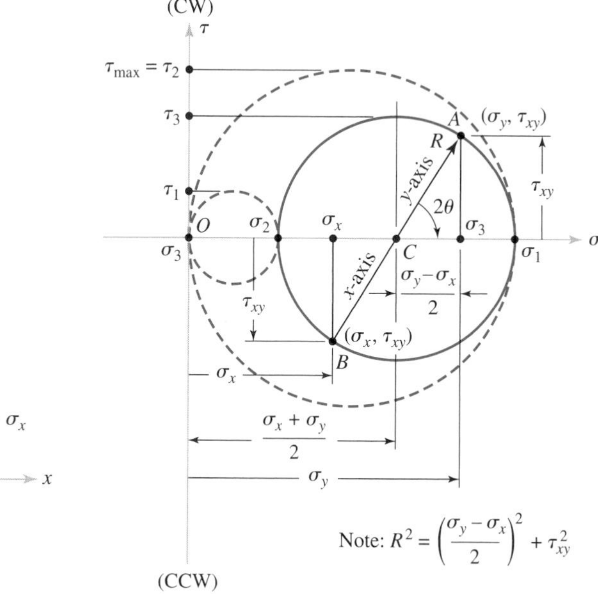

(*a*) Biaxial element at critical point.

(*b*) Mohr's circle for *xy* biaxial state of stress shown in (a); note that $\sigma_3 = 0$ for this biaxial case.

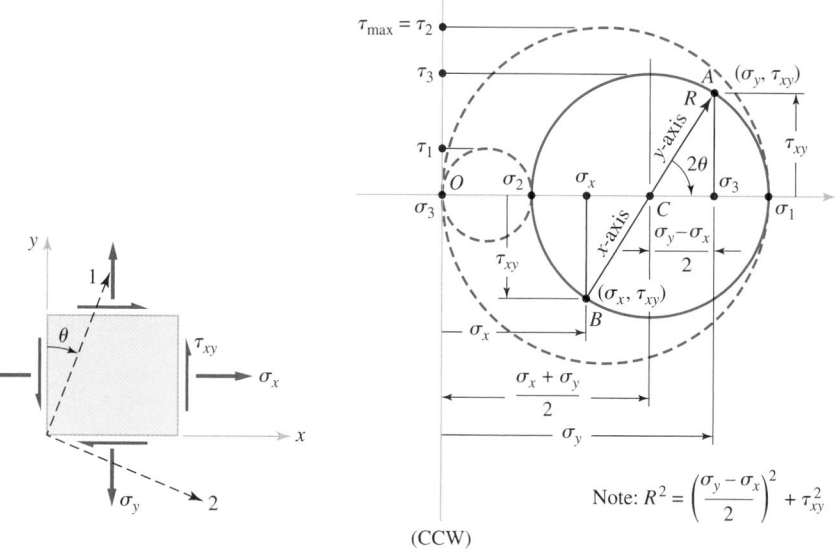

Figure 4.12
Mohr's circle for a biaxial state of stress in the *x-y* plane.

$$h = \frac{\sigma_x + \sigma_y}{2} \tag{4-68}$$

$$k = 0 \tag{4-69}$$

$$R = \sqrt{\left(\frac{\sigma_x - \sigma_y}{2}\right)^2 + \tau_{xy}^2} \tag{4-70}$$

Figure 4.12 shows such a plot. To successfuly utilize Mohr's circle as a stress analysis tool the following sign conventions should be adopted:

1. Normal stresses should be plotted as positive if tensile, and negative if compressive.
2. Shearing stresses should be plotted as positive if they produce a clockwise (CW) couple, and negative if they produce a counterclockwise (CCW) couple. (This convention is used *only* for Mohr's circle applications.)

The plotting procedure, illustrated in Figure 4.12, is to plot point A (tensile σ_y and CW τ_{xy} couple), then point B (tensile σ_x and CCW τ_{xy} couple), then pass a circle through A and B. The location of center C and magnitude of radius R may be established geometrically as shown, and principal stresses σ_1 and σ_2 pinpointed where the circle intersects the σ-axis (shear is 0 on the σ-axis). The principal shearing stress τ_3 may also be found as shown (see (4-62)). The third principal stress, $\sigma_3 = 0$, should not be ignored; plot σ_3 at the origin of the $\sigma - \tau$ plane and then construct two additional Mohr's circles, one through $\sigma_2 - \sigma_3$ and another through $\sigma_1 - \sigma_3$. From these circles τ_1 and τ_2 may be found as shown. If the maximum shearing stress for this state of stress is needed for failure prevention calculations, it will be the largest of τ_1, τ_2, and τ_3; for the case illustrated in Figure 4.12 this is $\tau_{max} = \tau_2$. Note specifically that τ_{max} is *not* τ_3, which is based on the initially plotted *solid* circle. If $\sigma_3 = 0$ had not been plotted, the *dashed* circles would have been missed, and, therefore, the true maximum shearing stress would have been missed.

It is also possible to find the orientation of the principal axes from the Mohr's circle construction, using the expression

$$2\theta = \tan^{-1}\frac{2\tau_{xy}}{\sigma_y - \sigma_x} \tag{4-71}$$

where the double angle 2θ is shown in Figure 4.12(b), measured *from* the ray *CA to* the principal stress ray σ_1. This corresponds to the rotation angle θ in Figure 4.12(a) measured in the same direction as 2θ, *from* the *y*-axis *to* the principal axis 1. A consequence of this procedure is that principal axis 2 will be perpendicular to axis 1 in the *x-y* plane, and principal axis 3 will be mutually perpendicular to axes 1 and 2.

Example 4.9 Mohr's Circle for Stress; Principal Stresses

Using the Mohr's circle analogy, solve the problem posed in Example 4.8 to find

a. The principal stresses

b. The principal shearing stresses

Solution

Reviewing the analysis of Example 4.8, the state of stress at the critical point *A* is shown in Figure E4.9A, repeated from Figure E4.8C.

Also, the magnitudes of σ_x, σ_y, and τ_{xy} were determined in Example 4.6 to be

$$\sigma_x = 117{,}200 \text{ psi} \tag{1}$$

$$\sigma_y = 0 \tag{2}$$

$$\tau_{xy} = 16{,}300 \text{ psi} \tag{3}$$

a. Using the Mohr's circle analogy to find principal stresses, the *xy* plane of Figure E4.9A may be redrawn as shown in Figure E4.9B. Mohr's circle may next be constructed as shown in Figure E4.9C by plotting points *A* and *B* as shown, passing a circle through them and finding principal stresses σ_1 and σ_2 semigraphically. The third principal stress, $\sigma_3 = 0$, is plotted at the origin.

From Figure E4.9C, the following values may be found graphically

$$\sigma_1 = \overline{C} + \overline{R} = \frac{117{,}200}{2} + 60{,}825 = 119{,}425 \text{ psi} \tag{4}$$

$$\sigma_2 = \overline{C} - \overline{R} = \frac{117{,}200}{2} - 60{,}825 = -2225 \text{ psi} \tag{5}$$

and for this *biaxial* state of stress

$$\sigma_3 = 0 \tag{6}$$

These values are the same as those found in Example 4.8 by solving the stress cubic equation.

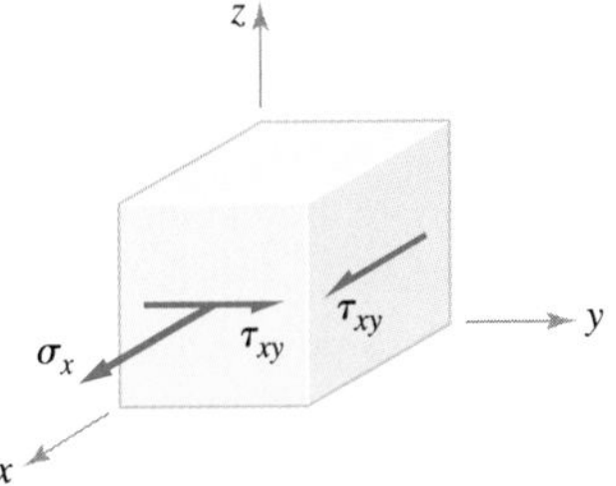

Figure E4.9A
State of stress at critical point *A*.

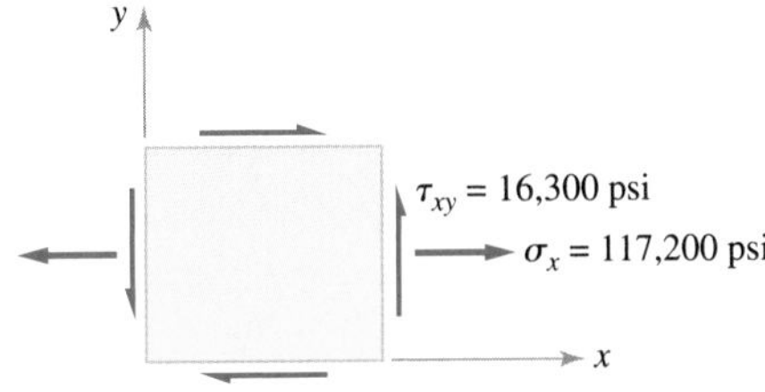

Figure E4.9B
Biaxial *xy* element from Figure E4.9A.

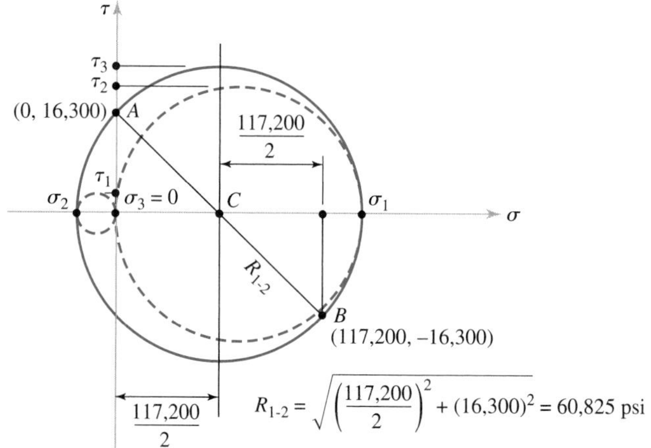

Figure E4.9C
Mohr's circle constructed for biaxial element shown in Figure E4.9B.

b. To find the principal shearing stresses, the remaining two (dashed) Mohr's circles are constructed, and the principal shearing stresses are found to be, from the radii of the three circles,

$$|\tau_1| = R_{2-3} = \frac{2225}{2} = 1113 \text{ psi} \tag{7}$$

$$|\tau_2| = R_{1-3} = \frac{119{,}425}{2} = 59{,}713 \text{ psi} \tag{8}$$

$$|\tau_3| = R_{1-2} = 60{,}825 \text{ psi} \tag{9}$$

These values are the same as those shown in Example 4.8. From these values it may be noted that

$$\tau_{max} = \tau_3 = 60{,}825 \text{ psi} \tag{10}$$

Strain Cubic Equation and Principal Strains

Just as stress is an important measure of loading severity, strain may also be an important parameter in assessing failure potential in a machine part. *Strain* is the term used to define the intensity and direction of the deformation at a given critical point, with respect to a specified plane or set of planes passing through the critical point. Analogous to state of stress, the state of strain at a point may be fully defined by six components of strain relative to any selected x-y-z coordinate system: three normal strain components and three shearing strain components. The normal strain components are usually designated ε_x, ε_y, and ε_z, and the shearing strain components γ_{xy}, γ_{yz}, and γ_{zx}. Just as the state of stress may be completely defined in terms of three principal stresses and their directions, the state of strain may be completely defined in terms of three principal strains and their directions. The principal strains may be determined from a general strain cubic equation analogous to the general stress cubic equation (4-59). The strain cubic equation is

$$\begin{aligned}\varepsilon^3 - \varepsilon^2[\varepsilon_x + \varepsilon_y + \varepsilon_z] + \varepsilon\left[\varepsilon_x\varepsilon_y + \varepsilon_y\varepsilon_z + \varepsilon_z\varepsilon_x - \frac{1}{4}(\gamma_{xy}^2 + \gamma_{yz}^2 + \gamma_{zx}^2)\right] \\ - \left[\varepsilon_x\varepsilon_y\varepsilon_z + \frac{1}{4}(\gamma_{xy}\gamma_{yz}\gamma_{zx} - \varepsilon_x\gamma_{yz}^2 - \varepsilon_y\gamma_{zx}^2 - \varepsilon_z\gamma_{xy}^2)\right] = 0\end{aligned} \tag{4-72}$$

This equation is formally the same as the stress cubic equation (4-59) except that the normal strains, and the shearing strains divided by two, replace the normal stresses and shearing stresses. The three solutions of (4-72) are the three principal normal strains ε_1, ε_2, and ε_3.

Mohr's Circle Analogy for Strain

In some circumstances, a designer may be interested in finding stresses or strains at particular points in an existing machine part. Such an interest may stem from the need to improve failure resistance, to verify a calculation procedure, or some other requirement. In these cases it is common to *bond* a strain gage or a strain rosette to the surface at one or more critical points.[21] Using the strain gage data to evaluate biaxial solutions of the strain cubic equation (4-72), the principal strains and their directions may be found. Use of a Mohr's strain circle to determine the principal strains is common.

Mohr's strain circles may be constructed using the techniques just described for Mohr's stress circles, except that normal and shearing strains are plotted on the cartesian axes ε and $\gamma/2$ rather than σ and τ. To utilize the resulting Mohr's strain circle solutions for principal strains, with the objective of estimating stresses at the critical point, relationships between stress and strain must be known.

Elastic Stress-Strain Relationships (Hooke's Law)

For a broad class of engineering materials it has been experimentally well established that a linear relationship exists between stress and strain so long as the material is not loaded beyond its elastic range. These linear elastic relationships were first presented in the seventeenth century by Robert Hooke, and have come to be known as the *Hooke's Law relationships*. They are

$$\varepsilon_x = \frac{1}{E}[\sigma_x - \nu(\sigma_y + \sigma_z)] \tag{4-73}$$

$$\varepsilon_y = \frac{1}{E}[\sigma_y - \nu(\sigma_z + \sigma_x)] \tag{4-74}$$

$$\varepsilon_z = \frac{1}{E}[\sigma_z - \nu(\sigma_x + \sigma_y)] \tag{4-75}$$

and

$$\gamma_{xy} = \frac{\tau_{xy}}{G} \tag{4-76}$$

$$\gamma_{yz} = \frac{\tau_{yz}}{G} \tag{4-77}$$

$$\gamma_{zx} = \frac{\tau_{zx}}{G} \tag{4-78}$$

where the elastic constants are

E = Young's modulus of elasticity
ν = Poisson's ratio
G = shear modulus of elasticity

[21] A strain gage is a device designed to measure the displacement between two points some known distance apart. They may be mechanical, optical, electroresistive, piezoresistive, capacitive, inductive, or acoustical in nature. The electrical-resistance strain gage is often employed on the free surface of a machine part or specimen to find stresses there. It is usual to use an arrangement of three gages to measure three strains at known angles to each other, from which principal surface strains and directions may be found. These three-gage arrays are called *rosettes*. To convert strains to stresses, Young's modulus of elasticity and Poisson's ratio must also be known.

Numerical values of these elastic constants are given in Table 3.9 for several engineering materials.

It should be noted that the Hooke's Law relationships are valid for any arbitrarily selected *x-y-z* cartesian coordinate system orientation, including the principal *1-2-3* orientation. The Hooke's Law relationships may also be inverted to give

$$\sigma_x = \frac{E}{(1 - \nu - 2\nu^2)}[(1 - \nu)\varepsilon_x + \nu(\varepsilon_y + \varepsilon_z)] \tag{4-79}$$

$$\sigma_y = \frac{E}{(1 - \nu - 2\nu^2)}[(1 - \nu)\varepsilon_y + \nu(\varepsilon_z + \varepsilon_x)] \tag{4-80}$$

$$\sigma_z = \frac{E}{(1 - \nu - 2\nu^2)}[(1 - \nu)\varepsilon_z + \nu(\varepsilon_x + \varepsilon_y)] \tag{4-81}$$

4.6 Combined Stress Theories of Failure

Predicting failure, or establishing a combination of material and geometry that will avert failure, is a relatively simple matter if the machine part is subjected to a *static uniaxial* state of stress. It is necessary only to have available the simple uniaxial stress-strain curve for the material of interest, which can be readily obtained from one or a few simple tension and compression experiments. For example, if yielding has been established as the governing failure mode for a *uniaxially* stressed machine part, failure of the part would be predicted when its maximum normal stress equals or exceeds the *uniaxial* yield strength of the material.

If a machine part is subjected to a *multiaxial* state of stress, the accurate prediction of failure becomes far more difficult. No longer can one accurately predict yielding, for example, when the maximum normal stress exceeds the *uniaxial* yield strength because the other two principal normal stress components at the critical point may also influence yielding behavior. Further, *multiaxial* yield strengths are generally unavailable because of the time and expense required to determine them experimentally. Therefore, when one desires to predict failure, or to pick a combination of material and geometry to avert failure, when the machine part is subjected to a multiaxial state of stress, it is usual to utilize an experimentaly validated theory that relates failure in the multiaxial state of stress to failure by the same mode in a simple uniaxial stress test. All such failure prediction theories are based on well-chosen *loading severity parameters* such as stress, strain, or strain energy density. Loading severity parameters must be *readily calculable* in the multiaxial state of stress and *readily measurable* in a simple uniaxial stress test. These theories, called *combined stress theories of failure*, all share a common postulate, namely, that *failure is predicted to occur when the maximum value of the selected loading severity parameter in the multiaxial state of stress becomes equal to or exceeds the value of the same loading severity parameter that produces failure in a simple uniaxial stress test using a specimen of the same material.*

Many combined stress failure theories have been proposed, but three have found wide acceptance because of their relatively good agreement with experimental results and reasonable simplicity in application. They are

1. Maximum normal stress theory
2. Maximum shearing stress theory
3. Distortion energy theory

Maximum Normal Stress Theory (Rankine's Theory)

In words, the maximum normal stress theory may be expressed as:

> Failure is predicted to occur in the multiaxial state of stress when the maximum principal normal stress becomes equal to or exceeds the maximum normal stress at the time of failure in a simple uniaxial stress test using a specimen of the same material. (4-82)

For a multiaxial state of stress, the maximum principal normal stress is the largest of the three roots to the stress cubic equation (4-59), that is, the largest of σ_1, σ_2, and σ_3. The maximum normal stress at the time of failure is equal to the uniaxial strength of the material corresponding to the governing failure mode. It should be noted that for some materials the failure strength under tensile loading may be different from the failure strength under compressive loading.

With these factors in mind, the word statement (4-82) may be expressed mathematically as *failure is predicted by the maximum normal stress theory to occur if (FIPTOI)*

$$\begin{aligned} \sigma_1 &\geq \sigma_{fail\text{-}t} \\ \sigma_2 &\geq \sigma_{fail\text{-}t} \\ \sigma_3 &\geq \sigma_{fail\text{-}t} \end{aligned} \tag{4-83}$$

or if

$$\begin{aligned} \sigma_1 &\leq \sigma_{fail\text{-}c} \\ \sigma_2 &\leq \sigma_{fail\text{-}c} \\ \sigma_3 &\leq \sigma_{fail\text{-}c} \end{aligned} \tag{4-84}$$

where $\sigma_{fail\text{-}t}$ is the uniaxial *tensile* (+) failure strength of the material and $\sigma_{fail\text{-}c}$ is the uniaxial *compressive* (−) failure strength of the material, corresponding to the governing failure mode (usually yielding or ultimate rupture if loading is static).

The maximum normal stress theory provides good results for brittle materials, as illustrated in Figure 4.13, but should not be used for ductile materials.

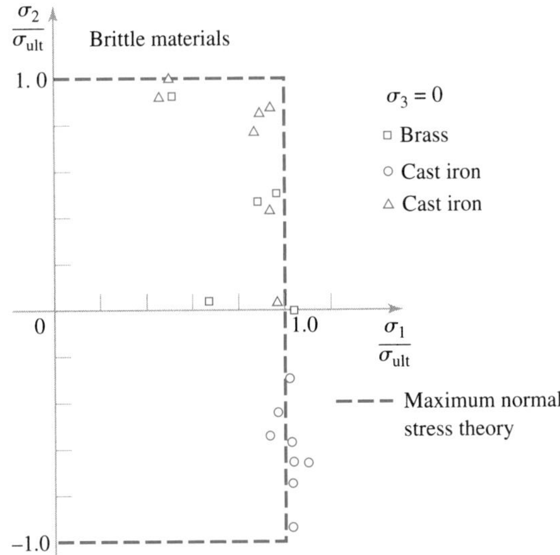

Figure 4.13
Comparison of biaxial ultimate strength data for brittle materials with the *maximum normal stress theory*.

Maximum Shearing Stress Theory (Tresca–Guest Theory)

In words, the maximum shearing stress theory may be expressed as

> Failure is predicted to occur in the multiaxial state of stress when the maximum shearing stress magnitude becomes equal to or exceeds the maximum shearing stress magnitude at the time of failure in a simple uniaxial stress test using a specimen of the same material. (4-85)

For a multiaxial state of stress the maximum shearing stress magnitude is the largest of the three principal shearing stresses τ_1, τ_2, and τ_3, given by (4-60), (4-61), and (4-62). For a *uniaxial* stress test, the only nonzero normal stress component is a principal stress component in the direction of the applied force. From (4-60), (4-61), and (4-62), in a uniaxial stress test at failure, two of the principal normal stresses are zero and the third is set equal to σ_{fail}, giving

$$\tau_{fail} = \frac{\sigma_{fail}}{2} \tag{4-86}$$

With these factors in mind, word statement (4-85) may be expressed mathematically as *failure is predicted by the maximum shearing stress theory to occur if (FIPTOI)*

$$\begin{aligned} |\tau_1| &\geq |\tau_{fail}| \\ |\tau_2| &\geq |\tau_{fail}| \\ |\tau_3| &\geq |\tau_{fail}| \end{aligned} \tag{4-87}$$

where τ_{fail} is the largest principal shearing stress at the time of failure in a *uniaxial* stress test, as given by (4-86).

The maximum shearing stress theory provides good results for ductile materials, as illustrated in Figure 4.14, but should not be used for brittle materials.

Distortion Energy Theory (Huber–von Mises–Hencky Theory)

In words, the distortion energy theory, sometimes called the von Mises theory, may be expressed as

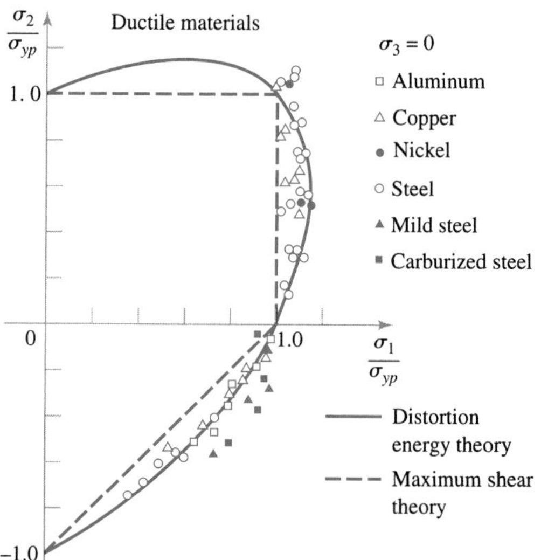

Figure 4.14
Comparison of biaxial yield strength data with the *maximum shearing stress theory* and the *distortion energy theory*.

> Failure is predicted to occur in the multiaxial state of stress when the distortion energy per unit volume becomes equal to or exceeds the distortion energy per unit volume at the time of failure in a simple uniaxial stress test using a specimen of the same material. (4-88)

The distortion energy theory, developed as an improvement over an earlier "total strain energy theory," is based on the postulate that the total strain energy U_T stored in a volume of stressed material may be divided into two parts: the energy associated solely with change in volume, U_v, termed *dilatation* energy, and the energy associated solely with change in shape, U_d, termed *distortion* energy. It was further postulated that failure, particularly under conditions of ductile behavior, is related *only* to the *distortion* energy. Thus

$$U_T = U_v + U_d \tag{4-89}$$

Dividing by the volume in each term, the distortion energy *per unit volume* may be expressed as

$$u_d = u_T - u_v \tag{4-90}$$

An expression for total strain energy per unit volume, u_T may be found[22] by calculating the work done in a triaxial state of stress by forces associated with σ_1, σ_2, and σ_3, acting over their respective areas, to induce the strains ε_1, ε_2, and ε_3, and their corresponding displacements.

Employing Hooke's Law, the resulting expression for total strain energy per unit volume becomes

$$u_T = \frac{1}{2E}[\sigma_1^2 + \sigma_2^2 + \sigma_3^2 - 2\nu(\sigma_1\sigma_2 + \sigma_2\sigma_3 + \sigma_3\sigma_1)] \tag{4-91}$$

Likewise, the dilatation (volume changing) energy per unit volume may be found[23] as

$$u_v = \frac{3(1 - 2\nu)}{2E}\left[\frac{\sigma_1 + \sigma_2 + \sigma_3}{3}\right]^2 \tag{4-92}$$

Then substituting (4-91) and (4-92) into (4-90), the distortion energy per unit volume is found to be

$$u_d = \frac{1}{2}\left[\frac{1 + \nu}{3E}\right][(\sigma_1 - \sigma_2)^2 + (\sigma_2 - \sigma_3)^2 + (\sigma_3 - \sigma_1)^2] \tag{4-93}$$

To find the distortion energy per unit volume at the time of failure, $u_{d\text{-}fail}$, (4-93) is evaluated under *uniaxial failure* conditions, when two of the principal stresses are equal to zero and the third one is equal to the uniaxial failure strength σ_{fail}. Thus

$$u_{d\text{-}fail} = \left[\frac{1 + \nu}{3E}\right]\sigma_{fail}^2 \tag{4-94}$$

With (4-93) and (4-94) at hand, word statement (4-88) may be expressed mathematically as *failure is predicted by the distortion energy theory to occur if (FIPTOI)*

$$\frac{1}{2}[(\sigma_1 - \sigma_2)^2 + (\sigma_2 - \sigma_3)^2 + (\sigma_3 - \sigma_1)^2] \geq \sigma_{fail}^2 \tag{4-95}$$

The distortion energy theory provides very good results for ductile materials, as illustrated in Figure 4.14, but generally should not be used for brittle materials.

[22] See equations (4-52), (4-59), (4-73), (4-74), and (4-75).

[23] For example, see ref. 6, p. 154.

The left side of (4-95) is sometimes defined as the square of the *von Mises stress*, the *effective stress*, or the *equivalent uniaxial stress*, σ_{eq}, giving the expression

$$\sigma_{eq} = \sqrt{\frac{1}{2}[(\sigma_1 - \sigma_2)^2 + (\sigma_2 - \sigma_3)^2 + (\sigma_3 - \sigma_1)^2]} \qquad (4\text{-}96)$$

Failure Theory Selection

Evaluation of the three failure theories just discussed in light of experimental evidence leads to the following observations:

1. For isotopic materials that fail by brittle fracture, the maximum normal stress theory is the best theory to use.
2. For isotropic materials that fail by yielding or ductile rupture, the distortion energy theory is the best theory to use.
3. For isotropic materials that fail by yielding or ductile rupture, the maximum shearing stress theory is almost as good as the distortion energy theory.
4. As a rule of thumb,[24] the maximum normal stress theory would be used for isotropic brittle materials (materials that exhibit a ductility of less than 5 percent elongation in 2 inches) and either the distortion energy theory or maximum shearing stress theory would be used for isotropic ductile materials (materials that exhibit a ductility of 5 percent or more in a 2-inch gage length). Where possible, a fracture mechanics analysis should be performed.

Example 4.10 Yielding Failure Prediction Under Static Multiaxial State of Stress

An aircraft wing flap actuator housing is made of cast magnesium alloy AZ63A-T4 (S_{yp} = 14,000 psi, S_u = 40,000 psi, e = 12 percent in 2 inches). At the suspected critical point it has been calculated that the state of stress is as shown in Figure E4.10. Would you predict failure of the part by yielding?

Solution

For the state of stress depicted in Figure E4.10, the stress cubic equation (4-59) becomes

$$\sigma^3 - \sigma^2(\sigma_x + \sigma_y + \sigma_z) + \sigma(\sigma_x\sigma_y + \sigma_y\sigma_z + \sigma_z\sigma_x - \tau_{yz}^2) - (\sigma_x\sigma_y\sigma_z - \sigma_x\tau_{yz}^2) = 0 \qquad (1)$$

or, substituting numerical values (in ksi),

$$\sigma^3 - \sigma^2(20 + 10 + 15) + \sigma[(20(10) + 10(15) + 15(20) - (5)^2] - [20(10)(15) - 20(5)^2] = 0 \qquad (2)$$

or

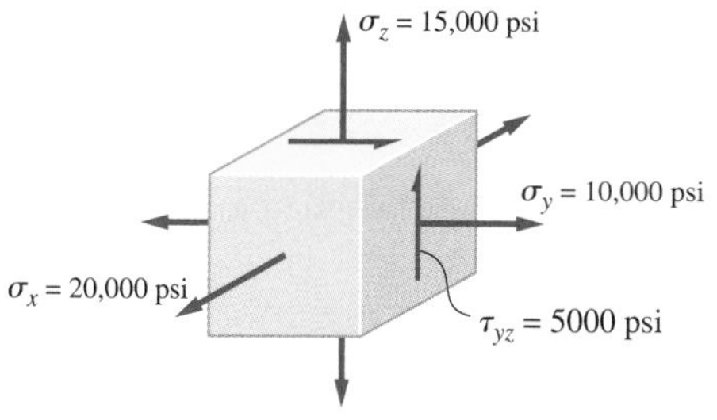

Figure E4.10
State of stress at suspected critical point.

[24]The 5 percent boundary between "brittle" and "ductile" behavior is an arbitrary but widely used rule of thumb.

Example 4.10 Continues

$$\sigma^3 - 45\sigma^2 + 625\sigma - 2500 = 0 \tag{3}$$

Since shearing stresses are zero on the x-plane in Figure E4.10, it is, by definition, a principal plane. Thus

$$\sigma_x = \sigma_1 = 20 \text{ ksi} \tag{4}$$

is one of the three solutions of (3). Next, dividing (3) by $(\sigma - 20)$ gives

$$\sigma^2 - 25\sigma + 125 = 0 \tag{5}$$

whence

$$\sigma = \frac{25 \pm \sqrt{(25)^2 - 4(125)}}{2} = 12.5 \pm 5.6 \tag{6}$$

giving the solutions

$$\sigma_2 = 18.1 \text{ ksi} \tag{7}$$

$$\sigma_3 = 6.9 \text{ ksi} \tag{8}$$

Since the elongation in 2 inches is given as e = 12 percent ($>$ 5 percent), the material may be regarded as ductile, and the distortion energy theory of failure, given in (4-95), is the best theory to use. Thus FIPTOI

$$\frac{1}{2}[(20 - 18.1)^2 + (18.1 - 6.9)^2 + (6.9 - 20)^2] \geq (14^2) \tag{9}$$

or, FIPTOI

$$150 \geq 196 \tag{10}$$

Since this condition is not met, *failure is not predicted* at the specified critical point.

4.7 Multiaxial States of Cyclic Stress and Multiaxial Fatigue Failure Theories

An extensive discussion of the prediction and prevention of fatigue failure and fatigue life under *uniaxial* states of cyclic stress was presented in 2.6. As mentioned in that discussion, most real design situations involve fluctuating loads that produce *multiaxial* states of cyclic stress. *A consensus has not yet been reached on the best approach to prediction of failure under multiaxial states of cyclic stress.* However, one technique used for ductile materials subjected to fluctuating multiaxial stresses is to utilize a cyclic adaptation of the *equivalent stress* expression of (4-96). For brittle materials subjected to fluctuating multiaxial stresses, the principal normal stress expressions are utilized.

An *equivalent uniaxial alternating stress* expression $\sigma_{eq\text{-}a}$ for ductile materials, based on the equivalent stress equation of (4-96), becomes

$$\sigma_{eq\text{-}a} = \sqrt{\frac{1}{2}[(\sigma_{1a} - \sigma_{2a})^2 + (\sigma_{2a} - \sigma_{3a})^2 + (\sigma_{3a} - \sigma_{1a})^2]} \tag{4-97}$$

An *equivalent uniaxial mean stress* expression $(\sigma_{eq\text{-}m})$ for ductile materials, based on the equivalent stress expression of (4-96), is

$$\sigma_{eq\text{-}m} = \sqrt{\frac{1}{2}[(\sigma_{1m} - \sigma_{2m})^2 + (\sigma_{2m} - \sigma_{3m})^2 + (\sigma_{3m} - \sigma_{1m})^2]} \tag{4-98}$$

The equations (4-97) and (4-98) may be substituted into any of the expressions discussed in 2.6 as *equivalents* for their uniaxial counterparts, σ_a and σ_m. For example, in using (2-42), if the loading conditions produce a *multiaxial* state of cyclic stress, the equations become

$$\sigma_{eq\text{-}CR} = \frac{\sigma_{eq\text{-}a}}{1 - (\sigma_{eq\text{-}m}/S_u)} \qquad \text{for } \sigma_{eq\text{-}m} \geq 0 \quad \text{and} \quad \sigma_{max} \leq S_{yp} \tag{4-99}$$

or

$$\sigma_{eq\text{-}CR} = S_{yp} \qquad \text{for } \sigma_{eq\text{-}m} \geq 0 \quad \text{and} \quad \sigma_{max} \geq S_{yp} \tag{4-100}$$

where

$$\sigma_{max} = \sigma_{eq\text{-}a} + \sigma_{eq\text{-}m} \tag{4-101}$$

For brittle materials subjected to multiaxial states of cyclic stress, if the convention $\sigma_1 \geq \sigma_2 \geq \sigma_3$ is adopted, the expressions σ_{1a} and σ_{1m} may be substituted into any of the uniaxial procedures discussed in 2.6. With brittle materials, for example, equations (2-42) would become, under multiaxial states of cyclic stress,

$$\sigma_{eq\text{-}CR} = \frac{\sigma_{1a}}{1 - (\sigma_{1m}/S_u)} \qquad \text{for } \sigma_{1m} \geq 0 \quad \text{and} \quad \sigma_{max} \leq S_u \tag{4-102}$$

Example 4.11 Fatigue Failure Prediction Under Fluctuating Multiaxial Stresses

A power transmission shaft of solid cylindrical shape is to be made of hot rolled 1020 steel with S_u = 65,000 psi, S_{yp} = 43,000 psi, e = 36 percent elongation in 2 inches, and fatigue properties as shown for 1020 steel in Figure 2.19. The shaft is to transmit 85 horsepower at a rotational speed of n = 1800 rpm, with no fluctuations in torque or speed. At the critical location, midspan between bearings, the rotating shaft is also subjected to a pure bending moment of 2000 in-lb, fixed in a vertical plane by virtue of a system of symmetrical external forces on the shaft. If the shaft diameter is 1.0 inch, what operating life would be predicted before fatigue failure occurs?

Solution

From the horsepower equation (4-32), the steady shaft torque is

$$T = \frac{63{,}025(hp)}{n} = \frac{63{,}025(85)}{1800} = 2976 \text{ in-lb (steady)} \tag{1}$$

and from the problem statement, the bending moment is completely reversed (due to shaft rotation), giving

$$M = 2000 \text{ in-lb (completely reversed)} \tag{2}$$

Since the maximum shearing stress due to torque T occurs at the surface, and cyclic bending stresses range from maximum to minimum and back to maximum at the surface with each rotation, all midspan surface points are equally critical.

A typical volume element at a midspan critical point is shown in Figure E4.11A. The steady torsional shearing stress may be calculated from (4-30) as

$$\tau_{xy} = \frac{Ta}{J} = \frac{16T}{\pi d^3} \tag{3}$$

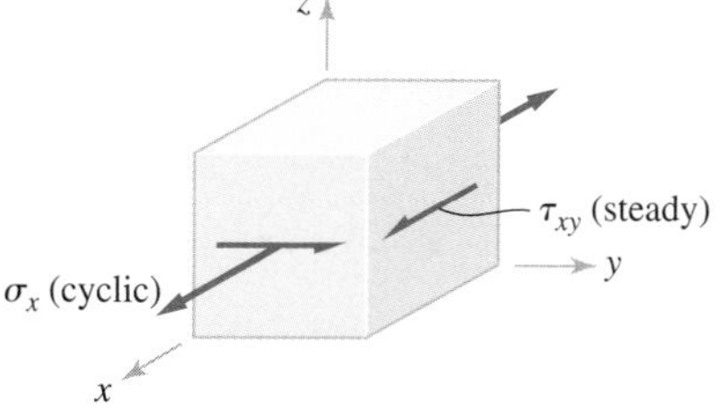

Figure E4.11A
State of stress at a typical midspan critical point.

Example 4.11 Continues

and cyclic bending stress from (4-5) as

$$\sigma_x = \frac{Mc}{I} = \frac{32M}{\pi d^3} \tag{4}$$

The stress cubic equation (4-59) for this state of stress reduces to

$$\sigma^3 - \sigma^2(\sigma_x) + \sigma(-\tau_{xy}^2) = 0 \tag{5}$$

or

$$\sigma(\sigma^2 - \sigma\sigma_x - \tau_{xy}^2) = 0 \tag{6}$$

The three roots (principal stresses) are

$$\sigma_1 = \frac{\sigma_x}{2} + \sqrt{\left(\frac{\sigma_x}{2}\right)^2 + \tau_{xy}^2} \tag{7}$$

$$\sigma_2 = 0 \tag{8}$$

$$\sigma_3 = \frac{\sigma_x}{2} - \sqrt{\left(\frac{\sigma_x}{2}\right)^2 + \tau_{xy}^2} \tag{9}$$

Substituting from (3) and (4) these become

$$\sigma_1 = \frac{16}{\pi d^3}\left[M + \sqrt{M^2 + T^2}\right] \tag{10}$$

$$\sigma_2 = 0 \tag{11}$$

$$\sigma_3 = \frac{16}{\pi d^3}\left[M - \sqrt{M^2 + T^2}\right] \tag{12}$$

Noting that $T_{max} = T_{min} = T_m = T = 2976$ in-lb, $M_{max} = +2000$ in-lb, $M_{min} = -2000$ in-lb, and $M_m = 0$, the torque-time and moment-time plots of Figure E4.11B may be sketched. From these plots, and known numerical values, (10), (11), and (12) may be used to write

$$\sigma_{1\text{-}max} = \frac{16}{\pi}\left[2000 + \sqrt{(2000)^2 + (2976)^2}\right] = 28{,}449 \text{ psi} \tag{13}$$

$$\sigma_{1\text{-}min} = \frac{16}{\pi}\left[-2000 + \sqrt{(-2000)^2 + (2976)^2}\right] = 8075 \text{ psi} \tag{14}$$

$$\sigma_{1m} = \frac{\sigma_{1max} + \sigma_{1min}}{2} = 18{,}262 \text{ psi} \tag{15}$$

$$\sigma_{1a} = \frac{\sigma_{1max} - \sigma_{1min}}{2} = 10{,}187 \text{ psi} \tag{16}$$

Also,

$$\sigma_{2max} = 0 \tag{17}$$

$$\sigma_{2min} = 0 \tag{18}$$

$$\sigma_{2m} = 0 \tag{19}$$

$$\sigma_{2a} = 0 \tag{20}$$

and

$$\sigma_{3max} = -8075 \text{ psi} \tag{21}$$

$$\sigma_{3min} = -28{,}449 \text{ psi} \tag{22}$$

$$\sigma_{3m} = -18{,}262 \text{ psi} \tag{23}$$

$$\sigma_{3a} = 10{,}187 \text{ psi} \tag{24}$$

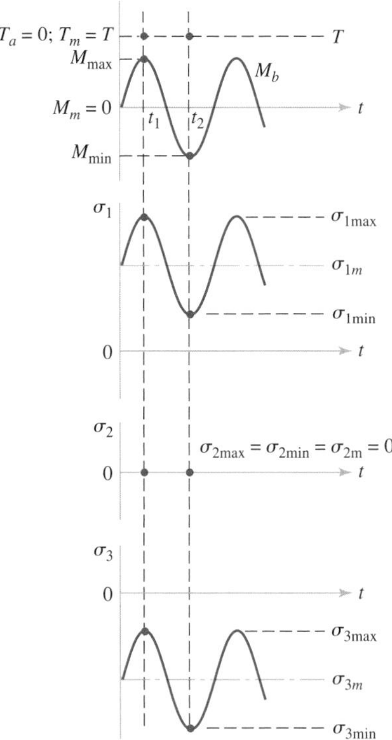

Figure E4.11B
Torques, moments, and principal stresses as a function of time *t*.

Note that subscripts *max* and *min* are determined by the *loading* cycle. Thus when the applied moment reaches its maximum value in Figure E4.11B at time t_1, the moment *and all principal stresses* are given the subscript *max.*

Since $e = 36$ percent, the material is ductile and the *equivalent stress* concepts of (4-97) and (4-98) may be used to write

$$\sigma_{eq\text{-}a} = \sqrt{\frac{1}{2}[(10{,}187 - 0)^2 + (0 - 10{,}187)^2 + (10{,}187 - 10{,}187)^2]} = 10{,}187 \text{ psi} \tag{25}$$

and

$$\sigma_{eq\text{-}m} = \sqrt{\frac{1}{2}[(18{,}262 - 0)^2 + (0 + 18{,}262)^2 + (-18{,}262 - 18{,}262)^2]} = 31{,}630 \text{ psi} \tag{26}$$

Using these expressions in (4-99), the condition on σ_{max} for using (4-99) may be checked by calculating

$$\sigma_{max} = 10{,}187 + 31{,}630 = 41{,}817 < S_{yp} = 43{,}000 \text{ psi} \tag{27}$$

Thus (4-99) is valid, and (4-99) gives

$$\sigma_{eq\text{-}CR} = \frac{10{,}187}{1 - (18{,}262/65{,}000)} = 14{,}167 \text{ psi} \tag{28}$$

Checking back to Figure 2.19, this value, $\sigma_{eq\text{-}CR} = 14{,}167$ psi, corresponds to infinite life for 1020 steel. No strength-influencing factors such as surface finish, corrosion, and so forth, have been considered. A more accurate prediction could be made by considering such factors, as was done in Example 2.9.

4.8 Stress Concentration

Failures in machines and structures virtually always initiate at sites of *local stress concentration* caused by geometrical or microstructural discontinuities. These stress concentrations, or *stress raisers*, often lead to local stresses many times higher than the nominal net section stress that would be calculated without considering stress concentration effects. An intuitive appreciation of the stress concentration associated with a geometrical discontinuity may be developed by thinking in terms of *force flow* through a member as it is subjected to external loads (see 4.3). The sketches of Figure 4.15 illustrate the concept. In Figure 4.15(a) the rectangular flat plate of width w and thickness t is fixed at its lower edge and subjected to a total vertical force F uniformly distributed along the upper edge. The dashed lines each represent a fixed quantum of force, and the local spacing between lines is therefore an indication of the local force intensity, or stress. In Figure 4.15(a) the lines are uniformly spaced throughout the plate, and the stress σ is uniform and calculable as

$$\sigma = \frac{F}{wt} \tag{4-103}$$

In the sketch of Figure 4.15(b) a flat rectangular plate of the same thickness is subjected to the same total force F, but the plate has been made wider and notched back to provide the same *net* section width w at the site of the notch. The lines of force flow may be visualized in very much the same way that streamlines would be visualized in the steady flow of a fluid through an obstructed channel with the same cross-sectional shape as the plate cross section. It may be noted in Figure 4.15(b) that no force can be supported across the notch, and therefore the lines of force flow must pass around the root of the notch, just as a flowing fluid must pass around the obstruction. In so doing, force-flow lines crowd together locally near the root of the notch, producing a higher force intensity, or stress, at the notch root. Even though the net section *nominal stress* is still properly calculated by (4-103), the *actual local stress* at the root of the notch may be many times higher than the calculated nominal stress.

Some common examples of stress concentration are illustrated in Figure 4.16. Discontinuities at the roots of gear teeth, at the corners of keyways in shafting, at the roots of screw threads, at the fillets of shaft shoulders, around rivet holes and bolt holes, and in the neighborhood of welded joints all constitute stress raisers that usually must be considered by a designer. The seriousness of the stress concentration depends on the type of loading, the type of material, and the size and shape of the discontinuity.

Stress Concentration Effects

For the purposes of studying stress concentration effects, stress raisers may be classified as being either *highly local* or *widely distributed*. Highly local stress raisers are those for

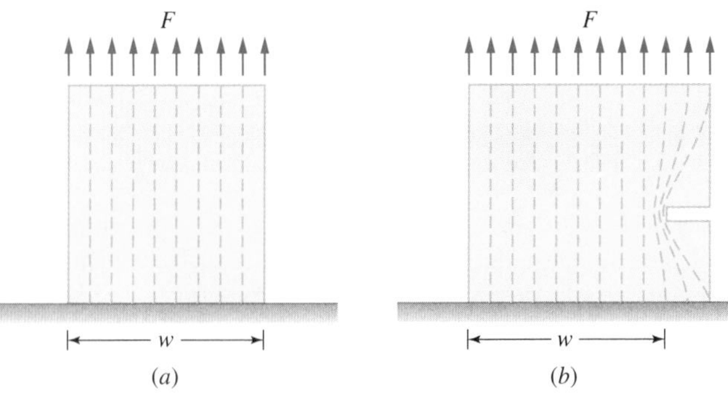

Figure 4.15
Intuitive concept of stress concentration. (a) Without stress concentration. (b) With stress concentration.

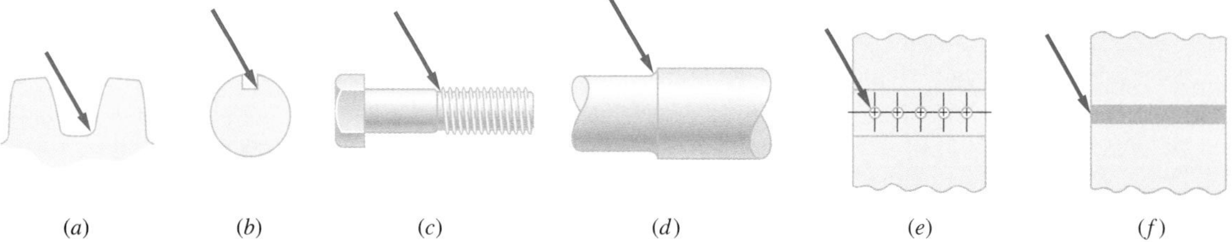

Figure 4.16
Some common examples of stress concentration. (a) Gear teeth. (b) Shaft keyway. (c) Bolt threads. (d) Shaft shoulder. (e) Riveted or bolted joint. (f) Welded joint.

which the volume of material containing the concentration of stress is a *negligibly small portion* of the overall volume of the stressed member. Widely distributed stress raisers are those for which the volume of material containing the concentration of stress is a *significant portion* of the overall volume of the stressed member. For the case of a highly local stress concentration, the overall size and shape of the stressed part would not be significantly changed by yielding in the region of the stress concentration. For a widely distributed stress concentration, the overall size and shape of the stressed part would be subject to significant changes by virtue of yielding in the region of stress concentration. Small holes and fillets would usually be regarded as highly local stress concentrations; curved beams, curved hooks, and eye-and-clevis joints would usually be categorized as cases of widely distributed stress concentration.

With the foregoing definitions made, stress concentration effects may be classified as shown in Table 4.7. From the table it may be noted that the effects of stress concentration must be considered for all combinations of geometry, loading, and material with the possible exception of one, namely, the case of a *highly local* stress concentration in a *ductile* material subjected to *static* loading. In this case the local yielding is usually negligible and a stress concentration factor of unity may often be used. All other cases must be analyzed for potential failure because of the effects of stress concentration. The final column in Table 4.7 lists K_t or K_f as stress concentration factors. The factor K_t is the theoretical elastic stress concentration factor, which is defined to be the ratio of the actual maximum local stress in the region of the discontinuity to the nominal net section stress calculated by simple theory as if the discontinuity exerted no stress concentration effect; that is,

$$K_t = \frac{\text{actual maximum stress}}{\text{nominal stress}} = \frac{\sigma_{act}}{\sigma_{nom}} \tag{4-104}$$

It should be noted that the value of K_t is valid only for stress levels within the *elastic* range, and it must be suitably modified if stresses are in the plastic range.

TABLE 4.7 Effects of Stress Concentration

Type of Stress Concentration	Type of Loading	Type of Material	Type of Failure	Stress Concentration Factor
Widely distributed	Static	Ductile	Widely distributed yielding	K_t (modified)
Widely distributed	Static	Brittle	Brittle fracture	K_t
Widely distributed	Cyclic	Any	Fatigue failure	K_f
Highly local	Static	Ductile	No failure (redistribution occurs)	$K_t \rightarrow 1$
Highly local	Static	Brittle	Brittle fracture	K_t
Highly local	Cyclic	Any	Fatigue failure	K_f

The factor K_f is the fatigue stress concentration factor, which is defined to be the ratio of the effective fatigue stress that actually exists at the root of the discontinuity to the nominal fatigue stress calculated as if the notch has no stress concentration effect. K_f may also be defined as the ratio of the fatigue strength of an *unnotched* member to the fatigue strength of the same member *with a notch*. Thus, the fatigue stress concentration factor may be defined as

$$K_f = \frac{\text{effective fatigue stress}}{\text{nominal fatigue stress}} = \frac{S_N(\text{unnotched})}{S_N(\text{notched})} \tag{4-105}$$

Stress concentration factors are determined in a variety of different ways, including direct measurement of strain, utilization of photoelastic techniques, application of the principles of the theory of elasticity, and finite element analysis. Numerical values for a wide variety of geometries and types of loading are presented in reference 9. A few of the more common cases drawn from the literature are reproduced in Figures 4.17 through 4.25. Data for the important case of screw threads are sparse, but reference 9 reports a range of values for K_t from about 2.7 to 6.7 at the thread root of standard threads. Most (but not all) threaded fastener failures tend to occur in the thread at the nut face.

Multiple Notches

Sometimes it will be found that one stress raiser is superimposed upon another, such as a notch within a notch or a notch in a fillet. Although accurate calculation of the overall stress concentration factor is difficult for such combinations, reasonable estimates can be made.[25] Figure 4.26(a) illustrates a large notch with a smaller notch at its root. To estimate the combined influence of these notches, the stress concentration factor K_{t1} for the large notch is determined as if the small notch does not exist. This permits an estimate of the stress σ'_n at the root of the large notch by multiplying the nominal stress σ_n by K_{t1} to give

$$\sigma'_n = K_{t1}\sigma_n \tag{4-106}$$

Now, assuming that the stress σ'_n occurs throughout the zone within the dashed line near the notch root of Figure 4.26(a), σ'_n becomes the *nominal* stress as far as the *small* notch

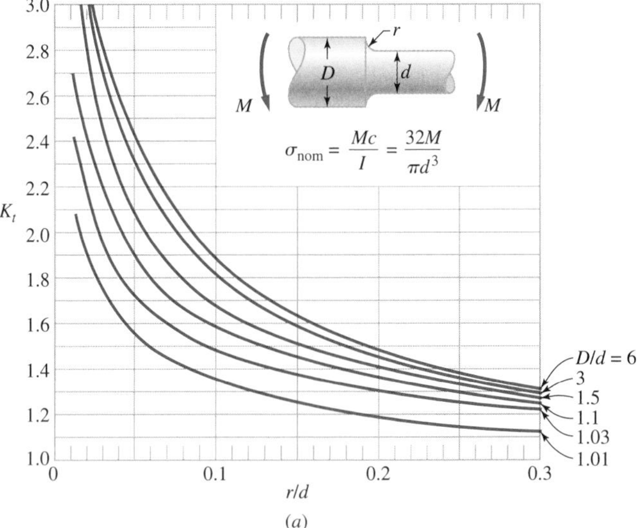

Figure 4.17
Stress concentration factors for a shaft with a fillet subjected to (a) bending, (b) axial load, or (c) torsion. (From ref. 9; adapted with permission of John Wiley & Sons, Inc.)

[25]See ref. 10.

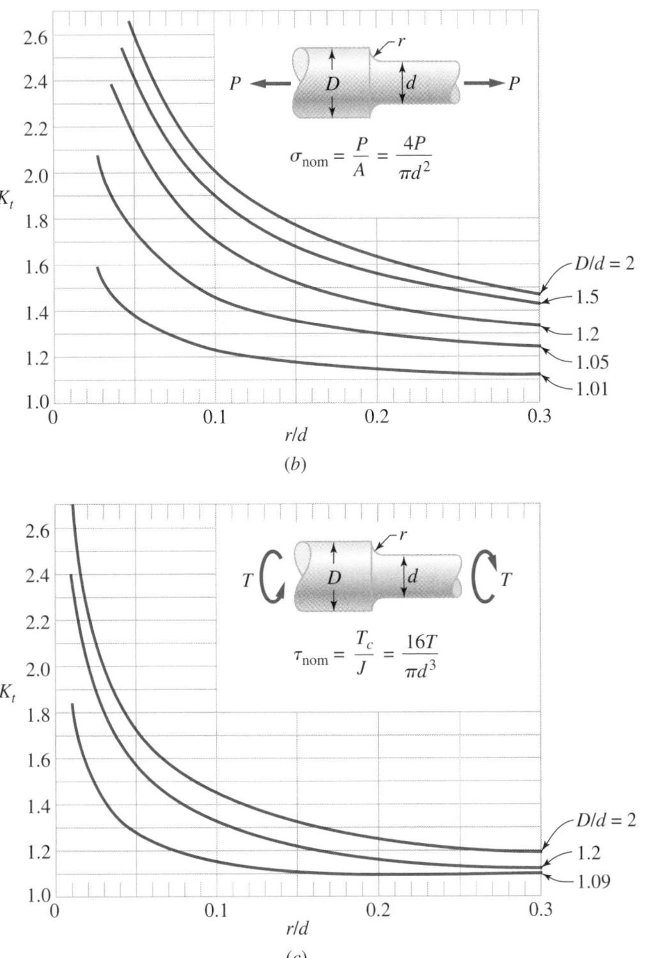

Figure 4.17
(Continued)

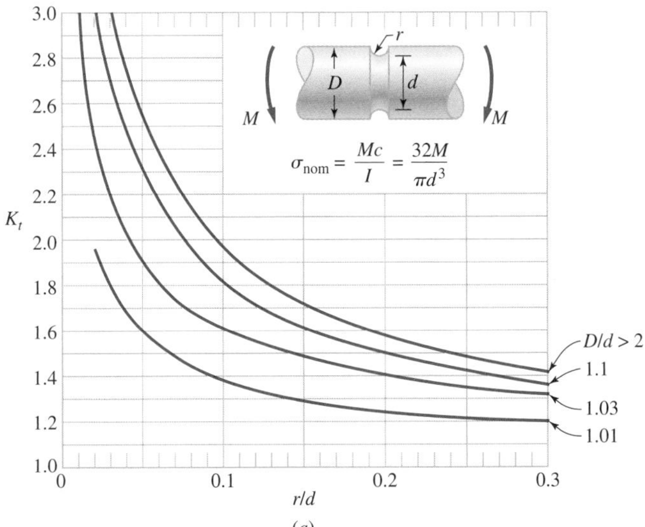

Figure 4.18
Stress concentration factors for a shaft with a groove subjected to (a) bending, (b) axial load, or (c) torsion. (From ref. 9; adapted with permission from John Wiley & Sons, Inc.)

Figure 4.18
(*Continued*)

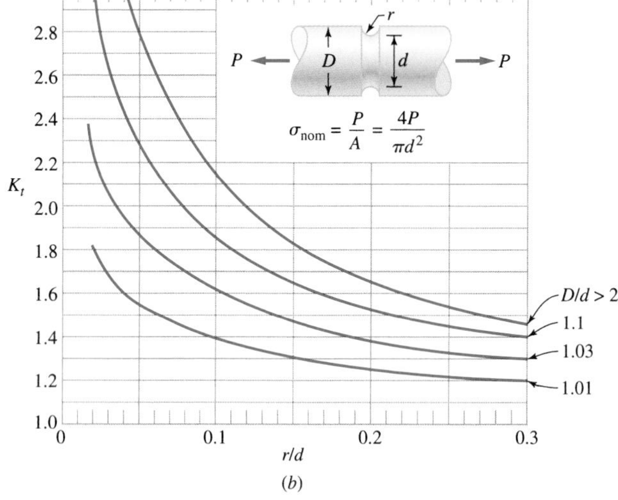

(*b*)

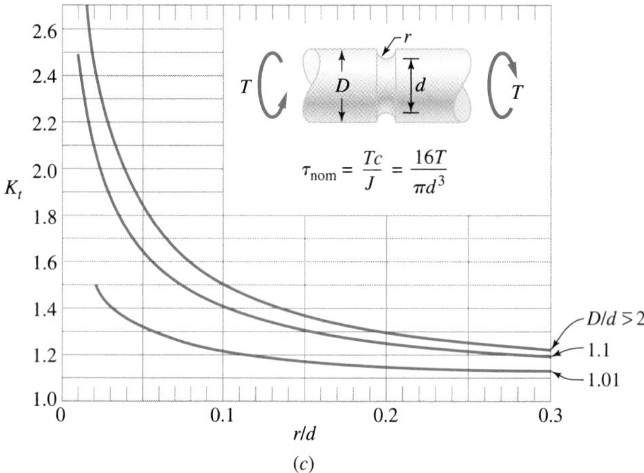

(*c*)

Figure 4.19
Stress concentration factors for (a) a shaft with a radial hole subjected to axial load, bending, or torsion (from ref. 4 with permission of the McGraw-Hill Companies), (b) a shaft with a straight parallel keyway subjected to torsion (from ref. 5, with permission of the McGraw-Hill Companies), and (c) a shaft with an eight-tooth spline subjected to torsion (from ref. 9, with permission from John Wiley & Sons, Inc.).

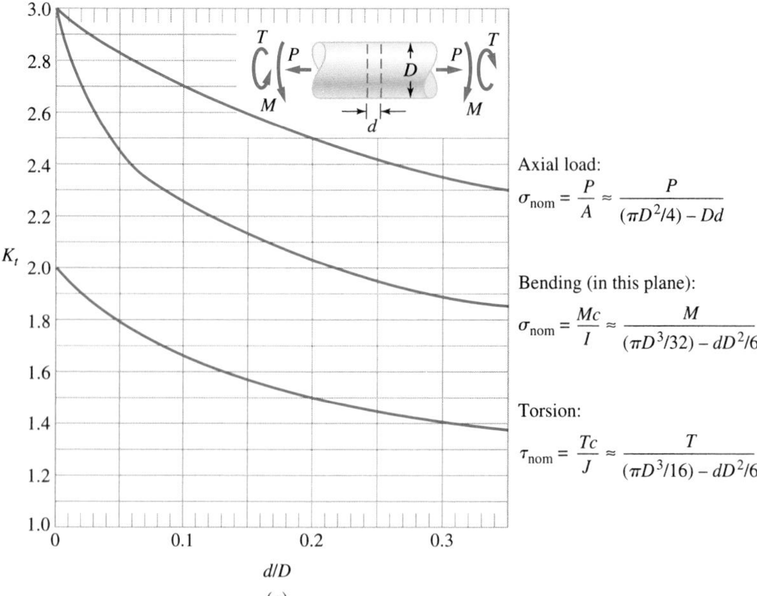

(*a*)

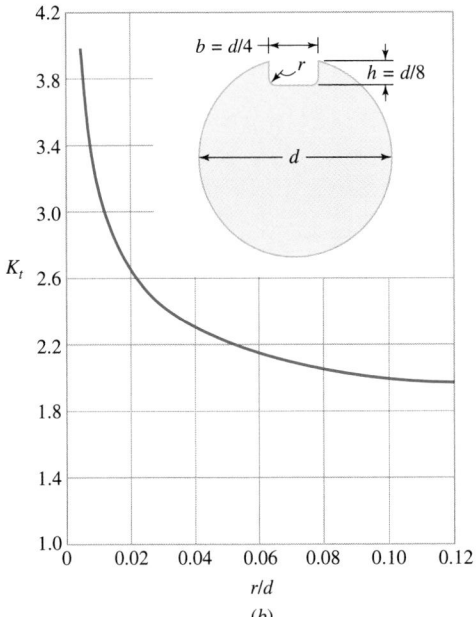

$b = d/4$ r $h = d/8$ d

K_t

r/d

(*b*)

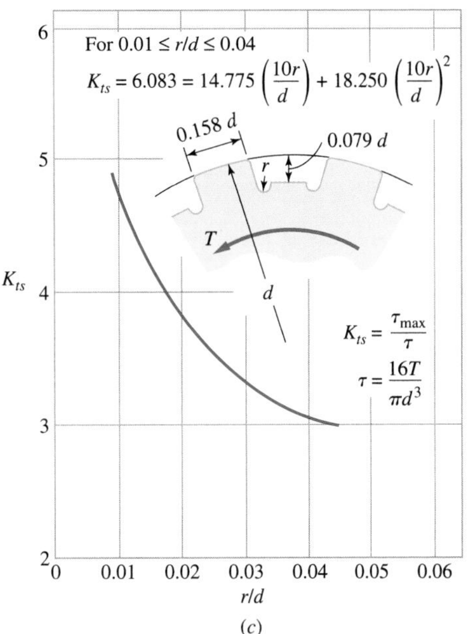

Figure 4.19
(*Continued*)

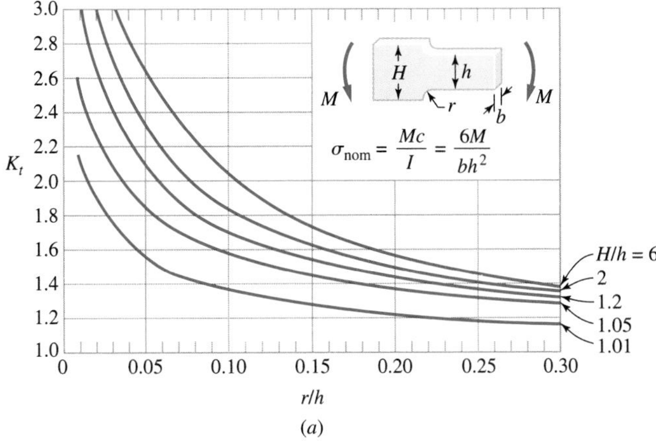

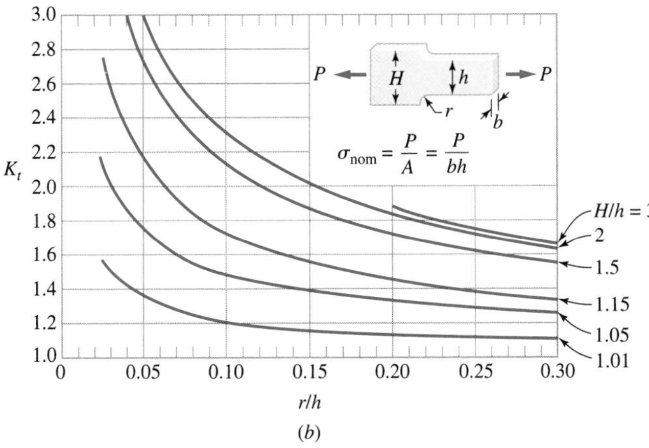

Figure 4.20
Stress concentration factors for a flat bar with a shoulder fillet subjected to (a) bending or (b) axial load. (From ref. 9; adapted with permission from John Wiley & Sons, Inc.)

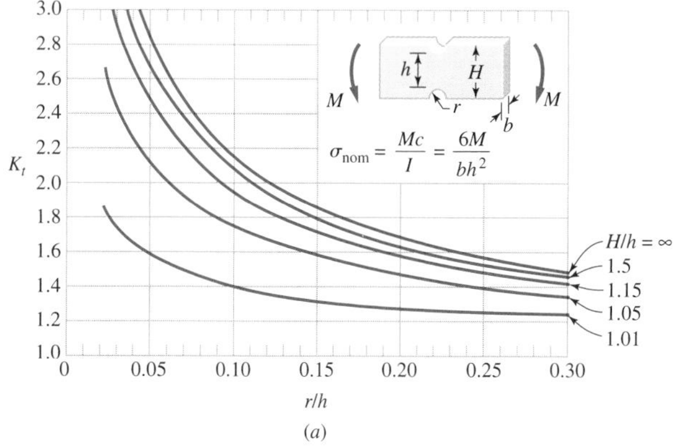

(a)

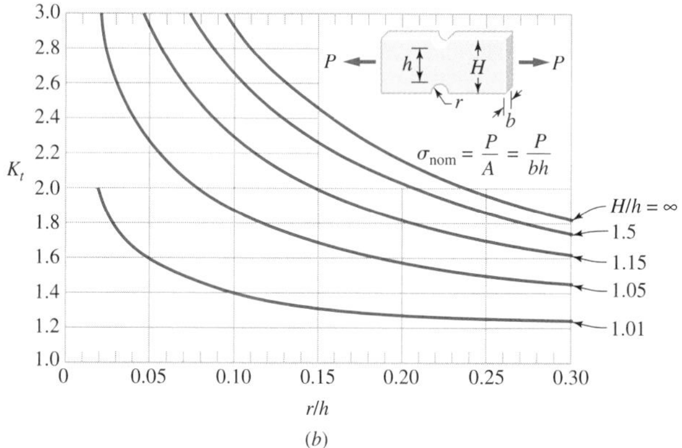

(b)

Figure 4.21
Stress concentration factors for a flat bar with a notch subjected to (a) bending or (b) axial load. (From ref. 9; adapted with permission from John Wiley & Sons, Inc.)

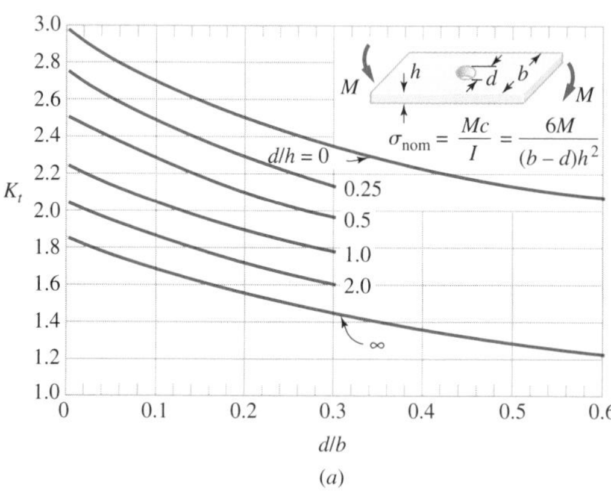

(a)

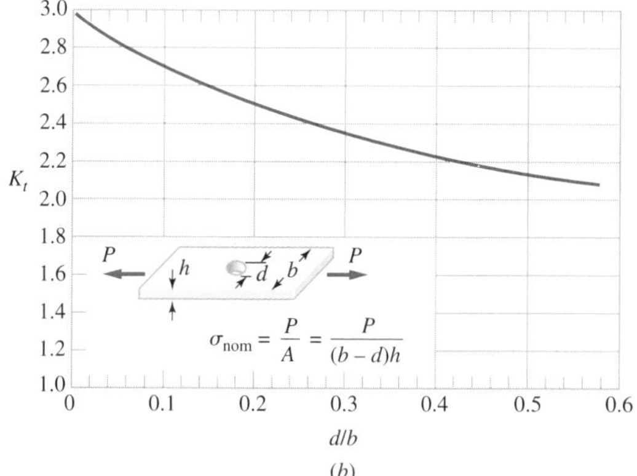

(b)

Figure 4.22
Stress concentration factors for a flat plate with a central hole subjected to (a) bending or (b) axial load. (From ref. 9; adapted with permission from John Wiley & Sons, Inc.)

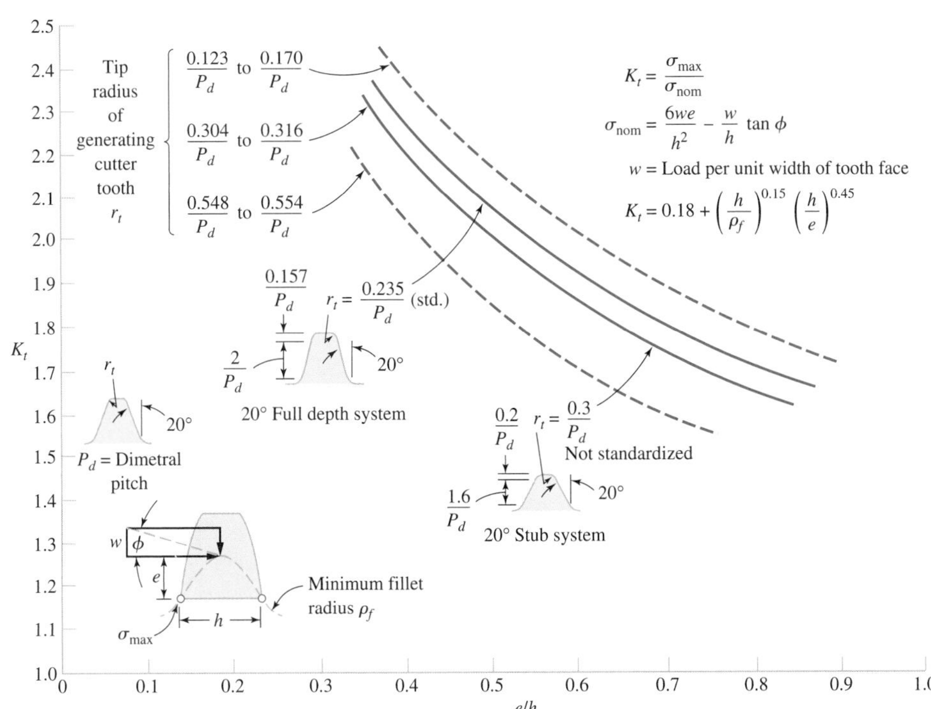

Figure 4.23
Theoretical stress concentration factor K_t for a tension-side gear tooth fillet. Pressure angle is $\varphi = 20°$. (From ref. 9; adapted with permission from John Wiley & Sons, Inc.)

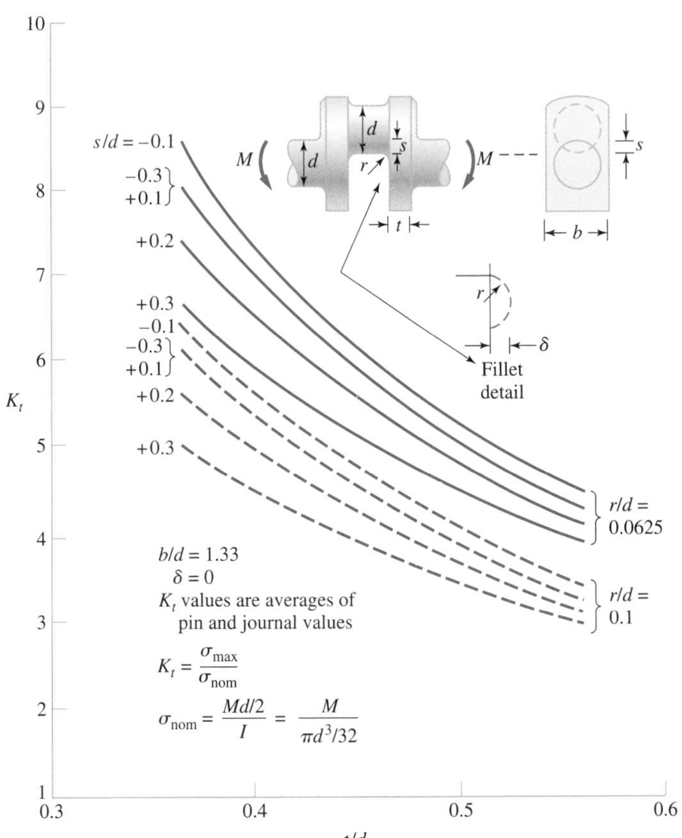

Figure 4.24
Stress concentration factors K_t for a crankshaft in bending. *Note:* When the inside of the crankpin and the outside of the journal are in line, $s = 0$ When the crankpin is closer, s is positive (as shown). When the crankpin's inner surface is farther away than $d/2$, s is negative. (From ref. 9; adapted with permission from John Wiley & Sons, Inc.)

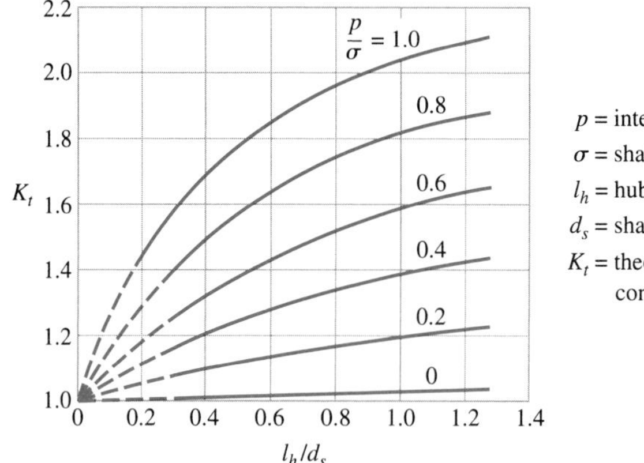

Figure 4.25
End-of-hub stress concentration factors for press fit assemblies subjected to bending moments. (From ref. 16; reprinted by permission of Pearson Education, Inc., Upper Saddle River, NJ.)

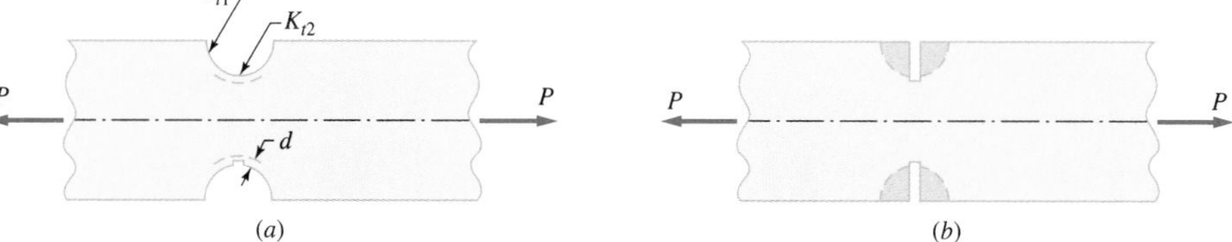

Figure 4.26
Stress concentration effects due to superimposed multiple notches. (After ref. 10; adapted with permission of The McGraw-Hill Companies.)

is concerned (because the entire small notch lies within the σ'_n field). Next, determining the stress concentration factor K_{t2} for the small notch acting alone, it may be multiplied times σ'_n to obtain the actual stress at the small notch root, or

$$\sigma_{actual} = K_{t2}\sigma'_n \tag{4-107}$$

Next, utilizing (4-106),

$$\sigma_{actual} = K_{t1}K_{t2}\sigma_n \tag{4-108}$$

Thus, the combined theoretical stress concentration factor K_{tc} for the multiple notch is the *product* of the stress concentration factors for the two notches considered individually, giving

$$K_{tc} = K_{t1}K_{t2} \tag{4-109}$$

This has been verified photoelastically.[26] The combined fatigue stress concentration factor K_{tc}, as for any other stress concentration factor, depends[27] on q and can be calculated from (4-111) by substituting K_{tc} for K_t in that equation. One technique for estimating a conservative value for K_{tc} is sketched in Figure 4.26(b). The technique assumes the notch of 4.26(a) to be filled in as shown by the crosshatched area. This leaves a single, deep, narrow notch, for which the theoretical stress concentration factor will always be *greater* than the stress concentration factor for the multiple notch.

Fatigue Stress Concentration Factors and Notch Sensitivity Index

Unlike the theoretical stress concentration factor K_t, the fatigue stress concentration factor K_f is a *function of the material*, as well as geometry and type of loading. To account for the influence of material characteristics, a *notch sensitivity index* q has been defined to re-

[26]Ibid.

[27]See (4-110).

late the actual effect of a notch on fatigue strength of a material to the effect that might be predicted solely on the basis of elastic theory. The definition of notch sensitivity index q is given by

$$q = \frac{K_f - 1}{K_t - 1} \tag{4-110}$$

where K_f = fatigue stress concentration factor
K_t = theoretical stress concentration factor
q = notch sensitivity index valid for high-cycle fatigue range

The magnitude of q ranges from *zero* for no notch effect to *unity* for full notch effect. The notch sensitivity index is a function of both material and notch radius, as illustrated in Figure 4.27 for a range of steels and an aluminum aloy. For finer-grained materials, such as quenched and tempered steels, q is usually close to unity. For coarser-grained materials, such as annealed or normalized aluminum alloys, q approaches unity if the notch radius exceeds about one-quarter inch. In view of these facts it is tempting to recommend the use of $K_f = K_t$ as a simplifying assumption. Doing so, however, would ignore several important notch sensitivity effects, including:

1. Under fatigue loading, an alloy steel with superior *static* properties will often be found *not* to have superior *fatigue* properties when compared to a plain carbon steel, because of the difference in notch sensitivities.
2. There is a tendency to improperly assess the effects of tiny scratches and cavities unless notch sensitivity effects are recognized.
3. Serious errors in applying the results from models to large structures may be made if notch sensitivity effects are not recognized.
4. In critical design situations, inefficiencies may accrue if notch sensitivity effects are not considered.

Based on (4-110), an expression for fatigue stress concentration factor may be written as

$$K_f = q(K_t - 1) + 1 \tag{4-111}$$

The theoretical elastic stress concentration factor K_t may be determined, on the basis of geometry and loading, from handbook charts such as those depicted in Figures 4.17

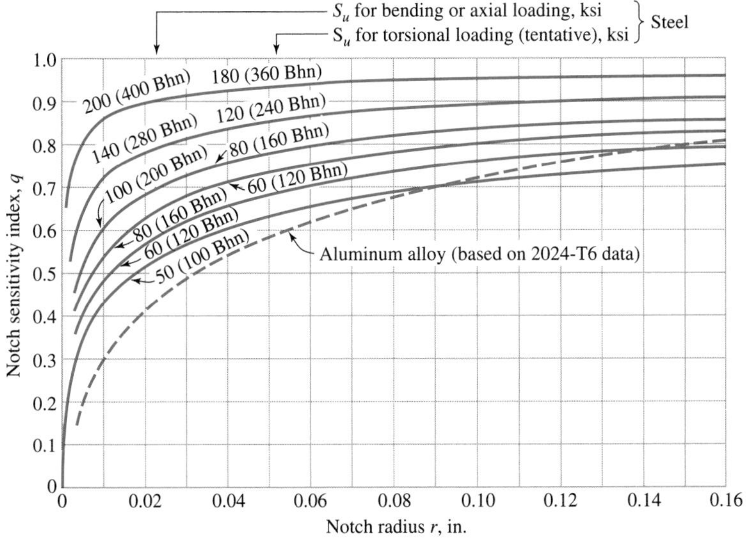

Figure 4.27
Curves of notch sensitivity index versus notch radius for a range of steels and an aluminum alloy subjected to axial, bending, and torsional loading. (After ref. 10; reprinted with permission of the McGraw-Hill Companies.)

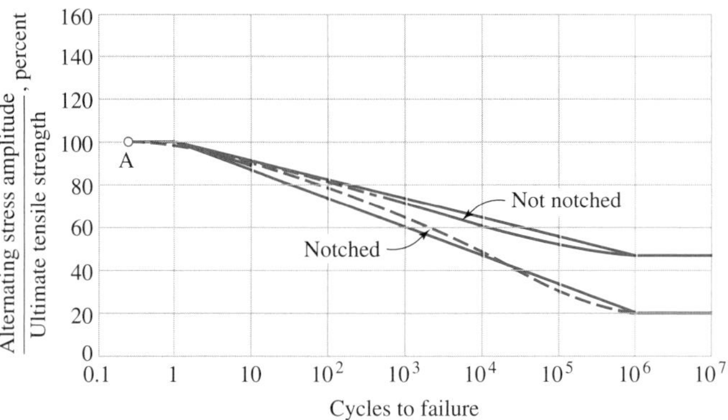

Figure 4.28
S-N curves for notched and unnotched specimens subjected to completely reversed axial loading. (After ref. 11.)

through 4.25. The notch sensitivity index q may also be read from charts, such as the one shown in Figure 4.27.

For *uniaxial* states of cyclic stress it is sometimes convenient to use K_f as a "strength reduction factor" rather than as a "stress concentration factor." This may be done by *dividing* the fatigue limit *by* K_f rather than *multiplying* the applied nominal cyclic stress *times* K_f. Although conceptually it is more correct to think of K_f as a stress concentration factor, computationally it is often simpler to use K_f as a strength reduction factor when the cyclic stresses are uniaxial. For multiaxial states of stress, however, K_f *must* be used as a stress concentration factor.

The fatigue stress concentation factor (or strength reduction factor) determined from (4-11) is strictly applicable only in the high-cycle fatigue range (lives of 10^5–10^6 cycles and greater). It has earlier been noted that for ductile materials and static loads, effects of stress concentration may usually be neglected. In the intermediate and low-cycle life range (from a quarter-cycle up to about 10^5–10^6 cycles), the stress concentration factor increases from unity to K_f, so the notched and unnotched *S-N* curves tend to converge toward the quarter-cycle point *A*, as shown in Figure 4.28.

Estimates for fatigue stress concentration factor can be made by constructing a straight line from the ultimate strength plotted at a life of 1 cycle to the unnotched fatigue strength divided by K_f plotted at a life of 10^6 cycles, on a semilogarithmic *S-N* plot.[28] The ratio of unnotched to notched fatigue strength at any selected intermediate life may be used as an estimate of fatigue stress concentration factor for that life.

Finally, experimental investigations have indicated that for evaluating fatigue of *ductile materials*, the fatigue stress concentration factor for any *nonzero mean* cyclic state of stress should be applied *only* to the *alternating* component of stress (and *not* to the *steady* component). In evaluating the fatigue loading of *brittle* materials, the stress concentration factor should be applied to *both* the alternating and the steady component.

Example 4.12 Fatigue Life Prediction Under Uniaxial Stress, Including Stress Concentration

A 2.25-inch-wide by 0.10-inch-thick rectangular bar of annealed 1040 steel has a 0.25-inch-diameter hole drilled through, as shown in Figure E4.12. The properties of the 1040 steel are $S_u = 54{,}000$ psi, $S_{yp} = 48{,}000$ psi, $e = 50$ percent in 2 inches, and $S_f = 27{,}000$ psi. The bar is to be subjected to a completely reversed alternating direct force of 2000 lb. Buckling is not a problem. What cyclic life would you predict for the bar if the unnotched *S-N* curve of Figure 4.28 is valid?

[28]See Figure 4.28.

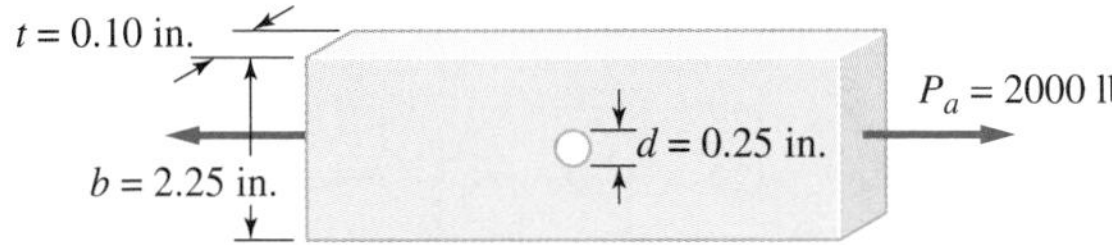

Figure E4.12
Steel bar with stress concentration subjected to cyclic loading.

Solution

The *actual* stress amplitude at the critical point adjacent to the hole is

$$(\sigma_a)_{act} = K_f(\sigma_a)_{nom} \tag{1}$$

The *nominal* stress amplitude may be calculated as

$$(\sigma_a)_{nom} = \frac{P_a}{A_{net}} = \frac{2000}{(2.0)(0.10)} = 10{,}000 \text{ psi} \tag{2}$$

From (4-111)

$$K_f = q(K_t - 1) + 1 \tag{3}$$

Using Figure 4.27, for steel with σ_u = 54,000 psi and r = 0.125 inch,

$$q = 0.76 \tag{4}$$

From Figure 4.22(b), with d/b = 0.25/2.25 = 0.11,

$$K_t = 2.65 \tag{5}$$

Hence (3) becomes

$$K_f = 0.76(2.65 - 1) + 1 = 2.25 \tag{6}$$

Using results from (2) and (6), (1) becomes

$$(\sigma_a)_{act} = 2.25(10{,}000) = 22{,}500 \text{ psi} \tag{7}$$

Calculating the ordinate ratio,

$$\frac{(\sigma_a)_{act}}{S_u} = \frac{22{,}500}{54{,}000} = 0.42 = 42\% \tag{8}$$

and using it to read into the "not notched" *S-N* curve of Figure 4.28, the predicted cyclic life of the bar is infinite.[29] It should be recognized, however, that there is virtually no margin of safety. It would be prudent, therefore, to impose an appropriate safety factor or reliability assessment before going further.

Example 4.13 Fatigue Life Prediction Under Multiaxial Stresses, Including Stress Concentration Effects

Refer to Figure E4.13A, in which a torsionally oscillating shaft of 1.25-inch diameter has a 0.25-inch diametral hole all the way through it. By the way the shaft is loaded, it is subjected to a *released* torsional moment of 8300 in-lb, and an in-phase *released* cyclic bending moment, in the plane of the through-hole axis, of 3700 in-lb. If the shaft is made of 4340 steel with S_u = 150,000 psi, S_{yp} = 120,000 psi, e = 15 percent in 2 inches, and the fatigue properties shown in Figure E4.13B, how many torsional oscillations would you expect could be completed before fatigue failure of the shaft takes place? (Assume that the critical point for bending and torsion coincide.)

[29]Because the ordinate ratio 0.42 lies below the *S-N* curve.

Example 4.13 Continues

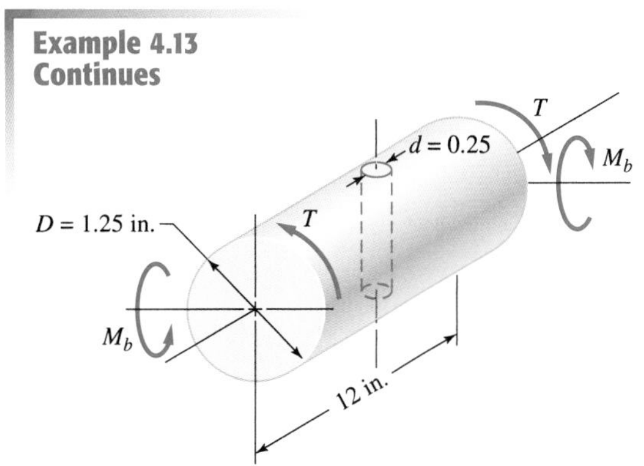

Figure E4.13A
Oscillating shaft with through hole.

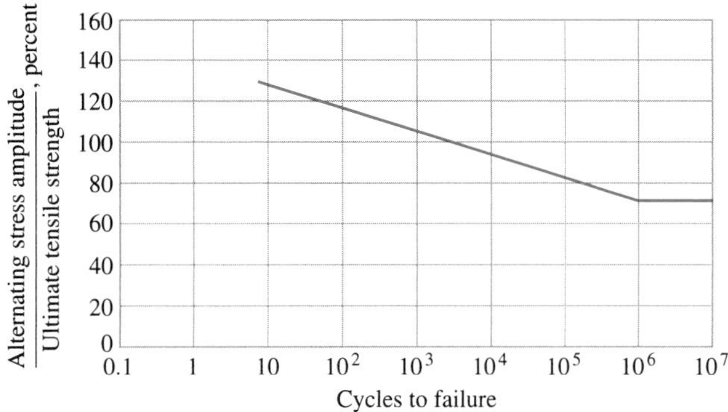

Figure E4.13B
Uniaxial fatigue properties of 4340 steel used in the torsionally oscillating shaft.

Solution

The critical points for bending and torsion are assumed to be at the same location, adjacent to the edge of the hole at top and botom surface elements. The state of stress at the critical point is shown in Figure E4.13C.

From Figure 4.19, the nominal bending and torsional stresses as well as bending and torsional stress concentration factors may be found. The nominal bending stress is

$$\sigma_{nom} = \frac{M}{\dfrac{\pi D^3}{32} - \dfrac{dD^2}{6}} = \frac{M}{\dfrac{\pi(1.25)^3}{32} - \dfrac{0.25(1.25)^2}{6}} = 7.90M \tag{1}$$

and the nominal torsional stress is

$$\tau_{nom} = \frac{T}{\dfrac{\pi D^3}{16} - \dfrac{dD^2}{6}} = \frac{T}{\dfrac{\pi(1.25)^3}{16} - \dfrac{0.25(1.25)^2}{6}} = 3.14T \tag{2}$$

Referring again to Figure 4.19, for

$$\frac{d}{D} = \frac{0.25}{1.25} = 0.2 \tag{3}$$

it may be found that

$$(K_t)_{tor} = 1.50 \tag{4}$$

and

$$(K_t)_{bend} = 2.03 \tag{5}$$

Also, from Figure 4.27, for S_u = 150,000 psi and r = 0.125,

$$q_{tor} = 0.94 \tag{6}$$

and

$$q_{bend} = 0.92 \tag{7}$$

Then from (4-111), and (4) through (7),

$$K_{f\text{-}tor} = 0.94(1.50 - 1) + 1 = 1.47 \tag{8}$$

and

$$K_{f\text{-}bend} = 0.92(2.03 - 1) + 1 = 1.95 \tag{9}$$

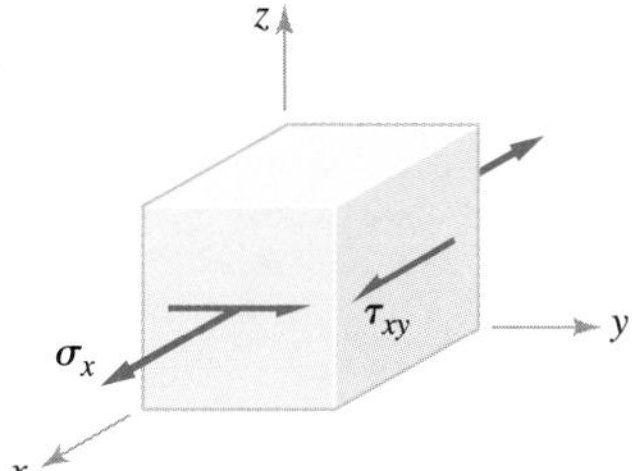

Figure E4.13C
State of stress at the critical point.

Then from (2) and (8)

$$\tau_{xy} = 1.47(3.14T) = 4.62T \tag{10}$$

and from (1) and (9)

$$\sigma_x = 1.95(7.90M) = 15.41M \tag{11}$$

For the state of stress depicted in Figure E4.13C, the stress cubic equation (4-59) becomes

$$\sigma^3 - \sigma^2\sigma_x + \sigma(-\tau_{xy}^2) = 0 \tag{12}$$

with roots

$$\sigma_1 = \frac{\sigma_x}{2} + \sqrt{\left(\frac{\sigma_x}{2}\right)^2 + \tau_{xy}^2} \tag{13}$$

$$\sigma_2 = 0 \tag{14}$$

$$\sigma_3 = \frac{\sigma_x}{2} - \sqrt{\left(\frac{\sigma_x}{2}\right)^2 + \tau_{xy}^2} \tag{15}$$

or, substituting (10) and (11),

$$\sigma_1 = \frac{15.41M}{2} + \sqrt{\left(\frac{15.41M}{2}\right)^2 + (4.62T)^2} \tag{16}$$

$$\sigma_2 = 0 \tag{17}$$

$$\sigma_3 = \frac{15.41M}{2} - \sqrt{\left(\frac{15.41M}{2}\right)^2 + (4.62T)^2} \tag{18}$$

Figure E4.13D shows the relationships among torque, moment, and these principal stresses, as a function of time. The subscripts *max* and *min* are determined by the peaks and valleys of the *loads*.

From known values of maximum torque and moment, the following cyclic principal stresses may be calculated:

$$\sigma_{1max} = \frac{15.41(3700)}{2} + \sqrt{\left[\frac{15.41(3700)}{2}\right]^2 + [4.62(8300)]^2} \tag{19}$$

giving

$$\sigma_{1max} = 28{,}509 + 47{,}782 = 76{,}291 \text{ psi} \tag{20}$$

$$\sigma_{1min} = 0 \tag{21}$$

$$\sigma_{1m} = \frac{76{,}291 + 0}{2} = 38{,}146 \text{ psi} \tag{22}$$

$$\sigma_{1a} = \left|\frac{76{,}291 - 0}{2}\right| = 38{,}146 \text{ psi} \tag{23}$$

$$\sigma_{2max} = \sigma_{2min} = \sigma_{2m} = \sigma_{2a} = 0 \tag{24}$$

$$\sigma_{3max} = 28{,}509 - 47{,}782 = -19{,}273 \text{ psi} \tag{25}$$

$$\sigma_{3min} = 0 \tag{26}$$

Example 4.13 Continues

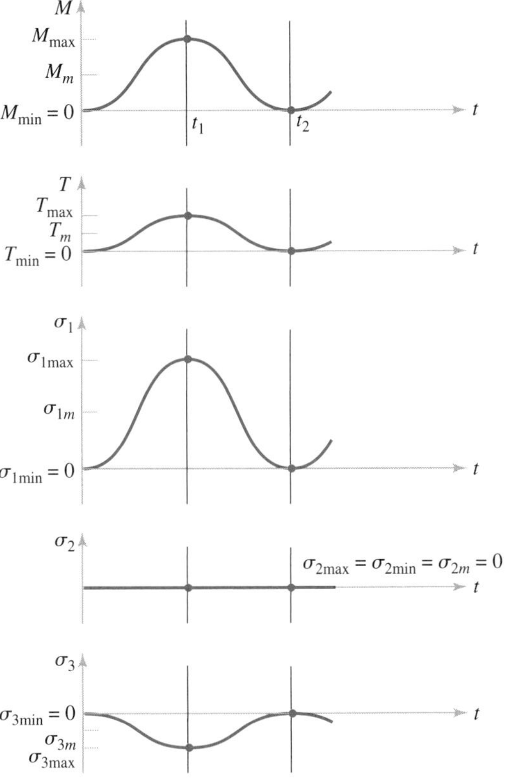

Figure E4.13D
Relationships among torque, moment, and principal stresses as a function of time.

$$\sigma_{3m} = \frac{-19{,}273 + 0}{2} = -9637 \text{ psi} \tag{27}$$

$$\sigma_{3a} = \left|\frac{-19{,}273 - 0}{2}\right| = 9637 \text{ psi} \tag{28}$$

Since $e = 15$ percent, the material is ductile, and the *equivalent stress* concepts of (4-97) and (4-98) may be used to give

$$\sigma_{ea} = \sqrt{\frac{1}{2}[(38{,}146 - 0)^2 + (0 - 9637)^2 + (9637 - 38{,}146)^2]} = 34{,}356 \text{ psi} \tag{29}$$

and

$$\sigma_{em} = \sqrt{\frac{1}{2}[(38{,}146 - 0)^2 + (0 + 9637)^2 + (-9637 - 38{,}146)^2]} = 43{,}768 \text{ psi} \tag{30}$$

Considering the expressions of (4-99) and (4-100), the condition on using them is based on the magnitude of maximum stress, which is

$$\sigma_{max} = 34{,}356 + 43{,}768 = 78{,}124 \text{ psi} \tag{31}$$

Based on the validity criteria

$$\sigma_{em} \geq 0 \tag{32}$$

and

$$\sigma_{max} = 78{,}124 < S_{yp} = 120{,}000 \tag{33}$$

(4-99) is chosen, with the result

$$\sigma_{eq\text{-}CR} = \frac{34{,}356}{1 - (43{,}768/150{,}000)} = 48{,}510 \text{ psi} \tag{34}$$

Since the fatigue limit in Figure E4.13B is shown as $S_f = 75{,}000$ psi, the smaller value $\sigma_{eq\text{-}CR} = 58{,}694$ psi is predicted to give infinite life (an infinite number of torsional oscillations) for this part. Strength-influencing factors such as surface finish, corrosion, and so on, should be reviewed, however, to improve the life assessment. (See Example 2.9.)

4.9 Bending of Initially Curved Beams

In 4.4, where bending of straight beams was discussed, it was noted that for *straight* beams the neutral axis of bending passes through the centroid of the beam cross section. A linear stress distribution from tensile outer fibers to compressive outer fibers results, as given by (4-4), or (4-7) and (4-8). If a beam is *initially curved*, however, the neutral axis of bending does *not* coincide with the centroidal axis, and the stress distribution becomes nonlinear. Examples of curved beams in practice include C-clamps, hooks, and C-shaped machine frames. The neutral axis is shifted *toward* the center of curvature of an initially curved beam by a distance e, as depicted in Figure 4.29. Based on equilibrium considerations it may be shown that[30]

$$e = r_c - \frac{A}{\displaystyle\int \frac{dA}{r}} \tag{4-112}$$

where A is area of the beam cross section, r is radius from center of initial curvature to differential area dA, and r_c is the radius of curvature of the centroidal surface of the curved beam. Evaluations of the integral $\int dA/r$ are given in Table 4.8 for several curved beam cross sections.[31]

The (hyperbolic) stress distribution from inner to outer curved beam surfaces is given by

$$\sigma = -\frac{My}{eA(r_n + y)} \tag{4-113}$$

where r_n is the radius of curvature of the neutral surface and a *positive* moment is defined as one tending to *straighten* the beam. The extreme values of stress at the inner and outer surfaces, where y is equal to c_i and c_o, respectively, are

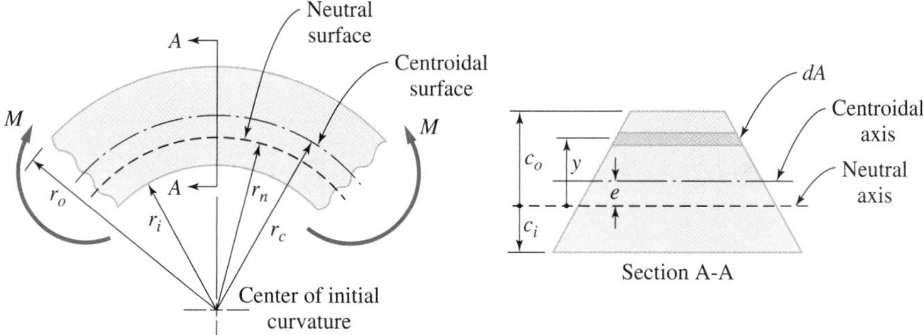

Figure 4.29
Initially curved beam subjected to pure bending moment.

[30]See, for example, ref. 12, pp. 107 ff.

[31]For other cross sections see, for example, refs. 4 and 13.

TABLE 4.8 Integral Evaluations and Widely Distributed Stress Concentration Factors k_i as a Function of $\frac{r_c}{c}$ for Curved Beams with Various Cross Sections[1]

Cross Section	$\int \frac{dA}{r}$	$\frac{r_c}{c}$	k_i
1. Rectangular	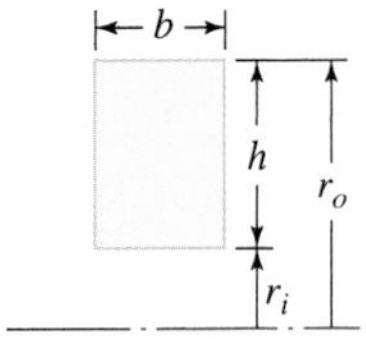 $b \ln \frac{r_o}{r_i}$	1.20	2.888
		1.40	2.103
		1.60	1.798
		1.80	1.631
		2.00	1.523
		3.00	1.288
		4.00	1.200
		6.00	1.124
		8.00	1.090
		10.00	1.071
2. Trapezoidal	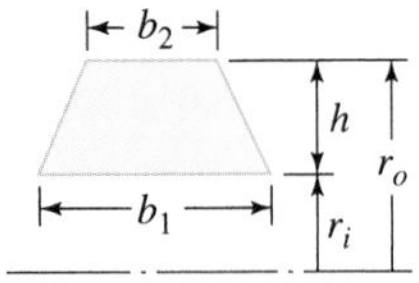 $\left(\frac{b_1 r_o - b_2 r_i}{h} \ln \frac{r_o}{r_i}\right) - b_1 + b_2$	when $b_2/b_1 = 1/2$	
		1.20	3.011
		1.40	2.183
		1.60	1.859
		1.80	1.681
		2.00	1.567
		3.00	1.314
		4.00	1.219
		6.00	1.137
		8.00	1.100
		10.00	1.078
3. Triangular (base inward)	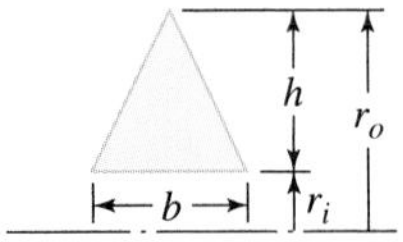 $\left(\frac{b r_o}{h} \ln \frac{r_o}{r_i}\right) - b$	1.20	3.265
		1.40	2.345
		1.60	1.984
		1.80	1.784
		2.00	1.656
		3.00	1.368
		4.00	1.258
		6.00	1.163
		8.00	1.120
		10.00	1.095
4. Circular	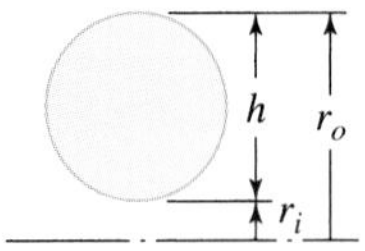 $2\pi\left\{\left(r_i + \frac{h}{2}\right) - \left[\left(r_i + \frac{h}{2}\right)^2 - \frac{h^2}{4}\right]^{1/2}\right\}$	1.20	3.408
		1.40	2.350
		1.60	1.957
		1.80	1.748
		2.00	1.616
		3.00	1.332
		4.00	1.229
		6.00	1.142
		8.00	1.103
		10.00	1.080

TABLE 4.8 (*Continued*)

Cross Section	$\int \frac{dA}{r}$	$\frac{r_c}{c}$	k_i
5. Inverted tee	$b_1 \ln \frac{r_i + h_1}{r_i} + b_2 \ln \frac{r_o}{r_i + h_1}$	(when $b_2/b_1 = 1/4$ and $h_1/(h_1 + h_2) = 1/4$)	
		1.20	3.633
		1.40	2.538
		1.60	2.112
		1.80	1.879
		2.00	1.731
		3.00	1.403
		4.00	1.281
		6.00	1.176
		8.00	1.128
		10.00	1.101

[1]Extracted from ref. 4. Note that k_i is to be applied to a nominal "straight-beam" bending stress in which $c_i = c_o = c$; hence the "straight-beam" parameter r_c/c is tabulated.

$$\sigma_i = \frac{Mc_i}{eAr_i} \tag{4-114}$$

and

$$\sigma_o = -\frac{Mc_o}{eAr_o} \tag{4-115}$$

Thus if a positive moment is defined as one tending to straighten the beam, tensile stresses result at the inner radius of the beam, r_i, and compressive stresses result at the outer radius r_o. Equations (4-114) and (4-115) are, strictly speaking, applicable only for the case of pure bending. If, as illustrated in Figure 4.30, the bending moment is produced by a system of forces whose resultant line of action does not pass through the centroid of the critical cross section, the moment of the forces, M, should be computed about the *centroidal axis*, shown as point a in Figure 4.30, not the neutral axis. Further, an additional direct stress component must be superposed on the bending stress to obtain the proper resultant stresses. In such cases, (4-114) and (4-115) are modified to give

$$\sigma_i = \frac{Mc_i}{eAr_i} + \frac{F}{A} \tag{4-116}$$

and

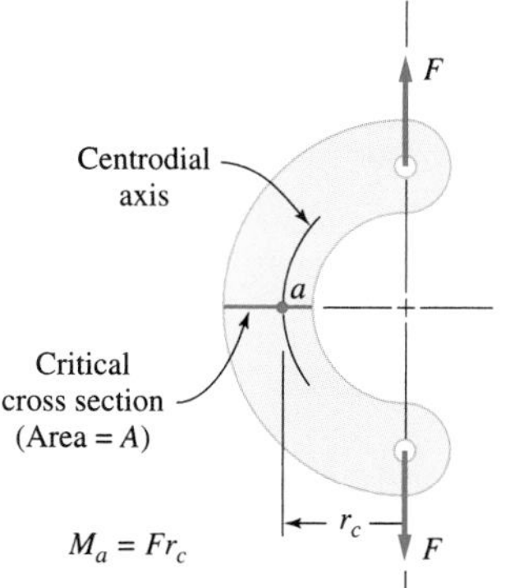

Figure 4.30
Curved beam subjected to a system of forces.

$$\sigma_o = -\frac{Mc_o}{eAr_o} + \frac{F}{A} \tag{4-117}$$

Finally, critical inner-fiber stresses are sometimes calculated as the result of widely distributed stress concentration, as suggested in the discussion of Table 4.7. Using this approach, the inner-fiber bending stress is calculated by multiplying the *nominal* bending stress, calculated as if the beam were straight, by a widely distributed stress concentration factor k_i (sometimes called "curvature factor"), giving

$$\sigma_i = k_i \sigma_{nom} = k_i\left(\frac{Mc}{I}\right) \tag{4-118}$$

Values of stress concentration factor k_i are given in Table 4.8 for several curved beam cross sections and curvature/depth ratios.[32] Generally, the results from using (4-116) are more accurate, but the use of Table 4.8 with (4-118) gives a quick estimate with little effort.

Example 4.14 Crane Hook (Curved Beam) Under Static Loading

A crane hook with the shape and dimensions shown in Figure E4.14 is loaded as shown when a 3-ton weight is very slowly lifted. Assuming the load P to be static, do the following analyses for critical section A-A.

a. Calculate the distance e that the neutral axis of bending is shifted toward the center of curvature from the centroidal axis, due to initial curvature.

b. Calculate stresses on section A-A at the inner radius r_i and outer radius r_o when load $P = 6000$ lb is applied to the hook.

c. Using the concept of widely distributed stress concentration, recalculate the stress at inner radius r_i of section A-A.

d. If the hook is made of 1020 steel with properties as given in Table 3.3, would you predict yielding of the hook?

Solution

a. The neutral axis shift, e, may be calculated using (4-112). From Table 4.8, case 1, the integral may be evaluated for this rectangular cross section as

$$\int \frac{dA}{r} = b\ln\frac{r_o}{r_i} = (1)\ln\frac{6}{2} = 1.10 \tag{1}$$

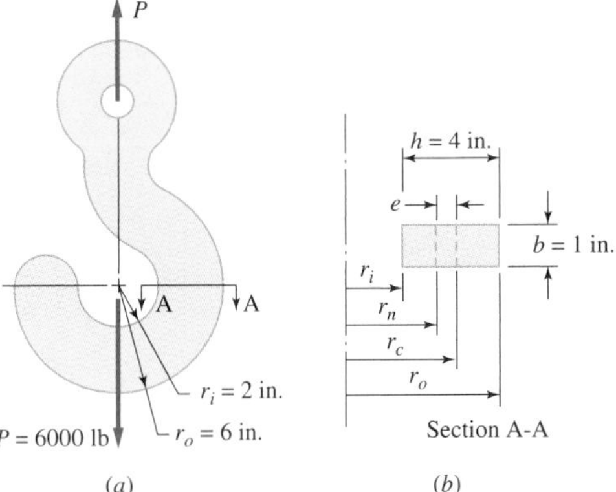

Figure E4.14
Crane hook with rectangular cross section.

[32]For other cross sections see, for example, ref. 4, Table 16. It is important to keep in mind that k_i values in Table 4.8 are to be applied to the nominal bending stress from a "straight-beam" calculation for which $c_i = c_o = c$. Hence the straight-beam parameter r_c/c is tabulated in Table 4.8.

and

$$A = 1(4) = 4 \text{ in}^2 \tag{2}$$

Hence, from (4-111),

$$e = r_c - \frac{A}{\int \frac{dA}{r}} = 4.00 - \frac{4.00}{1.10} = 0.36 \text{ inch} \tag{3}$$

b. Since the load on the curved beam (the hook) produces both a bending moment and a direct stress on section *A-A*, (4-115) and (4-116) are applicable. The bending moment produced is

$$M = Pr_c = 6000(4.00) = 24{,}000 \text{ in-lb} \tag{4}$$

Also,

$$c_o = r_o - r_n = 6.0 - (4.00 - 0.36) = 2.36 \text{ inches} \tag{5}$$

and

$$c_i = r_n - r_i = 3.64 - 2.0 = 1.64 \text{ inches} \tag{6}$$

Then from (4-116)

$$\sigma_i = \frac{(24{,}000)(1.64)}{0.36(4.0)(2.0)} + \frac{6000}{4.0} = 15{,}167 \text{ psi} \tag{7}$$

and from (4-117)

$$\sigma_o = -\frac{(24{,}000)(2.36)}{0.36(4.0)(6.0)} + \frac{6000}{4.0} = -5056 \text{ psi} \tag{8}$$

c. Using the concept of widely distributed stress concentration, the stress on section *A-A* at inner radius r_i, due to bending, is given by (4-118). The *nominal* bending stress, based on bending of a straight beam, is

$$\sigma_{nom} = \frac{Mc}{I} = \frac{6M}{bd^2} = \frac{6(24{,}000)}{(1)(4)^2} = 9000 \text{ psi} \tag{9}$$

The stress concentration factor k_i may be found in Table 4.8, case 1, using the ratio

$$\frac{r_c}{c} = \frac{4.0}{2.0} = 2.0 \tag{10}$$

The corresponding value of k_i is

$$k_i = 1.52 \tag{11}$$

Thus the bending stress at r_i becomes, from (4-118),

$$(\sigma_i)_b = 1.52(9000) = 13{,}680 \text{ psi} \tag{12}$$

Adding the direct stress

$$\sigma_i = 13{,}680 + \frac{6000}{4.0} = 15{,}180 \text{ psi} \tag{13}$$

This is in agreement with the result of (7).

d. From Table 3.3, the yield strength of 1020 steel is $S_{yp} = 51{,}000$ psi. Since

$$\sigma_i = 15{,}167 < S_{yp} = 51{,}000 \tag{14}$$

failure by yielding would not be predicted.

4.10 Stresses Caused by Curved Surfaces in Contact

In 4.4 the development of contact pressure between mating surfaces of joints, where loads are transmitted from one machine part to another, was briefly discussed. As noted in that discussion, the *Hertz contact stress* distributions at and below the surfaces within the contact region of the mating parts are typically *triaxial*. Because of the triaxiality, a failure theory must be utilized if it is desired to assess the potential for failure by the governing failure mode.

The *general* case of contact stress occurs when each of the two contacting bodies loaded against each other has two mutually perpendicular principal curvatures at the contact site, measured by $R_{1\,max}$ and $R_{1\,min}$ for body 1, and $R_{2\,max}$ and $R_{2\,min}$ for body 2.[33] The two more common *specific* cases are two spheres in contact (including a sphere on a sphere, a sphere on a flat plane, and a sphere in a spherical cup) and two parallel cylinders in contact (including a cylinder on a cylinder, a cylinder on a flat plane, and a cylinder in a cylindrical groove). Examples of machine elements having characteristics of such contact geometries include ball bearings, roller bearings, cams, and gear teeth.

For the case of solid spheres with diameters d_1 and d_2, pressed together by a force F, the "footprint" of the small contact area is circular, having radius a as shown in Figure 4.31. The radius of the circular contact area is given by

$$a = \sqrt[3]{\frac{3F\left[\left(\dfrac{1-\nu_1^2}{E_1}\right)+\left(\dfrac{1-\nu_2^2}{E_2}\right)\right]}{8\left(\dfrac{1}{d_1}+\dfrac{1}{d_2}\right)}} \tag{4-119}$$

where E_1, E_2 = moduli of elasticity for spheres 1 and 2, respectively
ν_1, ν_2 = Poisson's ratios for spheres 1 and 2, respectively

The maximum contact pressure p_{max}, at the center of the circular contact area, is

$$p_{max} = \frac{3F}{2\pi a^2} \tag{4-120}$$

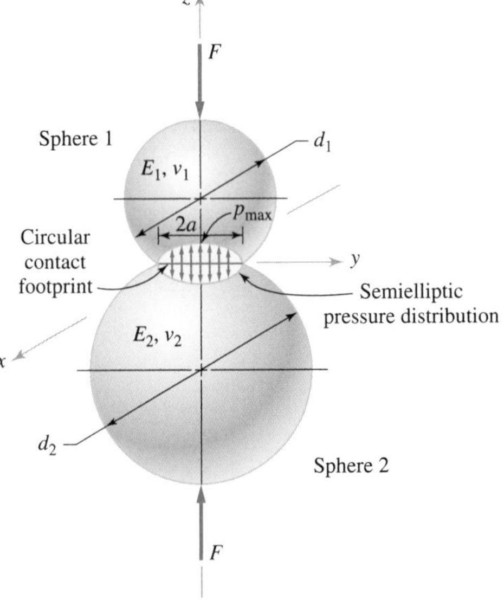

Figure 4.31
Two spheres in contact, loaded by force *F*.

[33]See, for example, ref. 14, pp. 581 ff., or ref. 4, p. 652.

Expressions (4-119) and (4-120) are equally valid for two spheres in contact (d_1 and d_2 positive), a sphere on a plane ($d = \infty$ for the plane), and a sphere in a spherical cup (d is negative for a cup).

The maximum stresses σ_x, σ_y, and σ_z are generated on the z-axis, where they are *principal* stresses. On the z-axis then, the principal stresses are

$$\sigma_x = \sigma_1 = \sigma_y = \sigma_2 = -p_{max}\left[(1+\nu)\left(1 - \frac{z}{a}\tan^{-1}\frac{a}{z}\right) - \frac{1}{2\left(\dfrac{z^2}{a^2}+1\right)}\right] \qquad (4\text{-}121)$$

and

$$\sigma_z = \sigma_3 = -p_{max}\left[\frac{1}{\left(\dfrac{z^2}{a^2}+1\right)}\right] \qquad (4\text{-}122)$$

Since $\sigma_1 = \sigma_2$, from (4-62) $|\tau_3| = 0$, and from (4-60) and (4-61)

$$|\tau_1| = |\tau_2| = \tau_{max} = \left|\frac{\sigma_1 - \sigma_3}{2}\right| \qquad (4\text{-}123)$$

Figure 4.32 depicts these stresses as a function of distance below the contact surfaces up to a depth of $3a$. It may be noted that the maximum shearing stress, τ_{max}, reaches a peak value slightly below the contact surface, as alluded to in the discussion of surface fatigue failure in 2.3.

When two parallel solid cylinders of length L, with diameters d_1 and d_2, are pressed together radially by a force F, the "footprint" of the narrow contact area is rectangular, having half-width b as shown in Figure 4.33.

The half-width of the narrow rectangular contact area may be calculated from

$$b = \sqrt{\frac{2F\left[\left(\dfrac{1-\nu_1^2}{E_1}\right) + \left(\dfrac{1-\nu_2^2}{E_2}\right)\right]}{\pi L\left(\dfrac{1}{d_1} + \dfrac{1}{d_2}\right)}} \qquad (4\text{-}124)$$

The maximum contact pressure p_{max}, along the centerline of the narrow rectangular contact area, is

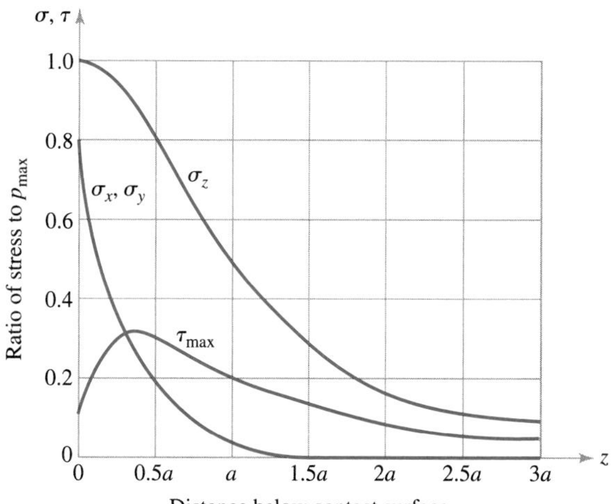

Figure 4.32
Magnitudes of σ_x, σ_y, σ_z, and τ_{max} as a function of maximum contact pressure between spheres for various distances z below the contact interface (for $\nu = 0.3$).

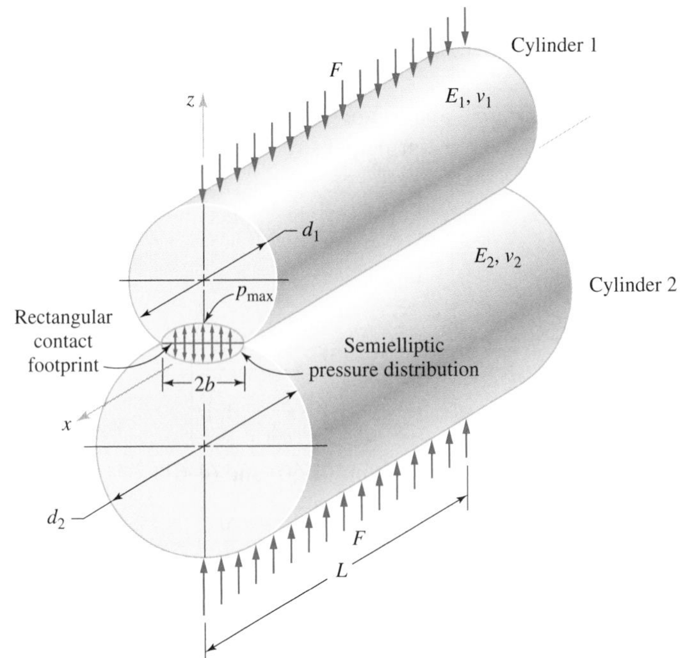

Figure 4.33
Two cylinders in contact, loaded by force F uniformly distributed over their length L.

$$p_{max} = \frac{2E}{\pi b L} \tag{4-125}$$

Expressions (4-124) and (4-125) are equally valid for two parallel cylinders in contact (d_1 and d_2 positive), a cylinder on a plane ($d = \infty$ for the plane), or a cylinder in a cylindrical groove (d is negative for the groove).

The maximum normal stresses (principal stresses) σ_x, σ_y, and σ_z occur on the z-axis. Thus the principal stresses are

$$\sigma_1 = \sigma_x = -2\nu p_{max}\left[\sqrt{\frac{z^2}{b^2} + 1} - \frac{z}{b}\right] \tag{4-126}$$

$$\sigma_2 = \sigma_y = -p_{max}\left[\left(2 - \frac{1}{\left(\frac{z^2}{b^2} + 1\right)}\right)\sqrt{\frac{z^2}{b^2} + 1} - 2\frac{z}{b}\right] \tag{4-127}$$

and

$$\sigma_3 = \sigma_z = -p_{max}\left[\frac{1}{\sqrt{\frac{z^2}{b^2} + 1}}\right] \tag{4-128}$$

From equations (4-60), (4-61), and (4-62)

$$|\tau_1| = \left|\frac{\sigma_2 - \sigma_3}{2}\right| \tag{4-129}$$

$$|\tau_2| = \left|\frac{\sigma_1 - \sigma_3}{2}\right| \tag{4-130}$$

and

$$|\tau_3| = \left|\frac{\sigma_1 - \sigma_2}{2}\right| \tag{4-131}$$

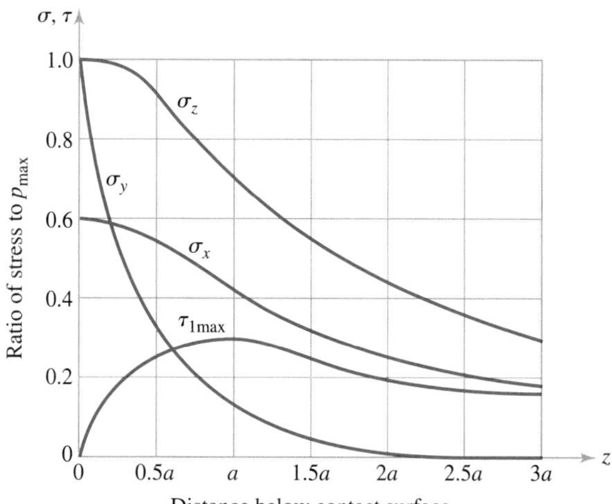

Figure 4.34
Magnitudes of σ_x, σ_y, σ_z, and τ_{max} as a function of maximum contact pressure between cylinders, for various distances z below the contact interface (for $\nu = 0.3$).

Of these three principal shearing stresses, τ_1 is a maximum at about $z = 0.75b$ below the surface, as shown in Figure 4.34, and is larger there than either τ_2 or τ_3 (although τ_1 is not largest for *all* values of z/b).

In some instances the "normal approach" (displacement of the centers of two contacting spheres or cylinders toward each other) caused by the load-induced elastic contact deformations will be of interest. For example, the *overall stiffness* of a machine assembly may be needed for certain design evaluations, requiring therefore that the stiffness be known for *each component* in the assembly. If bearings, cams, or gear teeth are components of an assembly, the normal approach due to contact deformation may be a very significant part of the overall deformation.

For the case of two *spheres* in contact, the normal approach Δ_s is given by[34]

$$\Delta_s = 1.04\sqrt[3]{F^2\left(\frac{1}{d_1} + \frac{1}{d_2}\right)\left[\left(\frac{1-\nu_1^2}{E_1}\right) + \left(\frac{1-\nu_2^2}{E_2}\right)\right]^2} \tag{4-132}$$

For the case of two *parallel cylinders* (of the same material) in contact, the normal approach Δ_c may be calculated, defining $\nu_1 = \nu_2 = \nu$ and $E_1 = E_2 = E$, as

$$\Delta_c = \frac{2F(1-\nu^2)}{\pi LE}\left(\frac{2}{3} + \ln\frac{2d_1}{b} + \ln\frac{2d_2}{b}\right) \tag{4-133a}$$

For a *cylinder in a cylindrical groove*

$$\Delta_c = \frac{2F(1-\nu^2)}{LE}\left(1 - 2\ln\frac{b}{2}\right) \tag{4-133b}$$

It should be noted that these displacements are highly nonlinear functions of the load; hence the *stiffness* characteristics of curved surfaces in contact are highly nonlinear.

Example 4.15 Cylindrical Surfaces in Direct Contact

Figure E4.15 shows the tentative concept for a special load transfer joint in a torque measuring fixture. It is important to accurately maintain distance L at a constant value. The joint is to be constructed using a 2-inch-long cylinder of ultrahigh strength AISI 4340 steel (see

[34]See, for example, ref. 4, Table 33.

Example 4.15 Continues

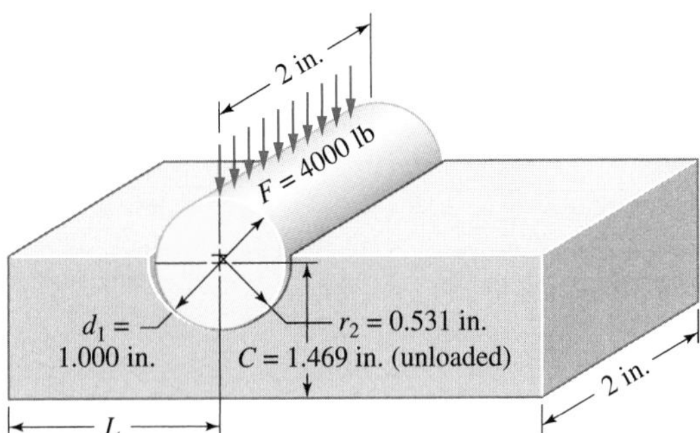

Figure E4.15
Special-load transfer joint, made of AISI 4340 steel.

Tables 3.3 and 3.9) with a diameter of 1.000 inch, residing in a 0.531-inch-radius semicircular groove in a 2-inch-wide block of the same material. It has been estimated that a total force of 4000 lb will be distributed across the cylinder as shown. Experimentation has indicated that for satisfactory service the maximum surface contact pressure should not exceed 250,000 psi. For the proposed design configuration:

a. Determine whether the surface contact pressure is below the specified limiting value.
b. Determine the magnitude of the maximum principal shearing stress in the contact region, and its depth below the surface.
c. Estimate the change in dimension C as the joint goes from the unloaded to the fully loaded condition.

Solution

a. For the parallel cylindrical surfaces in contact, the maximum contact pressure is given by (4-125) as

$$p_{max} = \frac{2F}{\pi b L} \tag{1}$$

where b is given by (4-124). To calculate b, we first find from Table 3.9 and the problem statement that

$$\begin{aligned} E_1 &= E_2 = 30 \times 10^6 \text{ psi} \\ v_1 &= v_2 = v = 0.3 \\ L &= 2.0 \text{ inches} \\ F &= 4000 \text{ lb} \end{aligned} \tag{2}$$

hence

$$b = \sqrt{\frac{2(4000)\left[2\left(\dfrac{1-0.3^2}{30 \times 10^6}\right)\right]}{\pi(2.0)\left[\dfrac{1}{1.000} + \dfrac{1}{-1.062}\right]}} = 0.013 \text{ inch} \tag{3}$$

and from (1)

$$p_{max} = \frac{2(4000)}{\pi(0.013)(2.0)} = 95{,}426 \text{ psi} \tag{4}$$

Since the limiting allowable contact pressure is given as 250,000 psi, the calculated value of p_{max} = 95,426 psi is acceptable.

b. To find the value of $\tau_{1\,max}$, the plot of Figure 4.30 may be utilized. Noting that $\tau_{1\,max}$ peaks at about $z = 0.75b$ (for $\nu = 0.3$), the corresponding ratio of $\tau_{1\,max}/p_{max}$ may be read as

$$\frac{\tau_{1\,max}}{p_{max}} = 0.3 \tag{5}$$

Thus

$$\tau_{1\,max} = 0.3(95{,}426) = 28{,}628 \text{ psi} \tag{6}$$

and it occurs at a depth

$$z = 0.75(0.013) = 0.010 \text{ inch} \tag{7}$$

below the surface. The shearing stress and its depth below the surface are approximately the same in both members since the material is the same for both.

c. The change in dimension C, when the joint is loaded as shown, may be calculated from (4-133b) as

$$\Delta_c = \frac{2(4000)(1 - 0.3^2)}{(2.0)30 \times 10^6}\left(1 - 2\ln\frac{0.013}{2}\right) = 0.0013 \text{ inch} \tag{8}$$

4.11 Load Sharing in Redundant Assemblies and Structures

Force-flow concepts[35] are useful in visualizing the paths taken by lines of force as they pass through a machine or structure from points of load application to points of structural support. If the structure is simple and *statically determinate*, the equations of static equilibrium given in (4-1) are sufficient to calculate all the reaction forces. If, however, there are *redundant* supports, that is, supports in addition to those required to satisfy the conditions of static equilibrium, equations (4-1) are no longer sufficient to explicitly calculate the magnitudes of *any* of the support reactions. Mathematically, this happens because there are more unknowns than equilibrium equations. Physically, it happens because each support behaves as a separate "spring," deflecting under load in proportion to its stiffness, so that the reactions are shared among the supports in an unknown way.

From a mathematical point of view, additional deflection equations must be written and combined with the equilibrium equations of (4-1) so that the number of unknowns matches the number of independent equations.

From a physical point of view it is important to appreciate how load sharing relates to relative stiffness. If a stiff spring, or stiff load path, is in *parallel* with a soft spring or soft load path, as illustrated in Figure 4.35, the stiff path carries more of the load.[36] If a stiff spring or stiff load path is in *series* with a soft spring or load path, as illustrated in Figure 4.36, the loads carried are equal but the stiff spring deflection is smaller than the soft spring deflection.[37] *The importance of these simple concepts cannot be overemphasized because all real machines and structures are combinations of springs in series and/or parallel.* Using these concepts, a designer may quantitatively determine the overall spring rate for any member or combination of members in a machine or structure. Consequently, load carrying and load sharing may be quantitatively evaluated at an early design stage.

Examples of machine and structural elements for which load-sharing concepts may be useful include bearings, gears, splines, bolted joints, screw threads, chain and sprocket

[35]Discussed in 4.3.

[36]May be verified using (4-139).

[37]May be verified using (4-142).

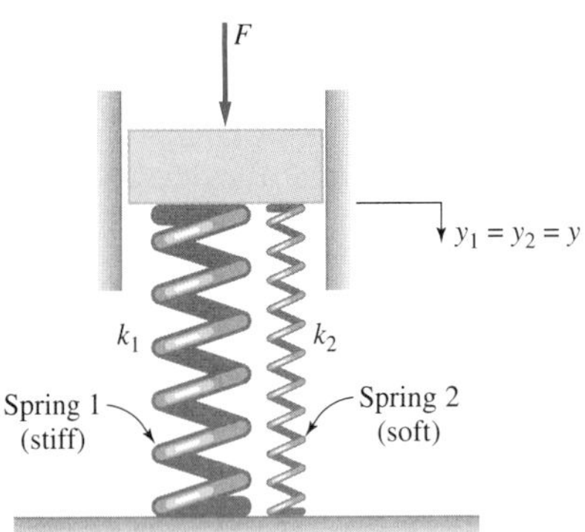

Figure 4.35
Parallel spring configuration.

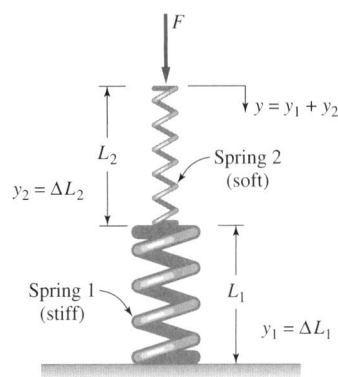

Figure 4.36
Series spring configuration.

drives, timing belt drives, multiple V-belt drives, filamentary composite parts, machine frames, housings, and welded structures. In certain applications, such as machine grips, hub-and-shaft assemblies, or stacked elements subjected to cyclic loads, load-sharing concepts may be used to design for local *strain-matching* at critical interfaces.[38]

Machine Elements as Springs

Although the rigid body assumption is a useful tool in force analysis, no real machine element (or material) is actually rigid. All real machine parts have finite stiffnesses, that is, finite spring rates. A uniform prismatic bar in axial tension as analyzed in 2.4, leads to (2-7), repeated here as

$$k_{ax} = \frac{AE}{L} \tag{4-134}$$

where k_{ax} is axial spring rate, A is cross-sectional area, E is modulus of elasticity, and L is bar length. Figure 2.2 depicts the linear force-deflection curve characterizing an axially loaded uniform bar in tension. Many, but not all, machine elements exhibit *linear* force-deflection curves. Some exhibit *nonlinear* curves, as sketched in Figure 4.37. For example, nonlinear *stiffening* curves are typified by Hertz contact configurations such as bearings, cams, and gear teeth, and nonlinear *softening* curves are associated with arches, shells, and conical-washer (Belleville) springs. In addition to axial loading, other configurations exhibiting linear spring rates include elements in torsion, bending, or direct shear.

Figure 4.37
Various force-deflection characteristics exhibited by springs or machine elements when loaded.

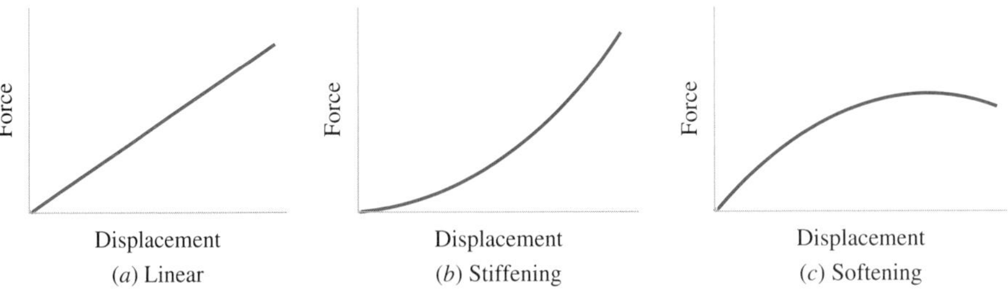

[38]If small-amplitude cyclic relative motions are induced between two contacting surfaces because of differential stiffness qualities of the two mating bodies, fretting fatigue or fretting wear may become potential failure modes. Reducing the amplitude of the cyclic sliding motion by strain-matching may reduce or eliminate fretting. See 2.11.

From earlier equations for deflection, given in 4.4, spring rate (force or torque per unit displacement) expressions may be developed.

For torsionally loaded members, from (4-37),

$$k_{tor} = \frac{T}{\theta} = \frac{KG}{L} \tag{4-135}$$

where k_{tor} is *torsional* spring rate, T is applied torque, θ is angular displacement, G is shear modulus, L is length, and K is a cross-sectional shape constant from Table 4.4.

For members in bending the concept is the same but details are more involved, requiring that members in bending be handled on a case-by-case basis. For example, case 1 of Table 4.1 would give

$$k_{bend\text{-}1} = \frac{P}{y_{load}} = \frac{48EI}{L^3} \tag{4-136}$$

where $k_{bend\text{-}1}$ is *bending* spring rate for case 1 of Table 4.1, P is concentrated center load, y_{load} is deflection at the loading point, E is modulus of elasticity, and L is length of the simply supported beam.

For simple members in direct shear,[39] the spring rate is

$$k_{dir\text{-}sh} = \frac{P}{\delta_P} = \frac{AG}{L} \tag{4-137}$$

where $k_{dir\text{-}sh}$ is the *direct shear* spring rate, δ_P is displacement of the loaded edge, and G is shear modulus of elasticity.

For nonlinear springs such as bearings or gear teeth (Hertzian contacts), the spring rate is not constant and cannot be simply described. A *linearized* spring rate approximation is often defined about the operating point, especially if vibrational characteristics are of interest.

As illustrated by Figures 4.35 and 4.36, spring elements may combine in either parallel or series arrangements. If springs combine in *parallel*, as shown in Figure 4.35, the displacements are equal but the total force F is split between spring 1 and spring 2. That is,

$$F = F_1 + F_2 \tag{4-138}$$

and since $y_1 = y_2 = y$,

$$\frac{F}{y} = \frac{F_1}{y_1} + \frac{F_2}{y_2} \tag{4-139}$$

or

$$k_p = k_1 + k_2 \tag{4-140}$$

where k_p is the combined spring rate for the springs in parallel.

On the other hand, if springs combine in *series*, as shown in Figure 4.36, the force F is the same on *both* springs but displacements for spring 1 and spring 2 add to give total displacement y. That is,

$$y = y_1 + y_2 \tag{4-141}$$

Since $F = F_1 = F_2$, (4-141) may be rewritten as

$$\frac{y}{F} = \frac{y_1}{F_1} + \frac{y_2}{F_2} \tag{4-142}$$

[39]For example, a short block of "length" L and cross-sectional "shear" area A, if fixed at one "end" and loaded at the other by a force P, illustrates the case of direct shear.

from which

$$k_s = \frac{1}{\dfrac{1}{k_1} + \dfrac{1}{k_2}} \tag{4-143}$$

where k_s is the combined spring rate for the springs in series.

These results may be extended to any number of spring elements in series, in parallel, or in any series-parallel network.

Example 4.16 Spring Rates and End Deflection of a Right-Angle Support Bracket

The steel right-angle support bracket with leg lengths $L_1 = 10$ inches and $L_2 = 5$ inches, as shown in Figure E4.16, is to be used to support the static load $P = 1000$ lb. The load is to be applied vertically at the free end of the cylindrical leg, as shown. Both bracket leg centerlines lie in the same horizontal plane. If the square leg has side $s = 1.25$ inches, and the cylindrical leg has diameter $d = 1.25$ inches,

a. Develop an expression for the overall spring rate k_o of the bracket in the vertical direction at the point of load application (point O).

b. Calculate the numerical value of k_o.

c. Find the deflection y_o for a load $P = 1000$ lb.

Solution

a. The overall spring rate at point O is

$$k_o = \frac{P}{y_o} \tag{1}$$

It may be determined by noting that effective "springs" in the bracket include

1. k_1, caused by bending of the square leg
2. k_2, caused by torsion of the square leg reflected to point O through rigid body rotation of cylinder leg length L_2
3. k_3, caused by bending of the cylindrical leg

Since these three springs are in series ($y = y_1 + y_2 + y_3$ and $P = P_1 = P_2 = P_3$), from (4-113)

$$k_o = \frac{1}{\dfrac{1}{k_1} + \dfrac{1}{k_2} + \dfrac{1}{k_3}} \tag{2}$$

Using case 8 of Table 4.1,

$$k_1 = \frac{P}{y_1} = \frac{3EI}{L_1^3} \tag{3}$$

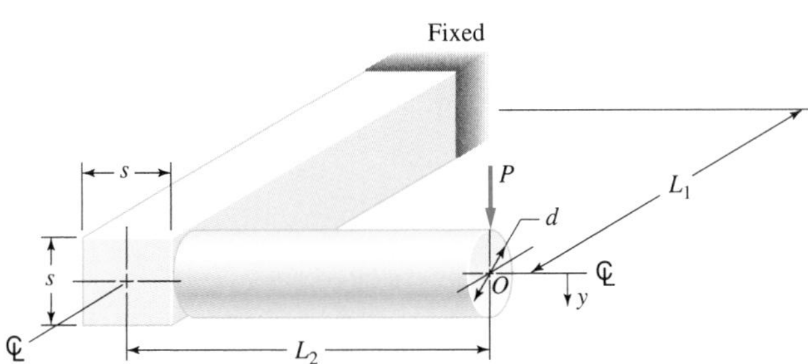

Figure E4.16
Right-angle support bracket.

For the square cross section

$$I = \frac{bd^3}{12} = \frac{s^4}{12} \tag{4}$$

and

$$k_1 = \frac{3Es^4}{12L_1^3} = \frac{Es^4}{4L_1^3} \tag{5}$$

Next, for torsion of the square cross section, using case 3 of Table 4.4 with (4-37), and recalling from geometry that $y_2 = L_2\theta$,

$$k_2 = \frac{P}{y_2} = \frac{P}{L_2\theta} = \frac{KG}{L_1L_2^2} \tag{6}$$

where, from Table 4.4,

$$K = 2.25\left(\frac{s}{2}\right)^4 \tag{7}$$

Hence

$$k_2 = \frac{0.14s^4G}{L_1L_2^2} \tag{8}$$

Finally, again using case 8 of Table 4.1, for bending of the cylindrical leg,

$$k_3 = \frac{P}{y_3} = \frac{3EI}{L_2^3} \tag{9}$$

For the circular cross section

$$I = \frac{\pi d^4}{64} \tag{10}$$

and

$$k_3 = \frac{3\pi Ed^4}{64L_2^3} \tag{11}$$

From (2) then

$$k_o = \frac{1}{\dfrac{4L_1^3}{Es^4} + \dfrac{L_1L_2^2}{0.14Gs^4} + \dfrac{64L_2^3}{3\pi Ed^4}} \tag{12}$$

b. Since the material is steel, $E = 30 \times 10^6$ psi and $G = 11.5 \times 10^6$ psi, and

$$k_o = \frac{1}{\dfrac{4(10)^3}{30 \times 10^6(1.25)^4} + \dfrac{(10)(5)^2}{0.14(11.5 \times 10^6)(1.25)^4} + \dfrac{64(5)^3}{3\pi(30 \times 10^6)(1.25)^4}}$$

$$= 7.70 \times 10^3 \frac{\text{lb}}{\text{in}} \tag{13}$$

c. From (1)

$$y_o = \frac{P}{k_o} = \frac{1000}{7.70 \times 10^3} = 0.13 \text{ inch} \tag{14}$$

Example 4.17 Load-Sharing in a Redundant Structure

The pinned-joint structure shown in Figure E4.17A is to be used to support the load P. The structural members are solid cylindrical steel bars. For the dimensions given, how would the load P be shared among the bars?

Solution

For this *statically indeterminate structure*, the load will be shared in proportion to stiffness among the three links (springs) in parallel. From Figure E4.17B the spring rate for link 1 is (in the vertical direction)

$$k_{1V} = \frac{F_{1V}}{y} = \frac{F_1 \cos\alpha}{y_1/\cos\alpha} = \frac{F_1 \cos^2\alpha}{y_1} \tag{1}$$

or, using (4-134),

$$k_{1V} = \frac{A_1 E}{L_1}\cos^2\alpha \tag{2}$$

For link 2

$$k_{2V} = \frac{F_{2V}}{y} = \frac{F_2}{y_2} = \frac{A_2 E}{L_2} \tag{3}$$

and for link 3

$$k_{3V} = \frac{F_{3V}}{y} = \frac{F_3 \cos\alpha}{y_3/\cos\alpha} = \frac{A_3 E}{L_3}\cos^2\alpha \tag{4}$$

Noting that

$$L_2 = L_3 = \frac{L_1}{\cos\alpha} \tag{5}$$

these spring rates may be written as

$$k_{1V} = \frac{\pi E}{4}\left(\frac{d_1^2 \cos^2\alpha}{L_2 \cos\alpha}\right) \tag{6}$$

$$k_{2V} = \frac{\pi E}{4}\left(\frac{d_2^2}{L_2}\right) \tag{7}$$

$$k_{3V} = \frac{\pi E}{4}\left(\frac{d_3^2 \cos^2\alpha}{L_2 \cos\alpha}\right) \tag{8}$$

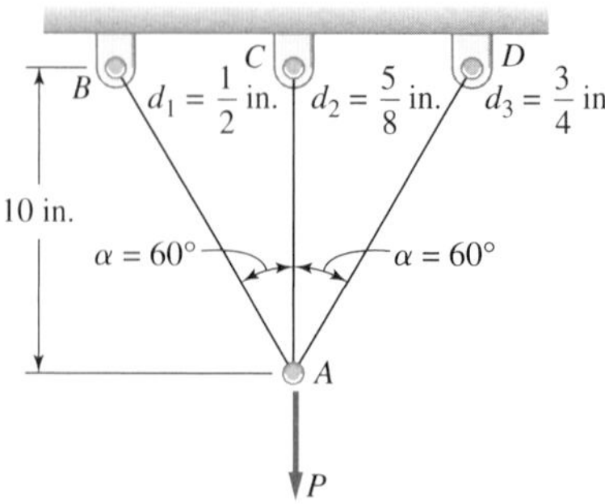

Figure E4.17A
Pinned-joint structure.

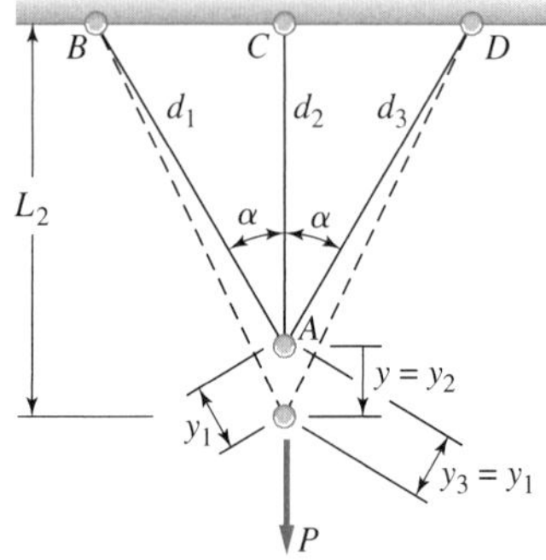

Figure E4.17B
Deflections of pinned joint structure.

Since the links are in parallel, the combined spring rate, vertically at point A, is

$$k_A = \frac{\pi E}{4L_2}(d_1^2 \cos\alpha + d_2^2 + d_3^2 \cos\alpha) \tag{9}$$

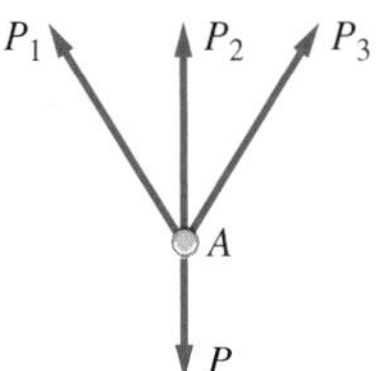

Figure E4.17C
Free-body diagram of joint A.

From Figure E4.17C the load-sharing proportions then are, for links 1, 2, and 3 respectively,

$$\frac{P_1}{P} = \frac{k_{1V}}{k_A} \tag{10}$$

$$\frac{P_2}{P} = \frac{k_{2V}}{k_A} \tag{11}$$

and

$$\frac{P_3}{P} = \frac{k_{3V}}{k_A} \tag{12}$$

Evaluating these three expressions numerically,

$$k_{1V} = \frac{\pi(30 \times 10^6)}{4(10)}(0.5)^2 \cos 60 = 2.95 \times 10^5 \frac{\text{lb}}{\text{in}} \tag{13}$$

$$k_{2V} = \frac{\pi(30 \times 10^6)}{4(10)}(0.625)^2 = 9.20 \times 10^5 \frac{\text{lb}}{\text{in}} \tag{14}$$

$$k_{3V} = \frac{\pi(30 \times 10^6)}{4(10)}(0.75)^2 \cos 60 = 6.63 \times 10^5 \frac{\text{lb}}{\text{in}} \tag{15}$$

whence

$$k_A = k_{1V} + k_{2V} + k_{3V} = (2.95 + 9.20 + 6.63)10^5 = 18.78 \times 10^5 \frac{\text{lb}}{\text{in}} \tag{16}$$

Using (10), (11), and (12) then gives

$$P_1 = \frac{2.95}{18.78}P = 0.157P \tag{17}$$

$$P_2 = \frac{9.20}{18.78}P = 0.490P \tag{18}$$

$$P_3 = \frac{6.63}{18.78}P = 0.353P \tag{19}$$

4.12 Preloading Concepts

Preloads may be induced in structures or machine assemblies by forcing two or more parts together or apart (intentionally or unintentionally), then clamping them in such a way that the tension in some parts equilibrates the compression in other parts. A consequence of preloading is that "built-in" stresses are produced within the assembled device, with no externally applied loads. As discussed in 4.11, parts that are preloaded against each other behave as integrated systems of springs in series and/or parallel. Proper preloading has many potential advantages, including elimination of unwanted clearance gaps between parts, increased stiffness of machine assemblies, and improved fatigue resistance of component parts. Examples of components and/or assemblies that may display significant improvements in performance as a result of proper preloading include bearing assemblies, gear trains, bolted joints, flange-and-gasket seals, and springs. Preloading may be used to increase the axial or radial stiffness of rolling element bearings,[40] to eliminate backlash from gear meshes, to avoid separation of bolted joints subjected to cyclic loading, to prevent separation of flange-and-gasket seals under fluctuating loads or pressures,[41] and to improve the dynamic response characteristics of cyclically loaded assemblies. It is important to remember, however, that when determining the dimensions of critical cross sections, the "built-in" stresses induced by preloading must always be *superposed* upon the stresses produced by operational loads.

Enhancement of fatigue resistance by initially preloading parts that are to be subjected to fluctuating loads is a particularly useful concept. The effect of tensile preloading on a completely reversed cyclically loaded member is to increase the *mean* stress from *zero* to a significant *tensile* value.[42] The preload typically has a small effect on the *maximum* cyclic stress. Thus (2-42) provides some insight into the improvement of fatigue resistance as a result of preloading. Rewriting (2-42) gives

$$\sigma_{eq\text{-}CR} = \frac{\sigma_a}{1 - \dfrac{\sigma_m}{S_u}} = \frac{\sigma_{max} - \sigma_m}{1 - \dfrac{\sigma_m}{S_u}} \quad \text{for} \quad \sigma_m \geq 0 \quad \text{and} \quad \sigma_{max} \leq S_{yp}$$

$$\sigma_{eq\text{-}CR} = S_{yp} \quad \text{for} \quad \sigma_m \geq 0 \quad \text{and} \quad \sigma_{max} \geq S_{yp} \tag{4-144}$$

If, for exmaple, the external load on a selected member induces a maximum stress value of, say, $\sigma_{max} = 0.75S_u$, and the *mean stress is zero*, from (4-144) the equivalent completely reversed cyclic stress amplitude is

$$\sigma_{eq\text{-}CR} = \sigma_{max} = 0.75S_u \tag{4-145}$$

If the cyclically loaded member is initially preloaded to produce a *tensile mean stress* of, say, $0.9\sigma_{max}$, and operational cyclic loading remains unchanged, (4-144) gives

$$\sigma_{eq\text{-}CR} = \frac{\sigma_{max} - 0.9\sigma_{max}}{1 - \dfrac{0.9\sigma_{max}}{S_u}} = 0.31\sigma_{max} = 0.23S_u \tag{4-146}$$

For this case, preloading would reduce the equivalent completely reversed cyclic stress from $0.75S_u$ to $0.23S_u$, approximately a threefold reduction in equivalent stress amplitude. The corresponding factor on *life improvement* would usually be even larger. Further discussions of fatigue life enhancement are presented in later chapters as they pertain to the specific machine element or assembly under consideration.

Example 4.18 System Stiffness as a Function of Preload

The "rigid" block shown in Figure E4.18A is supported on a "frictionless" surface between two linear compression springs. The outer end of each spring abutts a "fixed" wall,[43] and there are no gaps between the inner spring ends and the block. Also, there is no "attachment" of the block to the inner spring ends (hence, the block separates from the spring if tension is applied). For case 1 shown in Figure E4.18A, there is no preload. For case 2, the

[40] See 11.6.

[41] See 13.4.

[42] See 2.6.

[43] Of course there are in reality no *rigid* blocks, *frictionless* surfaces, or *fixed* walls. In any real application the validity of such assumptions would require evaluation. Nevertheless, the preloading concepts remain valid.

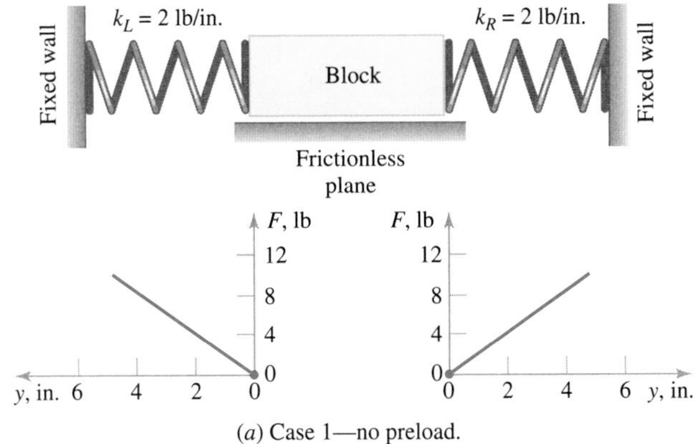

(*a*) Case 1—no preload.

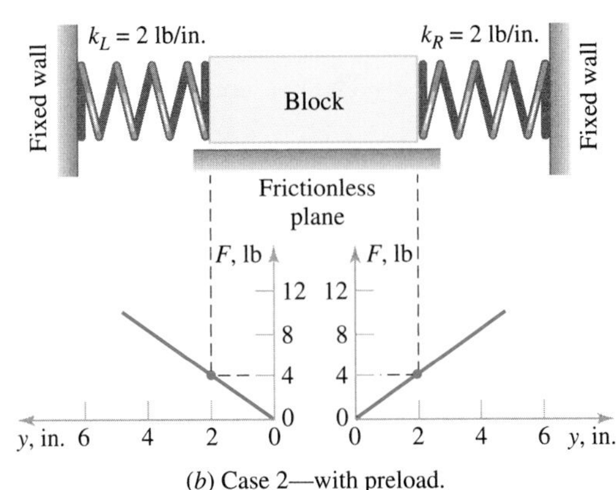

(*b*) Case 2—with preload.

Figure E4.18A
Comparison of behavior of preloaded spring system versus non-preloaded system.

fixed outer walls are relocated closer together, squeezing the springs compressively to produce a "built-in" 4-lb spring force (preload) in each spring. The linear force-deflection curve is shown in Figure E4.18A for each spring, both for the case of *no preload* and for the case of a *4-lb preload.*

Compare the system stiffness for the non-preloaded case with the system stiffness for the preloaded case.

Solution

For *non-preloaded* case 1, the rigid block may be taken as a free body, as shown in Figure E4.18B. Summing forces horizontally for case 1,

$$F_{L1} + F_1 = F_{R1} \tag{1}$$

Since there is no preload, any rigid block displacement y to the right causes separation of the block from the left spring, hence

$$F_{L1} = 0 \tag{2}$$

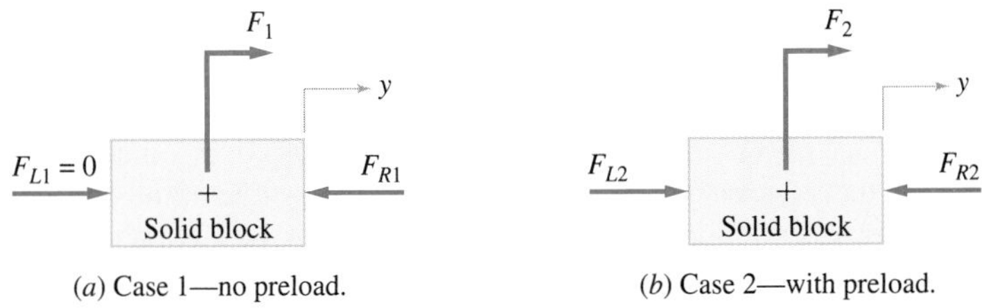

(*a*) Case 1—no preload. (*b*) Case 2—with preload.

Figure E4.18B
Rigid block from Figure E4.18A, taken as a free body.

Example 4.18 Continues

Thus (1) becomes

$$F_1 = F_{R1} = k_R y = 2y \tag{3}$$

and the spring rate k_{NP} for the *non-preloaded* system may be written as

$$k_{NP} = \frac{F_1}{y} = 2\frac{\text{lb}}{\text{in}} \tag{4}$$

For *preloaded* case 2, the rigid block may again be taken as a free body, as shown in Figure E4.18B. Summing forces horizontally for case 2,

$$F_{L2} + F_2 = F_{R2} \tag{5}$$

Since there is a preload of 4 lb initialy built into the system, for the zero-displacement equilibrium position shown in Figure E4.18A, for case 2 $F_2 = 0$, giving

$$F_{L2} = F_{R2} = 4 \text{ lb} \tag{6}$$

For displacements y to the right, reading the force-displacement diagrams for case 2, shown in Figure E4.18A,

$$F_{L2} = 4 - k_L y \qquad \text{for} \quad y \le \frac{4}{k_L} \tag{7}$$

and

$$F_{L2} = 0 \qquad \text{for} \quad y \ge \frac{4}{k_L} \tag{8}$$

Also for case 2,

$$F_{R2} = 4 + k_R y \tag{9}$$

From (5), with a 4-lb preload,

$$4 - k_L y + F_2 = 4 + k_R y \qquad \text{for} \quad y \le \frac{4}{k_L} \tag{10}$$

or

$$F_2 = (k_L + k_R)y \qquad \text{for} \quad y \le \frac{4}{k_L} \tag{11}$$

and

$$F_2 = F_{R2} = 4 + k_R y \qquad \text{for} \quad y > \frac{4}{k_L} \tag{12}$$

so the spring rate k_P for the *preloaded* system may be written as

$$k_P = \frac{F_2}{y} = k_L + k_R = 4\frac{\text{lb}}{\text{in}} \qquad \text{for} \quad y \le 2 \text{ inches} \tag{13}$$

and

$$k_P = \frac{F_2}{y} = \frac{4}{y} + k_R = \frac{4}{y} + 2\frac{\text{lb}}{\text{in}} \qquad \text{for} \quad y \ge 2 \text{ inches} \tag{14}$$

Thus the stiffness $k_{NP} = 2$ lb/in, calculated in (4) for the *non-preloaded* system, has been doubled to $k_P = 4$ lb/in, calculated in (13) for the *preloaded* system, as long as rigid block displacements are less than 2 inches. For displacements over 2 inches, a gap begins to open between the left spring and the block, and the spring rate of the preloaded system falls off. However, in a well-designed system no gaps would be permitted to open.

4.13 Residual Stresses

A usual premise in stress-analyzing machine elements is that *if the load is zero, the stress is zero*. However, in 4.12 the discussion of preloading introduces the idea that two or more parts may be clamped together in such a way that built-in tension in some parts opposes built-in compression in others. Consequently, *nonzero stresses* are induced in the parts by preloading, even when the *external loads are all zero*.

Additionally, when ductile machine parts have stress concentration sites (due to holes, fillets, etc.), or regions where load-induced stress gradients exist (due to bending, torsion, etc.), yield strength levels may be exceeded locally, producing highly local plastic flow. These small regions of high stress and highly local plastic flow produce little or no observable change in macroscopic dimensions or appearance of the loaded member.[44] When external loads are released, the elastic "springback" of the bulk of the material forces the small plastically deformed regions back into an *oppositely* stressed condition. These stresses, induced by nonuniform plastic deformation, are called *residual stresses* because they persist *after* the external loading has been *removed*. Residual stresses may be induced, either accidentally or intentionally, during processing, machining, and/or assembly operations, including heat treatment, shot-peening, rolling, forming, presetting, turning, milling, grinding, polishing, plastic bending or torsion, shrink-fit assembly processes, overpressurization, or overspeeding. The residual stresses induced may be either tensile or compressive, depending on how they are generated, and are typically difficult to detect, measure, or estimate.[45] Since residual stresses at any critical point add directly to the load-induced operational stresses at the same point, residual stresses, although unknown or even unsuspected, may greatly influence the potential for failure. That is, *residual stresses may be either very harmful or very beneficial*, depending upon how they combine with the operational stresses.

Another basic concept in generating beneficial residual stresses is that residual stress should tend to *oppose* operational stress at the critical point, so that the resultant stress is minimized. While the concept is true both for static and fluctuating loads, it is especially important for fluctuating loads because they may lead to fatigue failure. Since a very high proportion of fatigue cracks originate at surfaces of machine parts, significant improvement in fatigue resistance may often be made by treatments that strengthen only the *surface layers* by inducing a favorable (usually compressive) residual stress field there. As discussed in 2.6, "Factors That May Affect *S-N* Curves," the processes of shot-peening, cold-rolling, and presetting all may be used to intentionally induce compressive residual surface stresses. Figure 2.23 illustrates the effects of *shot-peening*, and Figure 2.24 the effects of *cold-rolling*. Figure 4.38 illustrates the effects of *axial static prestress* on the fatigue strength of 7075-T6 aluminum specimens. Note in Figure 4.38 that residual compressive stress at the notch root (induced by tensile prestress) significantly improves fatigue resistance, while residual tensile stress at the notch root (induced by compressive prestress) significantly diminishes fatigue strength. This is an example of a basic principle of effective prestressing, namely, *the imposition of an overload that causes local yielding will produce a residual stress field favorable to future loads in the same direction, and unfavorable to future loads in the opposite direction*. Thus, if prestressing is to be used, the prestressing force should always be imposed in the *same direction* as anticipated operational loads.

[44]See 4.8.

[45]See, for example, ref. 10, pp. 461–465, or ref. 13, p. 659 ff.

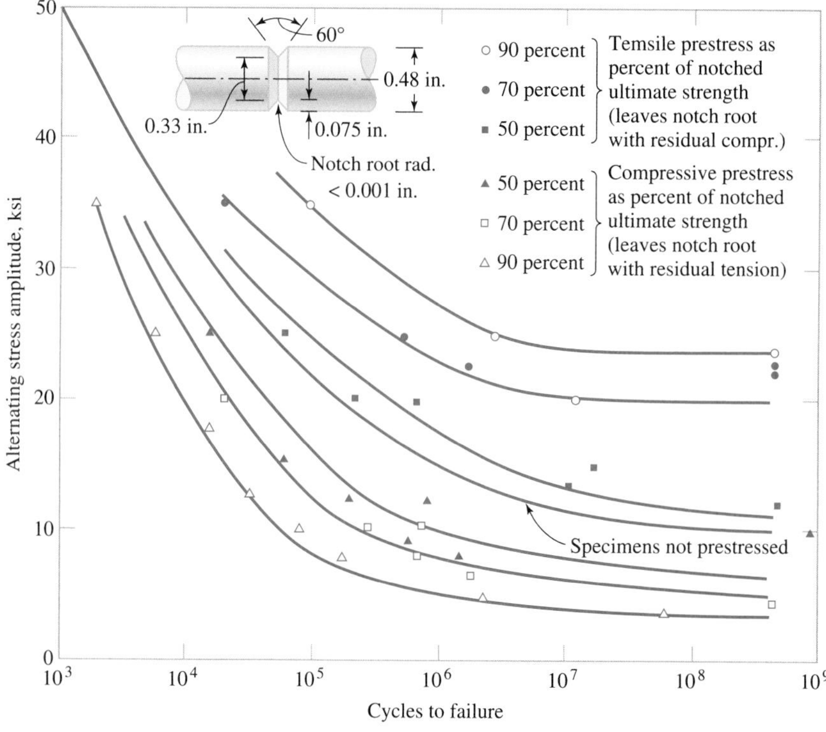

Figure 4.38
Effects of axial static prestress on the *S-N* curve of 7075-T6 specimens tested in rotating bending fatigue tests. (Data from ref. 15.)

Estimating Residual Stress

For simple loading and simple geometry, residual stresses may be estimated if the stress-strain properties are known for the material and stress concentration factors are available for the geometry of the part. For example, if the material behavior is idealized to have an elastic–perfectly plastic stress-strain curve, as shown in Figure 4.39, the residual stresses usually may be estimated. For instance, an axially loaded uniform straight cylindrical bar with a circumferential semicircular notch, such as the one shown in Figure 4.40(a), may be analyzed for residual stresses if the dimensions of the bar and the notch are known. The procedure for making such an estimate would be to

1. Define a specific (idealized) stress-strain curve for the material, in a format similar to Figure 4.39.
2. From loading and geometry determine stress concentration factor K_t, using the method discussed in 4.8 and a chart of the type shown in Figure 4.18.

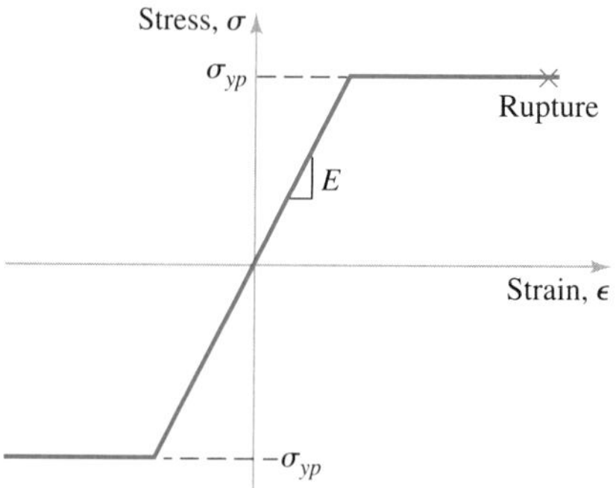

Figure 4.39
Stress-strain curve for idealized elastic–perfectly plastic material.

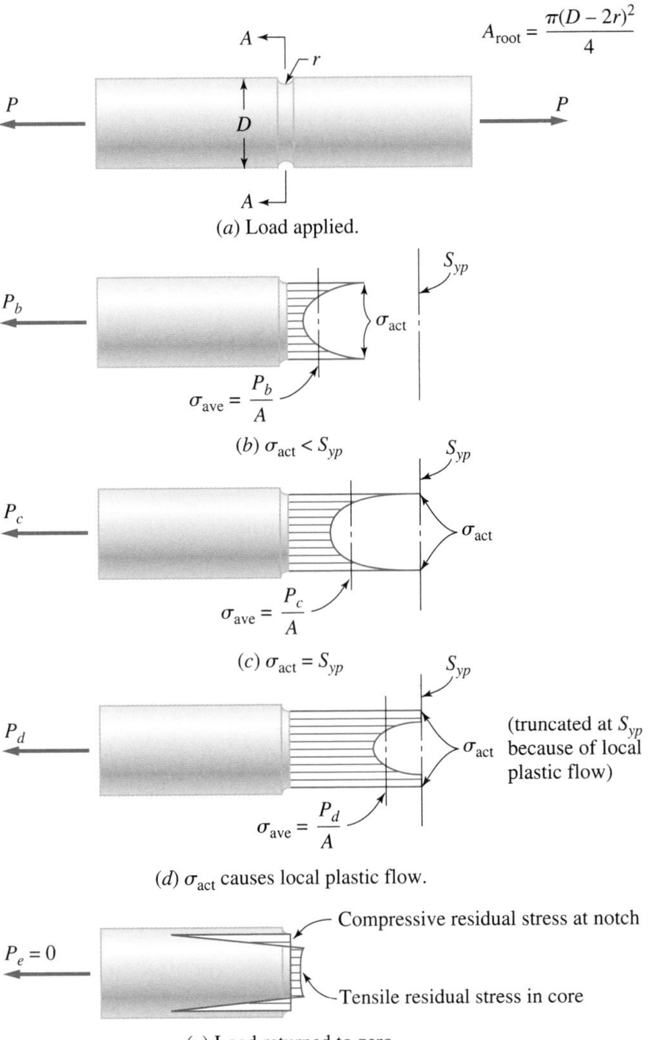

Figure 4.40
Stress distributions for selected sequence of applied axial load *P* to a notched tension bar.

3. Calculate $\sigma_{act} = K_t\sigma_{nom}$ as a function of axial load P, and visualize changes in stress distribution on cutting plane *A-A* using a series of free-body sketches related to increasing values of P, as shown in Figure 4.40. Note in Figure 4.40 that for free-body sketch (b), $\sigma_{act} < S_{yp}$, for (c) $\sigma_{act} = S_{yp}$, for (d) σ_{act} tries to exceed S_{yp} but cannot due to perfectly plastic behavior (consequently local plastic flow occurs to reestablish equilibrium), and for (e) load P is returned to zero, giving rise to elastic springback that forces a pattern of residual stresses, as shown. The compressive residual stress at the surface is typically about one-half the strain-hardened yield strength of the material when well-designed prestressing procedures are used. It is important to note that when subsequent operational loads are applied in the same direction as the preload force, the compressive residual stresses add algebraically to the tensile operational stresses. This results in a reduction of the peak resultant surface stress, as long as the new operational load *P does not exceed* P_c [see Figure 4.40(c)]. These basic concepts apply to bending and torsional prestressing configurations as well as to axial configurations.[46]

[46]See, for example, ref. 10, pp. 103 ff.

Example 4.19 Estimating Residual Stress Distribution

The simply supported beam in Figure E4.19A is subjected to a pure applied bending moment $M_a = 60{,}000$ in-lb, as shown. The steel material is ductile and has a yield strength of $S_{yp} = 70{,}000$ psi. Elastic–perfectly plastic behavior may be assumed, and buckling is not a problem.

a. Calculate the moment M_{fy} at which first yielding occurs, and sketch the stress distribution at midspan when M_{fy} is applied.

b. Calculate and sketch the midspan stress distribution when moment M_a is applied.

c. Find the residual stresses when the moment M_a is removed, and sketch the corresponding residual stress distribution.

Solution

a. Using (4-5) with $\sigma_{max} = S_{yp}$,

$$S_{yp} = \frac{M_{fy}c}{I} = \frac{6M_{fy}}{bd^2} \tag{1}$$

or

$$M_{fy} = \frac{S_{yp}bd^2}{6} = \frac{(70{,}000)(1)(2)^2}{6} = 46{,}667 \text{ in-lb} \tag{2}$$

is the moment that causes first yielding at the outer fibers. The stress distribution produced by this moment is shown in Figure E4.19B.

b. Since $M_a = 60{,}000$ in-lb exceeds $M_{fy} = 46{,}667$ in-lb, plastic flow occurs at the outer fibers, to some depth d_p, as indicated in Figure E4.19C. Noting that the moment of resultant forces F_p due to stresses in the plastic zones of the distribution plus the moment of resultant forces F_e due to stresses in the elastic zones must balance the applied moment M_a, we find

$$F_p(70{,}000)(d_p)(1) = 70{,}000d_p \tag{3}$$

$$F_e = \frac{1}{2}(70{,}000)(1 - d_p)(1) = 35{,}000(1 - d_p) \tag{4}$$

and by moment equilibrium

$$(70{,}000d_p)2\left(1 - \frac{d_p}{2}\right) + 35{,}000(1 - d_p)2\left[\frac{2}{3}(1 - d_p)\right] = M_a = 60{,}000 \tag{5}$$

or

$$d_p = 1 \pm 0.654 \tag{6}$$

Since d_p physically cannot exceed 1.0 inch,

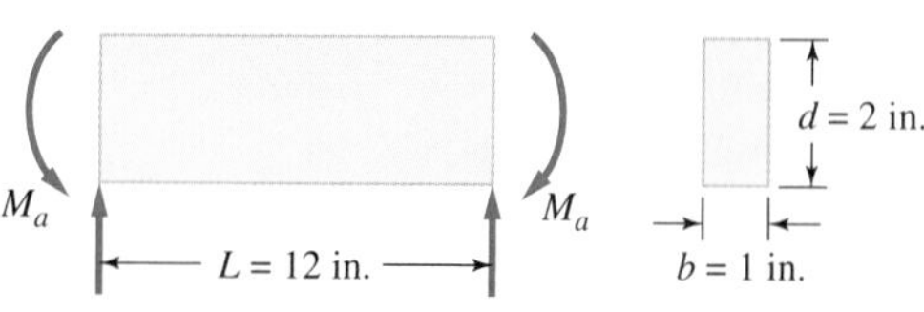

Figure E4.19A
Simple beam with rectangular cross section, subjected to pure bending moment.

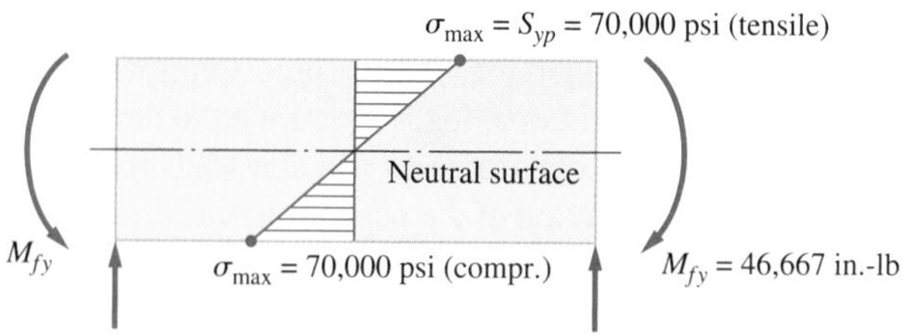

Figure E4.19B
Stress distribution when first yielding occurs.

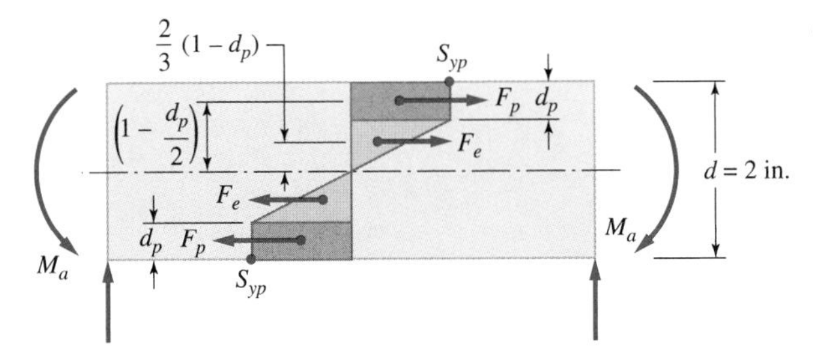

Figure E4.19C
Stress distribution when M_a = 60,000 in-lb is applied.

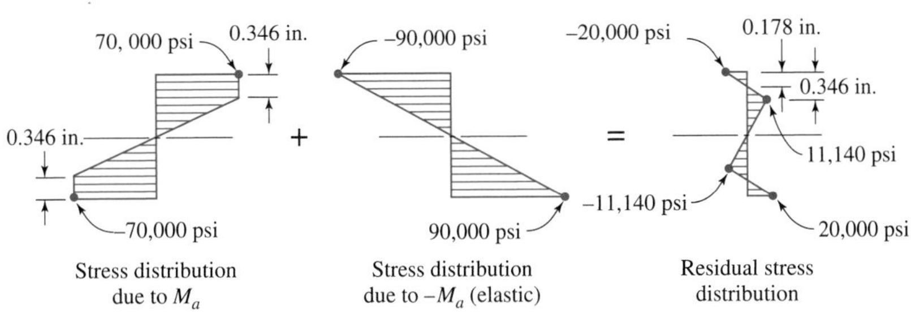

Figure E4.19D
Residual stress distribution in beam obtained by superposing stresses due to $+M_a$ upon stresses due to $-M_a$.

$$d_p = 0.346 \text{ inch} \tag{7}$$

Thus the midspan stress distribution is shown in Figure E4.19C, with d_p = 0.346 inch.

c. To find the residual stress distribution when M_d = 60,000 in-lb is released, it may be noted that releasing M_a is equivalent to superposing a moment of $-M_a$ upon the stress distribution of Figure E4.19C, where the superposed stresses will remain completely elastic (as long as all of the residual stresses, as displayed in Figure E4.19D, are within the elastic range). Thus, when $-M_a$ is applied, the outer fiber (elastic) stress $\sigma_{max\text{-}e}$ is

$$\sigma_{max\text{-}e} = -\frac{M_a c}{I} = -\frac{6M_a}{bd^2} = -\frac{6(60{,}000)}{1(2)^2} = -90{,}000 \text{ psi} \tag{8}$$

Superposing this stress distribution upon the distribution shown in Figure E4.19C gives the residual stress distribution in the unloaded beam, as indicated in Figure 4.19D.

Example 4.20 Fatigue Life Improvement Resulting from Shot-Peening (Favorable Residual Stresses)

A cylindrical support bar is to be subjected to a *released* cyclic tensile load of P_{max} = 23,000 lb. The critical section in the bar is at the root of a circumferential groove of semicircular profile, having the dimensions shown in Figure E4.20A. The material is to be a wrought carbon steel alloy with S_u = 76,000 psi, S_{yp} = 42,000 psi, and e (2 inches) = 18 percent when in hot rolled condition, and S_u = 70,000 psi, S_{yp} = 54,000 psi, and e (2 inches) = 15 percent

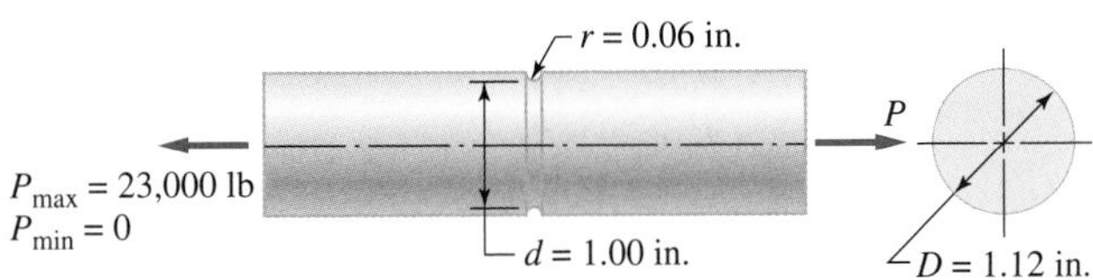

Figure E4.20A
Cyclically loaded cylindrical steel support bar with circumferential groove of semi-circular profile.

Example 4.20 Continues

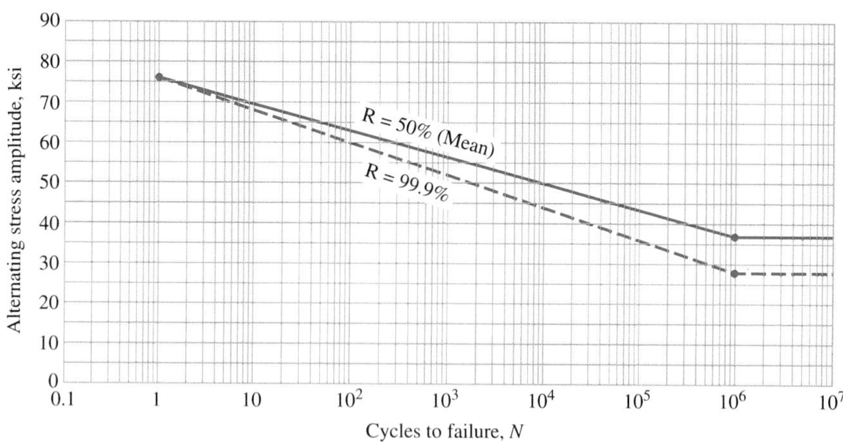

Figure E4.20B

Estimated *S-N* curves for fatigue strength reliability levels of 50% and 99.9% (repeat of Figure E2.6A). Note that S_f for R = 50% is 38,000 psi and S_f for R = 99.9% is 28,600 psi.

when in cold drawn condition. Fatigue properties are defined by the *S-N* curve from Example 2.6, and repeated here as Figure E4.20B.

a. Using a fatigue strength reliability level of 99.9 percent, estimate the life of the bar under the conditions given, if the material is in hot-rolled condition.

b. Estimate the improvement in life (if any) that might be expected from the hot-rolled bar if the groove is properly shot-peened, assuming that the shot-peening process results in a compressive residual surface stress of about 50 percent of the strain-hardened yield strength.

Solution

a. For the *released* loading applied, with P_{max} = 23,000 lb, the nominal cyclic stress ranges from

$$(\sigma_{min})_{nom} = 0 \tag{1}$$

to

$$(\sigma_{max})_{nom} = \frac{P_{max}}{A_{net}} = \frac{23{,}000}{\dfrac{\pi(1.00)^2}{4}} = 29{,}285 \text{ psi} \tag{2}$$

giving nominal values of stress amplitude and tensile mean stress of

$$(\sigma_a)_{nom} = \frac{29{,}285 - 0}{2} = 14{,}642 \text{ psi} \tag{3}$$

and

$$(\sigma_m)_{nom} = \frac{29{,}285 + 0}{2} = 14{,}642 \text{ psi} \tag{4}$$

Using methods developed in 4.8, since the material is ductile,

$$(\sigma_a)_{act} = K_f(\sigma_a)_{nom} \tag{5}$$

and

$$(\sigma_m)_{act} = (\sigma_m)_{nom} \tag{6}$$

Then from (4-110)

$$K_f = q(K_t - 1) + 1 \tag{7}$$

where q may be read from Figure 4.24 with $r = 0.06$ and $S_u = 76{,}000$ psi, as

$$q = 0.74 \tag{8}$$

From Figure 4.18(b), with $r/d = 0.06$ and $D/d = 1.12$, K_t may be read as

$$K_t = 2.18 \tag{9}$$

giving

$$K_f = 0.74(2.18 - 1) + 1 = 1.87 \tag{10}$$

From this, (5) and (6) become,

$$(\sigma_a)_{act} = 1.87(14{,}642) = 27{,}381 \text{ psi} \tag{11}$$

and

$$(\sigma_m)_{act} = 14{,}642 \text{ psi} \tag{12}$$

Therefore, *effective* values for σ_{max} and σ_{min} are

$$(\sigma_{max})_{eff} = 27{,}381 + 14{,}642 = 42{,}023 \text{ psi} \tag{13}$$

and

$$(\sigma_{min})_{eff} = 14{,}642 - 27{,}381 = -12{,}739 \text{ psi} \tag{14}$$

Now, using (2-42),

$$\sigma_{eq\text{-}CR} = \frac{(\sigma_a)_{act}}{1 - \dfrac{\sigma_m}{S_u}} = \frac{27{,}381}{1 - \dfrac{14{,}642}{76{,}000}} = 33{,}915 \text{ psi} \tag{15}$$

Taking this equivalent completely reversed stress amplitude to the 99.9 percent reliability *S-N* curve of Figure E4.20B, the estimated life is read as

$$N = 2 \times 10^5 \text{ cycles} \tag{16}$$

b. If the groove were shot-peened to give a residual compressive stress of 50 percent of the strain-hardened yield strength, the compressive residual stress at the surface would be

$$\sigma_{res} = -0.5(54{,}000) = -27{,}000 \text{ psi} \tag{17}$$

Superposing the residual stress on the effective operational stresses at the notch root, (13) and (14) would be changed to

$$(\sigma_{max})_{sp\text{-}eff} = 42{,}023 + (-27{,}000) = 15{,}023 \text{ psi} \tag{18}$$

and

$$(\sigma_{min})_{sp\text{-}eff} = -12{,}739 + (-27{,}000) = -39{,}739 \text{ psi} \tag{19}$$

so

$$(\sigma_a)_{sp\text{-}eff} = \frac{15{,}023 - (-39{,}739)}{2} = 27{,}381 \text{ psi} \tag{20}$$

and

$$(\sigma_m)_{sp\text{-}eff} = \frac{15{,}023 + (-39{,}739)}{2} = -12{,}358 \text{ psi} \tag{21}$$

For the shot-peened groove, then, since mean stress is compressive, using (2-43),

Example 4.20 Continues

$$\sigma_{eq\text{-}CR} = (\sigma_a)_{sp\text{-}eff} = 27{,}381 \text{ psi} \tag{22}$$

Taking the equivalent completely reversed stress amplitude to the 99.9 percent reliability *S-N* curve of Figure E4.20B, the estimated life is read as

$$N_{sp} = \infty \text{ cycles} \tag{23}$$

because

$$\sigma_{eq\text{-}CR} = 27{,}381 < S_f(R = 99.9\%) = 28{,}600 \text{ psi} \tag{24}$$

Shot-peening the groove in this case is estimated to increase fatigue life of the part from about 200,000 cycles to infinite life. Shot-peening is widely used to improve fatigue failure resistance.

4.14 Environmental Effects

When discussing materials selection in Chapter 3, the effect of environment upon material properties was an important consideration. In Table 3.1, for example, environment-related factors in material selection include strength at elevated temperature, long-term dimensional stability, dimensional stability under temperature fluctuation, resistance to chemically reactive environments, and resistance to nuclear radiation environments. Also, the discussion of potential failure modes in Chapter 2 addresses the influence of environment. Environmentally sensitive failure modes listed in 2.3 include temperature-induced elastic deformation, thermal fatigue, corrosion, corrosion fatigue, stress corrosion, erosion, abrasive wear, corrosive wear, creep, stress rupture, thermal shock, radiation damage, and creep buckling.

Apart from material selection and failure mode definition, another important design consideration involves the assessment of environmentally induced changes in loading, dimensions, or functional capability. For example, temperature, moisture, wind or flowing fluids, abrasives or contaminants, vibration, or earthquake activity may interfere with the ability of a machine or structure to perform its function. If a designer creates an otherwise superb machine, but fails to consider an important environmental excitation, a surprise failure may result. Addressing pertinent environmental factors often requires ingenuity and insight. "Standard" equations and approaches are not always available. The following brief discussion illustrates the concern.

Temperature changes may cause significant expansion or contraction of an element or a structure. For example, bridges, pipelines, or machine frames, in some cases, may develop serious operational problems if thermal expansions or contractions are not properly assessed. If expanding elements are constrained by stiff surrounding structure, high internal forces may be developed. Such forces may cause misalignment, binding, or excessive stress levels. If two different materials with two *different* thermal expansion coefficients are used for *parallel* structural elements, destructive internal forces may be generated. Even *small* temperature changes may induce destructive internal forces in some cases. If *temperature gradients* are induced in a machine element, warpage and distortion may cause intolerable changes in geometry or changes in fit. Carefully planned preloads, designed to increase stiffness or reduce susceptibility to cycling loads, may be relaxed or lost because of elevated temperature. The effects of beneficial residual stresses may be reduced or lost as well.

Water, oil, or other fluids may influence operational characteristics in some cases. For example, the coefficients of friction may be reduced, causing problems such as reduced torque capacities of brakes and clutches, or altered tightening torque requirements for bolted joints. Volumetric changes in polymeric or composite elements, because of fluid absorption, may cause unacceptable dimensional mismatches.

Abrasives or other contaminants may infiltrate joints, bearings, and gear boxes in applications such as mining machines, construction equipment, or farm machinery. Sand, shale, coal dust, limestone, and phosphate dust are examples of contaminants that should be considered when assessing potential lubrication requirements, as they relate to potential wear failure of critical elements or subassemblies.

Wind or other flowing-fluid environments may induce significant structural loads or deflections, may excite unanticipated vibratory behavior, or may carry abrasive particles or other contaminants to critical operating interfaces of bearings or gears. The vibration of smokestacks, submarine periscopes, off-shore drilling platforms, ship propellers, turbine blades, and the "galloping" of electrical transmission lines, are examples of *flow-induced vibration* that may cause failure. Civilian and military vehicles operating in desert environments provide an example of potential wind-driven sand and dust contamination of engine bearings and gear boxes.

Remote vibratory sources sometimes induce waves that propagate through foundations or the earth to interfere with operation of other machines located far from the vibrating source. Precision measuring equipment is especially vulnerable to such adverse effects. In some cases, a *statically* loaded machine element may be transformed into a *cyclically* loaded element because of a remote vibratory disturbance. Earthquake or *seismic* activity should also be considered by designers of certain equipment and structures scheduled for installation in seismically active areas.

It is important for a designer to assess the direct influence of environment on the integrity of machine elements and structures so that surprise failures are not generated. In most cases, environmentally induced changes in loading, dimensions, or functional behavior can be anticipated if environmental influences are carefully considered.

Problems

4-1. For the pliers shown in Figure P4.1, construct a complete free-body diagram for the pivot pin. Pay particular attention to moment equilibrium.

4-2. For the bolted bracket assembly shown in Figure P4.2, construct a free-body diagram for each component, including each bracket-half, the bolts, the washers, and the nuts. Try to give a qualitative indication of relative magnitudes of force vectors where possible. Indicate sources of the various force vectors you show.

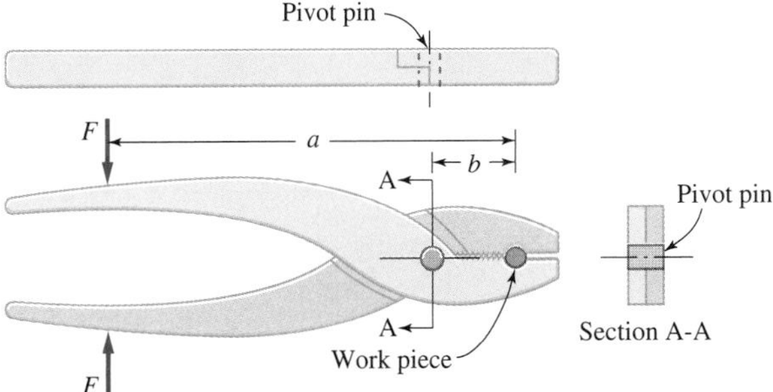

Figure P4.1
Sketch of a pair of pliers.

4-3. For the simple short-shoe block brake shown in Figure P4.3, construct a free-body diagram for the actuating lever and short block, taken together as the free body.

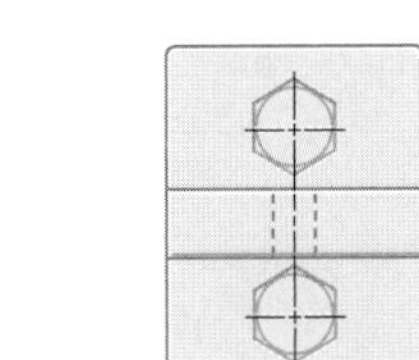

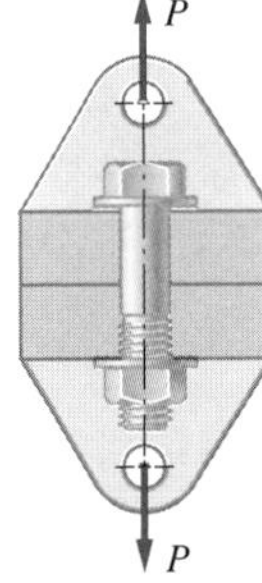

Figure P4.2
Bolted bracket assembly.

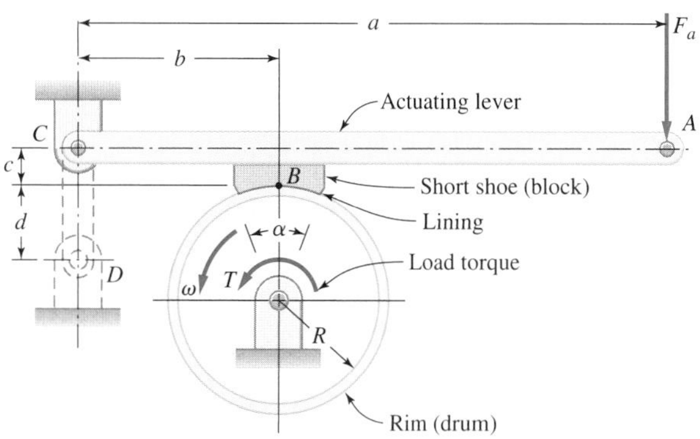

Figure P4.3
External short-shoe block brake.

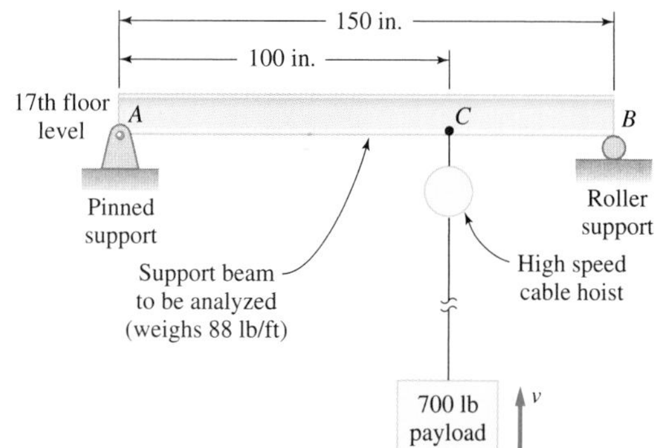

Figure P4.6
Support beam for high-speed cable hoist.

4-4. A floor-support beam of rectangular cross section supports a uniformly distributed load of w lb/ft over its full length, and its ends may be regarded as fixed.

a. Construct a complete free-body diagram for the beam.

b. Construct shear and bending moment diagrams for the beam.

4-5. The toggle mechanism shown in Figure P4.5 is to be used to statically load a helical coil spring so that it may be inspected for cracks and flaws while under load. The spring has a free length of 3.5 inches when unloaded, and a spring rate of 240 lb/in. When the static actuating force is applied, and the mechanism is at rest, the spring is compressed as shown in Figure P4.5, with dimensions as noted. Determine all the forces acting *on link 3*, and neatly draw a free-body diagram for *link 3*. Clearly show the numerical magnitudes and actual directions for all forces on *link 3*. Do only enough analysis to determine the forces on link 3, *not the entire mechanism*.

4-6. A simply supported beam is to be used at the 17th floor of a building under construction to support a high-speed cable hoist, as shown in Figure P4.6. This hoist brings the 700-pound payload quickly from zero velocity at ground level to a maximum velocity and back to zero velocity at any selected floor between the 10th and 15th floor levels. Under the most severe operating conditions, it has been determined that the *acceleration* of the payload from zero to maximum velocity produces a dynamic cable load of 1913 lb. Perform a force analysis of the beam under the most severe operating conditions. Show final results, including magnitudes and actual directions of all forces, on a neat free-body diagram of the beam.

4-7. As shown in Figure P4.7, a beam of solid rectangular cross section is fixed at one end, subjected to a downward vertical load (along the z-axis) of $V = 8000$ lb at the free end, and at the same time subjected to a horizontal load (along the y-axis) to the right, looking directly at the free end of the beam, of $H = 3000$ lb. For the beam dimensions shown, do the following:

a. Identify the precise location of the most serious critical point for this beam. Explain your logic clearly.

b. Calculate the maximum stress at the critical point.

c. Predict whether failure by yielding should be expected if the beam is made of AISI 1020 annealed carbon steel.

4-8. The electronic detector package for monitoring paper thickness in a high-speed paper mill scans back and forth along horizontal precision guide rails that are solidly supported at 24-inch intervals, as shown in Figure P4.8. The detector package fails to make acceptable thickness measurements if its vertical

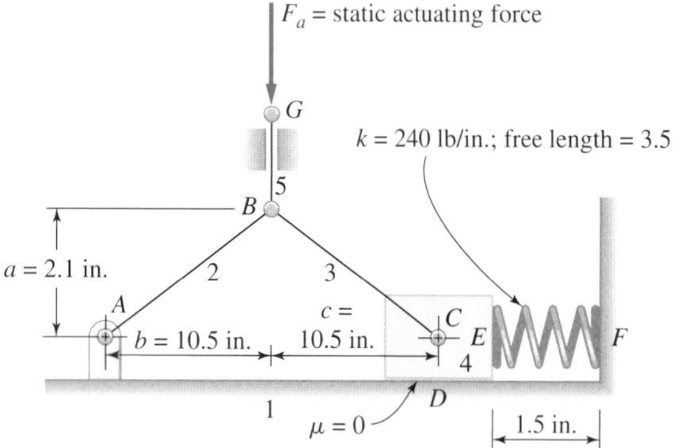

Figure P4.5
Toggle mechanism.

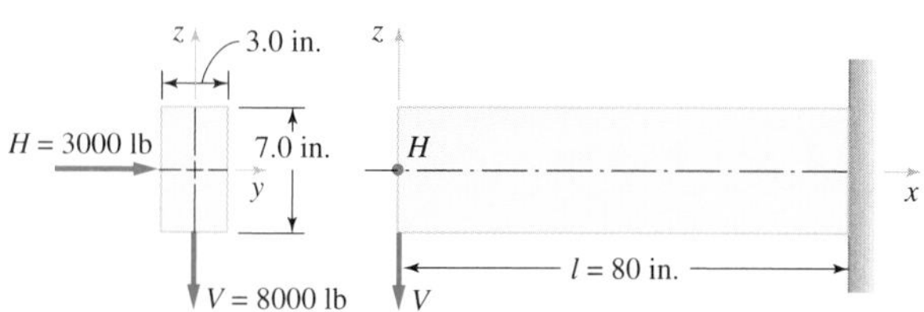

Figure P4.7
Cantilever beam subjected to simultaneous vertical and horizontal end-loads.

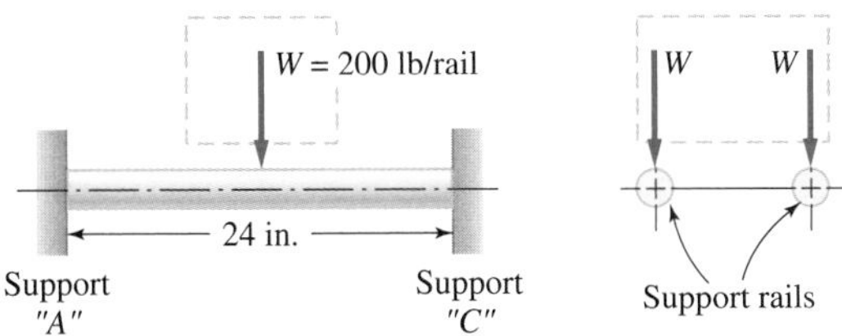

Figure P4.8
Simplified sketch of thickness-measuring scanner for paper mill application.

displacements exceed 0.005 inch as it moves along the guide rails during the scanning process. The total weight of the detector package is 400 lb and each of the two guide rails is a solid AISI 1020 cold drawn steel cylindrical bar ground to 1.0000 inch in diameter. Each of the support rails may be modeled as a beam with fixed ends and a midspan concentrated load. Half the detector weight is supported by each rail.

a. At a minimum, what potential failure modes should be considered in predicting whether the support rails are adequately designed?

b. Would you approve the design of the rails as proposed? Clearly show each step of your supporting analysis, and be complete in what you do.

c. If you *do not* approve the design, what recommendations would you make for specific things that might be done to improve the design specifications? Be as complete as you can.

4-9. A newly designed "model" is to be tested in a hot flowing gas to determine certain response characteristics. It is being proposed that the support for the model be made of Ti-6Al-4V titanium alloy. The titanium support is to be a rectangular plate, as shown in Figure P4.9, 3.00 inches in the flow direction, 20.00 inches vertically (in the load-carrying direction), and 0.0625 inch thick. A vertical load of 17,500 pounds must be carried at the bottom end of the titanium support, and the top end of the support is fixed for all test conditions by a special design arrangement. During the test the temperature is expected to increase from ambient (75°F) to a maximum of 400°F. The vertical displacement of the bottom end of the titanium support must not exceed 0.1250 inch, or the test will be invalid.

a. What potential failure modes should be considered in predicting whether this support is adequately designed?

b. Would you approve the proposed design for the titanium support? Support your response with clear, complete calculations.

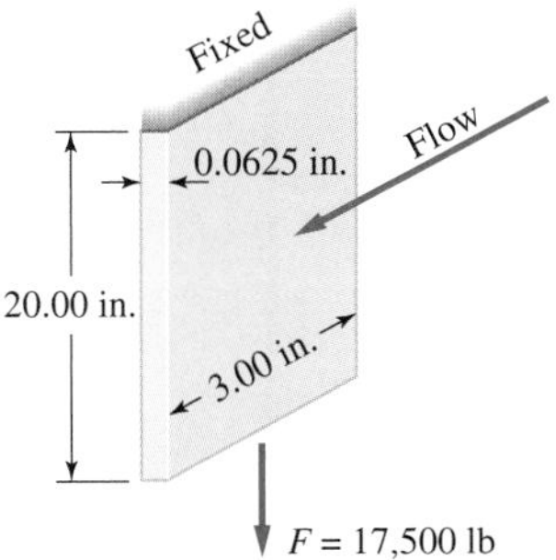

Figure P4.9
Titanium support for flow-testing a model.

4-10. A polar exploration team based near the south pole is faced with an emergency in which a very important "housing and supplies" module must be lifted by a special crane, swung across a deep glacial crevasse, and set down in a safe location on the stable side of the crevasse. The only means of supporting the 450-kN module during the emergency move is a 3.75-m-long piece of steel with a rectangular cross section of 4 cm thick by 25 cm deep with two small holes. The holes are both 3 mm in diameter, and are located at midspan 25 mm from the upper and the lower edges, as shown in Figure P4.10. These holes were drilled for some earlier use, and careful inspection has shown a tiny through-the-thickness crack, approximately 1.5 mm long, emanating from each hole, as shown. The support member may be modeled for this application as a 3.75-m-long simply supported beam that symmetrically supports the module weight at two points, located 1.25 m from each end, as shown. The material is known to be D6AC steel (1000°F temper). Ambient temperature is about −54°C.

If the beam is to be used only once for this purpose, would you approve its use? Support your answer with clearly explained calculations based on the most accurate techniques that you know.

4-11. The support towers of a suspension bridge, which spans a small estuary on a tropical island, are stabilized by anodized aluminum cables. Each cable is attached to the end of a cantilevered support bracket made of D6AC steel (tempered at 1000°F) that is fixed in a heavy concrete foundation, as shown in Figure P4.11. The cable load, F, may be regarded as static and has been measured to be about 200,000 lb, but under hurricane conditions may reach 500,000 lb due to wind loading.

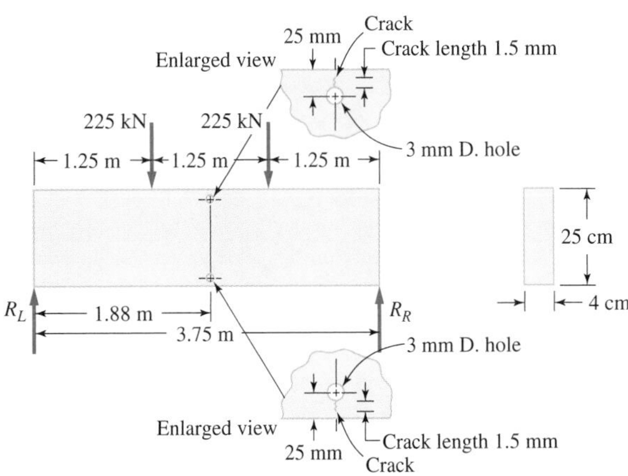

Figure P4.10
Module support beam with two small holes.

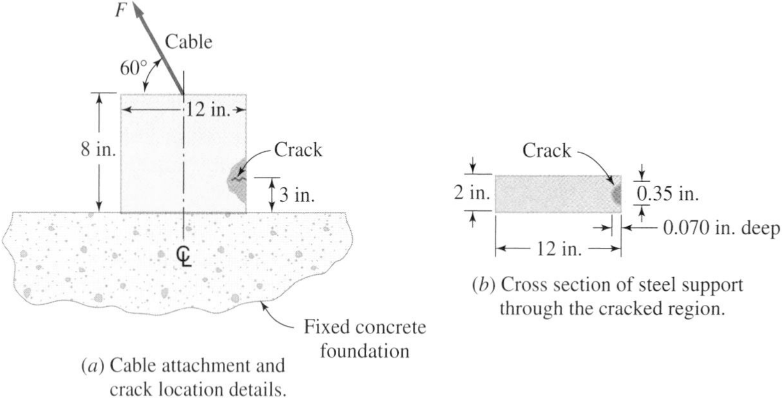

Figure P4.11
Cantilevered support bracket for anchoring bridge cables.

Inspection of the rectangular cross-section brackets has turned up a crack, with dimensions and location as shown in Figure P4.11. Assuming that fatigue is not a potential failure mode in this case, would you recommend that the cracked support bracket be replaced (a very costly procedure) or allow it to remain in service? (Repair procedures such as welding of the crack are not permitted by local construction codes.)

4-12. A horizontal cantilever beam of square cross section is 10 inches long, and is subjected to a vertical cyclic load at its free end. The cyclic load varies from a downward force of 1000 lb to an upward force of 3000 lb. Estimate the required cross-sectional dimensions of the square beam if the steel material has the following properties: $S_u = 95{,}000$ psi, $S_{yp} = 80{,}000$ psi, and $S_f = 50{,}000$ psi (note that S_f has already been corrected for the *influencing factors* of Table 2.2). Infinite life is desired. For this preliminary estimate, the issues of safety factor and stress concentration may both be neglected.

4-13. A short horizontal cantilever bracket of rectangular cross section is loaded vertically downward (z-direction) at the end by a force of $F = 85{,}000$ lb, as shown in Figure P4.13. The beam cross section is 3.0 inches by 1.5 inches, as shown, and the length is 1.2 inches. The beam is made of hot-rolled AISI 1020 steel.

a. Identify potential critical points other than the point directly under the force F.

b. For each identified critical point, show a small volume element including all nonzero stress components.

c. Calculate the magnitude of each stress component shown in (b). Neglect stress concentration effects.

d. Determine whether failure by yielding will occur, and if it does, state clearly *where* it happens. Neglect stress concentration effects.

4-14. The stubby horizontal cantilevered cylindrical boss shown in Figure P4.14 is loaded at the free end by a vertically downward force of $F = 575$ kN. The circular cross section has a diameter of 7.5 cm and a length of just 2.5 cm. The boss is made of cold-rolled AISI 1020 steel.

a. Identify clearly and completely the locations of all potential critical points that you believe should be investigated, and clearly explain why you have chosen these particular points. *Do not consider* the point where force F is concentrated on the boss.

b. For each potential critical point identified, neatly

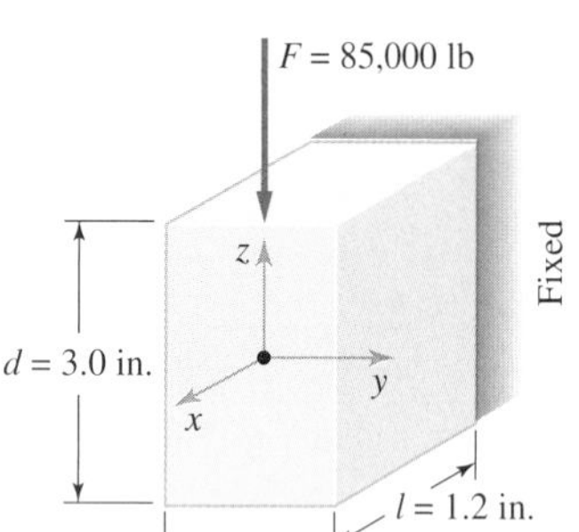

Figure P4.13
Short rectangular cantilever bracket with transverse end-load.

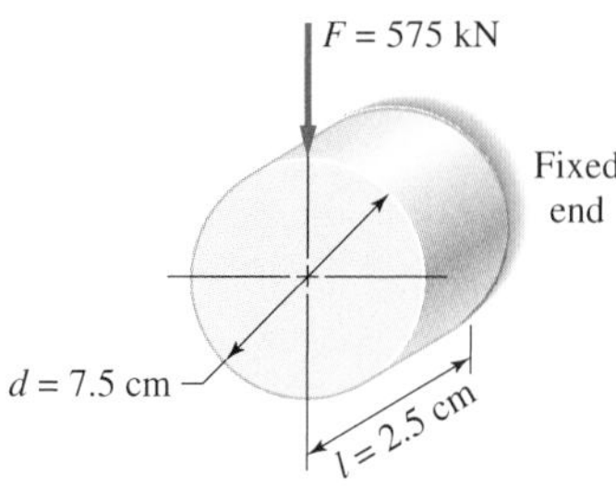

Figure P4.14
Stubby cylindrical cantilever boss with transverse end-load.

sketch a small-volume element showing all pertinent stress components.

c. Calculate a numerical value for each stress component shown in (b).

d. At each of the critical points identified, determine whether yielding should be expected to occur. Show calculation details for each case.

4-15. The short tubular cantilever bracket shown in Figure P4.15 is to be subjected to a transverse end-load of $F = 30{,}000$ lb. Neglecting possible stress concentration effects, do the following:

a. Specify precisely and completely the location of all potential critical points. Clearly explain why you have chosen these particular points. Do not consider the point where force F is applied to the bracket.

b. For each potential critical point identified, sketch a small-volume element showing all nonzero components of stress.

c. Calculate numerical values for each of the stresses shown in (b).

d. If the material is cold-drawn AISI 1020 steel, would you expect yielding to occur at any of the critical points identified in (a)? Clearly state which ones.

4-16. It is being proposed to use AISI 1020 cold-drawn steel for the shaft of a 22.5-horsepower electric motor designed to operate at 1725 rpm. Neglecting possible stress concentration effects, what minimum diameter should the solid steel motor shaft be made if yielding is the governing failure mode? Assume the yield strength in shear to be one-half the tensile yield strength.

4-17. It is desired to use a solid circular cross section for a rotating shaft to be used to transmit power from one gear set to another. The shaft is to be capable of transmitting 18 kilowatts at a speed of 500 rpm. If yielding is the governing failure mode and the shear yield strength for the ductile material has been determined to be 900 MPa, what should the minimum shaft diameter be to prevent yielding?

4-18. A solid steel shaft of square cross section is to be made of annealed AISI 1020 steel. The shaft is to be used to transmit power between two gearboxes spaced 10.0 inches apart. The shaft must transmit 75 horsepower at a rotational speed of 2500 rpm. Based on *yielding* as the governing failure mode, what minimum dimension should be specified for the sides of the square shaft to just prevent yielding? Assume the yield strength in shear to be one-half the tensile yield strength. There are no axial or lateral forces on the shaft.

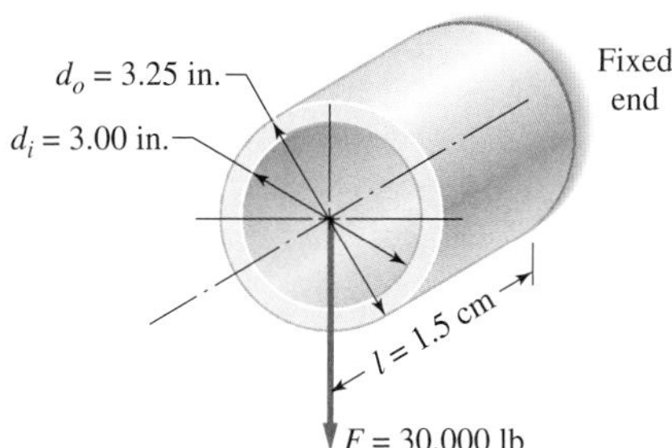

Figure P4.15
Short tubular cantilever bracket with transverse end-load.

4-19. It is necessary to use a solid equilateral triangle as the cross-sectional shape for a rotating shaft to transmit power from one gear reducer to another. The shaft is to be capable of transmitting 25 horsepower at a speed of 1500 rpm. Based on *yielding* as the governing failure mode, if the shear yield strength for the material has been determined to be 35,000 psi, what should the minimum shaft dimensions be to just prevent yielding?

4-20. a. Find the torque required to produce first yielding in a box-section torsion-bar built up from two equal-leg *L*-sections (structural angles), each $2\frac{1}{2} \times 2\frac{1}{2} \times \frac{1}{4}$ inch, welded together continuously at two places all along their full length of 3 feet. The material is hot-rolled AISI 1020 steel. Assume the yield strength in shear to be one-half the tensile yield strength. Neglect stress concentration effects.

b. For the box-section torsion-bar of (a), what torque would cause first yielding if the welder forgot to join the structural angles along their length? Compare with the result from (a).

4-21. A hollow square tube is to be used as a shaft to transmit power from an electric motor/dynamometer to an industrial gearbox which requires an input of 42 horsepower at 3400 rpm, continuously. The shaft material is annealed AISI 304 stainless steel. The dimensions of the square shaft cross section are 1.25 inches outside, the well thickness is 0.125 inch, and the shaft length is 20 inches. There are no significant axial or lateral loads on the shaft.

a. Based on yielding as a failure mode, what existing safety factor would you calculate for this shaft when it is operating under full power? Assume the yield strength in shear to be one-half the tensile yield strength.

b. What angle of twist would you predict for this shaft when operating under full power?

4-22. A horizontal steel cantilever beam is 10 inches long and has a square cross section that is one-inch on each side. If a vertically downward load of 100 pounds is applied at the midlength of the beam, 5 inches from the fixed end, what would be the vertical deflection at the free end if transverse shear is neglected. Use Castigliano's theorem to make your estimate.

4-23. a. Using the strain energy expression for torsion in Table 4.6, verify that if a prismatic member has a uniform cross section all along its length, and if a constant torque T is applied, the stored strain energy in the bar is properly given by (4-54).

b. Using Castigliano's method, calculate the angle of twist induced by the applied torque T.

4-24. The steel right-angle support bracket, with leg lengths $L_1 = 10$ inches and $L_2 = 5$ inches, as shown in Figure P4.24, is

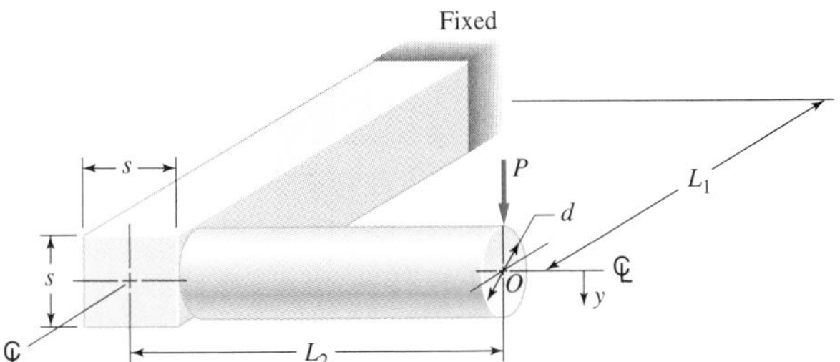

Figure P4.24
Right-angle support bracket with transverse end-load.

to be used to support the static load $P = 1000$ lb. The load is to be applied vertically downward at the free end of the cylindrical leg, as shown. Both bracket-leg centerlines lie in the same horizontal plane. If the square leg has sides $s = 1.25$ inches, and the cylindrical leg has diameter $d = 1.25$ inches, use Castigliano's theorem to find the deflection y_o under the load P.

4-25. Make two neat, clear sketches illustrating two ways of completely defining the state of stress at a point. Define all symbols used.

4-26. A solid cylindrical bar is fixed at one end and subjected to a pure torsional moment M_t at the free end, as shown in Figure P4.26. Find, for this loading, the principal normal stresses and the principal shearing stresses at the critical point, using the stress cubic equation (4-59).

4-27. Solve problem 4-26 using the Mohr's circle analogy.

4-28. A solid cylindrical bar of diameter d is fixed at one end and subjected to both a pure torsional moment M_t and a pure bending moment M_b at the free end. Using the stress cubic equation (4-59), find the principal normal stresses and principal shearing stresses at the critical point for this loading, in terms of applied moments and bar dimensions.

4-29. Solve problem 4-28 using the Mohr's circle analogy.

4-30. From the stress analysis for a machine part at a specified critical point, it has been found that $\sigma_z = 6$ MPa, $\tau_{xz} = 2$ MPa, and $\tau_{yz} = 5$ MPa. For this state of stress, determine the principal stresses and the maximum shearing stress at the critical point.

4-31. A solid cylindrical bar of 7075-T6 aluminum is 3 inches in diameter, and is subjected to a torsional moment of T_x = 75,000 in-lb, a bending moment of M_y = 50,000 in-lb, and a

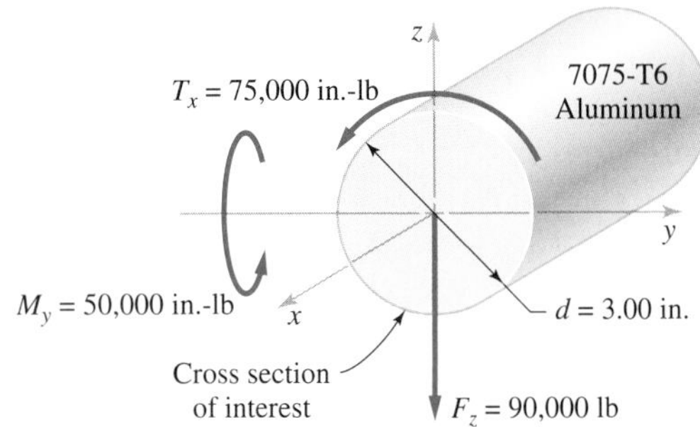

Figure P4.31
Solid cylindrical bar subjected to torsion, bending, and transverse shear.

transverse force of F_z = 90,000 lb, as shown in the sketch of Figure P4.31.

a. Clearly establish the location(s) of the potential critical point(s), giving logic and reasons why you have selected the point(s).

b. Calculate the magnitudes of the principal stresses at the selected critical point(s).

c. Calculate the magnitude(s) of the maximum shearing stress(es) at the critical point(s).

4-32. The square cantilever beam shown in Figure P4.32 is subjected to pure bending moments M_y and M_z, as shown. Stress concentration effects are negligible.

a. For the critical point, make a complete sketch depicting the state of stress.

b. Determine the magnitudes of the principal stresses at the critical point.

4-33. Equations (4-73), (4-74), and (4-75) represent Hooke's Law relationships for a *triaxial* state of stress. Based on these equations:

a. Write the Hooke's Law relationships for a *biaxial* state of stress.

b. Write the Hooke's Law relationships for a *uniaxial* state of stress.

c. Does a uniaxial state of stress imply a uniaxial state of strain? Explain.

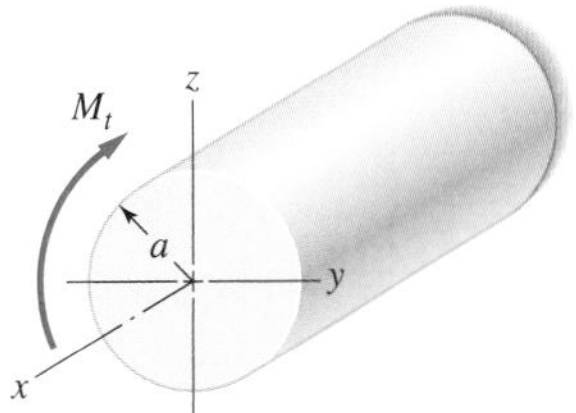

Figure P4.26
Solid cylindrical bar subjected to a pure torsional moment.

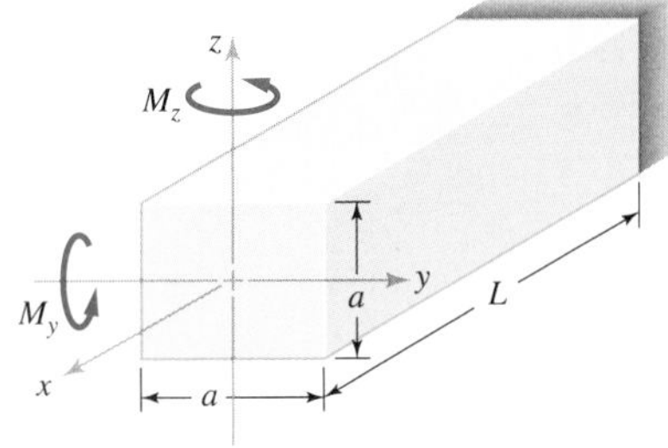

Figure P4.32
Cantilever beam subjected to bending about two axes.

4-34. It has been calculated that the critical point in a 4340 steel part is subjected to a state of stress in which $\sigma_x = 6000$ psi, $\tau_{xy} = 4000$ psi, and the remaining stress components are all zero. For this state of stress, determine the sum of the normal strains in the x, y, and z directions; that is, determine the magnitude of $\varepsilon_x + \varepsilon_y + \varepsilon_z$.

4-35. For the case of pure biaxial shear, that is, the case where τ_{xy} is the only nonzero component of stress, write expressions for the principal normal strains. Is this a biaxial state of strain? Explain.

4-36. Explain why it is often necessary for a designer to utilize a failure theory.

4-37. What are the essential attributes of any useful failure theory?

4-38. What is the basic assumption that constitutes the framework for all failure theories?

4-39. a. The *first strain invariant* may be defined as $I_1 \equiv \varepsilon_i + \varepsilon_j + \varepsilon_k$. Write in words a "first strain invariant" theory of failure. Be complete and precise.

b. Derive a complete mathematical expression for your "first strain invariant" theory of failure, expressing the final result in terms of *principal stresses* and *material properties*.

c. How could one establish whether or not this theory of failure is valid?

4-40. The solid cylindrical cantilever bar shown in Figure P4.40 is subjected to a pure torsional moment T about the x-axis, pure bending moment M_b about the y-axis, and pure tensile force P along the x-axis, all at the same time. The material is a ductile aluminum alloy.

a. Carefully identify the most critical point(s), neglecting stress concentration. Give detailed reasoning for your selection(s).

b. At the critical point(s), draw a cubic volume element, showing all stress vectors.

c. Carefully explain how you would determine whether or not to expect yielding at the critical point.

4-41. In the triaxial state of stress shown in Figure P4.41, determine whether failure would be predicted. Use the maximum normal stress theory for brittle materials and both the distortion energy theory and the maximum shearing stress theory for ductile materials:

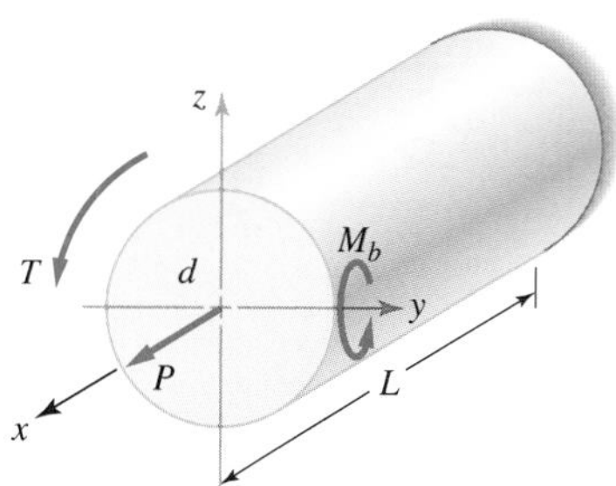

Figure P4.40
Cylindrical cantilever beam subjected to torsion, bending, and direct tension.

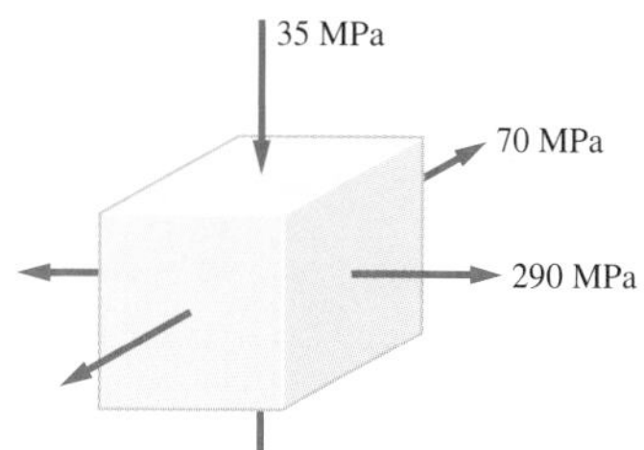

Figure P4.41
Triaxial state of stress.

a. For an element stressed as shown, made of 319-T6 aluminum (S_u = 248 MPa, S_{yp} = 165 MPa, e = 20 percent in 50 mm).

b. For an element stressed as shown, made of 518.0 aluminum, as cast (S_u = 310 MPa, S_{yp} = 186 MPa, e = 8 percent in 50 mm).

4-42. The axle of an electric locomotive is subjected to a bending stress of 25,000 psi. At the same critical point, torsional stress due to the transmission of power is 15,000 psi and a radial compressive stress of 10,000 psi results from the fact that the wheel is pressed onto the axle. Would you expect yielding at the selected critical point if the axle is made of AISI 1060 steel in the "as-rolled" condition?

4-43. A hollow tubular steel bar is to be used as a torsion spring subjected to a cyclic pure torque ranging from –5000 in-lb to +15,000 in-lb. It is desired to use a thin-walled tube with wall thickness t equal to 10 percent of the outside diameter d. The steel material has an ultimate strength of 200,000 psi, a yield strength of 180,000 psi, and an elongation of e = 15 percent in 2 inches. The fatigue limit is 95,000 psi. Find the minimum tube dimensions that should just provide infinite life. The polar moment of inertia for a thin-walled tube may be approximated by the expression $J = \pi d^3 t/4$.

4-44. Using the "force-flow" concept, describe how one would assess the relative severity of various types of geometrical discontinuities in a machine part subjected to a given set of external loads. Use a series of clearly drawn sketches to augment your explanation.

4-45. The support bracket shown in Figure P4.45 is made of permanent-mold cast-aluminum alloy 356.0, solution-treated and aged (see Tables 3.3 and 3.10), and subjected to a static pure bending moment of 850 in-lb. Would you expect the part to fail when the load is applied?

4-46. The machine part shown in Figure P4.46 is subjected to a completely reversed (zero mean) cyclic bending moment of ±4000 in-lb, as shown. The material is annealed 1020 steel with S_u = 57,000 psi, S_{yp} = 43,000 psi, and elongation in 2 inches of 25 percent. The S-N curve for ths material is given in Figure 2.19. How many cycles of loading would you estimate could be applied before failure occurs?

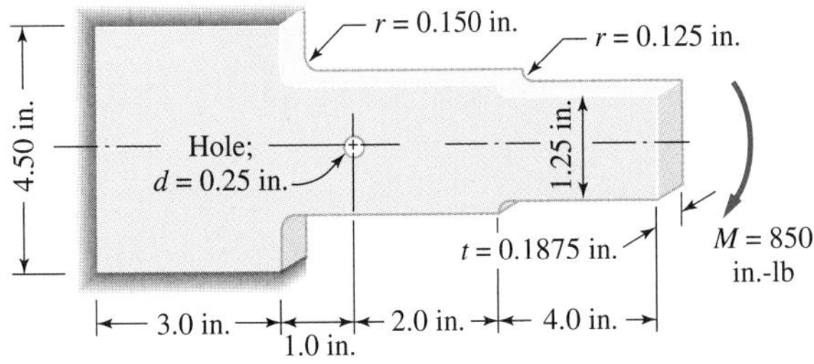

Figure P4.45
Cast-aluminum support bracket subjected to pure bending moment at the free end.

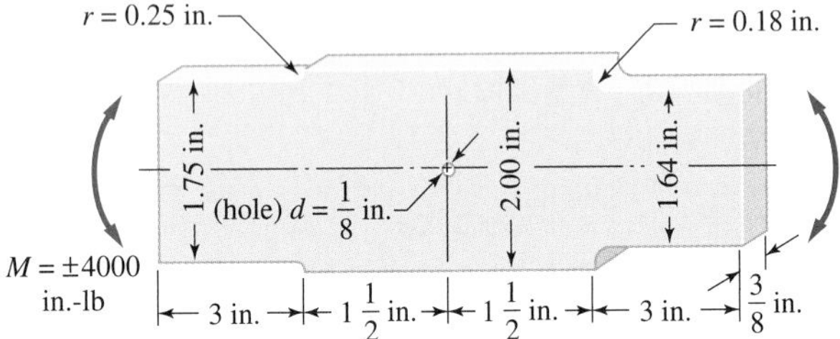

Figure P4.46
Machine part subjected to completely reversed cyclic bending moment.

4-47. a. The mounting arm shown in Figure P4.47 is to be made of Class 60 gray cast iron with ultimate strength of 60,000 psi in tension, and elongation in 2 inches of less than 0.5 percent. The arm is subjected to a static axial force of P = 50,000 lb and a static torsional moment of T = 18,000 in-lb, as shown. For the dimensions shown, could the arm support the specified loading without failure?

b. During a different mode of operation, the axial force P cycles repeatedly from 50,000 psi tension to 50,000 psi compression, and the torsional moment T remains zero at all times. What would you estimate the life to be for this cyclic mode of operation? (The S-N curve may be estimated using the methods described in 2.6.)

4-48. An S-hook, as sketched in Figure P4.48, is being proposed as a means of hanging unitized dumpster bins in a new state-of-the-art dip-style painting process. The maximum weight of a dumpster bin is estimated to be 300 pounds, and two hooks will typically be used to support the weight, equally split between two lifting lugs. However, the possibility exists that on some occasions the entire weight may have to be supported by a single hook. It is estimated that each pair of hooks will be loaded, unloaded, and then reloaded approximately every five minutes. The plant is to operate 24 hours per day, 7 days a week. The proposed hook material is commercially polished AM 350 stainless steel in age-hardened condition (see Table 3.3). Preliminary considerations suggest that both yielding and fatigue may be potential failure modes.

a. To investigate potential yielding failure, identify critical points in the S-hook, determine maximum stresses at each critical point, and predict whether the loads can be supported without failure by yielding.

b. To investigate potential failure by fatigue, identify critical points in the S-hook, determine pertinent cyclic stresses at each critical point, and predict whether a 10-year design life could be achieved with 99 percent reliability.

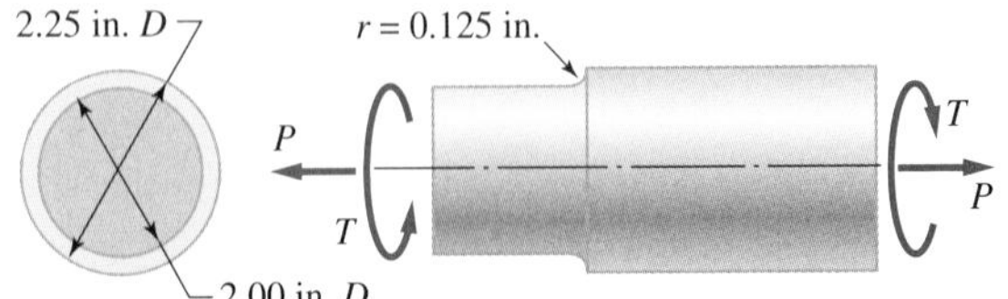

Figure P4.47
Cast-iron mounting arm subjected to tension and torsion.

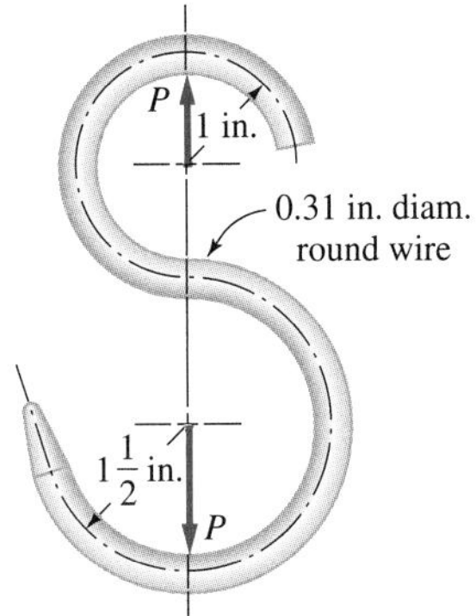

Figure P4.48
S-hook of circular cross section, supported at top and loaded at bottom by force *P*.

4-49. The support (shackle) at one end of a symmetrical leaf spring (e.g., see Figure 14.15) is depicted in Figure P4.49. The

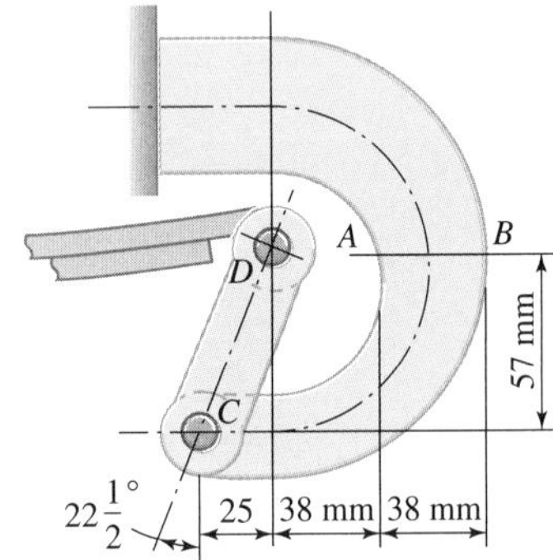

Figure P4.49
Support bracket for leaf spring attachment.

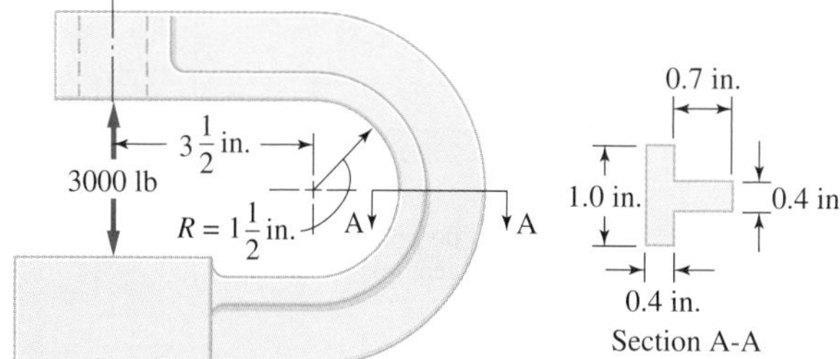

Figure P4.50
Cast-iron C-frame subjected to 3000-lb separating force.

cross section at A-B is rectangular, with dimensions of 38 mm by 25 mm thickness in and out of the paper. The total vertical force at the center of the leaf spring is 18 kN up on the spring.

a. Find the maximum stress at the critical point in the support.

b. Would it be reasonable to select ASTM A-48 (class 50) gray cast iron as a potential material candidate for the support? (See Table 3.3 for properties.)

4-50. A $1\frac{1}{2}$-ton hydraulic press for removing and reinstalling bearings in small to medium-size electric motors is to consist of a commercially available hydraulic cylinder mounted vertically in a C-frame, with dimensions as sketched in Figure P4.50. It is being proposed to use ASTM A-48 (Class 50) gray cast iron for the C-frame material. (See Table 3.3 for properties.) Predict whether the C-frame can support the maximum load without failure.

4-51. Your group manager tells you that he has heard that a sphere of AISI 1020 steel will produce plastic flow at the region of contact due to its own weight if placed on a flat plate of the same material. Determine whether the allegation may be true, and, if true, under what circumstances. Use properties for *hot-rolled* 1020 steel to make your determination.

4-52. Two mating spur gears (see Chapter 15) are 25 mm wide and the tooth profiles have radii of curvature at the line of contact of 12 mm and 16 mm, respectively. If the transmitted force between them is 180 N, estimate the following:

a. The width of the contact zone

b. The maximum contact pressure

c. The maximum subsurface shearing stress and its distance below the surface

4-53. The preliminary sketch for a device to measure the axial displacement (normal approach) associated with a sphere sandwiched between two flat plates is shown in Figure P4.53. The material to be used for both the sphere and the plates is AISI 4340 steel, heat treated to a hardness of R_C 56 (see Tables 3.3, 3.9, and 3.13). Three sphere diameters are of interest: d_s = 0.500 inch, d_s = 1.000 inch, and d_s = 1.500 inches.

a. To help in selecting a micrometer with appropriate measurement sensitivity and range, *estimate* the range of

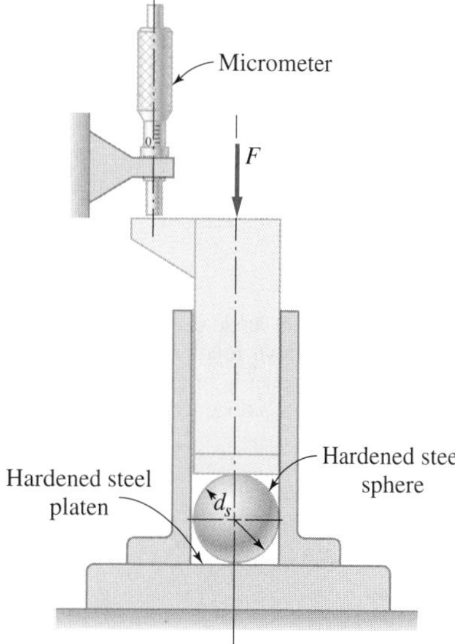

Figure P4.53
Device for measuring nonlinear deflection as a function of load for contact between a sphere and flat plate.

normal approach for each sphere size, corresponding to a sequence of loads from 0 to 3000 pounds, in increments of 500 pounds.

b. Plot the results.

c. Would you classify these force-deflection curves as linear, stiffening, or softening? (See Figure 4.37).

4-54. It is being proposed to use a single small gas turbine power plant to drive two propellers in a preliminary concept for a small vertical-stationary-take-off-and-landing (VSTOL) aircraft. The power plant is to be connected to the propellers through a "branched" system of shafts and gears, as shown in Figure P4.54. One of many concerns about such a system is that rotational vibrations between and among the propeller masses and the gas turbine mass may build up their vibrational amplitudes to cause high stresses and/or deflections that might lead to failure.

a. Identify the system elements (shafts, gears, etc.) that might be important "springs" in analyzing this rotational mass-spring system. Do not include the gas turbine or the propellers themselves.

b. For each element identified in (a), list the types of springs (torsion, bending, etc.) that might have to be analyzed to determine vibrational behavior of the rotational vibrating system.

4-55. a. A steel horizontal cantilever beam having the dimensions shown in Figure P4.55(a) is to be subjected to a vertical end-load of F = 100 lb. Calculate the spring rate of the cantilever beam referred to its free end (i.e., at the point of load application). What vertical deflection at the end of the beam would you predict?

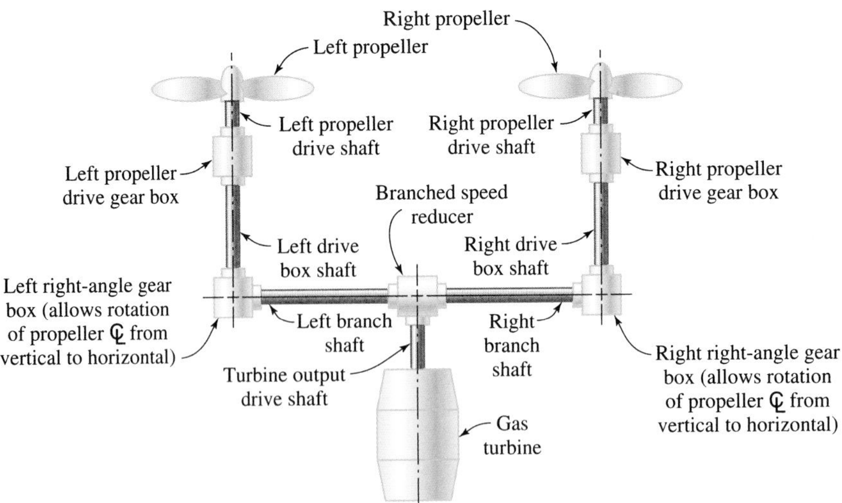

Figure P4.54
Branched drive system being proposed for a small VSTOL aircraft.

b. The helical coil spring shown in Figure P4.55(b) has been found to have a linear spring rate of k_{sp} = 300 lb/in. If an axial load of F = 100 lb is applied to the spring, what axial (vertical) deflection would you predict?

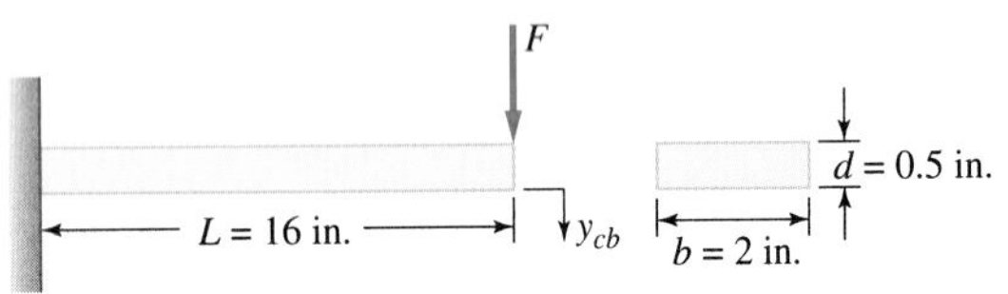

(*a*) Cantilever beam spring.

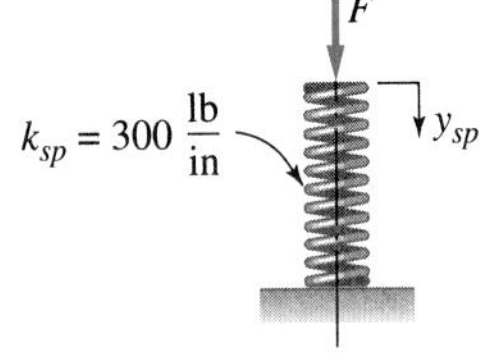

(*b*) Helical coil spring.

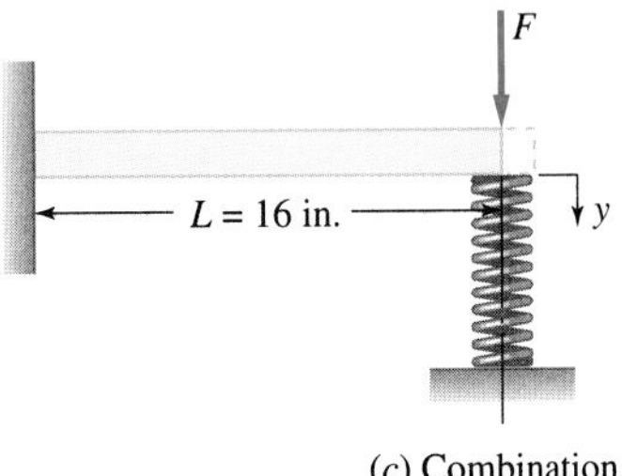

(*c*) Combination.

Figure P4.55
Combination of a cantilever beam and helical coil spring to support a load F.

c. In Figure P4.55(c), the helical coil spring of (b) is placed under the end of the cantilever beam of (a) with no clearance or interference between them, so that the centerline of the coil spring coincides with the end of the cantilever beam. When a vertical load of F = 100 lb is applied at the end of the beam, calculate the spring rate of the combined beam and coil spring assembly referred to the free end of the beam (i.e., at the point of load application). What vertical deflection of the end of the beam would you predict?

d. What portion of the 100-lb force F is carried by the cantilever beam, and what portion is carried by the coil spring?

4-56. To help assess the influence of bearing stiffness on lateral vibration behavior of a rotating steel shaft with a 100-lb steel flywheel mounted at midspan, you are asked to make the following estimates:

a. Using the configuration and the dimensions shown in Figure P4.56(a), calculate the static midspan deflection and spring rate, assuming that the bearings are infinitely stiff radially (therefore they have no vertical deflection under load), but support no moment (hence the shaft is simply supported).

b. Using the actual force-deflection bearing data shown in Figure P4.56(b) (supplied by the bearing manufacturer), calculate the static midspan deflection of the shaft bearing system and the midspan spring rate for the shaft bearing system.

c. Estimate the percent change in system stiffness attributable to the bearings, as compared to system stiffness calculated by ignoring the bearings. Would you consider this to be a *significant* change?

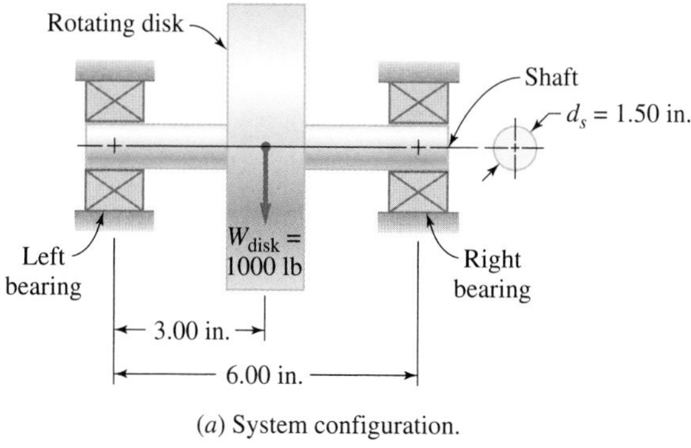

(*a*) System configuration.

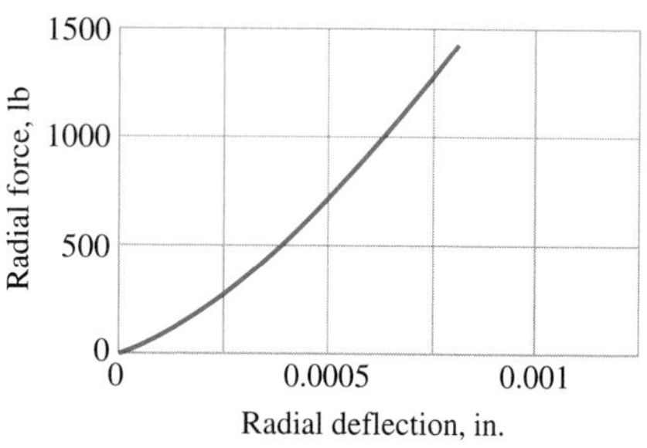

(*b*) Force-deflection curve for single bearing.

Figure P4.56
Flywheel and shaft supported by ball bearings.

4-57. A bolted joint of the type shown in Figure P4.57A employs a "reduced-body" bolt to hold the two flanged members together. The area of the reduced-body steel bolt at critical cross section A is 0.068 in^2. The steel bolt's static material properties are S_u = 60,000 psi and S_{yp} = 36,000 psi. The external force P cycles from zero to 1200 lb tension. The clamped flanges of the steel housing have been estimated to have an effective axial stiffness (spring rate) of three times the axial bolt stiffness over its effective length L = 1.50 inches.

a. Plot the cyclic force-time pattern in the reduced-body bolt if no preload is used.

b. Using the *S-N* curve of Figure P4.57B, estimate bolt life for the case of no preload.

c. Plot the cyclic force-time pattern in the reduced-body bolt if the nut is initially tightened to induce a preload force of F_i = 1000 lb in the bolt body (and a preload force of −1000 lb in the clamped flanges). A separate analysis has determined that when the 1000-lb preload is present, the peak external force of 1200 lb will not be enough to cause the flanges to separate. (See Example 13.1 for details.)

d. Estimate the bolt life for the case of an initial preload force of 1000 lb in the bolt, again using the *S-N* curve of Figure P4.57B.

e. Comment on the results.

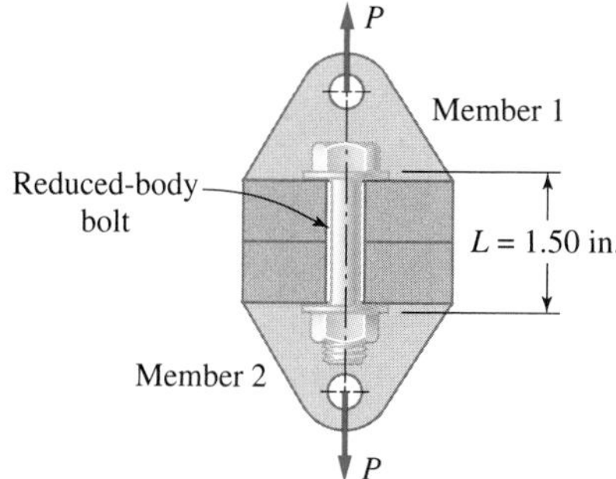

Figure P4.57A
Reduced-body bolt subjected to cyclic axial loading.

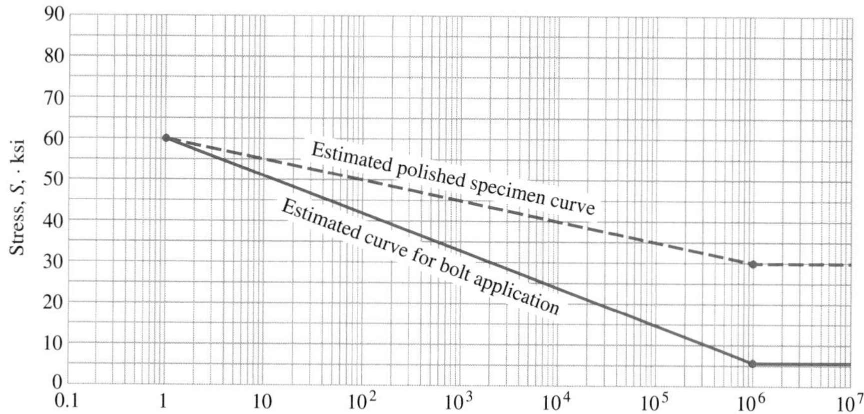

Figure P4.57B
Uniaxial *S-N* curve (a) for polished specimens and (b) for a bolt as used in this application.

4-58. Examining the rotating bending fatigue test data for 60° V-notched specimens depicted in Figure 4.38, respond to the following questions:

a. For notched specimens that have not been prestressed, if they are subjected to rotating bending tests that induce an applied alternating stress amplitude of 20,000 psi at the notch root, what mean life might reasonably be expected?

b. If similar specimens are first subjected to an axial *tensile* static preload level that produces local stresses of 90 percent of notched ultimate strength, then released and subjected to rotating bending tests that induce an applied alternating stress amplitude of 20,000 psi at the notch root, what mean life might reasonably be expected?

c. If similar specimens are first subjected to an axial *compressive* static preload level that produces local stresses of 90 percent of notched ultimate strength, then released and subjected to rotating bending tests that induce an applied alternating stress amplitude of 20,000 psi at the notch root, what mean life might reasonably be expected?

d. Do these results seem to make sense? Explain.

4-59. A notched rectangular bar of the type illustrated in Figure 4.21(b) is 1.15 inches wide, 0.50 inch thick, and is symmetrically notched on both sides by semicircular notches with radii of $r = 0.075$ inch. The bar is made of a ductile steel with yield strength of $S_{yp} = 50{,}000$ psi. Sketch the stress distribution across the minimum section for each of the following circumstances, assuming elastic–perfectly plastic behavior.

a. A tensile load of $P_a = 10{,}000$ lb is applied to the bar.

b. The 10,000-lb tensile load is released.

c. A tensile load of $P_c = 20{,}000$ lb is applied to a new bar of the same type.

d. The 20,000-lb load is released.

e. A tensile load of $P_e = 30{,}000$-lb is applied to another new bar of the same type.

f. Would the same or different results be obtained if the same bar were used for all three loads in sequence?

4-60. An initially straight and stress-free beam, similar to the one shown in Figure E4.19A, is 5.0 cm high and 2.5 cm wide. The beam is made of a ductile aluminum material with yield strength of $S_{yp} = 275$ MPa.

a. What applied moment is required to cause yielding to a depth of 10.0 mm if the material behaves as if it were elastic–perfectly plastic?

b. Determine the residual stress pattern across the beam when the applied moment of (a) is released.

Chapter 5

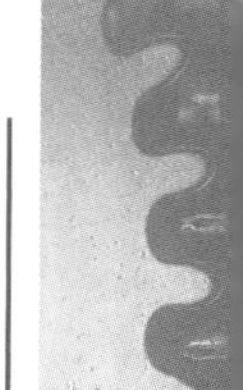

The Role of Safety Factors; Reliability Concepts

5.1 Purpose of Safety Factors in Design Calculations

Averting failure of a machine or structure by *predicting and correcting potential failure scenarios at the design stage*, before the machine is built, is a key design strategy. By identifying the probable governing failure mode and tentatively selecting the best available candidate material, the postulated failure prediction scenarios provide a basis for choosing shapes and sizes for all parts. Ideally, if loads, environments, and material properties were perfectly known, the shapes and sizes of machine parts could readily be determined by simply making sure that pertinent operating loads or stresses never exceed capacities or strengths at any of the critical points in the machine. Practically, *uncertainties* and *variabilities* always exist in design predictions. Loads are often variable and inaccurately known, strengths are variable and sometimes inaccurately known for certain failure modes or certain states of stress, calculation models embody assumptions that may introduce inaccuracies in determining dimensions, and other uncertainties may result from variations in quality of manufacture, operating conditions, inspection, and maintenance practices. These uncertainties and variabilities clearly complicate the design task.

Uncertainties in selecting shapes, sizes, and materials that will provide safe, reliable operation must be directly addressed. To accomplish this *failure prevention objective*, a designer has two choices: (1) *select a design safety factor* that will assure that minimum strength or capacity safely exceeds maximum stress or load for all foreseeable conditions, or (2) *statistically define* strength or capacity, stress or load, modeling errors, manufacturing variabilities, and variations in operational environment and maintenance, so that the *probability of failure* may be kept below a *preselected* level of acceptability. Because of the difficulty and cost of statistically defining many of these variables, a usual first choice is to select an appropriate *design safety factor*.

5.2 Selection and Use of a Design Safety Factor

In practice, the shapes and sizes of machine parts are usually determined by first defining a *design-allowable value* for whatever *loading severity parameter* is selected, whether it be stress, deflection, load, speed, or something else. To determine the design-allowable value, the *critical failure level* corresponding to the selected loading severity parameter is divided by a *design safety factor* (a number always greater than 1) to account for perceived uncertainties. Dimensions are then calculated so that maximum operating values of the selected loading severity parameters are less than design-allowable values. Mathematically, this may be expressed as

$$P_d = \frac{L_{fm}}{n_d} \tag{5-1}$$

where P_d is the design-allowable value of the loading severity parameter, L_{fm} is the critical failure level dictated by the governing failure mode, and n_d is the design safety factor *chosen by the designer* to account for all of the perceived uncertainties and variabilities. Usually (but not always) the selected loading severity parameter will be stress and the critical failure level will be *critical material strength corresponding to the governing failure mode*. Hence, a more usual form of (5-1) is

$$\sigma_d = \frac{S_{fm}}{n_d} \tag{5-2}$$

where σ_d is the design-allowable stress, S_{fm} is failure strength of the selected material corresponding to the governing failure mode, and n_d is the selected design safety factor. To assure a safe design, dimensions are calculated so that maximum operating stress levels are equal to or less than the design stress.

Selection of a design safety factor must be undertaken with care since there are unacceptable consequences associated with selected values that are either too low or too high. If the selected value is too small, the probability of failure will be too great. If the selected value is too large, the size, weight, or cost may be too high. Proper safety factor selection requires a good working knowledge of the limitations and assumptions in the calculation models or simulation programs used, the pertinent properties of the proposed materials, and operational details of the proposed application. *Design experience is extremely valuable* in selection of an appropriate design safety factor, but a rational selection may be made even with *limited experience*. The method suggested here breaks the selection down into a series of semiquantitative smaller decisions that may be weighted and empirically recombined to calculate an acceptable value for the design safety factor, tailored to the specific application. Even experienced designers find this approach valuable when faced with designing a new product or redesigning an existing product for a new application.

To implement the selection of a design safety factor, consider separately each of the following eight *rating factors*:

1. The accuracy with which the loads, forces, deflections, or other failure-inducing agents can be determined
2. The accuracy with which the stresses or other loading severity parameters can be determined from the forces or other failure-inducing agents
3. The accuracy with which the failure strengths or other measures of failure can be determined for the selected material in the appropriate failure mode
4. The need to conserve material, weight, space, or dollars
5. The seriousness of the consequences of failure in terms of human life and/or property damage
6. The quality of workmanship in manufacture
7. The conditions of operation
8. The quality of inspection and maintenance available or possible during operation

A semiquantitative assessment of these rating factors may be made by assigning a *rating number*, ranging in value from -4 to $+4$, to each one. These *rating numbers (RNs)* have the following meanings:

$RN = 1$	*mild need* to modify n_d
$RN = 2$	*moderate need* to modify n_d
$RN = 3$	*strong need* to modify n_d
$RN = 4$	*extreme need* to modify n_d

Further, if there is a perceived need to *increase* the safety factor, the selected rating number is assigned a *positive* (+) sign. If the perceived need is to *decrease* the safety factor, the selected rating number is assigned a *negative* (−) sign.

The next step is to calculate the algebraic sum, t, of the eight rating numbers, giving

$$t = \sum_{i=1}^{8} (RN)_i \tag{5-3}$$

Using the result from (5-3), the design safety factor, n_d, may be empirically estimated from

$$n_d = 1 + \frac{(10 + t)^2}{100} \quad \text{for } t \geq -6 \tag{5-4}$$

or

$$n_d = 1.15 \quad \text{for } t < -6 \tag{5-5}$$

Using this method, the design safety factor will never be less than 1.15, and rarely larger than 4 or 5. This range is broadly compatible with the usual list of suggested safety factors found in most design textbooks or handbooks,[1] but n_d is specifically determined for each application on a more rational basis.

Example 5.1 Determining Design Safety Factor

You have been asked to propose a value for the design safety factor to be used in determining the dimensions for the main landing gear support for a new executive jet aircraft. It has been determined that the application may be regarded as "average" in many respects, but the material properties are known a little better than for the average design case, the need to conserve weight and space is strong, there is a strong concern about threat to life and property in the event of a failure, and the quality of inspection and maintenance is regarded as excellent. What value would you propose for the design safety factor?

Solution

Based on the information given, the *rating numbers* assigned to each of the eight rating factors might be

Rating factor	Selected Rating Number (RN)
1. Accuracy of loads knowledge	0
2. Accuracy of stress calculation	0
3. Accuracy of strength knowledge	−1
4. Need to conserve	−3
5. Seriousness of failure consequences	+3
6. Quality of manufacture	0
7. Conditions of operation	0
8. Quality of inspection/maintenance	−4

[1]See, for example, refs. 1, 2, 3, or 4.

Example 5.1 Continues

From (5-3)

$$t = 0 + 0 - 1 - 3 + 3 + 0 + 0 - 4 = -5 \tag{1}$$

and since $t \geq -6$, from (5-4)

$$n_d = 1 + \frac{(10 - 5)^2}{100} = 1.25 \tag{2}$$

The recommended value for a design safety factor appropriate to this application would be, therefore, $n_d = 1.25$. To obtain the *design-allowable stress*, the material strength corresponding to the governing failure mode, probably fatigue, would be divided by $n_d = 1.25$. Buckling might also be a potential failure mode, in which case the *design-allowable axial load* would be determined by dividing the critical buckling load (see 2.7) by $n_d = 1.25$.

5.3 Determination of Existing Safety Factor in a Completed Design: A Conceptual Contrast

To avoid confusion in understanding the term "safety factor" in design deliberations, it is well to reflect upon the dual roles of a designer, one as a *synthesist* and the other as an *analyst*. As a synthesist, the designer must formulate safe, reliable new machines that have not yet been built. To do this, a *design* safety factor is *selected* by the designer, following the guidelines of 5.2, then materials and dimensions are determined in accordance with the *chosen* design safety factor. As an analyst, the designer is charged with the task of examining an existing machine, already built and perhaps in service, to find out what *existing* safety factor has already been built into the existing machine by someone else. In this case, the designer has no choice in the existing safety factor; it must be *calculated* from material properties, dimensions, and loads that exist. Thus, in contrast to the design safety factor n_d, which is selected by a designer to create a new design, the existing safety factor n_{ex} is calculated by a designer based upon examination of a completed design proposal or a machine already in service.

To calculate an existing safety factor, the critical failure level of the selected loading severity parameter, L_{fm}, is divided by the maximum operating value of the loading severity parameter, P_{max}, to give

$$n_{ex} = \frac{L_{fm}}{P_{max}} \tag{5-6}$$

In the more usual context, this may be expressed as

$$n_{ex} = \frac{S_{fm}}{\sigma_{max}} \tag{5-7}$$

where S_{fm} is the critical strength corresponding to the governing failure mode and σ_{max} is the maximum operating stress level at the critical point.

Although n_{ex} and n_d are clearly related, they are conceptually very different. Keeping the contrast in mind will help avoid confusion in design discussions.

Example 5.2 Assessment of Probable Failure Mode and Existing Safety Factor

An axially loaded straight cylindrical bar of diameter $d = 0.50$ inch is made of 2024-T4 aluminum with ultimate strength $S_u = 68{,}000$ psi, yield strength $S_{yp} = 48{,}000$ psi, and fatigue properties shown in Figure 2.19. The bar is subjected to completely reversed axial loading of 4000 lb maximum, and has a design life requirement of 10^7 cycles.

a. What potential failure modes should be investigated?
b. What is the existing factor of safety?

Solution

a. The two most probable candidates for governing failure mode are yielding and fatigue. Both should be calculated to determine which one governs.
b. For yielding,

$$(n_{ex})_{yp} = \frac{S_{yp}}{\sigma_{max}} \tag{1}$$

and

$$\sigma_{max} = \frac{P_{max}}{A} = \frac{4000}{\pi\left[\frac{(0.50)^2}{4}\right]} = \frac{4000}{0.196} = 20{,}408 \text{ psi} \tag{2}$$

The existing safety factor for yielding then is

$$(n_{ex})_{yp} = \frac{48{,}000}{20{,}408} = 2.35 \tag{3}$$

For fatigue

$$(n_{ex})_f = \frac{S_{N=10^7}}{\sigma_{max}} \tag{4}$$

From Figure 2.19,

$$S_{N=10^7} = 23{,}000 \text{ psi} \tag{5}$$

The existing safety factor for fatigue then is

$$(n_{ex})_f = \frac{23{,}000}{20{,}408} = 1.13 \tag{6}$$

Since $(n_{ex})_f < (n_{ex})_{yp}$, the *fatigue* failure mode governs; hence the existing safety factor for this member, as presently designed, is

$$n_{ex} = 1.13 \tag{7}$$

5.4 Reliability: Concepts, Definitions, and Data

A guiding principle followed by effective designers is to utilize as much available *quantitative* information as possible in making design decisions. Therefore, if probabilistic descriptions are available in the form of statistical data for describing strength distributions, loading distributions, or variations in environment, manufacturing, inspection, and/or maintenance practices, these data should be utilized to keep the *probability of failure* low, or, to state if differently, to keep the *relibility above a prescribed level of acceptability.*

Reliability may be defined as the probability that a machine or machine part will perform its intended function without failure for its prescribed design lifetime. If the probability of failure is denoted by $P\{failure\}$, the reliability, or probability of survival, is $R = 1 - P\{failure\}$. Thus reliability is a quantitative measure of survival success, typically based on distribution functions verified by experimental data.

Implementation of the *probabilistic design approach* requires that the *distribution function (probability density function)* be known or assumed for both the stress at the critical point (and all factors influencing stress) and the strength at the critical point (and all factors influencing strength).

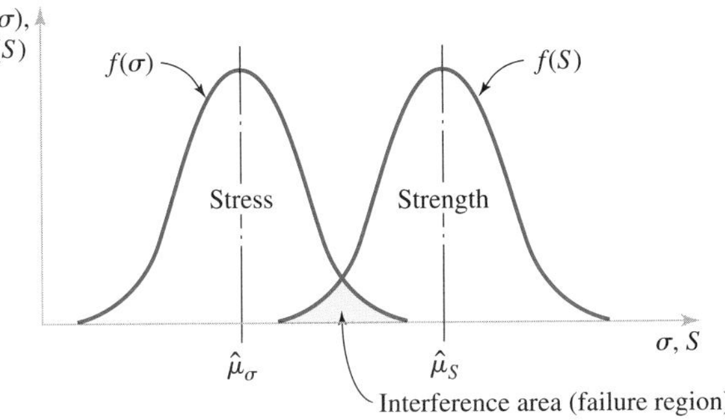

Figure 5.1
Probability density functions for stress and strength, showing interference area (failure region).

If the probability density function for stress, $f(\sigma)$, and the probability density function for strength, $f(S)$, are known, they may be plotted as shown in Figure 5.1. By definition, the reliability is the probability that the strength exceeds the stress, or

$$R = P\{S > \sigma\} = P\{S - \sigma > 0\} \tag{5-8}$$

This corresponds to the area in Figure 5.1 that lies *outside* the shaded *interference* area. The shaded area represents *probability of failure*, that is, $P\{\sigma \geq S\}$. The probability of failure is sometimes called *unreliability*.

Three probability density functions of particular interest to a designer are the *normal*, the *log-normal*, and the *Weibull* distribution functions. Only one of these, the normal distribution, will be illustrated here, but similar developments are available for the other two.[2] The probability density function $f(x)$ for the *normal* distribution is

$$f(x) = \frac{1}{\hat{\sigma}\sqrt{2\pi}} e^{-\frac{1}{2}\left(\frac{x-\hat{\mu}}{\hat{\sigma}}\right)^2} \quad \text{for} \quad -\infty < x < \infty \tag{5-9}$$

where x is a random variable such as stress or strength, $\hat{\mu}$ is the estimated *population* mean, and $\hat{\sigma}$ is the estimated *population* standard deviation, where

$$\hat{\mu} = \frac{1}{n}\sum_{i=1}^{n} x_i \tag{5-10}$$

and

$$\hat{\sigma} = \sqrt{\frac{1}{n-1}\sum_{i=1}^{n}(x_i - \mu)^2} \tag{5-11}$$

In these expressions, n equals the number of items in the population. As a matter of interest, the square of the standard deviation, $\hat{\sigma}^2$, is defined as the *variance*. Both variance and standard deviation are measures of dispersion, or scatter, of a distribution. Conventional notation for describing a normal distribution is

$$x \overset{d}{=} N(\hat{\mu}, \hat{\sigma}) \tag{5-12}$$

which is to be read "x is distributed normally with mean $\hat{\mu}$ and standard deviation $\hat{\sigma}$." As shown in Figure 5.2(a), a normal distribution has the well-known bell shape, is symmetrical about its mean, and tails off to infinity in both directions. The corresponding *normal cumulative distribution function*, $F(X)$, is plotted in Figure 5.2(b), where

$$F(X) = P\{X \leq X_0\} = \int_{-\infty}^{X} \frac{1}{\sqrt{2\pi}} e^{-\frac{t^2}{2}} dt \tag{5-13}$$

[2]See, for example, refs. 5, 6, 7, or 8.

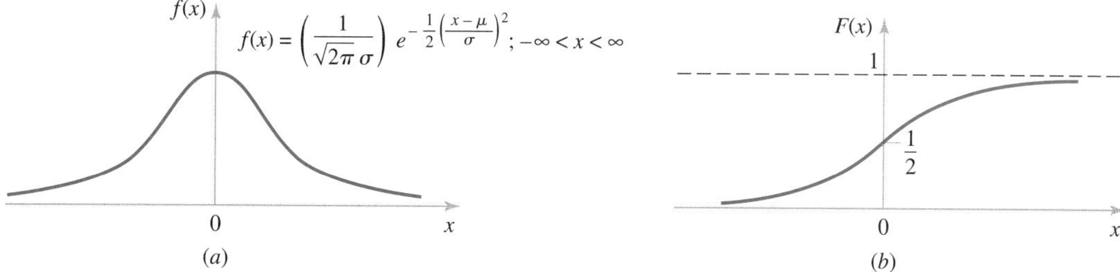

Figure 5.2
Plots of the probability density function and the cumulative distribution function for a *normal* distribution. (a) Probability density function. (b) Cumulative distribution function.

and

$$X = \frac{x - \hat{\mu}}{\hat{\sigma}} \tag{5-14}$$

is defined as the *standard normal variable*, which has a normal distribution with mean of 0 and standard deviation of 1. X_0 is any specified value of the random variable X. Any normal distribution with mean $\hat{\mu}$ and standard deviation $\hat{\sigma}$ can be transformed into a standard normal distribution using (5-14). Because all normal distributions can be fully defined by $\hat{\mu}$ and $\hat{\sigma}$, they are called *two-parameter* distributions. Table 5.1 gives values for the cumulative distribution function $F(X)$ for the standard normal distribution.

TABLE 5.1 Cumulative Distribution Function *F(X)* for the Standard Normal Distribution, Where
$$F(X) = \int_{-\infty}^{X} \frac{1}{\sqrt{2\pi}} e^{(-t^2/2)} dt$$

X	.00	.01	.02	.03	.04	.05	.06	.07	.08	.09
0	.5000	.5040	.5080	.5120	.5160	.5199	.5239	.5279	.5319	.5359
.1	.5398	.5438	.5478	.5517	.5557	.5596	.5636	.5675	.5714	.5753
.2	.5793	.5832	.5871	.5910	.5948	.5987	.6026	.6064	.6103	.6141
.3	.6179	.6217	.6255	.6293	.6331	.6368	.6406	.6443	.6480	.6517
.4	.6554	.6591	.6628	.6664	.6700	.6736	.6772	.6808	.6844	.6879
.5	.6915	.6950	.6985	.7019	.7054	.7088	.7123	.7157	.7190	.7224
.6	.7257	.7291	.7324	.7357	.7389	.7422	.7454	.7486	.7517	.7549
.7	.7580	.7611	.7642	.7673	.7704	.7734	.7764	.7794	.7823	.7852
.8	.7881	.7910	.7939	.7967	.7995	.8023	.8051	.8078	.8106	.8133
.9	.8159	.8186	.8212	.8238	.8264	.8289	.8315	.8340	.8365	.8389
1.0	.8413	.8438	.8461	.8485	.8508	.8531	.8554	.8577	.8599	.8621
1.1	.8643	.8665	.8686	.8708	.8729	.8749	.8770	.8790	.8810	.8830
1.2	.8849	.8869	.8888	.8907	.8925	.8944	.8962	.8980	.8997	.9015
1.3	.9032	.9049	.9066	.9082	.9099	.9115	.9131	.9147	.9162	.9177
1.4	.9192	.9207	.9222	.9236	.9251	.9265	.9279	.9292	.9306	.9319
1.5	.9332	.9345	.9357	.9370	.9382	.9394	.9406	.9418	.9429	.9441
1.6	.9452	.9463	.9474	.9484	.9495	.9505	.9515	.9525	.9535	.9545
1.7	.9554	.9564	.9573	.9582	.9591	.9599	.9608	.9616	.9625	.9633
1.8	.9641	.9649	.9656	.9664	.9671	.9678	.9686	.9693	.9699	.9706
1.9	.9713	.9719	.9726	.9732	.9738	.9744	.9750	.9756	.9761	.9767
2.0	.9772	.9778	.9783	.9788	.9793	.9798	.9803	.9808	.9812	.9817
2.1	.9821	.9826	.9830	.9834	.9838	.9842	.9846	.9850	.9854	.9857
2.2	.9861	.9864	.9868	.9871	.9875	.9878	.9881	.9884	.9887	.9890

TABLE 5.2 (*Continued*)

X	.00	.01	.02	.03	.04	.05	.06	.07	.08	.09
2.3	.9893	.9896	.9898	.9901	.9904	.9906	.9909	.9911	.9913	.9916
2.4	.9918	.9920	.9922	.9925	.9927	.9929	.9931	.9932	.9934	.9936
2.5	.9938	.9940	.9941	.9943	.9945	.9946	.9948	.9949	.9951	.9952
2.6	.9953	.9955	.9956	.9957	.9959	.9960	.9961	.9962	.9963	.9964
2.7	.9965	.9966	.9967	.9968	.9969	.9970	.9971	.9972	.9973	.9974
2.8	.9974	.9975	.9976	.9977	.9977	.9978	.9979	.9979	.9980	.9981
2.9	.9981	.9982	.9982	.9983	.9984	.9984	.9985	.9985	.9986	.9986
3.0	.9987	.9987	.9987	.9988	.9988	.9989	.9989	.9989	.9990	.9990
3.1	.9990	.9991	.9991	.9991	.9992	.9992	.9992	.9992	.9993	.9993
3.2	.9993	.9993	.9994	.9994	.9994	.9994	.9994	.9995	.9995	.9995
3.3	.9995	.9995	.9995	.9996	.9996	.9996	.9996	.9996	.9996	.9997
3.4	.9997	.9997	.9997	.9997	.9997	.9997	.9997	.9997	.9997	.9998
3.5	.9998	.9998	.9998	.9998	.9998	.9998	.9998	.9998	.9998	.9998
3.6	.9998	.9998	.9999	.9999	.9999	.9999	.9999	.9999	.9999	.9999
3.7	.9999	.9999	.9999	.9999	.9999	.9999	.9999	.9999	.9999	.9999
3.8	$.9^4 28$	$.9^4 31$	$.9^4 33$	$.9^4 36$	$.9^4 38$	$.9^4 41$	$.9^4 43$	$.9^4 46$	$.9^4 48$	$.9^4 50$
3.9	$.9^4 52$	$.9^4 54$	$.9^4 56$	$.9^4 58$	$.9^4 59$	$.9^4 61$	$.9^4 63$	$.9^4 64$	$.9^4 66$	$.9^4 67$
4.0	$.9^4 68$	$.9^4 70$	$.9^4 71$	$.9^4 72$	$.9^4 73$	$.9^4 74$	$.9^4 75$	$.9^4 76$	$.9^4 77$	$.9^4 78$
4.1	$.9^4 79$	$.9^4 80$	$.9^4 81$	$.9^4 82$	$.9^4 83$	$.9^4 83$	$.9^4 84$	$.9^4 85$	$.9^4 85$	$.9^4 86$
4.2	$.9^4 87$	$.9^4 87$	$.9^4 88$	$.9^4 88$	$.9^4 89$	$.9^4 89$	$.9^4 90$	$.9^5 00$	$.9^5 02$	$.9^5 07$

An important theorem of statistics states that when independent normally distributed random variables are summed, the resulting sum is itself normally distributed with mean equal to the sum of the individual distribution means and standard deviation equal to the square root of the sum of the individual distribution variances. Thus when $f(\sigma)$ and $f(S)$ are *both* normal probability density functions, the random variable $y = (S - \sigma)$ used in (5-8) is also normally distributed with a mean of

$$\hat{\mu}_y = \hat{\mu}_S - \hat{\mu}_\sigma \tag{5-15}$$

and standard deviation

$$\hat{\sigma}_y = \sqrt{\hat{\sigma}_S^2 + \hat{\sigma}_\sigma^2} \tag{5-16}$$

The reliability R then may be expressed as[3]

$$R = P\{y > 0\} = P\{S - \sigma > 0\} = \frac{1}{2\pi} \int_{-\frac{\hat{\mu}_S - \hat{\mu}_\sigma}{\sqrt{\hat{\sigma}_S^2 + \sigma_\sigma^2}}}^{\infty} e^{-\frac{t^2}{2}}\, dt \tag{5-17}$$

where

$$t = \frac{y - \hat{\mu}_y}{\hat{\sigma}_y} \tag{5-18}$$

is the standard normal variable [see (5-14) and Table 5.1].

Example 5.3 Existing Safety Factor and Reliability Level

An axially loaded, straight, cylindrical bar of diameter $d = 0.50$ inch is made of 2024-T4 aluminum. Experimental data for the material tested under conditions that closely correspond to actual operating conditions indicate that the yield strength is normally distributed with a mean value of 48,000 psi and standard deviation of 5000 psi. The static load on the

[3]See, for example, ref. 6.

bar has a nominal value of 7000 lb, but due to various operational procedures, and excitations from adjacent equipment, the load has been found to actually be a normally distributed random variable with standard deviation of 500 lb.

a. Find the existing factor of safety for the bar, based on yielding as a failure mode.

b. Find the reliability level of the bar, based on yielding as a failure mode.

Solution

a. Based on yielding, the existing safety factor is

$$n_{ex} = \frac{S_{yp-nom}}{\sigma_{nom}} \tag{1}$$

where

$$S_{yp-nom} = 48{,}000 \text{ psi} \tag{2}$$

and

$$\sigma_{nom} = \frac{P_{nom}}{A} = \frac{7000}{\pi\left[\frac{(0.5)^2}{4}\right]} = \frac{7000}{0.196} = 35{,}714 \text{ psi} \tag{3}$$

whence the existing safety factor is

$$n_{ex} = \frac{48{,}000}{35{,}714} = 1.34 \tag{4}$$

b. From given data,

$$P \stackrel{d}{=} N(7000 \text{ lb}, 500 \text{ lb}) \tag{5}$$

and

$$S_{yp} \stackrel{d}{=} N(48{,}000 \text{ psi}, 5000 \text{ psi}) \tag{6}$$

From (5) the estimated mean stress may be calculated as

$$\hat{\mu}_{\sigma} = \frac{P}{A} = \frac{7000}{0.196} = 35{,}714 \text{ psi} \tag{7}$$

and the estimated standard deviation of stress is

$$\hat{\sigma}_{\sigma} = \frac{500}{0.196} = 2550 \text{ psi} \tag{8}$$

hence

$$\sigma \stackrel{d}{=} N(35{,}714 \text{ psi}, 2550 \text{ psi}) \tag{9}$$

These calculations neglect statistical variability in dimensions used to calculate area.

From (6) and (9) the lower limit of the reliability integral in (5-17) may be calculated as

$$X = -\frac{\hat{\mu}_S - \hat{\mu}_{\sigma}}{\sqrt{\hat{\sigma}_S^2 + \hat{\sigma}_{\sigma}^2}} = -\frac{48{,}000 - 35{,}714}{\sqrt{5000^2 + 2550^2}} = -2.19 \tag{10}$$

From Table 5.1, since $R = 1 - P$, the reliability corresponding to $X = -2.19$ may be read as

$$R = 0.986 \tag{11}$$

Thus one would expect that 98.6 percent of all installations would function properly, but 14 of every 1000 installations would be expected to fail. A decision must be made by the designer about the acceptability of this failure rate for this application.

System Reliability, Reliability Goals, and Reliability Allocation

Efforts by a designer toward achieving the failure prevention objective at the design stage often include the tasks of determining the required reliability levels for individual components or subsystems that will assure the specified *reliability goal* for the whole machine assembly. The process of assigning reliability requirements to individual components to attain the specified system reliability goal is called *reliability allocation*. Although the allocation problem is complex, the principles may be demonstrated in a straighforward way by making certain *simplifying assumptions*.

As a practical matter, designers are well advised to work with a reliability engineering specialist at this stage. Reliability specialists can provide valuable insights in selecting distributions that are well suited to the application, data interpretation, experimental design, reliability allocation, and analytical and simulation techniques for reliability assessment and system optimization.

However, the task of setting appropriate reliability goals lies primarily in the province of the designer, in cooperation with engineering and company management. The daunting fact that a *reliability of 100 percent can never be achieved* forces careful consideration of potential failure consequences. Choosing reliability levels appropriate and acceptable for the particular machine application being proposed and selection of reliability goals are often difficult tasks.[4] It is alleged, for example, that engineers seeking an appropriate and acceptable reliability goal during early pioneering efforts in spacecraft design decided that the probability that a *toreador* can avoid being gored in a bull fight would also serve as an acceptable reliability goal for spacecraft design. The aerospace industry specifies a reliability of "five-nines" ($R = 0.99999$) in many cases, while the "standard" reliability of rolling element bearings is usually taken to be $R = 0.90$. Table 5.2 illustrates probability-of-failure criteria used by some industries. Clearly the selection of an appropriate reliability goal depends upon both the consequences of failure and the cost of achieving the goal.[5]

TABLE 5.2 Reliability-Based Design Goals as a Function of Perceived Hazard Level[1]

Potential Consequences of Failure	Designated Hazard Assessment Category	Suggested Criteria for Design-Acceptable Probability of Failure
Some reduction in safety margin or functional capability.	Minor	$\leq 10^{-5}$
Significant reduction in safety margin or functional capability.	Major	$10^{-5}-10^{-7}$
Device no longer performs its function properly; small numbers of serious or fatal injuries may occur.	Hazardous	$10^{-7}-10^{-9}$
Complete equipment failure, unable to correct the situation; multiple deaths and/or high collateral damage may occur.	Catastrophic	$\leq 10^{-9}$

[1]Excerpted in part from ref. 11.

[4]See also 1.8.

[5]For example, in the 1980s Motorola inaugurated the practice of setting allowable failure rates to a very low value, by design, and named the process *Six Sigma*. (Six sigma corresponds to a failure rate goal of no more than about 3.4 failures per million units. By contrast, U.S. industry as a whole operates around three sigma, or a failure rate of about 66,000 units per million.) Adopting the Six Sigma process, General Electric Company estimated that spending about $500 million in 1999 on Six Sigma would produce a return of approximately $2 billion. See ref. 10.

For the purpose of making a *simplified* reliability analysis at the design stage, the reliabilities of components and subsystems may be *assumed* constant over the design life of the machine. To assess the effects of component or subsystem failure on overall system performance, a *functional block diagram* is a helpful tool. For a simplified analysis, each block in the diagram represents a "black box" that is assumed to be in one of two states: *failed* or *functional*. Depending upon the arrangement of components in the machine assembly, the failure of a component will have different effects upon the reliability of the overall machine.

Figure 5.3 illustrates four possible arrangements of components in a machine assembly. The *series* configuration of Figure 5.3(a), in which *all components must function properly for the system to function*, is probably the most common arrangement encountered in practice. For a series arrangement the system reliability R_S is found to be

$$R_S = \prod_{i=1}^{n} R_i \tag{5-19}$$

where the right-hand side is the product of the individual component reliabilities. If it is assumed that the failure probability q is identical for all components, (5-19) may be rewritten as

$$R_S = (1 - q)^n \tag{5-20}$$

which may be expanded by the binomial theorem to give, neglecting higher-order terms,

$$R_S = 1 - nq \tag{5-21}$$

From this it may be noted that, for a series arrangement, the system reliability *decreases* rapidly as the number n of series components increases. In general, for series systems, the system reliability will always be less than or equal to the reliability of the *least reliable* component. This is sometimes called the "weakest link" failure model.

The *parallel* arrangement of Figure 5.3(b), in which the system *continues to function until all components have failed*, has a system reliability of

$$R_S = 1 - \prod_{i=1}^{p} (1 - R_i) \tag{5-22}$$

provided that all components are active in the system and that failures do not influence the reliabilities of the surviving components. In practice, these assumptions may be invalid if, for example, a standby redundant component is not activated unless the on-line component fails, or if the failure rate of surviving components increases as failures occur. If it is assumed that the failure probability is identical for all components, (5-22) may be rewritten as

$$R_S = 1 - q^p \tag{5-23}$$

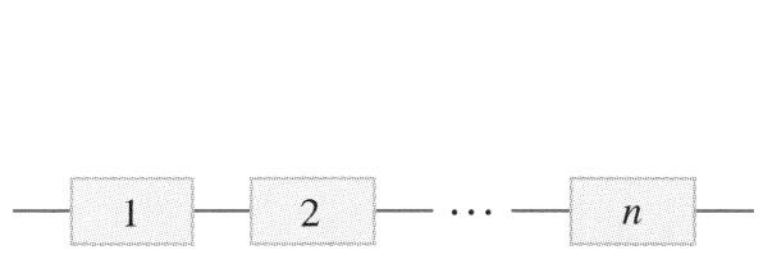

(*a*) Series arrangement of n components.

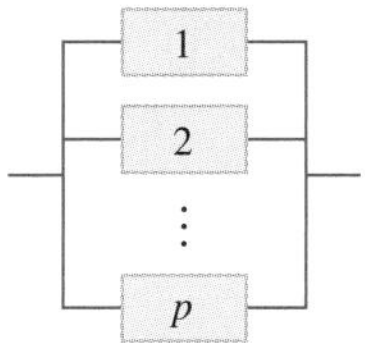

(*b*) Parallel arrangement of p components.

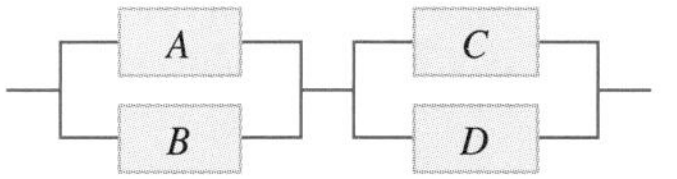

(*c*) Series-parallel arrangement of components (component level redundancy).

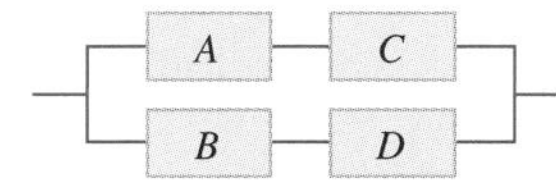

(*d*) Parallel-series arrangement of components (subsystem-level redundancy).

Figure 5.3
Block diagrams illustrating various design arrangements of components in a machine system, from a reliability analysis point of view.

From (5-23) it may be noted that, for a parallel arrangement, the system reliability *increases* as the number of parallel components increases. As a practical matter, however, designing a parallel system may be difficult and costly, and for a given component reliability, the gain in system reliability slows down rapidly as more parallel components are added. Nevertheless, the procedure of adding a second component or subsystem in parallel is the basis for the very important *fail-safe* design concept discussed in 1.8.

Virtually all arrangements of components in a machine assembly may be modeled as combinations of series and parallel configurations. Figure 5.3(c), for example, is a series arrangement of two parallel subsystems, and Figure 5.3(d) shows a parallel arrangement of two series subsystems. For Figure 5.3(c) there is *redundancy at the component level* while for Figure 5.3(d) there is *redundancy at the subsystem level*. System analysis has shown that for a system with a given number of components, component-level redundancy gives a higher system reliability than subsystem-level redundancy.

The task of simplified reliability allocation may be accomplished by utilizing a technique called *equal apportionment*. This technique models the system reliability as a series of m subsystems, each having equal reliability R_i. Each subsystem may include one or more components in series or parallel arrangements. The specified system reliability then is

$$(R_S)_{\substack{specified\\system}} = \prod_{i=1}^{m} R_i \tag{5-24}$$

and the required component reliability to achieve the specified system reliability is

$$(R_i)_{req'd} = \left[(R_S)_{\substack{specified\\system}}\right]^{1/m} \tag{5-25}$$

Example 5.4 Component Reliability Required to Meet Specified Overall Machine-Reliability Goals

A machine assembly of four components may be modeled as a series-parallel arrangement similar to that shown in Figure 5.3(c). It has been determined that a system reliability of 95 percent is necessary to meet design objectives.

a. Considering subsystems A-B and C-D, what subsystem reliability is required to meet the specified 95 percent reliability goal for the machine?

b. What component reliabilities would be required for A, B, C, and D to meet the 95 percent reliability specification for the machine?

Solution

a. Using (5-25), the required subsystem reliability is

$$R_{req'd} = (0.95)^{1/2} = 0.975 \tag{1}$$

b. For subsystem A-B, the components A and B are in parallel, hence (5-22) may be used to give

$$R_{A\text{-}B} = 1 - (1 - R_i)^2 \tag{2}$$

or

$$R_i = 1 - \sqrt{1 - R_{A\text{-}B}} = 1 - \sqrt{1 - 0.975} = 0.84 \tag{3}$$

Hence the system reliability goal of 0.95 can be met only if the *subsystem* reliabilities are at least 0.975. However, because of the redundancy within the subsystems, *components* having reliabilities as low as 0.84 can meet the reliability requirements of the system.

Reliability Data

A designer striving to achieve the failure prevention objective by implementing reliability methodology at the design stage quickly confronts the need for *pertinent*, *valid*, *quantitative* data relating to both strength and loading. Distributional data defining mean values and variances in all parameters that affect strength and loading are required. A sufficient body of pertinent distributional design data rarely exists. Valid and useful new data are costly and time consuming to collect. At the design stage it is typically necessary to rely on existing experimental material strength data interpreted statistically, and extrapolated statistical operational field data for loads and environments, gathered from similar service applications.

Historically, experimental material strength data have often been gathered without reporting information now known to be statistically important; in many cases archival information, or published histographic information, can be reinterpreted statistically[6] to recover the reliability indices of interest. In addition to probability of failure, other indices of use might include average time between failures, average duration of system "downtime," expected loss of revenue due to failure, or expected loss of manufacturing output due to failure. In the final analysis, to be useful to a designer at the early design stage, probabilistic descriptions of such indices must have the potential to be quantitatively linked to strength and/or loading distributions. In critical applications, to support design improvement activities, it may become necessary to generate specific statistical data during the development stage or during field service monitoring. A review of the fundamental design steps in Table 1.1 serves to reiterate this concept.

5.5 The Dilemma of Reliability Specification versus Design Safety Factor

The dilemma for a designer in attempting to implement the "best" failure prevention strategy often seems to rest upon the choice between the *reliability* approach and the *design safety factor* approach. The spotty availability of pertinent statistical data often leads a designer to simply ignore the data that *are* available. Thus the basic principle of utilizing as much quantitative information as possible when making design decisions is violated.

Another viewpoint sometimes taken is that when using the probabilistic design approach (reliability approach), the design safety factor should be abandoned. If, indeed, probabilistic descriptions were available for strength, load, environment, manufacturing, inspection, and maintenance practices, there would be no need for a design safety factor. However, such complete information is rarely, if ever, available to a designer.

Rather than choosing between the two approaches, a more productive viewpoint might be to *combine the best attributes* of each in making design decisions. Thus, where well-defined probabalistic data are available for describing strength, loading, manufacturing practice, or other design parameters, these quantitative probablistic data might be incorporated piecewise in the design safety factor approach of 5.2. As more precise probabilistic data are incorporated, the semiquantitative *rating numbers (RNs)* tend to be driven toward more negative values, since more precise information in these cases would prompt a perceived need to *lower* the design safety factor. The magnitude of the design safety factor calculated from (5-4) would, in such cases, be driven toward lower values, ultimately approaching the lower limiting value prescribed by (5-5). Because of using the available probabilistic information, this approach promotes more precision in making de-

[6]An excellent example is provided by C. R. Mischke in ref. 8, pp. 1–10.

sign decisions about materials and dimensions, and, at the same time, preserves the designer's ability to account for uncertainties and variabilities in parameters that are unsupported by statistical data. The approach has much to recommend it.

Example 5.5 Change in Selection of Design Safety Factor as a Result of Higher-Reliability Strength Data

Since the determination of a design safety factor of $n_d = 1.25$ in Example 5.1, an extensive body of experimental data has been discovered for fatigue strength of the material. This data bank indicates that the fatigue endurance limit S'_f is normally distributed. Since the data are extensive, accurate estimates of the mean and standard deviation may be determined with a high level of confidence. How would this newly discovered body of statistical data influence the value of the design safety factor determined in Example 5.1?

Solution

The only change in determining n_d from the rating factors would involve reevaluation of rating number 3, which was chosen as $RN = -1$ in Example 5.1. The new, much more accurate, data might suggest a change in value to $RN = -3$. Thus, from (5-3) for this case

$$t = 0 + 0 - 3 - 3 + 3 + 0 + 0 - 4 = -7 \tag{1}$$

and since $t < -6$, from (5.5)

$$n_d = 1.15 \tag{2}$$

Using the statistically significant new strength data therefore results in a design safety factor reduction of about 8 percent. It should be cautioned that an 8 percent reduction in design safety factor does not necessarily correspond to an 8 percent increase in the calculated design-allowable stress. The reason for this is that the fatigue endurance limit value used in the calculation of design-allowable stress would (because the probabilistic approach is being used to define fatigue strength) now depend upon the choice of an appropriate fatigue strength reliability level for the application (see Example 2.9).

Problems

5-1. Define the term "design-allowable stress," write an equation for design-allowable stress, define each term in the equation, and tell how or where a designer would find values for each term.

5-2. Your company desires to market a new type of lawn mower with an "instant-stop" cutting blade. (For more details about the application, see Example 16.1). You are responsible for the design of the actuation lever. The application may be regarded as "average" in most respects, but the material properties are known a little better than for the average design case, the need to consider threat to human health is regarded as strong, maintenance is probably a little poorer than average, and it is extremely important to keep the cost low. Calculate a proper safety factor for this application, clearly showing details of your calculation.

5-3. You are asked to review the design of the shafts and gears that are to be used in the drive mechanism for a new stair-climbing wheelchair for quadriplegic users. The wheelchair production rate is to be about 1200 per year. From the design standpoint the application may be regarded as "average" in many respects, but the need to consider threat to human health is regarded as extremely important, the loads are known a little better than for the average design project, there is a strong desire to keep the weight down, and a moderate desire to keep the cost down. Calculate a proper safety factor for this application, clearly showing all details of how you arrive at your answer.

5-4. A novel design is being proposed for a new attachment link for a chair lift at a ski resort. Carefully assessing the potential importance of all pertinent "rating factors," calculate a proper safety factor for this application, clearly showing the details of how you arrive at your answer.

5-5. Stainless-steel alloy AM 350 has been tentatively selected for an application in which a cylindrical tension rod must support an axial load of 10,000 lb. The ambient temperature is known to be 800°F. If a design safety factor of 1.8 has been selected for the application, what minimum diameter should the tension rod have? (*Hint:* Examine "materials properties" charts given in Chapter 3.)

5-6. It has been discovered that for the application described in problem 5-5, an additional design constraint must be satisfied, namely, the creep strain rate must not exceed 1×10^{-6} in/in/hr at the ambient temperature of 800°F. To meet the 1.8 safety factor requirement for this case, what minimum diameter should the tension rod have? (*Hint:* Examine "material properties" charts given in Chapter 3.)

5-7. A design stress of $\sigma_d = 32{,}000$ psi is being suggested by a colleague for an application in which 2024-T4 aluminum alloy has tentatively been selected. It is desired to use a design safety factor of 1.5. The application involves a solid cylindrical shaft continuously rotating at 120 revolutions per hour, simply supported at the ends, and loaded at midspan, downward, by a static load P. To meet design objectives, the aluminum shaft must operate without failure for 10 years. For 2024-T4 aluminum, $S_u = 68{,}000$ psi and $S_{yp} = 48{,}000$ psi. Some other properties data for 2024-T4 aluminum are shown in Figure 2.19. Would you agree with your colleague's suggestion for a design stress of $\sigma_d = 32{,}000$ psi? Explain.

5-8. A 304 stainless-steel alloy, annealed, has been used in a deflection-critical application to make the support rod for a test package suspended near the bottom of a deep cylindrical cavity. The solid stainless-steel support rod has a diameter of 0.750 inch and a precisely manufactured length of 16.000 feet. It is oriented vertically and fixed at the top end. The 6000-pound test package is attached at the bottom, placing the vertical bar in axial tension. The vertical deflection at the end of the bar must not exceed a maximum of 0.250 inch. Calculate the existing safety factor.

5-9. A very wide sheet of 7075-T651 aluminum plate, $^5/_{16}$ inch thick, has just been found to have a single-edge through-the-thickness crack, 1.0 inch long. The loading produces a gross nominal tensile stress of 10,000 psi perpendicular to the plane of the crack. Estimate the existing safety factor with the 1.0-inch crack in place.

5-10. a. Estimate and plot the *S-N* curve for ASTM A-48 (Class 50) gray cast iron, using the static properties data given in Table 3.3.

b. Based on the estimated *S-N* curve, what existing safety factor would you calculate for parts made from this cast-iron material if they are subjected to completely reversed uniaxial cyclic stresses of 30 ksi amplitude, and the desired design life is 10^4 cycles?

5-11. A vertical solid cylindrical steel bar is 2.0 inches in diameter and 12 feet long. Both ends are pinned and the top pinned end is vertically guided, as for the case shown in Figure 2.37(a). If a centered static load of $P = 5000$ pounds must be supported at the top end of the vertical bar, what is the existing safety factor?

5-12. A fatigue testing program has produced the life data given in Table P5.12 for 35 specimens, all from a single heat of aluminum alloy, all tested to failure at a completely reversed stress amplitude of 26,000 psi. Compute the estimated mean life and the estimated standard deviation on life for the population of material from which this sample was drawn, assuming the population distribution to be normal.

TABLE P5.12 Fatigue LIfe Data

Fatigue Life in Thousands of Cycles						
290	490	342	456	517	310	445
540	233	376	410	439	403	315
439	433	473	367	400	358	445
358	351	422	276	560	360	406
400	395	321	356	443	404	362

5-13. A supplier of 4340 steel material has shipped enough material to fabricate 100 fatigue-critical tension links for an aircraft application. As required in the purchase contract, he has conducted uniaxial fatigue tests on random specimens drawn from the lot of material, and has certified that the mean fatigue strength corresponding to a life of 10^6 cycles is 68,000 psi, that the standard deviation on strength corresponding to a life of 10^6 cycles is 3400 psi, and that the distribution of strength at a life of 10^6 cycles is normal.

a. Estimate how many tension links from the lot of 100 may be expected to fail when operated for 10^6 cycles if the applied operating stress amplitude is less than 60,000 psi.

b. Estimate how many tension links may be expected to fail when operated for 10^6 cycles at stress levels between 60,000 psi and 68,000 psi.

5-14. A lot of 4340 steel material has been certified by the supplier to have a fatigue strength distribution at a life of 10^7 cycles of

$$S_{N=10^7} \stackrel{d}{=} N(68{,}000 \text{ psi}, 2500 \text{ psi})$$

Experimental data collection over a long period of time indicates that operating stress levels at the critical point of an important component with a design life of 10^7 cycles have a stress distribution of

$$\sigma_{op} \stackrel{d}{=} N(60{,}000 \text{ psi}, 5000 \text{ psi})$$

Estimate the reliability level corresponding to a life of 10^7 cycles for this component.

5-15. It is known that a titanium alloy has a standard deviation on fatigue strength of 3000 psi over a wide range of strength levels and cyclic lives. Also, experimental data have been collected to indicate that operating stress levels at the critical point of an important ccmponent with a design life of 5×10^7 cycles have a stress distribution of

$$\sigma_{op} \stackrel{d}{=} N(50{,}000 \text{ psi}, 4000 \text{ psi})$$

If a reliability level of "five-nines" (i.e., $R = 0.99999$) is desired, what *mean* strength would the titanium alloy need to have?

5-16. Using the tabulated normal cumulative distribution function given in Table 5.1, verify the strength reliability factors listed in Table 2.3.

5-17. The main support shaft in a new 10-ton-hoist design project is under consideration. Clearly, if the shaft fails the falling 10-ton payload could inflict serious injuries, or even fatalities. Suggest a design-acceptable probability of failure for this potentially hazardous failure scenario.

5-18. A series-parallel arrangement of components consists of a series of n subsystems, each having p parallel components. If the probability of failure for each component is q, what would be the system reliability for the series-parallel arrangement described?

5-19. A parallel-series arrangement of components consists of p parallel subsystems, each having n components in series. If the probability of failure for each component is q, what would be the system reliability for the parallel-series arrangement described?

5-20. A critical subsystem for an aircraft flap actuation assembly consists of three components in series, each having a component reliability of 0.90.

a. What would the subsystem reliability be for this critical three-component subsystem?

b. If a second (redundant) subsystem of exactly the same series arrangement were placed in parallel with the first subsystem, would you expect a significant improvement in reliability? How much?

c. If a third redundant subsystem of exactly the same arrangement as the first two were also placed in parallel with them, would you expect a significant additional improvement in reliability? Make any comments you think appropriate.

d. Can you think of any reason why several redundant subsystems should not be used in this application in order to improve reliability?

5-21. A machine assembly of four components may be modeled as a parallel-series arrangement similar to that shown in Figure 5.3(d). It has been determined that a system reliability of 95 percent is necessary to meet design objectives.

a. Considering subsystems *A-C* and *B-D*, what subsystem reliability is required to meet the 95 percent reliability goal for the machine?

b. What component reliabilities would be required for *A, B, C,* and *D* to meet the 95 percent reliability specification for the machine?

Chapter 6

Geometry Determination

6.1 The Contrast in Objectives Between Analysis and Design

The objective of *analysis*, within the context of machine design, is to examine machines and/or machine elements for which sizes, shapes, and materials have already been proposed or selected, so that loading severity parameters (e.g., stresses) may be calculated and compared with critical capacities (e.g., strengths corresponding to governing failure modes) at each critical point.[1] Adjunct analyses might also be undertaken to calculate and compare such attributes as cost, life, weight, noise level, safety risks, or other pertinent performance parameters.

The objective of *design* (or *synthesis*) is to examine performance requirements associated with a particular design mission, then select the best possible material and determine the best possible shape, size, and arrangement, within specified constraints of life, cost, weight, safety, reliability, or other performance parameters.

The contrast is clear. *Analysis* can be undertaken only if shapes, sizes, and materials are already known. Analytical procedures are typified by calculating stresses from known loading and geometry, then comparing the calculated stresses with the strengths of known materials. On the other hand, *design* is undertaken to *create* shapes, *determine* sizes, and *select* materials so that the specified loading may be sustained without causing failure during the specified design lifetime. Design procedures embody the tasks of determining probable failure modes, selecting appropriate materials, selecting an appropriate design safety factor, calculating a design stress, and determining shapes and sizes so that maximum stresses are equal to the design stress. In a nutshell, the input requirements for analysis are the same as the output results for design.

Although there is a clear contrast between the objectives of analysis and design, the basic concepts used, the pertinent equations of mechanics, the mathematical models utilized, and the useful data sources are the same for both activities. In fact, one technique sometimes used in accomplishing the design objective is to solve a series of analysis problems, making appropriate changes in materials and/or geometry, until design requirements are met. In the final analysis a good designer must first be a good analyst. A good analyst must possess (1) the ability to reduce a complicated real problem to a pertinent but solvable mathematical model by making good simplifying assumptions, (2) the ability to carry through an appropriate solution by using techniques and data sources best suited to the task, and (3) the ability to interpret results through a basic understanding of the models, equations, and/or software used, and their limitations.

One task, however, that lies uniquely within the sphere of design activity is creation

[1]See 5.3 for additional discussion.

of the best possible shape for a proposed machine part. Basic guidelines for accomplishing this task are presented next.

6.2 Basic Principles and Guidelines for Creating Shape and Size

The basic principles for creating the shape of a machine part and determining its size are

1. Create a shape that will, as nearly as possible, result in a *uniform stress distribution throughout all of the material* in the part, and
2. For the shape so chosen, find dimensions that will produce *maximum operating stresses equal to the design stress.*

Interpreting these principles in terms of the five common stress patterns discussed in 4.4, a designer should, insofar as possible, choose shapes and arrangements that will produce direct axial stress (tension or compression), uniform shear, or fully conforming surface contact, and avoid bending, transverse shear, torsion, and Hertzian contact geometry. If bending, transverse shear, torsion, or Hertzian contact cannot be avoided, the designer should persist in developing shapes that minimize stress gradients and eliminate lightly stressed or "lazy" material. With these basic principles in mind, several guidelines may be stated[2] in terms of identifying desirable geometric choices for shapes and arrangements of machine parts. These *configurational guidelines* include:

1. Using direct load paths
2. Tailoring element shape to loading gradient
3. Incorporating triangular ot tetrahedral arrangements or shapes
4. Avoiding buckling-prone geometry
5. Utilizing hollow cylinders and I-beams to achieve near-uniform stress
6. Providing conforming surfaces at mating interfaces
7. Removing lightly stressed or "lazy" material
8. Merging different shapes gradually from one to another
9. Matching element surface strains at joints and contacting surfaces
10. Spreading loads at joints

Each of these goals is briefly discussed next.

Direct Load Path Guideline

The concept of *force-flow lines* was introduced in 4.3. To follow the direct load path guideline, the lines of *force flow* should be kept as direct and as short as possible. For example, Figure 6.1 shows two alternative configurations for a proposed machine part to transmit a direct tensile load from joint A to joint B. Proposal 1 is a U-shaped link and Proposal 2 is a straight link. To follow the direct load path guideline, Proposal 2 should be adopted because the lines of force flow are shorter and more direct. This choice eliminates undesirable bending stresses from the link shown as Proposal 1, and results in a more uniform stress distribution.

[2]These concepts and guidelines were first articulated by Professor Emeritus Walter L. Starkey of The Ohio State University in the 1950s, when he was developing an area of design activity that he termed *form synthesis*. The concepts were later expanded and published in ref. 1 by K. M. Marshek in 1987. See ref. 1.

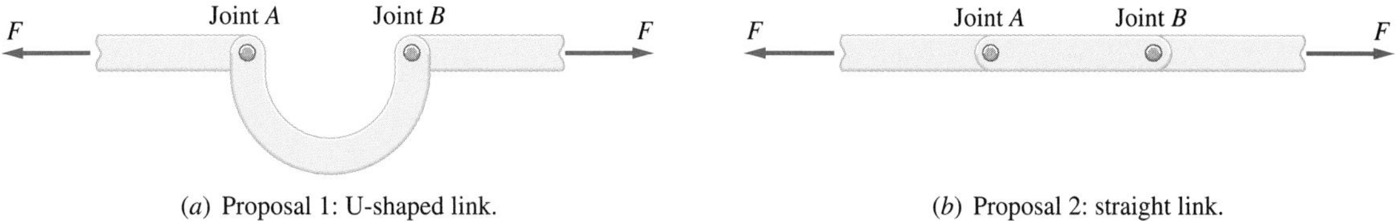

(a) Proposal 1: U-shaped link.

(b) Proposal 2: straight link.

Figure 6.1
Two alternative configurations demonstrating the direct load path guideline. Proposal 2 is preferred because it uses a direct load path.

Tailored-Shape Guideline

Under some circumstances, gradients in loading or stress lead to regions of lightly stressed material if a constant cross-sectional shape and size are maintained throughout the part. For example, a cylindrical bar in torsion exhibits maximum torsional shear stress at the outer fibers but zero stress at the center; a beam in bending reaches maximum stress at the outer fibers, but has zero stress at the neutral axis; a cantilever beam in bending under a transverse force at the free end experiences a maximum bending moment at the fixed end but zero moment at the free end. To follow the tailored-shape guideline, the shape of the machine part should be tailored in proportion to the gradient in such a way that stress level is kept as nearly constant as possible throughout the volume of the part. For example, Figure 6.2 shows two alternative configurations for a proposed cantilever beam that must support an end load P. Proposal 1 is a beam of constant rectangular cross section having width $b = b_1$ and depth $d = d_1$ for the whole length L of the beam. Proposal 2 is a beam of constant depth $d = d_1$, but width b varies linearly from b_1 at the fixed end to zero at the free end. To follow the tailored shape guideline, Proposal 2 should be chosen. The reasoning is as follows. The maximum bending stress at the outer fibers of a cantilever beam, at any distance x from the free end, is given by [see (4-5) and Table 4.1]

$$(\sigma_{max})_x = \frac{M_x c_x}{I_x} = \frac{6PLx}{bd^2} \tag{6-1}$$

For Proposal 1, since $b = b_1$ is constant,

$$(\sigma_{max})_{x-1} = \frac{6PLx}{b_1 d_1^2} \tag{6-2}$$

For Proposal 2, since b varies linearly with x,

$$(\sigma_{max})_{x-2} = \frac{6PLx}{\left(\dfrac{b_1}{L}\right)xd_1^2} = \frac{6PL^2}{b_1 d_1^2} \tag{6-3}$$

Comparing (6-2) with (6-3), the maximum outer-fiber stress for Proposal 1 varies from

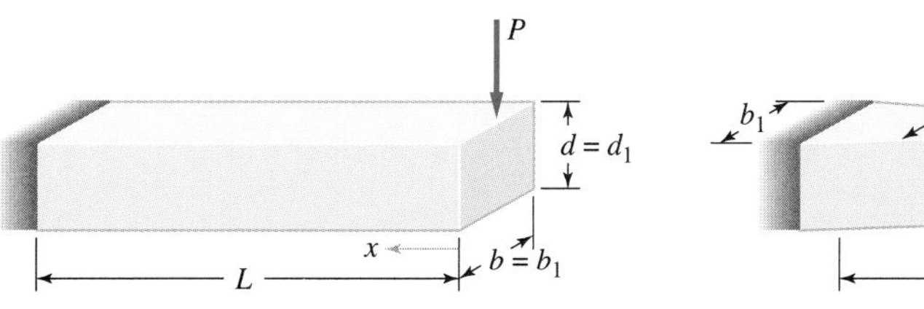

(a) Proposal 1: constant-width cantilever beam.

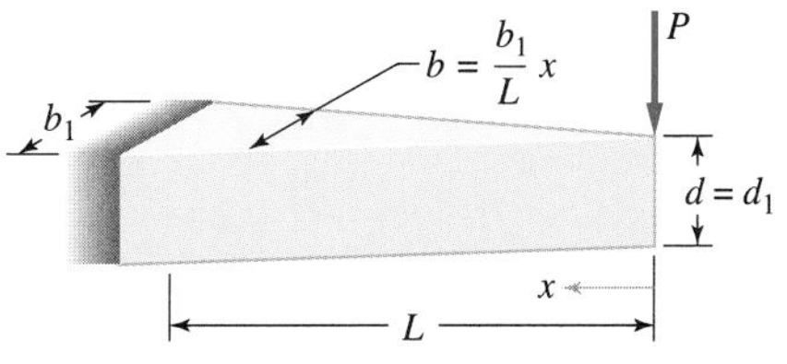

(b) Proposal 2: tapered-width cantilever beam.

Figure 6.2
Two alternative configurations demonstrating the tailored shape guideline. Proposal 2 is preferred because shape is tailored to the loading gradient.

$6PL/b_1 d_1^2$ at the fixed end, where $x = L$, down to zero at the free end, where $x = 0$. On the other hand, for Proposal 2 the maximum outer-fiber stress remains at a constant level of $6PL/b_1 d_1^2$ over the whole length of the beam.

Triangle-Tetrahedron Guideline

Analysis of a three-link planar, pinned-joint, triangular structure, such as the one shown in Figure 6.3(a), shows that any equilibrium force system *in the plane of the triangle*, when forces are applied solely at the joints, will result in uniformly distributed tensile or compressive stresses in each of the links. No bending stresses will be induced, assuming negligible friction in the pinned joints. If, for example, a vertical load P is to be supported by some type of bracket attached to a vertical wall a distance L away, two alternative configurations for the support bracket are proposed in Figure 6.3(b) and 6.3(c). Proposal 1 is a width-tapered, rectangular cross-section, cantilever beam, similar to the one shown in Figure 6.2(b). Proposal 2 is a pinned-joint triangular truss with links of solid compact cross section. To follow the triangle-tetrahedron guideline, Proposal 2 should be adopted. Even though the tailored-shape guideline illustrated in Figure 6.2 favors selection of a width-tapered cantilever beam when compared to a constant-width beam, the nonuniform stress distribution from top to bottom, as depicted in Figure 4.3, still leaves a substantial volume of lightly stressed material near the neutral axis of bending. To achieve a near-uniform stress distribution, the triangular truss of Proposal 2 is a better choice. A modified version of the truss is illustrated as Proposal 3, shown in Figure 6.3(d), where the pinned joint truss is replaced by a "cut-out" triangular plate that approximates the truss geometry. The stress distribution in the cut-out plate of Proposal 3 is clearly more complicated than the pinned-joint truss of Proposal 2, but may retain enough triangular truss "advantage" to justify eliminating the design complexities associated with pinned joints.

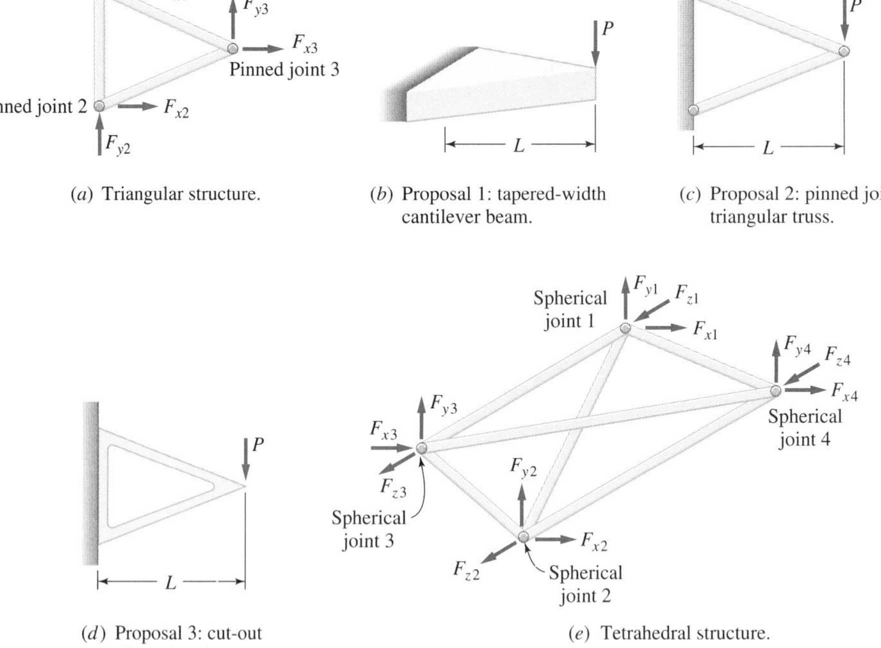

Figure 6.3
Various configurations illustrating the triangle-tetrahedron guideline. Proposal 2 is preferred to Proposal 1 because it results in more uniformly stressed materal. A tetrahedral structure must be used if loading is three-dimensional.

Finally, it is important to recognize that a *triangular pinned-joint structure is able to support force components only in the plane of the triangle*. To support out-of-plane force components, a three-dimensional structure must be provided. The three-dimensional structure corresponding in principle to a planar triangle is a tetrahedral truss constructed of six links joined by spherical joints, as shown in Figure 6.3(e). *Any* equilibrium force system, when forces are applied solely at the joints, will result in uniformly distributed tensile or compressive stresses in each of the links of the tetrahedral structure.

Buckling Avoidance Guideline

Euler's equation for critical buckling load, as given by (2-75) for column buckling, generally characterizes the effects of geometry on buckling resistance. Repeating the equation,

$$P_{cr} = \frac{\pi^2 EI}{L^2} \tag{6-4}$$

where P_{cr} is the critical buckling load. To make a column more resistant to buckling, the magnitude of P_{cr} must be increased. Hence, for a given material, the geometry should be modified toward larger values of area moment of inertia, I, and/or smaller column length, L. To achieve larger area moments of inertia, from the basic equation for area moment of inertia about the x-axis,[3]

$$I_x = \int y^2 \, dA \tag{6-5}$$

It may be noted from (6-5) that I becomes larger when the cross-sectional shape is configured to concentrate most of the material at relatively large distances y from the neutral axis of bending. To follow the buckling avoidance guideline, either the column length should be shortened (often difficult because of functional requirements) or column cross-sectional shape should concentrate material away from the neutral axis. For example, Figure 6.4 shows two alternative cross-sectional shapes for a proposed column design application. Both proposed cross sections have the same net cross-sectional area (same amount of material) but Proposal 1 is a compact, solid, circular cross section of diameter d_1, while Proposal 2 is a hollow annular cross section of inner diameter $d_{i2} = d_1$ and outer diameter $d_{o2} = \sqrt{2}d_1$ to give equal cross-sectional areas for both shapes. For these dimensions, the area moment of inertia for the compact solid cross section of Proposal 1 is

$$I_1 = \frac{\pi d_1^4}{64} \tag{6-6}$$

and for the expanded annular cross section of Proposal 2

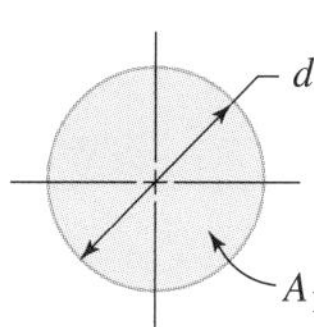

(*a*) Proposal 1: compact cross section of area A_1.

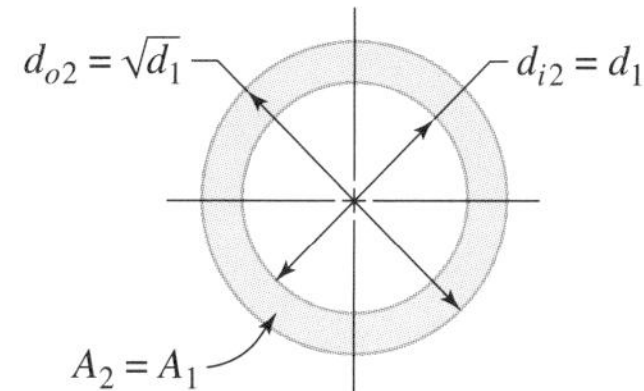

(*b*) Proposal 2: expanded cross section with area $A_2 = A_1$.

Figure 6.4
Two alternative cross-sectional shapes demonstrating the buckling avoidance guideline. Proposal 2 is preferred because it provides a larger area moment of inertia with the same amount of material.

[3]See, for example, ref. 2, Ch. 5.

$$I_2 = \frac{\pi[(\sqrt{2}d_1)^4 - d_1^4]}{64} = \frac{3\pi d_1^4}{64} \tag{6-7}$$

Comparing (6-7) with (6-6), the moment of inertia for the annular shape, I_2, is *three times as high* as I_1 for the compact solid shape. The critical buckling load for the annular column cross section is, therefore, three times as high as for the compact cross section using the same amount of material.

Hollow Cylinder and I-Beam Guideline

When torsional moments are applied to solid cylindrical members, the resulting torsional shearing stresses vary linearly from zero at the center to a maximum at the outer fibers, as illustrated in Figure 4.7. Likewise, when bending moments are applied to solid rectangular beams, the resulting bending stresses vary linearly from zero at the neutral axis to a maximum at the outer fibers, as illustrated in Figure 4.3. In both cases, to produce a more nearly uniform stress distribution, material near the central part of the cross section could be removed and concentrated near the outer fibers. For torsional moments applied to cylindrical members, this may be accomplished by converting from solid to hollow cylinders with the same net cross-sectional area, as shown in Figure 6.5(a). For bending moments applied to rectangular cross section beams, it may be accomplished by reconfiguring the rectangular shape into an I-section with the same net cross-sectional area as shown in Figure 6.5(b).[4] To follow the hollow cylinder guideline then, for members in torsion, a *hollow* cylinder (such as Case 2 of Figure 6.5) should be chosen over a solid cylinder (such as Case 1). To follow the I-beam guideline, an I-section (such as Case 4 of Figure 6.5), or a hollow box section, should be chosen over a solid rectangular section such as Case 3. For all cases, if the net cross-sectional area is maintained constant, moving material away from the central region results in larger moments of inertia and reduced stress levels.

Conforming Surface Guideline

When two *nonconforming* surfaces are pressed into contact, the *actual* areas of contact are typically very small; hence the local contact pressures are very high. The discussion of Hertz contact stresses in 4.10 illustrates this observation for contact between spheres and between cylinders. When two surfaces *exactly conform* to each other at the contact interface, the contact area is taken to be the "projected area," whether the contact geometry is curved or planar. Thus conforming surfaces give larger *actual* areas of contact, and consequently, lower contact pressures. To follow the conforming surface guideline, the con-

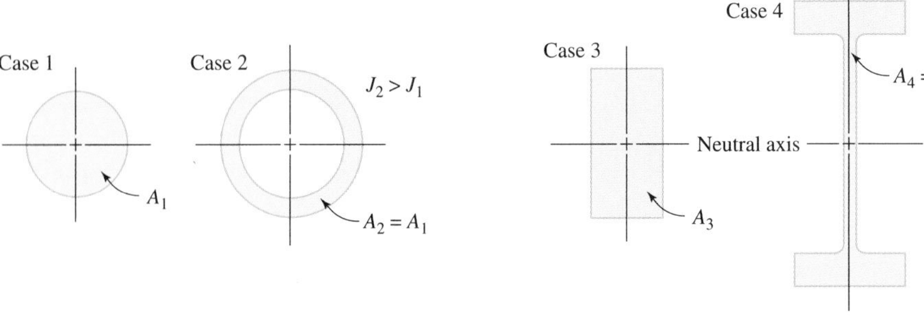

Figure 6.5
Alternative configurations for cross sections in torsion and bending. Case 2 is preferred to Case 1, and Case 4 is preferred to Case 3 because in both instances the preferred cases provide more uniformly stressed material.

[4]An equivalent hollow box section could also be used.

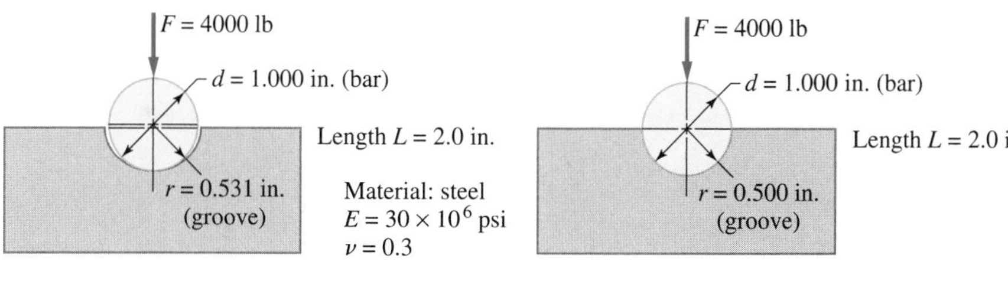

(*a*) Proposal 1: nonconforming contact.

(*b*) Proposal 2: conforming contact.

Figure 6.6
Two alternative configurations demonstrating the conforming surface guideline. Proposal 2 is preferred over Proposal 1 because the actual contact area is larger and the contact pressure is lower.

tact geometry at the interface between any two parts should be chosen to provide a large actual contact area by making contact surfaces conform to each other as closely as possible within the functional constraints on the device. For example, Figure 6.6 shows two alternative configurations for a proposed pivot joint made by placing a load-bearing solid cylindrical bar in a cylindrical groove. Proposal 1 is for a 2.0-inch-long steel cylinder of diameter 1.000 inch, placed in a 0.531-inch-radius semicircular groove in a 2.0-inch-wide block of the same material. Proposal 2 is a generally similar arrangement except the radius of the semicircular groove is 0.500 inch. Both proposals must support a total vertical downward force of 4000 lb. To follow the conforming surface guideline, Proposal 2 should be chosen since the dimensions of the circular profile of the cylindrical bar and the mating groove exactly conform. Contact pressures may be compared utilizing the Hertz contact pressure equation (4-124) for Proposal 1, and the "projected area" calculation for Proposal 2. Thus, for Proposal 1, (4-124) gives an interfacial contact width of

$$b = \sqrt{\frac{2(4000)\left[2\left(\frac{1-0.3^2}{30\times 10^6}\right)\right]}{\pi(2.0)\left[\frac{1}{1.000}+\frac{1}{-1.062}\right]}} = 0.013 \text{ inch} \tag{6-8}$$

and (4-125) then gives maximum contact pressure as

$$p_{max} = \frac{2(4000)}{\pi(0.013)(2.0)} = 95{,}426 \text{ psi} \tag{6-9}$$

For Proposal 2, the contact prssure is uniform and is calculated as force divided by projected area of contact, or

$$p_{max} = \frac{F}{dL} = \frac{4000}{(1.0)(2.0)} = 2000 \text{ psi} \tag{6-10}$$

and the choice of Proposal 2 is confirmed.

Lazy-Material Removal Guideline

When preliminary proposals for the geometry of a machine part are examined carefully, it may be found that certain regions of material are lightly stressed, or not stressed at all. To follow the lazy-material removal guideline, regions of lightly stressed (lazy) material should be removed in the interests of conserving weight, material, and dollars. For example, Figure 6.7 shows alternative proposals for a member of nonuniform cross section in tension, a beam in bending, and a bar in torsion. For each case, Proposal 1 is a simple solid or stepped section, while Proposals 2 and 3 represent modified geometries obtained by removing regions of material that support little or no stress when loaded. To follow the lazy-material removal guideline, Proposal 2 or 3 should be chosen for each case, since little

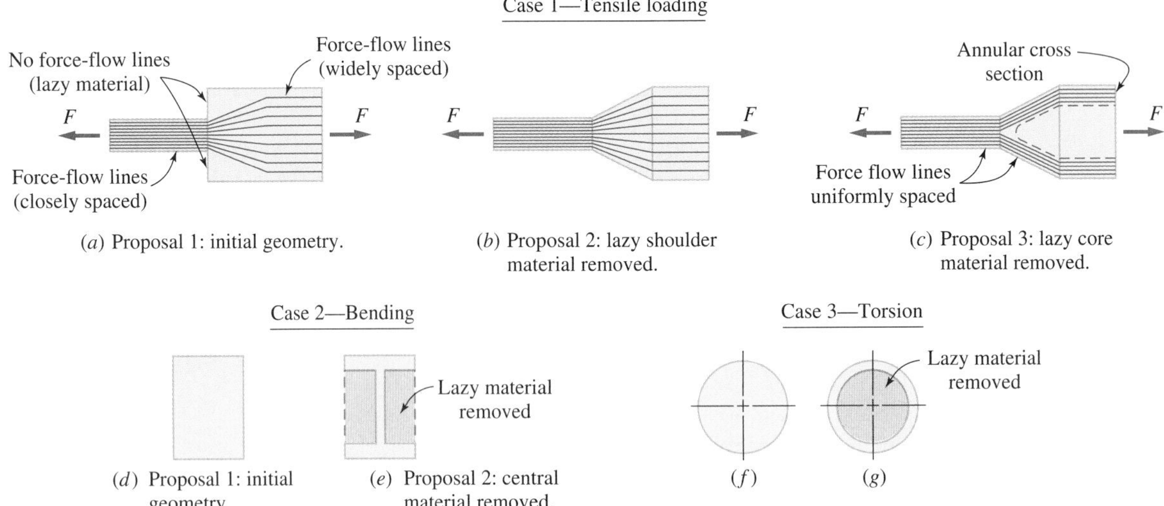

Figure 6.7
Various alternative configurations demonstrating the lazy-material removal guideline. Proposal 2 is preferred over Proposal 1 because "lazy" material has been removed; Proposal 3 is preferred over Proposal 2 because additional "lazy" material has been removed.

change in stress level is induced and significant material savings can be achieved. For the tension member of Case 1 in Figure 6.7(a), a stepped cylindrical bar is subjected to axial tension. The lines of force flow indicate that the shoulder region is virtually unstressed and could be removed with little consequence, as shown in Figure 6.7(b). It may be further recognized that the larger-diameter section of the bar is more lightly stressed than the smaller-diameter section, since tensile stress is proportional to area. Thus, if functional constraints permit, the third proposal of Figure 6.7(c) might be implemented by removing the central core of the larger-diameter section to increase the stress there to the same level as in the smaller-diameter section.

For Case 2 in Figure 6.7(d), a beam of solid rectangular cross section is subjected to a bending moment, giving a linear stress gradient from maximum tension at the top to maximum compression at the bottom [see Figure 4.3(b)]. Lazy material could be removed from the central portion of the beam (near the neutral axis) as shown in Figure 6.7(e), leaving most of the remaining material near the outer fibers where the stress is higher. Possible cross-sectional shapes that might result would include the I-shape as shown in 6.7(e), an equivalent hollow box section, or a less-desirable asymmetric section shape such as a channel, angle, or zee.

Case 3, shown in Figure 6.7(f), depicts a bar of solid circular cross section subjected to a torsional moment, giving a linear shear stress gradient from zero at the center to maximum at the outer fibers [see Figure 4.7(b)]. Lazy material could be removed from the central core of the cylinder as shown in Figure 6.7(g) to leave material only near the outer fibers, where the shearing stress is higher, resulting in a more efficient annular cross section.

Merging Shape Guideline

From the discussions of stress concentration in 4.8, it is evident that geometric discontinuities may lead to local stresses many times higher than those for a smooth, gradual transition from one section size or shape to another. For example, the stress concentration factor charts for a shouldered shaft in Figure 4.17, a grooved shaft in Figure 4.18, and a shouldered flat bar in Figure 4.20 all clearly show that larger fillet radii and smoother tran-

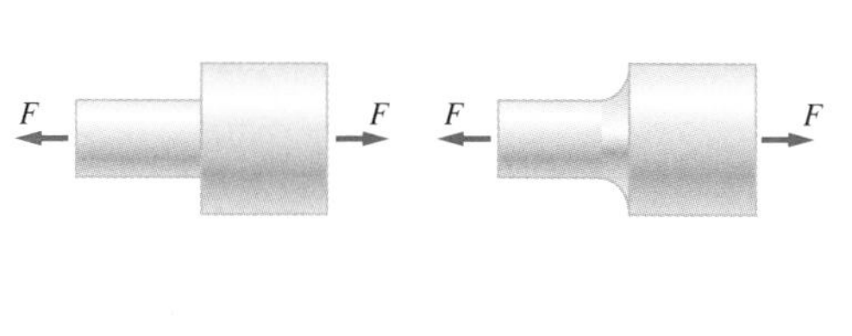

(a) Proposal 1: shaft with sudden step.

(b) Proposal 2: large fillet transition.

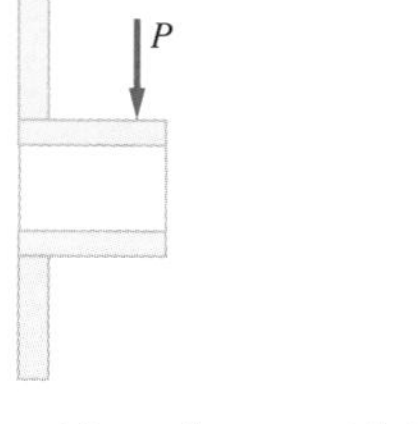

(c) Proposal 1: cantilever bearing support.

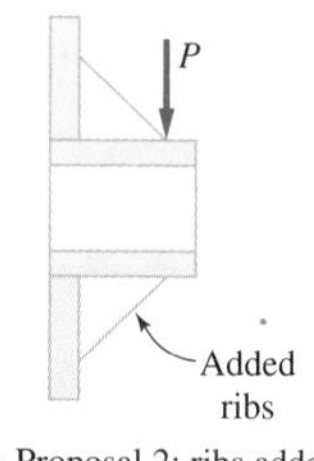

(d) Proposal 2: ribs added to merge geometry.

Figure 6.8
Alternative configurations demonstrating the merging shape guideline. Proposal 2 is preferred over Proposal 1 in both instances because gradual transitions in shape provide more uniform stresses and lower peak stresses.

sitions from one section to another result in lower peak stress values. To follow the merging shape guideline, sudden changes in size or shape should be avoided, and transitions in size or shape should be made by gradually merging from one to the other. For example, Figure 6.8 shows alternative proposals for a shouldered shaft in tension and for a cantilevered bearing support in bending. In each case Proposal 1 shows a sudden change in geometry from one region to another while Proposal 2 provides a more gradual transition. To follow the merging shape guideline, Proposal 2 should be chosen for each case. For the stepped shaft, a large, smooth fillet provides the merging transition, and for the bearing support, the addition of ribs to gradually distribute the load from the bearing to the supporting wall provides the desired merging effect.

Strain-Matching Guideline

When loads are transferred from one machine element to another, it is important that the lines of force flow be distributed as uniformly as possible across the entire contact interface between the elements. If local elastic strains induced in the contacting elements are not well matched at the interface, locally high stresses may result, or relative displacements (slip) along the contact interface may be produced. The consequence may be unacceptably high local stresses or, if cyclic loads are involved, the consequence may be slip-induced fretting damage. To follow the strain-matching guideline, elastic strains and relative displacements between mating element surfaces should be controlled by carefully specifying element shapes and dimensions that provide uniformly distributed load transfer across the interface. For example, Figure 6.9 shows two alternative proposals for the threaded interface between a tension support rod and its mating retention nut. In the conventional configuration of Proposal 1, shown in Figure 6.9(a), the lines of force flow tend to concentrate in threads 1, 2, and 3, with little of the load carried across threads 4, 5, and 6. This happens because threads 1, 2, and 3 represent the shorter, stiffer load paths; therefore, in accordance with the discussions of 4.11, most of the load is transferred across these stiffer thread paths. In Proposal 2, shown in Figure 6.9(b), the nut has been given a special tapered shape to make the nut axially more flexible in the region of threads 1, 2, and 3, shifting more of the total transferred load to threads 4, 5, and 6. To follow the strain matching guideline, Proposal 2 should be chosen.

Load-Spreading Guideline

At joints or contacting surfaces between two machine parts, the load should be transferred as gradually and as uniformly as possible from one part to the other. To follow the load-spreading guideline, the following techniques should be implemented when possible:

1. Use many small fasteners rather than a few large ones.
2. Use generous fillets or tapered sections rather than sharp reentrant corners or large square shoulders.

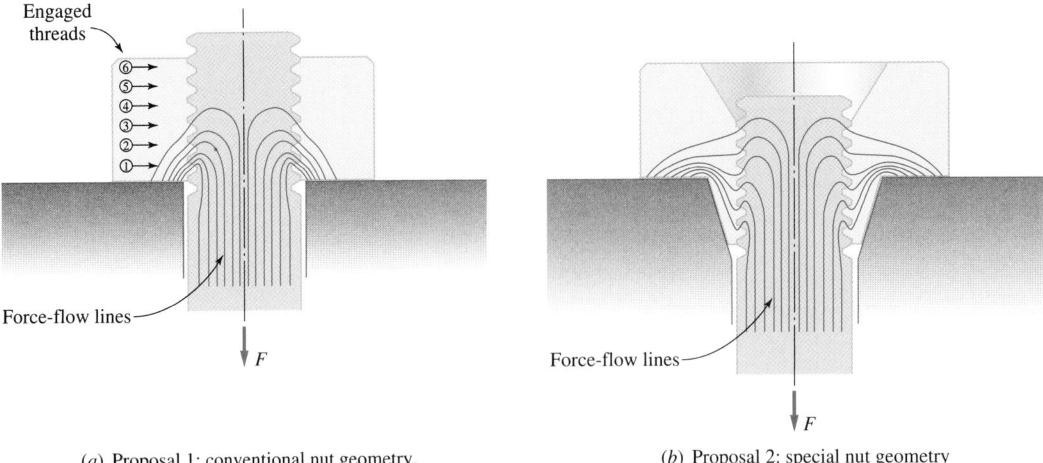

(*a*) Proposal 1: conventional nut geometry.

(*b*) Proposal 2: special nut geometry

Figure 6.9
Two alternative proposals for retention nut geometry demonstrating strain-matching guideline. Proposal 2 is preferred over Proposal 1 because strain matching lowers peak stresses by providing more uniform load transfer across the threaded interface. However, cost advantage may be an important argument for selecting Proposal 1.

3. Use washers under bolt heads and nuts.
4. Use ribs to distribute loads from lugs or feet to supporting walls.
5. Use raised pads on shafting to mount hubs, gears, or bearings.
6. Use innovative transition geometry to spread load or reduce stress concentration, especially in critical design situations.

Example 6.1 Creating Shapes by Implementing Configurational Guidelines

An L-shaped support bracket made of two members, one generally cylindrical and one generally rectangular, is being proposed to support the load as shown schematically in Figure E6.1A. The cylindrical member A is attached to the support wall at one end and press-fit into the rectangular member B at the other end. Without making any calculations, identify which of the configurational guidelines of 6.2 might be applicable in determining an appropriate shape for each member of this bracket, and, based on these guidelines, sketch an initial proposal for the shape of each member.

Figure E6.1A
Basic geometry of proposed two-member support bracket.

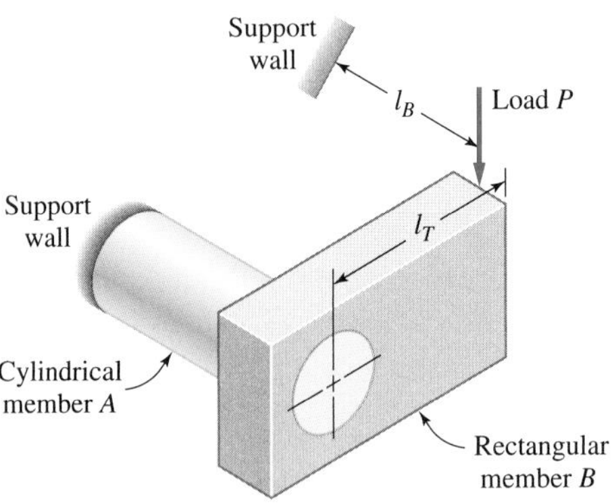

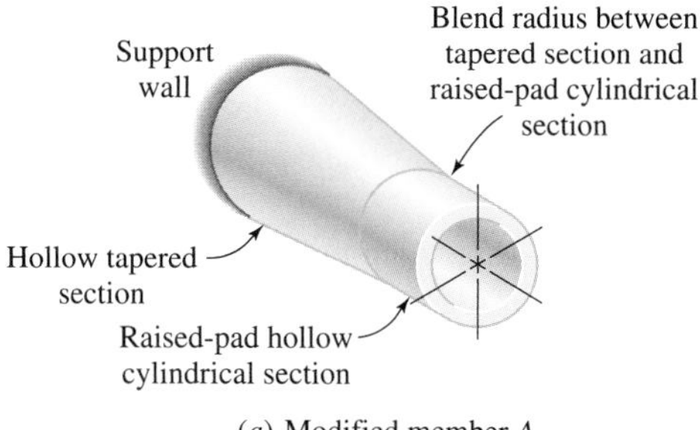

(*a*) Modified member *A*.

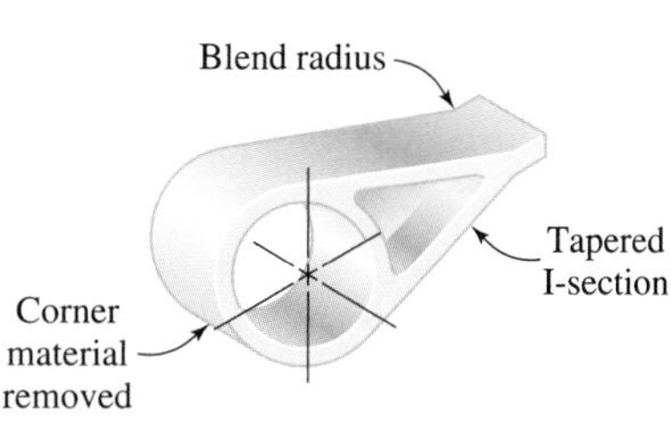

(*b*) Modified member *B*.

Figure E6.1B
Shape modifications based on configurational guidelines of 6.2 (as compared to Figure E6.1A).

Solution

The load P at the end of member 3 in Figure E6.1A will produce, primarily, *cantilever bending* in member B, and *cantilever bending plus torsion* in member A. *Transverse shear* is also present in both members. Reviewing the list of configurational guidelines in 6.2, the potentially applicable guidelines for member A would include:

2. Tailoring shape to loading gradient (bending)
5. Utilizing hollow cylinder shapes to achieve near-uniform stress (torsion)
6. Providing conforming surfaces at mating interfaces
7. Removing lazy material
8. Merging shapes gradually from one shape to another

For member B the applicable guidelines would include:

2. Tailoring shape to loading gradient (bending)
5. Utilizing I-beam shapes to achieve near-uniform stress (bending)
6. Providing conforming surfaces at mating interfaces
7. Removing lazy material
8. Merging shapes gradually from one shape to another

Incorporating these guidelines to refine the shapes of members A and B sketched in Figure E6.1A, the initial proposal for refined shapes might take the form sketched in Figure E6.1B.

6.3 Critical Sections and Critical Points

In the list of fundamental steps in the design of a machine, as outlined in Table 1.1 of Chapter 1, step VII outlines the important aspects of designing each part. This includes initial geometry determination, force analysis, identification of potential failure modes, material selection, selection of potential critical sections and critical points, safety factor selection, and use of appropriate stress and deflection analyses. Based on these considerations, proper dimensions of critical sections may be determined, usually through a process of iteration. With the exception of selecting critical sections and critical points, all of the important aspects of design just reiterated have been discussed in detail in Chapter 2 through this point in Chapter 6. Techniques for selecting critical sections and critical points will be addressed next.

Critical sections in a machine part are those cross sections that, because of the geometry of the part and the magnitude and orientation of forces and moments on it, may contain critical points. A *critical point* is a point within the part that has a high potential for failure because of high stresses or strains, low strength, or a critical combination of these. Typically, a designer first identifies the potential critical sections, then identifies the possible critical points within each critical section. Finally, appropriate calculations are made to determine the *governing* critical points so that the calculated dimensions will assure safe operation of the part over its prescribed design lifetime.

The number of potential critical points requiring investigation in any given machine part is directly dependent upon the *experience and insight* of the designer. A very *inexperienced* designer may have to analyze *many, many* potential critical points. A very *experienced* and insightful designer, analyzing the same part, may only need to investigate *one or a few* critical points because of ingrained knowledge about failure modes, how forces and moments reflect upon the part, and how stresses and strains are distributed across the part. In the end, the careful inexperienced designer and the experienced expert should both reach the same conclusions about where the governing critical points are located, but the expert designer typically does so with a smaller investment of time and effort.

Example 6.2 Critical Point Selection

It is desired to examine member *A* shown in Figure E6.1B of Example 6.1, with the objective of establishing critical sections and critical points in preparation for calculating dimensions and finalizing the shape of the part.

With this objective, select appropriate critical sections and critical points, give the rationale for the points you pick, and make sketches showing the locations of the selected critical points.

Solution

In the solution of Example 6.1, it was established that member *A* is subjected to cantilever bending, torsion, and transverse shear, giving rise to the proposed geometry shown in Figure E6.1B. Member *A* is again sketched in Figure E6.2 to show the *cross-sectional* geometry more clearly. Since cantilever bending produced by an end-load results in a maximum bending moment at the fixed end, as well as uniform transverse shear along the whole beam length, and since there is a constant torsional moment all along the beam length, critical section 1 *at the fixed end* is clearly a well-justified selection. Also, it may be noted that the annular wall is thinnest where the tapered section blends into the raised cylindrical mounting pad near the free end. At this location the bending moment is less than at the fixed end, transverse shear is the same, and torsional moment is the same as at critical section 1, but the wall is thinner and stress concentration must be accounted for; hence critical section 2 should also be investigated.

At critical section 1, four critical points may initially be chosen, as shown in Figure E6.2(b). At critical points *A* and *B*, bending and torsion combine, and at critical points *C* and *D*, torsion and transverse shear combine. In both cases potentially critical multiaxial stress states are produced. Since the state of stress at *A* is the same as at *B* (except that

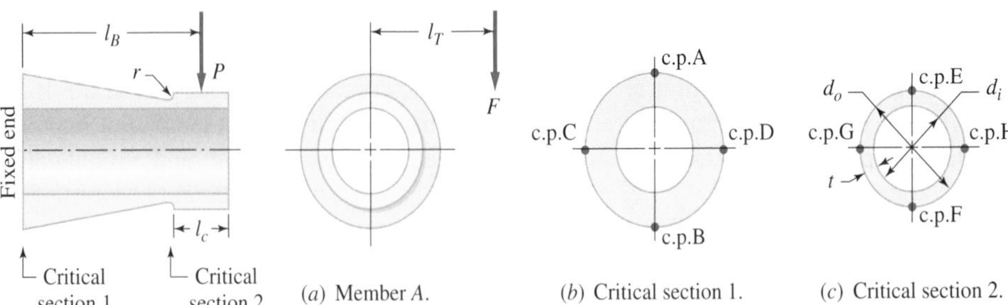

(*a*) Member *A*. (*b*) Critical section 1. (*c*) Critical section 2.

Figure E6.2 Potential critical sections and critical points in member *A* of the Example 6.1 bracket (c.p. = critical point).

bending produces tension at A and compression at B), investigation of critical point A is alone sufficient. Also, since torsional shear stress adds to transverse shear stress at D and subtracts at C, investigation of critical point D is alone sufficient. Therefore, it is concluded that critical points A and D should be investigated, with the knowledge that B and C are less serious.

A similar consideration of critical section 2 leads to the similar conclusion that critical points E and H should be investigated, knowing that A and C are less serious.

Summarizing, critical points A, D, E, and H should be investigated. If a designer has doubts about any other potential critical point within the member, it too should be investigated.

6.4 Transforming Combined Stress Failure Theories into Combined Stress Design Equations

The state of stress at a critical point is typically multiaxial; therefore, as discussed in 4.6, the use of a *combined stress theory of failure* is usually necessary in critical point analysis. Further, as discussed in 5.2, dimensions are usually determined by making sure that the maximum operating stress levels do not exceed the design-allowable stress at any critical point. A useful formulation may be obtained by transforming the combined stress failure theories given in (4-83), (4-84), (4-87), and (4-95) into *combined stress design equations*, from which required dimensions may be calculated at any critical point. Such transformations may be accomplished by using *only* the equal signs of the failure theory equations, and inserting the design-allowable stress in place of the failure strength in each equation. The resulting combined stress design equations then contain known loads, known material strength properties corresponding to the governing failure mode, known safety factor, and unknown dimensions. The unknown dimensions may be calculated by inverting the applicable combined stress design equation. Details of a solution may be complicated, often requiring iterative techniques. The rules for selecting the *applicable* design equation, based on material ductility, are the same as the rules for failure theory selection given in 4.6.

In more detail, if a material exhibits *brittle* behavior (elongation less than 5 percent in 2 inches), the maximum normal stress failure theory, given by (4-83) and (4-84), would be transformed into the *maximum normal stress design equations*

$$\sigma_{max} = \sigma_{d-t} \tag{6-11}$$

if σ_{max} is tensile, or

$$\sigma_{min} = -\sigma_{d-c} \tag{6-12}$$

if σ_{min} is compressive, where σ_{d-t} and σ_{d-c} are, respectively, tensile and compressive design stresses, σ_{max} is the largest of the principal stresses σ_1, σ_2, and σ_3 at the critical point, and σ_{min} is, algebraically, the smallest principal stress. The governing equation is the one that gives the largest required dimensions.

If a material exhibits *ductile* behavior (elongation of 5 percent or more in 2 inches), the maximum shearing stress failure theory given by (4-87) would be transformed into the *maximum shearing stress design equation*

$$|\tau_{max}| = \tau_d = \frac{\sigma_d}{2} \tag{6-13}$$

where $|\tau_{max}|$ is the largest magnitude of principal shearing stress at the critical point (see (4-60), (4-61) and (4-62)).

Also for *ductile* behavior, the distortion energy failure theory given by (4-95), and supplemented by (4-96), would be transformed into the *distortion energy design equation*

$$\sigma_{eq}^2 = \frac{1}{2}[(\sigma_1 - \sigma_2)^2 + (\sigma_2 - \sigma_3)^2 + (\sigma_3 - \sigma_1)^2] = \sigma_d^2 \tag{6-14}$$

where σ_1, σ_2, and σ_3 are the three principal normal stresses produced by the loading at the critical point. In all of the above equations known loads, known material strength properties, and known safety factor would be inserted and design dimensions (the only unknowns) would be found by solving the equation.

6.5 Simplifying Assumptions: The Need and the Risk

After the initial proposal for the shape of each of the parts and their arrangement in the assembled machine, and after all critical points have been identified, the critical dimensions may be calculated for each part. In principle, this task merely involves utilizing the stress and deflection analysis equations of Chapter 4 and the multiaxial stress design equations of 6.4. In practice, the complexities of complicated geometry, redundant structure, and implicit or higher-order mathematical models often require one or more *simplifying assumptions* in order to obtain a manageable solution to the problem of determining the dimensions.

Simplifying assumptions may be made with respect to loading, load distribution, support configuration, geometric shape, force flow, predominating stresses, stress distribution, applicable mathematical models, or any other aspect of the design task, to make possible a solution. The purpose of making simplifying assumptions is to reduce a complicated real problem to a pertinent but solvable mathematical model. The coarsest simplifying assumption would be to assume the "answer" with no analysis. For an experienced designer, a routine application, light loads, and minimal failure consequences, directly assuming the dimensions might be acceptable. The most refined analysis might involve very few simplifying assumptions, modeling the loading and geometry in great detail, possibly creating in the process very detailed and complicated mathematical models that require massive computational codes and large investments of time and effort to find the dimensions. For very critical applications, where loading is complicated, failure consequences are potentially catastrophic, and the nature of the application will reasonably support large investments, such detailed modeling might be acceptable (but usually would be used only after initially exercising simpler models).

Typically, a few well-chosen simplifying assumptions are needed to reduce the real design problem to one that can be tentatively solved with a reasonable effort. More accurate analyses may be made in subsequent iterations, if necessary. The *risk* of making simplifying assumptions must always be considered; if the assumptions are not true, the resulting model will not reflect the performance of the real machine. The resulting poor predictions might be responsible for premature failure or unsafe operation unless the analysis is further refined.

6.6 Iteration Revisited

Many details of the mechanical design process have been examined since design was first characterized in 1.4 as an *iterative* decision-making process. Now that the basic principles and guidelines for determining shape and size have been presented, and details of material selection, failure mode assessment, stress and deflection analysis, and safety factor determination have been discussed, it seems appropriate to briefly revisit the important role of iteration in design.

During the first iteration a designer typically concentrates on meeting functional performance specifications by selecting candidate materials and potential geometric arrangements that will provide strength and life adequate for the loads, environment, and potential failure modes governing the application. An appropriate safety factor is chosen to account for uncertainties, and carefully chosen simplifying assumptions are made to implement a manageable solution to the task of determining critical dimensions. A consideration of manufacturing processes is also appropriate in the first iteration. Integrating the selection of the manufacturing process with the design of the product is necessary if the advantages and economies of modern manufacturing methods are to be realized.

A second iteration usually establishes nominal dimensions and detailed material specifications that will safely satisfy performance, strength, and life requirements. Many loops may be embedded in this iteration.

Typically, a third iteration carefully audits the second iteration design from the perspectives of fabrication, assembly, inspection, maintenance, and cost. This is often accomplished by utilizing modern methods for global optimization of the manufacturing system, a process usually called *design for manufacture* (*DFM*).[5]

A final iteration, undertaken before the design is released, typically includes the establishment of fits and tolerances for each part, and final modifications based on the third-iteration audit. A final safety factor check is then usually made to assure that strength and life of the proposed design meet specifications without wasting materials or resources.

As important as understanding the *iterative* nature of the design process is understanding the *serial* nature of the iteration process. Inefficiencies generated by deeply embedded *early design decisions* may make cost reduction or improved manufacturability difficult and expensive at later stages. Such inefficiencies are being addressed in many modern facilities by implementing the *simultaneous engineering* approach. Simultaneous engineering involves on-line computer linkages among all activities, including design, manufacturing, testing, production, marketing, sales, and distribution, with early and continuous input and auditing throughout the design, development, and field service phases of the product. Using this approach, the various iterations and modifications are incorporated so rapidly, and communicated so widely, that inputs and changes from all departments are virtually simultaneous.

Example 6.3 Determining Dimensions at a Selected Critical Point

Continuing the examination of member A, already described in Examples 6.1 and 6.2, it is desired to find dimensions of the annular cross section shown in Figure E6.2 of Example 6.2, at critical section 2. The load P to be supported is 10,000 lb. The distance from the fixed wall to load P is $l_B = 10$ inches, and the distance from the centerline of member A to load P is $l_T = 8$ inches (see Figure E6.1A of Example 6.1). The tentative material selection for this first cut analysis is 1020 cold-drawn steel, and it has been determined that yielding is the probable failure mode. A preliminary analysis has indicated that a design safety factor of $n_d = 2$ is appropriate.

Determine the dimensions of member A at critical section 2.

Solution

From Figure E6.2, the dimensions to be determined at critical section 2 include outside diameter d_o, inside diameter d_i, wall thickness t, and fillet radius r, all unknown. The length of the raised pad, l_c, is also unknown, but is required for calculation of bending moment M_2 at critical section 2. The material properties of interest for 1020 CD steel are

[5] See 7.4.

Example 6.3 Continues

$$S_{yp} = 51{,}000 \text{ psi} \quad (\text{Table 3.3})$$
$$e(2 \text{ inches}) = 15\% \quad (\text{Table 3.10}) \tag{1}$$

From the solution to Example 6.2, the critical points to be analyzed for section 2 are c.p. E (bending and torsion) and c.p. H (torsion and transverse shear), as shown in Figure E6.2.

To start the solution, the following simplifying assumptions may be made:

1. The annular wall is thin, so assume

$$d_o = d_i = d \tag{2}$$

and

$$t = 0.1d \tag{3}$$

2. A common proportion for bearing surfaces is to make diameter equal to length, so assume

$$l_c = d \tag{4}$$

At critical section 2, the bending moment M_2 may be written as

$$M_2 = \frac{Pl_c}{2} = \frac{Pd}{2} = \frac{10{,}000d}{2} = 5000d \text{ in-lb} \tag{5}$$

and torsional moment T_2 may be written as

$$T_2 = Pl_T = 10{,}000(8) = 80{,}000 \text{ in-lb} \tag{6}$$

Examining c.p. E first, the elemental volume depicting the state of stress may be constructed as shown in Figure E6.3A.

The nominal axial stress $\sigma_{x\text{-}nom}$, caused by bending moment M_2, may be written as

$$\sigma_{x\text{-}nom} = \frac{M_2 c}{I} \tag{7}$$

and the nominal shearing stress $\tau_{xy\text{-}nom}$, caused by torsional moment T_2, may be written as

$$\tau_{xy\text{-}nom} = \frac{T_2 a}{J} \tag{8}$$

For thin annular sections,[6] the area moment of inertia, I, about the neutral axis of bending, and the polar moment of inertia, J, may be approximated as

$$I = \frac{\pi d^3 t}{8} \tag{9}$$

and

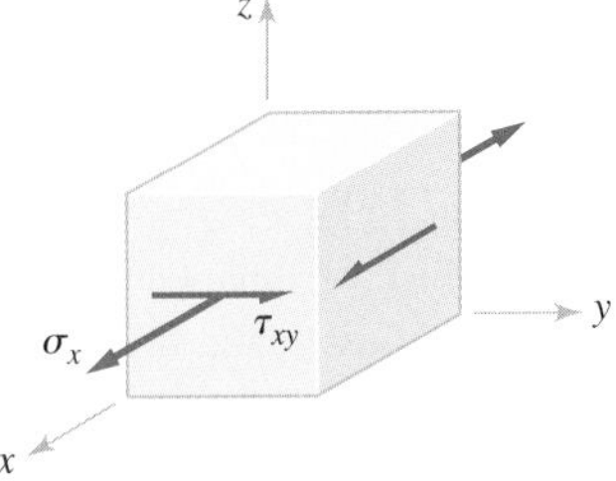

Figure E6.3A
State of stress at c.p. E.

[6]See ref. 2, Table 1, case 16, or use Table 4.2, case 5, to approximate I for a thin-walled tube by incorporating (2) and (3).

$$J = \frac{\pi d^3 t}{4} \tag{10}$$

Using (3), these become

$$I = \frac{\pi(0.1)d^4}{8} = 0.039d^4 \tag{11}$$

and

$$J = \frac{\pi(0.1)d^4}{4} = 0.079d^4 \tag{12}$$

Stress concentration factors due to the fillet may be approximated from Figure 4.17 by initially assuming the ratios $r/d = 0.05$ and $D/d = 1.1$, to give the approximate values

$$K_{tb} = 1.9 \qquad \text{(for bending)} \tag{13}$$

and

$$K_{t\tau} = 1.3 \qquad \text{(for torsion)} \tag{14}$$

From these values, the actual bending stress σ_x at c.p. E may be written as

$$\sigma_x = (\sigma_{x\text{-}nom})K_{tb} = \left(\frac{M_2 c}{I}\right)K_{tb} = \left[\frac{5000d(d/2)}{0.039d^4}\right]1.9 = \frac{1.22 \times 10^5}{d^2} \tag{15}$$

and the actual torsional shearing stress τ_{xy} at c.p. E may be written as

$$\tau_{xy} = (\tau_{xy\text{-}nom})K_{t\tau} = \left(\frac{T_2 a}{J}\right)K_{t\tau} = \left[\frac{80{,}000(d/2)}{0.079d^4}\right]1.3 = \frac{6.58 \times 10^5}{d^3} \tag{16}$$

Using the stress cubic equation, (4-59), the principal stresses at c.p. E. may then be found. The stress cubic equation becomes

$$\sigma^3 - \sigma^2\sigma_x + \sigma(-\tau_{xy}^2) = 0 \tag{17}$$

giving the principal stress solutions

$$\sigma_1 = \frac{\sigma_x}{2} + \sqrt{\left(\frac{\sigma_x}{2}\right)^2 + \tau_{xy}^2} \tag{18}$$

$$\sigma_2 = \frac{\sigma_x}{2} - \sqrt{\left(\frac{\sigma_x}{2}\right)^2 + \tau_{xy}^2} \tag{19}$$

$$\sigma_3 = 0 \tag{20}$$

Noting from (1) that the material is ductile, the distortion energy design equation (6-14) is chosen, giving

$$\sigma_{eq}^2 = \frac{1}{2}[(\sigma_1 - \sigma_2)^2 + (\sigma_2 - \sigma_3)^2 + (\sigma_3 - \sigma_1)^2] = \sigma_d^2 \tag{21}$$

Substitution of principal stress expressions (18), (19), and (20) into (21) gives

$$\sigma_x^2 + 3\tau_{xy}^2 = \sigma_d^2 \tag{22}$$

Also from (1), noting from the specifications that yielding is the probable failure mode and $n_d = 2$,

$$\sigma_d = \frac{S_{yp}}{n_d} = \frac{51{,}000}{2} = 25{,}500 \text{ psi} \tag{23}$$

Then from (15), (16), and (23), (22) becomes

$$\left(\frac{1.22 \times 10^5}{d^4}\right)^2 + 3\left(\frac{6.58 \times 10^5}{d^3}\right)^2 = (25{,}500)^2 \tag{24}$$

Example 6.3 Continues

or

$$d^6 - 22.92d^2 = 2000 \tag{25}$$

This may be solved for d using equation-solver software or may be iterated manually to find a solution. A starting point for a manual iteration might be found by calculating a diameter d_b from (15), based on bending only, and a diameter d_τ from (20), based on torsion only. Using the larger of these two diameters as a starting point, the combined stress value of d may be calculated from (25).

Setting $\sigma_x = \sigma_d$ in (15),

$$\frac{1.22 \times 10^5}{d_b^2} = 25{,}500 \tag{26}$$

or, for bending stress only,

$$d_b = 2.19 \text{ inches} \tag{27}$$

Setting $\tau_{xy} = \tau_d$ in (16), where $\tau_d = \sigma_d/\sqrt{3}$, based on the distortion energy theory,

$$\frac{6.58 \times 10^5}{d_\tau^3} = \frac{25{,}500}{\sqrt{3}} \tag{28}$$

or, for shearing stress only,

$$d_\tau = 3.55 \text{ inches} \tag{29}$$

Now, (25) may be iterated to a solution for d by starting with the initial value of $d =$ 3.55 inches. The iteration sequence is shown in Table E6.3A. The next step is to reanalyze c.p. E using

$$d_o = 3.63 \text{ inches} \tag{30}$$

with related values of

$$t = 0.36 \text{ inch} \tag{31}$$

and

$$d_i = d_o - 2t = 2.91 \text{ inches} \tag{32}$$

Also try

$$r = 0.18 \text{ inch} \tag{33}$$

With these dimensions,

$$\frac{r}{d_o} = \frac{0.18}{3.63} = 0.05 \tag{34}$$

and

$$\frac{D}{d} = \frac{d_o + 2r}{d_o} = \frac{3.63 + 2(0.18)}{3.63} = 1.1 \tag{35}$$

hence, from Figure 4.17,

$$K_{tb} = 1.87 \tag{36}$$

TABLE E6.3A Iteration Sequence to Find Diameter *d*

d, in	d^6	$22.92d^2$	Result
3.55	2001.57	288.85	1712.72 $<$ 2000
3.60	2176.78	297.43	1879.35 $<$ 2000
3.65	2364.60	305.35	2059.25 $>$ 2000
3.63	2287.91	302.01	1985.90 (close)

and

$$K_{t\tau} = 1.27 \tag{37}$$

Recalculating moments of inertia I and J more accurately,

$$I = \frac{\pi}{64}(d_o^4 - d_i^4) = \frac{\pi}{64}(3.63^4 - 2.91^4) = 5.00 \text{ in}^4 \tag{38}$$

and

$$J = 2I = 2(5.00) = 10.00 \text{ in}^4 \tag{39}$$

Also, using (4),

$$l_c = D = d_o + 2r = 3.63 + 2(0.18) = 4.0 \text{ inches} \tag{40}$$

Recalculating σ_x using (15),

$$\sigma_x = \left[\frac{(5000)4.0(3.63/2)}{5.00}\right]1.87 = 13{,}576 \text{ psi} \tag{41}$$

and recalculating τ_{xy} using (16),

$$\tau_{xy} = \left[\frac{80{,}000(3.63/2)}{10.00}\right]1.27 = 18{,}440 \text{ psi} \tag{42}$$

hence from (18), (19), and (20),

$$\sigma_1 = \frac{13{,}576}{2} + \sqrt{\left(\frac{13{,}576}{2}\right)^2 + (18{,}440)^2} \tag{43}$$

giving

$$\sigma_1 = 6788 + 19{,}650 = 26{,}438 \text{ psi} \tag{44}$$

$$\sigma_2 = 6788 - 19{,}650 = -12{,}862 \text{ psi} \tag{45}$$

$$\sigma_3 = 0 \tag{46}$$

Equation (21) then gives a basis for the question

$$\sigma_{eq}^2 = \frac{1}{2}[(26{,}438 + 12{,}862)^2 + (-12{,}862)^2 + (12{,}862)^2] \stackrel{?}{=} (25{,}500)^2 \tag{47}$$

or

$$\sigma_{eq} = \sqrt{\frac{1}{2}(1.54 \times 10^9 + 1.65 \times 10^8 + 1.65 \times 10^8)} \stackrel{?}{=} 25{,}500 \tag{48}$$

or

$$\sigma_{eq} = 30{,}622 \stackrel{?}{=} 25{,}500 = \sigma_d \tag{49}$$

Since the equality is not satisfied, and the left side *exceeds* the design stress on the right side of (49), an increase in moments of inertia will be required to reduce the operating stress. This might be accomplished by decreasing d_i, for example. Before pursuing this issue, however, c.p. H should be checked.

The elemental volume depicting the state of stress at c.p. H may be constructed as shown in Figure E6.3B.

The shearing stress due to torsion at c.p. H will be the same as at c.p. E; hence, from (42),

$$\tau_{xy\text{-}tor} = 18{,}440 \text{ psi} \tag{50}$$

and the shearing stress due to transverse shear at c.p. H, using Table 4.3, case 3, and (37), is

$$\tau_{xz\text{-}ts} = \left(2\frac{P}{A}\right)K_{t\tau} = \left[\frac{2(10{,}000)}{\frac{\pi}{4}(3.63^2 - 2.91^2)}\right]1.27 \tag{51}$$

Example 6.3 Continues

Figure E6.3B
State of stress at c.p. H.

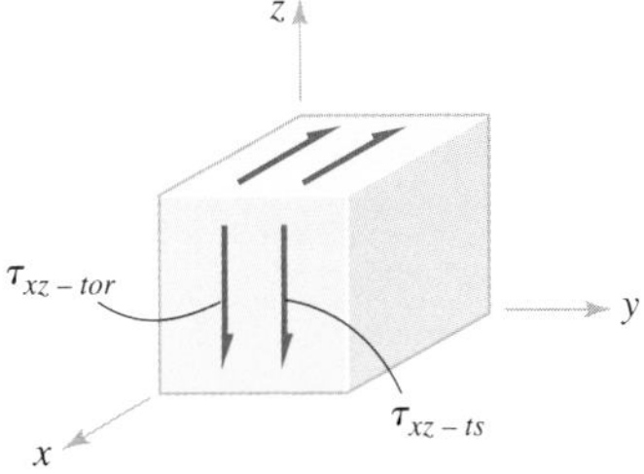

or

$$\tau_{xz\text{-}ts} = 6868 \text{ psi} \tag{52}$$

Since the shearing stresses add directly at c.p. H,

$$\tau_{xz} = 18{,}440 + 6868 = 25{,}308 \text{ psi} \tag{53}$$

At c.p. H the stress cubic, (4-59), reduces to

$$\sigma^3 - \sigma(-\tau_{xy}^2) = 0 \tag{54}$$

with principal stress solutions

$$\sigma_1 = \tau_{zx} = 25{,}308 \text{ psi} \tag{55}$$

$$\sigma_2 = -\tau_{zx} = -25{,}308 \text{ psi} \tag{56}$$

$$\sigma_3 = 0 \tag{57}$$

Using these values in the distortion energy design equation, (6-14), the question becomes

$$\sigma_{eq}^2 = \frac{1}{2}[(25{,}308 + 25{,}308)^2 + (-25{,}308)^2 + (-25{,}308)^2] \stackrel{?}{=} (25{,}500)^2 \tag{58}$$

or

$$\sigma_{eq} = \sqrt{\tfrac{1}{2}(2.56 \times 10^9 + 6.40 \times 10^8 + 6.40 \times 10^8)} \stackrel{?}{=} 25{,}500 \tag{59}$$

or

$$\sigma_{eq} = 43{,}800 \stackrel{?}{=} 25{,}500 = \sigma_d \tag{60}$$

Again, the equality is not satisfied, and this result is unacceptable because the left side is too large. Further, comparing (60) with (49), c.p. H is more critical than c.p. E and should be used for the next iteration.

To proceed, try an initial value of $d_i = 2.5$, and iterate until (60) is satisfied. Table E6.3B details the iteration sequence, keeping $d_0 = 3.63$, as before.

Based on these results, the dimensions recommended for critical section 2 of member *A* are as shown in Figure E6.3C.

Of course, many other dimensional combinations could be found for critical section 2 that would also be safe and acceptable. Further, it remains to find acceptable dimensions

TABLE E6.3B Iteration Sequence to Find Inside Diameter d_i

d_i	J, in^4	$\tau_{xz\text{-}tor}$	A, in^2	$\tau_{xz\text{-}ts}$	τ_{xz}	σ_{eq}
2.50	13.21	13,959	5.44	4671	18,630	32,268 > 25,500
2.25	14.53	12,691	6.37	3989	16,680	28,891 > 25,500
2.00	15.48	11,912	7.21	3524	15,436	26,736 > 25,500
1.75	16.13	11,432	7.94	3200	14,632	25,343 (close enough)

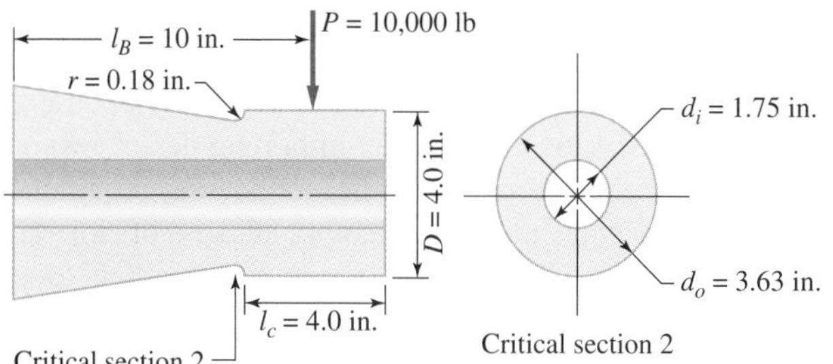

Figure E6.3C
Sketch showing recommended initial dimensions at critical section 2 of member *A* of the bracket shown in Example 6.1.

at critical section 1 before the initial design proposal for member A is completed. It should also be clear by now that writing or utilizing appropriate software to expedite the solution to an iterative design problem, such as the one just completed, is often justified.

6.7 Fits, Tolerances, and Finishes

All of the discussions so far in this chapter have dealt with determination of the "macrogeometry" of machine parts. In many cases the "microgeometry" of a machine part, or an assembly of parts, also has great importance in terms of proper function, prevention of premature failure, ease of manufacture and assembly, and cost. The important microgeometric design issues include: (1) the specification of the *fits* between mating parts to assure proper function, (2) the specification of *allowable variation* in manufactured part dimensions (*tolerances*) that will simultaneously guarantee the specified fit, expedite assembly, and optimize overall cost, and (3) the specification of *surface texture and condition* that will ensure proper function, minimize failure potential, and optimize overall cost. Some examples of machine parts and assemblies in which one or more of the microgeometric design issues may be important are:

1. The press fit connection between a flywheel hub and the shaft upon which it is mounted (see Chapters 9 and 18). The fit must be tight enough to assure proper retention, yet the stresses generated must be within the design-allowable range, and assembly of the flywheel to the shaft must be feasible. Both fits and tolerances are at issue.
2. The light interference fit between the inner race of a ball bearing and the shaft mounting pad upon which it is installed (see Chapter 11). The fit must be tight enough to prevent relative motion during operation, yet not so tight that internal interference between the balls and their races, generated by elastic expansion of the inner race when pressed on the shaft, shortens the bearing life. Premature failure due to fretting fatigue, initiated between the inside of the inner race and the shaft, might also be a consideration, as might be operational constraints on radial stiffness or the need to accommodate thermal expansion. Fits, tolerances, and surface textures are all important issues.
3. The radial clearance between a hydrodynamically lubricated plain bearing sleeve and the mating journal of a rotating shaft, as well as the surface roughnesses of the mating bearing surfaces (see Chapter 10). The clearance must be large enough to allow development of a "thick" film of lubricant between the bearing sleeve and the shaft journal, yet small enough to limit the rate of oil flow through the bearing clearance space so that hydrodynamic pressure can develop to support the load. The surface roughness of each member must be small enough so that roughness protuberances do not penetrate the lubricant film to cause "metal-to-metal" contact, yet large enough to allow ease of manufacture and a reasonable cost. Tolerances and surface texture are issues of importance.

Important design consequences hinge upon the decisions made about fits, tolerances, and surface textures, as illustrated by the three examples just cited. Specification of appropriate fits, tolerances, and surface textures is usually based upon experience with the specific application of interest. However, it is an important design responsibility to assure that "experience-based guidelines" meet specific application requirements such as preventing the loss of interference in a press fit assembly because of "tolerance stackup," preventing metal-to-metal contact in a hydrodynamic bearing due to excessive surface roughness, assuring that mating parts can be assembled and disassembled with relative ease, assuring that interference fits can sustain operating loads without separation or slip, assuring that differential thermal expansion does not excessively alter the fit, and ensuring that specified tolerances are neither so large that interchangeability is compromised nor so small that manufacturing cost is excessive. It is well established that increasing the *number* and *tightness* of specified tolerances causes a corresponding increase in cost and difficulty of manufacturing, as illustrated, for example, in Figure 6.10.

The design decisions on fits, tolerances, and surface texture must be accurately and unambiguously incorporated into detail and assembly drawings. In some cases, for example cylindrical fits between shafts and holes, extensive standards have been developed to aid in specification of proper fits and tolerances for a given application.[7] For reasons of cost effectiveness, primarily in manufacturing, the standards suggest lists of preferred *basic sizes* that should be chosen unless special conditions exist that prevent such a choice. Therefore, when nominal dimensions are calculated based on strength, deflection, or other

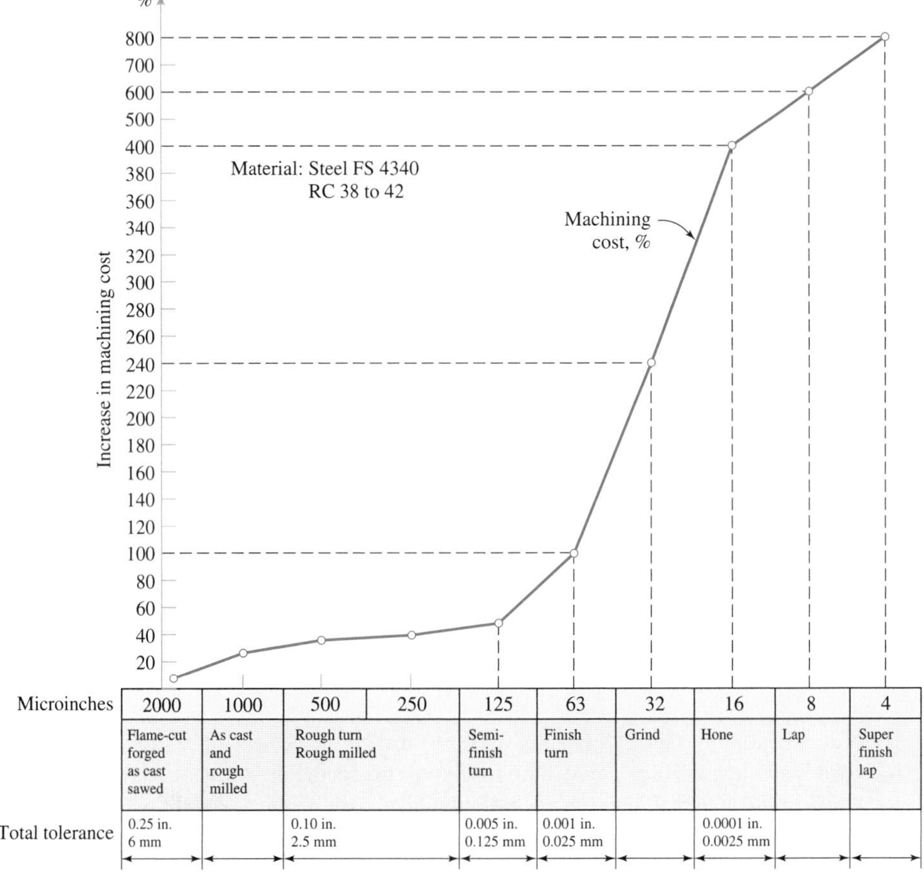

Figure 6.10
Increase in machining costs as a function of tighter tolerances and finer surface finishes. (Attributed to Association for Integrated Manufacturing Technology.)

[7]See refs. 3 and 4.

TABLE 6.1 Preferred Basic Sizes[1] (Fractional Inch Units)

$\frac{1}{64}$	0.015625	$\frac{7}{16}$	0.4375	$1\frac{3}{4}$	1.7500
$\frac{1}{32}$	0.03125	$\frac{1}{2}$	0.5000	2	2.0000
$\frac{1}{16}$	0.0625	$\frac{9}{16}$	0.5625	$2\frac{1}{4}$	2.2500
$\frac{3}{32}$	0.09375	$\frac{5}{8}$	0.6250	$2\frac{1}{2}$	2.5000
$\frac{1}{8}$	0.1250	$\frac{11}{16}$	06875	$2\frac{3}{4}$	2.7500
$\frac{5}{32}$	0.15625	$\frac{3}{4}$	0.7500	3	3.0000
$\frac{3}{16}$	0.1875	$\frac{7}{8}$	0.8750	$3\frac{1}{4}$	3.2500
$\frac{1}{4}$	0.2500	1	1.0000	$3\frac{1}{2}$	3.5000
$\frac{5}{16}$	0.3125	$1\frac{1}{4}$	1.2500	$3\frac{3}{4}$	3.7500
$\frac{3}{8}$	0.3750	$1\frac{1}{2}$	1.5000	4	4.0000

[1]Additional standard preferred basic fractional inch sizes up to 20 inches are given in ref. 3. Excerpted from ref. 3 by permission from American Society of Mechanical Engineers.

performance requirements, the closest preferred basic size should usually be chosen from Table 6.1 (fractional inch units), Table 6.2 (decimal inch units), or Table 6.3 (SI metric units), depending upon the application.[8]

The general term *fit* is used to characterize the range of "tightness" or "looseness" that may result from a specific combination of *allowances*[9] and *tolerances*[10] applied in the design of mating parts. Fits are of three general types: *clearance*, *transition*, and *interference*.

The designations of standard fits are usually conveyed by the following letter symbols:

RC	Running or sliding clearance fit
LC	Locational clearance fit
LT	Locational transition clearance or interference fit
LN	Locational interference fit
FN	Force or shrink fit

TABLE 6.2 Preferred Basic Sizes[1] (Decimal Inch Units)

0.010	0.08	0.60	2.40
0.012	0.10	0.80	2.60
0.016	0.12	1.00	2.80
0.020	0.16	1.20	3.00
0.025	0.20	1.40	3.20
0.032	0.24	1.60	3.40
0.040	0.30	1.80	3.60
0.05	0.40	2.00	3.80
0.06	0.50	2.20	4.00

[1]Additional standard preferred basic decimal inch sizes up to 20 inches are given in ref. 3. Excerpted from ref. 3 by permission from American Society of Mechanical Engineers.

[8]These tables are truncated versions of corresponding tables from the standards listed in ref. 3.

[9]Allowance is a prescribed difference between the maximum size-limit of an external dimension (shaft) and the minimum size-limit of a mating internal dimension (hole). It is the minimum clearance (positive allowance) or the maximum interference (negative allowance) between such parts.

[10]Tolerance is the total permissible variation of a size.

TABLE 6.3 Preferred Basic Sizes[1] (mm)

First Choice	Second Choice	First Choice	Second Choice	First Choice	Second Choice	First Choice	Second Choice
1		3		10		30	
	1.1		3.5		11		35
1.2		4		12		40	
	1.4		4.5		14		45
1.6		5		16		50	
	1.8		5.5		18		55
2		6		20		60	
	2.2		7		22		70
2.5		8		25		80	
	2.8		9		28		90

[1]Additional standard preferred basic metric sizes up to 1000 mm are given in ref. 3. Excerpted from ref. 3 by permission from American Society of Mechanical Engineers.

These letter symbols are used in conjunction with numbers representing the *class*[11] of fit; for example, FN 4 represents a class 4 force fit.

Standard *running and sliding* fits (clearance fits) are divided into nine classes,[12] designated RC 1 through RC 9, where RC 1 provides the smallest clearance and RC 9 the largest. Guidelines for selecting an appropriate fit for any given clearance application are shown in Table 6.4. Standard limits and clearances are tabulated in Table 6.5 for a selected range of nominal (design) sizes.

TABLE 6.4 Guidelines for Selecting Clearance Fits (Fractional and Decimal Inch)

Class of Fit	Intended Application
RC 1	*Close sliding fits*; intended for accurate location of parts that must assemble without perceptible play.
RC 2	*Sliding fits*; intended for accurate location but with greater clearance than class RC 1. Parts move and turn easily but are not intended to run freely. In larger sizes parts may seize as a result of small temperature changes.
RC 3	*Precision running fits*; intended for precision work at slow speeds and light loads. About the closest fit that can be expected to run freely. Not usually suitable if appreciable temperature changes are likely to be encountered.
RC 4	*Close running fits*; intended for running fits on accurate machinery at moderate speeds and loads. Provides accurate location and minimum play.
RC 5	*Medium running fits*; intended for higher speeds and/or higher loads.
RC 6	*Medium running fits*; intended for applications similar to RC 5 but where larger clearances are desired.
RC 7	*Free running fits*; intended for use where accuracy is not essential or where large temperature changes are likely to be encountered, or both.
RC 8	*Loose running fits*; intended for use where larger commercial (as-received) tolerances may be advantageous or necessary.
RC 9	*Loose running fits*; intended for applications similar to RC 8 but where even larger clearances may be desired.

[11]See, for example, Table 6.5 or 6.7.

[12]For fractional and decimal inch dimensions. Similar, but slightly different, guidelines for metric dimensions are available from ANSI B4.2, cited in ref. 3.

TABLE 6.5 Selected[1] Standard Limits and Clearances for Running and Sliding Fits, Using the Basic Hole System[2] (thousandths of an inch)

	Class RC 1			Class RC 3			Class RC 5			Class RC 7			Class RC 9		
	Limits of	Standard Limits		Limits of	Standard Limits		Limits of	Standard Limits		Limits of	Standard Limits		Limits of	Standard Limits	
Nom. Size Range (in.)	Clearance	Hole	Shaft	Clearance	Hole	Shaft	Clearance	Hole	Shaft	Clearance	Hole	Shaft	Clearance	Hole	Shaft
0–0.12	0.1	+0.2	−0.1	0.3	+0.4	−0.3	0.6	+0.6	−0.6	1.0	+1.0	−1.0	4.0	+2.5	−4.0
	0.45	0	−0.25	0.95	0	−0.55	1.6	0	−1.0	2.6	0	−1.6	8.1	0	−5.6
0.12–0.24	0.15	+0.2	−0.15	0.4	+0.5	−0.4	0.8	+0.7	−0.8	1.2	+1.2	−1.2	4.5	+3.0	−4.5
	0.5	0	−0.3	1.2	0	−0.7	2.0	0	−1.3	3.1	0	−1.9	9.0	0	−6.0
0.24–0.40	0.2	+0.25	−0.2	0.5	+0.6	−0.5	1.0	+0.9	−1.0	1.6	+1.4	−1.6	5.0	+3.5	−5.0
	0.6	0	−0.35	1.5	0	−0.9	2.5	0	−1.6	3.9	0	−2.5	10.7	0	−7.2
0.40–0.71	0.25	+0.3	−0.25	0.6	+0.7	−0.6	1.2	+1.0	−1.2	2.0	+1.6	−2.0	6.0	+4.0	−6.0
	0.75	0	−0.45	1.7	0	−1.0	2.9	0	−1.9	4.6	0	−3.0	12.8	0	−8.8
0.71–1.19	0.3	+0.4	−0.3	0.8	+0.8	−0.8	1.6	+1.2	−1.6	2.5	+2.0	−2.5	7.0	+5.0	−7.0
	0.95	0	−0.55	2.1	0	−1.3	3.6	0	−2.4	5.7	0	−3.7	15.5	0	−10.5
1.19–1.97	0.4	+0.4	−0.4	1.0	+1.0	−1.0	2.0	+1.6	−2.0	3.0	+2.5	−3.0	8.0	+6.0	−8.0
	1.1	0	−0.7	2.6	0	−1.6	4.6	0	−3.0	7.1	0	−4.6	18.0	0	−12.0
1.97–3.15	0.4	+0.5	−0.4	1.2	+1.2	−1.2	2.5	+1.8	−2.5	4.0	+3.0	−4.0	9.0	+7.0	−9.0
	1.2	0	−0.7	3.1	0	−1.9	5.5	0	−3.7	8.8	0	−5.8	20.5	0	−13.5
3.15–4.73	0.5	+0.6	−0.5	1.4	+1.4	−1.4	3.0	+2.2	−3.0	5.0	+3.5	−5.0	10.0	+9.0	−10.0
	1.5	0	−0.9	3.7	0	−2.3	6.6	0	−4.4	10.7	0	−7.2	24.0	0	−15.0
4.73–7.09	0.6	+0.7	−0.6	1.6	+1.6	−1.6	3.5	+2.5	−3.5	6.0	+4.0	−6.0	12.0	+10.0	−12.0
	1.8	0	−1.1	4.2	0	−2.6	7.6	0	−5.1	12.5	0	−8.5	28.0	0	−18.0
7.09–9.85	0.6	+0.8	−0.6	2.0	+1.8	−2.0	4.0	+2.8	−4.0	7.0	+4.5	−7.0	15.0	+12.0	−15.0
	2.0	0	−1.2	5.0	0	−3.2	8.6	0	−5.8	14.3	0	−9.8	34.0	0	−22.0
9.85–12.41	0.8	+0.9	−0.8	2.5	+2.0	−2.5	5.0	+3.0	−5.0	8.0	+5.0	−8.0	18.0	+12.0	−18.0
	2.3	0	−1.4	5.7	0	−3.7	10.0	0	−7.0	16.0	0	−11.0	38.0	0	−26.0

[1]Data for classes RC 2, RC 4, RC 6, and RC 8, and additional sizes up to 200 inches available from ref. 3.

[2]A basic hole system is a system in which the design size of the hole is the basic size and the allowance, if any, is applied to the shaft.

TABLE 6.6 Guidelines for Selecting Interference Fits (Fractional and Decimal Inch)

Class of Fit	Intended Application
FN 1	*Light drive fits*; require only light pressure to assemble mating parts to produce more-or-less permanent assemblies. Suitable for thin sections, long fits, or cast-iron external members.
FN 2	*Medium drive fits*; suitable for "ordinary" steel parts, or shrink fits for light sections.
FN 3	*Heavy drive fits*; suitable for heavier steel parts, or shrink fits for medium sections.
FN 4	*Force fits*; suitable for parts that can be highly stressed, or for shrink fits where the heavy pressing forces required for assembly would be impractical.
FN 5	*Force fits*; similar to FN 4 but for even higher interference pressures.

Standard *force* fits (interference fits) are divided into five classes,[13] FN 1 through FN 5, where FN 1 provides minimum interference and FN 5 provides maximum interference. Guidelines for selecting an appropriate fit for a given force fit application are shown in Table 6.6. Standard limits and interferences are tabulated in Table 6.7 for a selected range of nominal (design) sizes.

Standard locational fits for fractional and decimal inch dimensions[14] are divided into 20 classes: LC 1 through LC 11, LT 1 through LT 6, and LN 1 through LN 3. The (extensive) data for transitional fits are not included in this text, but are available in the standards cited in reference 3.

Details of dimensioning, although important, will not be discussed here since many excellent references on this topic are available in the literature.[15] In particular, the techniques of *true-position dimensioning* and *geometric dimensioning and tolerancing*, embody important concepts that not only ensure proper function of the machine but expedite manufacture and inspection of the product as well. Software packages have been developed for statistically analyzing tolerance "stackup" in complex two-dimensional and three-dimensional assemblies, and predicting the impact of design tolerances and manufacturing variations on assembly quality, before a prototype is built.[16]

Finally, Figure 6.11 is included to illustrate the range of expected surface roughnesses corresponding to various production processes. It is the designer's responsibility to strike a proper balance between a surface texture smooth enough to assure proper function, but rough enough to permit economy in manufacture. The roughness measure used in Figure 6.11 is the arithmetic average of deviation from the mean surface roughness height, in micrometers (microinches).

[13]For fractional and decimal inch dimensions. Similar but slightly different guidelines for metric dimensions are available from ANSI B4.2, cited in ref. 3.

[14]Similar but slightly different guidelines for metric dimensions are available from ANSI B4.2, cited in ref. 3.

[15]See, for example, refs. 5 and 6.

[16]See, for example, refs. 8 and 9.

TABLE 6.7 Selected[1] Standard Limits and Interferences for Force and Shrink Fits, Using the Basic Hole System[2] (thousandths of an inch)

	Class FN 1			Class FN2			Class FN3			Class FN4			Class FN5		
Nom. Size Range (in.)	Limits of Interference	Standard Limits Hole	Standard Limits Shaft	Limits of Interference	Standard Limits Hole	Standard Limits Shaft	Limits of Interference	Standard Limits Hole	Standard Limits Shaft	Limits of Interference	Standard Limits Hole	Standard Limits Shaft	Limits of Interference	Standard Limits Hole	Standard Limits Shaft
0–0.12	0.05	+0.25	+0.5	0.2	+0.4	+0.85				0.3	+0.4	+0.95	0.3	+0.6	+1.3
	0.5	−0	+0.3	0.85	−0	+0.6				0.95	−0	+0.7	1.3	−0	+0.9
0.12–0.24	0.1	+0.3	+0.6	0.2	+0.5	+1.0				0.4	+0.5	+1.2	0.5	+0.7	+1.7
	0.6	−0	+0.4	1.0	−0	+0.7				1.2	−0	+0.9	1.7	−0	+1.2
0.24–0.40	0.1	+0.4	+0.75	0.4	+0.6	+1.4				0.6	+0.6	+1.6	0.5	+0.9	+2.0
	0.75	−0	+0.5	1.4	−0	+1.0				1.6	−0	+1.2	2.0	−0	+1.4
0.40–0.56	0.1	+0.4	+0.8	0.5	+0.7	+1.6				0.7	+0.7	+1.8	0.6	+1.0	+2.3
	0.8	−0	+0.5	1.6	−0	+1.2				1.8	−0	+1.4	2.3	−0	+1.6
0.56–0.71	0.2	+0.4	+0.9	0.5	+0.7	+1.6				0.7	+0.7	+1.8	0.8	+1.0	+2.5
	0.9	−0	+0.6	1.6	−0	+1.2				1.8	−0	+1.4	2.5	−0	+1.8
0.71–0.95	0.2	+0.5	+1.1	0.6	+0.8	+1.9				0.8	+0.8	+2.1	1.0	+1.2	+3.0
	1.1	−0	+0.7	1.9	−0	+1.4				2.1	−0	+1.6	3.0	−0	+2.2
0.95–1.19	0.3	+0.5	+1.2	0.6	+0.8	+1.9	0.8	+0.8	+2.1	1.0	+0.8	+2.3	1.3	+1.2	+3.3
	1.2	−0	+0.8	1.9	−0	+1.4	2.1	−0	+1.6	2.3	−0	+1.8	3.3	−0	+2.5
1.19–1.58	0.3	+0.6	+1.3	0.8	+1.0	+2.4	1.0	+1.0	+2.6	1.5	+1.0	+3.1	1.4	+1.6	+4.0
	1.3	−0	+0.9	2.4	−0	+1.8	2.6	−0	+2.0	3.1	−0	+2.5	4.0	−0	+3.0
1.58–1.97	0.4	+0.6	+1.4	0.8	+1.0	+2.4	1.2	+1.0	+2.8	1.8	+1.0	+3.4	2.4	+1.6	+5.0
	1.4	−0	+1.0	2.4	−0	+1.8	2.8	−0	+2.2	3.4	−0	+2.8	5.0	−0	+4.0
1.97–2.56	0.6	+0.7	+1.8	0.8	+1.2	+2.7	1.3	+1.2	+3.2	2.3	+1.2	+4.2	3.2	+1.8	+6.2
	1.8	−0	+1.3	2.7	−0	+2.0	3.2	−0	+2.5	4.2	−0	+3.5	6.2	−0	+5.0
2.56–3.15	0.7	+0.7	+1.9	1.0	+1.2	+2.9	1.8	+1.2	+3.7	2.8	+1.2	+4.7	4.2	+1.8	+7.2
	1.9	−0	+1.4	2.9	−0	+2.2	3.7	−0	+3.0	4.7	−0	+4.0	7.2	−0	+6.0
3.15–3.94	0.9	+0.9	+2.4	1.4	+1.4	+3.7	2.1	+1.4	+4.4	3.6	+1.4	+5.9	4.8	+2.2	+8.4
	2.4	−0	+1.8	3.7	−0	+2.8	4.4	−0	+3.5	5.9	−0	+5.0	8.4	−0	+7.0

[1]Data for additional sizes up to 200 inches available from ref. 3.

[2]A basic hole system is a system in which the design size of the hole is the basic size and the allowance, if any, is applied to the shaft.

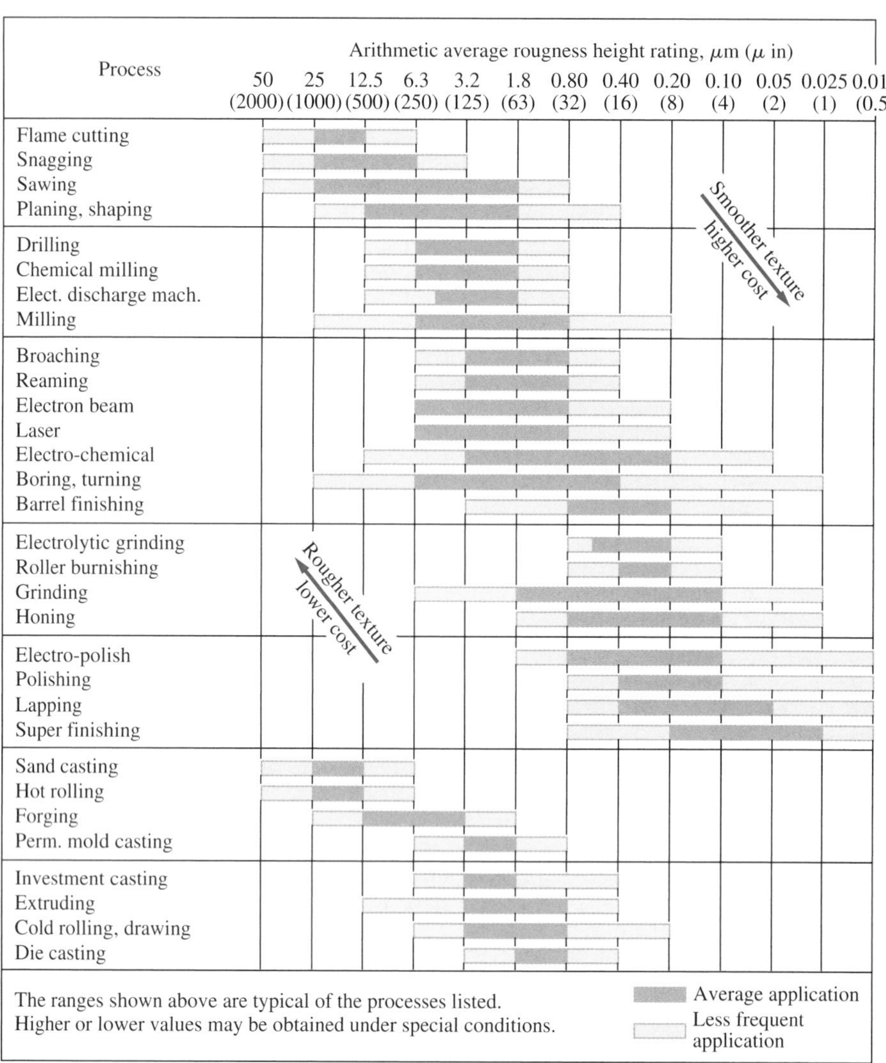

Figure 6.11
Surface roughness ranges produced by various manufacturing processes. (From ref. 7, by permission of the McGraw-Hill Companies.)

Problems

6-1. List the basic principles for creating the shape of a machine part and determining its size. Interpret these principles in terms of the five common stress patterns discussed in 4.4.

6-2. List 10 configurational guidelines for making good geometric choices for shapes and arrangements of machine parts.

6-3. In Proposal 1 shown in Figure 6.1(a), a "U-shaped" link is suggested for transferring direct tensile force F from joint A to joint B. Although the *direct load path guideline* clearly favors Proposal 2 shown in Figure 6.1(b), it has been discovered that a rotating cylindrical drive shaft, whose center lies on a virtual line connecting joints A and B, requires that some type of U-shaped link must be used to make space for the rotating drive shaft. Without making any calculations, identify which of the configurational guidelines of 6.2 might be applicable in determining an appropriate geometry for the U-shaped link, and, based on these guidelines, sketch an initial proposal for the overall shape of the link.

6-4. Referring to Figure 16.4, the brake system shown is actuated by applying a force F_a at the end of an *actuating lever*, as shown. The actuating lever is to be pivoted at point C. Without making any calculations, identify which of the configurational guidelines of 6.2 might be applicable in determining an appropriate shape for the actuating lever, and, based on these guidelines, sketch an initial proposal for the overall shape of the lever. Do not include the shoe, but provide for it.

6-5. Figure P6.5 shows a sketch of a proposed torsion bar spring, clamped at one end to a rigid support wall, supported by a bearing at the free end, and loaded in torsion by an attached lever arm clamped to the free end. It is being proposed to use a split-clamp arrangement to clamp the torsion bar to the fixed

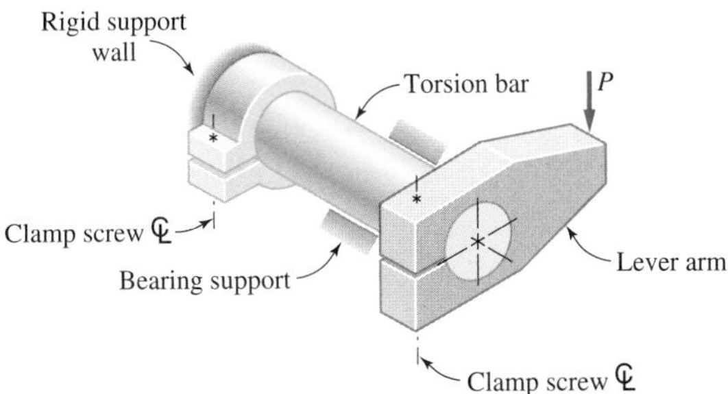

Figure P6.5
Sketch of a torsion bar spring arrangement.

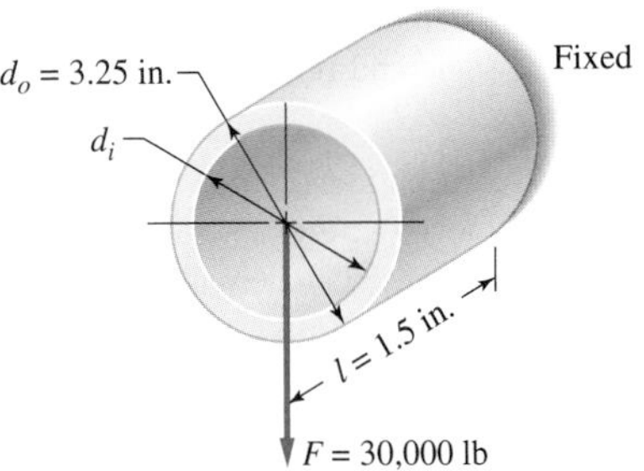

Figure P6.8
Short tubular cantilever bracket subjected to transverse end load.

support wall and also to use a split-clamp configuration to attach the lever arm to the free end of the torsion bar. Without making any calculations, and concentrating only on the torsion bar, identify which of the configurational guidelines of 6.2 might be applicable in determining an appropriate shape for this torsion bar element. Based on the guidelines listed, sketch an initial proposal for the overall shape of the torsion bar.

6-6. a. Referring to the free-body diagram of the brake actuating lever shown in Figure 16.4(b), identify appropriate critical sections in preparation for calculating dimensions and finalizing the shape of the part. Give your rationale.

b. Assuming that the lever will have a constant solid circular cross section over the full length of the beam, select appropriate critical points in each critical section. Give your reasoning.

6-7. a. Figure P6.7 shows a channel-shaped cantilever bracket subjected to an end load of $P = 8000$ lb, applied vertically downward as shown. Identify appropriate critical sections in preparation for checking the dimensions shown. Give your rationale.

b. Select appropriate critical points in each critical section. Give your reasoning.

c. Can you suggest improvements on shape or configuration for this bracket?

6-8. The short tubular cantilever bracket shown in Figure P6.8 is to be subjected to a transverse end load of $F = 30{,}000$ lb vertically downward. Neglecting possible stress concentration effects, do the following:

a. Identify appropriate critical sections in preparation for determining the unspecified dimensions.

b. Specify precisely and completely the location of all potential critical points in each critical section identified. Clearly explain why you have chosen these particular points. Do not consider the point where force F is applied to the bracket.

c. For each potential critical point identified, sketch a small volume element showing all nonzero components of stress.

d. If cold-drawn AISI 1020 steel has been tentatively selected as the material to be used, yielding has been identified as the probable governing failure mode, and a safety factor of $n_d = 1.20$ has been chosen, calculate the required numerical value of d_i.

6-9. The cross hatched critical section in a solid cylindrical bar of 2024-T3 aluminum, as shown in the sketch of Figure P6.9, is subjected to a torsional moment of $T_x = 8500$ N-m, a bending moment of $M_y = 5700$ N-m, and a vertically downward transverse force of $F_z = 400$ kN.

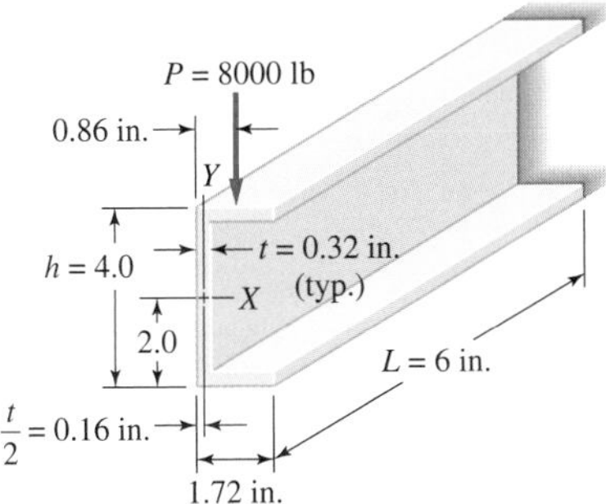

Figure P6.7
Channel-shaped cantilever bracket subjected to transverse end load.

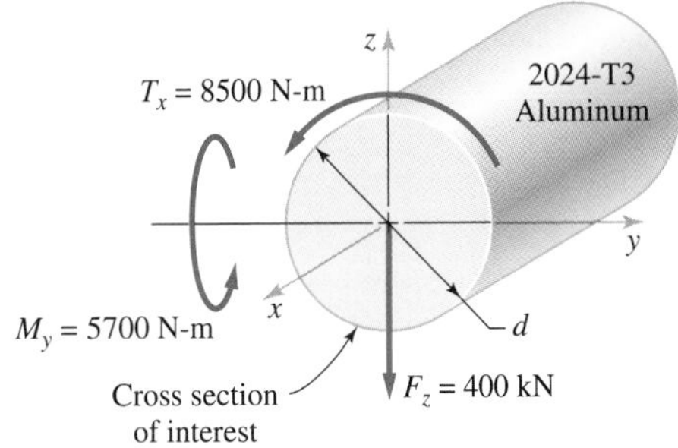

Figure P6.9
Solid cylindrical bar subjected to torsion, bending, and transverse shear.

a. Clearly establish the location(s) of the potential critical point(s), giving logic and reasons why you have selected the point(s).

b. If yielding has been identified as the probable governing failure mode, and a safety factor of 1.15 has been chosen, calculate the required numerical value of diameter d.

6-10. A fixed steel shaft (spindle) is to support a rotating idler pulley (sheave) for a belt drive system. The nominal shaft diameter is to be 2.00 inches. The sheave must rotate in a stable manner on the shaft, at relatively high speeds, with the smoothness characteristically required of accurate machinery. Write an appropriate specification for the limits on shaft size and sheave bore, and determine the resulting limits of clearance. Use the basic hole system.

6-11. A cylindrical bronze bearing sleeve is to be installed into the bore of a fixed cylindrical steel housing. The bronze sleeve has an inside diameter of 2.000 inches and nominal outside diameter of 2.500 inches. The steel housing has a nominal bore diameter of 2.500 inches and an outside diameter of 3.500 inches. To function properly, without "creep" between the sleeve and the housing, it is anticipated that a "medium drive fit" will be required. Write an appropriate specification for the limits on sleeve outer diameter and housing bore diameter, and determine the resulting limits of interference. Use the basic hole system.

6-12. For a special application, it is desired to assemble a phosphor bronze disk to a hollow steel shaft, using an interference fit for retention. The disk is to be made of C-52100 hot-rolled phosphor bronze, and the hollow steel shaft is to be made of cold-drawn 1020 steel. As shown in Figure P6.12, the proposed nominal dimensions of the disk are 10 inches for outer diameter and 3 inches for the hole diameter, and the shaft, at the mounting pad, has a 3-inch outer diameter and a 2-inch inner diameter. The hub length is 4 inches. Preliminary calculations have indicated that in order to keep stresses within an acceptable range, the interference between the shaft mounting pad and the hole in the disk must not exceed 0.0040 inch. Other calculations indicate that to transmit the required torque across the interference fit interface the interference must be at least 0.0015 inch. What class fit would you recommend for this application, and what dimensional specifications should be written for the shaft mounting pad outer diameter and for the disk hole diameter? Use the basic hole system for your specifications.

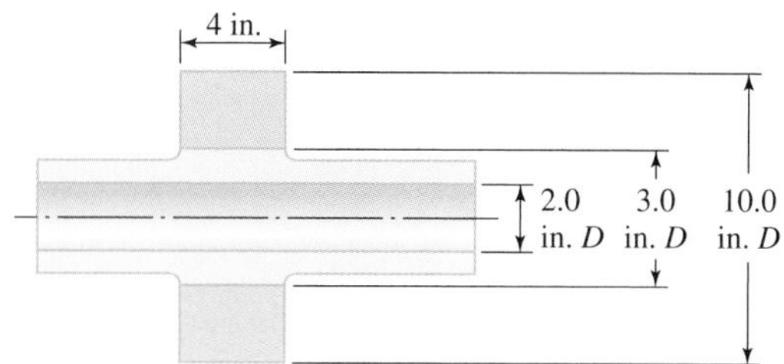

Figure P6.12
Sketch showing the nominal configuration of an interference fit assembly.

6-13. It is desired to design a hydrodynamically lubricated plain bearing (see Chapter 10) for use in a production line conveyor to be used to transport industrial raw materials. It has been estimated that for the anticipated operating conditions and the lubricant being considered, a minimum lubricant film thickness of $h_o = 0.0046$ inch can be sustained. Further, it is being proposed to *finish-turn* the bearing journal (probably steel) and *ream* the bearing sleeve (probably bronze). An empirical relationship has been found in the literature (see Chapter 10) that claims satisfactory wear levels can be achieved if

$$h_o \geq 5.0(R_j + R_b)$$

where R_j = arithmetic average asperity peak height above mean bearing *journal* surface, in

R_b = arithmetic average asperity peak height above mean bearing *sleeve* surface, in.

Determine whether bearing wear levels in this case would be likely to lie within a satisfactory range.

6-14. You have been assigned to a design team working on the design of a boundary-lubricated plain bearing assembly (see Chapter 10) involving a 4340 steel shaft heat-treated to a hardness of Rockwell C 40 (RC 40), rotating in an aluminum-bronze bushing. One of your colleagues has cited data that seem to indicate that a 20 percent improvement in wear life might be achieved by *grinding* the surface of the steel shaft at the bearing site, as opposed to a *finish-turning* operation, as currently proposed. Can you think of any reasons *not* to grind the shaft surface?

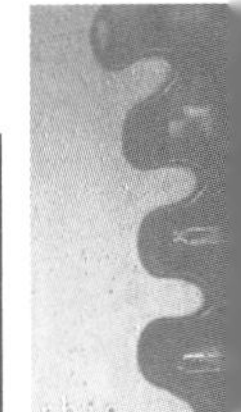

Chapter 7

Design-Stage Integration of Manufacturing and Maintenance Requirements

7.1 Concurrent Engineering

To avoid the potential penalties of locking in early design decisions, the strategy of *concurrent engineering* deserves careful consideration. The objective of concurrent engineering, or *concurrent design*, is to organize the information flow among all project participants, from the time marketing goals are established until the product is shipped. Information and knowledge about all of the design-related issues during the life cycle of the product are made as available as possible at all stages of the design process. Concurrent engineering strategy, especially in mass production industries, is typically implemented by using a team approach to involve engineers and others working on every phase of the entire life cycle of the product, to communicate changes as they develop. Participating groups may include design, tooling, fabrication, assembly, processing, maintenance, inspection, marketing, shipping, and recycling or disposal. For the concurrent design strategy to be effective, team members from downstream processes must be continuously and deeply involved in the discussions and decision making all along the way, starting at the preliminary design stage; company management must also support the strategy. Interactive computer systems, including CAD (computer-aided design) software for product data management, and solid modeling software form the cornerstones for implementation of concurrent engineering strategy. The technique allows on-line review and update of the current design configuration by any team member at any time.[1] Properly executed, this approach prevents the need for costly redesigns and capitalizes upon the availability of modern flexible manufacturing systems and automation technology.

Concurrent engineering strategy is sometimes referred to as *Design for "X" (DFX)* strategy, where "*X*" is the symbol for any of the engineering design issues associated with the product, including function, performance, reliability, manufacture, assembly, disassembly, maintenance, inspection,[2] and robustness.[3] The approach is to evaluate each of the issues, qualitatively and quantitatively, with the goal of optimizing performance, manufacturing, and maintenance requirements, as well as minimizing life cycle costs for the

[1]For example, *Windchill* software, a product of *Parametric Technology, Inc.*, automatically resizes products when one dimension is changed, and uses the Internet to link computerized design with purchasing, outsourcing, manufacturing, and long-term maintenance. (See ref. 14.)

[2]Inspection here refers not only to the ability to examine manufactured parts for compliance with specifications and tolerances, but, of equal importance, the ability to access and examine potential failure initiation sites throughout the life of the machine.

[3]*Robustness* is a term that refers to the ability of a product or system to perform properly in the presence of variations in the manufacturing process, variations in environmental operating conditions, or service-induced changes in geometry or material properties.

overall system. A brief discussion of some of the DFX issues follows next, with emphasis on the concept that input from as many downstream system activities as possible, as early as possible, is important in avoiding or minimizing expensive design changes.

7.2 Design for Function, Performance, and Reliability

The traditional design responsibilities of making sure that a proposed machine or system fulfills all of the specified functions, performs efficiently over the design lifetime, and does not fail prematurely are universally accepted. Chapters 1 through 6 of this textbook are devoted to detailed discussions of design procedures aimed at reaching these goals. It only remains to be stated that these design responsibilities must continue to be met as other DFX demands are introduced to implement and optimize downstream processes. An ongoing review of potential changes in functionality, performance, and reliability, generated by downstream process-improvement activity, is essential to the successful design, production, and marketing of a competitive end product. It is equally true that a careful ongoing review of material selection, part geometry, and overall configuration is essential to efficient cost-effective fabrication, assembly, and maintenance of the final product.

7.3 Selection of the Manufacturing Process

Changing the shape and size of available stock or bulk raw material into parts with the sizes, shapes, and finishes specified by the designer is the objective of any manufacturing process. It will often be true that more than one manufacturing method is available for producing a particular part. Selection of the best process may depend upon one or more of the following factors:

1. Type, form, and properties of raw material
2. Desired properties of finished part, including strength, stiffness, ductility, and toughness
3. Size, shape, and complexity of finished part
4. Tolerances required and surface finishes specified
5. Number of parts to be produced
6. Availability and cost of capital equipment required
7. Cost and lead time for tooling required
8. Scrap rate and cost of reworking
9. Time and energy requirements for the overall process
10. Worker safety and environmental impact

In essence, all manufacturing processes may be categorized as methods for changing shape or size by one of five basic means: (1) flow of molten material, (2) fusion of component parts, (3) plastic deformation of ductile solid material, (4) selectively removing material by machining or chip-forming action, or (5) sintering powdered metal particles. Attributes and examples of each of these process categories are briefly summarized in Table 7.1.

A designer should give consideration to the selection of an appropriate manufacturing process for each part, early in the design stage. Details of material selection, size and shape

TABLE 7.1 Attributes of Manufacturing Process Categories

Process Category	Process Category Symbol	Processing Power Required	Processing Time Required	Special Capital Equipment Costs	Special Tooling Costs	Relative Strength of Product	Examples of Process[1]
Flow of molten material (casting process)	C	relatively low	relatively low	relatively high	relatively low	generally poorest	Sand casting Shell mold casting Ceramic mold casting Permanent mold casting Die casting Centrifugal casting Investment casting Others
Fusion of component parts by local application of heat (welding process)	W	moderate	moderate	relatively low	relatively low	moderate	Arc welding Gas welding Resistance welding Electron beam welding Others
Plastic deformation of ductile solid material (forming process)	F	relatively high	relatively low	relatively high	relatively high	generally best	Hammer forging Press forging Rolling Drawing Extruding Bending Deep drawing Spinning Stretching Others
Material removal by chip-forming action (machining process)	M	moderate	relatively high	relatively low	relatively low	second best	Turning Facing Boring Milling Drilling Grinding Sawing Others
Sintering of powdered metal particles (sintering process)	S	moderate	moderate	moderate	relatively high		Diffusion bonding Liquid-phase sintering Spark sintering Hot isostatic processing (HIP)

[1]For detailed discussion of processes see, for example, refs. 1, 2, or 3.

of a part, number to be produced, and strength requirements all have an impact on selection of the manufacturing process best suited to a particular part. In turn, design decisions on details of shape, orientation, retention, or other features are, in many cases, dependent upon the selected manufacturing process. A designer is well advised to consult with man-

TABLE 7.2 Selection of Manufacturing Process Based on Application Characteristics

Characteristic	Application Description or Requirement	Process Category Generally Best Suited[1]
Shape	uniform, simple	M, F, S
	intricate, complex	C, W
Size	small	M, F, S
	medium	M, F, C, W
	large	C, M, W
Number to be produced	one or a few	M, W
	low mass production	M, F, C, S, W
	high mass production	F, C
Strength required	minimal	C, F, M, S
	average	F, M, W
	maximum available	F

[1]See Table 7.1 for definition of process category symbols.

ufacturing engineers early in the design stage to avoid later problems. Concurrent engineering strategy directly supports effective design decision making in this context.

Preliminary guidelines for selecting appropriate manufacturing processes are summarized in Table 7.2. Although these guidelines are very useful for a designer, it is emphasized that a team approach involving manufacturing engineers and materials engineers usually pays dividends. Table 3.17 should be checked to make sure the process selected is compatible with the proposed material.

Example 7.1 Tentative Selection of Manufacturing Process

The frame sketched in Figure 20.1(c) is to be used for an experimental fatigue testing machine that will operate in a laboratory environment. It is anticipated that three such machines will be constructed. Utilizing Tables 7.1 and 7.2, tentatively select an appropriate manufacturing process for producing the frame, assuming that low-carbon steel will be chosen as the preferred material.

Solution

Evaluating each of the "characteristics" included in Table 7.2 in terms of the corresponding assessment of the "application description" best describing the frame sketched in Figure 20.1(c), and using "process category" symbols defined in Table 7.1, the preliminary evaluation shown in Table E7.1 may be made. Tallying the results from column three of Table E7.1, the frequency of citation for "applicable process categories" may be listed as follows:

M:	3 times
F:	1 time
C:	2 times
S:	0 times
W:	4 times

Welding appears to be the most appropriate manufacturing process, and will tentatively be selected. From Table 3.17, this selection is compatible with low-carbon steel.

TABLE E7.1 Manufacturing Process Suitability

Characteristic	Application Description	Applicable Process Category
Shape	intricate, complex	C, W
Size	medium	M, F, C, W
Number to be produced	a few	M, W
Strength required	average	F, M, W

7.4 Design for Manufacturing (DFM)

After the materials have been selected and processes identified, and after the sizes and shapes have been created by the designer to meet functional and performance requirements, each part, and the overall machine assembly, should be scrutinized for compliance with the following guidelines for efficient manufacture.

1. The total number of individual parts should be minimized.
2. Standardized parts and components should be used where possible.
3. Modular components and subassemblies with standardized interfaces to other components should be used where possible.
4. Individual part geometry should accommodate the selected manufacturing process to minimize waste of material and time.
5. Near net shape manufacturing processes should be specified where possible to minimize the need for secondary machining and finishing processes.
6. Parts and component arrangements should be designed so that all assembly maneuvers may be executed from a single direction during the assembly process, preferably from the top down to capitalize on gravity-assisted feeding and insertion.
7. As far as possible, the function-dictated sizes, shapes, and arrangements of parts in the assembly should be augmented by geometric features that promote ease of alignment, ease of insertion, self-location, and unobstructed access and view during the assembly process. Examples of such features might include well-designed chamfers, recesses, guideways, or intentional asymmetry.
8. The number of separate fasteners should be minimized by utilizing assembly tabs, snap-fits, or other interlocking geometries, where possible.

Again, as the designer strives to comply with these guidelines, he or she would be well advised to engage in frequent consultations with manufacturing and materials engineers.

7.5 Design for Assembly (DFA)

The assembly process often turns out to be the most influential contributor in determining the overall cost of manufacturing a product, especially for higher production rates. For this reason the assembly process has been intensively studied over the past two decades, and several techniques, including both qualitative and quantitative approaches, have been developed for evaluating and choosing the best assembly method for a given product.[4] Basically, all assembly processes may be classified as either *manual* (performed by people) or

[4]See refs. 3 through 10 for detailed discussions.

automated (performed by mechanisms). Manual assembly processes range from *bench* assembly of the complete machine at a single station to *line* assembly, where each person is responsible for assembling only a small portion of the complete unit as it moves from station to station along a production line. Automated assembly may be subcategorized into either *dedicated* automatic assembly or *flexible* automatic assembly. Dedicated assembly systems involve the progressive assembly of a unit using a series of single-purpose machines, in line, each dedicated to (and capable of) only *one* assembly activity. In contrast, flexible assembly systems involve the use of one or more machines that have the capability of performing *many* activities, simultaneously or sequentially, as directed by computer-managed control systems.

The design importance of knowing early in the design stage which assembly process will be used lies in the need to configure parts[5] for the selected assembly process. Table 7.3 provides preliminary guidelines for predicting which assembly process will probably be used to best meet the needs of the application. Realistically, it is imporant to note that[6] only 10 percent of products are suitable for line assembly, only 10 percent of products are suitable for dedicated automatic assembly, and only 5 percent of products are suitable for flexible assembly. Clearly, manual assembly is by far the most widely used assembly process.

To facilitate manual assembly, the designer should attempt to configure each part so that it may be easily grasped and manipulated without special tools. To accomplish this, parts should not be heavy, sharp, fragile, slippery, sticky, or prone to nesting or tangling. Ideally, parts should be symmetric, both rotationally and end-to-end, so that orientation and insertion are fast and easy. For effective automatic assembly, parts should have the ability to be easily oriented, easily fed, and easily inserted. They should therefore *not* be very thin, very small, very long, very flexible, or very abrasive, and they should not be hard to grasp. In the final analysis, the designer would be well advised to engage in frequent dialogues with manufacturing engineers throughout the design process.

TABLE 7.3 Preliminary Guidelines for Selection of Assembly Process Based on Application Characteristics

Characteristic	Application Characteristic or Requirement	Assembly Method Generally Best Suited[1]
Total number of parts in one assembly	low	M
	medium	M, D, F
	high	D, F
Projected production volume	low	M, F
	medium	M, D, F
	high	D, F
Cost of available labor	low	M
	moderate	M, D, F
	high	D, F
Difficulty in handling (acquiring, orienting, and transporting parts) and insertion	little	D, F
	moderate	M, D, F
	great	M

[1]M = manual assembly
D = dedicated automatic assembly
F = flexible automatic assembly

[5]Assuming that the proposed part configuration also meets functional specifications.

[6]See ref. 4, p. 24.

7.6 Design for Critical Point Accessibility, Inspectability, Disassembly, Maintenance, and Recycling

In 1.8 the heavy dependence of both *fail-safe design* and *safe life design* upon regular *inspection of critical points* was emphasized. Nevertheless, designers have rarely considered inspectability of critical points at the design stage. To prevent failure, and minimize downtime, it is imperative that designers configure machine components, subassemblies, and fully assembled machines so that the critical points established during the functional design process are *accessible* and *inspectable*. Further, inspection should be possible with a minimum of disassembly effort. Also, maintenance and service requirements should be carefully examined by the designer as early as possible in the design stage to minimize downtime (especially unscheduled downtime) and maintain functionality throughout the life cycle.

As design calculations proceed, a list of governing critical points (see 6.3) should be compiled, prioritized, and posted. As subassembly and machine assembly layout drawings are developed, careful attention should be given to accessibility of the governing critical points to inspection. These considerations are especially important for high-performance machines and structures such as aircraft, spacecraft, high-speed rail vehicles, off-shore oil platforms, and other devices operating under highly loaded conditions or in adverse environments. To provide appropriate accessibility to critical point inspection it is important for a designer to have a working knowledge of nondestructive evaluation (NDE) techniques, and equipment[7] that might be utilized to implement the inspection process. Potential NDE techniques, which range from very simple to very complex, include:

1. Direct visual examination
2. Visual examination using inspection mirrors or optical magnification
3. Use of borescopes or fiber-optic bundles
4. Use of liquid or dye penetrant flaw-detection techniques
5. Use of electromagnetic flaw-detection techniques
6. Use of microwave techniques
7. Use of ultrasonic or laser ultrasonic examination
8. Use of eddy-current techniques
9. Use of thermographic techniques
10. Use of acoustic emission procedures

A designer would be well advised to consult, as early as is practical in the design process, specialists in NDE methods to help choose wise approaches for inspecting the governing critical points. When inspection methods have been selected, it is important to configure components, subassemblies, and machines so that supporting equipment can be easily maneuvered to each critical site with a minimum of effort and as little disassembly as possible. The provision of access ports, inspection plates, line-of-sight corridors, inspection mirror clearance, borescope access, and clearance for transducers or other supporting inspection equipment is a direct responsibility of the designer. Consideration of these requirements should begin as early as possible to avoid costly design changes later.

Life cycle maintenance and service requirements should also be examined by the designer with the objective of configuring components, subassemblies, and the overall assembly so that maintenance and service are as easy as is practical. In this context, virtual assembly and disassembly software[8] may be useful at the design stage to identify and cor-

[7]See, for example, ref. 11.

[8]See, for example, refs. 12 and 13.

rect potential problems—such as assembly and disassembly interference; insufficient clearances for wrenches, pullers, or presses; the need for special tools; or the need for assembly or disassembly sequence improvement—before hardware prototypes exist. Consideration should be given to assuring that expendable or recyclable maintenance items such as filters, wear plates, belts, and bearings are easily replaceable. Some additional guidelines in designing for more efficient maintenance include:

1. Providing appropriate access ports and inspection plates
2. Providing accessible gripping sites, jacking sites, recesses or slots for pullers, or other clearances to simplify the disassembly process
3. Providing bosses, recesses, or other features to facilitate pressing bearings in and out, pulling or pressing gears and seals off and on, or other required assembly and disassembly tasks
4. Using integral fasteners, such as studs or tabs, to replace loose parts that are easily lost
5. If possible, avoiding permanent or semipermanent fastening methods such as staking, welding, adhesive bonding, or irreversible snap-fits

Finally, in accord with the definition of mechanical design given in 1.4, design for resource conservation and minimization of adverse environmental impact are increasingly important responsibilities to be addressed at the design stage. Design for recycling, reprocessing, and remanufacturing can often be enhanced simply by considering the need during the design stage. Not only must a designer compete in the marketplace by optimizing the design with respect to performance, manufacturing, and maintenance requirements, but he or she must respond responsibly to the clear and growing obligation of the global technical community to conserve resources and preserve the earth's environment.

Problems

7-1. Define the term "concurrent engineering" and explain how it is usually implemented.

7-2. List the five basic methods for changing the size or shape of a work piece during the manufacturing process and give two examples of each basic method.

7-3. Explain what is meant by "near net shape" manufacturing.

7-4. Basically, all assembly processes may be classified as either *manual, dedicated automatic*, or *flexible automatic* assembly. Define and distinguish among these assembly processes, and explain why it is important to tentatively select a candidate process at an early stage in the design of a product.

7-5. Explain how "design for inspectability" relates to the concepts of *fail-safe design* and *safe life design* described in 1.8.

7-6. Give three examples from your own life-experience in which you think that "design for maintenance" could have been improved substantially by the designer or manufacturer of the part or machine being cited.

7-7. The gear support shaft depicted in Figure 8.1(a) is to be made of AISI 1020 steel. It is anticipated that 20,000 of these shafts will be manufactured each year for several years. Utilizing Tables 7.1 and 7.2, tentatively select an appropriate manufacturing process for producing the shafts.

7-8. It is being proposed to use AISI 4340 steel as the material for a high-speed flywheel such as the one depicted in Figure 18.10. It is anticipated that 50 of these high-speed flywheels will be needed to complete an experimental evaluation program. It is desired to achieve the highest practical rotational speeds. Utilizing Tables 7.1 and 7.2, tentatively select an appropriate manufacturing process for producing these high-speed rotors.

7-9. The rotating power screw depicted in Figure 12.1 is to be made of AISI 1010 carburizing-grade steel. A production run of 500,000 units is anticipated. Utilizing Tables 7.1 and 7.2, tentatively select an appropriate manufacturing process for producing the power screws.

7-10. Figure 8.1(c) depicts a flywheel drive assembly. Studying this assembly, and utilizing the discussion of 7.5, including Table 7.3, suggest what type of assembly process would probably be best. It is anticipated that 25 assemblies per week will satisfy market demand. The assembly operation will take place in a small midwestern farming community.

PART TWO

DESIGN APPLICATIONS

Chapter 8

Power Transmission Shafting; Couplings, Keys, and Splines

8.1 Uses and Characteristics of Shafting

Virtually all machines involve the transmission of power and/or motion from an input source to an output work site. The input source, usually an electric motor or internal combustion engine, typically supplies power as a rotary driving torque to the *input shaft* of the machine under consideration, through some type of a *coupling* (see 8.8.) A shaft is typically a relatively long cylindrical element supported by bearings (see Chapters 10 and 11), and loaded torsionally, transversely, and/or axially as the machine operates. The operational loads on a shaft arise from elements mounted on or attached to the shaft, such as gears (see Chapter 14 and 15), belt pulleys (see Chapter 17), chain sprockets (see Chapter 17), or flywheels (see Chapter 18), or from bearings mounted on the shaft that support other operational subassemblies of the machine. Some schematic examples of typical shafting configurations are shown in Figure 8.1.

Most power transmission shafts are cylindrical (solid or hollow), and often are stepped. In special applications, shafts may be square, rectangular, or some other cross-sectional shape. Usually the shaft rotates and is supported by bearings attached to a fixed frame or machine housing. Sometimes, however, the shaft is fixed to the housing, so that the bearings of *idler* gears, pulleys, or wheels may be mounted upon it. Short, stiff, fixed cantilever shafts, such as those used to support the nondriving wheels of an automobile, are usually called *spindles*.

Since power transmission shafting is so widely required in virtually all types of machinery and mechanical equipment, its design or selection may be the most frequently encountered design task. In most cases, the approximate positions of gears, pulleys,

Figure 8.1
Sketches of some typical shafting configurations.

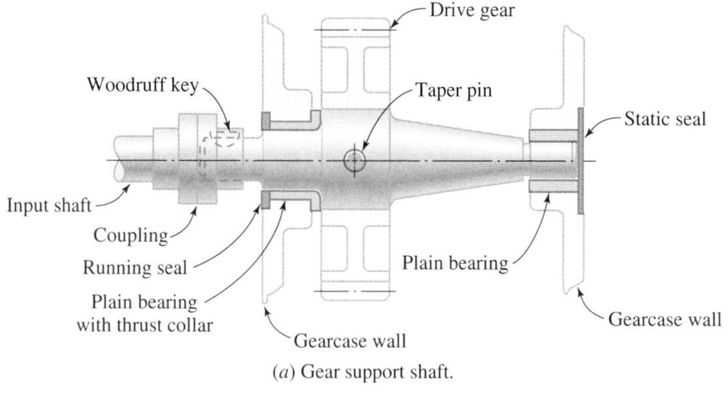

(*a*) Gear support shaft.

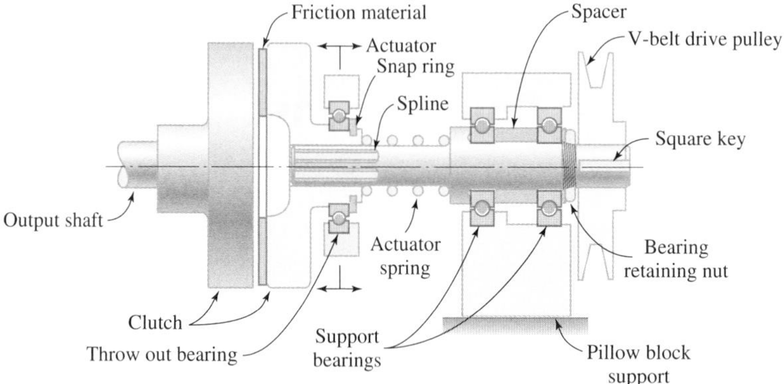

(*b*) Clutch drive shaft.

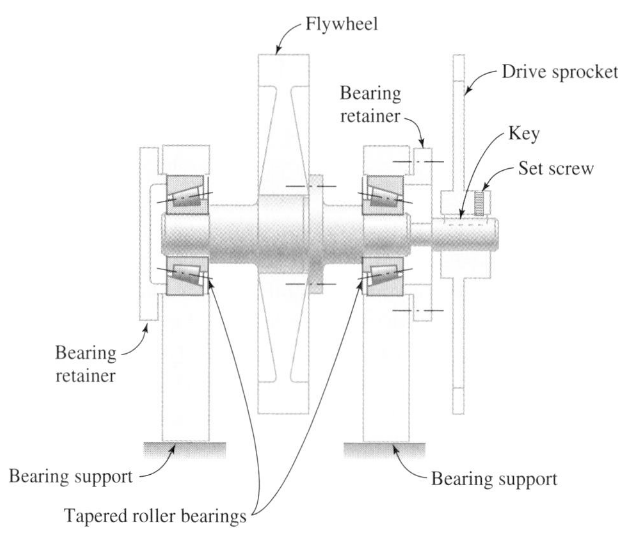

(*c*) Flywheel drive shaft.

sprockets, and supporting bearings along the shaft axis are dictated by the functional specifications for the machine. The initial position layout of these elements is the first step in the design of a shaft. Next, a conceptual sketch of the shaft configuration is made, showing the main features required for the mounting and positioning of the elements along the shaft. The need to consider locating shoulders for accurate axial positioning of bearings or

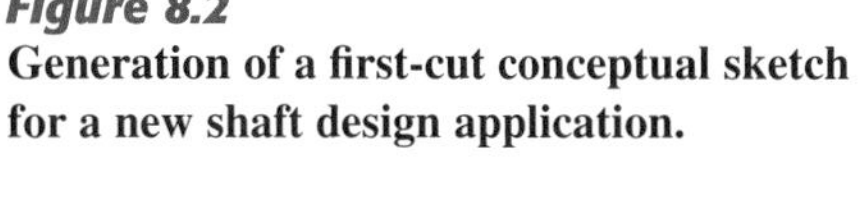

Figure 8.2
Generation of a first-cut conceptual sketch for a new shaft design application.

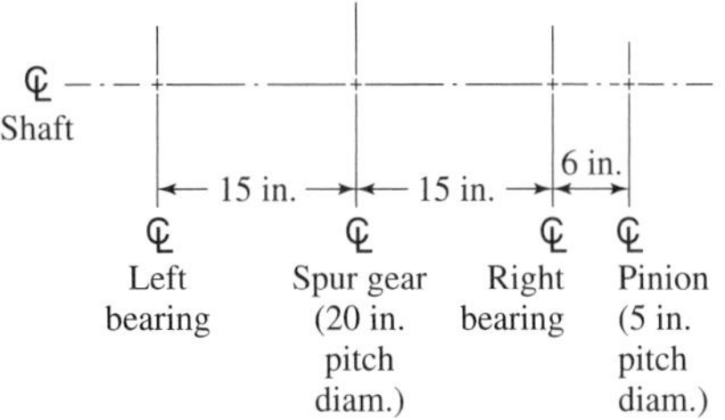

(*a*) Position layout of gears and bearings as dictated by functional requirements of machine.

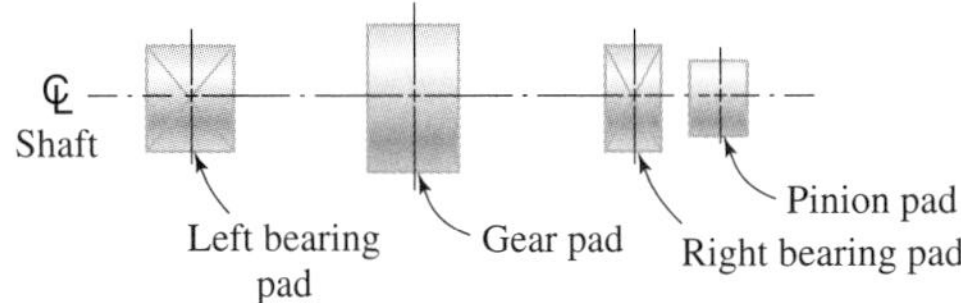

(*b*) Sketch of shaft mounting features required for gears and bearings.

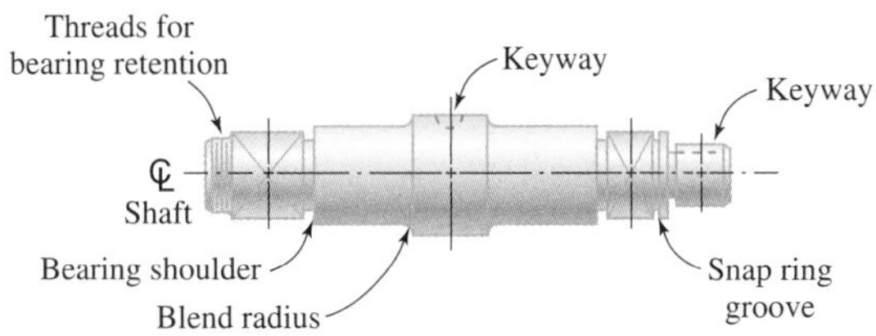

(*c*) Completion of first-cut shaft sketch by connecting the mounting features and adding shoulders, fillets, and retention features.

gears, raised mounting pads to facilitate pressing gears or bearings on or off the shaft, and progressive increases in shaft diameter inward from the two ends (to permit assembly) is important, even at this early stage. The contemplated use of other mounting or retention features, such as keys, splines, pins, threads, or retaining rings, may also be included in the conceptual sketch of the shaft, even before any design calculations have been made. The generation of a first conceptual sketch for a new shaft design application is illustrated in Figure 8.2.

8.2 Potential Failure Modes

Most shafts rotate. Transverse loads from gears, sprockets, pulleys, and bearings that are mounted upon a rotating shaft result in completely reversed *cyclic bending stresses*. In some cases, transverse loads may also result in completely reversed *transverse shearing stresses*. In addition, axial loads, such as those from helical gears or preloaded bearings, may produce both *axial stresses* and/or superposed *bending moments* that are usually steady, but sometimes fluctuating. Transmitted shaft torques induce *torsional shearing stresses*. These torsional shearing stresses are usually steady, but may sometimes fluctuate, depending on the application. From these observations, and the discussions of Chapter 2, it is clear that *fatigue* is a very important potential failure mode for power transmission shafting.

Furthermore, excessive *misalignments* in gear meshes, bearings, cams, sprockets, or seals may lead to failure of these elements to function properly. Bending *deflections* or shaft *slopes* that lead to excessive misalignment may be said to induce failure by *force-induced elastic deformation* (see Chapter 2). If plain bearing journals, gear teeth, splines, or cam lobes are integral features of a shaft, *wear* must also be considered as a potential failure mode (see Chapter 2).

Because shafts are nearly always part of a dynamic system of interacting springs and masses, it is important to examine the possibility that operation at certain *critical speeds* may excite intolerable vibrations. If not adequately damped, vibration amplitudes may suddenly increase and may destroy the system.[1] The *response* of any vibrating system to an exciting force or motion is dependent upon the *springs*, *masses*, and *dampers* in the system, and how they are supported and coupled. Since a shaft may be regarded as a spring element, both in flexure and torsion (and sometimes axially), with attached mass elements (such as gears, pulleys, flywheels, and the mass of the shaft itself), and possible damping due to friction, windage, or lubrication, the accurate modeling of shaft vibration response may become a complex task. A designer is well advised to consult with a specialist in mechanical vibrations to help analyze complex vibrational systems. However, simple preliminary estimates of the *fundamental natural frequency*[2] of the system usually can be made, and then compared with the system *forcing frequency*[3] to make sure that *resonance*[4] *(or near resonance) is avoided*. The *critical speed* of a shaft is usually defined to be the lowest shaft speed that excites a resonant condition in the system. Higher critical speeds may also exist, but are usually less serious.

It is an important design responsibility to assure that the shaft stiffness is great enough to keep the fundamental natural frequency well above the forcing frequency (related to shaft speed) so that excessive deflections (force-induced elastic deformations) are not induced.

Summarizing, the primary potential failure modes to be considered when designing a power transmission shaft are:

1. Fatigue failure
2. Force-induced elastic deformation failure (either misalignments or critical speed resonances)
3. Wear failure (when bearing journals, gear teeth, or cams are integral shaft features)

8.3 Shaft Materials

Following the material selection guidelines discussed in Chapter 3, and recognizing the potential failure modes suggested in 8.2, candidate materials for power transmission shafting typically should have good strength (especially fatigue strength), high stiffness, low cost, and, in some applications, good wear resistance. Steel materials meet the strength, stiffness, and cost criteria. If case hardening or through hardening alloys are utilized, such steel materials can meet the wear resistance criterion as well.

Most power transmission shafting is made of low- or medium-carbon steel, either hot-rolled or cold-drawn. Materials such as AISI 1010, 1018, 1020, or 1035 steel are commonly chosen for shafting applications. If higher strength is required, low-alloy steels such

[1]Almost no details about the mechanical vibrations of spring-mass systems are presented in this book. For vibration details see, for example, refs. 1, 2, or 3.

[2]*Fundamental natural frequency* is the lowest frequency assumed by an undamped vibrating system when displaced from its equilibrium position and released to oscillate naturally.

[3]System *forcing frequency* is the input frequency to the system provided by the time-varying applied loads.

[4]Resonance is the condition existing when the *forcing frequency* corresponds to a *natural frequency*.

as AISI 4140, 4340, or 8640 may be selected, using appropriate heat treatment to achieve the desired properties. For forged shafting, such as for automotive crankshafts, 1040 or 1045 are commonly chosen steels. If case hardening of selected surfaces is necessary to achieve acceptable wear resistance, carburizing steels such as 1020, 4320, or 8620 may be used to support case-hardened surfaces at the required sites.

In applications where special design conditions or environments must be accommodated, the methods of Chapter 3 should be used to select an appropriate shaft material. If conditions such as corrosive environment or elevated temperature are present, for example, materials such as stainless steel, titanium, or Inconel might be required in spite of higher cost or greater difficulty in fabrication.

8.4 Design Equations—Strength Based

As suggested in 8.2, the state of stress at a selected critical point on the surface of a rotating power transmission shaft may involve torsional shearing stress, transverse shearing stress, bending stress, or axial stress components, any or all of which may be fluctuating about zero or nonzero mean values. In general, therefore, shaft design equations must be based on multiaxial states of stress produced by fluctuating loads, as discussed in Chapters 2, 4, and 6. The procedure is illustrated in detail for a relatively simple shaft loading case in Example 4.11 of Chapter 4. As a practical matter, many shaft design cases involve a reasonably simple state of stress, often characterized as a *steady torsional shearing stress component* produced by a steady operating torque, and a *completely reversed bending stress component* produced by steady transverse forces on the rotating shaft. This simplified characterization has led to the traditional shaft design equations found in most machine design textbooks, and also forms the basis for the current ANSI/ASME standard for the design of power transmission shafting.[5] Before using any of these equations or standards, a designer is responsible for making sure that the state of stress *actually produced* in any specific application is well approximated by the simplified case of steady torsional shear together with completely reversed cyclic bending stresses. Otherwise, the more general methods for analyzing multiaxial states of stress under fluctuating loads must be utilized.

For the case of steady torsion and completely reversed bending stress, the methods of Chapters 2, 4, and 6 may be simplified substantially. This is because the state of stress reduces to a *biaxial* case, and therefore there are only *two* nonzero principal stresses. Still, a solution for shaft diameter cannot be extracted explicitly, and an iterative solution is ultimately necessary. To *avoid* a trial-and-error solution, various simplifying assumptions have been proposed to facilitate making an explicit estimate of shaft diameter.[6] One approach is to utilize an appropriate *static* combined stress design equation (see 6.4), assuming that it can be used with *cyclic* stress components. Since most shafting materials are ductile, the appropriate selection for a combined stress design equation would usually be either the *maximum shearing stress design equation* (6-13), or the *distortion energy design equation* (6-14). The distortion energy design equation is considered more accurate, so (6-14) will be used for the following development.

Based on the definitions illustrated in Figure 2.12, and utilizing (4-5),

$$(\sigma_x)_{max} = (\sigma_x)_m + (\sigma_x)_a = \frac{(M_m + M_a)c}{I} \tag{8-1}$$

[5]*Design of Transmission Shafting*, ANSI/ASME B106.1M-1985, American Society of Mechanical Engineers, 345 East 47th St., New York, NY 10017.

[6]See, for example, refs. 3, 4, and 5.

For the case of completely reversed bending $M_m = 0$, and for ductile materials a fatigue stress concentration factor, K_{fb}, is appropriately applied to the *alternating* component of bending stress, so (8-1) becomes, for a cylindrical shaft,

$$(\sigma_x)_{max} = \frac{K_{fb} M_a c}{I} = \frac{32 K_{fb} M_a}{\pi d^3} \tag{8-2}$$

By similar reasoning, and utilizing (4-30),

$$(\tau_{xy})_{max} = (\tau_{xy})_m + (\tau_{xy})_a = \frac{(T_m + T_a)a}{J} \tag{8-3}$$

For the case of steady torsion $T_a = 0$, consequently no stress concentration factor need be considered and (8-3) becomes, for a cylindrical shaft,

$$(\tau_{xy})_{max} = \frac{T_m a}{J} = \frac{16 T_m}{\pi d^3} \tag{8-4}$$

Continuing with the case of steady torsion and completely reversed bending, the stress cubic equation (4-59) may be solved for principal stresses[7] by setting all stress components equal to zero except for τ_{xy} and σ_x. The resulting expressions for σ_1, σ_2, and σ_3 may then be substituted into (6-14) to give

$$\sqrt{\sigma_x^2 + 3\tau_{xy}^2} = \sigma_d \tag{8-5}$$

Next, combining (8-2), (8-4), and (8-5),

$$\sqrt{\left(\frac{32 K_{fb} M_a}{\pi d^3}\right)^2 + 3\left(\frac{16 T_m}{\pi d^3}\right)^2} = \sigma_d \tag{8-6}$$

The design stress, σ_d, based on fatigue strength S_N, is

$$\sigma_d = \frac{S_N}{n_d} \tag{8-7}$$

Combining (8-6) and (8-7), and solving for shaft diameter d,

$$d = \sqrt[3]{\frac{32 n_d}{\pi S_N}\sqrt{(K_{fb} M_a)^2 + 0.75 T_m^2}} \tag{8-8}$$

where d = required shaft diameter for a life of N cycles
n_d = selected design safety factor
K_{fb} = fatigue stress concentration factor for bending case
M_a = alternating bending moment ($M_a = M_{max}$ for completely reversed case)
T_m = steady torsional moment ($T_m = T_{max}$ for steady case)
S_N = fatigue strength for a design life of N cycles ($S_N = S_f$ for infinite life)

Reiterating, (8-8) gives a valid estimate for shaft diameter only for the case of steady torque and completely reversed bending, applied to a solid rotating shaft of circular cross section and made of ductile material. For more complicated multiaxial states of stress the methods of Chapters 2, 4, and 6 should be used in an iterative solution for shaft diameter. In such cases (8-8) is sometimes used to calculate initial values of shaft diameter to *start* the iteration. Since most shafting applications involve stepped shafts with many different diameters (see Figures 8.1 and 8.2, for example) it is usually necessary to calculate diam-

[7]See Example 4.11.

eters at many different critical points along the length of a shaft, even during preliminary design calculations.

Example 8.1 Designing a Shaft for Strength

A proposed shaft is to be supported on two bearings spaced 30 inches apart. A straight spur gear with a pitch diameter[8] of 20 inches is to be supported midway between bearings, and a straight spur pinion having a pitch diameter of 5 inches is to be supported 6 inches to the right of the right-hand bearing. The 20° involute gears are to transmit 150 horsepower at a rotational speed of 150 rpm. The proposed shaft material is hot-rolled 1020 steel with $S_u = 65{,}000$ psi, $S_{yp} = 43{,}000$ psi, $e = 36$ percent elongation in 2 inches, and fatigue properties as shown for 1020 steel in Figure 2.19. Shoulders for gears and bearings are to be a minimum of 0.125 inch (0.25 inch on the diameter). Design the shaft using a design safety factor $n_d = 2.0$.

Solution

The first step is to make a first-cut conceptual sketch of the new shaft. Following the guidelines illustrated in Figure 8.2, the conceptual sketch of Figure E8.1A is proposed. Next, a coordinate system is established, and a *stick-sketch* made of the shaft, gears, and bearings, so that all forces and moments on the shaft (and the attached gears) may be shown.

For the initial set of design calculations, cross sections of the shaft at sections A, B, C, and D of Figure E8.1B will be selected as *critical sections.*

The next step is to perform a force analysis to determine the magnitudes and directions for all forces and moments on the shaft. Because *straight spur gears* are used in this proposed system, no axial (y-direction) forces are developed, so only x- and z-force components will exist, as shown in Figure E8.1B.

From (4-32) the transmitted torque may be calculated as

$$T = \frac{63{,}025(hp)}{n} = \frac{63{,}025(150)}{150} = 63{,}025 \text{ in-lb} \qquad (1)$$

This torque is transmitted from the driver pinion at B, through the shaft, to the driven gear at D.

Forces may be calculated as follows: The tangential force F_{Bz} (force at B in the z-direction) at the pitch radius is

$$F_{Bz} = \frac{T}{\left(\frac{D_p}{2}\right)} = \frac{63{,}025}{10} = 6302 \text{ lb} \qquad (2)$$

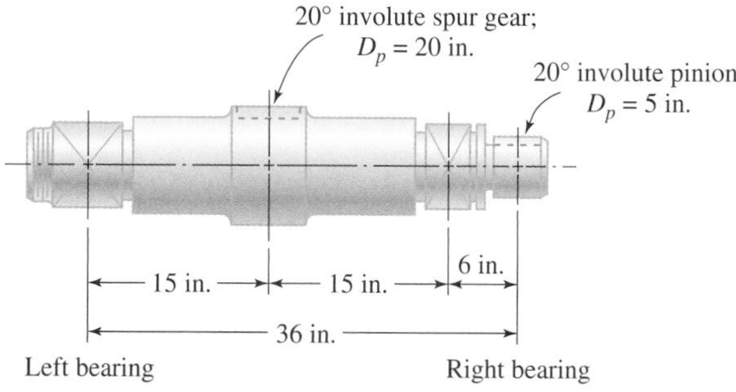

Figure E8.1A
First-cut conceptual sketch.

[8]See Chapter 15 for definitions and discussions relating to spur gear meshes.

Example 8.1 Continues

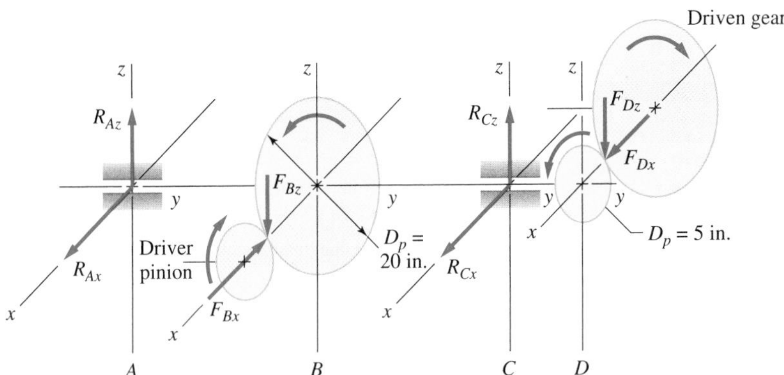

Figure E8.1B
Coordinate system and force notation.

Using (15-26) from Chapter 15, the radial separating force, F_{Bx}, for an involute tooth profile with a pressure angle $\varphi = 20°$, is given by

$$F_{Bx} = F_{Bz} \tan \varphi = 6302 \tan 20 = 2294 \text{ lb} \tag{3}$$

Similarly,

$$F_{Dz} = \frac{63{,}025}{2.5} = 25{,}210 \text{ lb} \tag{4}$$

and

$$F_{Dx} = 25{,}210 \tan 20 = 9176 \text{ lb} \tag{5}$$

Next, summing moments about the z-axis at A

$$15F_{Bx} - 30R_{Cx} - 36F_{Dx} = 0 \tag{6}$$

giving

$$R_{Cx} = \frac{15(2294) - 36(9176)}{30} = -9864 \text{ lb} \tag{7}$$

Summing forces in the x-direction

$$R_{Ax} + R_{Cx} + F_{Dx} - F_{Bx} = 0 \tag{8}$$

or

$$R_{Ax} = 2294 - 9176 + 9864 = 2982 \text{ lb} \tag{9}$$

Finally, summing moments about the x-axis at A gives

$$R_{Cz} = \frac{15(6302) + 36(25{,}210)}{30} = 33{,}403 \text{ lb} \tag{10}$$

and summing forces in the z-direction

$$R_{Az} = 6302 - 33{,}403 + 25{,}210 = -1891 \text{ lb} \tag{11}$$

Again stick-sketching a free-body diagram for the shaft and attached gears (based on Figure E8.1B), the actual force magnitudes, directions, and locations, based on the calculations just made, may be shown as indicated in Figure E8.1C. From Figure E8.1C, bending moments M and torques T may be calculated at each of the critical sections A, B, C, and D. Symmetry about the shaft axis allows vector addition of bending force components.

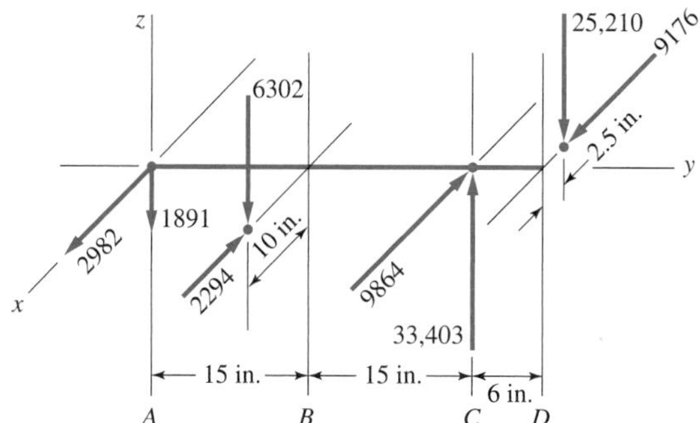

Figure E8.1C
Stick-sketch showing direction, magnitude, and point of application of forces on the proposed shaft.

Thus,

$$T_{A-B} = 0 \tag{12}$$

$$T_{B-D} = 63{,}025 \text{ in-lb} \tag{13}$$

$$M_A = 0 \tag{14}$$

$$M_B = 15\sqrt{(2982)^2 + (1891)^2} = 52{,}966 \text{ in-lb} \tag{15}$$

$$M_C = 6\sqrt{(25{,}210)^2 + (9176)^2} = 160{,}968 \text{ in-lb} \tag{16}$$

$$M_D = 0 \tag{17}$$

Applying shaft design equation (8-8) to each critical section in turn gives the following approximations for diameters at A, B, C, and D. Since diametral dimensions are still unknown, stress concentration factors cannot be accurately found from the charts of 4.8. However, values may be estimated for a first-cut calculation by examining Figures 4.17(a) and 4.18(a), since grooves and fillets are the shaft features of concern. Assuming that D/d values are close to unity, and r/d values tend to be small, an estimate of $K_t = 1.7$ seems appropriate. From Figure 4.27, for a steel with $S_u = 65{,}000$ psi, a q value around 0.8 seems reasonable.

Then, from (4-111), K_{fb} may be estimated as

$$K_{fb} = q(K_t - 1) + 1 = 0.8(1.7 - 1) + 1 \approx 1.6 \tag{18}$$

When shaft dimensions become better known, more accurate estimates of K_{fb} may be made.

At Critical Section *A*

Using information now known or estimated, at section A (8-8) becomes, for infinite life design ($S_N = S_f = 33{,}000$ psi),

$$d_A = \sqrt[3]{\frac{32(2)}{\pi(33{,}000)}\sqrt{[(1.6)(0)]^2 + 0.75[0]^2}} = 0 \tag{19}$$

Clearly this is not a reasonable result since the bearing at A must support the loads shown in Figure E8.1C, and a zero-diameter shaft cannot do that. Reexamining the assumptions leading to (8-8), transverse shear was assumed to be negligible; for critical section A the transverse shear is evidently *not* negligible, and must be introduced into (8-8). From case 2 of Figure 4.3,

$$\tau_{ts} = \frac{4}{3}\frac{F}{A} = \frac{16F}{3\pi d_A^2} \tag{20}$$

Substituting (20) back into (8-5) for τ_{xy} gives

$$\sqrt{3\left(\frac{16F}{3\pi d_A^2}\right)^2} = \sigma_d = \frac{S_N}{n_d} \tag{21}$$

Example 8.1 Continues

or

$$d_A = \sqrt{\frac{16\sqrt{3}n_d F}{3\pi S_N}} \tag{22}$$

where F is the resultant radial force at A. Thus, (22) gives

$$d_A = \sqrt{\frac{16\sqrt{3}(2.0)\sqrt{(2982)^2 + (1891)^2}}{3\pi(33{,}000)}} = 0.79 \text{ inch} \tag{23}$$

It is appropriate to again note that a designer is always responsible for making sure that the *assumptions used* in developing a design equation *are met* before using the equation. If the assumptions are not met, an appropriate modification must be made, or a new equation must be developed.

At Critical Section *B*

Using (8-8)

$$d_B = \sqrt[3]{\frac{32(2.0)}{\pi(33{,}000)}\sqrt{[(1.6)(52{,}966)]^2 + 0.75[63{,}025]^2}} = 3.96 \text{ inches} \tag{24}$$

At Critical Section *C*

Again using (8-8),

$$d_C = \sqrt[3]{\frac{32(2.0)}{\pi(33{,}000)}\sqrt{[(1.6)(160{,}968)]^2 + 0.75[63{,}025]^2}} = 5.46 \text{ inches} \tag{25}$$

At Critical Section *D*

First using (8-8), noting $M_D = 0$,

$$d_D = \sqrt[3]{\frac{32(2.0)}{\pi(33{,}000)}\sqrt{0 + 0.75[63{,}025]^2}} = 3.23 \text{ inches} \tag{26}$$

Since transverse shear is not included in (26), and since results at critical section A suggest that it may be significant, the expression of (20) may be introduced into (8-5) to combine the effects of torsion and transverse shear. Since $M_D = 0$,

$$\sqrt{3(\tau_m + \tau_{ts})^2} = \frac{S_N}{n_d} \tag{27}$$

or

$$\frac{16T_D}{\pi d_D^3} + \frac{16F_D d_D}{3\pi d_D^3} = \frac{S_N}{\sqrt{3}n_d} \tag{28}$$

or

$$d_D = \sqrt[3]{\frac{16\sqrt{3}n_d}{\pi S_N}\left(T_D + \frac{F_D d_D}{3}\right)} \tag{29}$$

or

$$d_D = \sqrt[3]{\frac{16\sqrt{3}(2.0)}{\pi(33{,}000)}\left(63{,}025 + \frac{\sqrt{(25{,}210)^2 + (9176)^2}d_D}{3}\right)} = \sqrt[3]{39.69 + 4.62d_D} \tag{30}$$

This implicit expression for d_D may be solved by iteration, using the result of (26) as an initial value, to obain

$$d_D = 3.86 \text{ inches} \tag{31}$$

In this case the estimate given by (26), which *does not* include transverse shear, results in a shaft diameter about 15 percent smaller than the estimate of (31), which *does* in-

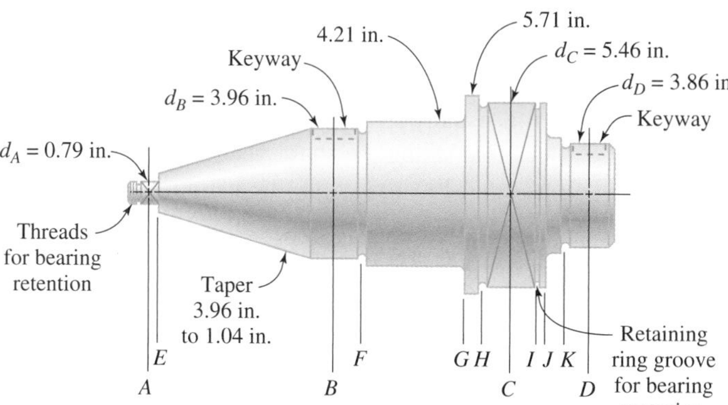

Figure E8.1D
Second-cut sketch of shaft showing principal dimensions (not to scale).

clude transverse shear. Again, a designer is responsible for deciding whether (8-8) is accurate enough for a first-cut calculation of shaft diameter at this critical section or whether a more refined estimate should be made.

Next, using the diameters just calculated at sections *A*, *B*, *C*, and *D*, and the shoulder restrictions from the problem statement, the first-cut sketch of Figure E8.1A may be updated and modified to obtain the second-cut sketch of Figure E8.1D.

With the updated sketch of Figure E8.1D, a new set of critical sections should be chosen (for example, *E, F, G, H, I, J, K*) for a second round of design calculations using the new shaft diameters, fillet radii, and groove dimensions from Figure E8.1D. Before starting this second round of design calculations, however, several other design issues should probably be addressed, including the following:

1. It is advisable to produce a scale drawing of the shaft sketch shown in Figure E8.1D. A scale drawing often helps a designer to "intuitively" identify potential trouble spots in the configuration. (Either CAD programs or manual drafting may be used at this stage.)
2. Tentative bearing selections should be made for sections *A* and *D*, using the calculated loads at those sites together with the specified shaft speed (see Chapters 11 and 12). In some cases, the required bearing size will force a significant change in shaft dimensions, overriding the shaft strength–based diameters. Such changes should be incorporated in the updated scale drawing of the shaft before second-round design calculations are made.
3. Tentative keyway dimensions should probably be established (see 8.8) so that more accurate stress concentration factors can be used in the second-round design calculations.
4. Tentative retaining ring selections should probably be made so that a more accurate stress concentration factor can be used at those critical points.
5. Potential pinion mounting problems at critical section *D* should be examined since a five-inch pitch-diameter pinion mounted on a d_D = 3.86-inch shaft leaves little radial space for a mounting hub on the pinion. If this can't be worked out, an *integral* pinion may be a viable option at this location. In turn, an integral pinion would probably require case hardening of tooth surfaces, and a review of shaft material.

The iterative nature of shaft design is evident in this example.

8.5 Design Equations—Deflection Based

As noted in 8.2, misalignments in gear meshes, bearings, cams, sprockets, seals, or other shaft-mounted elements may result in malfunction of these items due to nonuniform pressure distribution, interference, backlash, excessive wear, vibration, noise, or heat generation. Excessive *shaft bending deflection* or excessive *shaft slope* may therefore lead to failure of such shaft-mounted elements. Before continuing with the strength-based design refinements discussed in 8.4, it is usually advisable to estimate the shaft deflection (and/or slope) at deflection-critical locations along the shaft. Using the first-cut shaft dimensions based on strength calculations (e.g., Example 8.1), the bending deflection and slope calculations may be made (with whatever accuracy the designer desires). Rough approximations are often accurate enough at the early design stage, but more accurate calculations are usually required as the design is finalized.

A *first-cut* estimate of the bending deflection and slope of a *stepped shaft* may be made by first approximating the proposed stepped-shaft as a shaft of *uniform diameter* ("eye-balled" to be a little smaller than the average stepped-shaft diameter), then using or superposing the appropriate beam-bending equations (see Table 4.1) to find the deflections or slopes of interest.

If force components exist in two coordinate directions (as in x- and z-directions of Example 8.1), calculations must be made separately for the x- and the z-directions, and results combined vectorially. In some cases, it may be useful to make a somewhat more accurate estimate by assuming two or three diametral steps along the shaft.

Calculations of slopes and deflections for a stepped-shaft model are more complicated since both moment M and cross-sectional moment of inertia I change along the shaft length. These changing values require the use of graphical integration,[9] area-moment methodology,[10] numerical integration,[11] transfer matrix methods,[12] or, possibly, a finite element solution[13] (usually reserved until final design iterations are being made). Graphical and numerical integration methods are based on successive integration of the differential equation for the elastic beam-deflection curve given in (4-42). Thus, to find slope and deflection, the expressions

$$\frac{d^2y}{dx^2} = \frac{M}{EI} \tag{8-9}$$

$$\theta = \frac{dy}{dx} = \int \frac{M}{EI}\, dx \quad \text{(slope)} \tag{8-10}$$

$$y = \frac{d\theta}{dx} = \iint \frac{M}{EI}\, dx\, dx \quad \text{(deflection)} \tag{8-11}$$

are the working equations. The integrations may be performed either analytically, graphically, or numerically.

Torsional deflections may also be of interest in some applications. If a shaft has a uniform diameter over its whole length, the angular displacement θ may be readily calculated from (4-41), repeated here as

$$\theta = \frac{TL}{KG} \tag{8-12}$$

where θ is angular deflection in radians, T is applied torque, L is shaft length between torque application sites, G is shear modulus of elasticity, and K (see Table 4.4) is equal to the polar moment of inertia J for a circular cross section.

For a stepped shaft, the different-diameter shaft sections may be regarded as torsional springs in *series*. Hence, from (4-142) and (4-134), the angular deflection of a circular-sec-

[9]See, for example, ref. 6. [10]See, for example, ref. 8, pp. 552–556.

[11]See, for example, ref. 7, pp. 11.14–11.17, or ref. 9, pp. 347–352.

[12]See, for example, ref. 13. [13]See, for example, ref. 14, Ch. 7.

tion shaft with i steps, each of length L_i, having polar moments of inertia J_i, may be calculated from

$$\theta_{stepped} = \frac{T}{G}\left(\frac{L_1}{J_1} + \frac{L_2}{J_2} + \cdots + \frac{L_i}{J_i}\right) \tag{8-13}$$

Example 8.2 Estimating Shaft Deflection and Slope

From a strength-based analysis for a proposed steel shaft, the approximate dimensions have been determined as shown in Figure E8.2A. Estimate the maximum bending deflection of the shaft, and indicate approximately where along the shaft it occurs. Also estimate the shaft slope at each of the bearing sites.

Solution

A first estimate for maximum shaft deflection may be made by approximating the stepped shaft as a shaft of "equivalent" uniform diameter d_e, subjected to bending loads P_1 and P_2, and simply supported by bearings at sections A and F. Bending deflection may be calculated by using case 2 of Table 4.1 twice, once for load P_1 separately, and again for load P_2 separately, then superposing the results.

For P_1 acting alone, case 2 must be slightly reinterpreted so that $a > b$ (by inverting the shaft end-for-end to be analyzed). For this calculation then

$$\begin{aligned} a &= 28 \text{ in.} \\ b &= 8 \text{ in.} \\ L &= 36 \text{ in.} \end{aligned} \tag{1}$$

Selection of an "equivalent" diameter is easier with experience. The 4-inch diameter extends over nearly half the shaft length; therefore a reasonable guess for d_e may be about 3.5 inches. Using this value

$$I = \frac{\pi d_e^4}{64} = \frac{\pi(3.5)^4}{64} = 7.37 \text{ in}^4 \tag{2}$$

Then from case 2 of Table 4.1

$$y_{max\text{-}P_1} = \frac{(2000)(28)(8)(28 + 2\{8\})\sqrt{3(28)(28 + 2\{8\})}}{27(30 \times 10^6)(7.37)(36)} = 0.006 \text{ inch} \tag{3}$$

This deflection occurs at

$$x' = \sqrt{\frac{28(28 + 2\{8\})}{3}} = 20.26 \text{ inches (from right end of the inverted shaft)} \tag{4}$$

hence

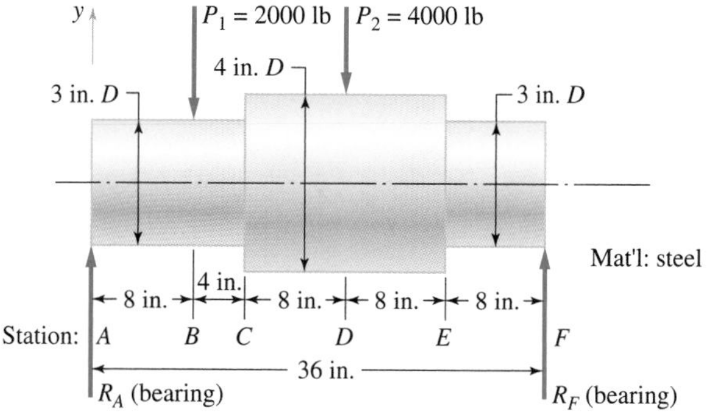

Figure E8.2A
Sketch of proposed shaft showing loads and nominal dimensions.

Example 8.2 Continues

$$x_{P_1} = 36 - 20.26 = 15.74 \text{ inches (from the left end of actual shaft)} \tag{5}$$

For P_2 acting alone, case 2 may be applied again with

$$\begin{aligned} a &= 20 \text{ in.} \\ b &= 16 \text{ in.} \\ L &= 36 \text{ in.} \end{aligned} \tag{6}$$

giving

$$y_{max\text{-}P_2} = \frac{(4000)(20)(16)(20 + 2\{16\})\sqrt{3(20)(20 + 2\{16\})}}{27(30 \times 10^6)(7.37)(36)} = 0.017 \text{ inch} \tag{7}$$

This deflection occurs at

$$x_{P_2} = \sqrt{\frac{20(20 + 2\{16\})}{3}} = 18.62 \text{ inches (from left end)} \tag{8}$$

Since x_{P_1} and x_{P_2} are rather close to each other, the maximum deflections for load P_1 acting alone and load P_2 acting alone will be superposed without further refinement. Thus

$$y_{max} = 0.006 + 0.017 = 0.023 \text{ inch (occurs around 17 inches from left bearing)} \tag{9}$$

Bearing *slopes* could also be estimated by superposing calculation results from P_1 and P_2, separately applied, to determine the slopes θ_A and θ_F. The calculation is deferred here in favor of a slightly more accurate estimate based on *graphical integration* of the stepped shaft, to be discussed next.

To implement the graphical integration approach, the three-stepped shaft of Figure E8.2A will be analyzed without dimensional modification. Equations (8-9), (8-10), and (8-11) will be written explicitly for this particular shaft, then successively integrated using the following graphical procedure.

1. Sketch the shaft *to scale*, showing all loads and dimensions, and display the arbitrarily selected drawing scale, k_z.
2. Calculate M, I, and M/EI at every point along the shaft where either moment M or moment of inertia I changes.
3. Plot to scale M/EI versus z along the shaft, using the already selected scale for k_z and arbitrarily selected scale for $k_{M/EI}$.
4. Graphically integrate the M/EI function plotted in step (3). To do this, perform the following steps:
   a. Divide the z-axis into a suitable (arbitrary) number of small z-increments, then project the z-midpoint M/EI ordinate from the curve to the M/EI axis, labeling the increment, and repeat for all increments.
   b. Establish a "pole" P at an arbitrarily selected distance along the z-axis to the *left* of the origin. Establishing a suitable scale, note the magnitude of the pole distance.
   c. Construct a *slope-ray* from the pole point P to each labeled ordinate projection described in (a).
   d. Construct, to scale, a new slope function, $\int M/EI\,dz$, by connecting end-to-end, in turn, line segments *parallel to the slope rays* drawn in step (c). The scale factor k_z remains the same as before, and the scale factor for the $\int M/EI$ axis may be calculated as

$$k_{\int M/EI} = P k_{M/EI} k_z \tag{10}$$

5. Graphically integrate the $\int M/EI$ function plotted in step (4), repeating, in sequence, steps (a) through (d) to obtain the $\int\int M/EI$ function. For this plot, k_z remains un-

changed, and for the $\int\int M/EI$ axis, the scale factor may be calculated as

$$k_{\int\int M/EI} = P k_{\int M/EI} k_z \tag{11}$$

6. Deflections at the bearing sites A and F are known to be zero. These boundary conditions are imposed by connecting points A and F on the $\int\int M/EI$ curve by a straight baseline. Actual shaft deflections at points along the z-axis are measured *vertically* from the baseline to the $\int\int M/EI$ curve, applying scale factor $k_{\int\int M/EI}$.
7. Using data from step (6), plot the shaft deflection y as a function of position z along the shaft axis.
8. Shaft slope may be estimated at any point as

$$\theta_i \approx \tan\theta_i = \frac{\Delta y_i}{\Delta z_i} \tag{12}$$

Using data from Figure E8.2A, summing moments at A gives

$$R_F = \frac{8(2000) + 20(4000)}{36} = 2667 \text{ lb} \tag{13}$$

and summing forces vertically

$$R_A = 2000 + 4000 - 2667 = 3333 \text{ lb} \tag{14}$$

With this information, calculations may be made at each of stations A through F for moment, area moment of inertia (two values at each shoulder), and the function M/EI (two values at each shoulder). Results of these calculations are shown in Table E8.2A.

Next, scale factors k_z and $k_{M/EI}$ are selected for plotting purposes as

$$k_z = 10 \tag{15}$$

$$k_{M/EI} = 10^{-4} \tag{16}$$

Pole distances are selected as $P = 1.0$ inch, for plotting both $\int M/EI$ and $\int\int M/EI$. Using (9) and (10)

$$k_{\int M/EI} = 1.0(10^{-4})(10) = 10^{-3} \tag{17}$$

and

$$k_{\int\int M/EI} = 1.0(10^{-3})(10) = 10^{-2} \tag{18}$$

With these data, the graphical integration steps (1) through (8), described above, may be accomplished as follows:

1. The stepped shaft is drawn to scale $k_z = 10$, as shown in Figure E8.2B(a), and loads are shown in their proper positions along the shaft.

TABLE E8.2A Tabulation of *M*, *I*, and *M/EI* at Each Station (for steel, $E = 30 \times 10^6$ psi)

Station	M, in-lb	I, in^4	M/EI, in^{-1}
A	0	3.98	0
B	26,664	3.98	2.23×10^{-4}
C_L	31,996	3.98	2.68×10^{-4}
C_R	31,996	12.57	0.85×10^{-4}
D	42,672	12.57	1.13×10^{-4}
E_L	21,336	12.57	0.57×10^{-4}
E_R	21,336	3.98	1.79×10^{-4}
F	0	3.98	0

Example 8.2 Continues

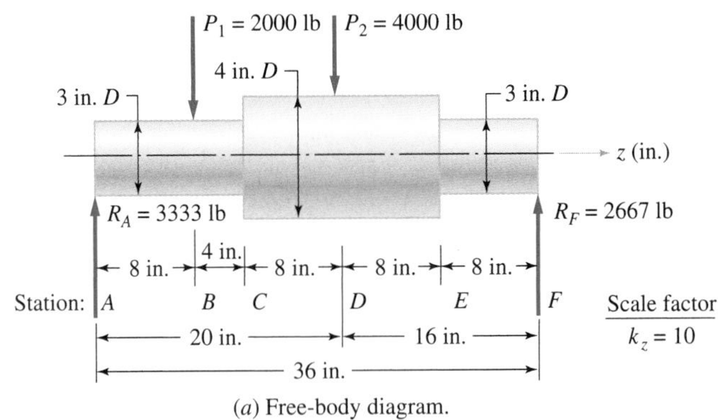

(a) Free-body diagram.

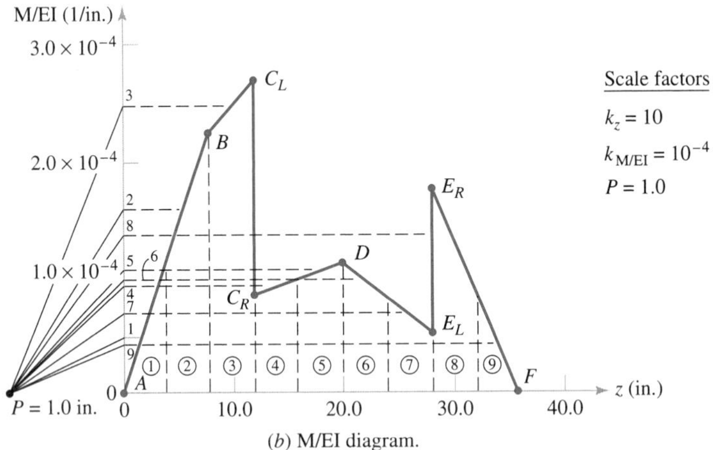

(b) M/EI diagram.

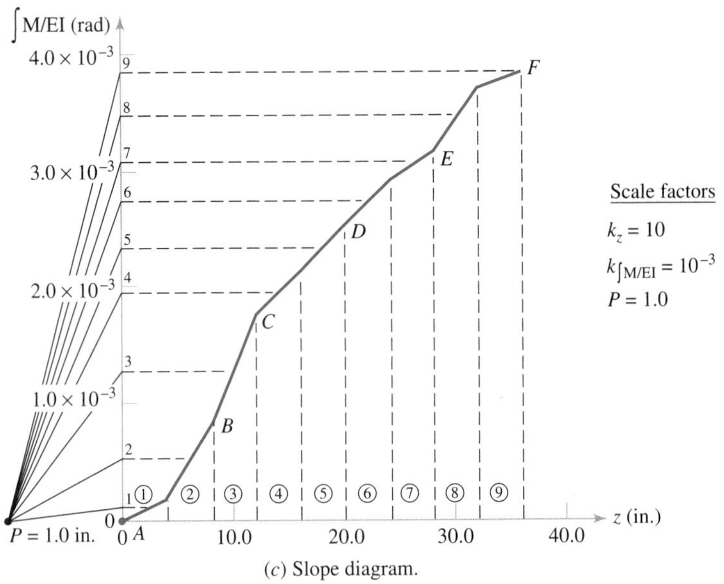

(c) Slope diagram.

Figure E8.2B

Determination of shaft deflection and slope by graphical integration.

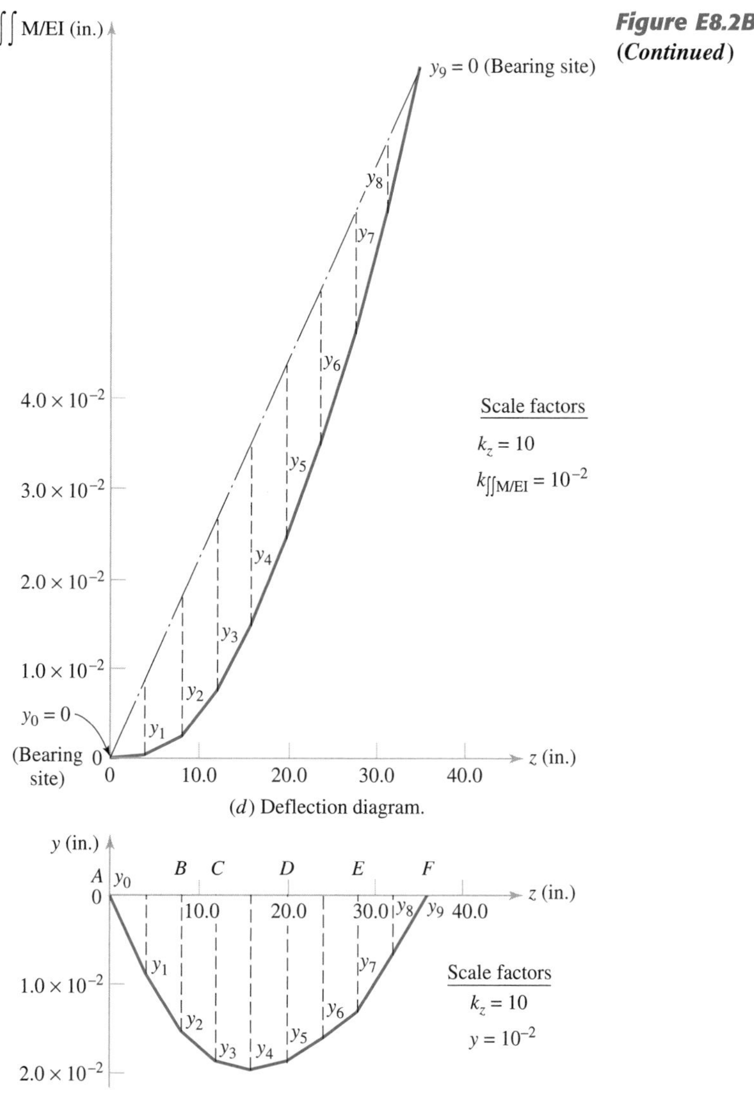

(d) Deflection diagram.

(e) Shaft deflection and slope.

Figure E8.2B **(*Continued*)**

2. Calculations for M/EI are summarized in Table E8.2A.
3. The M/EI function is plotted to scale $k_{M/EI} = 10^{-4}$, again using $k_z = 10$, in Figure E8.2B(b).
4. The M/EI function is graphically integrated to give the $\int M/EI$ curve plotted in Figure E8.2B(c), where $k_{\int M/EI} = 10^{-3}$ and $k_z = 10$.
5. The $\int M/EI$ function is graphically integrated to give the $\int\int M/EI$ curve plotted in Figure E8.2B(d), where $k_{\int\int M/EI} = 10^{-2}$ and $k_z = 10$.
6. In Figure E8.2B(d), the bearing sites are connected by a straight baseline (shaft centerline), and shaft deflections are found by measuring *vertically* between the baseline and the $\int\int M/EI$ curve at selected locations, then multiplying times the scale factor $k_{\int\int M/EI} = 10^{-2}$. Results are tabulated in Table E8.2B.
7. Using Table E8.2B, the shaft deflection curve is plotted in Figure E8.2B(e), using scale factors $k_y = 10^{-2}$ and $k_z = 10$. The maximum deflection is

Example 8.2 Continues

TABLE E8.2B Shaft Deflection at Ten Locations Along the Shaft

Location	Shaft Bending Deflection, in
0	0
1	0.009
2	0.015
3	0.018
4	0.019
5	0.018
6	0.015
7	0.014
8	0.007
9	0

$$y_{max} = 0.019 \text{ inch} \quad \text{(occurs around 16 inches from the left bearing)} \tag{19}$$

8. Shaft slopes may be calculated at bearing sites A and F using (12) to give (noting that the z-increment size is 4 inches),

$$|\theta_A| = \frac{y_1}{4} = \frac{0.009}{4} = 0.0023 \text{ rad} \tag{20}$$

and

$$|\theta_F| = \frac{y_8}{4} = \frac{0.007}{4} = 0.0018 \text{ rad} \tag{21}$$

Comparing the deflection estimates from (7) and (19), it may be noted that the straight-shaft approximation is reasonably close to the more accurate stepped-shaft analysis. The acceptability of a maximum shaft deflection of 0.019 inch must be evaluated in view of the specific application. One rule of thumb[14] suggests that shaft deflections generally should not exceed 0.001 inch per foot of shaft between bearing supports, which indicates that for *this* application, y_{max} should not exceed about 0.003 inch. A stiffer shaft might therefore be required here. Likewise, the shaft *slopes* at the bearing sites, given in (20) and (21), must be evaluated against allowable angular misalignments for the specific bearings chosen. For example, deep-groove ball bearings generally operate well up to about $\pm 0.25°$ angular misalignment, so the estimated shaft slopes should be acceptable for this application if deep-groove ball bearings are used.

8.6 Shaft Vibration and Critical Speed

As discussed in 8.2, rotating shafts are nearly always part of a dynamic system of springs and masses. Because of imperfections in the manufacture and assembly of shafts and mounted components, as well as static deflections caused by the weights of these elements, the center of mass does not coincide *exactly* with the center of rotation of the system. Consequently, as the rotational speed increases, the rotating centrifugal forces on the *eccentric*

[14]See ref. 6.

mass centers increase. In turn, the "whirling" eccentricities also increase, producing even higher rotating centrifugal forces. If the actual rotational shaft speed approaches any of the *critical speeds* of the system (see 8.2), a violent bending-mode vibration may result, and it may destroy the system. This *self-excited* vibrational phenomenon is called *shaft whirl*. The lowest (fundamental) critical speed for a simply supported shaft with i mounted masses may be estimated from the following equation based on Rayleigh's Energy Method[15]

$$n_{cr} = \frac{60}{2\pi}\sqrt{g\frac{\sum m_i y_i}{\sum m_i y_i^2}} \tag{8-14}$$

or

$$n_{cr} = 187.7\sqrt{\frac{\sum W_i y_i}{\sum W_i y_i^2}} \tag{8-15}$$

where n_{cr} = critical shaft speed, rpm
W_i = weight of ith mass, lb
y_i = static displacement of ith mass center from centerline of rotation, inch

Operating a rotating shaft at or near the critical speed(s) must be avoided. If possible, the operating shaft speed should be kept *safely below* the lowest (fundamental) critical speed (typically by a factor of 2 or 3). If, for some reason, the system cannot be operated below the lowest critical speed, it may be possible to accelerate *quickly through resonance* to avoid the destructive self-excited vibration at the fundamental critical speed. Operation at speeds above the lowest critical speed (but below higher resonant speeds) may then be satisfactory, but each shutdown and startup cycle must be carefully monitored to avoid damage as the system passes through resonance. A designer is well advised to consult with a specialist in vibration and modal analysis to avoid problems in these areas.

In some systems, torsional vibration of the shaft and its associated components may be excited by fluctuating torques on the rotating system. In principle, estimating the fundamental torsional resonant frequency for a simple system is not difficult, but most shafting systems involve many *torsional mass moments of inertia* and many *springs*. The spring rates and masses of stepped shafts, gear meshes, and bearings, in addition to other system components, must be properly modeled to produce a reasonable estimate of critical torsional frequencies. It is advisable to consult a vibration specialist when making these estimates.

Example 8.3 Shaft Critical Speed

The stepped shaft of Example 8.2 rotates at 1200 rpm. Using the results from Example 8.2, estimate the fundamental critical speed for the three-step shaft shown in Figure E8.2A, and comment on its acceptability.

Solution

The critical frequency of this shaft may be estimated from (8-15). To use the equation, it is necessary to calculate the weight of each stepped segment of the shaft, and find the static deflection of the shaft at the mass center of each segment. From Table 3.4 the specific weight of steel is 0.283 lb/in^3. Static deflections at the mass centers of the three shaft segments may be read from Figure E8.2B(e) or interpolated from Table E8.2B. These data are summarized in Table E8.3.

[15]See, for example, ref. 1.

Example 8.3 Continues

TABLE E8.3 Weight and Deflection Estimates for Three Stepped Shaft Segments

Item	Equation or Source	Segment A-C	Segment C-E	Segment E-F
Volume	$V_i = \pi D^2 L/4$, in^3	84.82	201.01	56.55
Weight	$W_i = 0.283 V_i$, lb	24.00	56.90	16.00
Location z_i of mass center from station A	Figure 8.2A, in	6.0	20.0	32.0
Deflection y_i of mass center	Table E8.2B, in	0.012	0.018	0.007

Entering the data from Table E8.3 into (8-15) gives the critical speed for this shaft as

$$n_{cr} = 187.7\sqrt{\frac{24(0.012) + 56.90(0.018) + 16(0.007)}{24(0.012)^2 + 56.90(0.018)^2 + 16(0.007^2)}} = 1488 \text{ rpm} \tag{1}$$

Comparing the critical shaft speed of 1488 rpm with the proposed operating speed of 1200 rpm raises a serious question about shaft stiffness since the ratio of critical to operating speed is

$$\frac{n_{cr}}{n_{op}} = \frac{1488}{1200} = 1.24 \tag{2}$$

instead of the recommended minimum of 2 or 3. Further, the stiffnesses of support bearings, usually an important consideration, have not been included in the estimate, nor have the masses of any of the attached components. A vibration specialist should be consulted in this case before proceeding with more refined stress analyses because the shaft size will probably need be increased to avoid problems with shaft whirl or lateral vibration.

8.7 Summary of Suggested Shaft Design Procedure; General Guidelines for Shaft Design

Based on the discussions of 8.1 through 8.6, the procedural steps in developing a successful shaft design may be summarized as follows:

1. Based on functional specifications and contemplated system configuration, generate a first conceptual sketch for the shaft (see Figure 8.2).
2. Identify potential failure modes (see 8.2).
3. Select a tentative shaft material (see 8.3).
4. Based on loads and geometry, determine the state of stress at each potential critical point (see Chapter 4 and 8.4).
5. Select an appropriate design safety factor (see Chapter 5).
6. Calculate a tentative shaft diameter for each critical section along the length of the shaft, based on stress, strength, and failure prevention considerations (see 8.4).
7. Check deflections and slopes of the tentative strength-based shaft design, to assure proper function of shaft-mounted components such as gears and bearings (see 8.5).
8. Check critical speed and vibrational characteristics to assure that failure does not occur due to shaft whirl, resonant vibration, excessive noise, or heat generation.

As a new shaft is designed by following these steps, the following experience-based guidelines may also be useful:

1. Try to make the shaft as short, stiff, and light as possible. This may imply hollow shafting in some cases and the use of high-modulus-of-elasticity materials such as steel.
2. Try to use a straddle-mounted bearing support configuration if possible (simply supported beam concept). If it is necessary to design a shaft that overhangs the bearing support (cantilever beam concept), minimize the length of the overhang.
3. Try to place shaft support bearings close to the lateral bending loads on the shaft.
4. Try to avoid the use of more than two support bearings for the shaft. Three or more bearings require high-precision alignment to avoid potentially large "built-in" displacement-induced bending moments on the shaft.
5. Try to configure the shaft so that stress concentration sites do not coincide with high *nominal* stress regions. The use of generous radii for fillets and relief grooves, smooth surfaces, and, in critical applications, the use of shot-peening or cold-rolling may be appropriate.
6. Consider potential failure problems due to excessive deflection or shaft slope at mounting sites for gears, bearings, sprockets, cams, and shaft seals. Typically, gear mesh separation should not exceed 0.005 inch and relative slope at the mesh site should not exceed 0.03 degree. For rolling element bearings the slope should be less than 0.04 degree unless the bearings are self-aligning. For plain bearings the shaft slope should be small enough that end-to-end radial displacement of the journal relative to the sleeve is small compared to lubricant film thickness. Specific manufacturing guidelines should be followed where available.

8.8 Couplings, Keys, and Splines

Power transmission to or from a rotating shaft is accomplished either by (1) *coupling* the rotating shaft end-to-end with a power source (such as an electric motor or internal combustion engine), or end-to-end with the input shaft of a power-dissipating system (such as a machine tool or automobile), or (2) by *attaching power input or output components* (such as pulleys, sprockets, or gears) to the shaft using keys, splines, or other retention devices.

The basic design requirements for both couplings and component retention devices are the same; rated shaft torque must be transmitted without slip, and premature failure must not be induced in any part of the operating machine. In the case of couplings, it may be necessary to accommodate shaft misalignments[16] to prevent premature failures.

Rigid Couplings

Mechanical couplings used to connect rotating shafts are typically divided into two broad categories: *rigid couplings* and *flexible couplings*. Rigid couplings are simple, inexpensive, and relatively easy to design, but require precise colinear alignment of the mating shafts. Also, the shafts must be well supported by bearings located near the coupling. Another advantage of a rigid coupling is that it provides high stiffness across the joint. This results in *small* relative torsional displacements, so the phase relationship between the power source and driven machine can be accurately preserved if necessary. Rigid couplings also provide higher *torsional critical speeds* for the system.

[16]Parallel centerline offset, nonparallel (angular) centerlines, or both may contribute to shaft misalignments.

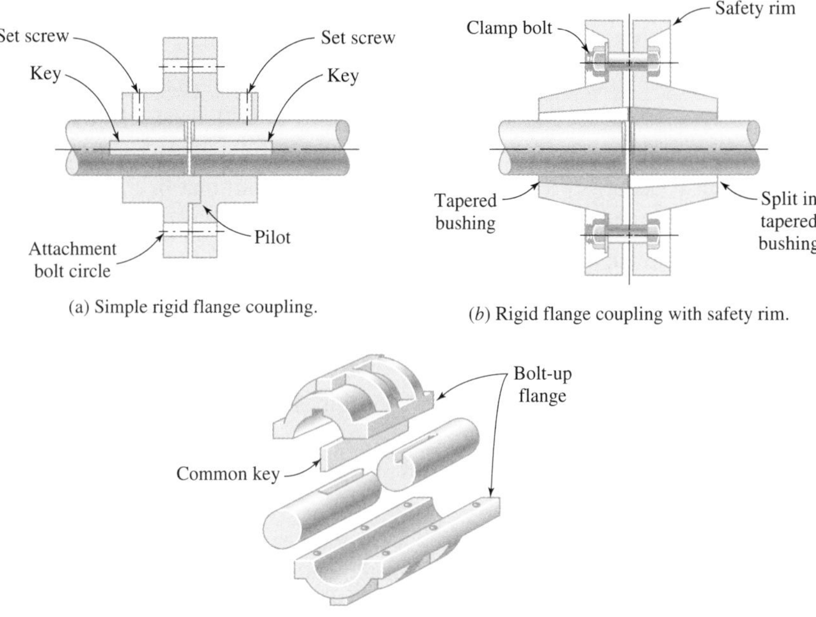

(a) Simple rigid flange coupling.

(*b*) Rigid flange coupling with safety rim.

(*c*) Rigid ribbed coupling.

Figure 8.3
Three types of rigid couplings for shafts.

The major disadvantage of installing a rigid coupling is that when the flange bolts are tightened, any misalignment between the two shafts has the potential to cause large forces and bending moments that may overload the coupling, the shafts, the bearings, or the support housing.

The typical geometry of a rigid coupling involves two similar halves, each with a hub to accommodate attachment to its respective shaft, a piloted boss for precise alignment, and attachment bolts on a bolt circle to lock the two halves together. Figure 8.3(a) illustrates the geometry of a simple, flange-type, rigid coupling. An outer protective rim is often added to the flanges to provide a safety shield for the bolt heads, as illustrated in Figure 8.3(b). The protective rim also provides a means of dynamic balancing after assembly (by drilling or removing small slugs of material to restore balance).

If transmitted torque is steady, and no vibration is induced, the probable governing failure mode for a rigid coupling is *yielding*. If torque fluctuates, or vibration is present, or if there is significant misalignment in the rotating shaft system, *fatigue* and *fretting fatigue* become the probable governing failure modes.

Attachment of the rigid coupling hubs to the shafts may be accomplished by using keys, taper-lock sleeves, or interference fits (requires pressing at assembly). Figures 8.3(a) and 8.3(c) illustrate the use of keys. Figure 8.3(c) shows a "ribbed" coupling in which one long common key is held in place along both shafts by a "coupler housing" bolted in place on the rotating shafts. Taper-lock sleeves are illustrated in the sketch of a compression-type rigid coupling shown in Figure 8.3(b). Each axially-split tapered sleeve is wedged against its shaft by tightening the flange bolts, in turn providing a *frictional* drive torque capability. Such couplings may be installed or removed easily but are limited to low or moderate torque applications.

Design of a rigid coupling such as the one shown in Figure 8.3(a) typically involves an investigation of the following potentially critical areas:

1. Shear and bearing in the key
2. Shear and bearing in the flange attachment bolts, including the influence of flange bolt *preloading* and/or bolt *bending*, if applicable

3. Bearing on the flange at attachment bolt interfaces
4. Shear in the flange at the hub

Flexible Couplings

In coupling applications, misalignment is the rule rather than the exception. To accommodate small misalignments betwen two shafts, *flexible* couplings are usually chosen. Since a wide variety of flexible couplings are commercially available, a designer typically *selects* a suitable coupling for the application by referring to manufacturer's catalogs, rather than designing one from scratch. A few of the many types of commercially available flexible couplings are illustrated in Figure 8.4. The couplings shown may be grouped into three basic categories, according to the way in which shaft misalignments are accommodated. These are:

1. Misalignment accommodated by an *interposed solid element* that slides, or introduces small clearances (backlash) beween shafts
2. Misalignment accommodated by one or more *interposed flexible metallic elements*
3. Misalignment accommodated by an *interposed flexible elastomeric element*

Sketches (a), (b), and (c) of Figure 8.4 represent examples of the first category. In the *sliding disk coupling* of Figure 8.4(a), two facing slotted flanges are coupled by an intermediate disk with mating cross-keys having enough clearance to allow sliding motion between the disk and the flanges. Such couplings, intended for low-speed high-torque drives, typically accommodate angular misalignments up to about $^1/_2$ degree, and shaft-centerline parallel misalignments up to $^1/_4$ inch. *Fretting fatigue* and *fretting wear* are potential failure modes.

The *gear coupling* of Figure 8.4(b) is probably the most widely used type of shaft coupling. It consists of two mounting hubs with external gear teeth that engage internal teeth in a captured sleeve that fits over both hubs. Curved teeth are often used to accommodate larger angular misalignments. Backlash in the engaged gears typically allows up to about one degree of angular misalignment for *straight* teeth, and up to three degrees if the hub teeth are *curved*. However, good shaft-centerline alignment is required.

Figure 8.4(c) illustrates a *roller chain coupling*, in which sprockets are attached at adjacent ends of the two abutting shafts, then wrapped together by a common roller chain segment that spans both sprockets. Clearance between the chain and the sprockets allows up to $1^1/_2$ degrees of angular shaft-centerline misalignment and up to about 0.010 inch parallel shaft-centerline misalignment. Roller chain couplings are low-cost, high-torque devices, but may be noisy. *Wear* or *fretting wear* are potential failure modes.

Figures 8.4(d), (e), and (f) illustrate three examples of flexible couplings where misalignments are accommodated by metallic flexure elements. *Spring couplings* typically accommodate angular misalignments up to about four degrees, and parallel shaft misalignments of up to about $^1/_8$ inch. *Flexible disk couplings* allow about one degree of angular misalignment, and around $^1/_{16}$ inch of parallel shaft misalignment. A *bellows coupling* may allow up to about nine degrees of angular misalignment, and $^1/_4$ inch of parallel shaft misalignment, but is typically limited to low-torque applications. *Fatigue* is the probable governing failure mode for all couplings in the flexible metallic element category.

Figures 8.4(g), (h), and (i) are examples of the third category of flexible couplings, in which an elastomeric element in compression, bending, or shear provides the means of accommodating misalignment. In Figure 8.4(g), two shaft-mounted flanges, each with concentric internal and external teeth, are coupled by a mating *elastomeric sleeve*. This arrangement gives high torsional flexibility, and tends to attenuate shock and vibration. Angular misalignment up to about one degree can be tolerated as well. The *bonded elastomeric disk coupling* of Figure 8.4(h) places the elastomeric disk in shear, and is typically limited

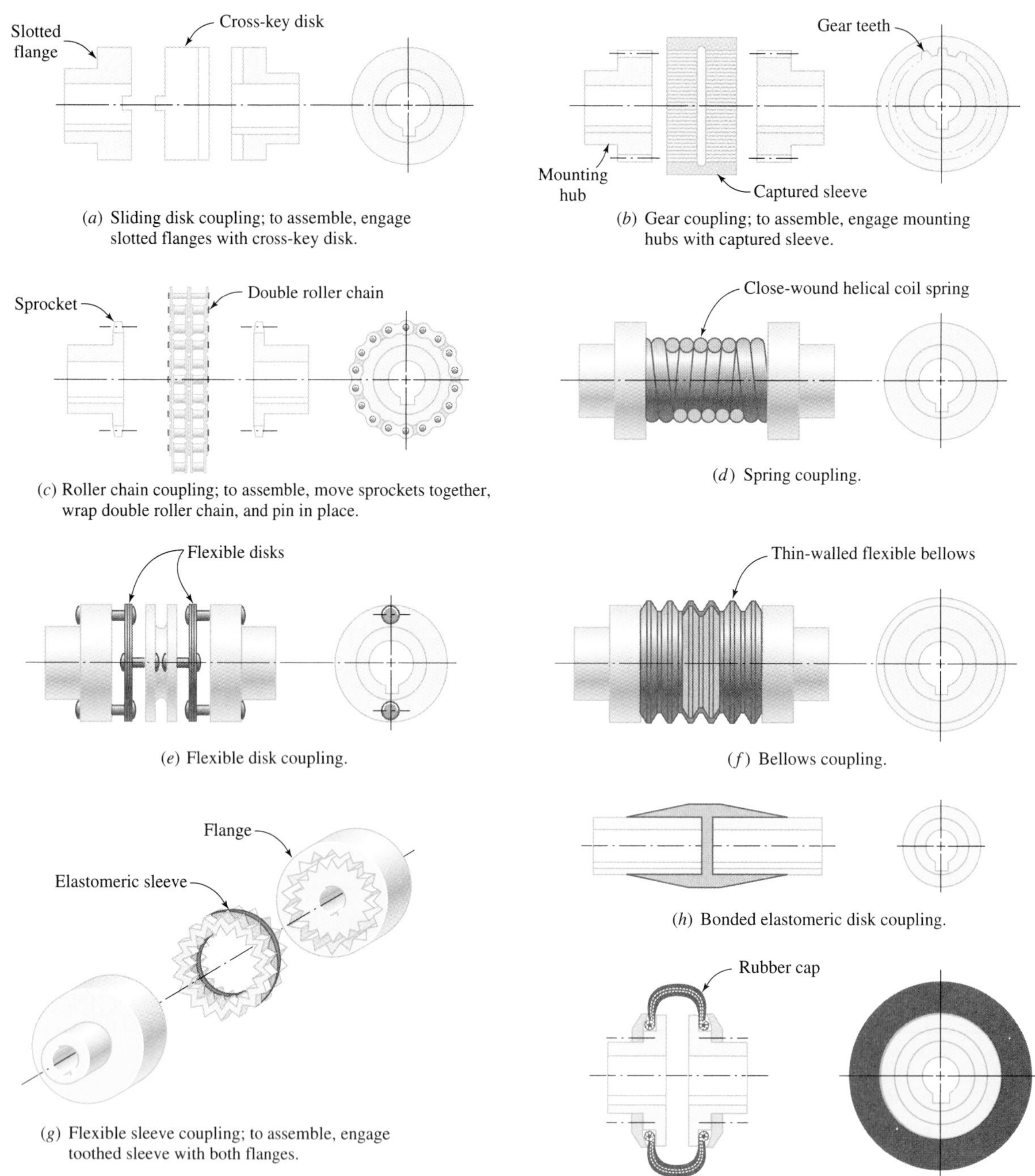

(*a*) Sliding disk coupling; to assemble, engage slotted flanges with cross-key disk.

(*b*) Gear coupling; to assemble, engage mounting hubs with captured sleeve.

(*c*) Roller chain coupling; to assemble, move sprockets together, wrap double roller chain, and pin in place.

(*d*) Spring coupling.

(*e*) Flexible disk coupling.

(*f*) Bellows coupling.

(*g*) Flexible sleeve coupling; to assemble, engage toothed sleeve with both flanges.

(*h*) Bonded elastomeric disk coupling.

(*i*) Rubber cap coupling.

Figure 8.4
Conceptual sketches of various types of flexible couplings for shafts.

to low-torque applications. The *rubber cap coupling* of Figure 8.4(i) captures the rubber element compressively in each flange cavity, and transmits the torque by shear of the elastomeric coupling element. Such couplings are available in high-torque capacities, and may accommodate up to one degree of angular misalignment and up to $^1/_4$-inch parallel shaft misalignment. The probable failure mode for elastomeric element couplings is *fatigue*.

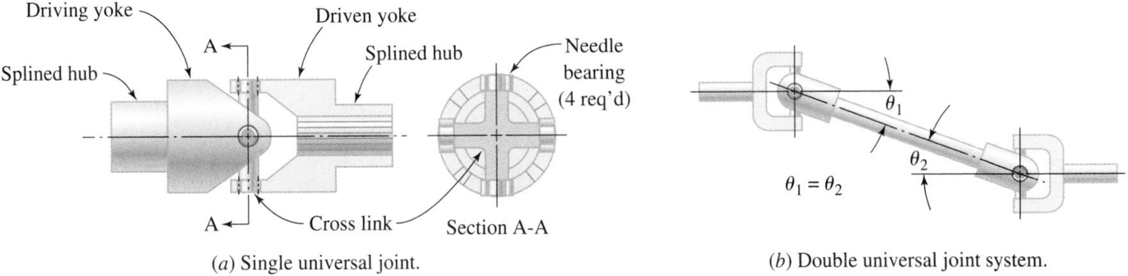

(*a*) Single universal joint.

(*b*) Double universal joint system.

Figure 8.5
Simple universal joint arrangements.

If power transmission shafts must intersect at angles larger than can be accommodated by flexible couplings (just described), *universal joints* (*U-joints*) may sometimes be used to couple the shafts. Figure 8.5(a) illustrates a single universal joint and 8.5(b) a double universal joint configuration. Typical applications for U-joints include automotive drive shafts, articulating farm tractors, industrial rolling mill drives, and mechanical control mechanisms, among others. The *single universal joint* configuration embodies a splined-hub *driving* yoke, a splined-hub *driven* yoke, and a cross-link (sometimes called a *spider*) connecting the two yokes through pivot bearings (usually needle bearings), as shown in Figure 8.5(a). Angular misalignments up to about 15 degrees between shaft centerlines are readily accommodated, and up to 30 degrees in special circumstances. One important observation is that the angular velocity ratio between the input shaft and output shaft of a simple U-joint is not constant, raising the potential for torsional vibration problems in the system. If a *double U-joint* system is used, such as the one shown in Figure 8.5(b), variations in angular velocity ratio are very slight because variables in the second joint tend to compensate for those in the first joint. Shaft-centerline *offset* can also be accommodated by the double U-joint arrangement. *Constant velocity ratio* universal joints (CV joints) have been developed and are commercially available, but are more expensive. Potential governing failure modes for universal joints include *fretting wear* in the needle bearings that connect the yokes to the cross-link (because of small-amplitude oscillatory motion) and *fretting fatigue* in the splined connections between each yoke and its mating shaft.

Keys, Splines, and Tapered Fits

When power is to be transmitted to, or supplied from, a rotating shaft, it is necessary to attach components such as pulleys, sprockets, or gears to the shaft. To prevent relative rotation between the shaft and an attached component, the connection between the component's hub and the shaft must be secured. Retention devices such as keys, splines, or tapered fits are commonly used to prevent such relative rotation. For lighter-duty applications, pins or setscrews may be used, sometimes in conjunction with retaining rings to provide axial constraint of the shaft-mounted component. Of the keys shown in Figure 8.6, the parallel *square key* and the *Woodruff key* are probably used more widely than the others. Recommendations for key size and keyway depth, as a function of shaft diameter, are provided by ASME/ANSI Standards B17.1-1967 and B17.2-1967, as illustrated in abridged form in Tables 8.1 and 8.2. For parallel keys it is common practice to use setscrews to prevent backlash between the key and keyway, especially if fluctuating torques occur during operation. Recommended setscrew sizes are included in Table 8.1. In some applications *two* setscrews are used, one bearing *directly on the key*, and a second one located at 90° *away from the keyway*, where it bears directly on the *shaft* (usually where a local shallow *flat* has been machined).

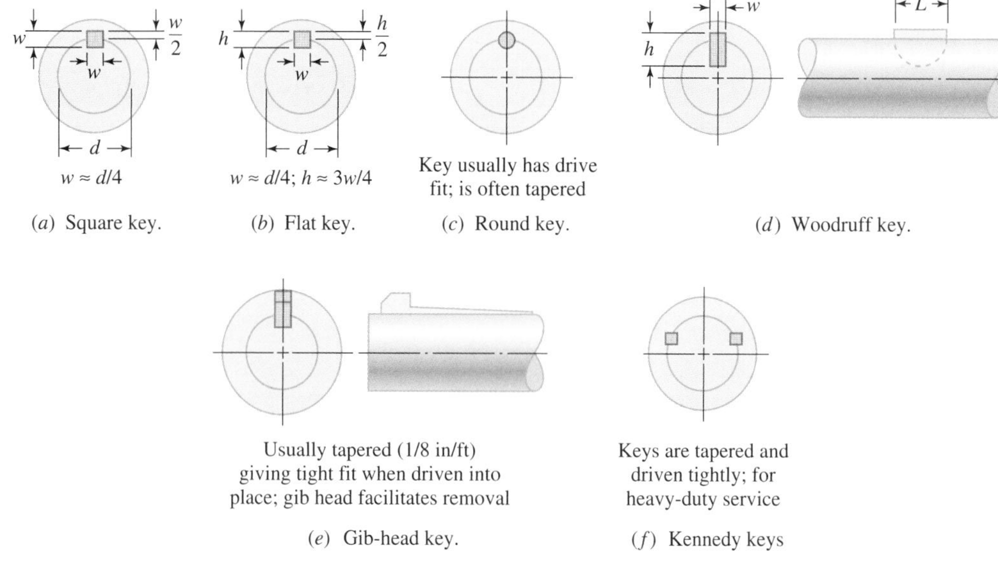

Figure 8.6
Various types of shaft keys.

The potential failure modes for keyed connections include *yielding* or *ductile rupture*, or, if fluctuating loads or torques are involved, *fatigue* or *fretting fatigue* of the key or the shaft near the end of the keyway. Parallel keyways machined in the shaft have either a "sled runner" geometry or a profiled geometry, as shown in Figure 8.7.

Keyways in hubs are machined all the way through the hub, usually by a broaching operation. Stress concentration factors for standard shaft keyways, when the shaft is subjected to *bending*, are about $K_{tb} \approx 1.8$ for a profiled keyway and $K_{tb} \approx 1.4$ for a sled run-

TABLE 8.1 Recommendations for Selection of Standard Square Parallel Keys (abridged from ANSI Standard B17.1-1967)

Shaft Diameter Range, in	Nominal Key Size, in	Nominal Set Screw Size, in
$^5/_{16}$–$^7/_{16}$	$^3/_{32}$	#10–32
$^1/_2$–$^9/_{16}$	$^1/_8$	$^1/_4$–20
$^5/_8$–$^7/_8$	$^3/_{16}$	$^5/_{16}$–18
$^{15}/_{16}$–$1^1/_4$	$^1/_4$	$^3/_8$–16
$1^5/_{16}$–$1^3/_8$	$^5/_{16}$	$^7/_{16}$–14
$1^7/_{16}$–$1^3/_4$	$^3/_8$	$^1/_2$–13
$1^{13}/_{16}$–$2^1/_4$	$^1/_2$	$^9/_{16}$–12
$2^5/_{16}$–$2^3/_4$	$^5/_8$	$^5/_8$–11
$2^{13}/_{16}$–$3^1/_4$	$^3/_4$	$^3/_4$–10
$3^5/_{16}$–$3^3/_4$	$^7/_8$	$^7/_8$–9
$3^{13}/_{16}$–$4^1/_2$	1	1–8
$4^9/_{16}$–$5^1/_2$	$1^1/_4$	$1^1/_8$–7
$5^9/_{16}$–$6^1/_2$	$1^1/_2$	$1^1/_4$–6

TABLE 8.2 Recommendations for Selection of Wodruff Keys (abridged from ANSI Standard B17.2-1967)

Shaft Diameter Range, in	Key Number[1]	Nominal Key Size $w \times L$, in	Nominal Key Height h, in
$^{7}/_{16}$–$^{1}/_{2}$	305	$^{3}/_{32} \times {}^{5}/_{8}$	0.250
$^{11}/_{16}$–$^{3}/_{4}$	405	$^{1}/_{8} \times {}^{5}/_{8}$	0.250
$^{13}/_{16}$–$^{15}/_{16}$	506	$^{5}/_{32} \times {}^{3}/_{4}$	0.312
1–$1^{3}/_{16}$	608	$^{3}/_{16} \times 1$	0.437
$1^{1}/_{4}$–$1^{3}/_{4}$	809	$^{1}/_{4} \times 1^{1}/_{8}$	0.484
$1^{7}/_{8}$–$2^{1}/_{2}$	1212	$^{3}/_{8} \times 1^{1}/_{2}$	0.641

[1]Last two digits give nominal key diameter in eighths. Digits preceding the last two give nominal width in thirty-seconds.

ner keyway.[17] If the shaft is subjected to *torsion*, the torsional stress concentration factor is usually around $K_{tT} = 1.7$ for both profiled and sled runner keyways. Stress concentration factors for Woodruff keyways in the shaft are similar to those for the sled runner keyway.

It is often desirable to "size" a key so that it will shear off by ductile rupture in the event an overload occurs in or on the machine. Using an inexpensive shaft key as a "mechanical safety fuse" in this way protects the shaft and other more expensive machine elements from damage. Typically, therefore, the selected key material is soft, ductile, low carbon steel, cold-rolled to standard keystock sizes, and custom cut to an appropriate length. Design of the key for such an application must assure that operational torques will be transmitted without failure, while torques produced by overload conditions, such as jamming or seizure of a system component, *do* cause the key to shear off across the shaft-hub interface.

A keyed connection between a shaft and a hub is illustrated in Figure 8.8. For the case shown the key width is w, its radial height is h, and its length is l. The torque-induced force F_s is trasmitted from the shaft through the key to the hub, where it is reacted by force F_h. Because the distance $h/2$ between force vectors F_s and F_h is small compared to radius $D/2$ for the shaft, it may be assumed that

$$F_s = F_h \equiv F \tag{8-16}$$

It may also be assumed that forces F act at the mean radius, $D/2$.

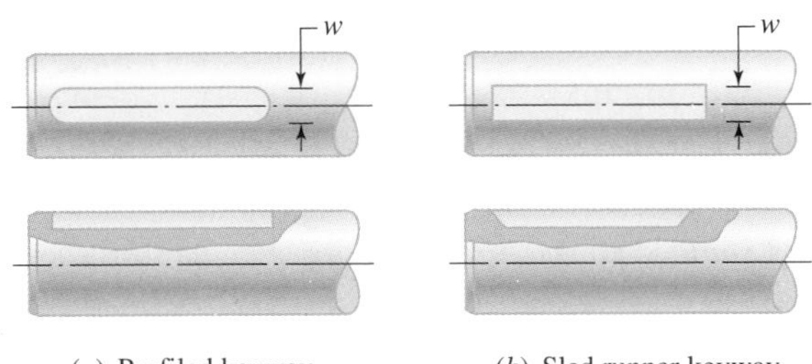

Figure 8.7
Keyway geometry. Keyways may be embedded in the shaft (as shown) to capture the key, or may run out at the end of the shaft.

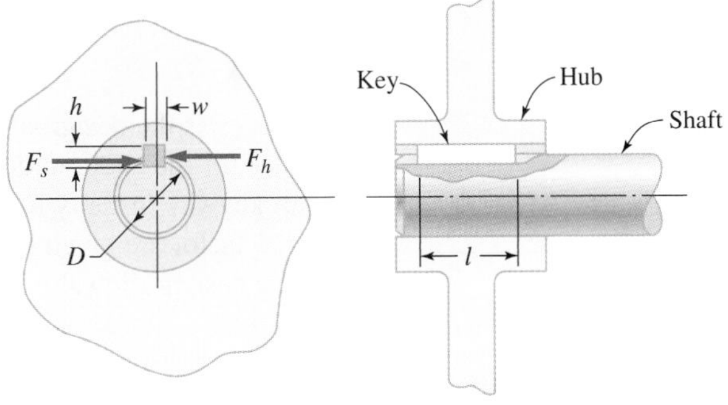

Figure 8.8
Sketch of a keyed connection between a shaft and a hub.

[17]See ref. 10, p. 245 ff.

For the key, potential critical sections include (1) the shear plane between shaft and hub at radius $D/2$, and (2) the contact planes between the sides of the key and the sides of the keyway. If the torque to be transmitted from shaft to hub is defined to be T, the torque-induced force F may be calculated as

$$F = \frac{T}{D/2} = \frac{2T}{D} \tag{8-17}$$

The area of the shearing plane of the key is

$$A_s = wl \tag{8-18}$$

so, the *average* shearing stress across the shearing plane may be written as

$$\tau_s = \frac{F}{A_s} = \frac{2T}{Dwl} \tag{8-19}$$

For the contact plane between the side of the key and the side of the keyway, the contact area is

$$A_c = \frac{h}{2}l \tag{8-20}$$

The compressive bearing stress on the contact plane therefore becomes

$$\sigma_c = \frac{F}{A_c} = \frac{4T}{Dhl} \tag{8-21}$$

For a *square* key, $h = w$, so from (8-21)

$$\sigma_c = \frac{4T}{Dwl} \tag{8-22}$$

Using the concepts of 4.5, it may be noted that the maximum principal stress σ_s on the shear plane is equal to $2\tau_s$, making (8-19) and (8-22) equally critical. Hence, only (8-19) need be used for further analysis.

The design-allowable torque for the *key* may be found from (8-19) to be

$$T_{key-allow} = \frac{\tau_{d-key} Dwl}{2} \tag{8-23}$$

where τ_{d-key} is the design-allowable shearing stress (corresponding to the governing failure mode) for the key.

The design-allowable torque for the *shaft*, based on (4-30), may be found from

$$T_{shaft-allow} = \frac{\pi D^3 \tau_{d-shaft}}{16 K_{t\tau}} \tag{8-24}$$

where $\tau_{d-shaft}$ is the design-allowable shearing stress (corresponding to the governing failure mode) for the shaft, and $K_{t\tau}$ is the theoretical shearing stress concentration factor for the shaft keyway (usually $K_{t\tau} \approx 1.7$ for *static* loading). A few values of *fatigue* stress concentration factor are given in Table 8.3.

If the design-allowable stress for the shaft and the key are the same,

$$\tau_{d-key} = \tau_{d-shaft} \equiv \tau_d \tag{8-25}$$

To provide *equal strength* for the shaft and the key under these assumptions, (8-23) and (8-24) may be equated and solved for the key length l, giving, for equal strength,

$$l_{eq-str} = \frac{\pi D^2}{8 w K_{t\tau}} \tag{8-26}$$

TABLE 8.3 Fatigue Stress Concentration Factors for Keyways[1]

	Annealed Steel		Hardened Steel	
Type of Keyway	Bending	Torsion	Bending	Torsion
Profile	1.6	1.3	2.0	1.6
Sled runner	1.3	1.3	1.6	1.6

[1]From ref. 11.

Recommended proportions[18] for *square* keys suggest that $w = D/4$, so for a square key (8-26) becomes

$$l_{eq-str} = \frac{\pi D}{2K_{tT}} \tag{8-27}$$

and if $K_{tT} = 1.7$ is a fair estimate of the shaft stress concentration factor at the keyway under torsion,[19] the key length for equal strength of shaft and key is

$$l_{eq-str} = \frac{\pi D}{2(1.7)} \approx 0.9D \tag{8-28}$$

If it is desired to use the key as a "mechanical fuse" to protect the shaft from failure, the *key length should be reduced* by an appropriate factor (perhaps to 80% of l_{eq-str}), giving

$$l_{fuse} \approx 0.7D \tag{8-29}$$

As usual, it is the designer's responsibility to assure that the assumptions made in developing (8-28) and (8-29) are sufficiently accurate for the particular application at hand, making appropriate modifications otherwise.

Example 8.4 Designing a Rigid Coupling

A proposed power drive shaft must transmit 130 horsepower at a speed of 1200 rpm. The shaft may be accurately approximated as a solid cylindrical member, well supported by bearings placed near each end of the shaft. The proposed material for the shaft is hot-rolled 1020 steel having $S_u = 65{,}000$ psi, $S_{yp} = 43{,}000$ psi, $e = 36$ percent elongation in 2 inches, and $S_f' = 33{,}000$ psi. It is desired to couple this drive shaft, end-to-end, to a gear transmission input shaft of the same diameter and same material. It is desired to use a simple, rigid flange coupling, similar to the one shown in Figure 8.3(a). It is further desired to use the keyed connection between the coupling hub and the transmission input shaft as a mechanical safety fuse to protect the transmission input shaft and internal transmission components. (The internal transmission components are known to have strengths equal to or greater than the transmission input shaft.) A design safety factor of *two* has been chosen for this application, no significant bending moments are expected on the coupling, and long life is desired. Design a simple, rigid flange coupling for this application.

Solution

Using the basic configuration of the rigid coupling sketched in Figure 8.3(a), Figure E8.4 shows the dimensions to be determined. As discussed earlier under "Rigid Couplings," the critical areas to be investigated include:

[18]Based on ANSI Standard B17.1-1967.

[19]Recall, however, that for static loads, ductile materials, and highly local stress concentration, the *actual* value of K_{tT} is reduced because of local plastic flow.

Example 8.4 Continues

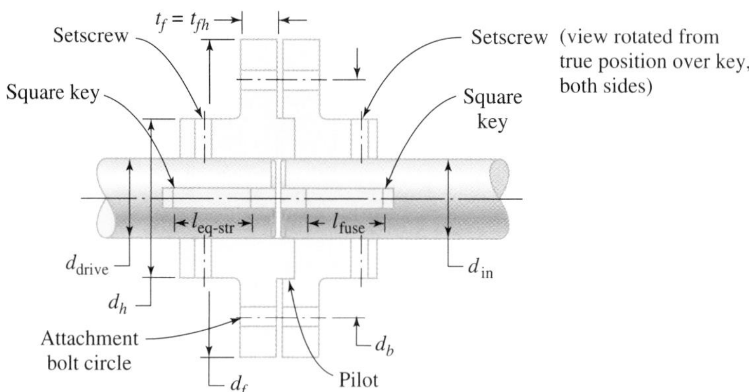

Figure E8.4
Sketch of proposed rigid flange coupling.

1. Shear and bearing in the keys
2. Shear and bearing in the flange attachment bolts
3. Bearing on the flange at attachment bolt interfaces
4. Shear in the flange at the hub

Using (4-32), and the specifications above,

$$T = \frac{63{,}025(hp)}{n} = \frac{63{,}025(130)}{1200} = 6828 \text{ in-lb} \tag{1}$$

and since there are no significant bending moments, the power drive shaft diameter may be calculated from (8-8) as

$$d_{drive} = \sqrt[3]{\frac{32(2.0)}{\pi(33{,}000)}\sqrt{0.75(6828)^2}} = 1.54 \text{ inches} \tag{2}$$

It is worth remembering here that (2) includes a torsional stress concentration factor of $K_{tT} = 1$, based on the concept that static load, ductile material, and highly local stress concentration permits local plastic flow that alleviates stress concentration effects without adverse consequences. To be more conservative, a larger stress concentration factor than unity could be included as a torsional shearing stress multiplier [a torque multiplier in (2)].

From Table 8.1, for a 1.54-inches-diameter shaft, a $^3/_8$-inch square key is recommended. Following this recommendation, selecting the same material for the key as specified for the shaft, and using (8-27), gives a key length for the drive shaft connection to the coupling hub of

$$l_{eq-str} = \frac{\pi(1.54)}{2(1.0)} = 2.4 \text{ inches} \tag{3}$$

The key between the coupling hub and the transmission input shaft is to be used as a mechanical safety fuse, so its length must be reduced by a selected factor, say to 80 percent of l_{eq-str}. Thus in Figure E8.4,

$$l_{fuse} = 0.8(2.4) = 1.9 \text{ inches} \tag{4}$$

Since d_{drive} is 1.54 inches, a reasonable value for shaft hub diameter d_h may be chosen from a scaled layout. The value chosen here is

$$d_h = 2.13 \text{ inches} \tag{5}$$

Also, a reasonable bolt circle diameter, d_B, would appear to be

$$d_B = 3.00 \text{ inches} \tag{6}$$

and a reasonable outer flange diameter, d_f, would seem to be

$$d_f = 4.00 \text{ inches} \tag{7}$$

A tentative choice of six $^5/_{16}$-inch-diameter bolts on the bolt circle also appears to be reasonable.

Based upon the above design decisions, a flange thickness t_f, at the bolt circle, may be calculated as follows.

The torque-induced force at the bolt circle is

$$F_B = \frac{2T}{d_B} = \frac{2(6828)}{3.00} = 4552 \text{ lb} \tag{8}$$

Assuming the force to be distributed among three of the six bolts (a judgment call), the compressive bearing stress between each contacting bolt and flange interface is

$$\sigma_{c-bf} = \frac{F_B}{3(A_c)} = \frac{4552}{3(0.313)(t_f)} = \frac{4848}{t_f} \tag{9}$$

and setting σ_{c-bf} equal to the design stress

$$\frac{43{,}000}{2.0} = \frac{4848}{t_f} \tag{10}$$

gives

$$t_f \approx 0.25 \text{ inch} \tag{11}$$

The required shearing area for the bolts, A_{sb}, may be found using (6-13) as

$$\tau_{sb} = \frac{4552}{A_{sb}} = \frac{\sigma_d}{2} = \frac{(43{,}000/2)}{2} \tag{12}$$

giving

$$A_{sb} = \frac{4552}{10{,}750} = 0.423 \text{ in}^2 \tag{13}$$

If three of the six bolts carry the load (as assumed above), the nominal *bolt* diameter should be

$$d_{bolt} = \sqrt{\frac{4(0.423/3)}{\pi}} = 0.42 \text{ inch} \tag{14}$$

hence, $^7/_{16}$-inch bolts should probably be used rather than $^5/_{16}$-inch bolts.

Finally, the flange thickness at the edge of the hub, t_{fh}, based on shear in the flange at the hub, may be checked. Again using (6-13),

$$\tau_{fh} = \frac{F_h}{A_{sh}} = \frac{\sigma_d}{2} = 10{,}750 \text{ psi} \tag{15}$$

Thus

$$A_{sh} = \frac{F_h}{10{,}250} = \frac{6828/(2.13/2)}{10{,}750} = 0.60 \text{ in}^2 \tag{16}$$

and

$$t_{fh} = \frac{0.60}{\pi(2.13)} = 0.09 \text{ inch} \tag{17}$$

The 0.25-inch flange thickness already chosen is therefore adequate.

Example 8.4 Continues

Summarizing, the following dimensional recommendations are made for the preliminary design dimensions of the rigid flange coupling shown in Figure E8.4.

$$d_{drive} = d_{in} = 1.54 \text{ inches}$$
$$\text{key size} = {}^{3}\!/_{8}\text{- inch square}$$
$$\text{setscrew size} = {}^{1}\!/_{2}\text{-13}$$
$$l_{eq-str} = 2.4 \text{ inches}$$
$$l_{fuse} = 1.9 \text{ inches}$$
$$d_h = 2.13 \text{ inches}$$
$$d_B = 3.00 \text{ inches}$$
$$d_f = 4.0 \text{ inches}$$
$$t_f = 0.25 \text{ inches}$$

flange bolts: use 6 bolts ${}^{7}\!/_{16}$-14, grade 1, equally spaced

In applications where higher torques must be transmitted, *keys* may not have enough capacity. In such cases *splines* are often used. In essence, splines are integral keys uniformly spaced around the outside of shafts or inside of hubs, as illustrated in Figure 8.9(a). Shaft splines are often cut into *raised pads* to reduce stress concentration effects. Splines may be either straight sided, as illustrated in Figure 8.9(b), or have involute tooth profiles, as shown in Figure 8.9(c).

Like keyed connections, the potential failure modes for splines include *yielding* for steady torque applications, or, if fluctuating loads or torques are involved, *fatigue* or *fretting fatigue*. In addition, *wear* or *fretting wear* may govern in some cases, since cyclic axial sliding is common in splined connections.

Three classes of fit for straight splines have been standardized as follows:

1. Class A fit: permanent connection—not to be moved after installation.
2. Class B fit: accommodates axial sliding when *no torque* is applied.
3. Class C fit: accommodates axial sliding with *load torque* applied.

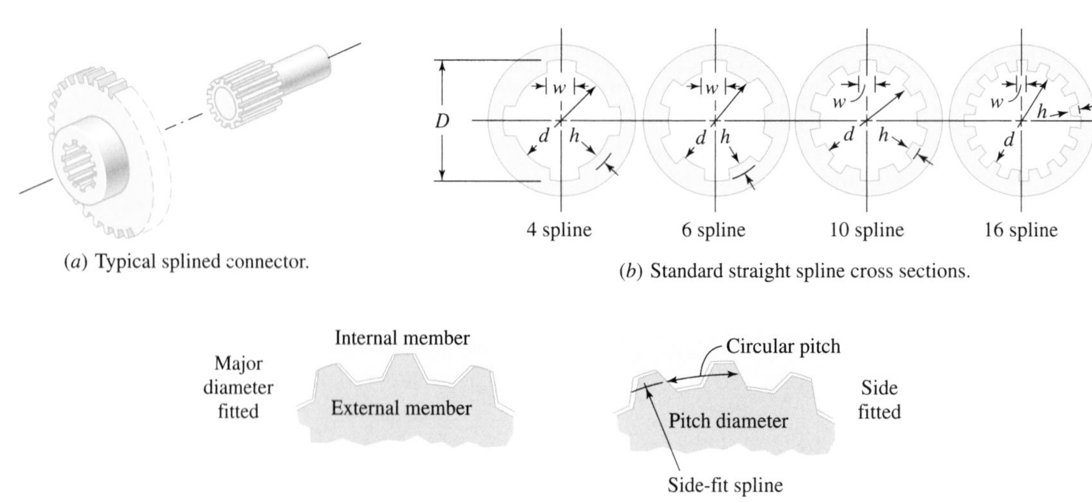

(*a*) Typical splined connector.

(*b*) Standard straight spline cross sections.

(*c*) Involute spline cross sections.

Figure 8.9
Spline geometry.

TABLE 8.4 Standard Dimensions for Straight Spline Teeth[1]

No. of Spline Teeth	w (for all fits)	Class *A* Fit (permanent)		Class B Fit (to slide *without* load)		Class C Fit (to slide *with* load)	
		h	d	h	d	h	d
Four	0.241D	0.075D	0.850D	0.125D	0.750D	—	—
Six	0.250D	0.050D	0.900D	0.075D	0.850D	0.100D	0.800D
Ten	0.156D	0.045D	0.910D	0.070D	0.860D	0.095D	0.810D
Sixteen	0.098D	0.045D	0.910D	0.070D	0.860D	0.095D	0.810D

[1]See Figure 8.9(b) for definitions of symbols.

Dimensions for *straight splines* operating under any of these conditions may be determined from Table 8.4.

For typical manufacturing tolerances, experience has shown that only about 25 percent of the teeth in a splined connection actually carry the load. Based on this assumption, the required spline engagement length, l_{sp}, to provide equal strength for spline and shaft, may be estimated in essentially the same way that equal-strength keyed connections were estimated [see (8-27)]. Thus the shearing stress τ_{sp} at the root of the spline may be equated to the design stress τ_d, giving

$$\tau_{sp} = \frac{F_s}{0.25A_s} = \frac{2T}{0.25d_p\left(\dfrac{\pi d_r}{2}\right)l_{sp}} = \tau_d \tag{8-30}$$

or, assuming root diameter and pitch diameter nearly equal,

$$d_r \approx d_p \equiv d \tag{8-31}$$

so the allowable torque for the spline becomes

$$T_{sp-allow} = \frac{\pi d^2 l_{sp}\tau_d}{16} \tag{8-32}$$

From (8-24) the allowable shaft torque, assuming the shaft diameter to be equal to the root diameter of the spline, is

$$T_{shaft-allow} = \frac{\pi d^3\tau_d}{16K_{t\tau}} \tag{8-33}$$

Equating (8-32) to (8-33) gives the spline engagement length l_{sp}, for equal strength of shaft and spline, as

$$l_{sp} = \frac{d}{K_{t\tau}} \tag{8-34}$$

Stress concentration factor data for an 8-tooth straight spline in torsion are given in Figure 4.19(c). For standard involute splines in torsion, the value of $k_{t\tau}$ seems to be about 2.8.[20] Recalling once again that for static loads, ductile materials, and highly local stress concentrations, local plastic flow reduces the *actual* stress concentration factor toward a value near unity, a designer may sometimes choose to set $K_{t\tau} = 1$.

Involute splines are widely used in modern practice. They are typically stronger, tend to be more self-centering, and are easier to cut and fit than straight splines. The teeth have an involute profile similar to gear teeth, usually with a 30° pressure angle (see Chapter 15), and half the depth of a standard gear tooth. Internal splines are typically machined by

[20]See ref. 10, pp. 248–249.

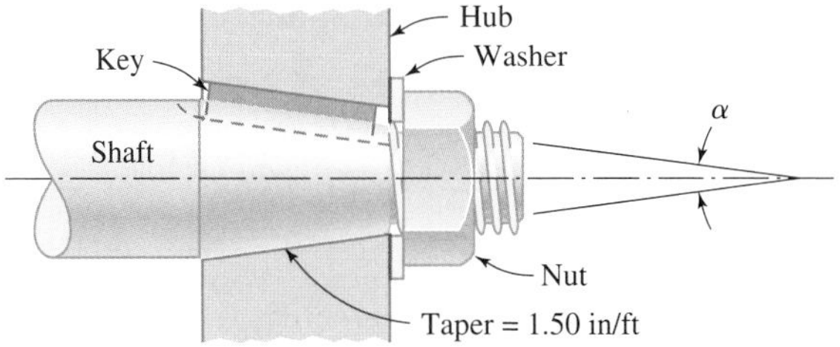

Figure 8.10
Typical tapered fit connection between shaft and hub.

broaching or shaping, and external splines by hobbing or shaping. Splines are "fitted" either at the major diameter or at the sides, as shown in Figure 8.9(c). Spline dimensions and tolerances have been standardized.[21]

Tapered fits are sometimes used to mount a power input or output component at the end of a shaft. Typically, a threaded stub and nut are used to secure the component by axially forcing the taper to tighten, as illustrated in Figure 8.10. Tapered fits provide good concentricity but only moderate torque transmission capacity. Torque capacity can be increased, however, by augmenting the tapered connection with a key. Precise axial location of a component on the shaft is not possible when a tapered connection is used. Typical tapers used in these connections are *self-locking* (i.e., $\alpha \leq 2\tan^{-1}\mu$, where α is the included angle of the tapered section, and μ is static coefficient of friction), so a designer should provide a means of inserting or attaching a "puller" to facilitate *disassembly* of the tapered joint if necessary.

In some applications, *interference fits* may be used to mount a power-input or -output component on a shaft to provide torque transfer by *friction* at the interface. The interference fit may be produced by axially *pressing* on to the shaft a component with a hub hole slightly smaller than the shaft mounting diameter, or by *heating* the hub, *cooling* the shaft, or *both*, to facilitate assembly. The shaft mounting diameter is usually raised above the overall shaft diameter to minimize stress concentration and permit precision machining of the mounting diameter. A shaft shoulder is usually provided to assure accurate axial positioning. The frictional torque transfer capacity, T_f, for such a joint, depends upon the interfacial pressure p, the diameter of the shaft d_s, the hub length l_h, and the friction coefficient μ. Transmitted torque is given by

$$T_f = \frac{F_f d_s}{2} = \frac{\mu p \pi d_s^2 l_h}{2} \tag{8-35}$$

The interfacial pressure p may be found from (9-48) or, in some cases, (9-49).

For lighter-duty applications, *setscrews* or *pins* may be used to transfer the torque between the mounted component and the shaft. Setscrews are threaded fasteners that are tightened into radial threaded holes in the hub, bearing against the outer surface of the shaft to provide frictional resistance to motion between shaft and hub. Often, a shallow *flat* is machined on the shaft to receive the setscrew point so that local burrs do not interfere with disassembly. Various types of points for setscrews are commercially available, as illustrated in Figure 8.11. The *cup point* is probably the most commonly used, primarily with *ductile* shafting. The *oval point* is used for similar applications, but typically requires a groove or keyway for spotting it on the shaft. The *cone point* is used when no adjustment of the hub position relative to the shaft is anticipated. When hardened-steel shafts are used, or if frequent adjustment is anticipated, a *flat point* setscrew would usually be used. The *half-dog point* is used for cases where relative position of the hub on the shaft must be preserved; a properly registered mating hole is drilled in the shaft to receive the dog point.

[21]See ANSI Standard B92.1 and B92.2M.

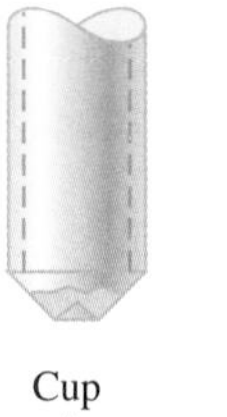

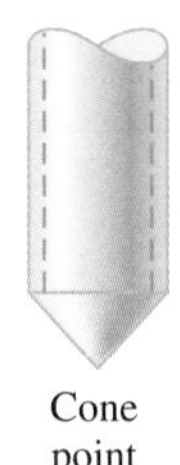

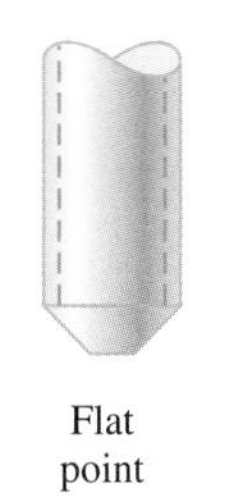

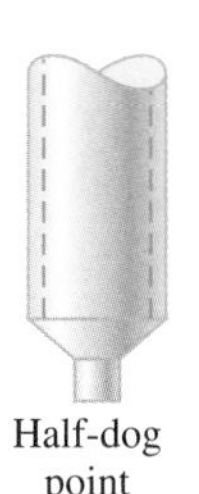

Figure 8.11
Common types of setscrew points.

Setscrew sizes are usually chosen to be about $^1/_4$ the mating shaft diameter, with a nominal length equal to about $^1/_2$ the shaft diameter. The *holding power* of a setscrew is the frictional resistance to slip (a tangential friction force) between shaft and hub, created by the tightened setscrew. Table 8.5 gives estimated values of holding power for cup point setscrews when they are installed using the seating torques indicated. Torque capacity can be increased substantially by utilizing two setscrews side by side. The frequently encountered problem of setscrew loosening under fluctuating or vibrational loads may be alleviated somewhat by using setscrews with deformable locking plastic inserts in the thread engagement region, or by tightening a second setscrew on top of the first to lock it in place.

Pins of various types may also be used in lighter-duty applications to provide torque transfer between mounted components and the shaft. To utilize any type of pin requires that a diametral hole be drilled through the shaft to accommodate the pin, creating a significant stress concentration at the site of the hole [see Figure 4.19(a)]. Several types of commercially available pins are sketched in Figure 8.12. *Clevis pins* are usually used in

TABLE 8.5 Holding Power of Cup-Point Socket Set Screws Against a Steel Shaft[1]

Size, in	Seating Torque, in-lb	Holding Power, lb
#0	1.0	50
#1	1.8	65
#2	1.8	85
#3	5	120
#4	5	160
#5	10	200
#6	10	250
#8	20	385
#10	36	540
$^1/_4$	87	1000
$^5/_{16}$	165	1500
$^3/_8$	290	2000
$^7/_{16}$	430	2500
$^1/_2$	620	3000
$^9/_{16}$	620	3500
$^5/_8$	1325	4000
$^3/_4$	2400	5000
$^7/_8$	5200	6000
1	7200	7000

[1]From ref. 5, p. 366.

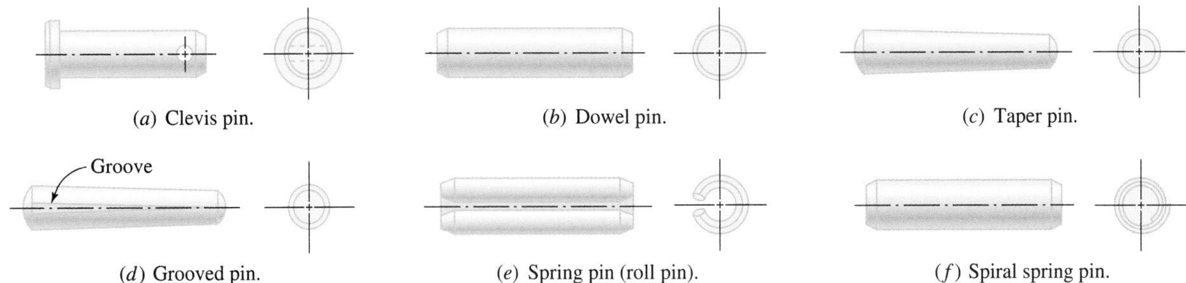

(*a*) Clevis pin. (*b*) Dowel pin. (*c*) Taper pin.

(*d*) Grooved pin. (*e*) Spring pin (roll pin). (*f*) Spiral spring pin.

Figure 8.12
Various types of commercially available pins.

cases where quick detachment of the mounted component is desired, and they may be made of ductile low-carbon steel to provide a safety shear pin that will protect the machine from damage. *Dowel pins* are usually hardened and ground for accurate dimensions and are used in applications where precise location is required. *Taper pins* are similar to dowel pins, but are ground to a slight taper that matches a taper-reamed, diametral hole through the shaft. Taper pins are self-locking and must be *driven out* to permit disassembly. *Grooved pins* are inexpensive, easy to install, and very popular. Many different styles of grooved pins are commercially available, in addition to the one shown in Figure 8.12(d). *Spring pins* (roll pins) are widely used and economical. They are easily installed, and hole tolerances are less critical than for solid pins. This is because the springy cross section deforms elastically to allow assembly. Frictional resistance created by the radial spring force keeps the spring pin in place. Of course the hollow cross section provides less shear strength than a solid pin with similar material properties, as shown in Table 8.6. *Spiral*

TABLE 8.6 Nominal Failure Loads for Commercially Available Pins in Double Shear (also see Figure 8.12)

	Nominal Failure Load in Double Shear,[1] lb					
Nominal Pin Diameter,[2] in	Clevis[3]	Dowel[4]	Taper[3]	Grooved[4]	Spring[4]	Spiral, Heavy Duty[4]
0.031	—	—	—	200	—	—
0.062	—	800	700	780	425	450
0.125	—	3200	2900	3120	2100	2100
0.188	6400	7200	6600	6990	4400	4400
0.250	11,330	12,800	11,800	12,430	7200	7700
0.375	25,490	28,700	26,400	27,950	17,600	17,600
0.500	45,320	51,000	46,900	49,700	25,800	30,000
0.625	70,810	79,800	73,400	—	46,000	46,000
0.750	101,970	114,000	104,900	—	66,000	66,000
0.875	138,780	156,000	143,500	—	—	—
1.000	181,270	204,000	187,700	—	—	—

[1]Tangential force at shaft–hub interface that will shear off the pin assuming the load to be equally distributed between the two shear areas.
[2]Other standard sizes available.
[3]Based on 1095 steel quenched and drawn to Rockwell C 42 (see Table 3.3). Other materials are available.
[4]Selected data excerpted from ref. 12.

spring pins have similar characteristics to the standard axially slotted spring pins, but have better shock load resistance and better fatigue resistance, and produce a tighter fit in the receiving drilled hole. Table 8.6 provides comparative data for a pin's *nominal* failure load, as a function of size (based on double shear), for the various pin types illustrated in Figure 8.12.

Problems

8-1. A drive shaft for a new rotary compressor is to be supported by two bearings spaced 8 inches apart. A V-belt system drives the shaft through a V-sheave (see Figure 17.9) mounted at midspan, and the belt is pretensioned to T_0 lb, giving a vertically downward force of $2T_0$ at midspan. The right end of the shaft is directly coupled to the compressor input shaft through a flexible coupling. The compressor requires a steady input torque of 5700 N-m. Make a first-cut conceptual sketch of a shaft configuration that would be appropriate for this application.

8-2. The drive shaft of a rotary coal grinding mill is to be driven by a gear reducer through a flexible shaft coupling, as shown in Figure P8.2. The main shaft of the gear reducer is to be supported on two bearings mounted 10 inches apart at A and C, as shown. A 1 : 3 spur gear mesh drives the shaft. The 20° spur gear is mounted on the shaft at midspan between the bearings, and has a pitch diameter of 9 inches. The pitch diameter of the drive pinion is 3 inches. The grinder is to be operated at 600 rpm and requires 100 horsepower at the input shaft. The shaft material is to be AISI 1060 cold-drawn carbon steel (see Table 3.3). Shoulders for gears and bearings are to be a minimum of $^1/_8$ inch ($^1/_4$ inch on the diameter). A design safety factor of 1.5 is desired. Do a *first-cut* design of the shaft, including a *second-cut sketch* showing principal dimensions.

8-3. A belt-driven jack-shaft is sketched schematically in Figure P8.3.

a. Construct load, shear, and bending moment diagrams for the shaft in both the horizontal and the vertical plane.

b. Develop an expression for the resultant bending moment on the shaft segment between the left pulley and the right bearing.

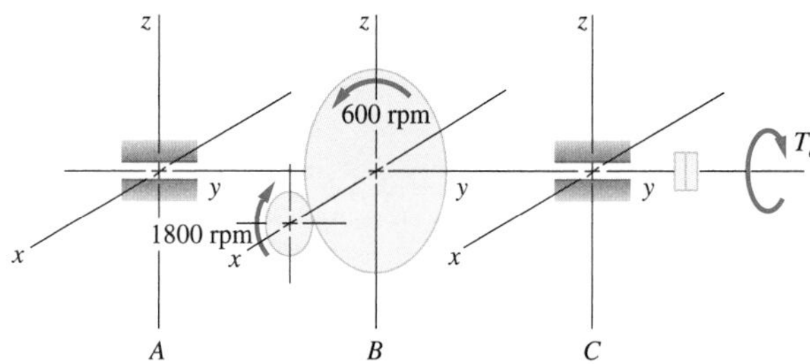

Figure P8.2
Schematic arrangement and coordinate system for gear reducer main shaft.

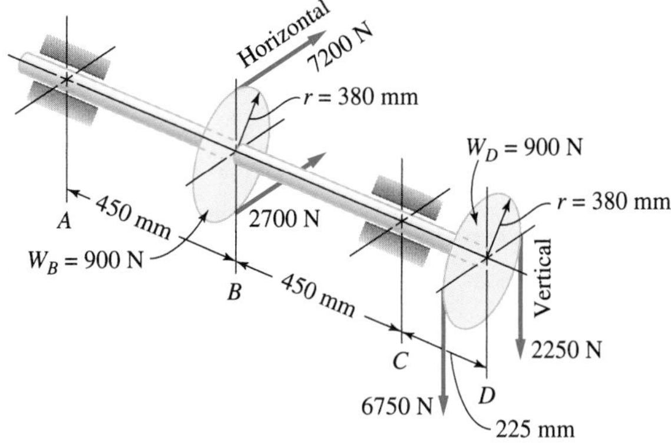

Figure P8.3
Schematic sketch of a belt-driven jack-shaft.

c. Find the location and magnitude of the minimum value of bending moment on the shaft segment between the left pulley and the right bearing.

d. Calculate the torque in the shaft segment between pulleys.

e. If the shaft is to be made of hot-rolled 1020 steel (see Figure 2.19), is to rotate at 1200 rpm, and a design safety factor of 1.7 is desired, what diameter would be required to provide infinite life?

8-4. Repeat problem 8-3 except that the shaft is to be made of AISI 1095 steel, quenched and drawn to Rockwell C 42 (see Table 3.3).

8-5. A pinion shaft for a helical gear reducer (see Chapter 15) is sketched in Figure P8.5, where the reaction forces on the pinion are also shown. The pinion shaft is to be driven at 1140 rpm by a motor developing 14.9 kW.

a. Construct load, shear, and bending moment diagrams for the shaft, in both the horizontal and the vertical plane. Also make similar diagrams for axial load and for torsional moment on the shaft, assuming that the bearing at the right end (nearest the gear) supports all thrust (axial) loading.

b. If the shaft is to be made of 1020 steel (see Figure 2.19), and a design safety factor of 1.8 is desired, what diameter would be required at location B to provide infinite life?

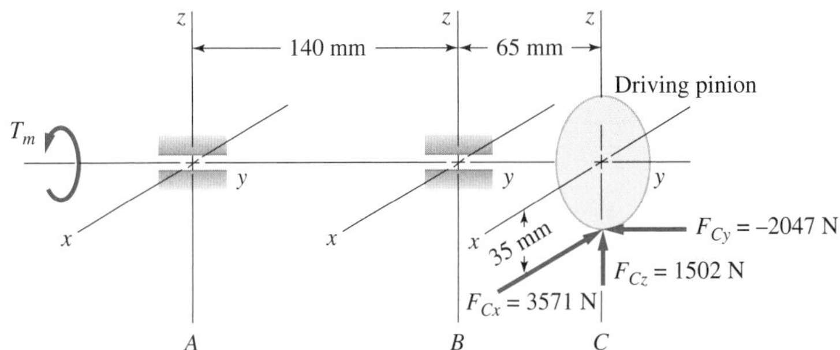

Figure P8.5
Pinion shaft for helical gear reducer.

8-6. A power transmission shaft of hollow cylindrical shape is to be made of hot-rolled 1020 steel with $S_u = 65{,}000$ psi, $S_{yp} = 43{,}000$ psi, $e = 36$ percent elongation in 2 inches, and fatigue properties as shown for 1020 steel in Figure 2.19. The shaft is to transmit 85 horsepower at a rotational speed of $n = 1800$ rpm, with no fluctuations in torque or speed. At the critical section, midspan between bearings, the rotating shaft is also subjected to a pure bending moment of 2000 in-lb, fixed in a vertical plane by virtue of a system of symmetrical external forces on the shaft. If the shaft outside diameter is 1.25 inches and the inside diameter is 0.75 inch, what operating life would be predicted before fatigue failure occurs?

8-7. A solid cylindrical power transmission shaft is to be made of AM 350 stainless steel for operation in an elevated temperature air environment of 1000°F (see Table 3.5). The shaft is to transmit 200 horsepower at a rotational speed of 3600 rpm, with no fluctuations in torque or speed. At the critical section, midspan between bearings, the rotating shaft is also subjected to a pure bending moment of 2500 in-lb, fixed in a vertical plane by a system of symmetrical external forces on the shaft. If the shaft diameter is 1.275 inches, predict a range within which the mean operational life would be expected to fall.

8-8. A shaft of square cross section, 2.0 inches by 2.0 inches, is being successfully used to transmit power in an application where the shaft is subjected to constant steady pure torsion only. If the same material is used and the same safety factor is desired, and for exactly the same application, what diameter should a solid *cylindrical* shaft be made for equivalent performance?

8-9. A shaft with a raised bearing pad, shown in Figure P8.9, must transmit 100 horsepower on a continuous basis at a rotational speed of 1725 rpm. The shaft material is annealed AISI 1020 steel. A notch-sensitivity index of $q = 0.7$ may be assumed for this material. Using the most accurate procedure you know, estimate how large the vertical midspan bearing force P can be and still maintain a safety factor of 1.3 based on infinite-life design requirements.

8-10. A solid circular cross-section shaft made of annealed AISI 1020 steel (see Figure 2.19) with an ultimate strength of 57,000 psi and yield strength of 43,000 psi is shouldered as shown in Figure P8.10. The shouldered shaft is subjected to a pure bending moment $M_b = 1600$ in-lb, and rotates at a speed of 2200 rpm. How many revolutions of the shaft would you predict before failure takes place?

8-11. A rotating solid cylindrical shaft must be designed to be as light as possible for use in an orbiting space station. A safety factor of 1.15 has been selected for this design, and the tentative material selection is Ti-150a titanium alloy (see Figure

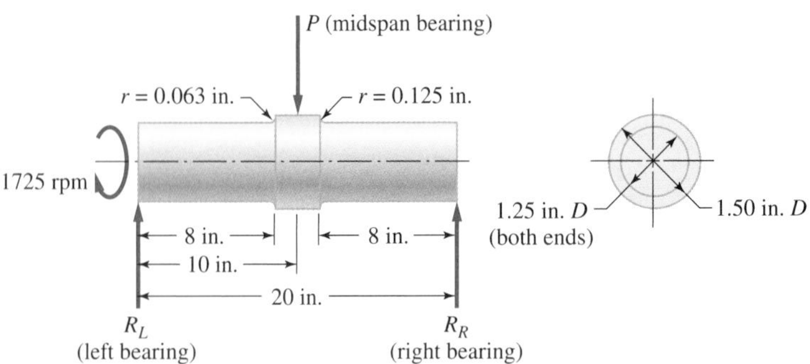

Figure P8.9
Shaft with a raised bearing pad.

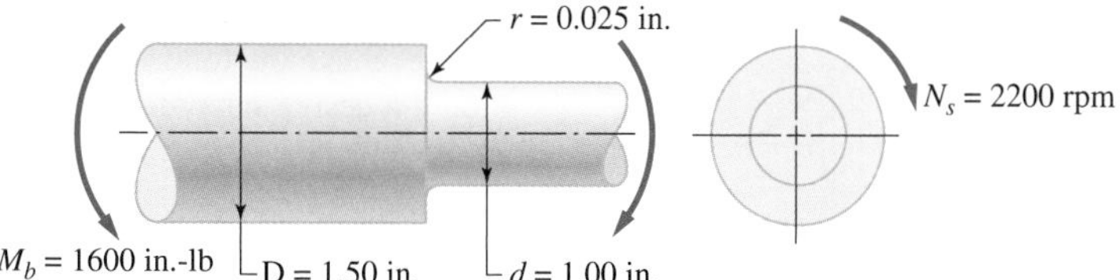

Figure P8.10
Shouldered shaft.

2.19). This shaft will be required to rotate a total of 200,000 revolutions during its design life. At the most critical section of the shaft, it has been determined from a force analysis that the rotating shaft will be subjected to a steady torque of 9000 in-lb and a bending moment of 11,000 in-lb. It is estimated that the fatigue stress concentration factors for this critical section will be 1.8 for bending and 1.4 for torsion. Calculate the required minimum shaft diameter at this critical section.

8-12. The sketch in Figure P8.12 shows a shaft configuration determined by using a now-obsolete ASME shaft code equation to estimate several diameters along the shaft. It is desired to check the critical sections along the shaft more carefully. Concentrating attention on critical section *E-E*, for which the proposed geometry is specified in Figure P8.12, a force analysis has shown that the bending moment at *E-E* will be 100,000 in-lb, and the torsional moment is steady at 50,000 in-lb. The shaft rotates at 1800 rpm. Tentatively, the shaft material has been chosen to be AISI 4340 ultra-high-strength steel (see Table 3.3). A safety factor of 1.5 is desired. Calculate the minimum diameter the shaft should have at location *E-E* if infinite life is desired.

8-13. Repeat problem 8-12 except that the shaft is to be made of AISI 1095 steel quenched and drawn to Rockwell C 42 (see Table 3.3).

8-14. A new application is being considered for the machine in which the shaft depicted in Figure P8.12 is to be used. All requirements and specifications for the shaft in the new application remain the same as described in problem 8-12 with one exception: The torsional moment at location *E-E*, instead of being a steady 50,000 in-lb, fluctuates once per revolution from a minimum of 25,000 in-lb to a maximum of 75,000 in-lb, and back. Determine the required minimum diameter at *E-E* to accommodate the fluctuating torque requirement of the new application while maintaining the safety factor of 1.5 based on the infinite-life requirement.

8-15. To obtain a quick-and-dirty estimate for maximum deflection of the loaded stepped steel shaft shown in Figure P8.15, it is being proposed to approximate the stepped shaft by an "equivalent" shaft of uniform diameter $d_{eq} = 4.0$ inches. The shaft may be assumed to be simply supported by bearings at sections *A* and *G*, and loaded by transverse loads of 8000 lb down at section *C* and 2000 lb up at section *F*, as shown.

a. Calculate maximum deflection of the equivalent 4.0-inch-uniform-diameter shaft.

b. Calculate slopes of the equivalent shaft at the bearing sites *A* and *G*.

8-16. For the stepped steel shaft shown in Figure P8.15, use the method of graphical integration to obtain a more accurate estimate of maximum bending deflection of the shaft and the location along the shaft where the maximum bending deflection occurs. Also, estimate the shaft slope at each of the bearing sites.

8-17. A rotating hollow shaft having a 5.00-cm outside diameter and a 6.0-mm-thick wall is to be made of AISI 4340 steel. The shaft is supported at its ends by bearings that are *very stiff*, both radially and in their ability to resist angular deflections caused by shaft bending moments. The support bearings are spaced 60 cm apart. A solid-disk flywheel weighing 450 N is mounted at midspan, between the bearings. What limiting maximum shaft

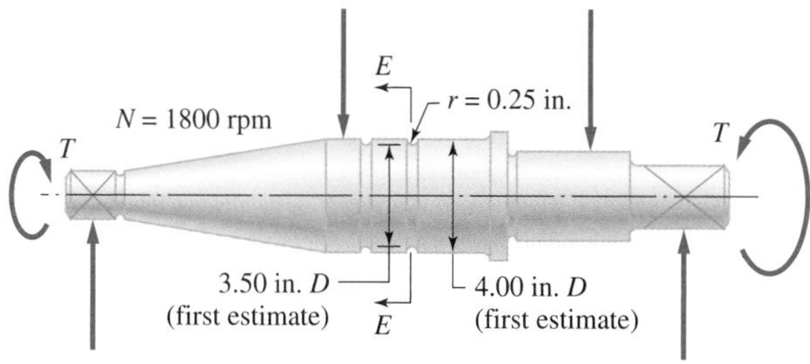

Figure P8.12
Shaft design configuration established using an "old" ASME code equation.

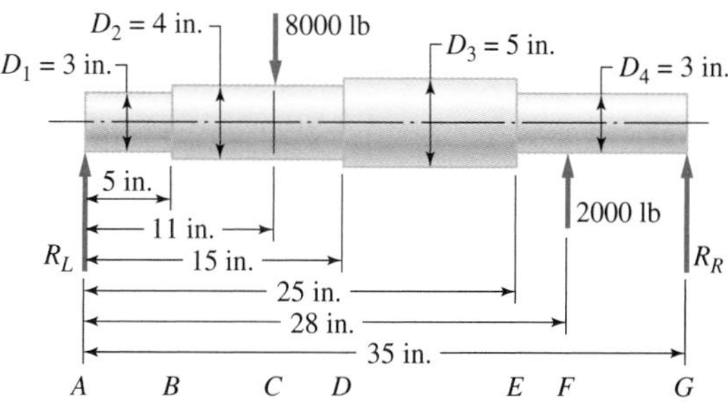

Figure P8.15
Simply supported stepped shaft with transverse loads.

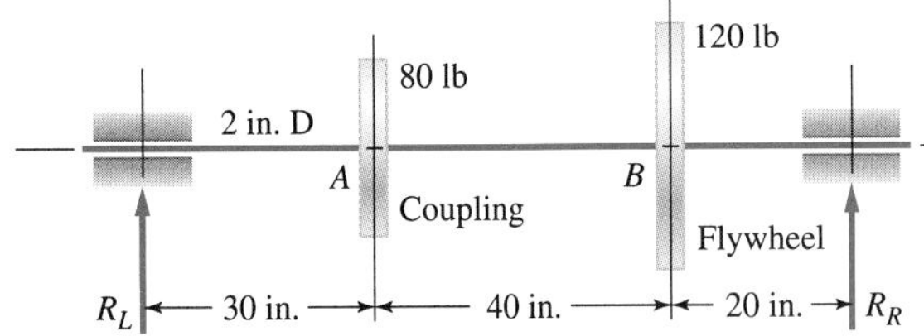

Figure P8.19
Rotating shaft system with attached masses.

speed would you recommend for this application, based on the need to avoid lateral vibration of the rotating system?

8-18. Repeat problem 8-17 using a *solid* shaft of the same outside diameter instead of the hollow shaft.

8-19. A 2-inch-diameter solid cylindrical 1020 steel shaft is supported on identical rolling-element bearings (see Chapter 11) spaced 90 inches apart, as sketched in Figure P8.19. A rigid coupling weighing 80 lb is incorporated into the shaft at location *A*, 30 inches from the left bearing, and a small solid-disk flywheel weighing 120 lb (see Chapter 18) is mounted on the shaft at location *B*, 70 inches from the left bearing. The shaft is to rotate at 240 rpm. The bearings are not able to resist any shaft bending moments.

a. *Neglecting* any radial elastic deflection in the support bearings, and *neglecting* the mass of the shaft, estimate the critical speed for lateral vibration of the rotating system shown. If this estimate of critical speed is correct, is the proposed design acceptable?

b. Reevaluate the critical speed estimate of (a) by *including* the mass of the shaft in the calculation. If this new estimate of critical speed is correct, is the proposed design acceptable?

c. Reevaluate the critical speed estimate of (b) if the radial elastic deflections of the bearings (the spring rate of each bearing has been provided by the bearing manufacturer as 5×10^5 lb-in) are included in the calculation. Does this new estimate of critical speed, if correct, support the postulate that the system is adequately designed from the standpoint of lateral vibration?

8-20. For the proposed rigid coupling sketched in Figure P8.20, evaluate the following aspects of the proposed configuration if a design safety factor of 2.0 is desired.

a. Shear and bearing in the keys

b. Shear and bearing in the flange attachment bolts

c. Bearing on the flange at attachment bolt interfaces

d. Shear in the flange at the hub

The input shaft has a nominal diameter of 2.25 inches, and supplies a steady input of 50 hp at 150 rpm. The bolt circle diameter is d_B = 6.0 inches. Cold-drawn AISI 1020 steel is being proposed as the material for the coupling components, including the bolts, and also the material for the key (see Table 3.3). Is the coupling design acceptable as proposed?

8-21. As a new engineer, you have been assigned the task of recommending an appropriate shaft coupling for connecting the output shaft of a 12-hp gear-motor drive unit, operating at 600 rpm, to the input shaft of a newly designed seed-corn cleaning machine ordered by a farm-supply depot. Based on the components selected for this device, and on the assembly capabilities within your company's production facility, it has been estimated that the parallel centerline misalignment between the motor drive shaft and the input shaft of the seed cleaning machine may be as much as $^1/_{32}$ inch, and the angular misalignment between shafts may be as much as 2°. What type of coupling would you recommend?

8-22. a. A chain drive (see Chapter 17) delivers 110 horsepower to the input shaft of an industrial blower in a paint manufacturing plant. The drive sprocket rotates at 1700 rpm, and has a bore diameter of 2.50 inches and a hub length of 3.25 inches. Propose an appropriate geometry for a standard square key, including width and length dimensions, if the key is to be made of 1020 cold-drawn steel having S_u = 61,000 psi and S_{yp} = 51,000 psi. The key material may be assumed to be weaker than either the mating shaft material or hub material. A design safety factor of 3 is desired.

b. Would it be possible to use a standard Woodruff key of the same material in this application?

8-23. Repeat problem 8-22, except that the drive sprocket rotates at 800 rpm.

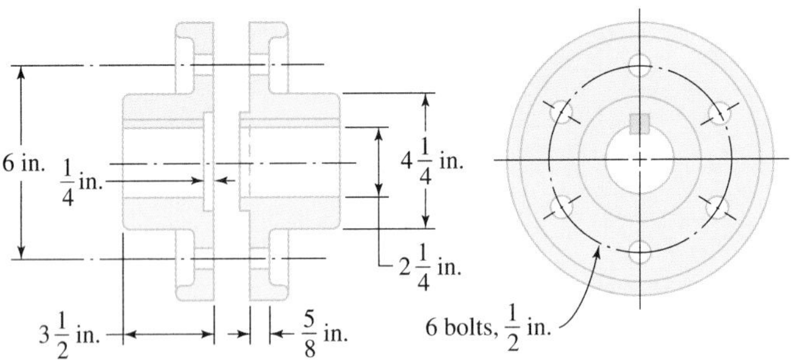

P8.20
Proposed shaft coupling configuration.

8-24. For the chain drive specifications given in problem 8-22, and for the same sprocket dimensions, select the minimum size of grooved pin that could be used to attach the sprocket to the shaft, assuming the grooved pin to be made of 1095 steel quenched and drawn to Rockwell C 42 (see Table 3.3).

8-25. A six-tooth spline is under consideration for a splined connection to a universal joint between the rotating power take-off shaft of a new medium-size farm tractor and the power input shaft of a towed, newly designed sod-removal machine. Because of the kinematic relationship of the hitch to the splined connection, the spline is required to slide under load as the machine operates. It is initially desired to design the joint so the strength of the splined section of the shaft is equal to the basic unsplined shaft strength. What spline engagement length should be used for this application if the nominal shaft diameter has been determined to be d = 1.25 inches and a spline root fillet radius of $^1/_{32}$ inch has been selected? [*Hint*: Assume that the stress concentration factors given for the eight-tooth spline of Figure 4.19(c) provide a satisfactory estimate for the six-tooth spline under consideration here.]

8-26. a. A V-pulley is to be mounted on the steel 1.0-inch-diameter engine drive-shaft of a lawn tractor. The pulley must transmit 14 horsepower to the drive-shaft at a speed of 1200 rpm. If a cup point setscrew were used to attach the pulley hub to the shaft, what size setscrew would be required? A design safety factor of 2 is desired.

b. What seating torque would be recommended to properly tighten the setscrew so that it will not slip when power is being transmitted?

Chapter 9

Pressurized Cylinders; Interference Fits

9.1 Uses and Characteristics of Pressurized Cylinders

Containment vessels for pressurized fluids may be spherical, ellipsoidal, toroidal, cylindrical, or combinations of these shapes. In spite of the fact that spherical vessels are more favorably stressed than the others, the practical advantage of easy fabrication makes the *cylindrical* pressure vessel, by far, the most widely used shape.

In practice, cylindrical pressure vessels may be either *internally* pressurized or *externally* pressurized, may be *thin-walled* or *thick-walled*, and may be *open* at the ends or *closed* at the ends. They may be fabricated as monolithic cylindrical shells, multilayered shrink-fit concentric cylindrical shells, filament-wound cylindrical shells, or wire-wrapped cylindrical shells. Filament-wound composite shells are made by wrapping a filamentary material, such as fiberglass, on a cylindrical mandrel and impregnating it with a polymer, curing the resulting *filamentary composite*, and removing the mandrel. Wire-wrapped cylindrical shells are made by tightly wrapping a high-strength steel wire around the cylindrical vessel for reinforcement.

Applications involving pressurized cylinders include tanks, hoops, pipes, gun barrels, rocket-engine thrust chambers, and hydraulic and pneumatic actuators, as well as the related applications embodying interference fits between a cylindrical hub and mating shaft.

Geometrical discontinuities in pressure vessels require special attention from the designer, just as all other stress concentration sites do. The juncture between the cylindrical vessel and the end closures, whether flat, semiellipsoidal, or hemispherical, must be evaluated. Openings (nozzles) in the vessel wall to allow inward and outward flow, manholes for inspection and cleaning, attached support flanges, weld seams, and thickness transition sites must also be carefully considered in the safe design of a pressure vessel. All of these discontinuities produce complicated states of stress and uncertainties in failure analysis. For these reasons, a designer is well advised to consult appropriate pressure vessel design *codes* when analyzing pressure vessels. In most states, conformance with the *ASME Boiler and Pressure Vessel Code*[1] is required by law. Code design procedures integrate generations of failure prevention experience with applicable theoretical analyses to maximize the reliability and safety of the structure.

9.2 Interference Fit Applications

The relationship between cylindrical pressure vessels and interference fits is direct. When the cylindrical hub of a shaft-mounted component is assembled to the shaft by *pressing it*

[1]See ref. 1.

in place, or when *differential thermal expansion* techniques are used for assembly (see 8.8), an interfacial pressure is generated between the shaft and the hub. The state of stress produced in the *hub* by the interference-fit-induced internal pressure is completely analogous to the state of stress induced in the wall of an *internally* pressurized cylindrical vessel. Likewise, if the shaft is hollow, the state of stress in the *shaft* is analogous to that in the wall of an *externally* pressurized cylindrical vessel. Because the friction torque transmitted by an interference fit assembly is directly related to interfacial pressure [see (8-35)], the diametral interference required to provide a specified friction torque may be calculated, and the resulting states of stress in the hub and the shaft may be determined as well. The strategy here will be to first develop expressions for stress distributions in pressurized hollow cylinders, as a function of pressure, dimensions, and material properties, then adapt these expressions to the case of interference fits.

9.3 Potential Failure Modes

Depending upon the application, the pressure level, the temperature level, the environment, and the composition of the pressurizing fluid, pressure vessels are vulnerable to potential failure by many possible modes. Reviewing the failure modes listed in 2.3, pressure vessel failure might occur by *yielding, ductile rupture, brittle fracture* (even if the material is nominally ductile), *fatigue* (including low-cycle, thermal, or corrosion fatigue), *stress-corrosion-cracking, creep*, or *combined creep and fatigue*. In interference fit applicaitons, such as a hub pressed on a shaft, failure modes might include *yielding, ductile rupture, brittle fracture* (perhaps induced by a poorly chosen diametral interference), *fretting fatigue* at the interface, or *fatigue* initiated at end-of-hub stress concentration sites. It is worth noting that interference fit failures are sometimes generated by designer inattention to the important task of specifying proper tolerances on the shaft diameter and mating hub bore. Usually, a designer must investigate both the tolerance-dictated *minimum interference* condition (to ensure that torque transfer requirements are met) and the tolerance-dictated *maximum interference* condition (to assure that stress levels are within safe ranges). Any differential expansion or contraction induced by operational temperature changes must also be evaluated in terms of consequential tolerance changes.

9.4 Materials for Pressure Vessels

Considering the potential failure modes suggested in 9.3, and utilizing the material selection guidelines of Chapter 3, it may be observed that candidate materials for pressurized cylinders should have good strength (including good fatigue strength), high ductility, good formability, good weldability, and low cost. In elevated temperature applications, the selected material should also have good strength at the operating temperature, good creep resistance, and in many cases, good corrosion resistance. Steel materials typically meet most of these criteria, and therefore steel materials are widely used in fabrication of pressure vessels. In special circumstances, other materials might be selected, however, especially if light weight, very high temperatures, or specific corrosive environments must be accommodated. In the case of interference fit applications, 8.3 provides additional material selection insights.

If, as suggested in 9.1, the ASME Boiler and Pressure Vessel Code is consulted, an extensive catalog of material specifications and related code-specified design stresses may be found for an array of pressure vessel and piping materials. This list of materials includes carbon and low-alloy steels, nonferrous alloys, high-alloy steels, cast iron, and fer-

Table 9.1 ASME Code Specifications for Selected Steel Pressure Vessel Plate Materials[1]

Type of Steel	Specification No.	Grade	Thickness, in	Tensile Strength, ksi	Min. Yield Strength, ksi	Elongation in 2 Inches, percent
Carbon	SA-285/ SA-285M	A	2 maximum	45–65	24	30
Carbon	"	B	"	50–70	27	28
Carbon	"	C	"	55–75	30	27
Mn-Si	SA-299/ SA-299M	—	up to 1	75–95	42	19
Mn-Si	"	—	1 to 8	75–95	40	19
Mn-Mo	SA-302/ SA 302M	A	0.25 to 8	75–95	45	19
Mn-Mo	"	B	"	80–100	50	18
Mn-Mo-Ni	"	C	"	80–100	50	20
Mn-Mo-Ni	"	D	"	80–100	50	20

[1]Hundreds of additional material specifications, both ferrous and nonferrous, are included in the ASME Boiler and Pressure Vessel Code. See ref. 1.

Table 9.2 Example Listing of ASME Code Specification for Maximum Allowable Stress (ksi) as a Function of Temperature (°F)

Spec. No.[1]	Grade	Maximum Allowable Stress, ksi −20 to 650°	up to 700°	up to 750°	up to 800°	up to 850°	up to 900°	up to 950°	up to 1000°
SA 285	A	11.3	11.0	10.3	9.0	7.8	6.5	—	—
"	B	12.5	12.1	11.2	9.6	8.1	6.5	—	—
"	C	13.8	13.3	12.1	10.2	8.4	6.5	—	—

[1]See Table 9.1.

ritic steels. Information is provided for plates, sheets, castings, forgings, weldments, end closures, hoops, and layered construction. Table 9.1 illustrates the type of information available in the code for a few selected steel pressure vessel plate materials. Table 9.2 illustrates the type of information available for code-specified design-allowable stress, as a function of metal operating temperature. Designers of pressure vessels are well advised to familiarize themselves with the scope of the code, even if code design is not *legally* required.

9.5 Principles from Elasticity Theory

When designers investigate stresses and strains in a machine element, they base their analyses either upon a *strength of materials* model or a *theory of elasticity* model. In using the strength of materials approach, *simplifying assumptions are made about the stress distributions and strain distributions* within the body. The familiar assumptions that (1) planes remain plane, (2) bending stress is proportional to distance away from the neutral axis, or (3) the hoop stress in the wall of a *thin-walled* pressurized cylinder is *uniform* across the wall thickness are cases in point. On the other hand, use of the more complicated (but more accurate) theory of elasticity model permits a *determination* of the *distributions of stresses and strains* within the body. In cases where the simplifying assumptions of the strength of materials approach are too inaccurate, such as for a *thick-walled* pressurized cylinder, the theory of elasticity approach is recommended.

The basic relationships from elasticity theory[2] needed to analyze thick-walled cylinders include:

1. Differential equations of force equilibrium
2. Force-displacement relationships, for example, Hooke's Law
3. Geometrical compatibility relationships
4. Boundary conditions

The requirements of force equilibrium are familiar (see 4.2). Likewise, Hooke's Law is familiar (see 4.5). Geometrical compatibility requires material continuity; that is, no voids may be torn open and no material may "bunch up." Boundary conditions require that forces associated with stress distributions within the body must be in *equilibrium* with the external forces on its boundaries. The use of these four *principles of elasticity* will be demonstrated in developing the thick-walled cylinder equations in 9.7.

9.6 Thin-Walled Cylinders

For purposes of analysis, pressurized cylinders are often categorized as either *thin-walled* or *thick-walled*. To qualify as a thin-walled cylindrical pressure vessel, the wall must be thin enough to satisfy the assumption that the radial stress component in the wall, σ_r, is *negligibly small* compared to the *tangential (hoop) stress* component σ_t, and that σ_t is *uniform* across the thickness. Typically, if the thickness t is 10 percent of the diameter d, or less, the pressure vessel may be accurately analyzed as a thin-walled vessel. For thick-walled pressure vessels, *both* σ_r and σ_t vary (nonlinearly) across the wall thickness. Of course, any pressure vessel may be correctly analyzed as thick-walled, but if the thin-walled assumptions are valid, a significant simplification results in the analysis.

Referring to Figure 9.1(a), the through-the-thickness volume element shown is sub-

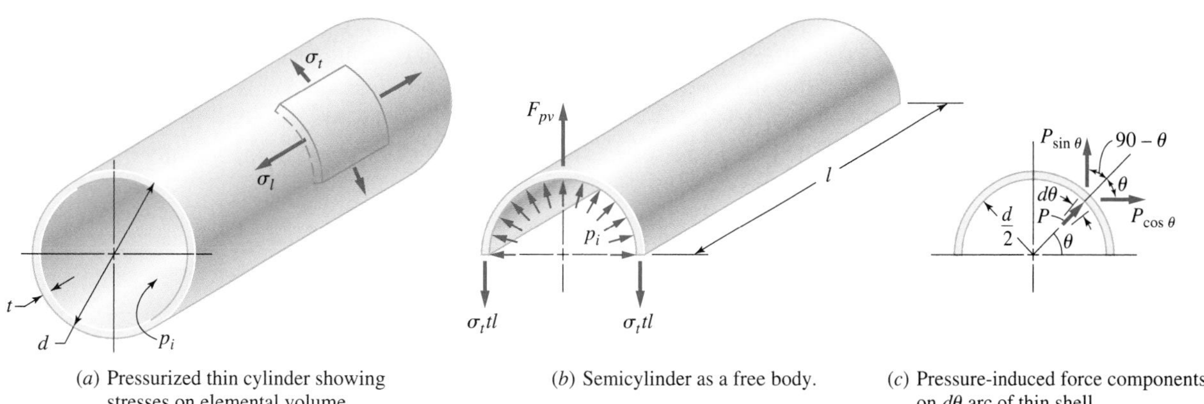

(*a*) Pressurized thin cylinder showing stresses on elemental volume.

(*b*) Semicylinder as a free body.

(*c*) Pressure-induced force components on $d\theta$ arc of thin shell.

Figure 9.1
Analysis of a thin-walled cylinder.

[2]See, for example, ref. 2.

jected to a *biaxial* state of stress, because by the thin-walled assumption[3]

$$\sigma_r = 0 \tag{9-1}$$

Taking the upper half of the internally pressurized thin-walled cylinder as a free body, as shown in Figure 9.1(b), vertical force resolution gives

$$F_{pv} - 2(\sigma_t tl) = 0 \tag{9-2}$$

The vertical force component F_{pv} produced by the pressure p_i may be found by integrating the vertical component of pressure-induced force, $p_i dA$, over the semicircular surface. Thus (9-2) may be rewritten, based on Figure 9.1(c), as

$$2\int_0^{\pi/2}\left[p_i\left(\frac{d}{2}d\theta\right)l\right]\sin\theta - 2\sigma_t tl = 0 \tag{9-3}$$

giving

$$p_i dl - 2\sigma_t tl = 0 \tag{9-4}$$

The product $(d)(l)$ is defined as *projected area*, a concept also used in analyzing plain bearings (see Chapter 10) and certain other machine elements. From (9-4) it may be found that the *tangential* stress component is

$$\sigma_t = \frac{p_i d}{2t} \tag{9-5}$$

If the pressure vessel is closed at the ends, the *longitudinal* stress component σ_l may be found from a *free-body diagram*, constructed by passing a cutting plane through the cylinder, perpendicular to the cylinder axis, and writing an axial force resolution to give

$$p_i\left(\frac{\pi d^2}{4}\right) = \sigma_l(\pi dt) \tag{9-6}$$

From (9-6), the *longitudinal* stress component is

$$\sigma_l = \frac{p_i d}{4t} \tag{9-7}$$

Comparing (9-5) with (9-7), it may be noted that the tangential (hoop) stress σ_t in a thin-walled pressure vessel is twice the longitudinal stress σ_l. Because a *biaxial state of stress* exists, it is necessary to utilize an appropriate failure theory.[4] Also, it is important to understand that the stress calculations just developed are valid only in the cylindrical wall at a *sufficient distance away*[5] from joints with end closures, nozzles, or supports, as discussed in 9.1. Local stresses must be separately determined in the vicinity of any stress concentration site.

9.7 Thick-Walled Cylinders

If thin-walled assumptions are *not* valid, it becomes necessary to employ the principles of elasticity from 9.5 to properly determine the stress *distributions* for σ_r and σ_t. Figure

[3]Furthermore, radial pressures are nearly always negligibly small compared to the tangential and longitudinal stresses.

[4]See 4.6.

[5]Based on *St. Venant's principle*, which, in essence, states that if a system of forces acts on a small local portion of an elastic body, the stresses are changed only locally in the immediate vicinity of the force application, and remain virtually *unchanged* in other parts of the body.

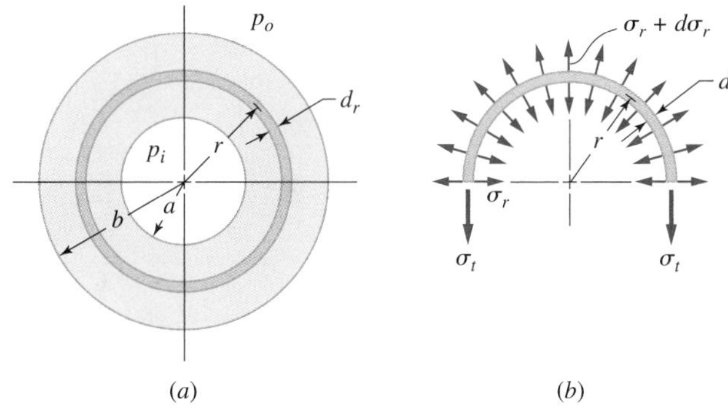

Figure 9.2
Analysis of a thick-walled cylinder.

9.2(a) shows a cross-sectional view of a *thick-walled* cylinder of length l. It will be initially assumed that there are no axial forces on the cylinder; hence the state of stress at any point within the wall is *biaxial*. If axial forces *do* exist, the associated longitudinal stress component σ_l may be superposed later.

Because a cylinder is *symmetrical* about a central axis, plane sections remain plane when the cylinder is pressurized, automatically satisfying the *geometrical compatibility* requirement. The thick-walled cylinder shown in Figure 9.2(a) is subjected to both an *internal* pressure p_i and an *external* pressure p_o. An elemental ring of radial thickness dr may be defined as shown in Figure 9.2(a). The ring has a mean radius r that lies between the *inner* cylinder-wall radius a and the *outer* cylinder-wall radius b. Taking the upper half of the elemental ring as a free body, the forces on the chosen free body may be deduced from the discussions of 9.6 and the stress state shown in Figure 9.2(b). The *differential equation of force equilibrium* may be written by taking a vertical force resolution to give (using the projected area concept of 9.6)

$$\sigma_r(2r)l - (\sigma_r + d\sigma_r)[2(r + dr)]l + 2\sigma_t l dr = 0 \tag{9-8}$$

or

$$\sigma_t - \sigma_r - r\frac{d\sigma_r}{dr} = 0 \tag{9-9}$$

To solve for the *two* unknowns, σ_r and σ_t, requires another independent relationship. For linear elastic material behavior, the *force-displacement relationships*, expressed as Hooke's Law, satisfy this need. Thus (4-73), (4-74), and (4-75) may be simplified for a biaxial state of stress ($\sigma_t = 0$), to give

$$\varepsilon_r = \frac{1}{E}[\sigma_r - v\sigma_t] \tag{9-10}$$

$$\varepsilon_t = \frac{1}{E}[\sigma_t - v\sigma_r] \tag{9-11}$$

$$\varepsilon_l = \frac{1}{E}[-v(\sigma_r + \sigma_t)] \tag{9-12}$$

where ε_r, ε_t, and ε_l are, respectively, normal radial strain, normal tangential strain, and normal longitudinal strain. It is worth noting that ε_l is *not* zero, in spite of the fact that σ_l *is* zero.

Rewriting (9-12),

$$\sigma_r + \sigma_t = 2C_1 \tag{9-13}$$

where

$$2C_1 \equiv -\frac{E\varepsilon_l}{\nu} \tag{9-14}$$

Subtracting (9-13) from (9-9) gives

$$-2\sigma_r = r\frac{d\sigma_r}{dr} - 2C_1 \tag{9-15}$$

Next, multiplying by r and rearranging,

$$r^2\frac{d\sigma_r}{dr} + 2r\sigma_r = 2rC_1 \tag{9-16}$$

or

$$\frac{d}{dr}(r^2\sigma_r) = 2rC_1 \tag{9-17}$$

Integrating (9-17) gives

$$r^2\sigma_r = r^2C_1 + C_2 \tag{9-18}$$

where C_2 is a constant of integration.

Solving (9-18) for σ_r

$$\sigma_r = C_1 + \frac{C_2}{r^2} \tag{9-19}$$

Combining (9-19) with (9-13) then gives

$$\sigma_t = C_1 - \frac{C_2}{r^2} \tag{9-20}$$

Boundary conditions of interest may be formulated from Figure 9.2 as

$$\sigma_r = -p_i \quad \text{at} \quad r = a \tag{9-21}$$

$$\sigma_r = -p_o \quad \text{at} \quad r = b \tag{9-22}$$

Inserting these boundary conditions, one at a time, into (9-19) gives

$$-p_i = C_1 + \frac{C_2}{a^2} \tag{9-23}$$

and

$$-p_o = C_1 + \frac{C_2}{b^2} \tag{9-24}$$

These two equations may be solved simultaneously for C_1 and C_2, then the values of C_1 and C_2 inserted into (9-19) and (9-20) to give, finally,

$$\sigma_r = \frac{p_ia^2 - p_ob^2 + \left(\frac{a^2b^2}{r^2}\right)(p_o - p_i)}{b^2 - a^2} \tag{9-25}$$

and

$$\sigma_t = \frac{p_ia^2 - p_ob^2 - \left(\frac{a^2b^2}{r^2}\right)(p_o - p_i)}{b^2 - a^2} \tag{9-26}$$

In these equations (+) stress is tension and (−) stress is compression.

In many applications, the pressure vessel is subjected to *internal pressure only* (p_o = 0), giving

$$\sigma_r = \frac{a^2 p_i}{b^2 - a^2}\left[1 - \frac{b^2}{r^2}\right] \tag{9-27}$$

$$\sigma_t = \frac{a^2 p_i}{b^2 - a^2}\left[1 + \frac{b^2}{r^2}\right] \tag{9-28}$$

For *internally* pressurized vessels, the peak stress *magnitudes* of σ_r and σ_t *both occur at the inner radius*[6] $r = a$, where

$$\sigma_r|_{r=a} = -p_i \tag{9-29}$$

$$\sigma_t|_{r=a} = p_i\left(\frac{b^2 + a^2}{b^2 - a^2}\right) \tag{9-30}$$

The forms of the stress distributions for σ_r and σ_t for the case of an internally pressurized cylinder are sketched in Figure 9.3(a).

If an internally pressurized thick-walled cylindrical pressure vessel is closed at both ends, the resulting longitudinal stress σ_l may be found by adapting the equilibrium concepts leading to (9-6) as

$$\sigma_l = p_i\left(\frac{a^2}{b^2 - a^2}\right) \tag{9-31}$$

It should be noted that the pressure vessel stresses σ_r, σ_t, and σ_l are all *principal stresses*.

In some applications, it may be found that the pressure vessel is subjected to *external pressure only* (p_i = 0), giving

$$\sigma_r = \frac{-b^2 p_o}{b^2 - a^2}\left(1 - \frac{a^2}{r^2}\right) \tag{9-32}$$

$$\sigma_t = \frac{-b^2 p_o}{b^2 - a^2}\left(1 + \frac{a^2}{r^2}\right) \tag{9-33}$$

For *externally* pressurized vessels the peak stress magnitudes of σ_r and σ_t *do not occur at the same radius*. The peak magnitude of σ_r occurs at the inner radius, $r = a$. The

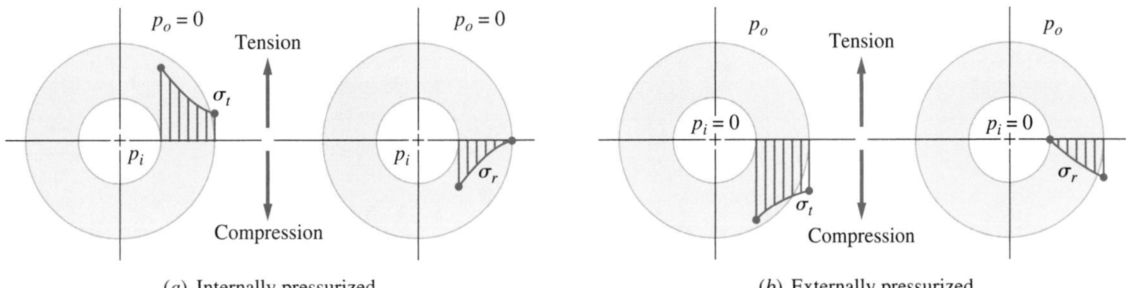

(*a*) Internally pressurized.

(*b*) Externally pressurized.

Figure 9.3
Radial and tangential stress distributions across the wall thickness of internally and externally pressurized thick-walled cylinders.

[6]Since the maximum tensile tangential stress $(\sigma_t)_{max}$ occurs at the *inner* surface of the pressure vessel wall, the more likely crack initiation sites are on the *inside* surface of a thick-walled cylindrical pressure vessel. This demands special attention to the task of choosing an appropriate inspection technique for crack detection. See 7.6.

forms of the stress distributions for σ_r and σ_t for the case of an externally pressurized cylinder are sketched in Figure 9.3(b).

Example 9.1 Designing the Pressurized Shell of a Thick-Walled Cylindrical Hydraulic Actuator

A single-acting vertical-lift hydraulic actuator is to be designed to operate at an ambient temperature of 900°F. The proposed material for the cylindrical outer shell is 1020 steel. As shown in Figure E9.1, the actuator must have the ability to vertically lift a static maximum load of 14,000 pounds, and hold it in place. The available pressure supply provides 2000 psi fluid pressure to the actuator. A pressure release valve allows the load to retract the actuator when desired. The outer diameter of the steel cylinder must not exceed 4.0 inches because of clearance restrictions, but the smaller the better. Perform the initial design calculations to determine pertinent dimensions for the cylindrical shell, disregarding for now the complications of stress concentration. A design safety factor of 2.5 is desired.

Solution

The pertinent dimensions required are inner diameter d_i and outer diameter d_o. The solution strategy will be to first find the piston diameter d_i required to support the load, and then find the outer diameter d_o to provide a design safety factor of $n_d = 2.5$ at the critical point.

Since the maximum load is $W = 14{,}000$ pounds, and the supply pressure is 2000 psi, the required piston area A_i may be found as

$$A_i = \frac{W}{p_i} = \frac{14{,}000}{2000} = 7.0 \text{ in}^2 \tag{1}$$

whence the inside diameter required to lift the load is

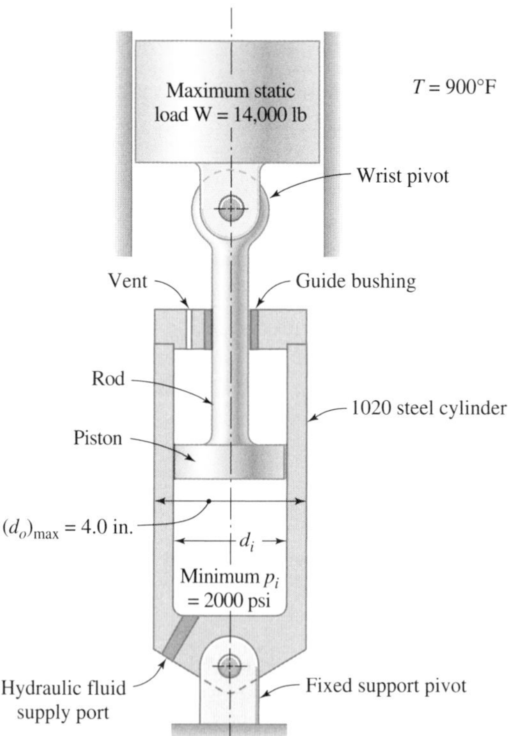

Figure E9.1
Conceptual sketch of high-temperature hydraulic actuator.

$$d_i = \sqrt{\frac{4A_i}{\pi}} = \sqrt{\frac{4(7.0)}{\pi}} = 3.0 \text{ inches} \tag{2}$$

Examining Figure E9.1, it may be noted from the concept configuration that the cylinder walls carry no axial load because the pressure-induced axial force on the piston is transmitted directly to the load as the rod slides through the guide bushing. Hence for this design calculation

$$\sigma_l = 0 \tag{3}$$

resulting in a *biaxial* state of stress throughout the cylinder wall. Since this is a case of internal pressure only, the critical points all lie on the inner wall where the radial and tangential stresses both peak in magnitude, as given by (9-29) and (9-30). Further, there are no shearing stress components on volume elements at radius $r_i = d_i/2$, making σ_r and σ_t principal stresses *by definition*.

Since the 1020 steel material is ductile, the state of stress is multiaxial (biaxial), and the loading is static, the distortion energy design equation (6-14) is appropriate for use. Reducing (6-14) to the biaxial case, we desire

$$\frac{1}{2}[(\sigma_t - \sigma_r)^2 + \sigma_r^2 + \sigma_t^2] = \sigma_d^2 \tag{4}$$

The governing failure mode for this case is judged to be yielding at an ambient temperature of 900°F. From Table 3.5, the yield strength of 1020 steel at 900°F is found to be

$$S_{yp-900} = 24{,}000 \text{ psi} \tag{5}$$

hence the design stress becomes

$$\sigma_d = \frac{S_{yp-900}}{n_d} = \frac{24{,}000}{2.5} = 9600 \text{ psi} \tag{6}$$

At the inner radius, $r_i = 1.5$ inches, so (9-29) and (9-30) give

$$\sigma_r|_{r=1.5} = -2000 \text{ psi} \tag{7}$$

and

$$\sigma_t|_{r=1.5} = 2000\left[\frac{r_o^2 + 1.5^2}{r_o^2 - 1.5^2}\right] \tag{8}$$

Inserting (6), (7), and (8) into (4) gives

$$\frac{1}{2}\left[\left(2000\left\{\frac{r_o^2 + 1.5^2}{r_o^2 - 1.5^2}\right\} - \{-2000\}\right)^2 + (-2000)^2 + \left(2000\left\{\frac{r_o^2 + 1.5^2}{r_o^2 - 1.5^2}\right\}\right)^2\right] = (9600)^2 \tag{9}$$

Perhaps the most expedient way to solve this fourth-order implicit equation for r_o is by trial and error. Since the physical constraints limit r_o to a maximum value of 4.0 inches, we may try this value first. From (9) we may iterate as follows, first dividing both sides by the term $(2000)^2/2$.

From Table E9.1, the initial design recommendations are

Table E9.1 Iteration Summary

r_0	r_0^2	Left side	Right side	Comment
2.0	4.0	34.65	46.08	no solution
1.8	3.24	74.59	46.08	no solution
1.9	3.61	47.75	46.08	close enough

Example 9.1 Continues

$$d_i = 3.0 \text{ inches} \tag{10}$$

$$d_o = 2r_o = 3.8 \text{ inches} \tag{11}$$

As a matter of interest, the ratio of wall thickness to mean diameter is

$$R = \frac{t}{d_m} = \frac{0.4}{3.4} = 0.12 \tag{12}$$

so, from the discussions of 9.6, the solution of this design problem using the thin-walled assumptions would have been marginal to unsatisfactory in terms of making an accurate assessment of the wall-thickness requirement.

9.8 Interference Fits: Pressure and Stress

As discussed in 9.2, the stress distributions in the hub and the shaft of an interference fit assembly may be readily found from (9-29) and (9-30) for the hub, and from (9-32) and (9-33) for the shaft. To find the interfacial pressure p as a function of materials and geometry of the shaft-hub assembly, the four basic principles from elasticity theory, presented in 9.5, may again be utilized. Considering the shaft and hub to be concentric cylinders, assembled one upon the other with an interference fit, as illustrated in Figure 9.4, the requirements of *geometrical compatibility* may be written as

$$|f - a| + |d - f| = \frac{\Delta}{2} \tag{9-34}$$

where

$$\Delta \equiv \text{diametral interference} \tag{9-35}$$

The radii of shaft and hub before and after assembly, as shown in Figure 9.4, are defined as follows:

$a =$ hub inner radius before assembly

$b =$ hub outer radius before assembly

$c =$ shaft inner radius before assembly

$d =$ shaft outer radius before assembly

$e =$ shaft inner radius after assembly

$f =$ common interfacial radius after assembly

$g =$ hub outer radius after assembly

Figure 9.4
Sketches showing dimensions of a hollow shaft and a hub before and after assembly.

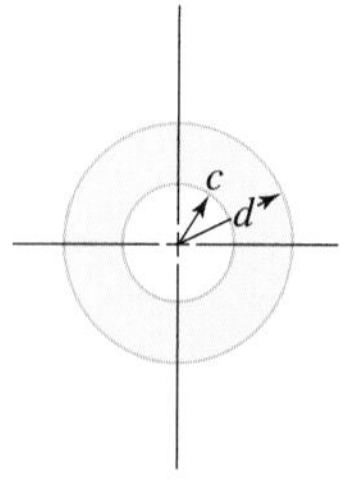

(*a*) Hollow shaft before assembly.

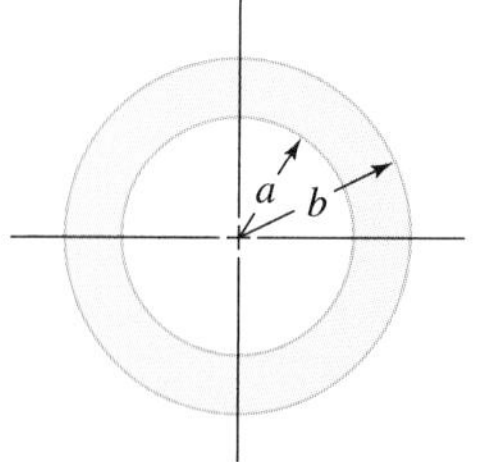

(*b*) Hub before assembly.

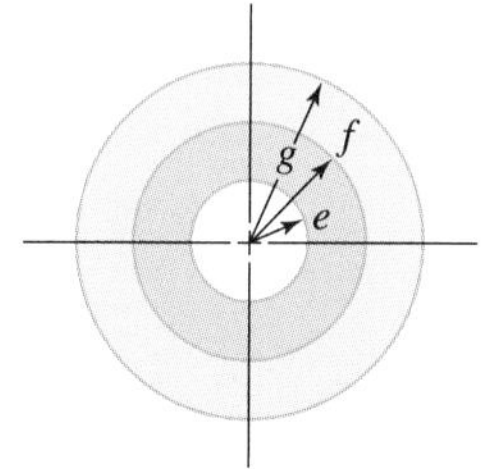

(*c*) Shaft and hub after assembly.

The following subscript notation will be utilized to expedite development of an expression for interfacial pressure p:

h = hub

s = shaft

t = tangential

r = radial

l = longitudinal

To proceed, the magnitudes of tangential (circumferential) strains in the hub and the shaft, respectively, at radius f in the assembly, may be expressed as

$$|\varepsilon_{th}| = \left|\frac{2\pi f - 2\pi a}{2\pi a}\right| = \left|\frac{f - a}{a}\right| \tag{9-36}$$

and

$$|\varepsilon_{ts}| = \left|\frac{2\pi d - 2\pi f}{2\pi d}\right| = \left|\frac{d - f}{d}\right| \tag{9-37}$$

Next, writing *Hooke's Law* equations (valid for both hub and shaft at their common radius f),

$$\varepsilon_{th} = \frac{1}{E_h}[\sigma_{th} - \nu_h(\sigma_{rh} + \sigma_{lh})] \tag{9-38}$$

and

$$\varepsilon_{ts} = \frac{1}{E_s}[\sigma_{ts} - \nu_s(\sigma_{rs} + \sigma_{ls})] \tag{9-39}$$

Also, *by equilibrium*, at radius f

$$\sigma_{rh} = \sigma_{rs} \tag{9-40}$$

and

$$\sigma_{lh} = \sigma_{ls} = 0 \tag{9-41}$$

assuming the length of shaft and hub to be equal. (Usually they are not equal in length and it becomes necessary to introduce an end-of-hub stress concentration factor.)

Rewriting (9-30) for the hub,

$$\sigma_{th} = p\left(\frac{b^2 + a^2}{b^2 - a^2}\right) \tag{9-42}$$

and from (9-29)

$$\sigma_{rh} = -p \tag{9-43}$$

where p is the interfacial pressure resulting from the interference fit assembly of the hub on the shaft.

Likewise, for the shaft, (9-33) gives

$$\sigma_{ts} = -p\left(\frac{d^2 + c^2}{d^2 - c^2}\right) \tag{9-44}$$

and from (9-32)

$$\sigma_{rs} = -p \tag{9-45}$$

Inserting (9-36) and (9-37) into (9-34) gives

$$|\varepsilon_{th}a| + |\varepsilon_{ts}d| = \frac{\Delta}{2} \tag{9-46}$$

Substituting (9-42), (9-43), (9-44), and (9-45) into (9-38) and (9-39), then placing the resulting expressions into (9-46), gives

$$\left|\frac{a}{E_h}\left[p\left(\frac{b^2+a^2}{b^2-a^2}\right)-(-p)\nu_h\right]\right|+\left|\frac{d}{E_s}\left[-p\left(\frac{d^2+c^2}{d^2-c^2}\right)-(-p)\nu_s\right]\right|=\frac{\Delta}{2} \tag{9-47}$$

which may be solved for interfacial pressure p as

$$p=\frac{\Delta}{2\left[\frac{a}{E_h}\left(\frac{b^2+a^2}{b^2-a^2}+\nu_h\right)+\frac{d}{E_s}\left(\frac{d^2+c^2}{d^2-c^2}-\nu_s\right)\right]} \tag{9-48}$$

By using the value of p from (9-48), the radial and tangential stresses in the hub and in the shaft may be found using (9-27), (9-28), (9-29), and (9-33).

If the shaft and hub are made of the same material, $E_s = E_h \equiv E$, and $\nu_s = \nu_h \equiv \nu$. Also, if the shaft is solid, $c = 0$ and $a = d$, to good approximation. Under these circumstances (9-48) simplifies to

$$p=\frac{E\Delta}{4a}\left[1-\frac{a^2}{b^2}\right] \tag{9-49}$$

In cases where the hub length is shorter than the shaft (the usual case) stress concentration occurs at the hub ends due to the sudden change in geometry. Figure 4.7 gives end-of-hub stress concentration factors for this condition.[7] As a practical matter, a raised pad, with generous blending fillet radii, is nearly always specified for the shaft at the site where the interference fit hub is mounted. This practice greatly reduces stress concentration effects at the hub ends and expedites assembly (the hub slips easily over the shaft until it arrives at the raised diameter).

Example 9.2 Designing an Interference Fit Assembly

For a special application, it is desired to assemble a phosphor bronze disk to a hollow steel shaft, using an interference fit for retention. The disk is to be made of C-52100 hot-rolled phosphor bronze, and the hollow steel shaft is to be made of cold-drawn 1020 steel. As shown in Figure E9.2, the proposed nominal dimensions of the disk are 10 inches for outer diameter and 3 inches for the hole diameter, and for the shaft at the mounting pad, 3 inches outer diameter and 2 inches inner diameter. The hub length is 4 inches. The decision has been made that the maximum stress in the disk should not exceed one-half the tensile yield strength of the disk material.

a. What is the maximum diametral interference that should be specified for the fit between the phosphor bronze disk and the steel shaft?

b. If the coefficient of friction between phosphor bronze and steel is about 0.34 dry or 0.17 greasy (see Appendix Table A.1), what torque would you estimate could be transferred from shaft to disk with no slippage?

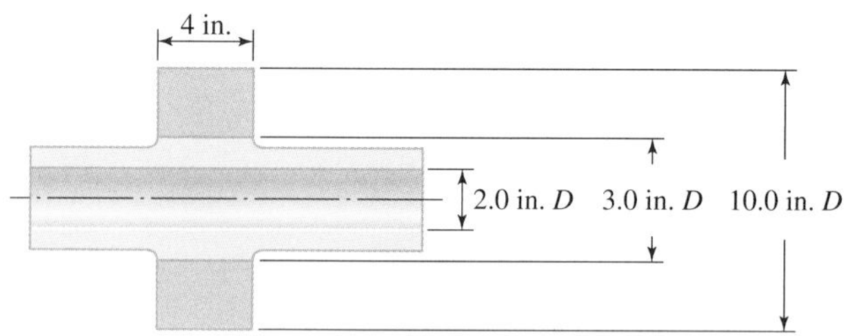

Figure E9.2
Sketch showing nominal configuration of the interference fit assembly.

[7]See also ref. 3, ¶5.5.

c. Approximately what hydraulic press capacity would you estimate might be needed to press the shaft out of the disk after it had been assembled using the interference specified in (a)?

Solution

The material properties of interest may be found in Tables 3.3, 3.9, and 3.10. They are, for commercial bronze (52100):

$$S_u = 55{,}000 \text{ psi}$$
$$S_{yp} = 30{,}000 \text{ psi}$$
$$E = 16 \times 10^5 \text{ psi}$$
$$\nu = 0.35$$
$$e(2 \text{ inches}) = 70 \text{ percent}$$

and for cold-drawn 1020 steel:

$$S_u = 61{,}000 \text{ psi}$$
$$S_{yp} = 51{,}000 \text{ psi}$$
$$E = 30 \times 10^6 \text{ psi}$$
$$\nu = 0.30$$
$$e(2 \text{ inches}) = 25 \text{ percent}$$

a. For the phosphor bronze disk, from (9-29) and (9-30),

$$\sigma_r|_{r=1.5} = -p \tag{1}$$

and

$$\sigma_t|_{r=1.5} = p\left[\frac{(5)^2 + (1.5)^2}{(5)^2 - (1.5)^2}\right] = 1.20p \tag{2}$$

Since the state of stress is biaxial at the inner radius of the disk, and the material is ductile, the distortion energy design equation (6-14) reduces to

$$\frac{1}{2}[(\sigma_t - \sigma_r)^2 + \sigma_r^2 + \sigma_t^2] = \left(\frac{30{,}000}{2}\right)^2 \tag{3}$$

Substituting (1) and (2) into (3) gives

$$p = 7860 \text{ psi} \tag{4}$$

Then using (9-48)

$$7860 = \frac{\Delta}{2\left[\dfrac{15}{17 \times 10^6}\left(\dfrac{5^2 + 1.5^2}{5^2 - 1.5^2} + 0.35\right) + \dfrac{1.5}{30 \times 10^6}\left(\dfrac{1.5^2 + 1.0^2}{1.5^2 - 1.0^2} - 0.30\right)\right]} \tag{5}$$

which gives

$$\Delta = 0.004 \text{ inch} \tag{6}$$

as the *maximum* recommended diametral interference to satisfy the specified requirements.

b. The friction torque capacity of an interference fit joint is given by (8-35) as

$$T_f = \frac{\mu p \pi d_s^2 l_h}{2} \tag{7}$$

Since it is common practice to lubricate the shaft and hub surfaces before pressing them together, the "greasy" friction coefficient $\mu = 0.17$ should be used in (7) to give

Example 9.2 Continues

$$T_f = \frac{(0.17)(7860)\pi(3.0)^2(4.0)}{2} = 75{,}560 \text{ in-lb} \tag{8}$$

c. The axial force required to press the shaft out of the hub is

$$F_{ax} = \mu N = \mu p A = \mu p \pi d_s l_h \tag{9}$$

So as not to underestimate the hydraulic press capacity requirement, the "dry" friction coefficient $\mu = 0.34$ should be used in (9), giving

$$F_{ax} = (0.34)(78{,}560)\pi(3.0)(4.0) = 100{,}750 \text{ lb} \tag{10}$$

If the assembly has been in service, the axial force required for removal might even be higher because of fretting or corrosion.

To be conservative, a hydraulic press capacity of 100 tons is estimated for disassembly.

9.9 Design for Proper Interference

Interference fits are obtained by machining the bore of a hub to a slightly *smaller* diameter than the mating shaft, then assembling the two parts. Assembly may be accomplished by forcefully pressing them together (press fitting), or by heating the hub or cooling the shaft (shrink fitting). The counterposing elastic deflections of shaft and hub create potentially large normal interfacial pressures and slip-resisting frictional forces between them (see 9.8). The question becomes, How much diametral interference should be specified for proper performance? The factors to be considered by a designer in answering this question include:

1. The torque transfer requirement or frictional resistance requirement to meet design specifications
2. The state of stress at the critical point in the shaft-hub assembly, under operating conditions
3. The materials being used
4. The probable governing failure mode
5. Specified machining tolerances for both the shaft mounting pad diameter and the hub bore diameter

Usually, the torque transfer requirement and the nominal shaft and hub sizes are predetermined by power, speed, loading, and configuration specifications for the application. From these, the required interfacial pressure may be calculated by using (8-35) or some similar relationship, and the desired diametral interference Δ may be found from (9-48) or (9-49).

It often turns out that the specification of proper tolerances on shaft and bore diameters is just as important as specification of the diametral interference. This is true because the "spring rates" of the shaft and hub are so stiff that the calculated interference Δ is often of the same order as typical machining tolerances for carefully controlled fits. Consequently, "in-tolerance" shafts and hubs may produce variations in assembly interference that result in significant numbers of field failures. The failures may range from slippage or fretting because of excessively *loose* fits, to premature fatigue failure or hub yielding because of excessively *tight* fits. A designer is well advised to carefully examine the potential failure consequences of inadequate tolerance specifications but to remain mindful of the cost of unnecessarily restrictive tolerancing.

Example 9.3 Assessing Failures in a Press Fit Assembly

A small manufacturer of light industrial conveyors has been experiencing numerous field failures at the input drive sprocket of a new conveyor. About 1 of every 5 installations seems to exhibit rapid galling at the press fit interface between the drive sprocket hub and the shaft, and the sprocket then twists loose from the shaft after only a few hours of operation. You are asked to evaluate the cause of failure and recommend corrective action.

Information supplied for your use in as follows:

> Both the shaft and hub are made of 1020 steel. The mounting diameter of the hole is specified as 1.2500 inches and the mating shaft diameter is specified as 1.2510 inches. Based on power and speed, the torque to be transmitted has been calculated as 1250 in-lb. The outer diameter of the hub is 2.00 inches, and hub length is 1.25 inches.
>
> The project design engineer reports that he has used a factor of safety of *two* on required torque transfer at the interference fit, in calculating the specified diametral interference $\Delta = 0.0010$ inch, yet these galling failures are occurring.

Check the design engineer's calculations, suggest possible causes of failure, and suggest corrective actions.

Solution

To check the design engineer's calculation, first impose the safety factor of *two* on the operational torque transfer requirement to find the design value of friction torque as

$$T_{fd} = 2(1250) = 2500 \text{ in-lb} \tag{1}$$

The required interfacial pressure then is, using (8-35),

$$p = \frac{2(2500)}{(0.11)\pi(1.25)^2(1.25)} = 7410 \text{ psi} \tag{2}$$

where coefficient of friction $\mu = 0.11$ has been taken as the "typical" value for lubricated mild steel on mild steel, from Appendix Table A.1. Since shaft and hub are both made of 1020 steel, (9-45) gives required diametral interference Δ as

$$\Delta = 7410\left[\frac{4\left(\frac{1.25}{2}\right)}{30 \times 10^6\left(1 - \frac{1.25^2}{2.0^2}\right)}\right] = 0.0010 \text{ inch} \tag{3}$$

Thus the design engineer's specification of 0.0010 in interference appears to be correct, giving a hole diameter of 1.2500 and shaft diameter of 1.2510.

Next, the hub stress levels may be checked by using (9-25) and (9-30) to give

$$\sigma_{rh} = -7410 \text{ psi} \tag{4}$$

and

$$\sigma_{th} = \left[\frac{2.0^2 + 1.25^2}{2.0^2 - 1.25^2}\right]7410 = 16{,}910 \text{ psi} \tag{5}$$

Assuming that yielding is the governing failure mode, the yield strength for 1020 steel may be read from Table 3.3 as

$$S_{yp} = 51{,}000 \text{ psi} \tag{6}$$

which gives a design stress, using a design safety factor of 2, of

Example 9.3 Continues

$$\sigma_d = \frac{51{,}000}{2} = 25{,}500 \text{ psi} \tag{7}$$

Since the material is ductile, the distortion energy design equation (6-14) is appropriate for investigating this biaxial state of stress. From (6-14), check to see if

$$\frac{1}{2}[(16{,}910 - \{-7410\})^2 + (-7410)^2 + (16{,}910)^2] \overset{?}{\leq} (25{,}500)^2 \tag{8}$$

or, if

$$4.66 \times 10^8 \overset{?}{\leq} 6.5 \times 10^8 \tag{9}$$

The condition is satisfied, so the hub stress levels *are acceptable*.

What then might be the reason for the field failures? The potential influence of *tolerances* on the fit is worth examining. The title block on the shaft and hub drawings includes the following tolerance specifications:

Unless otherwise noted, dimensions are in inches and tolerances are:
.XX . . . . ±0.03
.XXX . . . . ±0.005
.XXXX . . . . ±0.0005

With these tolerances, the allowable ranges of hole and shaft diameters are found to be

$$d_h = 1.2495 \text{ to } 1.2505 \quad \text{(hole)} \tag{10}$$

$$d_s = 1.2515 \text{ to } 1.2505 \quad \text{(shaft)} \tag{11}$$

Examining diametral interference for the smallest shaft paired with the largest hole gives the *minimum potential interference* as

$$\Delta_{min} = 1.2505 - 1.2505 = 0.0000 \text{ inch} \tag{12}$$

and pairing the largest shaft with the smallest hole gives the *maximum potential interference* as

$$\Delta_{max} = 1.2515 - 1.2495 = 0.0020 \text{ inch} \tag{13}$$

Clearly, $\Delta_{min} = 0$ is unacceptable since no friction torque can be transferred under such circumstances.

As a matter of further concern, the state of stress under maximum interference conditions should be examined. For $\Delta_{max} = 0.0020$ inch, adapting (3),

$$p = 14{,}800 \text{ psi} \tag{14}$$

giving, from (9-29) and (9-30),

$$\sigma_{rh} = -14{,}800 \text{ psi} \tag{15}$$

and

$$\sigma_{th} = 33{,}820 \text{ psi} \tag{16}$$

Substituting these values into (8), check to see if

$$\frac{1}{2}[(33{,}820 - \{-14{,}800\})^2 + (-14{,}800)^2 + (33{,}820)^2] \overset{?}{\leq} (25{,}500)^2 \tag{17}$$

or if

$$1.86 \times 10^9 \overset{?}{\leq} 6.50 \times 10^8 \tag{18}$$

The condition is *not* satisfied, so hub stress levels *exceed* design stress levels (and, in fact, *exceed yield strength* levels as well). Consequently, $\Delta_{max} = 0.0020$ inch is *unacceptable* since yielding leads to loss of fit and possible inability to transfer operating torques.

Summarizing, inadequate tolerance specification in this application leads to the potential for failure at both the minimum interference condition and the maximum interference condition. Tighter tolerances are therefore recommended, in spite of increased production costs.

Before making final recommendations it would be well to explore potential effects of other possible variables, including the range of operating temperatures (not likely to be a problem since shaft and hub are the same material) and the validity of the value of friction coefficient used in the calculations.

Example 9.4 Feasibility of Using an Interference Fit to Transmit Torque

The second-cut sketch of the shaft design configuration developed in Example 8.1 is reproduced as Figure E9.4.

Consideration is being given to using an *interference fit* to mount the 5-inch pitch diameter, 20° involute, straight spur pinion at location *D*, where the nominal diameter is shown as 3.86 inches. It has been separately determined that the pinion may be well approximated as a toothless disk of 1020 steel with outer diameter of 4.67 inches and axial dimension (length) of 2.5 inches. Make a recommendation regarding the feasibility of using an interference fit if a design safety factor of 1.5 is desired.

Solution

From (1) of Example 8.1, the torque to be transferred across the proposed interference fit joint at *D* is

$$T = 63{,}025 \text{ in-lb} \tag{1}$$

Imposing the design safety factor of 1.5 on this torque value, the friction torque required at *D* is

$$T_{fd} = 1.5(63{,}025) = 94{,}540 \text{ in-lb} \tag{2}$$

The required interfacial pressure may be determined from (8-35) as

$$p = \frac{2T_{fd}}{\mu \pi d_s^2 l_h} = \frac{2(94{,}540)}{(0.11)\pi(3.86)^2(2.5)} = 14{,}690 \text{ psi} \tag{3}$$

where the coefficient of friction $\mu = 0.11$ has been taken as the "typical" value for lubricated mild steel on mild steel, from Appendix Table A.1. Since the shaft and the hub are

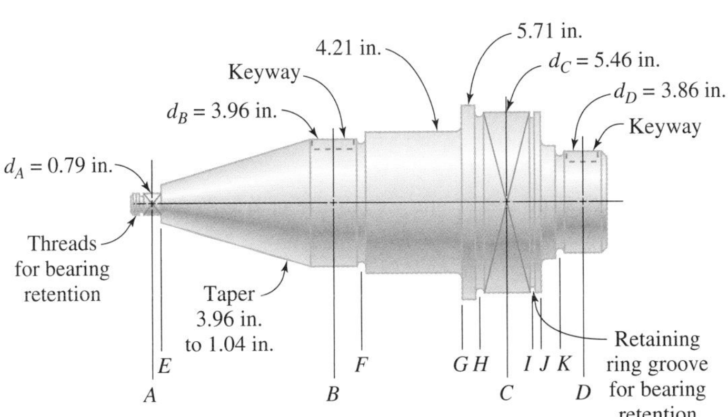

Figure E9.4
Second-cut sketch of shaft, showing principal dimensions (not to scale).

Example 9.4 Continues

both made of the same material, (9-49) may be utilized to find the diametral interference Δ as

$$\Delta = p\frac{4a}{E\left[1 - \dfrac{a^2}{b^2}\right]} \tag{4}$$

and since (3) gives the *minimum* allowable pressure to meet design torque transfer requirements, the *minimum* allowable diametral interference is

$$\Delta_{min} = (14{,}690)\frac{4\left(\dfrac{3.86}{2}\right)}{30 \times 10^6\left[1 - \dfrac{3.86^2}{4.67^2}\right]} = 0.120 \text{ inch} \tag{5}$$

Checking stresses in the hub after assembly, using (9-29) and (9-30) at the interface,

$$\sigma_{rh} = -14{,}690 \text{ psi} \tag{6}$$

and

$$\sigma_{th} = \left[\frac{4.67^2 + 3.86^2}{4.67^2 - 3.86^2}\right]14{,}690 = 78{,}050 \text{ psi} \tag{7}$$

Assuming that yielding is the governing failure mode, the yield strength for 1020 steel may be read from Table 3.3 as

$$S_{yp} = 51{,}000 \text{ psi} \tag{8}$$

This, in turn, gives a design stress of

$$\sigma_d = \frac{S_{yp}}{n_d} = \frac{51{,}000}{1.5} = 34{,}000 \text{ psi} \tag{9}$$

Since the material is ductile, the distortion energy design equation (6-14) is appropriate for investigating this biaxial state of stress. From (6-14), check to see if

$$\frac{1}{2}[(\sigma_t - \sigma_r)^2 + \sigma_r^2 + \sigma_t^2] \overset{?}{\leq} \sigma_d^2 \tag{10}$$

or if

$$\frac{1}{2}[(78{,}050 - \{-14{,}690\})^2 + (-14{,}690)^2 + (78{,}050)^2] \overset{?}{\leq} (34{,}000)^2 \tag{11}$$

or if

$$10.32 \times 10^9 \overset{?}{\leq} 1.16 \times 10^9 \tag{12}$$

Since the expression is *not* satisfied, the *stresses are too high* and the minimum required value of Δ_{min} causes *yielding* of the hub, so $\Delta_{min} = 0.120$ inch *cannot* be sustained.

One possibility for reducing the high stress might be to increase the outer diameter of the toothless disk used to approximate the gear mounted at location D (see Figure E9.4). Equations (9), (9-29), and (9-30) may be solved together to find a new outer diameter for the gear disk.

To solve for the required outer radius r_o of the gear disk, (9-29) and (9-30) may first be used, with $r_i = 3.86/2$, to find

$$\sigma_r|_{r=3.86/2} = -14{,}690 \text{ psi} \tag{13}$$

$$\sigma_t|_{r=3.86/2} = 14{,}690\left[\frac{r_o^2 + \left(\dfrac{3.86}{2}\right)^2}{r_o^2 - \left(\dfrac{3.86}{2}\right)^2}\right] \tag{14}$$

Since the material is ductile, the distortion energy design equation (6-14) is appropriate for investigating this biaxial state of stress. Accordingly, using (9), it may be written that

$$\frac{1}{2}\left[\left(14{,}690\left\{\frac{r_o^2 + 3.725}{r_o^2 - 3.725}\right\} - \{-14{,}690\}\right)^2 + (-14{,}690)^2 + \left(14{,}690\left\{\frac{r_o^2 + 3.725}{r_o^2 - 3.725}\right\}\right)^2\right] = (34{,}000)^2 \tag{15}$$

As discussed in Example 9.1, (15) may be solved iteratively to give, approximately,

$$d_o = 2r_o = 2(3.9) = 7.8 \text{ inches} \tag{16}$$

To meet both the requirements of torque transfer and acceptable stress levels, the pinion diameter would have to be around 7.8 inches. This represents about a 60 percent increase in pinion diameter over the 5-inch-diameter pinion specified in Example 8.1, and implies that the mating gear (probably larger than the pinion) will also grow by about 60 percent. Whether this significant size increase is acceptable must be carefully evaluated by the designer before proceeding.

Problems

9-1. When stresses and strains in a machine element or a structure are investigated, analyses are based on either a "strength of materials" approach or a "theory of elasticity" model. The theory of elasticity model facilitates *determining* the distributions of stresses and strains within the body rather than *assuming* the distributions as required by the strength of materials approach. List the basic relationships from elasticity theory needed to determine the distributions of stresses and strains within elastic solids subjected to externally applied forces and displacements.

9-2. Equations for stresses in thin-walled cylinders are less complicated than equations for stresses in thick-walled cylinders because of the validity of two *simplifying assumptions* made when analyzing thin-walled cylinders. What are these two assumptions?

9-3. a. A *thin-walled* cylindrical pressure vessel with closed ends is to be subjected to an *external* pressure of p_o with an internal pressure of zero. Starting with the generalized Hooke's Law equations, develop expressions for radial strain, tangential (hoop) strain, and longitudinal (axial) strain in the cylindrical vessel wall as a function of pressure p_o, diameter d, wall thickness t, Young's modulus E, and Poisson's ratio ν.

b. If the vessel material is AISI 1018 HR steel [$\sigma_u = 58{,}000$ psi, $\sigma_{yp} = 32{,}000$ psi, $\nu = 0.3$, and e (2 inches) = 25 percent] and if the external pressure is $p_o = 3000$ psi, what is the *change in length* of the vessel if it has an outer diameter of 5.00 inches, wall thickness of 0.25 inch, and length of 15 inches?

c. Does it shorten by this much or lengthen by this much?

d. How much, if any, would the thickness change?

e. If it changes, would it get thicker or thinner?

f. Would you predict yielding of the vessel wall? (Neglect stress concentration and clearly support your prediction with appropriate calculations.)

9-4. Based on the concepts utilized to derive expressions for the stresses in the wall of a thin-walled *cylindrical* pressure vessel, derive expressions for the stresses in the wall of a thin-walled *spherical* pressure vessel.

9-5. A steel hydraulic cylinder, closed at the ends, has an inside diameter of 3.00 inches and an outside diameter of 4.00 inches. The cylinder is internally pressurized by an oil pressure of 2000 psi. Calculate (a) the maximum tangential stress in the cylinder wall, (b) the maximum radial stress in the cylinder wall, and (c) the maximum longitudinal stress in the cylinder wall.

9-6. A cylindrical pressure vessel made of AISI hot-rolled steel plate is closed at the ends. The cylindrical wall has an outside diameter of 12.0 inches and an inside diameter of 8.0 inches. The vessel is internally pressurized to a gage pressure of 15,000 psi.

a. Determine, as accurately as you can, the magnitudes of maximum radial stress, maximum tangential stress, and maximum longitudinal stress in the cylindrical pressure vessel wall.

b. Making the "usual" thin-walled assumptions, and using a mean cylindrical wall diameter of 10.0 inches for your calculations, again determine the magnitude of the maximum radial stress, the maximum tangential stress, and the maximum longitudinal stress in the cylindrical pressure vessel wall.

c. Compare results of (a) and (b) by calculating percentage errors as appropriate, and comment on the results of your comparison.

9-7. Calculate the maximum tangential stress in the steel hub of a press fit assembly when pressed upon the mounting diameter of a hollow steel shaft. The unassembled hub dimensions are 3.0000 inches for the inside diameter and 4.00 inches for the outside diameter, and the unassembled dimensions of the shaft at the hub mounting site are 3.0300 inches outside diameter and 2.00 inches for the inside diameter. Proceed by first calculating the interfacial pressure at the mating surfaces caused by the press fit, then calculating the hub tangential stress caused by this pressure.

9-8. In the design of a jet cargo aircraft, the tail stabilizer, a horizontal flight-control surface, is to be mounted high up on the tail rudder structure, and is to be controlled by two actuator units. The aft unit is to provide the major large-amplitude movement, while the forward unit is to provide the trim action. The forward actuator consists essentially of a power-screw (see Chapter 12) driven by an electric motor, with, for dual-unit safety purposes, an alternative drive consisting of a hydraulic motor that can also drive the screw. In addition, a hand drive is provided in case both the electric drive and the hydraulic drive unit should fail.

The screw consists of a hollow tube of high-strength steel with threads turned on the outer surface, and, for fail-safe dual-load-path purposes, a titanium tube is to be shrink-fitted inside of the hollow steel tube. For preliminary design purposes, the screw may be considered to be a tube of 4 inches inside diameter and $\frac{1}{2}$-inch wall thickness. The proposed titanium tube would have a 4-inch nominal outside diameter and 1-inch-thick wall. The tubes are to be assembled by the hot-and-cold shrink assembly method. The linear coefficient of thermal expansion for the steel material is 6.5×10^{-6} in/in/°F, and the linear coefficient of thermal expansion for the titanium is 3.9×10^{-6} in/in/°F.

a. Determine the actual dimensions at 70°F that should be specified if the diametral interference must never, at any temperature within the expected range, be less than 0.002 inch. Expected temperatures range between the extremes of −60°F and 145°F. Also, the steel tube must not exceed a tangential stress level of 120,000 psi at either temperature extreme.

b. Determine the temperature to which the screw must be heated for assembly purposes if the titanium tube is cooled by liquid nitrogen to −310°F, and if the diametral clearance between tubes for assembly purposes should be about 0.005 inch.

9-9. A steel gear is to be shrink-fitted over a mounting diameter on a solid steel shaft, and its hub abutted against a shoulder to provide axial location. The gear hub has a nominal inside diameter of $1\frac{1}{2}$ inches and a nominal outer diameter of 3 inches. The nominal shaft diameter is $1\frac{1}{2}$ inches. To transmit the torque, it has been estimated that a class FN5 force fit (see Table 6.7) will be required. Stresses in the hub must not exceed the yield strength of the hub material, and a design safety factor of at least 2, based on yielding, is desired.

Two ductile candidate steel materials have been proposed for the gear: AISI 1095 steel quenched and drawn to a hardness of Rockwell C 42, and AISI 4620 hot-rolled steel (with case-hardened teeth). Evaluate these two materials for the proposed application, and decide which material to recommended (see Table 3.3).

9-10. A steel gear has a hub with a nominal bore diameter of 1.0 inch, outer hub diameter of 2.0 inches, and hub length of 1.5 inches. It is being proposed to mount the gear on a steel shaft of 1.0-inch diameter using a class FN4 force fit (see Table 6.7).

a. Determine the maximum tangential and radial stresses in the hub and the shaft for the condition of *loosest* fit.

b. Determine the maximum tangential and radial stresses in the hub and the shaft for the condition of *tightest* fit.

c. Estimate the maximum torque that could be transmitted across the press fit connection before slippage would occur between the gear and the shaft.

Chapter 10

Plain Bearings and Lubrication

10.1 Types of Bearings

Bearings are machine elements that *permit directed relative motion between two parts*, while transmitting forces from one part to the other *without allowing motion in the direction of the applied loads*. For example, each of the shafting applications shown in Figure 8.1 requires a set of bearings to stably support the shaft and its applied loads, while at the same time permitting the shaft to rotate freely. The applied loads on a shaft are typically generated by shaft-mounted gears, belts, chain drives, flywheels, or other specialized elements. Every mechanical device with moving parts requires at least one bearing of some type to permit the desired relative motion while providing the necessary constraints and load-carrying capability.

All bearings may be broadly classified as one of two types:

1. Plain (sliding) bearings
2. Rolling element (antifriction) bearings

Plain bearings, characterized by the sliding of one moving surface upon another, are discussed in detail in the following sections of this chapter. Rolling element bearings, characterized by interposed elements such as balls or rollers between the moving surfaces, are discussed in Chapter 11.

10.2 Uses and Characteristics of Plain Bearings

Plain bearing applications include reciprocating sliders, rotating or oscillating cylindrical members sliding in annular sleeves, and rotating or rotationally oscillating disks sliding on mating disks. These various types of plain bearings are illustrated conceptually in Figure 10.1. Advantages of plain bearings over rolling element bearings, when properly designed, include:[1]

1. Lower first cost
2. Simple design of both shaft and housing
3. Small radial space required
4. Quiet operation
5. Not very sensitive to dust or grit

[1]For advantages of rolling element bearings, see Chapter 11.

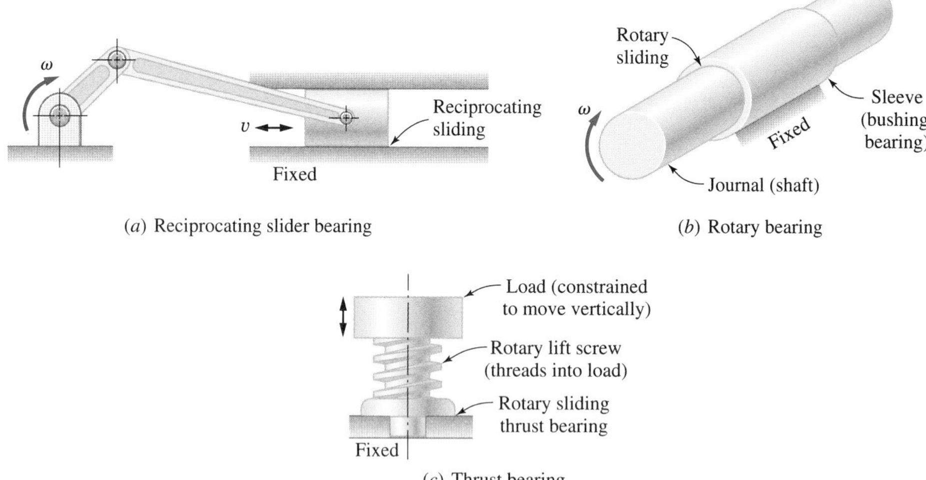

(a) Reciprocating slider bearing

(b) Rotary bearing

(c) Thrust bearing

Figure 10.1
Conceptual sketches of various types of plain bearings.

6. Less subject to fatigue failure
7. Less subject to fretting fatigue failure when small-amplitude, cyclic, relative motion is involved
8. Relatively lightweight
9. Easy to replace

Most plain bearing applications involve the use of a lubricant at the sliding interface to reduce frictional drag and power loss, support the transmitted load (sometimes), and help dissipate the heat generated. In fact, plain bearings are usually subclassified with respect to the types of lubricating conditions that prevail at the sliding interface. As discussed in detail in 10.5, plain bearing lubrication categories include hydrodynamic lubrication, boundary lubrication, hydrostatic lubrication, and solid film lubrication.

10.3 Potential Failure Modes

Depending upon the specific application, the character of the relative sliding motion, the type of lubrication regime at the sliding interface, and the environment, plain bearings may be vulnerable to failure by several possible modes. From the list of possible failure modes described in 2.3, plain bearing failure might occur by yielding, corrosion, adhesive wear, abrasive wear, corrosive wear, surface fatigue wear, fretting wear, creep, or galling and seizure. For example, if the radial bearing loads are high on a *journal and sleeve* type of plain bearing, especially when at rest, significant yielding might possibly occur at the contact site (see discussion of Hertz contact stresses in 4.10). Acid formation during oxidation of a lubricant, or outside contamination, may induce unacceptable corrosion of bearing surfaces. If the lubricant film is so thin that metal-to-metal contact occurs at the bearing interface (often the case during initial startup), adhesive wear may damage the bearing surfaces. Foreign particles from the operating environment, or oxidized wear particles, may produce damaging abrasive wear that could, in serious cases, result in galling and seizure. Depending upon the materials used, the corresponding friction-generated heating at the bearing interface, and the heat dissipation capability of the design configuration, resulting elevated temperatures could induce failures such as creep, lubricant breakdown, or unacceptable temperature-induced changes in dimensions or clearances. If loads or relative sliding motions are cyclic, and the sliding amplitudes are small, surface fatigue wear or fretting wear may damage the mating surfaces, ultimately leading to failure.

10.4 Plain Bearing Materials

Keeping in mind the potential failure modes highlighted in 10.3, and the material selection guidelines of Chapter 3, candidate material pairs for plain bearings should have, in the aggregate, adequate compressive strength, good fatigue strength, low shear strength (to facilitate smoothing of surface asperities), low elastic modulus (to accommodate misalignment and deflection), good ductility and low hardness (to promote embedding of small abrasive foreign particles), high thermal conductivity (to dissipate friction-induced heat), and compatible thermal expansion coefficients (to minimize binding due to differential thermal expansion). Under most circumstances, a bearing pair consists of a hard material, such as a steel shaft, sliding against a soft, ductile material, such as a bronze sleeve (often regarded as a replaceable wear element).

Shafting material selection was discussed in 8.3. Materials typically used for the softer wear element include bronze bearing alloys (e.g. leaded bronze, tin bronze, aluminum bronze, beryllium copper), babbitt metal (lead based, tin based), sintered porous metals (bronze, tin, aluminum), and self-lubricating nonmetallic materials (teflon, nylon, acetal, phenolic, or polycarbonate, any of which may be filled with graphite or molybdenum disulfide). Silver is occasionally used as a bearing surface, usually as a very thin layer plated on a higher-strength substrate. In special applications, such as for water-immersed bearings, fluted rubber or other elastomers are sometimes chosen.

10.5 Lubrication Concepts

Ideally, *lubrication* of plain bearings involves supplying a sufficient quantity of clean, uncontaminated lubricant (usually oil) to the sliding interface in the hope of separating the two rubbing surfaces enough so that *no asperity contacts* occur during operation. If the lubricated bearing system is able to operate under the applied loads with no asperity contacts between the two moving surfaces, the lubricant (oil) film is called a *thick film* or a *full film*. If asperities of one surface *do* contact asperities of the other surface *through the lubricant film*, it is called a *thin film* or a *partial film*. If circumstances are such that *no lubricant at all* can find its way into the contacting interface, a *zero film* condition is said to exist.

If a thick film can be developed and maintained during the operation of a bearing, no wear is expected. For this case, the predicted bearing life would be infinite (at least, very long). In practice, even when a bearing system is designed for thick film operation, periods of thin film (or zero film) operation may exist. For example, when a shaft first starts to rotate from rest, or when periodic high load peaks occur, a brief thin film event nearly always occurs. Under thin film conditions *bearing damage* usually occurs, and bearing life is usually shortened.

The lubricant film between two sliding bearing surfaces may be developed either by providing a sufficient quantity of *externally pressurized* lubricant to the interface, so that the surfaces separate, or by taking advantage of the *viscous*[2] *character* of the oil, the *relative motion* of the surfaces, and the *geometry of the interface* to produce pressure locally that can separate the surfaces. When externally pressurized lubrication is utilized, no rela-

[2]Viscous fluids exhibit a resistance to shear; hence, in this application, the oil is "dragged" into the interfacial region by the relative motion (see 10.7).

tive motion is required to separate the surfaces, and the bearing is classified as *hydrostatic*. When a viscous lubricant is "pumped" into the wedge-shaped zone of a journal bearing by the rotating shaft (see Figure 10.5), the bearing is classified as *hydrodynamic*.

A special subcategory of hydrodynamic lubrication, called *elastohydrodynamic lubrication*, or *squeeze film lubrication*, has been recognized in recent years as an important phenomenon. The potential for elastohydrodynamic lubrication exists when highly loaded, nonconforming, lubricated surfaces, such as rolling element bearings, gear teeth, or cams, pass through zones of rolling contact. In these circumstances, the lubricated surfaces rapidly approach each other to "squeeze" the lubricant film. Due to the extremely high pressures generated in the squeezed film, elastic deformations of the bearing surfaces may become significant, and sometimes the effective oil viscosity increases in the region of contact by as much as *20 orders of magnitude*. For such cases, asperity contacts through the oil film may not occur.[3]

If a plain bearing has a surface area that is too small or too rough, or if the relative velocity is too low, or if the volume of lubricant supplied is too small, or if temperatures increase too much (so that viscosity is lowered too much), or if loads become too high, asperity contacts may be induced. In this case, load *sharing* between the solid asperity contact sites and the pressurized lubricant film may change the lubrication behavior significantly. Such a lubrication scenario is often referred to as *boundary lubrication*.

An additional lubrication category, *solid film lubrication*, is usually designated for cases where dry lubricants, such as graphite or molybdenum disulfide, or self-lubricating polymers, such as Teflon or nylon, are used.

10.6 Boundary-Lubricated Bearing Design

If loads are light, it is often possible to utilize a simple, low-cost, boundary-lubricated bearing to meet design requirements. Electric motor shaft bearings, office machinery bearings, electric fan bearings, power screw support bearings, and home appliance bearings are examples of successful boundary-lubricated bearing designs. The lubricant in such cases may be supplied by hand oilers, drip or wick oilers, grease fittings, oil-impregnated porous-metal bearings, graphite- or molybdenum disulfide–impregnated (or filled) bearings, or self-lubricated polymeric bearings.

Design of boundary lubricated bearings is largely an empirical process based on *documented* user experience. The primary design parameters are unit bearing load P, sliding velocity V, and operating temperature Θ at the sliding interface. Conceptually, the equilibrium operating temperature may be calculated by equating the *heat generated per unit time* at the sliding interface to the *heat dissipated per unit time* to the surroundings (by conduction, convection, and/or radiation), then solving for the temperature rise $\Delta\Theta$ at the sliding interface. As a practical matter, wide variations in configuration and operating conditions make estimating the heat dissipation extremely complex, and nearly impossible to reduce to a simple procedure. The heat balance equation takes the general form

$$\Delta\Theta = f(\text{geometry, material, friction coefficient})(PV) \qquad (10\text{-}1)$$

From this, it may be observed that the PV product serves as an *index to temperature rise* at the sliding interface, and it is widely used as a design parameter for boundary-lubricated bearings. Table 10.1 lists experience-based design-limiting values of PV for many materials, along with design-limiting values for P, V, and Θ. For an acceptable bearing design configuration, operating values of P, V, and PV must all be less than the limiting values

[3]See, for example, ref. 9.

Table 10.1 Design Limits for Boundary-Lubricated Bearings Operating in Contact with Steel Shafts (Journals)[1]

Bearing Sleeve Material	Max. Allowable Unit Load P_{max}, psi	Max. Allowable Sliding Velocity V_{max}, ft/min	Max. Alowable PV Product $(PV)_{max}$, psi-ft/min	Approx. Max. Allowable Op. Temp.[2] Θ_{max}, °F
Porous bronze	2000	1200	50,000	450
Porous Lead-bronze	800	1500	60,000	450
Porous iron	3000	400	30,000	450
Porous Bronze-iron	2500	800	35,000	450
Porous Lead-iron	1000	800	50,000	450
Porous Aluminum	2000	1200	50,000	250
Phenolics	6000	2500	15,000	200
Nylon	2000	600	3,000	200
Teflon	500	50	1,000	500
Filled Teflon	2500	1000	10,000	500
Teflon fabric	60,000	150	25,000	500
Polycarbonate	1000	1000	3,000	220
Acetal	2000	600	3,000	200
Carbon-Graphite	600	2500	15,000	750
Rubber	50	4000	—	150
Wood	2000	2000	12,000	160

[1]See ref. 1.

[2]If lubricant breakdown temperature is lower, it should be used as limiting temperature.

shown in Table 10.1. The parameters P, V, and PV may be calculated as

$$P = \frac{\text{bearing load}}{\text{projected area}} = \frac{W}{dL}\ \text{psi} \tag{10-2}$$

$$V_{cont} = \text{sliding velocity for continuous motion} = \frac{\pi d N}{12}\ \frac{\text{ft}}{\text{min}} \tag{10-3a}$$

or

$$V_{osc} = \text{average sliding velocity for oscillatory motion} \tag{10-3b}$$

$$= \frac{\pi d}{12}\left(\frac{\varphi}{2\pi}\right)f = \frac{\varphi f d}{24}\ \frac{\text{ft}}{\text{min}}$$

and

$$PV = (P)(V)\ \frac{\text{psi-ft}}{\text{min}} \tag{10-4}$$

The unusual *mixed units* for PV have become traditional. Variables in (10-2), (10-3), and (10-4) are defined as follows:

W = total radial bearing load, lb

d = journal diameter, inches

L = sleeve length, inches

N = journal speed relative to sleeve, rpm

$\varphi =$ total angle swept each oscillation, rad

$f =$ oscillating frequency, osc/min

One preliminary design procedure that may be utilized to *initially* size a boundary-lubricated plain bearing is:

1. Calculate the journal (shaft) diameter d using a strength-based analysis
2. Find the resultant radial bearing load W by performing a system force analysis
3. Select tentative materials for journal and sleeve
4. Calculate sliding velocity V from (10-3), and compare to V_{max} from Table 10.1. If $V > V_{max}$, redesign is required
5. For selected materials find limiting value $(PV)_{max}$ from Table 10.1
6. Using results of steps 4 and 5, calculate P and compare to P_{max} from Table 10.1. If $P > P_{max}$, redesign is required
7. Using the result of step 6, and known values of W and d, calculate L from (10-2)
8. It is useful to observe that most successful bearing configurations have length/diameter ratios in the range

$$\frac{1}{2} \leq \frac{L}{d} \leq 2 \tag{10-5}$$

If the result of step 7 lies within this range, the preliminary design is complete. If it does not lie within the range, a design decision must be made as to whether additional redesign is required.

Meeting all of the criteria just discussed is typically an iterative process, and, at best, provides only a *preliminary* design configuration. Experimental verification of any new design for a boundary-lubricated bearing application is recommended.

Example 10.1 Preliminary Design of a Boundary-Lubricated Plain Bearing

From a strength-based analysis, the shaft diameter at one of the bearing sites has been found to be 0.79 inch (for reference, see Figure E8.1D of Example 8.1). The resultant radial load on this bearing was found to be $W = 3530$ pounds, the axial load was zero, and the shaft rotates at 150 rpm. No torque is transmitted through this portion of the shaft. The shaft is to be made of hot-rolled 1020 steel.

It is desired to design a plain bearing for this location. A porous metal sleeve impregnated with oil is the first choice of management since low cost and low maintenance have been included as requirements in the design specifications.

1. Conduct a preliminary design investigation of the feasibility of using such a bearing in this application.
2. If necessary, propose alternatives.

Solution

Since this is a case of *continuous* rotation, the sliding velocity V may be calculated from (10-3a) as

$$V = \frac{\pi d N}{12} = \frac{\pi(0.79)(150)}{12} = 31.0 \text{ fpm} \tag{1}$$

This is well below the limiting velocity $V_{max} = 1200$ fpm (from Table 10.1) for a porous bronze sleeve. Also from Table 10.1, for a porous bronze sleeve,

$$(PV)_{max} = 50{,}000 \frac{\text{psi-ft}}{\text{min}} \tag{2}$$

From (10-4) then

$$P = \frac{50{,}000}{31.0} = 1615 \text{ psi} \tag{3}$$

This is below the limiting unit load $P_{max} = 2000$ psi (from Table 10.1) for a porous bronze sleeve and therefore acceptable.

Finally, from (10-2),

$$L = \frac{W}{Pd} = \frac{3530}{(1615)(0.79)} = 2.77 \text{ inches} \tag{4}$$

From this, the L/d ratio is

$$\frac{L}{d} = \frac{2.77}{0.79} = 3.5 \tag{5}$$

This value for L/d lies well outside the recommended range defined by (10-5). To adjust the sleeve configuration, the upper bound may be selected using (10-5), to give

$$\frac{L}{d} = 2 \tag{6}$$

From this

$$d = \frac{L}{2} = \frac{2.77}{2} = 1.39 \text{ inches} \tag{7}$$

Rechecking (10-2) with these values,

$$P = \frac{3530}{1.39(2.77)} = 917 \text{ psi} \tag{8}$$

This is still well below $P_{max} = 2000$ psi.

Rechecking (10-3a),

$$V = \frac{\pi(1.39)(150)}{12} = 54.6 \text{ fpm} \tag{9}$$

still well below $V_{max} = 1200$ fpm.

Rechecking (10-4),

$$PV = (917)(54.6) = 50{,}000 \frac{\text{psi-ft}}{\text{min}} \tag{10}$$

an acceptable value, as shown in Table 10.1.

The preliminary design recommendations may be summarized as follows:

1. Increase the local shaft diameter at this bearing site to approximately $d = 1.39$ inches.
2. Make the bearing length at this location approximately $L = 2.77$ inches.
3. Make the bearing sleeve of porous bronze, oil impregnated.

10.7 Hydrodynamic Bearing Design

In contrast to the design of boundary-lubricated bearings, as just discussed, hydrodynamically lubricated bearings depend upon the development of a *thick* film of lubricant between the journal and the sleeve so that surface asperities do not make contact through the fluid film. Since there is *no asperity contact* in ideal hydrodynamic lubrication, the material properties of the journal and the sleeve become relatively unimportant, but the properties of the lubricant in the clearance space between the journal and sleeve become *very* important.[4] Discussion of lubricant properties and their measurement and detailed development of hydrodynamic lubrication theory are readily available in a wide variety of textbooks.[5] Only the essential concepts needed for the *design* of hydrodynamic bearings will be included here.

Ultimately, the basic objectives of the designer are to *choose* the bearing *diameter* and *length*, *specify* the surface *roughness* and *clearance* requirements between journal and sleeve, and *determine* acceptable *lubricant properties* and *flow rates* that will assure support of specified design loads and minimize frictional drag. The final design specifications must also produce an acceptable design life without premature failure, while meeting cost, weight, and space envelope requirements.

Lubricant Properties

A *lubricant* is any substance that tends to separate the sliding surfaces, reduce friction and wear, remove friction-generated heat, enhance smooth operation, and provide an acceptable operating lifetime for bearings, gears, cams, or other sliding machine elements. Most lubricants are liquids (e.g., petroleum-based mineral oils), but may be solids or gasses. *Additives* are often used to tailor lubricating oils to specific applications by retarding oxidation, dispersing contaminants, improving viscosity, or modifying the pour point.[6]

The lubricant property of greatest importance in designing hydrodynamically lubricated bearings is *absolute* (*dynamic*) *viscosity* η, usually just called *viscosity*. Viscosity is a measure of the internal shear resistance of a fluid (analogous to shear modulus of elasticity for a solid). When a layer of fluid is interposed between a solid flat plate moving with steady velocity U, and a fixed flat plate, as illustrated in Figure 10.2, experimental results indicate that when the rate of shear du/dy is plotted versus applied shearing stress τ, curves of different shapes result for different lubricants. If a resulting curve is linear, the fluid is said to be *Newtonian* (most lubricating oils are Newtonian). Based on Figure 10.2, Newtonian behavior may be modeled as

$$\tau = \frac{F}{A} = \eta\frac{du}{dy} = \eta\frac{U}{h} \tag{10-6}$$

where F = friction force, lb
η = viscosity of the lubricant, lb-sec/in^2 (reyns)
U = steady velocity of the moving plate, in/sec
h = lubricant film thickness, in
A = area of moving plate in contact with oil film, in^2
τ = shearing stress, psi

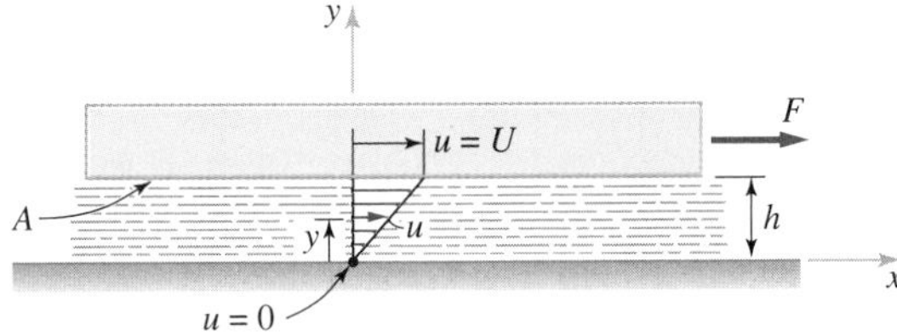

Figure 10.2
Illustration of linear velocity gradient across the film thickness in a Newtonian fluid layer between parallel flat plates.

[4]However, for nonconforming bearing surfaces, such as rolling element bearings, cams, and meshing gear teeth, *elastohydrodynamic* lubrication sometimes prevails. For elastohydrodynamic lubrication, the elastic deformations (hence, the material properties) of the lubricated surfaces become significant, in addition to the properties of the lubricant film.

[5]See, for example, refs. 2, 3, 4, or 5.

[6]The temperature at which an oil will just start to flow under prescribed conditions.

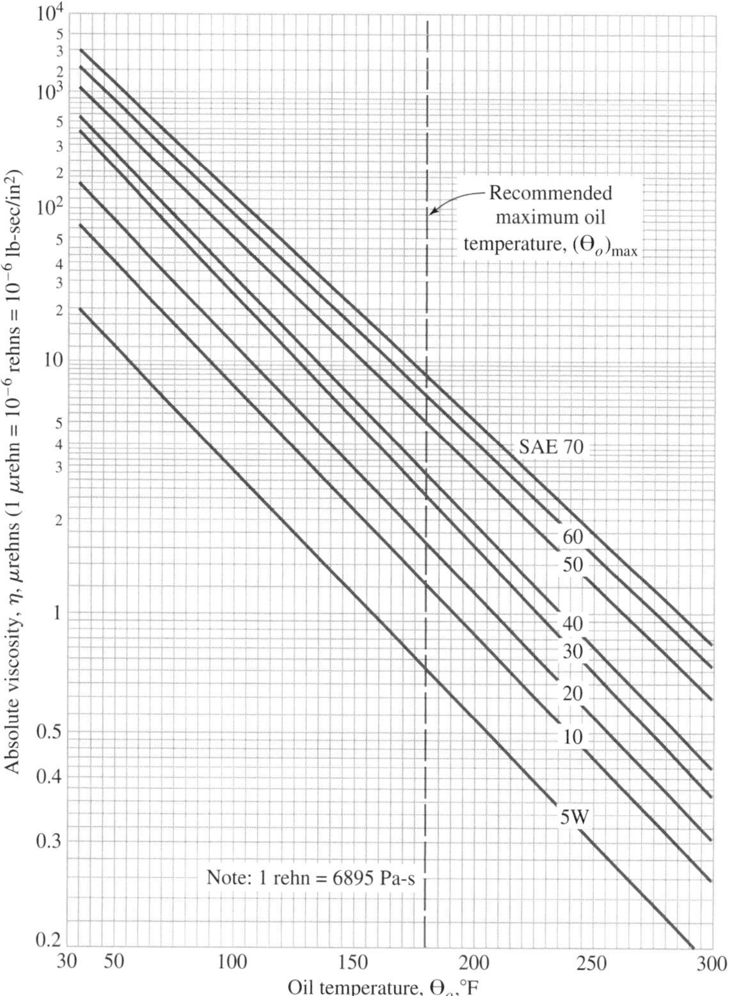

Figure 10.3
Viscosity as a function of temperature for several SAE numbered oils. (From ref. 6.)

The viscosity η fully characterizes a Newtonian fluid, and is a constant for any fixed combination of temperature and pressure. Typically, changes in viscosity for lubricating oils are negligibly small with pressure change, but very significant with temperature change. Figure 10.3 shows a widely published plot of viscosity as a function of temperature for several standard SAE (Society of Automotive Engineers) numbered oils.[7]

Loading, Friction, and Lubricant Flow Relationships

The viscous drag equation (10-6) for the case of parallel flat plates may be adapted to the hypothetical case of a cylindrical journal under no load, rotating concentrically within a bearing sleeve. This is accomplished, conceptually, by "wrapping" the plates into a cylindrical geometry for which the moving plate becomes the rotating journal, the fixed plate becomes the fixed sleeve, and the film thickness h becomes the uniform oil-filled radial clearance c, as illustrated in Figure 10.4. If the resulting journal has diameter $d = 2r$ and length L, the film contact area A becomes

$$A = \pi dL = 2\pi rL \tag{10-7}$$

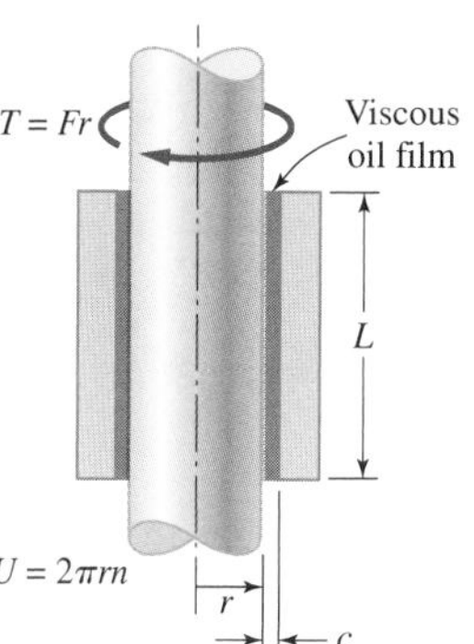

Figure 10.4
Illustration of viscous friction torque for nonloaded concentric journal bearing (Petroff's analysis).

[7]From ref. 6.

Substituting (10-7) and $h = c$ into (10-6), and solving for F, gives

$$F = \frac{2\pi L\eta U r}{c} \tag{10-8}$$

where F is the tangential friction force. If the tangential friction force *per unit bearing length* is defined as $F_1 = F/L$, (10-8) may be rewritten as

$$\left(\frac{F_1}{\eta U}\right)\left(\frac{c}{r}\right) = 2\pi \tag{10-9}$$

This equation, known as *Petroff's equation*, while valid only for the hypothetical case of *zero radial load* and *zero eccentricity* (see Figure 10.9), establishes two important dimensionless parameters useful in bearing design.

Early experiments by *Tower*[8] revealed that *loaded* hydrodynamic journal bearings experience a *pressure buildup* within the lubricant film when the journal is rotating, and that the average pressure multiplied times the projected bearing area is equal to the supported bearing load.

Tower's experimental results were analyzed by *Reynolds*[9] under the assumption that the rotating journal tends to move the viscous fluid into a narrow wedge-shaped zone of lubricant[10] between the journal and sleeve, as depicted in Figure 10.5(c). The development of the wedge-shaped zone is depicted by the sequence of events shown in Figure 10.5. First, the journal is shown at rest, and in metal-to-metal contact with the sleeve. Next, as the journal starts to rotate, it climbs up the sleeve surface, sustained for a while by dry friction. Finally, the journal slips back, shifts center position, and reaches a new equilibrium position associated with the operating conditions for the application. When running at equilibrium, a wedge-shaped zone of oil with a film thickness of h_0 at its apex is established, as shown in Figure 10.5(c). To write equilibrium equations for an elemental volume of oil in the wedge-shaped zone *dx-dy-dz*, the forces on the element may be expressed as shown in Figure 10.5(d). The resulting differential equation of equilibrium reduces to

$$\frac{dp}{dx} = \frac{\partial \tau}{\partial y} \tag{10-10}$$

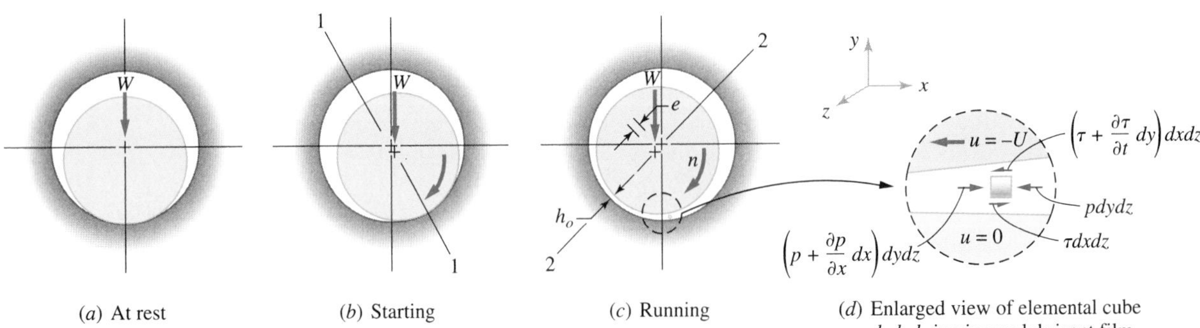

Figure 10.5
Schematic sketch showing relative position change of shaft in the sleeve, starting from rest and increasing to steady-state rotational speed.

[8] See ref. 7. [9] See ref. 8.

[10] A lubricant film between two *parallel* flat plates, or two *concentric* cylindrical surfaces, will not support a bearing load. To support a load, the plates must be slightly *nonparallel*, or the journal center must be slightly *eccentric* with respect to the sleeve to form a wedge-shaped zone. See ref. 6.

This equation was first developed by Reynolds[11] using the following assumptions:[12]

1. The lubricant is a Newtonian fluid
2. Inertia forces produced by the moving fluid are negligible
3. The lubricant is an incompressible fluid
4. The lubricant viscosity is constant throughout the fluid film
5. There is no pressure variation in the axial direction
6. There is no lubricant flow in the axial direction
7. The film pressure is constant in the y-direction (pressure is a function of x only)
8. Lubricant particle velocity is a function of x and y only (no side flow)

Using these assumptions, (10-10) may be integrated, evaluating constants of integration by imposing appropriate boundary conditions, to yield the expression

$$\frac{d}{dx}\left(\frac{h^3}{\eta}\frac{dp}{dx}\right) = -6U\frac{dh}{dx} \tag{10-11}$$

This differential equation is known as the *classical Reynolds equation for one-dimensional flow*, neglecting side leakage. In a similar development, Reynolds utilized a reduced form of the general *Navier-Stokes equations*[13] to derive a differential equation, similar to (10-11), that *does* include the side leakage in the z-direction. This equation is

$$\frac{\partial}{\partial x}\left(\frac{h^3}{\eta}\frac{\partial p}{\partial x}\right) - \frac{\partial}{\partial z}\left(\frac{h^3}{\eta}\frac{\partial p}{\partial z}\right) = -6U\frac{\partial h}{\partial x} \tag{10-12}$$

Other than for the case of a journal bearing of infinite length, analytical solutions of (10-12) have not been found. Various *approximate* solutions have been developed, including an important solution by *Sommerfeld*[14] and a computer-based finite difference solution by *Raimondi* and *Boyd*.[15] Selected results plotted from reference 6 are included here as Figures 10.6 through 10.14. These plots will be used as the basis for hydrodynamic bearing design in this textbook. Data are included here only for 360° full journal bearings and 180° partial bearings. Additional data for 120° and 60° partial bearings may be found in reference 6. The design curves in Figures 10.6 through 10.14 are plotted for L/d ratios (length/diameter ratios) of ∞, 1.0, 0.5, and 0.25, and take into account the possibility that rupture of the lubricant film may occur due to subatmospheric pressure developed in certain portions of the film.

If values of any of the load functions, friction functions, or flow functions plotted in Figures 10.6 through 10.14 are desired for L/d ratios other than the four plotted, the following interpolation equation may be used for any value i of L/d between $\frac{1}{4}$ and ∞.[16]

$$f_i = \frac{1}{\left(\frac{L}{d}\right)^3}\left[-\frac{1}{8}\left(1-\frac{L}{d}\right)\left(1-\frac{2L}{d}\right)\left(1-\frac{4L}{d}\right)f_\infty + \frac{1}{3}\left(1-\frac{2L}{d}\right)\left(1-\frac{4L}{d}\right)f_{1.0}\right.$$
$$\left. - \frac{1}{4}\left(1-\frac{L}{d}\right)\left(1-\frac{4L}{d}\right)f_{0.5} + \frac{1}{24}\left(1-\frac{L}{d}\right)\left(1-\frac{2L}{d}\right)f_{0.25})\right] \tag{10-13}$$

where f_∞, $f_{1.0}$, $f_{0.5}$, and $f_{0.25}$ are the values of the function read from the pertinent plot for L/d ratios of ∞, 1.0, 0.5, and 0.25, respectively. The various functions in Figures 10.6

[11]See ref. 8. [12]See ref. 2, pp. 32 ff, or ref. 10, pp. 487 ff. [13]See for example, ref. 11. [14]See ref. 12.
[15]See ref. 6. [16]See ref. 6.

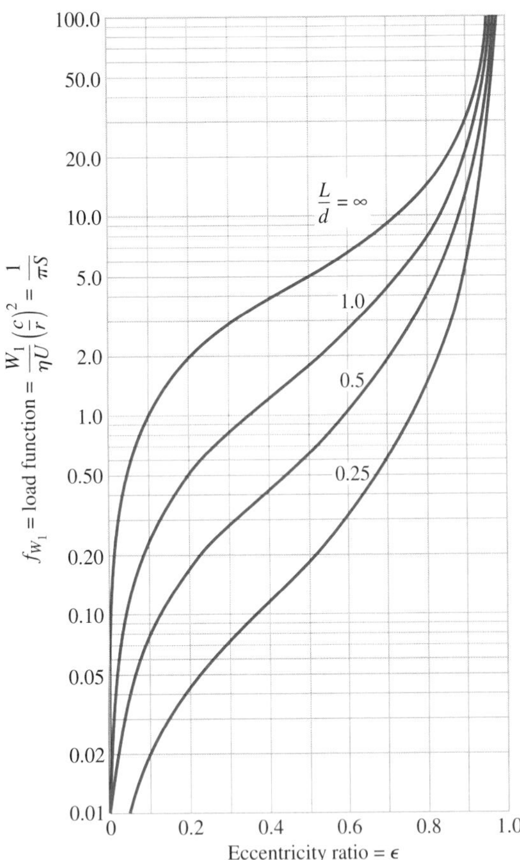

Figure 10.6
Load characteristics of 180° journal bearings. (Plotted from data tabulated in ref. 6.)

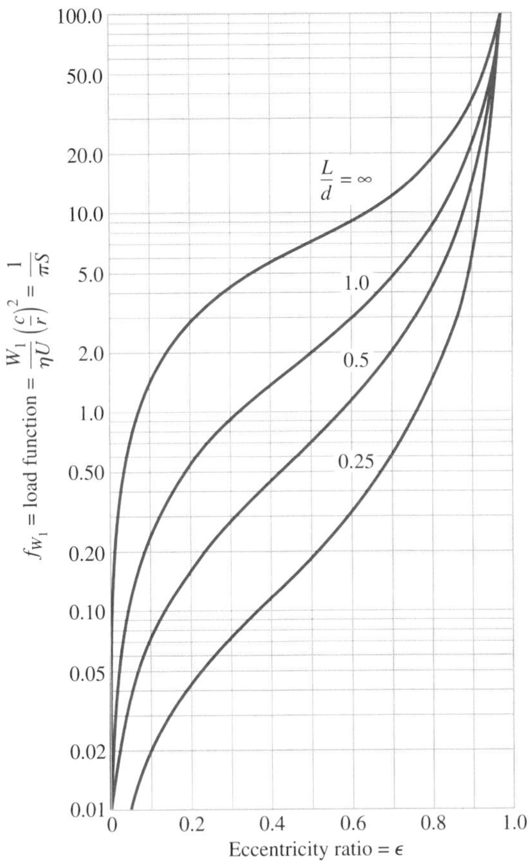

Figure 10.7
Load characteristics of 360° journal bearings. (Plotted from data tabulated in ref. 6.)

through 10.14 are plotted versus the eccentricity ratio ε, where

$$\varepsilon = \frac{e}{c} \tag{10-14}$$

and

e = eccentricity (see Figure 10.5)
c = radial thickness of uniform oil-filled clearance space between concentric journal and sleeve

The load function f_{W_1} and friction function f_{F_1} in Figures 10.6 through 10.9 may be plotted in terms of basic bearing design parameters, or, alternatively, in terms of the *Sommerfeld Number*, *S*, also called *bearing characteristic number*, where

$$S = \left(\frac{r}{c}\right)^2 \frac{\eta n}{P} \tag{10-15}$$

with

r = journal radius, in
c = uniform radial clearance, in
η = viscosity, rehns

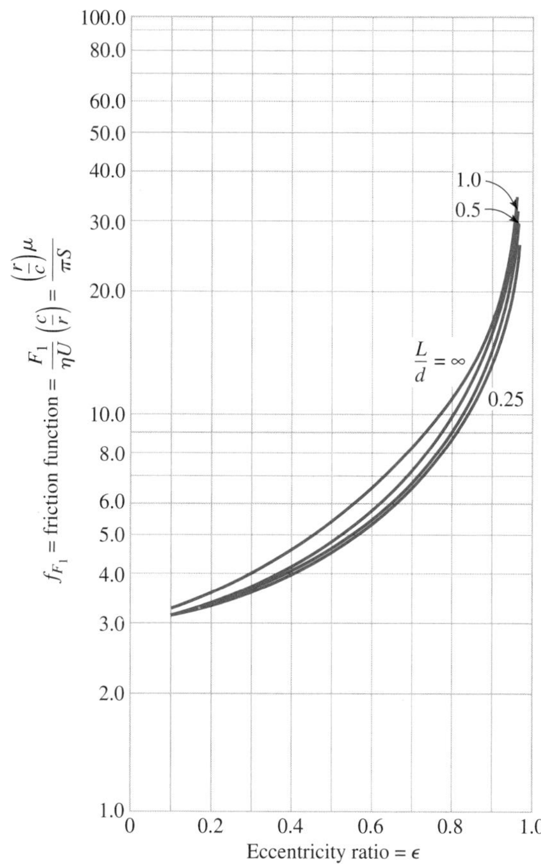

Figure 10.8
Friction characteristics of 180° journal bearings. (Plotted from data tabulated in ref. 6.)

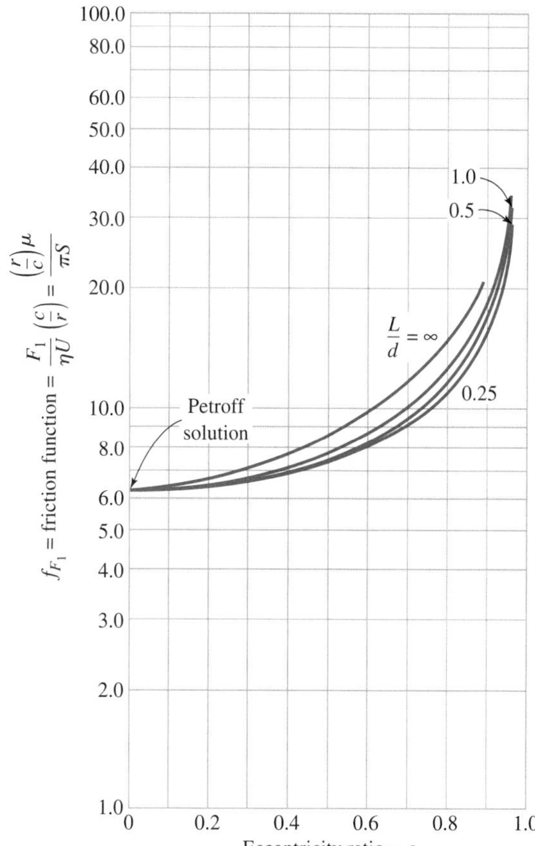

Figure 10.9
Friction characteristics of 360° journal bearings. (Plotted from data tabulated in ref. 6.)

$n =$ journal rotational speed, rev/sec
$P = \dfrac{W}{2rL} =$ unit bearing load, psi
$W =$ radial bearing load, lb
$L =$ bearing length, in

For design purposes, f_{W_1} and f_{F_1} are more conveniently defined in terms of basic bearing parameters; hence they will be used here in the form

$$f_{W_1} = \frac{W_1}{\eta U}\left(\frac{c}{r}\right)^2 \tag{10-16}$$

and

$$f_{F_1} = \frac{F_1}{\eta U}\left(\frac{c}{r}\right) \tag{10-17}$$

where $W_1 = W/L =$ bearing load per unit length, lb/in
$U = 2\pi rn =$ relative surface speed, in/sec
$F_1 = F/L =$ tangential friction force per unit length, lb/in
$F =$ tangential friction force, lb

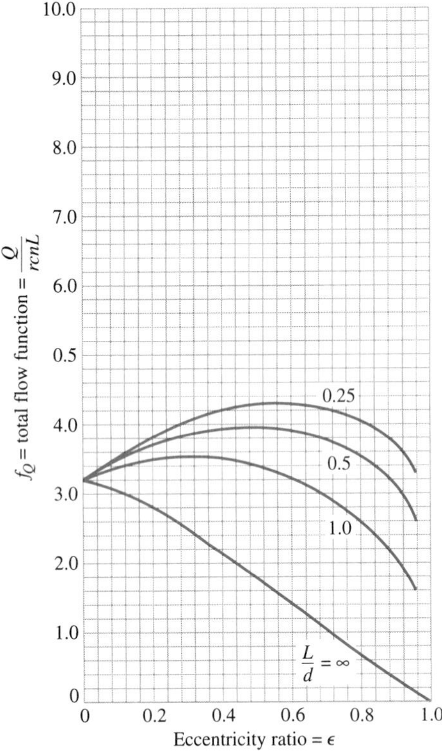

Figure 10.10
Total flow characteristics of 180° journal bearings. (Plotted from data tabulated in ref. 6.)

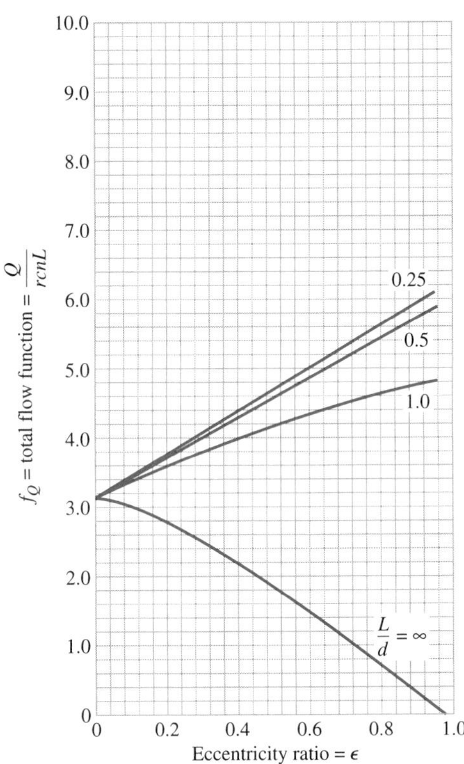

Figure 10.11
Total flow characteristics of 360° journal bearings. (Plotted from data tabulated in ref. 6.)

The relationship between eccentricity ratio ε and *Sommerfeld Number S* is shown in Figure 10.14 for 360° journal bearings. An *optimal design region* is defined in Figure 10.14 between a *minimum friction loss* boundary (left boundary) and a *maximum load capacity* boundary (right boundary). Depending upon the priorities in a specific application, and for a particular length-to-diameter ratio, the eccentricity ratio for minimum friction loss, $\varepsilon_{min\text{-}fr}$, may be read from the chart, as well as the eccentricity ratio for maximum load-carrying capacity, $\varepsilon_{max\text{-}load}$. An optimal value of eccentricity ratio may then be chosen (between $\varepsilon_{min\text{-}fr}$ and $\varepsilon_{max\text{-}load}$), depending upon the relative importance of low friction loss compared to maximum load capacity.

Thermal Equilibrium and Oil Film Temperature Rise

Because lubricant viscosity changes significantly with temperature, and because most commonly used mineral oils tend to break down rapidly above about 180°F, it is important to be able to estimate oil film temperature Θ_o at equilibrium operating conditions. Although external cooling coils may be used in demanding applications, it is more usual for a bearing to be "self-cooling," dissipating its friction-generated heat to the ambient air, primarily by convection from the exposed metal surface area of the bearing (and sump) housing.

Under equilibrium operating conditions, the rate at which heat is generated by fluid friction drag within the rotating bearing must be equal to the rate at which heat is dissipated to the ambient atmosphere. The equilibrium oil film temperature at which this balance is reached must be satisfactory in terms of both equilibrium viscosity and limiting

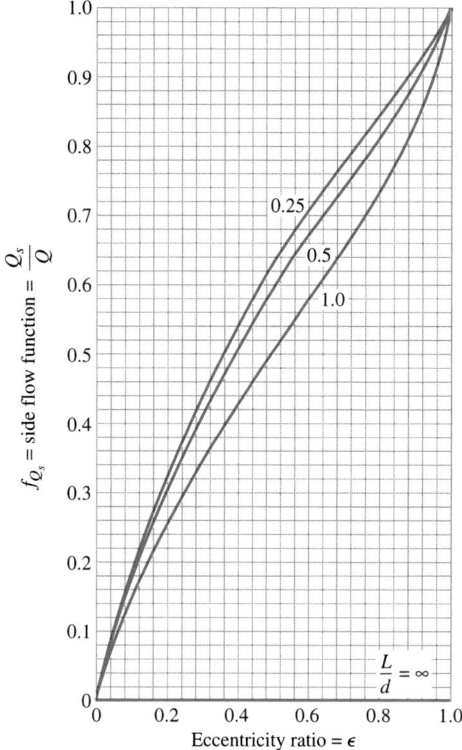

Figure 10.12
Side flow characteristics of 180° journal bearings. (Plotted from data tabulated in ref. 6.)

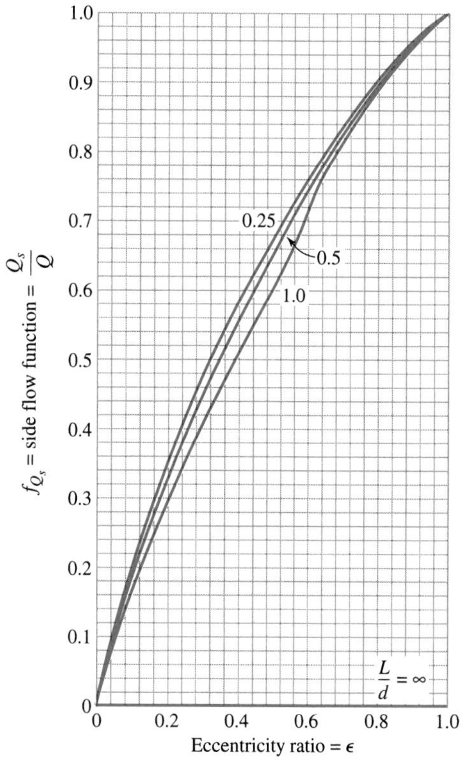

Figure 10.13
Side flow characteristics of 360° journal bearings. (Plotted from data tabulated in ref. 6.)

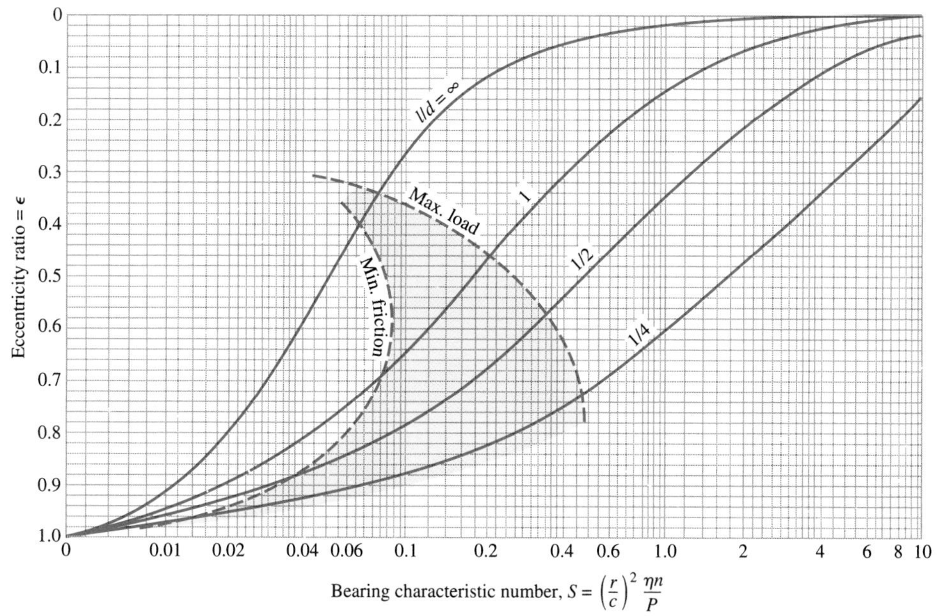

Figure 10.14
Definition of *optimal design region* between curved boundary on the left corresponding to minimum friction loss and curved boundary on the right corresponding to maximum load-carrying capacity. Chart is for 360° journal bearings. (Plotted from data tabulated in ref. 6.)

maximum oil temperature. It is relatively easy to estimate the friction-generated heating rate, but accurate estimation of the heat dissipation rate is difficult, and the results are uncertain. Although the estimates presented here are widely used for preliminary design purposes, experimental validation of the proposed bearing configuration is recommended.

The friction-generated heating rate, H_g, may be calculated as

$$H_g = \frac{2\pi r(60n)(F_1L)}{J_\Theta} = \frac{60UF_1L}{J_\Theta}\ \frac{\text{Btu}}{\text{min}} \tag{10-18}$$

where U = journal surface speed, in/sec
n = journal rotational speed, rev/sec
F_1 = tangential friction force per unit length, lb/in
L = bearing length, in
J_Θ = mechanical equivalent of heat = 9336 in-lb/Btu (10^3 N-m/W-s)

The heat dissipation rate, H_d, may be estimated from Newton's law of cooling[17] as

$$H_d = k_1A_h(\Theta_s - \Theta_a)\frac{\text{Btu}}{\text{min}} \tag{10-19}$$

where k_1 = modified overall heat transfer coefficient, Btu/min-in^2-°F
A_h = exposed surface area of housing, in^2
Θ_s = housing surface temperature, °F
Θ_a = ambient air temperature, °F

At equilibrium it must be true that

$$H_g = H_d \tag{10-20}$$

or

$$\frac{60UF_1L}{J_\Theta} = k_1A_h(\Theta_s - \Theta_a) \tag{10-21}$$

To utilize this equilibrium relationship, it is necessary to evaluate or estimate k_1, A_h, and $(\Theta_s - \Theta_a)$. Typical preliminary design estimates for k_1 are, for "standard" ambient air, and for 500-feet-per-minute forced air flow, respectively,

$$(k_1)_{std} = 2\frac{\text{Btu}}{\text{hr-ft}^2\text{-°F}} = 2.31 \times 10^{-4}\frac{\text{Btu}}{\text{min-in}^2\text{-°F}}\left(11.3\frac{\text{W}}{\text{m}^2\text{-°C}}\right) \tag{10-22}$$

and

$$(k_1)_{500\,fpm} = 3\frac{\text{Btu}}{\text{hr-ft}^2\text{-°F}} = 3.86 \times 10^{-4}\frac{\text{Btu}}{\text{min-in}^2\text{-°F}}\left(17.0\frac{\text{W}}{\text{m}^2\text{-°C}}\right) \tag{10-23}$$

The exposed surface area of the housing may be roughly estimated as a function of the projected area of the bearing, giving

$$A_h = C_hA_p = C_h(\pi dL) \tag{10-24}$$

where the surface area factor C_h ranges from about 5 to about 15, depending upon the design configuration.

The mean temperature rise of the housing surface is often estimated to be about one-half the mean temperature rise of the oil film, giving

$$(\Theta_s - \Theta_a) = \frac{1}{2}(\Theta_o - \Theta_a) \tag{10-25}$$

[17]See ref. 2, p. 56 ff.

Clearly, the values given in (10-22) and (10-23) are only ballpark estimates at best, making experimental validation of the resulting design configuration important.

Design Criteria and Assumptions

To implement a rational design procedure for hydrodynamic journal bearings, it is useful to clearly identify *performance criteria*, *life criteria*, and *assumptions made*. Based on the discussions just presented, these may be summarized as follows:

I. Performance criteria:

A. Support the load

$$\frac{W_1}{\eta U}\left(\frac{c}{r}\right)^2 = f_{W_1}\left(\varepsilon, \frac{L}{d}\right) \tag{10-26}$$

B. Minimize frictional drag

$$\frac{F_1}{\eta U}\left(\frac{c}{r}\right) = f_{F_1}\left(\varepsilon, \frac{L}{d}\right) \tag{10-27}$$

C. Meet space, weight, and cost constraints.

II. Life criteria:

A. Prevent overheating

$$\Theta_o \leq 180°\text{F} \tag{10-28}$$

as determined by the heat balance equation

$$\frac{60UF_1L}{J_\Theta} = k_1A_h(\Theta_s - \Theta_a) \tag{10-29}$$

B. Maintain acceptable viscosity

$$\eta = f(\text{oil}, \Theta_o) \tag{10-30}$$

C. Prevent excessive wear; maintain a minimum film thickness that is greater than combined surface asperity heights of journal and sleeve, by setting

$$h_o = c(1 - \varepsilon) \geq \rho_1 R_j + \rho_2 R_b \tag{10-31}$$

where h_o = minimum film thickness, in
c = radial clearance, in
ε = eccentricity ratio
R_j = arithmetic average asperity peak height above mean *journal* surface, in (see Figure 6.11)
R_b = arithmetic average asperity peak height above mean *bearing* surface, in (see Figure 6.11)
ρ_1,ρ_2 = constants relating "predominant peak roughness" heights to arithmetic average peak heights; take both equal to 5.0 unless more accurate information is available

III. Assumptions:

A. Overall heat transfer coefficient, depending upon ambient conditions, is either

$$(k_1)_{\text{std}} = 2.31 \times 10^{-4} \frac{\text{Btu}}{\text{min-in}^2\text{-°F}}\left(11.3 \frac{\text{W}}{\text{m}^2\text{-°C}}\right) \tag{10-32}$$

or

$$(k_1)_{500\,\text{fpm}} = 3.86 \times 10^{-4} \frac{\text{Btu}}{\text{min-in}^2\text{-°F}}\left(17.0 \frac{\text{W}}{\text{m}^2\text{-°C}}\right) \tag{10-33}$$

B. Housing surface area factor

$$C_h = 8 \qquad \text{(typical range is 5 to 15)} \tag{10-34}$$

C. Housing surface temperature rise

$$\Theta_s - \Theta_a = \frac{1}{2}(\Theta_o - \Theta_a) \tag{10-35}$$

Based on these criteria and assumptions, a step-by-step design procedure may be defined.

Suggested Design Procedure

To meet the basic objectives of hydrodynamic bearing design, namely, choosing *bearing diameter* and *length*, specifying *surface roughness* and *clearance requirements* between journal and sleeve, and defining acceptable *lubricant properties* and *lubricant flow rates*, the following procedure may be followed:

1. Choose an L/d ratio between $\frac{1}{2}$ and 2.
2. Select an appropriate c/r ratio for the design application, utilizing Table 10.2 as a guideline.

Table 10.2 Recommended Bearing Clearances for Various Design Applications

	Running Clearance, inch, for Shaft Diameter Under				
	$\frac{1}{2}$ inch	1 inch	2 inches	$3\frac{1}{2}$ inches	$5\frac{1}{2}$ inches
Precision spindle practice—hardened and ground spindle lapped into bronze bushing. Below 500 fpm and 500 psi.	0.00025 to 0.00075	0.00075 to 0.0015	0.0015 to 0.0025	0.0025 to 0.0035	0.0035 to 0.005
Precision spindle practice—hardened and ground spindle lapped into bronze bushing. Above 500 fpm and 500 psi.	0.0005 to 0.001	0.001 to 0.002	0.002 to 0.003	0.003 to 0.0045	0.0045 to 0.0065
Electric motor and generator practice—ground journal in broached or reamed babbitt bushing.	0.0005 to 0.0015	0.001 to 0.002	0.0015 to 0.0035	0.002 to 0.004	0.003 to 0.006
General machine practice (continuous rotating motion)—turned steel or cold-rolled steel journal in bored and reamed bronze or poured and reamed babbitt bushing.	0.002 to 0.004	0.0025 to 0.0045	0.003 to 0.005	0.004 to 0.007	0.005 to 0.008
General machine practice (oscillating motion)—journal and bearing materials as above	0.0025 to 0.0045	0.0025 to 0.0045	0.003 to 0.005	0.004 to 0.007	0.005 to 0.008

Table 10.2 (*Continued*)

	Running Clearance, inch, for Shaft Diameter Under				
	$\frac{1}{2}$ inch	1 inch	2 inches	$3\frac{1}{2}$ inches	$5\frac{1}{2}$ inches
Rough machine practice—turned steel or cold-rolled steel journal in poured babbitt bearing	0.003 to 0.006	0.005 to 0.009	0.008 to 0.012	0.011 to 0.016	0.014 to 0.020
Automotive crankshaft—					
Babbit bearing	—	—	0.0015	0.0025	—
Cadmium-silver-copper bearing	—	—	0.0020	0.0030	—
Copper-lead bearing	—	—	0.0014	0.0035	—

3. Choose an appropriate eccentricity ratio ε for high load-carrying capacity (large ε), low friction drag (small ε), or a suitable compromise between load-carrying capacity and frictional drag (intermediate ε). See Figure 10.14 to match the choice of ε to application priorities.
4. Select tentative machining processes for journal and sleeve, and determine corresponding arithmetic average roughness peak heights R_j and R_b from Figure 6.11.
5. Write expressions for L, W_1, U, A_h, and c, all in terms of d.
6. Read values of f_{W_1} and f_{F_1} from Figures 10.6 through 10.9, for selected values of ε and L/d. If necessary, use interpolation function f_i from (10-13).
7. Use the load function equation (10-25) to determine
$$d = f_1(\eta) \tag{10-36}$$
8. Use the friction drag equation (10-27), together with life criteria equations (10-28), (10-29), and (10-30), to determine
$$d = f_2(\eta, \Theta_o) \tag{10-37}$$
9. Combine (10-36) and (10-37) to obtain
$$\eta = f_3(\Theta_o) \tag{10-38}$$
10. Refer to Figure 10.3, which provides another (graphical) relationship for
$$\eta = f_{gr}(\Theta_o, \text{oil}) \tag{10-39}$$
11. Select from Figure 10.3 one or more candidate oils, and find (by trial and error) combinations of η and Θ_o that simultaneously satisfy (10-38) and (10-39), specifying, finally, the resulting combination of selected oil, η, and Θ_o.
12. Calculate required bearing diameter d from either (10-36) or (10-37), and bearing length L from step 1. Check the value of diameter d against the corresponding recommended clearance values in Table 10.2, and again iterate steps 1 through 12 if necessary.
13. Impose wear criterion (10-31) to assure that h_o exceeds the combined predominant peak roughness height by an acceptable margin.
14. Calculate lubricant flow rate requirement, and corresponding side leakage flow rate, using Figures 10.10 through 10.13.
15. Summarize results, specifying d, L, Θ_o, oil specification, Q, and Q_s.

Example 10.2 Preliminary Design of a Hydrodynamically Lubricated Plain Bearing

It is desired to design a hydrodynamically lubricated, 180° partial-plain-bearing for use in a production line conveyor to be used to transport industrial raw materials. The vertical bearing load is essentially constant at 500 pounds downward, and the journal is to rotate at a constant speed of 600 rpm. High load-carrying capacity is regarded as more important than low friction loss in the bearing. Conduct a preliminary design investigation to determine a combination of dimensions and lubricant parameters suitable for the proposed application.

Solution

Following the suggested design procedure step by step, and using engineering judgment where necessary, the following sequence of calculations may be made:

1. Choose

$$\frac{L}{d} = 1.0 \tag{1}$$

2. Referring to Table 10.2, the proposed industrial conveyor falls into the category "general machine practice (continuous rotating motion)." For this category, clearance values c from Table 10.2 are shown in Table E10.2A as a function of shaft diameter, together with corresponding c/r ratio values.

   From the options given in Table E10.2A, first try:

$$\frac{c}{r} = 0.003 \tag{2}$$

3. From Figure 10.14, for $L/d = 1.0$, read values of $\varepsilon_{min\text{-}fr} = 0.7$ and $\varepsilon_{max\text{-}load} = 0.47$. Since high load-carrying capacity has a high priority in this application, first try an eccentricity ratio of

$$\varepsilon = 0.5 \tag{3}$$

4. Make a tentative design decision to *turn* the journal (probably steel) and *ream* the bearing sleeve (probably bronze). Using midrange average surface roughness values from Figure 6.11,

$$R_j = 63\ \mu\text{-in} \tag{4}$$

   and

$$R_b = 63\ \mu\text{-in} \tag{5}$$

5. Writing all pertinent variables in terms of diameter d:

$$L = \left(\frac{L}{d}\right)d = (1.0)d = d \tag{6}$$

Table E10.2A Conveyor Bearing Clearance Data

d, in	r, in	c, in	c/r
$\frac{1}{2}$	0.25	0.0020–0.0040	0.008–0.016
1	0.50	0.0025–0.0045	0.005–0.009
2	1.00	0.003–0.005	0.003–0.005
$3\frac{1}{2}$	1.75	0.004–0.007	0.002–0.004
$5\frac{1}{2}$	2.75	0.005–0.008	0.002–0.003

$$W_1 = \frac{W}{L} = \frac{500}{d}\ \frac{\text{lb}}{\text{in}} \tag{7}$$

$$U = \pi d n = \pi d\left(\frac{600}{60}\right) = 31.42d\,\frac{\text{in}}{\text{sec}} \tag{8}$$

$$A_h = C_h(\pi d L) = 8\pi d^2\ \text{in}^2 \tag{9}$$

$$c = \left(\frac{c}{r}\right)\left(\frac{d}{2}\right) = \frac{0.003d}{2} = 0.0015d \tag{10}$$

6. Using $L/d = 1.0$ and $\varepsilon = 0.5$, read values of f_{W_1} and f_{F_1} for a 180° partial bearing from Figures 10.6 and 10.8, giving

$$f_{W_1} = 1.65 \tag{11}$$

and

$$f_{F_1} = 4.80 \tag{12}$$

7. From (10-26) and (11)

$$\frac{W_1}{\eta U}\left(\frac{c}{r}\right)^2 = 1.65 \tag{13}$$

or

$$\frac{(500/d)}{\eta(31.42d)}(0.003)^2 = 1.65 \tag{14}$$

giving

$$d = \sqrt{\frac{500(0.003)^2}{1.65(31.42)\eta}} = \sqrt{\frac{8.68 \times 10^{-5}}{\eta}} \tag{15}$$

8. Combining (10-29), (10-32), (10-35), (6), (8), and (9), and assuming the ambient factory air temperature to be $\Theta_a = 85°\text{F}$,

$$F_1 = \frac{k_1 A_h(\Theta_s - \Theta_a)J_\Theta}{60UL} = \frac{2.31 \times 10^{-4}(8\pi d^2)\left(\dfrac{\Theta_o - 85}{2}\right)9336}{60(31.42d)(d)}$$

$$= 1.44 \times 10^{-2}(\Theta_o - 85) \tag{16}$$

Also, from (10-27) and (12)

$$F_1 = \frac{f_{F_1}\eta U}{\left(\dfrac{c}{r}\right)} = \frac{4.80\eta(31.42d)}{0.003} = 5.03 \times 10^4 \eta d \tag{17}$$

Equating (16) and (17)

$$1.44 \times 10^{-2}(\Theta_o - 85) = 5.03 \times 10^4 \eta d \tag{18}$$

whence

$$d = \frac{2.86 \times 10^{-7}}{\eta}(\Theta_o - 85) \tag{19}$$

9. Equating (15) and (19)

$$\sqrt{\frac{8.68 \times 10^{-5}}{\eta}} = \frac{2.86 \times 10^{-7}}{\eta}(\Theta_o - 85) \tag{20}$$

gives

$$\eta = 9.42 \times 10^{-10}(\Theta_o - 85)^2 \tag{21}$$

Example 10.2 Continues

Table E10.2B Oil Film Temperature and Viscosity Data

Oil Specification	Θ_o,°F	η, rehns from (21)	η, rehns from Fig. 10.3	Comment
SAE 10	130	1.91×10^{-6}	2.45×10^{-6}	No Good
SAE 10	135	2.36×10^{-6}	2.23×10^{-6}	Close
SAE 10	134	2.26×10^{-6}	2.30×10^{-6}	Good

10. From Figure 10.3, the graphical data provide the relationship

$$\eta = f_{gr}(\Theta_o, \text{oil}) \tag{22}$$

11. To solve (21) and (22) simultaneously, select a candidate oil (repeat for other oils if necessary) and compile Table E10.2B by iteration until the selected oil and the mean oil film temperature Θ_o simultaneously satisfy (21) and (22). From this table, tentatively select SAE 10 oil, which has a viscosity of $\eta = 2.3 \times 10^{-6}$ reyns at a mean oil film temperature of $\Theta_o = 134$°F.
12. Required bearing diameter d may be calculated from either (15) or (19). Using (15),

$$d = \sqrt{\frac{8.68 \times 10^{-5}}{2.3 \times 10^{-6}}} = 6.14 \text{ inches} \tag{23}$$

Checking this value against the recommended clearance data of Table E10.2A, c is at the boundary of the tabled data, but will be judged acceptable for now. Also, from (6) and (23)

$$L = d = 6.14 \text{ inches} \tag{24}$$

13. From (10-31), (4), (5), and (10), check

$$h_o = 0.0015(6.14)(1 - 0.5) \overset{?}{\geq} 5(63 \times 10^{-6} + 63 \times 10^{-6}) \tag{25}$$

or

$$0.0046 \overset{?}{\geq} 0.0006 \tag{26}$$

Thus the minimum film thickness h_o is more than seven times the predominant peak roughness height of the journal and sleeve combined, an acceptable value.
14. From Figure 10.10, for $\varepsilon = 0.5$ and $L/d = 1.0$,

$$f_Q = 3.4 = \frac{Q}{rcnL} = \frac{Q}{\left(\frac{6.14}{2}\right)(0.009)(10)(6.14)} = \frac{Q}{1.70} \tag{27}$$

whence

$$Q = 3.4(1.70) = 5.77 \frac{\text{in}^3}{\text{sec}} \tag{28}$$

is the required oil flow that must be supplied to the bearing clearance space. Also, from Figure 10.12, for $\varepsilon = 0.5$ and $L/d = 1.0$,

$$f_{Q_s} = 0.5 = \frac{Q_s}{Q} \tag{29}$$

or

$$Q_s = 0.5(5.77) = 2.88 \frac{\text{in}^3}{\text{sec}} \tag{30}$$

This means that about half the oil flow into the bearing clearance space leaves the bearing as side flow.

15. Summarizing, the preliminary design investigation indicates that an acceptable configuration for this bearing design project should result if the following combination of dimensions and lubricant parameters are specified:

$$d = 6.14 \text{ inches}$$

$$L = 6.14 \text{ inches}$$

$$c = 0.009 \text{ inch (nominal)}$$

$$\textit{surface roughness} \leq 63 \text{ microinches (same for both journal and sleeve)}$$

$$\text{oil} = \text{SAE 10}$$

$$Q = 5.77 \text{ in}^3/\text{sec (about 1.5 gal/min)}$$

$$\Theta_o = 134°F$$

To emphasize it again, many approximations were used in reaching these values, so experimental validation is recommended before finalizing the design. Also, it is well to recognize that the solution presented here is not unique; many other potentially satisfactory combinations of dimensions and lubricants could be found.

10.8 Hydrostatic Bearing Design

As just discussed, in hydrodynamic bearing lubrication the motion of the journal rotating within the sleeve creates a pressurized thick oil film that can support the bearing load without surface asperity contact. In some applications, it may not be possible to sustain a hydrodynamic oil film between the moving elements of the bearing. For example, the motion may be very slow, reciprocating, or oscillating, or loads may be very high, or it may be desired to have very low frictional drag at startup. In such cases it may become necessary to utilize *hydrostatic* lubrication, where lubricant from a positive displacement pump is forced into a recess, or *pool*, in the bearing surface. The externally pressurized lubricant lifts the load, even when the journal is not rotating, allowing the lubricant to flow out of the recess through the clearance space between the separated surfaces, returning to a sump for recirculation.

The main elements of a hydrostatic bearing system are a positive displacement pump, an oil reservoir, and a supply manifold to supply oil to all the bearing pad recesses in the system. Flow control orifices are used to meter the flow so that balanced pressure may be achieved among the supporting pools to allow each pad to lift its share of the load. When operating at equilibrium, the pump-supplied lubricant flow just matches the total outflow from the supporting pools, producing a film thickness h between bearing surfaces great enough so that surface asperities do not make contact through the oil film.

Although the principles of hydrostatic bearing design are relatively simple, the details are highly empirical and experience based. For example, the number, shape, and size of the recesses required for stability and adequate load capacity, the clearances and tolerances, the surface finishes, the pumping capacity and power required, and system cost can best be determined by specialists in hydrostatic bearing design. Further details of hydrostatic bearing design are beyond the scope of this text.[18]

[18]For additional information, see, for example, ref. 13.

Problems

10-1. Plain bearings are often divided into four categories, according to the prevailing type of lubrication at the bearing interface. List the four categories, and briefly describe each one.

10-2. From a strength-based shaft design calculation, the shaft diameter at one of the bearing sites on a steel shaft has been found to be 1.50 inches. The radial load at this bearing site is 150 lb, and the shaft rotates at 500 rpm. The operating temperature of the bearing has been estimated to be about 200°F. It is desired to use a length-to-diameter ratio of 1.5 for this application. Based on environmental factors, the material for the bearing sleeve has been narrowed down to a choice between nylon and filled Teflon (see Table 10.1). Which material would you recommend?

10-3. It is being proposed to use a nylon bearing sleeve on a fixed steel shaft to support an oscillating conveyor tray at equal intervals along the tray, as shown in Figure P10.3. Each bearing bore is to be 0.50 inch, bearing length is to be 1.00 inch, and it is estimated that the maximum load to be supported by each bearing is about 500 lb. Each bearing rotates ±10 degrees per oscillation on its fixed steel journal, at a frequency of 60 oscillations per minute. Would the proposed nylon bearing sleeve be acceptable for this application?

10-4. A local neighborhood organization has become interested in replicating a waterwheel-driven grist mill of the type that had been used in the community during the nineteenth century, but they have not been able to locate any detailed construction plans. One of their concerns is with the bearings needed to support the rotating waterwheel. To give an authentic appearance, they would like to use an oak bearing on each side of the waterwheel to support a cast-iron waterwheel shaft. The waterwheel weight, including the residual weight of the retained water, is estimated to be about 12,000 lb, and the wheel is to rotate at about 30 rpm. It has been estimated on the basis of strength that the cast-iron shaft should be no less than 3 inches in diameter. The bearings need to be spaced about 36 inches apart. Propose a suitable dimensional configuration for each of the two proposed oak bearings so that bearing replacement will rarely be needed. It is anticipated that 68°F river water will be used for lubrication.

10-5. From a strength-based analysis, a shaft diameter at one of its support bearing sites must be at least 1.50 inches. The maximum radial load to be supported at this location is estimated to be about 150 lb. The shaft rotates at 500 rpm. It is desired to use a nylon bearing sleeve at this location. Following established design guidelines for boundary-lubricated bearings, and keeping the bearing diameter as near to the 1.50-inch minimum as possible, propose a suitable dimensional configuration for the bearing.

10-6. A preliminary result obtained as a possible solution for problem 10-5 indicates that the smallest acceptable bearing diameter for the specifications given is about 3.3 inches. Engineering management would prefer to have a bearing diameter of about 1.50 inches (the minimum based on shaft strength requirements), and they are asking whether it would be possible to find another polymeric bearing material that might be satisfactory for this application. Using Table 10.1 as your resource, can you find a polymeric bearing material other than nylon that will meet established design guidelines and function properly with a diameter of 1.50 inches?

10-7. A plain bearing is to be designed for a boundary-lubricated application in which a 3.00-inch-diameter steel shaft rotating at 1750 rpm must support a radial load of 225 lb. Using established design guidelines for boundary-lubricated bearings, and using Table 10.1 as your resource, select an acceptable bearing material for this application.

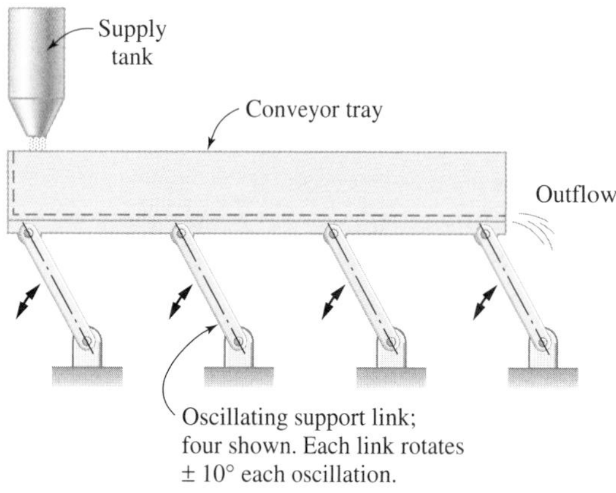

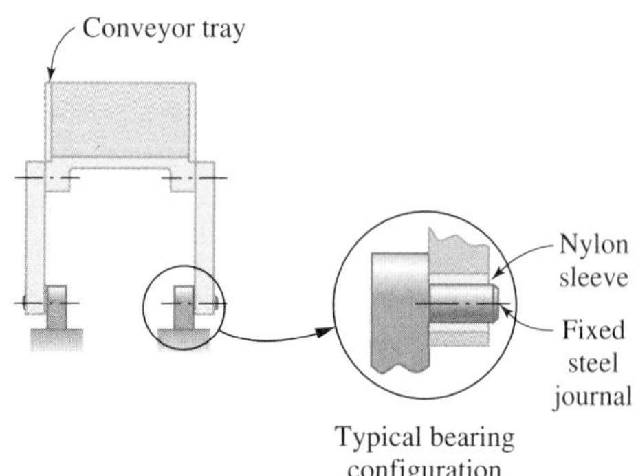

Figure P10.3
Schematic arrangement of an oscillating conveyor.

10-8. A plain bearing is to be designed for a boundary-lubricated application in which a 0.5-inch-diameter steel shaft rotating at 1800 rpm must support a radial load of 75 lb. Using established design guidelines for boundary-lubricated bearings, and using Table 10.1 as your resource, select an acceptable bearing material for this application if the operating temperature is estimated to be about 350°F.

10-9. A proposed flat belt drive system (see Chapter 17) is being considered for an application in which the driven steel shaft is to rotate at a speed of 1440 rpm, and the power to be transmitted is 800 W. As shown in Figure P10.9, the power is transmitted to the 10-mm-diameter (driven) shaft by a flat belt running on a shaft-mounted pulley. The pulley has a nominal pitch diameter of 60 mm, as sketched in Figure P10.9. It is desired to support the driven shaft using two grease-lubricated plain bearings, one adjacent to each side of the pulley (see Figure P10.9). The two bearings share the belt load equally. It has been determined that the initial belt tension, T_0, should be 150 N (in each side of the belt) to achieve optimum performance, and it may reasonably be assumed that the *sum* of tight side and slack side belt tension will remain approximately equal to $2T_0$ for all operating conditions. Select satisfactory plain bearings for this application, including their diameter, their length, and an acceptable material from which to make them (see Table 10.1).

10-10. It is desired to use a hydrodynamically lubricated 360° babbitt-metal plain bearing for use in supporting the crankshaft (see Chapter 19) of an automotive-type internal combustion engine for an agricultural application. Based on strength and stiffness calculations, the minimum nominal journal diameter is 2 inches, and a length-to-diameter ratio of 1.0 has been chosen. The maximum radial load on the bearing is estimated as 700 lb, and the journal rotates in the bearing sleeve at 1200 rpm. High load-carrying ability is regarded as much more important than low friction loss. Tentatively, an SAE 30 oil has been chosen, and the average bearing operating temperature has been determined to be about 150°F. Estimate the power loss due to bearing friction.

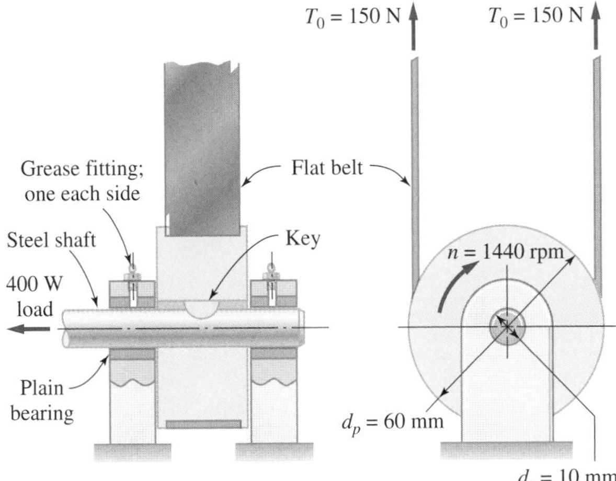

Figure P10.9
Flat belt drive system.

10-11. In an automotive crankshaft application, a hydrodynamic full 360° journal bearing must be 2 inches in nominal diameter based on strength requirements, and the bearing length being considered is 1.0 inch. The journal is to be made of steel and the bearing sleeve is to be made of a copper-lead alloy (see Table 10.2). The bearing must support a radial load of 1000 lb, and the journal rotates at 3000 rpm. The lubricant is to be SAE 20 oil, and the average operating temperature at the bearing interface has been estimated to be about 130°F. Load-carrying ability and low friction loss are regarded as about equally important.

a. Find the minimum *film thickness* required for this application.

b. What *manufacturing processes* would you recommend for finishing the journal and the sleeve to provide hydrodynamic lubrication at the bearing interface? Justify your recommendations. (*Hint*: Examine Figure 6.11.)

c. Estimate the *power loss* resulting from bearing friction.

d. What *oil flow rate* must be supplied to the bearing clearance space?

10-12. A hydrodynamic journal bearing rotates at 3600 rpm. The bearing sleeve has a 32-mm diameter and is 32 mm long. The bearing radial clearance space is to be 20 μm, and the radial load on the bearing is to be 3 kN. The lubricant chosen is SAE 10 oil supplied at an average temperature of 60°C. Estimate the friction-generated heating rate in this bearing if the *eccentricity ratio* has been determined to be 0.65.

10-13. It is desired to design a hydrodynamically lubricated 360° plain bearing for a special factory application in which a rotating steel shaft must be at least 3.0 inches nominal diameter and the bushing (sleeve) is to be bronze, reamed to size. The radial bearing load is to be 1000 lb. The desired ratio of length to diameter is 1.5. The shaft is to rotate at a speed of 1000 rpm. It has been estimated that an eccentricity ratio of 0.5 should be a good starting point for designing the bearing, based on an evaluation of the optimal design region of Figure 10.14 for a length-to-diameter ratio of 1.5.

Conduct a preliminary design investigation to determine a combination of dimensions and lubricant parameters suitable for the proposed application.

10-14. For the design result you found in solving problem 10-13,

a. Find the friction drag torque.

b. Find the power dissipated as a result of friction drag.

10-15. A hydrodynamically lubricated 360° plain bearing is to be designed for a machine tool application in which a rotating steel spindle must be at least 1.00 inch nominal diameter, the bushing is to be bronze, and the steel spindle is to be lapped into the bronze bushing. The radial bearing load is 40 lb, and the spindle is to rotate at 2500 rpm. The desired ratio of length to diameter is 1.0. Conduct a preliminary design study to determine a combination of dimensions and lubricant parameters suitable for this application.

10-16. For your proposed design result found in solving problem 10-15,

a. Find the friction drag torque.

b. Find the power dissipated as a result of friction drag.

10-17. A hydrodynamically lubricated 360° plain bearing is to be designed for a conveyor-roller support application in which the rotating cold-rolled steel shaft must be at least 100 mm nominal diameter and the bushing is to be made of poured babbitt, reamed to size. The radial bearing load is to be 18.7 kN. The desired ratio of length to diameter is 1.0. The shaft is to rotate continuously at a speed of 1000 rpm. Low friction drag is regarded as more important than high load-carrying capacity. Find a combination of dimensions and lubricant parameters suitable for this conveyor application.

10-18. For your proposed design result found in solving problem 10-17, find the friction drag torque.

Chapter 11

Rolling Element Bearings

11.1 Uses and Characteristics of Rolling Element Bearings

As for the case of plain bearings (see Chapter 10), rolling element bearings are designed to permit relative motion between two machine parts, usually a rotating shaft and a fixed frame, while supporting the applied loads. In contrast to the sliding interface that characterizes plain bearings, for rolling element bearings the rotating shaft is separated from the fixed frame by interposed rolling elements, so that *rolling friction* prevails rather than sliding friction. Consequently, both startup torque and operational friction losses are typically much lower than for plain bearings.

Rolling element bearing applications range from tiny instrument bearings with bore diameters of only a few millimeters to huge special bearings such as drag-line trunnion bearings with bores of 20 feet.

A wide array of sizes and types between these extremes is commercially available, and a large percentage of consumer products embody rolling element bearings. In addition to the more common rotational configurations, rolling element bearing concepts have been extended to linear, linear-rotary, and curved-path applications by using recirculating loops, circuits, or guideways for the rolling elements.[1]

Advantages of rolling element bearings include (see Chapter 10 for advantages of plain bearings):

1. High reliability with minimum maintenance
2. Minimum lubrication required. Lubricant can often be sealed in for the "lifetime" of the bearing
3. Good for low-speed operation
4. Low starting friction and low power loss due to frictional drag
5. Can readily support radial, thrust, or combined radial and thrust loads
6. Small axial space required
7. Nearly universal interchangeability among manufacturers due to industry-wide standardized sizes and closely controlled tolerances
8. Can be preloaded to eliminate internal clearances, improve fatigue life, or increase bearing stiffness (see 4.12 and Figure 11.7)
9. Increase in operational noise level warns of impending failure

[1]Applications include ball splines, linear ball bearings, flat-way linear bearings, and ball screws. See, for example, ref. 1.

11.2 Types of Rolling Element Bearings

Rolling element bearings may be broadly classified as either ball bearings (spherical rolling elements) or roller bearings (nominally cylindrical rolling elements). Within each of these broad categories there are a host of geometrical configuraitons commercially available. Rolling element bearings are almost universally standardized[2] (cooperatively) by the *American Bearing Manufacturers Association*[3] *(ABMA)*, American National Standards Institute (ANSI), and the International Standards Organizaiton (ISO).

As illustrated in Figure 11.1 the typical structure for rolling element bearings involves an *inner race*, an *outer race, rolling elements* captured between the races, and a *separator* (or *cage*) used to space the rolling elements so they do not rub each other during operation. The nomenclature shown in Figure 11.1 is widely used for both ball and roller bearings, except that the outer race of a tapered roller bearing is usually called a *cup* and the inner race a *cone*. Other exceptions include thrust bearings (typical configuration is not consistent with the nomenclature of Figure 11.1) and needle bearings (small-diameter rollers, often retained in a thin outer shell, and sometimes used with the needle rollers directly in contact with a hardened steel shaft surface).

Figure 11.2 illustrates a few selected examples from the wide variety of commercially available ball bearing configurations. Most bearing manufacturers publish engineering manuals containing extensive descriptions of the various types and sizes available, and procedures for making appropriate selections.[4] The *single-row deep-groove (Conrad) ball bearing*, shown in Figure 11.2(a), is probably the most widely utilized rolling element bearing because it can support both radial loads and moderate thrust loads in either direction. This is made possible by using continuous shoulders uninterrupted by *filling notches*. Further, these bearings are relatively inexpensive and operate smoothly over a wide speed range. Deep-groove ball bearings are manufactured by moving the inner race to an eccentric position, inserting as many balls as possible, repositioning the inner race to a concentric location, spacing the balls uniformly, and assembling the separator to maintain ball spacing with minimum frictional drag. To increase the *radial* load capacity, a *filling notch* is sometimes cut into one side of each bearing race, permitting more balls to be inserted than is possible for the Conrad design. This maximum-capacity design, illustrated in Fig-

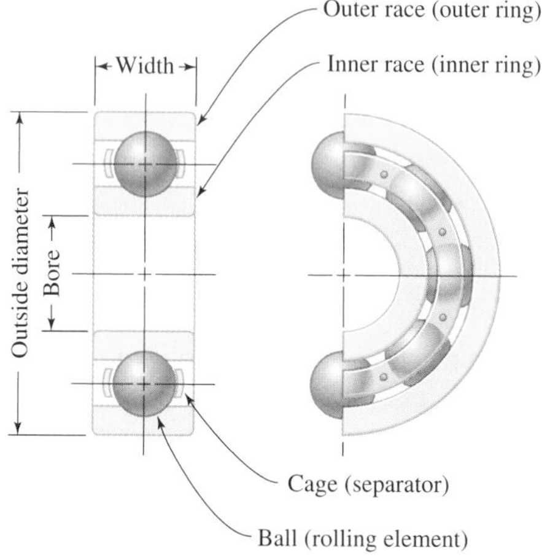

Figure 11.1
Bearing nomenclature. A ball bearing is illustrated, but roller bearing nomenclature is substantially the same.

[2]See, for example, ANSI/AFBMA standards 9-1990 for ball bearings (ref. 2) and 11-1990 for roller bearings (ref. 3).

[3]Formerly Antifriction Bearing Manufacturers Association (AFBMA).

[4]See, for example, ref. 5.

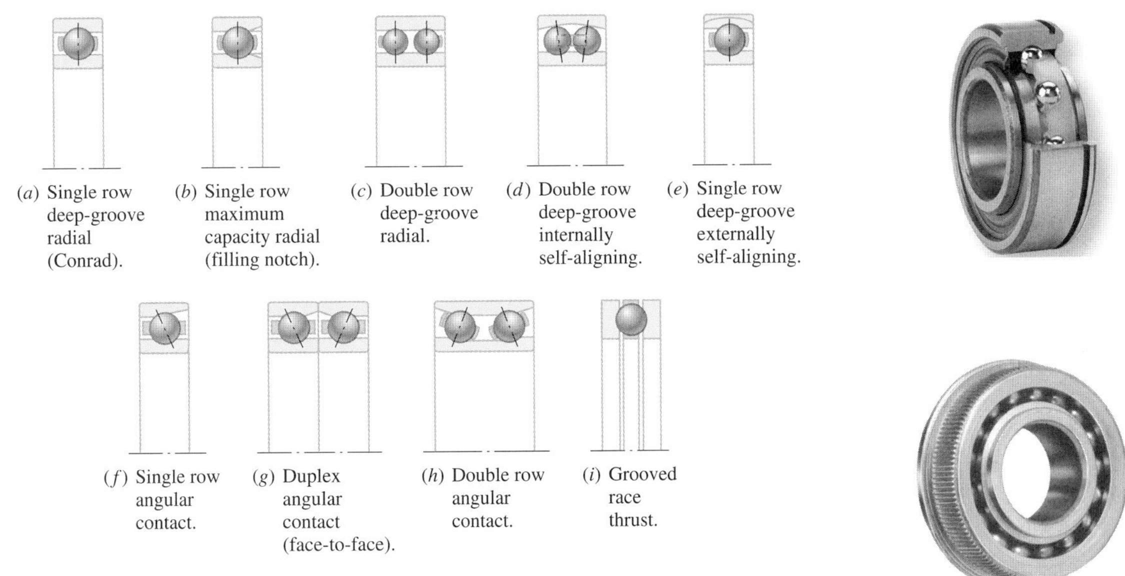

Figure 11.2
Selected examples from the many types of commercially available ball bearings.

ure 11.2(b), achieves the *maximum radial capacity* possible for a single-row ball bearing, but thrust capacity is reduced by the shoulder discontinuities caused by the filling notches.

Double-row deep-groove ball bearings, shown in Figure 11.2(c), can support heavier combinations of radial and thrust loads, and, by grinding a spherical raceway surface in the outer race, they can be made *internally self-aligning*, as shown in Figure 11.2(d). *Externally self-aligning bearings*, having two-piece outer races with matched spherical interfaces, are also commercially available [see Figure 11.2(e)].

Higher thrust capacity *in one direction* is achieved in single-row *angular contact ball bearings* by configuring the grooves so that the centerline of contact between balls and raceways is at an angle with respect to a plane perpendicular to the axis of rotation, as illustrated in Figure 11.2(f). One side of the outer race is counter bored to remove most of the raceway shoulder, while a higher shoulder is provided on the opposing thrust face. Thermal expansion of the outer race is utilized to assemble a maximum ball complement, resulting in a nonseparable bearing after cooling. Angular contact ball bearings are usually used in pairs, arranged face-to-face [shown in Figure 11.2(g)], back-to-back, or in tandem with controlled axial preloading to obtain desired bearing stiffness (see 11.7). *Double-row angular contact ball bearings* with "built-in" preload (one row of balls is axially preloaded against the other) are available as shown in Figure 11.2(h). *Ball thrust bearings* are commercially available, as illustrated in Figure 11.2(i).

Some examples of bearings.

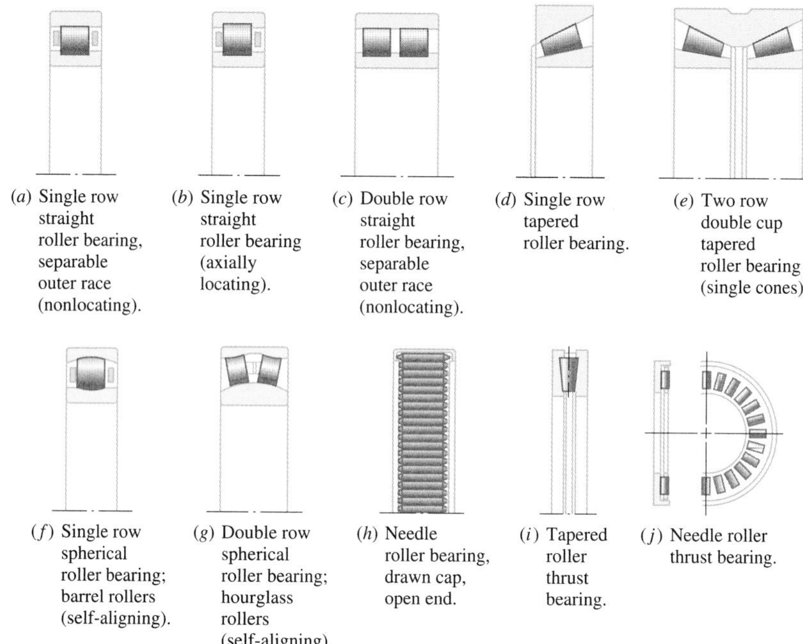

(a) Single row straight roller bearing, separable outer race (nonlocating).
(b) Single row straight roller bearing (axially locating).
(c) Double row straight roller bearing, separable outer race (nonlocating).
(d) Single row tapered roller bearing.
(e) Two row double cup tapered roller bearing (single cones).
(f) Single row spherical roller bearing; barrel rollers (self-aligning).
(g) Double row spherical roller bearing; hourglass rollers (self-aligning).
(h) Needle roller bearing, drawn cap, open end.
(i) Tapered roller thrust bearing.
(j) Needle roller thrust bearing.

Figure 11.3
Selected examples from the many types of commercially available roller bearings.

In addition to the basic bearing types, many specialty and custom-designed bearings can be ordered, including bearings with two-piece inner races, diametrally split ball bearings for difficult mounting configurations, extended inner race design with locking collars, integral mounting flange configurations, and others. Bearings with seals, shields, or both, installed on one or both sides, are available in many cases, as well as snap ring grooves for outer race retention.

In contrast with ball bearings, *straight roller bearings* are unable to support thrust loads, and are usually made so that the races can be axially separated, as shown in Figure 11.3(a). They can, therefore, readily accommodate small axial displacements of a shaft relative to the housing. Such a displacement might occur, for example, because of differential thermal expansion between the shaft and the housing. A small axial thrust capacity can be incorporated into straight roller bearings for locational purposes [by incorporating shoulders as shown in Figure 11.3(b)], and heavier radial load support can be achieved with *double-row roller bearings*, as illustrated in Figure 11.3(c).

If combined radial and thrust loads must be accommodated, *tapered roller bearings*, such as the one shown in Figure 11.3(d), may be used. The conically angled raceways allow tapered roller bearings to carry combinations of radial and thrust loads. By the same token, even when a purely radial external load is applied to a tapered roller bearing, a thrust (axial) load is induced within the bearing because of the taper. Hence, as for the case of angular contact ball bearings discussed earlier, tapered roller bearings are usually used in pairs (or sometimes in four-row configurations) to resist the thrust reaction or to provide desired stiffness. Such bearings can be obtained in various two- and four-row arrangements, such as the two-row double-cup single-cone assembly illustrated in Figure 11.3(e).

Self-aligning (spherical) roller bearings are available, as illustrated in Figure 11.3(f) and (g), where barrel rollers with a spherical outer race or hourglass rollers with a spherical inner race provide the self-aligning capability.

When radial space is limited, *needle roller bearings* are sometimes used. One of many needle bearing configurations is shown in Figure 11.3(h). Various types of roller, tapered roller, and needle *thrust bearings* are also available as shown, for example, in Figures 11.3(i) and (j).

11.3 Potential Failure Modes

Rolling element bearings may be vulnerable to failure by any of several possible modes, depending upon the specified operating loads and speeds. Under typical operating conditions, *surface fatigue failure* is the most likely failure mode (see 2.3). The cyclic subsurface Hertzian shear stresses produced by the curved surfaces in rolling contact may initiate and propagate minute cracks that ultimately dislodge particles and generate surface pits. Typically, the *raceways* pit first, generating progressively intensifying noise, vibration, and heat. If bad bearings are not replaced, races may fracture, bearings may jam, and serious damage to other machine elements may result. In some cases, static loads on a bearing during idle nonrotating segments of a duty cycle may cause *brinelling* of the races, resulting in subsequent generation of noise, vibration, and heat as the rolling elements pass the local discontinuities in the raceway (see 2.3). When selecting bearings, designers must routinely examine both the ability of a bearing to resist surface fatigue failure (basic dynamic load rating) and the ability of a bearing to resist brinelling failure (basic static load rating). Which of these two failure modes governs will depend upon the specific details of the duty cycle.

It is worth noting that for *lightly loaded* bearings the initiation of surface fatigue crack nuclei may not occur, and very long operating lives may be achieved. This is true since, in effect, the cyclic Hertzian stresses are below the fatigue endurance limit of the ferrous bearing alloys (see 2.6 and Figure 2.18). At least one bearing manufacturer[5] now provides *fatigue load limit* data, for which fatigue load limit is defined as the bearing *load* below which fatigue failure will not occur for that particular bearing, assuming contaminant-free, ideal lubrication between rolling elements and raceways.

In some applications, *fretting wear* may be the governing failure mode (see 2.3). Notably, if bearings are operated with small-amplitude reversing cyclic motion, even when loads are light, the ensuing fretting conditions may produce significant local fretting wear, and ultimately result in unacceptable noise, vibration, or heat, as well as contamination of the bearing lubricant from accumulated fretting debris. The failure of needle bearings in automotive universal joints provides a common example of the potential for fretting wear failure of a needle roller bearing.

As for power transmission shafting (see 8.2), rolling element bearings are usually an integral part of a dynamic system of springs and masses. Hence, it is important to examine the possibility that operation at certain critical speeds may excite destructive vibrational behavior. Experience has shown that typical spring rates of rolling element bearings are low (soft) compared to many other machine elements, and therefore *bearing* stiffness, in some cases, may dominate the vibrational behavior of a machine (see 4.11). It is an important design responsibility to assure that installed bearing stiffness is great enough to keep the fundamental natural frequency of the system well above the forcing frequency[6] so that excessive deflection (*force-induced elastic deformation failure*) does not result. Proper *preloading* is a means of improving the failure resistance of rolling element bearings to force-induced elastic deformation (see 11.6).

11.4 Bearing Materials

Rolling element bearings are probably the most highly refined, carefully engineered, precisely produced, statistically tested devices of all the machine elements. For these reasons the responsibility for achieving the keystone design objectives (selection of best material and determination of best geometry) lies with the specialized designer working in the bearing industry. Most designers need only be concerned with the *selection* of an appropriate bearing from com-

[5]See ref. 5.

[6]See 8.2 for definitions.

mercially available inventory. This discussion of bearing materials is, accordingly, brief, since in most cases bearing manufacturers will have already selected the optimum materials.

Most *ball bearings*, including both balls and races, are made from vacuum-degassed high-carbon chrome steel, hardened and tempered for optimum strength and toughness, sometimes stabilized for precise dimensional control. Stainless-steel alloys are sometimes used when corrosion resistance is needed or moderately elevated temperatures must be accommodated. For higher-temperature applications, cobalt base alloys may be used. AISI 52100 steel is widely used for both races and balls, and is typically through-hardened to Rockwell C61–C64. Stainless-steel alloys 400 C or 440 CM may be used for special applications. Separator materials include phenolic, bronze, phosphor bronze, or alloy steels such as AISI 4130.

In the case of *roller bearings*, races and rollers are usually manufactured from low-carbon carburizing-grade, electric furnace, vacuum-degassed alloy steels. They are then case-carburized and heat-treated to produce a hard, fatigue-resistant case encompassing a tough ductile core. In this context, case-hardened AISI 3310, 4620, and 8620 are often used. Specialty steels may be used for critical applications.

11.5 Bearing Selection

In selecting an appropriate bearing for any given application, design decisions must be made about bearing type, size, space allocation, mounting methods, and other details. Typically, a *unique* bearing choice will *not* be possible for a given application because of overlapping bearing capabilities. As usual, designer experience plays an important role in the selection process. Most bearing manufacturers provide technical assistance in this area. Table 11.1 provides overall guidance in selecting an appropriate bearing type, but to use the table a designer must first summarize the specific design requirements at the bearing

Table 11.1 Bearing Characteristics That May Influence the Selection of Bearing Type[1]

Bearing Type	Radial Capacity	Thrust Capacity	Limiting Speed	Radial Stiffness	Axial Stiffness
Deep-groove ball	moderate	moderate—both directions	high	moderate	low
Maximum-capacity ball	moderate (plus)	moderate—one direction	high	moderate (plus)	low (plus)
Angular contact ball	moderate	moderate (plus)—one direction	high (minus)	moderate	moderate
Cylindrical roller	high	none	moderate (plus)	high	none
Spherical roller	high	moderate—both directions	moderate	high (minus)	moderate
Needle roller	moderate to high	none	moderate to very high	moderate to high	none
Single-row tapered roller	high (minus)	moderate (plus)—one direction	moderate	high (minus)	moderate
Double-row tapered roller	high	moderate—both directions	moderate	high	moderate
Four-row tapered roller	high (plus)	high—both directions	moderate (minus)	high (plus)	high
Ball thrust	none	high—one direction	moderate (minus)	none	high
Roller thrust	none	high (plus)—one direction	low	none	high (plus)
Tapered roller thrust	locational only	high (plus)—one direction	low	none	high (plus)

[1]Adapted from Torrington bearing catalog data.

site in terms of radial loading spectrum, thrust loading spectrum, operating speeds, design life requirements, stiffness requirements, and operating environment.

After a preliminary selection of the *type* of bearing to be used, determination of the bearing *size* may be accomplished by following the procedures outlined in the following paragraphs. These procedures are representative of those recommended by many bearing manufacturers.

Before detailing bearing selection procedures, however, it is important to understand the concepts of *basic dynamic load rating*, C_d, and *basic static load rating*, C_s, universally used by bearing manufacturers in characterizing the load-carrying capacity of all types of rolling element bearings. Also of interest are the data-based relationship between the load and number of rotations to failure, and the influence on the selection process of reliability requirements higher than the standard 90 percent.

Basic Load Ratings

Basic load ratings have been standardized[7] by the bearing industry as a uniform means of describing the ability of any given rolling element bearing to resist failure by (1) surface fatigue and (2) brinnelling (see 11.3). The basic dynamic load rating, C_d, is a measure of resistance to *surface fatigue failure*. The basic static load rating, C_s, is a measure of resistance to failure by *brinnelling*. Typically, values of C_d and C_s are published for each bearing listed in a manufacturer's catalog.[8]

Basic dynamic radial load rating $C_d(90)$ is *defined to be the largest constant stationary radial load that 90 percent of a group of apparently identical bearings will survive for 1 million revolutions (inner race rotating, outer race fixed) with no evidence of failure by surface fatigue*. Based on extensive experimental data, the relationship between radial bearing load P and bearing life L (revolutions to failure) for any given bearing is

$$\frac{L}{10^6} = \left(\frac{C_d}{P}\right)^a \tag{11-1}$$

where $a = 3$ for ball bearings and $a = {}^{10}\!/_3$ for roller bearings.[9]

Basic static radial load rating C_s is *defined to be the largest stationary radial load that will produce significant brinnelling evidence at the most heavily loaded rolling element contact site*. "Significant brinnelling evidence" is defined to be any plastic deformation diameter (or width) greater than 0.0001 of the diameter of the rolling element.

Reliability Specifications

Values of dynamic load rating corresponding to a reliability of $R = 90$ percent, $C_d(90)$, are routinely published by all bearing manufacturers. While 90 percent reliability is acceptable for a wide variety of industrial applications, higher reliabilities are sometimes desired by a designer. *Reliability-adjustment factors*, based on actual failure rate data, allow a designer to select bearings for reliability levels higher than 90 percent.[10] Although it is possible to express reliability adjustment as either a *load* factor or a *life* factor, current practice is to use a reliability *life*-adjustment factor, K_R. Thus the basic bearing rating life, $L_{10} = 10^6$ revolutions (corresponds to $P = 10$ or $R = 90$), may be adjusted to any chosen higher reliability R using

$$L_P = K_R L_{10} \tag{11-2}$$

[7]See refs. 2 and 3.

[8]See, for example, Tables 11.5, 11.6, and 11.7.

[9]See refs. 2 and 3; also note that some manufacturers recommend the value $a = {}^{10}\!/_3$ for both ball and roller bearings.

[10]Recall that reliability R in percent is equal to $100 - P$, where P is probability of failure in percent.

Table 11.2 Reliability Life-Adjustment Factor K_R for Bearing Reliabilities Different from $R = 90\%$

Reliability R, percent	Probability of Failure P, percent	K_R
50	50	5.0
90	10	1.0
95	5	0.62
96	4	0.53
97	3	0.44
98	2	0.33
99	1	0.21

where the value of K_R is read from Table 11.2. L_P is called the *reliability-adjusted* rating life (revolutions). The L_{10} life is sometimes called the B_{10} life.

Suggested Selection Procedure for Steady Loads

The procedure followed by a designer when selecting an appropriate bearing for a specific application usually involves the following steps:

1. First, design the shaft by taking it as a *free body*, perform a complete *force analysis*, and employ appropriate *shaft design equations* to determine a tentative strength-based shaft diameter at the bearing site (See Chapter 8).
2. From the force analysis, calculate the radial load F_r and axial *thrust load* F_a to be supported by the proposed bearing.
3. Determine the *design life requirement*, L_d, for the bearing.
4. Determine the *reliability* R appropriate to the application, and select the corresponding *life-adjustment factor*, K_R.
5. Assess the severity of any shock or impact associated with the application so that an *impact factor*, *IF*, may be determined (see Table 11.3 and review 2.8 of Chapter 2).
6. Tentatively select the *type of bearing* to be used (see Table 11.1).
7. Calculate a *dynamic equivalent radial load* P_e from the following empirical relationship[11]

$$P_e = X_d F_r + Y_d F_a \tag{11-3}$$

Table 11.3 Estimated Impact Factors for Various Applications[1]

Type of Application	Impact Factor *IF*
Uniform load, no impact	1.0–1.2
Precision gearing	1.1–1.2
Commercial gearing	1.1–1.3
Toothed belts	1.1–1.3
Light impact	1.2–1.5
Vee belts	1.2–2.5
Moderate impact	1.5–2.0
Flat belts	1.5–4.5
Heavy impact	2.0–5.0

[1]See also 2.8 in Chapter 2.

[11]Some sources recommend the use of a "race rotation" factor that increases P_e for any case in which the *outer* race rotates, instead of the inner race. However, a race rotation factor is not used here, because the bearing manufacturers supplying the selection data for Tables 11.5, 11.6, and 11.7 are not currently using it.

Table 11.4 Approximate Radial Load Factors for Selected Bearing Types[1]

Bearing Type	Dynamic				Static			
	X_{d_1}	Y_{d_1}	X_{d_2}	Y_{d_2}	X_{s_1}	Y_{s_1}	X_{s_2}	Y_{s_2}
Single-row radial ball bearing	1	0	0.55	1.45	1	0	0.6	0.5
Single-row angular contact ball bearing (shallow angle)	1	0	0.45	1.2	1	0	0.5	0.45
Single-row angular contact ball bearing (steep angle)	1	0	0.4	0.75	1	0	0.5	0.35
Double-row radial ball bearing	1	0	0.55	1.45	1	0	0.6	0.5
Double-row angular contact ball bearing (shallow angle)	1	1.55	0.7	1.9	1	0	1	0.9
Double-row angular contact ball bearing (steep angle)	1	0.75	0.6	1.25	1	0	1	0.65
Self-aligning single-row ball bearing[2]	1	0	0.4	$0.4\cot\alpha$	1	0	0.5	$0.2\cot\alpha$
Self-aligning double-row ball bearing	1	$0.4\cot\alpha$	0.65	$0.65\cot\alpha$	1	0	1	$0.45\cot\alpha$
Straight roller bearing[2] ($\alpha = 0$); (cannot take thrust)	1	0	—	—	1	0	1	0
Single-row roller bearing[3] ($\alpha \neq 0$)	1	0	0.4	$0.4\cot\alpha$	1	0	0.5	$0.2\cot\alpha$
Double-row roller bearing ($\alpha \neq 0$)	1	$0.45\cot\alpha$	0.65	$0.65\cot\alpha$	1	0	1	$0.45\cot\alpha$
Self-aligning single-row roller bearing	1	0	0.4	$0.4\cot\alpha$	1	0	0.5	$0.2\cot\alpha$
Self-aligning double-row roller bearing	1	$0.45\cot\alpha$	0.65	$0.65\cot\alpha$	1	0	1	$0.45\cot\alpha$

[1]For more accurate values, see refs. 2 and 3.
[2]Nominal contact angle α is the angle between a plane perpendicular to the bearing axis and the nominal line of action of the resultant force transmitted from a bearing race to a rolling element.
[3]These values may be used for preliminary selection of tapered roller bearings, but mounting details and preloading have significant influence; manufacturer's catalogs should be consulted. See also Table 11.7.

where X_d = dynamic radial load factor, based on bearing geometry (see Table 11.4)
Y_d = dynamic axial (thrust) load factor, based on bearing geometry (see Table 11.4)

The combinations X_{d_1}, Y_{d_1}, and X_{d_2}, Y_{d_2} should both be calculated, and whichever combination gives the larger value for P_e should be used.

8. Using (11-1) as a basis, calculate the basic dynamic load rating *requirement* corresponding to the reliability level selected in step 4, as

$$[C_d(R)]_{req} = \left[\frac{L_d}{K_R(10^6)}\right]^{1/a}(IF)P_e \tag{11-4}$$

where $[C_d(R)]_{req}$ = required radial dynamic load rating to give a bearing reliability of R percent
L_d = design life, revolutions
K_R = reliabiliy adjustment factor from Table 11.2
IF = application impact factor from Table 11.3
P_e = equivalent radial load from (11-3)
a = exponent equal to 3 for ball bearings or $^{10}/_3$ for roller bearings

9. With the result from (11-4), enter a basic load rating table for the type of bearing selected in step 6 and tentatively select the smallest bearing with a basic radial load rating C_d of at least $[C_d(R)]_{req}$. Such tables may be found for all types of bearings in bearing manufacturers' catalogs.[12] Illustrative examples of bearing selection tables are included here as Tables 11.5, 11.6, and 11.7.

[12]See, for example, refs. 5 and 6.

Table 11.5 Dimensions and Load Ratings for Selected Single-Row Radial Deep-Groove (Conrad Type) Ball Bearings: Series 60, 62, and 63 (Data courtesy of SKF® USA Inc., Norristown, PA)

Bearing Number	Bore		Outside Diameter		Width		Max Fillet Radius[1]		Min Shaft Abutment Diameter[2]		Basic Load Rating[3] C_d		C_s		Approximate Fatigue Load Limit[4] P_f		Limiting Speed,[5] 10^3 rpm
	mm	in	mm	in	mm	in	mm	in	mm	in	kN	10^3 lbf	kN	10^3 lbf	kN	10^3 lbf	With Grease/ Oil
6000	10	0.3937	26	1.0236	8	0.3150	0.3	0.012	12	0.472	4.62	1.04	1.96	0.44	0.08	0.019	30/36
6200			30	1.1811	9	0.3543	0.6	0.024	14	0.551	5.07	1.14	2.36	0.53	0.10	0.023	24/30
6300			35	1.3780	11	0.4331	0.6	0.024	14	0.551	8.06	1.81	3.40	0.76	0.14	0.032	20/26
6002	15	0.5906	32	1.2598	9	0.3543	0.3	0.012	17	0.669	5.59	1.26	2.85	0.64	0.12	0.027	22/28
6202			35	1.3780	11	0.4331	0.6	0.024	19	0.748	7.80	1.75	3.75	0.84	0.16	0.036	19/24
6302			42	1.6535	13	0.5118	1	0.039	20	0.787	11.40	2.56	5.40	1.21	0.23	0.051	17/20
6004	20	0.7874	42	1.6535	12	0.4724	0.6	0.024	24	0.945	9.36	2.10	5.00	1.12	0.21	0.048	17/20
6204			47	1.8504	14	0.5512	1	0.039	25	0.984	12.70	2.86	6.55	1.47	0.28	0.063	15/18
6304			52	2.0472	15	0.5906	1	0.039	26.5	1.043	15.90	3.57	7.80	1.75	0.34	0.075	13/16
6005	25	0.9843	47	1.8504	12	0.4724	0.6	0.024	29	1.142	11.20	2.52	6.55	1.47	0.28	0.062	15/18
6205			52	2.0472	15	0.5906	1	0.039	30	1.181	14.00	3.15	7.80	1.75	0.34	0.075	12/15
6305			62	2.4409	17	0.6693	1	0.039	31.5	1.240	22.50	5.06	11.60	2.61	0.49	0.110	11/14
6006	30	1.1811	55	2.1654	13	0.5118	1	0.039	35	1.378	13.30	2.99	8.30	1.87	0.36	0.080	12/15
6206			62	2.4409	16	0.6299	1	0.039	35	1.378	19.50	4.38	11.20	2.52	0.48	0.107	10/13
6306			72	2.8346	19	0.7480	1	0.039	36.5	1.437	28.10	6.32	16.00	3.60	0.67	0.151	9/11
6007	35	1.3780	62	2.4409	14	0.5512	1	0.039	40	1.575	15.90	3.57	10.20	2.29	0.44	0.099	10/13
6207			72	2.8346	17	0.6693	1	0.039	41.5	1.634	25.50	5.73	15.30	3.44	0.66	0.147	9/11
6307			80	3.1496	21	0.8268	1.5	0.059	43	1.693	33.20	7.46	19.00	4.27	0.82	0.183	8.5/10
6008	40	1.5748	68	2.6772	15	0.5906	1	0.039	45	1.772	16.80	3.78	11.60	2.61	0.49	0.110	9.5/12
6208			80	3.1496	18	0.7087	1	0.039	46.5	1.831	30.70	6.90	19.00	4.27	0.80	0.180	8.5/10
6308			90	3.5433	23	0.9055	1.5	0.059	48	1.890	41.00	9.22	24.00	5.40	1.02	0.229	7.5/9

6009	45	1.7717	75	2.9528	16	0.6299	1	0.039	50	1.969	20.80	4.68	14.60	3.28	0.64	0.144	9/11
6209			85	3.3465	19	0.7480	1	0.039	51.5	2.028	33.20	7.46	21.60	4.86	0.92	0.206	7.5/9
6309			100	3.9370	25	0.9843	1.5	0.059	53	2.087	52.70	11.90	31.50	7.08	1.34	0.301	6.7/8
6010	50	1.9685	80	3.1496	16	0.6299	1	0.039	55	2.165	21.60	4.86	16.00	3.60	0.71	0.160	8.5/10
6210			90	3.5433	20	0.7874	1	0.039	56.5	2.224	35.10	7.89	23.20	5.22	0.98	0.220	7/8.5
6310			110	4.3307	27	1.0630	2	0.079	59	2.323	61.80	13.90	38.00	8.54	1.60	0.360	6.3/7.5
6011	55	2.1654	90	3.5433	18	0.7087	1	0.039	61.5	2.421	28.10	6.32	21.20	4.77	0.90	0.202	7.5/9
6211			100	3.9370	21	0.8268	1.5	0.059	63	2.480	43.60	9.80	29.00	6.52	1.25	0.281	6.3/7.5
6311			120	4.7244	29	1.1417	2	0.079	64	2.520	71.50	16.10	45.00	10.10	1.90	0.427	5.6/6.7
6012	60	2.3622	95	3.7402	18	0.7087	1	0.039	66.5	2.618	29.60	6.65	23.20	5.22	0.98	0.220	6.7/8
6212			110	4.3307	22	0.8661	1.5	0.059	68	2.677	47.50	10.70	32.50	7.31	1.40	0.315	6/7
6312			130	5.1181	31	1.2205	2	0.079	71	2.795	81.90	18.40	52.00	11.70	2.20	0.495	5/6
6016	80	3.1496	125	4.9213	22	0.8661	1	0.039	86.5	3.406	47.50	10.70	40.00	8.99	1.66	0.373	5.3/6.3
6216			140	5.5118	26	1.0236	2	0.079	89	3.504	70.20	15.80	55.00	12.40	2.20	0.495	4.5/5.3
6316			170	6.6929	39	1.5354	2	0.079	91	3.583	124.00	27.90	86.50	19.50	3.25	0.731	3.8/4.5
6020	100	3.9370	150	5.9055	24	0.9449	1.5	0.059	108	4.252	60.50	13.60	54.00	12.10	2.04	0.459	4.3/5
6220			180	7.0866	34	1.3386	2	0.079	111	4.370	124.00	27.90	93.00	20.90	3.35	0.753	3.4/4
6320			215	8.4646	47	1.8504	2.5	0.098	113	4.449	174.00	39.10	140.00	31.50	4.75	1.070	3/3.6
6030	150	5.9055	225	8.8583	35	1.3780	2	0.079	161	6.339	125.00	28.10	125.00	28.10	3.90	0.877	2.6/3.2
6230			270	10.6299	45	1.7717	2.5	0.098	163	6.417	174.00	39.10	166.00	37.30	4.90	1.100	2/2.6
6330			320	12.5984	65	2.5591	3	0.118	166	6.535	276.00	62.10	285.00	64.10	7.80	1.750	1.9/2.4
6040	200	7.8740	310	12.2047	51	2.0079	2	0.079	211	8.307	216.00	48.60	245.00	55.10	6.40	1.440	1.9/2.4
6240			360	14.1732	58	2.2835	3	0.118	216	8.504	270.00	60.70	310.00	69.70	7.80	1.750	1.7/2
6340			420	16.5354	80	3.1496	4	0.157	220	8.661	377.00	84.80	465.00	105.0	11.2	2.520	1.5/1.8

[1]Maximum allowable fillet radius at shaft (and housing) abutment.

[2]For housing dimensions see manufacturer's catalogs.

[3]C_d is basic dynamic radial load rating for $R = 90\%$; C_s is basic static radial load rating.

[4]Equivalent radial load below which infinite life may be expected; analogous to fatigue endurance limit; contact SKF® for more accurate values.

[5]*Absolute* rotational speed of the *inner race relative to the outer race*.

Table 11.6 Dimensions and Load Ratings for Selected Single-Row Cylindrical Roller Bearings: Series 20, 22, and 30 (Data courtesy of SKF® USA Inc., Norristown, PA)

Bearing Number	Bore		Outside Diameter		Width		Max Fillet Radius[1]		Min Shaft Abutment Diameter[2]		Basic Load Rating[3] C_d		C_s		Approximate Fatigue Load Limit[4] P_f		Limiting Speed,[5] 10^3 rpm
	mm	in	mm	in	mm	in	mm	in	mm	in	kN	10^3 lbf	kN	10^3 lbf	kN	10^3 lbf	With Grease/Oil
202	15	0.5906	35	1.3780	11	0.4331	0.6	0.024	17	0.669	12.5	2.81	10.2	2.29	1.22	0.27	18/22
302			42	1.6535	13	0.5118	1	0.039	19	0.748	19.4	4.36	15.3	3.44	1.86	0.42	16/19
204	20	0.7874	47	1.8504	14	0.5512	1	0.039	24	0.945	25.1	5.64	22.0	4.95	2.75	0.62	13/16
2204			47	1.8504	18	0.7087	1	0.039	24	0.945	29.7	6.68	27.5	6.18	3.45	0.78	13/16
304			52	2.0472	15	0.5906	1	0.039	24	0.945	30.8	6.92	26.0	5.85	3.25	0.73	12/15
205	25	0.9843	52	2.0472	15	0.5906	1	0.039	29	1.142	28.6	6.43	27.0	6.07	3.35	0.75	11/14
2205			52	2.0472	18	0.7087	1	0.039	29	1.142	34.1	7.67	34.0	7.64	4.25	0.96	11/14
305			62	2.4409	17	0.6693	1	0.039	31.5	1.240	40.2	9.04	36.5	8.21	4.55	1.02	9.5/12
206	30	1.1811	62	2.4409	16	0.6299	1	0.039	34	1.339	38.0	8.54	36.5	8.21	4.55	1.02	9.5/12
2206			62	2.4409	20	0.7874	1	0.039	34	1.339	48.4	10.9	49.0	11.0	6.10	1.37	9.5/12
306			72	2.8346	19	0.7480	1	0.039	36.5	1.437	51.2	11.5	48.0	10.8	6.20	1.39	9/11
207	35	1.3780	72	2.8346	17	0.6693	1	0.039	39	1.535	48.4	10.9	48.0	10.8	6.10	1.37	8.5/10
2207			72	2.8346	23	0.9055	1	0.039	39	1.535	59.4	13.4	63.0	14.2	8.15	1.83	8.5/10
307			80	3.1496	21	0.8268	1.5	0.059	41.5	1.634	64.4	14.5	63.0	14.2	8.15	1.83	8/9.5
208	40	1.5748	80	3.1496	18	0.7087	1	0.039	46.5	1.831	53.9	12.1	53.0	11.9	6.70	1.51	7.5/9
2208			80	3.1496	23	0.9055	1	0.039	46.5	1.831	70.4	15.8	75.0	16.9	9.65	2.17	7.5/9
308			90	3.5433	23	0.9055	1.5	0.059	48	1.890	80.9	18.2	78.0	17.5	10.2	2.29	6.7/8
209	45	1.7717	85	3.3465	19	0.7480	1	0.039	51.5	2.028	60.5	13.6	64.0	14.4	8.15	1.83	6.7/8

2209			85	3.3465	23	0.9055	1	0.039	51.5	2.028	73.7	16.6	81.5	18.3	10.6	2.38	6.7/8
309			100	3.9370	25	0.9843	1.5	0.059	53	2.087	99.0	22.3	100.0	22.5	12.9	2.90	6.3/7.5
210	50	1.9685	90	3.5433	20	0.7874	1	0.039	56.5	2.224	64.4	14.5	69.5	15.6	8.8	1.98	6.3/7.5
2210			90	3.5433	23	0.9055	1	0.039	56.5	2.224	78.1	17.6	88.0	19.8	11.4	2.56	6.3/7.5
310			110	4.3307	27	1.0630	2	0.079	59	2.323	110.0	24.7	112.0	25.2	15.0	3.37	5/6
211	55	2.1654	100	3.9370	21	0.8268	1.5	0.059	61.5	2.421	84.2	18.9	95.0	21.4	12.2	2.74	6/7
2211			100	3.9370	25	0.9843	1.5	0.059	61.5	2.421	99.0	22.3	118.0	26.5	15.3	3.44	6/7
311			120	4.7244	29	1.1417	2	0.079	64	2.520	138.0	31.0	143.0	32.2	18.6	4.18	4.8/5.6
212	60	2.3622	110	4.3307	22	0.8661	1.5	0.059	68	2.677	93.5	21.0	102.0	22.9	13.4	3.01	5.3/6.3
2212			110	4.3307	28	1.1024	1.5	0.059	68	2.677	128.0	28.8	153.0	34.4	20.0	4.50	5.3/6.3
312			130	5.1181	31	1.2205	2	0.079	71	2.795	151.0	34.0	160.0	36.0	20.8	4.68	4.3/5
216	80	3.1496	140	5.5118	26	1.0236	2	0.079	89	3.504	138.0	31.0	166.0	37.3	21.2	4.77	4/4.8
2216			140	5.5118	33	1.2992	2	0.079	89	3.504	187.0	42.0	245.0	55.1	31.0	6.97	4/4.8
316			170	6.6929	39	1.5354	2	0.079	91	3.583	260.0	58.5	290.0	65.2	36.0	8.09	3.2/3.8
220	100	3.9370	180	7.0866	34	1.3386	2	0.079	111	4.370	251.0	56.4	305.0	68.6	36.5	8.21	3.2/3.8
2220			180	7.0866	46	1.8110	2	0.079	111	4.370	336.0	75.5	450.0	101.0	54.0	12.1	3.2/3.8
320			215	8.4646	47	1.8504	2.5	0.098	113	4.449	391.0	87.9	440.0	98.9	51.0	11.5	2.4/3
230	150	5.9055	270	10.6299	45	1.7717	2.5	0.098	163	6.417	446.0	100.0	600.0	135.0	64.0	14.4	1.9/2.4
2230			270	10.6299	73	2.8740	2.5	0.098	163	6.417	627.0	141.0	930.0	209.0	100.0	22.5	1.9/2.4
330			320	12.5984	65	2.5591	3	0.118	166	6.535	781.0	176.0	965.0	217.0	100.0	22.5	1.7/2
240	200	7.8740	360	14.1732	58	2.2835	3	0.118	216	8.504	765.0	172.0	1060.0	238.0	106.0	23.8	1.5/1.8
2240			360	14.1732	98	3.8583	3	0.118	216	8.504	1230.0	277.0	1900.0	427.0	190.0	42.7	1.5/1.8
340			420	16.5354	80	3.1496	4	0.157	220	8.661	990.0	223.0	1320.0	297.0	125.0	28.1	1.3/1.6

[1]Maximum allowable fillet radius at shaft (and housing) abutment shoulder.

[2]For housing dimensions see manufacturer's catalogs.

[3]C_d is basic dynamic radial load rating for R = 90%; C_s is basic static radial load rating.

[4]Equivalent radial load below which infinite life may be expected; analogous to fatigue endurance limit; contact SKF® for more accurate values.

[5]*Absolute* rotational speed of the *inner race relative to the outer race*.

Table 11.7 Dimensions and Load Ratings for Selected Single-Row Tapered Roller Bearings: Series 302, 303, and 323[1] (Data Source: Timken Company®, Canton, Ohio)

Bearing Number	Bore mm	Bore in	Outside Diameter mm	Outside Diameter in	Width mm	Width in	Max Fillet Radius[2] mm	Max Fillet Radius[2] in	Min Shaft Abutment Diameter[3] mm	Min Shaft Abutment Diameter[3] in	Basic Load Rating[4] C_d kN	Basic Load Rating[4] C_d 10^3 lbf	Basic Load Rating[4] C_s kN	Basic Load Rating[4] C_s 10^3 lbf	Dynamic Axial Load Factor[5] Y_{d_2}	Thrust Ratio[6] $\frac{(F_a)_i}{F_r}$
30204	20	0.7874	47	1.8504	15.25	0.6004	1	0.039	25.5	1.004	28.3	6.37	29.2	6.56	1.74	0.28
30304			52	2.0472	16.25	0.6398	1.5	0.059	27	1.063	35.6	8.01	34.5	7.76	2.00	0.24
32304			52	2.0472	22.25	0.8760	1.5	0.059	28	1.102	46.4	10.4	48.3	10.9	2.00	0.24
30205	25	0.9843	52	2.0472	16.25	0.6398	1	0.039	30.5	1.201	31.6	7.11	34.4	7.74	1.60	0.30
30305			62	2.4409	18.25	0.7185	1.5	0.059	32.5	1.280	50.2	11.3	50.1	11.3	2.00	0.24
32305			62	2.4409	25.25	0.9941	1.5	0.059	35	1.328	67.0	15.1	72.3	16.3	2.00	0.24
30206	30	1.1811	62	2.4409	17.25	0.6791	1	0.039	36	1.417	40.5	9.12	43.8	9.86	1.60	0.30
30306			72	2.835	20.75	0.8169	1.5	0.059	38	1.496	59.5	13.40	60.6	13.6	1.90	0.25
32306			72	2.835	28.75	1.1319	1.5	0.059	39.5	1.555	81.1	18.2	89.8	20.2	1.90	0.25
30207	35	1.3780	72	2.835	18.25	0.7185	1.5	0.059	42.5	1.673	53.4	12.0	59.2	13.3	1.60	0.30
30307			80	3.150	22.75	0.8957	2	0.079	44	1.732	80.4	18.1	85.6	19.2	1.90	0.25
32307			80	3.150	32.75	1.2894	2	0.079	46	1.811	107.0	24.1	123.0	27.7	1.90	0.25
30208	40	1.5748	80	3.150	19.75	0.7776	1.5	0.059	48	1.890	59.9	13.5	65.4	14.7	1.60	0.30
30308			90	3.543	25.25	0.9941	2	0.079	50	1.969	91.4	20.6	102.0	23.0	1.74	0.28
32308			90	3.543	35.25	1.3878	2	0.079	55	2.165	123.0	27.7	150.0	33.7	1.74	0.28
30209	45	1.7717	85	3.346	20.75	0.8169	1.5	0.059	53	2.087	64.8	14.6	74.7	16.8	1.48	0.33
30309			100	3.937	27.25	1.0728	2	0.079	56	2.205	114.0	25.6	130.0	29.2	1.74	0.28
32309			100	3.937	38.25	1.5059	2	0.079	57	2.244	143.0	32.1	172.0	38.6	1.74	0.28
30210	50	1.9685	90	3.346	21.75	0.8563	1.5	0.059	59	2.323	73.6	16.6	87.5	19.7	1.43	0.34
30310			110	4.331	29.25	1.1516	2.5	0.098	62	2.441	131.0	29.4	150.0	33.8	1.74	0.28
32310			110	4.331	42.25	1.6634	2.5	0.098	65	2.559	173.0	38.9	211.0	47.5	1.74	0.28
30211	55	2.1654	100	3.937	22.75	0.8957	2	0.079	64	2.520	94.9	21.3	114.0	25.7	1.48	0.33
30311			120	4.724	31.5	1.2402	2.5	0.098	68	2.677	150.0	33.7	172.0	38.6	1.74	0.28
32311			120	4.724	45.5	1.7913	2.5	0.098	70	2.756	211.0	47.4	265.0	59.7	1.74	0.28
30212	60	2.3622	110	4.331	23.75	0.9350	2	0.079	70	2.756	99.1	22.3	117.0	26.2	1.48	0.33
30312			130	5.118	33.5	1.3189	3	0.118	74	2.913	186.0	41.9	221.0	49.8	1.74	0.28
32312			130	5.118	48.5	1.9094	3	0.118	78	3.071	244.0	55.0	310.0	69.8	1.74	0.28
30216	80	3.1496	140	5.512	28.25	1.1122	2.5	0.098	92	3.543	151.0	34.0	187.0	42.0	1.43	0.34
30220	100	3.9370	180	7.087	37	1.4567	3	0.118	119	4.685	278.0	62.6	375.0	84.2	1.43	0.34

[1]First digit, 3, indicates tapered roller bearing; second and third digits refer to width series and diameter series.
[2]Maximum allowable fillet radius at shaft (and housing) abutment shoulder.
[3]For housing dimensions see manufacturer's catalogs.
[4]C_d is basic dynamic radial load rating for $R = 90\%$; C_s is basic static radial load rating.
[5]See also Table 11.4.
[6]Ratio of induced axial thrust force to applied radial load for radially loaded bearings, assuming 180° roller contact zone when radially loaded.

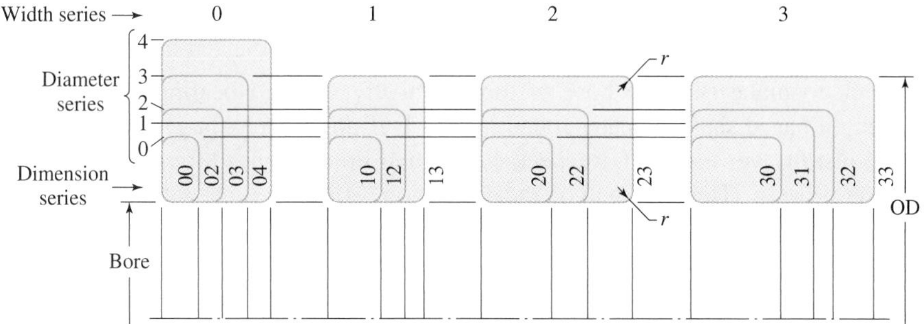

Figure 11.4
Illustration of selected examples from ABMA standard boundary dimensions for ball bearings, and for straight and spherical roller bearings. Boundaries shown are *not* valid for tapered roller bearings or for "inch-series" bearings.

Tables 11.5, 11.6, and 11.7 include not only basic dynamic and static load ratings, but boundary dimensions for the listed bearings as well. The ABMA has formulated a plan[13] for *standardized boundary dimensions* for bearings, including standard combinations of bore, outside diameter, width, and fillet sizes for shaft and housing shoulders. For a given bore, an assortment of standard widths and outside diameters are available, giving great flexibility in choosing a bearing geometry suitable to most applications. Figure 11.4 illustrates the ABMA plan for a selection of bearing boundary dimension specifications. Bearings are identified by a two-digit number called the *dimension-series code*. The *first* digit is from the *width series* (0, 1, 2, 3, 4, 5, or 6). The *second* digit is from the (*outside*) *diameter series* (8, 9, 0, 1, 2, 3, or 4). Figure 11.4 shows several samples of bearing profiles that may be obtained for a given bore size. Tables 11.5 and 11.6 illustrate the resulting dimensions for several selected series of ball and roller bearings, and Table 11.7 provides similar information for tapered roller bearings.

Not only have bearing dimensions been standardized, but dimensional *tolerances* have also been standardized for shaft and housing fits at several levels of precision.[13] These range from the "usual" ABEC-1 ball bearing tolerances and RBEC-1 roller bearing tolerances recommended for most applications to the ABEC-9 and RBEC-9 tolerances available (at higher cost) for exceptionally high-precision applications such as machine tool spindles.

Similar (but not exactly the same) precision tolerance levels for tapered roller bearings are provided by six tolerance *classes*.[14] To establish appropriate precision tolerance levels, bearing manufacturers' catalogs should be consulted.

10. Check to make sure operating speed lies below *limiting speed* for the tentatively selected bearing. If not, select another bearing that meets *both* the dynamic load rating requirement and the limiting speed requirement. Remember that the *speed* is the *absolute rotational speed of the inner race relative to the outer race.*

11. Calculate a static equivalent radial load rating P_{se} from the following empirical relationship

$$P_{se} = X_s F_{sr} + Y_s F_{sa} \tag{11-5}$$

where F_{sr} = static radial load
F_{sa} = static axial load
X_s = static radial load factor, based on bearing geometry (see Table 11.4)
Y_s = static axial (thrust) load factor, based on bearing geometry (see Table 11.4)

The combination X_{s_1}, Y_{s_1} or X_{s_2}, Y_{s_2} should both be calculated, and whichever combination gives the larger value of P_{se} should be used.

[13]See ref. 4.

[14]See ref. 6.

12. Check bearing selection tables to make sure that P_{se} does not exceed the basic static radial load rating C_s for the bearing that was tentatively selected in step 10.
13. Check to make sure the bore of the tentatively selected bearing will fit over the strength-based shaft diameter found in step 1. If not, select a larger bearing bore that will just fit over the shaft. Recheck C_d, C_s, and limiting speed for the newly selected larger bearing. (Depending upon the dimension series, bearings with larger bores may have lower basic load ratings.)
14. Using the final bearing selection, increase the shaft diameter at the bearing site to the nominal bearing bore size, using proper mounting dimensions and tolerances, as specified by the bearing manufacturer.

Example 11.1 Rolling Element Bearing Selection for Steady Loading

A gleaning-cylinder support shaft for a proposed "new concept" agricultural combine has been stress-analyzed to find that the minimum required strength-based shaft diameter is 1.60 inches at one of the proposed bearing sites. From the associated force analysis and other design specifications, the following information has been assembled:

1. Radial bearing load $F_r = 370$ lb (steady).
2. Axial bearing load $F_a = 130$ lb (steady).
3. Shaft speed $n = 350$ rpm.
4. Design life specification is 10 years of operation, 50 days/year, 20 hr/day.
5. Design reliability specification is $R = 95$ percent.
6. The shaft is to be V-belt driven.

Select a rolling element bearing for this application.

Solution

To select appropriate candidate bearings, the 14-step selection procedure just outlined may be utilized as follows:

1. Strength-based shaft diameter is (from problem statement)

$$(d_s)_{str} = 1.60 \text{ inches} \tag{1}$$

2. From problem specifications, the radial and axial applied load components are

$$F_r = 370 \text{ lb} \tag{2}$$

$$F_a = 130 \text{ lb} \tag{3}$$

3. Design life requirement is

$$L_d = 10 \text{ yr}\left(50\frac{\text{days}}{\text{yr}}\right)\left(20\frac{\text{hr}}{\text{day}}\right)\left(60\frac{\text{min}}{\text{hr}}\right)350\frac{\text{rev}}{\text{min}} = 2.1 \times 10^8 \text{ revolutions} \tag{4}$$

4. Using Table 11.2 for the specified reliability of $R = 95$ percent,

$$K_R = 0.62 \tag{5}$$

5. Using Table 11.3, since a V-belt drive is specified, a midrange impact factor may be selected as

$$IF = 1.9 \tag{6}$$

6. By evaluating the characteristics of this application (radial load, thrust load, low speed), it may tentatively be deduced, from Table 11.1, that appropriate bearing types would be

a. single-row, deep-groove ball bearing

b. single-row, tapered roller bearing (probably paired with another similar bearing at other end of the shaft)

7. Considering first a deep-groove ball bearing, the dynamic equivalent radial load P_e may be calculated from (11-3) as

$$P_e = X_d F_r + Y_d F_a \tag{7}$$

where, from Table 11.4, for a single-row radial ball bearing,

$$X_{d_1} = 1 \qquad Y_{d_1} = 0 \tag{8}$$

and

$$X_{d_2} = 0.55 \qquad Y_{d_2} = 1.45 \tag{9}$$

Hence

$$(P_e)_1 = (1)(370) + (0)(130) = 370 \text{ lb} \tag{10}$$

and

$$(P_e)_2 = (0.55)(370) + (1.45)(130) = 392 \text{ lb} \tag{11}$$

since $(P_e)_2 > (P_e)_1$,

$$P_e = (P_e)_2 = 392 \text{ lb} \tag{12}$$

8. The basic dynamic radial load rating requirement for $R = 95$ percent may be calculated from (11-4) as

$$[C_d(95)]_{req} = \left[\frac{2.1 \times 10^8}{0.62(10^6)}\right]^{1/a} (1.9)(392) \tag{13}$$

For ball bearings $a = 3$; hence

$$[C_d(95)]_{req} = \left[\frac{2.1 \times 10^8}{0.62(10^6)}\right]^{1/3} (1.9)(392) = 5192 \text{ lb} \tag{14}$$

9. Entering Table 11.5, either bearing number 6300 or 6002 would be appropriate choices for the "smallest" candidate, since 6300 has the smallest OD (42 mm) and 6002 has the smallest bore (10 mm).
10. The limiting speed for both bearings is far above the required operating speed of 350 rpm.
11. The static equivalent rating load for this application may be calculated from (11-5) as

$$P_{se} = X_s F_{sr} + Y_s F_{sa} \tag{15}$$

where, from Table 11.4,

$$X_{s_1} = 1.0 \qquad Y_{s_1} = 0 \tag{16}$$

$$X_{s_2} = 0.6 \qquad Y_{s_2} = 0.5 \tag{17}$$

hence

$$(P_{se})_1 = 1(370) + 0(130) = 370 \text{ lb} \tag{18}$$

$$(P_{se})_2 = 0.6(370) + 0.5(130) = 287 \text{ lb} \tag{19}$$

and therefore

$$P_{se} = 370 \text{ lb} \tag{20}$$

12. Checking P_{se} against the basic static load ratings in Table 11.5 for candidate bearings 6300 ($C_s = 764$ lb) and 6002 ($C_s = 641$ lb), both have acceptable static load ratings.
13. Checking bore diameters of bearings 6300 ($d = 0.3937$ inch) and 6002 ($d = 0.5906$ inch), it is clear that neither can be mounted on the minimum shaft diameter of 1.60

Example 11.1 Continues

inches specified in step 1. Entering Table 11.5 again, the smallest acceptable available bore is 45 mm, exhibited by bearing numbers 6009, 6209, and 6309. Of these, the smallest acceptable value of basic dynamic radial load rating (greater than 5192 lb) is bearing number 6209. The basic static load rating for this bearing is also acceptable ($4860 > 287$), and limiting speed is satisfactory as well.

14. The final selection of a ball bearing candidate, therefore, is bearing number 6209. The nominal shaft mounting diameter at this bearing site should be increased to

$$d = 1.7717 \text{ inches} \tag{21}$$

and manufacturers' catalogs should be consulted to determine appropriate tolerances. The minimum abutment diameter for the shaft shoulder at this bearing site may also be found in Table 11.5 as 2.087 inches, and the maximum fillet radius where shaft mounting diameter meets the abutment shoulder is $r = 0.039$ inch.

Going back now to step 6, a similar selection procedure may be carried through for a tapered roller bearing candidate as follows:

1–6. Same as steps 1–6 just above for the deep-groove ball bearing.

7. Considering a tapered roller bearing (single row), the dynamic equivalent radial load P_e may be calculated from (11-3) as

$$P_e = X_d F_r + Y_d F_a \tag{22}$$

where, from Table 11.4, for a single-row roller bearing ($\alpha \neq 0$)

$$X_{d_1} = 1 \qquad Y_{d_1} = 0 \tag{23}$$

and

$$X_{d_2} = 0.4 \qquad Y_{d_2} = 0.4 \cot\alpha \tag{24}$$

The value of Y_{d_2} is a function of α, unknown until the bearing has been selected. Hence an iterative process is indicated. One approach is to utilize Y_{d_2} values given in Table 11.7, noting that for all bearings in the size range of interest,

$$Y_{d_2} \approx 1.5 \tag{25}$$

Hence,

$$(P_e)_1 = (1)(370) + (0)(130) = 370 \text{ lb} \tag{26}$$

and

$$(P_e)_2 = (0.4)(370) + (1.5)(130) = 343 \text{ lb} \tag{27}$$

8. The basic dynamic load rating requirement for $R = 95$ percent may be calculated from (11-4), using $a = {}^{10}\!/_3$, as

$$[C_d(95)]_{req} = \left[\frac{2.1 \times 10^8}{0.62(10^6)}\right]^{3/10}(1.9)(370) = 4036 \text{ lb} \tag{28}$$

9. Entering Table 11.7, it may be noted that all bearings in the table meet the basic dynamic load rating requirements.
10. No tabled values of limiting speed are available. Generally, manufacturers' catalogs should be consulted. In this particular application, the operating speed of 350 rpm is so low that no problems would be anticipated. (Rough estimates of limiting speed may also be extracted from Table 11.6 for straight roller bearings.)
11. The static equivalent radial load, using (11-5) and Table 11.4 (noting $Y_{s_2} = Y_{d_2}/2$), is

$$(P_{se})_1 = 1(370) + 0(130) = 370 \text{ lb} \tag{29}$$

$$(P_{se})_2 = 0.5(370) + 0.75(130) = 283 \text{ lb} \tag{30}$$

hence

$$P_{se} = 370 \text{ lb} \tag{31}$$

12. Checking P_{se} against the basic static load rating in Table 11.7, all bearings in the table are acceptable.

13. As for the ball bearing candidate, 45 mm (1.7717 in) is the smallest acceptable value for the available bore sizes given in Table 11.7.[15] Hence bearing number 30209 would be the tentative choice. Comparing $Y_{d_2} = 1.48$ for bearing number 30209 with the assumed value of $Y_{d_2} = 1.5$ from (25), the difference in $(P_e)_2$ is small, $(P_e)_1$ still governs, as shown in (26), and no recalculation is necessary.

14. The final selection for a tapered roller bearing candidate therefore is bearing number 30209. The nominal shaft mounting diameter at the bearing site should be increased to

$$d = 1.7717 \text{ inches} \tag{32}$$

and manufacturers' catalogs should be consulted to determine appropriate tolerances. From Table 11.7, the minimum shaft-shoulder-abutment diameter is given as 2.087 inches, and the maximum fillet radius between shaft mounting diameter and abutment shoulder is $r = 0.059$ inch. The bearing OD is 3.3465 inches and bearing width is 0.6853 inch.

The final choice between the ball bearing candidate and the tapered roller bearing candidate would typically depend upon cost, availability, and other factors such as stiffness requirements or system vibrational characteristics.

Suggested Selection Procedure for Spectrum Loading

If a bearing is subjected to a *spectrum* of different applied load levels during each duty cycle in the operation of a machine, one approach to bearing selection would be to assume that the largest load is applied to the bearing every revolution, even when the *actual load* may be smaller during some segments of operation. Then the 14-step selection procedure outlined above would be followed. However, in high-performance design applications, where weight and space are at a premium, the cumulative damage concepts of 2.6 may be utilized to obtain a more efficient bearing selection. From (11-1) the bearing life L_i under applied load P_i, for a given bearing, is related to the basic dynamic radial load rating as follows:

$$10^6[C_d(90)]^a = L_i P_i^a = L_1 P_1^a = L_2 P_2^a = \cdots \tag{11-6}$$

If a duty cycle, such as the one illustrated in Figure 11.5, contains a total of n cycles, of which n_1 occur under applied load P_1, n_2 occur under P_2, and so on, then

$$n_1 + n_2 + \cdots + n_i = n \tag{11-7}$$

and the fraction of cycles at each applied load level P_i may be defined as

$$\begin{aligned} \alpha_1 &= \frac{n_1}{n} \equiv \text{fraction of cycles at } P_1 \\ \alpha_2 &= \frac{n_2}{n} \equiv \text{fraction of cycles at } P_2 \\ &\vdots \\ \alpha_i &= \frac{n_i}{n} \equiv \text{fraction of cycles at } P_i \end{aligned} \tag{11-8}$$

[15]Actually, from complete catalog listings (see ref. 6) a bearing bore of 1.6137 could be selected.

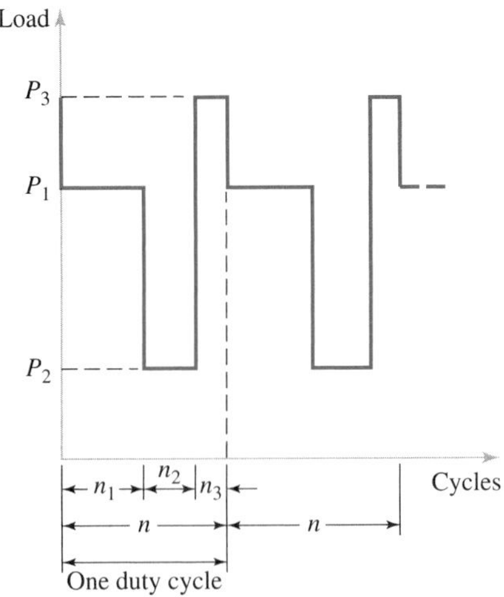

Figure 11.5
Example of spectrum loading on a rolling element bearing.

Also, by definition,

$$\begin{aligned} L_1 &= \text{cycles to failure if all cycles were loaded at } P_1 \\ L_2 &= \text{cycles to failure if all cycles were loaded at } P_2 \\ &\vdots \\ L_i &= \text{cycles to failure if all cycles were loaded at } P_i \end{aligned} \tag{11-9}$$

Based on (11-8), the total cycles of operation at load P_i may be expressed in terms of design life L_d as

$$\alpha_i L_d = \text{ total cycles at } P_i \tag{11-10}$$

Utilizing the Palmgren-Miner linear damage rule of (2-47), incipient failure is predicted if

$$\sum \frac{\alpha_i L_d}{L_i} = 1 \tag{11-11}$$

However, from (11-6),

$$L_i = \frac{10^6[C_d(90)]^a}{P_i^a} \tag{11-12}$$

so (11-11) becomes

$$\sum \frac{(\alpha_i L_d)P_i^a}{10^6[C_d(90)]^a} = 1 \tag{11-13}$$

or, incipient failure is predicted if

$$\alpha_1 P_1^a + \alpha_2 P_2^a + \cdots + \alpha_i P_i^a = \frac{10^6[C_d(90)]^a}{L_d} \tag{11-14}$$

Utilizing the same concepts of equivalent radial loads, service impact factors, and reliability-adjustment factors, (11-14) may be solved to give the basic radial load rating requirement $[C_d(R)]_{req}$ as

$$[C_d(R)]_{req} = \left[\frac{L_d}{K_R(10^6)}\right]^{1/a} \sqrt[a]{\sum \alpha_i[(IF)_i(P_e)_i]^a} \tag{11-15}$$

This basic radial load rating requirement may be used for bearing selection under spectrum loading conditions in the same way as for the case of steady loading, as described earlier.

Example 11.2 Rolling Element Bearing Selection for Spectrum Loading

From a complete force analysis of a newly proposed rotating shaft, stresses have been determined from a strength-based calculation that indicates the shaft diameter at one bearing site must be at least 1.63 inches. Also, from the force analysis and other design specifications, a duty cycle is well approximated by three segments, each segment having the characteristics defined in Table E11.2A.

The total design life for the bearing is to be 10^7 revolutions and the desired reliability is 97 percent. A single-row deep-groove ball bearing is preferred.

a. Select an appropriate bearing for this application using the spectrum loading procedure.
b. Compare the result of (a) with a bearing selection for this site using the steady load procedure, under the assumption that a constant load equal to the largest spectrum load remains on the bearing throughout all segments of its operation.

Solution

a. Using the specifications just given, and applying the concepts developed earlier in this chapter, the tabulation in Table E11.2B may be made.

Also, from Table 11.2, for $R = 97$ percent,

$$K_R = 0.44 \tag{1}$$

and the design life is specified as

$$L_d = 10^7 \text{ revolutions} \tag{2}$$

From (11-15), for a ball bearing ($a = 3$),

$$[C_d(97)]_{req} = \left[\frac{10^7}{0.44(10^6)}\right]^{1/3} \sqrt[3]{0.32[1.4(2000)]^3 + 0.65[1.8(3450)]^3 + 0.03[1.8(5000)]^3} \tag{3}$$

or

$$\begin{aligned}[C_d(97)]_{req} &= [2.83]\sqrt[3]{7.02 \times 10^9 + 156.0 \times 10^9 + 0.022 \times 10^9} \\ &= 2.83(5462) = 15{,}460 \text{ lb}\end{aligned} \tag{4}$$

From Table 11.5, the smallest acceptable bearing is number 6311. This bearing has a bore of 55 mm (2.164 inches), an OD of 120 mm (4.7244 inches), and a width of 29 mm (1.1417 inch). Checking the limiting speeds for bearing number 6311, it would be acceptable if oil lubrication were used (limiting speed 6700 rpm), but grease lubrication could *not* be used (limiting speed 5600 rpm). The basic static load rating of 10,100 lb is acceptable (10,100 > 5000). Also, the bore diameter of bearing number 6311, 2.1654 inches, is acceptable because it *will* go over the strength-based minimum shaft diameter of 1.63 inches.

Table E11.2A Duty Cycle Definition

Variable	Segment 1	Segment 2	Segment 3
F_r, lb	2000	1000	5000
F_a, lb	500	900	0
IF	light shock	moderate shock	moderate shock
n_i/duty cycle	1000	2000	100
N_{op}, rpm (rotational speed)	3600	7200	900

Example 11.2 Continues

Table E11.2B Segment-by-Segment Data

Variable	Segment 1	Segment 2	Segment 3
F_r	2000	1000	5000
F_a	500	2000	0
X_{d_1}	1	1	1
Y_{d_1}	0	0	0
X_{d_2}	0.55	0.55	0.55
Y_{d_2}	1.45	1.45	1.45
X_{s_1}	1	1	1
Y_{s_1}	0	0	0
X_{s_2}	0.6	0.6	0.6
Y_{s_2}	0.5	0.5	0.5
$(P_e)_1 = F_r$	2000	1000	5000
$(P_e)_2 = 0.55F_r + 1.45F_a$	1825	3450	2750
P_e	2000	3450	5000
$(P_{se})_1 = F_r$	2000	1000	5000
$(P_{se})_2 = 0.6F_r + 0.5F_a$	1450	1600	3000
P_{se}	2000	1600	5000
n_i	1000	2000	100
$\alpha_i = n_i/3100$	0.32	0.65	0.03
IF	1.4	1.8	1.8

The spectrum loading procedure therefore results in the selection of single-row ball bearing candidate number 6311. The nominal shaft mounting diameter at this bearing site should be increased to

$$d = 2.1654 \text{ inches} \tag{5}$$

and manufacturers' catalogs should be consulted to determine appropriate tolerances. Shaft-abutment-shoulder dimensions from Table 11.5 are 2.520 inches for minimum shoulder diameter and $r = 0.079$ inch, maximum, for shaft-shoulder fillet radius.

b. Using the *simplified method* (largest steady load), and choosing segment 3 loading data from Tables E11.2A and E11.2B, (11-4) gives

$$[C_d(97)]_{req} = \left[\frac{10^7}{0.44(10^6)}\right]^{1/3}(1.8)(5000) = 2.83(9000) = 25{,}470 \text{ lb} \tag{6}$$

From Table 11.5 the smallest acceptable bearing is number 6316, based on basic dynamic load rating. This bearing has a bore of 80 mm (3.1496 inches), an OD of 170 mm (6.6929 inches), and a width of 39 mm (1.5354 inches). Checking the limiting speed, even with oil lubricaiton the bearing's limiting speed (4500 rpm) does not meet the segment 2 speed requirement of 6700 rpm, and segment 2 represents about 63 percent of bearing operation.

Clearly, this (oversimplified) approach, for the case at hand, results in a much larger, heavier bearing and it doesn't meet limiting speed requirements. Consultation with a bearing manufacturer would probably be necessary to resolve the lubrication/speed issue.

The superiority of using the spectrum loading approach to bearing selection is evident for this particular case.

Lubrication

Until recently, metal-to-metal contact and boundary lubrication were assumed to characterize all rolling element bearing applications. However, in the development of elastohydrodynamic theory[16] in the 1960s and 1970s, it was postulated that for operational rolling element bearings, the squeeze film thickness, h, often may be the same order of magnitude (microinches) as surface roughness, making total separation of the surfaces by the film a feasible goal. Early investigators examined the influence on L_{10} bearing life of a dimensionless *film parameter* Λ, sometimes called the *lambda ratio*. It is defined as

$$\Lambda \equiv \frac{h_{min}}{(R_a^2 + R_b^2)^{1/2}} \tag{11-16}$$

where h_{min} = minimum elastohydrodynamic film thickness
R_a = relative roughness (arithmetic average roughness height) of bearing surface a (see Figure 6.11)
R_b = relative roughness (arithmetic average roughness height) of bearing surface b (see Figure 6.11)

Experimental results for ball bearings[17] and roller bearings[18] are shown in Figure 11.6. These results illustrate the *improvement* in ABMA L_{10} life that may be achieved by maintaining an appropriate minimum film thickness, h_{min}. The L_{10} life *degradation* that may result for small values of Λ is also shown. A magnitude of $\Lambda \approx 3$ may be regarded as near optimum when elastohydrodynamic conditions prevail. The minimum film thickness is a complicated function of the contact geometry, load, speed, and material properties. Its calculation is beyond the scope of this text, but is available in the literature.[19] Additional re-

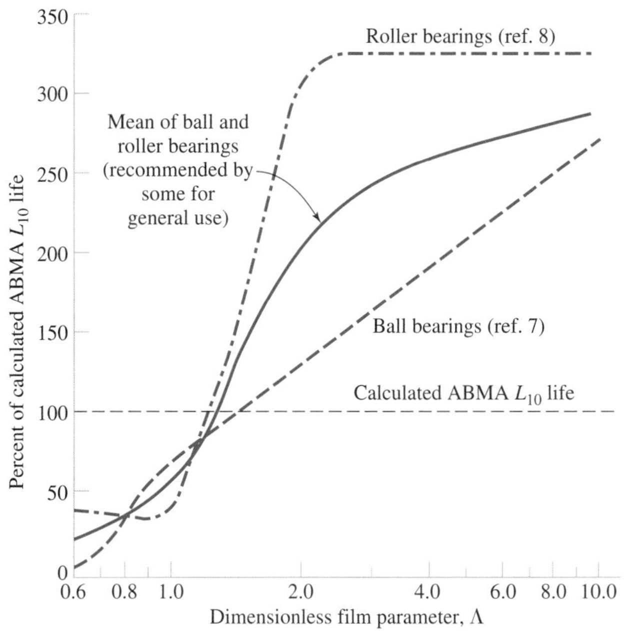

Figure 11.6
Influence of elastohydrodynamic film parameter on the ABMA L_{10} bearing life of ball and roller bearings.

[16]See 10.5.
[17]See ref. 7.
[18]See ref. 8.
[19]See ref. 9.

search is being conducted in this area. Bearing manufacturers' catalogs also provide useful information about the lubrication of rolling element bearings.[20]

11.6 Preloading and Bearing Stiffness

Although many design applications require no special attention to stiffness or deflection, *high-performance dynamic systems* may require high stiffness, high natural frequency, low deflection, and/or low noise levels. Bearings (which behave as springs; see 4.11) are usually in *series* with other machine components, such as shafts and support structure. If bearings are "soft" springs compared to the other components in series, they may dominate the combined spring rate and significantly lower the system stiffness, as evidenced by (4-143). Further, because rolling element bearing deflection under load is predominantly governed by Hertz contact behavior (see 4.10), the typical force-displacement curve for any such bearing is *nonlinear* and *stiffening* with increasing loads, as shown in Figures 4.33(b) and 11.7. That is, under *light loads* a bearing behaves as a *soft spring*, but at *higher loads* it behaves as a *stiff spring*. For this reason it is common practice to *preload* certain types of bearings to eliminate the "soft" lower branch of the force-displacement curve. As illustrated in Figure 11.7, the effect of preloading is to minimize the operational bearing deflection and increase effective stiffness. When it is desired to utilize preloading to increase effective bearing stiffness, it is first necessary to select bearings that can accommodate both axial and radial loads. Angular contact ball bearings, deep-groove ball bearings, and tapered roller bearings are all good candidates for preloading. They would typically be mounted in matched pairs, face-to-face or back-to-back, so they can be preloaded axially against each other. Achieving the desired preload is usually accomplished by using a pair of bearings with a controlled "stickout" between inner and outer race faces, by tightening a nut on the shaft or in the housing to produce an axial displacement of one race relative to the other, or by using thrust washers or springs. When the *axial* preload is induced in such bearings, a *radial* preload also results (because of the opposing contact angles of the

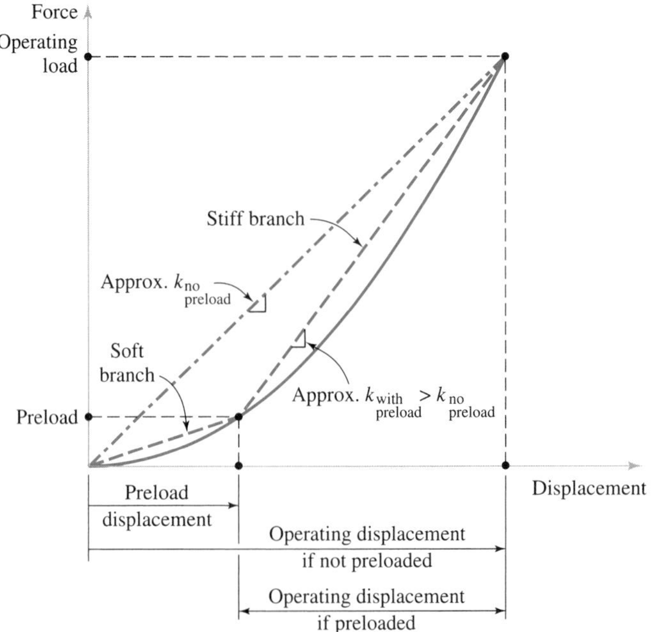

Figure 11.7
Linearized approximation of typical bearing force-displacement curve, showing the stiffening effect of preloading.

[20]See refs. 5 and 6.

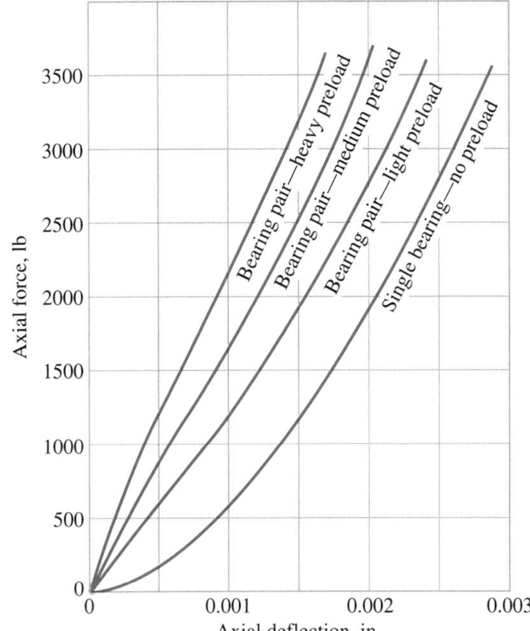

Figure 11.8
Sample *axial* force-deflection curves for ball bearings installed with various levels of axial preload. (*Data Source:* New Departure Hyatt[21] Catalog BC-7, 1977.)

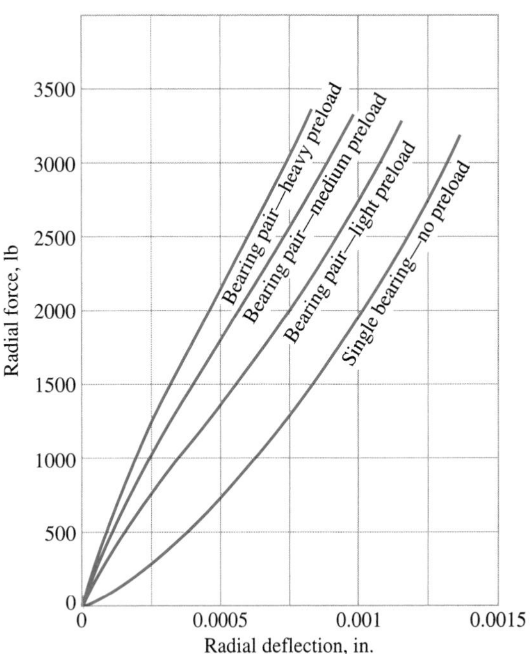

Figure 11.9
Sample *radial* force-deflection curves for ball bearings installed with various levels of axial preload; see Figure 11.8. (*Data Source:* New Departure Hyatt[22] Catalog BC-7, 1977.)

paired bearings). For example, Figure 11.8 shows the axial force-deflection curve for selected pairs of ball bearings installed *without* preload, as compared to paired ball bearings *with* three different levels of initial preload. For a single bearing of this type, with *no preload*, a 600-lb externally applied axial load produces an axial deflection of nearly 0.001 inch. Using a light (standard) preload reduces this deflection to about 0.0005 inch for the same 600-lb axial load, effectively *doubling* the bearing stiffness in this loading range. Medium and heavy preloads reduce the deflection under a 600-lb external load to 0.0003 and 0.0002 inch, respectively. The *radial* force-deflection behavior of these same bearings, with and without axial preload, is illustrated in Figure 11.9. It is clear that the *radial stiffness is significantly increased* when *axial preloads* are used.

Example 11.3 Influence of Bearing Stiffness and Preloading on System Behavior

A rotating steel disk, 35 inches in diameter and 2 inches thick, is to be mounted at midspan on a 1020 hot-rolled solid steel shaft having $S_u = 65{,}000$ psi, $e = 36$ percent elongation in 2 inches, and fatigue properties shown in Figure 2.19. A reliability of 99 percent is desired for the shaft and the bearings. The shaft length between symmetrical bearing centers [see (b) below for proposed bearings] is to be 6.0 inches. The operating speed of the rotating system is 7200 revolutions per minute, and a design life of 10^8 cycles is desired. When the system operates at steady-state full load, it has been estimated that about one horsepower of input to the rotating shaft is required.

[21]No longer in business.

[22]No longer in business.

Example 11.3 Continues

a. Estimate the required shaft diameter and the critical speed for the rotating system, assuming that the support bearings and the frame are rigid. The bending fatigue stress concentration factor has been estimated as $K_{fb} = 1.8$, and the composite strength-influencing factor, k_{10^8}, defined in (2-28), has been estimated as 0.25. A design safety factor of $n_d = 2.1$ is desired. Is the estimated critical speed acceptable?

b. Make a second estimate for the critical speed of the rotating system, this time including the bearing stiffness (elasticity). A design task group working under your direction has tentatively selected a single-row deep-groove ball bearing, number 6309 (see Table 11.5), with oil lubrication, for the application, using the procedure outlined in Example 11.1. In addition, the group has experimentally verified that the force-deflection data shown in Figures 11.8 and 11.9 are approximately correct for the tentatively selected bearing.

c. Compare results from (a) and (b), and judge the influence of bearing stiffness, as well as the acceptability of the design from a vibrational viewpoint.

d. Make a third estimate for critical speed of the rotating system if a standard (light) axial preload is induced by the way the bearings are mounted. Comment on the influence of preloading in this particular case, and suggest alternatives if necessary.

Solution

a. Using (8-8), the strength-based shaft diameter $(d_s)_{str}$ is given by

$$(d_s)_{str} = \sqrt[3]{\frac{32n_d}{\pi S_N}\sqrt{(K_{fb}M_a)^2 + 0.75(T_m)^2}} \tag{1}$$

Based on information provided in the problem statement, the following preliminary determinations and calculations may be made:

From Figure 2.19, the fatigue strength corresponding to $N = 10^8$ cycles is $S_{10^8} = 33{,}000$ psi. Using (2-26) therefore gives

$$S_N = S_{10^8} = k_{10^8}S'_{10^8} = 0.25(33{,}000) = 8250 \text{ psi} \tag{2}$$

The disk weight W_D is

$$W_D = 0.283\left[\frac{\pi(35)^2}{4}\right](2.0) = 544.6 \text{ lb} \tag{3}$$

From Table 4.1, case 1, the maximum moment at midspan is

$$M_{max} = \frac{W_D L}{4} = \frac{544.6(6.0)}{4} = 817 \text{ in-lb} \tag{4}$$

The reaction (radial) at each bearing site is

$$R_R = R_L = \frac{W_D}{2} = \frac{544.6}{2} = 272 \text{ lb} \tag{5}$$

From (4-32), the torque on the shaft is

$$T = \frac{63{,}025(1)}{7200} = 8.75 \text{ in-lb} \tag{6}$$

From (1) then

$$(d_s)_{str} = \sqrt[3]{\frac{32(2.1)}{\pi(8250)}\sqrt{(1.8(817))^2 + 0.75(8.75)^2}} = 1.56 \text{ inches} \tag{7}$$

Comparing this strength-based *minimum* shaft diameter *requirement* with the bore size of the selected bearing (number 6309; see Table 11.5), the bearing bore is seen to be larger than the shaft. The shaft diameter must therefore be increased locally to match the 1.7717-inch bore size at the mounting site, and, correspondingly, the abutment shoulder diameter should be made 2.087 inches. For *stiffness* calculations, the shaft diameter will be *assumed* to be 2.087 inches, uniform along the length.

From Table 4.1, case 1, the midspan (maximum) deflection is

$$y_m = \frac{WL^3}{48EI} = \frac{(544.6)(6.0)^3}{48(30 \times 10^6)\left(\dfrac{\pi(2.087)^4}{64}\right)} = 0.00009 \text{ inch} \tag{8}$$

The critical shaft frequency, assuming the bearings and housing to be infinitely stiff, then may be estimated from (8-15) as

$$(n_{cr})_{\substack{shaft\\only}} = 187.7\sqrt{\frac{1}{0.00009}} = 19{,}785\frac{\text{rev}}{\text{min}} \tag{9}$$

and

$$\frac{(n_{cr})_{\substack{shaft\\only}}}{n_{op}} = \frac{19{,}785}{7200} = 2.8 \tag{10}$$

According to the guidelines of 8.6, this is well within the acceptable range.

b. Using Figure 11.9 as the basis, and using the radial bearing reaction of 272 lb calculated from (5), the radial deflection for a single bearing with no preload may be read as

$$(y_{brg})_{no\text{-}pre} = 0.00025 \text{ inch} \tag{11}$$

so the total midspan lateral displacement of the disk center from the unloaded shaft centerline becomes

$$(y_m)_{no\text{-}pre} = 0.00009 + 0.00025 = 0.00034 \text{ inch} \tag{12}$$

Again using (8-15)

$$(n_{cr})_{no\text{-}pre} = 187.7\sqrt{\frac{1}{0.00034}} = 10{,}179\frac{\text{rev}}{\text{min}} \tag{13}$$

and

$$\frac{(n_{cr})_{no\text{-}pre}}{n_{op}} = \frac{10{,}179}{7200} = 1.4 \tag{14}$$

c. This ratio is far below the recommended guidelines of 8.6, and would be regarded as a risky design, requiring improvement or experimental investigation.

d. Again using Figures 11.8 and 11.9 as a basis, when a light preload is induced, the radial deflection of the bearing pair, each carrying a radial reaction force of 272 lb, may be read as

$$(y_{brg})_{light} = 0.00008 \text{ inch} \tag{15}$$

so the total midspan displacement of the disk center becomes

$$(y_m)_{light} = 0.00009 + 0.00008 = 0.00017 \text{ inch} \tag{16}$$

and from (8-15)

$$(n_{cr})_{light} = 187.7\sqrt{\frac{1}{0.00017}} = 14{,}396\frac{\text{rev}}{\text{min}} \tag{17}$$

giving

$$\frac{(n_{cr})_{light}}{n_{op}} = \frac{14{,}396}{7200} = 2.0 \tag{18}$$

This lies within the acceptable range.

Example 11.3 Continues

In this particular case, ignoring the bearing elasticity [as in (a)] gives an *erroneous conclusion* about the acceptability of the design. This becomes clear in (b) because when the bearing elasticity (stiffness) *is* included, the design becomes risky from the system vibration standpoint. A light axial preload on the ball bearings [as in (c)] brings the system stiffness back into the accpetable range. In other circumstances it might be necessary to consider heavier preloads or other types of bearings, such as double-row ball bearings or tapered roller bearings, to improve system stiffness.

11.7 Bearing Mounting and Enclosure

Because so many different arrangements are possible for mounting and enclosing bearings, a designer is well advised to consult bearing manufacturers' catalogs and handbooks for information relative to mounting options for the specific bearing selected and the specific application of interest. The discussion of mounting practice included here is brief, and intended only as broad guidance.

The operational clearances within *mounted* bearings that are running at steady-state operating temperatures should be *virtually zero* to provide acceptable life, vibration behavior, and noise levels. It is important to understand that the internal clearance of a mounted bearing is typically *smaller* than the internal clearance of the *same* bearing before it is mounted. Designers should follow bearing manufacturers' guidelines for shaft and housing fits and tolerances to obtain optimum benefits from the precision level of the bearings selected. In general, the *rotating* ring of a bearing should have an *interference fit* with its mating member (shaft or housing) and the *nonrotating* ring should have a *close push fit* with its mating member. This practice typically assures that the rotating ring will not slip or turn with respect to its mating shaft or housing, and that the nonrotating (unclamped) ring can accommodate axial sliding to avoid undesirable thermally induced thrust loads.

The majority of applications involve rotating shafts and stationary housings. For such a configuration, an approved interference fit should be specified between the shaft at the mounting site and the inner race bore, and a close push fit should be specified between the bearing's outer diameter and the housing bore. If housings are made of lower modulus-of-elasticity, higher thermal-coefficient-of-expansion materials, such as aluminum or magnesium, slightly tighter push fits should be used. If shafts are hollow, slightly tighter interference fits should be used.

Interference fits alone usually do not provide adequate precision for the axial location of a bearing ring. Some positive means of axially locating and securing the ring is usually required, such as an abutment shoulder, a spacer sleeve, a collar, or a snap ring, in conjunction with retention devices such as shaft nuts and lock washers, housing end caps, or threaded rings. Appropriate abutment dimensions must be specified for proper bearing operation (see, for example, Tables 11.5, 11.6, and 11.7, or bearing manufacturers' catalogs).

The basic arrangement for typical applications is to support a rotating shaft on *two* bearings, one near each end of the shaft. One (locating type) bearing is axially secured to both shaft and housing to provide accurate axial location and the ability to support both radial and thrust loads. The other (nonlocating type) bearing provides radial support only, allowing axial displacements to accommodate differential thermal expansion between shaft and housing without developing undesired thrust overloads. Axial displacements may, in some cases, take place within the bearing itself (e.g., cylindrical roller bearings) or by slip between the outer ring and housing bore. A typical arrangement is shown in Figure 11.10. Three or more bearings on a shaft are to be avoided unless absolutely necessary because

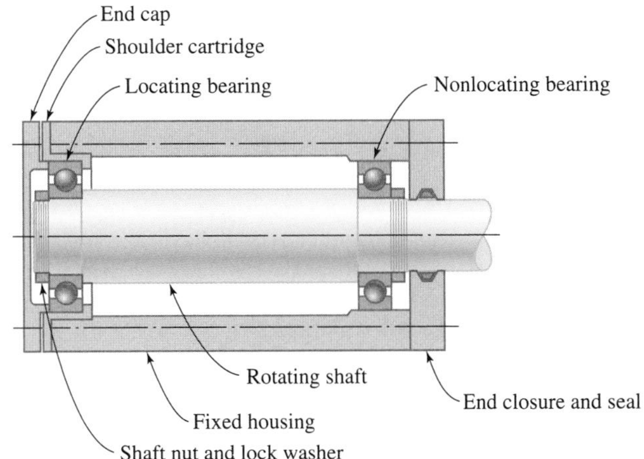

Figure 11.10
Typical mounting arrangement with locating bearing at left end and nonlocating bearing at right end. The locating bearing is fixed axially. The nonlocating bearing is free to slide axially so that neither temperature changes nor tightening procedures will overload the bearings in the axial direction.

centerline mismatch from bearing to bearing may produce "built-in" bending moments and unknown radial bearing loads that sometimes lead to premature failures.

It is a designer's responsibility to assure that bearings can physically be installed and that surrounding parts can be assembled without damaging the bearings. Proper access and clearance must be provided for mechanical, hydraulic, or thermal mounting and dismounting tools, so that no forces are ever transmitted across the rolling elements during assembly or disassembly. Access ports for inspection and cleaning must be provided so that these maintenance functions may be accomplished with a minimum of disassembly and downtime.

To prevent solid contaminants and moisture from getting into the bearings, and to retain lubricant, enclosures and/or seals are usually necessary. Seals may be an integral part of the bearing or they may be separate from the bearing. Selection of a suitable sealing arrangement may depend upon the type of lubricant (oil or grease), relative surface speed at the sealing interface, potential misalignment, available space, potential environmental influences, or other factors.

If integral seals are to be used, bearing manufacturers' catalogs should be consulted. Typically, bearings can be supplied with seals or shields at one or both sides. If separate seals are to be used, seal manufacturers' catalogs should be consulted. Separate seals may be either *nonrubbing seals*, such as gap-type seals or labyrinth seals, or *rubbing seals*, such as radial lip seals, V-ring seals, or felt seals. It is common practice to place a nonrubbing seal in front of the rubbing seal to protect it from damage by solid contaminants.

Problems

11-1. For each of the following applications, select two possible types of rolling element bearings that might make a good choice.

a. High-speed flywheel (see Chapter 18) mounted on a shaft rotating about a horizontal centerline.

b. High-speed flywheel mounted on a shaft rotating about a vertical centerline.

c. Low-speed flywheel mounted on a shaft rotating about a vertical centerline.

11-2. A single-row radial ball bearing has a basic dynamic load rating of 11.4 kN for an L_{10} life of 1 million revolutions. Calculate its L_{10} life if it operates with an applied radial load of 8.2 kN.

11-3. a. Determine the required basic dynamic load rating for a bearing mounted on a shaft rotating at 1725 rpm if it must carry a radial load of 1250 lb and the desired design life is 10,000 hours.

b. Select a single-row radial ball bearing from Table 11.5 that will be satisfactory for this application if the outside diameter of the bearing must not exceed 4.50 inches.

11-4. A single-row radial ball bearing must carry a radial load of 500 lb, and no thrust load. If the shaft that the bearing is mounted on rotates at 1175 rpm, and the desired L_{10} life of the bearing is 20,000 hr, select the smallest suitable bearing from Table 11.5 that will satisfy the design requirements.

11-5. In a preliminary design calculation, a proposed deep-groove ball bearing has been tentatively selected to support one end of a rotating shaft. A mistake has been discovered in the load calculation, and the correct load turns out to be about 25 percent higher than the earlier incorrect load used to select the ball bearing. To change to a larger bearing at this point means that a substantial redesign of all the surrounding components will probably be necessary. If no change is made in the original bearing selection, estimate how much reduction in bearing life would be expected.

11-6. A number 6005 single-row radial deep-groove ball bearing is to rotate at a speed of 1750 rpm. Calculate the expected bearing life in hours for radial loads of 400, 450, 500, 550, 600, 650, and 700 lb, and make a plot of life versus load. Comment on the reuslt.

11-7. Repeat problem 11-6, except use a number 205 single-row cylindrical roller bearing instead of the 6005 radial ball bearing.

11-8. A number 207 single-row cylindrical roller bearing has tentatively been selected for an application in which the design life corresponding to 90 percent reliability (L_{10} life) is 7500 hr. Estimate what the corresponding lives would be for reliabilities of 50 percent, 95 percent, and 99 percent.

11-9. Repeat problem 11-8, except use a number 6007 single-row radial ball bearing instead of the 207 roller bearing.

11-10. A solid steel spindle shaft of circular cross section is to be used to support a ball bearing idler pulley as shown in Figure P11.10. The shaft may be regarded as simply supported at the ends and the shaft does not rotate. The pulley is to be mounted at the center of the shaft on a single-row radial ball bearing. The pulley must rotate at 1725 rpm and support a load of 800 lb, as shown in the sketch. A design life of 1800 hours is required and a reliability of 90 percent is desired. The pulley is subjected to moderate shock loading conditions.

a. Pick the smallest acceptable bearing from Table 11.5 if the shaft at the bearing site must be at least 1.63 inches in diameter.

b. Again using Table 11.5, select the smallest bearing that would give an infinite operating life, if you can find one. If you find one, compare its size with the 1800-hour bearing.

11-11. A helical idler gear (see Chapter 15) is to be supported at the center of a short hollow circular shaft using a single-row radial ball bearing. The inner race is pressed on the fixed non-rotating shaft, and the rotating gear is attached to the outer race of the bearing. The gear is to rotate at a speed of 900 rpm. The forces on the gear produce a resultant radial force on the bearing of 400 lb, and a resultant thrust force on the bearing of 325 lb. The assembly is subjected to light shock loading conditions. Based on a preliminary stress analysis of the shaft, it must have at least a 2.0-inch outside diameter. It is desired to use a bearing that will have a life of 3000 hours with 99 percent reliability. Select the smallest acceptable bearing (bore) from Table 11.5.

11-12. An industrial punching machine is being designed to operate 8 hours per day, 5 days per week, at 1750 rpm. A 10-year design life is desired. Select an appropriate *Conrad* type single-row radial ball bearing to support the drive shaft if bearing loads have been estimated as 1.2 kN radial and 1.5 kN axial, and light impact conditions prevail. Standard L_{10} bearing reliability is deemed to be acceptable for this application.

11-13. The shaft shown in Figure P11.13 is to be supported by two bearings, one at location *A* and the other at location *B*. The shaft is loaded by a commercial-quality driven helical gear (see Chapter 15) mounted as shown. The gear imposes a radial load of 7000 lb and a thrust load of 2500 lb applied at a pitch radius of 3 inches. The thrust load is to be fully supported by bearing *A* (bearing *B* takes no thrust load). It is being proposed to use a single-row tapered roller bearing at location *A*, and another one at location *B*. The device is to operate at 350 rpm, 8 hours per day, 5 days per week, for 3 years before bearing replacement is necessary. Standard L_{10} reliability is deemed acceptable. A strength-based analysis has shown that the minimum shaft

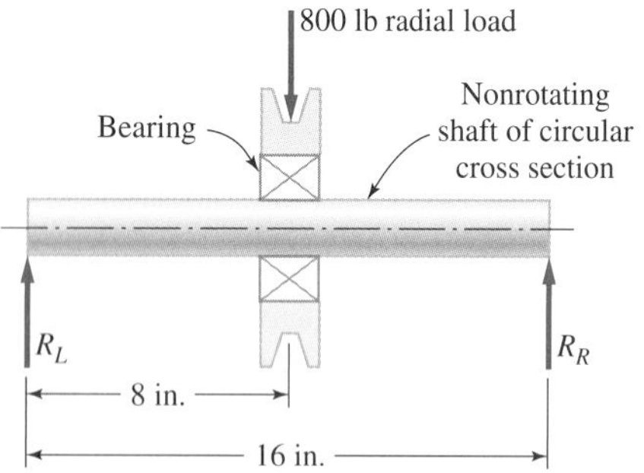

Figure P11.10
Idler pulley supported by nonrotating shaft.

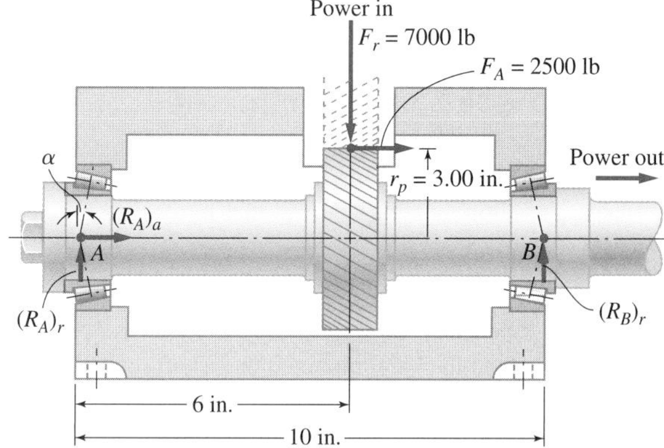

Figure P11.13
Support bearing arrangement for gear-driven shaft.

diameter must be 1.375 inches at both bearing sites. Select suitable bearings for both location A and location B.

11-14. From a stress analysis of a rotating shaft, it has been determined that the shaft diameter at one particular bearing site must be at least 80 mm. Also, from a force analysis and other design specifications, a duty cycle is well approximated by three segments, each segment having the characteristics defined in Table P11.14.

The total design life for the bearing is to be 40,000 hours and the desired reliability is 95 percent. A single-row deep-groove ball bearing is preferred.

a. Select an appropriate bearing for this application, using the spectrum loading procedure.

b. Compare the result of (a) with a bearing selection for this site using the steady load procedure, assuming that a *constant* radial load (and corresponding axial load) is applied to the bearing throughout all segments of its operation.

11-15. A preliminary stress analysis of the shaft for a rapid-return mechanism has established that the shaft diameter at a particular bearing site must be at least 0.70 inch. From a force analysis and other design specifications, one duty cycle for this device lasts 10 seconds, and is well approximated by two segments, each segment having the characteristics defined in Table P11.15.

The total design life for the bearing is to be 3000 hours. A single-row tapered roller bearing is preferred, and a standard L_{10} reliability is acceptable.

a. Select an appropriate bearing for this application, using the spectrum loading procedure.

b. Compare the result of (a) with a bearing selection for this site using the steady load procedure, assuming that a *constant* radial load equal to the largest spectrum load (and corresponding axial load) is applied to the bearing throughout the full duty cycle.

Table P11.14 Duty Cycle Definition

Variable	Segment 1	Segment 2	Segment 3
F_r, kN	7	3	5
F_a, kN	3	0	0
IF	light impact	heavy impact	moderate impact
n_i per duty cycle	100	500	300
N_{op}, rpm	500	1000	1000

Table P11.15 Duty Cycle Definition

Variable	Segment 1	Segment 2
F_r, lb	800	600
F_a, lb	400	0
IF	light impact	steady load
Operating time per cycle, sec	2	8
N_{op}, rpm	900	1200

11-16. A preliminary analysis of bearing A in Figure P11.13 has indicated that a 30209 tapered roller bearing will provide a satisfactory L_{10} bearing life of 3 years (operating at 350 rpm for 8 hours per day, 5 days per week) before bearing replacement is necessary. A lubrication consultant has suggested that if an ISO/ASTM viscosity-grade-46 petroleum oil is sprayed into the smaller end of the bearing (tapered roller bearings provide a geometry-based natural pumping action, inducing oil flow from their smaller ends toward their larger ends), a minimum elastohydrodynamic film thickness (h_{min}) of 10 microinches can be maintained. If the bearing races and the tapered rollers are all lapped to a surface roughness height of 4 microinches (see Figure 6.11), estimate the bearing life of the 30209 tapered roller bearing under these elastohydrodynamic lubrication conditions.

11-17. A rotating steel disk, 40 inches in diameter and 4 inches thick, is to be mounted at midspan on a 1020 hot-rolled solid steel shaft, having $S_u = 65{,}000$ psi, $e = 36$ percent elongation in 2 inches, and fatigue properties as shown in Figure 2.19. A reliability of 90 percent is desired for the shaft and the bearings, and a design life of 5×10^8 cycles has been specified. The shaft length between symmetrical bearing centers [see (b) below for proposed bearings] is to be 5 inches. The operating speed of the rotating system is 4200 revolutions per minute. When the system operates at steady-state full load, it has been estimated that about three horsepower of input to the rotating shaft is required.

a. Estimate the required shaft diameter and the critical speed for the rotating system, assuming that the support bearings and the frame are rigid in the radial direction. The bending fatigue stress concentration factor has been estimated as $K_{fb} = 1.8$, and the composite strength-influencing factor, $k_{5\times10^8}$, used in (2-28), has been estimated as 0.55. A design safety factor of 1.9 has been chosen. Is the estimated critical speed acceptable?

b. Make a second estimate for the critical speed of the rotating system, this time including the bearing stiffness (elasticity). Based on the procedure outlined in Example 11.1, a separate study has suggested that a single-row deep-groove ball bearing number 6209 (see Table 11.5), with oil lubrication, may be used for this application. In addition, an experimental program has indicated that the force-deflection data shown in Figures 11.8 and 11.9 are approximately correct for the tentatively selected bearing. Is your second estimate of critical speed acceptable? Comment on your second estimate, and if not acceptable, suggest some design changes that might make it acceptable.

c. Make a third estimate for critical speed of the rotating system if a *medium preload* is induced by the way the bearings are mounted. Comment on your third estimate.

Chapter 12

Power Screw Assemblies

12.1 Uses and Characteristics of Power Screws

Power screws, sometimes called *jack screws*, *lead screws*, or *linear actuators*, are machine elements that transform rotary motion into translational motion, or amplify a small tangential force moving (rotationally) through a large distance into a large axial force moving through a small distance. Geometrically, a power screw is a threaded shaft with an attached *thrust collar* at one end, engaged into a mating *nut*. With suitable constraint, either the nut may be rotated to cause axial translation of the threaded shaft (screw), or the screw may be rotated to cause axial translation of the nut. Common examples include screw jacks for load lifting, C-clamps, vises, lead screws for precision lathes or other machine tools, positive positioners for control rod drives in nuclear power reactors, and compaction drives for home garbage compactors.

The basic configuration for a load-lifting application is illustrated in Figure 12.1. Because large mechanical advantages are possible (the threads are essentially inclined planes,

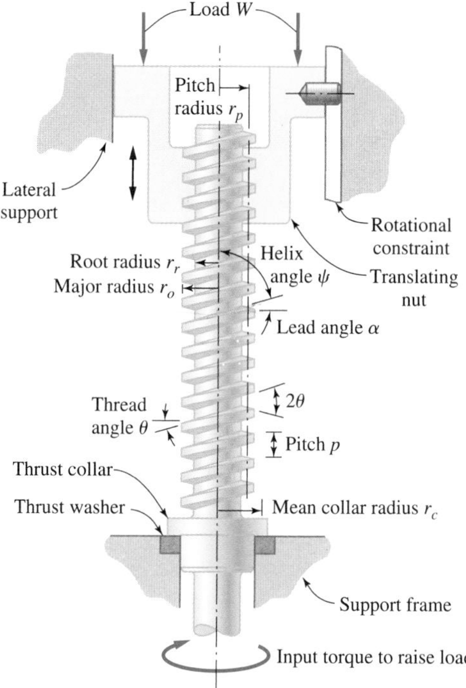

Figure 12.1
Basic configuration of load-raising power screw.

Some examples of power screws.

wrapped helically around the shaft), thread forms are chosen to maximize axial load-carrying capacity and to minimize frictional drag. The most useful thread forms that may be used for power screws are the *square* thread, the *modified square* thread, the *Acme* thread, and (for unidirectional loads) the *buttress* thread, all sketched in Figure 12.2. The square thread provides the best strength and efficiency but is difficult to manufacture. The 5° *thread angle*, θ, (10° included angle) of the modified square thread improves its manufacturability. The Acme thread illustrated in Figure 12.1, with its 29° included angle ($\theta = 14\frac{1}{2}°$) is easy to manufacture and allows the use of a split nut that can be squeezed radially to adjust for wear. The buttress thread provides greater strength for unidirectional loads. Stress analysis of power screw threads is discussed in 12.6. Power screw thread

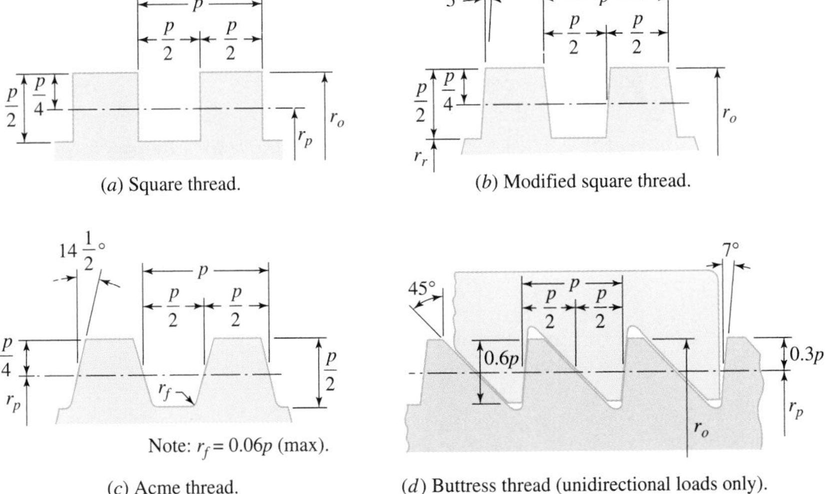

Figure 12.2
Common thread forms used for power screw applications.

Table 12.1 Selected Data for Standardized Power Screw Thread Forms[1]

Nominal Major Outside Diameter, d_o, in	Threads per Inch		
	Square and Modified Square	Acme	Buttress
1/4	10	16	—
5/16	—	14	—
3/8	—	12	—
3/8	8	10	—
7/16	—	12	—
7/16	—	10	—
1/2	6 1/2	10	16
5/8	5 1/2	8	16
3/4	5	6	16
7/8	4 1/2	6	12
1	4	5	12
1 1/2	3	4	10
2	2 1/4	4	8
2 1/2	2	3	8
3	1 3/4	2	6
4	1 1/2	2	6
5	—	2	5

[1]See refs. 1, 2, 3 for more complete data.

standards have been developed[1] for most practical thread forms, as illustrated by the selected data shown in Table 12.1. Metric power screw threads are rarely used in the United States, but are often used elsewhere.[2]

As shown in Figure 12.1 the *pitch* is defined as the axial distance along an element of the pitch cylinder measured from a reference point on one thread to a corresponding point on the next adjacent thread. The pitch is the reciprocal of the number of threads per inch. The *major* (outer) diameter is $d_o = 2r_o$. The *lead angle* (complement of the helix angle), α, is the angle between a plane drawn tangent to the pitch helix of a square thread and a plane drawn normal to the screw axis.[3] The *lead* l is the axial displacement of the nut relative to the screw for one rotation of the screw. If a *single-threaded* configuration is used, the lead is *equal* to the pitch. Single threads are used in most applications. If a *double-* or *triple-threaded* configuration is used, the lead is equal to *twice* the pitch or *three times* the pitch, respectively. Therefore, the increased lead of a multiple thread provides the advantage of rapid axial advancement. For double- or triple-threaded configurations, either two or three parallel helical threads are machined side by side around the screw. If n is the number of parallel threads,

$$l = np \tag{12-1}$$

By "developing" one full turn of a thread onto a plane, it may be deduced that

[1]See, for example, refs. 1, 2, and 3.

[2]See, for example, ref. 4.

[3]Ambiguities in terminology exist in the literature; some authors call α the *lead angle* while others call α the *helix angle*. In this text, the lead angle is consistently measured from a plane *normal* to the screw axis while helix angle is measured from the screw axis itself, as shown in Figure 12.1. Hence, $\alpha + \psi = 90°$.

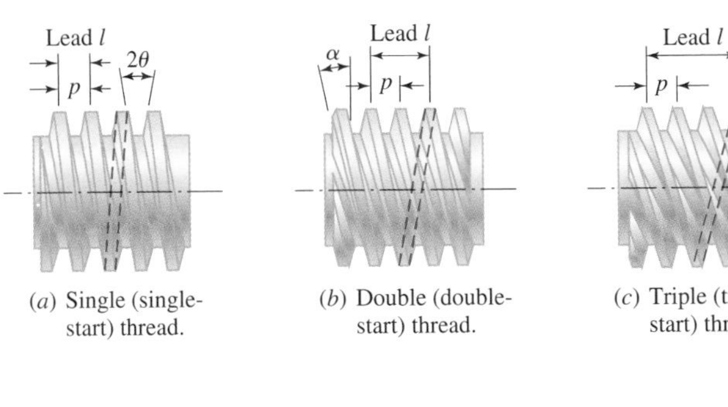

Figure 12.3
Multiple thread configurations.

$$\tan\alpha = \frac{l}{2\pi r_p} = \frac{np}{2\pi r_p} \tag{12-2}$$

Single, double, and triple threads are illustrated in Figure 12.3. Usual practice is to utilize right-hand threads, but for special applications left-hand threads may be used.

In some special applications, where it is very important to minimize thread-friction drag, the sliding friction between screw and nut threads may be replaced by rolling friction through use of a *ball screw*. In ball screws a continuous stream of balls flows along between threadlike nominally semicircular grooves in the screw and nut. Each ball, after rolling to the end of the nut groove, enters a return tube on the outside of the nut to be fed back into the circuit at the entry point. A sketch of a ball screw is shown in Figure 12.4. Such devices are commonly used in automotive steering assemblies, aircraft flap actuators, thrust reversers, landing gear retractors, door closers, and other special applications.

12.2 Potential Failure Modes

Examination of the typical configuration of a power screw assembly, as illustrated for example in Figure 12.1, suggests that power screws embody certain characteristics of plain

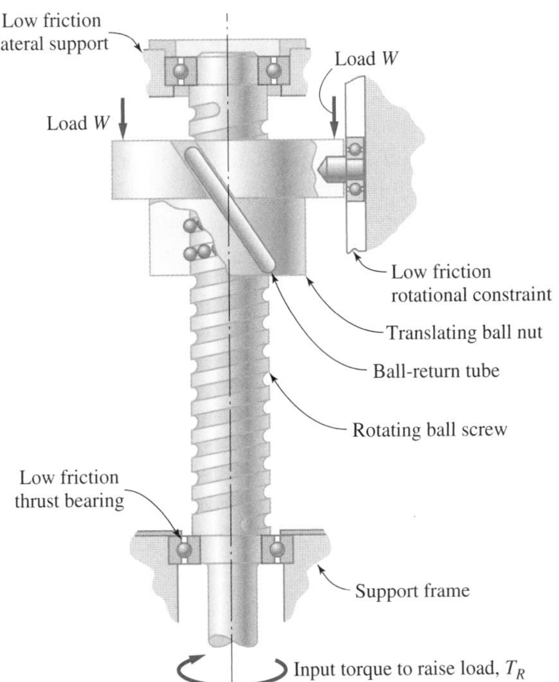

Figure 12.4
Sketch of a *ball screw* version of the load-raising power screw shown in Figure 12.1. Because of inherently low friction in a ball screw, a brake is usually required to prevent the load from driving the screw in reverse when not under power, a condition called "overhauling."

bearings (Chapter 10), power transmission shafting (Chapter 8), axially loaded columns (2.7), and (for ball screw versions) rolling element bearings (Chapter 11). The potential failure modes for a power screw assembly therefore include candidates from each of these applications, including failure by yielding, brinnelling, fatigue, corrosion, surface fatigue wear, adhesive wear, abrasive wear, corrosive wear, fretting wear, galling, seizure, and possibly, for screws with long unsupported spans, failure by buckling or elastic instability. (See 2.3 for failure mode descriptions.)

Design to prevent failure by certain ones of these failure modes is further complicated by the indeterminate nature of the thread contact region between the screw and the nut. Theoretically, all threads in engagement should share the load, but inaccuracies in thread profile and spacing cause virtually all of the load to be carried by the first few threads. This uncertainty has relevance to the determination of proper thread-bearing contact area, to selection of acceptable lubricants, and to minimization of wear, galling, or seizure (see 10.6). It also impacts thread stress analyses required to prevent failure by fatigue, yielding, or brinnelling (see 12.6). As always, the designer is responsible for assessing which of the potential failure modes must be considered in each particular case, then for making the necessary judgments and simplifying assumptions needed to design the assembly and avoid failure. In the final analysis, experimental testing and development are virtually always required to achieve the desired performance with acceptable failure rates. If the application is critical, periodic inspection of the power screw assembly, at specified intervals, is essential to the avoidance of catastrophic failure.[4]

12.3 Materials

From the potential failure modes listed in 12.2 and the material selection guidelines of Chapter 3, candidate material pairs for power screws and mating nuts should have good compressive strength, good fatigue strength, good ductility, good thermal conductivity, and good material compatibility between mating surfaces of the screw and nut. For the case of ball screws, material surfaces should have high surface hardness, usually a hard fatigue-resistant case over a tough ductile core.

In most circumstances, a threaded power screw would be manufactured from low-carbon carburizing-grade alloy steel, case carburized and heat treated. AISI 1010, 3310, 4620, or 8620 might be used, depending upon the specific application. The mating nut would typically be made of a soft ductile material such as bronze, leaded bronze, or aluminum bronze, or sintered porous versions of these materials, impregnated with lubricant. Thrust collar material selection would follow the guidelines for plain bearings in Chapter 10, or rolling element bearings in Chapter 11. Ball screw applications usually involve *selection* of an appropriate stock assembly from a manufacture's catalog, so "material" issues usually will have already been resolved by the ball screw manufacturer.

12.4 Power Screw Torque and Efficiency

The input torque T_R required to operate a power screw when *raising* the load, as illustrated in Figure 12.1, may be estimated as the sum of the torque T_W required to lift the load, the torque T_{tf} required to overcome friction between the contacting threads of the screw and the nut, and the torque T_{cf} required to overcome friction between the thrust collar and the

[4]Even when periodic inspections *are* performed, catastrophic failures may sometimes occur. A recent case in point is the crash of Alaskan Airlines Flight 261 on January 31, 2000. It has been hypothesized that the crash may have been caused by the failure of a jack-screw-and-nut assembly used to operate the horizontal stabilizer of the aircraft (see ref. 5).

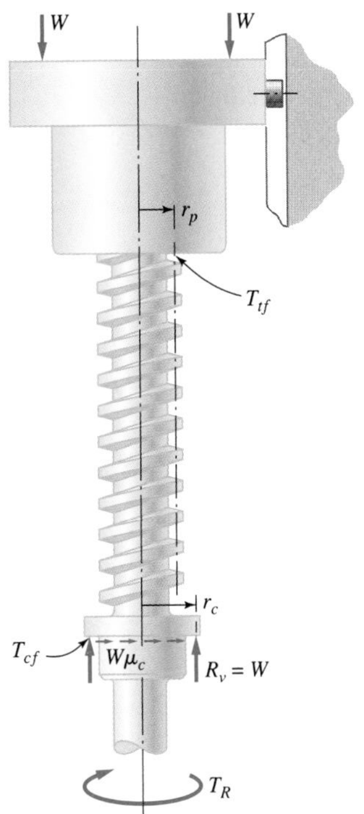

Figure 12.5
Screw and nut of Figure 12.1 taken together as a free body. The load is being raised.

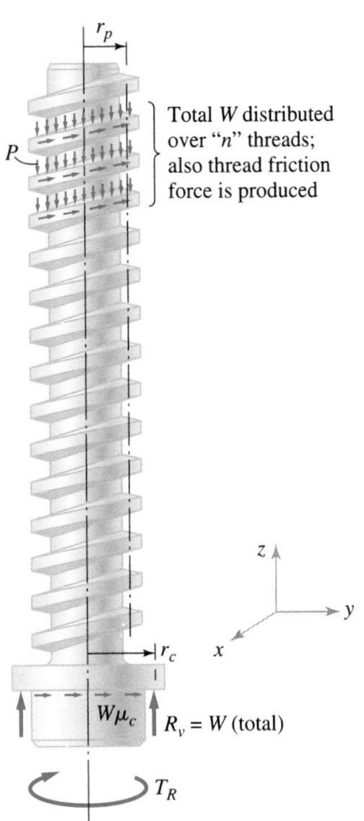

Figure 12.6
Screw of Figure 12.5 taken alone as a free body when the load is being raised.

supporting frame. Thus

$$T_R = T_W + T_{tf} + T_{cf} \tag{12-3}$$

To evaluate these terms, the equilibrium conditions of (4-1) are applied to the screw taken as a free body, illustrated in Figures 12.5, 12.6, and 12.7. From Figure 12.6 it may

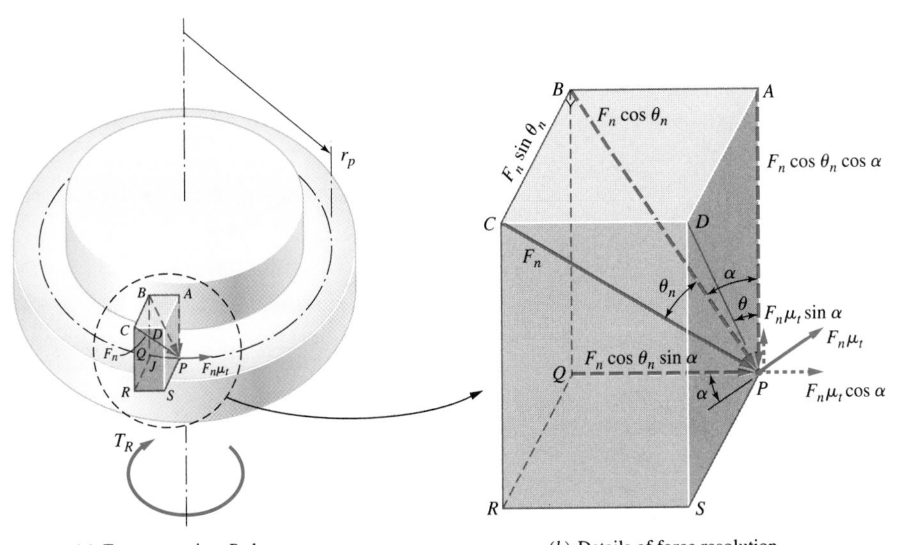

(*a*) Forces at points *P* along pitch helix of Figure 12.6.

(*b*) Details of force resolution.

Figure 12.7
Force resolution at a typical point in the thread contact zone when the load is being raised.

be noted that $\Sigma F_x = 0$ and $\Sigma F_y = 0$ are satisfied by symmetry, and $\Sigma M_x = 0$ and $\Sigma M_y = 0$ are identically satisfied.

From the remaining two conditions of (4-1), utilizing Figures 12.6 and 12.7,

$$\Sigma F_z = W + F_n\mu_t \sin\alpha - F_n \cos\theta_n \cos\alpha = 0 \tag{12-4}$$

and

$$\Sigma M_z = T_R - W\mu_c r_c - r_p F_n \cos\theta_n \sin\alpha - r_p F_n \mu_t \cos\alpha = 0 \tag{12-5}$$

From (12-4)

$$F_n = \frac{W}{\cos\theta_n \cos\alpha - \mu_t \sin\alpha} \tag{12-6}$$

Combining 12.6 with 12.5,

$$T_R = Wr_p\left[\frac{\cos\theta_n \sin\alpha + \mu_t \cos\alpha}{\cos\theta_n \cos\alpha - \mu_t \sin\alpha}\right] + Wr_c\mu_c \tag{12-7}$$

The first term of this input torque expression is the torque required to raise the load W and overcome thread friction. The second term is the torque required to overcome collar friction. The expression for input torque required to operate the power screw when *lowering* the load, T_L, may be found by rewriting (12-4) and (12-5), noting that the sense of the input torque is *reversed* when lowering the load. The directions of the thread friction force vectors and collar friction force vectors are reversed as well, giving

$$T_L = Wr_p\left[\frac{-\cos\theta_n \sin\alpha + \mu_t \cos\alpha}{\cos\theta_n \cos\alpha + \mu_t \sin\alpha}\right] + Wr_c\mu_c \tag{12-8}$$

The expressions for T_R and T_L may be written in a more convenient form by noting, from the geometry of Figure 12.7, that

$$\tan\theta_n = \tan\theta \cos\alpha \tag{12-9}$$

and for small lead angles (the usual case) cos $\alpha \approx 1$, giving from (12-9)

$$\theta_n = \theta \tag{12-10}$$

Dividing the bracketed expressions of (12-7) and (12-8) by cos α in both numerator and denominator, and evaluating tan α from (12-2), expressions (12-7) and (12-8) may be rewritten in terms of lead l and thread angle θ as

$$T_R = Wr_p\left[\frac{l\cos\theta + 2\pi r_p\mu_t}{2\pi r_p\cos\theta - l\mu_t}\right] + Wr_c\mu_c \tag{12-11}$$

and

$$T_L = Wr_p\left[\frac{-l\cos\theta + 2\pi r_p\mu_t}{2\pi r_p\cos\theta + l\mu_t}\right] + Wr_c\mu_c \tag{12-12}$$

For the case of square threads, cos $\theta = 1$. For the case of a thrust collar that embodies a rolling element bearing, the collar friction coefficient, μ_c, is very low, and the second term (collar friction term) may usually be neglected. For the case of a ball screw, the thread friction coefficient is usually very small, and for the limiting case of *zero* thread friction (never actually achievable),

$$T_R = -T_L = \frac{Wl}{2\pi} \tag{12-13}$$

Depending upon friction coefficients and thread geometry, the load being raised may or may not *back-drive* the screw if the applied torque T_R is removed. That is, axial load W may or may not be able lower itself by causing the screw to rotate in reverse. If the screw *cannot* be rotated in reverse by any magnitude of axial load W, the assembly is said to be *self-locking*. If applying an axial load on the nut *can* cause the screw to rotate, the assembly is said to be *overhauling*. If collar friction is negligibly small, the condition that represents the boundary between self-locking and overhauling is that $T_L = 0$ in (12-12). Since the denominator of the bracketed expression in (12-12) cannot be zero, for the screw to be *self-locking*

$$\mu_t > \frac{l\cos\theta}{2\pi r_p} \tag{12-14}$$

and for the screw to be *overhauling*

$$\mu_t < \frac{l\cos\theta}{2\pi r_p} \tag{12-15}$$

Design applications exist for each category. For example, it is essential that screw jacks and threaded fasteners be self-locking. It is equally essential that linear actuators to be designed so that an axial force on the nut will cause rotation of the screw be overhauling.

Efficiency of a power screw assembly, e, may be defined as the ratio of driving torque $T_{\mu=0}$ (theoretical torque to raise the load if there were no friction in either collar or thread contact zone) to T_R (with collar and thread friction). Utilizing (12-7) to obtain an expression for the theoretical torque assuming zero friction,

$$T_{\mu=0} = Wr_p\tan\alpha \tag{12-16}$$

whence

$$e = \frac{T_{\mu=0}}{T_R} = \frac{Wr_p\tan\alpha}{Wr_p\left[\dfrac{\cos\theta_n\sin\alpha + \mu_t\cos\alpha}{\cos\theta_n\cos\alpha - \mu_t\sin\alpha}\right] + Wr_c\mu_c} \tag{12-17}$$

This may be rewritten as

$$e = \frac{1}{\left[\dfrac{\cos\theta_n + \mu_t\cot\alpha}{\cos\theta_n - \mu_t\tan\alpha}\right] + \mu_c\dfrac{r_c}{r_p}\cot\alpha} \tag{12-18}$$

If collar friction is negligibly small, and (12-10) is utilized, (12-18) simplifies to

$$e_{\mu_c=0} = \frac{\cos\theta - \mu_t\tan\alpha}{\cos\theta + \mu_t\cot\alpha} \tag{12-19}$$

Efficiency as a function of lead angle α has been plotted in Figure 12.8 for Acme threads, using several values of thread friction coefficient and assuming collar friction to be negligibly small. It may be noted that efficiencies are very low for lead angles approaching 0° or 90°. Near-maximum efficiencies are achieved using lead angles in the range of 30° to 60° and low thread friction coefficients. The lead angle α for standard Acme power screws ranges from about 2° to 5°, and practical values of μ_t are around 0.1 for dry surfaces or 0.03 for lubricated surfaces (see Appendix Table A.1). Therefore, for Acme threads, operating efficiencies tend to be rather low, ranging around 20 percent for larger sizes and dry surfaces to around 70 percent for smaller sizes and lubricated surfaces. A similar plot for ball screws with lead angles in the range 2° to 5° would typically exhibit efficiencies of 90 percent or higher.

Example 12.1 Power Screw Torque and Efficiency

A power screw assembly, of the type illustrated in Figure 12.1, is to be used to cyclically raise and lower a load of 2500 lb at an axial displacement rate of 3 in/sec. It is being proposed that a single-start modified square thread form, with outside diameter of 1.50 inches,

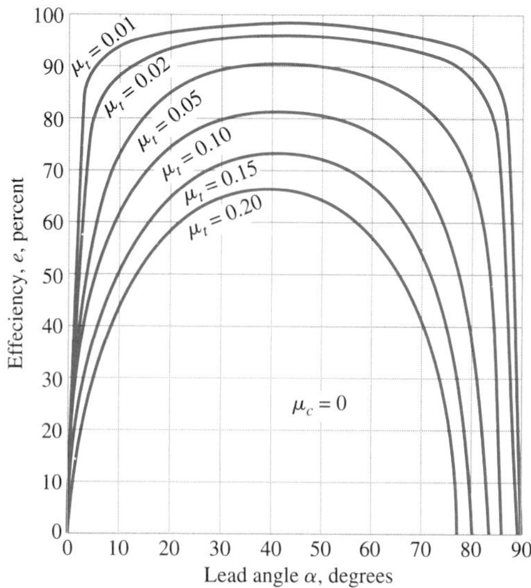

Figure 12.8
Efficiencies for Acme screws as a function of lead angle and thread friction coefficient. Collar friction is not included.

Example 12.1 Continues

be used for the application. Tentatively, a steel screw and bronze nut are being proposed, and a bronze thrust washer is to be secured to the frame at the thrust collar contact site. The proposed mean collar radius is 0.88 inch. It is anticipated that both the screw threads and the thrust collar will be oil lubricated.

a. If these proposals are adopted, find the pitch, lead, pitch radius, and lead angle for this configuration.

b. Find the starting torque and running torque required to raise and lower the load.

c. Would you expect the screw assembly to be self-locking or overhauling?

d. Calculate the power screw efficiency of the proposed assembly. Does collar friction contribute significantly to loss of efficiency?

e. If a special gear motor is to be used to drive this power screw assembly, what would be the minimum power and torque requirements for the drive motor?

f. Using Figure 12.8 as the basis, compare the thread efficiency of the proposed modified square thread with the thread efficiency if a standard Acme thread form were used.

Solution

a. From Table 12.1, a standard modified square thread with $d_o = 1.50$ inches has 3 threads per inch; hence the pitch is

$$p = \frac{1}{3} = 0.33 \text{ inch} \tag{1}$$

and from (12-1), for this single-start thread, the lead is

$$l = p = 0.33 \text{ inch} \tag{2}$$

From Figure 12.2 the pitch radius is

$$r_p = r_o - \frac{p}{4} = \frac{1.50}{2} - \frac{0.33}{4} = 0.67 \text{ inch} \tag{3}$$

and using (12-2), the lead angle is

$$\alpha = \tan^{-1}\frac{l}{2\pi r_p} = \tan^{-1}\frac{0.33}{2\pi(0.67)} = 4.5° \tag{4}$$

b. The torque to raise the load is given by (12-11), and to lower the load given by (12-12). From Figure 12.2 the thread angle θ is 5°. The only difference between starting and running torque is the difference between static and sliding friction coefficients, which are, from Table A.1, for lubricated steel on bronze,

$$\mu_{t\text{-}stc} = \mu_{c\text{-}stc} = 0.06 \tag{5}$$

$$\mu_{t\text{-}sld} = \mu_{c\text{-}sld} = 0.03 \tag{6}$$

Thus to *raise* the load, using (12-11),

$$\begin{aligned}(T_R)_{start} &= 2500(0.67)\left[\frac{0.33\cos 5 + 2\pi(0.67)(0.06)}{2\pi(0.67)\cos 5 - 0.33(0.06)}\right] + 2500(0.88)(0.06)\\ &= 233.3 + 132.0 = 365.3 \text{ in-lb}\end{aligned} \tag{7}$$

and

$$\begin{aligned}(T_R)_{run} &= 2500(0.67)\left[\frac{0.33\cos 5 + 2\pi(0.67)(0.03)}{2\pi(0.67)\cos 5 - 0.33(0.03)}\right] + 2500(0.88)(0.03)\\ &= 182.2 + 66.0 = 248.2 \text{ in-lb}\end{aligned} \tag{8}$$

To *lower* the load, using (12-12),

$$\begin{aligned}(T_L)_{start} &= 2500(0.67)\left[\frac{-0.33\cos 5 + 2\pi(0.67)(0.06)}{2\pi(0.67)\cos 5 + 0.33(0.06)}\right] + 132.0\\ &= -30.3 + 132.0 = 101.7 \text{ in-lb}\end{aligned} \tag{9}$$

and

$$\begin{aligned}(T_L)_{run} &= 2500(0.67)\left[\frac{-0.33\cos 5 + 2\pi(0.67)(0.03)}{2\pi(0.67)\cos 5 + 0.33(0.03)}\right] + 132.0\\ &= -80.7 + 132.0 = 51.3 \text{ in-lb}\end{aligned} \tag{10}$$

c. Using (12-14), the thread will be self-locking if

$$\mu_t > \frac{l\cos\theta}{2\pi r_p} = \frac{0.33\cos 5}{2\pi(0.67)} \tag{11}$$

or if

$$\mu_t > 0.08 \tag{12}$$

From (5) and (6), the static and sliding values of μ_t are both less than 0.08, so the screw would be classified as overhauling, not self-locking. From (b), however, positive torques must be applied both to start and run the screw, even when the load is being lowered, implying that the assembly behaves as if it were self-locking. The reason is that the collar friction acts as a brake to prevent overhauling of the screw.

d. From (12-18) the efficiency of the power screw assembly is

$$e = \frac{1}{\left[\dfrac{\cos 5 + 0.03\cot 4.5}{\cos 5 - 0.03\tan 4.5}\right] + 0.03\left(\dfrac{0.88}{0.67}\right)\cot 4.5} = 0.53 \tag{13}$$

Without collar friction, from (12-19),

$$e_{\mu_c=0} = \frac{\cos 5 - 0.03\tan 4.5}{\cos 5 + 0.03\cot 4.5} = 0.72 \tag{14}$$

Example 12.1 Continues

Collar friction reduces efficiency from about 72 percent to about 53 percent, a significant reduction. The use of a *rolling element* thrust bearing rather than the bronze thrust washer should be considered.

e. The minimum torque requirement is the *starting* torque needed to raise the load, as calculated in (7). Thus

$$T_{min} = 365.3 \text{ in-lb} \tag{15}$$

The minimum *power* requirement may be calculated from (4-32), using the *running* torque to raise the load, as calculated in (8). The rotational speed of the screw may be calculated as

$$n = \frac{3\dfrac{\text{in}}{\text{sec}}\left(60\dfrac{\text{sec}}{\text{min}}\right)}{0.33\dfrac{\text{in}}{\text{rev}}} = 540\frac{\text{rev}}{\text{min}} \tag{16}$$

whence

$$(hp)_{min} = \frac{Tn}{63{,}025} = \frac{(248.2)(540)}{63{,}025} = 2.13 \text{ horsepower} \tag{17}$$

An appropriate *safety factor* on minimum torque and power requirements would need to be applied when selecting the drive motor.

f. If an Acme thread form of the same outside diameter $d_o = 1.50$ inches were used, Table 12.1 indicates that the standard specification would be 4 threads per inch, giving a pitch of 0.25 and lead angle of, from (4),

$$\alpha = \tan^{-1}\frac{0.25}{2\pi(0.67)} = 3.4° \tag{18}$$

From Figure 12-8, using $\mu_t = 0.03$ and $\alpha = 3.4$

$$(e)_{Acme} = 66\% \tag{19}$$

as compared with 72 percent for the modified square thread form.

12.5 Suggested Power Screw Design Procedure

As for any other machine component, the primary objectives for design of a power screw are to select the best material and conceive the best geometry to satisfy the design specifications. Following the guidelines of 6.2, the screw should be designed so as to place it in *tension*, if possible. If it must support *compression* due to the axial load, buckling must be considered when determining its diameter. Determination of thread bending, bearing, and shearing stresses within the thread engagement zone is important in preventing potential failures due to wear or fatigue. Stress concentration at the roots of the more heavily loaded threads may be an important factor, since *cyclic loading* of the threads is usual in power screw operation. Typically, power screw thread standards limit the length of engagement to twice the major diameter of the screw.[5] However, due to manufacturing tolerances and load-induced pitch changes, the first one, two, or three threads in the engagement zone carry most of the load, a fact important when estimating thread stresses and bearing pressures. At certain critical points the nominal screw stresses may combine with local thread stresses, so an appropriate combined stress design theory (see 6.4) may be required. With these observations in mind, the following procedure is suggested.

[5]See, for example, ref. 1.

1. Tentatively select the materials for screw and nut, including heat treatment, and select the desired thread form, consulting pertinent thread standards or Figure 12.2. The type of thrust collar bearing should also be tentatively selected.
2. If the selected design configuration places the screw in compression, or if off-center loading or spurious external bending moments are induced, estimate a tentative screw diameter based on buckling (see 2.7). If buckling is not a potential failure mode, estimate the tentative screw diameter based on direct stress. Use of the resulting diameter from these estimates as the root diameter usually provides a slightly conservative result.
3. Using the tentative root diameter from step 2, and the selected thread form, examine Figure 12.2 and Table 12.1 to determine tentative thread pitch, lead, and other dimensions.
4. Utilizing (12-11) and/or (12-12), together with design specifications, calculate the required driving torque for the power screw assembly.
5. Identify applicable critical sections and critical points, calculate nominal direct stresses and torsional shear stresses in the screw, and calculate thread bending stresses, thread shearing (stripping) stresses, and thread bearing pressures in the engagement zone. Where pertinent, calculate principal stresses so that combined stress design theories or failure theories may be used to assess the acceptability of the design at each critical point.
6. Using an appropriate *design factor of safety*, for each potential failure mode, calculate the *design stress* for each candidate material (see 5.2).
7. Using the principal stresses determined in step 5, pertinent design stresses from step 6, and pertinent combined stress design theories (see 6.4), adjust dimensions as necessary to provide an acceptable design. Acceptability of the bearing pressure between screw and nut threads in the engagement zone may be based on the plain bearing criteria of Table 10.1.

12.6 Critical Points and Thread Stresses

Three critical points in the thread engagement zone of a power screw are of importance. These critical points are illustrated in Figure 12.9, where an elemental volume showing the state of stress at each of the critical points is also shown.

At critical point A, the probable governing failure mode is wear. The allowable bearing pressure between engaged threads may be selected from Table 10.1. The bearing pressure p on the thread surface may be estimated using the effective projected thread contact area. Thus

$$\sigma_B = p_B = \frac{W}{A_p} = \frac{W}{\pi(r_o^2 - r_r^2)n_e} \tag{12-20}$$

where W = load, lb
r_o = major (outer) thread radius, in
r_r = root radius, in
n_e = effective number of threads in the engagement zone that carry the load

At critical point B the probable governing failure modes are yielding and fatigue. The stress components at the critical point include torsional shear stress in the body of the screw, direct stress in the body of the screw, and transverse shear stress due to thread bending, as shown on the sketch of elemental volume B in Figure 12.9. Torsional shear stress in the root cross section of the screw may be calculated using (4-26) and (12-3), to give

$$\tau_s = \frac{(T_R - T_{cf})r_r}{J_r} = \frac{(T_R - T_{cf})r_r}{\left(\dfrac{\pi r_r^4}{4}\right)} = \frac{4(T_R - T_{cf})}{\pi r_r^3} \tag{12-21}$$

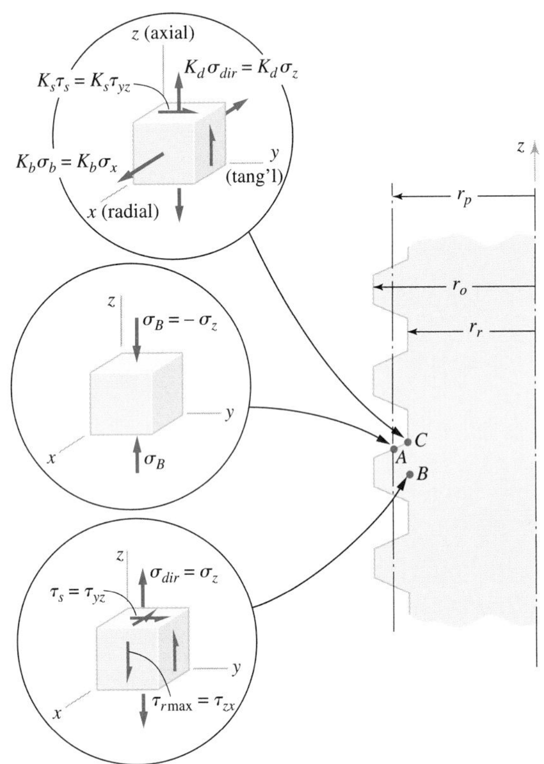

Figure 12.9
Critical points and stress components in power screws.

where T_R = torque to raise load, in-lb
T_{cf} = torque to overcome collar friction, in-lb
r_r = root radius of screw, in

Direct stress in the root cross section is given by

$$\sigma_{dir} = \frac{W}{A_r} = \frac{W}{\pi r_r^2} \tag{12-22}$$

Considering the thread as a cantilever beam with a rectangular root cross section, the maximum transverse shearing stress is (see Table 4.3)

$$\tau_{r\text{-}max} = \frac{3\,W}{2A_{tr}} = \left(\frac{3}{2}\right)\frac{W}{2\pi r_r\left(\dfrac{p}{2}\right)n_e} = \frac{3W}{2\pi r_r p n_e} \tag{12-23}$$

where p is thread pitch in inches.

At critical point C, the probable governing failure modes are yielding or fatigue, with brittle fracture possible but less likely. The stress components at this critical point include torsional shear in the body of the screw [see (12-21)], direct stress in the body of the screw [see (12-22)], and thread bending stress, as shown on the sketch of elemental volume C in Figure 12.9.

The nominal thread bending stress may again be estimated by considering the thread as a cantilever beam with an "end" load distributed along the pitch line. Using (4-5) gives

$$\sigma_b = \frac{Mc}{I} = \frac{6M}{bd^2} = \frac{6[W(r_p - r_r)]}{(2\pi r_r n_e)\left(\dfrac{p}{2}\right)^2} = \frac{12W(r_p - r_r)}{\pi r_r n_e p^2} \tag{12-24}$$

where p = thread pitch, in

An appropriate stress concentration factor should be applied to each stress component at this critical point, as shown in Figure 12.9.

Example 12.2 Preliminary Design and Stress Analysis for a Power Screw

Consideration is being given to the use of a power screw assembly as the driving mechanism for a mechanical compactor to be used in a sheet-metal fabrication plant. The device is to be used for compacting sheet-metal scrap into small cylindrical "slugs" for shipment to a recycling facility. The proposed configuration is sketched in Figure E12.2. The estimated unsupported length of the screw when fully extended (corresponds to maximum axial load) is 72 inches, and the end supports of the nut and platen may be considered to be pinned at each location. The rotating drive nut is powered by a V-belt, engaged with the drive pulley shown. The rotating nut, supported on a low-friction rolling element thrust bearing, causes the screw (and attached platen) to translate up and down. The screw is constrained against rotation by a low-friction rolling element follower running in a vertical groove in the support frame. When the metal scraps in the compaction chamber are compressed enough to generate a resisting force on the compaction platen of 10,000 lb, a force sensor activates a reversing switch on the drive, retracting the screw to its fully up position.

Make a preliminary determination of a satisfactory thread form and screw size for the power screw to be used in this device. A design safety factor of 1.5 has been specified by engineering management, and it is anticipated that the threads will be lubricated.

Solution

Using the suggested step-by-step procedure in 12.5, a preliminary design configuration for the screw may be found as follows:

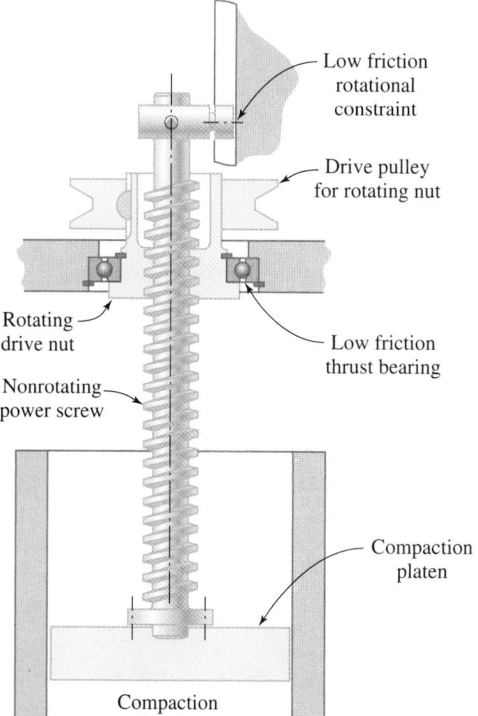

Figure E12.2
Sketch of proposed arrangement for scrap sheet-metal compactor.

Example 12.2 Continues

1. A steel screw and bronze nut will tentatively be selected. For preliminary design purposes, material properties will be taken as those for 1020 steel for the screw and C-22000 bronze for the nut (see Table 3.3). This combination provides good wear resistance (harder steel on softer bronze; see Table 3.13), good buckling resistance (high modulus of elasticity for steel screw), and relatively low coefficient of thread friction (see Appendix Table 2.1). The Acme thread form will tentatively be selected. As shown in Figure E12.2, the thrust collar embodies a rolling element bearing, so collar friction may be neglected at the preliminary stage. "Follower" friction may be neglected as well.

2. Since the screw is in compression, a tentative root diameter for the screw may be estimated on the basis of buckling resistance. Using Euler's equation (2-76), with $L_e = L$ since the column has been judged to be pinned-pinned (see problem statement), gives

$$P_{cr} = \frac{\pi^2 EI}{L^2} \tag{1}$$

Also, since the safety factor is prescribed to be $n_d = 1.5$, from (5-1)

$$P_d = \frac{P_{cr}}{1.5} \tag{2}$$

The maximum compressive force (design value) on the platen is given as 10,000 lb, hence from step 2

$$(P_{cr})_{req'd} = 1.5(10{,}000) = 15{,}000 \text{ lb} \tag{3}$$

Thus from step 1, for the steel screw of length 72 inches,

$$I_{req'd} = \frac{\pi d_r^4}{64} = \frac{P_{cr}L^2}{\pi^2 E} = \frac{15{,}000(72)^2}{\pi^2(30 \times 10^6)} = 0.263 \text{ in}^4 \tag{4}$$

So the tentative root diameter d_r is

$$d_r = \left[\frac{(0.263)(64)}{\pi}\right]^{1/4} = 1.52 \text{ inches} \tag{5}$$

3. From Table 12.1, the nearest standard-size Acme thread appears to be the 2-inch nominal major diameter thread. For this selection

$$\begin{aligned} d_o &= 2.0 \text{ inches} \\ p &= 0.25 \text{ inch (4 threads/in)} \\ d_r &= d_o - 2\left(\frac{p}{2}\right) = 1.75 \text{ inches} \\ l &= p = 0.25 \text{ inch (single start)} \end{aligned} \tag{6}$$

4. Utilizing (12-11), the driving torque for this device may be calculated for maximum loading. It should be noted that the rolling element thrust collar bearing permits the collar friction term to be neglected, and from Appendix Table A.1 the coefficient of friction between the steel screw and bronze nut may be read as

$$(\mu_t)_{dry} = 0.08 \quad \text{(sliding)} \tag{7}$$

or

$$(\mu_t)_{lubricated} = 0.03 \quad \text{(sliding)} \tag{8}$$

For the selected 2-inch Acme thread, (12-11) may be written, for lubricated threads, as

$$T_R = 10{,}000\left(\frac{1.0 + 0.88}{2}\right)\left[\frac{0.25\cos 14.5 + 2\pi(0.94)(0.03)}{2\pi(0.94)\cos 14.5 - 0.25(0.03)}\right]$$
$$= 10{,}000(0.94)[0.07] = 690 \text{ in-lb} \tag{9}$$

This is the peak driving torque required at the end of the compaction cycle (screw fully extended).

5. Critical points to be investigated are those shown in Figure 12.9 as A, B, and C.

For *critical point* A, in the contact zone between screw threads and nut threads, it will be assumed that 3 threads carry the full load. Using (12-20),

$$p_B = \frac{10{,}000}{\pi(1.0^2 - 0.88^2)(3)} = 4703 \text{ psi} \tag{10}$$

From Table 10.1, for steel (screw) on bronze (nut), the maximum allowable pressure is only 2000 psi, so a larger projected thread contact area is required. Trying the next larger standard Acme thread size in Table 12.1,

$$d_o = 2.5 \text{ inches}$$
$$p = 0.33 \text{ inch (3 threads/in)}$$
$$d_r = 2.17 \text{ inches}$$
$$l = p = 0.33 \text{ inch}$$

hence, from (12-20)

$$p_B = \frac{10{,}000}{\pi(1.25^2 - 1.08^2)(3)} = 2679 \text{ psi} \tag{11}$$

This bearing pressure is still a little high, so a 3-inch Acme thread should be chosen. For a 3-inch thread, from Table 12.1,

$$d_o = 3 \text{ inches}$$
$$p = 0.5 \text{ inch}$$
$$d_r = 2.5 \text{ inches}$$
$$l = p = 0.5 \text{ inch}$$

Again using (12-20), reducing n_e to $2\frac{1}{2}$ threads carrying the load (larger and stiffer threads),

$$p_B = \frac{10{,}000}{\pi(1.5^2 - 1.25^2)2.5} = 1852 \text{ psi} \tag{12}$$

This configuration makes critical point A acceptable.

The torque calculation of step 9 must be modified for this new screw size to

$$T_R = 10{,}000\left(\frac{1.5 + 1.25}{2}\right)\left[\frac{0.5\cos 14.5 + 2\pi(1.38)(0.03)}{2\pi(1.38)\cos 14.5 - 0.5(0.03)}\right]$$
$$= 10{,}000(1.38)[0.09] = 1226 \text{ in-lb} \tag{13}$$

For *critical point* B at the thread bending neutral axis, using (12-21), (12-22), and (12-23), the nominal stress components are torsional shear in the screw,

$$\tau_s = \frac{4(1226)}{\pi(1.25)^3} = 799 \text{ psi} = 0.80 \text{ ksi} \tag{14}$$

direct stress in the screw,

Example 12.2 Continues

$$\sigma_{dir} = -\frac{10{,}000}{\pi(1.25)^2} = -2037 \text{ psi} = -2.04 \text{ ksi} \tag{15}$$

and maximum transverse shear stress due to thread bending,

$$\tau_{r\text{-}max} = \frac{3}{2}\left[\frac{10{,}000}{2\pi(1.25)(0.25)(2.5)}\right] = 3056 \text{ psi} = 3.06 \text{ ksi} \tag{16}$$

Using the stress cubic equation, (4-59), the principal stresses for critical point 3 may be determined as follows:

$$\sigma^3 - \sigma^2(-2.04) + \sigma(-0.80^2 - 3.06^2) = 0 \tag{17}$$

or

$$\sigma^3 + 2.04\sigma^2 - 9.45\sigma = 0 \tag{18}$$

giving principal stress solutions as

$$\sigma_1 = 2.22 \text{ ksi} \tag{19}$$

$$\sigma_2 = 0 \tag{20}$$

$$\sigma_3 = -4.26 \text{ ksi} \tag{21}$$

For *critical point C* at the thread root, using (14), (15), and (12-24) for nominal stresses,

$$\tau_s = 0.80 \text{ ksi} \tag{22}$$

$$\sigma_{dir} = -2.04 \text{ ksi} \tag{23}$$

and

$$\sigma_b = \frac{12(10{,}000)(1.38 - 1.25)}{\pi(1.25)(2.5)(0.5)^2} = 6356 \text{ psi} = 6.36 \text{ ksi} \tag{24}$$

At this critical point, appropriate stress concentration factors should be applied to each of the stress components, since fatigue is a potential failure mode. Specific stress concentration factor data could not be found. The shouldered shaft data of Figure 4.17 may be utilized for approximation purposes. For the 3.0-inch Acme thread under investigation,

$$d_o = 3 \text{ inches}$$

$$d_r = 2.5 \text{ inches}$$

$$p = 0.5 \text{ inch}$$

$$r_f = 0.06(0.5) = 0.03 \text{ inch [see Figure 12.2(c)]}$$

From Figure 4.17, then, with $r_f/d_r = 0.012$ and $r_o/r_r = 1.20$,

$$K_b = 2.6 \tag{25}$$

$$K_d = 2.9 \text{ (extrapolated)} \tag{26}$$

$$K_s = 2.3 \tag{27}$$

Based on the stress flow concepts discussed in 4.8, and the thread geometry illustrated in Figure 12.9, these stress concentration values are probably conservative.

From Figure 4.27 it may be found that for 1020 steel ($S_u = 61{,}000$ psi) the notch sensitivity index for $r_f = 0.03$ inch is in the range 0.60–0.65 for tension and bending, and 0.65–0.70 for torsion. Hence it may be assumed that

$$q_b = q_d = 0.65 \tag{28}$$

and

$$q_s = 0.70 \tag{29}$$

Using (4-111) then

$$K_{fb} = 0.65(2.6 - 1) + 1 = 2.0 \tag{30}$$

$$K_{fd} = 0.65(2.9 - 1) + 1 = 2.2 \tag{31}$$

$$K_{fs} = 0.70(2.3 - 1) + 1 = 1.9 \tag{32}$$

and combining these with (22), (23), and (24),

$$K_{fb}\sigma_b = (2.0)(6.36) = 12.72 \text{ ksi} \tag{33}$$

$$K_{fd}\sigma_{dir} = (2.2)(-2.04) = -4.49 \text{ ksi} \tag{34}$$

$$K_{fs}\tau_s = (1.9)(0.80) = 1.52 \text{ ksi} \tag{35}$$

Again using the stress cubic equation (4-59), the principal stresses for critical point C may be determined as follows:

$$\sigma^3 - \sigma^2[12.72 + (-4.49)] + \sigma[12.72(-4.49) - (1.52)^2] - [-12.72(1.52)^2] = 0 \tag{36}$$

or

$$\sigma^3 - 8.23\sigma^2 - 59.42\sigma + 29.39 = 0 \tag{37}$$

Since shearing stresses are zero on the x-plane in Figure 12.9 at critical point C, it is, by definition, a principal plane. Therefore,

$$\sigma_x = \sigma_1 = 12.72 \text{ ksi} \tag{38}$$

is one of the three solutions of (37). Next, dividing (37) by $(\sigma - 12.72)$ gives

$$\sigma^2 + 4.49\sigma - 2.31 = 0 \tag{39}$$

whence

$$\sigma = \frac{-4.49 \pm \sqrt{(4.49)^2 - 4(-2.31)}}{2} = -2.25 \pm 2.71 \tag{40}$$

giving

$$\sigma_2 = 0.46 \text{ ksi} \tag{41}$$

$$\sigma_3 = -4.96 \text{ ksi} \tag{42}$$

as principal stress solutions.

6. At critical point A, where the probable governing failure mode is wear, the safety factor is already included in the *allowable pressure* data of Table 10.1. Hence the design stress (design allowable pressure) of 2000 psi requires no further modification.

At critical points B and C, where the probable governing failure mode is fatigue, the design stress may be determined as follows: In step 1 of this solution the material properties chosen were those of 1020 steel, as specified in Chapter 3. These are

$$S_u = 61{,}000 \text{ psi}$$

$$S_{yp} = 51{,}000 \text{ psi}$$

$$e(2 \text{ inches}) = 15\%$$

Taking the fatigue endurance limit of the material to be 0.5 S_u,

$$S'_{N=\infty} = 0.5(61{,}000) = 30{,}500 \text{ psi} \tag{43}$$

Using Table 2.1 as a guideline, for this application (infinite life assumed) the value of k_∞ may be estimated as[6]

$$k_\infty = (1.0)(1.0)(1.0)(0.65)(0.9)(1.0)(1.0)(0.9)(1.0)(0.9) = 0.47 \tag{44}$$

[6]Stress concentration factors are included in *stress* calculations.

Example 12.2 Continues

Hence, for infinite life, the fatigue endurance limit *of the lead screw* may be estimated using (2-26) as

$$S_{N=\infty} = 0.47(30{,}500) = 14{,}450 \text{ psi} \tag{45}$$

and, using the specified design safety factor, the design stress becomes

$$\sigma_d = \frac{S_{N=\infty}}{n_d} = \frac{14{,}450}{1.5} = 9630 \text{ psi} \tag{46}$$

for the steel screw.

7. As the compactor is actuated, the force increases to 10,000 pounds, then reverses and retracts, the load dropping to approximately zero. The stresses at critical points B and C consequently are *released cyclic stresses*. The principal stresses, given as (19), (20), and (21) for critical point B, and (38), (41), and (42) for critical point C, are *peak* cyclic stress values; each returns to zero stress when the load is released. Following the concepts of 4.7, the equivalent alternating stress amplitude for critical point B may be estimated from (4-97) as

$$\sigma_{eq\text{-}a} = \sqrt{\frac{1}{2}\left[(\sigma_{1a} - \sigma_{2a})^2 + (\sigma_{2a} - \sigma_{3a})^2 + (\sigma_{3a} - \sigma_{1a})^2\right]} \tag{47}$$

where

$$\sigma_{1a} = \frac{2.22 - 0}{2} = 1.11 \text{ ksi}; \qquad \sigma_{1m} = \frac{2.22 + 0}{2} = 1.11 \text{ ksi} \tag{48}$$

$$\sigma_{2a} = 0; \qquad \sigma_{2m} = 0 \tag{49}$$

$$\sigma_{3a} = \frac{0 - (-4.26)}{2} = 2.13 \text{ ksi}; \qquad \sigma_{3m} = \frac{0 + (-4.26)}{2} = -2.13 \text{ ksi} \tag{50}$$

hence

$$(\sigma_{eq\text{-}a})_B = 1.85 \text{ ksi} \tag{51}$$

Using (4-98), the equivalent mean stress at critical point B may be estimated as

$$(\sigma_{eq\text{-}m})_B = \sqrt{\frac{1}{2}\left[(\sigma_{1m} - \sigma_{2m})^2 + (\sigma_{2m} - \sigma_{3m})^2 + (\sigma_{3m} - \sigma_{1m})^2\right]} = 2.85 \text{ ksi} \tag{52}$$

Similarly, for critical point C,

$$\sigma_{1a} = \frac{12.72 - 0}{2} = 6.36 \text{ ksi}; \qquad \sigma_{1m} = \frac{12.72 + 0}{2} = 6.36 \text{ ksi} \tag{53}$$

$$\sigma_{2a} = \frac{0.46 - 0}{2} = 0.23 \text{ ksi}; \qquad \sigma_{2m} = \frac{0.46 + 0}{2} = 0.23 \text{ ksi} \tag{54}$$

$$\sigma_{3a} = \frac{0 - (-4.96)}{2} = 2.48 \text{ ksi}; \qquad \sigma_{3m} = \frac{-4.96 + 0}{2} = -2.48 \text{ ksi} \tag{55}$$

and from (47)

$$(\sigma_{eq\text{-}a})_C = 5.37 \text{ ksi} \tag{56}$$

The equivalent mean stress at critical point C is

$$(\sigma_{eq\text{-}m})_C = \sqrt{\frac{1}{2}\left[(6.36 - 0.23)^2 + (0.23 - \{-2.48\})^2 + (-2.48 - 6.36)^2\right]} = 7.84 \text{ ksi} \tag{57}$$

Since the steel material is ductile (e = 15 percent), the equivalent completely reversed alternating stress amplitude at C may be calculated using (4-99) or (4-100).

For critical point B, from (4-101),

$$\sigma_{max} = 1.85 + 2.85 = 4.70 < S_{yp} = 51.0 \text{ ksi} \tag{58}$$

hence (4-99) was the proper equation to use, so

$$\sigma_{eq\text{-}CR} = \frac{1.85}{1 - \left(\frac{4.70}{61.0}\right)} = 2.00 \text{ ksi} \tag{59}$$

Similarly, for critical point C,

$$\sigma_{max} = 5.37 + 7.84 = 13.21 < S_{yp} = 51.0 \text{ ksi} \tag{60}$$

and again using (4-99),

$$\sigma_{eq\text{-}CR} = \frac{5.37}{1 - \left(\frac{7.84}{61.0}\right)} = 6.16 \text{ ksi} \tag{61}$$

From (59) and (61) it may be noted that critical point C governs.

Finally, comparing (61) with the design stress from (46), the question becomes

$$\sigma_d \overset{?}{\geq} \sigma_{eq\text{-}CR} \tag{62}$$

or

$$9630 \overset{?}{\geq} 6160 \tag{63}$$

Since the inequality is satisfied, the design stress exceeds the actual stress and the design configuration is acceptable.

It may be observed from these design calculations that the *overall governing failure mode is wear.*

To summarize, the preliminary recommendation is to use a standard 3.0-inch Acme single-start thread for this application. This choice appears to satisfy all design criteria and avoid potential failure by buckling, wear, and fatigue. A hollow screw should probably be considered to save material and reduce weight. The nut should be separately examined.

Problems

12-1. Figures 12.5, 12.6, and 12.7 depict a power screw assembly in which the rotating screw and nonrotating nut will *raise* the load W when the torque T_R is applied in the direction shown (CCW rotation of screw if viewed from bottom end). Based on a force analysis of the power screw system shown in the three figures cited, the torque required to raise the load is given by (12-7).

a. List the changes that must be made in the free-body diagrams shown in Figures 12.6 and 12.7 if the load is to be *lowered* by reversing the sense of the applied torque.

b. Derive the torque equation for *lowering* the load in this power screw assembly. Compare your result with (12-8).

12-2. The power lift shown in Figure P12.2 utilizes a motor-driven Acme power screw to raise the platform, which weighs a maximum of 3000 lb when loaded. Note that the nut, which is fixed to the platform, does not rotate. The thrust collar of the power screw presses against the support structure, as shown, and the motor drive torque is supplied to the drive shaft below the thrust collar, as indicated. The thread is a $1\frac{1}{2}$-inch Acme with 4 threads per inch. The thread coefficient of friction is 0.40. The mean collar radius is 2.0 inches, and the collar coefficient of friction is 0.30. If the rated power output of the motor drive unit is 7.5 hp, what maximum platform lift speed (ft/min) could be specified without exceeding the rated output power of the motor drive unit? (Note any approximations used in your calculations.)

12-3. A power lift similar to the one shown in Figure P12.2 uses a single-start square-thread power screw to raise a load of 50 kN. The screw has a major diameter of 36 mm and a pitch of 6 mm. The mean radius of the thrust collar is 40 mm. The static thread coefficient of friction is estimated as 0.15 and the static collar coefficient of friction as 0.12.

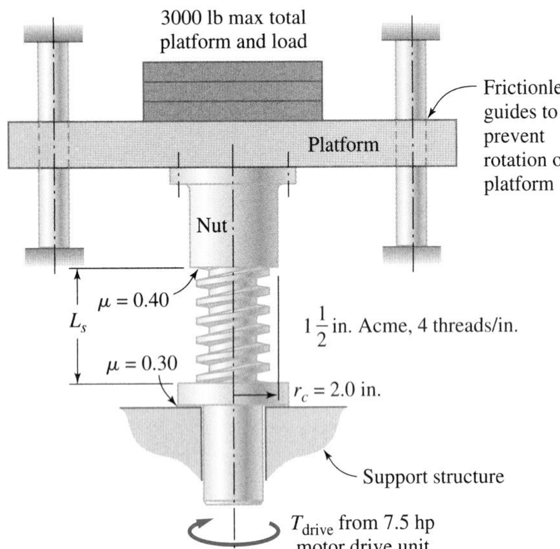

Figure P12.2
Power lift driven by an Acme power screw.

a. Calculate the thread depth.

b. Calculate the lead angle.

c. Calculate the helix angle.

d. Estimate the starting torque required to raise the load.

12-4. In a design review of the power lift assembly shown in Figure P12.2, a consultant has suggested that buckling of the screw might become a problem if the lift height (screw length) becomes "excessive." He also suggests that for buckling considerations, the lower end of the steel screw, where the collar contacts the support structure, may be regarded as fixed, and at the upper end where the screw enters the nut, the screw may be regarded as pinned but guided vertically. If a safety factor of 2.2 is desired, what would be the maximum acceptable lift height L_s?

12-5. Replot the family of efficiency curves shown in Figure 12.8, except do the plot for *square threads* instead of Acme threads. Use the same array of friction coefficients, and again assume the collar friction to be negligibly small.

12.6. A one-inch Acme power screw with 5 threads per inch is driven by a one-horsepower drive unit at a speed of 20 rpm. The thrust is taken by a rolling element bearing, so collar friction may be neglected.

a. What is the maximum load that can be lifted without stalling the drive if the thread coefficient of friction is 0.20?

b. Estimate the efficiency of this power screw.

c. Would you expect this power screw to "overhaul" under maximum load if the power source is disconnected? (Justify your answer with appropriate calculations.)

12-7. A standard $1\frac{1}{2}$-inch rotating power screw with triple square threads is to be used to lift a 4800-lb load at a lift speed of 10 ft/min. Friction coefficients for both the threads and the collar have been experimentally determined to be 0.12. The mean thrust collar friction diameter is 2.75 inches.

a. What horsepower would you estimate to be required to drive this power screw assembly?

b. What motor horsepower would you recommend for this installation?

12-8. Repeat problem 12-7 if everything remains the same except that the power screw has double square threads.

12-9. Repeat problem 12-7 if everything remains the same except that the power screw has a single square thread.

12-10. Find the torque required to drive a $\frac{5}{8}$-inch square-thread power screw with $5\frac{1}{2}$ threads per inch. The load to be lifted is 800 lb. The collar has a mean friction diameter of 1.0 inch, and a coefficient of friction of 0.13. The coefficient of thread friction is 0.15.

12-11. A mild-steel C-clamp has a standard single-start $\frac{1}{2}$-inch Acme thread and mean collar radius of $\frac{5}{16}$ inch. Estimate the force required at the end of a 6-inch handle to develop a 300-lb clamping force. (*Hint:* See Appendix Table A.1 for friction coefficients.)

12.12. Design specifications for a power screw lifting device require a single-start square thread having a major diameter of 0.75 inch and six threads per inch. The load to be lifted is 4000 lb, and it is to be lifted at a rate of 0.5 inch per second. The coefficient of friction for both the threads and the collar is estimated to be about 0.15. The mean collar diameter is 1.0 inch. Calculate the required rotational speed of the screw and the power required to drive it.

12-13. A 20-mm power screw for a hand-cranked arbor press is to have a single-start square thread with a pitch of 4 mm. The screw is to be subjected to an axial load of 5 kN. The coefficient of friction for both threads and collar is estimated to be about 0.09. The mean friction diameter for the collar is to be 30 mm.

a. Find the nominal thread width, thread height, mean thread diameter, and the lead.

b. Estimate the torque required to "raise" the load.

c. Estimate the torque required to "lower" the load.

d. Estimate the efficiency of this power screw system.

12-14. Based on design specifications and loads, a standard single-start 2-inch Acme power screw with 4 threads per inch has tentatively been chosen. Collar friction is negligible. The screw is in tension and the torque required to raise a load of 12,000 lb at the specified lift speed has been calculated to be 2200 in-lb. Concentrating your attention on *critical point B* shown in Figure 12.9, calculate the following:

a. Nominal torsional shearing stress in the screw.

b. Nominal direct stress in the screw.

c. Maximum transverse shearing stress due to thread bending. Assume that three threads carry the full load.

d. Principal stresses at critical point B.

e. If the screw is to be made of 1020 cold-drawn steel (see Table 3.3), the governing failure mode is specified to be yielding, stress concentration may be ignored, and a design safety factor of 2.3 has been chosen, would the state of stress at critical point B be acceptable?

12-15. For the power screw application described in problem 12-14, concentrate your attention on critical point C shown in Figure 12.9. Since cold-drawn steel is ductile (see Table 3.10), and yielding is the governing failure mode (by specification), stress concentration may be ignored. Calculate the following for critical point C:

a. Torsional shearing stress in the screw.

b. Direct stress in the screw.

c. Bending stress in the thread. Assume that 3 threads carry the full load.

d. Principal stresses at critical point C.

e. If the screw is to be made of 1020 cold-drawn steel (see Tables 3.3 and 3.10), the governing failure mode is specified to be yielding, and a design safety factor of 2.3 has been chosen, would the state of stress at critical point C be acceptable?

12-16. A special square-thread single-start power screw is to be used to raise a 10-ton load. The screw is to have a mean thread diameter of 1.0 inch, and four threads per inch. The mean collar radius is to be 0.75 inch. The screw, the nut, and the collar are all to be made of mild steel, and all sliding surfaces are lubricated. (See Appendix Table A.1 for typical coefficients of friction.) It is estimated that three threads carry the full load. The screw is in tension.

a. Calculate the outside diameter of this power screw.

b. Estimate the torque required to raise the load.

c. Estimate the torque required to lower the load.

d. If a rolling element bearing were installed at the thrust collar (gives negligible collar friction), what would be the minimum coefficient of thread friction needed to prevent overhauling of the fully loaded screw?

e. Calculate, for the conditions of (d), the nominal values of torsional shearing stress in the screw, direct axial stress in the screw, thread bearing pressure, maximum transverse shearing stress in the thread, and thread bending stress.

12-17. A power screw lift assembly is to be designed to lift and lower a heavy cast-iron lid for a 10-foot-diameter pressure cooker used to process canned tomatoes in a commercial canning factory. The proposed lift assembly is sketched in Figure P12.17. The weight of the cast-iron lid is estimated to be 4000 lb, to be equally distributed between two support lugs as shown in Figure P12.17. It may be noted that the screw is in *tension*, and it has been decided that a standard Acme thread form should be used. Preliminary calculations indicate that the nominal tensile stress in the screw should not exceed a design stress of 8000 psi, based on yielding. Stress concentration and safety factor have both been included in the specification of the 8000 psi design stress. Fatigue may be neglected as a potential failure mode

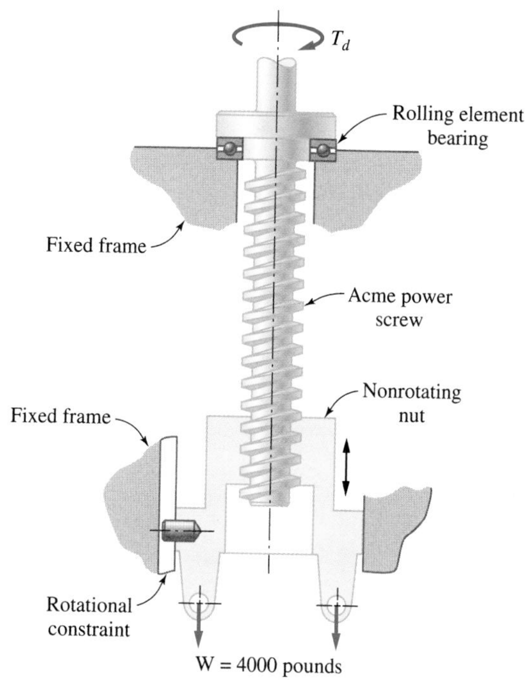

Figure P12.17
Power screw lift assembly for canning factory pressure-cooker lid.

because of the infrequent use of the lift assembly. The rotating steel screw is supported on a rolling element bearing (negligible friction), as shown, and the nonrotating nut is to be made of porous bronze (see Table 10.1). The coefficient of friction between the screw and the nut has been estimated to be 0.08.

a. Estimate the tentative minimum root diameter for the screw, based on yielding due to direct tensile load alone as the governing failure mode.

b. From the result of (a), what Acme thread specification would you suggest as a first-iteration estimate for this application?

c. What would be the maximum driving torque, T_d, for the Acme thread specified in (b)?

d. What torsional shearing stress would be induced in the root cross-section of the suggested power screw by driving torque T_R.

e. Identify the critical points that should be investigated in the Acme thread power screw.

f. Investigate the contact zone between screw threads and nut threads, and resize the screw if necessary. Assume that the full load is carried by three threads. If resizing *is* necessary, recalculate the driving torque for the revised screw size.

g. What horsepower input would be required to drive the screw, as sized in (f), if it is desired to raise the lid 18 inches in no more than 15 seconds?

Chapter 13

Machine Joints and Fastening Methods

13.1 Uses and Characteristics of Joints in Machine Assemblies

Virtually all machines and structures, both large and small, comprise an assemblage of individual parts, separately manufactured, and joined together to produce the complete article. The joints and connections between parts must be given special attention by the designer because they always represent geometrical discontinuities that tend to disrupt uniform force flow. Consequently, either the stresses at a joint are high (because of stress concentration), or a "bulbous" geometry must be used to prevent the high local stresses. Also, joints may involve adverse interactions between two different materials in contact, may sometimes contribute to damaging misalignments, nearly always constitute potential failure sites, and often represent more than half of the cost of the machine (if joint analysis, design, and assembly costs are included). The configurational guidelines for determining size and shape, given in 6.2, should be followed as closely as possible when designing joints, but are sometimes difficult to implement. The basic challenge is to design the joint so that the components may be assembled and secured economically, with maximum joint integrity.

Where possible, joints and fasteners *should be eliminated*. However, to facilitate manufacture, to accommodate machine delivery through standard door openings, to provide for repair or replacement of parts, to accommodate shipping and handling, and to permit disassembly for maintenance procedures, joints become a necessity. No standards have been developed for uniform joint configurations. The more usual structural joint configurations include butt joints, lap joints, flange joints, sandwich joints, face joints or connections, movable joints, or combinations of these. Most joint types permit choices among a variety of different permanent and removable fastening techniques. Factory assembly of joined components is usually preferable, especially for permanent joints, but field assembly may be required or desired in some circumstances. The potenial hazard of in-service joint loosening must also be addressed by the designer. Figure 13.1 illustrates several basic types of joints.

13.2 Selection of Joint Type and Fastening Method

Selection of the type of joint to be used and the method of fastening depends upon many factors, including the loading direction, magnitude, and spectrum characteristics, whether the load is symmetric or eccentric, whether materials to be joined are the same or different, the sizes, thicknesses, geometries, and weights of the parts to be joined, the precision of alignment and dimensional tolerances required, whether the joint is to be permanent or detach-

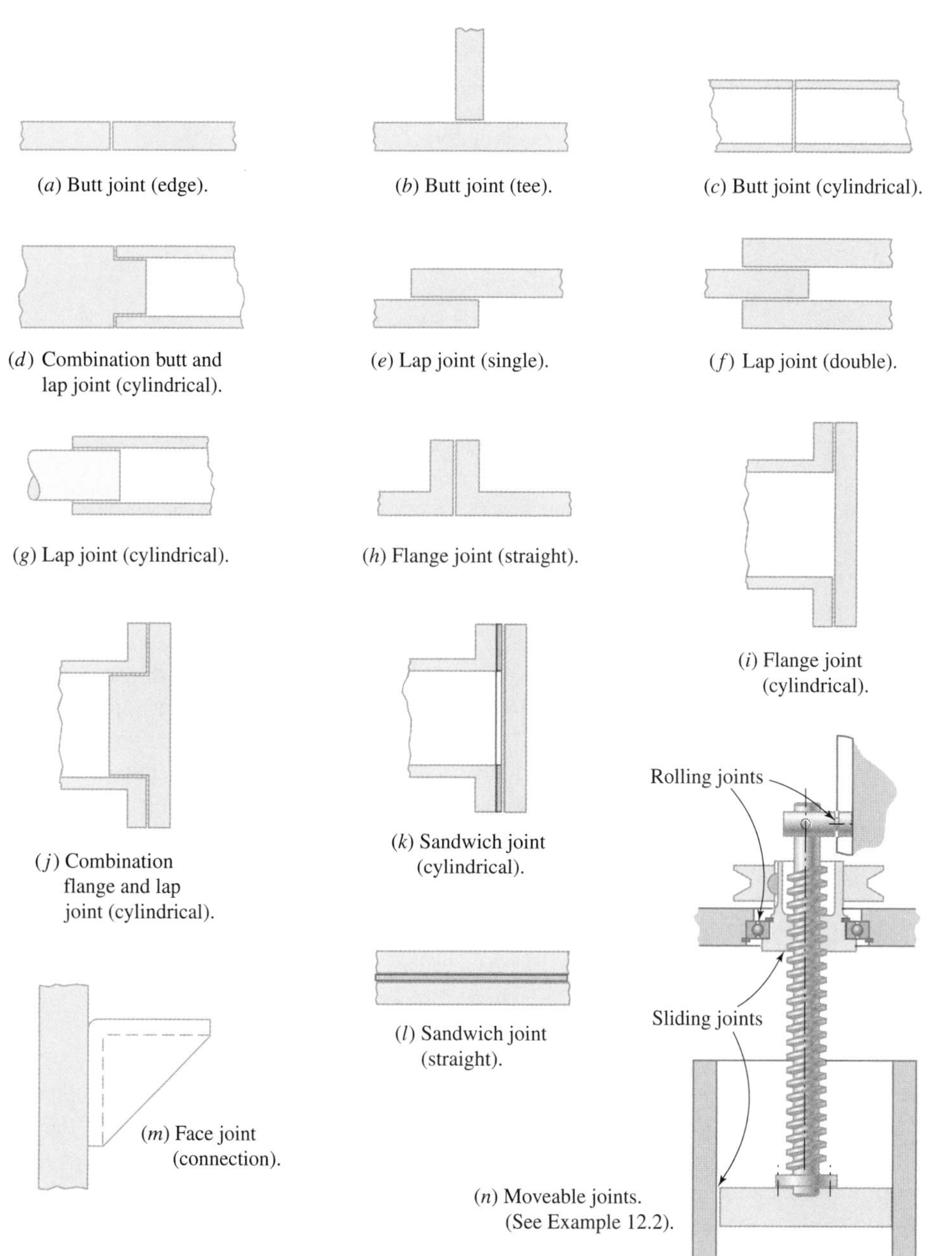

Figure 13.1
Variations on basic joint types.

able, whether the joint must be pressure sealed, and the cost of assembly. As usual, designer experience is a valuable asset in making the best choice. Evaluation of each proposed joint interface in terms of the factors just listed and the various basic joint types available (illustrated in Figure 13.1) usually narrows the choice to one or two satisfactory joint options.

Potential fastening methods include interference fits, threaded fasteners, welding, bonding (brazing, soldering, adhesive bonding), crimping, staking, clamping, or the use of pins, retaining rings, clips, or other specialty fasteners.[1] Which method is selected by the designer depends largely upon the choice of basic joint type, the forces to be transmitted, whether a detachable fastener is required or desired, and the cost of assembly. Threaded

[1]See, for example, ref. 1.

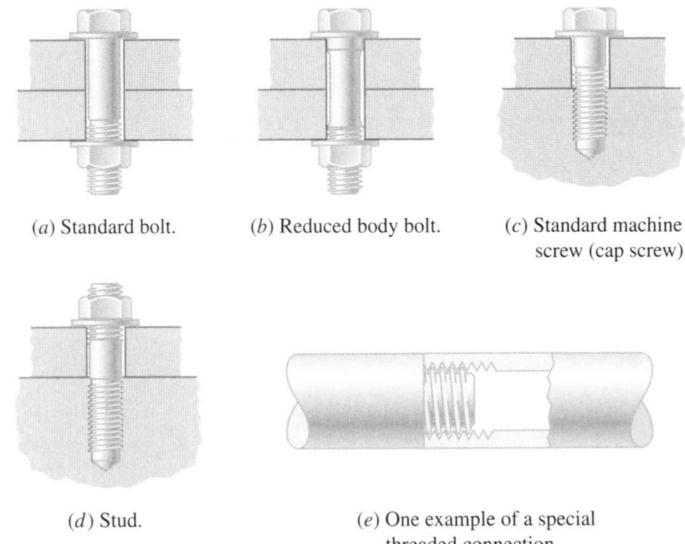

Figure 13.2
Various threaded fastening arrangements.

fasteners, including bolts, machine screws (cap screws), and studs, are the most widely selected fastener types, both for detachable and permanent joints. Special threaded connections are also used in some applications. Various threaded fastening arrangements are illustrated in Figure 13.2.

For permanent joints, welding is widely used, but rivets are also frequently chosen. Various bonding techniques are used in suitable applications. Pins, retaining rings, and set screws are chosen for special applications, such as retaining shaft-mounted components (see 8.8), and interference fits are widely used for mounting rolling element bearings (see 11.7), gears, flywheels, and other similar components. Design considerations related to threaded fasteners, rivets, welding, and bonding are discussed in later sections of this chapter.

13.3 Potential Failure Modes

Because joints, by definition, involve at least two components, an interface between them, and fasteners or fastening media to hold them together (except for movable joints), the identification of potential failure modes is more complicated than usual. This is because failure may occur in either of the components near the contact surface, in the fasteners or the fastening medium, or failure may be initiated primarily because of interfacial contacts. Examination of potential failure modes from this perspective, utilizing 2.2 and 2.3 as guidelines, suggests that potential governing failure modes may be *elastic deformation* failure (leading to unacceptable loss of fastener preload or joint separation under load), *plastic deformation* failure (leading to permanent distortion or displacement that interferes with machine function, *ductile rupture* or *brittle fracture* in components, fasteners, or fastening media, *corrosion* (especially galvanic corrosion due to dissimilar metals in contact), *fatigue* in components, fasteners, or fastening media, *fretting fatigue* generated by small-amplitude cyclic relative motions at the interface, *stress corrosion*, *corrosion fatigue*, and, for elevated temperature environments, *creep*, *thermal relaxation*, or *stress rupture*. For the case of a moveable joint, the failure modes discussed in 10.3 and 11.3 should also be considered.

Some joints involve multiple bolts, rivets, or weld-runs, often in a repeating pattern, within or along the joint. This complicates the designer's task of failure prevention because the analysis of redundant load paths in multiply fastened joints requires either the use of indeterminate structural analysis techniques (see 4.11) or realistic simplifying assumptions by the designer to determine how the loads are shared among the bolts, rivets, or weld-runs. Ultimately, experimental testing and development are usually necessary to achieve required joint integrity with acceptable failure rates.

13.4 Threaded Fasteners

Because threaded fasteners offer so many distinct advantages, they are more widely used than any other means for joining components and assembling machines or structures. Threaded fasteners are commercially available in a wide range of standardized styles, sizes, and materials, throughout the industrialized world. They may be used to join components of the same or different materials, for simple or complex joint configurations, in factories, or in the field. They may be readily and safely installed with standard hand tools or power tools, and if maintenance or repair is required they may be removed or replaced just as easily.

As illustrated in Figure 13.2, the basic threaded fastening system consists of an external (male) threaded element such as a bolt, machine screw, or stud, assembled to a mating internal (female) threaded element, such as a nut, threaded insert, or tapped hole. Nearly all of the basic joint types shown in Figure 13.1 may be secured by using appropriate threaded fastening techniques. Because threaded fasteners are so widely standardized, interchangeability and low unit cost are virtually guaranteed, irrespective of manufacturer.

Many different threaded fastener styles are commercially available off the shelf, and some manufacturers also supply *special-order designs* for nonstandard applications. Various standard head styles (see Figure 13.3) and thread configurations (see "Screw Thread Standards and Terminology," below) are readily available. A wide variety of materials and grades (standardized strength levels) are also available (see "Threaded Fastener Materials," below). In addition, many different types of nuts, locknuts, washers, and lock washers may be obtained (see Figure 13.4).

Threads are manufactured either by *cutting* or *forming*.[2] Smaller sizes may be produced by utilizing cutting tools called *taps* for internal threads or *dies* for external threads. Larger sizes are usually lathe-turned. Thread *forming* is accomplished by rolling a "blank" between hardened dies that cause the metal to flow radially into the desired thread shape. This *cold-rolling* process produces favorable residual stresses (see 4.13) that enhance fatigue strength, and results in a smoother, harder, more wear-resistant surface. High-production-rate automatic screw machines utilize the thread-forming process.

Heads are typically formed by a cold-forming process called "upsetting" in which the fastener blank is forced to flow plastically into a head die of the desired shape. Slots or flats may be machined in a subsequent operation.

Reduced body bolts [see Figure 13.2(b)] are sometimes used to enhance fatigue performance by reducing stress levels within the thread zone, especially at the critical first thread. Since for reduced body bolts the minor (root) thread diameter is *larger* than the bolt

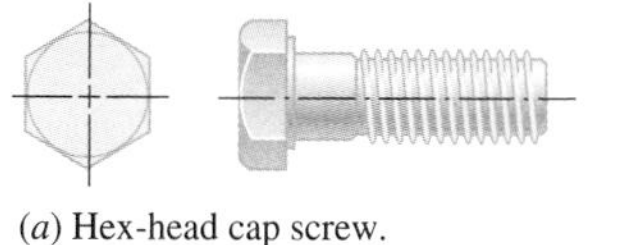

(*a*) Hex-head cap screw.

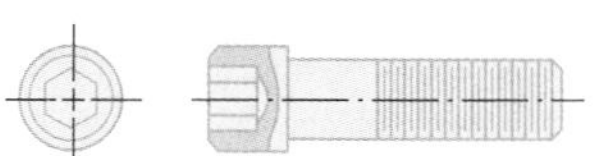

(*b*) Hexagonal socket-head screw.

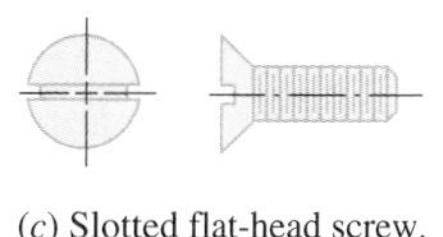

(*c*) Slotted flat-head screw.

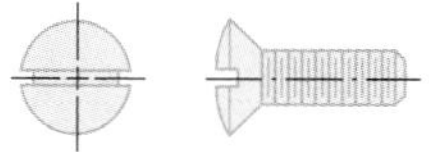

(*d*) Slotted oval-head screw.

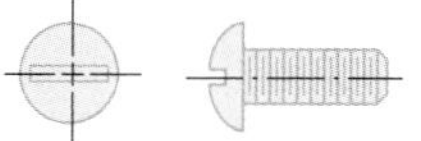

(*e*) Slotted round-head screw.

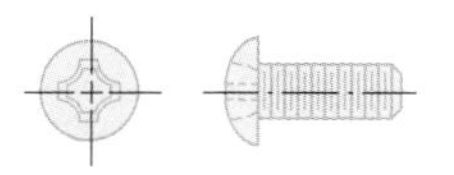

(*f*) Phillips round-head screw.

Figure 13.3
A sample of readily available standard head styles for bolts and screws. Many other styles are available.

[2]See, for example, ref. 2.

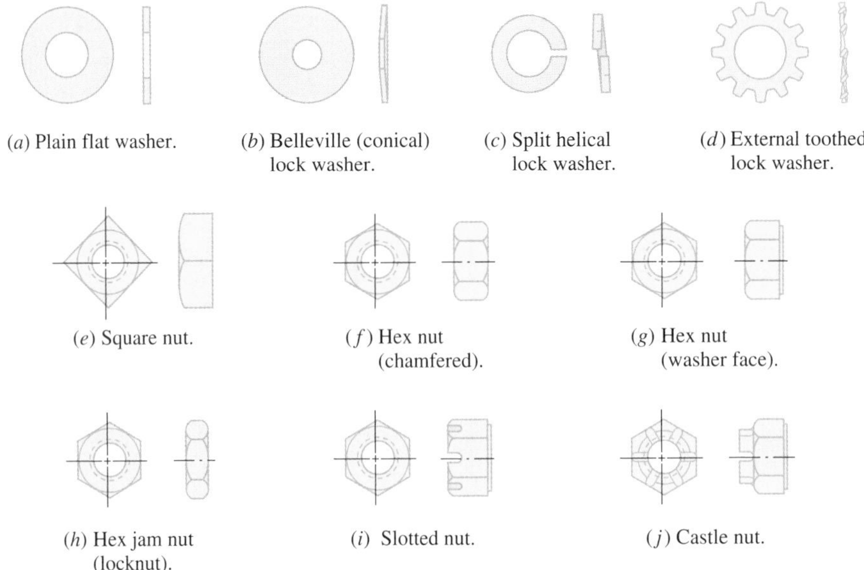

Figure 13.4
A sample of readily available standard washers and nuts. Many other styles are available.

body diameter, the *nominal stress is lower* at the thread root for a given load. Also, generous fillets between the reduced body and the threads lead to *lower* stress concentration factors. The combination of lower nominal stress and lower stress concentration factor often provides a remarkable improvement in resistance to fatigue failure. Strain matching between threads of the bolt and nut, as illustrated in Figure 6.9, is also utilized in some cases to improve the fatigue resistance of a threaded fastener system.

Screw Thread Standards and Terminology

The terminology used in discussing threaded fasteners parallels the power screw terminology defined in 12.1. Thus, as illustrated in Figure 13.5, the *pitch*, p, is the axial distance between corresponding points on adjacent threads. In Unified Inch standard threads (see ref. 3), the pitch is the reciprocal of the number of threads per inch. The *major diameter*, d, is the largest diameter of the (male) thread and the *minor diameter*, d_r, is the smallest (root) diameter of the thread. The *lead*, l, is the axial displacement of the mating nut for one nut rotation. For a *single thread* the lead is equal to the pitch. For *double threads* (two adjacent parallel threads) or *triple threads* (three adjacent parallel threads), the lead is equal to twice the pitch or three times the pitch, respectively (see Figure 12.3). Although multiple threads

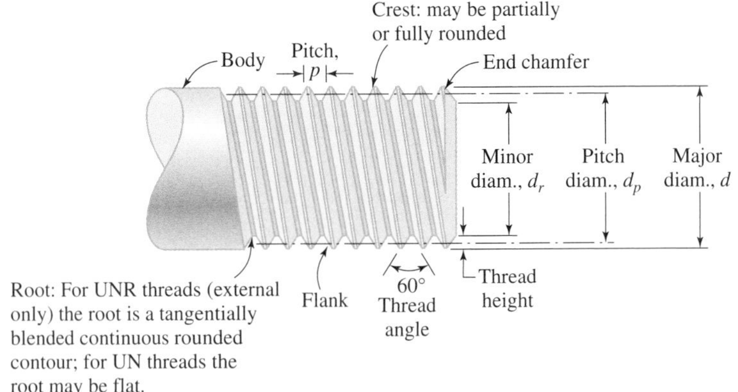

Figure 13.5
Screw thread profile and terminology (see ref. 3, 4, or 5 for standard details).

provide the advantage of more rapid nut advancement, they are rarely used in threaded fastener applications because they tend to loosen more easily due to the steeper lead angle.

The *size* of a screw thread refers to the nominal major diameter, or, for a male thread, the nominal diameter of the stock on which the helical thread is cut. In the Unified Inch system, screw thread sizes smaller than 1/4-inch diameter are numbered from No. 12 (0.216 inch) down to No. 0 (0.060 inch), and even smaller for special applications. Selected dimensional information is shown in Table 13.1 for Unified Inch standard screw threads, and Table 13.2 for Metric standard threads.

Basic thread form profiles (see, for example, Figure 13.5) are given the standard designations UN and UNR for *Unified Inch* screw threads, and M and MS for *Metric* threads.[3]

Table 13.1 Dimensions of Unified Inch Standard Screw Threads[1]

		Coarse Thread Series—UNC/UNRC				Fine Thread Series—UNF/UNRF			
Size	Major Diam. d, in	Threads per Inch	Minor Diam. d_r, in	Minor Diam. Area A_r, in^2	Tensile Stress Area[2], in^2	Threads per Inch	Minor Diam. d_r, in	Minor Diam. Area A_r, in^2	Tensile Stress Area[2], in^2
0	0.0600	—	—	—	—	80	0.0465	0.0015	0.0018
1	0.0730	64	0.0561	0.0022	0.0026	72	0.0580	0.0024	0.0028
2	0.0860	56	0.0667	0.0031	0.0037	64	0.0691	0.0034	0.0039
3	0.0990	48	0.0764	0.0041	0.0049	56	0.0797	0.0045	0.0052
4	0.1120	40	0.0849	0.0050	0.0060	48	0.0894	0.0057	0.0066
5	0.1250	40	0.0979	0.0067	0.0080	44	0.1004	0.0072	0.0083
6	0.1380	32	0.1042	0.0075	0.0091	40	0.1109	0.0087	0.0101
8	0.1640	32	0.1302	0.0120	0.0140	36	0.1339	0.0129	0.0147
10	0.1900	24	0.1449	0.0145	0.0175	32	0.1562	0.0175	0.0200
12	0.2160	24	0.1709	0.0206	0.0242	28	0.1773	0.0226	0.0258
1/4	0.2500	20	0.1959	0.0269	0.0318	28	0.2113	0.0326	0.0364
5/16	0.3125	18	0.2524	0.0454	0.0524	24	0.2674	0.0524	0.0581
3/8	0.3750	16	0.3073	0.0678	0.0775	24	0.3299	0.0809	0.0878
7/16	0.4375	14	0.3602	0.0933	0.1063	20	0.3834	0.1090	0.1187
1/2	0.5000	13	0.4167	0.1257	0.1419	20	0.4459	0.1486	0.1600
9/16	0.5625	12	0.4723	0.1620	0.1819	18	0.5024	0.1888	0.2030
5/8	0.6250	11	0.5266	0.2018	0.2260	18	0.5649	0.2400	0.2560
3/4	0.7500	10	0.6417	0.3020	0.3345	16	0.6823	0.3513	0.3730
7/8	0.8750	9	0.7547	0.4193	0.4617	14	0.7977	0.4805	0.5095
1	1.000	8	0.8647	0.5510	0.6057	12	0.9098	0.6245	0.6630
1 1/8	1.1250	7	0.9704	0.6931	0.7633	12	1.0348	0.8118	0.8557
1 1/4	1.2500	7	1.0954	0.8898	0.9691	12	1.1598	1.0237	1.0729
1 3/8	1.3750	6	1.1946	1.0541	1.1549	12	1.2848	1.2602	1.3147
1 1/2	1.5000	6	1.3196	1.2938	1.4053	12	1.4098	1.5212	1.5810
1 3/4	1.7500	5	1.5335	1.7441	1.8995				
2	2.0000	4.5	1.7594	2.3001	2.4982				
2 1/4	2.2500	4.5	2.0094	3.0212	3.2477				
2 1/2	2.5000	4	2.2294	3.7161	3.9988				
2 3/4	2.7500	4	2.4794	4.6194	4.9340				

[3]See refs. 3, 4, and 5.

Table 13.1 ***(Continued)***

		Coarse Thread Series—UNC/UNRC				Fine Thread Series—UNF/UNRF			
Size	Major Diam. d, in	Threads per Inch	Minor Diam. d_r, in	Minor Diam. Area A_r, in^2	Tensile Stress Area, in^2	Threads per Inch	Minor Diam. d_r, in	Minor Diam. Area A_r, in^2	Tensile Stress Area, in^2
3	3.0000	4	2.7294	5.6209	5.9674				
$3\frac{1}{4}$	3.2500	4	2.9794	6.720	7.0989				
$3\frac{1}{2}$	3.5000	4	3.2294	7.918	8.3286				
$3\frac{3}{4}$	3.7500	4	3.4794	9.214	9.6565				
4	4.0000	4	3.7294	10.608	11.0826				

[1]Reprinted from ref. 3 by permission of The American Society of Mechanical Engineers. All rights reserved.
[2]Tensile stress area is based on using the average of the root diameter d_r and pitch diameter d_p.

Table 13.2 Dimensions of Selected Metric Standard Screw Threads[1] (preferred combinations of diameter and pitch; other standard options are available)

	Coarse Thread Series				Fine Thread Series			
Major Diam. d, mm	Pitch p, mm	Minor Diam. d_r, mm	Minor Diam. Area A_r, mm^2	Tensile Stress Area[2], mm^2	Pitch p, mm	Minor Diam. d_r, mm	Minor Diam. Area A_r, mm^2	Tensile Stress Area[2], mm^2
3.0	0.50	2.459	4.75	5.18				
3.5	0.60	2.850	6.38	6.98				
4.0	0.70	3.242	8.25	9.05				
5.0	0.80	4.134	13.4	14.6				
6.0	1.00	4.917	19.0	20.7				
8.0	1.25	6.647	34.7	37.6	1.00	6.917	38.0	40.0
10.0	1.50	8.376	55.1	59.5	1.25	8.647	58.7	62.5
12.0	1.75	10.106	80.2	86.3	1.25	10.647	89.0	93.6
14.0	2.00	11.835	110	118	1.50	12.376	120	127
16.0	2.00	13.835	150	160	1.50	14.376	162	170
18.0	2.50	15.294	184	197	1.50	16.376	211	219
20.0	2.50	17.294	235	250	1.50	18.376	265	275
22.0	2.50	19.294	292	309	1.50	20.376	326	337
24.0	3.00	20.752	338	360	2.00	21.835	374	389
27.0	3.00	23.752	443	468	2.00	24.835	484	501
30.0	3.50	26.211	540	571	2.00	27.835	609	628
33.0	3.50	29.211	670	705	2.00	30.835	747	768
36.0	4.00	31.670	788	831	3.00	32.752	842	876
39.0	4.00	34.670	944	992	3.00	35.752	1004	1041
42.0	4.50	37.129	1083	1140	2.00	39.835	1246	1274
48.0	5.00	42.587	1424	1498	2.00	45.835	1650	1681
56.0	5.50	50.046	1967	2062	2.00	53.835	2276	2313
64.0	6.00	57.505	2591	2716	2.00	61.835	3003	3045
72.0	6.00	65.505	3370	3505	2.00	69.835	3830	3878
80.0	6.00	73.505	4243	4395	2.00	77.835	4758	4811
90.0	6.00	83.505	5477	5648	2.00	87.835	6059	6119
100.0	6.00	93.505	6867	7059	2.00	97.835	7518	7584

[1]Reprinted from ref. 4 by permission of The American Society of Mechanical Engineers. All rights reserved.
[2]Tensile stress area is based on using the average of the root diameter d_r and pitch diameter d_p.

The Inch-series UNR and Metric-series MS threads have flank-blended continuous rounded root contours (external threads only), which reduce stress concentration and enhance resistance to fatigue failure.

Screw thread standards have identified useful groups of diameter-pitch combinations called *thread series*. *Coarse-series* threads are advantageous where rapid assembly or disassembly is required, or to reduce the likelihood of "cross-threading" where screws are inserted into softer materials such as cast iron, aluminum, or plastics. *Fine-series* threads are used where higher bolt strength is important, since smaller thread depth and larger root diameter provides higher basic tensile strength. Fine threads have less tendency to loosen under vibration than coarse threads because the lead angle is smaller. An *extra-fine series* may be used in special cases where more precise adjustments must be made (e.g., bearing retaining nuts) or for thin-wall tubing applications. *Constant pitch series* (4, 6, 8, 12, 16, 20, 28, and 32 threads per inch) have also been standardized.[4]

Thread *classes* distinguish standard specified ranges of dimensional tolerance and allowance. Classes 1A, 2A, and 3A apply to *external threads* only, and classes 1B, 2B, and 3B apply to *internal threads* only. Tolerances decrease (higher precision) as class number increases. Classes 2A and 2B are the *most commonly used.*

Screw threads are specified by designating in sequence the nominal size, pitch, series, class, and hand. For example,

$$^1\!/_4\text{-28 UNF-2A}$$

defines a 0.250-inch-diameter thread with 28 threads per inch, unified fine thread series, class 2 fit, external right-hand thread. For a left-hand thread the designation LH would be appended.

An example of a metric thread specification would be

$$\text{MS } 10 \times 1.5$$

which defines a 10-mm-diameter thread with thread pitch of 1.5 mm, rounded root (MS profile), external thread (since rounded root profiles are used only on external threads).

Threaded Fastener Materials

As for any other machine element, the important task of selecting an appropriate fastener material may be accomplished by applying the materials selection methodology described in Chapter 3. Because of the many advantages of selecting a steel material (see 3.4), carbon steels and steel alloys are by far the most widely used materials for threaded fastener applications. Common steels used for threaded fasteners include 1010 (no critical strength requirements), 1020 (bright cap screws, other special items), 1038 (high-strength fasteners), 1045 (special high-strength requirements), and 1100 resulfurized (usually for nuts).

To improve both economy and reliability, several standards-making organizations have established well-defined specifications for threaded fastener materials and strength levels. A system of *head markings* that identify material *class* or *grade* and minimum strength level for each individual fastener has also been standardized. (*Imported* fasteners may not have head markings.) Figure 13.6 illustrates head-marking specifications established by the Society of Automotive Engineers (SAE) and American Society for Testing and Materials (ASTM). Tables 13.3, 13.4, and 13.5 provide corresponding material and property information. In these tables the *proof of strength* is defined as the minimum tensile stress that must be sustained by the fastener without significant deformation or failure, and generally corresponds to about 85 percent of the yield strength.

In addition to the standard steel materials, fasteners may be made of alloys of aluminum, brass, copper, nickel, stainless steel, or beryllium, from plastics, and, for high-tem-

[4]See ref. 3.

Figure 13.6
Standard head markings for Unified Inch and Metric bolts and cap screws.

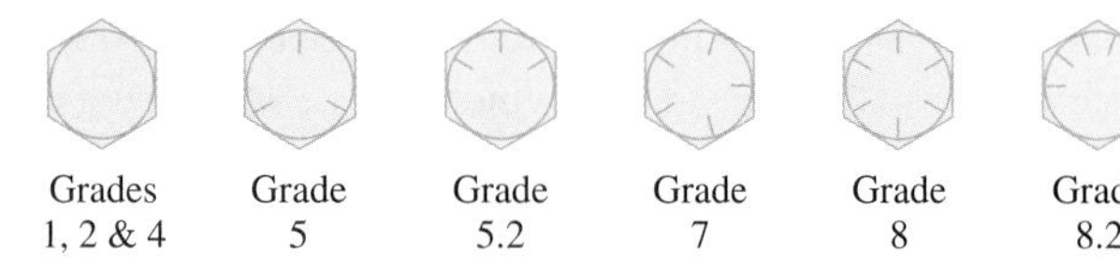

(*a*) SAE head markings. (See Table 13.3 for properties.)

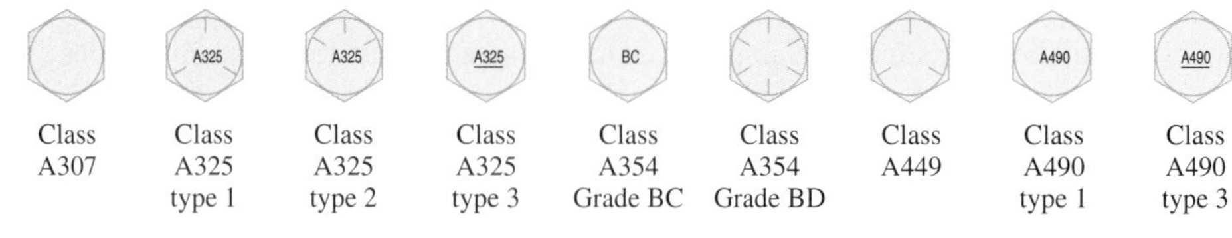

(*b*) ASTM head markings. (See Table 13.4 for properties.)

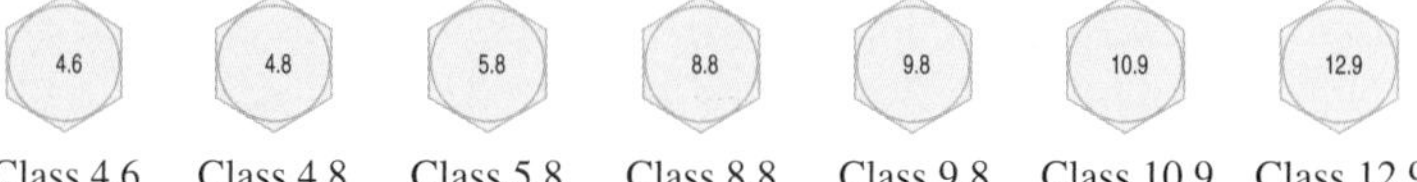

(*c*) Metric head markings. (See Table 13.5 for properties.)

Table 13.3 SAE Grade Specifications for Steel Bolts

SAE Grade	Material	Size Range, in	Minimum Tensile Strength, ksi	Minimum Yield Strength, ksi	Minimum Proof Strength, ksi
1	Low or medium carbon	$\frac{1}{4}$–$1\frac{1}{2}$	60	36	33
2	Low or medium carbon	$\frac{1}{4}$–$\frac{3}{4}$	74	57	55
		$\frac{7}{8}$–$1\frac{1}{2}$	60	36	33
4	Medium carbon, cold drawn	$\frac{1}{4}$–$1\frac{1}{2}$	115	100	65
5	Meidum carbon, Q&T	$\frac{1}{4}$–1	120	92	85
		$1\frac{1}{8}$–$1\frac{1}{2}$	105	81	74
5.2	Low-carbon martensite, Q&T	$\frac{1}{4}$–1	120	92	85
7	Medium-carbon alloy, Q&T	$\frac{1}{4}$–$1\frac{1}{2}$	133	115	105
8	Medium-carbon alloy, Q&T	$\frac{1}{4}$–$1\frac{1}{2}$	150	130	120
8.2	Low-carbon martensite, Q&T	$\frac{1}{4}$–1	150	130	120

Table 13.4 ASTM Class Specifications for Steel Bolts

ASTM Class	Material	Size Range, in	Minimum Tensile Strength, ksi	Minimum Yield Strength, ksi	Minimum Proof Strength, ksi
A307	Low carbon	$\frac{1}{4}$–$1\frac{1}{2}$	60	36	33
A325, type 1	Medium carbon,	$\frac{1}{2}$–1	120	92	85
	Q&T	$1\frac{1}{8}$–$1\frac{1}{2}$	105	81	74
A325, type 2	Low-carbon	$\frac{1}{2}$–1	120	92	85
	martensite, Q&T	$1\frac{1}{8}$–$1\frac{1}{2}$	105	81	74

Table 13.4 (*Continued*)

ASTM Class	Material	Size Range, in	Minimum Tensile Strength, ksi	Minimum Yield Strength, ksi	Minimum Proof Strength, ksi
A325, type 3	Weathering steel, Q&T	$\frac{1}{2}$–1	120	92	85
		$1\frac{1}{8}$–$1\frac{1}{2}$	105	81	74
A354, grade BC	Alloy steel, Q&T	$\frac{1}{4}$–$2\frac{1}{2}$	125	109	105
A354, grade BD	Alloy steel, Q&T	$\frac{1}{4}$–4	150	130	120
A449	Medium carbon, Q&T	$\frac{1}{4}$–1	120	92	85
		$1\frac{1}{8}$–$1\frac{1}{2}$	105	81	74
		$1\frac{3}{4}$–3	90	58	55
A490, type 1	Alloy steel, Q&T	$\frac{1}{2}$–$1\frac{1}{2}$	150	130	120
A490, type 3	Weathering steel, Q&T				

Table 13.5 Metric Class Specifications for Steel Bolts

Property Class[1]	Material	Size Range	Minimum Tensile Strength, MPa	Minimum Yield Strength, MPa	Minimum Proof Strength, MPa
4.6	Low or medium carbon	M5–M36	400	240	225
4.8	Low or medium carbon	M1.6–M16	420	340	310
5.8	Low or medium carbon	M5–M24	520	420	380
8.8	Medium carbon, Q&T	M16–M36	830	660	600
9.8	Medium carbon, Q&T	M1.6–M16	900	720	650
10.9	Low-carbon martensite, Q&T	M5–M36	1040	940	830
12.9	Alloy, Q&T	M1.6–M36	1220	1100	970

[1]Number to left of decimal point designates approximate minimum tensile strength, S_u, in hundreds of megapascals; approximate yield strength, S_{yp}, in each case, is obtained by multiplying S_u times decimal fraction to the right of (and including) the decimal point [e.g., for class 4.6, S_u = 400 MPa and S_{yp} = 0.6(400) = 240 MPa].

perature applications. Hastelloy, Inconel, or Monel may be used as well. Coatings or special finishes are sometimes used to improve appearance, improve corrosion resistance, or provide lubricity.

Critical Points and Thread Stresses

As for power screw threads (see 12.6), there are three potential critical points in the thread engagement zone, illustrated in Figure 13.7 as *A*, *B*, and *C*. In addition, a fourth critical point in the bolt at the joint parting line, shown as *D* in Figure 13.7, must be considered if the joint is subjected to shear loading.

Unlike the case of power screw threads, critical point *A* is usually ignored when analyzing *fastener* threads because there is no relative motion between the threads of the bolt or

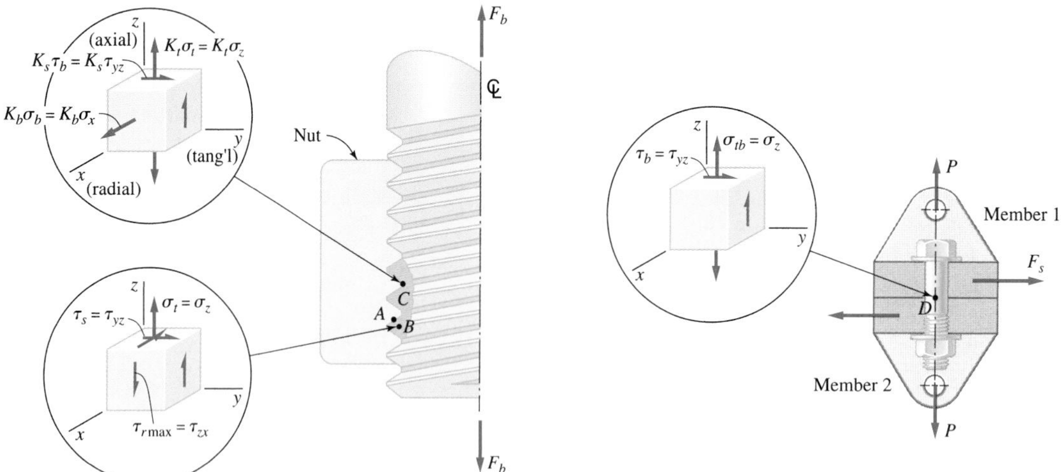

Figure 13.7
Critical points and stress components in threaded fasteners.

screw and the nut after assembly; hence wear is not an applicable failure mode. Potential fastener loosening, however, may involve an analysis of thread friction at this critical point.

At critical point B the analysis closely parallels the power screw thread analysis except that much of the torsion is released after completion of the initial tightening process. Thus the potential failure modes are *yielding* (*stripping*) and *fatigue*, and the pertinent stress components are torsional shear, τ_s, due to tightening, direct tensile stress, σ_t, due to tensile force F_b on the bolt (including preload), and transverse shearing stress, $\tau_{tr\text{-}max}$, due to thread bending. To avoid stripping, the minimum nut length is typically made long enough so that the axial force on the bolt will produce a tensile failure before the threads are stripped. Stripping of *external* threads occurs when male threads fail in shear at the *minor* diameter, and stripping of *internal* threads occurs when female threads fail in shear at the *major* diameter. For standard threads, *if bolt and nut materials are the same*, a nut length of at least one-half the nominal thread diameter produces tensile failure before stripping. Standard nuts have a length of approximately $^7/_8$ of the nominal thread diameter.

Adapting (12-21) to the threaded fastener case

$$\tau_s = \frac{32(T_t - T_{nf})}{\pi d_r^3} \tag{13-1}$$

where T_t = tightening torque to produce desired preload, in-lb
T_{nf} = torque to overcome friction at contact face of nut (if turned) or head (if turned), in-lb
d_r = root diameter of screw thread, in

Adapting (12-22) to the threaded fastener case

$$\sigma_t = \frac{4F_b}{\pi d_r^2} \tag{13-2}$$

where F_b = tensile force on bolt (including preload), lb.

Adapting (12-23) to the threaded fastener case,

$$\tau_{tr-max} = \frac{3F_b}{\pi d_r p n_e} \tag{13-3}$$

where p = thread pitch, in
n_e = effective number of threads in the engagement zone that carry the bolt load F_b

At critical point C, the analysis again parallels the power screw thread analysis. Thus the potential failure modes are *yielding* or *fatigue* (brittle fracture is possible, but less likely). The pertinent stress components are torsional shearing stress, τ_s (see 13-1), direct tensile stress, σ_t (see 13-2), and thread bending stress, σ_b. Adapting (12-24) to the threaded fastener case,

$$\sigma_b = \frac{12F_b(d_p - d_r)}{\pi d_r N_e p^2} \tag{13-4}$$

where d_p = pitch diameter (mean diameter) of screw thread, inches.

As in the case of power screw analysis, an appropriate stress concentration factor should be properly applied to *each stress component* at this critical point if fatigue is a potential failure mode. As noted in 4.8, thread root *theoretical* stress concentration factors, K_t, range from about 2.7 to about 6.7. Typical values of notch sensitivity index range around 0.7 for mild steel (Grade 1), so *fatigue* stress concentration factors typically range from about 2.2 to about 5. A value of $K_f = 3.5$ is often chosen for preliminary calculations. If loading is static, and the bolt or screw is made of a ductile material (usual case), a stress concentration factor of *unity* is usually appropriate (see Table 4.7).

At critical point D the potential failure modes are *yielding* (*shearing*) and *fatigue*, and the pertinent stress components are direct tensile stress in the body, σ_{tb}, and shearing stress in the body, τ_b. The direct tensile stress is given by

$$\sigma_{tb} = \frac{4F_b}{\pi d_b^2} \tag{13-5}$$

where d_b = body diameter of bolt, inches.

The shearing stress τ_b is usually calculated as *direct* (*pure*) shear, although, as a practical matter, there is virtually always *some* bolt bending, and therefore *transverse* shearing stress is a more accurate characterization than pure shear. In *high-performance* design situations, these bolt-bending and transverse-shearing stresses should be carefully investigated. Assuming that the bolted joint is tightly clamped and symmetrically loaded, as shown in Figure 13.7, the shearing stress τ_b may be estimated as

$$\tau_b = \frac{4F_s}{\pi d_b^2} \tag{13-6}$$

where F_s = shear force on bolt, lb.

This estimate assumes that the bolt body is snug in the holes and the joint is tightly clamped, but the *entire shear load*, F_s, is transferred by the *bolt* alone, *none* of it by friction at the joint interface (an overconservative assumption in some cases). If more than one fastener is used to secure the joint, and the joint loading is in any way asymmetric, additional torsion-like or bending-like stress components may require assessment, as described below in "Multiply Bolted Joints; Symmetric and Eccentric Loading."

Preloading Effects; Joint Stiffness and Gasketed Joints

When a bolted joint, such as the one shown in Figure 13.7, is tightened to clamp members 1 and 2 together, the tightening process induces tension in the bolt and compression in the clamped flanges of members 1 and 2. The consequence of the tightening process is that the bolt is *preloaded* (in tension) against the clamped flanges (in compression), producing internal stresses without any *externally* applied load. As discussed in 4.11, such a joint behaves as a system of *springs*; the bolt is in parallel with the flanges, which are in series with each other (see Figures 4.35 and 4.36). Properly preloaded bolted joints of this type may be used to eliminate the development of clearance gaps under operational loads, to enhance resistance to potential shear failure, to enhance resistance to fastener loosening, and to enhance resistance to fatigue failure by reducing the cyclic stress amplitude carried by the bolt (see 4.12).

As discussed in 4.11, a preloaded bolted joint constitutes a statically indeterminate elastic system; hence the axial tensile force in the bolt is a function both of the initial preload force, F_i, due to tightening, and the subsequently applied operating force, P, which tends to separate the clamped members. Considering the bolt and the clamped members to be linear springs (usually a valid assumption), a bolted joint such as the one shown in Figure 13.7 may be modeled as the parallel-series spring system shown in Figure 13.8. Such a system requires both force-equilibrium and force-displacement relationships to explicitly calculate forces in the bolt and in the clamped members.

From Figure 13.8,

$$k_b = \frac{P_b}{y_b} \tag{13-7}$$

and

$$k_m = \frac{P_m}{y_m} \tag{13-8}$$

By force equilibrium,

$$P = P_b + P_m \tag{13-9}$$

where F_b = force on the bolt, lb
F_m = force on the clamped members, lb

By geometric compatibility,

$$y_b = y_m \tag{13-10}$$

Combining (13-10) with (13-7) and (13-8),

$$\frac{P_b}{k_b} = \frac{P_m}{k_m} \tag{13-11}$$

so (13-9) may be rewritten as

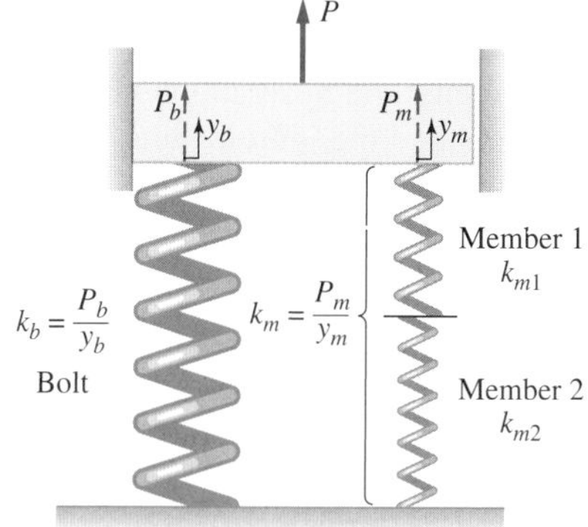

Figure 13.8
Parallel–series model of preloaded bolted joint of type shown in Figure 13.7.

$$P = P_b + \frac{k_m}{k_b} P_b = \left(1 + \frac{k_m}{k_b}\right) P_b \tag{13-12}$$

From this,

$$P_b = \left(\frac{k_b}{k_b + k_m}\right) P \tag{13-13}$$

and

$$P_m = \left(\frac{k_m}{k_b + k_m}\right) P \tag{13-14}$$

Assuming that an initial bolt preload force F_i ($-F_i$ in the clamped members) has been induced by tightening the fastener enough so that joint separation never occurs under operating load P, the initial preload force and the operating forces add to give resultant forces F_b and F_m on the bolt and on the clamped members, respectively. These forces are

$$F_b = P_b + F_i = \left(\frac{k_b}{k_b + k_m}\right) P + F_i \tag{13-15}$$

and

$$F_m = P_m - F_i = \left(\frac{k_m}{k_b + k_m}\right) P - F_i \tag{13-16}$$

To reiterate, the initial preload force induced by bolt tightening must be large enough so the clamped members do not separate under operating loads; otherwise the bolt is forced to carry the entire applied load P. It is common practice to specify an initial preload (or tightening torque) that brings the fastener to approximately 75 percent of yield strength, or, equivalently, 85 percent of proof stress (see Tables 13.3, 13.4 and 15.5).

Example 13.1 Bolt Preloading

A bolted joint of the type shown in Figure 13.7 has been tightened by applying torque to the nut to induce an initial axial preload force of $F_i = 1000$ lb in the $^3/_8$-16 UNC Grade 1 steel bolt. The external (separating) force on the bolted assembly is to be $P = 1200$ lb. The clamped flanges of the steel housing, each 0.75 inch thick, have been estimated to have an axial stiffness of three times the bolt stiffness in the axial direction.

a. Find the resultant tension in the bolt and compression in the clamped members when the external load of 1200 lb is fully applied to the preloaded joint.

b. Will the members separate or remain in contact under full load?

Solution

a. Utilizing (13-15),

$$\begin{aligned} F_b &= \left(\frac{k_b}{k_b + 3k_b}\right) 1200 + 1000 = 0.25(1200) + 1000 \\ &= 1300 \text{ lb} \qquad \text{(tension in bolt)} \end{aligned} \tag{1}$$

and, from (13-16),

$$\begin{aligned} F_m &= \left(\frac{3k_b}{4k_b}\right) 1200 - 1000 = 0.75(1200) - 1000 \\ &= -100 \text{ lb} \qquad \text{(compression in clamped members)} \end{aligned} \tag{2}$$

b. Since there remains a compressive force in the clamped members, they do not separate under the fully applied operating load of 1200 pounds.

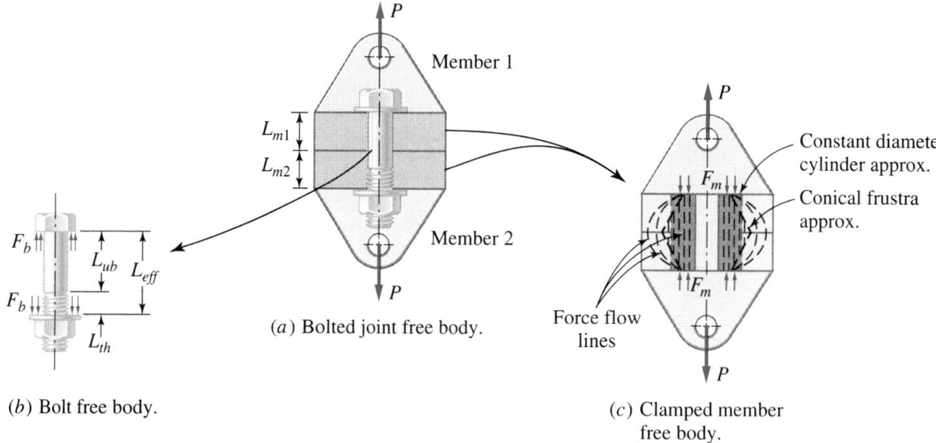

(*a*) Bolted joint free body.

(*b*) Bolt free body.

(*c*) Clamped member free body.

Figure 13.9
Free bodies of bolt, and of clamped members, defined to augment the estimation of spring rates k_b and k_m.

The bolted joint shown in Figure 13.7, modeled as the spring system shown in Figure 13.8, may be broken down into a separate free body for the bolt and another for the clamped members, as shown in Figure 13.9. The axial spring rate for the bolt, k_b, is relatively easy to obtain from Figure 13.9(b) using (4-134) to give

$$k_b = \frac{A_{ub}E_b}{L_{eff}} = \frac{\pi d_b^2 E_b}{4L_{eff}} \tag{13-17}$$

where A_{ub} = effective (unthreaded) cross-sectional area of bolt, in^2
L_{eff} = effective load-carrying length, in
E_b = modulus of elasticity for bolt material, psi
d_b = nominal bolt diameter, in

If more precision is desired, the bolt body and bolt threads may be separately modeled [see Figure 13.9(b)] and combined in series to obtain the spring rate k_b. That is, utilizing (4-143),

$$k_b = \frac{1}{\dfrac{1}{k_{ub}} + \dfrac{1}{k_{th}}} \tag{13-18}$$

where k_{ub} = spring rate of unthreaded body

and

k_{th} = spring rate of threaded length

These spring rates may be calculated as

$$k_{ub} = \frac{A_{ub}E_b}{L_{ub}} = \frac{\pi d_b^2 E_b}{4L_{ub}} \tag{13-19}$$

and

$$k_{th} = \frac{A_t E_b}{L_{th}} = \frac{\pi\left(\dfrac{d_r + d_p}{2}\right)^2 E_b}{4L_{th}} \tag{13-20}$$

To estimate the axial spring rate for the clamped members, lines of force flow (see 4.3 and 4.8) may be visualized as shown, for example, in Figure 13.9(c). Although it is not possible to sketch *precise* lines of force flow, it is clear that they flow from the area of bolt head contact, diverge as they spread into the flange material, then converge back to the area of nut and washer contact. Various methods of defining the boundaries of the force-flow region have been suggested. The two more practical approaches[5] are (1) to define a hollow constant-diameter cylindrical approximation to the force-flow region, or (2) to define two hollow conical frustra, back-to-back, to approximate the force-flow region, as shown in Figure 13.9(c). All such approximations require judgment, and accuracy of the estimate is enhanced by designer experience. If the design is critical, a more accurate modeling approach, such as finite element analysis, or an experimental setup to determine k_m may be advisable. If the hollow conical frustrum approximation is utilized, the fixed cone angle between an element of the surface of the cone and the bolt centerline, starting at the edge of the bolt head or nut and washer, is usually taken to be approximately 30 degrees.[6] For less critical cases, it may be enough to know that the effective outside diameter, d_m, of a constant-diameter hollow-cylinder model usually lies in the range from 1.5 to 2.5 times the nominal bolt diameter. A "typical" value of 2 times the bolt diameter is often chosen for preliminary calculations. If the constant-diameter hollow-cylinder model is chosen, the spring rate k_m for the clamped member may be calculated from a series spring combination of member 1 and member 2 (see Figure 13.8), giving

$$k_m = \frac{1}{\dfrac{1}{k_{m1}} + \dfrac{1}{k_{m2}}} \tag{13-21}$$

where

$$k_{m1} = \frac{A_{m1}E_{m1}}{L_{m1}} = \frac{\pi(d_m^2 - d_{ub}^2)E_{m1}}{4L_{m1}} \tag{13-22}$$

and

$$k_{m2} = \frac{A_{m2}E_{m2}}{L_{m2}} = \frac{\pi(d_m^2 - d_{ub}^2)E_{m2}}{4L_{m2}} \tag{13-23}$$

with

d_m = assumed diameter of hollow constant-diameter cylinder approximation to the force-flow region

If design conditions require a pressure-sealed bolted joint, a *gasket* must usually be installed at the interface between the bolted members. If the gasket is clamped between the flanges, as illustrated in Figure 13.10(a), the spring rate of the gasket material, k_g, must be included in the spring rate of the clamped members, given in (13-21), by adding k_g in series with k_{m1} and k_{m2}. This gives

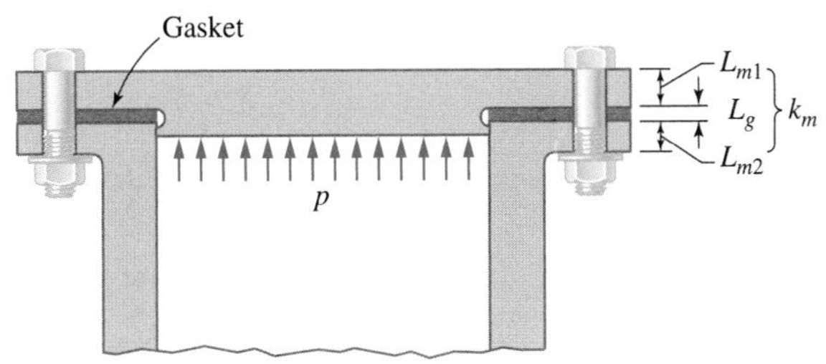

(*a*) Gasket sandwiched between flanges.

O-ring
L_{m1}
L_{m2}
k_m
p

(*b*) O-ring constrained in groove; flanges in solid contact.

Figure 13.10
Pressure-sealed joints.

[5]See also refs. 6 and 7. [6]See ref. 12.

Table 13.6 Modulus of Elasticity and Temperature Limits for Selected Gasket Materials

Material	Temperature Limit, °F (°C)	Modulus of Elasticity, psi (Mpa)
Stainless steel	1000–2100 (538–1149)	29×10^6 (20×10^4)
Copper	900 (482)	18.7×10^6 (12.9×10^4)
Aluminum	800 (427)	10×10^6 (6.9×10^4)
Asbestos (compressed)	800 (427)	70×10^3 (480)
Spiral wound	800 (427)	40×10^3 (300)
Teflon	250 (121)	40×10^3 (300)
Rubber	250 (121)	10×10^3 (69)
Cork	250 (121)	12×10^3 (86)

$$k_m = \frac{1}{\dfrac{1}{k_{m1}} + \dfrac{1}{k_g} + \dfrac{1}{k_{m2}}} \tag{13-24}$$

In many cases the modulus of elasticity of the gasket material is several orders of magnitude softer than for the flanges of the clamped members. When this is true it may be seen from (13-24) that the soft gasket "spring" dominates and:

$$k_m \approx k_g \tag{13-25}$$

The moduli of elasticity for a few gasket materials are shown in Table 13.6. If the soft gasket spring rate does dominate the clamped members' stiffness, the advantages of bolt preloading are largely lost. To avoid this potential problem, *captured* gaskets or O-rings are often used, as illustrated in Figure 13.10(b). For this arrangement the soft gasket material does not participate as a series spring sandwiched between the flanges of the clamped members.

As a general rule, gaskets should be made as thin as possible, but thick enough to provide an effective seal. Thin gaskets are not only stiffer springs [see (4-134)] but also result in reduced thermal relaxation and the associated loss of preload, provide higher resistance to "blowout," and promote better heat transfer.

Example 13.2 Cyclic Loading of Preloaded and Gasketed Joints

It is desired to further investigate the bolted joint discussed in Example 13.1, where it was found that a preload force of 1000 lb in the $^3/_8$-16 UNC Grade 1 bolt was sufficient to keep the joint from separating under the peak operating load of 1200 lb. It has now been learned that the external load is actually not static, but a released cyclic force ranging from a minimum of 0 to a maximum of 1200 lb at 3000 cycles per minute.

a. Plot the cyclic force-time pattern in the bolt if no preload is used.
b. Estimate bolt life for the case of no preload.
c. Plot the cyclic force-time pattern in the bolt if the initial bolt preload force is 1000 lb, as for the case of Example 13.1.
d. Estimate bolt life for the case of $F_i = 1000$ lb.
e. If a $^3/_{32}$-inch thick gasket made of compressed asbestos were sandwiched between member 1 and member 2, in the manner shown in Figure 13.10(a), plot the resulting cyclic force-time pattern in the bolt for the preloaded case of $F_i = 1000$ lb.
f. Estimate bolt life for the gasketed joint with $F_i = 1000$ lb.

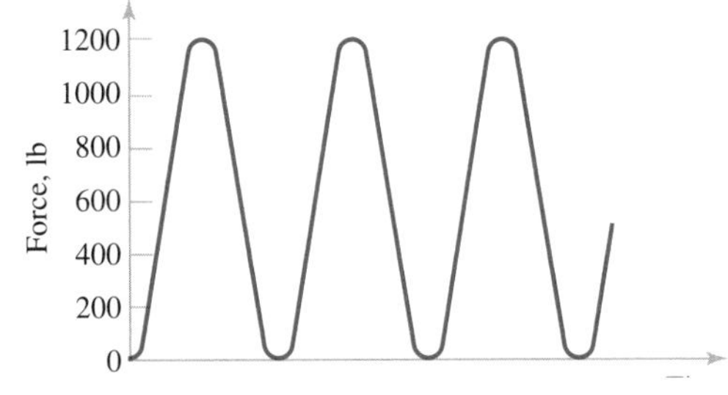

Figure E13.2A
Cyclic force-time pattern for the case of no initial preload.

Solution

a. With no preload the bolt is subjected to the full operational cyclic force, ranging from $F_{min} = 0$ to $F_{max} = 1200$ lb. This force-time pattern is shown in Figure E13.2A.

b. To estimate bolt life, the fatigue strength properties must be obtained or estimated for the Grade 1 bolt material in its operating environment, and the stress-time pattern at the thread root must be defined. From Table 13.3, Grade 1 bolts in this size range have the properties

$$S_u = 60{,}000 \text{ psi}$$
$$S_{yp} = 36{,}000 \text{ psi} \tag{1}$$

As suggested in "Critical Points and Thread Stresses" above, the fatigue stress concentration factor K_f will be taken as 3.5. Using the techniques outlined in 2.6 for estimating *S-N* curves, the polished specimen mean *S-N* curve for the material may be plotted as shown by the dashed line in Figure E13.2B. Utilizing (2-25), (2-27), and Table 2.2, and choosing a 90 percent reliability level for material properties,

$$S_f = k_\infty S'_f \tag{2}$$

with

$$k_\infty = (1.0)(1.0)\left(\frac{1}{3.5}\right)(0.8)(1.0)(1.0)(1.0)(1.0)(1.0)(0.9) = 0.21 \tag{3}$$

Hence

$$S_f = 0.21(30{,}000) = 6170 \text{ psi} \tag{4}$$

Utilizing this value, the solid *S-N* curve for the bolt, as used in this application and corresponding to 90 percent reliability, may be plotted as shown in Figure E13.2B.

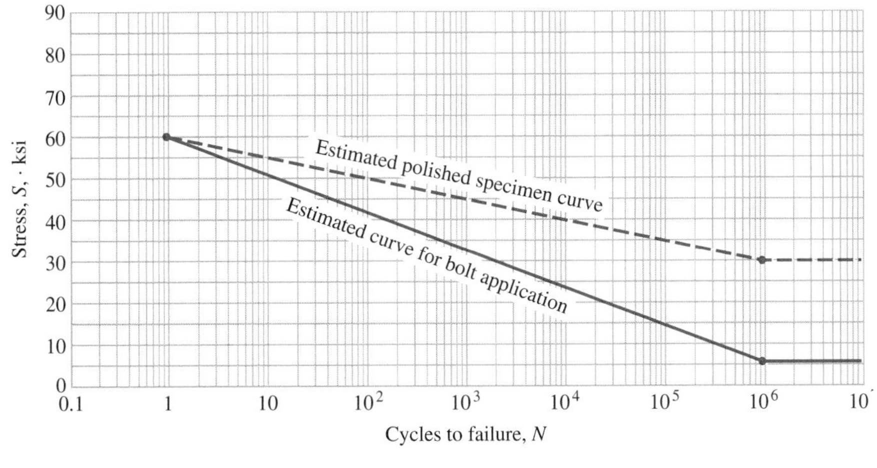

Figure E13.2B
Estimated uniaxial *S-N* curves for Grade 1 material, (a) for polished specimens, and (b) for bolts as used in this application with a reliability $R = 90$ percent.

Example 13.2 Continues

From Table 13.1, the root area of $^3/_8$-16 UNC bolt is $A_r = 0.068$ in^2. Hence the nominal stresses at the thread root are

$$\sigma_{max} = \frac{F_{max}}{A_r} = \frac{1200}{0.068} = 17{,}650 \text{ psi} \tag{5}$$

and

$$\sigma_{min} = 0 \tag{6}$$

Accordingly,

$$\sigma_m = \frac{17{,}650 + 0}{2} = 8825 \text{ psi} \tag{7}$$

Using (4-144), since $\sigma_{max} < \sigma_{yp}$,

$$\sigma_{eq-CR} = \frac{\sigma_{max} - \sigma_m}{1 - \left(\dfrac{\sigma_m}{S_u}\right)} = \frac{17{,}650 - 8825}{1 - \left(\dfrac{8825}{60{,}000}\right)} = 10{,}350 \text{ psi} \tag{8}$$

Reading into the *bolt S-N* curve (solid curve in Figure E13.2B) with $\sigma_{eq-CR} = 10{,}350$ psi, the estimated bolt life may be read out as approximately 300,000 cycles, or about 100 minutes of operating time.

c. With an initial preload of $F_i = 1000$ lb, the maximum and minimum cyclic tensile forces in the bolt, as calculated in Example 13.1, are $F_{max} = 1300$ lb and $F_{min} = 1000$ lb. This force-time pattern is plotted in Figure E13.2C.

d. Using the same approach as in (b),

$$\sigma_{max} = \frac{F_{max}}{A_r} = \frac{1300}{0.068} = 19{,}120 \text{ psi} \tag{9}$$

and

$$\sigma_{min} = \frac{F_{min}}{A_r} = \frac{1000}{0.068} = 14{,}705 \text{ psi} \tag{10}$$

Accordingly,

$$\sigma_m = \frac{19{,}120 + 14{,}705}{2} = 16{,}910 \text{ psi} \tag{11}$$

Again using (4-144),

$$\sigma_{eq-CR} = \frac{\sigma_{max} - \sigma_m}{1 - \left(\dfrac{\sigma_m}{S_u}\right)} = \frac{19{,}120 - 16{,}910}{1 - \left(\dfrac{16{,}910}{60{,}000}\right)} = 3080 \text{ psi} \tag{12}$$

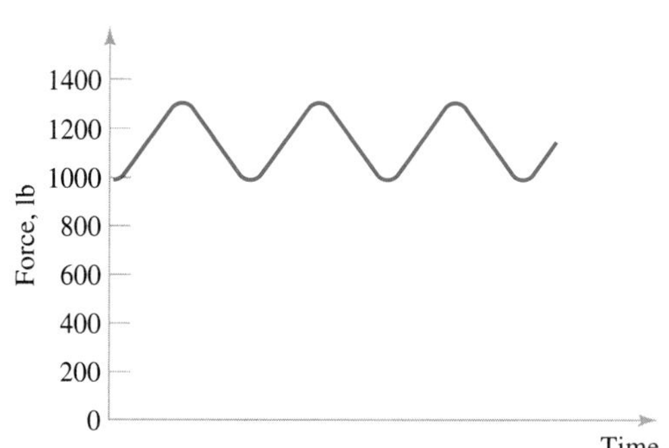

Figure E13.2C
Cyclic force-time pattern for the case of initial preload of $F_i = 1000$ lb.

Reading into the bolt *S-N* curve of Figure E13.2B again, with $\sigma_{eq-CR} = 3080$ psi, the bolt life is found to be infinite.

The result of preloading in this case is to improve bolt life from about 100 minutes to infinite life.

e. If a gasket is sandwiched between members 1 and 2, the spring rate k_m is changed, and forces in the bolt and in the housing, when the initial preload force is applied, must be recomputed.

Assuming the effective outer diameter of the gasket to be d_m, just as for the clamped members, the *gasket* spring rate may be calculated in the manner of (13-22) as

$$k_g = \frac{\pi(d_m^2 - d_b^2)E_g}{L_g} \tag{13}$$

From Table 13.6, the value of E_g for compressed asbestos is 70×10^3 psi, and from the problem statement $L_g = {}^3/_{32}$ inch. Hence (13) becomes, assuming $d_m = 2d_b$,

$$k_g = \frac{\pi[(2d_b)^2 - d_b^2](70 \times 10^3)}{(3/32)} = \frac{\pi[3(0.375)^2](70 \times 10^3)}{0.094} = 9.90 \times 10^5 \frac{\text{lb}}{\text{in}} \tag{14}$$

For the 0.75-inch-thick steel flanges

$$k_{m1} = k_{m2} = \frac{\pi[3(0.375)^2](30 \times 10^6)}{0.75} = 5.30 \times 10^7 \frac{\text{lb}}{\text{in}} \tag{15}$$

Using (13-26)

$$k_m = \frac{1}{\dfrac{1}{5.3 \times 10^7} + \dfrac{1}{9.9 \times 10^5} + \dfrac{1}{5.3 \times 10^7}} = \frac{1}{1.05 \times 10^{-6}} = 9.54 \times 10^5 \frac{\text{lb}}{\text{in}} \tag{16}$$

It is worth noting that in this case the gasket dominates the spring rate k_m. The bolt spring rate k_b may be calculated from (13-17) as

$$k_b = \frac{\pi(0.375)^2(30 \times 10^6)}{4(1.5)} = 2.21 \times 10^6 \frac{\text{lb}}{\text{in}} \tag{17}$$

Now, using (13-15) and (13-16) to calculate the force in the bolt and in the clamped members, respectively,

$$F_b = \left(\frac{2.21 \times 10^6}{2.21 \times 10^6 + 9.54 \times 10^5}\right)1200 + 1000 = 0.70(1200) = 1838 \text{ lb} \tag{18}$$

and

$$F_m = \left(\frac{9.54 \times 10^5}{3.16 \times 10^6}\right)1200 - 1000 = 0.30(1200) - 1000 = -640 \text{ lb} \tag{19}$$

For the gasketed joint with an initial preload of $F_i = 1000$ lb therefore, the maximum and minimum cyclic tensile forces in the bolt are $F_{max} = 1838$ lb and $F_{min} = 1000$ lb. This force-time pattern is plotted in Figure E13.2D.

f. Using the same approach as for (b) and (d),

$$\sigma_{max} = \frac{F_{max}}{A_r} = \frac{1838}{0.068} - 27{,}030 \text{ psi} \tag{20}$$

and

$$\sigma_{min} = \frac{F_{min}}{A_r} = \frac{1000}{0.068} - 14{,}705 \text{ psi} \tag{21}$$

Example 13.2 Continues

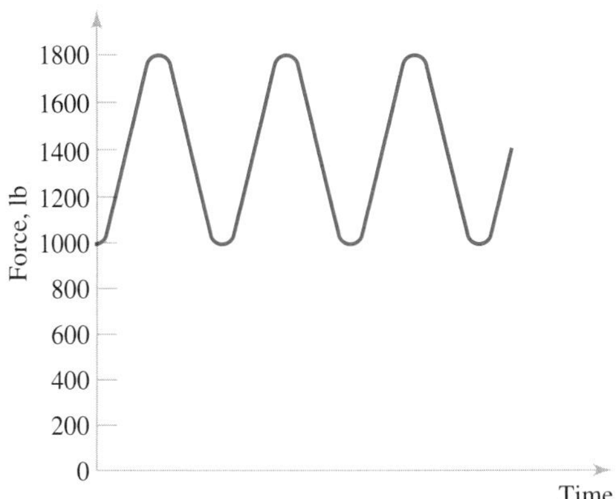

Figure E13.2D
Cyclic force-time pattern for the case of *gasketed* joint with initial preload of F_i = 1000 lb.

Accordingly,

$$\sigma_m = \frac{27{,}030 + 14{,}705}{2} = 20{,}870 \text{ psi} \tag{22}$$

Again using (4-144),

$$\sigma_{eq-CR} = \frac{27{,}030 - 20{,}870}{1 - \left(\dfrac{20{,}870}{60{,}000}\right)} = 9445 \text{ psi} \tag{23}$$

Reading into the bolt S-N curve of Figure E13.2B again with $\sigma_{eq-CR} = 9445$ psi, the estimated bolt life is approximately 300,000 cycles (about the same as for the case of no preload). This life, about 100 minutes, is likely unacceptable, and a different gasketing arrangement or an upgraded bolt fatigue strength should be considered.

Tightening Torque; Fastener Loosening

To obtain the maximum benefits of preloading, it is essential that the designer-specified initial preload *actually be induced* in the bolt by the tightening process. The most accurate method of inducing the desired axial preload involves direct measurement of the axial force-induced elastic deformation, δ_f, in the bolt, by using a micrometer or an electronic device for precisely measuring bolt elongation. From (2-3), the *required* value of bolt elongation, $(\delta_f)_{req'd}$, corresponding to the specified preload force, may need to be calculated, then the bolt tightened until the measured value δ_f is equal to $(\delta_f)_{req'd}$. This method produces accurate preloads but is impractical in most applications because it is slow and expensive. A more convenient (but less accurate) method of inducing the desired preload is to apply a measured torque to the nut or screw head, using a *torque wrench*. The torque needed to develop a desired axial preload may be calculated by adapting the power screw equation (12-11) to the bolt-tightening scenario, giving

$$T_i = F_i \frac{d_p}{2}\left[\frac{l\cos\theta + \pi d_p \mu_t}{\pi d_p \cos\theta - l\mu_t}\right] + F_i \frac{d_c}{2}\mu_c \tag{13-26}$$

where T_i is the tightening torque required to produce initial preload force F_i in the bolt. Often, the pitch diameter, d_p, is approximated by the nominal bolt diameter, d_b, and the mean collar (contacting washer-face) diameter, d_c, approximated as $(d_b + 1.5d_b)/2$. Using these estimates, (13-26) may be approximated as

$$T_i \approx F_i \frac{d_p}{2}\left[\frac{l\cos\theta + \pi d_b \mu_t}{\pi d_b \cos\theta - l\mu_t}\right] + F_i \frac{(1.25d_b)}{2}\mu_c \tag{13-27}$$

or

$$T_i \approx \left[\frac{1}{2}\left(\frac{l\cos\theta + \pi d_b \mu_t}{\pi d_b \cos\theta - l\mu_t}\right) + 0.625\mu_c\right] F_i d_b \tag{13-28}$$

The *bracketed* coefficient is usually defined as the *torque coefficient*, K_T. It is often assumed that the thread friction coefficient, μ_t, and the collar friction coefficient, μ_c, for an average application, are both nominally equal to 0.15. For this assumption, it turns out that the nominal value of torque coefficient is

$$K_T \approx 0.2 \tag{13-29}$$

for all bolt sizes and for both coarse and fine thread pitches. It is cautioned, however, that values of K_T ranging from 0.07 to 0.3 have been reported, depending upon finishes and lubricants.[7] Thus, for an average application, (13-28) may be approximated as

$$T_i \approx 0.2F_i d_b \tag{13-30}$$

A more modern method is to use a "smart" torque wrench which continuously monitors wrench torque and nut rotation, sending these data to a controlling computer program. When the onset of yielding is detected by the computer, the wrench is disengaged, leaving a tensile preload approaching the proof load[8] for the fastener being tightened.

Under less demanding conditions, for lower performance designs, or when torque wrenches are not available, an alternative technique called the *turn-of-the-nut* method is sometimes used. The nut is "seated" to a finger-tight condition, then subjected to a specified additional rotation to induce the preload. More accurate methods should be used when practical.

Initial preload is often relaxed over time by virtue of creep, corrosion, wear, or other processes that slowly reduce the axial elastic deformation induced by the initial preloading (tightening torque). Elevated temperature or cyclic temperature fluctuations may accelerate the loss of preload.

Loss of preload not only has the potential for diminishing fatigue resistance of the joint (see Example 13.2), but also may contribute to *loosening* of the threaded fastener, possibly allowing the joint to separate. Ease of disassembly has often been cited as an advantage of threaded fasteners, allowing maintenance or replacement of parts, but this very characteristic represents a disadvantage in terms of allowing threaded assemblies to loosen and separate during operation. It is common practice to specify a *retightening schedule* for threaded fasteners, to prevent excessive loosening and to periodically reestablish the proper preload.

Multiply Bolted Joints; Symmetric and Eccentric Loading

In the section "Critical Points and Thread Stresses," the analyses were based on the presumption that shear and tensile loads on the bolt were known. More often than not, bolted joints involve *many* bolts placed in a specified pattern to improve the strength and stability of the joint. In *multiply* bolted joints, it becomes important to define the *distribution* of applied forces *among* the bolts and the stresses that result in each bolt. Several different

[7]See ref. 1. [8]See Tables 13.3, 13.4, and 13.5 for proof strength values.

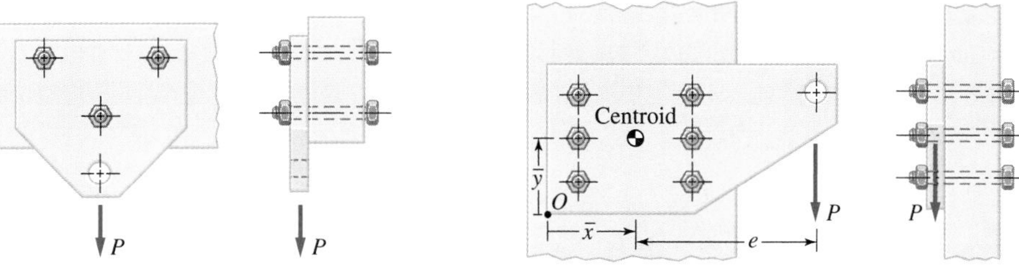

(a) Symmetric loading on bolt. (b) Eccentric loading on bolt.

Figure 13.11
Multiply bolted joints in shear, with symmetric and eccentric loading illustrated.

configurations of multiply bolted joints, some subjected to symmetric loads and others to eccentric loads, some under shear loads and others under tension loads, are illustrated in Figures 13.11 and 13.12.

If *shear loading* is *symmetrically* applied to a multiply bolted joint, as illustrated in Figure 13.11(a), it is usually acceptable to assume that the total shear is uniformly distributed over all bolts. (It should be noted, however, that if there are more than *two* bolts in the pattern, a more accurate assessment of the loading distribution may be required for critical design applications.) If the shear is assumed to be equally distributed over all bolts, and if it is assumed that clamping friction provides no shear resistance, equation (13-6) may be adapted as

$$\tau_b = \frac{P}{\sum_{i=1}^{n_b} A_i} \tag{13-31}$$

where A_i = area of the ith bolt, n_b = number of bolts, and shearing stress τ_b has the sense of P.

If *shear loading* is *eccentrically* applied to a multiply bolted joint, as illustrated in Figure 13.11(b), there will be an additional "torsion-like" shearing stress developed because of the couple Pe. This torsion-like shearing stress component may be estimated by transferring the eccentric shear force P to the *centroid* of the bolt pattern as a direct shear force P and a torsion-like couple Pe, as shown in Figure 13.13. Again neglecting any shear resistance due to clamping friction, the torsion-like shear stress may be calculated from

$$\tau_{ti} = \frac{(Pe)r_i}{J_j} \tag{13-32}$$

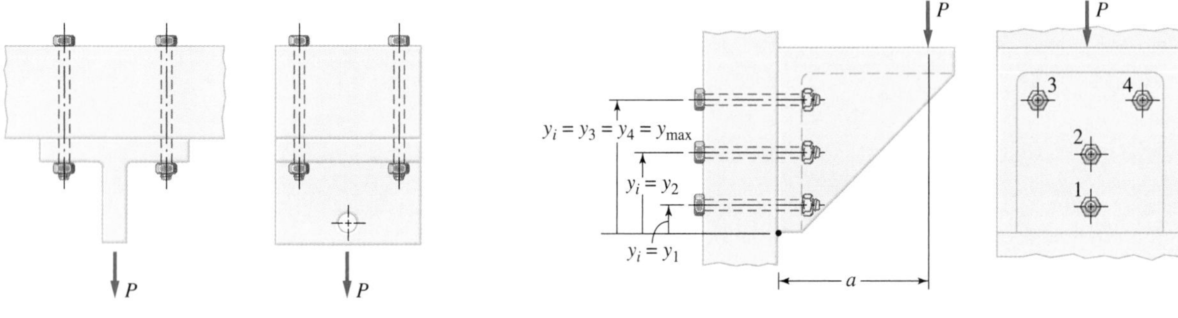

(a) Symmetric loading on bolts. (b) Eccentric loading on bolts.

Figure 13.12
Multiply bolted joints in tension, with symmetric and eccentric loading illustrated.

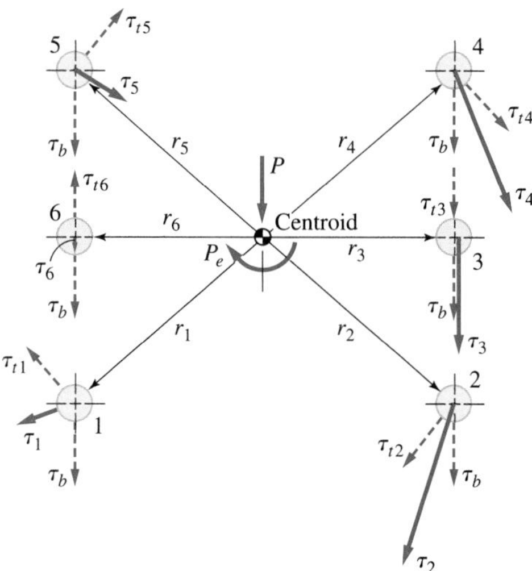

Figure 13.13
Illustration of resultant shearing stress τ_i caused by an eccentric shear force on a multiply bolted joint such as shown in Figure 13.11(b).

where τ_{ti} = torsion-like shear stress on ith bolt with direction perpendicular to radius r_i and sense of Pe

r_i = radius from centroid to ith bolt

$J_j = \sum_{i=1}^{n_b} A_i r_i^2$ = "joint" polar moment of inertia

The location of the centroid may be found with respect to any convenient x-y coordinate origin from

$$\bar{x} = \frac{\sum_{i=1}^{n_b} A_i x_i}{\sum_{i=1}^{n_b} A_i} \tag{13-33}$$

and

$$\bar{y} = \frac{\sum_{i=1}^{n_b} A_i y_i}{\sum_{i=1}^{n_b} A_i} \tag{13-34}$$

where $\bar{x},\bar{y}$ = coordinates of centroid with respect to selected coordinate origin

x_i, y_i = coordinates of ith bolt with respect to selected coordinate origin

n_b = number of bolts

The resultant shearing stress (assuming no frictional shear resistance) on the ith bolt may then be found as the *vector* sum of (13-31) and (13-32). Figure 13.13 illustrates the concept graphically.

When bolts are preloaded, combined stress failure theories or design theories must be appropriately applied.

If *tension loading* is *symmetrically* applied to a multiply bolted joint, as illustrated in Figure 13.12(a), it is usually acceptable to assume that the external load P is uniformly distributed among all bolts, so, for preloaded bolts, (13-15) may be adapted as

$$F_b = \left(\frac{k_b}{k_b + k_m}\right)\frac{P}{n_b} + F_i \tag{13-35}$$

If *tensile loading* is *eccentrically* applied to a multiply bolted joint, as illustrated in

Figure 13.12(b), an additional "bending-like" load distribution may be produced in the bolts. To find the load-sharing *distribution* for such a case, as illustrated in Figure 13.12(b), the load in any given bolt may be written as

$$F_k = A_k \sigma_{tk} = A_i \left[\frac{M(y_k)}{I_j} \right] \tag{13-36}$$

from which the external force on the most highly loaded bolt in the group may be written as

$$(F_b)_{max} = A_b \left[\frac{(Pa)(y_k)_{max}}{I_j} \right] \tag{13-37}$$

where A_b = area of most heavily loaded bolt
P = total applied load
a = perpendicular distance from hinge point H to applied load line of action (see Figure 13.12)
$(y_k)_{max}$ = distance from reference plane through H and perpendicular to line of action of P, for the most heavily loaded bolt

and

$$I_j = \sum_{k=1}^{n_b} A_k y_k^2 = \text{“joint” moment of inertia} \tag{13-38}$$

For eccentrically loaded joints having preloaded bolts then, (13-37) and (13-38) may be combined with (13-15) to give the tensile force in the most heavily loaded bolt as

$$(F_b)_{max} = \left(\frac{k_b}{k_b + k_m} \right) A_b \left[\frac{(Pa)(y_k)_{max}}{\sum_{k=1}^{n_b} A_k y_k^2} \right] + F_i \tag{13-39}$$

Again, if the resulting state of stress is *multiaxial*, combined stress failure theories or design equations should be used. As usual, however, appropriate simplifying assumptions may be made for preliminary design calculations.

Example 13.3 Multiply Bolted Joints and Tightening Torques

An L-shaped 1020 steel support bracket must support a static load of $P = 3000$ lb, as shown in Figure E13.3A. A bolt pattern using three bolts, with locations as shown in Figure E13.3A, has been suggested. If the bolts are to be *SAE Grade 1* (see Table 13.3) with unified standard coarse threads, and if a design safety factor of 2 is desired, recommend an appropriate bolt size and tightening torque for this application. As a practical matter, all three bolts are to be the same size. Make enough simplifying assumptions to expedite the calculations, since this is a first-cut preliminary design estimate.

Solution

From the sketch of Figure E13.3A, it may be deduced that the bolts are subjected to direct shear, torsion-like shear, and bending-like tension, due to the eccentrically applied external load P. By specification, an initial preload tension (tightening torque) is also to be applied. Because of the complexity of finding a proper bolt size explicitly, an iterative process may be more manageable. To do this, an initial bolt size will be assumed, stress components calculated and combined using the distortion energy design equation (6-14), and the bolt size adjusted if necessary. The process must be repeated until a satisfactory bolt size is found. The (conservative) assumption will also be made that all of the shear loading is supported by the bolts; none of it by clamping friction. In more refined calculations, this assumption should probably be reassessed.

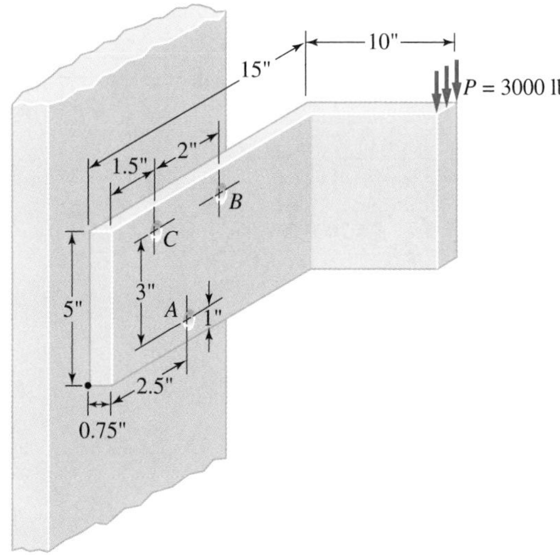

Figure E13.3A
Sketch of L-shaped support bracket, showing proposed bolt pattern.

An initial guess for required bolt area will be made by arbitrarily doubling the area needed to support direct shear alone, as calculated from (13-31).

Thus, for three bolts of the same size,

$$3(A_b)_0 = \frac{P}{\tau_d} = \frac{P}{\left(\dfrac{\tau_{yp}}{n_d}\right)} = \frac{P}{\left(\dfrac{S_{yp}}{2n_d}\right)} \tag{1}$$

whence

$$(A_b)_0 = \frac{2n_d P}{3S_{yp}} = \frac{2(2)(3000)}{3(36{,}000)} = 0.11 \text{ in}^2 \tag{2}$$

Doubling the value,

$$(A_b)_1 = 2(0.11) = 0.22 \text{ in}^2 \tag{3}$$

This estimate will be used as the initial assumption for bolt area.

Based on $(A_b)_1 = 0.22$ in^2, and a design safety factor of 2, the initial *allowable* preload force may be estimated as

$$(F_i)_1 = \left(\frac{36{,}000}{2}\right)(0.22) = 3960 \text{ lb} \tag{4}$$

(Remember that these are crude first estimates.)

Using (13-31), the *direct* shear stress components (equal in magnitude and vertically downward in direction) may be calculated as

$$(\tau_b)_1 = \frac{3000}{3(0.22)} = 4545 \text{ psi} \tag{5}$$

The *torsion-like* shear stress components may be calculated using (13-32), but first the centroid of the bolt pattern must be found. This may be done by arbitrarily selecting a coordinate origin, O, as illustrated in Figure E13.3B, then utilizing (13-33) and (13-34) to find

$$\bar{x} = \frac{0.22(1.5 + 2.5 + 3.5)}{3(0.22)} = 2.5 \tag{6}$$

and

Example 13.3 Continues

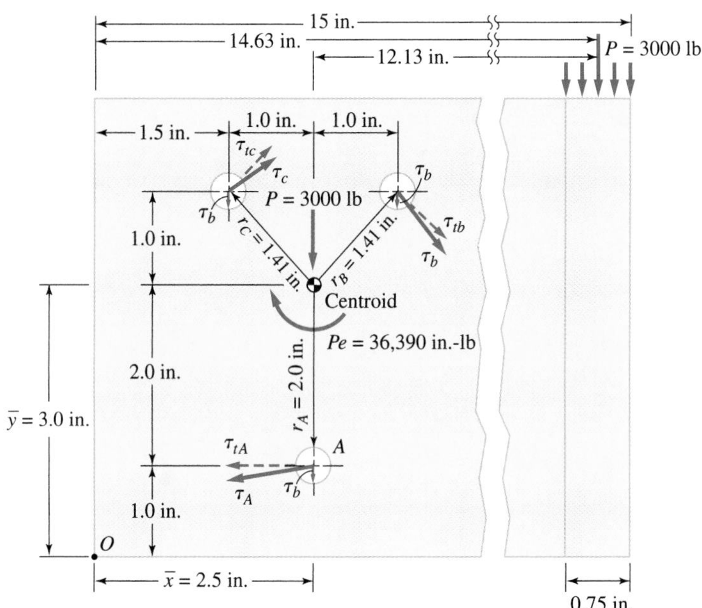

Figure E13.3B
Centroidal location and resultant shearing stresses for the bolt pattern illustrated in Figure E13.3A.

$$\bar{y} = \frac{0.22(4 + 1 + 4)}{3(0.22)} = 3.0 \tag{7}$$

Based on these coordinates, the centroidal position is as shown in Figure E13.3B.

Next, as also illustrated in Figure E13.3B, the radii from the centroid to each bolt center may be determined. Then, from (13-32),

$$(\tau_{tA})_1 = \frac{[(3000)(12.13)](2.0)}{0.22(2.0^2 + 1.41^2 + 1.41^2)} = 41{,}350 \text{ psi} \tag{8}$$

$$(\tau_{tB})_1 = \frac{36{,}390(1.41)}{1.76} = 29{,}155 \text{ psi} \tag{9}$$

$$(\tau_{tC})_1 = \frac{36{,}390(1.41)}{1.76} = 29{,}155 \text{ psi} \tag{10}$$

Each of these shearing stress vectors is perpendicular to its centroidal radius, with the *sense* of moment Pe.

To find resultant shearing stresses on each bolt, the direct shear must be *vectorially* added to the torsion-like shear. In this particular problem, it may be determined by inspection (Figure E13.3B) that bolt A is most critical, and the two shear components are perpendicular; hence

$$(\tau_A)_1 = \sqrt{4445^2 + 41{,}350^2} = 41{,}590 \text{ psi} \tag{11}$$

It is of interest to note that the direct shear component is negligibly small compared to the torsion-like shear, and will be dropped from further iterations.

Before going on to include the bending-like tensile component and the preload component, (11) should be evaluated with respect to the shear design stress, which is, based on (6-14),

$$\tau_d = 0.577\left(\frac{\sigma_{yp}}{n_d}\right) = 0.577\left(\frac{36{,}000}{2}\right) = 10{,}390 \text{ psi} \tag{12}$$

Since $(\tau_A)_1$ far exceeds τ_d (by a factor of about 4), a new (larger) assumed bolt area must be tried.

Applying a factor of 4 to $(A_b)_1 = 0.22$, anticipating that some additional area will be required to support the preload (and bending-like tension), and using a standard bolt size, a new iteration will be based on a standard $1\frac{1}{8}$-inch-diameter bolt, giving

$$(A_b)_2 = \frac{\pi(1.125)^2}{4} = 0.99 \text{ in}^2 \tag{13}$$

Repeating the iteration process for this new assumed area of 0.99 in^2, (5) will become negligible based on the observations about (11), and (8), (9), and (10) become

$$(\tau_{tA})_2 = \frac{[(3000)(12.13)](2.0)}{0.99(2.0^2 + 1.41^2 + 1.41^2)} = 9215 \text{ psi} \tag{14}$$

and

$$(\tau_{tB})_2 = (\tau_{tC})_2 = \frac{36{,}390(1.41)}{7.90} = 6495 \text{ psi} \tag{15}$$

Since the direct shear is negligible, these expressions represent the *resultant* shearing stresses as well. Thus

$$(\tau_A)_2 = 9215 \text{ psi} \tag{16}$$

and

$$(\tau_B)_2 = (\tau_C)_2 = 6495 \text{ psi} \tag{17}$$

Next, the bending-like bolt tension may be evaluated for the bolt pattern of Figure E13.3A by adapting (13-39) to this case. Taking the fulcrum axis H to be the lower edge of the clamped face, as shown in Figure E13.3A, assuming the clamped members to have an effective stiffness k_m of three times the axial bolt stiffness (see Figure 13.9 and related discussion), and using $(A_b)_2 = 0.99$ in^2, (13-39) may be adapted for bolt A as

$$F_{bA} = 0.25(0.99)\left[\frac{(3000)(10)(1.0)}{0.99(1^2 + 4^2 + 4^2)}\right] + F_i = 0.25[918.3] + F_i = 229.6 + F_i \tag{18}$$

and for bolt B or C as

$$F_{bB} = F_{bC} = 0.25\left[\frac{(3000)(10)(4)}{0.99(1^2 + 4^2 + 4^2)}\right] + F_i = 0.25[3673.2] + F_i = 918.3 + F_i \tag{19}$$

For purposes of establishing a *minimum* preload to prevent joint separation, (13-16) may be used to assure that the clamping force F_m on the members always remains compressive (minus sign). Thus for bolt A

$$F_{mA} = 0.75[918.3] - F_i = 689 - F_i \tag{20}$$

and for bolts B and C

$$F_{mB} = F_{mC} = 0.75[3673.2] - F_i = 2755 - F_i \tag{21}$$

Hence if all three bolts are to be initially tightened to the same preload F_i, F_i must be *no smaller* than 2755 lb in order to prevent separation of the joint. At the other extreme, the preload should not produce a stress greater than σ_d when the bolt is tightened. That is

$$(F_i)_{max} = \sigma_d A_b = \frac{36{,}000}{2}(0.99) = 17{,}820 \text{ lb} \tag{22}$$

(This is a little lower than the "85 percent of proof stress" rule of thumb would give for this $1\frac{1}{8}$-inch bolt.)

Based on (21) and (22) then, the range of F_i should be

$$2755 < F_i < 17{,}820 \tag{23}$$

with higher values preferred from the standpoint of loosening.

Example 13.3 Continues

For purposes of the preliminary assessment, a midrange value for preload, $F_i = 10{,}000$ lb, will be tried. Then using (18), the tensile stress in the bolt may be calculated for bolt A as

$$(\sigma_A)_2 = \frac{F_{bA}}{A_b} = \frac{229.6 + 10{,}000}{0.99} = 10{,}335 \text{ psi} \tag{24}$$

and for bolts B and C, using (19), as

$$(\sigma_B)_2 = (\sigma_C)_2 = \frac{918.3 + 10{,}000}{0.99} = 11{,}030 \text{ psi} \tag{25}$$

Critical point D in Figure 13.7(b) illustrates the state of stress being investigated in this example. For such a state of stress, the results from above may be summarized as follows: For bolt A,

$$\begin{aligned} \sigma_A &= 10{,}335 \text{ psi} \\ \tau_A &= 9215 \text{ psi} \end{aligned} \tag{26}$$

and for bolt B or C,

$$\begin{aligned} \sigma_{B/C} &= 11{,}030 \text{ psi} \\ \tau_{B/C} &= 6495 \text{ psi} \end{aligned} \tag{27}$$

The stress cubic equation (4-59) may be used to find the principal stresses for bolt A as (see also Example 4.8)

$$\begin{aligned} \sigma_1 &= \frac{\sigma_A}{2} + \sqrt{\left(\frac{\sigma_A}{2}\right)^2 + \tau_A^2} = \frac{10{,}335}{2} + \sqrt{\left(\frac{10{,}335}{2}\right)^2 + (9215)^2} \\ &= 5168 + \sqrt{(5168)^2 + (9215)^2} = 5168 + 10{,}565 = 15{,}735 \text{ psi} \end{aligned} \tag{28}$$

$$\sigma_2 = 0 \tag{29}$$

$$\sigma_3 = 5168 - 10{,}565 = -5397 \text{ psi} \tag{30}$$

and for bolts B and C as

$$\sigma_1 = \frac{11{,}030}{2} + \sqrt{\left(\frac{11{,}030}{2}\right)^2 + (6495)^2} = 5515 + 8520 = 14{,}035 \text{ psi} \tag{31}$$

$$\sigma_2 = 0 \tag{32}$$

$$\sigma_3 = 5515 - 8520 = -3005 \text{ psi} \tag{33}$$

With these principal stresses, the distortion energy design equation (6-14) may be utilized to assess the choice of a $1\frac{1}{8}$-inch bolt. Using (6-14) for bolt A, the question is

$$\sqrt{\frac{1}{2}\{[15{,}735 - 0]^2 + [0 - (-5397)]^2 + [-5397 - 15{,}735]^2\}} \overset{?}{=} \frac{36{,}000}{2} \tag{34}$$

or

$$19{,}015 \overset{?}{=} 18{,}000 \tag{35}$$

This is close, but slightly under the specified design safety factor of 2. Due to the preliminary nature of these calculations we would probably accept this result as satisfactory, especially since no credit has been given for frictional resistance to shear loading. For bolts B and C, using (6-14), the question is

$$\sqrt{\frac{1}{2}\{[14{,}035 - 0]^2 + [0 - (-3005)]^2 + [-3005 - 14{,}035]^2\}} \overset{?}{=} \frac{36{,}000}{2} \tag{36}$$

or

$$15{,}755 \overset{?}{=} 18{,}000 \tag{37}$$

Bolts B and C are acceptable.

Since the usual recommendation is to limit initial preload to 85 percent of proof stress (see Table 13.3), for a $1^1/_8$-7 UNC Grade 1 bolt (see Table 13.1), the proof load would be

$$F_p = \sigma_p A_t = 33{,}000(0.763) = 25{,}180 \text{ lb} \tag{38}$$

and

$$(F_i)_{limit} = 0.85(25{,}180) = 21{,}400 \text{ lb} \tag{39}$$

The selected preload of $F_i = 10{,}000$ lb used in the above calculations is only about half of the limiting value, so thread stresses from the tightening process should be satisfactory.[9]

The preliminary recommendations then are to use

$$1^1/_8\text{-7 UNC-2A Grade 1 bolts} \tag{40}$$

and, based on (13-30), a tightening torque of

$$T_i = 0.2(10{,}000)(1.125) = 2250 \text{ in-lb } (188 \text{ ft-lb}) \tag{41}$$

13.5 Rivets

For *permanent* joints, rivets have been widely used to secure both *structural joints*, for which strength is an important design consideration, and lower-performance *industrial joints*, for which strength requirements are modest but production costs and assembly time are key factors. Rivet use is virtually always restricted to *lap joints* (see Figure 13.1) because rivets are *efficient in shear* but inefficient in tension. Traditionally, *structural* rivets have been used in civil engineering structures such as buildings, bridges, and ships, as well as many mechanical engineering applications including pressure vessels, automotive applications, and aircraft structures. Although the development of modern welding equipment and techniques has reduced the importance of riveting for many of these uses, it remains important in high-performance applications such as stressed-skin aircraft structures. Rivets are cost-effective, do not change the material properties of the members fastened, do not warp the joined parts, and serve as fatigue crack stoppers.

For lower-performance *industrial* joints, such as used in the assembly of appliances, electronic devices, business machines, furniture, and other similar applications, the potentially lower costs and higher assembly speeds often make riveting a wise choice. Additional advantages of riveting are that rivets do not tend to shake loose, can be assembled blind (fully installed from one side), can be used to join dissimilar materials in various thicknesses, and are simple and safe to install.

Four common rivet types are illustrated in Figure 13.14. Many other variations are commercially available,[10] including self-piercing rivets, compression rivets, and other special cold-headed configurations.

Rivet Materials

Rivets may be made of virtually any ductile material with acceptable shear strength for the application. The selection of an appropriate material for a given high-performance structural application should follow the guidelines given in Chapter 3.

[9]Interestingly, if thread stresses due to tightening were actually calculated and combined for the thread critical points shown in Figure 13.7(a), they would be found to far exceed yield strength. In practice, however, local yielding redistributes the stresses and acceptable performance may be expected if tightening preload is limited to 85 percent of the proof stress.

[10]See, for example, ref. 1.

Figure 13.14
Various types of rivets.

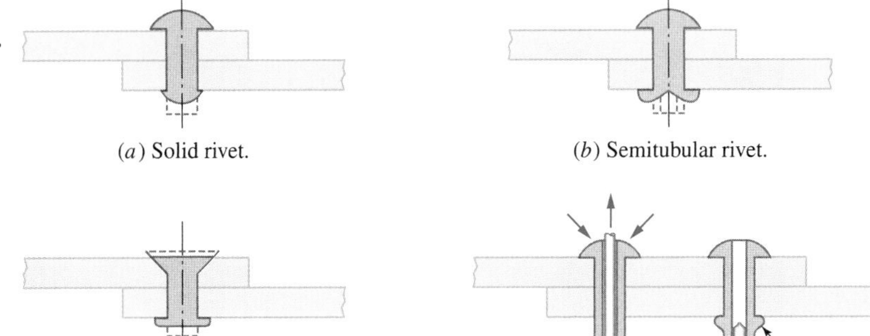

(*a*) Solid rivet.

(*b*) Semitubular rivet.

Blind side upset

(*c*) Countersunk flush rivet.

(*d*) Closed end break-mandrel blind rivet.

Low- and medium-carbon steel rivets are most widely used because of low cost, high strength, and good ductility (formability). Low-carbon grades (1006 to 1015) are preferred because of better formability (if strength requirements are adequate). Aluminum rivets are also widely used, notably in aircraft applications, but also are used in consumer products such as tubular furniture, storm windows and doors, and automotive parts. For consumer products, almost any soft grade of aluminum may be used. For aircraft applications, 2024, 5052, or other ductile high-strength aluminum alloys may be utilized for rivets. Other ductile materials, including brass or copper alloys, stainless steel, and even precious metals, may be used in special environments if warranted by the application.

Critical Points and Stress Analysis

In the elementary analysis of riveted joints and connections, it is usually assumed that bending and tension in the rivets may be neglected, that friction between parts does not contribute to *force transfer* across the joint, and that residual stresses are negligible. Further, shear in the rivets is assumed to be uniform and equally shared among rivets. Potential critical sections, therefore, may be identified as

1. Tensile failure of net plate cross section between rivets
2. Shearing of rivet cross section
3. Compressive bearing failure between rivet and plate
4. Edge shear-out at rivet hole
5. Edge tearing at rivet hole

These various critical sections are sketched in Figure 13.15. For elementary riveted-joint analysis, the stresses associated with each critical section may be estimated as described next.

For *tensile failure of the plate* between rivets [see Figure 13.15(a)], the plate tensile stress, σ_t, is

$$\sigma_t = \frac{F_s}{(b - N_r D_h)t} \tag{13-40}$$

where F_s = total shear load, lb
b = gross plate width, in
t = plate thickness, in
D_h = hole diameter, in (slightly larger than rivet diameter)
N_r = number of load-carrying rivets

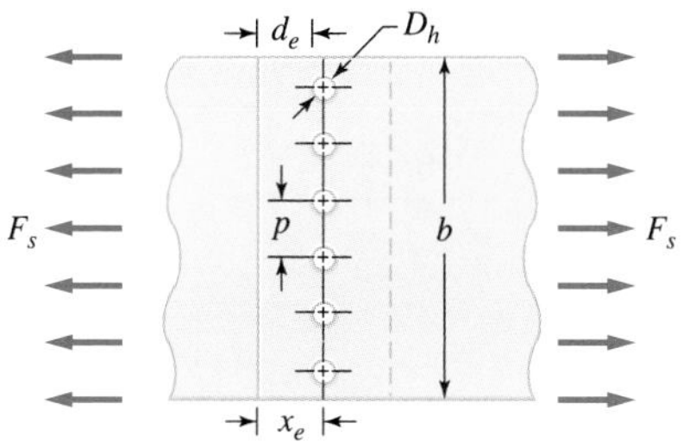

(a) Tensile stress σ_t on net plate cross section.

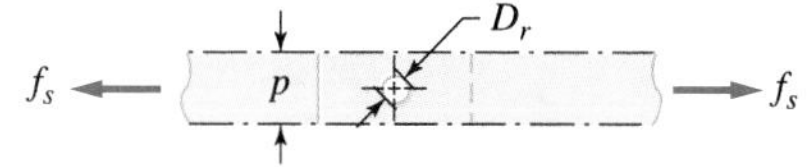

(b) Shear stress τ_s on rivet cross section.

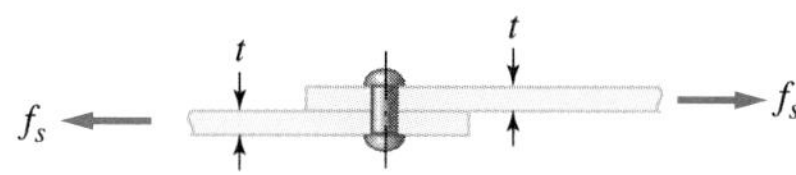

(c) Compressive bearing stress σ_c between plate and rivet.

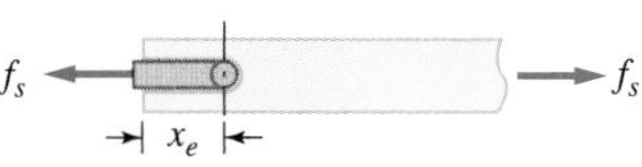

(d) Edge shear-out stress τ_e in plate.

(e) Edge tearing stress σ_e in plate.

Figure 13.15
Potential critical sections in a riveted lap joint.

For *rivet shearing* [see Figure 13.15(b)], the shearing stress, τ_s, is

$$\tau_s = \frac{f_s}{A_s} = \frac{\left(\dfrac{F_s}{N_r}\right)}{\left(\dfrac{\pi D_r^2}{4}\right)} = \frac{4F_s}{\pi D_r^2 N_r} \tag{13-41}$$

where A_s = shear area of rivet, in^2
f_s = shear force per rivet, lb
D_r = rivet diameter, in

For compressive *bearing* failure [see Figure 13.15(c)], the compressive bearing stress, σ_c, is

$$\sigma_c = \frac{f_s}{tD_r} = \frac{F_s}{tD_r N_r} \tag{13-42}$$

For *edge shear-out* of the rivet to the plate edge [see Figure 13.15(d)], the shear-out stress, τ_e, is

$$\tau_e = \frac{f_s}{2x_e t} = \frac{F_s}{2x_e t N_r} \tag{13-43}$$

where x_e = distance from rivet hole center to edge of plate (recommended minimum values for x_e are $2D_r$ for protruding head rivets and $2.5D_r$ for countersunk flush rivets)

For *edge tearing* failure [modeled as a center-loaded fixed-fixed beam in bending; see Figure 13.15(e)], the edge-tearing tensile stress is

$$\sigma_e = \frac{Mc}{I} = \frac{6M}{td_e^2} = \frac{6\left(\dfrac{f_s p}{8}\right)}{td_e^2} = \frac{3F_s p}{4td_e^2 N_r} \tag{13-44}$$

where p = pitch (recommended minimum values are shown in Table 13.7; if two rows are used, the row spacing should be such that any two rivets in the two rows are spaced at least p apart)

It should be emphasized that selecting rivet materials and sizes to keep these elementary stress estimates below the design stress provides only a *preliminary* configuration. For

Table 13.7 Recommended Minimum Rivet Pitch (Spacing)[1]

Rivet diameter D_r, in	$1/8$	$5/32$	$3/16$	$1/4$
Minimum pitch p for protruding head rivets, in	$1/2$	$9/16$	$11/16$	$7/8$
Minimum pitch p for countersunk flush rivets, in	$11/16$	$27/32$	$1\,1/32$	$1\,1/4$

[1]See ref. 8.

critical applications, experimental verifications should be conducted to qualify the joint for actual operating conditions.

If eccentric loading is applied to multiply riveted joints, the concepts of eccentrically loaded multiply *bolted* joints may be directly applied, as discussed in detail in 13.4.

13.6 Welds

When permanent joints are an appropriate design choice, *welding* is often an economically attractive alternative to threaded fasteners or rivets. Most industrial welding processes involve local *fusion* of the parts to be joined, at their common interfaces, to produce a *weldment*. Heat is supplied by a controlled electric arc passing from an *electrode* to the *workpiece*, or by use of an oxyfuel gas *flame*. Typically, an *inert gas* or a *flux* is used to shield the weld zone from the atmosphere during the welding process, and a *filler metal* is introduced so that sound, uncontaminated welds result. Table 13.8 lists the more frequently used industrial welding processes and their abbreviated letter designations.

Shielded metal arc welding (SMAW), sometimes called *stick welding*, is the common manual process used for repair work, and for welding large steel structures, where a consumable flux-coated electrode (welding rod) is fed into the molten zone as the welder traverses the joint. *Gas metal arc welding (GMAW)*, sometimes called *metal–inert gas (MIG) welding*, is a common automated process for producing high-quality welds at high welding speeds. The GMAW process may be used for most metals. For production welding, the goal is to select a process that provides the desired quality and strength at the lowest avail-

Table 13.8 Frequently Used Welding Processes[1]

Letter Designation	Welding Process
SMAW	shielded metal arc welding
SAW	submerged arc welding
GMAW	gas metal arc welding
FCAW	flux cored arc welding
GTAW	gas tungsten arc welding
PAW	plasma arc welding
OFW	oxyfuel gas welding
EBW	electron beam welding
LBW	laser beam welding
RSW	resistance spot welding
RSEW	resistance seam welding

[1]For a detailed description of these processes, see ref. 9.

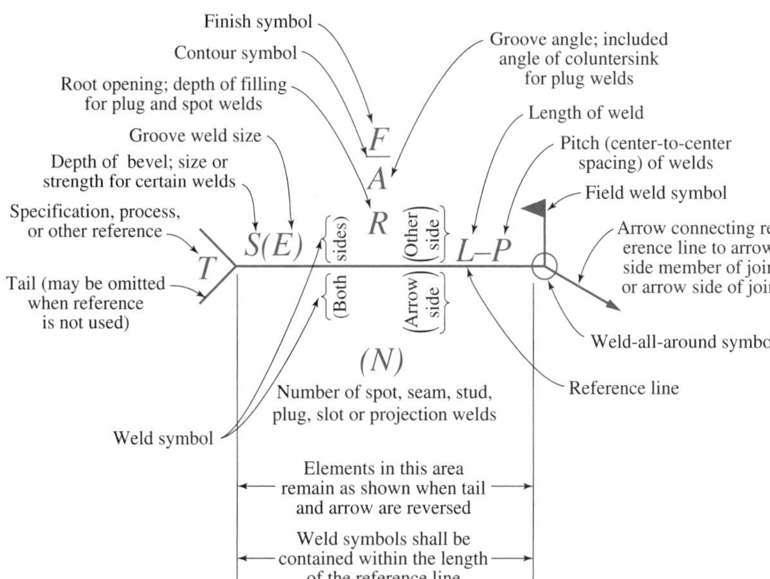

Figure 13.16
Standard location of the elements of the welding symbol. (Reprinted from ref. 13 with permission from American Welding Society.)

able cost. A designer is well advised to consult with a welding specialist when selecting the welding process to be used in a production environment.

Specification of welds has been standardized by the American Welding Society (AWS) through the use of a *basic welding symbol*,[11] shown in Figure 13.16. The basic welding symbol consists of a *reference line*, which carries precise information about type, size, edge preparation, contour, finish, and any other pertinent welding data, and an *arrow* which points to one side of the joint to be welded, designated as the *arrow side* (as opposed to the *other side*). As indicated in Figure 13.16, all welding specifications for the *arrow side* are placed *below* the reference line, and welding specifications for the *other side* are placed *above* the reference line. Standard symbols for edge preparation, weld type, and joint finishing, shown in Figure 13.17, are placed as needed on or near the reference line of Figure 13.16.

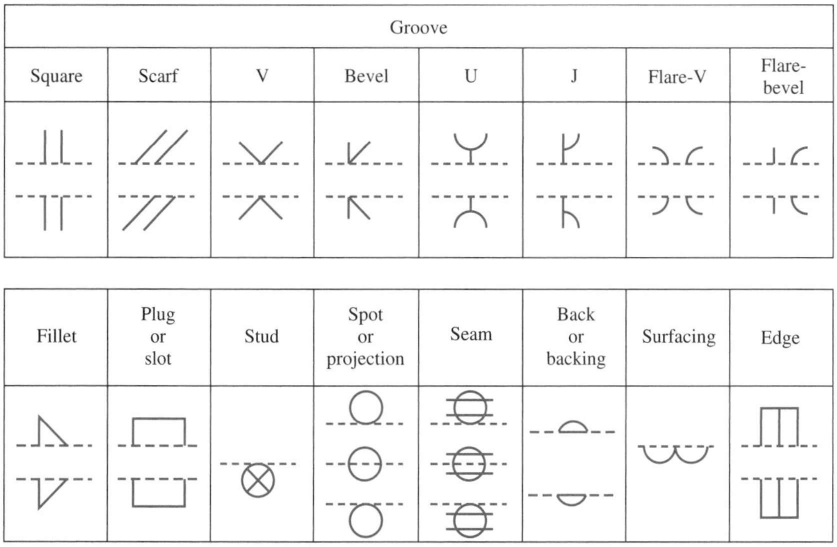

Groove							
Square	Scarf	V	Bevel	U	J	Flare-V	Flare-bevel

Fillet	Plug or slot	Stud	Spot or projection	Seam	Back or backing	Surfacing	Edge

Note: The reference line is depicted as a dashed line fir illustrative purposes.

Figure 13.17
Standard symbols for edge preparation, weld type, and joint finishing, defined for use with AWS basic welding symbol shown in Figure 13.16. Resulting welds are sketched in Figure 13.18. (Adapted from ref. 13 with permission from American Welding Society.)

[11]See ref. 9, Ch. 6.

Figure 13.17
(*Continued*)

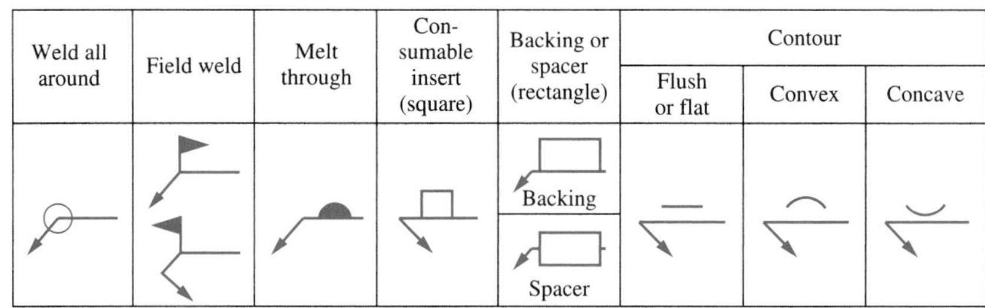

Weld all around	Field weld	Melt through	Con-sumable insert (square)	Backing or spacer (rectangle)	Contour		
					Flush or flat	Convex	Concave
				Backing Spacer			

Sketches of various types of weld joints are shown in Figure 13.18. It is estimated that about 85 percent of all industrial welds are *fillet welds*, about 10 percent are *butt welds*, and *special welds* account for the remaining 5 percent. Figure 13.19 provides preliminary guidance for appropriate *weld selection* for various basic joint types. Final selection of the best type of weld for a given application should be made in consultation with a welding specialist.

No matter what welding technique is used, the weld joint must be regarded as a metallurgically nonhomogeneous zone ranging from *unheated parent metal* through the *heat-affected zone (HAZ)* to the *weld metal zone*, a region of *cast* material. The gradient in

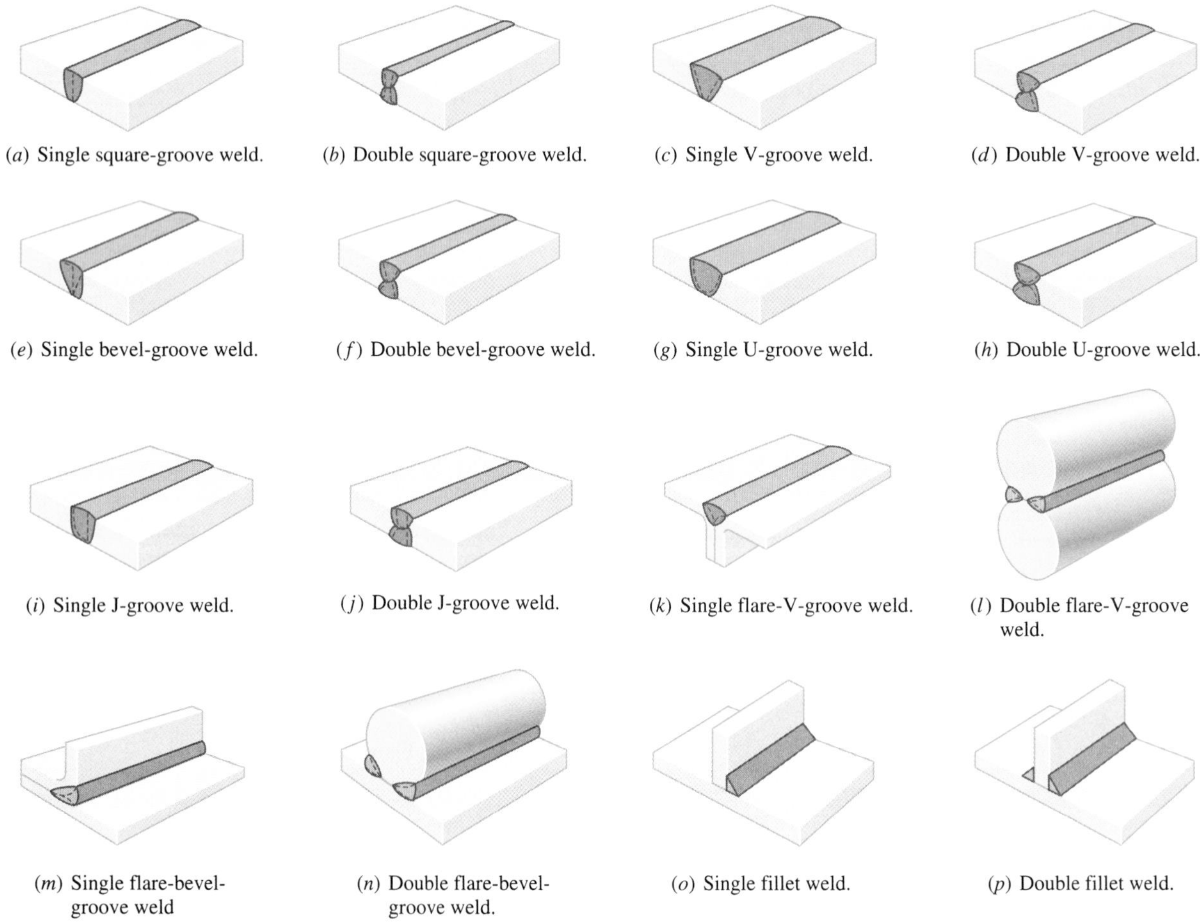

(*a*) Single square-groove weld.

(*b*) Double square-groove weld.

(*c*) Single V-groove weld.

(*d*) Double V-groove weld.

(*e*) Single bevel-groove weld.

(*f*) Double bevel-groove weld.

(*g*) Single U-groove weld.

(*h*) Double U-groove weld.

(*i*) Single J-groove weld.

(*j*) Double J-groove weld.

(*k*) Single flare-V-groove weld.

(*l*) Double flare-V-groove weld.

(*m*) Single flare-bevel-groove weld

(*n*) Double flare-bevel-groove weld.

(*o*) Single fillet weld.

(*p*) Double fillet weld.

Figure 13.18
Sketches of various types of welds and their edge preparation. (Excerpted from ref. 9, with permission from American Welding Society.)

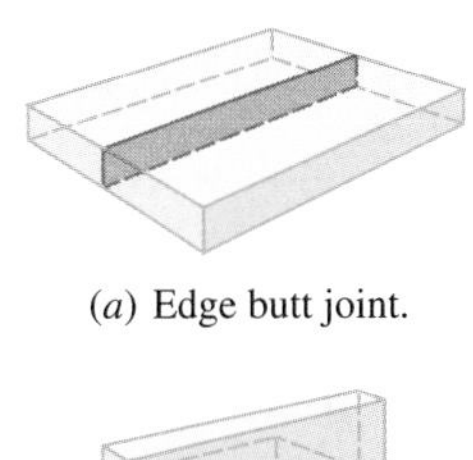

(*a*) Edge butt joint.

Applicable Welds	
Square-groove	J-groove
V-groove	Flare-V-groove
Bevel-groove	Flare-bevel-groove
U-groove	

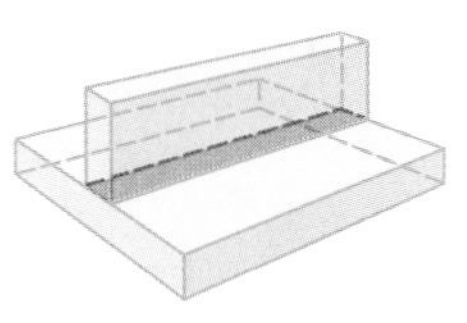

(*b*) Tee butt joint.

Applicable Welds	
Fillet	J-groove
Plug	Flare-bevel-groove
Slot	Spot
Square-groove	Projection
Bevel-groove	Seam

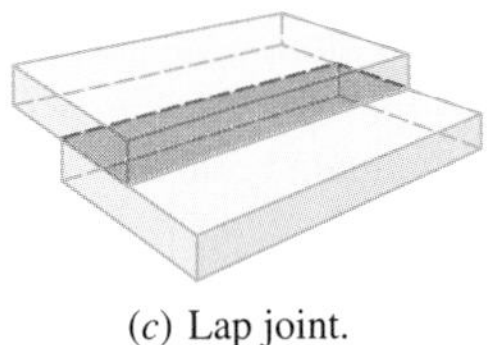

(*c*) Lap joint.

Applicable Welds	
Fillet	J-groove
Plug	Flare-bevel-groove
Slot	Spot
Bevel-groove	Projection
Seam	

Figure 13.19
Applicable weld selections for various basic joint types (see Figure 13.1). Some of these welds are sketched in Figure 13.8. (Excerpted from ref. 9, with permission from American Welding Society.)

homogeneity, together with any geometrical discontinuities, gives rise to *stress concentration* that must be considered by the designer, especially if fluctuating loads are applied to the weldment. Fatigue stress concentration factors commonly used for various weld-zone critical points are given in Table 13.9.

A significant design consideration in specifying welded construction is that *unfavorable residual stresses* (see 4.13) of high magnitude may result from postcooling shrinkage stresses, usually greatest in the transverse direction. These residual stresses not only have the potential for adversely influencing fatigue resistance, but may also result in unacceptable *distortion* of the weldment. *Preheating* and/or *postweld thermal treatments* are often used to reduce the residual stresses and the distortions to acceptable levels. *Mechanical cold-working* treatments of the completed welds, such as *shot-peening, cold-rolling*, or *cold-stretching*, are also used to produce *favorable* residual stresses at the weld-zone critical points (see 2.6).

Base Metals, Filler Materials, and Weldability

Most metals can be welded by selecting appropriate processes, electrodes, and shielding media. The term *weldability* refers to a qualitative description of the weld quality that can be achieved by using specified materials and processes. Table 13.10 provides weldability characteristics for several metals and alloys, using approved welding practices.[12]

Table 13.9 Fatigue Stress Concentration Factors K_f for Various Weld-Zone Critical Points

Location	K_f
HAZ[1] of reinforced butt weld (even if ground flush)	1.2
Toe of transverse fillet weld	1.5
End of parallel fillet weld	2.7
Toe of tee-butt weld joint	2.0

[1]Heat-affected zone.

[12]See ref. 9.

Table 13.10 Weldability of Various Metals (G = good, F = fair, U = unacceptable)

Metal	Arc	Gas	Metal	Arc	Gas
Carbon steel			Magnesium alloys	U	G
Low and medium carbon	G	G			
High carbon	G	F	Copper and copper alloys		
Tool steel	F	F	Deoxidized copper	F	G
			Pitch, electrolytic, and lake	G	F
Cast steel, plain carbon	G	G	Commercial bronze, red brass, and low brass	F	G
			Spring, admiralty, yellow, and commercial brass	F	G
Gray and alloy cast iron	F	G	Muntz metal, naval brass, and magnesium bronze	F	G
			Phosphor bronze, bearing bronze, and bell metal	G	G
Malleable iron	F	F	Aluminum bronze	G	F
			Beryllium copper	G	—
Low-alloy high-strength steels					
Ni-Cr-Mo and Ni-Mo	F	F	Nickel and nickel alloys	G	G
Most others	G	G			
			Lead	U	G
Stainless steel					
Chromium	G	F			
Chromium-nickel	G	G			
Aluminum and Al alloys					
Commercially pure	G	G			
Al-Mn alloys	G	G			
Al-Mg-Mn and Al-Si-Mg alloys	G	F			
Al-Cu-Mg-Mn alloys	F	U			

Electrode materials usually have compositions similar to the base metals being welded, but often have slightly higher tensile strengths. Specifications for electrodes have been developed and published as ANSI/AWS standards. For example, references covering electrode specifications for the SMAW and GMAW welding processes, briefly discussed above, are shown for various materials in Tables 13.11 and 13.12, respectively. (For other processes, see ref. 9.)

Table 13.11 AWS Specifications for Covered Electrodes Used in the SMAW Process

Base Material Type	AWS Specification
Carbon steel	A5.1
Low-alloy steel	A5.5
Corrosion-resistant steel	A5.4
Cast iron	A5.15
Aluminum and Al alloys	A5.3
Copper and copper alloys	A5.6
Nickel and nickel alloys	A5.11
Surfacing	A5.13 and A5.21

Table 13.12 AWS Specifications for Electrode Wire Used in the GMAW Process

Base Material Type	AWS Specification
Carbon steel	A5.18
Low-alloy steel	A5.28
Aluminum alloys	A5.10
Copper alloys	A5.7
Magnesium	A5.19
Nickel alloys	A5.14
300 series stainless steel	A5.9
400 series stainless steel	A5.9
Titanium	A5.16

Table 13.13 Minimum Properties for Carbon Steel and Low-Alloy Steel Welding Electrodes

ANSI/AWS Electrode	Tensile Strength, ksi (MPa)	Yield Strength, ksi (MPa)	Elongation in 2 in, percent
E 60	62 (427)	50 (345)	17–25
E 70	70 (482)	57 (393)	22
E 80	80 (551)	67 (462)	19
E 90	90 (620)	77 (531)	14–17
E 100	100 (689)	87 (600)	13–16
E 120	120 (827)	107 (737)	14

In ANSI/AWS A5.1, *Specification for Covered Carbon Steel Arc Welding Electrodes* (see Table 13.11), a simple numbering system is specified for electrode classification. The system uses a prefix E (for *electrode*), followed by four (or five) digits. The first two (or three) digits give the *minimum tensile strength* (ksi) of undiluted weld metal in as-welded condition. The next digit indicates *welding position suitability* (1: all positions, 2: flat and horizontal fillets, 3: flat only). The last digit refers to *electrode covering* and *current* type. *Carbon steel* electrodes are available in strength level series 60 and 70. *Low-alloy steel* electrodes are available in strength levels from 70 to 120 ksi in 10-ksi increments. Nominal weld metal properties for some of these electrodes are given in Table 13.13.

Design calculations for welded structures may be based on either electrode material strength or base metal strength, depending upon configuration, and design safety factors are chosen using the methods of Chapter 5. In certain applications, however, where public safety is at issue, safety factors (allowable design stresses), as well as other specifications, are prescribed by law. Designers are constrained to follow the applicable construction codes in these cases.[13]

Butt Welds

If approved welding techniques are used, and a sound full-penetration weld is achieved, the stress analysis of a *butt-welded critical section* may be conducted using the methods of Chapter 4 for analyzing stress patterns in *monolithic* machine elements. To use this approach, the weld zone is assumed to have the properties and characteristics of the parent metal in the welded plates. An appropriate stress concentration factor should be used if *fluctuating* loads are imposed on the weld zone. If loads are *static*, and materials are *ductile*, a stress concentration factor is not usually necessary (see Table 4.6). If a simple butt-welded strap, such as the one shown in Figure 13.20, is subjected to a direct tensile load P, for example, the weld-zone tensile stress may be calculated as

$$\sigma = K_f \sigma_{nom} = K_f\left(\frac{P}{tL_w}\right) \tag{13-45}$$

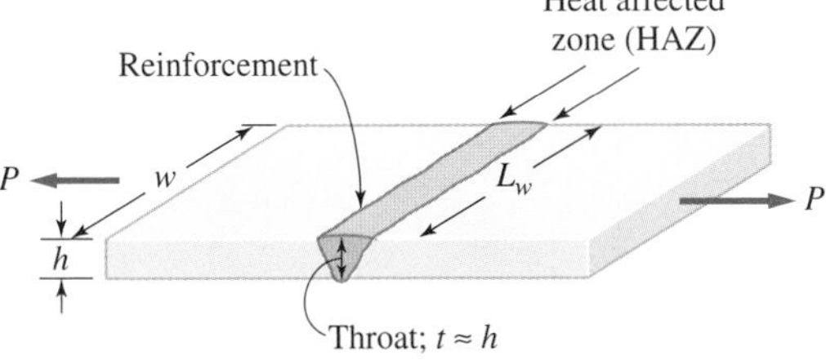

Figure 13.20
Butt-welded strap subjected to tensile load *P*. [Critical section is usually the heat affected zone (HAZ) at the edge of the weld.]

[13]For example, see American Institute of Steel Construction (AISC) *Manual of Steel Construction*, and American Society of Mechanical Engineers (ASME) *Boiler and Pressure Vessel Code*.

where K_f = fatigue stress concentration factor (see Table 13.9)
P = tensile load, lb
t = weld throat (plate thickness), in
L_w = effective weld length (usually taken as plate width, w, but a deduction may be made for imperfections at weld-run ends caused by starting and stopping the weld seam)

Typically, no load-carrying credit is given to the reinforcement, since failure usually initiates in the heat-affected zone. If the potential for fatigue failure exists, the reinforcement is often ground flat, but a fatigue stress concentration factor should still be applied because of weld-zone imperfections.

Fillet Welds

Actual stress distributions in loaded fillet welds are nonlinear and difficult to estimate accurately.[14] It has become common practice to base the required size of fillet welds upon the *average shearing stress* across the throat. As shown in Figure 13.21, the fillet weld *throat* is the height t of an equilateral triangle whose legs s are equal to the fillet weld *size*. Thus

$$t = 0.707s \tag{13-46}$$

and the shear area of a fillet weld is

$$A_s = 0.707sL_w \tag{13-47}$$

where L_w = effective weld length.

Thus for symmetrically loaded welds subjected either to cross shear or longitudinal shear, the average shearing stress on the weld throat may be calculated as

$$\tau_w = \frac{P}{A_s} = \frac{P}{0.707sL_w} \tag{13-48}$$

For fluctuating loads, an appropriate fatigue stress concentraiton factor should also be used (see Table 13.9).

Just as for the case of multiply *bolted* joints, or multiply *riveted* joints, if multiply *welded* joints are subjected to eccentric loading, as illustrated in Figure 13.22, an additional torsion-like shearing stress, τ_t, will be developed in the welds. This torsion-like shearing stress must be *vectorially* added to the direct shear, τ_w, at the weld critical points. To estimate the torsion-like shearing stress, it may be assumed that the base metal parts are *rigid*, with only the *welds* behaving as linear elastic (spring) elements.

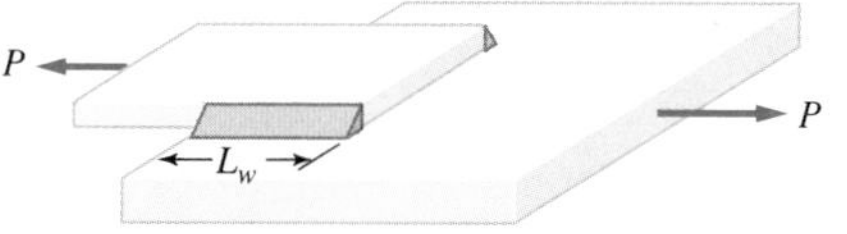

(*a*) Parallel fillet welds subjected to longitudinal shear. (Load is parallel to weld run.)

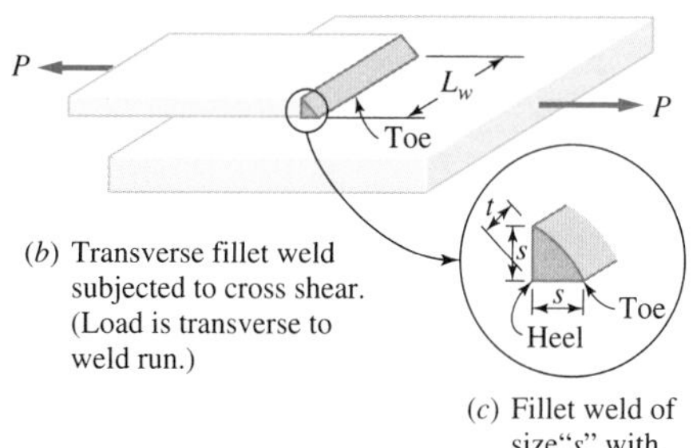

(*b*) Transverse fillet weld subjected to cross shear. (Load is transverse to weld run.)

(*c*) Fillet weld of size "*s*" with throat "*t*".

Figure 13.21
Fillet welds subjected to loads parallel and transverse to the weld run.

[14]See, for example, ref. 10.

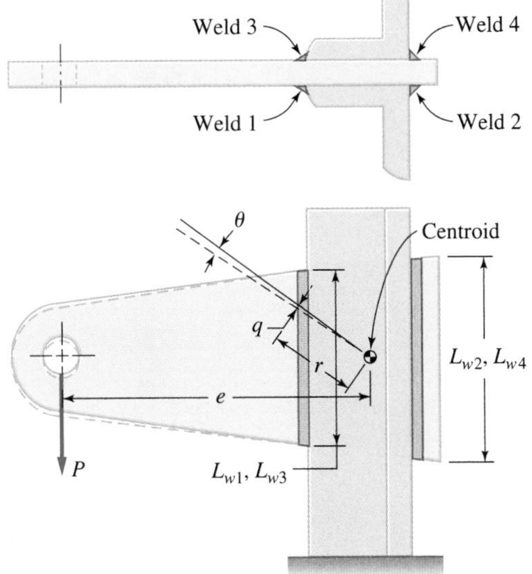

Figure 13.22
Multiply fillet-welded joint subjected to eccentric loading.

First, the centroid of the *weld group* may be found using equations of the form of (13-35) and (13-36), except that the A_i's are the weld *throat* areas for the various welds in the group, and x_i, y_i are individual weld centroid coordinates for each of the i welds.

Next, the basic principles of elasticity may be applied (see 9.5). By geometrical compatibility, the rigid body angle of rotation, θ, about the weld group centroid, must be the same for all points in the weld-zone throat areas, as illustrated in Figure 13.22. Hence for any dA in any weld throat area, the tangential displacement, q, is

$$q = r\theta \tag{13-49}$$

Considering each elemental area, dA, as a tiny spring with *spring rate k per unit area*, the differential force-displacement relationship may be written as

$$df = kq\,dA \tag{13-50}$$

where df = elemental spring force
q = tangential displacement
dA = elemental throat shear area

Based on moment equilibrium

$$T = Pe = \int_{joint} r\,df \tag{13-51}$$

Noting that the torsion-like shearing stress, τ_t may be written for the elemental area dA as

$$\tau_t = \frac{df}{dA} \tag{13-52}$$

The elemental spring force, df, may be expressed by combining (13-49) and (13-50), to give

$$df = kr\theta\,dA \tag{13-53}$$

so (13-52) may be written for any point as

$$\tau_{ti} = \frac{kr_i\theta\,dA}{dA} = kr_i\theta \tag{13-54}$$

Also, (13-51) may be rewritten as

$$T = \int_{joint} r(kr\theta\, dA) = k\theta \int_{joint} r^2\, dA \tag{13-55}$$

whence

$$k\theta = \frac{T}{\displaystyle\int_{joint} r^2\, dA} = \frac{T}{J_j} \tag{13-56}$$

and (13-56) becomes

$$\tau_{ti} = \frac{Tr_i}{J_j} \tag{13-57}$$

where $T = Pe$ = torsion-like couple on weld joint
r_i = radius from the joint *centroid* to the ith critical point of interest
J_j = weld-joint polar moment of inertia

The weld-joint polar moment of inertia may be calculated by finding the polar moment of inertia of each weld in the group about its own centroid, adding the *transfer term* to find its moment of inertia about the joint centroid, and summing for all n welds in the group. Thus

$$J_j = \sum_{i=1}^{n} [J_{oi} + A_i \hat{r}_i^2] \approx \sum_{i=1}^{n} \left[A_i \left(\frac{L_{wi}^2}{12} + \hat{r}_i^2 \right) \right] \tag{13-58}$$

where J_{oi} = polar moment of inertia of ith weld about its own centroid, $\approx \dfrac{A_i L_{wi}^2}{12}$
$\hat{r}_i$ = radius from *joint* centroid to *weld* centroid for the ith weld

Just as for bolted joints, illustrated in Figure 13.13, the torsion-like shearing stresses, τ_{ti}, must be added *vectorially* to the direct shearing stress, τ_w, at each potential critical point.

With these expressions at hand, a suggested fillet weld design procedure may be established as follows:

1. Try to configure the proposed weldment so that welds can be placed where they will be most effective. For example, symmetric placement of weld runs with respect to the weld group centroid will tend to maximize the value of J_j [see (13-58)], and therefore minimize the torsion-like shearing stress.
2. Sketch the proposed weld joint, showing the proposed location and the length of each weld run. Also, show external loads and supports. (This may require three views if three-dimensional weld-run geometry or three-dimensional loading is involved.)
3. Estimate or assume a tentative weld size, s.
4. Find the joint centroid and the joint polar moment of inertia, J_j, about the centroid of the weld group.
5. Select potential critical points. Usually these occur at the ends or edges of welds that are farthest from the joint centroid (maximum value of r_i).
6. Using (13-48) and (13-57), calculate tentative values for τ_w and τ_{ti}, and add them vectorially to obtain a tentative *resultant* shearing stress, τ_{si}, at each potential critical point. If loads are *fluctuating*, appropriate fatigue stress concentration factors should be employed.
7. Compare the largest *resultant* shearing stress, τ_{s-max}, with the predefined design-allowable shearing stress, τ_d. If $\tau_{s-max} < \tau_d$, decrease the weld size and repeat the pro-

cedure. If $\tau_{s-max} > \tau_d$, increase the weld size, and repeat the procedure. Continue to iterate until $\tau_{s-max} \approx \tau_d$.

Example 13.4 Determining Fillet Weld Sizes

It is desired to weld a 1020 steel side plate to a 1020 steel column in accordance with the specifications of Figure E13.4A. The side plate is to support a 6000-lb static vertical load applied at a horizontal distance of 5 inches away from the edge of the column. Determine the size of the fillet weld that should be used if all welds are to be the same size and a design safety factor of 2 has been selected.

Solution

From Figure E13.4A, it may be noted that eccentric load P produces both *direct shear* and *torsion-like shear* in the two fillet welds specified. Following the suggested weld design procedure, the following results are obtained:

1. Weld placement is predetermined by problem statement.
2. Sketch of weld joint is predetermined by problem statement.
3. A tentative weld size must be assumed or estimated. A crude estimate may be made, for example, by *arbitrarily* basing the weld size on twice the *direct* shearing stress. Since the load is static, the probable failure mode is yielding. From Table 13.13, the yield strength for an E6012 electrode is

$$S_{yp} = 50{,}000 \text{ psi} \tag{1}$$

Using the specified safety factor of 2 and the distortion energy theory,

$$\tau_d = \frac{\tau_{yp}}{2} = \frac{0.577 S_{yp}}{2} = 0.29(50{,}000) = 14{,}500 \text{ psi} \tag{2}$$

Using (13-48) we may, by assumption, set

$$\tau_d = 2\tau_w = \frac{2P}{0.707 s L_w} \tag{3}$$

Solving for weld size s,

$$s = \frac{2(6000)}{0.707(6+6)14{,}500} = 0.1 \text{ inch} \tag{4}$$

From the result of (4), a $^1/_8$-inch fillet weld size will be assumed for the first iteration.

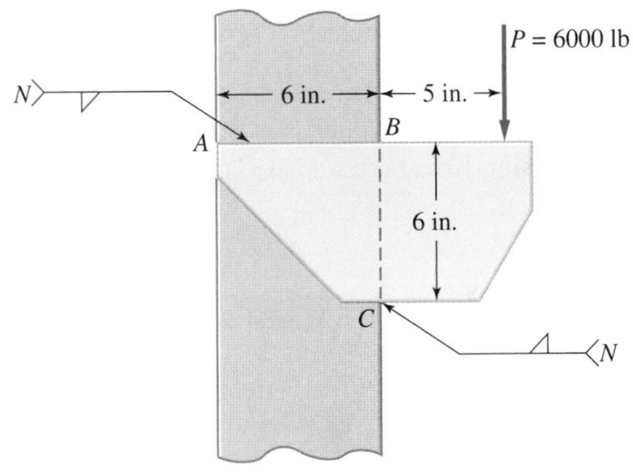

Figure E13.4A
Sketch of proposed side plate installation.

Example 13.4 Continues

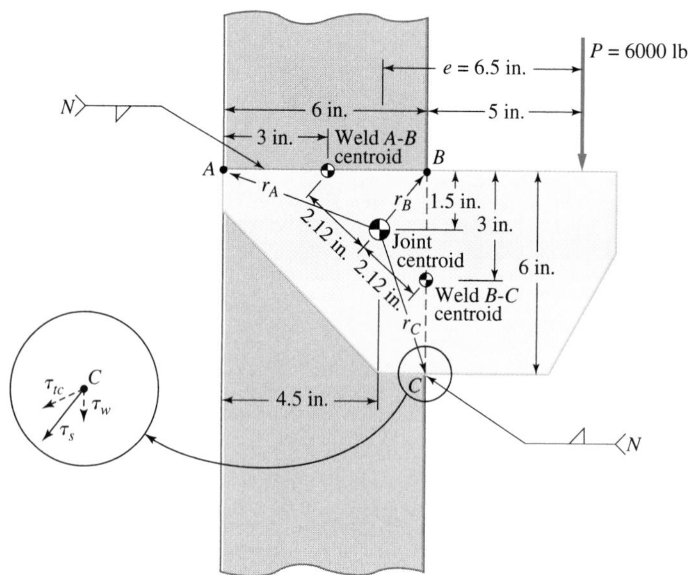

Figure E13.4B
Centroidal position and critical point locations.

4. From Figure E13.4B, the weld group centroidal coordinates relative to A are

$$\bar{x} = \frac{(6s)3 + (6s)6}{12s} = \frac{(18 + 36)s}{12s} = 4.5 \text{ in} \quad \text{(right)} \tag{5}$$

and

$$\bar{y} = \frac{(6s)0 + (6s)3}{12s} = \frac{(0 + 18)s}{12s} = 1.5 \text{ in} \quad \text{(down)} \tag{6}$$

Using (13-58), the polar moment of inertia about the *joint* centroid is, for a $^1\!/_8$-inch weld,

$$J_j = 6\left(\frac{1}{8}\right)\left(\frac{6^2}{12} + 2.12^2\right) + 6\left(\frac{1}{8}\right)\left(\frac{6^2}{12} + 2.12^2\right) = 2(5.62) = 11.24 \text{ in}^4 \tag{7}$$

5. Checking potential critical points, it may be noted that

$$r_A = r_C = 4.74 \text{ inches} > r_B = 2.12 \text{ inches} \tag{8}$$

which eliminates B from further consideration.

Because $T = Pe$ is clockwise, and P is vertically down, the direct shear at A *tends to oppose* the torsion-like shear, and at C they *tend to add*. Hence, C is the critical point to be examined.

6. From (13-48)

$$\tau_W = \frac{6000}{0.707\left(\frac{1}{8}\right)6} = 11{,}315 \text{ psi} \quad \text{(vertically down)} \tag{9}$$

and from (13-57)

$$\tau_{tC} = \frac{[6000(6.5)]4.74}{11.24} = 16{,}447 \text{ psi} \quad \text{(perpendicular to } r_C \text{ and CW)} \tag{10}$$

Adding (9) and (10) vectorially, as shown graphically in the inset of Figure E13-4B, the vertical and horizontal components of τ_{tc} are

$$\tau_{tCV} = 16{,}447\left(\frac{1.5}{4.74}\right) = 5205 \text{ psi} \tag{11}$$

and

$$\tau_{tCH} = 16{,}447\left(\frac{4.5}{4.74}\right) = 15{,}615 \text{ psi} \tag{12}$$

giving vertical and horizontal components of resultant shearing stress τ_s as

$$\tau_{sV} = 5205 + 11{,}315 = 16{,}520 \text{ psi} \tag{13}$$

and

$$\tau_{sH} = 15{,}615 + 0 = 15{,}615 \text{ psi} \tag{14}$$

hence

$$(\tau_s)_{max} = \sqrt{(16{,}520)^2 + (15{,}615)^2} = 22{,}730 \text{ psi} \tag{15}$$

7. Comparing (15) with (2),

$$(\tau_s)_{max} = 22{,}730 > \tau_d = 14{,}500 \tag{16}$$

Therefore the assumed weld size of $^1/_8$ inch is too small.

For the next iteration a $^1/_4$-inch weld will be tried.

Since (5) and (6) are not functions of weld size, the position of the *weld group centroid* does not change. Therefore C remains the critical point and r_C is unchanged. Further, (7) is a linear function of weld size; hence for the $^1/_4$-inch weld

$$J_j = \frac{(1/4)}{(1/8)}(11.24) = 22.48 \text{ in}^4 \tag{17}$$

Also, (9) and (10) become

$$\tau_W = \frac{6000}{0.707(1/4)6} = 5660 \text{ psi} \quad \text{(vertically down)} \tag{18}$$

and

$$\tau_{tC} = \frac{[(6000)(6.5)]4.74}{22.48} = 8225 \text{ psi} \quad \text{(perpendicular to } r_C \text{ and } CW) \tag{19}$$

Adding (18) and (19) vectorially (see Figure E13.4B), the vertical and horizontal components of τ_{tC} now become

$$\tau_{tCV} = 8225\left(\frac{1.5}{4.74}\right) = 2603 \text{ psi} \tag{20}$$

and

$$\tau_{tCH} = 8225\left(\frac{4.5}{4.74}\right) = 7808 \text{ psi} \tag{21}$$

so vertical and horizontal components of τ_s become

$$\tau_{sV} = 2603 + 5660 = 8263 \text{ psi} \tag{22}$$

$$\tau_{sH} = 7808 + 0 = 7808 \text{ psi} \tag{23}$$

and

$$(\tau_s)_{max} = \sqrt{(8263)^2 + (7808)^2} = 11{,}370 \text{ psi} \tag{24}$$

Example 13.4 Continues

comparing (24) with (2),

$$(\tau_s)_{max} = 11{,}370 < \tau_d = 14{,}500 \tag{25}$$

This is acceptable for static loading. *A $^1/_4$-inch fillet weld is recommended for this application.* (It is worth noting that if the loading were cyclic, a fatigue stress concentration factor would be required and a larger weld size would likely be necessary.)

13.7 Adhesive Bonding

Adhesive bonding of structural parts has become more attractive as a design option with the development of improved adhesive formulations. Designers have found that, in some cases, structural adhesives produce reliable joints at lower total cost than other fastening methods. An adhesively bonded joint is actually a *composite* structure in which the two elements being joined (the *adherends*) are held together by surface attachment provided by the *adhesive*. Thin adhesive layers (ideally 0.005 inch or less) and intimate adherend surface contact are essential for strong bonds, and joint configurations that load the adhesive bonding material in *shear* have the highest probability of success.

Advantages of adhesive bonding over other joining techniques, such as bolting, riveting, or welding, include:

1. Loads transferred across the joint are distributed over the entire area of the joint's faying surfaces (rather than the smaller discrete shear areas of bolts, rivets or welds), allowing the successful use of lower specific strength adhesive agents.
2. Stress concentrations are minimized because there are no holes, as with bolts or rivets, and no warping, residual stresses, or local discontinuities, as with welding. In some cases, this may result in thinner sections and lighter weight.
3. Smoother unbroken surfaces (no protruding fasteners) may provide better appearance, easier finishing, and/or more streamlined fluid flow characteristics.
4. Adhesive bonding is well suited to joining *dissimilar* materials, and may be used to electrically insulate them from each other (to avoid galvanic corrosion), or seal them against air or liquid leakage.
5. The viscoelastic properties and flexibility of the adhesive layer may accommodate differential thermal expansion between the adherends, provide hysteretic damping to attenuate vibration and sound transmission, reduce impact loading, or contribute to improved fatigue resistance of the joint.

Joint Design

Joint type (see Figure 13.1) must be carefully considered if adhesive bonding is anticipated, since bonded joints are most effective when the adhesive layer is loaded in shear. *Tensile* loading of the adhesive layer should generally be *avoided* in structural applications, especially if nonuniform loading distributions tend to produce *cleavage* or *peeling* [see Figure 13.23(c)]. Typically, lap joints are preferred; butt joints are not usually practical. It is important to note, however, that *single* lap joints, as shown in Figure 13.24(a), develop *bending moments* because applied loads are not colinear. *Double* lap joints may be required in some cases to avoid potential cleavage or peeling. A hybrid version, called a *scarf* joint, is sometimes used, but is expensive to machine and is impractical for thin sections. A scarf joint is illustrated in Figure 13.24.

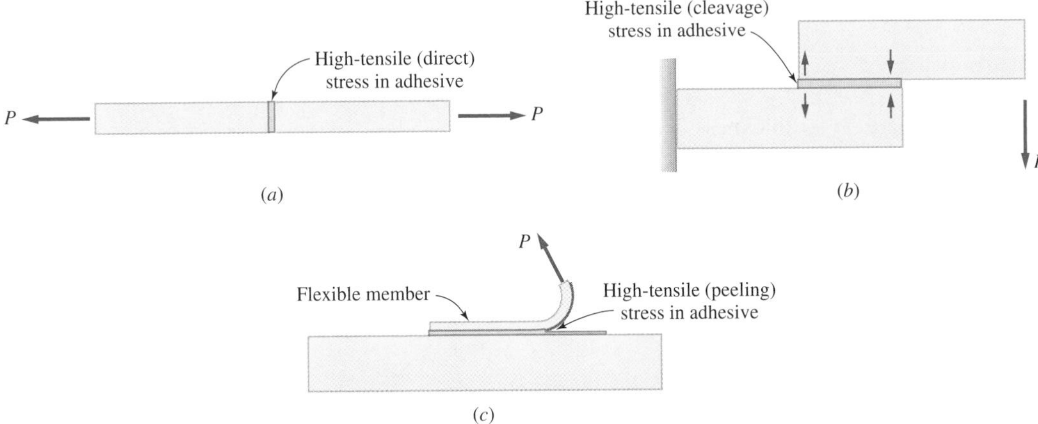

Figure 13.23
Illustration of potential failures of the adhesive layer due to (a) direct tension, (b) cleavage, and (c) peeling. *Such designs are to be avoided.*

It has been shown[15] that the shearing stress distribution in the adhesive layer of a lap joint loaded as shown in Figure 13.24(a) is highly nonlinear, reaching maximum values at the adhesive boundaries that may be more than *twice* the *average* shearing stress across the joint. For preliminary design purposes, the *average* shearing stress in the adhesive layer may be calculated, then multiplied by a *stress distribution factor,*[16] K_s. Likewise, a stress distribution factor may be used for scarf joints [see Figure 13.24(b)]. Specific values for K_s are not readily available.

The maximum shearing stress, τ_{max}, in the adhesive layer of a lap joint, such as the one shown in Figure 13.24(a), may be calculated as

$$\tau_{max} = K_s \tau_{ave} = K_s\left(\frac{P}{bL_L}\right) \tag{13-59}$$

where K_s = stress distribution factor (use K_s = 2 in the absence of specific data)
P = applied load
b = width of adhesive layer
L_L = length of adhesive layer

The maximum shearing stress, τ_{max}, in the adhesive layer of a scarf joint, such as the one shown in Figure 13.2(b), may be calculated as

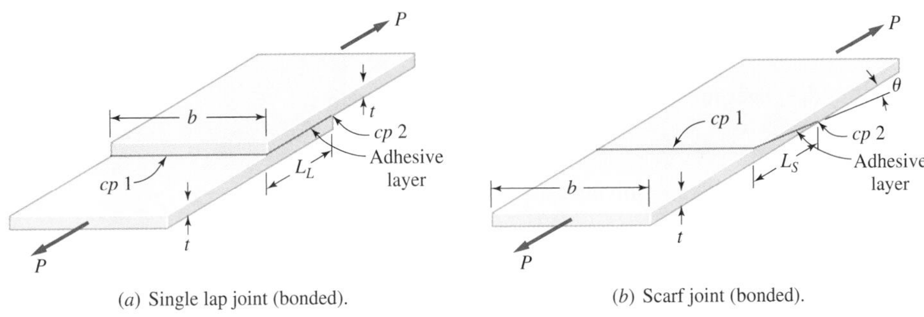

Figure 13.24
Adhesively bonded (a) lap joint and (b) scarf joint.

[15]See, for example, ref. 11.

[16]Similar to a stress concentration factor but with smaller stress gradient.

$$\tau_{max} = K_s\tau_{ave} = K_s\left[\frac{P\cos\theta}{\left(\dfrac{bt}{\sin\theta}\right)}\right] = K_s\left(\frac{P}{bt}\sin\theta\cos\theta\right) \tag{13-60}$$

where t = thickness of adherend members

θ = scarf angle $= \tan^{-1}\dfrac{t}{L_s}$

Typically, the design objective for adhesively bonded joints is to select an adhesive material and joint dimensions so that the load-carrying capability of the adhesive layer is equal to the load-carrying ability of the members being joined.

Structural Adhesive Materials

Selection of an appropriate structural adhesive material depends upon many factors, including strength requirements, type of joint, operational environment (temperature, humidity, solvents), substrates to be bonded (porosity, coatings, contaminants), available manufacturing facilities, and cost factors. Table 13.14 gives a brief summary of properties and characteristics of several commonly used structural adhesive materials.

Example 13.5 Adhesive Bonding

It is desired to investigate adhesive bonding as a means of joining two Ti-6A1-4V titanium plates having thicknesses of 0.060 inch and widths of 4 inches.

a. If a lap joint of the type shown in Figure 13.24(a) were used, and an epoxy adhesive selected, what would be the required length of joint overlap?

b. If a scarf joint of the type shown in Figure 13.24(b) were used, and an epoxy adhesive selected, what would be the required length L_s of the scarf joint?

c. Which type of joint would you recommend?

Solution

a. The basic design criterion is to achieve a *joint strength equal to the plate strength*. That is, the tensile load-carrying capability of the titanium plates should equal the ability of the adhesive to carry the load in shear. Thus

$$P_{t-pl} = P_{s-adh} \tag{1}$$

Using (13-59) for the adhesive layer in shear, at incipient failure,

$$P_{s-adh} = \frac{bL_L\tau_{max-f}}{K_s} \tag{2}$$

For the plates in tension, at incipient failure by yielding,

$$P_{t-pl} = S_{yp}bt \tag{3}$$

Combining these and solving for L_L,

$$L_L = K_s t\left(\frac{S_{yp}}{\tau_{max-f}}\right) \tag{4}$$

From Table 3.3, S_{yp} = 128,000 psi for T_i-6AI-4V titanium alloy, and from Table 13.14, τ_{max-f} = 2200 psi for a typical epoxy adhesive. Using K_s = 2 and a plate thickness of t = 0.060 inch, (4) gives

Table 13.14 Properties and Characteristics of Selected Structural Adhesive Materials

	Epoxies	Urethanes	Cyanoacrylates	Acrylics	Anaerobics	Hot Melts	Silicones
Typical RT[1] shear strength, psi	2200	2200	2700	3700	2500	500	200–250
Max. continuous operating temp., °F (°C)	390 (200)	210 (100)	175 (80)	210 (100)	390 (200)	390 (200)	500 (260)
Impact resistance	Poor	Excellent	Poor	Good	Fair	Fair	Excellent
Surfaces bonded	Metals, glass, plastics, ceramics	Metals, glass, plastics, ceramics, rubber	Metals, plastics, ceramics, rubber, wood	Metals, glass, ceramics, plastics, wood	Metals, glass, ceramics, some plastics	All	All
Gap limitations, in	None	None	0.010	0.030	0.025	None	0.25
Substrate preparation required	Yes—careful cleaning required	Varies with substrate	Yes	Some can bond oily surfaces	Yes	Yes	Yes—careful cleaning required
Oily substrate bonding	Fair–good	Fair	Poor–fair	Good	Fair	Poor	Poor
Moisture resistance	Excellent	Fair	Poor	Good	Good	Fair–good	Good
Solvent resistance	Excellent	Good	Good	Good	Excellent	Fair–good	Fair
Typical cure (hr. at °F)	2–24 @ 70 0.1–4 @ 300	4–12 @ 70 0.2–0.5 @ 150	0.5–5 @ 70	0.5–1 @ 70 (for handling)	1–12 @ 70 0.05–2 @ 250	seconds	24 @ 77
Fastest RT[1] cure to handleable state	5 min	5 min	< 10 sec	2 min	5 min	seconds	2 hr
Fastest RT[1] full cure	< 24 hr	< 24 hr	< 2 hr	< 12 hr	< 12 hr	—	2–5 days
Odor/toxicity	Mild/moderate	Mild/moderate	Irritating/low	Strong/low–moderate	Mild/low	Mild/low	Mild–irritating/low
Shelf life	6 mo–1 yr	6 mo–1 yr	1 yr	6 mo–1 yr	> 1 yr	> 1 yr	6 mo–1 yr
Remarks	Brittle; good for dissimilar materials; low shrinkage	Excellent for low temps and dissimilar materials	Brittle; low shrinkage; thin films best	Good for dissimilar materials	Thin glue lines best	Flexible; softens at high temps	Excellent for extreme temp's and dissimilar materials; provides sealing

[1]Room temperature.

Example 13.5 Continues

$$L_L = 2(0.060)\left(\frac{128{,}000}{2200}\right) = 6.98 \text{ inches} \tag{5}$$

b. Using the same design criterion for the scarf joint and utilizing (13-60),

$$P_{s\text{-}adh} = \frac{bt\tau_{max-f}}{K_s \sin\theta\cos\theta} \tag{6}$$

where

$$\theta = \tan^{-1}\frac{t}{L_s} \tag{7}$$

Combining (1), (3), and (6) and solving for scarf angle θ,

$$\theta = \frac{1}{2}\sin^{-1}\left[\frac{2}{K_s}\left(\frac{\tau_{max-f}}{S_{yp}}\right)\right] \tag{8}$$

or

$$\theta = \frac{1}{2}\sin^{-1}\left[\frac{2}{2}\left(\frac{2200}{128{,}000}\right)\right] = 0.487^\circ \tag{9}$$

Since

$$\theta = \tan^{-1}\frac{t}{L_s} = 0.487^\circ \tag{10}$$

the length of the scarf joint, L_s, may be found as

$$L_s = \frac{t}{\tan 0.487} = \frac{0.060}{\tan 0.487} = 7.06 \text{ inches} \tag{11}$$

c. The required length of the scarf joint and the lap joint are both approximately 7 inches. However, machining a 7-inch scarf in 0.060-inch-thick titanium is not very practical. A *7-inch lap joint* is recommended for this application.

Problems

13-1. You have been assigned the task of examining a number of large flood gates installed in 1931 for irrigation control at a remote site on the Indus River in Pakistan. Several large steel bolts appear to have developed cracks, and you have decided that they should be replaced to avert a potentially serious failure of one or more of the flood gates. Your Pakistani assistant has examined flood gate specifications, and has found that the original bolts may be well characterized as 32-mm medium-carbon quenched and tempered steel bolts, of property class 8.8. You have brought with you only a limited number of replacement bolts in this size range, some of which are SAE Grade 7, and others that are ASTM Class A325, type 3. Which, if either, of these replacement bolts would you recommend as a substitute for the cracked originals? Justify your recommendation.

13-2. A high-speed "closing machine" is used in a tomato canning factory to install lids and seal the cans. It is in the middle of the "pack" season and a special bracket has separated from the main frame of the closing machine because the $\frac{3}{8}$-24 UNF-2A hex-head cap screws used to hold the bracket in place have failed. The head markings on the failed cap screws consist of the letters BC in the center of the head. No cap screws with this head marking can be found in the storeroom. The $\frac{3}{8}$-24 UNF-2A cap screws that can be found in the "high-strength" bin have five equally spaced radial lines on the heads. Because it is so important to get up-and-running immediately to avoid spoilage, you are being asked, as an engineering consultant, whether the available cap screws with head markings of five radial lines can be safely substituted for the broken originals. How do you respond? Justify your recommendation.

13-3. A cylindrical flange joint requires a total clamping force between the two mating flanges of 9000 lb. It is desired to use six equally spaced cap screws around the flange. The cap screws pass through clearance holes in the top flange and thread into tapped holes in the bottom flange.

a. Select a set of suitable cap screws for this applicaiton.

b. Recommend a suitable tightening torque for the cap screws.

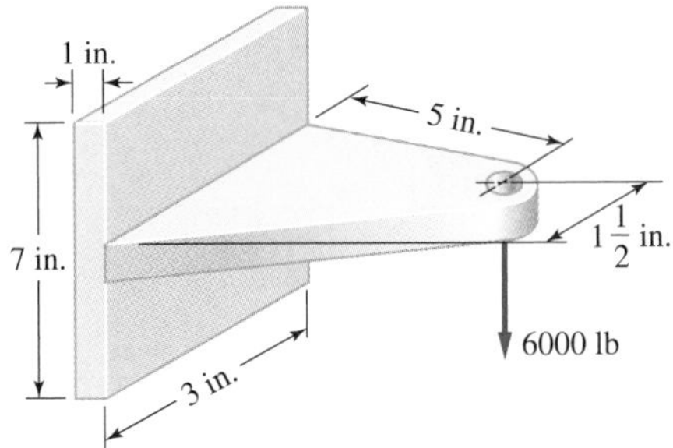

Figure P13.4
Steel support bracket to be bolted to a stiff steel column.

13-4. It is desired to use a set of four bolts to attach the bracket shown in Figure P13.4 to a stiff steel column. For purposes of economy, all bolts are to be the same size. It is desired to use ASTM Class A307 low-carbon steel material and standard UNC threads. A design safety factor of 2.5 has been selected, based on yielding as the governing failure mode.

a. What bolt-hole pattern would you suggest and what bolt specification would you recommend?

b. What tightening torque would you recommend if it is desired to produce a preload force in each bolt equal to 85 percent of the minimum proof strength?

13-5. Estimate the nominal size of the smallest SAE Grade 1 standard UNC bolt that will not yield under a tightening torque of 1000 in-lb. Neglect stress concentration.

13-6. A standard fine-thread metric machine screw made of steel has a major diameter of 8.0 mm and a head marking of 9.8. Determine the tensile proof force (kN) for this screw. It may be assumed that the coefficient of friction is about 0.15 for both the threads and the collar.

13-7. A standard coarse-thread metric cap screw made of steel has a major diameter of 10.0 mm. If a torque wrench is used to tighten the cap screw to a torque of 35 N-m, estimate the axial preload force induced in the cap screw. It may be assumed that the coefficient of friction is about 0.15 for both threads and collar.

13-8. Engineering specifications for a machine tool bracket application call for a nonlubricated M30 × 2 threaded fastener of property class 8.8 to be tightened to 100 percent of proof load. Calculate the torque required to accomplish this. It may be assumed that the coefficient of friction is about 0.15 for both the threads and the collar.

13-9. A $^3/_4$-16 SAE Grade 2 steel bolt is to be used to clamp two 1.00-inch-thick steel flanges together with a $^1/_{16}$-inch-thick special lead-alloy gasket between the flanges, as shown in Figure P13.9. The effective load-carrying area of the steel flanges and of the gasket may be taken as 0.75. sq. in. Young's modulus for the gasket is 5.3×10^6 psi. If the bolt is initially tightened to induce an axial preload force in the bolt of 6000 lb, and if an external force of 8000 lb is then applied as shown,

a. What is the force on the bolt?

b. What is the force on each of the steel flanges?

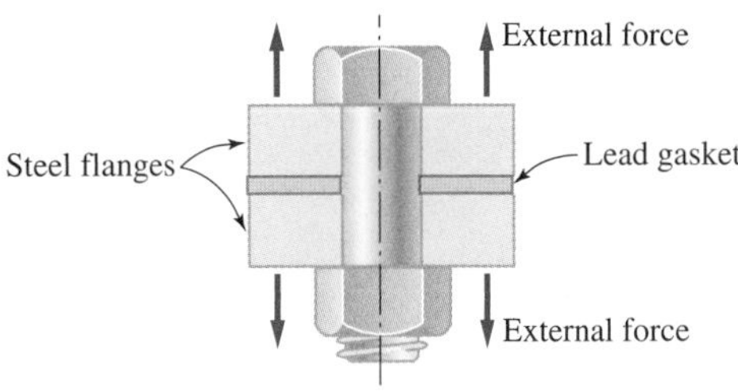

Figure P13.9
Bolted flange joint with interposed gasket.

c. What is the force on the gasket?

d. If the stress concentration factor for the bolt thread root is 3.0, would local yielding at the thread root be expected?

13-10. A special reduced-body bolt is to be used to clamp two $^3/_4$-inch-thick steel flanges together with a $^1/_8$-inch-thick copper-asbestos gasket between the flanges in an arrangement similar to the one shown in Figure P13.9. The effective area for both the steel flanges and the copper-asbestos gasket may be taken as 0.75 square inch. Young's modulus of elasticity for the copper-asbestos gasket is 13.5×10^6 psi. The special bolt has $^3/_4$-16 UNF threads but the body of the bolt is reduced to 0.4375 inch in diameter and generously filleted, so stress concentration may be neglected. The bolt material is AISI 4620 cold-drawn steel.

a. Sketch the joint, showing the reduced-body bolt, and the loading.

b. If the bolt is tightened to produce a preload in the joint of 5000 lb, what external force P_{sep} could be applied to the assembly before the joint would start to separate?

c. If the external load P fluctuates from 0 to 5555 lb at 3600 cycles per minute, and the desired design life is 7 years of continuous operation, would you predict failure of the bolt by fatigue?

13-11. A typical bolted joint of the type shown in Figure 13.9 uses a $^1/_2$-13 UNC bolt, and the length of the bolt and length of the housing is the same. The threads stop immediately above the nut. The bolt is steel with $S_u = 101{,}000$ psi, $S_{yp} = 85{,}000$ psi, and $S_f = 50{,}000$ psi. The thread stress concentration factor is 3. The effective area of the steel housing is 0.88 in^2. The load fluctuates cyclically from 0 to 2500 lb at 2000 cpm.

a. Find the existing factor of safety for the bolt if no preload is present.

b. Find the minimum required value of preload to prevent loss of compression in the housing.

c. Find the existing factor of safety for the bolt if the preload in the bolt is 3000 lb.

13-12. A $^1/_2$-20 UNF-2A SAE Grade 2 steel cap screw is being considered for use in attaching a cylinder head to an engine block made of 356.0 cast aluminum (see Table 3.3). It is being proposed to engage the cap screw into an internally threaded hole tapped directly into the aluminum block. Estimate the re-

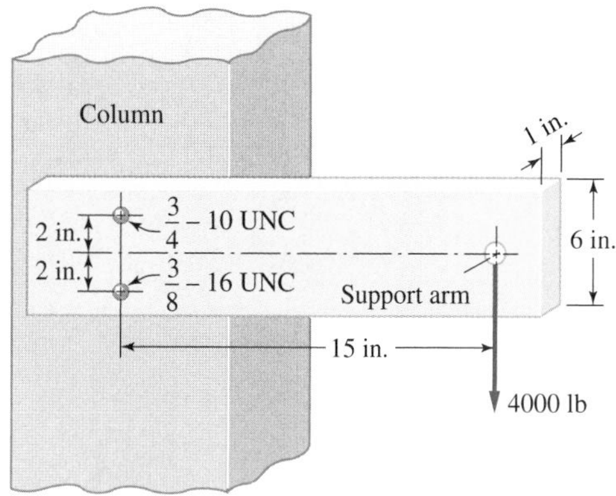

Figure P13.13
Support arm bolted to a rigid column.

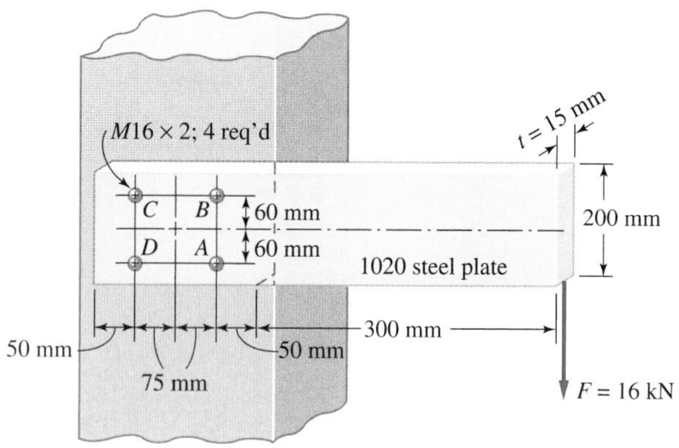

Figure P13.15
Cantilevered support plate bolted to a steel column.

quired length of thread engagement that will ensure tensile failure of the cap screw before the threads are stripped in the aluminum block. Assume that all engaged threads participate equally in carrying the load. Base your estimate on direct shear of the aluminum threads at the major thread diameter, and use the distortion energy theory of failure to estimate the shear yield strength for the aluminum block.

13-13. A support arm is to be attached to a rigid column using two bolts located as shown in Figure P13.13. The bolt at A is to have a $^3/_4$-10 UNC thread specification and the bolt at B is to have a $^3/_8$-13 UNC specification. It is desired to use the same material for both bolts, and the probable governing failure mode is yielding. If a design safety factor of 1.8 has been selected, what minimum tensile yield strength is required of the bolt material? No significant preload is induced in the bolts as a result of the tightening process, and it may be assumed that friction between the arm and the column does not contribute to supporting the 4000-lb load.

13-14. A steel side plate is to be bolted to a vertical steel column as shown in Figure P13.14, using $^3/_4$-10 UNC SAE Grade 8 steel bolts.

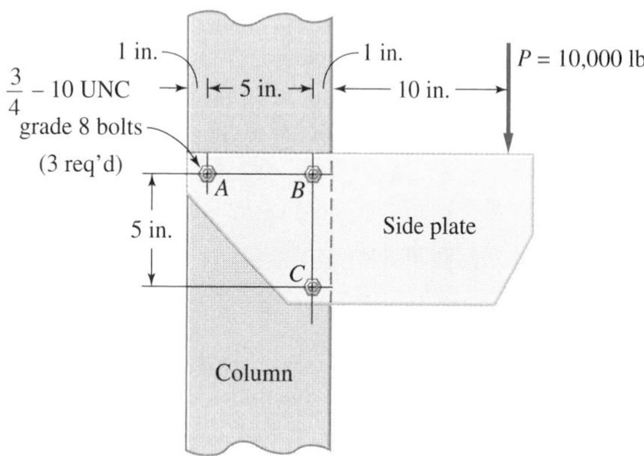

Figure P13.14
Steel side plate bolted to a steel column.

a. Determine and clearly indicate the magnitude and direction of the direct shearing stress for the most critically loaded bolt.

b. Determine and clearly indicate the magnitude and direction of the torsion-like shearing stress for the most critically loaded bolt.

c. Determine the existing safety factor on yielding for the most critically loaded bolt, assuming that no significant preload has been induced in the bolt due to tightening.

13-15. A 1020 hot-rolled steel cantilevered support plate is to be bolted to a stiff steel column using four M16 × 2 bolts of Property Class 4.6, positioned as shown in Figure P13.15. For the 16-kN static load and the dimensions given, and assuming that none of the load is supported by friction, do the following:

a. Find the resultant shear force on each bolt.

b. Find the magnitude of the maximum bolt shearing stress, and its location.

c. Find the maximum bearing stress and its location.

d. Find the maximum bending stress in the cantilevered support plate, and identify where it occurs. Neglect stress concentration.

e. Determine whether yielding would be expected at any point within the installed support plate under the 16-kN applied load.

13-16. For the eccentrically loaded riveted joint shown in Figure P13.16, do the following:

a. Vertify the location of the centroid for the joint.

b. Find the location and magnitude of the force carried by the most heavily loaded rivet. Assume that the force taken by each rivet depends linearly on its distance from the joint centroid.

c. Find the maximum rivet shearing stress if $^3/_4$-inch rivets are used.

d. Find the location and magnitude of the maximum bearing stress if $^5/_{16}$-inch thick plate is used.

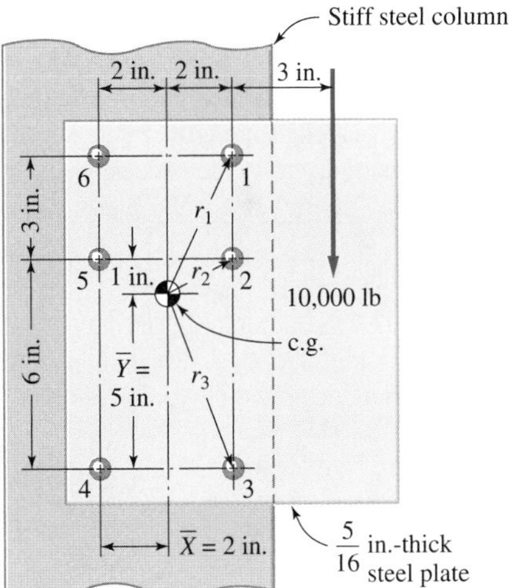

Figure P13.16
Eccentrically loaded riveted joint.

13-17. For a bracket riveted to a large steel girder, as sketched in Figure P13.17, perform a complete stress analysis of the riveted joint. If S_{yp} for the plate is 40,000 psi, and S_{yp} for the rivet is 50,000 psi, what are the existing safety factors on yielding for each potential type of failure for the riveted joint? Assume the rivet centerline is 1.5 times the rivet diameter away from the edge of the plate and protruding head rivets are used. The plate is $^1/_4$ inch thick, and the girder is much thicker.

13-18. A simple butt-welded strap, similar to the one shown in Figure 13.20, is limited by surrounding structure to a width of 4 inches. The material of the strap is annealed AISI 1020 steel (see Table 3.3), and an E 6012 welding electrode has been recommended for this application. The applied load P fluctuates from a minimum of 0 to a maximum of 25,000 lb and back, continuously.

a. If a safety factor of 2.25 has been selected, k_∞ is approximately 0.8 [see (2-27)], and infinite life is desired, what thickness should be specified for the butt-welded strap?

b. If any fatigue failures *do* occur when these welded straps are placed in service, at what location would you expect to see the fatigue cracks initiating?

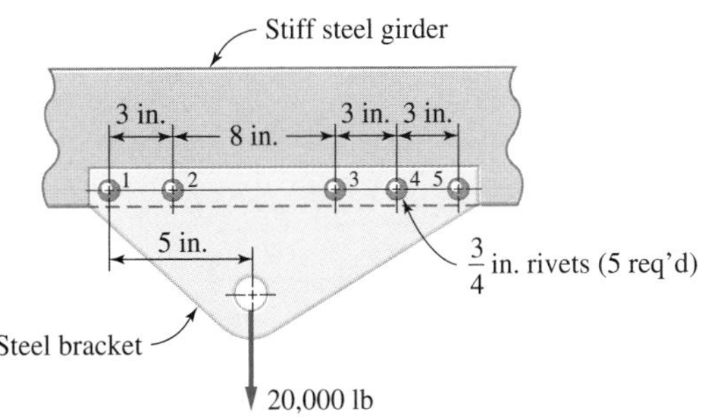

Figure P13.17
Steel bracket riveted to stiff steel girder.

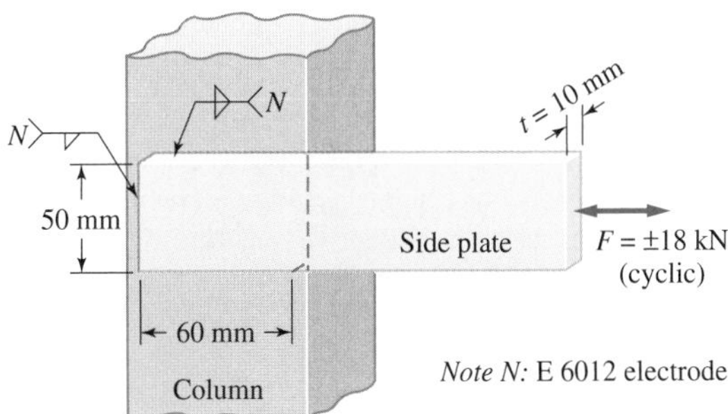

Figure P13.19
Steel side plate welded to a stiff steel column.

13-19. A horizontal side plate made of 1020 steel (see Figure 2.19) is to be welded to a stiff steel column using an E 6012 electrode, as specified in Figure P13.19. If the horizontally applied load F fluctuates cyclically from +18 kN (tension) to −18 kN (compression) each cycle, k_∞ is approximately 0.75 [see (2-27)], and a design safety factor of 2.5 is desired, what fillet weld size would your recommend if all fillet welds are to be the same size? Infinite life is desired.

13-20. A steel side plate is to be welded to a vertical steel column according to the specifications in Figure P13.20. Neglecting stress concentration effects, calculate the magnitude and clearly indicate the direction of the resultant shearing stress at the critical point. In selecting the critical point be sure to consider effects of both torsion-like shear and direct shear.

13-21. A proposed double lap-joint [see Figure 13.1(f)] is to be symmetrically loaded in tension, parallel to the plane of the straps to be joined. Adhesive bonding is being considered as a means of joining the straps. The single center strap is titanium and the double outer straps are medium-carbon steel. This aerospace application involves continuous operation at a temperature of about 350°F, moderate impact loading, and occasional exposure to moisture. What types of structural adhesives would

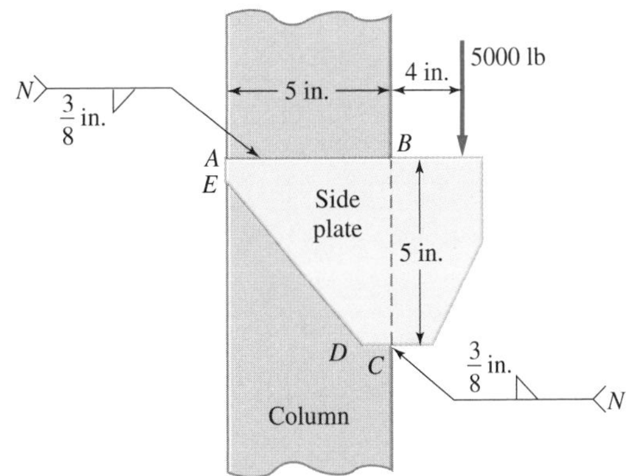

Figure P13.20
Side-plate welded to a vertical steel column.

you recommend as good candidates for bonding this double lap joint?

13-22. In an adhesively bonded lap joint (see Figure 13.24) made of two sheets of metal, each having thickness t, it has been found from an experimental testing program that the maximum shearing stress in the adhesive may be estimated as

$$(\tau_{max})_{adh} = \frac{2P}{bL_L}$$

where P = total in-plane tensile force on the joint (perpendicular to b)

b = sheet width

L_L = overlap dimension of joint.

If the adhesive material has a shear failure strength of τ_{sf}, and the metal sheets have a yield strength of S_{yp}, derive an equation for overlap dimension L_L for which adhesive shear failure and metal sheet yielding failure are equally likely.

13-23. It is being proposed to use a lap joint configuration (see Figure 13.24) to adhesively bond two 20-gage strips (0.036 inch thick) of 2024-T3 aluminum (see Table 3.3) using an epoxy adhesive. Assuming a stress distribution factor of $K_s = 2$, what bonded overlap length would you recommend?

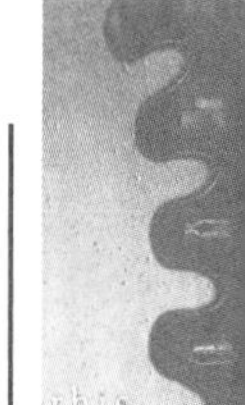

Chapter 14

Springs

14.1 Uses and Characteristics of Springs

Springs may be broadly defined as structures or devices that exhibit elastic deformation when loaded, and recover their initial configuration when the load is removed. In most applications *linear* spring rates are desired (see 2.4). Because every *real* material has a *finite* modulus of elasticity, *machine elements of all types* necessarily behave as "springs." When analyzing *load sharing* in redundant structures or in preloaded systems,[1] or when investigating *vibrational response* to operating loads and frequencies,[2] the spring properties of machine elements must be considered.

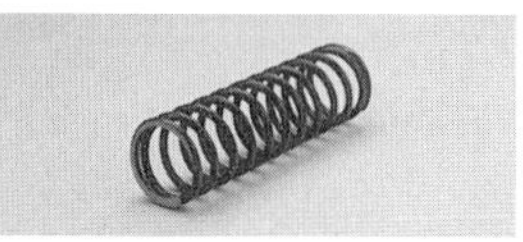

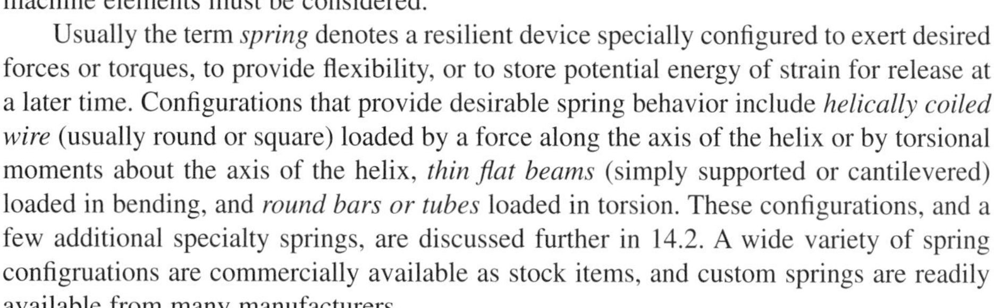

Usually the term *spring* denotes a resilient device specially configured to exert desired forces or torques, to provide flexibility, or to store potential energy of strain for release at a later time. Configurations that provide desirable spring behavior include *helically coiled wire* (usually round or square) loaded by a force along the axis of the helix or by torsional moments about the axis of the helix, *thin flat beams* (simply supported or cantilevered) loaded in bending, and *round bars or tubes* loaded in torsion. These configurations, and a few additional specialty springs, are discussed further in 14.2. A wide variety of spring configruations are commercially available as stock items, and custom springs are readily available from many manufacturers.

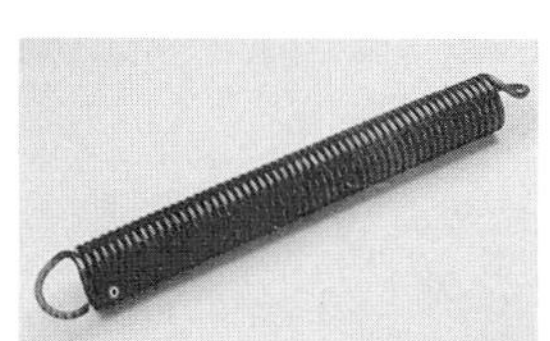

14.2 Types of Springs

Helical-coil springs are probably more widely used than any other type. As illustrated in Figure 14.1, helical-coil springs may be used to support compressive loads (pushing), tensile loads (pulling), or torsional moments (twisting). In addition to the standard helical-coil compression spring shown in Figure 14.1(a), several nonlinear configurations designed to solve special problems are shown in Figure 14.1(b) through (e). A typical helical-coil extension spring is illustrated in Figure 14.1(f), and a helical torsion spring is shown in Figure 14.1(h).

Beam springs (leaf springs) of various types are illustrated in Figure 14.2. Leaf springs may be either single or multileaf cantilever beams subjected to transverse end-loads, as shown in Figures 14.2(a) and (b), or single or multileaf simply supported beams subjected to center loads, as shown in Figures 14.2(c) and (d). Multileaf springs are usually proportioned to approximate *constant-strength beams* (see 14.7).

Some examples of springs.

Torsion bar springs, as illustrated in Figure 14.3, may be solid or hollow bars with cir-

[1]See 4.11 and 4.12. [2]See 8.6 and 11.6.

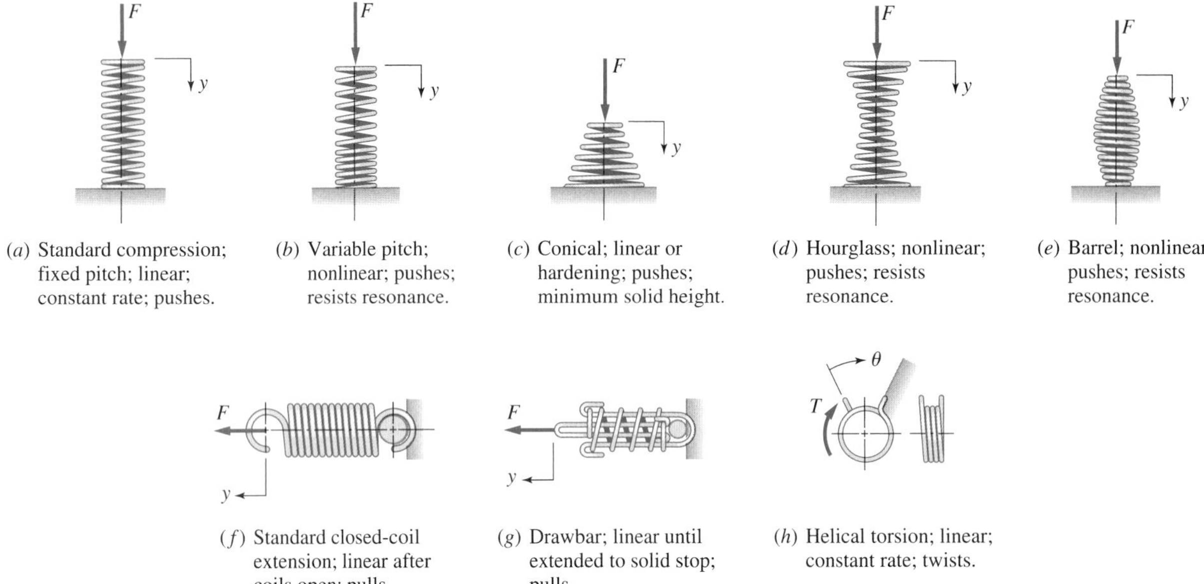

(*a*) Standard compression; fixed pitch; linear; constant rate; pushes.

(*b*) Variable pitch; nonlinear; pushes; resists resonance.

(*c*) Conical; linear or hardening; pushes; minimum solid height.

(*d*) Hourglass; nonlinear; pushes; resists resonance.

(*e*) Barrel; nonlinear; pushes; resists resonance.

(*f*) Standard closed-coil extension; linear after coils open; pulls.

(*g*) Drawbar; linear until extended to solid stop; pulls.

(*h*) Helical torsion; linear; constant rate; twists.

Figure 14.1
Various helical-coil spring configurations.

cular cross sections subjected to torsional moments that induce angular displacements. End attachments for torsion bar springs require special design attention to minimize stress concentration problems. Occasionally, torsion bar springs may be made with noncircular cross sections for special applications, but circular cross sections are more efficient.

Many other specialty springs have been devised. A few of these are shown in Figures 14.3(c) through (h). The *volute spring*, shown in 14.3(c), may be used when *high* friction damping is desired. *Rubber springs*, such as the one shown in 14.3(d), provide high damping as well, and have been successfully used as "shock insulators" for mounting heavy equipment such as automotive engines. *Pneumatic springs*, such as the two-convolution bellows arrangement shown in Figure 14.3(e), are basically columns of confined gas, properly contained so that the compressibility of the gas provides the desired displacement be-

Figure 14.2
Various beam spring (leaf spring) configurations.

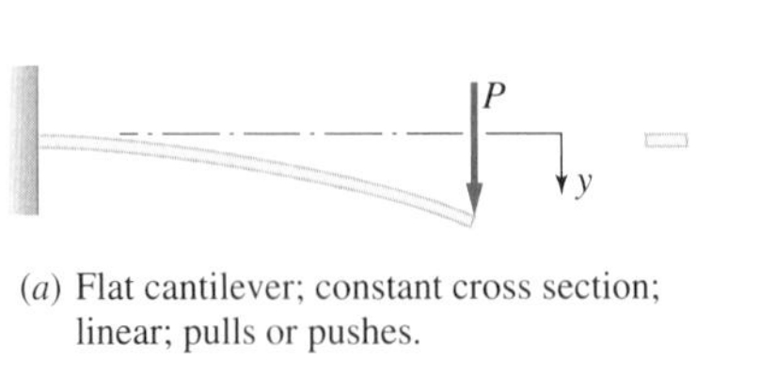

(*a*) Flat cantilever; constant cross section; linear; pulls or pushes.

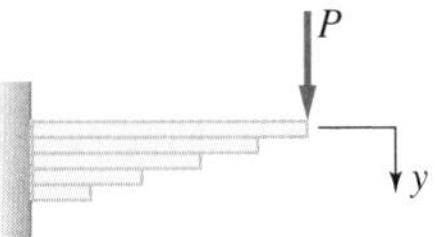

(*b*) Multileaf cantilever; approximates uniform strength cantilever; linear; pulls or pushes (if properly oriented).

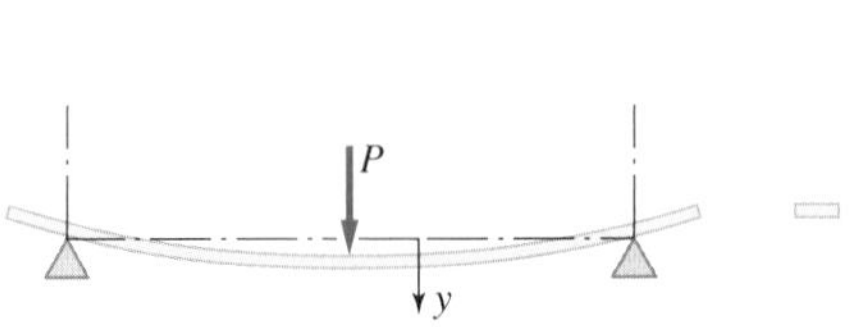

(*c*) Flat simply supported constant cross section; linear; pulls or pushes.

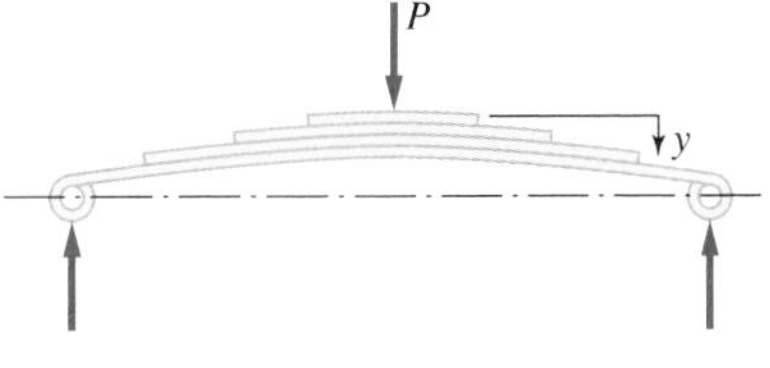

(*d*) Multileaf simply supported; positive arc height (camber); approximates uniform strength beam; linear; pushes or pulls (if properly oriented).

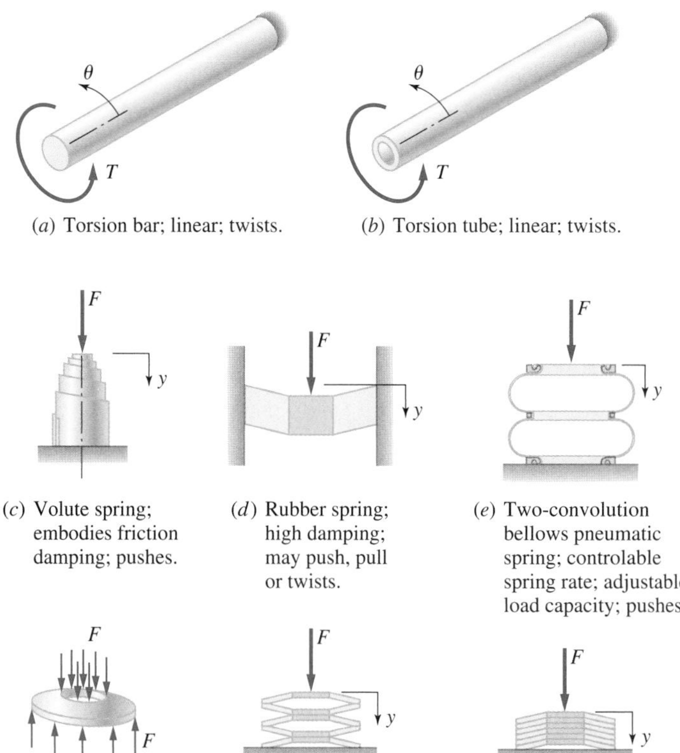

θ
T
θ
T
(a) Torsion bar; linear; twists.
(b) Torsion tube; linear; twists.
F
y
F
y
F
y
(c) Volute spring; embodies friction damping; pushes.
(d) Rubber spring; high damping; may push, pull or twists.
(e) Two-convolution bellows pneumatic spring; controlable spring rate; adjustable load capacity; pushes.
F
F
F
y
F
y
(f) Bellville washer; high loads; nonlinear; pushes.
(g) Belleville washers stacked in series.
(h) Belleville washers stacked in parallel.

Figure 14.3
Torsion bar springs and a few other speciality springs.

havior. *Belleville washers (coned-disk springs)*, such as the one shown in 14.3(f), may be used where space is limited and high loads with small deflections are required. By varying the dimensions of the coned disks, or by stacking them in *series* or *parallel*, as shown in Figures 14.3(g) and (h), the spring rate may be made approximately linear, nonlinear hardening, or nonlinear softening (see Figure 4.37). Many other types of special-purpose springs are commercially available.

14.3 Potential Failure Modes

Springs of all types are expected to operate over long periods of time without significant changes in dimensions, displacements, or spring rates, often under fluctuating loads. Based on these requirements and the various spring configurations discussed in 14.2, the potential failure modes (see 2.3) include yielding, fatigue, corrosion fatigue, fretting fatigue, creep, thermal relaxation, buckling, and/or force-induced elastic deformation (in the guise of resonant response or "surging"). To clarify, when springs are deflected under full load, the stresses induced must not exceed the yield strength of the material. If they do, the resulting permanent dimensional changes may interfere with the spring's ability to provide required forces or deliver stored strain energy essential to subsequent operation. Similarly, creep may lead to unacceptable long-term dimensional changes, even under static loading, a condition sometimes referred to as "set." If operating conditions include elevated temperatures, thermal relaxation must not produce unacceptable changes in dimensions or reduction of load-supporting capability. Fluctuating loads, often applied to springs, may lead to fatigue failure. Corrosive environments may make matters even worse, giving rise to accelerated failures due to corrosion fatigue. Fretting conditions *between* leaves of multileaf springs, *between* torsion bars and attached lever-arms, *between* washers of stacked Belleville springs, and any other configurations where cyclic strains induce small-amplitude sliding *between* contacting surfaces of spring elements may lead to fretting fatigue failures. Like columns, axially loaded open-coil compression springs may buckle if they are too slender or exceed critical deflections. When cyclic operating frequencies are close to the resonant frequency[3] of a spring, erratic force-displacement behavior may be induced because of wave propagation phenomena, sometimes called "surging." The prevention of surging is especially important for helical-coil spring applications.

As always, it is an important design responsibility to identify probable failure modes at the design stage, for the particular application at hand, and select an appropriate material and geometry to minimize the likelihood of potential failures.

14.4 Spring Materials

Material selection guidelines established in Chapter 3, together with failure mode discussions of 14.3, suggest that candidate materials for springs should have high strength (ultimate, yield, and fatigue), high resilience, good creep resistance, and, in some applications, good corrosion resistance and/or resistance to elevated temperatures. Materials meeting these criteria include carbon steel, alloy steel, stainless steel, spring brass, phosphor bronze, beryllium copper, and nickel alloys. Any of these spring materials may be formed into bars, wire, or strip by various hot-forming or cold-forming processes.

Cold-formed spring wire is produced by cold drawing the material through carbide dies to produce the desired size, surface finish, dimensional accuracy, and mechanical properties. Spring wire may be obtained in annealed, hard-drawn, or pretempered conditions. Strength properties of many materials are strongly *size dependent*, as illustrated in Figure 14.4.

Flat wire is produced by passing round wire through the rolls of a flattening mill, then quenching and tempering the flat strand to obtain the desired properties. *Spring steel strip* is produced by subjecting hot-rolled strip to a cleaning operation, followed by a combination of cold-rolling and thermal treatments to obtain the properties desired.

After the wire or strip material is coiled or formed, a shot-peening, strain-peening, or presetting operation is sometimes used to enhance fatigue resistance.[4] Corrosion resistance may be enhanced by coating, plating, or painting the spring.

Spring wire materials that are widely used by the spring industry include[5]:

1. Music wire (highest quality; highest strength; widely used)
2. Oil-tempered steel valve spring wire (high quality; high strength; limited sizes)
3. Oil-tempered steel spring wire (good quality; good strength; often used)
4. Hard-drawn steel wire (inexpensive; modest strength; used for static loads)
5. Alloy steel wire (for elevated temperatures to 230°C; high quality; high strength; e.g., chrome-vanadium, chrome-silicon)
6. Stainless-steel wire (good corrosion resistance for elevated temperatures to 260°C; high quality; high strength)
7. Beryllium copper wire (good conductivity; high strength; excellent fatigue resistance)
8. Nickel alloy wire (good corrosion resistance for elevated temperatures to 600°C; e.g., Inconel X-750; high cost)

Ultimate strength properties for many of these materials may be closely approximated by the empirical expression

[3]See 8.2 and 8.6. [4]See 2.6 and 4.13. [5]See ref. 1.

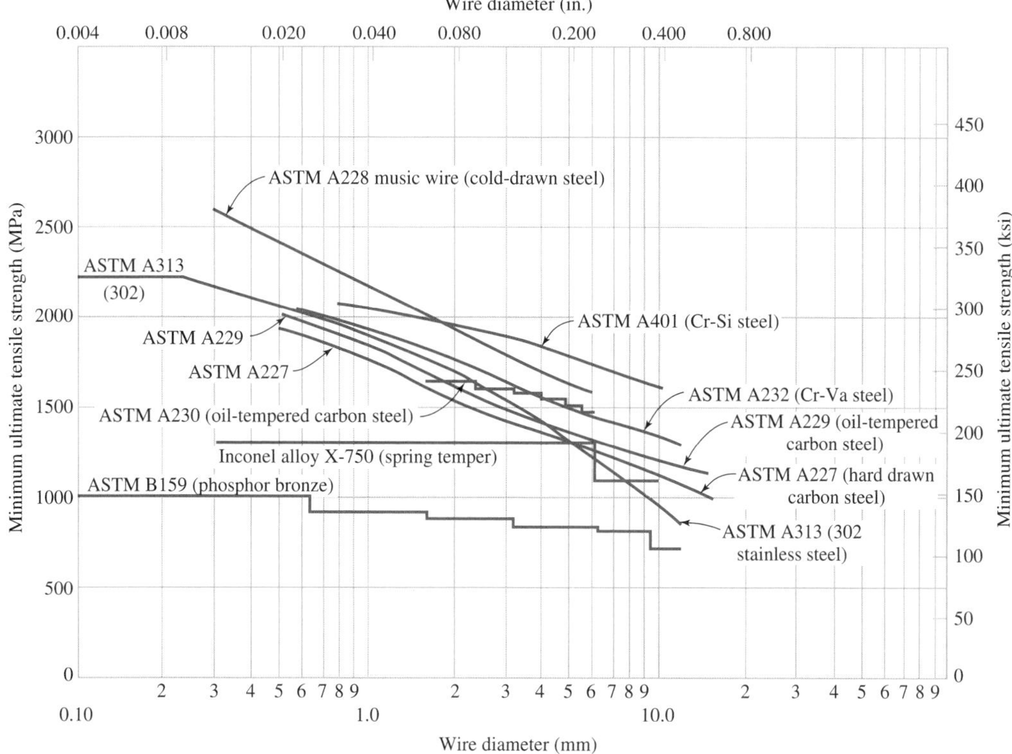

Figure 14.4
Minimum ultimate tensile strengths of several spring wire materials, as a function of wire diameter. (From ref. 2 with permission of Associated Spring, Barnes Group, Inc., Bristol, CT.)

$$S_{ut} = Bd^{a} \tag{14-1}$$

where S_{ut} = ultimate strength in tension
d = wire diameter
a = exponent
B = coefficient

Using data from Figure 14.4, the exponent a and coefficient B may be evaluated for five of the materials, as shown in Table 14.1. For materials not included in Table 14.1, ultimate strengths may be read directly from Figure 14.4. Commonly available stock wire sizes in the diameter ranges shown in Table 14.1 are listed in Table 14.2. Springs are usually cold-formed when wire diameters are less than 10 mm ($^3/_8$ inch), and hot-wound when wire diameters exceed 16 mm ($^5/_8$ inch).

TABLE 14.1 Values of *a* and *B* for Five of the Materials Shown in Figure 14.4

Material	Diametral Range of Validity, in (mm)	Exponent a	Coefficient B, ksi (MPa)
Music wire	0.010–0.250 (0.25–6.5)	−0.1625	184.6 (2153.5)
Oil-tempered steel	0.020–0.625 (0.5–16)	−0.1833	146.8 (1831.2)
Hard-drawn steel	0.020–0.625 (0.5–16)	−0.1822	141.0 (1753.3)
Cr-Va alloy steel	0.020–0.500 (0.5–13)	−0.1453	173.1 (1909.9)
Cr-Si alloy steel	0.031–0.437 (0.8–11)	−0.0934	220.8 (2059.2)

TABLE 14.2 Commonly Available Spring Wire Stock Diameters[1]

in	mm	in	mm
0.010	0.25	0.092	
0.012	0.30	0.098	2.50
0.014	0.35	0.105	
0.016	0.40	0.112	2.80
0.018	0.45	0.125	
0.020	0.50	0.135	3.50
0.022	0.55	0.148	
0.024	0.60	0.162	4.00
0.026	0.65	0.177	4.50
0.028	0.70	0.192	5.00
0.030	0.80	0.207	5.50
0.035	0.90	0.225	6.00
0.038	1.00	0.250	6.50
0.042	1.10	0.281	7.00
0.045		0.312	8.00
0.048	1.20	0.343	9.00
0.051		0.362	
0.055	1.40	0.375	
0.059		0.406	10.0
0.063	1.60	0.437	11.0
0.067		0.469	12.0
0.072	1.80	0.500	13.0
0.076		0.531	14.0
0.081	2.00	0.562	15.0
0.085	2.20	0.625	16.0

[1]Custom wire sizes also available at extra cost.

Flat steel spring stock is usually made of cold-rolled AISI 1050, 1065, 1074, or 1095 steel, typically available in annealed, $\frac{1}{4}$ hard (pretempered), $\frac{1}{2}$ hard, $\frac{3}{4}$ hard, or full hard condition. Automotive leaf springs have been made from various other fine-grained alloy steels such as SAE 9260, SAE 6150, and SAE 5160. In all cases, *hardenability* must be adequate to ensure a fully martensitic microstructure throughout the entire spring cross section. Properties for flat spring steel strip and other materials of interest are shown in Table 14.3, and preferred widths and thicknesses are given in Table 14.4.

TABLE 14.3 Typical Properties of Selected Spring Alloy Flat Strip[1]

Material	S_{ut}, ksi (MPa)	Rockwell Hardness	Elongation (2 in), percent
Spring steel	246 (1700)	C50	2
Stainless 302	189 (1300)	C40	5
Monel 400	100 (690)	B95	2
Monel K500	174 (1200)	C34	40
Inconel 600	151 (1040)	C30	2
Inconel X-750	152 (1050)	C35	20
Beryllium Copper	189 (1300)	C40	2
Phosphor bronze	100 (690)	B90	3

[1]From ref. 2.

TABLE 14.4 Preferred Widths and Thicknesses for Flat Strip Leaf Spring Stock Cross Sections[1]

Width		Thickness					
in	mm	in	mm	in	mm	in	mm
1.57	40.0	0.28	7.10	0.52	13.20	0.98	25.00
1.77	45.0	0.30	7.50	0.55	14.00	1.04	26.50
1.97	50.0	0.31	8.00	0.59	15.00	1.10	28.00
2.20	56.0	0.33	8.50	0.63	16.00	1.18	30.00
2.48	63.0	0.35	9.00	0.67	17.00	1.24	31.50
2.95	75.0	0.37	9.50	0.71	18.00	1.32	33.50
3.54	90.0	0.39	10.00	0.75	19.00	1.40	35.50
3.94	100.0	0.42	10.60	0.79	20.00	1.48	37.50
4.92	125.0	0.44	11.20	0.83	21.20	—	—
5.91	150.0	0.46	11.80	0.88	22.40	—	—
		0.49	12.50	0.93	23.60	—	—

[1]See American National Standard ANSI Z17.1; Custom sizes available at extra cost.

14.5 Axially Loaded Helical-Coil Springs; Stress, Deflection, and Spring Rate

The design of helical-coil springs involves selection of a material, and determination of the wire diameter, d, mean coil radius, R, number of active coils, N, and other spring parameters (see Figure 14.5) so that the desired force-deflection response is obtained, without exceeding the design stress under the most severe operating conditions.

For *open-coil* helical springs, loaded along the coil axis, whether the load produces

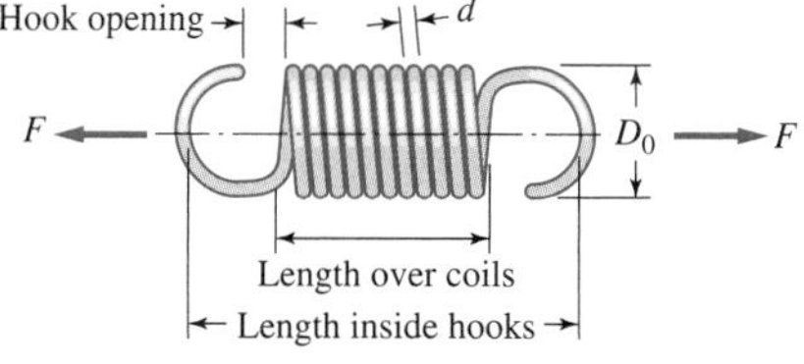

(*a*) Closely wound helical-coil extension spring with half hook at each end.

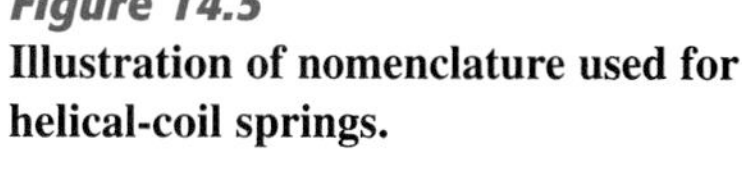

Figure 14.5
Illustration of nomenclature used for helical-coil springs.

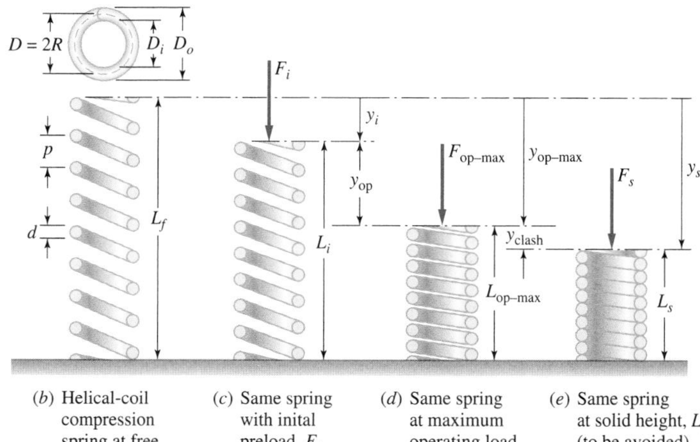

(*b*) Helical-coil compression spring at free length, L_f; no load.

(*c*) Same spring with inital preload, F_i.

(*d*) Same spring at maximum operating load.

(*e*) Same spring at solid height, L_s (to be avoided).

extension or compression, the primary stress developed in the wire is *torsion*. For tightly coiled and preloaded extension springs, torsion is also the primary stress in the wire, but it remains constant until the external extension force exceeds the built-in preload. Thus, a helical-coil spring may be thought of as a coiled torsion bar. To illustrate this, a *straight* bar (wire) of diameter d and length L may be loaded *torsionally* by attaching levers at each end as shown in Figure 14.6(a). If the levers have length R, and are loaded by equal but opposite force-couples as shown, the bar is subjected to the torsional moment

$$T = FR \tag{14-2}$$

The same bar may be formed into a *helix* of N coils, each having a mean radius R, then subjected to opposing columnar forces F, as shown in Figure 14.6(b). The *free body* that results when a section is cut through the helically coiled bar, as shown in Figure 14.6(c), indicates that the bar experiences a torsional moment $T = FR$, just as for the straight torsion bar, and in addition, a transverse shear force F. Investigations have shown that the *torsional shearing stresses* induced in the wire are *the primary stresses, but transverse shearing stresses* in helical-coil springs *are important enough* to consider. From (4-30) and (4-28), the maximum torsional shearing stress over the surface of the wire[6] is

$$\tau_{tor-max} = \frac{16FR}{\pi d^3} \tag{14-3}$$

where F = axial load on the spring
R = mean coil radius (radius from coil centerline to wire centerline)
d = wire diameter

The transverse shearing stress has been shown[7] to reach a maximum value at the mid-height of the wire cross section, and to have the magnitude

$$\tau_{ts-max} = 1.23\frac{F}{A_w} \tag{14-4}$$

where $A_w = \dfrac{\pi d^2}{4}$ = area of wire cross section

Figure 14.6
Torsion and transverse shear in helical-coil springs.

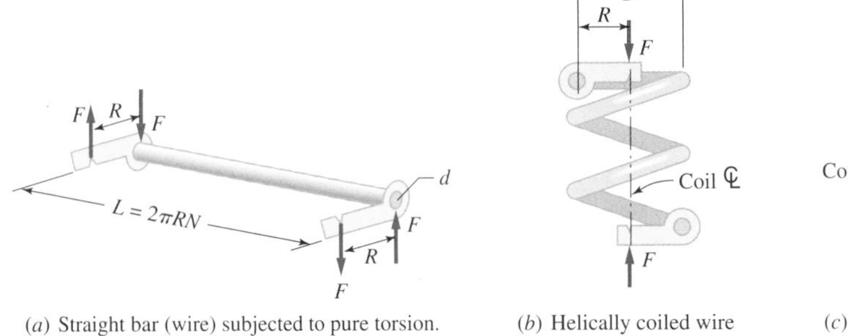

(a) Straight bar (wire) subjected to pure torsion.

(b) Helically coiled wire subjected to axial forces.

(c) Free body obtained by cutting through wire at any cross section.

Coil ℄

$\tau_{tor-max}$ (outer surface)

τ_{ts-max} (wire mid-height)

Critical point at inner coil radius where $\tau_{tor-max}$ and τ_{ts-max} add

(d) State of stress at the critical point.

[6]Not including curvature effects. [7]See ref. 3.

As shown in Figure 14.6(d), the torsional shear and transverse shear *oppose* at the *outer* coil radius but *add* at the *inner* coil radius.

Further, because of coil curvature, a slightly larger shearing strain is produced (by the torsion) at the inner (shorter) fiber of the coil than at the outer (longer) fiber, inducing a slightly higher torsional shearing stress at the *inner* fiber. This stress-increasing *curvature factor,* K_c, may be estimated as[8]

$$K_c = \frac{4c - 1}{4c - 4} \tag{14-5}$$

where the *spring index, c,* is defined to be

$$c \equiv \frac{2R}{d} = \frac{D}{d} \tag{14-6}$$

The curvature factor is often neglected when springs are *statically* loaded.

The maximum shearing stress, therefore, occurs at the mid-height of the wire at the inner coil radius, and may be estimated by combining (14-3), (14-4), and (14-5), as follows:

$$\tau_{max} = K_c(\tau_{tor-max}) + \tau_{ts-max} \tag{14-7}$$

or

$$\tau_{max} = \left(\frac{4c - 1}{4c - 4}\right)\left(\frac{16FR}{\pi d^3}\right) + 1.23\left(\frac{F}{A_w}\right) \tag{14-8}$$

which may be rewritten as

$$\tau_{max} = \left(\frac{4c - 1}{4c - 4}\right)\left(\frac{16FR}{\pi d^3}\right) + 1.23\left(\frac{F}{\pi(d^2)/4}\right)\left(\frac{R}{R}\right)\left(\frac{d}{d}\right) \tag{14-9}$$

or

$$\tau_{max} = \left(\frac{4c - 1}{4c - 4} + \frac{0.615}{c}\right)\frac{16FR}{\pi d^3} \tag{14-10}$$

Defining

$$K_w \equiv \left(\frac{4c - 1}{4c - 4} + \frac{0.615}{c}\right) \equiv \textit{Wahl factor} \tag{14-11}$$

the expression (14-10), for maximum shearing stress, becomes

$$\tau_{max} = K_w\left(\frac{16FR}{\pi d^3}\right) \tag{14-12}$$

Values of Wahl factor, K_w, as a function of spring index, c, are shown in Table 14.5.

Experience has shown that springs proportioned to give values of c less than about *4* are difficult to manufacture, and if values of c exceed about *12*, springs tend to buckle in compression, and tangle when stored in bulk. From Table 14.5, it may be noted that the range of K_w is from about 1.4 when c is 4 to about 1.1 when c is 12. A midrange value is around $K_w = 1.2$. This midrange value is often used as an initial estimate for K_w in (inherently iterative) spring design calculations.

The *end loops* of helical-coil extension springs must be separately analyzed because of *stress concentration* at the point where the end loop is "bent up" [see Figure 14.7(b)], and at the inner radius of the hook [see Figure 14.7(a)]. At critical section B, where torsion is the primary stress, the stress concentration factor is often approximated as

$$k_{iB} = \frac{r_{mB}}{r_{iB}} \tag{14-13}$$

[8]See ref. 3.

TABLE 14.5 Values of Wahl Factor K_w as a Function of Spring Index c

$c \rightarrow$ ↓	0.0	0.1	0.2	0.3	0.4	0.5	0.6	0.7	0.8	0.9
2	2.058	1.975	1.905	1.844	1.792	1.746	1.705	1.669	1.636	1.607
3	1.580	1.556	1.533	1.512	1.493	1.476	1.459	1.444	1.430	1.416
4	1.404	1.392	1.381	1.370	1.360	1.351	1.342	1.334	1.325	1.318
5	1.311	1.304	1.297	1.290	1.284	1.278	1.273	1.267	1.262	1.257
6	1.253	1.248	1.243	1.239	1.235	1.231	1.227	1.223	1.220	1.216
7	1.213	1.210	1.206	1.203	1.200	1.197	1.195	1.192	1.189	1.187
8	1.184	1.182	1.179	1.177	1.175	1.172	1.170	1.168	1.166	1.164
9	1.162	1.160	1.158	1.156	1.155	1.153	1.151	1.150	1.148	1.146
10	1.145	1.143	1.142	1.140	1.139	1.138	1.136	1.135	1.133	1.132
11	1.131	1.130	1.128	1.127	1.126	1.125	1.124	1.123	1.122	1.120
12	1.119	1.118	1.117	1.116	1.115	1.114	1.113	1.113	1.112	1.111
13	1.110	1.109	1.108	1.107	1.106	1.106	1.105	1.104	1.103	1.102
14	1.102	1.101	1.100	1.099	1.099	1.098	1.097	1.097	1.096	1.095
15	1.095	1.094	1.093	1.093	1.092	1.091	1.091	1.090	1.090	1.089
16	1.088	1.088	1.087	1.087	1.086	1.086	1.085	1.085	1.084	1.084
17	1.083	1.083	1.082	1.082	1.081	1.081	1.080	1.080	1.079	1.079
18	1.078	1.078	1.077	1.077	1.077	1.076	1.076	1.075	1.075	1.074

where r_{mB} = mean bend radius at section B
r_{iB} = inner radius at section B

Combining this with (14-3) gives

$$\tau_{max\,B} = k_{iB}\tau_{tor-max} = \frac{r_{mB}}{r_{iB}}\left(\frac{16FR}{\pi d^3}\right) \tag{14-14}$$

At critical section A, where bending stress and direct tension add, this critical point occurs at the inner radius of the hook. It is common practice to approximate this stress concentration factor as

$$k_{iA} = \frac{r_{mA}}{r_{iA}} \tag{14-15}$$

The stress at the inner wire radius of section A then may be written as

Figure 14.7
Extension spring end-loop stresses. (See also Figure 14.10.)

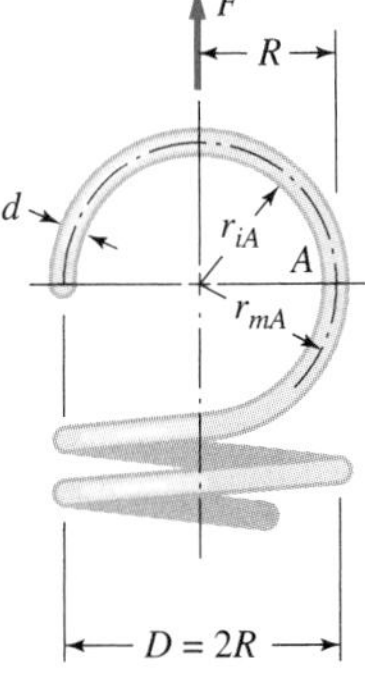

(*a*) Bending and tension at critical section A of end loop; $k_{iA} \approx \dfrac{r_{mA}}{r_{iA}}$.

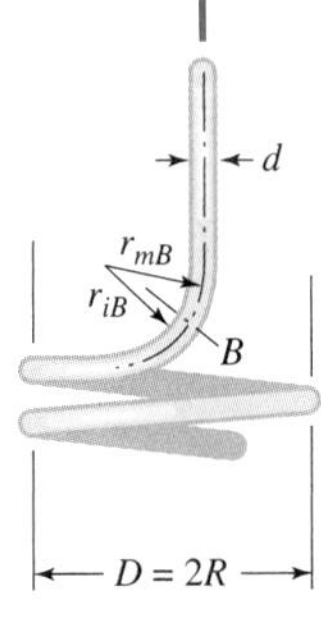

(*b*) Torsion at critical section B of end loop; $k_{iB} \approx \dfrac{r_m B}{r_i B}$.

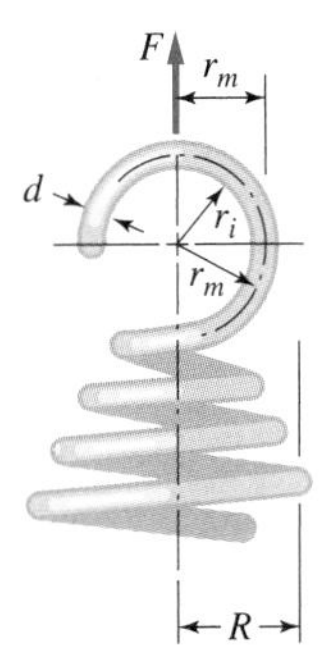

(*c*) Coned-end design for reducing stresses in end loops.

$$\sigma_{max A} = k_i\left(\frac{Mc}{I}\right) + \frac{F}{A} = \frac{r_{mA}}{r_{iA}}\left(\frac{32FR}{\pi d^3}\right) + \frac{4F}{\pi d^2} \tag{14-16}$$

If desired, the hook may be more accurately analyzed as a *curved beam* using (4-118) together with data from Table 4.8. It is usually recommended that k_i not be permitted to exceed about 1.25, corresponding to the requirement that $r_i \geq d$. In some cases, the end-loop maximum stress may be reduced by "coning" the end coils to shorten the bending moment arm from R to r_m, as illustrated in Figure 14.7(c).

Deflection and Spring Rate

Axial deflection of a helical-coil spring under *axial* load may be found by first determining the magnitude of the relative *angular* rotation between two adjacent cross sections of the wire, spaced a distance dL apart, produced by an applied torque $T = FR$.

Referring to Figure 14.8, consider for the moment that only a small segment $ABCD$ of the wire is flexible, and the rest of the spring wire is infinitely stiff. From (4-37), the angle of twist, $d\varphi$, of a circular bar of length dL, subjected to a torque FR, may be written as

$$d\varphi = \frac{FR\,dL}{JG} \tag{14-17}$$

Using this equation, the rotation of wire section CD may be calculated with respect to section AB. As shown in Figure 14.8, this angular displacement permits calculation of the *axial* displacement, dy, of a point E located 90° away. Because of the *rigid body* assumption for everything but segment $ABCD$, the axial displacement of point F, at the center of the helix, is the same as the axial displacement of E. The axial displacement of point F, therefore, is

$$dy = R d\varphi = \frac{FR^2 dL}{JG} \tag{14-18}$$

Now, if the *entire spring* is permitted to be flexible, the total axial deflection y may be found by integrating (14-18) over the entire active wire length, L, to give

$$y = \frac{FR^2 L}{JG} \tag{14-19}$$

If the spring has N active coils, its *active length* may be expressed as

$$L = 2\pi RN \tag{14-20}$$

and (14-19) may be rewritten as

$$y = \frac{FR^2(2\pi RN)}{\left(\frac{\pi d^4}{32}\right)G} = \frac{64FR^3N}{d^4G} \tag{14-21}$$

From this, an expression for spring rate k [see 2.4 and equation (2-1)] may be written as

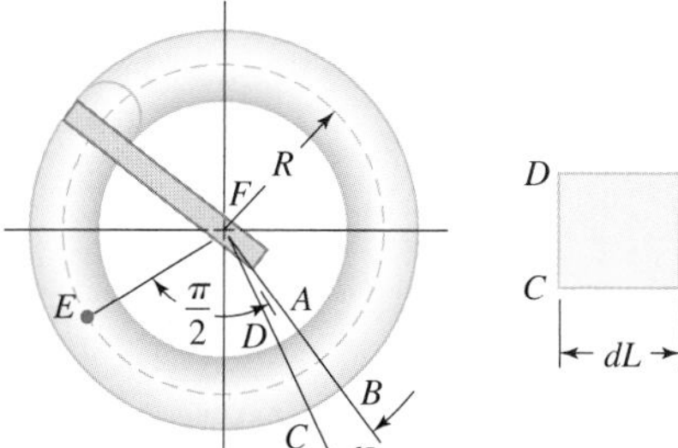

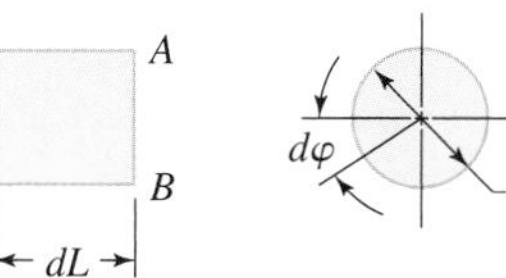

Figure 14.8
Deflections in helical-coil springs under axial load.

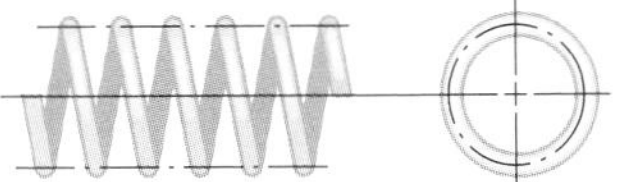
(*a*) Plain ends; no inactive coils unless special end fitting used.

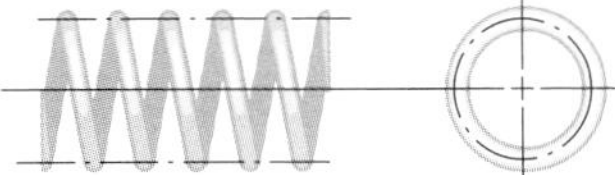
(*b*) Plain ends ground; one-half inactive coil each end.

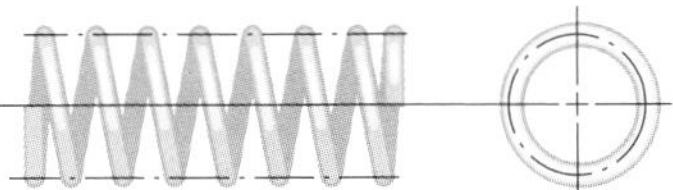
(*c*) Closed ends; one inactive coil each end; right-hand helix.

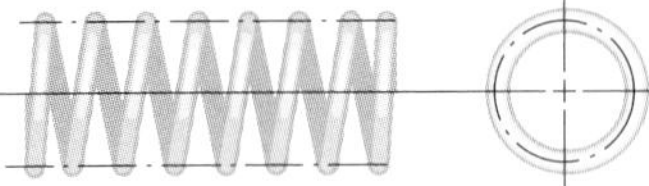
(*d*) Closed ends; one inactive coil each end; left-hand helix.

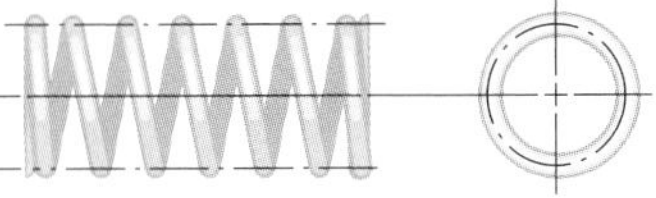
(*e*) Closed and ground ends; one inactive coil each end.

Figure 14.9
Various helical-coil compression spring configurations, and the influence of end coils on spring rate.

$$k = \frac{F}{y} = \frac{d^4 G}{64 R^3 N} \tag{14-22}$$

When writing expressions for spring rate or deflection in terms of the number of *active coils, N,* it is important to note that the *end coil* configuration often locally influences overall spring flexibility. End coil configuration should be accounted for in determining the *total* number of coils, N_t, used in winding the spring. The total number of coils, N_t, is the total number of turns from *end of wire to end of wire*; the number of active coils, N, is the number or turns over which the *wire twists* under load, and therefore contributes to axial spring deflection. For helical-coil compression springs, N_t is determined by *adding* the number of inactive coils, N_i, to the number of active coils N, to obtain

$$N_t = N + N_i \qquad \text{(compression springs)} \tag{14-23}$$

Figure 14.9 gives the number of *inactive* coils associated with each of the more common end conditions used for helical-coil compression springs.

For the case of extension springs, the end loops contribute *additional* elasticity to the springs, depending upon the end-loop configuration. Referring to Figure 14.5(a), the effective number of active coils, N, is obtained by *adding* the *equivalent coils for both end loops*, N_e, to the number of coils in the *length over coils*, N_c, to obtain

$$N = N_c + N_e \qquad \text{(extension springs)} \tag{14-24}$$

The equivalent numbers of coils for a few common end loops used for extension springs are given in Figure 14.10.

Figure 14.10
Various helical-coil extension spring end configurations. Many other variations are possible.

(*a*) Half hook over center; adds axial deflection equivalent to about 0.1 coil each end.

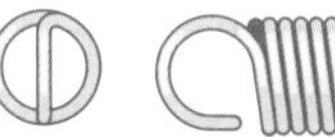
(*b*) Half loop over center; adds axial deflection equivalent to about 0.1 coil each end.

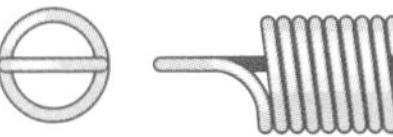
(*c*) Single full loop over center; adds axial deflection equivalent to about 0.5 coil each end.

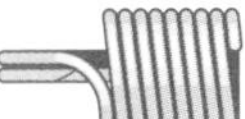
(*d*) Double full loop over center.

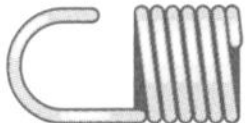
(*e*) Long round-end hook over center.

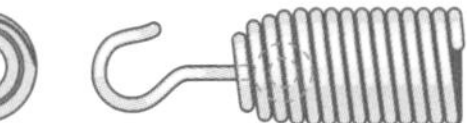
(*f*) Coned end with swivel hook; coned-end design tends to decrease stress concentration in hook.

Buckling and Surging

Two other design issues that may be important in some cases are: (1) elastic instability, or *buckling* of long, skinny helical-coil springs under compressive loads, and (2) resonant frequency response, or surging, if axial operating frequencies approach the axial natural frequency of the helical-coil spring (see 8.2).

Just as long thin columns subjected to axial compression become elastically unstable and *buckle* when loads become too large (see 2.7), helical-coil springs loaded in compression will buckle if axial deflections become too large. Equations for prediction of critical buckling deflection, y_{cr}, developed in a manner similar to column-buckling equations[9], may be expressed in terms of free length L_f, mean coil radius R, and the method of constraining the ends of the spring (see Table 14.6). Figure 14.11 shows *critical deflection ratio* plotted versus *slenderness ratio* for springs with both ends hinged ($\alpha = 1$) and springs with both ends fixed ($\alpha = 0.5$). Values of α for other end constraints included in Table 14.5 may be used to plot additional critical deflection curves if needed. After a spring has been tentatively designed to achieve the desired spring rate within acceptable stress levels, potential buckling failure should be checked using the curves of Figure 14.11. If the deflection ratio for the proposed spring exceeds the critical value read from the curve, the spring should be redesigned.

An alternative buckling solution sometimes used is to contain the spring inside a closely fitting guide cylinder, or to insert an internal cylindrical guide mandrel to prevent buckling, but friction and wear may produce failures of another type if this solution is chosen. If guide cylinders or mandrels are used in this way, a diametral clearance of approximately 10 percent of the cylinder or mandrel diameter is commonly used to avoid rubbing between the spring and its guide.

If springs are used in high-speed cyclic applications, the resonant response, called *surging*, should be investigated. If the material and geometry of the axially reciprocating spring are such that its *axial natural frequency* is close to the operating frequency, a traveling *displacement wave front* is propagated and reflected along the spring with about the same frequency as the exciting force.[10] This condition results in local *condensations* and *rarefactions*[11] propagating along the spring. These phenomena may produce high stresses and/or erratic forces locally, with consequent loss of control of the spring-loaded object. The often-cited example of an automotive valve spring illustrates the condition; surging of a valve spring may allow the valve to open erratically when it should be closed, or vice versa.

TABLE 14.6 End Constraint Constants for Compressively Loaded Helical-Coil Springs

Type of End Constraint	End Constraint Constant[1] α
Both ends pivoted (hinged)	1
One end supported by flat surface perpendicular to spring axis (fixed); other end pivoted (hinged)	0.7
One end clamped; other end free	2
Spring supported between flat parallel surfaces (fixed ends)	0.5

[1]See column end constraint constants shown in Table 2.4, for comparison.

[9]See ref. 3. [10]See for example, refs. 4 and 5.

[11]Local changes in pitch wherein a few coils may momentarily be very close (condensation) or significantly separated (rarefaction).

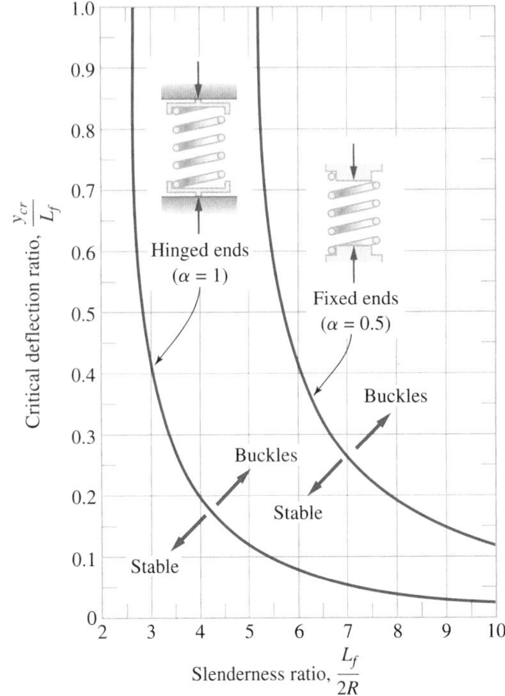

Figure 14.11
Critical deflection curves defining onset of buckling of helical-coil springs loaded in axial compression. (α = end condition constant; see Table 14.6.)

The axial natural frequency, f_n, of a helical-coil spring (using fixed-fixed end conditions) has been developed[12] from the basic equation

$$f_n = \frac{\omega_n}{2\pi} = \frac{\pi\sqrt{\frac{kg}{W}}}{2\pi} = \frac{1}{2}\sqrt{\frac{kg}{W}} \tag{14-25}$$

where f_n = fundamental axial natural frequency, Hz
ω_n = axial circular frequency, rad/sec
k = spring rate
W = weight of spring's active coils
g = gravitational constant

For a helical-coil spring this may be rewritten as

$$f_n = \frac{d}{2\pi R^2 N}\sqrt{\frac{Gg}{32w}} \tag{14-26}$$

where G = shear modulus of elasticity of spring material
w = specific weight of spring wire

For the common case where steel wire is used to wind the spring, $G = 11.5 \times 10^6$ psi, and $w = 0.283$ lb/in^3, so (14-22) becomes

$$(f_n)_{steel} = \frac{3510d}{R^2N} \tag{14-27}$$

It is important to note that these equations give the natural frequency of the *spring* itself, *not* the natural frequency of a spring-mass system where a mass W_1 is attached to a

[12] See ref. 3.

spring of much smaller mass. In some applications higher harmonics[13] may be excited, giving potential resonant responses at frequencies much higher than the fundamental natural frequency of the spring. Harmonics up to about the *thirteenth* have been reported to be significant in some applications.[14] Based on such evidence, it is usually recommended that cyclically loaded springs should have a fundamental natural frequency at least *fifteen times* the operating frequency to avoid resonances at the higher harmonics.

14.6 Summary of Suggested Helical-Coil Spring Design Procedure, and General Guidelines for Spring Design

Spring design is an inherently iterative process in which, at a minimum, the stress requirements and force-deflection requirements must be independently satisifed. Other factors, such as envelope constraints and environment, may also play a role. The following steps may be utilized to develop a successful spring design, but many other approaches could also be used.

1. Based on functional specifications and contemplated system configuration, generate a first conceptual sketch for the proposed spring, approximately to scale. Incorporate any restrictions imposed by the spring space envelope, including maximum outer coil diameter, minimum inner coil diameter, limits on free length or solid length, or limits on operating displacements (see Figure 14.5).
2. Identify potential failure modes (see 14.3).
3. Select a tentative spring material (see 14.4).
4. Select an appropriate design safety factor (see Chapter 5).
5. Calculate the design stress σ_d using (5-2), repeated here for reference.

$$\sigma_d = \frac{S_{fm}}{n_d} \qquad \text{(5-2) (repeated)}$$

or, by extension,

$$\tau_d = \frac{\tau_{fm}}{n_d} \qquad (14\text{-}28)$$

where S_{fm} = failure strength of the selected material corresponding to the governing failure mode

n_d = design safety factor

τ_d = design shearing stress

τ_{fm} = shear failure strength of the selected material corresponding to the governing failure mode

In spring design applications, three complicating factors are often encountered when attempting to find an appropriate value for failure strength S_{fm} or τ_{fm}; they are

a. Spring wire *strength* is often a function of wire size, d, as shown in (14-1), but at the outset d is unknown.

b. Yield strength data are sparse for spring wire materials, and relationships between ultimate strength and yield strength, especially in torsion, can only be defined empirically and approximately. (Recall that torsion is the primary state of stress at

[13]Harmonics are integral multiples of the fundamental natural frequency. [14]See ref. 6.

TABLE 14.7 Approximate Torsional Shear Yield Strength τ_{yp} as a Function of Ultimate Tensile Strength S_{ut} for Helical-Coil Compression Springs Under Static Loading[1]

Material	τ_{yp}/S_{ut}
Carbon steel (cold drawn)	0.45
Music wire	0.40
Carbon and low-alloy steel (hardened and tempered)	0.50
Stainless steel	0.35
Nonferrous alloys	0.35

[1]See ref. 2.

the critical points of helical-coil springs). Table 14.7 includes some data that may be helpful in relating τ_{yp} to S_{ut}.

c. Cyclically loaded springs are usually subjected to nonzero-mean loading cycles. Since torsion is a *multiaxial* state of stress, the critical points in cyclically loaded helical-coil springs must be analyzed as multiaxial states of nonzero-mean cyclic stress, as discussed in 4.7. As noted in 2.6, however, if a designer is fortunate enough to find *multiaxial nonzero-mean* cyclic *strength* data that match the operating conditions of the application, these data should be used *directly* rather than using the approximations of 4.7. The spring industry has developed a small body of data of this type for the special (but common) case of released loading ($R = 0$), as shown in Table 14.8.

6. Determine wire diameter, d, mean coil radius, R, and number of active coils, N, so that shearing stress τ_{max} from (14-12) and deflection y from (14-21) are *independently* satisfied in such a way that strength, life, and spring rate all meet design specifications (from step 1). If design performance specifications are *not* met, *changes must be made* in d, R, and/or N, iteratively, until specifications *are* satisfied.

   Usually, stress requirements are more easily satisfied *first* in the iteration process, since τ_{max} is a function of only two spring parameters, d and R. Then, deflection requirements may be satisfied by determining N. Space envelope constraints, reasonable values of spring index, and designer judgment are the basis for initial selections of d and R to start the iteration process.

TABLE 14.8 Approximate Fatigue Shearing Strength τ_f for Round Wire Helical-Coil Compression Springs Under Released Cyclic Loading Conditions[1] **($R = 0$) as a Function of Ultimate Tensile Strength S_{ut} (see Figure 14.4)**

	τ_f/S_{ut}			
	Austenitic Stainless and Nonferrous Alloys		Steel Alloys ASTM A230 and A232	
Fatigue Life N, cycles	Unpeened	Peened	Unpeened	Peened
10^5	0.36	0.42	0.42	0.49
10^6	0.33	0.39	0.40	0.47
10^7	0.30	0.36	0.38	0.46

[1]Room temperature, benign environment, typical spring wire sizes and surfaces (see ref. 2).

7. Using the tentative values of d, R, and N from step 6, determine spring rate, k, and check to assure that it also meets any *other* functional requirements for k.
8. Select an appropriate end configuration (see Figures 14.9 and 14.10) and determine the number of inactive coils (for compression springs) or equivalent extra coils (for extension springs). Calculate total number of coils in the spring using (14-23) or (14-24).
9. Determine solid height, free height, and operational deflection to make sure that no design requirements are violated.
10. Check potential buckling of compression springs (see Figure 14.11).
11. Check potential surging of the proposed design [see (14-25)].
12. Continue to iterate until all design requirements are satisfied. Summarize the final specifications for material, heat treatment, and dimensions. As the spring design is developed using these steps, the following experience-based guidelines may be found useful:
    a. If the spring is guided (to prevent buckling), allow a minimum diametral clearance between the spring and the guiding hole or mandrel, or approximately 10 percent of the coil diameter.
    b. Use material strength data from Tables 14.7 and 14.8, with (14-1), if applicable. Otherwise, use procedures of 4.7 for multiaxial states of nonzero-mean cyclic stress to determine failure strength.
    c. For initial selection of wire diameter, d, and mean coil radius, R, try to proportion the spring so that spring index c will lie in the range between *4* and *12*; an *initial* assumption of *7* or *8* usually produces a well-proportioned spring. Any space envelope constraints should also be invoked at this first iteration. Commonly available wire sizes, as shown in Table 14.2, should be selected unless there is an important reason to select other sizes.
    d. If designing a *compression* spring, *closed and ground ends* [see Figure 14.9(e)] are usually a good choice, unless minimum cost is essential. If designing an *extension* spring, the use of *full end loops*, over center (see Figure 14.7), is usually a good choice, unless minimum cost is essential.
    e. Allow a clearance *between coils* when the spring is at its maximum operating deflection, y_{op-max} (see Figure 14.5), to avoid coil-to-coil contact (*clash*). A *clash allowance*, y_{clash}, of at least 10 percent of the maximum operating defleciton, is usually recommended.
    f. For the usual case of highly stressed springs, the use of favorable residual stress fields should be considered as a means of improving life and reliability, especially under nonzero-mean cyclic loading. Shot-peening (see 2.6) and presetting (see 4.13) are often used to induce favorable residual stresses. *Presetting* is accomplished for a compression spring by manufacturing the spring to be *longer* than the actual length desired, then *overloading* it in the direction it will be loaded during operation, causing it to *yield*. The objective is to achieve the desired free length upon removal of the overload. An extension spring would be preset from a manufacturing length *shorter* than the free length desired.

Example 14.1 Helical-Coil Spring Design for Static Loads

A helical-coil compression spring is to be designed to exert a static force of 100 lb when the spring is compressed 2.0 inches from its free length, and it must fit inside a cylindrical hole 2.25 inches in diameter. The environment is laboratory air. Design a spring suitable

Example 14.1 Continues

for this application. Only five such springs are to be manufactured. A design safety factor of 2 is desired.

Solution

1. Following the spring design procedure just discussed, a conceptual sketch may be made, as shown in Figure E14.1. To provide the 10 percent diametral clearance suggested in guideline 12.a above,

$$1.10D_o = 1.10(2R + d) \leq D_h \tag{1}$$

or, for hole diameter $D_h = 2.25$ inches, as specified,

$$D_o = 2R + d \approx 2.0 \text{ inches} \tag{2}$$

2. The most probable failure mode is *yielding*, since the load is static and buckling is prevented by the guide hole.
3. Since it is widely available, has excellent properties, and can be cold-formed, *music wire* will be tentatively selected for the spring material.
4. By specification, the design safety factor (see Chapter 5) is to be

$$n_d = 2 \tag{3}$$

5. From (14-28), the shear design stress is

$$(\tau_d)_{static} = \frac{\tau_{yp}}{n_d} \tag{4}$$

From Table 14.7,

$$\tau_{yp} = 0.40S_{ut} \tag{5}$$

Using (14-1) and the data from Table 14.1, the ultimate tensile strength, S_{ut}, may be written as a function of wire diameter, d, as

$$S_{ut} = 184.6d^{-0.1625} \text{ ksi} \tag{6}$$

Combining (3), (4), (5), and (6),

$$(\tau_d)_{static} = 36.9d^{-0.1625} \text{ ksi} \tag{7}$$

6. Using guideline 12.c, the spring index may be tentatively selected as

$$c = 8 \tag{8}$$

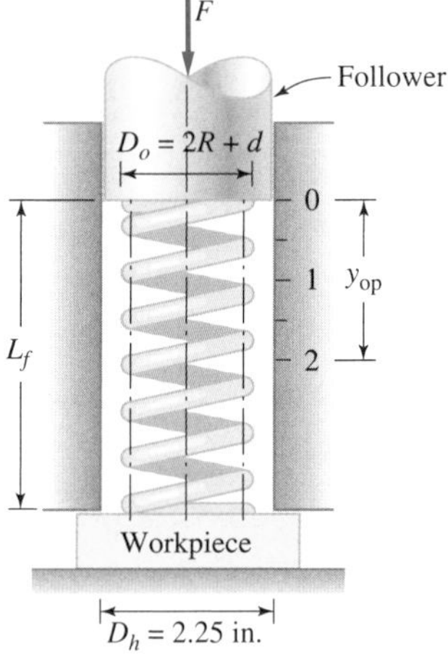

Figure E14.1
Conceptual sketch of proposed spring.

and from (14-6), using (8),

$$2R = 8d \tag{9}$$

but from (2)

$$2R = 2.0 - d \tag{10}$$

so (9) gives

$$d = 0.222 \text{ inch} \tag{11}$$

whence,

$$R = \frac{2.0 - 0.222}{2} = 0.889 \text{ inch} \tag{12}$$

Using these dimensions, (7) gives

$$(\tau_d)_{static} = 36.9(0.222)^{-0.1625} = 47.12 \text{ ksi} = 47{,}120 \text{ psi} \tag{13}$$

From (14-12) then, using K_w = 1.18 from Table 14.4,

$$\tau_{max} = 1.18\left(\frac{16(100)(0.889)}{\pi(0.222)^3}\right) = 48{,}830 \text{ psi} \tag{14}$$

Thus τ_{max} slightly exceeds the design stress τ_d, but they are close. Before making any changes, other parameters should be examined, and before proceeding further, a tentative "standard" wire size should be selected from Table 14.2.

Rewriting (14-21),

$$N = \frac{d^4Gy}{64FR^3} = \frac{(0.222)^4(11.5 \times 10^6)(2.0)}{64(100)(0.889)^3} = 12.42 \text{ active coils} \tag{15}$$

7. No additional "functional" spring rate requirements are specified.
8. Using guideline 12.d for stability, closed and ground ends will be selected. From Figure 14.9(e), one inactive coil for each end should be allowed for this end coil configuration. Using (14-23), the total number of coils for the proposed spring is

$$N_t = 12.42 + 2(1) = 14.42 \text{ coils} \tag{16}$$

9. Approximate solid height (see Figure 14.5), L_s, may be calculated as

$$L_s = N_t d = 14.42(0.222) = 3.20 \text{ inches} \tag{17}$$

From guideline 12.e, an appropriate clash allowance would be

$$y_{clash} = 0.10(2.0) = 0.20 \text{ inch} \tag{18}$$

The free height, therefore, should be

$$L_f = y_{op-max} + y_{clash} + L_s = 2.0 + 0.20 + 3.20 = 5.4 \text{ inches} \tag{19}$$

10. Although this spring is guided, its buckling tendency may be checked by using Figure 14.11. For this spring the slenderness ratio is

$$SR = \frac{(5.4)}{2(0.889)} = 3.04 \tag{20}$$

Thus from Figure 14.11, the *critical* deflection ratio is so large that the deflection ratio *for this spring*

$$DR = \frac{2.0}{5.4} = 0.37 \tag{21}$$

will clearly not cause buckling. (Actually, this spring will not buckle at any physically achievable deflection.)

Example 14.1 Continues

To refine these calculations, a "standard" wire size of $d = 0.225$ inch will be selected from Table 14.2. Then, from (7),

$$(\tau_d)_{static} = 36.9(0.225)^{-0.1625} = 47{,}000 \text{ psi} \tag{22}$$

and using

$$c = \frac{2R}{d} = \frac{2(0.889)}{0.225} = 7.9 \tag{23}$$

From Table 14.4,

$$K_w = 1.19 \tag{24}$$

and (14-12) becomes

$$\tau_{max} = 1.19\left(\frac{16(100)(0.889)}{\pi(0.225)^3}\right) = 47{,}300 \text{ psi} \tag{25}$$

Thus, $\tau_{max} \approx \tau_d$, and the safety factor requirement is met.

The number of active coils, revising (15) slightly, is

$$N = \frac{(0.225)^4(11.5 \times 10^6)(2.0)}{64(100)(0.889)^3} = 13.1 \text{ active coils} \tag{26}$$

This gives a revised number of total coils, N_t, of

$$N_t = 13.1 + 2 = 15.1 \text{ coils} \tag{27}$$

The revised solid height becomes

$$L_s = 15.1(0.225) = 3.40 \text{ inches} \tag{28}$$

and the free height becomes

$$L_f = 2.0 + 0.2 + 3.4 = 5.6 \text{ inches} \tag{29}$$

Summarizing the results, the design recommendations for this spring are:

1. Use ASTM A228 music wire, cold forming the spring from standard stock diameter $d = 0.225$-inch wire.
2. Use closed and ground ends.
3. Wind the spring to a mean coil radius of $R = 0.889$ inch ($D = 1.776$ inches).
4. Wind a total number of 15.1 coils, tip-of-wire to tip-of-wire.
5. Wind the spring to a finished free height of $L_f = 5.6$ inches. If it is necessary to adjust the manufactured free height using plastic deformation to achieve the value 5.6 inches, do so by winding the spring slightly longer and then *compressively* overloading to obtain the finished free height desired.

Example 14.2 Helical-Coil Spring Design for Fluctuating Loads

It is desired to investigate the spring design generated in Example 14.1 to find out if it could survive 10^7 cycles if the load *cycled* repeatedly from zero to a peak of 100 lb. If the design of Example 14.1 is *not* satisfactory, explain how to redesign the spring to obtain a life of 10^7 cycles.

Solution

To determine whether the spring design of Example 14.1 will survive 10^7 cycles of released load ($R = 0$), one need only compare the shear design stress in *fatigue*, corresponding to 10^7 cycles of released load, to the *static yielding* design stress calculated in equation (7) of Example 14.1. Thus the *fatigue* design stress is

$$(\tau_d)_{10^7} = \frac{\tau_{10^7@R=0}}{n_d} = \frac{0.38S_{ut}}{2} = 35.1d^{-0.1625} \text{ ksi} \tag{1}$$

The value $0.38S_{ut}$ is read from Table 14.8. Comparing (1) above with (7) of Example 14.1.

$$(\tau_d)_{10^7} = 35.1d^{-0.1625} < (\tau_d)_{static} = 36.9d^{-0.1625} \tag{2}$$

therefore the design is a little less than satisfactory for 10^7 cycles of released loading.

Several options are available to the designer in attempting to redesign the spring for a longer life:

1. The spring could be *shot-peened*. From Table 14.7, it may be seen that shot peening would increase $\tau_{10^7@R=0}$ from $0.38S_{ut}$ to $0.46S_{ut}$, more than enough to produce a satisfactory design stress.
2. The *safety factor* could be reviewed to see if the small decrease in n_d (from 2.0 to 1.9[15]) could be accepted.
3. The spring could be *redesigned*, using the approach of Example 14.1, to modify d, R, and/or N until an acceptable design revision is achieved.

14.7 Beam Springs (Leaf Springs)

As illustrated in Figure 14.2, thin flat beams may be used as springs in some applications. Thin simply supported beams subjected to transverse center loads, or thin cantilever beams subjected to transverse end loads, are the more common configurations used. From (4-5), together with Tables 4.1 and 4.2, the bending stresses and deflections may be written for both types of beam springs. For the case of a *cantilever* spring of constant rectangular cross section, loaded at the free end as illustrated in Figure 14.12(a), the bending stress at

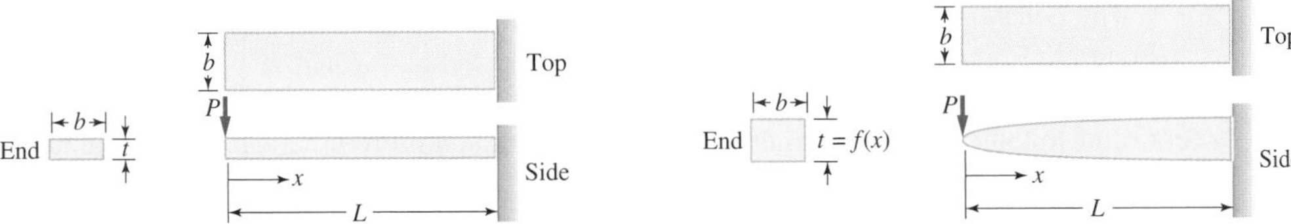

(*a*) End-loaded single-leaf cantilever with constant cross section.

(*b*) End-loaded uniform fiber strength cantilever with constant width (parabolic thickness profile).

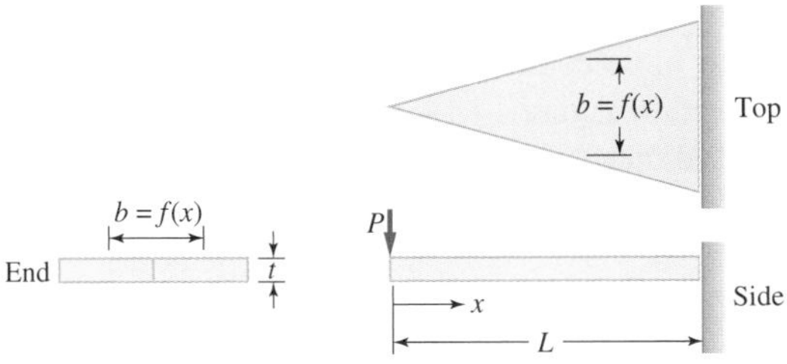

(*c*) End-loaded uniform fiber strength cantilever with constant thickness (triangular planform).

Figure 14.12
Various shapes used as end-loaded cantilever springs.

[15]Obtained from the coefficient ratio (35.1/36.9) from (2), multiplied times $n_d = 2$.

the outer fibers, σ_x, as a function of position x along the beam, is

$$\sigma_x = \frac{M_x c}{I} = \frac{(Px)\left(\frac{t}{2}\right)}{\left(\frac{bt^3}{12}\right)} = \left(\frac{6P}{bt^2}\right)x \tag{14-29}$$

where P = transverse end load
b = beam width
t = beam thickness
x = distance from free end of beam

By inspection, the maximum outer-fiber stress occurs at the fixed end where $x = L$, so from (14-29)

$$\sigma_{max} = \frac{6PL}{bt^2} \tag{14-30}$$

At the free end, where $x = 0$, the maximum outer fiber stress is zero. To use material more efficiently, it is common practice to employ the *tailored-shape guideline*[16] to eliminate the lightly stressed material near the free end. This may be accomplished by prescribing that stress σ_x in (14-29) remain constant, and equal to design stress σ_d all along the beam. To obtain constant outer-fiber stress all along the beam, either b or t may be varied with x in such a way that

$$\sigma_x = \sigma_d \tag{14-31}$$

From (14-29), this requirement may be satisfied by prescribing either that

$$t^2 = \left(\frac{6P}{b\sigma_d}\right)x \tag{14-32}$$

or that

$$b = \left(\frac{6P}{t^2\sigma_d}\right)x \tag{14-33}$$

The *constant-width* beam of *parabolic thickness-profile*, resulting from (14-32), is illustrated in Figure 14.12(b). The *constant-thickness* beam of *linearly varying width* (triangular planform), resulting from (14-33), is illustrated in Figure 14.12(c). Of these two, the constant-thickness beam of triangular planform is much more practical to manufacture. To make it more compact, it is usually manufactured as an equivalent *multiple leaf spring*, as shown in Figure 14.13. In the multiple leaf arrangement, each leaf approximates two symmetric half-strips cut from a constant-strength triangular planform, then welded together lengthwise. Shorter leaves are then stacked on top of longer leaves to form a bundle of n leaves, each having width b_1, as shown in the side view of Figure 14.13. Similar reasoning holds for the more common case of the center-loaded simply supported multiple leaf spring illustrated in Figure 14.14, in which a diamond-shaped planform is used to eliminate lightly stressed material near the supports. Noting that the beam width at any position x is z_x, the area moment of inertia is

$$I_x = \frac{z_x t^3}{12} \tag{14-34}$$

By geometric similarity,

$$\frac{z_x}{nb_1} = \frac{x}{(L/2)} \qquad 0 \le x \le L/2 \tag{14-35}$$

[16]See 6.2.

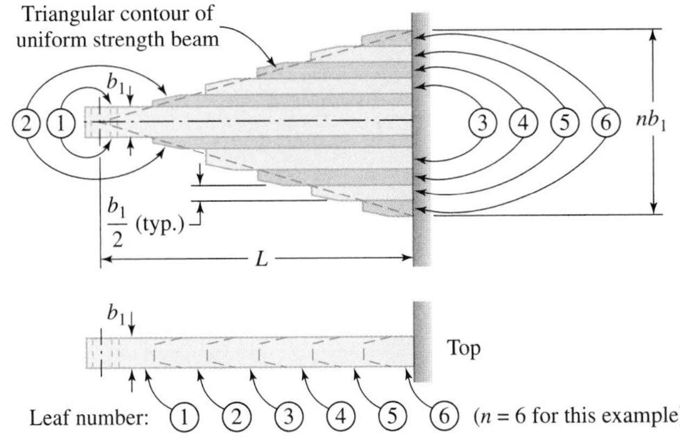

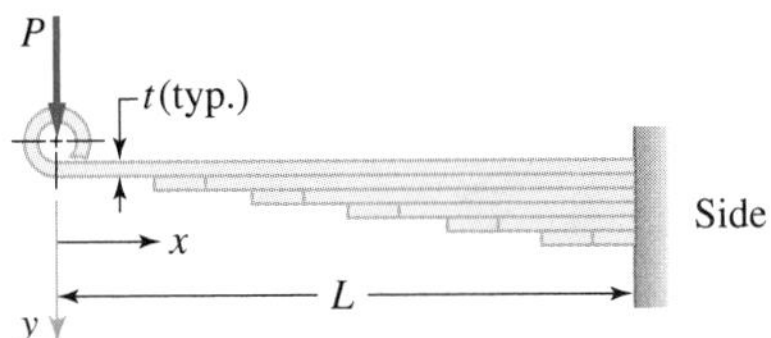

Figure 14.13
Uniform (outer-fiber) strength cantilever spring approximated by a multileaf cantilever spring. This is sometimes called a "quarter-elliptic" spring.

so (14-34) may be written as

$$I_x = \left(\frac{nb_1t^3}{6L}\right)x \qquad 0 \le x \le L/2 \tag{14-36}$$

Note also that

$$c = \frac{t}{2} \tag{14-37}$$

and

$$M_x = \left(\frac{P}{2}\right)x \qquad 0 \le x \le L/2 \tag{14-38}$$

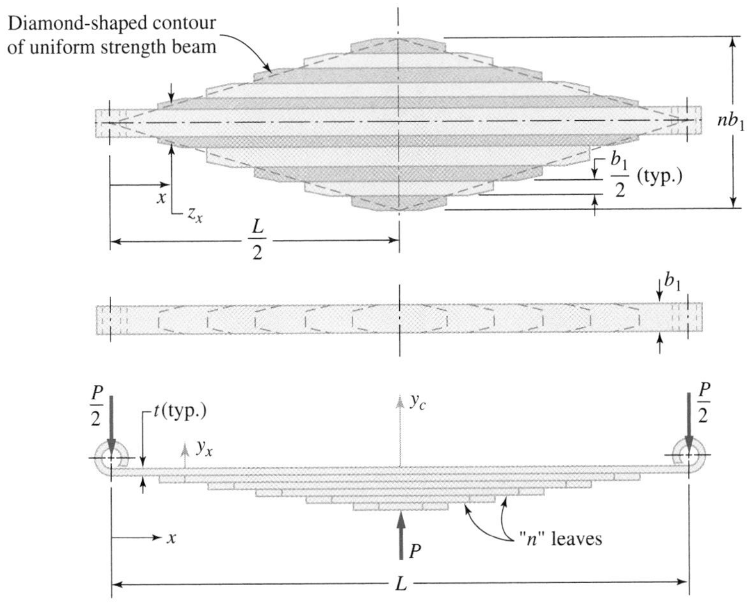

Figure 14.14
Uniform (outer-fiber) strength simply supported spring approximated by a simply supported multileaf spring. This is sometimes called a "semielliptic" spring.

By adapting (14-29), and utilizing symmetry,

$$\sigma_x = \frac{\left(\frac{Px}{2}\right)\left(\frac{t}{2}\right)}{\left(\frac{nb_1t^3x}{6L}\right)} = \frac{3PL}{2nb_1t^2} \qquad 0 \leq x \leq L \tag{14-39}$$

where σ_x (constant over the whole length of the spring) may be equated to the design stress σ_d.

The deflection y_x may be obtained by utilizing (4-46) to give

$$\frac{d^2y_x}{dx^2} = -\frac{M}{EI} = -\frac{\left(\frac{Px}{2}\right)}{E\left(\frac{nb_1t^3x}{6L}\right)} = -\frac{3PL}{Enb_1t^3} \tag{14-40}$$

Integrating once

$$\frac{dy_x}{dx} = -\frac{3PL}{Enb_1t^3}x + C_1 \tag{14-41}$$

and again

$$y_x = -\frac{3PL}{2Enb_1t^3}x^2 + C_1x + C_2 \tag{14-42}$$

The boundary conditions are

$$\frac{dy_x}{dx} = 0 \quad @ \quad x = L/2 \tag{14-43}$$

and

$$y_x = 0 \quad @ \quad x = 0 \tag{14-44}$$

First, using (14-43), (14-41) gives

$$C_1 = \frac{3PL^2}{2Enb_1t^3} \tag{14-45}$$

and using (14-44), (14-42) gives

$$C_2 = 0 \tag{14-46}$$

From (14-42) and (14-45) then

$$y_x = -\frac{3PLx^2}{2Enb_1t^3} + \frac{3PL^2x}{2Enb_1t^3} \tag{14-47}$$

The deflection under the center load P, y_c, may be obtained by setting $x = L/2$ to give

$$y_c = \frac{3PL^3}{8Enb_1t^3} \tag{14-48}$$

and the spring rate may then be written as

$$k = \frac{P}{y_c} = \frac{8Enb_1t^3}{3L^3} \tag{14-49}$$

In addition to selecting the spring material, and determining dimensional parameters (thickness, width, and number of leaves) to satisfy stress, life, deflection, and spring rate requirements, other design issues of importance must be resolved. The configuration of leaf *edges* and *ends* must be specified. Load transfer features such as *eyes*, supporting *shackles*, and *center clamping* details must be determined. Alignment and retention features, overall configuration to meet space envelope constraints, and initial curvature of the

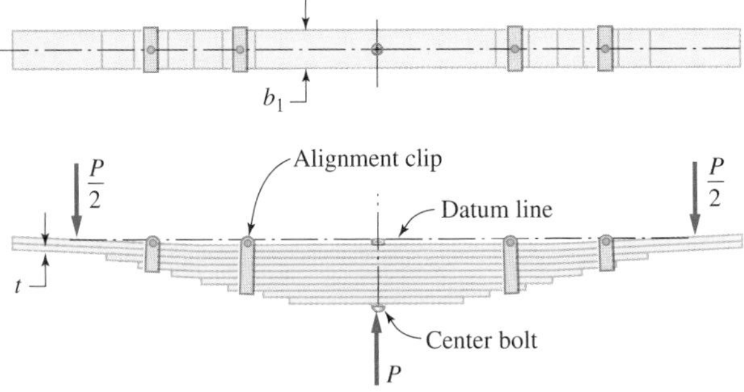

(*a*) Simply supported multileaf spring with plain ends and near-zero arc height (small initial curvature).

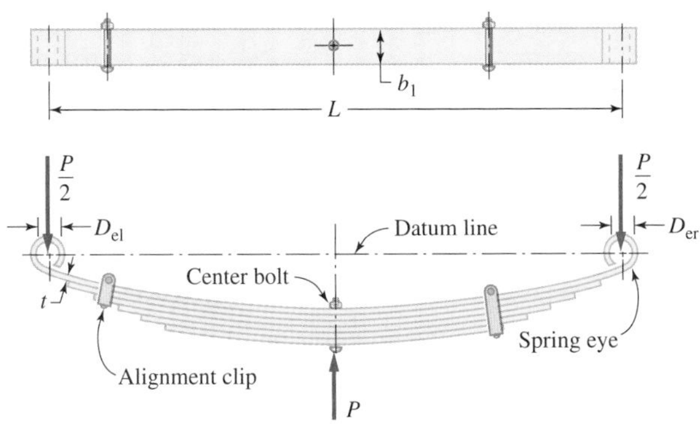

(*b*) Simply supported multileaf spring with upturned eye at each end and negative arc height (negative camber).

Figure 14.15
Typical leaf spring arrangements.

whole spring assembly (up or down), must be specified. A detailed consideration of these issues is beyond the scope of this text, but a few pertinent examples are illustrated here.[17] Figure 14.15 shows two of many such possible multileaf spring configurations.

Leaf edges are usually *rounded* to a convex arc with radius of curvature between 65 and 85 percent of the leaf thickness. Springs are often *preset* or *shot-peened* to improve fatigue resistance. *Protective finishes* or coatings such as grease, oil, paint, or plastic are sometimes used. Various *leaf end* configurations may be used, such as those illustrated in Figures 14.16(a) and (b). Spring ends must be supported by hinged *load transfer* devices,

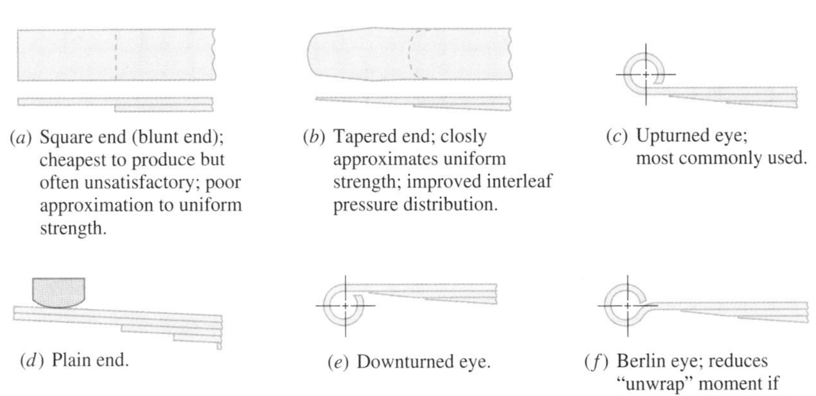

(*a*) Square end (blunt end); cheapest to produce but often unsatisfactory; poor approximation to uniform strength.

(*b*) Tapered end; closly approximates uniform strength; improved interleaf pressure distribution.

(*c*) Upturned eye; most commonly used.

(*d*) Plain end.

(*e*) Downturned eye.

(*f*) Berlin eye; reduces "unwrap" moment if longitudinal loads exists, sometimes welded if longitudinal forces are high.

Figure 14.16
Selected examples of leaf ends and spring eyes and ends. (After ref. 1.)

[17]For more detailed discussions, see ref. 1.

Figure 14.17
Selected examples of spring shackles and leaf alignment clips. (After ref. 1.)

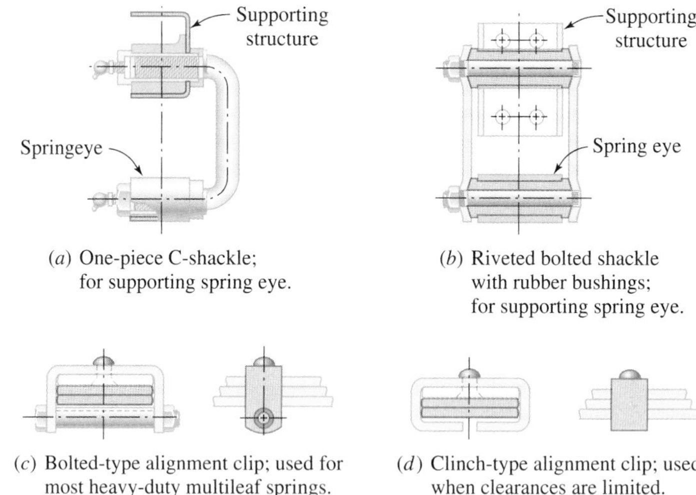

(*a*) One-piece C-shackle; for supporting spring eye.

(*b*) Riveted bolted shackle with rubber bushings; for supporting spring eye.

(*c*) Bolted-type alignment clip; used for most heavy-duty multileaf springs.

(*d*) Clinch-type alignment clip; used when clearances are limited.

usually called *shackles*, that permit small longitudinal end displacements as the spring center deflects laterally under load. Two spring shackle arrangements are illustrated in Figure 14.17, and various types of main-leaf spring eyes are illustrated in Figures 14.16(c) through (f). *Center load transfer* usually involves a clamping arrangement of U-bolts and contoured pads arranged to firmly attach the spring to its seat, as illustrated, for example, in Figure 14.18.[18] Alignment clips, often used to maintain the orientation and integrity of the leaf stack, are illustrated in Figures 14.17(c) and (d). Much more detailed descriptions of these features and others may be found in the literature.[19]

14.8 Summary of Suggested Leaf Spring Design Procedure

The following steps for developing a successful leaf spring design closely parallel the steps for design of a helical-coil spring outlined in 14.6.

1. Based on functional specifications and contemplated system configuration, generate a first conceptual sketch for the proposed spring, approximately to scale. Incorporate

Figure 14.18
Selected examples of center bolt and center clamping arrangements. (After ref. 1.)

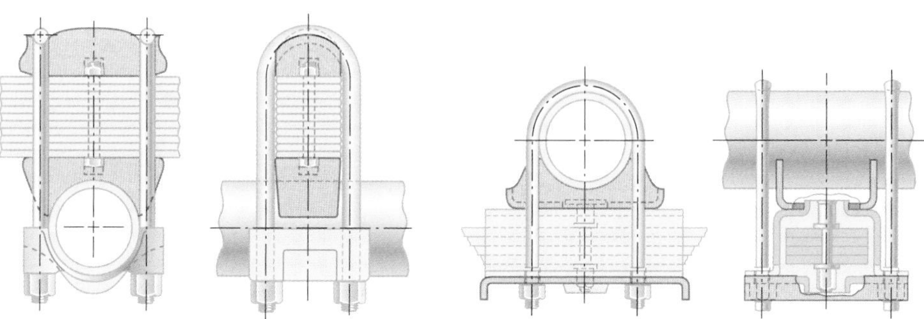

(*a*) Center clamping arrangement for overslung spring (inactive clamped length must be accounted for in design of spring).

(*b*) Center clamping arrangement for underslung spring (inactive clamped length must be accounted for in design of spring).

[18]The effects on spring behavior of the *inactive* clamped length must be taken into account.

[19]See, for example, ref. 1.

any restrictions imposed by the spring space envelope, including length, width, overall height, curvature, or limits on operating displacements.

2. Identify potential failure modes (see 14.3).
3. Select a tentative spring material (see 14.4).
4. Select an appropriate design safety factor (see Chapter 5).
5. Calculate the design stress σ_d using (5-2), repeated here for reference:

$$\sigma_d = \frac{S_{fm}}{n_d} \qquad \text{(5-2)(repeated)}$$

where S_{fm} = failure strength of the selected material corresponding to the governing failure mode

n_d = design safety factor

6. Determine thickness and width of each leaf, and the number of leaves to be used, so that spring rate requirements from (14-49) and bending stress requirements from (14-39) are independently satisfied. If design performance specifications are *not* met, changes must be made to t, b_1, and/or n, iteratively, until specifications *are* satisfied. Usually, *spring rate* requirements are satisfied *first* in the iteration process by selecting initial values for n and b_1, then calculating t from (14-39). Next, the stress requirements are satisfied. Space envelope constraints and designer judgment are the basis for *initial* selections for n and b_1. Preferred leaf widths and thicknesses, as shown in Table 14.4, should be selected unless there is an important reason to select other sizes.
7. Using the tentative values of t, b_1, and n, from step 6, determine spring rate k and check to assure that it meets all functional requirements. Check to make sure that there are no space envelope violations for any anticipated operating condition.
8. Select appropriate leaf edge and leaf end geometry, spring eye and spring shackle configuration, and center bolt and center clamping arrangement. Determine and account for inactive spring length arising from the selected center clamping arrangement.
9. Continue to iterate until all design requirements are satisfied. Summarize the final material and dimensional specifications.

Example 14.3 Beam Spring (Leaf Spring) Design for Fluctuating Loads

A proposed small, new, hydrocarbon-fuel-cell-driven passenger car is estimated to weigh 2500 lb fully loaded, with each wheel supporting an equal share of the load. It is desired to design a symmetrical semielliptic leaf spring for each of the rear wheel locations. Space envelope estimates indicate that a center-loaded spring, of approximately 48 inches length between shackles, can be accommodated. To avoid side interference, the leaf width should not exceed about 2 inches. Supporting design studies suggest that the spring deflection under static load will not exceed about 2.25 inches, and the peak deflection under dynamic operating conditions will be no more than about twice the static deflection, but an additional 0.50-inch clearance is desired before metal-to-metal contact occurs.[20] Estimates of k_∞ have also been made,[21] leading to a specified value of $k_\infty = 0.65$. The material is to be spring steel and a design safety factor of 1.3 has been selected. Propose a preliminary configuration for a semielliptic (simply supported multileaf) spring for this application that will provide infinite life.

[20]Actually, as a practical matter, elastomeric "snubbers" should probably be used to prevent metal-to-metal contact in case occasional high load spikes are encountered.

[21]See 2.6.

Example 14.3 Continues

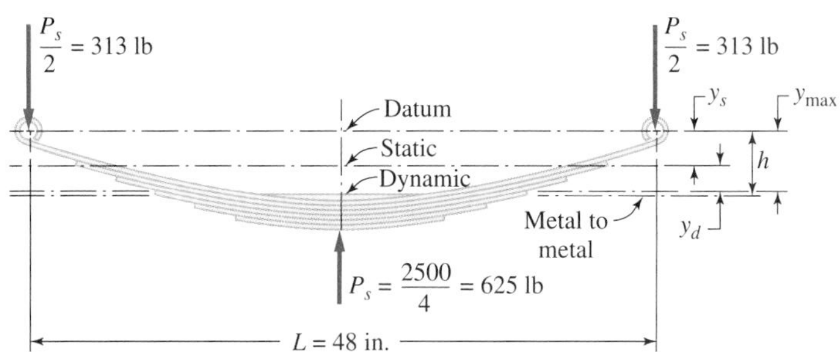

Figure E14.3
Conceptual sketch of proposed spring.

Solution

1. Following the spring design procedure of 14.8, a conceptual sketch may be constructed, as shown in Figure E14.3.
2. Based on the application, and because a *dynamic* deflection specification is given, the most probable failure mode is *fatigue*.
3. By specification, spring steel is the selected material (see Table 14.3).
4. By specification, the design safety factor is to be

$$n_d = 1.3 \tag{1}$$

5. From (5-2) the design stress is

$$\sigma_d = \frac{S_{fm}}{n_d} \tag{2}$$

Since the governing failure mode is fatigue, infinite life is specified, and the loading cycle has a nonzero mean,[22]

$$S_{fm} = S_{max-N=\infty} \tag{3}$$

Because fatigue properties for this material are not at hand, the methods of 2.6 will be used to estimate the fatigue properties. From Table 14.3, the ultimate tensile strength for spring steel is given as

$$S_{ut} = 246{,}000 \text{ psi} \tag{4}$$

and, since

$$S_{ut} > 200{,}000 \text{ psi} \tag{5}$$

the estimated value of S'_f for carefully prepared specimens of this material, under completely reversed loading, is

$$S'_f = 100{,}000 \text{ psi} \tag{6}$$

Since $k_\infty = 0.65$ has been specified, (2-25) gives

$$S_f = k_\infty S'_f = 0.65(100{,}000) = 65{,}000 \text{ psi} \tag{7}$$

From the deflection specifications it may be deduced that this is a case of non zero-mean cyclic loading. Since this is a linear spring, stress is proportional to load,

[22]See 2.6.

and deflection is proportional to load. From Figure E14.3 it may be further deduced that *static* deflection and load correspond to the *mean* of a released loading cycle. Maximum *dynamic* deflection and load correspond to the *peak* (maximum) of a released loading cycle. For the case at hand

$$R_t = \frac{y_m}{y_{max}} = \frac{P_m}{P_{max}} = \frac{\sigma_m}{\sigma_{max}} = 0.5 \tag{8}$$

and

$$m_t = \frac{S_u - S_f}{S_u} = \frac{246{,}000 - 65{,}000}{246{,}000} = 0.74 \tag{9}$$

as defined in (2-40). Using (2-40) to modify the completely reversed fatigue strength to account for the released loading conditions,

$$S_{max-N=\infty} = \frac{S_f}{1 - m_t R_t} = \frac{65{,}000}{1 - (0.74)(0.5)} = 103{,}170 \text{ psi} \tag{10}$$

Because this material is brittle (2% elongation in 2 inches, as shown in Table 14.3), $S_{yp} \approx S_{ut}$, the validity criteria of (2-40) are met, and (10) is valid.

Combining (10), (3), (2), and (1),

$$(\sigma_d)_{\substack{released\\load}} = \frac{103{,}170}{1.3} = 79{,}360 \text{ psi} \tag{11}$$

6. From Table 14.4, a standard preferred width that will accommodate the specified side clearances is

$$(b_1)_0 = 1.97 \text{ inches} \tag{12}$$

An initial selection for number of leaves will be taken as

$$(n)_0 = 6 \tag{13}$$

Based on specifications related to static load, and assumptions (12) and (13), (14-49) may be written as

$$(k)_0 = \frac{P_s}{y_s} = \frac{625}{2.25} = 277.8 = \frac{8(30 \times 10^6)(6)(1.97)t^3}{3(48)^3} \tag{14}$$

whence

$$t = \sqrt[3]{\frac{277.8(3)(48)^3}{8(30 \times 10^6)(6)(1.97)}} = 0.32 \text{ inch} \tag{15}$$

Since specification require that

$$y_s \leq 2.25 \text{ inches} \tag{16}$$

a standard preferred thickness, selected from Table 14.4, should be no less than 0.32 inch. Therefore the selection made is

$$t_0 = 0.33 \text{ inch} \tag{17}$$

From (14-48), using this thickness, the center deflection under the static load would be

$$(y_c)_0 = \frac{3(625)(48)^3}{8(30 \times 10^6)(6)(1.97)(0.33)^3} = 2.03 \text{ inches} \tag{18}$$

Since this is a significant departure from the specification value of 2.25 inches, the next smaller preferred width will be tried. It is

$$(b_1)_1 = 1.77 \text{ inches} \tag{19}$$

Example 14.3 Continues

From (18)

$$(y_c)_1 = \left(\frac{1.97}{1.77}\right)(y_c)_0 = (1.11)(2.03) = 2.26 \text{ inches} \tag{20}$$

This is an acceptable value. For these dimensions, the spring rate, as calculated from (14), becomes

$$(k)_1 = \frac{625}{2.26} = 277\frac{\text{lb}}{\text{in}} \tag{21}$$

Checking stresses, (14-39) may be solved for t_{min} by first setting

$$\sigma_x = (\sigma_d)_{\substack{released\\load}} = 79{,}360 \text{ psi} \tag{22}$$

and noting

$$P_{max} = \frac{y_{max}}{y_s}(P_s) = 2(625) = 1250 \text{ lb} \tag{23}$$

Thus

$$t_{min} = \sqrt{\frac{3P_{max}L}{2nb_1\sigma_x}} = \sqrt{\frac{3(1250)(48)}{2(6)(1.77)(79{,}360)}} = 0.33 \text{ inch} \tag{24}$$

This is compatible with the selection of (17), so the iteration is complete.

7. No additional constraints are known, except that a clearance of 0.5 inch is desired between $y_d = 2(2.26) = 4.52$ inches and any supporting structure, to avoid metal-to-metal contact. Initial curvature should be specified to provide a distance between the unloaded datum line and metal-to-metal contact of

$$h = 4.52 + 0.5 = 5.02 \text{ inches} \tag{25}$$

This value, h, usually called *opening* or *overall height* (depending upon how the spring is attached), is illustrated in Figure E14.3.

8. Rounded leaf edges, tapered ends, and upturned eyes are suggested for this application. Details of spring shackles, center bolt, and center clamping will be deferred to a later review.

9. Recommendations are summarized as follows:

   a. Use spring steel *leaves* with rounded edges and standard preferred cross-sectional dimensions, $t = 0.33$ inch by $b_1 = 1.77$ inches.

   b. Use 6 leaves, with the main leaf 48 inches long between centers of the upturned eyes. Use a center bolt and alignment clips, to be selected later.

   c. Use an initially curved spring of overall height $h = 5.02$ inches, generally configured as shown in Figure E14.3.

14.9 Torsion Bars and Other Torsion Springs

Torsion springs may be used in a wide variety of applications, ranging from precision instruments, balance springs, and window shades to automotive and military-tank suspension springs. Any application in which it is desired to apply a torque, or to store rotational energy, is a candidate for the use of a torsion spring.

Perhaps the simplest of all torsion springs is the *torsion bar* (or *torsion tube*) illustrated in Figure 14.19. The torsion bar shown in Figure 14.19(a) is typical of those used in suspension systems. Such torsion bars are usually loaded in one direction only, and are frequently shot-peened and/or preset to enhance fatigue resistance. End connections must be

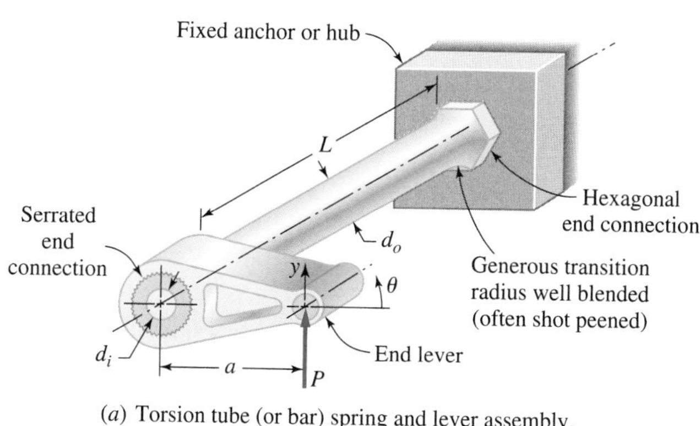

(*a*) Torsion tube (or bar) spring and lever assembly.

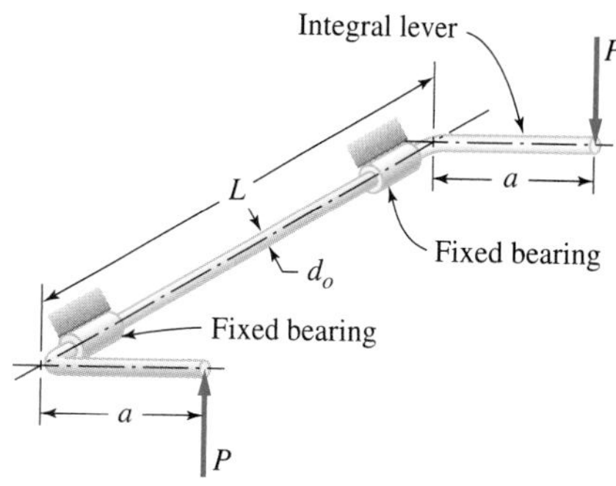

(*b*) Rod with bent ends used as single-piece torsion bar spring.

Figure 14.19
Examples of torsion bar springs.

carefully considered in torsion bar systems since they represent stress concentration sites at which premature failures may initiate. Typically, *serrated ends* permit the smallest end diameter; however, experience dictates that the diameter of the serrated end should be at least 1.15 times the body diameter of the torsion bar. The length of serration is usually made about 0.4 of the body diameter of the bar.[23] Serration *roots* are often shot-peened.

Hexagonal end connections are sometimes used, especially when high-volume production justifies the use of *upsetting* machines that can produce the larger hexagonal ends with no further machining required. One-piece torsion bars, such as the one sketched in Figure 14.19(b), may be used in such applications as counterbalancing car hoods, trunk lids, or truck tilt cabs, or as stabilizer bars (antiroll bars) in automotive wheel suspension systems. When the one-piece configuration is used, it is often necessary to include the bending flexibility of the integral levers in the spring rate determination.

The basic torsion bar equations for stress, τ, angular deflection, θ, and torsional spring rate, k_{tor}, as developed earlier in (4-29), (4-30), (4-37), and (4-135), are

$$\tau_{max} = \frac{T(d_o/2)}{J} = \frac{16Td_o}{\pi(d_o^4 - d_i^4)} \text{ psi} \tag{14-50}$$

$$\theta = \frac{TL}{JG} = \frac{32TL}{\pi(d_o^4 - d_i^4)G} \text{ rad} \tag{14-51}$$

$$k_{tor} = \frac{T}{\theta} = \frac{JG}{L} = \frac{\pi(d_o^4 - d_i^4)G}{32L} \frac{\text{in-lb}}{\text{rad}} \tag{14-52}$$

When helical-coil springs are used as torsion springs, they are wound in the same manner as for the compression and extension springs discussed in 14.5, except that the wire ends are configured to transmit torque [see Figure 14.20(a)]. Helical-coil torsion springs are usually mounted around a shaft or a mandrel for stability, and are typically made of round wire, close-wound. As shown in Figure 14.20(a), when load P is applied to produce the torsional moment Pa, the spring wire is subjected to a *bending* stress over its whole length. Due to coil curvature, this bending stress is maximum at the inner coil radius (see 4.9). The *bending stress concentration factor* at the inner coil radius, k_i, has been shown (by *Wahl*) to be well approximated by the *torsional* curvature factor,[24] K_c, given in (14-5). The maximum bending stress, σ_{max}, that occurs at the inner coil radius of a helical-coil torsion spring is given by

[23] See ref. 2. [24] See ref. 3.

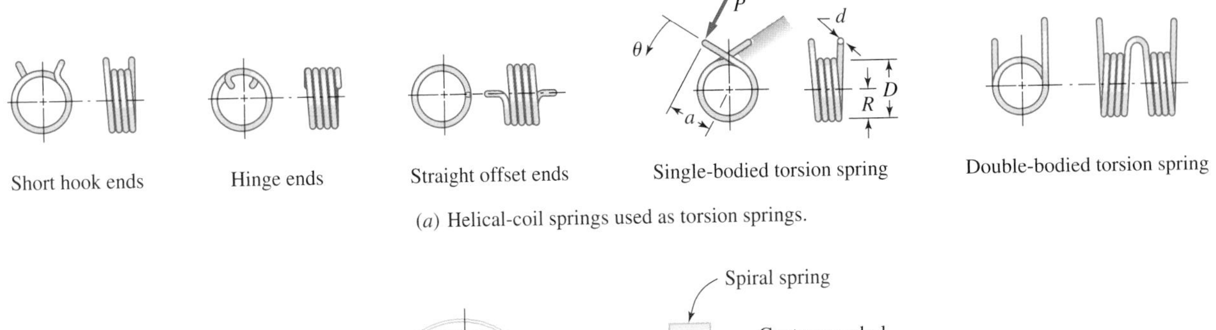

(*a*) Helical-coil springs used as torsion springs.

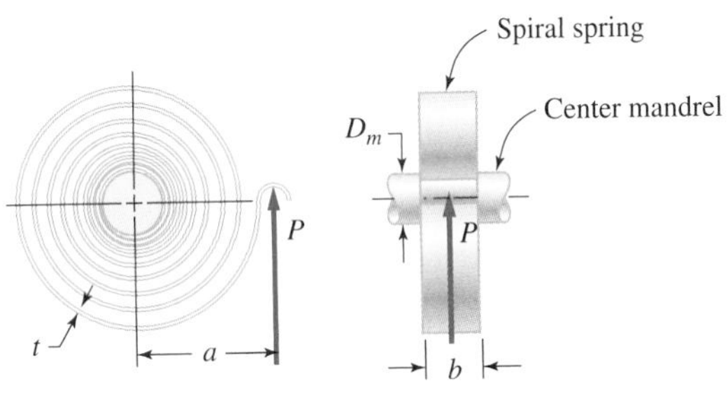

(*b*) Spiral torsion spring.

Figure 14.20
Various torsion spring configurations.

$$\sigma_{max} = k_i \frac{Mc}{I} = \left(\frac{4c - 1}{4c - 4}\right)\frac{32Pa}{\pi d^3} \tag{14-53}$$

Using Table 4.1, case 10, the angular deflection, θ, may be written as

$$\theta = \frac{ML}{EI} = \frac{64PaL}{\pi d^4 E} \tag{14-54}$$

and the torsional spring rate, k_{tor}, is

$$k_{tor} = \frac{T}{\theta} = \frac{\pi d^4 E}{64L} \tag{14-55}$$

For *spiral* torsion springs, such as the one shown in Figure 14.20(b), the same basic equations apply, but must be modified slightly because of the *rectangular* cross section of the spirally wound strip. For the spiral torsion spring

$$\sigma_{max} = k_i \frac{Mc}{I} = k_i\left(\frac{6Pa}{bt^2}\right) \tag{14-56}$$

where k_i = stress concentration factor at inner curved surface of the spring strip (ref. curved beam data of Table 4.8, case 1)

For example, when the spring is tightly wound on a mandrel of diameter D_m, typical values of k_i as a function of $(1 + D_m/t)$ are:

$$k_i \approx 1.5 \quad \text{for} \quad (1 + D_m/t) = 2.0$$
$$k_i \approx 1.2 \quad \text{for} \quad (1 + D_m/t) = 4.0$$
$$k_i \approx 1.1 \quad \text{for} \quad (1 + D_m/t) > 6.0$$

and

$$\theta = \frac{ML}{EI} = \frac{12PaL}{bt^3 E} \tag{14-57}$$

where L = total active length of spiral spring strip.

The overall torsional spring rate for the spiral spring strip is

$$k_{tor} = \frac{T}{\theta} = \frac{bt^3E}{12L} \tag{14-58}$$

14.10 Belleville (Coned-Disk) Springs

When space is limited and forces are high, *Belleville spring washers*, as sketched in Figures 14.3(f), (g), and (h), may often be used to good advantage. By appropriate selection of basic dimensions and stacking sequence, spring rates may be made approximately linear, nonlinear hardening, or nonlinear softening.[25] The cross section of a typical single-disk Belleville spring is sketched in Figure 14.21, with loads applied uniformly around inner and outer circumferential edges. Development of the nonlinear force-deflection equations, and expressions for critical stresses for Belleville washers, are complicated and beyond the scope of this text.[26] However, it has been shown that *radial stresses are negligible* and the potential *circumferential stress critical points* occur at A (compressive), B (tensile), and C (tensile; usually highest), as indicated in Figure 14.21.

An expression for the force F_{flat}, the force required to deflect the spring to a flat configuration, is[27]

$$F_{flat} = \frac{4E}{1-\nu^2}\left(\frac{ht^3}{K_1 D_o^2}\right) \tag{14-59}$$

where ν = Poisson's ratio
h = inside height at no load
t = thickness
D_o = outside washer diameter
D_i = inside washer diameter
K_1 = constant based on diameter ratio D_o/D_i (see Table 14.9)

For any particular Belleville spring, calculation of the washer flattening force, F_{flat}, from (14-59), together with the nondimensional plot of Figure 14.22, permits determination of the complete force-deflection curve for that spring. Force-deflection curves for stacked Belleville washers may be deduced from basic concepts of *series* and *parallel* spring arrangements.[28]

Further, the expression for circumferential tensile stress at critical point C (usually the *highest* tensile stress) is[29]

$$\sigma_C = \frac{4EtyD_i}{K_1 D_o^3(1-\nu^2)}\left[K_3 + (2K_3 - K_2)\left(\frac{h}{t} - \frac{y}{2t}\right)\right] \tag{14-60}$$

where y = spring deflection
K_1, K_2, K_3 = constants based on diameter ratio D_o/D_i (see Table 14.9)

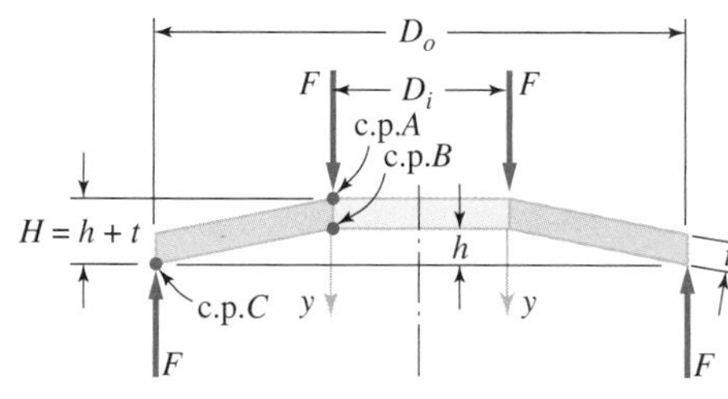

Figure 14.21
Cross section of a Belleville spring.

[25]See Figure 4.37. [26]See ref. 2 for more detail. [27]Ibid. [28]See 4.11. [29]See ref. 1 for more detail.

TABLE 14.9 Values of Constants[1] Used for Estimating Washer Flattening Force F_{flat} and Tensile Stress σ_c at Critical Point C (see Figure 14.22)

Diameter Ratio D_o/D_i	K_1	K_2	K_3
1.4	0.46	1.07	1.14
1.6	0.57	1.12	1.22
1.8	0.65	1.17	1.30
2.0	0.69	1.22	1.38
2.2	0.73	1.26	1.45
2.4	0.75	1.31	1.53
2.6	0.77	1.35	1.60
2.8	0.78	1.39	1.67
3.0	0.79	1.43	1.74
3.2	0.79	1.46	1.81
3.4	0.80	1.50	1.87
3.6	0.80	1.54	1.94
3.8	0.80	1.57	2.00
4.0	0.80	1.60	2.01

[1]Adapted from ref. 1, reprinted with permission from SAE Publication AE-21 © 1996, Society of Automotive Engineers, Inc.

14.11 Energy Storage in Springs

The capacity of a spring to store energy, especially if the spring must function within a limited space envelope, is often an important factor in the spring selection process. In accordance with the basic principles for creating shape and size, as discussed in 6.2, and the application of (4-43) to the case of a linear spring, the stored strain energy, U, is equal to

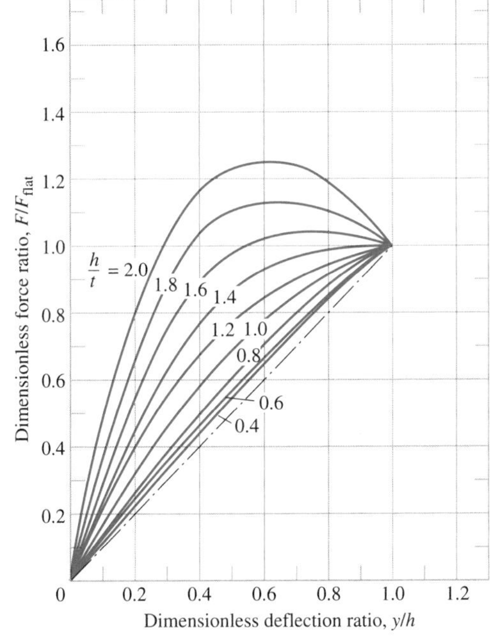

Figure 14.22
Nondimensional force-deflection characteristics of single-disk Belleville washers, as a function of *h/t* (see Figure 14.21). (Adapted from ref. 1, reprinted with permission from SAE Publication AE-21 © 1996, Society of Automotive Engineers, Inc.)

the work done during the loading process. Thus, for an initially unloaded spring

$$U = F_{ave}y = \left(\frac{0 + F}{2}\right)y = \frac{Fy}{2} \tag{14-61}$$

For example, the stored strain energy in a helical-coil compression spring, when deflected by axial load F, may be written, using (14-61) and (14-21), as

$$U_{\substack{helical\\compr}} = \frac{F\left(\dfrac{64FR^3N}{d^4G}\right)}{2} = \frac{32FR^3N}{d^4G} \text{ in-lb} \tag{14-62}$$

For maximum energy storage efficiency, as much of the spring material as possible should be stressed to the maximum allowable stress for the material. Thus the use of a direct tension bar as a spring would provide maximum efficiency because *all* of the material is equally stressed. Such a configuration is usually not very useful, however, because of the inherent high stiffness, large loads, and small deflections associated with energy storage in direct tension members. Thin tubes in torsion have good energy storage efficiency because the shearing stress is relatively uniform all across the wall. *Solid* members in bending or torsion have much lower energy storage efficiency because the *central* material is stressed to much lower levels (see Figures 4.3 and 4.7).

For comparison purposes, springs may be divided into *E springs* (loading produces primarily *normal* stress) and *G springs* (loading produces primarily shearing stress.[30]) A form coefficient, C_F, may be defined to characterize the *influence of geometry* when writing expressions for stored strain energy per unit volume. Using this approach,

$$u_v = C_F\left(\frac{\sigma_{max}^2}{2E}\right) \quad \text{(for E springs)} \tag{14-63}$$

and

$$u_v = C_F\left(\frac{\tau_{max}^2}{2G}\right) \quad \text{(for G springs)} \tag{14-64}$$

Form coefficients for most of the springs discussed in this chapter are shown in Table 14.10. Typical amounts of strain energy that can be stored within the spring *space envelope* are also shown in Table 14.10.[31]

TABLE 14.10 Form Coefficients and Energy Storage Capacity for Various Springs

Spring Type	Category	C_F	Typical Amounts of Energy Stored in Spring Space Envelope,[1] $\frac{\text{ft-lb}}{\text{in}^3}\left(\frac{\text{J}}{\text{mm}^3}\right)$
Tension bar	E	1.0	
Torsion tube; circular; thin wall	G	Approx. 0.90	
Torsion bar; circular; solid	G	0.50	

[30]See ref. 7. [31]See ref. 2.

TABLE 14.10 (*Continued*)

Spring Type	Category	C_F	Typical Amounts of Energy Stored in Spring Space Envelope,[1] $\frac{\text{ft-lb}}{\text{in}^3}\left(\frac{\text{J}}{\text{mm}^3}\right)$
Helical coil; round wire; compression or extension	G	Approx. 0.36	1.8–18 ($1.5–15\times10^{-4}$)
Spiral torsion; flat strip	E	Approx. 0.28	12–20 ($10–17\times10^{-4}$)
Helical-coil torsion; round wire	E	Approx. 0.20	1.2–6 ($1.0–5\times10^{-4}$)
Cantilever or simply supported beam spring; rectangular cross section	E	0.11	
Cantilever or simply supported constant-strength multileaf spring; flat strip	E	Approx. 0.38	
Belleville spring	E	Approx. 0.05–0.20	0.6–6 ($0.5–5\times10^{-4}$)

[1]See also ref. 1; for reference, the energy stored per unit volume of space envelope for a typical lead-acid battery is about 3000-4000 ft-lb/in^3 ($2500-3300\times10^{-4}$ J/mm^3).

Example 14.4 Energy Storage in Torsion Bar Springs

The value of the form coefficient given in Table 14.10 for a torsion bar of solid circular cross section is $C_F = 0.50$.

a. Explain how this numerical value has been obtained.

b. Explain why the value of C_F for a thin-walled circular torsion tube is so much larger than for a solid torsion bar.

Solution

a. Based on an adaptation of (14-61) to torsional loading, and using (14-51), the total stored strain energy, U, in a torsion bar is

$$U = T_{ave}\theta = \frac{T\theta}{2} = \frac{16T^2L}{\pi d_o^4 G} \tag{1}$$

The volume of material in a solid cylindrical torsion bar is

$$v = \frac{\pi d_o^2 L}{4} \tag{2}$$

Thus the strain energy stored per unit volume, u_v, is

$$u_v = \frac{U}{v} = \frac{16T^2L(4)}{\pi d_o^4 G \pi d_o^2 L} = \frac{64T^2}{\pi^2 d_o^6 G} \tag{3}$$

From (14-50)

$$\tau_{max} = \frac{16T}{\pi d_o^3} \quad (4)$$

Equating (3) to (14-64), and inserting (4),

$$C_F = \frac{\left(\dfrac{64T^2}{\pi^2 d_o^6 G}\right)}{\left(\dfrac{16T}{\pi d_o^3}\right)^2 \dfrac{1}{2G}} = 0.50 \quad (5)$$

b. C_F for a thin-walled tube is much larger because the "lazy" low-stress center material is eliminated, producing a nearly uniform high stress throughout the wall thickness.

Problems

14-1. You are asked, as a consultant, to determine a procedure for finding a "best estimate" for the *design stress* to be used in designing the helical-coil springs for a new off-the-road vehicle. The only *known* information is:

1. The spring material is a ductile high-strength ferrous alloy with known ultimate strength, S_u, and known yield strength, S_{yp}.
2. Spring deflection during field operation is estimated to range from a maximum of δ_{max} to a minimum of $\delta_{min} = 0.30\delta_{max}$.
3. Very long life is desired.

Based on the known information, write a concise step-by-step procedure for determining a "best estimate" value for the design stress.

14-2. An open-coil helical-coil compression spring has a spring rate of 80 lb/in. When loaded by an axial compressive force of 30 lb, its length was measured to be 0.75 inch. Its solid height has been measured as 0.625 inch.

a. Calculate the axial force required to compress the spring from its free length to its solid height.

b. Calculate the free length of the spring.

14-3. An open-coil helical-coil compression spring has a free length of 76.2 mm. When loaded by an axial compressive force of 100 N, its length is measured as 50.8 mm.

a. Calculate the spring rate of this spring.

b. If this spring, with a free length of 76.2 mm, were loaded by an axial *tensile* force of 100 N, what would you predict its corresponding length to be?

14-4. A helical-coil compression spring has an outside diameter of 1.100 inches, a wire diameter of 0.085 inch, and has closed and ground ends. The solid height of this spring has been measured as 0.563 inch.

a. Calculate the inner coil radius.

b. Calculate the spring index.

c. Estimate the Wahl factor.

d. Calculate the approximate total number of coils, end-of-wire to end-of-wire, in this spring.

14-5. An existing helical-coil compression spring has been wound from 3.50-mm peened music wire into a spring having an outside diameter of 22 mm and 8 active coils. What maximum stress and deflection would you predict if an axial static load of 27.5 N were applied?

14-6. An existing helical-coil compression spring has been wound from unpeened music wire of 0.105-inch diameter into a spring with mean coil radius of 0.40 inch. The applied axial load fluctuates continuously from zero to 25 lb, and a design life of 10^7 cycles is desired. Determine the existing safety factor for this spring as used in this application.

14-7. A round wire helical-coil compression spring with closed and ground ends must work inside a 2.25-inch-diameter hole. During operation the spring is subjected to a cyclic axial load that ranges from a minimum of 150 lb to a maximum of 550 lb. The spring rate is to be approximately 150 lb/in. A life of 200,000 load cycles is required. Initially, assume $k_{N=2\times10^5} = 0.85$. A design safety factor of 1.2 is desired. Design the spring.

14-8. A helical-coil spring with plain ends, ground, is to be used as a return spring on the cam-driven valve mechanism shown in Figure P14-8. The 1.50-inch-diameter rod must pass freely

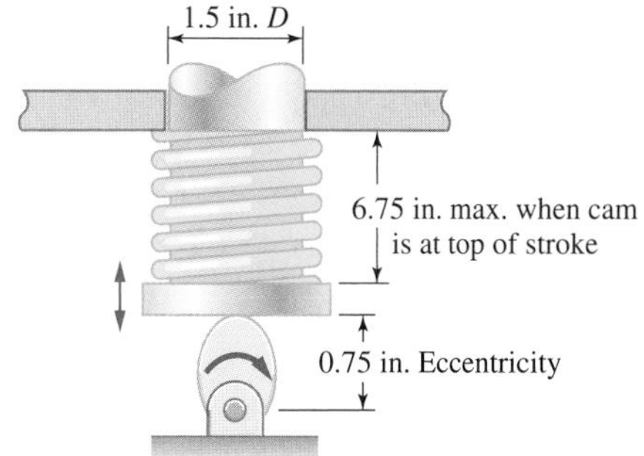

Figure P14.8
Cam-driven valve mechanism with preloaded return spring.

through the spring. The cam eccentricity is 0.75 inch (i.e., the total stroke is 1.50 inches). The height of the compressed spring when the cam is at the head-end-dead-center (HEDC) position is 3.0 inches, as shown. The spring must exert a force of 300 lb when at the HEDC position shown in the sketch, and must exert a force of 150 lb when at crank-end-dead-center (CEDC) at the bottom of the stroke. That is, the spring is preloaded into the machine. The spring is to be made of a patented spring steel wire that has a 200,000-psi ultimate tensile strength, 190,000-psi tensile yield strength, and 90,000-psi fatigue endurance limit. A safety factor of 1.25 is desired, based on infinite life design. Determine the following:

a. Mean coil radius, R

b. Wire diameter, d

c. Number of active coils, N

d. Spring rate, k

e. Free length of the spring, L_f

14-9. A proposed helical-coil compression spring is to be wound from standard unpeened music wire of 0.038-inch diameter, into a spring with outer coil diameter of $^7/_{16}$ inch and $12^1/_2$ total turns from end-of-wire to end-of-wire.
The ends are to be closed. Do the following:

a. Estimate torsional yield strength of the music wire.

b. Determine the maximum applied axial load that could be supported by the spring without initiating yielding in the wire.

c. Determine the spring rate of this spring.

d. Determine the deflection that would be produced if the incipient yielding load calculated in (b) above were applied to the spring.

e. Calculate the solid height of the spring.

f. If no permanent change in free height of the spring can be tolerated, determine the free height that should be specified so that when the spring is compressed to solid height and then released, the free height remains unchanged.

g. Determine the maximum operating deflection that should be recommended for this spring if no preload is anticipated.

h. Determine whether buckling of this spring might be a potential problem.

14-10. A helical-coil compression spring is to be designed for a special application in which the spring is to be initially assembled in the mechanism with a preload of 10 N, and exert a force of 50 N when it is compressed an additional 140 mm. Tentatively, it has been decided to use music wire, to use closed ends, and to use the smallest standard wire diameter that will give a satisfactory performance. Also, it is desired to provide a clash allowance of approximately 10 percent of the maximum operating deflection.

a. Find a standard wire diameter and corresponding mean coil radius that will meet the desired specifications.

b. Find the solid height of the spring.

c. Find the free height of the spring.

14-11. Two steel helical-coil compression springs are to be nested about a common axial centerline. The outer spring is to have an inside diameter of $1^1/_2$ inches, a standard wire diameter of 0.125 inch, and 10 active coils. The inner spring is to have an outside diameter of $1^1/_4$ inches, a standard wire diameter of 0.092 inch, and 13 active coils. Both springs are to have the same free length. Do the following:

a. Calculate the spring rate for each spring.

b. Calculate the axial force required to deflect the nested spring assembly a distance of 1.0 inch.

c. At an assembly deflection of 1.0 inch, as in (b), which spring will be more highly stressed?

14-12. A round wire helical-coil tension spring has end loops of the type shown in Figures 14.7(a) and (b). The wire diameter of the spring is 0.042 inch, and the mean coil radius is 0.28 inch. Pertinent end-loop dimensions are (ref. Figure 14.7) r_{iA} = 0.25 inch and r_{iB} = 0.094 inch. An applied axial static tension force of F = 5.0 lb is to be applied to the spring.

a. Estimate the maximum stress in the wire at critical point A.

b. Estimate the maximum stress in the wire at critical point B.

c. If the spring wire is ASTM A227 material, and a safety factor of 1.25 is desired, would the stresses at critical points A and B be acceptable?

14-13. For the helical-coil tension spring of problem 14-12, calculate the maximum stress in the main body of the spring, away from the ends, and identify where the critical point occurs. If the wire material is ASTM A227, would failure of the spring wire be expected?

14-14. A round wire helical-coil tension spring is to be used as a return spring on a cam-driven lever, as shown in Figure P14.14. The spring must be pretensioned to exert a 10-lb force at the "bottom" of its stroke, and should have a spring rate of 20 lb/inch. The peak-to-peak operating deflection for this spring is 1 inch. The spring is to be made of a patented steel-alloy spring wire which has a 200,000-psi ultimate strength, 190,000-psi yield strength, and 90,000-psi fatigue endurance limit. It is desired to have spring index of 8, and a safety factor of 1.5 has been selected. Design a lightweight spring for this application,

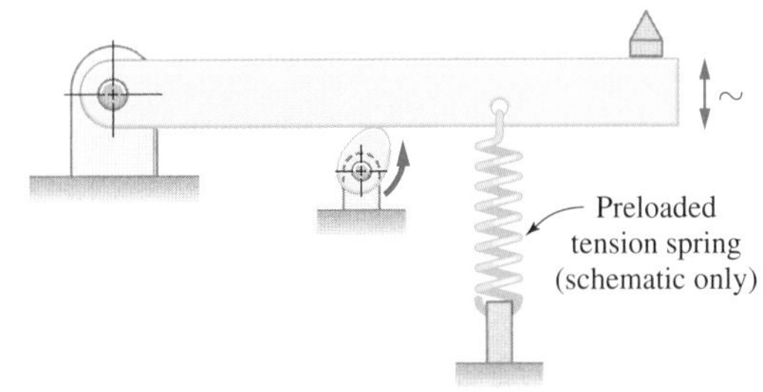

Figure P14.14
Cam-driven lever with return spring.

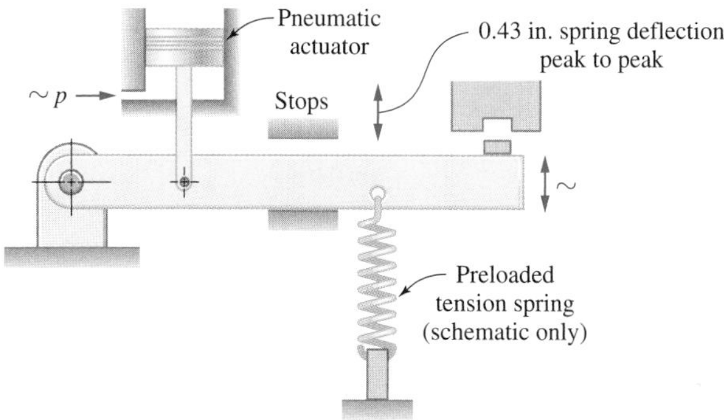

Figure P14.15
Return spring for pneumatically actuated lever.

specifically determining the following:

a. Wire diameter, d

b. Mean coil radius, R

c. Number of active coils, N

14-15. The round wire helical-coil tension spring shown in Figure P14.15 is to be used to make a return spring for a pneumatically actuated lever that operates between fixed stops, as shown. The spring must be pretensioned to 25 lb at the bottom stop (minimum load point), and operates through a total spring deflection of 0.43 inch, where it is halted by the upper stop, then returns to the lower stop and repeats the cycle. The spring is made of No. 12 wire (0.105-inch diameter), has a mean coil radius of 0.375 inch, and has been wound with 15 full coils plus a turned-up half loop on each end for attachment. Material properties for the spring wire are S_u = 200,000 psi, S_{yp} = 185,000 psi, e (2 inches) = 9 percent, S_f = 80,000 psi, $E = 30 \times 10^6$ psi, $G = 11.5 \times 10^6$ psi, and $\nu = 0.30$. Compute the existing safety factor for this spring, based on an infinite-life design criterion.

14-16. A round wire open-coil helical-coil spring is wound using No. 5 patented steel wire (d = 0.207 inch), with a mean coil radius of 0.65 inch. The spring has 15 active coils, and its free height is 6.0 inches. The material properties for the spring wire are given in Figure P14.16. The spring is to be used in an application where it is axially deflected 1.0 inch from its free height into tension, then 1.0 inch from its free height into compression during each cycle, at a frequency of 400 cycles per min.

a. Estimate the expected life in cycles before this spring fails.

b. Would *buckling* of this open-coil spring be expected?

c. Would you expect *surging* of the spring to be a problem in this application?

d. How much energy would be stored in the spring at maximum deflection?

14-17. A helical-coil spring is to be wound using No. 9 gage wire (d = 0.148 inch) made from a proprietary ferrous alloy for which only the static properties are known. These properties are S_u = 250,000 psi, S_{yp} = 230,000 psi, e (2 inches) = 7 percent, $E = 30 \times 10^6$ psi, $G = 11.5 \times 10^6$ psi, and $\nu = 0.35$. It is desired to use the spring in a cyclic loading situation where the axial load on the spring during each cycle ranges form 100-lb tension to 100-lb compression. The spring deflection at maximum load must be 2.0 inches. A spring with 18 active coils is being proposed for the application.

a. Compute the existing safety factor for this spring based on an infinite-life design criterion. Comment on your result.

b. If the spring is wound so that when it is in unloaded condition the space between coils is the same as the wire diameter, would you expect buckling to be a problem? (Support your answer with appropriate calculations.)

c. Approximately what *maximum* operating frequency should be specified for this mechanism?

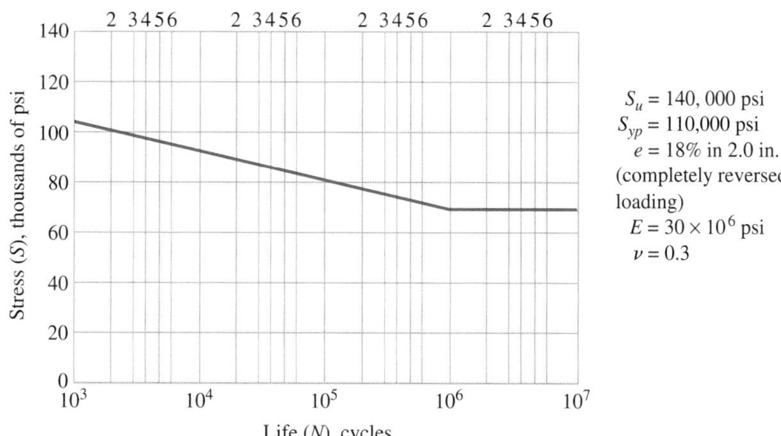

Figure P14.16
Material properties for patented steel spring wire.

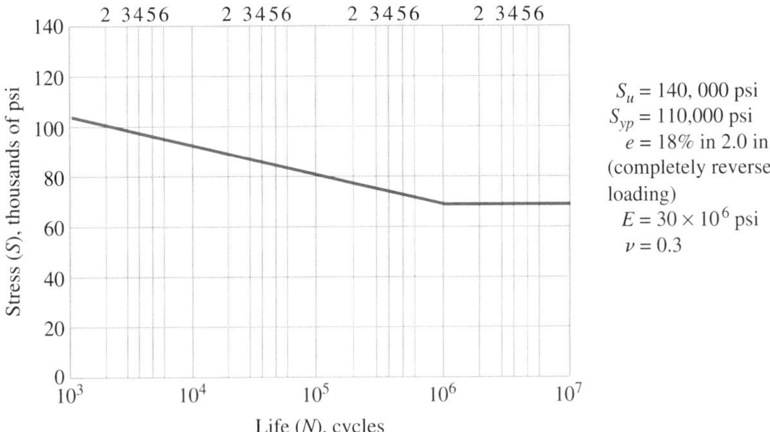

Figure P14.22
Material properties for multileaf spring.

14-18. For a simply supported flat-beam spring of rectangular cross section, loaded at midspan, answer the following questions:

a. What is the primary stress pattern in the beam spring?

b. Is it uniaxial or multiaxial?

c. If there are secondary stresses to be considered, what are they?

14-19. Derive an equation for the *maximum stress* in an end-loaded multileaf cantilever spring with n leaves, each having a width of b_1 and thickness t.

14-20. Derive an equation for the *spring rate* of an end-loaded multileaf cantilever spring with n leaves, each having a width of b_1 and thickness t.

14-21. A horizontal cantilever-beam spring of constant rectangular cross section is loaded vertically across the free end by a force that fluctuates cyclically from 1000 lb down to 5000 lb up. The beam is 5.0 inches wide and 10 inches long. The material is a ferrous alloy with S_u = 140,000 psi, S_{yp} = 110,000 psi, and S_f = 70,000 psi. If a safety factor of 1.5 is desired, and the cantilever-beam spring is to be designed for infinite life, what thickness should it be made? Neglect stress concentration.

14-22. A horizontal simply supported multileaf spring is to be subjected to a cyclic midspan load that fluctuates from 2500 lb down to 4500 lb down. The spring is to have 8 leaves, each 3.0 inches wide. The distance between shackles (simply supports) is to be 22.0 inches. Properties of the selected spring material are given in Figure P14.22.

a. Neglecting stress concentration effects, how thick should the leaves be made to provide infinite life, with a design safety factor of 1.2?

b. What would be the spring rate of this spring?

14-23. A multileaf simply supported truck spring is to be designed for each rear wheel, using AISI 1095 steel (see Table 3.3) with a hardness of Rockwell C-42. The truck weight is 3500 lb, with 65 percent of the weight on the rear wheels. The static midspan deflection is 4 inches and the maximum deflection in operation is 8 inches. The loading may be considered to be *released* cyclic loading. The length of the spring between supports must be between 48 inches and 66 inches. It has been decided that a safety factor of 1.3 should be used. Design a leaf spring to meet the requirements if infinite life is desired.

14-24. A single-piece torsion bar spring of the type sketched in Figure P14.24(a) is being considered by a group of students as a means of supporting the hood of an experimental hybrid vehicle being developed for intercollegiate competition. The maximum length L that can be accommodated is 48 inches. Figure P14.24(b) illustrates their concept. They plan to install the counterbalancing torsion bar spring along the hood-hinge centerline, with one of the 3-inch integral end-levers in contact with the hood, as shown. The hood will then be raised until it contacts the 45° hood stop; the 3-inch support lever on the opposite end will be rotated until the hood is just held in contact with the hood stop without any other external lifting force on the hood. The support lever will next be given an additional rotation to lightly preload the hood against the stop, and then clamped to the supporting structure.

a. Determine the diameter d_o of a solid-steel torsion bar that would counterbalance the hood weight and provide a 10 ft-lb torque to hold the hood against the stop shown, if the design stress in shear for the material is τ_d = 60,000 psi.

b. At what angle, with respect to a horizontal datum, should the clamp for the 3-inch support lever be placed to provide the desired 10 ft-lb preload torque? Neglect bending of the integral levers.

c. Make a plot showing gravity-induced torque, spring torque, and net torque, all plotted versus hood-opening angle.

d. Is the operating force, F_{op}, required to open or close the installed hood, reasonable for this design configuration?

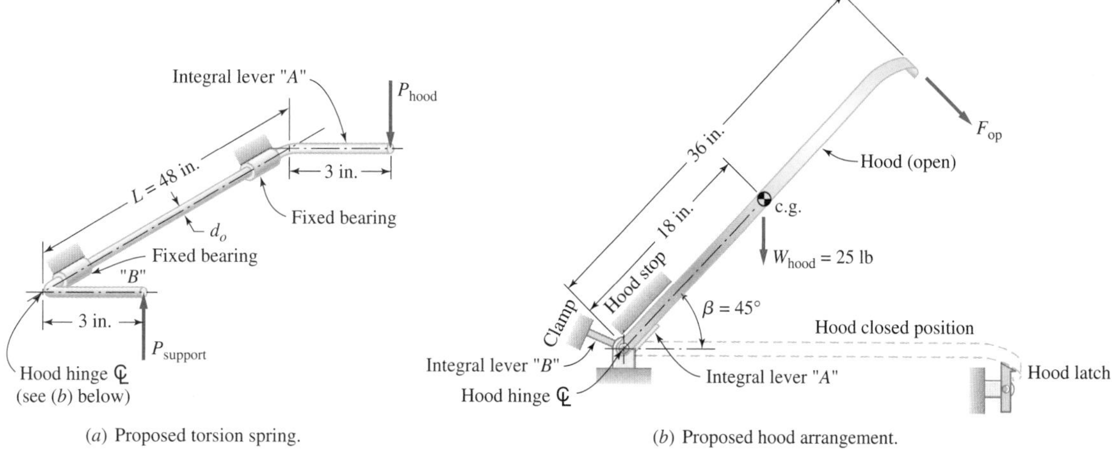

Figure P14.24
Proposed torsion bar spring for supporting the hood of an experimental vehicle.

What other potential problems can you foresee with the arrangement?

14-25. A single-bodied helical-coil torsion spring (see Figure 14.20) has a wire diameter of 0.038 inch and an outside coil diameter of 0.375 inch. The tightly coiled spring has 9.5 coils, with an end extension of a = 0.50 inch from the coil center to the point of load application at each end. The spring material is ASTM A401 Cr-Si steel.

a. Calculate the torsional spring rate for this spring.

b. If a torque of 1.0 in-lb were applied to this spring, what angular deflection (in degrees) would be expected? Neglect the contribution of end extensions.

c. What maximum stress would be predicted in the spring wire under the 1.0 in-lb torque?

d. If a design safety factor of 1.5, based on minimum ultimate tensile strength, is desired, would the spring design be acceptable as proposed?

14-26. A matched pair of torsional helical-coil springs, one left-hand and the other right-hand, is scheduled for use to counterbalance the weight of a residential overhead garage door. The arrangement is sketched in Figure P14.26. The 1-inch-diameter rotating shaft is supported on three bearings near the top of the door, one bearing at each end, and one at midspan. A small wire rope is wrapped around each of the pulleys to sym-

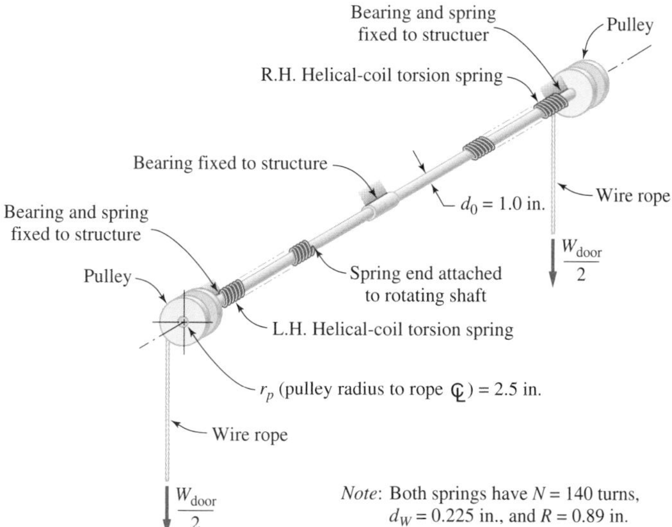

Figure P14.26
Sketch of a pair of helical-coil torsion springs used to counterbalance an overhead garage door.

metrically support the weight of the door. The 2.5-inch pulley radii are measured to the wire rope centerlines. Each spring is wound from standard oil-tempered steel wire having a wire diameter of 0.225 inch, a mean coil radius of 0.89 inch, and 140 turns, closely coiled. The total length of wire-rope excursion from spring-unloaded position to door-closed position is 85 inches.

a. Calculate the maximum bending stress in the springs when the door is closed, and tell where it occurs.

b. What would be the heaviest garage door that could be counterbalanced using the arrangement of Figure P14.26 and this matched pair of springs?

c. Estimate the free length and the weight of each spring.

14-27. As a diversion, a "machine design" professor has built a new rough-sawn cedar screened porch as an attachment to the back of his house. Instead of using a traditional "screen-door spring" (closed-coiled helical-coil tension spring) to keep his screen door closed, he has decided to use a helical-coil *torsion* spring of the single-bodied type shown in Figure 14.20(a). He plans to install the spring so its coil centerline coincides with the door-hinge centerline. The distance from the hinge centerline to the pull-handle is to be 32 inches, and his goal is to provide a handle-pull of 1 lb when the door is closed and a pull of 3 lb after the door has rotated open through an angle of 180° about its hinge centerline. Tentatively, a standard no. 6 music wire has been chosen ($d = 0.192$ inch) for the spring. A design stress of $\sigma_d = 165{,}000$ psi has been calculated, based on yielding as the probable governing failure mode.

a. Calculate the required mean coil radius for the spring.

b. Find the initial angular *preload* displacement of the spring that will produce a 1-lb pull at the handle to open the door from its closed position.

c. What would be the required number of active coils for the spring?

14-28. An unlabeled box of steel *Belleville spring washers* (all washers in the box are the same) has been found in a company storeroom. As a summer-hire, you have been asked to analytically evaluate and plot the force-deflection characteristics of the spring washers. The dimensions of the washers in the unlabeled box, with reference to the sketch of Figure 14.21, are as follows:

$D_o = 115$ mm

$D_i = 63.9$ mm

$t = 2.0$ mm

$h = 3.0$ mm

Do the following:

a. Estimate the force required to just "flatten" one of the Belleville washers.

b. Plot a force-deflection curve for one of the washers using force magnitudes ranging from *zero* up to the full *flattening force*. Characterize the curve as linear, nonlinear hardening, or nonlinear softening.

c. Plot a force-deflection curve for two of the washers stacked together in parallel. Characterize the curve as linear, nonlinear hardening, or nonlinear softening.

d. Plot a force-deflection curve for two of the washers stacked together in series. Characterize the curve as linear, nonlinear hardening, or nonlinear softening.

e. Calculate the magnitude of the highest tensile stress that would be expected in the single washer of (b) above at the time that the applied load just "flattens" the washer.

14-29. Considering a solid square bar of steel with side-dimension s and length-dimension L, would you predict that more elastic strain energy could be stored in the bar (without yielding) by using it as a *direct tension spring* axially loaded in the L-direction, or by using it as a *cantilever-bending spring* loaded perpendicular to the L-direction? Make appropriate calculations to support your prediction.

14-30. a. Write the equations from which the form coefficient C_F may be found for a simply supported center-loaded multileaf spring.

b. Find the numerical value of C_F for this type of spring and compare it with the value given in Table 14.10.

14-31. Repeat problem 14-30, except for the case of a simply supported beam spring of rectangular cross section.

14-32. Repeat problem 14-30, except for the case of an end-loaded cantilever-beam spring of rectangular cross section.

14-33. Repeat problem 14-30, except for the case of a spiral flat-strip torsion spring.

14-34. Repeat problem 14-30, except for the case of a round wire helical-coil torsion spring.

14-35. Repeat problem 14-30, except for the case of a round-wire helical-coil compression spring.

Chapter 15

Gears and Systems of Gears

15.1 Uses and Characteristics of Gears

When it is desired to transmit or transfer power or motion from one rotating shaft to another, many alternatives are available to a designer, including flat belts, V-belts, toothed timing belts, chain drives, friction wheel drives, and gear drives.[1] If smooth slip-free uniform motion, high speed, light weight, precise timing, high efficiency, or compact design are important design criteria, the selection of an appropriate system of gears will, in nearly all cases, fulfill these criteria better than any of the other alternatives. On the other hand, belt drives and chain drives are often less costly, and may be used to advantage when input and output shafts are widely spaced.

Figure 15.1
Friction wheel drives and analogous spur gear drive geometry.

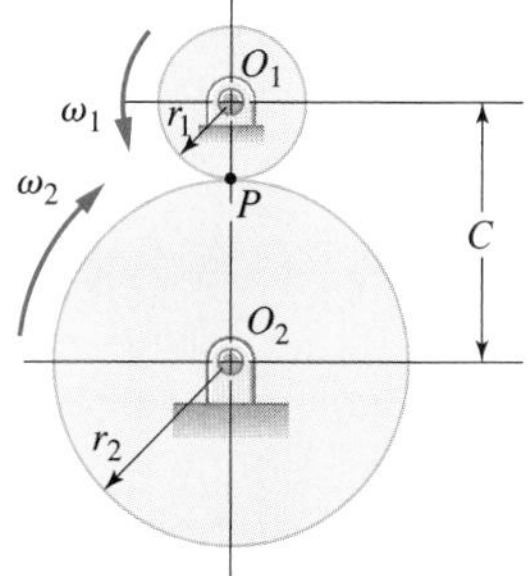

(*a*) Simple external friction drive.

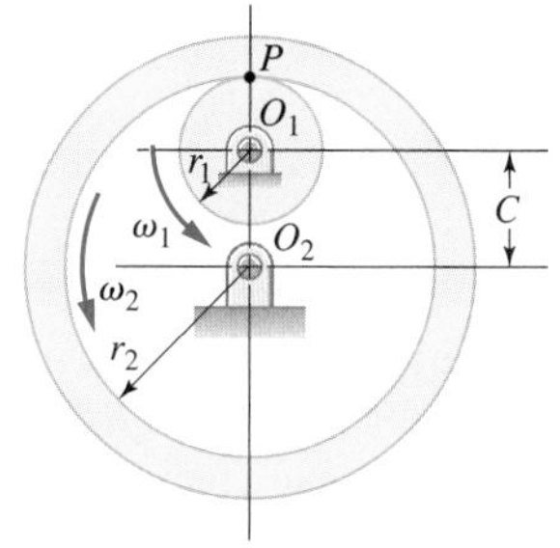
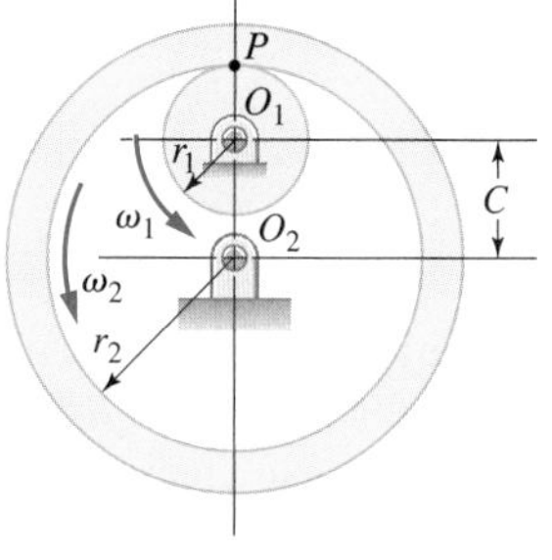

(*b*) Simple internal friction drive.

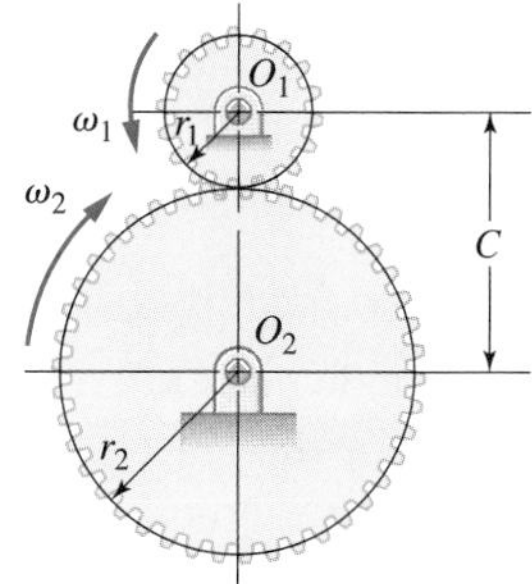

(*c*) Straight tooth spur gear drive with same angular velocity ratio as friction drive shown in (*a*).

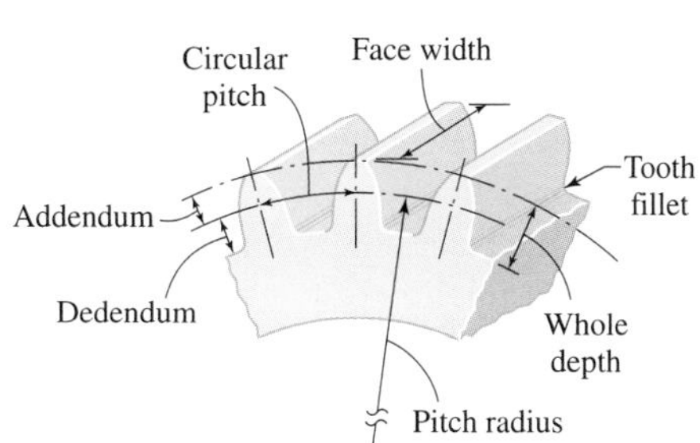

(*d*) Basic shape and terminology for straight tooth spur gear teeth, such as those in (*c*).

[1]See also Chapter 17.

Simple friction wheel drives, such as the external and internal drives sketched in Figures 15.1(a) and (b), may provide a smooth transmission of power from input cylinder 1 (driving) to output cylinder or annulus 2 (driven) if no slip occurs at contact site P. For the case of *no slip*, the magnitudes of tangential velocity are equal for the two contacting members; hence

$$|r_1\omega_1| = |r_2\omega_2| \tag{15-1}$$

where r_1, r_2 = radius of member 1, 2, respectively

ω_1, ω_2 = angular velocity of member 1, 2, respectively

For the *external drive* shown in Figure 15.1(a) the direction of ω_2 is *opposite* ω_1; for the *internal drive* shown in Figure 15.1(b) the direction of ω_2 is the *same* as ω_1. Conventional practice is to attach a minus sign when directions are opposite and a plus sign when directions are the same. For an external drive, therefore, the velocity ratio is taken as negative and for an internal drive it is taken as positive.

For light loads that *do not cause slip*, a friction drive may provide acceptable performance. If operating conditions *do tend to cause slip*, or if precise phasing (constant angular velocity ratio and absolutely no slip) is required between input and output shafts, a series of precise *intermeshing teeth* may be incorporated on each of the mating friction surfaces to ensure no-slip operation. If, as shown for the *external* drive of Figure 15.1(c), the teeth are arranged to extend both inside (*dedendum*) and outside (*addendum*) of the *pitch circles*[2] and are extended parallel to the cylinder axes (*face width*), the resulting toothed wheels become *straight-tooth spur gears* with pitch radii r_1 and r_2. Likewise, teeth may be incorporated on mating members of an *internal* drive. The smaller of two meshing gears is usually called the *pinion* and the larger is called the *gear*. Typically, the pinion is the driver (input) and the gear is the driven member (output).

The teeth must be carefully shaped so they do not interfere with each other as the gears rotate, and so that the *angular velocity ratio between the driving pinion and the driven gear neither increases nor decreases at any instant* as successive teeth pass through the *mesh* (region of tooth contact). *If these conditions are met, the gears are said to fulfill the fundamental law of gearing*. The kinematic conditions for the fundamental law of gearing and the process of determining profiles of meshing teeth that satisfy these conditions have been widely published[3], and are briefly summarized in 15.6. The basic nomenclature associated with spur gear teeth is shown in Figure 15.1(d) and discussed more fully in 15.6.

15.2 Types of Gears; Factors in Selection

Selection of the best type of gearing for a particular design scenario depends upon many factors, including geometric arrangement proposed for the machine, reduction ratio required, power to be transmitted, speeds of rotation, efficiency goals, noise-level limitations, and cost constraints.

In general, three shafting arrangements are encountered by the designer when contemplating the transmission of power or motion from one rotating shaft to another. They are (1) applications in which the *shaft axes are parallel*, (2) applications in which *shaft axes intersect*, and (3) applications in which *shaft axes are neither parallel nor do they intersect*.

Types of gears for use when the shaft axes are parallel are sketched in Figure 15.2. *Straight-tooth spur gears*, as shown in Figures 15.2(a) and (b), are relatively simple to de-

[2]The pitch circles correspond to the peripheries of the friction cylinders of Figure 15.1(a).

[3]See, for example, refs. 1, 2, 3, or 4.

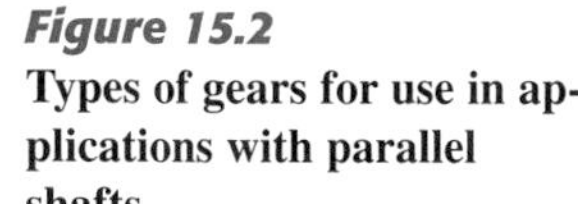

Figure 15.2
Types of gears for use in applications with parallel shafts.

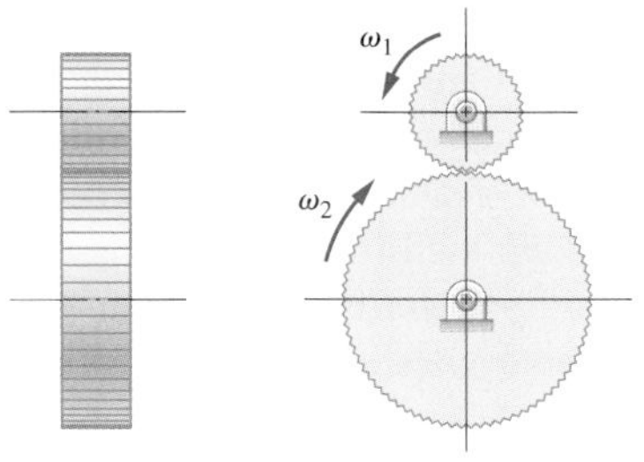

(*a*) External straight tooth spur gears.

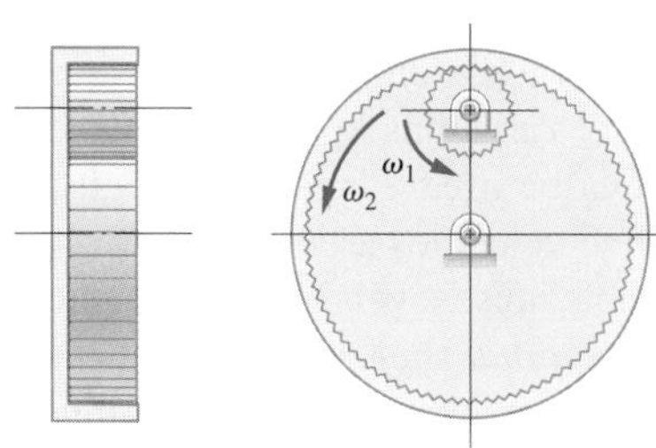

(*b*) Internal straight tooth spur gears.

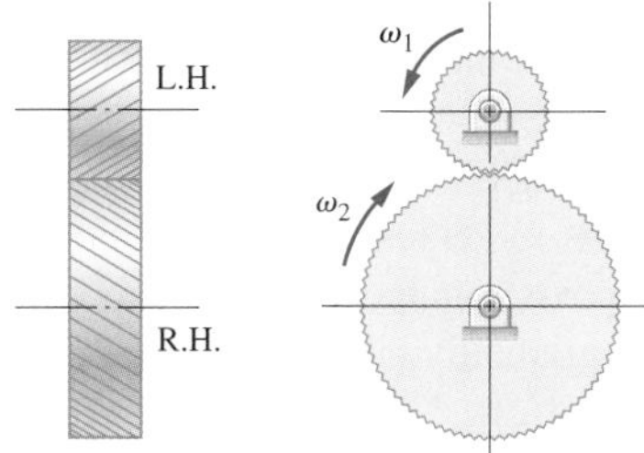

(*c*) External helical gears.

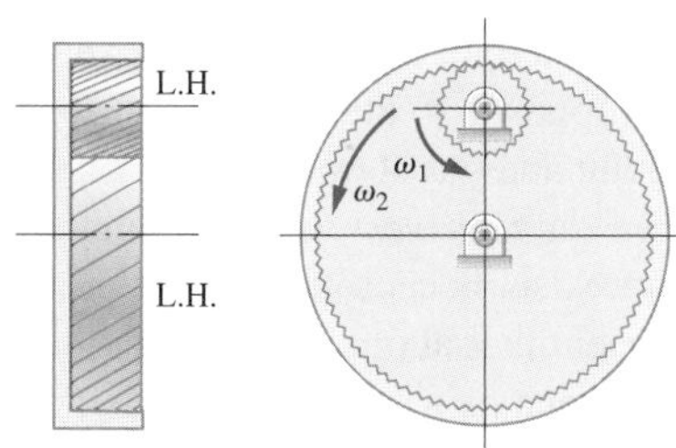

(*d*) Internal helical gears.

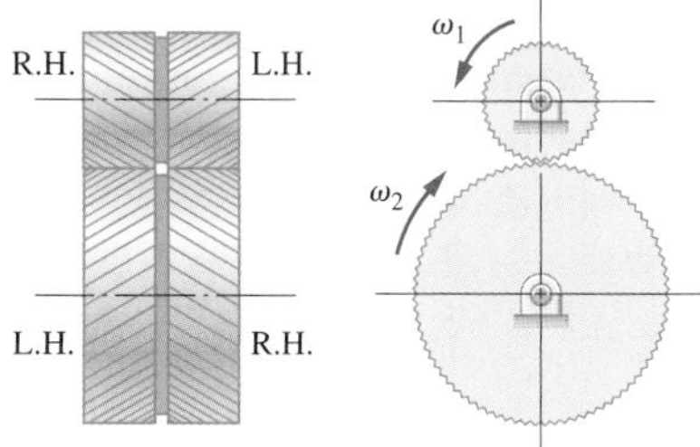

(*e*) Double helical gears (herringbone gears).

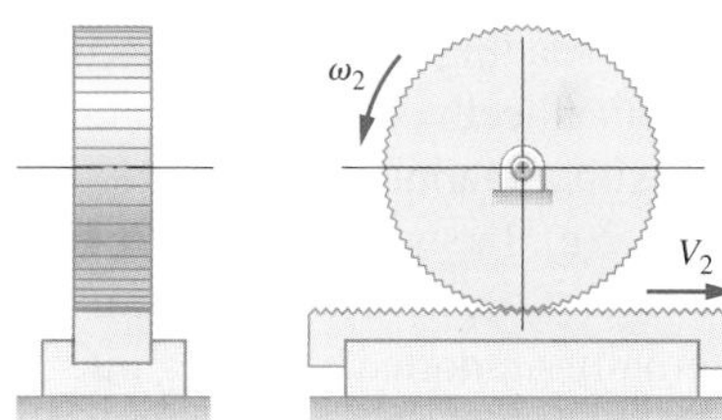

(*f*) Straight tooth spur rack and pinion (can also be helical).

Some examples of gears.

sign, manufacture, and check for precision, and are relatively inexpensive. They impose only radial loads on supporting bearings. Tooth profiles are ordinarily *involute*[4] in shape, and small variations in center distance are therefore usually well tolerated. Although spur gears can be used at speeds as high as other types of gears, they are usually limited to pitch-line velocities around 20 m/s (4000 fpm) to avoid high-frequency vibration and unacceptable noise levels[5]. *External* spur gears are usually preferred when design constraints permit; *internal* spur gears are sometimes used to achieve a short center distance, and are a necessity in most epicyclic[6] gearing arrangements.

Helical gears, as shown in Figures 15.2(c) and (d), are very similar to spur gears except that their teeth are angled with respect to the axis of rotation to form parallel helical spirals. The helixes for the two mating external gears must be manufactured to have the same helix angle, but for parallel shaft applications the hand of the helix on the pinion must be opposite to the hand of the gear. Because of the angled teeth, helical gears impose both radial and thrust (axial) loads on supporting bearings. Tooth profiles are ordinarily *involute in the transverse section*[7] and small variations in center distance are not a serious problem. Also, because of the angled teeth, as the gears rotate each tooth pair comes into engagement first at one end, with contact spreading gradually along a diagonal path across the tooth face as rotation continues [See Figure 15.33(c)]. This gradual-engagement pattern produces smoother and quieter operation than with straight-tooth spur gears. Pitch-line velocities in helical gearing applications have been permitted to exceed 50 m/s (10,000 fpm) in some cases without violating design criteria for vibration, noise, and design life. Internal helical gears, as sketched in Figure 15.2(d), are sometimes used. To eliminate thrust loading on support bearings when high horsepower must be transmitted, two sets of helical teeth of opposite hand are integrally manufactured on a single pinion, then meshed with a mating set of double-helical teeth on a single-gear wheel, as illustrated in Figure 15.2(e). In such *double-helical* or *herringbone* gears, the thrust loads generated by corresponding opposite-hand helical teeth (equal in magnitude and opposite in direction) are internally supported, inducing internal axial direct stresses in the pinion and in the gear, but eliminating thrust loads on supporting bearings.

A special case of gearing used for parallel shafts is the *rack-and-pinion* drive, illustrated in Figure 15.2(f). The straight rack may be considered as a segment of a gear having infinite pitch radius. The rotary motion of the pinion is converted into linear motion of the straight rack in this arrangement. Rack-and-pinion drives may be manufactured with either straight spur or helical teeth.

Types of gears for use when the *shaft axes intersect* are sketched in Figure 15.3. *Straight bevel gears*, as shown in Figure 15.3(a), represent the simplest type of gearing used for intersecting shafts. Usually the shaft axes intersect at an angle of 90° but almost any angle can be accommodated. The pitch surfaces for straight bevel gears are tangent conical frustra, as compared to tangent cylindrical surfaces for straight spur gears [see Figure 15.1(a)]. Bevel gear teeth are tapered in both tooth thickness and height, from a larger tooth profile at one end to a smaller tooth profile at the other end. Bevel gears impose both radial and thrust loads on supporting bearings. Tooth profiles closely resemble an involute curve in a section normal to the tooth axis. Bevel gears must be accurately mounted at the proper axial distance from the pitch cone apex for proper meshing. To provide smoother operation, *Zerol*[8] bevel gears are sometimes used. As shown in Figure 15.3(b), Zerol bevel

[4]See 15.6.

[5]Substantially higher pitch-line velocities may be allowed in certain specialized high-performance applications, such as helicopter gearing.

[6]See 15.3. [7]Cross section perpendicular to the gear axis of rotation.

[8]Zerol is a registered trademark of the Gleason Works, Rochester, NY.

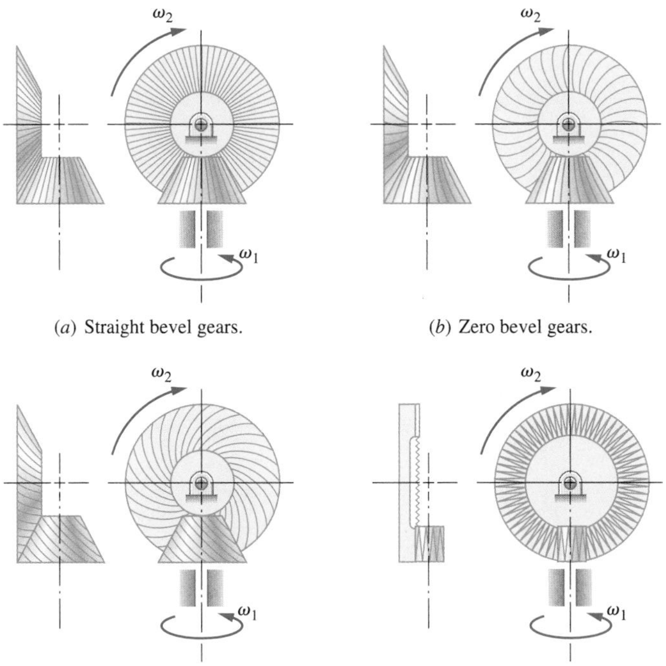

(*a*) Straight bevel gears.

(*b*) Zero bevel gears.

(*c*) Spiral bevel gears.

(*d*) Spur face gears
(can also be helical).

Figure 15.3
Types of gears for use in applications with intersecting shafts. (Zerol is a registered trademark of the Gleason Works, Rochester, NY.)

gears are similar to straight bevel gears except that they have teeth curved in their lengthwise direction, which provides a slight engagement overlap to produce smoother operation than straight bevel gears.

Spiral bevel gears, illustrated in Figure 15.3(c), are related to straight bevel gears in much the same way that helical gears are related to straight-tooth spur gears, thus providing the advantages of gradual engagement along the tooth face. Because of the conical geometry (as opposed to cylindrical), spiral bevel teeth do not have a true helical spiral, but have an appearance similar to that of a helical gear. Tooth profiles somewhat resemble an involute curve.

Face gears, which are functionally similar to bevel gears, have gear teeth cut on an annular ring at the outer edge of a gear "face," as sketched in Figure 15.3(d). A face gear is matched with a spur pinion mounted on an intersecting shaft (usually at 90°) as shown. Face gear teeth change shape from one end to the other. Pinion teeth need no special attributes to mesh with a face gear, and may be made either spur or helical. Pinion support bearings carry primarily radial loads but face gear support bearings carry both radial and thrust loads.

Finally, types of gears for use when shaft axes are neither parallel nor intersecting are sketched in Figure 15.4.

Hypoid gears resemble spiral bevel gears, but there is a small to moderate *offset* between their axes, as shown in Figure 15.4(a). The offset is measured along the perpendicular common to the two axes. If hypoid gears had zero offset, they would be spiral bevel gears.

Spiroid gears involve a face gear with teeth spirally curved along their length, mating with a tapered pinion, as sketched in Figure 15.4(b). Offsets for spiroid gearing are larger than for hypoid gearsets, and the pinion somewhat resembles a worm (see below). High reduction ratios are achievable in a compact space envelope and load-carrying capacity is good.

Figure 15.4
Types of gears for use in applications where shafts are neither parallel nor intersecting.

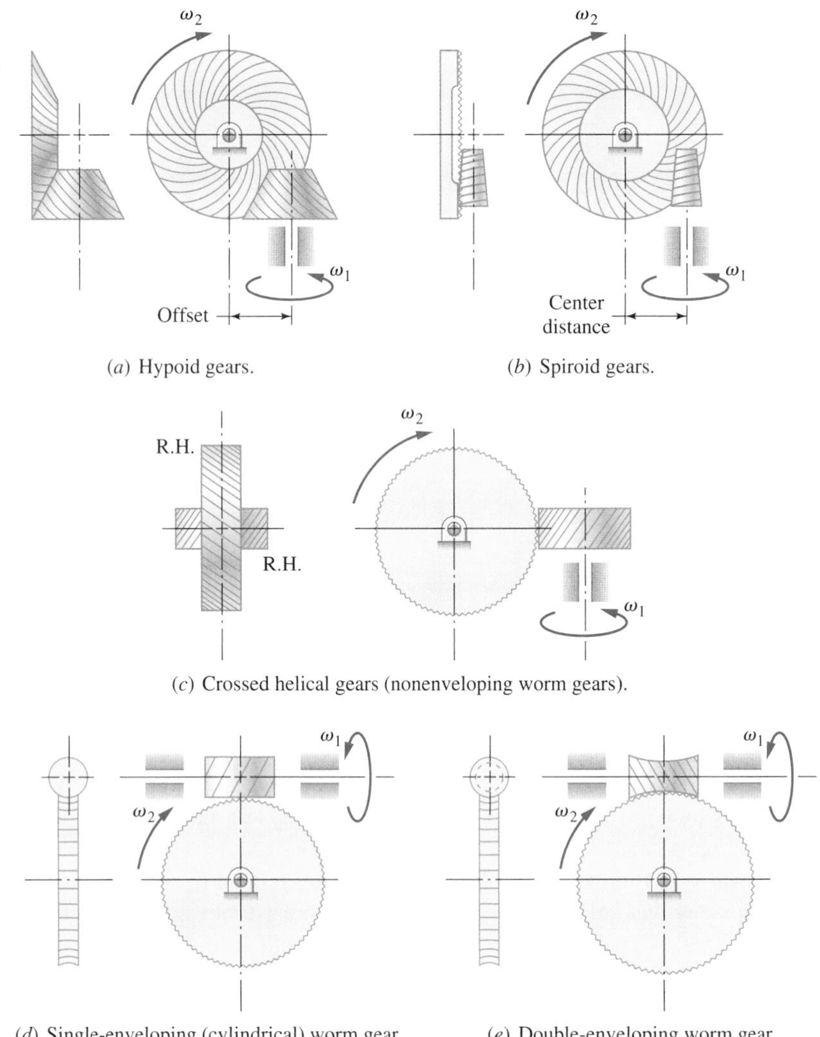

(*a*) Hypoid gears.

(*b*) Spiroid gears.

(*c*) Crossed helical gears (nonenveloping worm gears).

(*d*) Single-enveloping (cylindrical) worm gear.

(*e*) Double-enveloping worm gear.

Crossed-helical gears[9] are illustrated in Figure 15.4(c). Both members of a crossed-helical gearset have a cylindrical shape, as opposed to worm gearsets, in which one or both members are *throated* [see Figure 15.4(d) and (e)]. The nonintersecting axes of crossed-helical gears often are 90° to each other, but almost any angle can be accommodated. As the gears rotate, their meshing teeth first meet only in a *point* contact which travels across the tooth face along a sloping line; hence basic load-carrying capacity is small. After a wear-in period, the point contact evolves to a sloping *line of contact*, somewhat increasing load-carrying capacity. Usually a crossed-helical gear of one hand is meshed with a gear of the same hand, as illustrated in Figure 15.4(c), but it is possible to mesh gears of opposite hand if the shaft angle is properly set.

Worm gearsets, as illustrated in Figures 15.4(d) and (e), are characterized by a worm with "teeth" similar to the threads of a power screw (see Chapter 12), in mesh with a worm gear (worm wheel) having teeth similar to those of a helical gear, except that they are contoured to envelop the worm. Wormsets provide the easiest way to obtain large reduction ratios, but friction losses can be high because the worm threads slide sideways along the

[9]Sometimes called *nonenveloping* worm gears.

gear teeth. When only the worm wheel teeth are contoured (throated) to envelop a cylindrical worm, the wormset is called *single enveloping* or *cylindrical*, as sketched in Figure 15.4(d). When the worm profile is also throated to envelop the gear, as shown in Figure 15.4(e), the wormset is called *double enveloping*. Double-enveloping wormsets have more mesh contact area than single-enveloping wormsets and, hence greater load-carrying capacity; however, precision alignment is more critical. Worms may have single threads or multiple threads.[10] Worm shaft support bearings are typically subjected to radial loads and high thrust loads. Worm wheel bearings must support both radial and thrust loads.

15.3 Gear Trains; Reduction Ratios

A *pair* of meshing gears, such as the ones shown in Figure 15.1(c), is usually called a *gearset*, the simplest form of a *gear train*. A gear train is a sequence of several meshing gears arranged in such a way that desired output speed, torque, and/or direction of rotation are achieved using specified input conditions.

Various arrangements and sequences of gears may be used to achieve design goals. It is important to be able to readily determine the magnitude and direction of angular velocity of the output gear, given the input pinion velocity, for any gear train arrangement. Gear trains may be classified as *simple, compound*, or *epicyclic*.

A *simple* gear train such as the one shown in Figure 15.5(a) is an arrangement in which each shaft carries only one simple gear. Gear 2 in the simple gear train shown in Figure 15.5(a) serves the function of changing the *direction* of rotation of output gear 3 (as compared with a direct mesh of 1 and 3) but does not change the *magnitude* of angular velocity ratio between output gear 3 and input pinion 1. For this reason, gear 2 is called an *idler*. There is usually little justification for more than three gears in a simple train because no matter what their size, angular velocity ratios of additional idlers merely cancel each other and no contribution is made to the overall train reduction ratio.

When two concentric gears are rigidly mounted on a common shaft so that they are constrained to rotate together in the same direction with the same velocity, the resulting element is called a *compound gear*. Any gear train embodying at least one compound gear is called a compound gear train. Figures 15.5(b) and (c) illustrate compound gear trains in which the compound gear is 2–3. The compound gear train in Figure 15.5(b) is further subcategorized as *nonreverted* because the input shaft and output shaft *are not* colinear. Figure 15.5(c) illustrates a *reverted* compound gear train since the input shaft and output shaft *are* colinear. Reverted gear trains may be desirable or necessary in some applications (e.g., automotive transmissions) but are a little more complicated to design because of equal-center-distance constraints.

Borrowing the results from (15-15), it can be shown[11] that the train reduction ratio for a gear train in which all gears have space-fixed axes of rotation may be written as

$$\frac{\omega_{out}}{\omega_{in}} = \frac{n_{out}}{n_{in}} \pm \left| \frac{\text{product of number of teeth on driver gears}}{\text{product of number of teeth on driven gears}} \right| \tag{15-2}$$

where the algebraic sign is determined by inspection, in accordance with the convention that external mesh ratios are $(-)$ and internal mesh ratios are $(+)$. Appropriately applied and interpreted, (15-2) is equally valid for simple, compound, internal, external, spur, helical, bevel, or worm gear trains.

The train ratio will usually be less than 1 because most gear trains are speed *reducers*. In some cases, however, when the gear train is a speed *increaser*, the train ratio will be greater than 1.

[10]See Figure 12.3. [11]See, for example, ref. 3 or 4.

Figure 15.5
Various types of gear trains.

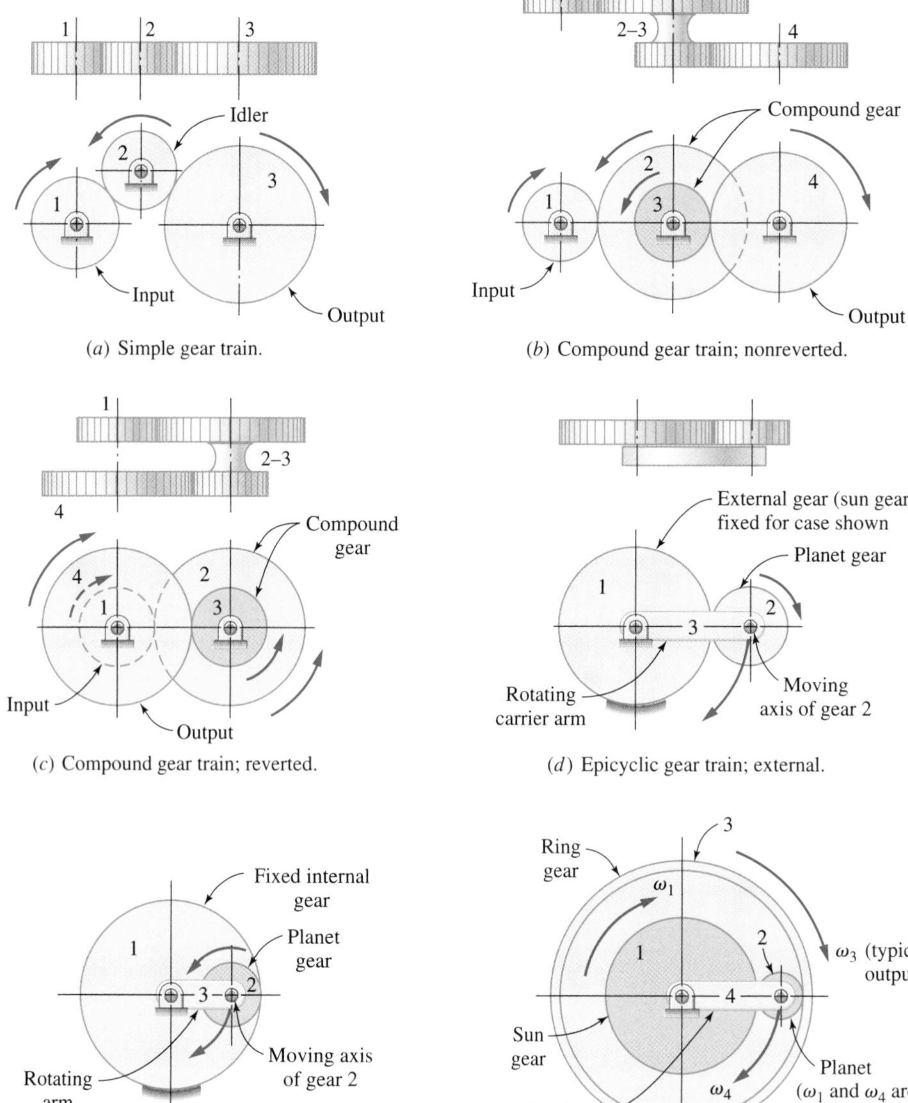

(*a*) Simple gear train.

(*b*) Compound gear train; nonreverted.

(*c*) Compound gear train; reverted.

(*d*) Epicyclic gear train; external.

(*e*) Epicyclic gear train; internal.

(*f*) Epicyclic gear train with ring gear.

For the simple and compound gear trains just discussed, the *axes of all gears are fixed* with respect to the frame, so these trains have *one degree of freedom.*[12] In contrast, an epicyclic (planetary) gear train is arranged so that one or more gears[13] are carried on a rotating arm (carrier), which *itself* rotates about a fixed center, as illustrated in Figures 15.5(d) and (e). Thus the *planet* gear not only rotates about its own center, but at the same time its

[12]Degrees of freedom (dof) are the number of independent parameters required to fully specify the position of every link (gear) relative to the fixed frame.

[13]Most *actual* epicyclic gear trains incorporate two or more planets, equally spaced around the sun gear, to balance forces acting on the sun gear, ring gear, and carrier arm. However, for the purpose of analyzing planetary train *speed ratios*, it is more convenient to use a single-planet format, and the result is the same.

center rotates about another center (that of the arm). Such a planetary gear train has *two degrees of freedom*, making determination of the train reduction ratio more complicated. In spite of the complication, planetary gear trains often provide an attractive design option because higher train reduction ratios may be obtained in smaller space envelopes, high efficiency is possible, *shifting* to obtain larger or smaller speed reduction ratios is feasible by programmed control of two inputs,[14] and a reverted train is characteristic.

If inputs are imposed upon the *sun gear* and *carrier arm* of the simple planetary arrangement of Figure 15.5(d), the only gear remaining to act as output is the *planet gear*, an impractical arrangement since its axis orbits the sun gear. For this reason, a *ring gear* is usually added to the train, in mesh with the planet and centered on the same axis as the sun gear and carrier arm, as shown in Figure 15.5(f). By selectively providing inputs (including fixity) to various combinations of elements in the planetary train, many planetary gear train variations are possible.[15]

In contrast to the relative ease with which power flow and rotational direction may be visualized for gear trains with all axes space fixed, it is very difficult to correctly visualize the behavior of a planetary gear train. Several methods have been developed for determining planetary gear ratios.[16] One method is to utilize the basic kinematic condition

$$\omega_{gear} = \omega_{arm} + \omega_{gear/arm} \tag{15-3}$$

where ω_{gear} = angular velocity of gear relative to fixed frame
ω_{arm} = angular velocity of arm relative to fixed frame
$\omega_{gear/arm}$ = angular velocity of gear relative to arm

Combining (15-3) with (15-2), it may be shown that

$$\frac{\omega_L - \omega_{arm}}{\omega_F - \omega_{arm}} = \pm \left| \frac{\text{product of number of teeth on driver gears}}{\text{product of number of teeth on driven gears}} \right| \tag{15-4}$$

where ω_F = angular velocity of first gear in the train (arbitrarily designated at either end)
ω_L = angular velocity of last gear in the train (at the other end)
ω_{arm} = angular velocity of arm

and the algebraic sign is determined in accordance with the convention that external mesh ratios are (−) and the internal mesh ratios are (+). Additional conditions on use of the equation are that the gears designated as *first* and *last* in the train must rotate about axes fixed to the frame (not orbiting) and there must be an uninterrupted sequence of meshes connecting them. Under these conditions, specification of any two angular velocities (known or specified inputs) allows determination of the third. One common arrangement, for example, is to fix the sun gear, drive the carrier arm by a chosen power source, and take the output from the ring gear, but other arrangements are also used.

Example 15.1 Compound and Planetary Gear Trains

Two preliminary gear train proposals have been submitted for evaluation. One is the four-gear train shown in Figure E15.1A and the other is the planetary gear train shown in Figure E15.1B. The primary design criteria are to achieve a train reduction ratio of 0.30, within 3 percent, and have the rotational direction of the output the same as the direction of the input.

[14]Because of two degrees of freedom. [15]See, for example, ref. 3, pp. 464–465.
[16]See, for example, refs. 3, 4, and 5.

Example 15.1 Continues

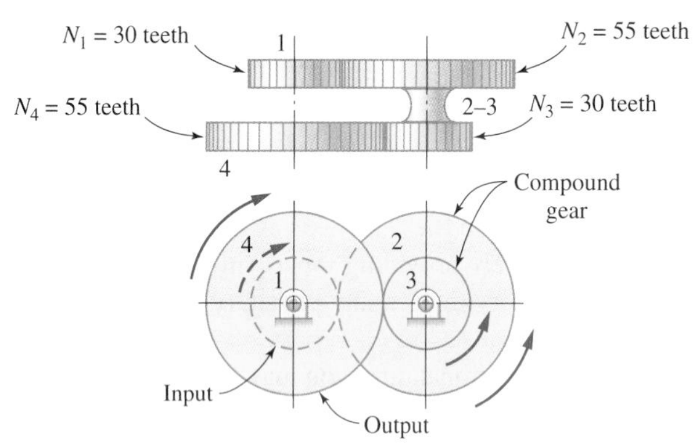

Figure E15.1A
Proposed four-gear train.

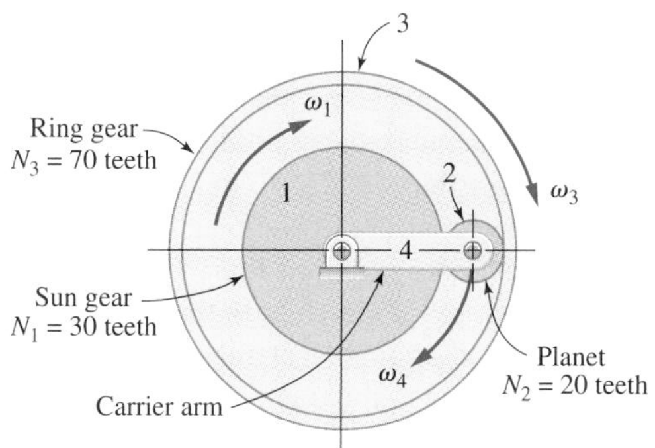

Figure E15.1B
Proposed planetary train. It is worth noting that the number of teeth cannot be *arbitrarily* specified for *all* of the gears in this type of gear train because of the physical space constraint that the sun-and-planet gears must fit properly *inside* the ring gear.

a. Does the four-gear train of Figure E15.1A meet the stated criteria?

b. Can the criteria be met using the planetary train of Figure E15.1B by fixing one element, driving a second element as the input, and using a third element as the output? If so, clearly specify an arrangement that satisfies the criteria.

Solution

a. Examining first the *four-gear* train, with ω_1 as input and ω_4 as output, (15-2) may be applied to give

$$\frac{\omega_{out}}{\omega_{in}} = \left(-\frac{N_1}{N_2}\right)\left(-\frac{N_3}{N_4}\right) = +\left|\frac{(30)(30)}{(55)(55)}\right| = +0.298 \tag{1}$$

The direction of the output is the same as the direction of the input (+), so one criterion is satisfied. The percent deviation of the train reduction ratio from the specified value of 0.30 is

$$e = \left(\frac{0.30 - 0.298}{0.30}\right)100 = 0.67\% \tag{2}$$

Thus *the proposed four-gear train meets the design criteria.*

b. To examine the proposed *planetary* train, several cases must be investigated. Each element of the planetary train, except for the planet gear, may be fixed in turn, and (15-4) used to evaluate the reduction ratio for each case. (Fixing the planet gear would cause the whole device to rotate as a rigid body with reduction ratio of unity.)

If carrier arm 4 were fixed, the proposed planetary train would degenerate to a simple three-gear train with the planet gear acting as an idler that meshes externally with the sun gear (−) and internally with the ring gear (+) to give a (−) train reduction ratio. The ring gear output would therefore rotate in the opposite direction to the sun gear input, violating the stated criterion of same-direction rotation. (The reduction ratio would also be wrong.)

If sun gear 1 were fixed ($\omega_1 = 0$), ring gear 3 used as input ($\omega_3 = \omega_{in}$), and carrier arm 4 used as output ($\omega_4 = \omega_{out}$), and if first and last gears in the train are taken as sun gear 1 and ring gear 3, respectively, (15-4) becomes

$$\frac{\omega_3 - \omega_4}{\omega_1 - \omega_4} = \frac{\omega_3 - \omega_4}{0 - \omega_4} = \left(-\frac{30}{20}\right)\left(+\frac{20}{70}\right) = -0.43 \tag{3}$$

whence

$$\frac{\omega_4}{\omega_3} = \frac{\omega_{out}}{\omega_{in}} = +0.70 \tag{4}$$

Thus the same-direction criterion is satisfied but the 0.70 ratio deviates far too much from the specified values of 0.30.

If ring gear 3 were fixed ($\omega_3 = 0$), sun gear 1 used as input ($\omega_1 = \omega_{in}$), and carrier arm 4 used as output ($\omega_4 = \omega_{out}$), and if first and last gears in the train are again taken as sun gear 1 and ring gear 3, (15-4) becomes

$$\frac{\omega_3 - \omega_4}{\omega_1 - \omega_4} = \frac{0 - \omega_4}{\omega_1 - \omega_4} = -0.43 \tag{5}$$

whence

$$\frac{\omega_4}{\omega_1} = \frac{\omega_{out}}{\omega_{in}} = +0.30 \tag{6}$$

This satisfies both the same-direction criterion and the specified reduction ratio value of 0.30. Thus the planetary train can meet the design criterion if

1. Ring gear 3 is fixed to the frame.
2. Sun gear 1 is attached to the input shaft.
3. Carrier arm 4 is attached to the output shaft.

15.4 Potential Failure Modes

Depending upon specific loads, speeds, type of gearing selected, gear materials and processing, manufacturing accuracy, mounting details, bearing and shaft characteristics, lubrication, and environmental factors, gears may be vulnerable to failure by any of several potential failure modes.[17] The basic design requirements imposed by the fundamental law of gearing, combined with the complexities of variable loading, combined rolling and sliding contact between curved tooth surfaces, and system interactions with other machine components, make the assessment of potential failure modes for gears more complicated than for any other machine element so far discussed.

Most gears rotate in one direction, subjecting each tooth on the gear to a *one-way* bending load each time it passes through the mesh (except for idler and planetary gears). This gives rise to potential fatigue failure (typically *high cycle fatigue* but may be low cycle fatigue in certain circumstances), due to nonzero mean (released) cyclic bending stresses at the root fillet of each tooth. The root fillet is also a site of stress concentration. The result of a root-fillet fatigue failure is ultimately to dislodge the tooth from the gear wheel.

Likewise, cyclic Hertz contact stresses are generated between the curved tooth surfaces as they repeatedly pass through the mesh, making *surface fatigue* a significant potential failure mode. The cyclic subsurface Hertzian shear stresses produced by the curved surfaces in repetitive contact may initiate and propagate minute cracks that ultimately dislodge particles and generate surface pits. The *smaller* pinion usually pits first because each tooth passes through the mesh more often than teeth on the *larger* gear. The pitting frequently is concentrated in a band along the pitch line (sometimes called pitch-line pitting) because the Hertz contact stresses tend to be higher in that region. Occasionally, tooth fracture may initiate from a surface pit.

Because of the kinematics of the contact between meshing gear teeth as they pass through the contact zone, a component of sliding motion exists in virtually all cases. Depending upon the tooth profile and type of gearing, the sliding component ranges from

[17]See 2.3 for list of potential failure modes.

small (e.g., spur gear meshes) to large (e.g., worm gear meshes). As a result, both *adhesive wear* and *abrasive wear* must be considered as potential failure modes. In addition, especially if lubrication is not used or is not adequate, adhesive wear behavior may degenerate to *galling*. If any of these wear phenomena significantly change the tooth profile, failure may occur because the angular velocity of the output gear becomes erratic (due to unacceptable clearances between mating teeth), resulting in unaccounted-for impact forces, or because intolerable levels of vibration or noise are generated.

Force-induced elastic deformation should also be considered as a potential failure mode because gear trains are nearly always an integral part of a dynamic system of springs and masses that may be excited to vibrate at resonant frequencies (see 8.2). Experience has shown that *spring rates* associated with the elastic deformation of typical gear-tooth contact geometry *are low* (soft) compared to many other machine elements, and therefore gear mesh stiffness in some cases may dominate, or at least significantly contribute to, the vibrational behavior of a machine (see 4.11). It is an important design responsibility to assure that installed gear mesh stiffness is great enough and constant enough to keep the fundamental natural frequency of the system well above the forcing frequency[18] so that excessive deflection (*force-induced elastic deformation failure*) does not result.

Vibrational behavior may also be the source of noise generation in the mesh, or the source of transmitted vibrations that excite other components, especially housing panels, that may behave as "speakers." *By definition*, if noise levels become unacceptable, failure has occurred (by force-induced elastic deformation). Several excitation sources for noise generation in gearing systems have been identified,[19] including cyclic changes in stiffness as teeth rotate through the mesh, and transmission error.[20] In high-precision gearing, unloaded profiles are sometimes modified during manufacture to provide *more nearly perfect output* under operating loads and speeds, thus reducing transmission error and consequential noise generation.

Other less frequently observed failure modes in gearing include root-fillet *yielding*, *brittle fracture-* or *ductile rupture* by an unanticipated external "overload," *brinnelling* of gear teeth due to impact overloading at mesh contact sites, and "subcase fatigue" cracking and spallation in case-hardened gears. The subcase fatigue nuclei are initiated in the strength-gradient transition zone between *low-strength* core material and *high-strength* case material. Finally, in occasional circumstances, a gear set may be "parked" in fixed position over a period of time. If operational vibrations are transmitted to the parked tooth contact sites, the small-amplitude cyclic relative motions may lead to failure by *fretting wear* or *fretting fatigue*.

The complexity of potential failure in gearing systems has led to an organized effort by the American Gear Manufacturers Association (AGMA) to categorize the various kinds of observed gear failures and establish a standardized nomenclature to describe them. Most of the failure modes just described are included in the AGMA failure "classes" specifically defined[21] by the terms *wear, scuffing, plastic deformation, contact fatigue, cracking, fracture*, and *bending fatigue*. Figure 15.6 illustrates, qualitatively, the operating regions in which each of the more likely failure modes may be expected to occur.[22]

15.5 Gear Materials

Based on the material selection guidelines of Chapter 3, and the failure mode discussion of 15.4, candidate materials for gears should have good strength (especially fatigue

[18]See 8.2 for definitions. [19]See ref. 7.

[20]Transmission error is defined as the deviation of the actual position of the output gear from the position it would occupy if the gear drive were perfect.

[21]See ref. 8, Table 1. [22]See ref. 9.

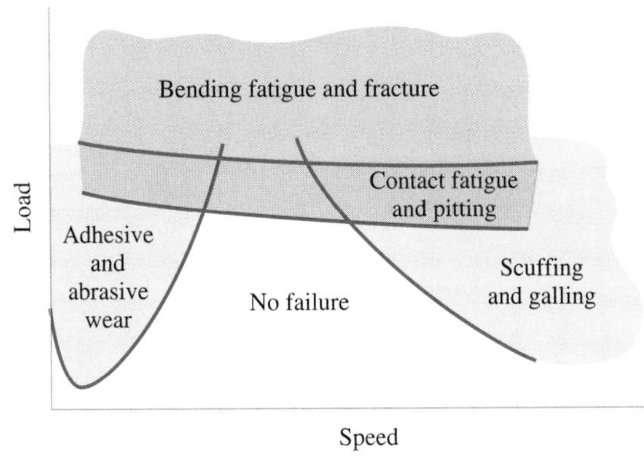

Figure 15.6
Failure mode likelihood as a function of load and speed.

strength), high stiffness, good wear resistance, high resistance to surface fatigue, high resilience, good damping ability, good machinability, and, in some applications, good corrosion resistance, as well as reasonable cost. Using the evaluation indexes of Table 3.2, materials meeting these criteria include steels and steel alloys, gray and alloy cast irons, brass, bronze, and certain polymeric materials, although other materials may be used for special applications. Steel gears are widely used because of high strength, good resilience, and moderate cost, but because of wear-resistance requirements steel gears are usually heat treated to produce a hard surface on the teeth. Thus a through-hardenable alloy such as 4140 or 4340 may be chosen, or a carburizing steel such as 4620 or 4320 may be used with an appropriate carburizing and case-hardening procedure, followed by heat treatment to produce a hard wear-resistant tooth surface. The nitriding process is also utilized to develop tooth surfaces with high hardness, using alloys such as 4140 or 4340 for the core material. Local surface heating by an induction coil or oxyacetylene flame, followed by rapid quenching (*induction hardening* or *flame hardening*), are also used with alloys such as 4140 or 4340 to produce hard tooth surfaces.

Cast-iron gears are inexpensive and have high hysteretic damping capacity, which tends to produce relatively quiet operation. ASTM Grade 20 is widely used, but higher-strength grades are used in more demanding applications. Steel pinions are often paired with cast-iron gears to obtain reasonable strength with quiet operation.

When nonferrous alloys are used to avoid corrosion problems, or in applications such as worm gear meshes where high sliding velocities must be accommodated and wear-in enhanced mesh conformability is advantageous, bronze alloys are often chosen. In the wormset application, a bronze gear is usually paired with a hardened steel worm.

Polymeric gears may be used in lightly loaded applications to provide quiet operation, effective load sharing due to low stiffness, and reasonable cost. Polymers may themselves be lubricative, or may be *filled* with solid lubricants in some cases to allow the mesh to run "dry." Extruded nylon gears and molded cloth-reinforced phenolic gears have been meshed with cast-iron or steel pinions in many cases.

Although the methods of 2.6 for defining a fatigue strength value appropriate to the application are as valid for gears as for any other machine element, the extensive specialized gearing database, developed within the gearing industry over many decades, makes it prudent to *first* search this database. Data for gear strength as a function of material, heat treatment, governing failure mode, reliability requirements, surface conditions, environment, speed, or other factors may be matched to the proposed gearing application. Any factors not included in the database strength values must, as usual, be accounted for by the designer. Examples of available specialized gear strength data are shown in Tables 15.10 and 15.12, based on gear tooth failure by bending fatigue, and in Table 15.15 and 15.16 for surface fatigue failure. Correction factors for other design lives, other temperatures, other hardnesses, or other reliability requirements[23] need to be applied to the data of Tables 15.10, 15.12, 15.15, and 15.16.

15.6 Spur Gears; Tooth Profile and Mesh Geometry

If the fundamental law of gearing is to be satisfied (see 15.1), the gear-tooth profile geometry must produce an exactly constant angular velocity ratio between the driver and driven gears at every position as successive teeth rotate through the mesh. If a meshing pair of gear teeth have profiles that satisfy this requirement, they are said to be *conjugate* profiles. Theoretically, it is possible to arbitrarily select any profile for one tooth, then find a profile for the meshing tooth that will produce conjugate action. As a practical matter, however, because of relative ease of manufacture and insensitivity of angular velocity ratio to minor variations in center distance, the only profile of importance in current gearing practice is the involute of a circle. (Figure 15.7).[24]

The *kinematic* requirement for conjugate action is that as the gears rotate, *the common normal to the curved tooth surfaces at their point of contact must, for all gear positions, intersect the line of centers at a fixed point P called the pitch point.*[25] This kinematic requirement is satisfied by meshing gear teeth that have conjugate involute profiles.

Involute Profiles and Conjugate Action

When any curved surface pushes on another, as shown in Figure 15.8, at contact point C the two surfaces are tangent and the force transmitted from one to the other acts along the common normal, ab. Line ab is called, therefore, *the line of action* or the *pressure line*. As illustrated in Figure 15.8, the line of action ab intersects the line of centers, O_1O_2, at the pitch point P. The angle φ between the pressure line and a reference line *through* the pitch point, *perpendicular* to the line of centers, and *tangent to* the base circle, is called the *pressure angle*. At the instantaneous position shown in Figure 15.8, circles may be drawn through pitch point P about each center O_1 and O_2. These circles are called *pitch circles* (see also Figure 15.1 and related discussion) and their radii, r_A and r_B, are called *pitch*

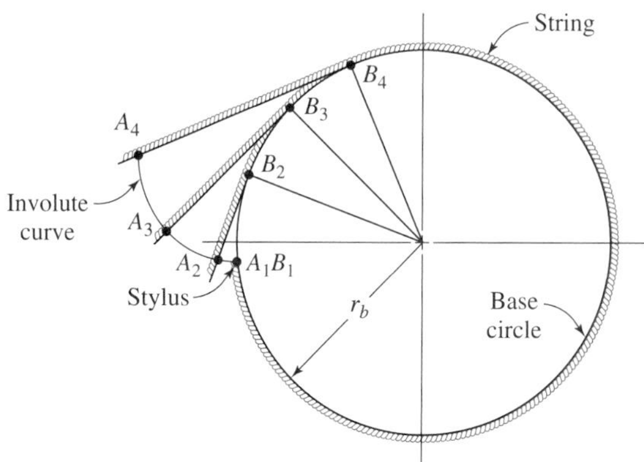

Figure 15.7
Generation of involute curve. The stylus attached to the string traces the curve $A_1 - A_2 - A_3 - A_4$ as a taut string unwraps. Corresponding tangents to the base circle (A_1B_1, A_2B_2, A_3B_3, A_4B_4) represent instantaneous radii of curvature of the involute. (Adapted from ref. 3, reprinted by permission of Pearson Education, Inc., Upper Saddle River, N. J.)

[23]See ref.10.

[24]The involute of a circle may be generated by unwrapping from a cylinder a taut string with an end stylus (see Figure 15.7). The planar trace of the end stylus as the string unwraps is an *involute* curve and the circular section of the cylinder is called the *base circle*.

[25]See, for example, ref. 3 or 4.

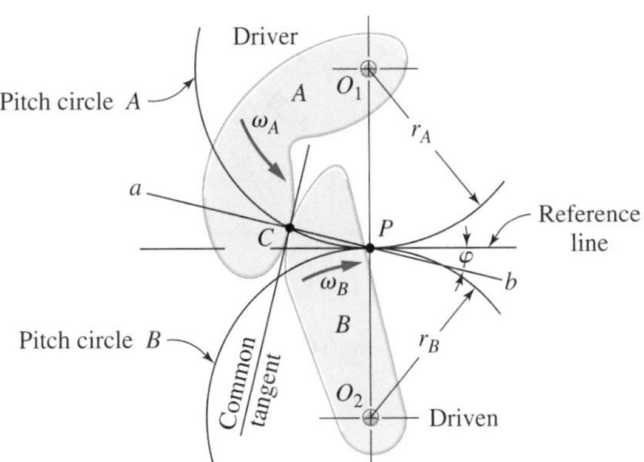

Figure 15.8
Transmission of force and motion from one rotating member to another across contacting curved surfaces.

radii. The angular velocity ratio of driven member B to driving member A is inversely proportional to their pitch radius ratio [see also (15-1)], giving

$$\frac{\omega_B}{\omega_A} = -\frac{r_A}{r_B} \tag{15-5}$$

Reiterating, a constant angular velocity ratio requires a constant pitch radius ratio. Therefore, lines of action for every instantaneous contact point between mating teeth as they rotate through the mesh must pass through the same (fixed) pitch point P.

Using the involute properties illustrated in Figure 15.7, a common *involute generator string* may be conceptually wrapped tightly in opposite directions about two adjacent base circles of different radii, centered at O_1 and O_2, as illustrated in Figure 15.9. If the base circles are rotated in opposite directions, keeping the string tight, a stylus point g on the common string will simultaneously trace an involute profile cd attached to member 1 and ef attached to member 2. Point g represents the point of contact. Line ab, the generating line, is always normal to both involutes (see Figure 15.7), and therefore represents the line of action. Thus the contact point moves along the line of action, but the line of action does not change position because it is always tangent to *both* base circles. Accordingly, the two conjugate involute profiles satisfy the fundamental law of gearing.

Figure 15.10 provides a basis for understanding the relationship between pitch circles and base circles. From (15-5) it is clear that the ratio of radii of the pitch circles is determined by the required angular velocity ratio for the design application. (Actual magnitudes for the radii depend upon other design factors, as discussed later.) If, as shown in Figure 15.10, the pitch circles are laid out to scale on centers O_1 and O_2 using the required radius

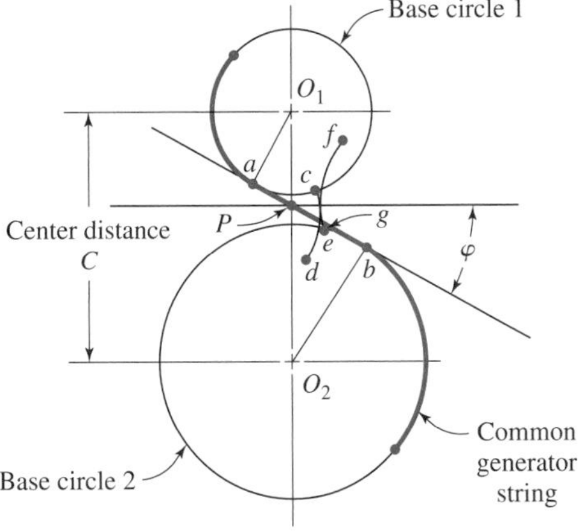

Figure 15.9
Simultaneous generation of involutes on two adjacent base circles using a common counterwrapped generator string.

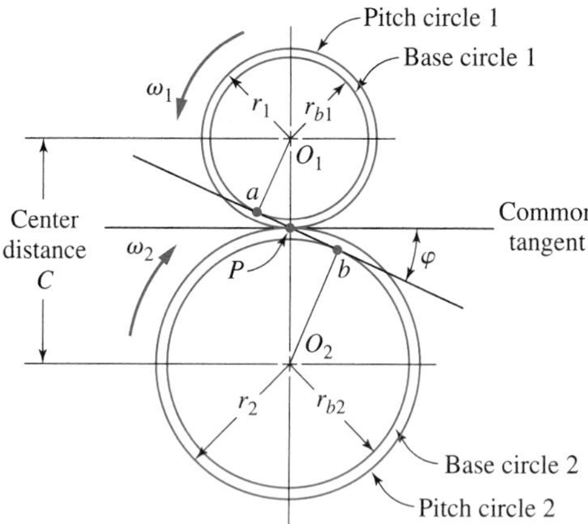

Figure 15.10
Relationship between pitch circles and base circles.

ratio, and in contact at the pitch point P, and the line of action ab is constructed through the pitch point using the selected pressure angle φ, the circles centered on O_1 and O_2 and tangent to the line of action at a and b are the base circles. Using these base circles, involute profiles may be generated to form the basic tooth profile for both sides of the involute gear teeth.

Gearing Nomenclature; Tooth Shape and Size

The distance from any selected reference point on one tooth to a corresponding point on the next adjacent tooth, measured along the pitch circle, is called the *circular pitch* p_c [see Figure 15.1(d)]. The circular pitch may be calculated as an arc length equal to the circumference of the *pitch* circle divided by the number of teeth N, or

$$p_c = \frac{\pi d}{N} = \frac{2\pi r}{N} \tag{15-6}$$

Similarly, the *base pitch*, p_b, is defined as an arc length equal to the circumference of the *base* circle divided by the number of teeth N. Since $r_b = r\cos\varphi$ (see Figure 15.10), it follows from (15-6) that

$$p_b = \frac{2\pi r_b}{N} = p_c \cos\varphi \tag{15-7}$$

The *circular tooth thickness* and *tooth space* are also measured along the pitch circle, each being nominally equal to $p_c/2$. As a practical matter, the tooth *space* is usually made slightly larger than the tooth *thickness*, to provide a small amount of *backlash*[26] for smooth nonbinding operation, as discussed later.

To define the remaining boundaries of the gear teeth, *addendum circles* and *dedendum circles* may be constructed centered on O_1 and O_2, as illustrated in Figure 15.11. The addendum circle is established by adding the addendum a (tooth height outside the pitch circle) to the pitch radius r, and the dedendum circle by subtracting the dedendum (tooth dimension inside the pitch circle) from the pitch radius. It is important to recognize that the *involute* tooth profile can extend inside the pitch circle only as far as the base circle; the portion of the tooth inside the base circle cannot participate in conjugate action and therefore must be shaped to provide tip clearance for the teeth of the mating gear. The true shape of this noninvolute portion of the tooth profile, often visualized as a straight radial line, is actually determined by the gear manufacturing process (see 15.7). From Figure 15.10 it may be deduced that larger pressure angles result in smaller base circles, conse-

[26]Nominally, backlash is the difference between tooth space and tooth thickness.

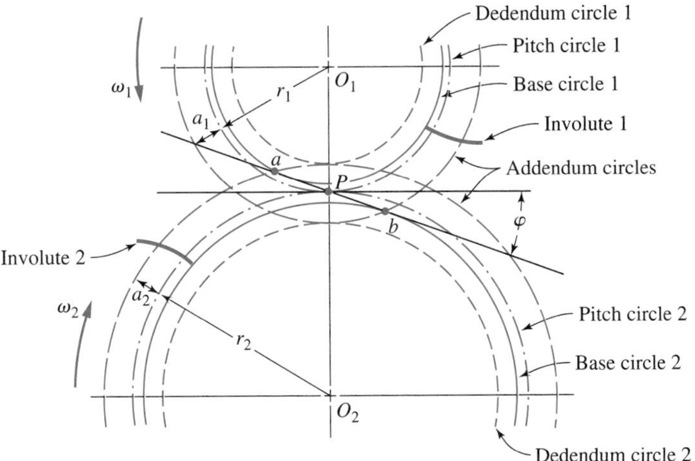

Figure 15.11
Defining boundaries for gear teeth.

quently extending conjugate action deeper into the dedundum flank of the gear tooth. The dedendum is typically made slightly larger than the addendum to provide positive clearance between the tooth tip of one gear and the bottom land of the tooth space of the mating gear. A root fillet is used to blend the tooth flank into the bottom land of the adjacent tooth space. Although the circular pitch is a measure of tooth size, more common indices of tooth size are *diametrical pitch*, P_d (used only with *U.S.* specification gears), or *module*, m (used only with *SI* or metric specifications). The diametrical pitch, defined as the number of teeth divided by pitch diameter in inches, is given by

$$P_d = \frac{N}{d} \tag{15-8}$$

and module, defined as pitch diameter in millimeters divided by number of teeth, is

$$m = \frac{d}{N} \tag{15-9}$$

Equations for *center distance C* may be written using (15-8) and (15-9), respectively, as

$$C = r_1 + r_2 = \frac{N_1}{2P_d} + \frac{N_2}{2P_d} = \frac{(N_1 + N_2)}{2P_d} \tag{15-10}$$

and

$$C = r_1 + r_2 = \frac{mN_1}{2} + \frac{mN_2}{2} = \frac{m}{2}(N_1 + N_2) \tag{15-11}$$

Also, from (15-8) and (15-9) other useful relationships may be developed, including

$$p_c P_d = \pi \qquad (p_c = \text{ inches; } \; P_d = \text{ teeth/inch}) \tag{15-12}$$

$$\frac{p_c}{m} = \pi \qquad (p_c = \text{ mm; } \; m = \text{ mm/tooth}) \tag{15-13}$$

and

$$m = \frac{25.4}{P_d} \tag{15-14}$$

Gear teeth on mating gears can mesh properly only if their *diametral pitches* or their *modules* are the same. Hence (15-1) may be combined with either (15-8) or (15-9) to give

$$\frac{\omega_2}{\omega_1} = \frac{n_2}{n_1} = \pm\frac{r_1}{r_2} = \pm\frac{d_1}{d_2} = \pm\frac{N_1}{N_2} \tag{15-15}$$

where the algebraic sign is determined by inspection, and in accordance with the convention that external mesh ratios are (−) and internal mesh ratios are (+).

Although there are theoretically no restrictions on tooth size, as a practical matter sizes are limited by available gear-cutting facilities. For standard involute spur gears to be fully interchangeable, they must have the same diametral pitch (or module), the same pressure angle, and the same addendum,[27] and the circular tooth thickness must be nominally equal to one-half the circular pitch.

The magnitude of the pressure angle significantly affects the shape of a gear tooth, as illustrated in Figure 15.12. As may be noted, the 25° teeth have thicker bases and larger radii of curvature at the pitch line, giving them higher bending-load-carrying capacity as compared to 20° teeth. The 25° teeth tend to generate more operating noise, however, due to their lower contact ratio (see below). As discussed next, standard tooth systems typically incorporate pressure angles of 20° or 25°, but if conditions warrant, a compromise pressure angle of $22\frac{1}{2}°$ is sometimes used.

Gear-Tooth Systems[28]

Standardized *tooth systems* have been widely adopted to facilitate interchangeability and availability by specifying selected pressure angles, then defining addendum, dedendum, working depth, whole depth, minimum tip clearance, and circular tooth thickness as functions of diametral pitch (or module). The most commonly used pressure angle is 20°, both

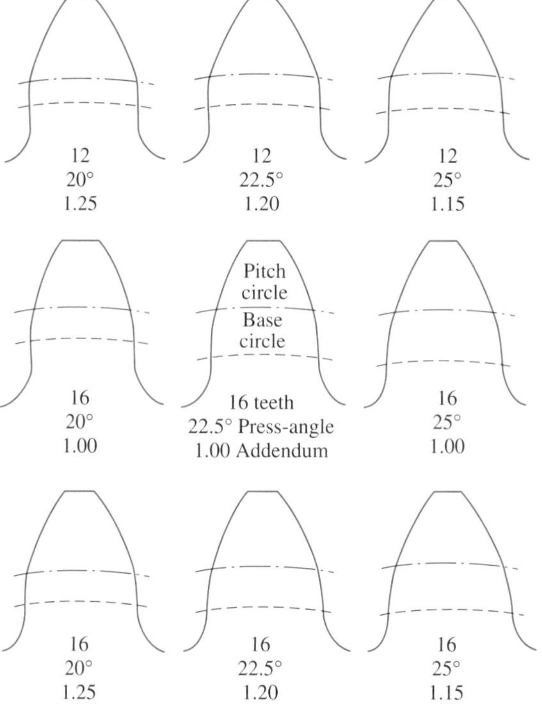

Figure 15.12
Comparison of gear tooth shape as a function of pressure angle. (Extracted from ref. 1. *Source:* Dudley Engineering Company, San Diego, CA.)

[27]For nonstandard systems, such as the *extended center distance system* and *long and short addendum system*, pinion and gear addenda may differ, as discussed in "Mesh Interactions" below.

[28]See, for example, refs. 11 and 12.

TABLE 15.1 Standard Proportions for AGMA Full-Depth Gear Teeth[1] (inch system)

	Coarse Pitch	Fine Pitch
	$(P_d < 20)$	$(P_d \geq 20)$
Pressure angle	20° or 25°	20°
Addendum	$1.000/P_d$	$1.000/P_d$
Dedendum	$1.250/P_d$	$(1.200/P_d) + 0.002$ (min)
Whole depth	$2.250/P_d$	$(2.200/P_d) + 0.002$ (min)
Working depth	$2.000/P_d$	$2.000/P_d$
Clearance (basic)	$0.250/P_d$	$(0.200/P_d) + 0.002$ (min)
Clearance (shaved or ground teeth)	$0.350/P_d$	$(0.350/P_d) + 0.002$ (min)
Circular tooth thickness	$1.571/P_d$	$1.571/P_d$

[1]See refs. 11 and 12.

for U.S. and SI gears, but a 25° pressure angle is also common in U.S. coarse-pitch gears ($P_d < 20$). Other pressure angles may be used for special applications. Tooth proportions for AGMA standard full-depth spur gear teeth are specified in Table 15.1, and the approximate actual sizes of standard involute spur gear teeth are shown in Figure 15.13 for a range of commonly used diametral pitches. Table 15.2 gives the range of tooth sizes (in terms of diametral pitch or module) generally available from gear manufacturers.

In addition to the standard tooth proportions defined by Table 15.1, the minimum root-fillet radius ρ_f typically is about $0.35/P_d$ or $m/3$, depending upon the manufacturing method and edge radius of the cutter. The face width b specified for U.S. and SI gears, respectively, is usually in the range

$$\frac{9}{P_d} \leq b \leq \frac{14}{P_d} \tag{15-16}$$

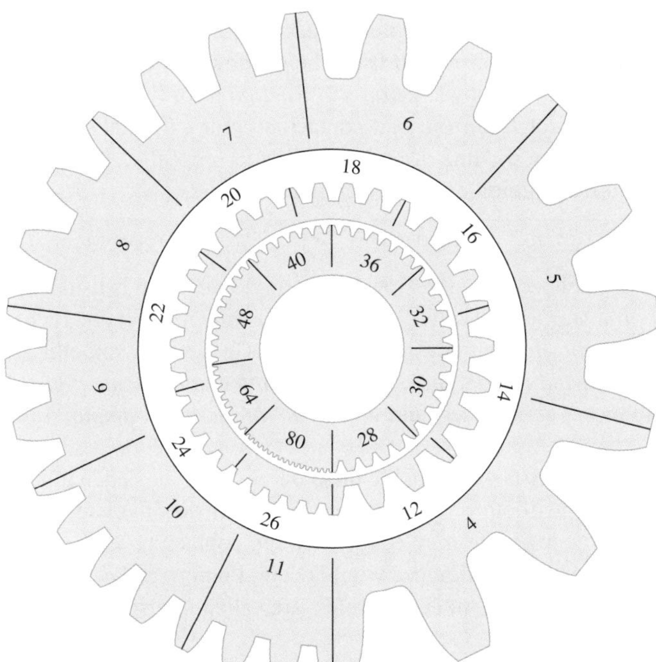

Figure 15.13
Approximate actual sizes of gear teeth with various standard diametral pitches. Generally, if $P_d \geq 20$, the gears are called *fine-pitch* gears; if $P_d < 20$ the gears are called *coarse-pitch* gears.

TABLE 15.2 Commonly Used Tooth Sizes

Diametral Pitch		Module			
Coarse	Fine	Preferred		Second Choice	
2	20	1	12	1.125	14
$2\frac{1}{4}$	24	1.25	16	1.375	18
$2\frac{1}{2}$	32	1.5	20	1.75	22
3	40	2	25	2.25	28
4	48	2.5	32	2.75	36
6	64	3	40	3.5	45
8	80	4	50	4.5	
10	96	5		5.5	
12	120	6		7	
16	150	8		9	
	200	10		11	

or

$$9m \leq b \leq 14m \tag{15-17}$$

Wider face widths typically result in *nonuniform* contact pressure across the face width, and narrower face widths may result in excessively *high* contact pressures.

When a designer specifies the use of gears manufactured to standard system specifications, the gears are usually available as stock items. On the other hand, the availability of modern computer-controlled gear manufacturing facilities allows the specification of special nonstandard *optimized* gears, if warranted by a specific application.

Mesh Interactions

From the time a pair of gear teeth enter contact, until they leave contact after rotating through the mesh, force is transmitted from the driving gear to the driven gear along the line of action. Figure 15.14 shows the *angle of action* θ for both driver and driven gears, corresponding to their rotation from first contact at point a to final separation at point b.

The distance Z along the line of action from a to b is called the *length of action* and may be calculated from pinion-and-gear geometry as[29]

$$Z = \sqrt{(r_p + a_p)^2 - (r_p \cos\varphi)^2} + \sqrt{(r_g + a_g)^2 - (r_g \cos\varphi)^2} - C\sin\varphi \tag{15-18}$$

where symbols are defined in Figure 15.15. The angle of action is usually subdivided into an *angle of approach* α and an *angle of recess* β as shown in Figure 15.14. Three important design tasks associated with designing teeth that will rotate smoothly through the angle of action are to *avoid interference*, to *assure uninterrupted contact* of at least one tooth pair (preferable more) at all times, and to *provide an optimum amount of backlash* in the mesh.

Since an involute exists only outside the base circle, if the dedendum circle is smaller than the base circle, the tooth profile between circles is not involute, tooth interaction is not conjugate, and interference may occur between contacting teeth. Figure 15.15 illustrates conditions for which interference would occur. Points c and d, the points of tangency between the line of action and the base circles, are called *interference points*. If an adden-

[29]For derivation see, for example, ref. 3.

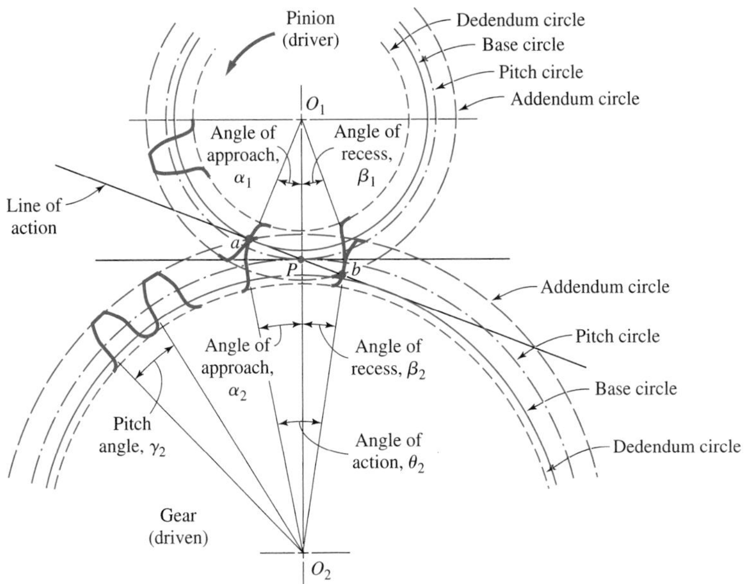

Figure 15.14
Interaction between contacting teeth as they rotate through the mesh.

dum circle, which defines tooth tip path, intersects the line of action *outside* the interference points, as illustrated, for example, by points *a* and *b*, interference will occur. One way to relieve the interference would be to remove the crosshatched tooth tip, as shown in Figure 15.15, creating a *stub-tooth* gear (seldom used in current practice). *Undercutting* the tooth flank of the mating gear also relieves interference but weakens the undercut teeth in the region of the already-critical root fillets.

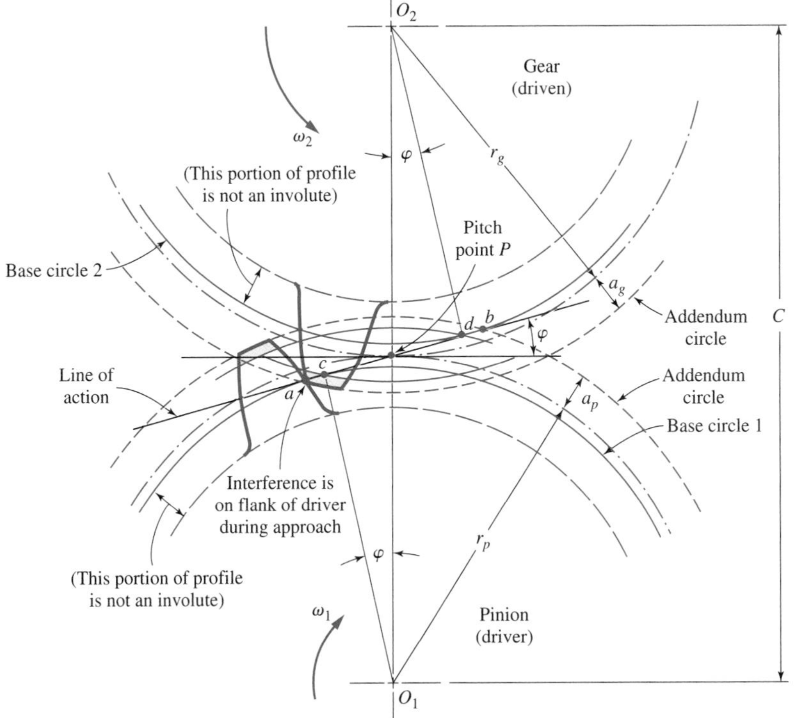

Figure 15.15
Interference between mating spur gear teeth.

Nonstandard spur gear systems are sometimes used to eliminate interference and undercutting. Two nonstandard systems that have been successful, and allow the generation of gear teeth using standard cutters, are the *extended center distance system* and the *long and short addendum system*. In the first system, when the pinion is being cut, the cutter is withdrawn from the blank a chosen amount so the tip of the basic rack cutter[30] passes through the interference point on the pinion. This eliminates undercutting, increases tooth thickness, and decreases tooth space. When the resulting nonstandard pinion is mated with its gear, its operating center distance will have increased because of the smaller tooth space, and the pressure angle will also be larger. The process may be applied to both pinion and gear if conditions call for this.

In the *long and short addendum system*, the cutter is advanced into the gear blank by the same amount it is withdrawn from the pinion blank during manufacture. This results in a *longer-than-standard* addendum for the pinion and *shorter-than-standard* addendum for the gear. The dedenda for both pinion and gear also change correspondingly to give a standard working depth, standard center distance, and standard pressure angle in the operating gear set. Since the longer addenda of the *driving* pinion teeth also shift the arc of action toward a *longer arc of recess* and a *shorter arc of approach*[31], such gearing is sometimes called *recess action gearing*.

In addition to eliminating interference and undercutting, *recess action gearing* tends to benefit from the fact that there is a *lower* coefficient of friction during recess than during approach. This results from the fact that the friction force component *aids the driving force* during recess[32], and surface smoothing and polishing occur during recess as compared with roughening and distress during approach.[33] Because of these beneficial factors, recess action gearing tends to exhibit less wear and quieter operation. Typically, it is recommended that the pinion tooth addendum be increased no more than 25 percent over the standard addendum. In some applications, however, longer addenda *are* used, sometimes extending to the extreme case where a 200 percent addendum is used for the driving pinion and 0 percent addendum for the gear to give *full recess action* and no approach action.[34]

Other steps useful for avoiding interference in standard gearing are to make sure that the pressure angle is not too small, that the number of pinion teeth is not too small, and that the number of gear teeth is not too large. Table 15.3 lists the minimum number of standard 20° full-depth involute pinion teeth that should be specified to avoid interference and undercutting for various mating gear sizes.

To maintain smooth operation and constant angular velocity ratio, it is necessary that at least one pair of teeth remain in contact at all times. This will be assured if the profile contact ratio m_p, defined as *the angle of action θ divided by the pitch angle γ*, or, alternatively, *the length of action Z divided by the base pitch p_b*, exceeds a value of unity. Referring to Figure 15.14, then, the profile contact ratio m_p may be written as (see Figures 15.14 and 15.15 for definitions)

$$m_p = \frac{\theta}{\gamma} = \frac{\alpha + \beta}{\gamma} \tag{15-19}$$

or, using the alternative definition and (15-7), (15-12), and (15-18),

[30]See Figure 15.17.

[31]See Figure 15.14 for definition of *arc of approach* and *arc of recess*. Approach action takes place from the time of first contact of an involute driving pinion tooth flank with a driven conjugate gear tooth profile, continuing as the point of contact moves deeper and deeper into the tooth space, until it reaches the pitch point. From here it withdraws (recess action) until the pinion tooth tip disengages from the gear.

[32]It opposes the driving force during approach.

[33]See ref. 30, p. 2. [34]More usual for worm gearing. See 15.20.

TABLE 15.3 Minimum Number of Pinion Teeth to Avoid Interference Between 20° Full-Depth Pinion and Mating Full-Depth Gears of Various Sizes

Minimum Number of Pinion Teeth	Number of Gear Teeth
13	16
14	26
15	45
16	101
17	∞

$$m_p = \frac{Z}{p_b} = \frac{P_d\left[\sqrt{(r_p + a_p)^2 - (r_p\cos\varphi)^2} + \sqrt{(r_g + a_g)^2 - (r_g\cos\varphi)^2} - C\sin\varphi\right]}{\pi\cos\varphi} \tag{15-20}$$

The profile contact ratio may be thought of as the average number of teeth in contact, and in practice typically ranges from about 1.4 to 1.8. As spur gears roll through the mesh, however, there are either two pairs of teeth in contact, or one pair. A gear set with a profile contact ratio of 1.5 never really has 1.5 teeth in contact. It has one pair of teeth in contact at all times and two pairs of teeth in contact 50 percent of the time. Larger profile contact ratios usually imply some *load sharing* among the teeth as well as smoother operation. Smaller teeth (larger P_d) and larger pressure angles tend to result in larger contact ratios.

An additional implication of load sharing when the profile contact ratio is greater than 1.0 is that the total transmitted bending force is supported by a single tooth pair until a second tooth pair comes into contact. The second tooth pair picks up its share of the load *before the force vector reaches the tip* of the single tooth pair. The bending moment on the single tooth therefore builds with rotation until the second tooth comes into contact, then drops off due to load sharing. This point is called the *highest point of single tooth contact (HPSTC)* and is used as the *moment arm* in certain bending stress calculations.[35]

Backlash is the amount by which the sum of the circular thicknesses of two meshing gear teeth is less than the circular pitch, or the gap distance between the noncontacting surfaces of adjacent teeth as they roll through the mesh. Backlash is often thought of as the freedom of one gear to move while the mating gear is held fixed. *Some* backlash is usually desirable because it tends to eliminate binding and to promote proper lubrication, but if the gear set is subjected to *reversing* torques, potential problems of impact, vibration, and noise may be encountered. In such cases, it may be necessary to preload[36] the mesh to eliminate the backlash, or increase the mesh stiffness in the torsional mode. Since conjugate action for involute profiles is insensitive to small changes in center distance, the backlash in a gear set is sometimes adjusted to a desirable amount by adjusting the center distance between the two gear shafts.

15.7 Gear Manufacturing; Methods, Quality, and Cost

To meet functional design requirements, maintain specified reliability, and achieve reasonable cost, a designer should have a general understanding of the various gear manu-

[35]See discussion in 15.9. [36]See 4.12.

facturing methods and their limitations. This brief discussion provides only a summary of the methods used to make gear teeth; much more detailed accounts are available in the literature.[37] Designers responsible for gearing used in *high-performance* applications are well advised to consult with gearing specialists or precision-gearing manufacturers.

Methods commonly used for producing gears include casting, forming, sintering, and machining processes (see Tables 7.1 and 7.2). *Cast gears* are limited to low-speed, low-precision applications. *Formed gears* such as die cast (e.g., zinc, aluminum, or brass alloys), extruded (e.g. aluminum or copper alloys), or injection-molded gears (e.g., thermoplastics such as nylon, acetal, or polycarbonate) may be used in applications where moderate accuracy is acceptable and high production quantities are required. *Sintered* powered-metal gears (e.g., iron, mild steel, or stainless steel) may be a good option if small gears of higher precision are required in larger production quantities. For power transmission, high-speed applications, high-load applications, and/or high-precision applications, gears made by *machining* teeth on a cast, forged, or wrought metal blank is by far the most usual manufacturing method.

Machining methods commonly used to *cut* gear teeth on a blank include *hobbing, shaping, milling*, and *broaching*. When there is a need for high-precision gear teeth, *finishing* operations, such as *shaving, rolling, grinding*, or *honing*, may follow the initial gear-cutting process to obtain more accurate teeth and better surface finish.

Gear Cutting

Gear teeth may be cut with either *form cutters* or *generating cutters*. When *form cutting*, the tooth space takes the exact shape of the cutter. When *generating*, a tool having a shape different from the tooth profile is moved relative to the gear blank in such a way that the *proper tooth profile is cut.*

Hobbing is a generating method in which the cutting tool (called a *hob*), shaped like a straight-sided worm thread with periodic relief grooves, and sharpened to provide cutting edges, is fed across the gear blank as both the gear blank and the hob are synchronously rotated. Spur, helical, and crossed-helical gears can be produced by feeding the hob *across the face width* of the gear, as shown in Figure 15.16(a). Worm gears can be produced by feeding the hob *tangentially* across the gear blank or *radially* into the gear blank, as illustrated in Figures 15.16(b) and (c). A high degree of profile accuracy is attainable by the hobbing process, excellent surface finish can be achieved, and the same cutter can be

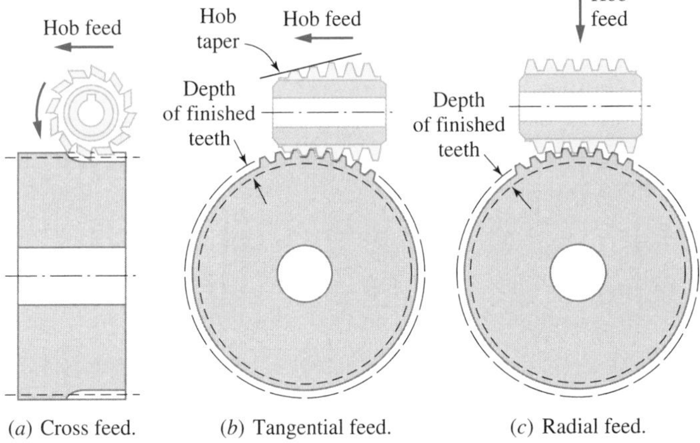

Figure 15.16
Formation of gear teeth by the hobbing method. (Extracted from ref. 1. *Source:* Dudley Engineering Company, San Diego, CA.)

[37]See, for example, ref. 1.

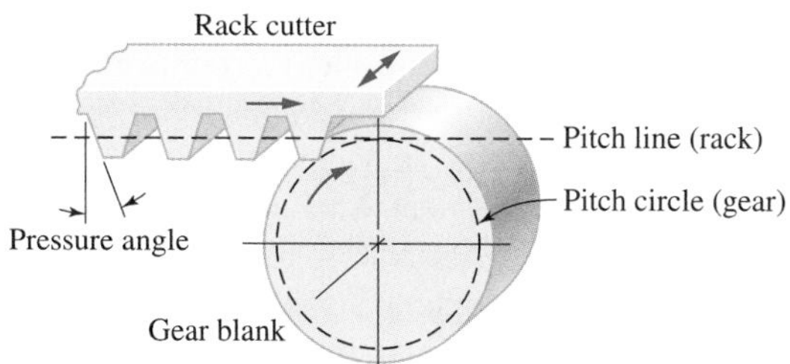

Figure 15.17
Rack generation of involute gear teeth. (Adapted from ref. 3, reprinted by permission of Pearson Education, Inc., Upper Saddle River, N. J.)

used for gears of *like pitch* but different numbers of teeth. High production rates can be achieved using the hobbing process.

Gear shaping, using either a *rack cutter* or a *pinion cutter*, is also a generating method. Since a rack is a gear with infinite pitch radius, involute rack teeth have a straight-sided profile with side slopes equal to the pressure angle. To cut gear teeth with a *rack cutter*, the pitch line of the rack is first aligned to be tangent to the pitch circle of the gear to be cut, as shown in Figure 15.17. Then the rack cutter, with sharpened tooth edges, is given a reciprocating motion across the face of the gear blank. As the gear blank is slowly rotated, the rack is synchronously fed parallel to its pitch line, as illustrated in Figure 15.17. Rack cutters cut only a few teeth during one generating cycle, then must be indexed back to pick up the next tooth to be cut. Rack cutters may be used to cut either spur or helical gears.

The shaping operation using a *pinion cutter* is basically the same as that using a rack cutter; the pinion-shaped cutting tool is reciprocated back-and-forth across the face of the gear blank as both pinion cutter and gear blank are synchronously rotated. Spur, helical, and face gears can be cut using a pinion cutter, as well as worms, and either internal or external gears can be cut. Like hobbing, either rack or pinion generation may be used to produce highly accurate profiles, and either tool can be used to cut gears with any number of teeth of the same pitch.

When cutting gear teeth by the *milling method*, a milling cutter having the exact shape of the tooth space is moved across the face of the gear blank, cutting out one complete tooth space. The gear is then rotated one circular pitch and the next tooth space milled; the process is repeated until all spaces have been cut, forming the gear. Worms, spur gears, and helical gears may be produced by the milling process. Theoretically, a different milling cutter must be used not only for different pitches but also for different numbers of teeth. As a practical matter, milling cutters are usually manufactured to provide the theoretically correct tooth shape for the gear with the smallest number of teeth in each of eight *assigned ranges*, for each standard pitch. Profile errors result for larger gears in the ranges but are acceptable for many applications. The milling method would not be used for high-speed or highly loaded gearsets.

Small internal gears (under 3 inches) can be cut with one pass of a broach[38] as long as the internal teeth are not in a blind hole. Larger internal gears are sometimes made by using a broach to make only a few teeth at a time, then indexing for repeated passes to complete the gear. Either spur or helical gears can be formed by broaching.

Gear Finishing

In high-performance applications, where gears operate at high speeds or under heavy loads, it is often necessary to enhance tooth surface quality by following the gear-cutting

[38] A broach is a straight tool with accurately shaped cutting teeth, increasing in size along its axis, to make a progressive series of cuts to the finished size and shape with a single pass of the tool, when pushed or pulled through the workpiece.

operation with a *finishing* operation. Finishing operations use highly accurate *gear-like tools, dies*, or *grinders* (usually computer controlled) to remove minute amounts of material to improve dimensional accuracy, surface finish, or surface hardness. *Gear shaving* is a corrective process of scraping away tiny shavings of material to improve profile accuracy and surface finish prior to heat treatment. Shaving is accomplished by either the *rotary shaving* method or the *rack shaving* method. In the rotary shaving method a hardened pinion-like cutter, with serrated teeth ground to a conjugate profile, is run against the gear to be finished. In the rack shaving method a hardened rack with serrated teeth is run against the gear.

The *rolling* process finishes the teeth by pressing a hardened tool with precisely ground teeth (called a *die*) against the gear to be finished using a high force, then rolling the die and the gear together. Gear rolling, generally applied *prior* to heat treatment, produces local small-scale yielding, compressive residual surface stresses, and a smooth burnished surface finish.

The *grinding* process of gear finishing usually is performed after carburizing and heat treating the already-cut gear teeth to a high surface hardness. Medium-hardness gears are also ground in some cases. Several grinding methods are used, generally paralleling the cutting methods just discussed. *Form grinders* utilize an abrasive disk, with an involute shape dressed into its sides, to grind both sides of the space between two teeth. *Generating grinders* either use a *single abrasive wheel* dressed to the shape of a basic rack tooth and reciprocated along the face width as the gear to be ground is rolled through the mesh, or use *two saucer-shaped wheels* concave toward the tooth flank with their narrow rims generating the involute as the gear to be finished is rolled through the grinding area.[39] Threaded grinding wheels are also used in some applications, in a process similar to hobbing.

In the *honing* process a fine-grit abrasive gear-shaped tool, called a *hone*, is rolled against the gear to be honed using a very light force to hold them in contact. Honing tends to "average" the surface irregularities, removing local bumps and scale. High-speed gears are often honed *after* finish grinding to produce a very smooth surface finish. This enhances load sharing among meshing teeth and reduces vibration and noise.

Cutter Path Simulation, Mesh Deflection, and Profile Modification

With the advent of electronic computers, not only has it become possible to produce gears of improved precision and quality; it has become possible to model the changes in profile geometry induced by operating loads and speeds, then specify profile modifications that may be *manufactured into unloaded gear teeth* to produce an optimum *running* profile. Although these methods are still under development, good progress has been made in linking interactive graphics, finite element and boundary element analysis, solid modeling, and experimental databases developed for gear cutters and gearset performance,[40] allowing the production of higher-performance gearing in a more condensed time frame.

For example, the computer generation of a theoretical involute profile,[41] as illustrated in Figure 15.18, provides an accurate basis of comparison for internal and external spur and helical gear-tooth profiles. Composite cutter path simulations, such as the one shown in Figure 15.19 for a particular hobbing cutter, allow comparisons of the cut profile with the theoretical involute profile to expedite decisions about important details such as *tip relief, transition from involute flank to root radius*, and *material allowances* for finishing processes.

[39]This type of grinder is made only by the Maag Gear Wheel Company of Zurich, Switzerland; hence the process is known as Maag grinding.

[40]See, for example, ref. 13. [41]See Figure 15.7 and related discussion.

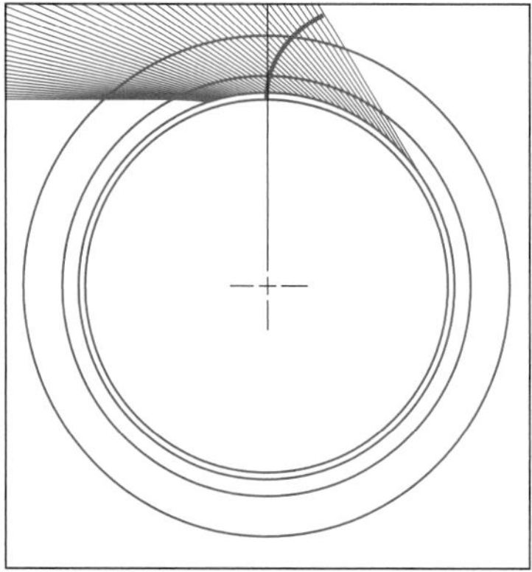

Figure 15.18
Computer-generated theoretical involute profile. (From ref. 13, reprinted with permission from SAE Publication AE-15 © 1990, Society of Automotive Engineers, Inc.)

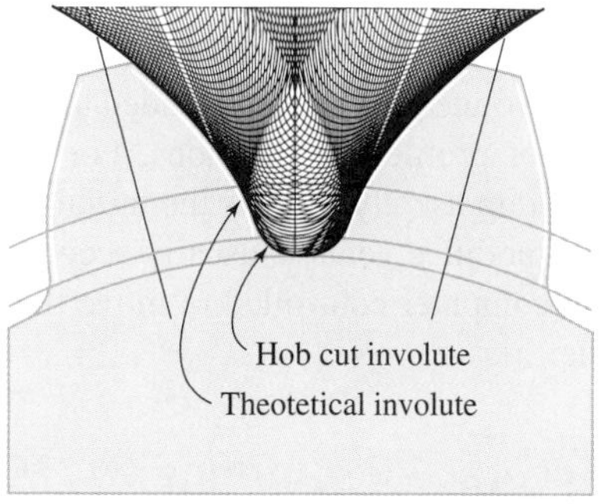

Figure 15.19
Comparison of composite cutter path simulation with theoretical involute profile. (From ref. 13, reprinted with permission from SAE Publication AE-15 © 1990, Society of Automotive Engineers, Inc.)

Tooth deflections caused by bending and Hertz contact deformations may cause *interference* with mating teeth rolling through the mesh, as shown in Figure 15.20(b). Interference may occur both in the zone of first contact at a and in the zone of last contact at b. Profile modifications are often used to avoid interference and provide better load distribution in these zones, as illustrated, for example, in Figure 15.20(c). Such profile modifica-

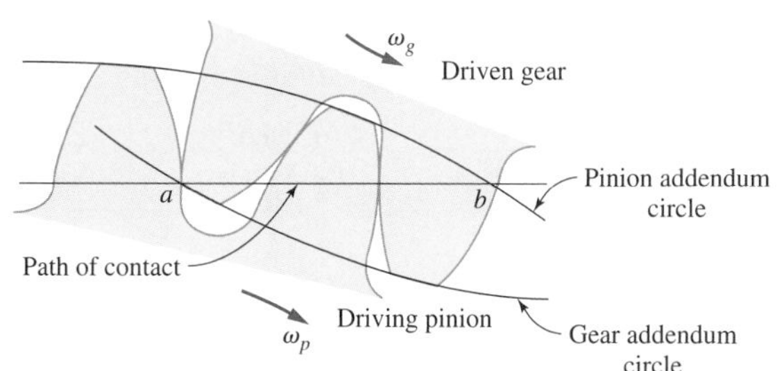

(*a*) Theoretical engagement geometry for involute teeth with *no load.*

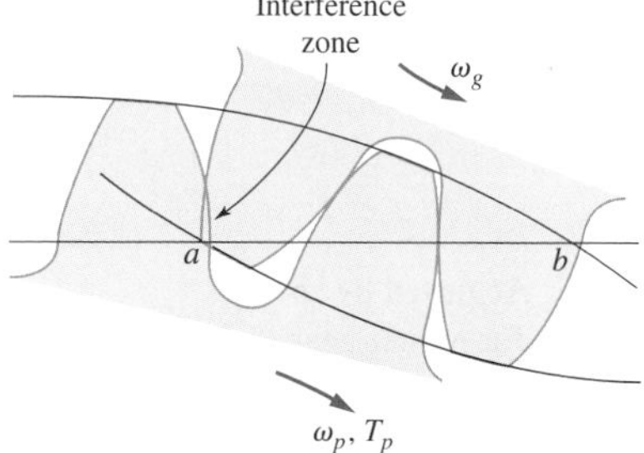

(*b*) Approximate engagement geometry under *operating load,* showing deflection-induced interference zone.

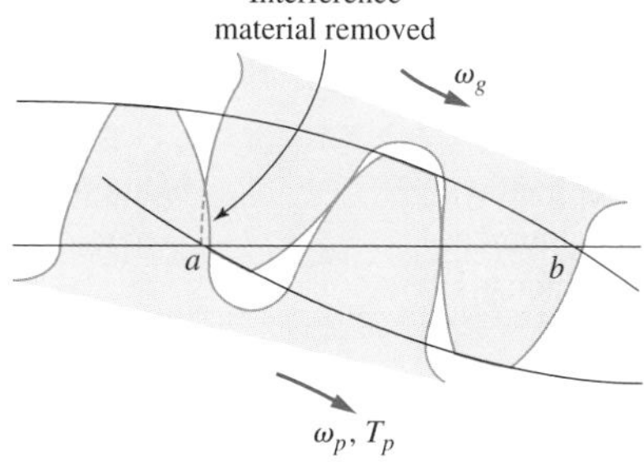

(*c*) Gear-tooth profile modification (tip relief) to remove interference material shown in (*b*).

Figure 15.20
Depiction of unloaded (theoretical) mesh geometry, tooth deformation under load, and a sample profile modification (tip relief), designed to provide smoother operation. (Courtesy of Gear Dynamics and Gear Noise Research Laboratory, Department of Mechanical Engineering, The Ohio State University.)

tions result in smoother, quieter operation, and reduce the risk of failure by surface fatigue (pitting), adhesive wear, galling, and tooth fracture. Computer simulations can be used to obtain three-dimensional tooth deflection data, and then generate plots of deviation from an involute profile as the loaded teeth roll through the mesh. Such plots form the initial basis for profile modifications. For high-precision gearing, a trial-and-error experimental program usually follows the initial profile modification proposal. The manufacture of profile-modified gears is usually accomplished using generating-type grinding machines that are computer controlled to move the grinding wheels shallower or deeper along the profile.

Accuracy Requirements, Measurement Factors, and Manufacturing Cost Trends

Casting the concepts of Chapter 1 into gear design strategy, the gear designer is responsible for achieving required load-carrying capacity and specified reliability while meeting gear speed, life, and noise requirements at a reasonable cost. High loads, high speeds, long lives, and quiet operation require high-quality gears manufactured to close tolerances; costs for such high-precision gears are high. When loads and speeds are less demanding, and more moderate application constraints with wider tolerances are permissible, accuracy requirements may be relaxed to give costs that are more moderate. Table 15.4 provides an experience-based guide to accuracy-level requirements for various applications.[42]

To assess the quality of a manufactured gear it is necessary to *measure* the finished product with an appropriate level of accuracy. Generally, it is necessary to have available

TABLE 15.4 Accuracy Levels Typical of Various Gearing Applications

		Approximate *Standard Quality Ranges*	
Accuracy Level	Dudley Designation[1]	AGMA[2] Q_v Value	DIN[3] Value
Highest possible accuracy. Achieved by special toolroom methods. Used for master gears, unusually critical high-speed gears, or when *both* highest load capacity and highest reliability are needed.	AA Ultra-high accuracy	14 or 15	2 or 3
High accuracy. Achieved by grinding or shaving with first-rate machine tools, and utilizing skilled operators. Widely used for turbine gearing and aerospace gearing. Sometimes used for critical industrial gears.	A High accuracy	12 or 13	4 or 5
Relatively high accuracy. Achieved by grinding or shaving with emphasis on production rate rather than highest quality. May be achieved by hobbing or shaping with best equipment under favorable conditions. Used for medium-speed industrial gears and critical vehicle gears.	B Medium-high accuracy	10 or 11	6 or 7
Good accuracy. Achieved by hobbing or shaping with first-rate machine tools and skilled operators. May be obtained in high-production grinding or shaving. Typically used for vehicle gears and electric motor industrial gears running at slower speeds.	C Medium accuracy	8 or 9	8 or 9

[42]Adapted from ref. 1. *Source:* Dudley Engineering Company, San Diego, CA.

TABLE 15.4 (*Continued*)

Accuracy Level	Dudley Designation[1]	Approximate *Standard Quality Ranges* AGMA[2] Q_v Value	DIN[3] Value
Nominal accuracy. Can be achieved by hobbing or shaping using older machine tools and less-skilled operators. Typically used for low-speed gears that *wear in* to yield a reasonable fit. (Lower hardness promotes wear-in.)	D Low accuracy	6 or 7	10 or 11
Minimal accuracy. For gears used at slow speeds and light loads. Teeth may be cast or molded in small sizes. Typically used in toys and gadgets. May be used for low-hardness gears when limited life and lower reliability are acceptable.	E Very low accuracy	4 or 5	12

[1]From ref. 1. *Source:* Dudley Engineering Company, San Diego, CA.
[2]See ref. 14.
[3]See ref. 15.

measuring devices to accurately check the prime variables of involute profile: *helix across the face width, tooth spacing, tooth finish,* and *tooth action when meshed with a master gear*[43] to obtain a *composite check* of individual tooth action and total gear runout. A plot of the cost trends of manufacturing gear teeth by different methods, as a function of required accuracy, is shown in Figure 15.21.

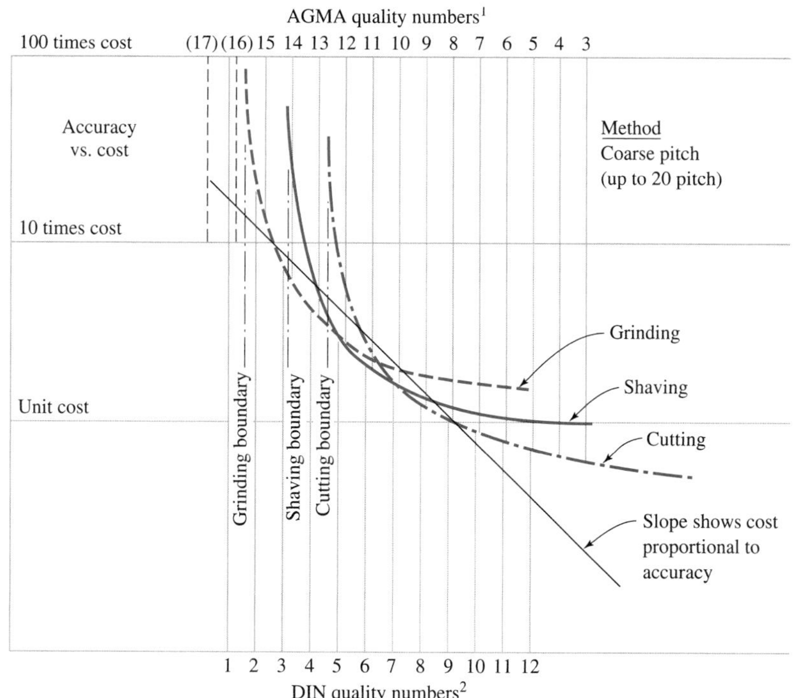

Figure 15.21
Cost trends for manufacturing gear teeth by different methods. (Extracted from ref. 1. *Source:* Dudley Engineering Company, San Diego, CA.)

[43]Carefully maintained, high-precision gear.

15.8 Spur Gears; Force Analysis

In Figure 15.22(a), a spur *pinion* (1) centered at O_1, rotating clockwise at a speed of n_1 rpm, drives a spur *gear* (2) centered at O_2, rotating counterclockwise at a speed of n_2 rpm. From Figure 15.8 and the related discussion, the normal force F_n transmitted from the driving pinion to the conjugate gear is known to be directed along the space-fixed line of action, which passes through the pitch point P. If an external power source produces a steady operating torque T_1 on the pinion, and the pinion and gear are each taken as separate free bodies, the result may be depicted as shown in Figures 15.22(b) and (c). R_1 and R_2 are resultant bearing support reactions. Examining the pinion free body, for example, it is clear that the resultant normal force F_n acting on the pinion at the pitch point P, along the line of action, may be resolved into a tangential component F_t and a radial component F_r, as illustrated in Figure 15.22(d). Bearing reaction forces may also be resolved into components if desired. The tangential component F_t is directed along the common tangent (hence perpendicular to pitch radius r_1), so by moment equilibrium about the axis of rotation

$$T_1 = F_t r_1 \tag{15-21}$$

Since F_t is directly related to power transmission, it is the "useful" force component. By contrast, the radial component F_r is simply a residue of tooth geometry, serving no useful purpose, tending only to push the gears apart. Often F_t is called *transmitted force* and F_r is called *separating force*. A similar force resolution could be made for the driven gear.

Transmitted force F_t may be calculated by using the power transferred from pinion to gear, as given by (4-32) or (4-34). From (4-32), for example,

$$hp = \frac{T(2\pi r n)}{12(33{,}000)} = \frac{F_t r_1(2\pi n_1)}{12(33{,}000)} = \frac{F_t r_2(2\pi n_2)}{12(33{,}000)} \tag{15-22}$$

based on the fact that transmitted power remains constant throughout a gear train if friction losses are neglected.

The pitch-line velocity V (ft/min) may be calculated as

$$V = \frac{2\pi r_1 n_1}{12} = \frac{2\pi r_2 n_2}{12} \tag{15-23}$$

Figure 15.22
Force analysis of spur gearset.

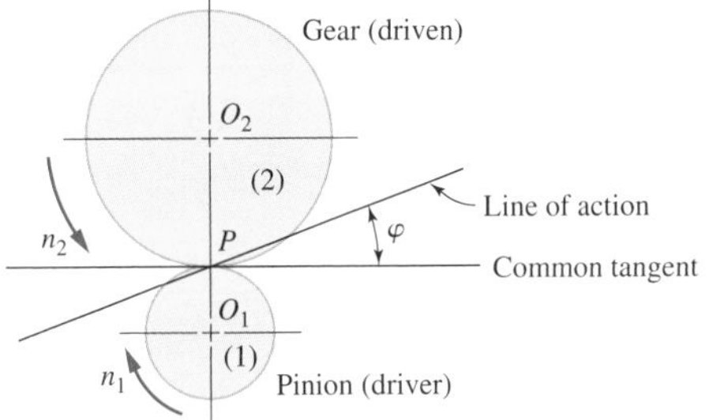

(*a*) Typical spur gearset layout showing pitch circles and line of action.

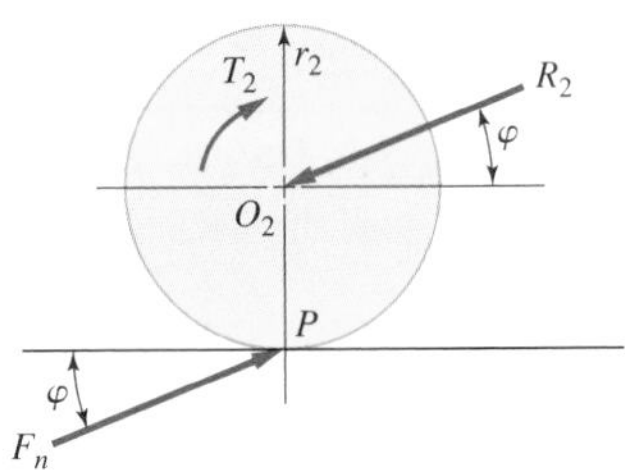

(*b*) Free-body diagram of driven gear.

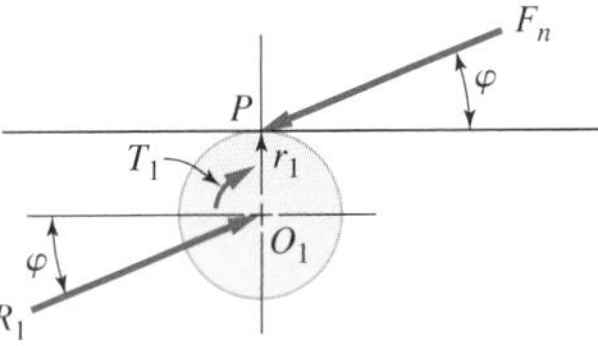

(*c*) Free-body diagram of driving pinion.

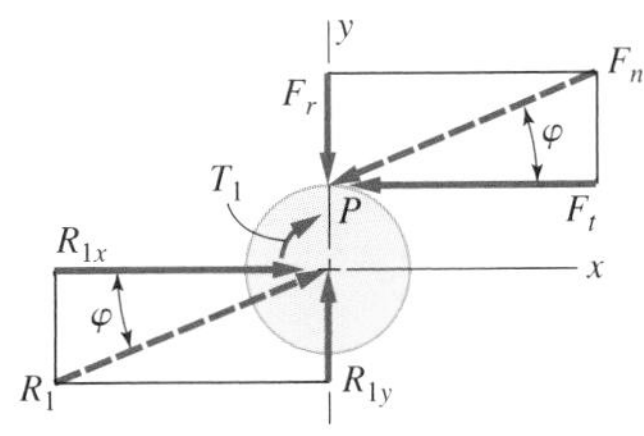

(*d*) Force resolution for pinion.

where r_1 and r_2 are pitch radii in inches, and n_1 and n_2 are rotational speeds in rpm. Combining (15-22) and (15-23) gives a convenient form of the horsepower equation for gearing applications as

$$hp = \frac{F_t V}{33{,}000} \qquad (15\text{-}24)$$

which allows calculation of required transmitted force as

$$F_t = \frac{33{,}000(hp)}{V} \qquad (15\text{-}25)$$

The power required and the pitch-line velocity V are usually known for a specific application.

If the pressure angle φ is known, the geometry of Figure 15.22(c) allows calculation of the separating force as

$$F_r = F_t \tan\varphi \qquad (15\text{-}26)$$

It is also of interest to note that

$$F_t = F_n \cos\varphi \qquad (15\text{-}27)$$

Finally, consideration of (15-21) and (15-22) leads to a useful extension of (15-15), namely

$$\frac{\omega_2}{\omega_1} = \frac{n_2}{n_1} = \pm\frac{r_1}{r_2} = \pm\frac{N_1}{N_2} = \pm\frac{T_1}{T_2} \qquad (15\text{-}28)$$

where ω = angular velocity
n = rotational speed
r = pitch radius
N = number of gear teeth
T = applied torque

15.9 Spur Gears; Stress Analysis and Design

Of the potential failure modes discussed in 15.4, the two most likely to govern a typical gearset design are *bending fatigue* failure at the root fillet of a tooth, or *surface fatigue* (*pitting*) failure generated by the cyclic Hertz contact stresses produced by repeatedly meshing gear teeth. *In principle*, gear-tooth design procedure is no different from the design procedure for any other machine element;[44] *loading severity parameters* (e.g., stresses) are calculated and compared with *critical capacities* (e.g., strengths corresponding to governing failure modes), adjusting materials and geometry until an appropriate safety factor or reliability level is achieved.[45] *In practice*, the results obtained by applying fundamental principles are typically refined by using a series of experience-based modifying factors (sometimes called *derating factors*) to account for manufacturing variabilities, operational dynamics, strength variabilities, environmental variabilities, and mounting and assembly variabilities. A widely accepted set of procedures for this purpose, with an extensive body of supporting data, has been standardized and published by the AGMA.[46] A designer responsible for gearing applications, especially high-performance gearing, is well advised to consult the most up-to-date AGMA standards (these standards are continuously being updated).

[44]See Chapter 6.

[45]See Chapter 5.

[46]American Gear Manufacturers Association, 1500 King Street, Suite 201, Alexandria, VA 22314.

The approach in this text will be to present a simplified analysis first, followed by a brief synopsis of the more refined AGMA standard procedures for analyzing tooth bending and surface durability.

Tooth Bending: Simplified Approach

A gear tooth may be idealized as a cantilever beam subjected to an end load,[47], as illustrated in Figure 15.23. Approximations first introduced by Lewis[48] in 1893 still are used to formulate the expression for nominal bending stress at the tensile-side root fillet designated as point a in Figure 15.23. Lewis observed that a constant-strength parabola[49] inscribed within an involute tooth profile, drawn tangent to the root fillets, and with apex located at the intersection of the normal force vector F_n and the tooth centerline, lies everywhere *within the tooth profile*. Assuming such a parabolic tooth shape provides a conservative geometric description for making a cantilever beam estimate of gear-tooth bending stresses. For a tooth of face width b, the nominal tensile bending stress at critical point a due to transmitted force F_t is[50]

$$(\sigma_b)_a = \frac{6F_tL}{bh^2} \tag{15-29}$$

A small (favorable) compressive stress component at a, produced by the separating force F_r, is usually neglected; the small *increase* in compressive stress at b, already compressive, is also usually neglected because cyclic *compressive* stressing is typically much less serious than cyclic *tensile* stressing at a.

To specialize the stress calculation of (15-29) to an involute gear tooth, as introduced in the Lewis development, a similar triangle proportionality in Figure 15.23 gives

$$\frac{h/2}{x} = \frac{L}{h/2} \tag{15-30}$$

whence

$$x = \frac{h^2}{4L} \tag{15-31}$$

Next, (15-29) is rearranged to the form

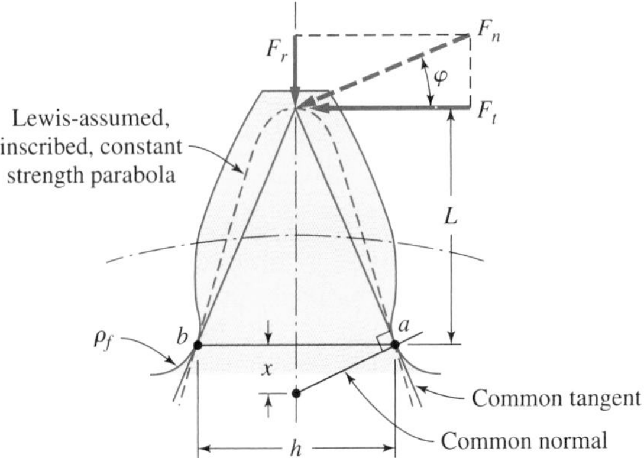

Figure 15.23
Cantilever bending model of an *assumed parabolic* spur gear tooth. Bending is produced by force F_n acting at the *apex* (tip) of the (Lewis) inscribed constant-strength parabola. The intersection of vector F_n with the tooth centerline defines the location of the apex.

[47]The load may actually be applied short of the end for precision gearing with a contact ratio greater than 1.0 because of *load sharing* with other teeth. See discussion of *highest point of single tooth contact* under "Mesh Interactions" in 15.6.

[48]See ref. 16. [49]See Figure 14.12(b) and (14-32). [50]See (14-30).

$$(\sigma_b)_{cp-a} = \frac{F_t}{b}\left(\frac{1}{h^2/4L}\right)\left(\frac{1}{4/6}\right) \tag{15-32}$$

Combining (15-31) with (15-32) and multiplying by (p_c/p_c),

$$(\sigma_b)_{cp-a} = \frac{F_t}{b}\left(\frac{1}{x}\right)\left(\frac{3}{2}\right)\left(\frac{p_c}{p_c}\right) = \frac{F_t}{bp_c y} \tag{15-33}$$

where

$$y \equiv \frac{2x}{3p_c} \tag{15-34}$$

is defined as the *Lewis form factor*, gives the original version of the Lewis equation.

Alternative versions written in terms of *diametral pitch* or *module* are usually preferred in modern practice. Using (15-12) and (15-13) these are, for U.S. units,

$$(\sigma_b)_{cp-a} = \frac{F_t P_d}{\pi b y} = \frac{F_t P_d}{bY} \tag{15-35}$$

and, for SI units,

$$(\sigma_b)_{cp-a} = \frac{F_t}{\pi m b y} = \frac{F_t}{mbY} \tag{15-36}$$

where $Y = \pi y$ is a Lewis form factor based on diametral pitch or module. Both Y and y are functions of tooth shape (but not size) and therefore vary with the number of teeth in the gears. Values of Y are given in Table 15.5 for various numbers of gear teeth.

To continue with the simplified approach, a fatigue stress concentration factor K_f should be applied to obtain the estimated actual stress at the tensile critical point a, using the methods discussed in 4.8.[51] Theoretical stress concentration factor K_t is given in Figure 4.23 for 20° involute gear teeth as a functionof diametrical pitch P_d. Recalling (4-111) then

$$K_f = q(K_t - 1) + 1 \tag{15-37}$$

and

TABLE 15.5 Values for Lewis Form Factor *Y* for 20° Full-Depth Involute Teeth

Number of Teeth	Y	Number of Teeth	Y
12	0.245	28	0.353
13	0.261	30	0.359
14	0.277	34	0.371
15	0.290	38	0.384
16	0.296	43	0.397
17	0.303	50	0.409
18	0.309	60	0.422
19	0.314	75	0.435
20	0.322	100	0.447
21	0.328	150	0.460
22	0.331	300	0.472
24	0.337	400	0.480
26	0.346	Rack	0.485

[51]Stress concentration was unknown at the time of the Lewis development for nominal stress at the root fillet.

$$(\sigma_b)_{act} = K_f(\sigma_b)_{nom} \tag{15-38}$$

Now, using (15-35), at critical point a,

$$(\sigma_b)_{act} = K_f\left(\frac{F_t P_d}{bY}\right) \tag{15-39}$$

The final step in the simplified approach is to obtain an appropriate value of equivalent completely reversed alternating stress amplitude σ_{eq-CR} corresponding to the actual mean and alternately stresses for the application.[52]

Comparing σ_{eq-CR} with fatigue strength S_N or fatigue limit S_f for the selected material under projected operating conditions[53] then gives the basis for an estimate of existing safety factor or reliability level for the proposed design. For example, the existing safety factor n_{ex}, based on (5-7), may be written for this case as

$$n_{ex} = \frac{S_N}{\sigma_{eq-CR}} \tag{15-40}$$

Example 15.2 Simplified Approach to Spur Gear Bending Fatigue

It is desired to further examine the proposed four-gear train shown in Figure E15.1A of Example 15.1, to establish preliminary estimates of the resistance to failure at the root fillets of gear no. 1, based on potential failure there due to tooth bending fatigue. Additional specifications for this gear train are as follows:

1. The gear train is to drive a large industrial blender (stirring machine) which requires 5 horsepower at a steady-state operating speed of 1000 rev/min at output gear 4.
2. This preliminary calculation may be based on the assumption that friction losses are negligible for all bearings and gear meshes.
3. It is desired to use Grade 1 AISI 1020 low-carbon steel, if possible, for the gear blanks.
4. Gear teeth are to be standard 20° spur full-depth involute teeth.
5. An in-house gearing specialist has suggested that a diametral pitch of about 10 would probably be a good initial assumption for the iterative calculation.

If a very long life is desired for this gear, and a reliability of 99 percent is required, propose an acceptable configuration for gear 4 based on tooth bending fatigue.

Solution

From Example 15.1, the reduction ratio is

$$\frac{n_{out}}{n_{in}} = \frac{n_4}{n_1} = 0.4 \tag{1}$$

so the speed of gear 1 is

$$n_1 = \frac{1000}{0.4} = 2500 \text{ rpm} \tag{2}$$

Also, since losses are negligible

$$hp_1 = hp_4 = 5 \text{ horsepower} \tag{3}$$

[52]Refer to equation (2-42) or (2-43) for a uniaxial state of stress, or (4-97) through (4-102) for multiaxial states of stress.

[53]Refer to (2-25), (2-26), (2-27), and (2-28).

Initially selecting $P_d = 10$ as recommended (this may require adjustment later), the pitch radius r_1 may be calculated from (15-8) as

$$r_1 = \frac{N_1}{2P_d} = \frac{20}{2(10)} = 1.0 \text{ inch} \tag{4}$$

Next from (15-23), the pitch line velocity V is

$$V = \frac{2\pi(1.0)(2500)}{12} = 1309 \text{ ft/min} \tag{5}$$

whence, from (15-25)

$$F_t = \frac{33{,}000(5.0)}{1309} = 126 \text{ lb} \tag{6}$$

To calculate the *nominal* bending stress at the tension-side root fillet from (15-35), a Lewis form factor Y may be read from Table 15.5, corresponding to $N_1 = 20$, as

$$Y = 0.322 \tag{7}$$

and a midrange face width b may be tentatively chosen, using (15-16), as

$$b = \frac{11.5}{P_d} = \frac{11.5}{10} = 1.15 \text{ inches} \tag{8}$$

Using these values in (15-35), the nominal bending stress at critical point a may be calculated as

$$(\sigma_b)_{nom} = \frac{F_t P_d}{bY} = \frac{(126)(10)}{1.15(0.322)} = 3403 \text{ psi} \tag{9}$$

To calculate the actual maximum (static) bending stress at critical point a, a fatigue stress concentration factor may be calculated using (15-37). The value of K_t may be estimated from Figure 4.23 and the value of q may be read from Figure 4.27.

To make the estimate of K_t from Figure 4.23, the moment arm e will be taken as the working depth (see Table 15.1) and tooth thickness h at the root will be assumed equal to base thickness (circular thickness multiplied by cosine of pressure angle); hence

$$\frac{e}{h} \approx \frac{2.000/P_d}{(1.571/P_d)\cos 20} = 1.35 \tag{10}$$

This ratio lies off the chart in Figure 4.23, so the estimate of K_t (crude at best) seems to be about

$$K_t \approx 1.5 \tag{11}$$

Using an estimate of root fillet radius ρ_f

$$\rho_f = \frac{0.35}{P_d} = \frac{0.35}{10} = 0.035 \text{ inch} \tag{12}$$

the value of q for 1020 steel read from Figure 4.27 is

$$q = 0.65 \tag{13}$$

so (15-37) gives

$$K_f = (0.65)(1.5 - 1) + 1 = 1.3 \tag{14}$$

permitting the evaluation of (15-38) for this proposed design as

$$(\sigma_b)_{act} = (1.3)(3403) = 4420 \text{ psi} \tag{15}$$

Referring to Table 3.3 the static strength properties for 1020 steel are

Example 15.2 Continues

$$S_u = 57{,}000 \text{ psi}$$
$$S_{yp} = 43{,}000 \text{ psi} \tag{16}$$

and from Figure 2.19, the fatigue limit for small, polished, 1020 steel specimens may be read as

$$S_f' = 33{,}000 \text{ psi} \tag{17}$$

Using (2-25) and (2-27) and the methods of Example 2.9, the value of k_∞ has been estimated to be about 0.65 for this application, including the 99 percent reliability adjustments. Thus

$$S_f = k_\infty(S_f') = 0.65(33{,}000) = 21{,}400 \text{ psi} \tag{18}$$

Finally, the pinion teeth experience *one-way* bending, producing *released tension* at the critical root fillet.[54]

Using (2-42) then,

$$\sigma_{eq-CR} = \frac{\sigma_a}{1 - \sigma_m/S_u} = \frac{(4420/2)}{1 - \dfrac{(4420/2)}{57{,}000}} = 2300 \text{ psi} \tag{19}$$

Based on (15-40) the existing safety factor at the tension-side root fillet for this proposed design would be

$$n_{ex} = \frac{S_f}{\sigma_{eq-CR}} = \frac{21{,}400}{2300} = 9.3 \tag{20}$$

Most designers would regard a safety factor of 9.3 as too large and would probably attempt to redesign (typicallly by selecting a new value for P_d) to better utilize material and space. It should be recognized, however, that this calculation has considered neither the possible dynamic effects of impact loading (see 2.8) nor the influence of deflection of shafts, bearings, and gear teeth upon tooth loading. These two influences could easily reduce n_{ex} by a factor of 2 or 3. Before redesigning on the basis of (20) therefore, it would be prudent to first make a preliminary evaluation of this design based on resistance to failure by surface fatigue wear (pitting), then proceed to more refined design calculations. Examples 15.3, 15.4, and 15.5 illustrate the procedure.

Tooth Bending: Synopsis of AGMA Refined Approach[55]

The AGMA approach to designing gear teeth to resist bending fatigue failure, while based on the idealized Lewis equation, involves an extensive list of empirical adjustment factors (sometimes called derating factors) to account for the influence of various manufacturing, assembly, geometric, loading, and material variabilities. Many pages of charts and graphs of supporting data are published by the AGMA. The presentation in this text represents only a *synopsis* of the AGMA procedure, with an *abridged* selection of supporting data to demonstrate the basic approach; any designer responsible for gearing design or development is well advised to consult the most up-to-date AGMA standards.

The basic AGMA bending stress[56] equation may be written as[57]

$$\sigma_b = \frac{F_t P_d}{bJ} K_a K_v K_m K_I \tag{15-41}$$

[54] See Figure 2.12. [55] See ref. 10.

[56] Referred to as *bending stress number* in AGMA standard 2001-C95.

[57] Other factors are available from AGMA standards to account for *overloads* (system vibration, overspeed events, changing process, load conditions), *size* effects, and *thin-rim* deflection effects.

where K_a = application factor (see Table 15.6)
K_m = mounting factor (see Table 15.7)
K_v = dynamic factor (see Figure 15.24)
K_I = idler factor = 1.42 for idler teeth with *two-way* bending; = 1.0 for gear teeth with *one-way* bending.
J = geometry factor (see Tables 15.8 and 15.9)
P_d = diametral pitch
b = face width
F_t = tangential load

TABLE 15.6 Application Factor, K_a

Prime Mover Characteristic	Driven Machine Characteristic: Uniform	Moderate Shock	Heavy Shock
Uniform (e.g., electric motor, turbine)	1.00	1.25	1.75 or higher
Light shock (e.g., multicylinder engine)	1.25	1.50	2.00 or higher
Medium shock (e.g., single-cylinder engine)	1.50	1.75	2.25 or higher

TABLE 15.7 Mounting Factor, K_m

Support Properties and Gear Quality	Face Width, in: 0 to 2	6	9	≥ 16
Accurate mountings, small bearing clearances, minimum deflections, precision gears	1.3	1.4	1.5	1.8
Less rigid mountings, more bearing clearance, less accurate gears, contact across full face	1.6	1.7	1.8	2.2
Combinations of mounting properties and gearing precision that produce less than full face contact	2.2 or higher			

TABLE 15.8 AGMA Geometry Factor J for Bending of 20° Full-Depth Involute Teeth Under *Tip Loading* (used for lower precision gearing)

Gear Teeth	Pinion Teeth: 12		14		17		21		26		35		55		135	
	P[1]	G	P	G	P	G	P	G	P	G	P	G	P	G	P	G
12	U[2]	U														
14	U	U	U	U												
17	U	U	U	U	U	U										
21	U	U	U	U	U	U	0.24	0.24								
26	U	U	U	U	U	U	0.24	0.25	0.25	0.25						
35	U	U	U	U	U	U	0.24	0.26	0.25	0.26	0.26	0.26				
55	U	U	U	U	U	U	0.24	0.28	0.25	0.28	0.26	0.28	0.28	0.28		
135	U	U	U	U	U	U	0.24	0.29	0.25	0.29	0.26	0.29	0.28	0.29	0.29	0.29

[1]P = pinion; G = gear.
[2]U indicates a combination that produces undercutting.

TABLE 15.9 AGMA Geometry Factor *J* for Bending of 20° Full-Depth Involute Teeth Under *HPSTC*[1] *Loading* (used for higher-precision gearing)

Gear Teeth	Pinion Teeth 12		14		17		21		26		35		55		135	
	P[2]	G	P	G	P	G	P	G	P	G	P	G	P	G	P	G
12	U[3]	U														
14	U	U	U	U												
17	U	U	U	U	U	U										
21	U	U	U	U	U	U	0.33	0.33								
26	U	U	U	U	U	U	0.33	0.35	0.35	0.35						
35	U	U	U	U	U	U	0.34	0.37	0.36	0.38	0.39	0.39				
55	U	U	U	U	U	U	0.34	0.40	0.37	0.41	0.40	0.42	0.43	0.43		
135	U	U	U	U	U	U	0.35	0.43	0.38	0.44	0.41	0.45	0.45	0.47	0.49	0.49

[1]*Highest Point of Single Tooth Contact*. See discussion under "Mesh Interactions" in 15.6.
[2]P = pinion; G = gear.
[3]U indicates a combination that produces undercutting.

The *application factor* K_a is used to account for shock or impact loading characteristics of the prime mover and the driven machine. The *dynamic factor* K_v is used to estimate the effect of dynamic loading when a detailed dynamic system analysis is not available. The gearing quality level, Q_v in Figure 15.24, reflects gearing accuracy, primarily based on transmission error. Guidance in selecting an appropriate quality-level curve may be found in Table 15.4. Pitch-line velocity for use in Figure 15.24 may be calculated using (15-23). The *mounting factor* K_m is used to account for nonuniform load distribution across the tooth face due to manufacturing variabilities, bearing clearances, and support deflections. The *idler factor* K_I reflects the difference in fatigue resistance of a gear tooth when subjected to completely reversed stressing (idler) as compared to released tension (typical gear tooth). The AGMA *geometry factor J* embodies the Lewis form factor, stress concentration effects, and load-sharing effects in a single parameter. Table 15.8 gives selected values of *J* for 20° full-depth teeth when the tangential load F_t is applied at the *tooth tip*, and Table 15.9 gives corresponding data for load applied at the *highest point of single tooth contact (HPSTC)*. If gearing precision is high, and load sharing occurs (for contact

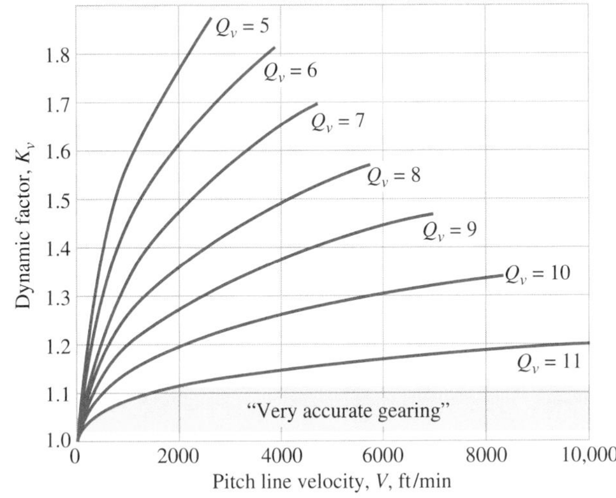

Figure 15.24
Dynamic factor K_v. The gearing quality values Q_v are a function of gearing accuracy (primarily based on transmission error). See Table 15.4 for guidance. (Adapted from ANSI/AGMA Standard 2001-C95, with the permission of the publisher, American Gear Manufacturers Association, 1500 King Street, Suite 201, Alexandria, VA 22314.)

ratios between 1 and 2), the factors in Table 15.9 for HPSTC loading are appropriate. If gearing precision is low, errors in profile and spacing may prevent effective load sharing, and a single tooth may be subjected to the *full load at its tip*; the factors in Table 15.8 are appropriate in this case.

On the strength side, published AGMA data for tooth bending fatigue strength S'_{tbf}, as illustrated in Tables 15.10 and 15.11 for steel gears, and Table 15.12 for iron and bronze gears, are all indexed to a reliability of 99 percent for a life of 10^7 cycles of one-way loading. For design lives other than 10^7 cycles, the strength S'_{tbf} may be multiplied times a *life adjustment factor* Y_N, read from Figure 15.28 for steel gears (values for Y_N for materials other than steel are not readily available). For reliability requirements other than 99 percent, the strength S'_{tbf} may be multiplied by a gearing reliability factor, R_g, read from Table 15.13.

Thus the tooth bending fatigue strength, S_{tbf}, corresponding to desired reliability and design life, becomes

$$S_{tbf} = Y_N R_g S'_{tbf} \tag{15-42}$$

Just as σ_{eq-CR} was compared with fatigue strength S_N in the simplified approach discussed earlier, in using the AGMA refined approach the AGMA bending stress σ_b is compared

TABLE 15.10 ANSI/AGMA Tooth Bending Fatigue Strength S'_{tbf} for Steel Gears[1]

		Metallurgical Quality[2]					
		Grade 1		Grade 2		Grade 3	
Material	Heat Treatment	Min. Surf. Hardness	S'_{tbf}, ksi	Min. Surf. Hardness	S'_{tbf}, ksi	Min. Surf. Hardness	S'_{tbf}, ksi
Steel	Through hardened	See Figure 15.25				—	—
	Flame or induction hardened, including root	—	45	28 R_C[3]	55	—	—
	Flame or induction hardened, except root	—	22	—	22	—	—
	Carburized and hardened (Min. core hard.)	55–64 R_C (21 R_C)	55	58–64 R_C (25 R_C)	65	58–64 R_C (30 R_C)	75
AISI 4140, AISI 4340 steel	Nitrided and through hardened	83.5 R_{15N}	see Figure 15.26	83.5 R_{15N}	see Figure 15.26	83.5 R_{15N}	—
Nitralloy 135 M, Nitralloy N, and 2.5% chrome (no aluminum)	Nitrided	87.5 R_{15N}	see Figure 15.27	87.5 R_{15N}	see Figure 15.27	87.5 R_{15N}	see Figure 15.27

[1]From ref. 10. Adapted from ANSI/AGMA Standard 2001-C95, with the permission of the publisher, American Gear Manufacturers Association, 1500 King Street, Suite 201, Alexandria, VA 22314.
[2]See Table 15.11 for quality characteristics of specified grade.
[3]R_C = Rockwell C scale.
R_{15N} = Rockwell 15N scale.

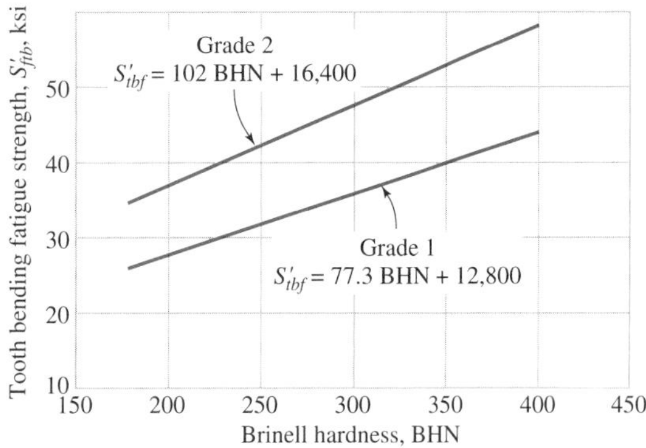

Figure 15.25
ANSI/AGMA tooth bending fatigue strength S'_{tbf} for through-hardened gears. Curves are based on 99 percent reliability for a life of 10^7 cycles of one-way loading. See Table 15.11 for quality characteristics of specified grade. (Adapted from ANSI/AGMA Standard 2001-C95, with the permission of the publisher, American Gear Manufacturers Association, 1500 King Street, Suite 201, Alexandria, VA 22314.)

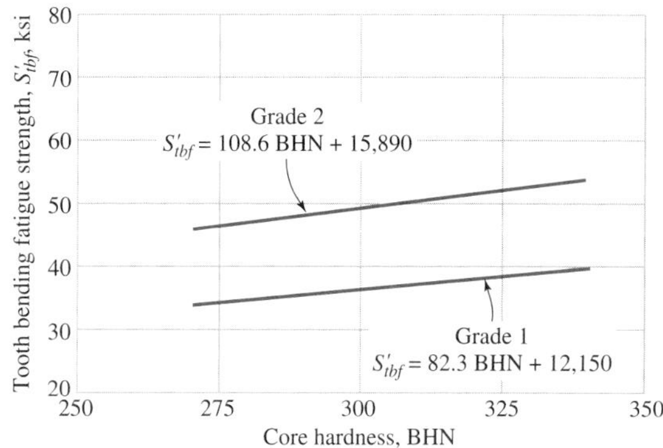

Figure 15.26
ANSI/AGMA tooth bending fatigue strength for nitrided through-hardened gears (AISI 4140 and 4340). Curves are based on 99 percent reliability for a life of 10^7 cycles of one-way loading. See Table 15.11 for quality characteristics of specified grade. (Adapted from ANSI/AGMA Standard 2001-C95, with the permission of the publisher, American Gear Manufacturers Association, 1500 King Street, Suite 201, Alexandria, VA 22314.)

Figure 15.27
ANSI/AGMA tooth bending fatigue strength for nitriding-steel gears. Curves are based on 99 percent reliability for a life of 10^7 cycles of one-way loading. See Table 15.11 for characteristics of specified grade. (Adapted from ANSI/AGMA Standard 2001-C95, with the permission of the publisher, American Gear Manufacturers Association, 1500 King Street, Suite 201, Alexandria, VA 22314.)

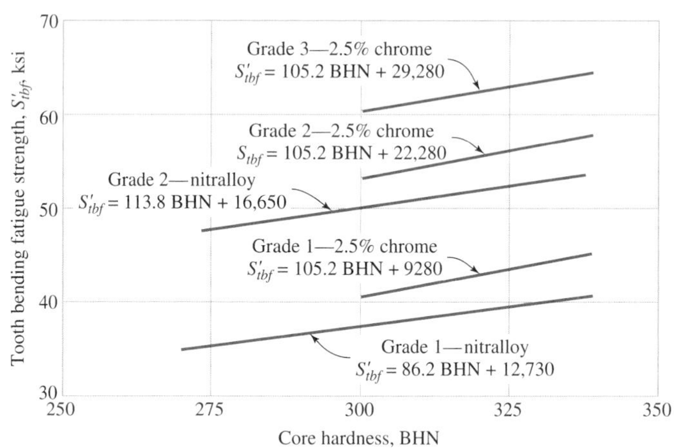

TABLE 15.11 Quality Characteristics of AGMA Steel Material Grades[1] (tentative)

Quality Grade	Characteristics
0 (ordinary quality)	No gross defects; no close control of quality items.[2]
1 (good quality)	Modest control of most important quality items; typical industrial practice.
2 (premium quality)	Close control of all critical quality items; results in improved performance but increases material cost.
3 (super quality)	Absolute control of all critical quality items; results in ultimate performance but high material cost; rarely required.

[1]See ref. 1.

[2]Quality items include surface hardness, core hardness, case structure, core structure, steel cleanliness, flank surface condition, root-fillet surface condition, grain size, and nonuniform hardness or structure.

TABLE 15.12 ANSI/AGMA Tooth Bending Fatigue Strength S'_{tbf} for Iron and Bronze Gears[1]

Material	Material Designation	Heat Treatment	Typical Min. Surface Hardness[2]	S'_{tbf}, ksi
ASTM A48 gray cast iron	Class 20	As cast	—	5
	Class 30	As cast	174 BHN	8.5
	Class 40	As cast	201 BHN	13
ASTM A536 ductile iron (nodular)	Grade 60-40-18	Annealed	140 BHN	22–33
	Grade 80-55-06	Q & T[3]	179 BHN	22–33
	Grade 100-70-03	Q & T	229 BHN	27–40
	Grade 120-90-02	Q & T	269 BHN	31–44
Bronze		Sand cast	Min. tensile strength, 40 ksi	5.7
	ASTM B-148 Alloy 954	Heat treated	Min. tensile strength, 90 ksi	23.6

[1]From ref. 10. Adapted from ANSI/AGMA Standard 2001-C95, with the permission of the publisher, American Gear Manufacturers Association, 1500 King Street, Suite 201, Alexandria, VA 22314.
[2]BHN = brinnell hardness number. [3]Quenched and tempered.

TABLE 15.13 AGMA Reliability Adjustment Factor R_g for Gears[1]

Desired Reliability, percent	R_g
99.99	0.67
99.9	0.80
99	1.0
90	1.18
50	1.43

[1]From ref. 10.

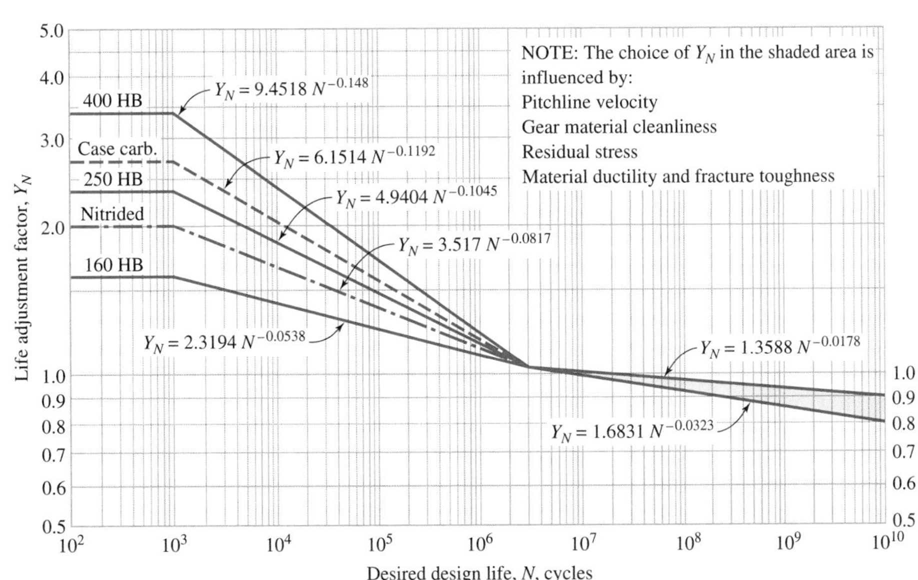

Figure 15.28
Life adjustment factor for adjusting the ANSI/AGMA bending fatigue strength to a life other than 10^7 cycles. Valid only for steel materials. (Adapted from ANSI/AGMA Standard 2001-C95, with the permission of the publisher, American Gear Manufacturers Association, 1500 King Street, Suite 201, Alexandria, VA 22314.)

with the AGMA bending fatigue strength S_{tbf} to determine whether the proposed design configuration is acceptable. If

$$\sigma_b \leq S_{tbf} \tag{15-43}$$

the design is regarded as *acceptable*.

Example 15.3 AGMA Refined Approach to Spur Gear Bending Fatigue

It is desired to compare the results of the simplified approach to designing gear teeth that will resist failure by bending fatigue, illustrated in Example 15.2, with results using the AGMA refined approach for the same gear (gear no. 1 of the gear train shown in Figure E15.1A of Example 15.1).

a. Make the necessary calculations using the AGMA refined approach.

b. Compare results of (a) with results of Example 15.2

Solution

a. From Example 15.2:

Material: AISI 1020 steel; Grade 1

S_u = 57,000 psi (Table 3.3)

S_{yp} = 43,000 psi (Table 3.3)

BHN = 121 (Table 3.13)

Reliability requirement: 99 percent
Life requirement: Very long
Tooth system: 20° full-depth involute

where $n_1 = 2500$ rpm
$hp_1 = 5$ horsepower
$r_1 = 1.0$ inch
$P_d = 10$
$N_1 = 20$ teeth
$N_2 = 32$ teeth (mating gear)
$V_1 = 1309$ ft/min
$F_{t1} = 126$ lb
$b_1 = 1.15$ inches
$n_{ex-1} = 9.3$

Using (15-41)

$$\sigma_b = \frac{F_t P_d}{bJ} K_a K_v K_m K_I \tag{1}$$

the factors may be evaluated as

$K_a = 1.0$ (Table 15.6)
$K_v = 1.15$ (Figure 15.24 with $Q_v = 10$ and $V = 1309$ ft/min)
$K_m = 1.6$ (Table 15.7 with typical mounting conditions)
$K_I = 1.0$ (one-way bending)
$J = 0.33$ (Table 15.9; HPSTC assumed valid)
$b = 1.15$ inches
$P_d = 10$
$F_t = 126$ lb

Substituting these values in (1)

$$\sigma_b = \frac{(126)(10)}{1.15(0.33)}(1.0)(1.15)(1.6)(1.0) = 6109 \text{ psi} \tag{2}$$

Using (15-42)

$$S_{tbf} = Y_N R_g S'_{tbf} \tag{3}$$

and the factors may be evaluated as

Y_N = 0.8 (from Figure 15.28, based on 8 years operation, 24 hr/day; approx. 10^{10} cycles)
R_g = 1.0 (Table 15.13)
$(S'_{tbf})_{N=10^{10}}$ = 22,000 psi (extrapolated from Figure 15.25, assuming Grade 1, BHN = 121; see Table 3.13)

whence (15-42) gives

$$(S_{tbf})_{N=10^{10}} = (0.8)(1.0)(22{,}000) = 17{,}600 \text{ psi} \tag{4}$$

Based on (15-43) then, for gear 1,

$$n_{ex-1} = \frac{(S_{tbf})_{N=10^{10}}}{\sigma_b} = \frac{17{,}600}{6109} = 2.9 \tag{5}$$

b. From (5) the calculated existing safety factor obtained by using the AGMA refined approach is about 2.9 compared with the calculated existing safety factor using the simplified method of about 9.3. This result is compatible with the notion that more accurate quantitative information about materials and operating conditions leads to less "random uncertainty" and a lower calculated existing safety factor (which is, by definition, associated with "random uncertainty").

Surface Durability: Hertz Contact Stresses and Surface Fatigue Wear

As discussed in 4.10, when curved surfaces, such as meshing gear teeth, are pressed together, the triaxial stress distributions at and below the surfaces of the contacting bodies may be described by their pertinent Hertz contact stress equations. Equations (4-126) through (4-131), developed for parallel cylinders in contact and graphically illustrated in Figure 4.33, have already been discussed. Since contacting gear teeth emulate cylinders in contact, the governing contact stress equation for surface fatigue failure of gear teeth may be adapted from the Hertz cylindrical contact stress model. For this case the maximum normal Hertz stress component σ_z, along the centerline as shown in Figure 4.33, is given by (4-128). Reexpressing (4-128) using gear tooth terminology as summarized in Table 15.14, the surface fatigue wear equation may be written as[58]

$$\sigma_{sf} = \sqrt{\frac{F_t\left(\dfrac{2}{d_p \sin\varphi} + \dfrac{2}{d_g \sin\varphi}\right)}{\pi b \cos\varphi\left(\dfrac{1 - v_p^2}{E_p} + \dfrac{1 - v_g^2}{E_g}\right)}} \tag{15-44}$$

[58]Based on the work of Earle Buckingham; see ref. 17.

TABLE 15.14 Gearing Notation Equivalents for Hertz Contact Stress Equations (4-124) and (4-125)

Hertz Equation Symbol	Gearing Notation Equivalent	Source or Definition
p_{max}	σ_{sf}	*Surface fatigue wear* stress (maximum Hertz contact stress)
F	$F_t/\cos\varphi$	Normal force (see 15-27)
L	b	Face width
d_1	$d_p \sin\varphi$	Basic involute geometry
d_2	$d_g \sin\varphi$	Basic involute geometry
ν_1, ν_2	ν_p, ν_g	Poisson's ratio of pinion, gear
E_1, E_2	E_p, E_g	Young's modulus for pinion, gear
	d_p, d_g	Pitch diameter of pinion, gear
	F_t	Tangential force at contact site
	φ	Pressure angle

Surface fatigue strength data, as shown in Figure 15.29 for case-hardened spur gear teeth, are plotted by calculating the value of σ_{sf} from (15-44) that corresponds to surface fatigue failure (significant degradation of function because of pitting or spalling). Surface fatigue strength data are not widely available.[59] It is worth noting that influencing factors such as sliding friction, lubrication, and thermal stresses, which typify gearing applications, probably account for the lower *S-N* curve in Figure 15.29, so it is important to use the *gearing S-N curve* for gearset design.

Finally, it is important to recognize that as gear teeth roll through the angle of action, the relative motion between contacting teeth consists of *rolling plus sliding*, except at the pitch point where it is *pure rolling*. The relative sliding velocity between contacting tooth surfaces is proportional to the distance between the contact point and the pitch point, and it reverses direction as a tooth pair rolls through the pitch point. These sliding velocity components may produce adhesive or abrasive wear and, in some cases, significant frictional heating, which may require special attention to lubrication and cooling.

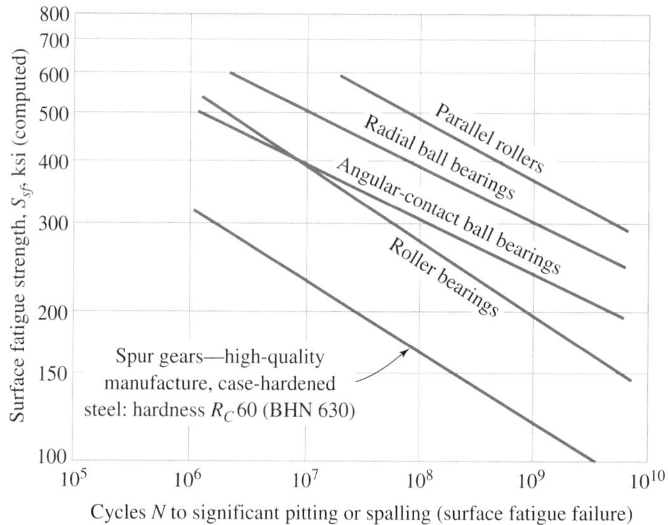

Figure 15.29
Contact stress *S-N* curves corresponding to 90 percent reliability. (Adapted from ref. 19 with permission of the McGraw-Hill Companies; also see ref. 18.)

[59]See ref. 19, however, for additional data.

Example 15.4 Simplified Approach to Surface Durability of Spur Gear Teeth

Continuing with the preliminary design of Examples 15.1, 15.2, and 15.3, it is desired to examine the resistance to surface fatigue wear of the teeth (see Figure E15.1A). The tentative material properties, dimensions, and operating configuration are summarized in Example 15.3. In addition, it is proposed to carburize and case harden the surfaces of the gear teeth to a hardness of approximately Rockwell C 60 (R_C 60) to improve resistance to failure by surface fatigue. Recalling that a very long life is desired, assess the adequacy of the teeth in terms of resisting failure by surface fatigue wear.

Solution

Utilizing (15-44) together with the data summarized in Examples 15.2 and 15.3,

$$\sigma_{sf} = \sqrt{\frac{126\left(\dfrac{2}{2.0\sin 20} + \dfrac{2}{3.2\sin 20}\right)}{\pi(1.15)\cos 20\left(\dfrac{1-0.3^2}{30\times 10^6} + \dfrac{1-0.3^2}{30\times 10^6}\right)}} = \sqrt{\frac{598.65}{2.06\times 10^{-7}}} = 53{,}910 \text{ psi} \qquad (1)$$

From Figure 15.29, the 90 percent reliability surface fatigue strength S_{sf} at a life of $N = 10^{10}$ cycles, for steel gears case hardened to R_C 60, may be read as (by extrapolation)

$$(S_{sf})_{N=10^{10}} = 90{,}000 \text{ psi } (R = 90\%) \qquad (2)$$

Adjusting the 90 percent reliability strength data of Figure 15.29 to the required 99 percent reliability level, Table 2.3 may be used to give

$$(S_{sf})_{N=10^{10}} = \left(\frac{0.81}{0.90}\right)90{,}000 = 81{,}000 \text{ psi } (R = 99\%) \qquad (3)$$

Based on (5-7), the existing safety factor for resistance to failure by surface fatigue wear is

$$n_{ex} = \frac{(S_{sf})_{N=10^{10}}}{\sigma_{sf}} = \frac{81{,}000}{53{,}910} = 1.5 \qquad (4)$$

Surface Durability: Synopsis of AGMA Refined Approach

The AGMA approach to designing gear teeth to resist surface fatigue (pitting) failure is based on the idealized Hertz contact stress equation (15-44) modified by a list of adjustment factors (derating factors) to account for the influence of various manufacturing, assembly, geometric, loading, and material variabilities, using a procedure very similar to that used for assessing the seriousness of tooth bending fatigue. The basic AGMA surface-pitting-resistance assessment is made by comparing the surface fatigue contact stress with the allowable surface fatigue strength.[60] The development proceeds by first rewriting (15-44) as

$$\sigma_{sf} = C_p\sqrt{\frac{F_t}{bd_pI}} \qquad (15\text{-}45)$$

where $C_p \equiv$ elastic coefficient $= \sqrt{\dfrac{1}{\pi\left(\dfrac{1-\nu_p^2}{E_p} + \dfrac{1-\nu_g^2}{E_g}\right)}}$

[60]Referred to as *allowable contact stress number* in AGMA standard 2001-C95.

$$I \equiv \text{geometry factor} = \frac{\sin\varphi\cos\varphi}{2}\left(\frac{m_G}{m_G + 1}\right)$$

$$m_G = \text{gear ratio} = \frac{d_g}{d_p} = \frac{N_g}{N_p}\ (\text{always} \geq 1.0)$$

then inserting the modifying factors to give[61]

$$\sigma_{sf} = C_p\sqrt{\frac{F_t}{bd_pI}K_aK_vK_m} \tag{15-46}$$

Generally, the *application factor* K_a, *dynamic factor* K_v, and *mounting factor* K_m are the same as for the tooth bending fatigue analysis, and may be read from Table 15.6, Figure 15.24, and Table 15.7 respectively.

On the strength side, published AGMA data for surface fatigue strength S'_{sf}, as shown in Table 15.15 for steel gears, Table 15.16 for iron and bronze gears, and Figure 15.30 for

TABLE 15.15 ANSI/AGMA Surface Fatigue Strength (Pitting Resistance) S'_{sf} for Steel Gears[1]

		Metallurgical Quality[2]					
		Grade 1		Grade 2		Grade 3	
Material	Heat Treatment	Min. Surf. Hardness[3]	S'_{sf}, ksi	Min. Surf. Hardness	S'_{sf}, ksi	Min. Surf. Hardness	S'_{sf}, ksi
Steel	Through hardened	See Figure 15.30				—	—
	Flame or induction hardened	50 R_C 54 R_C	170 175	— —	190 195	— —	— —
	Carburized and hardened (Min. core hard.)	55–64 R_C (21 R_C)	180	58–64 R_C (25 R_C)	225	58–64 R_C (30 R_C)	275
AISI 4140, AISI 4340 steel	Nitrided (through hardening)	83.5 84.5 R_{15N}	150 155	83.5 84.5 R_{15N}	163 168	83.5 84.5 R_{15N}	175 180
2.5% chrome (no aluminum)	Nitrided	87.5 R_{15N}	155	87.5 R_{15N}	172	87.5 R_{15N}	189
Nitralloy 135M	Nitrided	90.0 R_{15N}	170	90.0 R_{15N}	183	90.0 R_{15N}	195
Nitralloy N	Nitrided	90.0 R_{15N}	172	90.0 R_{15N}	188	90.0 R_{15N}	205
2.5% chrome (no aluminum)	Nitrided	90.0 R_{15N}	176	90.0 R_{15N}	196	90.0 R_{15N}	216

[1]From ref. 10. Adapted from ANSI/AGMA Standard 2001-C95, with the permission of the publisher, American Gear Manufacturers Association, 1500 King Street, Suite 201, Alexandria, VA 22314.

[2]See Table 15.11 for quality characteristics of specified grade.

[3]R_C = Rockwell C scale. R_{15N} = Rockwell 15N scale.

[61]Other factors are available from AGMA standards to account for *load distribution* (manufacturing, assembly, alignment, and mounting variabilities), *overloads* (system vibration, overspeed events, changing process-load conditions), *size*, and *surface condition* effects.

TABLE 15.16 ANSI/AGMA Surface Fatigue Strength S'_{sf} (Pitting Resistance) for Iron and Bronze Gears[1]

Material	Material Designation	Heat Treatment	Typical Min. Surface Hardness[2]	S'_{sf} ksi
ASTM A48 gray cast iron	Class 20	As cast	—	50–60
	Class 30	As cast	174 BHN	65–75
	Class 40	As cast	201 BHN	75–85
ASTM A536 ductile iron (nodular)	Grade 60-40-18	Annealed	140 BHN	77–92
	Grade 80-55-06	Q & T[3]	179 BHN	77–92
	Grade 100-70-03	Q & T	229 BHN	92–112
	Grade 120-90-02	Q & T	269 BHN	103–126
Bronze		Sand cast	Min. tensile strength, 40 ksi	30
	ASTM B-148 Alloy 954	Heat treated	Min. tensile strength, 90 ksi	65

[1]From ref. 10. Adapted from ANSI/AGMA Standard 2001-C95, with the permission of the publisher, American Gear Manufacturers Association, 1500 King Street, Suite 201, Alexandria, VA 22314.
[2]BHN = brinnell hardness number. [3]Quenched and tempered.

through-hardened steel gears, are all indexed to a reliability of 99 percent for a life of 10^7 cycles of one-way loading. For design lives other than 10^7 cycles, the strength S'_{sf} may be multiplied times a *life adjustment factor* Z_N, read from Figure 15.31 for steel gears. Values of Z_N for materials other than steel are not readily available. For reliability requirements other than 99 percent, the strength S'_{sf} may be multiplied by a gearing reliability factor R_g read from Table 15.13.

Thus the surface fatigue strength S_{sf} corresponding to desired reliability and design life becomes[62]

$$S_{sf} = Z_N R_g S'_{sf} \tag{15-47}$$

As usual, if

$$\sigma_{sf} \leq S_{sf} \tag{15-48}$$

the design is regarded as acceptable.

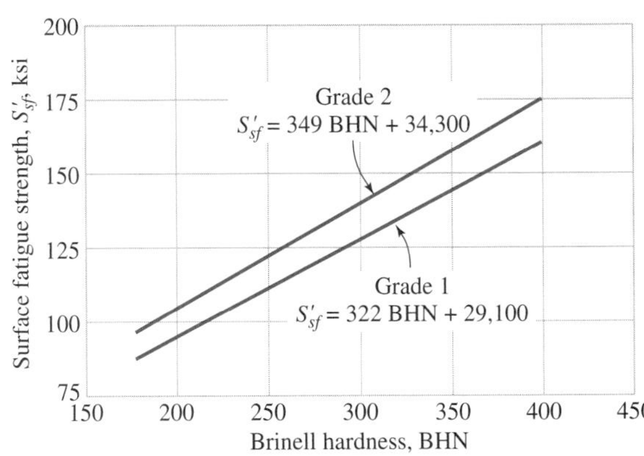

Figure 15.30
ANSI/AGMA surface fatigue strength S'_{sf} for through-hardened steel gears. Curves are based on 99 percent reliability for a life of 10^7 cycles of one-way loading. See Table 15.10 for quality characteristics of specified grade. (From ref. 10. Adapted from ANSI/AGMA Standard 2001-C95, with the permission of the publisher, American Gear Manufacturers Association, 1500 King Street, Suite 201, Alexandria, VA 22314.)

[62]A *hardness ratio factor* is also available from AGMA standards to account for small improvements in *gear* teeth (*not* pinion teeth) due to cold working by repetitive contacts with a harder pinion.

Example 15.5 AGMA Refined Approach to Surface Durability of Spur Gear Teeth

It is desired to refine the preliminary assessment of surface durability conducted in Example 15.4 by utilizing the AGMA approach. How would such a refinement be accomplished?

Solution

To utilize the AGMA approach, (15-46) is first evaluated. To do this the elastic coefficient may be calculated as

$$C_p = \sqrt{\frac{1}{\pi\left(\dfrac{1-0.3^2}{30\times 10^6}+\dfrac{1-0.3^2}{30\times 10^6}\right)}} = 2.29 \times 10^3 \tag{1}$$

The geometry factor may be evaluated as

$$I = \frac{\sin 20 \cos 20}{2}\left(\frac{1.6}{1.6+1}\right) = 0.10 \tag{2}$$

where

$$m_G = \frac{N_g}{N_p} = \frac{32}{20} = 1.6 \qquad \text{(from data in Example 15.3)}$$

Also from data in Example 15.3,

$$\begin{aligned} F_t &= 126 \text{ lb} \\ b &= 1.15 \text{ inches} \\ d_p &= 2r_1 = 2(1.0) = 2.0 \text{ inches} \\ K_a &= 1.0 \\ K_v &= 1.15 \\ K_m &= 1.6 \end{aligned}$$

so from (15-46)

$$\sigma_{sf} = 2{,}29 \times 10^3 \sqrt{\frac{126}{(1.15)(2.0)(0.10)}(1.0)(1.15)(1.6)} = 72{,}705 \text{ psi} \tag{3}$$

Evaluating the strength, using (15-47), the surface fatigue strength of carburized and hardened steel (Grade 1) may be read from Table 15.15 as

$$S'_{sf} = 180{,}000 \text{ psi} \tag{4}$$

The reliability factor for the original 99 percent reliability specification may be read from Table 15.13 as

$$R_g = 1.0 \tag{5}$$

and the life adjustment factor may be read from Figure 15.31 for a life of 10^{10} cycles as

$$Z_{N=10^{10}} = 0.67 \tag{6}$$

Using these values, (15-47) gives

$$(S_{sf})_{N=10^{10}} = 0.67(1.0)(180{,}000) = 120{,}600 \text{ psi} \tag{7}$$

Based on (5-7) then, the existing safety factor for resistance to failure by surface fatigue (pitting) is

$$n_{ex} = \frac{S_{sf}}{\sigma_{sf}} = \frac{120{,}600}{72{,}705} = 1.7 \tag{8}$$

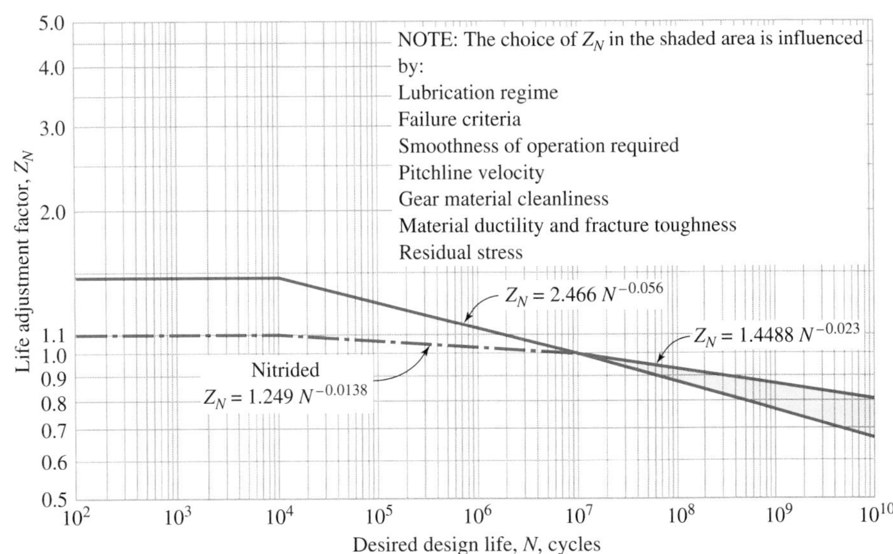

Figure 15.31
Life adjustment factor to adjust ANSI/AGMA surface fatigue strength to a life other than 10^7 cycles. Valid only for steel materials. (From ref. 10. Adapted from ANSI/AGMA Standard 2001-C95, with the permission of the publisher, American Gear Manufacturers Association, 1500 King Street, Suite 201, Alexandria, VA 22314.)

15.10 Lubrication and Heat Dissipation

As discussed in 15.9, the relative motion between contacting spur gear teeth as they roll through the angle of action consists of rolling plus sliding, except at the pitch point, where it is pure rolling. The relative sliding velocity between the contacting tooth surfaces is proportional to the distance between the contact point and the pitch point. Because these sliding velocity components may produce adhesive or abrasive wear, and, in some cases, significant frictional heating, adequate lubrication and cooling capacity are important for smooth operation and acceptable gear life. The lubrication and cooling issues become even more important for helical and bevel gears because of additional components of relative sliding motion, and, for worm gear sets, frictional heating is so significant that pumped-oil lubrication systems and external oil coolers are often required to prevent oil breakdown and system failure.

For spur gear meshes, power losses typically range from less than 0.5 percent of the transmitted power up to about 1 percent, depending upon materials, tooth system, surface characteristics, pitch-line velocity, and method of lubrication. For helical and bevel gear meshes, somewhat higher power losses are typical, often ranging from 1 to 2 percent. A common rule of thumb for spur, helical, and bevel gear meshes is to assume each mesh, including gears and supporting bearings, incurs a 2 percent power loss (98 percent efficiency). For worm gears, power loss may be estimated on the basis of a calculated *efficiency*, using the expression developed for power screws,[63] as discussed later in 15.21.

For light loads, low speeds, low power transmission, and intermittent operation, *unenclosed* gearing may be lubricated by using an oil can, drip oiler, or periodically brushed grease. When gears operate in an *enclosed* housing, or *gear case*, splash lubrication is widely used for gearing subjected to moderate loads, speeds, and transmitted power levels. In this case, one of the gears in a pair dips into an oil supply sump at the bottom of the gear case and carries the oil to the mesh. For high speeds and high-capacity gearing systems, positive oil circulation systems are often required, using a separate pump to draw oil from the sump and deliver it at a controlled rate to the meshing teeth, sometimes passing

[63] See 12.4.

the oil through an external heat exchanger to maintain an acceptable oil temperature (usually less than 180°F.[64]

In developing an acceptable lubrication system for gearing, the basic concepts presented for plain bearing lubrication[65] may be applied, and, depending upon the application, it may be appropriate either to strive for *thick film* lubrication in the mesh, or to accept *thin film* operation. Based on the discussion of elastohydrodynamic lubrication as related to rolling element bearings,[66] three potential gear mesh lubrication regimes may be identified. They are:[67]

Regime I: thin EHD oil film; essentially full metal-to-metal contact

Regime II: thin EHD oil film; partial metal-to-metal contact

Regime III: thick EHD oil film; no metal-to-metal contact

Regime I is characterized by slow speeds, high loads, and rough surface finishes, such as found in gearing for hand winches, food presses, and jacking devices. *Regime II*, with substantial, but not total, separation of tooth surfaces by the oil film, is typified by medium speeds, moderate to high loads, and good surface finishes, such as found in automotive, truck, and tractor transmission gearing. *Regime III* is representative of well-designed, carefully manufactured, high-speed gearing such as used in aerospace gearboxes and high-speed turbine gearing. As for rolling element bearings,[68] the elastohydrodynamic film parameter (lambda ratio) should have a value of approximately 3 for optimum lubrication performance.[69] Figure 15.32 depicts the approximate regions of operation associated with Regimes I, II, and III, as a function of elastohydrodynamic minimum film thickness and effective surface roughness of the gear teeth.

As was discussed for the design of hydrodynamic bearings in 10.7, the rate at which heat is generated by friction in the operating gear meshes (as well as in the associated sup-

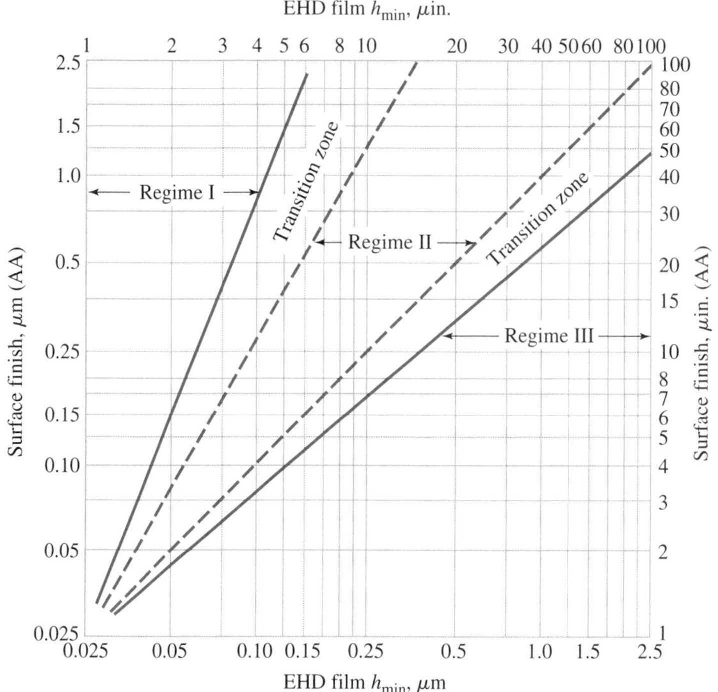

Figure 15.32
Probable lubrication regimes for operating gear teeth as a function of elastohydrodynamic (EHD) minimum film thickness and effective surface finish. See also (11-16). (From ref. 1. *Source:* Dudley Engineering Company, San Diego, CA.)

[64]See Figure 10.3. [65]See 10.5. [66]See (11-16) and associated discussion. [67]See ref. 1.
[68]See (11-16) and associated discussion. [69]See ref. 9 for calculation of EHD film thickness, h_{min}.

port bearings) must be equal to the rate at which heat is dissipated to the ambient atmosphere. The oil temperature at which this balance is reached must be satisfactory in terms of both equilibrium viscosity and limiting maximum temperature. The friction-generated heating rate, H_g, may be estimated by using thumb rules or efficiency calculations as just discussed, and the heat dissipation rate H_d for a gear transmission housing may be estimated from Newton's Law of cooling,[70] repeated here for convenience as

$$H_d = k_1 A_h(\Theta_s - \Theta_a) \text{ Btu/min} \tag{15-49}$$

where k_1 = modified overall heat transfer coefficient, Btu/min-in^2-°F [see (10-22) and (10-23) for approximate values]
A_h = exposed surface area of housing, in^2
Θ_s = housing surface temperature, °F
Θ_a = ambient air temperature, °F

To achieve equilibrium requires that H_d equal H_g and the resulting oil temperature must be acceptable for the lubricant chosen. External oil cooling may be separately considered, if necessary.

15.11 Spur Gears; Summary of Suggested Design Procedure

The design of gears is typically an iterative process in which potential failure modes are identified, tentative material candidates are selected, and initial gearing geometry is assumed. The initial proposal is then iterated until the operational loads and speeds imposed by design specifications can be applied for the specified design lifetime, or until the desired design safety factor has been achieved (see Chapter 5). The following design procedure is suggested for establishing a successful gearing design (many other approaches have been presented in the literature):[71]

1. Based on functional specifications and contemplated system configuration, generate a first conceptual sketch of the gear train, including required reduction ratio, proposed number of teeth for each gear (see Table 15.3 to avoid undercutting), gear shaft centerlines and bearing locations, and any other geometrical constraints (see Example 15.1).
2. Identify potential failure modes (see 15.4).
3. Select tentative gear materials (see 15.5).
4. Select tentative manufacturing and finishing methods appropriate to the application (see Table 15.4).
5. Tentatively select a tooth system, then, using specified power and speed requirements, perform a complete force analysis to determine torques, speeds, and transmitted loads at each mesh in the gear train (see Example 15.2).
6. Select a tentative diametral pitch (or module), based on either experience or expert advice, or by arbitrarily picking one from Figure 15.13 that appears to be consistent with the proposed design layout. Experience indicates that a balance between failure by tooth bending fatigue and surface fatigue durability usually occurs when the diametral pitch is chosen to be about 8 or 10. Coarser teeth tend to fail by surface fatigue and finer teeth by bending fatigue.
7. For each gear, calculate or estimate the tentative pitch radius, pitch-line velocity, face width, and transmitted load (see Example 15.2).

[70]See (10-19). [71]See, for example, ref. 1.

8. For each gear, calculate the safety factor, based on tooth bending fatigue, that would exist if the proposed (initial) selections for tooth geometry and material properties (including reliability specifications) were used for the design. Tooth bending fatigue stresses may be calculated using either the *simplified approach* (see Example 15.2) or the *AGMA refined approach* (see Example 15.3).
9. Compare the tentative *existing safety factor* with the specified *design safety factor*. If they are approximately equal, no further iteration is required. If they are *not* equal, another iteration *is* required, usually involving a better selection for diametral pitch. In some cases a change in selected material may be necessary. Continue iterating until the specified design safety factor is achieved. Repeat for each gear, as required.
10. Using the results of step 9, calculate the existing safety factor based on surface durability, using either the simplified Hertz contact surface fatigue approach (see Example 15.4) or the AGMA refined approach (see Example 15.5). If the existing safety factor does not agree with the specified design safety factor, additional iterations may be required.
11. Estimate friction losses, heat generation, and lubrication requirements to maintain acceptable temperature levels, noise levels, and design lives (see 15.10).

15.12 Helical Gears; Nomenclature, Tooth Geometry, and Mesh Interaction

As illustrated in Figures 15.2(c) and (d), helical gear drives share many of the attributes of straight-tooth spur gears when used to transmit power or motion between parallel shafts. The distinguishing geometrical difference is that spur gear teeth are straight and aligned with the axis of rotation [as shown in Figure 15.33(a)], while helical gear teeth are angled with respect to the axis of rotation at an angle ψ, called the *helix angle*, measured at the surface of the pitch cylinder (see Figure 15.34). A straight spur gear may be thought of as helical gear with a helix angle of zero.

The contact zone and mesh interactions of a helical gearset are much more difficult to visualize than the contact zone and mesh interactions for straight spur gears. Figure 15.33 provides one way to visualize the contact progression in a helical gear mesh by comparing it first with a straight spur gear mesh, then a stepped spur gear mesh. For the straight spur gear of Figure 15.33(a), when mating teeth engage, the line of contact is at once established all the way across the face width at or near the tip of the *driven* gear tooth. This contact line moves from the site of first contact, smoothly along the involute profile to the site of final contact at or near the root of the driven gear tooth, where separation occurs as the gears continue to rotate. Depending upon the profile contact ratio, as defined by (15-20), the total contact length *jumps from one face width to two face widths* and back,[72] as shown in the rotation diagram of Figure 15.33(a).

Next, visualize the straight spur gear of Figure 15.33(a) as being sliced perpendicular to its axis of rotation into several thin gears [four shown in Figure 15.33(b)], then visualize the small relative rotation of each thin gear slice, progressively, until a stepped-spur gear cluster is formed, as shown in Figure 15.33(b). With rotation of the stepped gear, a full-width line of contact is first established across the face of one gear slice, then, as the gear cluster rotates further, across the second, third, and fourth slices, successively, to generate the *stepped contact length* curve shown in the rotation diagram of Figure 15.33(b). It may be noted that the engagement of the stepped gear is, on average, more gradual than the engagement of the straight spur gear.

[72] Assuming $1 \leq m_c \leq 2$.

Figure 15.33
Comparison of contact patterns for *driven* spur, stepped, and right-hand helical gears. For clarity, the (overlapping engagement) contact patterns for adjacent teeth are omitted.

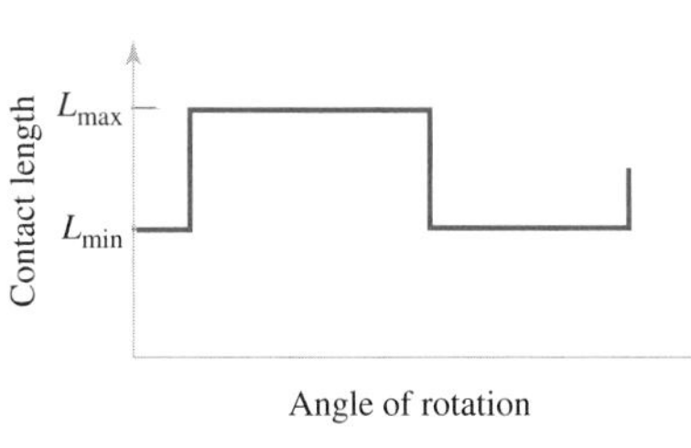

(*a*) Straight tooth spur gear contact pattern; also contact length as a function of rotation.

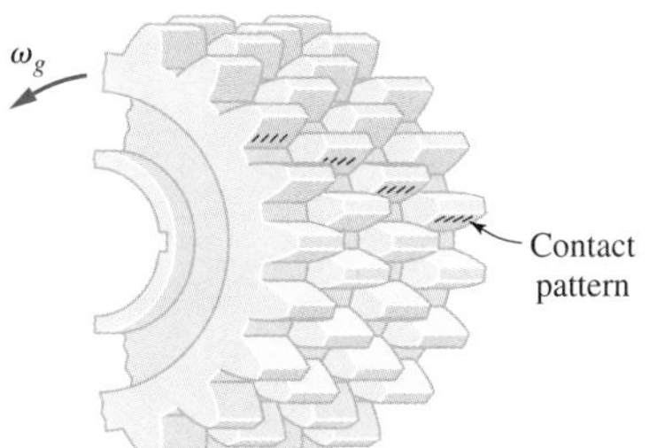

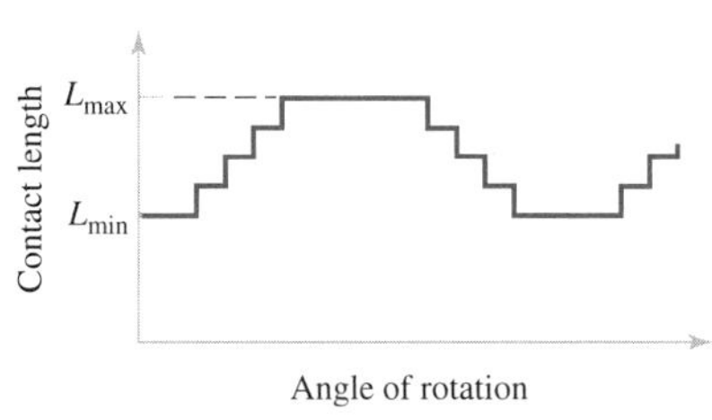

(*b*) Stepped spur gear contact pattern; also contact length as a function of rotation.

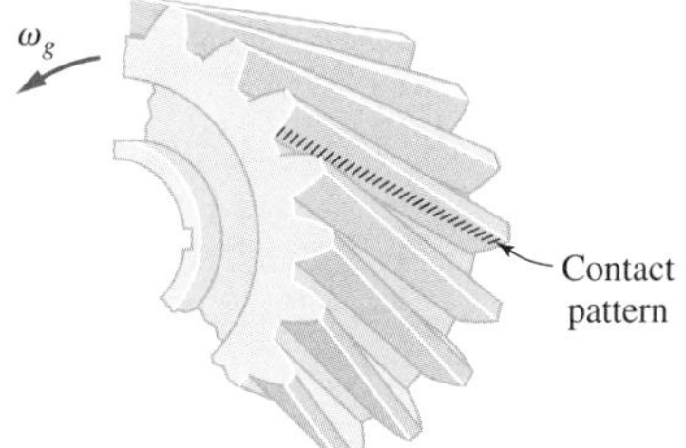

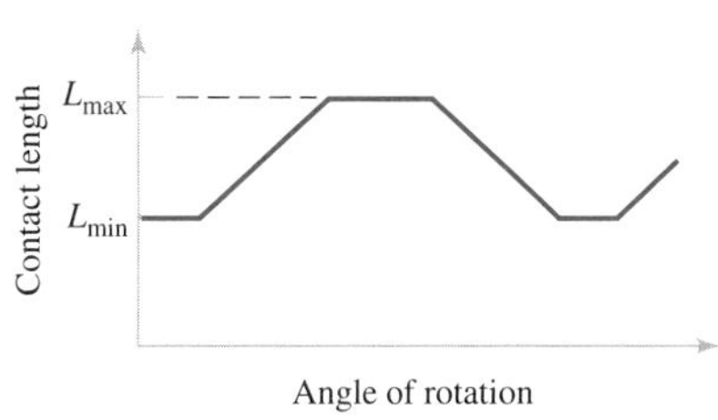

(*c*) Right hand helical gear contact pattern; also contact length as a function of rotation.

If the gear slices of Figure 15.33(b) are increased in number and decreased in thickness to maintain a fixed face width, and if each slice is progressively rotated through a slight angle with respect to the former slice, in the limit the stepped teeth give way to parallel continuous smooth teeth spiraling about the axis of rotation. Such a gear is called a *helical gear*. As illustrated in Figure 15.33(c), the rotation diagram for a helical gear *ramps smoothly from minimum to maximum contact length* and back, as the gears rotate. If the *face contact ratio*[73] is an *integer*, the minimum contact length, L_{min}, and the maximum contact length, L_{max}, are equal, so the contact length remains constant.

The basic terminology describing the features of helical gears is substantially the same as for straight-tooth spur gears. This may be verified by comparing Figure 15.34(a) with Figure 15.1(d). However, because of the angled teeth, some additional parameters are required to describe certain aspects of helical gearing. Figures 15.34(b) and (c) show sketches of a helical rack to assist in defining these additional terms. Figure 15.34(c) shows the top view of a basic right-hand[74] helical rack with helix angle ψ and face width

[73]See (15-60).

[74]One easy way to quickly determine the "hand" of a helical rack or a helical gear by observation is to align one's body with the axis of gear rotation and then raise whichever arm will parallel the teeth. The hand of the gear agrees with the hand raised. Helix angle is measured from the axis of rotation.

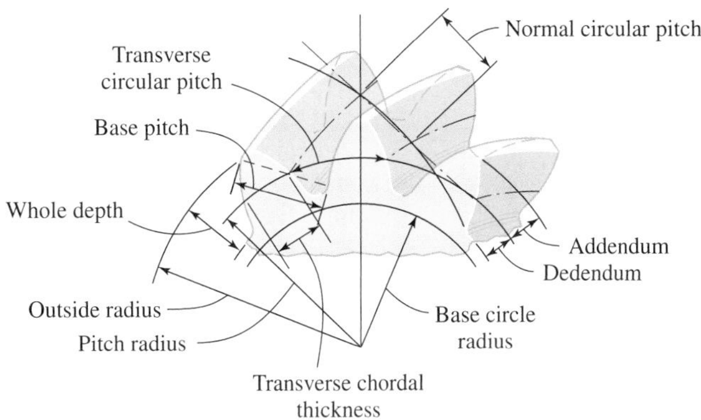

(*a*) Conceptual sketch of helical gear showing basic shape and terminology.

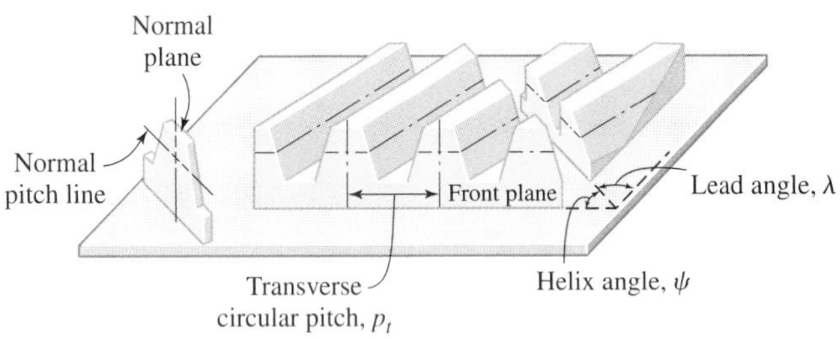

(*b*) Conceptual sketch of helical rack showing basic shape and terminology.

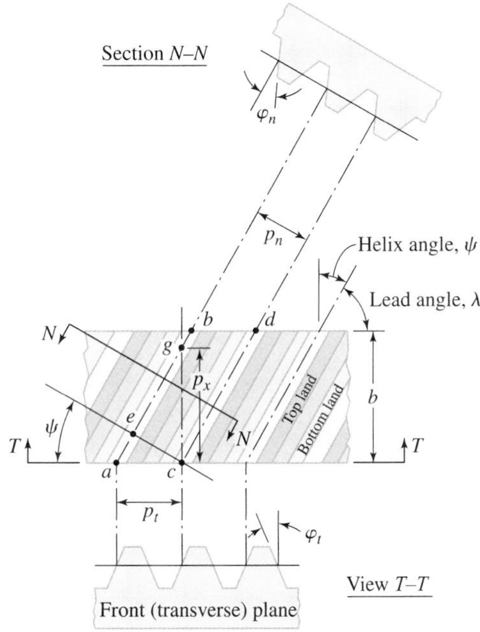

(*c*) Depiction of helical rack geometry showing relationships among transverse, normal, and axial parameters.

Figure 15.34
Geometry and nomenclature for helical gears and racks.

b. Lines *ab* and *cd* are reference lines lying on the surfaces of two adjacent helical teeth where the pitch plane intersects the tooth profiles. The angle made between either of these pitch plane reference lines and an intersecting line parallel to the axis of rotation of a mating gear is the helix angle ψ. In addition, *transverse planes* (normal to the axis of rotation), *normal planes* (perpendicular to the teeth), and *axial planes* (parallel to the axis of rotation) are of interest. In Figure 15.34(c), *T-T* is a transverse plane, *N-N* is a normal plane, and *cg* lies in an axial plane.

The additional parameters to be defined for helical gears are related to angles and distances measured on transverse, normal, and axial planes. Thus the *transverse circular pitch* p_t (equal to p_c) and *transverse pressure angle* φ_t (equal to φ) are measured on the *transverse* plane, the *normal circular pitch* p_n *and normal pressure angle* φ_n are measured on the *normal* plane, and the *axial pitch* p_x is measured on the axial plane. By definition, the transverse pitch is the same as circular pitch defined for straight-tooth spur gears. Thus from (15-6),

$$p_t = p_c = \frac{\pi d}{N} \tag{15-50}$$

Then from triangles *ace* and *acg*, respectively,

$$p_n = p_t \cos\psi \tag{15-51}$$

and

$$p_x = \frac{p_t}{\tan\psi} = \frac{p_n}{\sin\psi} \tag{15-52}$$

As for spur gears, diametral pitch is a more common index for sizing helical gear teeth. Based on (15-12)

$$P_t p_t = \pi \tag{15-53}$$

and similarly,

$$P_n p_n = \pi \tag{15-54}$$

where P_t is diametral pitch in the transverse plane (same as P_d for spur gears) and P_n is diametral pitch in the normal plane. From (15-53) and (15-54) it may be deduced that

$$P_t = P_n \cos\psi \tag{15-55}$$

It may also be deduced geometrically from Figure 15.34(c) that

$$\tan\varphi_t = \tan\varphi = \frac{\tan\varphi_n}{\cos\psi} \tag{15-56}$$

Whether φ_t or φ_n is taken as the standard value defining the helical tooth system depends on the gear-cutting method chosen.[75] Standard helical tooth proportions mimic spur gear tooth proportions (see Table 15.1) but are based on P_n and φ_n as illustrated in Table 15.17.

Of course, for high-performance applications, meeting design criteria for a high-quality gearset outweighs the importance of using standard available cutters, so in some applications special tooling is required.

Combining (15-8) with (15-55), the pitch diameter d of a helical gear may be calculated as

$$d = 2r = \frac{N}{P_d} = \frac{N}{P_n \cos\psi} \tag{15-57}$$

where N = number of teeth
ψ = helix angle
P_n = normal diametral pitch

The equation for center distance C for a helical gear pair may be written by combining (15-10) and (15-55) as

$$C = r_1 + r_2 = \frac{(N_1 + N_2)}{2P_t} = \frac{(N_1 + N_2)}{2P_n \cos\psi} \tag{15-58}$$

TABLE 15.17 Standard Proportions for AGMA Full-Depth Helical Gear Teeth (U.S. units)

	Coarse Pitch ($P_n < 20$)	Fine Pitch ($P_n \geq 20$)
Addendum	$1.000/P_n$	$1.000/P_n$
Dedendum	$1.250/P_n$	$(1.200/P_n) + 0.002$ (min)
Whole depth	$2.250/P_n$	$(2.200/P_n) + 0.002$ (min)
Working depth	$2.000/P_n$	$2.000/P_n$
Clearance (basic)	$0.250/P_n$	$(0.200/P_n) + 0.002$ (min)
Clearance (shaved or ground teeth)	$0.350/P_n$	$(0.350/P_n) + 0.002$ (min)
Circular tooth thickness	$1.571/P_n$	$1.571/P_n$

[75]For example, hobbing machines cut in the normal plane, so φ_n and P_n, the normal pressure angle and diametral pitch, are specified for the hobbing cutter. If a gear shaper is used, φ_t and P_t usually are used to select or design the cutting tool.

The face width b is typically made large enough so that for a given helix angle ψ, a positive overlap of adjacent teeth will occur in the axial direction. Referring to Figure 15.34(c), to achieve an effective overlap in the axial direction, it is often recommended that

$$b \geq \left[2.0p_x = \frac{2.0p_t}{\tan\psi} = \frac{2.0p_n}{\sin\psi} \right] \tag{15-59}$$

Face widths smaller than 1.15 p_x are not recommended. These guidelines give continuous contact in the axial plane as the gears rotate.

The (transverse) contact ratio m_p, sometimes called profile contact ratio,[76] remains a valid definition for helical gears. The positive helical overlap objective embodied in (15-59) can also be thought of in terms of a contact ratio m_f in the axial (face) direction. This is sometimes called the *face contact ratio*, where

$$m_f = \frac{b}{p_x} = \frac{P_t b \tan\psi}{\pi} = \frac{P_n b \sin\psi}{\pi} \tag{15-60}$$

Just as a larger profile contact ratio corresponds to better load sharing among multiple teeth simultaneously in contact, a larger face contact ratio corresponds to distribution of tooth loading along a greater contact length (wider face and/or larger helix angle). Typically, helix angles ranging from about 10° to about 35° are selected to balance smoother, quieter operation at larger helix angles with lower axial thrust loads produced by smaller helix angles. The sum of the profile contact ratio plus the face contact ratio is called the *total contact ratio* m_T, a measure of overall load sharing among the helical teeth in contact.

In addition to smoother, quieter operation, the teeth of a helical gear, geometrically defined in the normal plane, are thicker and stronger in the transverse plane [plane of transmitted torque; see (15-25)] than teeth of a straight spur gear with the same normal pitch, pitch diameter, and number of teeth. This may be verified from Figure 15.35, where

$$t_t = \frac{t_n}{\cos\psi} \tag{15-61}$$

Also from Figure 15.35, it may be noted that the normal plane *N-N* intersects the pitch cylinder in an elliptical trace whose radius r_e at point P is, from analytical geometry,

$$r_e = \frac{r}{\cos^2\psi} \tag{15-62}$$

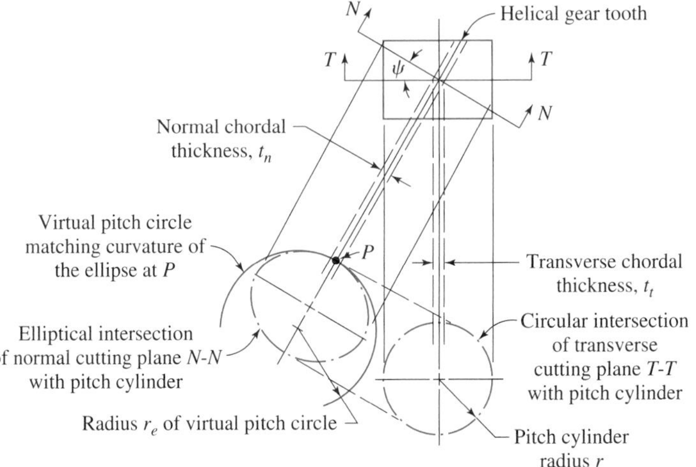

Figure 15.35
Sketch of right-hand helical gear, illustrating the concept of *virtual spur gear* and virtual pitch circle, based on elliptical intersection between normal plane and pitch cylinder.

[76]Defined for spur gears in (15-20).

The profile shape of the tooth in the normal plane is approximately (but not exactly) the same as the shape of a straight spur gear tooth for a pitch radius equal to instantaneous radius r_e of the ellipse at tangency point P, where the curvature of the ellipse and curvature of the *virtual pitch circle* are matched. Adapting (15-6) to the virtual pitch circle, and utilizing (15-62),

$$N_e = \frac{2\pi r_e}{p_n} = \frac{2\pi r}{p_n \cos^2\psi} \tag{15-63}$$

Incorporating (15-51) into (15-63),

$$N_e = \frac{N}{\cos^3\psi} \tag{15-64}$$

where N_e = *virtual*[77] number of teeth on virtual spur gear of pitch radius r_e and the same *actual* number of teeth as the helical gear
N = *actual* number of teeth on the helical gear
ψ = helix angle

The larger number of virtual teeth reduces the tendency for undercutting, allowing the use of a smaller minimum number of teeth for helical gears than for spur gears. When using (15-36) to calculate the critical bending stress for a helical gear tooth, the Lewis factor Y should be selected from Table 15.5 *using N_e as the number of teeth.* When using (15-41) to calculate the critical bending stress for a *helical gear*, the J factor should be selected from Figures 15.37 and 15.38, and the I factor for critical surface durability of a helical gear should be calculated as for spur gears.[78] These matters are discussed more fully in 15.14.

15.13 Helical Gears; Force Analysis

Free-body diagrams for a straight-tooth spur gearset are shown in Figure 15.22, where the normal (resultant) force F_n, known to be directed along the line of action, is resolved into a tangential component F_t, associated with useful power transmission, and a radial force component F_r, which serves no useful purpose. For helical gears, because the teeth are angled across the face of the gear, the normal (resultant) force F_n is not only inclined by pressure angle φ_t with respect to the pitch plane, but also inclined by helix angle ψ with respect to the transverse plane, as illustrated in Figure 15.36. Therefore, the resolution of F_n re-

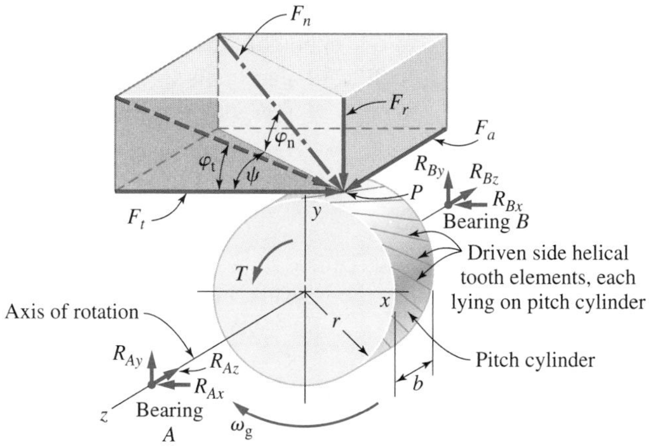

Figure 15.36
Force analysis of right-hand driven helical gear. Compare with Figure 15.22 for spur gears.

[77]Also called *equivalent* number of teeth or *formative* number of teeth.

[78]See (15-45) and associated definitions.

quires not only components F_t and F_r, as for straight spur gears, but also an axial thrust component F_a, as shown in Figure 15.36. Similar to spur gears, F_t is directly related to power transmission and torque developed, and radial component F_r is a (useless) separating force between the driving pinion and driven gear. The axial force F_a, generated by the inclined helical teeth, dictates that the support bearings (shown as A and B in Figure 15.36) must have the ability to resist not only the force components F_t and F_r in the transverse plane, but also the axial thrust component F_a. Also, as for spur gears, (15-25) allows calculation of transmitted force F_t, based on power transmission requirements, as

$$F_t = \frac{33{,}000(hp)}{V} \tag{15-65}$$

where pitch-line velocity V may be calculated using (15-23). Using the geometric interrelationships depicted in Figure 15.36, if the normal pressure angle φ_n and helix angle ψ are known (usual case), the resulting force components of interest may be calculated as functions of F_t, also known from (15-65). These results are

$$F_r = F_t \tan\varphi_t \tag{15-66}$$

$$F_a = F_t \tan\psi \tag{15-67}$$

$$F_n = \frac{F_t}{\cos\varphi_n \cos\psi} \tag{15-68}$$

The transverse pressure angle φ_t may be calculated as a function of normal pressure angle φ_n and helix angle ψ (both *chosen* by the designer), utilizing (15-56).

15.14 Helical Gears; Stress Analysis and Design

The AGMA refined approaches to the design evaluation of tooth bending and surface durability of spur gears, presented in 15.9, are also applicable to helical gears, with minor adjustments to account for geometry differences. Thus, for critical tooth bending stress σ_b in *helical gears*, based on (15-41),

$$\sigma_b = \frac{F_t P_t}{bJ} K_a K_v K_m K_I \tag{15-69}$$

where all terms are defined in the same way as for (15-41), except that helical gear geometry factor J is found by using Figures 15.37 and 15.38, secured from other literature sources, or calculated directly.[79]

For helical gears, a slightly modified version of (15-46) may be used to calculate surface fatigue contact stress σ_{sf}. The modification accounts for the increased total length of tooth contact, because of helical geometry, which may be estimated as[80] $(b/\cos\psi)(m_p)$, giving for helical gears

$$\sigma_{sf} = C_p \sqrt{\frac{F_t}{bd_p I}\left(\frac{\cos\psi}{m_p}\right) K_a K_v K_m} \tag{15-70}$$

Terms in (15-70) are defined or calculated in the same way as for (15-46). Angle ψ and m_p are helix angle and profile contact ratio, respectively. For the design to be acceptable based on tooth bending[81] it must be true that

[79]Procedures are available for *calculating* J for helical gears, but are beyond the scope of this text; e.g., see ref. 20 or 22.

[80]AGMA recommends that only 95 percent of this total helical contact length be used. See (15-20) for calculating m_p.

[81]See (15-43).

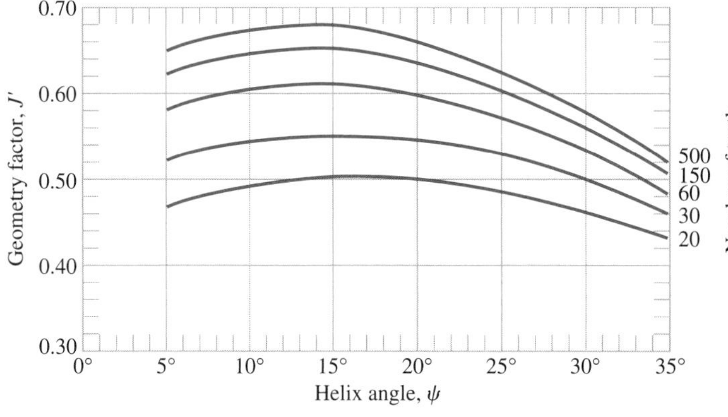

Figure 15.37
AGMA geometry factor J' (for tooth bending) for standard (equal) addendum helical gears with $\varphi_n = 20°$, meshing with a 75 tooth gear. For mating gears with other than 75 teeth, multiply modifying factor M_J times J' to obtain geometry factor J (see Figure 15.38). (From ref. 22. Adapted from ANSI/AGMA Standard 6021-G89, with the permission of the publisher, American Gear Manufacturers Association, 1500 King Street, Suite 201, Alexandria, VA 22314.)

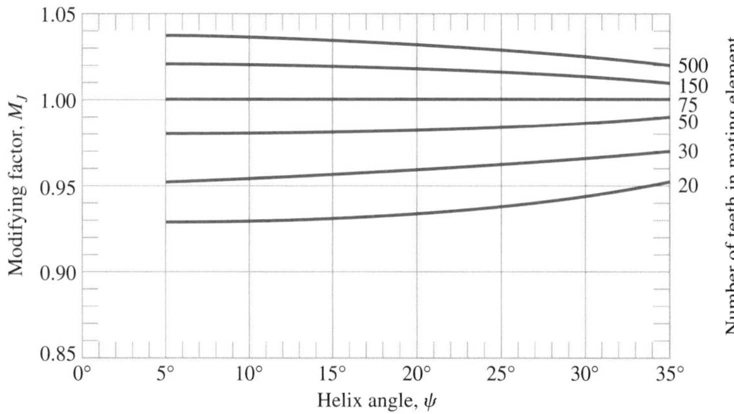

Figure 15.38
Modifying factor M_J to be multiplied times geometry factor J' (see Figure 15.37) when mating helical gear ($\varphi_n = 20°$) has other than 75 teeth. (From ref. 22. (Adapted from ANSI/AGMA Standard 6021-G89, with the permission of the publisher, American Gear Manufacturers Association, 1500 King Street, Suite 201, Alexandria, VA 22314.)

$$\sigma_b \leq S_{tbf} \tag{15-71}$$

where bending fatigue strength values may be based on Table 15.10, properly modified for the application. Likewise, for the design to be acceptable based on surface fatigue[82] it must be true that

$$\sigma_{sf} \leq S_{sf} \tag{15-72}$$

and surface fatigue strength values may be based on Table 15.14, properly modified for the application.

15.15 Helical Gears; Summary of Suggested Design Procedure

The design procedure for helical gears follows the same pattern as for spur gears (see 15.11) except for the complications introduced by the inclined teeth. Thus, a successful procedure for designing helical gear sets would usually include the following steps:[83]

1. Based on functional specifications and contemplated system configuration, generate a first conceptual sketch of the gear train, including required reduction ratio, proposed number of teeth for each helical gear, shaft centerline and bearing locations, and any other geometrical constraints (see Example 15.1).
2. Identify potential failure modes.
3. Select tentative gear materials (see 15.5).

[82]See (15-48). [83]Compare with spur gear procedure in 15.11.

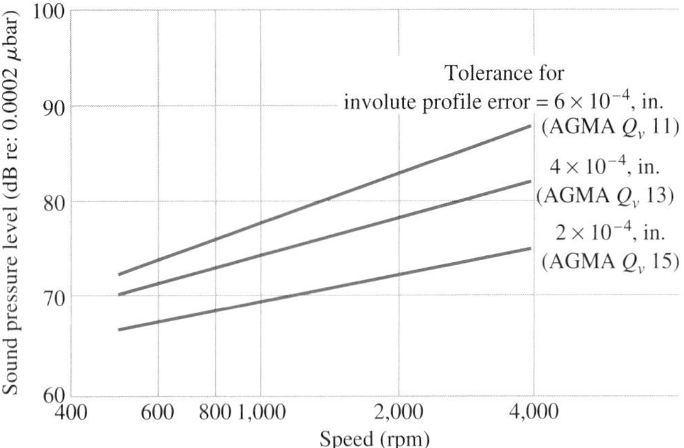

Figure 15.39
Illustrative example of noise generation produced by a particular gearset, operating over a range of speeds, as a function of manufacturing precision (gear quality). See Table 15.4 to interpret AGMA quality numbers Q_v. (Reprinted from ref. 21, with permission from *Machine Design*, a Penton Media publication.)

4. Select tentative manufacturing and gear finishing methods appropriate to the application (see Table 15.4), keeping in mind that higher quality and tighter tolerances tend to improve load sharing, reduce dynamic load amplification, reduce vibration, and reduce mesh noise generation (all at a higher cost). Figure 15.39, for example, illustrates the reduction in noise level[84] of a particular gearset as a function of tighter tolerances on the tooth profile,[85] for a range of operating speeds.
5. Tentatively select a helix angle and a tooth system; then, using application-specific power and speed requirements, perform a complete force analysis to determine torques, speeds, transmitted loads, separating forces, and axial thrust loads at each mesh in the gear train. When selecting an appropriate helix angle for the application, the following factors should be considered:
   a. Higher helix angles tend to result in smoother engagement and quieter operation (up to about $\psi = 35°$). For example, Figure 15.40 illustrates the reduction in noise level[86] for a particular gearset as a function of increasing helix angle.

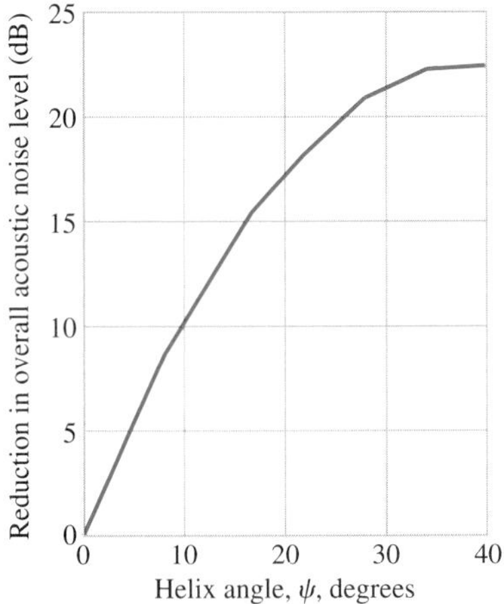

Figure 15.40
Illustrative example of noise reduction in a particular gearset as a function of helix angle. (Reprinted from ref. 21, with permission from *Machine Design*, a Penton Media publication.)

[84]A noise level of 90 dB (decibels) roughly corresponds to a shouting voice, or to a vaneaxial fan operating at 1500 cfm. A noise level of 70 dB roughly corresponds to a conversational voice level. For a more complete interpretation, a gear noise specialist should be consulted.

[85]However, effective noise reduction, as illustrated in Figure 15.40, can be achieved only if preliminary tooth design, including profile modification, is first nominally optimized for quiet operation.

[86]A noise-level reduction of 20 dB corresponds to a factor of 10 reduction in sound pressure. For a more complete interpretation, a gear noise specialist should be consulted.

b. Higher helix angles result in larger axial thrust loads. Usually, these thrust loads must be supported by the bearings, in addition to the radial loads on the bearings (see Chapters 10 and 11).

c. Maximum tooth strength for helical gears is typically achieved in the range of helix angles between 10 and 20 degrees.

d. For a given face width, higher helix angles result in larger face contact ratios.

e. For a given number of teeth with a specified normal diametral pitch, higher helix angles result in larger pitch diameters.

6. Select a tentative normal diametral pitch (or module), either based on experience, expert advice, or arbitrary selection of one from Figure 15.13 that appears to be consistent with the proposed design layout. Experience shows that a balance between failure by tooth bending fatigue and surface fatigue durability usually occurs when the normal diametral pitch is chosen to be about 8 or 10. Coarser teeth tend to fail by surface fatigue and finer teeth by bending fatigue.
7. For each gear, calculate or estimate the tentative pitch radius, pitch-line velocity, face width, and transmitted load.
8. For each gear, calculate the safety factor, based on tooth bending fatigue, that would exist if the proposed (initial) selections for helix angle, tooth geometry, and material properties (including reliability specifications) were used for the design. Tooth bending fatigue stresses may be calculated using (15-69).
9. Compare the tentative *existing* safety factor with the specified *design* safety factor. If they are approximately equal, no further iteration is required. If they are not equal, another iteration *is* required, usually involving a better selection for helix angle and/or normal diametral pitch. In some cases, a change of material may be necessary. Continue iterating until the specified design safety factor is achieved.
10. Using the results of step 9, calculate the existing safety factor based on surface durability, using (15-70). If the existing safety factor does not agree with the specified design safety factor, additional iterations may be required.
11. Check bearing size and availability to properly support radial and thrust loads produced by the operating helical gears (see Chapters 10 and 11).
12. Estimate friction losses, heat generation, and lubrication requirements to maintain acceptable temperature levels, noise levels, and design lives (see 15.10).

Example 15.6 Helical Gearset Design

Because a group of the industrial blending machines discussed in Examples 15.2 through 15.5 may be installed in the same large room, a question has been raised about the possibility of unacceptable noise levels. Engineering management has requested the evaluation of *helical* gears as a step toward noise reduction, even though the proposed design is still at a preliminary stage.

a. Using the spur gear data (see Example 15.3 for a summary) as a starting point, make a preliminary design proposal for a pair of parallel helical gears with the same center distance and the same angular velocity ratio as the currently proposed spur gear set. To do this, determine the helix angle, the outside diameter, and the face width of the new gears. Assume that the helical gears will be cut by a 10-pitch, 20° full-depth hob.

b. Compare existing safety factors for tooth bending and surface fatigue durability for the proposed helical gears with safety factors already found for the spur gears (2.6 for tooth bending fatigue[87] and 1.7 for surface fatigue durability[88]).

[87]See (5) of Example 15.3. [88]See (8) of Example 15.5.

Example 15.6 Continues

Solution

a. Pertinent spur gear data extracted from Example 15.3 are:

Material: AISI 1020 steel
S_u= 57,000 psi (Table 3.3)
S_{yp}= 43,000 psi (Table 3.3)
BHN = 121 (Table 3.13)
Reliability requirement: 99 percent
Life requirement: Very long
Tooth system: 20° full-depth involute spur gears

$$n_1 = 2500 \text{ rpm}$$
$$hp_1 = 5 \text{ horsepower}$$
$$r_1 = 1.0 \text{ inch}$$
$$P_d = 10$$
$$N_1 = 20 \text{ teeth}$$
$$N_2 = 32 \text{ teeth (mating gear)}$$
$$V_1 = 1309 \text{ ft/min}$$
$$F_{t1} = 126 \text{ lb}$$
$$b_1 = 1.15 \text{ inches}$$
$$n_{ex-1} = 9.3$$

Based on these data, (15-10) gives

$$C = \frac{(20 + 32)}{2(10)} = \frac{52}{20} = 2.600 \text{ inches} \tag{1}$$

and from (15-15)

$$\frac{\omega_2}{\omega_1} = -\left(\frac{20}{32}\right) = -0.625 \text{ (first mesh only)} \tag{2}$$

For the proposed helical gears:

$$P_n = 10 \quad \text{(matches hob)} \tag{3}$$

hence

$$P_t < 10 \quad [\text{see } (15 - 55)] \tag{4}$$

Again from (15-10)

$$P_t = \frac{N_1 + N_2}{2C} \tag{5}$$

By specification

$$N_2 = \frac{32}{20}N_1 \tag{6}$$

hence (5) becomes

$$P_t = \frac{N_1 + \left(\frac{32}{20}\right)N_1}{2\left(\frac{52}{20}\right)} = \frac{\left(\frac{52}{20}\right)N_1}{2\left(\frac{52}{20}\right)} = \frac{N_1}{2} \tag{7}$$

Iterating to find a combination of N_1, N_2, and P_t that maintains the same center distance and angular velocity ratio as the spur gear set, the results in Table E15.6 may be obtained. Hence, for the proposed gearset, tentatively select

$$N_1 = 15$$
$$N_2 = 24 \tag{8}$$

TABLE E15.6 Iteration to Find Compatible Set of Helical Gear Parameters

N_1	N_2	P_t	Remarks
20	32.0	10	Original spur gears
19	30.4	9.50	N_2 not a whole number[1]
18	28.8	9.00	N_2 not a whole number
17	27.2	8.50	N_2 not a whole number
16	25.6	8.00	N_2 not a whole number
15	24.0	7.50	OK[2]

[1]It should be clear that every operating gear must have an integral number of teeth (no partial teeth).
[2]Other combinations of numbers of teeth and helix angle exist, but this one should give the *smallest* acceptable helix angle.

Using (15-55),

$$\cos\psi = \frac{P_t}{P_n} = \frac{7.50}{10} = 0.75 \tag{9}$$

From (9)

$$\psi = 41.41° \tag{10}$$

This helix angle exceeds the recommended range of 10° to 35°, and should be reevaluated as the design progresses.

The pitch diameters of the pinion and gear may be calculated using (15-57), as

$$d_1 = \frac{N_1}{P_n \cos\psi} = \frac{15}{10\cos 41.41} = 2.000 \text{ inches} \tag{11}$$

and

$$d_2 = \frac{N_2}{P_n \cos\psi} = \frac{24}{10\cos 41.41} = 3.200 \text{ inches} \tag{12}$$

Nominal outside diameters of the two gears are (see Table 15.17)

$$\begin{aligned} d_{o1} &= 2r_{o1} = d_1 + 2a_1 = 2.000 + 2\left(\frac{1.000}{P_n}\right) \\ &= 2.000 + 2\left(\frac{1}{10}\right) = 2.200 \text{ inches} \end{aligned} \tag{13}$$

and

$$d_{o2} = d_2 + 2a_2 = 3.200 + 2\left(\frac{1}{10}\right) = 3.400 \text{ inches} \tag{14}$$

Using (15-59) and (15-54), the face width may be estimated as

$$b = \frac{2.0p_n}{\sin\psi} = \frac{2.0\pi}{P_n \sin\psi} = \frac{2.0\pi}{10\sin 41.41} = 0.95 \text{ inch} \tag{15}$$

b. The bending fatigue safety factor for the proposed helical gearset may be found using (15-69) and (15-71) together with K-values from Example 15.3. Thus

$$\sigma_b = \frac{F_t P_t}{bJ} K_a K_v K_m K_I = \frac{(126)(10\cos 41.41)}{(0.95)(J)}(1.0)(1.15)(1.6)(1.0) \tag{16}$$

The geometry factor J may be evaluated using Figures 15.37 and 15.38 to find

$$J = (J')M_J = (0.37)(0.97) = 0.36 \tag{17}$$

Example 15.6 Continues

It should be noted that *extrapolation* is required (because of the large helix angle) in both Figures 15.37 and 15.38 to obtain numerical values. Extrapolation is always undesirable, but will be utilized for this preliminary design. Ultimately, J should be calculated directly using methods of ref. 20 or 22. Combining (16) and (17)

$$\sigma_b = 5084 \text{ psi} \tag{18}$$

Utilizing (15-71) and using the same strength value $S_{tbf} = 16{,}000$ psi that was determined in Example 15.3, the existing safety factor, based on *tooth bending fatigue* in gear 1, is

$$n_{ex-1} = \frac{S_{tbf}}{\sigma_b} = \frac{16{,}000}{5084} = 3.2 \tag{19}$$

This represents an increase of over 20 percent in the value of tooth bending fatigue safety factor found for spur gears in Example 15.3.

Finally, the *surface fatigue* safety factor may be found for the proposed helical gearset using (15-70) and (15-72), together with the following data calculated in Example 15.5:

$$C_p = 2.29 \times 10^3$$
$$F_t = 126 \text{ lb}$$
$$K_a = 1.0$$
$$K_v = 1.15$$
$$K_m = 1.6$$

Also, from (10), (11), (12), and (15)

$$r_p = 1.000 \text{ inch}$$
$$r_g = 1.600 \text{ inches}$$
$$a_1 = a_2 = 0.200 \text{ inch}$$
$$\psi = 41.41°$$
$$b = 0.95 \text{ inch}$$

The profile contact ratio m_p may be calculated from (15-20) as

$$m_p = \frac{P_t\left[\sqrt{(r_p + a_p)^2 - (r_p\cos\varphi_t)^2} + \sqrt{(r_g + a_g)^2 - (r_g\cos\varphi_t)^2} - C\sin\varphi_t\right]}{\pi\cos\varphi_t} \tag{20}$$

From (15-55), or Table E15.6,

$$P_t = P_n\cos\psi = 10\cos 41.41 = 7.50 \tag{21}$$

and from (15-56)

$$\varphi_t = \tan^{-1}\frac{\tan 20}{\cos 41.41} - 25.9° \tag{22}$$

Using these values, (20) may be evaluated as

$$m_p = \frac{7.50\sqrt{(1.000 + 0.200)^2 - (1.000\cos 25.9)^2}}{\pi\cos 25.9} + \frac{7.50\sqrt{(1.600 + 0.200)^2 - (1.600\cos 25.9)^2}}{\pi\cos 25.9} - \frac{7.50(2.600\sin 25.9)}{\pi\cos 25.9} = \frac{7.50(0.79 + 1.08 - 1.14)}{2.83} = 1.93 \tag{23}$$

From (15-45)

$$I = \frac{\sin\varphi_t \cos\varphi_t}{2}\left(\frac{m_G}{m_G + 1}\right) \tag{24}$$

where

$$m_G = \frac{N_g}{N_p} = \frac{24}{15} = 1.60 \tag{25}$$

and

$$I = \frac{\sin 25.9 \cos 25.9}{2}\left(\frac{1.60}{1.60 + 1}\right) = 0.12 \tag{26}$$

Using these values in (15-70)

$$\sigma_{sf} = 2.29 \times 10^3 \sqrt{\frac{126}{(0.95)(2.00)(0.12)}\left(\frac{\cos 41.41}{1.93}\right)(1.0)(1.15)(1.6)}$$
$$= 45{,}521 \text{ psi} \tag{27}$$

Utilizing (15-72) and the same strength value $(S_{sf})_{N=10^{10}} = 120{,}600$ psi used in Example 15.5,

$$n_{ex-1} = \frac{(S_{sf})_{N=10^{10}}}{\sigma_{sf}} = \frac{120{,}600}{45{,}521} = 2.7 \tag{28}$$

This represents an increase of about 60 percent in surface fatigue safety factor over the value found for spur gears in Example 15.5.

Assuming these safety factors to be acceptable to engineering management, the following preliminary recommendations are made:

Material: AISI 1020 steel carburized and case hardened to Rockwell C 60
Reduction ratio: $m_G = 1.60$
Helical pinion: $N_p = 15$ teeth
Mating helical gear: $N_g = 24$ teeth
Helix angle: $\psi = 41.41°$ (try to reduce below 35° in final design; hand of pinion must be opposite to hand of gear)
Pinion pitch diameter: $d_p = 2.000$ inches
Gear pitch diameter: $d_g = 3.200$ inches
Center distance: $C = 2.600$ inches
Pinion outside diameter: $d_{op} = 2.200$ inches
Gear outside diameter: $d_{og} = 3.400$ inches
Face width: $b = 0.95$ inches
Governing safety factor (surface fatigue): $(n_{sf})_{N=10^{10}} = 2.7$

15.16 Bevel Gears; Nomenclature, Tooth Geometry, and Mesh Interaction

As discussed in 15.2 and illustrated in Figure 15.3(a), when gears are chosen for use in *intersecting* shaft applications, *straight bevel gears* represent the simplest (and most widely used) type of gearing available. In contrast to the *cylindrical* pitch surfaces of spur and helical gears (see Figures 15.1 and 15.2), the pitch surfaces of meshing bevel gears are *conical*, as sketched in Figure 15.41. The two contacting conical pitch surfaces roll together without slipping. They share a common apex, Q, at the point of shaft-centerline intersection. When

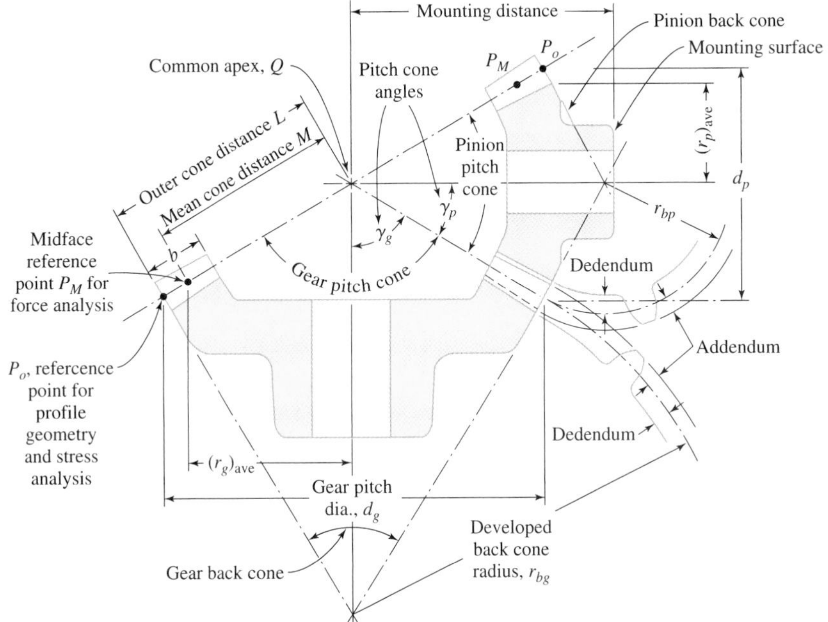

Figure 15.41
Basic geometry and nomenclature for straight-tooth bevel gears.

teeth are cut on the conical pitch surfaces they must be tapered, both in tooth thickness and height, from a large tooth profile at one end to a smaller tooth profile at the other end.

It is standard practice to define the size and shape of the tooth profile at the *larger end*. Referring again to Figure 15.41, the face width b is usually restricted to about 0.25 to 0.30 of the outer cone distance L because of the inherent difficulty in cutting very small teeth closer to the apex. Thus

$$b \leq 0.3L = 0.3\left(\frac{d}{2\sin\gamma}\right) \tag{15-73}$$

It is also recommended that the face width be limited to

$$b \leq 10/P_d \tag{15-74}$$

Practical values of diametral pitch lie in the range of 1 to 64.

Since all pitch cone elements are of equal length, a *circumscribing sphere* may be visualized, centered at the common apex Q, and in contact with the outer edges (bases) of all the conical pitch surfaces, as illustrated in Figure 15.42. Another *smaller concentric sphere* may be visualized at the inner ends of the teeth. Thus the conical frustra of all the pitch cones lie within the spherical shell bounded by these two concentric spheres. Based

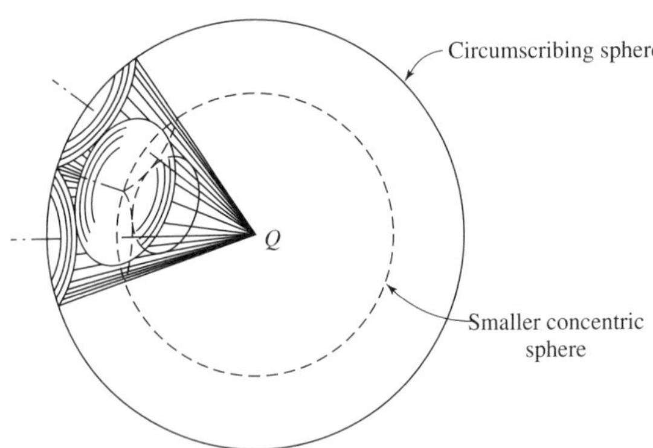

Figure 15.42
Frustra of conical bevel gear pitch surfaces enclosed between bounding concentric spheres. (From ref. 25.)

on the very successful use of the involute profile for spur and helical gearing applications, it is natural to consider the involute profile for bevel gear teeth. However, because of the spherical nature of bevel gear geometry, as illustrated in Figure 15.42, the involute shape would have to be generated on the *surface of a sphere* to secure *conjugate action*. Such a *spherical involute* tooth profile is impractical to manufacture, primarily because it would require the use of a single-point cutting tool.

Two other tooth profiles that are more practical (more easily manufactured) for use with straight bevel gears are the *octoid* tooth profile and the *Revacyle*[89] tooth profile. The octoid tooth form is widely used for *generated* bevel gear teeth because a simple reciprocating tool with a straight cutting edge can be used to cut the teeth. The *Revacycle* tooth profile is usually *close to circular* and is cut using a generating broach-type cutter; high production rates are possible if the Revacycle tooth profile is chosen.[90]

The increases in geometrical complexity just described make design calculations for bevel gears more complicated and more empirical than for spur or helical gears, but the underlying concepts are the same. The discussion of bevel gear design in this text is relatively brief, and data are included only for 20° *coniflex*[91] straight bevel gears. In any critical design project, the designer is well advised to consult the appropriate ANSI/AGMA standard[92] or seek the advice of a gearing specialist. As already noted, the size and shape of bevel gear teeth are defined at the larger end where they intersect the *back cones*, as illustrated in Figure 15.41. Therefore, *by definition*, both the diametral pitch P_d and pitch diameter d are based on the pitch circle at the large end of the teeth. Because tooling for bevel gears is not standardized according to pitch, the diametral pitch need not be an integer. Tooth profiles resemble spur gear tooth profiles of *virtual spur gears* having pitch radii equal to the *developed* pinion back cone radius r_{bp}, and the *developed* gear back cone radius r_{bg}. The *virtual numbers of teeth* in these imaginary spur gears are, for the pinion

$$N'_p = 2r_{bp}P_d = \frac{N_p}{\cos\gamma_p} \tag{15-75}$$

and for the gear

$$N'_g = 2r_{bg}P_d = \frac{N_g}{\cos\gamma_g} \tag{15-76}$$

Characterizing the size and shape of bevel gear teeth in terms of a virtual spur gear on the developed back cone is known as *Tredgold's approximation for bevel gears*.[93]

Bevel gears are generally not interchangeable because tooth form is closely tied to the method used for producing the gears, and a different value of addendum is used for each gear ratio. In practice, the pinion has a long addendum while the gear has a short addendum (see Table 15.18) to avoid undercutting. Hence bevel gears are nearly always made and replaced as *matched sets*. Tooth proportions for standard bevel gears are given in Table 15.18. Suggested *minimum* numbers of bevel pinion teeth are listed in Table 15.19.

The gear ratio for bevel gears can be determined from the number of teeth, the pitch diameters, or the pitch cone angles[94] as

$$m_G = \frac{\omega_p}{\omega_g} = \frac{N_g}{N_p} = \tan\gamma_g = \cot\gamma_p \quad (\text{always} > 1.0) \tag{15-77}$$

[89]Registered trade name of Gleason Works, Rochester, NY. [90]See ref. 24.

[91]Coniflex is a registered trade name of Gleason Works, Machine Division, Rochester, NY. The side-faces of Coniflex teeth are slightly crowned in their lengthwise direction, providing some tolerance to misalignment without concentrating tooth contact at the ends of the teeth.

[92]For example, ANSI/AGMA 2005-C96 (see ref. 23).

[93]See, for example, ref. 3. This procedure is similar to that leading to (15-64) for helical gears, also attributed to Tredgold.

[94]See Figure 15.41.

TABLE 15.18 Tooth Proportions[1] for Standard Full-Depth Straight Bevel Gears with 90° Shaft Angle (U.S. units)

Pressure angle	20°
Working depth	$2.000/P_d$
Whole depth	$(2.188/P_d) + 0.002(\text{min})$
Clearance	$(0.188/P_d) + 0.002(\text{min})$
Gear addendum	$[0.540 + 0.460(N_p/N_g)^2]/P_d$
Pinion addendum	$(2.000/P_d) - [0.540 + 0.460(N_p/N_g)^2]/P_d$
Gear dedendum	$[(2.188/P_d) + 0.002(\text{min})] - [0.540 + 0.460(N_p/N_g)^2]/P_d$
Pinion dedendum	$[(0.188/P_d) + 0.002(\text{min})] - [0.540 + 0.460(N_p/N_g)^2]/P_d$
Circular thickness	(see ref. 25)

[1]Defined at larger end of tooth; see, for example, ref. 26.

TABLE 15.19 Minimum Number of Bevel Gear Teeth[1] to Avoid Interference Between Standard 20° Full-Depth Pinion and Mating Bevel Gears of Various Sizes

Number of Bevel Pinion Teeth	Minimum Number of Mating Bevel Gear Teeth
16	16
15	17
14	20
13	31

[1]Extracted from ref. 1.

15.17 Bevel Gears; Force Analysis

As for all types of conjugate profile gearing, the normal (resultant) force F_n on a bevel gear tooth at the site of contact is directed along the line of action. Because the pitch surfaces of bevel gears are conical, the teeth are inclined with respect to the axis of rotation, as shown in Figure 15.41. Similar to the case for helical gears, the resolution of normal force F_n results in three mutually perpendicular components: a tangential component F_t in the plane of rotation, associated with useful power transmission, a radial separating force F_r, and an axial thrust component F_a. Neither F_r nor F_a serves a useful purpose. The midpoint of the face width b of the inclined teeth is usually assumed to be the point of application of resultant force F_n, where the pitch cone radius is r_{ave}, as shown in Figure 15.41. The average pitch cone radius r_{ave} to point P_M may be calculated from the geometry of Figure 15.41 as

$$r_{ave} = r - \frac{b}{2}\sin\gamma \tag{15-78}$$

As with spur gears and helical gears,[95] the transmitted force (tangential force) in pounds may be calculated as

$$F_t = \frac{33{,}000(hp)}{V_{ave}} \tag{15-79}$$

where hp is operational horsepower.

[95]See (15-25) and (15-65).

Using (15-78), the average pitch-line velocity at midface point P_M may be written, using (15-23), as

$$V_{ave} = \frac{2\pi(r_p)_{ave}n_p}{12} = \frac{2\pi(r_g)_{ave}n_g}{12} \tag{15-80}$$

where $(r_p)_{ave}$ and $(r_g)_{ave}$ are pitch radii in inches to midpoints of the meshing tooth faces, and n_p and n_g are rotational speeds in rpm; V_{ave} has units of ft/min.

Based on the selected pressure angle φ and pitch cone angle γ, the force components F_r and F_a may be geometrically determined as

$$F_r = F_t \tan\varphi \cos\gamma \tag{15-81}$$

and

$$F_a = F_t \tan\varphi \sin\gamma \tag{15-82}$$

To avoid a potential point of confusion when calculating *bending fatigue and surface fatigue stresses*, it should be carefully noted that the force components F_t, F_r and F_a, just calculated using (15-79), (15-81), and (15-82), are all referred to the *midface* reference point P_M in Figure 15.41. Because of the standard custom of defining bevel gear profile geometry and pitch radius r at P_o, on the *outer end* of the tooth (see Figure 15.41), *virtual force components* F_{to}, F_{ro}, and F_{ao}, if calculated from transmitted torque, speed, and horsepower, would have different values from F_t, F_r, and F_a. Specifically, it may be deduced from (15-21) that

$$T = F_t r_{ave} = F_{to}\left(\frac{d}{2}\right) \tag{15-83}$$

where T = torque on gear induced by power and speed requirements of the application
F_t, r_{ave} = tangential force and radius referred to *midface* reference point P_M
$F_{to}, \frac{d}{2}$ = virtual tangential force and standard pitch radius of gear referred to *tooth end* reference point P_o at larger (outer) end of the tooth

To avoid confusion, AGMA standard calculations for *bending fatigue* and *surface fatigue durability* (*pitting*) for bevel gears are currently formulated using *torque* in the expressions for σ_b and σ_{sf}[96] rather than *tangential force*, as commonly used for spur and helical gears.[97] That clarifying modification will be adopted in this text. Further, *pinion* torque, T_p, has been chosen as the basis for standard calculations. From (15-22) then, it becomes convenient to write

$$T_p = \frac{63{,}025(hp)}{n_p} \tag{15-84}$$

where hp = horsepower
n_p = pinion speed, rpm

15.18 Bevel Gears; Stress Analysis and Design

Calculations of bevel gear tooth bending stress σ_b and surface fatigue contact stress σ_{sf} are essentially the same as presented for spur[98] and helical[99] gears, except for minor adjustments associated with bevel gear geometry and the use of *pinion torque* in these ex-

[96]See (15-85) and (15-86). [97]See (15-41), (15-46), (15-69), and (15-70). [98]See (15-41) and (15-46).
[99]See (15-69) and (15-70).

pressions to avoid confusion (as discussed in 15.17). Specifically, for tooth bending fatigue[100]

$$\sigma_b = \frac{2T_p P_d}{d_p b J} K_a K_v K_m \tag{15-85}$$

and, for surface fatigue durability

$$\sigma_{sf} = (C_p)_{bevel}\sqrt{\frac{2T_p}{bd_p^2 I} K_a K_v K_m} \tag{15-86}$$

where P_d = diametral pitch (defined at larger ends of teeth)
T_p = pinion torque, in-lb
d_p = pinion pitch diameter (defined at larger ends of teeth)
b = face width, inch
J = bevel gear geometry factor for tooth bending (includes adjustments to account for conically inclined and tapered teeth, load sharing, location of most critical loading, stress concentration, and standard practice of defining P_d and d_p at larger ends of teeth). See Figure 15.44 for numerical values
K_a = application factor (see Table 15.6)
K_v = dynamic factor (see Figure 15.24)
K_m = bevel gear mounting factor (see Figure 15.43)

$$(C_p)_{bevel} = \sqrt{\frac{3}{2\pi\left(\frac{1-\nu_p^2}{E_p} + \frac{1-\nu_g^2}{E_g}\right)}}$$

= elastic coefficient for bevel gears. Comparing with equation (15-45) for spur gears, note that the bevel gear multiplying factor $\sqrt{3/2}$ reflects the *more nearly spherical* contact geometry of crowned bevel gears as compared to the *more nearly cylindrical* contact geometry of spur gears
I = bevel gear geometry factor for surface fatigue contact stress (includes similar adjustments to those noted for J above). See Figure 15.45 for numerical values

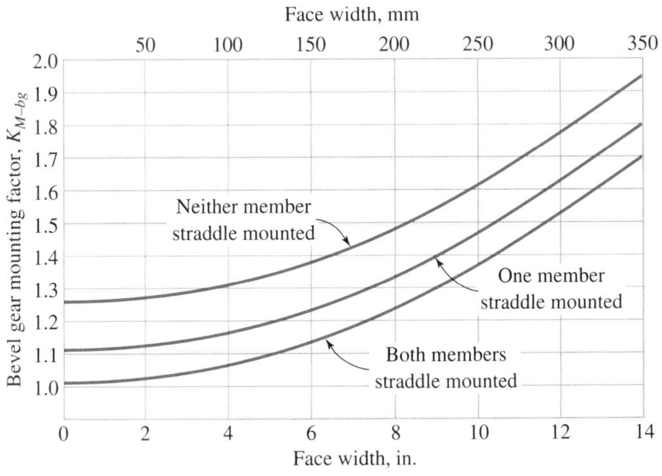

Figure 15.43
Bevel gear mounting factor, K_{m-bg}, for 20° crowned straight bevel gear teeth (90° shaft angle). Do not extrapolate. (From ref. 27. Adapted from ANSI/AGMA Standard 2003-B97, with the permission of the publisher, American Gear Manufacturers Association, 1500 King Street, Suite 201, Alexandria, VA 22314.)

[100] As for spur and helical gears, *additional* K-factors are available to improve accuracy.

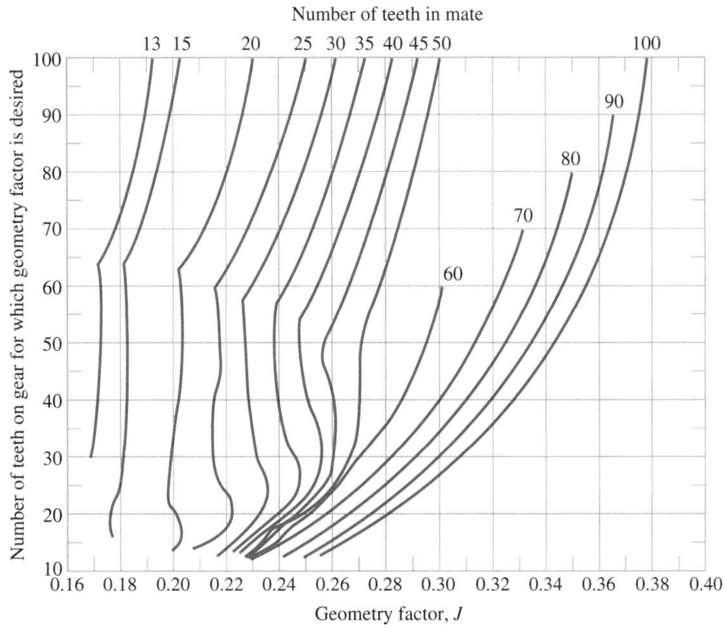

Figure 15.44
Geometry factor *J* (for tooth bending) for 20° *Coniflex*® straight bevel gears (90° shaft angle). (From ref. 27. Adapted from ANSI/AGMA Standard 2003-B97, with the permission of the publisher, American Gear Manufacturers Association, 1500 King Street, Suite 201, Alexandria, VA 22314.)

15.19 Bevel Gears; Summary of Suggested Design Procedure

Design procedure for bevel gears follows the same basic pattern established in 15.11 for straight spur gears and 15.15 for helical gears. For bevel gears, the design steps would usually include:

1. If functional specifications and contemplated system configuration include shafts at an angle of 90° to each other,[101] generate a first conceptual sketch of the gear train, including required reduction ratio, input pinion torque, gear shaft centerlines, bearing locations, and any other geometrical constraints.
2. Identify potential failure modes (see 15.4).

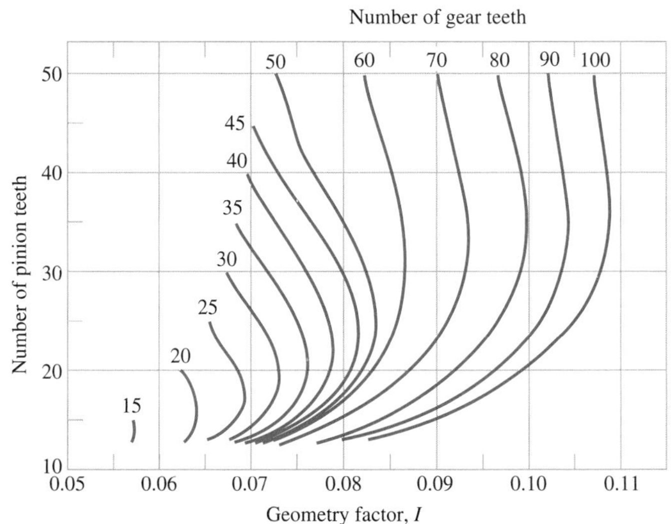

Figure 15.45
Geometry factor *I* (for surface fatigue durability) for 20° *Coniflex*® straight bevel gears (90° shaft angle). (From ref. 27. Adapted from ANSI/AGMA Standard 2003-B97, with the permission of the publisher, American Gear Manufacturers Association, 1500 King Street, Suite 201, Alexandria, VA 22314.)

[101]Actually, almost any angle can be accommodated but angles other than 90° are beyond the scope of this text. See ref. 23.

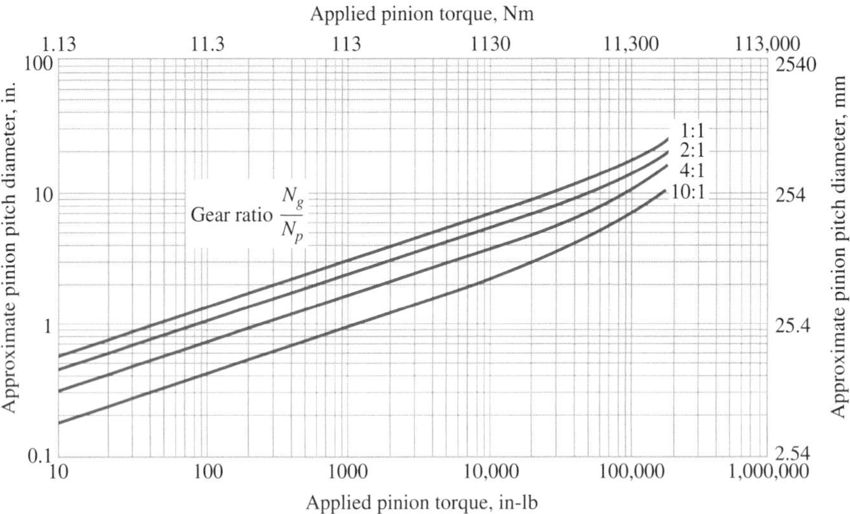

Figure 15.46
Approximate *pinion pitch diameter requirement* as a function of applied pinion torque and gear reduction ratio for 20° *Coniflex*® straight bevel gears, based on *tooth bending fatigue* (90° shaft angle). (From ref. 23. Adapted from ANSI/AGMA Standard 2005-C96, with the permission of the publisher, American Gear Manufacturers Association, 1500 King Street, Suite 201, Alexandria, VA 22314.)

3. Select tentative gear materials (see 15.5).
4. Select tentative manufacturing methods appropriate to the application (see Table 15.4).
5. Tentatively select a tooth system; then, using specified power and speed requirements, perform an analysis to determine pertinent *torques* and *speeds*.
6. From the *calculated* bevel pinion torque requirement T_p and *specified* gear ratio m_G, estimate the pinion pitch diameter d_p, using Figure 15.46 and 15.47 to obtain an acceptable *initial* selection of d_p; select the numbers of teeth for pinion and gear utilizing the recommendations of Figure 15.48.
7. Based on the definition of diametral pitch P_d given by (15-8), the approximate diametral pitch may be estimated (at the larger end of the teeth) from the results of step 6. For bevel gears, the diametral pitch need not be "standard" and any diametral pitch between 1 and 64, including nonintegral values, is acceptable.
8. Using the experience-based recommendations of (15-73) and (15-74), determine face width b by adopting the smaller of the two calculated values. Also, calculate average

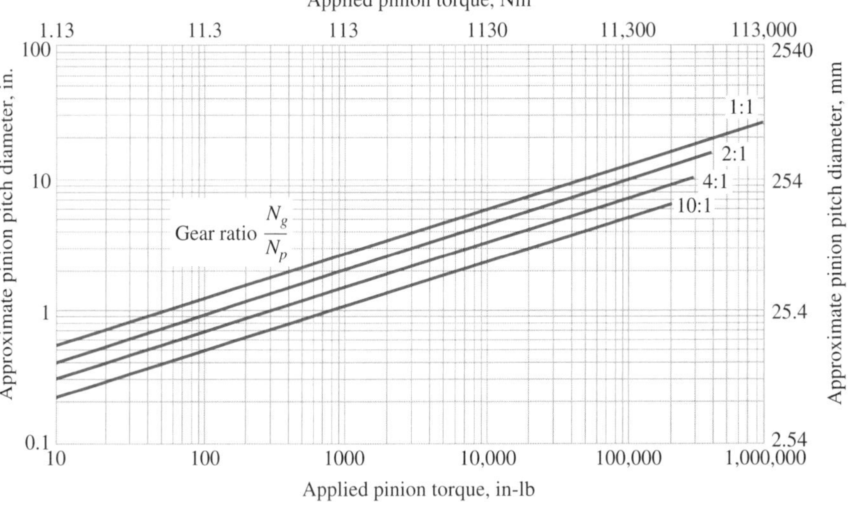

Figure 15.47
Approximate *pinion pitch diameter requirement* as a function of applied pinion torque and gear reduction ratio for 20° *Coniflex*® straight bevel gears, based on *surface fatigue durability* (90° shaft angle). (From ref. 23. Adapted from ANSI/AGMA Standard 2005-C96, with the permission of the publisher, American Gear Manufacturers Association, 1500 King Street, Suite 201, Alexandria, VA 22314.)

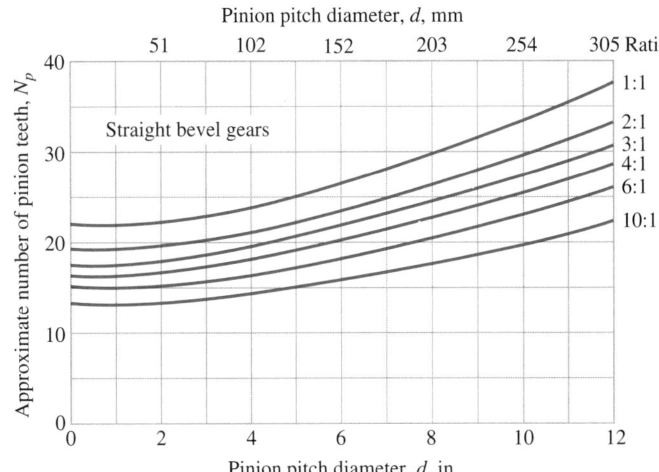

Figure 15.48
Recommended *number of pinion teeth* as a function of pinion pitch diameter and gear reduction ratio for 20° *Coniflex*® straight bevel gears (90° shaft angle). (From ref. 23. Adapted from ANSI/AGMA Standard 2005-C96, with the permission of the publisher, American Gear Manufacturers Association, 1500 King Street, Suite 201, Alexandria, VA 22314.)

pitch-line velocity from (15-80), then tangential, radial, and axial force components (to be reacted by support bearings).

9. For each bevel gear, calculate the tentative *existing safety factor*, based on *tooth bending fatigue*, that would exist if the proposed selections for tooth geometry and material properties (including reliability specifications) were used for the design. Tooth bending fatigue stresses may be calculated from (15-85) and material properties may be found in 15.9).[102]
10. Compare the tentative *existing safety factor* with the specified *design safety factor*. If they are approximately equal, no further iteration is required. If they are not equal, another iteration *is* required. Continue iterating until the specified design safety factor is achieved. Repeat for each bevel gear, as necessary.
11. Using the results of step 10, calculate the *existing safety factor* based on *surface fatigue durability*, from (15-86). If *this* existing safety factor, based on surface fatigue, does not equal or exceed the specified *design safety factor*, additional iterations may again be required.
12. Estimate friction losses, heat generation, and lubrication requirements to maintain acceptable temperature levels, noise levels, and design lives (see 15.10).

Example 15.7 Bevel Gearset Design

In the preliminary design of a special-purpose industrial honing machine, it is being proposed to use a straight bevel gearset to transform the rotation about a *vertical shaft axis* (of a motor-driven pinion) into rotation about an intersecting *horizontal shaft axis* (of the output gear). It is estimated that 7 horsepower will be required to drive the hone at a rotational speed of 350 rpm. The drive motor tentatively chosen runs under full power at 1150 rpm.

Additional specifications for this straight bevel gearset are:

1. These preliminary calculations may be based on the assumption that friction losses are negligible for all gears and bearings.
2. It is desired to use AISI 1020 low-carbon steel for the gear blanks, if possible; then, after machining, it is desired to carburize and case harden the tooth surfaces to R_C 60.

[102]However, AGMA standards include special tables of material properties for bevel gears (see ref. 27) that should be used in any critical design project.

Example 15.7 Continues

A design life of $N = 10^{10}$ cycles and a reliability of 99 percent are desired. Properties for AISI 1020 steel carburized and case hardened to R_C 60 have already been found in Examples 15.3 and 15.5, and may be summarized as follows:

a. Core properties are:

$$S_u = 57{,}000 \text{ psi}$$
$$S_{yp} = 43{,}000 \text{ psi}$$
$$(S_{tbf})_{N=10^{10}} = 16{,}000 \text{ psi}$$
$$\text{Hardness} = \text{BHN } 121$$

b. Case properties are:

$$(S_{sf})_{N=10^{10}} = 120{,}600 \text{ psi}$$

c. Gear teeth are to be standard 20° straight bevel gear teeth.

d. Total production is estimated to be 50 machines.

Propose an acceptable configuration that meets the specifications and provides a design safety factor of approximately $n_d = 1.5$.

Solution

a. Based on the specifications, a first conceptual layout of the bevel gearset is sketched as shown in Figure E15.7.

b. Potential primary failure modes appear to be tooth bending fatigue and surface fatigue pitting.

c. Since this is a low-production item, Coniflex® straight bevel gears are tentatively selected.

d. Using specified shaft speed requirements, the gear ratio m_G may be calculated from (15-77) as

$$m_G = \frac{\omega_p}{\omega_g} = \frac{n_p}{n_g} = \frac{1150}{350} = 3.29 \tag{1}$$

e. Operating pinion torque may be calculated using (15-84) as

$$T_p = \frac{63{,}025(hp)}{n_p} = \frac{63{,}025(7)}{1150} = 384 \text{ in-lb} \tag{2}$$

f. The approximate pinion pitch diameter requirement may be read from Figure 15.46

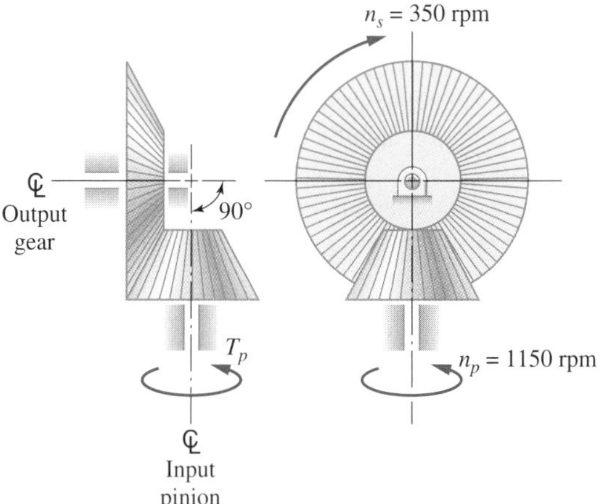

Figure E15.7
First conceptual sketch of proposed straight bevel gearset.

(based on tooth bending fatigue) or from Figure 15.47 (based on surface fatigue durability, which is often more critical). Using Figure 15.47 together with (1) and (2),

$$d_p \approx 1.5 \text{ inches} \tag{3}$$

(About the same result would be obtained from Figure 15.46.)

g. Next, from Figure 15.48, the recommended number of pinion teeth would be

$$N_p = 18 \text{ teeth} \tag{4}$$

and for the gear, using (1) and (15-77),

$$N_g = m_G N_p = 3.29(18) = 59.22 \approx 59 \text{ teeth} \tag{5}$$

Note that using the (required) *integral* number of gear teeth slightly changes the gear ratio calculated in (1) to m_G = (59/18) = 3.28.

h. The appropriate diametral pitch may next be estimated using (15-8) as

$$P_d = \frac{N_p}{d_p} = \frac{18}{1.5} = 12.0 \tag{6}$$

(Note that the integral value of P_d is fortuitous. Fractional values are acceptable.)

i. Using (15-73), face-width estimates may be made as

$$b_L \leq 0.3\left(\frac{d_p}{2\sin\gamma_p}\right) \tag{7}$$

where γ_p is determined from (15-77) as

$$\gamma_p = \cot^{-1} m_G = \cot^{-1} 3.28 = 17.0^\circ \tag{8}$$

Thus (7) gives

$$b_L \leq 0.3\left(\frac{1.5}{2\sin 17.0}\right) = 0.77 \text{ inch} \tag{9}$$

Also, from (15-74)

$$b_{P_d} \leq \frac{10}{P_d} = \frac{10}{12} = 0.83 \tag{10}$$

Taking the smaller of (9) or (10),

$$b = 0.77 \text{ inch} \tag{11}$$

j. Average pitch-line velocity V_{ave} may be calculated using (15-80). First, from (15-78)

$$(r_p)_{ave} = \frac{d_p}{2} - \frac{b}{2}\sin\gamma_p = \frac{1.5}{2} - \left(\frac{0.77}{2}\right)\sin 17.0 = 0.64 \text{ inch} \tag{12}$$

whence

$$V_{ave} = \frac{2\pi(r_p)_{ave} n_p}{12} = \frac{2\pi(0.64)(1150)}{12} = 385 \text{ ft/min} \tag{13}$$

k. Using (15-79), (15-81), and (15-82),

$$F_t = \frac{33{,}000(hp)}{V_{ave}} = \frac{33{,}000(7)}{385} = 600 \text{ lb} \tag{14}$$

then

$$(F_r)_p = F_t \tan\varphi \cos\gamma_p = (600)(\tan 20)\cos 17.0 = 209 \text{ lb} \tag{15}$$

and

$$(F_a)_p = F_t \tan\varphi \sin\gamma_p = 600 \tan 20 \sin 17.0 = 64 \text{ lb} \tag{16}$$

Example 15.7 Continues

Noting from (15-77) that

$$\gamma_g = \tan^{-1} m_G = \tan^{-1} 3.28 = 73.0° \tag{17}$$

the radial and axial components of force on the gear are

$$(F_r)_g = 600 \tan 20 \cos 73.0 = 64 \text{ lb} \tag{18}$$

and

$$(F_a)_g = 600 \tan 20 \sin 73.0 = 209 \text{ lb} \tag{19}$$

It may be noted that these force magnitudes are consistent with free-body equilibrium of the bevel gearset pictured in Figure E15.7.

l. The *bending fatigue safety factor* for the proposed bevel gearset may be found using (15-85). To facilitate the calculation, the known data may be summarized as

$$\begin{aligned} T_p &= 384 \text{ in-lb} \\ P_d &= 12.0 \\ d_p &= 1.5 \text{ inches} \\ b &= 0.77 \text{ inch} \\ N_p &= 18 \text{ teeth} \\ N_g &= 59 \text{ teeth} \end{aligned} \tag{20}$$

Next

K_a = 1.0 (from Table 15.6; uniform loading for both the drive source and the driven machine)
K_v = 1.15 (from Figure 15.24 using Q_v = 8 or 9 from Table 15.4, based on *good accuracy* gearing)
K_m = 1.1 (from Figure 15.43 for face width of 0.77 inch with one member straddle mounted; see proposed layout in Figure E15.7.
J = 0.24 (from Figure 15.44, for the pinion teeth, using N_p = 18 and N_g = 59)

Placing these values in (15-85)

$$\sigma_b = \frac{2(384)(12)}{(1.5)(0.77)(0.24)}(1.0)(1.15)(1.1) = 42{,}060 \text{ psi} \tag{21}$$

and the existing bending fatigue safety factor for this first design proposal would be

$$n_{ex-1} = \frac{(S_{tbf})_{N=10^{10}}}{\sigma_b} = \frac{16{,}000}{42{,}060} \approx 0.4 \tag{22}$$

A safety factor less than unity is clearly unacceptable.

To increase the safety factor on tooth bending fatigue, it is necessary to make the teeth stronger (thicker) in bending. As a second iteration, try increasing the pitch diameter to[103]

$$d_p = 2.0 \text{ inches} \tag{23}$$

Then, following the same calculation sequence just used, iterations on d_p are continued until an *existing safety factor* calculation results in a value approximately equal to the specified *design safety factor* value of n_d = 1.5. Table E15.7 illustrates the process.

m. The *surface fatigue safety factor* for the proposed bevel gearset may be found using (15-86), and it too must be 1.5 or greater, by specification. In addition to the third-it-

[103]Experience dictates that, when iterating at an *early stage*, larger jumps in the value of the iteration parameter tend to be more efficient in the long run.

TABLE E15.7 Iteration Process to Find Acceptable Pinion Specifications (Iterations 1–3 are based on tooth bending fatigue; iteration 4 is required by the result of (27) for surface fatigue.)

Iteration	d_p, in	P_d	b_L, in	b_{P_d}, in	b, in	$(r_p)_{ave}$, in	V_{ave}, ft/min	F_t, lb	K_a	K_V	K_m	J	σ_b, psi	$(n_{ex})_{bending}$
1	1.5	12.0	0.77	0.83	0.77	0.64	385	600	1.0	1.15	1.1	0.24	42,060	0.4
2	2.0	9.0	1.03	1.11	1.03	0.85	512	451	1.0	1.15	1.1	0.24	18,220	0.9
3	2.5	7.2	1.28	1.39	1.28	1.06	638	362	1.0	1.25	1.1	0.24	9,900	1.6[1]
4	3.0	6.0	1.54	1.67	1.50	1.28	987	234	1.0	1.25	1.1	0.24	5,870	2.7[2]

[1]This safety factor on bending fatigue is close enough to the specified value $n_d = 1.5$.
[2]This larger safety factor on *bending fatigue* is precipated by the need for an adequate safety factor on *surface fatigue* (more critical).

eration values from Table E15.7, the evaluation of (15-86) requires values for $(C_p)_{bevel}$ and I. From definitions following (15-86),

$$(C_p)_{bevel} = \sqrt{\frac{3}{2\pi\left[\left(\frac{1-\nu_p^2}{E_p}\right)+\left(\frac{1-\nu_g^2}{E_g}\right)\right]}}$$

$$= \sqrt{\frac{3}{2\pi\left[\left(\frac{1-0.3^2}{30\times 10^6}\right)+\left(\frac{1-0.3^2}{30\times 10^6}\right)\right]}} = 2805 \tag{24}$$

and from Figure 15.45,

$$I = 0.083 \tag{25}$$

Evaluating (15-86) then

$$\sigma_{sf-3} = 2805\sqrt{\frac{2(384)}{1.28(2.5)^2(0.083)}(1.0)(1.25)(1.1)} = 111{,}860 \text{ psi} \tag{26}$$

Thus the surface fatigue safety factor for the iteration-3 design proposal would be

$$n_{ex-3} = \frac{120{,}600}{111{,}860} \approx 1.1 \tag{27}$$

Because this existing safety factor is less than the specified design safety factor, additional iteration is required. A fourth iteration is therefore recorded in Table E15.7, based on surface fatigue requirements. Using these values, and repeating an updated calculation for (26),

$$\sigma_{sf-4} = 2805\sqrt{\frac{2(384)}{1.50(3.0)^2(0.083)}(1.0)(1.25)(1.1)} = 86{,}110 \text{ psi} \tag{28}$$

whence

$$n_{ex-4} = \frac{120{,}600}{86{,}110} \approx 1.4 \tag{29}$$

This safety factor is relatively close to $n_d = 1.5$, and could be iterated again if desired, to increase its value. It will be regarded as acceptable for now, however.

For this design project, surface fatigue durability ($n_{ex-4} = 1.4$) is more critical than tooth bending fatigue ($n_{ex-4} = 2.7$). Such results are typical. It may therefore be expedient (but not necessary) to make design calculations based on surface fatigue durability *first*, followed by calculations based on tooth bending fatigue.

15.20 Worm Gears and Worms; Nomenclature, Tooth Geometry, and Mesh Interaction

Worm gearsets have been briefly described in 15.2 and sketched in Figure 15.4. Only *cylindrical* worm gearing [see Figure 15.41(d)] will be discussed here[104] Worm gearing is used to transmit power and motion between nonintersecting shafts, usually at an angle of 90° to each other, as shown in Figure 15.49. The *worm* resembles a power screw (see Chapter 12) and the *worm gear* resembles a helical gear, except that it is *throated* to partially envelop the worm. If a rotating worm is constrained axially by thrust bearings, the worm gear is caused to rotate *about its own axis*, as the worm threads *slide sidewise* against the gear teeth. This simulates the kinematic behavior of a linearly advancing rack[105] engaged with a conjugate spur or helical gear. Worm gearing is often chosen to obtain large reduction ratios.[106] Ratios ranging from $3^1/_2:1$ to $100:1$ are common but worm gearsets can be produced to provide ratios ranging from $1:1$ to $360:1$. Because of the screw action, worm gearsets are quiet, vibration free, and produce a constant output speed free of pulsations.

Lines of contact between the worm and worm gear teeth progress *from the tips toward the roots* of the worm gear teeth as they pass through the mesh. Figure 15.50 illustrates the contact lines at a given instant for the case where three teeth are in contact, giving *three lines of contact*, all progressing toward the roots of the gear teeth as the worm rotates. The progression of these lines of contact sweeping across each worm gear tooth results in an *area of contact*. Thus the state of stress in the contact zone between the worm and the gear is influenced not only by *cyclic Hertz contact forces*, but also by *sliding friction forces* generated by the rotating worm, and by *load distribution* among the contacting teeth.

It is common practice to impose a gradual *run-in* period of new worm gear drives before operating them at full load. One recommended run-in process is to operate the gearset at half-load for a few hours, then increase the load in at least two stages to full load. This process results in smoother work-hardened surfaces, and a reduction in friction forces of up to 15 percent.

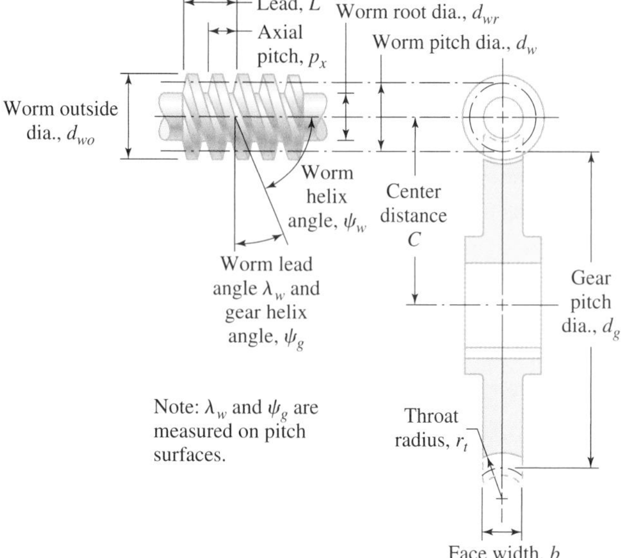

Figure 15.49
Sketch of worm gear arrangement for cylindrical (single-enveloping) worm, giving standard nomenclature. Shown for double-threaded worm (see Figure 12.3).

[104]See ref. 1 for discussion of *nonenveloping* and *double-enveloping* worm gearsets.

[105]The worm, when rotated, simulates a series of rack profiles being continuously advanced along its axis.

[106]Reduction ratio is the ratio of angular velocity of the worm to angular velocity of the gear.

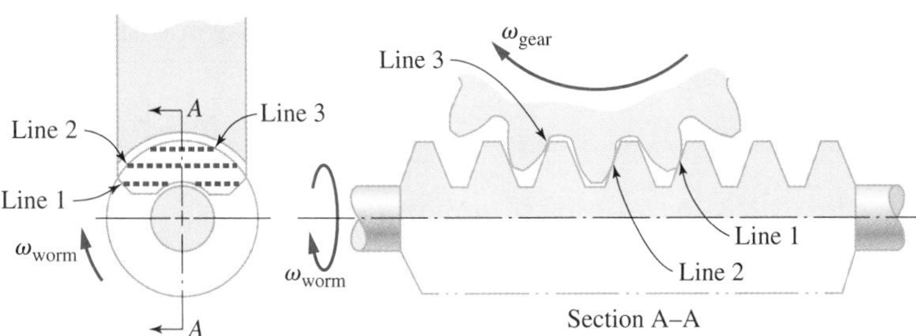

Figure 15.50
Instantaneous lines of contact on worm gear teeth. (From ref. 28. Adapted from ANSI/AGMA Standard 6022-C93, with the permission of the publisher, American Gear Manufacturers Association, 1500 King Street, Suite 201, Alexandria, VA 22314.)

The profile, or shape, of the worm threads is determined by the method used to manufacture them.To insure proper contact between the worm and the gear teeth, the *gear hob* (or other tooling used to produce the worm gear) typically has the same basic cutter shape used for the *worm*. Figure 15.51 depicts three common profile forms used in practice.

Customary tooth dimensions for worms and worm gears may be given as a function of diametral pitch P_d of the gears, as shown in Table 15.20, or as a function of axial pitch p_x.[107]

A basic requirement of a properly designed worm gearset is that the axial pitch p_x of the worm[108] must be equal to the circular pitch p_c of the gear.[109] Thus

$$p_x = \frac{p_n}{\cos \lambda_w} = p_c = \frac{p_n}{\cos \psi_g} = \frac{\pi d_g}{N_g} \tag{15-87}$$

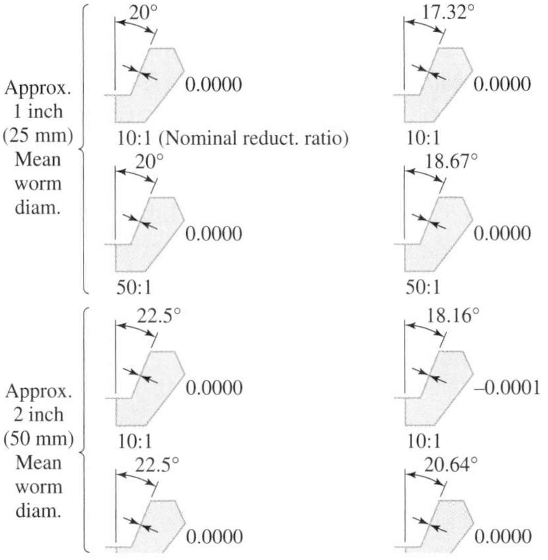

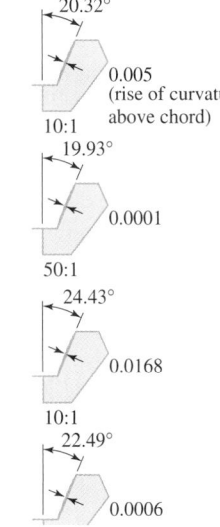

(*a*) Straight axial profile; cut using a straight-sided cutter in the axial plane.

(*b*) Straight normal profile; cut using a straight-sided cutter in the normal plane.

(*c*) Involute helicoid profile; cut using a straight-sided cutter or grinding wheel tilted to the worm lead angle and inclined to the chosen pressure angle.

Figure 15.51
Three common worm-thread profile forms. (Worm profile forms are not standardized; other forms are also used.) (From ref. 28. Selectively adapted from ANSI/AGMA Standard 6022-C93, with the permission of the publisher, American Gear Manufacturers Association, 1500 King Street, Suite 201, Alexandria, VA 22314.)

[107]It is common practice to express these dimensions in terms of axial pitch p_x. The expressions of Table 15.20 may be rewritten in terms of axial pitch by substituting $P_d = \pi/p_x$.

[108]See Figure 15.49.

[109]Same definition as for other gearing; e.g., see (15-6). Similar to the development of (15-51) and (15-52) for helical gears, the *worm gear* normal circular pitch is given by $p_n = p_c \cos\psi_g$, and the *worm* normal circular pitch is given by $p_n = p_x \sin\psi_w = p_x \sin(90 - \lambda_w) = p_x \cos\lambda_w$.

TABLE 15.20 Typical Tooth Proportions for Worms and Worm Gears[1] (U.S. units)

Addendum	$1.000/P_d$
Dedendum	$1.157/P_d$ for $P_d < 20$
	$(1.200 + 0.002)/P_d$ for $P_d \geq 20$
Whole depth	$2.157/P_d$
Working depth	$2.000/P_d$
Clearance	$0.157/P_d$
Root diameter of worm	$d_w - (2.314/P_d)$ for $P_d < 20$
	$d_w - (2.400 + 0.002)$ for $P_d \geq 20$
Outside diameter of worm	$d_w + (2.000/P_d)$
Root diameter of gear	$d_g - (2.314/P_d)$ for $P_d < 20$
	$d_g - (2.400 + 0.002)/P_d$ for $P_d \geq 20$
Throat diameter of gear	$d_g + (2.000/P_d)$
Outside diameter of gear (approx.)	$d_w + (3.000/P_d)$

[1]Compiled from ref. 28. Excerpted from ANSI/AGMA Standard 6022-C93, with the permission of the publisher, American Gear Manufacturers Association, 1500 King Street, Suite 201, Alexandria, VA 22314.

and, as for other types of gearing, the diametral pitch P_d (or module m) of the gear is

$$P_d = \frac{N_g}{d_g}\left(\text{or} \quad m = \frac{d_g}{N_g}\right) \tag{15-88}$$

Worms may be manufactured with a single thread or with multiple threads, similar to power screws.[110] The number of threads N_w on the worm is treated as if it were the number of "teeth" in the worm. The number of *threads* in the worm is often referred to as the number of *starts*.[111] Typically, reduction ratios above 30 : 1 utilize a *single* worm thread, and below 30 : 1 *multiple* threads are more common.

The number of teeth in the worm gear is determined by the required reduction ratio, together with the minimum number of teeth dictated by good design practice. Table 15.21 gives the suggested minimum number of worm gear teeth as a function of normal pressure angle.

TABLE 15.21 Suggested Minimum Number of Worm Gear Teeth[1] as a Function of Normal Pressure Angle φ_n

Normal Pressure Angle, φ_n	Minimum Number of Worm Gear Teeth
$14\frac{1}{2}$	40
$17\frac{1}{2}$	27
20	21
$22\frac{1}{2}$	17
25	14
$27\frac{1}{2}$	12
30	10

From ref. 28. Excerpted from ANSI/AGMA Standard 6022-C93, with the permission of the publisher, American Gear Manufacturers Association, 1500 King Street, Suite 201, Alexandria, VA 22314.

[110]See Figure 12.3.

[111]Looking at the end of a worm one can count the number of parallel threads that *start* at the end and wind helically along the cylindrical worm.

The pitch diameter of a worm is *not* a function of its number of threads. The *reduction ratio* of a worm gearset, therefore, is *not* equal to the ratio of gear and worm pitch diameters; the reduction ratio is a function of the *number of teeth* on the worm and on the gear. Hence

$$\frac{\omega_w}{\omega_g} = \frac{N_g}{N_w} \tag{15-89}$$

Usual practice is to select the worm pitch diameter d_w to lie in the range[112]

$$\frac{C^{0.875}}{3.0} \leq d_w \leq \frac{C^{0.875}}{1.6} \tag{15-90}$$

where C is center distance (inches). The gear pitch diameter may be calculated as

$$d_g = 2C - d_w \tag{15-91}$$

Typically, the worm gear face width should not exceed two-thirds the worm pitch diameter,[113] or

$$b \leq 0.67 d_w \tag{15-92}$$

This face-width restriction, based on experience, reflects the adverse effect of increasingly nonuniform load distribution along the contact thread length, as face width is increased.[114]

Similar to the case of power screws (Chapter 12), the *lead* is the axial advance of any point on the worm thread for one revolution of the worm, or the axial pitch of the worm multiplied times the number of threads on the worm. Thus

$$L_w = N_w p_x \tag{15-93}$$

The mean worm lead angle[115] λ_w is given by

$$\lambda_w = \tan^{-1} \frac{L_w}{\pi d_w} \tag{15-94}$$

For the usual case, where the worm shaft is 90° to the gear shaft, the *lead angle of the worm* is equal to the *helix angle of the worm gear*, or

$$\lambda_w = \psi_g \tag{15-95}$$

The *lead angle* of the worm (complement of the worm's helix angle) is the customary specification for worm threads (as opposed to helix angle). Usually the worm has a *helix angle* of more than 45° and the worm gear has a helix angle less than 45°. The "hand" of the worm must be the *same* as the hand of the gear.

Customary pressure angles for worm gearing are, in degrees, $14\frac{1}{2}$, $17\frac{1}{2}$, 20, $22\frac{1}{2}$, 25, $27\frac{1}{2}$, and 30. Either *normal* pressure angle φ_n or *transverse* pressure angle φ_t may be specified. They are related by

[112]Experience-based empiricism.

[113]More accurate guidelines may be found in ref. 28.

[114]Wider faces result in larger cantilever deflections of the worm threads near the extremities of the contact zone, concentrating more of the transmitted load toward the center of contact. This results in high local bending stresses and increases the potential for tooth bending fatigue failure at the root fillet. See ref. 30, p. 13.

[115]The lead angle varies significantly from the worm outside diameter to its root diameter, so the *mean* worm diameter is used as the basis for specifying lead angle. It is also worth noting that ambiguities in terminology exist in the literature that have the potential to create confusion: Some authors call λ the *lead angle* while others call λ the *helix angle*. In this text, the lead angle is consistently measured from a plane *normal* to the axis of rotation while helix angle is measured from the axis of rotation itself. Hence, the lead and helix angles are complements.

TABLE 15.22 Maximum Worm Lead Angle for Selected Pressure Angles

Pressure Angle φ_n, degrees	Maximum Lead Angle λ_w, degrees
$14\frac{1}{2}$	15
20	25
25	35
30	45

$$\varphi_n = \tan^{-1}\varphi_t \cos\lambda_w \tag{15-96}$$

Higher pressure angles may be selected when higher worm gear tooth strength is required, but consequences of selecting higher-pressure angles include fewer teeth in contact, higher bearing reaction forces, larger shaft bending stresses, and larger deflections of the worm shaft. To avoid interference, pressure angles are usually related to the worm lead angle, as shown, for example, in Table 15.22.

15.21 Worm Gears and Worms; Force Analysis and Efficiency

Force equilibrium and free-body analyses for a meshing worm and gear parallel the force analyses for a power screw assembly.[116] Conceptually, the worm is analogous to the power screw, and the worm gear is analogous to the nut. The force resolution for a power screw, as depicted in Figure 12.7, may therefore be directly applied to the case of a worm by observing that power screw lead angle α is equivalent to worm lead angle λ_w, and power screw normal thread angle θ_n is equivalent to normal pressure angle φ_n for the worm gearset. Making these equivalency substitutions in Figure 12.7, a *tangential* summation of forces on the worm, *when the worm is driving the gear*, gives

$$F_{wt} = F_n(\cos\varphi_n \sin\lambda_w + \mu\cos\lambda_w) \tag{15-97}$$

The normal force F_n may therefore be expressed as

$$F_n = \frac{F_{wt}}{\cos\varphi_n \sin\lambda_w + \mu\cos\lambda_w} \tag{15-98}$$

where μ is the sliding coefficient of friction between the worm threads and meshing gear teeth.[117]

Writing an *axial* force summation for the worm

$$F_{wa} = F_n(\cos\varphi_n \cos\lambda_w - \mu\sin\lambda_w) = F_{wt}\frac{\cos\varphi_n \cos\lambda_w - \mu\sin\lambda_w}{\cos\varphi_n \sin\lambda_w + \mu\cos\lambda_w} \tag{15-99}$$

Also, summing forces in the radial (separating) direction for the worm

$$F_{wr} = F_n \sin\varphi_n = F_{wt}\frac{\sin\varphi_n}{\cos\varphi_n \sin\lambda_w + \mu\cos\lambda_w} \tag{15-100}$$

Forces on the gear are related by equilibrium to forces on the worm as

[116] See 12.4.

[117] Coefficients of friction are shown in Appendix Table A.1. Empirical data show, however, that in worm gear applications the coefficient of friction is a function of sliding velocity. See ref. 29 for detailed data.

$$F_{gt} = F_{wa} \qquad [\text{see (15-99)}] \tag{15-101}$$

$$F_{ga} = F_{wt} \qquad [\text{see (15-97)}] \tag{15-102}$$

$$F_{gr} = F_{wr} \qquad [\text{see (15-100)}] \tag{15-103}$$

Power and speed requirements are typically utilized as the bases for calculating tangential force. Equation (15-25) may be used to calculate the tangential force on either the gear, if output power and rotational gear speed are known, or the worm, if input power and rotational worm speed are known. Since significant power losses result from sliding friction between worm and gear, the input power $(hp)_{in}$ and output power $(hp)_{out}$ often differ significantly. Thus[118]

$$F_{wt} = \frac{2T_w}{d_w} = \frac{33{,}000(hp)_{in}}{V_w} \tag{15-104}$$

and

$$F_{gt} = \frac{2T_g}{d_g} = \frac{33{,}000(hp)_{out}}{V_g} \tag{15-105}$$

where T_w = torque on worm, in-lb
T_g = gear torque, in-lb
d_w = worm pitch diameter, inches
d_g = gear pitch diameter, inches
V_w = pitch-line velocity of worm, fpm
V_g = pitch-line velocity of gear, fpm

The *efficiency* e for a worm gearset may be defined as the ratio of driving torque on the worm assuming no friction in the mesh, to driving torque on the worm when friction is included; efficiency may also be expressed as the ratio of output power at the gear shaft to input power at the worm shaft.[119] Thus, using (15-104) and (15-105)

$$e = \frac{(hp)_{out}}{(hp)_{in}} = \frac{F_{gt}V_g}{F_{wt}V_w} \tag{15-106}$$

Based on the pitch line speed of the gear, sliding velocity V_s is

$$V_s = \frac{V_g}{\sin\lambda_w} \tag{15-107}$$

and based on the pitch line speed of the worm

$$V_s = \frac{V_w}{\cos\lambda_w} \tag{15-108}$$

Hence

$$\frac{V_g}{V_w} = \frac{V_s\sin\lambda_w}{V_s\cos\lambda_w} = \tan\lambda_w \tag{15-109}$$

and (15-106) may be rewritten, using (15-97), (15-99), (15-101), and (15-109), as

$$e = \left(\frac{\cos\varphi_n\cos\lambda_w - \mu\sin\lambda_w}{\cos\varphi_n\sin\lambda_w + \mu\cos\lambda_w}\right)\tan\lambda_w = \frac{\cos\varphi_n - \mu\tan\lambda_w}{\cos\varphi_n + \mu\cot\lambda_w} \tag{15-110}$$

[118]AGMA practice is to use *mean diameter* d_m instead of pitch diameter d. Mean diameter is defined as twice the radius measured to the midpoint of the working depth (at the central plane of the gear). If there is a significant difference between d_m and d, d_m should be used.

[119]Assuming the worm to be driving the gear.

This is consistent with the efficiency expression for power screws given in (12-19); hence the efficiency trends, as a function of lead angle λ_w (power screw lead angle α) and friction coefficient, as depicted in Figure 12.8, are applicable to worm gears as well as power screws.

It is also important to remember that the efficiency of a wormset is an index to the *heat generated* in the mesh due to sliding between the worm thread and the gear teeth.[120] Further, the heat generated must be dissipated to the ambient atmosphere at such a rate that the lubricant sump temperature, under steady-state operation, does not exceed about 180°F. As discussed more fully in 15.10, various means of cooling the housing or the oil may be used, including the possible use of an external heat exchanger. Experience shows that if adequate oil pumps, heat exchangers, and mesh-directed oil jets are used, virtually any wormset can be operated at its full mechanical potential. Ultimately, experimental testing should be conducted to verify the acceptability of the proposed design.

Depending upon coefficient of friction and wormset geometry, the worm gear may or may not be able to *back-drive* the worm when worm torque is released. If the worm gear *cannot* back-drive the worm, the wormset is said to be *self-locking*. If the worm gear *can* back-drive the worm, the wormset is said to be *overhauling*. To determine whether or not a gearset is self-locking, the force resolution at the mesh site[121] must be reexamined for the case of the *gear driving the worm*. When the gear attempts to drive the worm (in the opposite direction from the *worm driving the gear*), the mesh sliding direction reverses and therefore the friction force also reverses direction. The resulting expression for tangential force on the worm under these circumstances becomes[122]

$$F_{wt} = F_n(\cos\varphi_n \sin\lambda_w - \mu\cos\lambda_w) \tag{15-111}$$

The gear can backdrive the worm only if F_{wt} is negative. If F_{wt} is positive or zero, the gearset is self-locking. Hence it may be deduced from (15-111) that *self-locking* conditions exist if

$$-\cos\varphi_n \sin\lambda_w + \mu\cos\lambda_w \geq 0 \tag{15-112}$$

or if

$$\mu \geq \cos\varphi_n \tan\lambda_w \tag{15-113}$$

Conversely, *overhauling* conditions exist if

$$\mu < \cos\varphi_n \tan\lambda_w \tag{15-114}$$

It should be noted that even if the wormset meets the self-locking criterion of (15-113), vibration or reversing cyclic torques may still cause it to "creep" backward. It is usually recommended that a lead angle no greater than 5 or 6 degrees be used if self-locking is desired, and that a separate braking device be incorporated if *no* creeping action can be tolerated. The low lead angles required for effective self-locking usually imply the use of a single-thread worm and produce low efficiencies.[123]

15.22 Worm Gears and Worms; Stress Analysis and Design

In principle, the design procedure for worm gearing follows the same pattern that has been described for spur, helical, and bevel gears, augmented by the additional consideration of adhesive/abrasive wear[124] that may result from the very significant sliding velocities gen-

[120]Wormset heat generation is not trivial. Consider that a 100-hp-capacity wormset with an efficiency of 75 percent (see Figure 12.8) generates an amount of heat equivalent to 25 hp, or nearly 20 kW.

[121]Refer again to Figure 12.7. [122]As compared to the *worm driving* case of (15-97). [123]See Figure 12.8.

[124]See 2.10.

erated between worm threads and worm gear teeth. *In practice*, because of complexities in profile geometry, the primitive state of design procedure for estimating component life when wear is the governing failure mode, and the sparsely populated wear databases available, the design of worm gearing is *almost wholly empirical.*

If a designer wished to pursue fundamental principles in the rational design of worm gearing, one good approach would be to assess the design based on bending fatigue stresses at the root fillets of the worm and worm gear teeth, surface fatigue durability based on cyclic Hertz contact stresses between worm threads and gear teeth, and adhesive/abrasive wear depths generated by the relatively high sliding velocities (resulting in large distances slid during the design lifespan of the unit) under relatively large contact pressures.[125] Such an assessment would be frustrated, however, because of sparse data for wear constants,[126] poorly defined design criteria for allowable adhesive/abrasive wear depth, and interactions among the potential failure modes (poorly understood and difficult to formulate). For these reasons an empirical approach is usually taken in wormset design.

A common empirical design approach is to assume that tooth bending fatigue is less serious than surface durability when defining an acceptable design.[127] Further, the surface durability criterion typically employed in practice is assumed to embody both surface fatigue and adhesive/abrasive wear parameters.[128]

Since a meaningful calculation has not yet been developed for an appropriate surface *stress* parameter, an alternative *loading severity parameter*[129] for wormset design is the tangential gear force F_{gt}. The *critical capacity* for this application, $(F_{gt})_{allow}$, is empirically calculated from the experience-based expression[130]

$$(F_{gt})_{allow} = d_g^{0.8} b K_s K_m K_v \tag{15-115}$$

where d_g = worm gear pitch diameter, inches
b = face width, inches ($\leq 0.67d_w$)
d_w = worm pitch diameter, inches
K_s = material factor (see expression below)
K_m = ratio correction factor (see expression below)
K_v = velocity factor (see expression below)

The material factor K_s for a *static chill-cast* or *forged* bronze gear, running against a *steel worm with surface hardness of Rockwell C 58* or higher, is given by[131]

$$(K_s)_C = 720 + 10.37C^3 \qquad \text{(for } C < 3.0 \text{ inches)} \tag{15-116}$$

$$(K_s)_{d_g} = 1000 \qquad \text{(for } d_g < 8.0 \text{ inches)} \tag{15-117}$$

$$(K_s)_{d_g} = 1411.6518 - 455.8259\log_{10} d_g \qquad \text{(for } d_g \geq 8.0 \text{ inches)} \tag{15-118}$$

K_s is taken as the largest of (15-116), (15-117), or (15-118).

The ratio correction factor K_m, which depends on gear ratio $m_G = N_g/N_w$, is given by

[125]See, for example, equations (2-119) and (2-121) of Chapter 2 for estimating adhesive and abrasive wear depth.

[126]Such as given in Tables 2.9, 2.10, and 2.11.

[127]Based on experience in the industry, this is usually, but not always, the case. See also (15-92) and related discussion.

[128]However, since cyclic Hertz contact stresses are not explicitly included, the criterion is more reflective of adhesive/abrasive wear than surface fatigue.

[129]See Chapter 6.

[130]Based on experience, this allowable value for tangential force on the worm gear should produce a nominal design life of 25,000 hours of operation. See ref. 29.

[131]These factors are defined by AGMA in ref. 29, which includes extensive tables of values and also (informative) empirical algorithms for them. Only selected excerpts are included in this text.

$$K_m = 0.0200(-m_G^2 + 40m_G - 76)^{0.5} + 0.46 \qquad (\text{for } 3 \leq m_G \leq 20) \qquad (15\text{-}119)$$

$$K_m = 0.0107(-m_G^2 + 56m_G + 5154)^{0.5} \qquad (\text{for } 20 \leq m_G \leq 76) \qquad (15\text{-}120)$$

$$K_m = 1.1483 - 0.00658m_G \qquad (\text{for } m_G > 76) \qquad (15\text{-}121)$$

The velocity factor K_v, which depends on sliding velocity V_s, is given by

$$K_v = 0.659e^{-0.0011V_s} \qquad (\text{for } 0 \leq V_s \leq 700 \text{ ft/min}) \qquad (15\text{-}122)$$

$$K_v = 13.31V_s^{-0.571} \qquad (\text{for } 700 \leq V_s \leq 3000 \text{ ft/min}) \qquad (15\text{-}123)$$

$$K_v = 65.52V_s^{-0.774} \qquad (\text{for } V_s > 3000 \text{ ft/min})^{132} \qquad (15\text{-}124)$$

Using these factors, (15-115) may be evaluated. For a design to be acceptable

$$F_{gt} \leq (F_{gt})_{allow} \qquad (15\text{-}125)$$

Adapting (15-105)

$$F_{gt} = \frac{12(33{,}000)(hp)_{out}}{\pi d_g n_g} = \frac{126{,}050(hp)_{out}}{d_g n_g} \qquad (15\text{-}126)$$

Incorporating (15-115) and (15-126) into (15-125), an acceptable design may be achieved if

$$\frac{126{,}050(hp)_{out}}{d_g n_g} \leq d_g^{0.8} b K_s K_m K_v \qquad (15\text{-}127)$$

One approach that may be used to iterate toward an acceptable design configuration is to isolate face width b from (15-127) to give, for an acceptable design

$$b \geq \frac{126{,}050(hp)_{out}}{d_g^{1.8} n_g K_s K_m K_v} \qquad (15\text{-}128)$$

If the wormset design proposal meets this face-width criterion, and also meets the condition imposed by (15-92), repeated here as

$$b \leq 0.67d_w \qquad (15\text{-}129)$$

an acceptable solution has been reached. If not, additional iterations are required.

15.23 Worm Gears and Worms; Suggested Design Procedure

The design procedure for worm gearsets follows the same basic guidelines that have been suggested for spur, helical, and bevel gears. Many alternative procedures are discussed in the literature, and variations are common depending upon design specifications. For worm gearing, the design steps would usually include the following:

1. If functional specifications and contemplated system configuration include nonintersecting shafts at an angle of 90° to each other, a high reduction ratio, and a compact space envelope, tentatively propose a worm gearset. Then, generate a first conceptual sketch of the gearset, specifying the required reduction ratio, input worm torque or power, output gear torque or power, shaft centerlines, bearing locations, operating speeds, efficiency limits, self-locking requirements, and any other geometrical constraints.
2. Identify potential failure modes (see 15.4).
3. Select tentative materials (see 15.5).

[132]Sliding velocities greater than 6000 ft/min are not recommended.

4. Select tentative manufacturing methods appropriate to the application, giving consideration to existing facilities and inventory of cutters (See Table 15.4.)
5. Select the *number of worm threads* (starts) to be used. Guidelines for making an initial selection are:
   a. Pick a *single start* if the desired reduction ratio is *greater* than 30 : 1.
   b. Pick *multiple starts* if the desired reduction ratio is *less* than 30 : 1.
6. From the specified ratio and N_w, calculate the required *number of worm gear teeth*, N_g.
7. Using (15-90) and known[133] center distance C, calculate a tentative *worm pitch diameter* d_w.
8. Using (15-91), calculate a *worm gear pitch diameter* d_g.
9. Using (15-87) or (15-88), calculate the *axial pitch* p_x or *diametral pitch* P_d. Tooth proportions for worms and worm gears, as a function of P_d or p_x, are shown in Table 15.20.
10. Using (15-93), calculate the *lead* of the worm, then from (15-94) calculate the *lead angle* λ_w.
11. Select[134] an appropriate normal *pressure angle* φ_n for the application, following guidelines given in Tables 15.21 and 15.22.
12. Determine the power or torque required to drive the load, then calculate transmitted load from (15-104) or (15-105).
13. Calculate tangential, axial, and radial force components acting on the worm threads and on the gear teeth, in the zone of contact, using (15-99) through (15-103).
14. Calculate the efficiency of the wormset using (15-110), or read an approximate efficiency from Figure 12.8 as a function of lead angle λ_w (power screw lead angle α) and friction coefficient μ.
15. If specifications require the wormset to be *self-locking*, check to make sure the unit meets the criterion given in (15-113). If specifications require the wormset to be *overhauling*, check to make sure the unit meets the criterion given in (15-114).
16. Using (15-128), and adopting the equal sign, calculate the minimum allowable face width b_{min}.
17. With the result from step 16, check the criterion for maximum allowable face width given by (15-129). It satisfied, the tentative design is acceptable. If not satisfied, additional iterations may be necessary to arrive at an acceptable design configuration.
18. From the efficiency e (see step 14), estimate heat generation and lubricant requirements to maintain acceptable temperature levels and design lives (see 15.10).
19. Laboratory testing of a prototype unit is highly recommended.

Example 15.8 Worm Gearset Design

It is desired to design a simple conveyor-lift such as the one crudely sketched in Figure E15.8A. As shown, the concept is to utilize two parallel roller chains[135] to support containment "buckets" between them, using pinned connections between the buckets and chains. The chains are to be driven by a pair of driving sprockets at the bottom (ground

[133]Center distance may be independently specified if reduction ratio can be *approximate*. If an *exact* reduction ratio is required, and it is desired to utilize an existing hob, it may be necessary to *calculate* the center distance needed to accommodate these specifications.

[134]An initial selection of $\varphi_n = 20°$ is often used.

[135]See Chapter 17.

Example 15.8 Continues

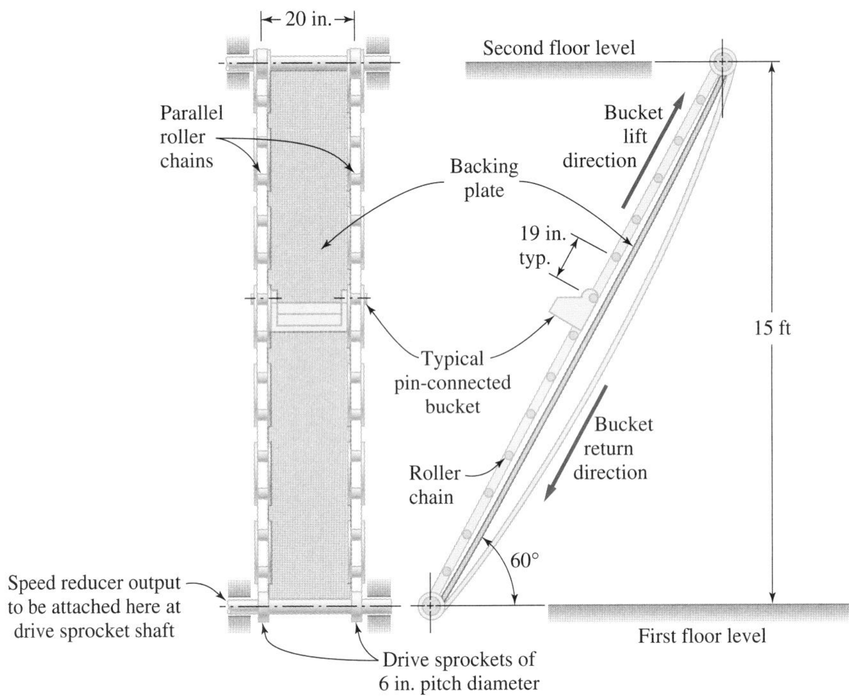

Figure E15.8A
Crude sketch of a proposed conveyor-lifting device.

floor) and supported at the top (second floor) by a pair of idler sprockets. A backing plate is proposed to enhance stability as the buckets slide up the steeply inclined backing plate from the first floor to the second foor, 15 feet above. Each bucket is to carry a payload of about 25 pounds, and a bucket must arrive at the second floor for unloading every 2 seconds. It may be assumed that the tare weights of ascending and descending buckets balance each other. The buckets are to be spaced 19 inches apart along the roller chains. It is initially suggested that a 1725-rpm electric motor be used to drive the lift at the first floor sprocket shaft. For safety reasons it is specified that electrical power failure at the motor must not result in self-driven conveyor reversal. With these specifications in mind, propose, and initially size, an appropriate speed reducer for this application.

Solution

a. From the discussions of 15.2, it would at first seem that either spur, helical, or worm gearing arrangements might be used for this application. To help decide among these possibilities, an estimate of required reduction ratio may be made. The reducer input speed from the electric motor is specified to be 1725 rpm. The reducer output speed (applied to drive sprocket shaft) must deliver a conveyor bucket to the second floor every 2 seconds (30 per minute). Hence, the drive sprocket speed in rpm may be estimated as follows. Tentatively selecting a 6-inch-diameter drive sprocket, the distance that the chain advances during each revolution of the drive sprocket is

$$L_{adv} = \frac{\pi(6)}{12} = 1.57\,\frac{\text{ft}}{\text{rev}} \tag{1}$$

The approximate length of chain supporting *loaded* buckets between the first and second floor is

$$L_{ch} = \frac{15}{\sin 60} = 17.32\text{ ft} \tag{2}$$

For the 19-inch bucket spacing, the speed of the sprocket drive shaft required to deliver 30 buckets per minute is

$$n_{spr} \approx \left(\frac{30 \text{ buckets}}{\text{min}}\right)\left(\frac{(19/12) \text{ ft}}{\text{bucket}}\right)\left(\frac{1 \text{ rev}}{1.57 \text{ ft}}\right) = 30.3 \frac{\text{rev}}{\text{min}} \tag{3}$$

Thus the reduction ratio must be, approximately,

$$m_G = \frac{1725}{30.3} = 56.9 \tag{4}$$

This relatively high reduction ratio suggests that worm gearing should be a primary candidate. For example, if spur or helical gears were used, and a single reduction were considered, a 2-inch drive pinion would require a 9-ft-diameter gear to give the ratio: a clearly impractical arrangement. If multiple reductions in series were considered, and each mesh reduction were to lie in the range of 3 to 4, a triple reduction unit would be required. While such an arrangement should probably be investigated, it is more complicated and may be more costly than a worm gearset. Worm gearing will be investigated here as a *first* step.

b. Following the design procedure given in 15.23, a conceptual sketch of the proposed wormdrive reduction unit is shown in Figure E15.8B.

c. Reviewing 15.4, the primary failure mode candidates appear to be tooth bending fatigue, surface fatigue pitting, and adhesive/abrasive wear.

d. Materials for worm gear applications are, commonly, a steel worm with surface hardness of Rockwell C 58 or higher in mesh with a chill-cast or forged bronze gear. These material choices will be adopted for this preliminary investigation.

e. No specific information is available about manufacturing facilities or cutter inventory, so it will be assumed that a hobbing method will be used for this wormset, and an accuracy level of AGMA 10 or 11 will be chosen, in accordance with the guidance of Table 15.4.

f. Since the required reduction ratio exceeds 30 : 1, a single start worm thread will be chosen, tentatively.

g. Using (15-89) and (4), the required number of worm gear teeth may be calculated as

$$N_g = (56.9)(1) = 57 \text{ teeth} \tag{5}$$

since N_g must have an integral value. For $N_w = 1$ and $N_g = 57$, the resulting ratio m_G becomes, again using (15-89),

$$m_G = 57 \tag{6}$$

h. From Figure E15.8B, the tentative center distance desired is 4.0 inches. Hence, from (15-90), the tentative worm pitch diameter is estimated as

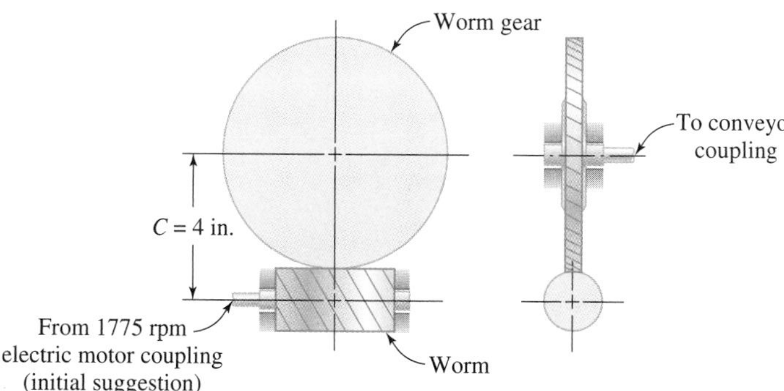

Figure E15.8B
First conceptual sketch of worm gearset layout.

Example 15.8 Continues

$$d_w = \frac{C^{0.875}}{2.3} = \frac{(4)^{0.875}}{2.3} = 1.5 \text{ inches} \tag{7}$$

which lies *midrange* of the (15-90) requirement.

i. Using (15-91), the gear pitch diameter may be calculated as

$$d_g = 2(4.0) - 1.5 = 6.5 \text{ inches} \tag{8}$$

j. From (15-88)

$$P_d = \frac{57}{6.5} = 8.77 \tag{9}$$

and from (15-87)

$$p_x = \frac{\pi}{P_d} = \frac{\pi}{8.77} = 0.36 \text{ inch} \tag{10}$$

k. From (15-93) the lead is

$$L_w = (1)(0.36) = 0.36 \text{ inch} \tag{11}$$

and from (15-94) the lead angle is

$$\lambda_w = \tan^{-1}\frac{0.36}{\pi(1.5)} = 4.6^\circ \tag{12}$$

l. A normal pressure angle of $\varphi_n = 20^\circ$ will be tentatively chosen (common initial choice). This is consistent with Tables 15.21 and 15.22.

m. An estimate of power required to drive the conveyor may be made by calculating power required to lift the load (change in potential energy per unit time), then adding the power required to overcome frictional sliding of the loaded buckets up the backing plate. This sum will correspond to minimum power output required from the worm gear drive shaft.

The power required to lift the load may be estimated as

$$(hp)_L = \left[\left(\frac{17.32 \text{ ft}}{1.58 \text{ ft/bucket}}\right)\left(\frac{25 \text{ lb}}{\text{bucket}}\right)\left(\frac{1.57 \sin 60 \text{ ft vertically}}{\text{rev}}\right) \times \left(\frac{30.3 \text{ rev}}{\text{min}}\right)\left(\frac{1 \text{ hp}}{33{,}000 \text{ ft-lb/min}}\right)\right] = 0.34 \text{ horsepower} \tag{13}$$

The power required to overcome frictional resistance of the buckets sliding up the inclined backing plate may be estimated by calculating the tensile force in the returning chain required to overcome frictional sliding (assuming no lubrication between buckets and backing plate).[136] Thus

$$F_{ch} = \mu F_n = 0.35\left(\frac{17.32 \text{ ft}}{1.58 \text{ ft/bucket}}\right)\left(\frac{25 \text{ lb}}{\text{bucket}}\right)\sin 30 = 48.3 \text{ lb} \tag{14}$$

The frictional drag induces a tangential force on the drive sprocket at a radius of 3.25 inches, producing a friction torque of

$$T_{fr} = (3.25)(48.3) = 157 \text{ in-lb} \tag{15}$$

and using (4-30)

$$(hp)_{fr} = \frac{(157)(30.1)}{63{,}025} = 0.07 \text{ horsepower} \tag{16}$$

[136]See Appendix Table A.1 for unlubricated mild steel sliding on mild steel (typical value, $\mu = 0.35$).

Thus, the minimum total power required by the conveyor (provided by the worm gear output shaft) is

$$(hp)_{out} = (hp)_{req'd} = (hp)_L + (hp)_{fr} = 0.34 + 0.07 = 0.41 \text{ horsepower} \tag{17}$$

n. Using (15-105)

$$F_{gt} = \frac{33{,}000(0.41)}{\pi\left(\frac{6}{12}\right)(30.1)} = 286 \text{ lb} \tag{18}$$

o. Using (15-101)

$$F_{wa} = F_{gt} = 286 \text{ lb} \tag{19}$$

then from (15-99)[137]

$$F_{wt} = \frac{286}{\left(\frac{\cos 20 \cos 4.6 - 0.09 \sin 4.6}{\cos 20 \sin 4.6 + 0.09 \cos 4.6}\right)} = 51 \text{ lb} \tag{20}$$

and from (15-102)

$$F_{ga} = F_{wt} = 51 \text{ lb} \tag{21}$$

Also, from (15-100) and (15-103),

$$F_{wr} = F_{gr} = 51\left(\frac{\sin 20}{\cos 20 \sin 4.6 + 0.09 \cos 4.6}\right) = 106 \text{ lb} \tag{22}$$

p. The efficiency of the wormset may be calculated from (15-110) as

$$e = \frac{\cos 20 \cos 4.6 - 0.09 \tan 4.6}{\cos 20 + 0.09 \cot 4.6} = 0.45 = 45\% \tag{23}$$

This value may be confirmed by consulting Figure 12.8.

q. To check the self-locking requirement using (15-113), to be self-locking

$$\mu \geq \cos 20 \tan 4.6 = 0.08 \tag{24}$$

Since

$$\mu = 0.09 > 0.08 \tag{25}$$

the wormset *is* self-locking by a narrow margin. An auxiliary braking device should be recommended to assure safety.

r. Using (15-128) with the "equal" sign

$$b_{min} = \frac{126{,}050(0.41)}{(6.5)^{1.8}(30.1)K_s K_m K_v} \tag{26}$$

The *material* factor K_s may be found from (15-117) as

$$K_s = 1000 \tag{27}$$

The *ratio correction* factor may be found from (15-120) as

$$K_m = 0.0107[-(57.3)^2 + 56(57.3) + 5145]^{0.5} = 0.76 \tag{28}$$

Finally, the *velocity* factor may be found from (15-122), using (15-107) first, to find sliding velocity as

[137]See typical value for dry steel on bronze of $\mu = 0.09$ given in Appendix A.1 (may be conservative).

Example 15.8 Continues

$$V_s = \frac{(30.1)\pi(6.5/12)}{\sin 4.5} = 639 \text{ ft/min} \tag{29}$$

Thus

$$K_v = 0.659e^{-0.0011(639)} = 0.33 \tag{30}$$

From (26) then

$$b_{min} = \frac{126{,}050(0.41)}{(6.5)^{1.8}(30.1)(1000)(0.76)(0.33)} = 0.24 \text{ inch} \tag{31}$$

s. Using (15-129)

$$b_{max} = 0.67(1.5) = 1.0 \text{ inch} \tag{32}$$

Based on the bracketing results of (31) and (32), a tentative face width will be specified as

$$b = 0.25 \text{ inch} \tag{33}$$

t. From (17) and (23) the minimum input horsepower from the electric motor may be estimated as

$$(hp)_{in}(e) = (hp)_{out} \tag{34}$$

or

$$(hp)_{in} = \frac{0.41}{0.45} = 0.92 \text{ horsepower} \tag{35}$$

Thus a one-horsepower 1725-rpm electric motor is required, at a minimum. Starting torque capacity should also be investigated since static coefficients of friction are higher than sliding coefficients of friction.

u. The approximate heat generation in the wormset is, from (35),

$$H_g = (hp)_{in}(1 - e) = 0.91(1 - 0.45) = 0.50 \text{ horsepower} \tag{36}$$

or[138]

$$H_g = (0.50)(42.41) = 21.2 \text{ Btu/min} \tag{37}$$

and utilizing (10-19)

$$H_d = H_g = k_1 A_h(\Theta_s - \Theta_a) \tag{38}$$

where,[139] for dissipation to still air,

$$(k_1)_{std} = 2.31 \times 10^{-4} \frac{\text{Btu}}{\text{min} - \text{in}^2 - {}^\circ\text{F}} \tag{39}$$

From (38) and (39) therefore, if a maximum steady-state temperature difference (between the housing and ambient air) of 100°F is deemed acceptable,

$$21.2 = 2.31 \times 10^{-4} A_h(100) \tag{40}$$

or

$$(A_h)_{req'd} \approx 918 \text{ in}^2 \tag{41}$$

Thus the surface area of the housing should be at least 918 square inches. If a cube-shaped housing were used, with each side about 10 inches, the housing surface area would meet this criterion. It appears feasible therefore to design the unit using an oil sump and housing (without any special external cooling provisions). Laboratory testing of a prototype unit is recommended.

[138] 1 horsepower $= 0.7068 \frac{\text{Btu}}{\text{sec}} = 42.41 \frac{\text{Btu}}{\text{min}}$

[139] See (10-22).

The preliminary design proposal may be summarized in the following specifications:

1. Use a wormset with a reduction ratio of 57 : 1.
2. Use a single start steel worm of pitch diameter d_w = 1.50 inches, surface hardened to Rockwell C 58 or harder, with a lead angle of 4.6 degrees.
3. Use a chill-cast bronze gear of pitch diameter d_g = 6.50 inches and face width of b = 0.25 inch with 57 teeth (p_x = 0.36 inch) and normal pressure angle $\varphi_n = 20°$.
4. Use a 4.0-inch center distance.
5. Use a splash lubrication system in a steel housing with a minimum of about 918 in^2 of surface area. No special external cooling measures are required.
6. Drive the unit with a 1-hp, 1725-rpm electric motor. (A starting torque investigation should be conducted before purchasing the motor.)
7. Plan to build and test a prototype unit.

Problems

15-1. For each of the design scenarios presented, suggest one or two types of gears that might make good candidates for further investigation in terms of satisfying the primary design requirements.

a. In the design of a new-concept agricultural hay-conditioner, it is necessary to transmit power from one rotating parallel shaft to another. The input shaft is to rotate at a speed of 1200 rpm and the desired output shaft speed is 350 rpm. Low cost is an important factor. What types(s) of gearing would you recommend? State your reasons.

b. In the design of a special speed reducer for a laboratory test stand, it is necessary to transmit power from one rotating shaft to another. The centerlines of the two shafts intersect. The driver shaft speed is 3600 rpm and the desired speed of the output shaft is 1200 rpm. Quiet operation is an important factor. What types(s) of gearing would you recommend? State your reasons.

c. It is desired to use a 1-hp, 1725-rpm electric motor to drive a conveyer input shaft at a speed of approximately 30 rpm. To give a compact geometry, the motor drive shaft is to be oriented at 90 degrees to the conveyor input shaft. The shaft centerline may either intersect or not, depending on designer judgment. What type(s) of gearing would you recommend? State your reasons.

15-2. The compound helical gear train sketched in Figure P15.2 involves three simple helical gears (1,2,5) and one compound helical gear (3,4). The number of teeth on each gear is indicated in the sketch. If the input gear (1) is driven clockwise at a speed of n_1 = 1725 rpm, calculate the speed and direction of the output gear (5).

15-3. The sketch of Figure P15.3 shows a two-stage reverted gear reducer that utilizes two identical pairs of gears to enable making the input shaft and output shaft colinear. If a 1-kw,

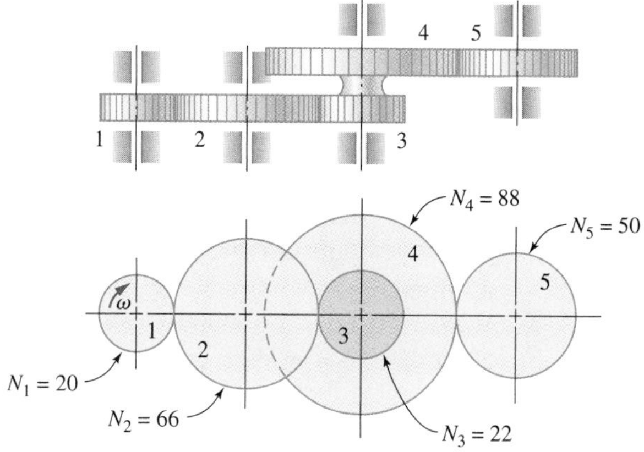

Figure P15.2
Helical gear train.

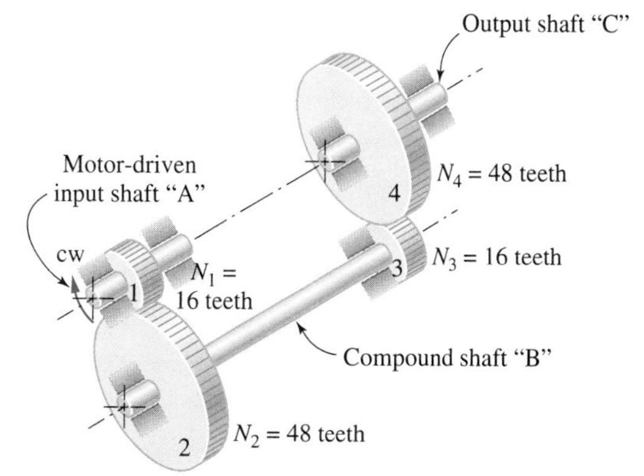

Figure P15.3
Two-stage reverted gear reducer.

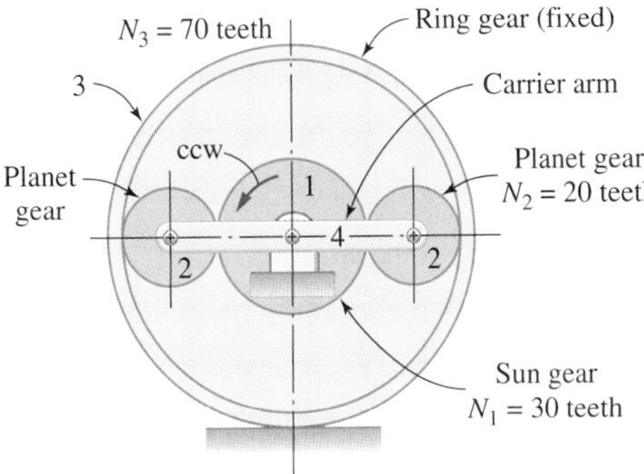

Figure P15.4
Two-planet epicyclic gear train.

1725-rpm motor operating at full rated power is used to drive the input shaft in the CW direction, do the following:

a. Determine the speed and direction of the compound shaft.

b. Determine the speed and direction of the output shaft.

c. Assuming a 98 percent efficiency of each gear mesh, calculate the torque available for driving the load at the output shaft.

15-4. A two-planet epicyclic gear train is sketched in Figure P15.4. If the ring gear is fixed, the sun gear is driven at 1200 rpm in the CCW direction, and the carrier arm is used as output, what would be the speed and direction of rotation of the carrier arm?

15-5. A special reverted planetary gear train is sketched in Figure P15.5. The planet gears (2-3) are connected together (compound) and are free to rotate together on the carrier shaft. In turn, the carrier shaft is supported by a symmetrical one-piece pair of carrier arms attached to an output shaft (5) that is colinear with the input shaft (1). Gear 4 is fixed. If the input shaft (1) is driven at 250 rpm in the CW direction, what would be the output shaft (5) speed and direction?

15-6. What are the kinematic requirements that must be met to satisfy the "fundamental law of gearing"?

15-7. Define the following terms, using a proper sketch where appropriate.

a. Line of action

b. Pressure angle

c. Addenum

d. Dedendum

e. Pitch diameter

f. Diametral pitch

g. Circular pitch

h. Pitch point

15-8. Describe what is meant by a "gear tooth system."

15-9. A straight-tooth spur gearset is being considered for a simple-speed reduction device at an early stage in the design process. It is being proposed to use standard 20° involute full-depth gear teeth with a diametral pitch of 4 and a 16-tooth pinion. A reduction ratio of 2.50 is needed for the application. Find the following:

a. Number of teeth on the driven gear

b. Circular pitch

c. Center distance

d. Radii of the base circles

e. Would you expect "interference" to be a problem for this gear mesh?

15-10. Repeat problem 15-9, except use a diametral pitch of 8 and 14-tooth pinion.

15-11. Repeat problem 15-9, except use a reduction ratio of 3.50.

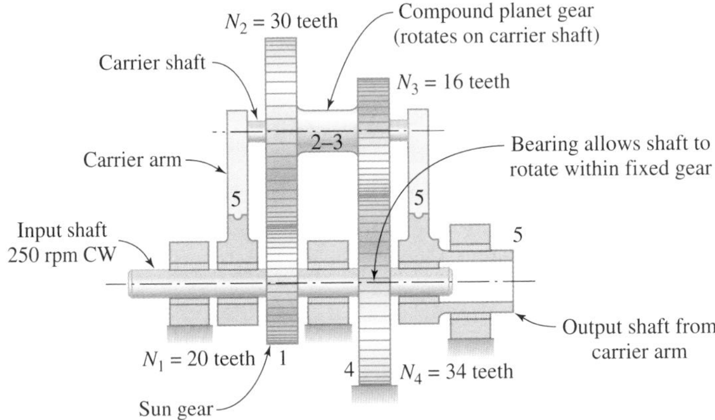

Figure P15.5
Reverted compound planetary gear train.

15-12. Repeat problem 15-9, except use a diametral pitch of 12 and a reduction ratio of 7.50.

15-13. Repeat problem 15-9, except use a diametral pitch of 12, a 17-tooth pinion, and a reduction ratio of 7.50.

15-14. A straight-tooth spur gearset has a 19-tooth pinion that rotates at a speed of 1725 rpm. The driven gear is to rotate at a speed of approximately 500 rpm. If the gear teeth have a module of 2.5, find the following:

a. Number of teeth on the driven gear

b. Circular pitch

c. Center distance

15-15. Repeat problem 15-14, except for a pinion that rotates at 3450 rpm.

15-16. Repeat problem 15-14, except for a module of 5.0.

15-17. Repeat problem 15-14, except for a driven gear that rotates at approximately 800 rpm.

15-18. A pair of 8-pitch straight-tooth spur gears is being proposed to provide a 3:1 speed *increase*. If the gears are mounted on 6-inch centers, find the following:

a. Pitch diameter of each gear.

b. Number of teeth on each gear.

c. If power supplied to the driving pinion at full load is 10 hp, and power loss at the gear mesh is negligible, what is the power available at the output gear shaft?

15-19. Repeat problem 15-18, except for a 3:1 speed *decrease*.

15-20. A proposed straight full-depth spur gear mesh is to consist of a 21-tooth pinion driving a 28-tooth gear. The proposed diametral pitch is to be 3, and the pressure angle is 20°. Determine the following, and where possible, show each feature on a simple scale drawing of the gear mesh.

a. Pitch circle for the pinion

b. Pitch circle for the gear

c. Pressure angle

d. Base circle for the pinion

e. Base circle for the gear

f. Addendum circle for the pinion

g. Dedendum circle for the pinion

h. Addendum circle for the gear

i. Dedendum circle for the gear

j. Circular pitch

k. Tooth thickness

l. One typical pinion tooth

m. One typical gear tooth

n. Length of action

o. Base pitch

p. Profile contact ratio

15-21. A proposed straight full-depth spur gear mesh is to have a reduction ratio of 4:1 and a center distance of 7.50 inches. The proposed diametral pitch is to be 3, and the pressure angle is 20°. Determine the following, and, where possible, show each feature on a simple scale drawing of the gear mesh.

a. Pitch circle for the pinion

b. Pitch circle for the gear

c. Pressure angle

d. Base circle for the pinion

e. Base circle for the gear

f. Addendum circle for the pinion

g. Dedendum circle for the pinion

h. Addendum circle for the gear

i. Dedendum circle for the gear

j. Circular pitch

k. Tooth thickness

l. Interference point locations

m. Whether interference will exist

15-22. Preliminary design calculations have suggested that design objectives may be met by a straight spur gearset using standard full-depth $2\frac{1}{2}$-pitch involute gear teeth, and a 21-tooth pinion meshing with a 28-tooth gear. A 25° pressure angle has been selected for this application, and the gear teeth are to be shaved to AGMA quality number $Q_v = 8$. Find the following:

a. Addendum

b. Dedendum

c. Clearance

d. Circular pitch

e. Circular tooth thickness

f. Base pitch

g. Length of action

h. Profile contact ratio

i. Module

15-23. Repeat problem 15-22, except for a 20° pressure angle.

15-24. A straight-tooth one-stage spur gear reducer has been used in a high-production home appliance for many years. The gear pair constitutes about one-half the \$50 production cost of the appliance. Consumer complaints about gear noise have grown over the years and sales are declining. One young engineer has found data that suggest the noise level would be significantly reduced if the AGMA quality number could be increased from its current value of $Q_v = 8$ for hobbed gears to a value of $Q_v = 11$, achieved by shaving the gears.

a. Estimate the increase in production cost of the appliance if gear shaving were used to achieve $Q_v = 11$.

b. Can you suggest any other approach that might accomplish the noise-reduction goal without resorting to shaving the gears?

15-25. A straight-toothed full-depth involute spur pinion with a pitch diameter of 100 mm is mounted on an input shaft driven by an electric motor at 1725 rpm. The motor supplies a steady torque of 225 N-m.

a. If the involute gear teeth have a pressure angle of 20°, determine the transmitted force, the radial separating force, and the normal resultant force on the pinion teeth at the pitch point.

b. Calculate the power being supplied by the electric motor.

c. Calculate the percent difference in resultant force if the pressure angle were 25° instead of 20°.

d. Calculate the percent difference in resultant force if the pressure angle were $14\frac{1}{2}°$ instead of 20°.

15-26. Referring to the two-stage gear reducer sketched in Figure P15.3, concentrate attention on the first-stage mesh between the pinion (1) and the gear (2). The pinion is being driven by a 1-kw, 1725-rpm electric motor, operating steadily at full capacity. The tooth system has a diametrical pitch of 8 and a pressure angle of 25°. Do the following for the first-stage gear mesh:

a. Sketch the gearset comprised of pinion 1 (driver) and gear 2 (driven) taken together as a free body, and assume that shaft support bearings are symmetrically straddle mounted about the gear on each shaft. Show all external forces and torques on the free body, speeds and directions of the two gears (refer to Figure P15.3), and the line of action.

b. Sketch the pinion, taken alone as a free body. Show all external forces and torques on the pinion, including the driving torque, tangential force, separating force, and bearing reaction forces. Give numerical values.

c. Sketch the driven gear, taken alone as a free body. Show all external forces and torques on the gear, including resisting torque, tangential force, separating force, and bearing reaction forces. Give numerical values.

15-27. In the two-stage gear reducer sketched in Figure P15.27, concentrate attention on the compound shaft "B," with attached gears 2 and 3, taken together as the free body of interest. The gears have standard 20° involute full-depth teeth, with a diametral pitch of 6. The motor driving the input shaft "A" is a 20-hp, 1725-rpm electric motor operating steadily at full rated power. For the chosen free body, do the following:

a. Clearly sketch a top view of shaft "B," and show all horizontal components of the loads and reactions.

b. Sketch a front view (elevation) of shaft "B," and show all vertical components of the loads and reactions.

c. If shaft "B" is to have a uniform diameter over its whole length, identify potential critical points that should be investigated when designing the shaft.

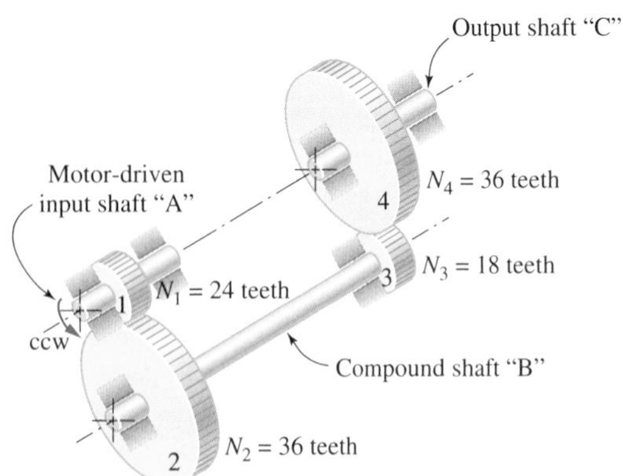

Figure P15.27
Two-stage gear reducer.

15-28. Referring again to Figure P15.4, note that the ring gear is fixed, the sun gear is driven at 1200 rpm in the CCW direction with a torque of 20 N-m, and the two-planet carrier arm is used as output. The 20° involute gears have a module of 2.5. Do the following:

a. Determine the circular pitch.

b. Determine the pitch diameter of each gear in the train, and verify that they are physically compatible in the assembly.

c. Find the center distance between planets on the 2-planet carrier arm.

d. Sketch each member of the train as a free body, showing numerical values and directions of all forces and torques on each free body.

e. Calculate the output torque.

f. Calculate the output shaft speed and determine its direction.

g. Calculate the nominal radial load on each of the bearings in the assembly, neglecting gravitational forces.

15-29. A 10-pitch 20° full-depth involute gearset with a face width of 1.25 inches is being proposed to provide a 2:1 speed reduction for a conveyor drive unit. The 18-tooth pinion is to be driven by a 15-hp, 1725-rpm electric motor operating steadily at full rated power. A very long life is desired for this gearset, and a reliability of 99 percent is required. Do the following:

a. Using the *simplified approach*, estimate the *nominal bending stress* at the tension-side root fillet of the *driving pinion*.

b. Estimate the *fatigue stress concentration factor* for the tension-side root fillet of the *driving pinion*.

c. Calculate the *actual bending stress* at the tension-side root fillet of the *driving pinion*.

d. Repeat (c) for the tension-side root fillet of the *driven gear*.

e. Based on the recommendation of an in-house materials specialist, Grade 1 AISI 4620 hot-rolled steel is to be used for both the pinion and the gear (see Tables 3.3 and 3.13), and the value of k_∞ [see (2-27)] has been estimated for this application to be 0.75, including the 99 percent reliability requirement but not including stress concentration effects. Estimate the existing safety factor at the tension-side root fillet of whichever of the gears is more critical, based on tooth bending fatigue.

15-30. For the gearset specifications of problem 15-29, do the following:

a. Using the *simplified approach*, estimate the *surface fatigue wear stress* for the meshing gear teeth.

b. If the Grade 1 4620 gear teeth are carburized and case hardened (not including the root fillet) to a hardness of approximately R_C 60, maintaining the 99 percent reliability requirement, and recalling that very long life is desired, determine the *surface fatigue strength* of the case-hardened teeth.

c. Estimate the *existing safety factor* based on surface fatigue wear failure.

15-31. Using the simplified approach (do *not* refine results by using AGMA equations), design a single-reduction straight spur gear unit to operate from a 5.0-hp electric motor running at 900 rpm to drive a rotating machine operating at 80 rpm. The motor is to operate steadily at full rated power. Near-infinite life is desired. A reliability of 90 percent is acceptable for this application. It is proposed to use ASTM A-48 (class 40) gray cast-iron material for both gears. Using $k_\infty = 0.70$, properties for this material may be based on Chapter 3 data, except for surface fatigue strength, which may be taken as $(S_{sf})_{N=10^8} = 28{,}000$ psi. A safety factor of 1.3 is desired. As part of the design procedure, select or determine the following so as to satisfy design specifications:

a. Tooth system

b. Quality level

c. Diametral pitch

d. Pitch diameters

e. Center distance

f. Face width

g. Number of teeth on each gear

Do not attempt to evaluate heat generation.

15-32. A 10-pitch 20° full-depth involute gearset, with $Q_v = 10$ and face width of 1.25 inches, is being proposed to provide a 2:1 speed reduction for a conveyor drive unit. The 18-tooth pinion is to be driven by a 15-hp, 1725-rpm electric motor operating steadily at full-rated power. A very long life is desired for this gearset, and a reliability of 99 percent is required. Do the following:

a. Using the *AGMA refined approach*, calculate the *tooth bending stress* at the tension-side root fillet of the *driving pinion*.

b. Repeat (a) for the tension-side root fillet of the *driven gear*.

c. If the proposed material for both gears is AISI 4620 through-hardened to BHN 207, estimate the existing safety factor at the tension-side root fillet of whichever gear is more critical, based on tooth bending fatigue.

15-33. For the gearset specifications of problem 15-32, do the following:

a. Using the *AGMA refined approach*, calculate the *surface fatigue contact stress* for the meshing gear teeth.

b. If the proposed material for both gears is AISI 4620 steel, and the teeth are carburized and case hardened (not including the root fillet) to a hardness of approximately R_C 60, maintaining the 99 percent reliability requirement, determine the *AGMA surface fatigue strength* (pitting resistance) for the carburized and case-hardened gear teeth.

c. Estimate the existing safety factor based on surface fatigue (pitting).

15-34. Using the *AGMA refined approach*, design a high-precision ($Q_v = 12$) single-reduction straight spur gear unit to operate from a 50-hp electric motor running at 5100 rpm to drive a rotating machine operating at 1700 rpm. The motor operates steadily at full rated power. A life of 10^7 pinion revolutions is desired, and a reliability of 99 percent is required. It is proposed to use Grade 2 AISI 4620 steel carburized and case hardened to R_C 60 for both gears. An important design constraint is to make the unit as *compact* as practical (i.e., use the minimum possible number of pinion teeth without undercutting). A safety factor of 1.3 is desired. Select or determine the following so as to satisfy the design specifications:

a. Tooth system

b. Diametral pitch

c. Pitch diameters

d. Center distance

e. Face width

f. Number of teeth on each gear

15-35. A right-hand helical gear, found in storage, has been determined to have a transverse circular pitch of 26.594 mm and a 30° helix angle. For this gear, calculate the following:

a. Axial pitch

b. Normal pitch

c. Module in the transverse plane

d. Module in the normal plane

15-36. The preliminary design proposal for a helical gearset running on parallel shafts proposes a left-hand 18-tooth pinion meshing with a 32-tooth gear. The normal pressure angle is 20°, the helix angle is 25°, and the normal diametral pitch is 10. Find the following:

a. Normal circular pitch

b. Transverse circular pitch

c. Axial pitch

d. Transverse diametral pitch

e. Transverse pressure angle

f. Pitch diameters of pinion and gear

g. Whole depth for pinion and gear

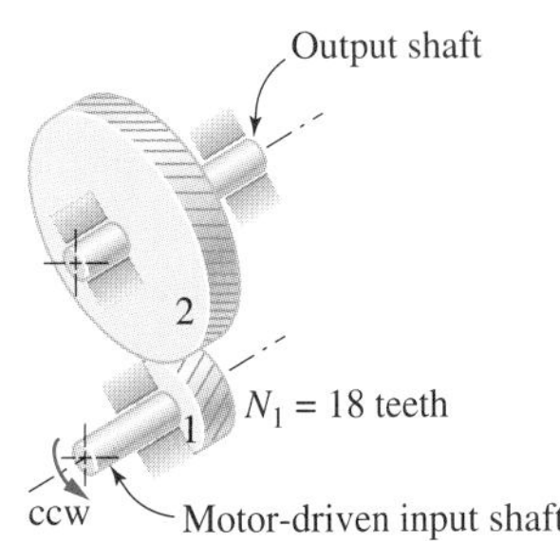

Figure P15.39
One-stage helical gear reducer.

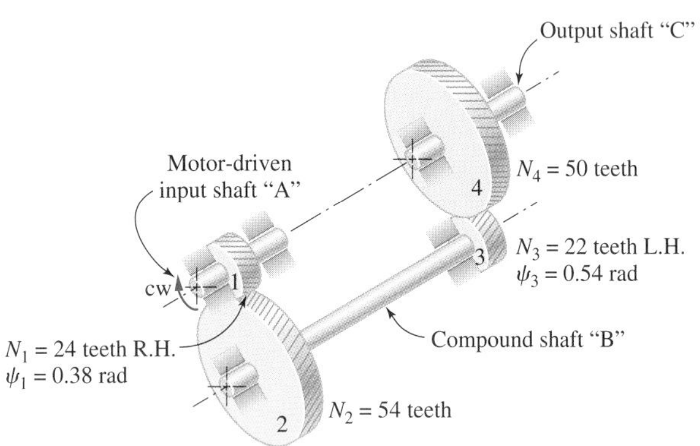

Figure P15.42
Two-stage reverted helical gear reducer.

15-37. Repeat problem 15-36, except that the 18-tooth helical pinion is right-hand.

15-38. Repeat problem 15-36, except that the normal diametral pitch is 16.

15-39. The sketch of Figure P15.39 shows a one-stage gear reducer that utilizes helical gears with a normal diametral pitch of 14, normal pressure angle of 20°, and helix angle of 30°. The helix of the 18-tooth drive pinion 1 is left-hand. The input shaft is to be driven in the direction shown (CCW) by a $^1/_2$-hp, 1725-rpm electric motor operating steadily at full rated power, and the desired output shaft speed is 575 rpm. Determine the following:

a. Transverse pressure angle

b. Transverse diametral pitch

c. Pitch diameter of pinion (1)

d. Pitch diameter of gear (2)

e. Number of teeth on gear (2)

f. Center distance

g. Pitch-line velocity

h. Numerical values and directions of tangential, radial, and axial force components *on the pinion* while operating at full rated motor horsepower

i. Minimum recommended face width

15-40. Repeat problem 15-39, except that the 18-tooth helical pinion is right-hand.

15-41. A parallel-shaft helical gearset is driven by an input shaft rotating at 1725 rpm. The 20° helical pinion is 250 mm in diameter and has a helix angle of 30°. The drive motor supplies a steady torque of 340 N-m.

a. Calculate the transmitted force, radial separating force, axial thrust force, and normal resultant force on the pinion teeth at the pitch point.

b. Calculate the power being supplied by the electric motor.

15-42. The sketch of Figure P15.42 shows a proposed two-stage reverted gear reducer that utilizes helical gears to provide quiet operation. The gears being suggested have a module of 4 mm in the normal plane, and a normal pressure angle of 0.35 rad. The input shaft is driven in the direction shown by a 22-kw, 600-rpm electric motor. Do the following:

a. Determine the speed and direction of the compound shaft.

b. Determine the speed and direction of the output shaft.

c. Sketch a free-body diagram of the 54-tooth gear (2), showing numerical values and directions of all force components applied to the gear (2) by the 24-tooth pinion (1).

d. Sketch a free-body diagram of the 22-tooth pinion (3), showing numerical values and directions of all force components applied to the pinion (3) by the 50-tooth gear (4).

15-43. An existing parallel-shaft single-reduction spur gear speed reducer is made up of a 21-tooth 8-pitch input pinion driving a 73-tooth gear mounted on the output shaft. The center distance between pinion and gear is 5.875 inches. The input shaft is driven by a 15-hp, 1725-rpm electric motor operating steadily at full rated capacity. To reduce vibration and noise, it is desired to substitute a helical gearset that can operate on the same center distance and provide approximately the same angular velocity ratio as the existing spur gearset. Study this request and propose a helical gearset that can perform the function satisfactorily at a reliability level of 99 percent for a very long lifetime. Assume that the helical gears will be cut by an 8-pitch 20° full-depth involute hob. The probable material is through-hardened Grade 1 steel with a hardness of BHN 350. Determine the following:

a. Using the spur gear data as a starting point, make a preliminary design proposal for a pair of helical gears with the same center distance and approximately the same angular velocity ratio as the existing spur gearset. Specifically, determine a combination of transverse diametral pitch, number of teeth on the pinion, and number of teeth on the mating gear that will satisfy specifications on center distance and angular velocity ratio.

b. Determine the helix angle. Does it lie within the recommended range of values?

c. Determine the pitch diameter for the pinion and the gear.

d. Determine the nominal outside diameter of the pinion and the gear.

e. Estimate an appropriate face width for the helical gear pair.

f. Calculate the existing safety factor for the tentatively selected helical gear pair, based on *tooth bending fatigue* as a potential failure mode.

g. Calculate the existing safety factor for the tentatively selected helical gear pair, based on *surface fatigue pitting wear* as a potential failure mode.

h. Comment on the governing safety factor.

15-44. Repeat problem 15-43, except assume that the helical gears will be cut using a 12-pitch 20° full-depth involute hob.

15-45. A newly proposed numerically controlled milling machine is to operate from a helical-gear speed reducer designed to provide 65 horsepower at an output shaft speed of 1150 rpm. It has been suggested by engineering management that a 3450-rpm electric motor be used to drive the speed reducer. An in-house gearing consultant has suggested that a normal diametral pitch of 12, a normal pressure angle of 20°, a helix angle of 15°, an AGMA quality number of 10, and a safety factor of 1.7 would be an appropriate starting point for the design. Design the gears.

15-46. Repeat problem 15-45, except that the suggested helix angle is 30°.

15-47. A pair of straight bevel gears, similar to those shown in Figure 15.41, has been incorporated into a right-angle speed reducer (shaft centerlines intersect at 90°). The straight bevel gears have a diametral pitch of 8 and a 20° pressure angle. The gear reduction ratio is 3:1, and the number of teeth on the bevel pinion is 16. Determine the following:

a. Pitch cone angle for the pinion

b. Pitch cone angle for the gear

c. Pitch diameter for the pinion

d. Pitch diameter for the gear

e. Maximum recommended face width

f. Average pitch cone radius for the pinion, assuming face width is maximum recommended value

g. Average pitch cone radius for the gear, assuming face width is maximum recommended value

h. Pinion addendum

i. Gear addendum

j. Pinion dedendum

k. Gear dedendum

15-48. Repeat problem 15-47, except use a diametral pitch of 12 and a reduction ratio of 4:1.

15-49. It is being proposed to use a Coniflex® straight bevel gearset to provide a 3:1 speed reduction between a 15-tooth pinion rotating at 300 rpm and a meshing gear mounted on a shaft whose centerline intersects the pinion shaft centerline at a 90° angle. The pinion shaft is driven steadily by a 3-hp source operating at full rated power. The bevel gears are to have a diametral pitch of 6, a 20° pressure angle, and a face width of 1.15 inches. Do the following:

a. Calculate the number of teeth in the driven gear.

b. Calculate the input torque on the pinion shaft.

c. Calculate average pitch-line velocity.

d. Calculate the transmitted (tangential) force.

e. Calculate the radial and axial forces on the pinion.

f. Calculate the radial and axial forces on the gear.

g. Determine whether the force magnitudes calculated for the pinion and the gear are consistent with the free-body equilibrium of the bevel gearset (see Figure 15.41 for geometric arrangement).

15-50. Repeat problem 15-49, except for a diametral pitch of 10.

15-51. Repeat problem 15-49, except for a diametral pitch of 16.

15-52. For the Coniflex® bevel gearset described in problem 15-49, the following information has been tabulated or calculated:

$$T_p = 630 \text{ in-lb}$$
$$P_d = 6$$
$$d_p = 2.50 \text{ inches}$$
$$b = 1.15 \text{ inches}$$
$$N_p = 15 \text{ teeth}$$
$$N_g = 45 \text{ teeth}$$

Further, it is proposed to use Grade 2 AISI 4140 steel nitrided and through-hardened to BHN 305 for both the pinion and the gear. Other known design information includes the following items:

1. Input power is supplied by an electric motor.
2. AGMA quality $Q_v = 8$ is desired.
3. The gear is straddle mounted with a closely positioned bearing on each side, but the pinion overhangs its support bearing.
4. A design life of 10^9 cycles has been specified.

A reliability of 99 percent is required, and a design safety factor of at least 1.3 is desired. Do the following:

a. Calculate the tooth bending fatigue *stress* for the more critical of the pinion or the gear.

b. Determine the tooth bending fatigue *strength* for the proposed AISI 4140 steel material corresponding to a life of 10^9 cycles.

c. Calculate the existing safety factor for the proposed design configuration, based on tooth bending fatigue as the governing failure mode. Compare this with the desired design safety factor, and make any comments you think appropriate.

15-53. Based on the specifications and data for the Coniflex® bevel gearset given in problem 15-52, do the following:

a. Calculate the surface fatigue durability *stress* for the Coniflex® bevel gearset under consideration.

b. Determine the surface fatigue durability *strength* for the proposed nitrided and hardened AISI 4140 steel material corresponding to a life of 10^9 cycles.

c. Calculate the existing safety factor for the proposed design configuration based on surface fatigue durability as the governing failure mode. Compare this with the desired design safety factor, and make any comments you think appropriate.

15-54. A Coniflex®, straight-tooth bevel gearset is supported on shafts with centerlines intersecting at a 90° angle. The gear is straddle mounted between closely positioned bearings, and the pinion overhangs its support bearing. The 15-tooth pinion rotates at 900 rpm, driving the 60-tooth gear, which has a diametral pitch of 6, pressure angle of 20°, and face width of 1.25 inches. The material for both gears is through-hardened Grade 1 steel with a hardness of BHN 300 (see Figure 15.25). It is desired to have a reliability of 90 percent, a design life of 10^8 cycles, and a governing safety factor of 2.5. Estimate the maximum horsepower that can be transmitted by this gear reducer while meeting all of the design specifications given.

15-55. Repeat problem 15-54, except change the material to through-hardened Grade 2 steel with a hardness of BHN 350 (see Figure 15.25).

15-56. It is desired to design a long-life right-angle straight bevel gear speed reducer for an application in which an 850-rpm, 5-hp internal combustion engine, operating at full power, drives the pinion. The output gear, which is to rotate at approximately 350 rpm, drives a heavy-duty industrial field conveyor. Design the gearset, including the selection of an appropriate material, if a reliability of 95 percent is desired.

15-57. Repeat problem 15-56, except use an 850-rpm, 10-hp internal-combustion engine, operating at full power, to drive the pinion.

15-58. A proposed worm gearset is to have a single-start worm with a pitch diameter of 1.250 inches, a diametral pitch of 10, and a normal pressure angle of $14\frac{1}{2}°$. The worm is to mesh with a worm gear having 40 teeth and a face width of 0.625 inch. Calculate the following:

a. Axial pitch

b. Lead of the worm

c. Circular pitch

d. Lead angle of the worm

e. Helix angle of the worm gear

f. Addendum

g. Dedendum

h. Outside diameter of the worm

i. Root diameter of the worm

j. Pitch diameter of the gear

k. Center distance

l. Velocity ratio

m. Root diameter of the gear

n. Approximate outside diameter of the gear

15-59. A double-start worm has a lead of 60 mm. The meshing worm gear has 30 teeth, and has been cut using a hob having a module of 8.5 in the *normal* plane. Do the following:

a. Calculate the pitch diameter of the worm.

b. Calculate the pitch diameter of the worm gear.

c. Calculate the center distance and determine whether it lies in the range of usual practice.

d. Calculate the reduction ratio of the worm gearset.

e. Calculate the diametral pitch of the gearset.

f. Calculate the outside diameter of the worm (mm).

g. Calculate the approximate outside diameter of the worm gear (mm).

15-60. A triple-start worm is to have a pitch diameter of 4.786 inches. The meshing worm gear is to be cut using a hob having a diametral pitch of 2 in the *normal* plane. The reduction ratio is to be 12:1. Do the following:

a. Calculate the number of teeth in the worm gear.

b. Calculate the lead angle of the worm.

c. Calculate the pitch diameter of the worm gear.

d. Calculate the center distance and determine whether it lies in the range of usual practice.

15-61. It is proposed to drive an industrial crushing machine, designed to crush out-of-tolerance scrap ceramic bearing liners, with an in-stock 2-hp, 1200-rpm electric motor coupled to an appropriate speed reducer. The crushing machine input shaft is to rotate at 60 rpm. A worm gear speed reducer is being considered to couple the motor to the crushing machine. A preliminary sketch of the wormset to be used in the speed reducer proposes a double-start right-hand worm with axial pitch of 0.625 inch, a normal pressure angle of $14\frac{1}{2}°$, and a center distance of 5.00 inches. The proposed material for the worm is steel with a minimum surface hardness of Rockwell C 58. The proposed gear material is forged bronze.

Calculate or determine the following, assuming the friction coefficient between worm and gear to be 0.09, and that the motor is operating steadily at full rated power:

a. Number of teeth on the gear

b. Lead angle of the worm

c. Sliding velocity between worm and gear

d. Tangential force on the worm

e. Axial force on the worm

f. Radial force on the worm

g. Tangential force on the gear

h. Axial force on the gear

i. Radial force on the gear

j. Power delivered to the crushing machine input shaft

k. Whether the wormset is self-locking

15-62. A worm gear speed reducer has a right-hand triple-threaded worm made of hardened steel, a normal pressure angle of 20°, an axial pitch of 0.25 inch, and a center distance of 2.375 inches. The gear is made of forged bronze. The speed reduction from input to output is 15:1. If the worm is driven by a $\frac{1}{2}$-hp, 1200-rpm electric motor operating steadily at full rated power, determine the following, assuming the coefficient of friction between worm and gear to be 0.09:

a. Number of teeth in the gear

b. Pitch diameter of the gear

c. Lead angle of the worm

d. Relative sliding velocity between worm and gear

e. Tangential force on the worm

f. Tangential force on the gear

g. An acceptable range for face width that should allow a nominal operating life of 25,000 hours. (*Hint:* See footnote 130 relating to equation (15-115).)

15-63. It is desired to utilize a worm gearset to reduce the speed of a 1750-rpm motor driving the worm down to an output gear shaft speed of approximately 55 rpm, and provide $1\frac{1}{2}$ horsepower to the load. Design an acceptable worm gearset, and specify the nominal required horsepower rating of the drive motor.

Chapter 16

Brakes and Clutches

16.1 Uses and Characteristics of Brakes and Clutches

Conceptually, brakes and clutches are nearly indistinguishable. Functionally, a *clutch* is a device for making a smooth, gradual connection between two separate elements rotating at different speeds about a common axis, bringing the two elements to a *common* angular velocity after the clutch is actuated. A *brake* serves a similar function, except that one of the elements is fixed to the frame, so the common angular velocity is *zero* following actuation of a brake. For example, each of the two rotating elements shown in Figure 16.1(a) has its own polar *mass moment of inertia* and its own *angular velocity*. Actuation of the brake/clutch package brings the rotating friction surfaces into tangential sliding contact, initiating a frictional drag torque that works to gradually reduce the angular velocity *difference* between the rotating elements to zero. When the *relative* frictional sliding velocity is reduced to zero, both elements have the *same* angular velocity. This use of the package makes it function as a clutch.

Figure 16.1
Functional comparison of brakes and clutches.

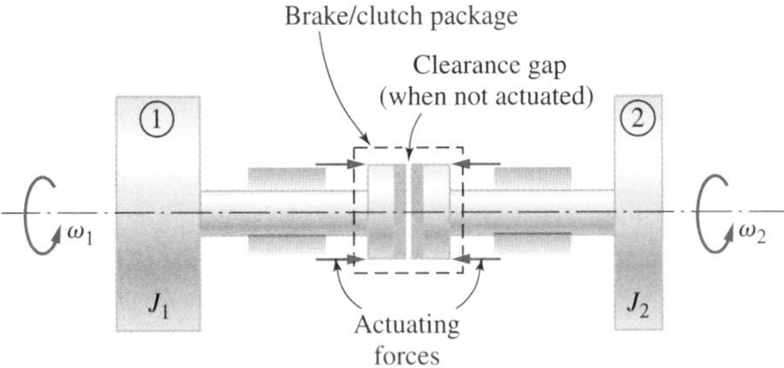

(*a*) Brake/clutch package used as a clutch.

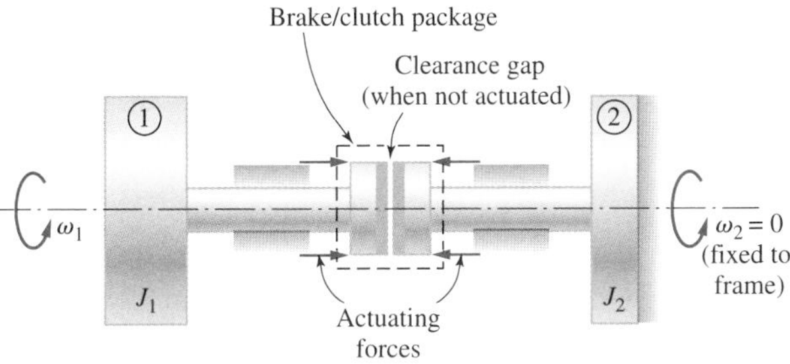

(*b*) Brake/clutch package used as a brake.

In Figure 16.1(b) the same device is shown, except that element 2 is fixed to the frame, so its angular velocity is always zero. In this case, actuation of the brake/clutch package brings the friction surfaces into contact as before, but the frictional drag torque works to gradually reduce the final angular velocity to *zero*. Therefore, for this case, the package functions as a brake.

Brakes and clutches are well known for their use in automotive applications, but are also extensively used in a wide variety of industrial machines including winches, hoists, excavators, tractors, mills, and elevators, and consumer products such as mowers, washing machines, garden tractors, chain saws, farm tractors, combines, and hay balers. Although a variety of different types of brakes and clutches have been devised, *friction clutches and brakes* are most common. In spite of the adage "friction is always against you," for friction clutches and brakes it is an essential design ingredient. Selection of a good *lining*[1] material usually implies selection of a material with a high coefficient of friction that remains essentially unchanged over a wide range of operating conditions. Only *friction* brakes and clutches will be discussed in detail in this text.[2]

16.2 Types of Brakes and Clutches

Designing or selecting a brake or a clutch for a specific application usually involves answering two basic questions: (1) What physical principles and basic arrangements for transferring energy from one element to the other seem best for this application? (2) What actuation method is appropriate?

Potential actuation methods include mechanical linkages, pneumatic or hydraulic actuators, electrically energized actuators, or dynamic force actuation at preselected speeds. *Fail-safe* actuation systems are sometimes used, in which the brake or clutch is *normally held in contact* by a spring load, and when release (separation) of the friction surfaces is desired, an actuator is energized to *retract* the spring. This concept[3] results in automatic actuation of the clutch or brake if actuator power is lost, preventing a potential runaway condition.

Several basic arrangements for friction-type and positive-contact brakes and clutches are sketched in Figure 16.2. The external contracting and internal expanding *shoe-and-rim (drum)* devices shown in Figure 16.2(a) and (b) are usually used as brakes.[4] The *band* type and *disk* type sketched in Figures 16.2(c) and (d) may be used as either brakes[5] or clutches.[6] The *cone* type device sketched in Figure 16.2(e) also may be used as either a clutch or a brake, but cone *clutches* are more common. The *positive contact* types shown in Figure 16.2(f) have mechanically interlocking surfaces that form a rigid mechanical junction when engaged. These devices are virtually always used as clutches, and can be engaged only when the relative angular velocity between the two rotating elements is near zero. Not shown is a specialty class of clutches known as *overrunning* or *one-way*

[1]Typically, one of the contacting friction surfaces is metal and the other is a high-friction material called a *lining*. (See also 16.4.)

[2]For other types of brakes and clutches, see, for example, ref. 1.

[3]First conceived by George Westinghouse for railroad brakes.

[4]Often used, for example, as rear wheel brakes in automotive applications.

[5]Caliper-type disk brakes, in which a stationary *caliper* mechanism straddles the edge of a rotating metal disk (rotor), are actuated by squeezing two opposing attached pads (linings) against the rotor (see also Figure 16.13). Typically, the pads have a rather short arc length of contact, and often the rotor embodies air-cooling vanes.

[6]Band-type devices are sometimes used as clutches in agricultural and construction equipment.

Figure 16.2
Various common types of brakes and clutches.

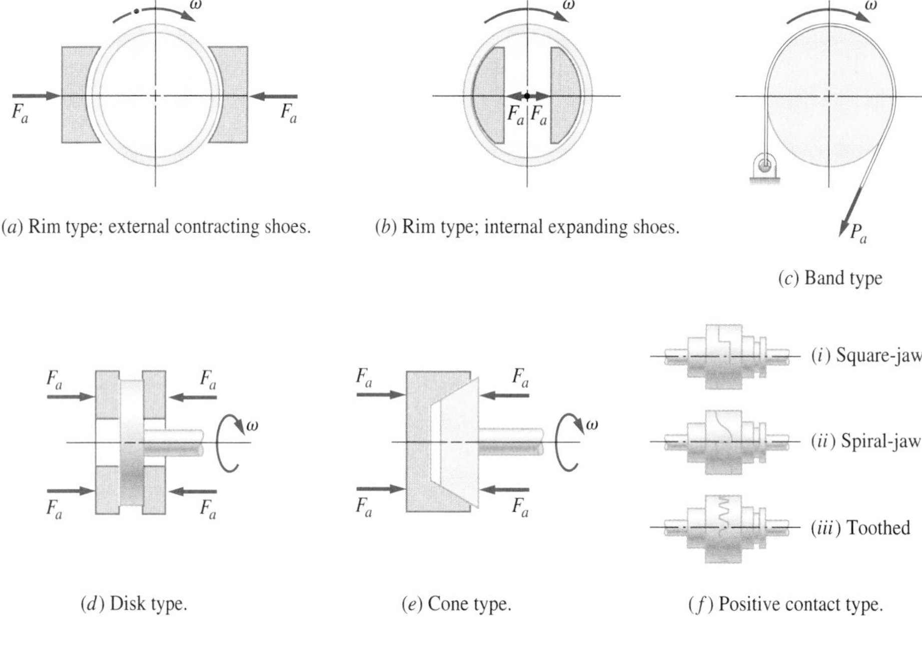

(*a*) Rim type; external contracting shoes.
(*b*) Rim type; internal expanding shoes.
(*c*) Band type
(*d*) Disk type.
(*e*) Cone type.
(*f*) Positive contact type.

clutches.[7] These devices allow relative rotation in only one direction, locking up if the relative rotation attempts to reverse.

For all types of friction clutches and brakes, a designer typically finds it necessary to *calculate* or *estimate:*

1. *Transmitted torque requirements* of the application, and *torque capacity* for the proposed brake/clutch package.
2. Actuating force requirements for the proposed configuration.

An example of a caliper disk brake.

[7]Examples are the *sprag clutch*, the *roller-ramp clutch*, and the *torsion spring clutch*. See ref. 1.

3. Limitations of the proposed device because of pressure, temperature, wear, or strength.
4. Energy generation, energy dissipation capability, and expected temperature rise in the device, especially at the friction interface.

All of these factors will be considered in developing the design procedure for friction clutches and brakes, as discussed below.

16.3 Potential Failure Modes

When clutches and brakes are actuated, two or more surfaces, moving with different velocities, are brought into contact by the actuating forces. Relatively high interfacial pressures may result. Significant pressure and relative sliding are the basic parameters that presage *wear* as a potential failure mode (see 2.3). Depending on the materials chosen and the operating environment, clutches and brakes may fail by *adhesive wear,*[8] by *abrasive wear*[9] or by *corrosive wear.*

Also, when friction clutches and brakes are actuated they *generate heat* at the rubbing interface, often at very high rates. Depending upon the ability of the clutch/brake package to dissipate the heat to the environment, the temperature may rise at the rubbing interface, sometimes to very high values. Thus, many of the temperature-related failure modes listed in 2.3 become potential failure modes for clutch or brake applications. These include temperature-induced elastic deformation (warpage), yielding, thermal fatigue, corrosion fatigue, creep, and in extreme cases, thermal shock. Other potential failure modes are high-cycle fatigue, low-cycle fatigue, corrosion fatigue, corrosion by direct chemical attack, or temperature-induced changes in material properties.

Of all the potential failure modes suggested, the preliminary design of a clutch or brake usually is based on wear, thermal fatigue, and change in material properties. If the friction lining wears away at an excessive rate, the lining may be entirely lost before an acceptable design life has been achieved. Because the *metallic side* of the rubbing pair may be subjected to temperature *spikes* if the brake or clutch is repeatedly actuated and released, these spikes may lead to *thermal fatigue*, sometimes causing a system of surface cracks called "heat checks." Because the *lining side* of the rubbing pair is often a molded or woven composite material with a polymeric binder, if the local interfacial temperature exceeds the binder breakdown temperature, the material properties of the lining may be transformed from a high-friction material to a *lubricative* material, giving rise to a potentially dangerous degradation in capacity. For many lining materials the coefficient of friction tends to decrease with increasing temperature, a condition known as *fading*. An example of this is loss of braking ability when driving on steep mountain grades, if brakes are constantly applied to control down-grade velocity. It is also important to consider potential failure modes for the brake or clutch *actuation system*. Material selection is especially important to assure acceptable brake and clutch performance.

16.4 Brake and Clutch Materials

Reviewing the list of potential failure modes just discussed in 16.3, and the material selection guidelines of Chapter 3, candidate materials for brake and clutch applications should have high resistance to wear, scoring, and galling, a high and stable coefficient of

[8]See (2-119). [9]See (2-121).

friction, the ability to withstand high temperatures, good thermal conductivity (to dissipate friction-induced heat), the ability to remain stable under variations in environment (such as the presence of moisture, dirt, or oil), high thermal fatigue resistance, good resistance to warpage (i.e., high thermal conductivity, low thermal coefficient of expansion, and high yield strength), and high resilience (to promote favorable distribution of interfacial pressure), and should be cost effective.

Under most circumstances, a brake/clutch package consists of a *metallic element* on one side, which when actuated is pressed against a mating element with an attached *lining of solid, molded, woven, or sintered material* having a high coefficient of friction, good temperature resistance, and good resilience. The metallic element (disk, rim, or drum) is commonly made of gray cast iron or steel.

The friction linings are usually made of composite materials in which particles or fibers are embedded in a matrix of thermosetting or elastomeric binder material, or of sintered metal (sometimes containing ceramic particles). *Molded* linings are most common and least costly. For many years the reinforcing fibers in brake and clutch linings were made of asbestos, but recognition that asbestos fibers are potentially carcinogenic has promoted the use of other materials, such as fiberglass or *coated* asbestos fibers. Metallic particles of brass or zinc are sometimes added to the matrix to improve thermal conductivity and wear resistance. Sintered metal linings may be used (at a higher cost) in applications where higher strength and higher temperature resistance are required. In some cases ceramic particles are added before sintering to form a *cermet* with higher temperature resistance. Table 16.1 gives brake/clutch lining properties for several commonly used materials.

Table 16.1 Coefficient of Friction and Design Limits for Commonly Used Brake/Clutch Friction Lining Materials Operating in Contact with Smooth Cast Iron or Steel[1]

Lining Material[2]	Approx. Friction Coefficient, μ	Max. Allowable Pressure, p_{max}, psi	Max. Allowable Temperature, Θ_{max}, °F		Max. Allowable Velocity, V_{max}, ft/min	Typical Applications
			Instantaneous	Continuous		
Carbon-graphite (dry)	0.25	300		1000		High-performance brakes
Carbon-graphite (in oil)	0.05–0.1	300		1000		High-performance brakes
Cermet	0.32	150	1500	750		Brakes and clutches
Sintered metal (dry)	0.15–0.45	400	1020	660	3600	Clutches and caliper disk brakes
Sintered metal (in oil)	0.05–0.08	500	930	570	3600	Clutches
Rigid molded-asbestos (dry)	0.31–0.49	100	750	350	3600	Drum brakes and clutches
Rigid molded-asbestos (in oil)	0.06	300	660	350	3600	Industrial clutches
Semirigid molded-asbestos	0.37–0.41	100	660	300	3600	Clutches and brakes
Flexible molded-asbestos	0.39–0.45	100	750	350	3600	Clutches and brakes
Wound asbestos yarn and wire	0.38	100	660	350	3600	Vehicle clutches
Woven asbestos yarn and wire	0.38	100	500	260	3600	Industrial clutches and brakes

Table 16.1 (*Continued*)

Lining Material[2]	Approx. Friction Coefficient, μ	Max. Allowable Pressure, p_{max}, psi	Max. Allowable Temperature, Θ_{max}, °F		Max. Allowable Velocity, V_{max}, ft/min	Typical Applications
			Instantaneous	Continuous		
Rigid molded-nonasbestos	0.33–0.63	150		500–750	4800–7500	Clutches and brakes
Woven cotton	0.47	100	230	170	3600	Industrial clutches and brakes
Resilient paper (in oil)	0.09–0.15	400	300		$pV < 500{,}000$ psi-ft/min	Clutches and transmission bands

[1]Adapted from ref. 1, with permission of The McGraw-Hill Companies. Brake/clutch manufacturers should be consulted for more accurate data.
[2]Typically, the fibrous, wound, woven, or paper materials are saturated with a suitable polymeric resin binder and cured under heat and pressure to form the lining.

16.5 Basic Concepts for Design of Brakes and Clutches

The procedure for analysis or design of a brake/clutch package is fundamentally the same for all types of brakes and clutches. The suggested steps include:

1. Select a proposed *type* of brake or clutch that seems best suited to the application,[10] and sketch a proposed configuration, including any dimensional or geometric constraints imposed by design specifications.
2. Based on probable governing failure modes, specified operating conditions, and required response times for executing the braking or clutching action, select an appropriate material pair.[11]
3. Estimate the torque required to accelerate or decelerate the proposed device to the desired common speed within the specified response time. This usually involves a dynamics analysis of the proposed machine, including inertial effects of all significant rotating and translating masses, reflected to the rotational speed of the brake/clutch package.[12]
4. Estimate the energy to be dissipated as frictional heat in the brake or clutch contact zone. This may be accomplished by summing changes in the kinetic energy of translation, changes in the kinetic energy of rotation, and any changes in potential energy due to elevation changes during the response period.
5. Estimate the pressure distribution over the contacting friction surfaces, by either calculation or assumption.
6. Determine the pressure at any point in the friction interface as a function of the *maximum* local pressure produced when the brake/clutch package is actuated.
7. Apply the principles of static equilibrium to determine the actuating force, friction torque, and support reactions. Determine suitable materials and dimensions that will provide design-specified values (or ranges) for friction torque and actuating force, including any specified factors of safety.
8. Iterate the proposed design configuration until it meets the functional specifications

[10]See Figure 16.2, for example [11]See 16.3, 16.4, and Example 16.1.

[12]The *effective inertia* of a connected mass operating at a speed different from the brake/clutch package is proportional to the square of the ratio of the speed of the connected mass to the speed of the brake/clutch.

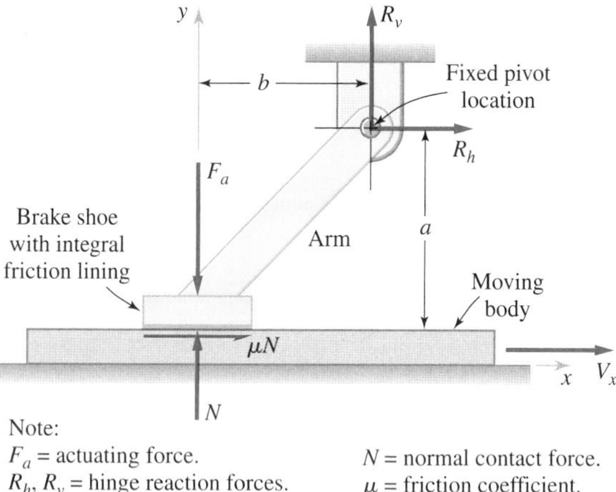

Figure 16.3
Simple fixed-hinge brake pad acting against a moving planar body.

and operates reliably for the specified design lifetime. Estimate potentially damaging temperature rise induced when the friction-generated heating rate exceeds the cooling rate. This is an important factor in the iteration process.

Several basic concepts may be illustrated by considering a simple, short, rigid, fixed-hinge brake shoe with integral lining, acting against the surface a planar translating body, as illustrated in Figure 16.3. As shown, the translating body has a velocity V_x to the right (x-direction), and the brake shoe is constrained by the hinged connection to remain stationary in the x-y plane. Since the brake shoe is *short* and *rigid*, it is reasonable to assume a uniform pressure distribution over the entire contacting friction surface. Hence,

$$p = p_{max} \tag{16-1}$$

where p = pressure at any specified point
p_{max} = maximum pressure in the contact zone

Since the pressure distribution is uniform, the normal force N may be expressed as

$$N = p_{max} A \tag{16-2}$$

where A = contact area of friction surface

If the brake-shoe-and-arm unit is taken as a free body, and moments are summed about the hinge pin, as depicted in Figure 16.3,

$$F_a b - Nb + \mu Na = 0 \tag{16-3}$$

From (16-3), the required actuating force F_a is

$$F_a = \frac{N(b - \mu a)}{b} = \frac{p_{max} A(b - \mu a)}{b} \tag{16-4}$$

Also, a hinge reaction force, R_h, in the x (horizontal) direction, may be found from a horizontal force summation *on the free body*, as

$$R_h + \mu p_{max} A = 0 \tag{16-5}$$

giving

$$R_h = -\mu p_{max} A \qquad \text{(i.e., directed to the left)} \tag{16-6}$$

The hinge reaction force, R_v, in the y (vertical) direction may be found from a vertical force summation as

$$R_v = p_{max} A \left[\frac{b - \mu a}{b} - 1 \right] \tag{16-7}$$

Thus, for the simple case illustrated in Figure 16.3, the design procedure would involve the selection of an appropriate friction lining material,[13] and iterative determination of dimensions a, b, and contact area A to meet the functional specifications of the application. At the same time, it is necessary to stay within pressure, temperature, and strength limitations of the lining material, as dictated by the governing failure mode.

Two additional concepts, *self-locking* and *self-energizing*, may also be understood by considering the simple device shown in Figure 16.3. Examining (16-4), it may be observed that if the coefficient of friction and geometric arrangement are such that

$$\frac{b - \mu a}{b} \leq 0 \tag{16-8}$$

no external actuation force is needed to actuate the brake; initial contact between the moving friction surfaces causes the brake to be actuated immediately and fully by the friction-generated moment. This condition, called *self-locking*, is usually not desirable[14] because the braking action is sudden and uncontrollable when the brake "grabs." On the other hand, it is often advantageous to capitalize upon the friction-generated moment to *assist* in applying the brake, reducing the magnitude of the required actuating force F_a to a smaller (but not zero) value. Such an arrangement is often called a *self-energizing* configuration. One way of determining a configuration that will provide self-energization without self-locking is to utilize a *pseudo–friction coefficient,* μ', in (16-4) when relating the hinge location to the required actuating force. The value of μ' is often taken as

$$\mu' \approx (1.25\mu \text{ to } 1.5\mu) \tag{16-9}$$

where μ is the maximum value of friction coefficient specified by the lining material manufacturer.

16.6 Rim (Drum) Brakes with Short Shoes

A more detailed sketch of the rim (drum) type external shoe (block) brake depicted in Figure 16.2(a) is shown in Figure 16.4. Such brakes may be categorized as *external* or *internal*, with *short shoes* (angle α subtending contact arc is 45° or less) or *long shoes*. Figure 16.4 shows a simple one-shoe, external, short block brake. As a practical matter, two *diametrically opposing shoes*, as depicted in Figure 16.2(a), are usually used to reduce drum bearing reactions. However, basic design concepts can be demonstrated in a simpler way by first considering a *single-shoe* short block brake.

Because the shoe sketched in Figure 16.4 is short and stiff, it is reasonable to assume a *uniform pressure distribution* over the whole contact interface; hence

$$p = p_{max} \tag{16-10}$$

and

$$N = p_{max} A \tag{16-11}$$

Although the normal force N and friction force μN are distributed continuously over the contacting surfaces of the rim and shoe, for a *short* shoe the assumption can usually be made that these forces are concentrated at the center of contact, shown as point B.

[13] See, for example, Table 16.1.

[14] Except when a ratcheting device is desired, in which relative motion in one direction is prevented by self-locking while motion in the opposite direction is unfettered.

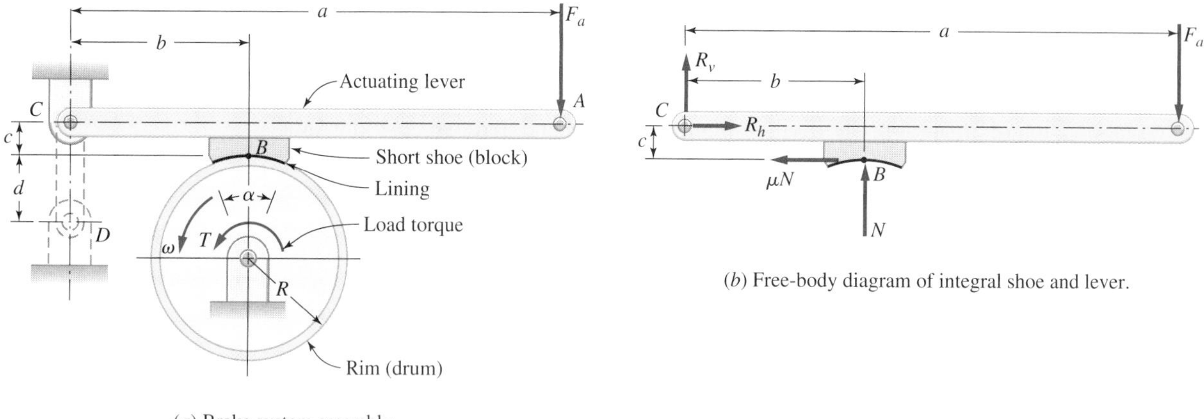

(a) Brake system assembly.

(b) Free-body diagram of integral shoe and lever.

Figure 16.4
Simple external short-shoe block brake.

Summing moments about hinge location C, and assuming a counterclockwise (CCW) direction of drum rotation (as shown), gives

$$Nb - \mu Nc - (F_a)_C a = 0 \tag{16-12}$$

whence

$$(F_a)_C = \frac{N(b - \mu c)}{a} \tag{16-13}$$

Based on the discussion of self-locking and self-energizing systems given in 16.5, the short block brake with hinge location C will be self-locking for the indicated CCW drum rotation if[15]

$$\frac{b - \mu c}{a} \leq 0 \tag{16-14}$$

The friction-based *braking torque capacity* for hinge location C, $(T_f)_C$, may be calculated utilizing (16-13) by summing moments on the drum about its axis of rotation O to obtain

$$(T_f)_C = R\mu N = R\mu\left(\frac{(F_a)_C a}{b - \mu c}\right) = (F_a)_C\left[R\mu\left(\frac{a}{b - \mu c}\right)\right] \tag{16-15}$$

Further, the *design* value of friction torque capacity must be large enough to safely bring the moving mass of the operating system from full operational velocity to a final velocity of zero within a specified response time t_r, and the energy stored in the moving mass must be safely dissipated by the brake package as the mass is decelerated to zero velocity.

If the hinge pin location were changed from C to D [shown by dashed lines in Figure 6.4(a)], the friction moment for the CCW drum rotation would tend to "unseat" the shoe (hinder actuation), and the expression for actuating force would become[16]

$$(F_a)_D = \frac{N(b + \mu c)}{a} \tag{16-16}$$

For hinge location D, therefore, the required actuating force $(F_a)_D$ would always be positive, self-locking would not occur, and self-energization could not be utilized to assist in applying the brake.

[15]Compare with (16-8). [16]Compare with (16-13) based on hinge location C.

The horizontal and vertical reaction forces for pin location C (i.e., self-energization is possible) may be found from horizontal and vertical force summations as

$$(R_h)_C = \mu N \tag{16-17}$$

and

$$(R_v)_C = (F_a)_C - N \tag{16-18}$$

The torque T (in-lb) required to decelerate the moving system to zero velocity within a specified response time t_r may be estimated as[17]

$$T = J_e\alpha = J_e\frac{(\omega_{op} - 0)}{t_r} = \frac{Wk_e^2\omega_{op}}{gt_r} \tag{16-19}$$

where J_e = effective polar mass moment of inertia (in-lb-sec^2), including inertial effects of all connected components[18] referred to the drum operating at angular velocity ω_{op}

α = angular acceleration, rad/sec^2

$\omega_{op} = \dfrac{2\pi n_{op}}{60}$ = angular velocity of drum operating at its full speed, rad/sec

n_{op} = rotational operating speed of drum, rpm

t_r = response time (braking time) to reduce the angular velocity ω_{op} to zero, sec

W = weight of rotating mass, lb

k_e = effective radius of gyration, inches

g = 386 in/sec^2

The energy E_d to be dissipated (by transfer of friction-generated heat to the surroundings) during the braking period t_r may be estimated as the sum of the kinetic energy of rotation, the kinetic energy of translation, and the potential energy of elevation change and/or elastic deformation.[19] For the simple case where *the only significant mass in the system is the rotating drum*, the energy to be dissipated is just the kinetic energy of rotation, or

$$E_d = \frac{1}{2}J_e\omega_{op}^2 \tag{16-20}$$

For any proposed design configuration, the friction-based braking torque *capacity*, as given in (16-15), must exceed, by a suitable design safety factor, the torque *required* to decelerate the moving system to zero velocity within the specified response time t_r. Thus, utilizing (16-19),

$$(T_f)_C = n_d T = \frac{n_d J_e \omega_{op}}{t_r} = \frac{2\pi n_d W k^2 n_{op}}{gt_r} \tag{16-21}$$

Combining (16-21) with (16-15), the design value of the *required* actuating force, $(F_a)_C$, may be found as[20]

$$(F_a)_C = \frac{2\pi n_d W k^2 n_{op}(b - \mu c)}{a t_r R \mu a} \tag{16-22}$$

When the brake is actuated, the energy content of the moving system must be wholly dissipated by converting it to frictional heat at the rubbing contact interface between the drum and the shoe. Since the metallic drum typically has a much higher thermal conduc-

[17]See any good undergraduate textbook on dynamics (e.g., ref. 2).

[18]In actual systems there may be several rotating components operating at different speeds but connected to the rotating drum through gearing or other speed changers. The contribution of an ith connected mass, rotating at speed n_i, is proportional to the square of the speed ratio, or $J_{ei} = J_i(n_i/n_{op})^2$.

[19]See, for example, ref. 2.

[20]As in all calculations, care must be taken to use compatible units for all terms in the equation.

tivity than the friction lining of the shoe, most of the heat generated will usually be conducted to the rim mass. An important issue is whether the proposed brake configuration will be able to dissipate the friction-generated heat without producing a local temperature rise so high that braking function is impaired or damage is inflicted upon the lining or the drum. Unfortunately, the friction-generated temperature rise is not easily calculated because of the uncertainties associated with temperature distributions, effective heat transfer areas, effective heat transfer coefficients, and environmental factors. If brakes are applied *infrequently and for brief periods*, so that the brake drum can conduct and radiate the friction-generated heat away before the next braking cycle begins, there will be *little or no residual temperature rise*. If braking periods are *too frequent* and/or *too lengthy*, the *initial* temperature for each braking cycle *will be higher than for the preceding cycle and the equilibrium temperature of the brake package may ratchet to a much higher level.*

For infrequent short-duration brake actuations, it is commonly assumed that the friction-generated heat will be absorbed by the mass of material adjacent to the frictional interface, often assumed to be the rim mass for the case of a rim-type brake. For such assumptions, the temperature rise $\Delta\Theta$ may be estimated, based on classic heat capacity concepts, as

$$\Delta\Theta = \frac{H_f}{CW} \tag{16-23}$$

where

$$H_f = \frac{E_d}{J_\Theta} = \frac{J_e \omega_{op}^2}{2J_\Theta} \tag{16-24}$$

and

J_e = polar mass moment of inertia of rotating drum
$\omega_{op} = 2\pi n_{op}$ = operating angular velocity of drum
n_{op} = rotational speed of drum
J_Θ = mechanical equivalent of heat (e.g., 9336 in-lb/Btu)
C = specific heat (e.g., 0.12 Btu/lb-°F for steel or cast iron)
W = weight of heat-absorbing mass (e.g., mass of brake rim, or weight of a selected *judgment-based thinner portion* of the brake rim)

When brake actuations are more frequent, of longer duration, or continuously applied, Newton's law of cooling[21] may be utilized to estimate the temperature rise $(\Theta_s - \Theta_a)$. Utilizing (10-19),

$$(\Theta_s - \Theta_a) = \frac{H_f}{k_1 A_s} \tag{16-25}$$

where Θ_s = brake package surface temperature
Θ_a = ambient temperature
A_s = exposed surface area of brake package
k_1 = modified overall heat transfer coefficient [see (10-22) and (10-23) for approximate values]

The temperature rises calculated from (16-23) or (16-25) must not heat the proposed brake materials to temperatures above the maximum allowable temperatures for the materials.[22]

Finally, when adhesive and/or abrasive wear are the probable governing failure modes, the wear concepts leading to (2-119) and (2-121) may be utilized to estimate the

[21]Discussed earlier in connection with plain bearings and gears. See, for example, (10-19) or (15-49). Also, see Table 16.1

[22]See, for example, Table 16.1

Table 16.2 Maximum Allowable Values of *pV* for Industrial Shoe Brakes

Operating Conditions	pV, psi-ft/min
Continuous braking, poor heat dissipation	30,000
Occasional braking, long rest periods, poor heat dissipation	60,000
Continuous braking, good heat dissipation (as into an oil bath)	85,000

wear depth normal to the friction surface of a brake lining, called *normal wear depth*, d_n, as

$$d_n = k_w p L_s = k_w p V t_{contact} \tag{16-26}$$

where k_w = material pair constant[23]
p = contact pressure
L_s = total distance slid, between drum and shoe, over the design lifetime
V = relative sliding velocity between lining and drum
$t_{contact}$ = time of drum/shoe contact during the design lifetime of the brake/clutch assembly

The common tactic of assuming normal wear *rate* δ_n to be proportional to the product of pressure p times sliding velocity V is supported by (16-26) since

$$\delta_n = \frac{d_n}{t_{contact}} = k_w(pV) \tag{16-27}$$

where, for a *short* block brake, wear rate δ_n is constant over the whole contact surface. Typically, most of the wear occurs on the *brake lining element*, which should be designed to be replaced when wear interferes with its ability to function properly. The brake drum, or rim, is usually designed to sustain little or no measurable wear.

In preliminary design scenarios, an experience-based alternative approach is to use the pV product as a preliminary criterion. It is common practice to express the pV product as pressure p(lb/in^2) times velocity V(ft/min). Based on this mixed-unit pV product, Table 16.2 gives approximate limiting values for several design scenarios.

Example 16.1 Preliminary Design of a Short-Shoe Drum-Brake

To meet new safety regulations, a manufacturer of push-type rotary lawn mowers is proposing to modify an existing mower design by incorporating a means of quickly bringing the rotating blade to a complete stop if the operator's hands are removed from the mower handle bar. The objective is to prevent injuries from the rotating blade; for example, if an operator looses control of the mower on a steep hillside, the blade should stop before the mower can roll back over the operator.

The basic concept being proposed is to utilize a spring-loaded lever[24] arrangement at the mower handle bar, attached by flexible cables to a clutch and a separate brake. Figure E16.1A shows a side view of the existing mower with a proposed commercial clutch inserted between the one-cylinder engine and the flywheel-and-blade assembly.[25] Axial

[23]The material pair constant may be available from the friction material manufacturer; otherwise, it must be determined experimentally.

[24]Sometimes called a "dead-man" lever because if for any reason the operator looses the ability to hold the lever down, the blade is brought to an immediate stop.

[25]Flywheels are used to smooth variations in speed caused by torque fluctuations, due to fluctuating operational loads or fluctuating prime-mover torque, over each cycle (as in the case of a one-cylinder internal combustion engine, for example). The rotating blade itself also produces a "flywheel effect." For more details, see Chapter 19.

Example 16.1 Continues

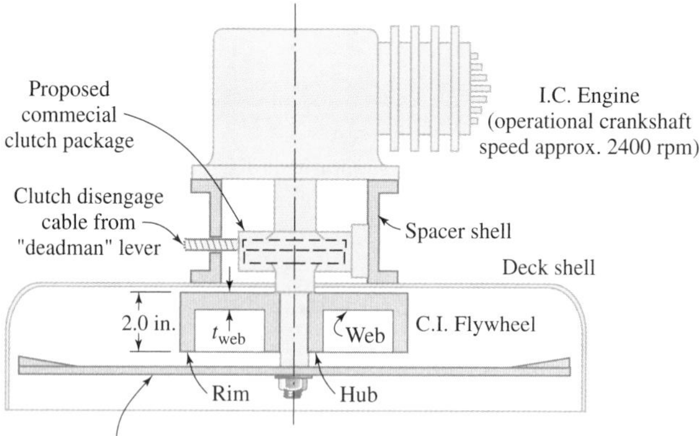

Figure E16.1A
Sketch of proposed mower modification, with clutch package and spacer inserted between engine and deck. (See Figure E16.1B for bottom view.)

space for the proposed clutch package is to be provided by installing a spacer shell between the engine and the deck. A preliminary design evaluation of the clutch has already been completed, and indicates that commercially available disk clutch packages with acceptable torque characteristics, actuation characteristics, and space envelopes can be purchased from stock.

The most serious potential sequence of events is envisioned as follows:

a. Operator is mowing uphill, slips on wet grass, falls (thereby releasing dead-man lever), and mower rolls back downhill over the operator.

b. When the spring-loaded dead-man lever is released, the clutch cable disengages the clutch, and the brake cable actuates the brake, thereby quickly stopping the rotary blade before the mower rolls over the operator.

You have been asked, as a consultant, to propose a *brake* design that will meet the design objective of stopping the rotating blade quickly enough upon release of the dead-man lever to prevent operator injury from the rotating blade. Estimates have been made that indicate the rotating blade must be reduced to zero speed in 10 revolutions (approximately 0.5 sec) to properly protect the operator from injury. Engineering management desires that you first consider the use of a *short-shoe block brake* arrangement that utilizes the existing cast-iron flywheel, by machining a smooth outer surface that can serve as a brake drum.

Propose a preliminary design that meets these specifications if a design safety factor of 1.5 is desired for all aspects of the design.

Solution

Utilizing the basic concepts outlined in 16.5 and 16.6, a preliminary brake package proposal may be developed as follows:

1. Management has specified that a *short-shoe drum brake* should be considered first. One tentative short-shoe arrangement that seems to have the potential for meeting management directives and design specifications is sketched in Figure E16.1B. This arrangement preserves much of the existing design, proposes to smooth the outer flywheel surface to act as the brake drum, and adds an actuation spring, lever, and short brake shoe to accomplish the braking task.

2. From Table 16.1, tentatively select a rigid, molded, nonasbestos lining for this appli-

cation; this lining has the following properties when used in contact with smooth cast iron:

$$\begin{aligned} \mu &= 0.33 \\ p_{max} &= 150 \text{ psi} \\ \Theta_{max} &= 500°\text{F} \\ V_{max} &= 4800-7500 \text{ ft/min} \end{aligned} \tag{1}$$

3. Utilizing (16-19), estimate the torque T required to decelerate the rotating cast-iron flywheel and steel blade from 2400 rpm to 0 rpm in 0.5 sec, as

$$T = J_e \frac{\omega_{op}}{t_r} \tag{2}$$

When the clutch disengages, the effective mass moment of inertia, J_e, includes only the flywheel and the blade. For the cast-iron[26] flywheel, utilizing Appendix Table A.2, and representing the flywheel by the combination of a solid cylindrical mass (hub and shaft), a thin ring (rim), and a thin disk (web),

$$J_{fw} = J_{hub} + J_{rim} + J_{web} \tag{3}$$

Using case 2 of Table A.2, and dimensions from Figures E16.1A and E16.1B,

$$\begin{aligned} J_{hub} &= \frac{W_{hub}}{2g} R_{hub}^2 = \frac{0.270[\pi R_{hub}^2 L_{hub}]}{2(32.2 \times 12)} R_{hub}^2 \\ &= \frac{0.270[\pi(1.0)^2(2.0 - 0.5)]}{2(32.2 \times 12)} (1.0)^2 = 0.0017 \text{ in-lb-sec}^2 \quad \text{(includes shaft)} \end{aligned} \tag{4}$$

Using case 4 of Table A.2,

$$\begin{aligned} J_{rim} &= \frac{W_{rim}}{g} R_{rim}^2 = \frac{0.270[2\pi(R_{mean})_{rim} t_{rim} L_{rim}]}{32.2 \times 12} R_{rim}^2 \\ &= \frac{0.270[\pi(3.75)(0.5)(2.0 - 0.5)]}{32.2 \times 12} (3.75)^2 = 0.1736 \text{ in-lb-sec}^2 \end{aligned} \tag{5}$$

And using case 3 of Table A.2,

$$\begin{aligned} J_{web} &= \frac{W_{web} R_{web}^2}{2g} = \frac{0.270[\pi R_{web}^2 t_{web}]}{2(32.2 \times 12)} R_{web}^2 \\ &= \frac{0.270[\pi(4.0)^2(0.5)]}{2(32.2 \times 12)} (4.0)^2 = 0.1405 \text{ in-lb-sec}^2 \end{aligned} \tag{6}$$

Thus, from (3),

$$J_{fw} = 0.0017 + 0.1786 + 0.1405 = 0.2810 \text{ in-lb-sec}^2 \tag{7}$$

For the steel blade,[27] case 2 of Table A.2 may be utilized by assuming the blade to be a long, thin ($D \approx 0$) member, giving[28]

$$J_{blade} = \frac{W_{blade} L_{blade}^2}{12g} = \frac{0.283[(0.25)(2.0)(20)](20)^2}{12(32.2 \times 12)} = 0.2441 \text{ in-lb-sec}^2 \tag{8}$$

and adding (7) to (8),

$$J_e = 0.2810 + 0.2441 = 0.525 \text{ in-lb-sec}^2 \tag{9}$$

[26]Weight density for cast iron is 0.270 lb/in^3 (see Table 3.4).

[27]Weight density for steel is 0.283 lb/in^3.

[28]Be careful to note in case 2 of Table A.2 that $J_{blade} = J_x = J_y \neq J_z$, so with $R \approx 0$, $J_{blade} = J_x = (m/12)(0 + h^2)$, where $h = L_{blade}$.

Example 16.1 Continues

Also, the operating angular velocity is

$$\omega_{op} = \frac{2\pi n_{op}}{60} = \frac{2\pi(2400)}{60} = 251.3 \frac{\text{rad}}{\text{sec}} \tag{10}$$

and, from specifications,

$$t_r = 0.5 \, sec \tag{11}$$

Using these values, (2) gives

$$T = \frac{0.525)251.3}{0.5} = 264 \text{ in-lb} \tag{12}$$

Next, using (16-21), the friction-based braking torque capacity must be

$$(T_f)_C = n_d T = (1.5)(264) = 396 \text{ in-lb} \tag{13}$$

Utilizing (16-15), the spring must exert an actuating force $(F_a)_C$ of

$$(F_a)_C = \frac{T_f(b - \mu c)}{R\mu a} \tag{14}$$

Substituting from (13) and (1),

$$(F_a)_C = \frac{(396)(b - 0.33c)}{(4.0)(0.33)a} = 300\left(\frac{b - 0.33c}{a}\right) \tag{15}$$

The actuating force $(F_a)_C$ that must be supplied by the brake actuation spring shown in Figure E16.1B is a function of dimensions a, b, and c, still to be determined. From (15), and referring to Figure E16.1B, the required actuating force may be reduced by selecting larger values of a, larger values of c, and/or smaller values of b, but limits are placed on these values by the geometric and dimensional constraints of the existing mower design.

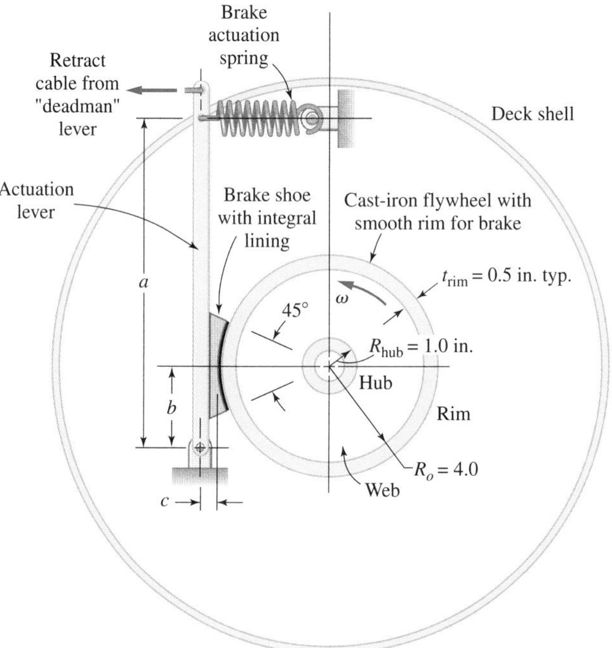

Figure E16.1B
Sketch of proposed short-shoe block brake arrangement for quickly stopping the rotary blade. View shown is bottom view of mower. (See Figure E16.1A for side view.)

As a first iteration, the following assumed values appear to be consistent with the mower geometry:

$$\begin{aligned} a_1 &= 12 \text{ inches} \\ b_1 &= 3 \text{ inches} \\ c_1 &= {}^3\!/_4 \text{ inch} \end{aligned} \tag{16}$$

Tentatively adopting these proposed "first-iteration" dimensions, (15) may be utilized to calculate the required actuating force as

$$(F_a)_C = 300\left[\frac{3 - 0.33(0.75)}{12}\right] = 69 \text{ lb} \tag{17}$$

It appears that a preloaded tension spring could reasonably provide such an actuating force. To develop a tentative spring design, see Chapter 14.

4. The energy to be dissipated by the brake package during the 0.5-sec brake stopping time may be estimated from (16-20), using (9) and (10), as

$$E_d = \frac{1}{2}(0.525)(251.3)^2 = 16{,}590 \text{ in-lb} \tag{18}$$

5. For a short-shoe brake (16-10) holds, so using data from (1),

$$p = p_{max} = 150 \text{ psi} \tag{19}$$

6. From (16-11), the required brake contact area $A_{req'd}$ is

$$A_{req'd} = \frac{N}{p_{max}} \tag{20}$$

7. The normal force N may be calculated from (16-13), using data from (16) and (17), as

$$N = \frac{F_a a}{b - \mu c} = \frac{69(12)}{3 - 0.33(0.75)} = 300 \text{ lb} \tag{21}$$

whence, from (20),

$$A_{req'd} = \frac{300}{150} = 2.00 \text{ in}^2 \tag{22}$$

In the proposed configuration sketched in Figure E16.1B, the contact length of the brake lining extends over a 45° ($\pi/4$ rad) segment of the cast-iron rim, giving a proposed contact length L_c of

$$L_c = R_o\left(\frac{\pi}{4}\right) = (4.0)\left(\frac{\pi}{4}\right) = 3.1 \text{ inches} \tag{23}$$

Using (22) and (23), the required lining contact with w_c is

$$w_c = \frac{A_{req'd}}{L_c} = \frac{2.00}{3.1} = 0.65 \text{ inch} \tag{24}$$

This width requirement is compatible with the existing 2.0-inch rim width shown in Figure E16.1A.

To estimate the temperature rise at the braking interface, (16-23) may be utilized, giving

$$\Delta\Theta = \frac{H_f}{CW} = \frac{(E_d/J_\Theta)}{CW} \tag{25}$$

Example 16.1 Continues

where

$$E_d = 16{,}590 \text{ in-lb [see (18)]}$$
$$J_\Theta = 9336 \text{ in-lb/Btu [see (16-24)]} \tag{26}$$
$$C = 0.12 \text{ Btu/lb-°F [see (16-24)]}$$

The weight of the heat-absorbing mass will be *estimated* as the weight of the *outer 10 percent*[29] of the cast-iron flywheel rim; hence

$$W = 0.270(2\pi)(4.0)[(0.10)(0.5)] = 0.34 \text{ lb} \tag{27}$$

Therefore, from (25), the estimated temperature rise would be

$$\Delta\Theta = \frac{(16{,}590/9336)}{(0.12)(0.34)} \approx 44°\text{F (above ambient temperature)} \tag{28}$$

Thus a maximum temperature around 114°F would be estimated. This is well within the acceptable range [see (1)]. It would probably be advisable to also investigate lower-cost linings such as woven cotton or resilient paper for this application.

Finally, the maximum relative tangential velocity between the lining and the cast-iron rim is

$$V_{max} = \frac{2\pi(4.0)(2400)}{12} = 5027 \text{ ft/min} \tag{29}$$

This value of maximum velocity is within the range of 4800–7500 ft/min, and will be regarded as acceptable, but should be checked experimentally.

Pivot pin reactions on the free body may be calculated from (16-17) and (16-18) as

$$R_h = \mu N = 0.33(300) = 99 \text{ lb} \tag{30}$$

and

$$R_v = F_a - N = 69 - 300 = -231 \text{ lb} \tag{31}$$

The resultant reaction force on the pin, therefore, is

$$R_{pin} = \sqrt{R_h^2 + R_v^2} = \sqrt{(99)^2 + (231)^2} = 251 \text{ lb} \tag{32}$$

The preliminary design recommendations may be summarized as follows:

1. Adopt the configuration sketched in Figure 16.13.
2. Tentatively adopt rigid molded-nonasbestos lining bonded to a shoe extending over a 45° circumferential segment of the 4.0-inch outer radius of the drum. The lining width should be approximately 0.65 inch, or more.
3. The lever and pivot location dimensions in Figure E16.1B should be approximately

   $a = 12$ inches
   $b = 3$ inches
   $c = 0.75$ inch

   and the brake actuation spring should be selected and preloaded to provide an actuation force of 69 lb.
4. Experimental testing should be conducted to validate all aspects of the proposed design before it is placed into production.

[29]This is purely a judgment call. The temperature rise should ultimately be experimentally validated before releasing the design to production engineering.

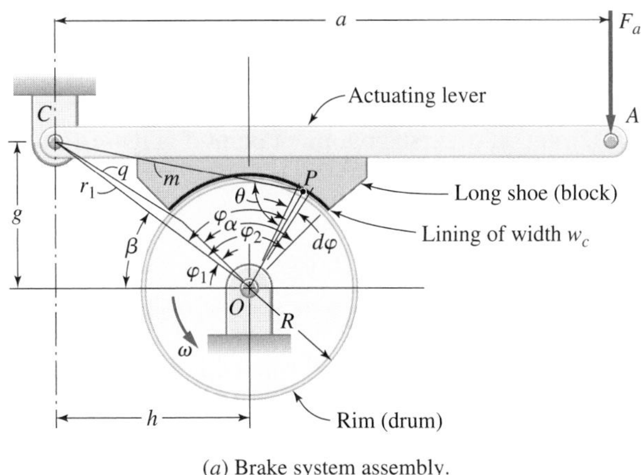

(a) Brake system assembly.

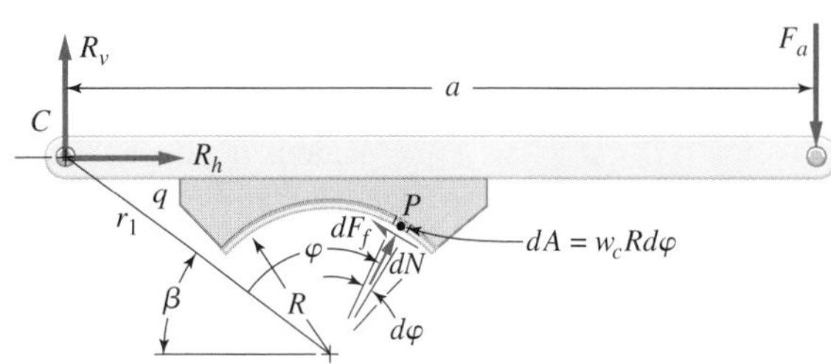

(b) Free body diagram of integral shoe and lever.

Note: w_c is contact width of brake shoe lining in and out of the paper.

Figure 16.5
External long-shoe block brake.

16.7 Rim (Drum) Brakes with Long Shoes

If the angle of shoe contact α [see Figure 16.5(a)] exceeds about 45°, the *short-shoe assumption* (that contact pressure is uniformly distributed) may introduce significant errors into the equations developed in 16.6.[30] To develop equations appropriate for *long shoes*, a more accurate estimate of pressure distribution over the shoe contact surface is necessary. One way of estimating long-shoe pressure distribution is based on the *normal wear* pattern expected for the lining as wear progresses, causing the shoe and actuating lever to rotate through angle γ about pivot point C. This rotation is shown in Figure 16.6 for an *external* shoe.[31] Utilizing (16-26), the *normal wear depth* at any arbitrary point P along the contact surface may be written as

$$(d_n)_P = k_w p_P V_P t_{contact} \tag{16-28}$$

and since the tangential velocity V and contact time $t_{contact}$ are constant over the whole rubbing contact surface,

$$(d_n)_P = (k_w V t_{contact}) p_P = K_{LS} p_P \tag{16-29}$$

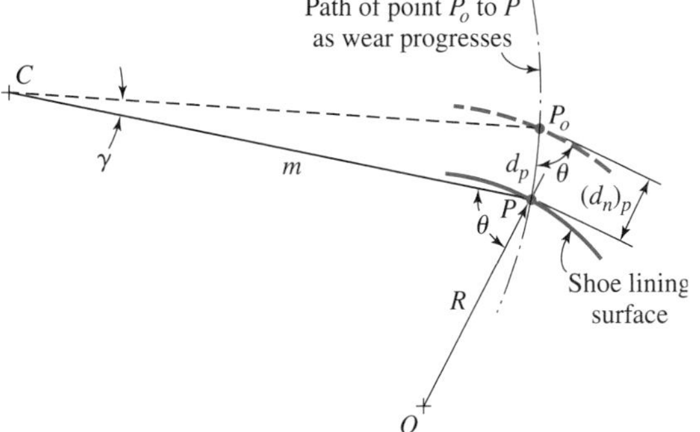

Figure 16.6
Displacement *d* of an arbitrary point on the contact surface as the lining wears and the shoe-and-arm unit rotates about hinge point *C*. (See also Figure 16.5.)

[30]Typically, however, the angle of contact for long-shoe brakes rarely exceeds about 120°.

[31]A similar argument may be made for an *internal* shoe.

where K_{LS} = constant for a material pair selected for long-shoe brakes operating at tangential sliding velocity V for a time of contact $t_{contact}$

p_P = local pressure at point P

Kinematically, as the shoe wears, rotation occurs about pivot point C. Figure 16.6 illustrates that the resultant wear d_P is

$$d_P = \gamma m \tag{16-30}$$

and the normal wear $(d_n)_P$, perpendicular to the surface at P, is

$$(d_n)_P = d_P \sin\theta = \gamma m \sin\theta \tag{16-31}$$

Constructing q through point C and perpendicular to OP, as illustrated in Figure 16.5(a),

$$q = r_1 \sin\varphi = m \sin\theta \tag{16-32}$$

whence, from (16-31),

$$(d_n)_P = \gamma r_1 \sin\varphi \tag{16-33}$$

Combining (16-29) with (16-33)

$$K_{LS} p_P = \gamma r_1 \sin\varphi \tag{16-34}$$

From this it may be deduced that the point of maximum pressure corresponds to the point where $\sin\varphi$ reaches a maximum value, so

$$K_{LS} p_{max} = \gamma r_1 (\sin\varphi)_{max} \tag{16-35}$$

Dividing (16-34) by (16-35)

$$p_P = \frac{\sin\varphi}{(\sin\varphi)_{max}} p_{max} \tag{16-36}$$

It should be noted in Figure 16.5(a) that

$$\begin{aligned} (\sin\varphi)_{max} &= \sin\varphi_2 && \text{if} \quad \varphi_2 \leq 90^\circ \\ (\sin\varphi)_{max} &= 1 && \text{if} \quad \varphi_2 > 90^\circ \end{aligned} \tag{16-37}$$

To sum moments about hinge pin location C, it should be noted from 16.5(b) that both the normal force components and the friction force components contribute, just as for a short-shoe brake, but for the long-shoe case the moment of normal forces dN and the moment of friction forces dF_f must be *integrated over the entire arc of contact surface* between the lining and the drum. Considering the elemental contact area $dA = w_c R d\varphi$ [see Figure 16.5(b)], the elemental normal force at P may be written as

$$dN = p_P w_c R d\varphi \tag{16-38}$$

and the elemental friction force at P as

$$dF_f = \mu dN = \mu p_P w_c R d\varphi \tag{16-39}$$

With these expressions, for counterclockwise drum rotation, the moment of *normal forces*[32] about hinge pin C may be written as [see 16.5(b)]

$$M_N = \int_{\varphi_1}^{\varphi_2} q dN = \int_{\varphi_1}^{\varphi_2} r_1 \sin\varphi dN = \int_{\varphi_1}^{\varphi_2} r_1 p_P w_c R \sin\varphi d\varphi \tag{16-40}$$

Substituting from (16-36),

$$M_N = \int_{\varphi_1}^{\varphi_2} r_1 \left[\frac{\sin\varphi}{(\sin\varphi)_{max}} p_{max}\right] w_c R \sin\varphi d\varphi = \frac{w_c R r_1 p_{max}}{(\sin\varphi)_{max}} \int_{\varphi_1}^{\varphi_2} \sin^2\varphi d\varphi \tag{16-41}$$

[32]Assuming counterclockwise rotation about C to be positive.

Integrating (16-41),

$$M_N = \frac{w_c R r_1 p_{max}}{4(\sin\varphi)_{max}}(2\alpha - \sin 2\varphi_2 + \sin 2\varphi_1) \tag{16-42}$$

Similarly, for counterclockwise drum rotation, the moment of *friction forces* about hinge pin C may be written as

$$M_f = \int_{\varphi_1}^{\varphi_2}(R - r_1\cos\varphi)\,dF_f = \int_{\varphi_1}^{\varphi_2}(R - r_1\cos\varphi)\,\mu p_P w_c R\,d\varphi \tag{16-43}$$

Again substituting from (16-36),

$$M_f = \frac{\mu w_c R p_{max}}{(\sin\varphi)_{max}}\int_{\varphi_1}^{\varphi_2}(R\sin\varphi - r_1\sin\varphi\cos\varphi)\,d\varphi \tag{16-44}$$

and integrating (16-44),

$$M_f = \frac{\mu w_c R p_{max}}{4(\sin\varphi)_{max}}\left[\frac{r_1}{2}(\cos 2\varphi_2 - \cos 2\varphi_1) - R(\cos\varphi_2 - \cos\varphi_1)\right] \tag{16-45}$$

Because of the sign convention used in developing these expressions for M_N and M_f, if numerical evaluation results in a *positive* moment, it is *counterclockwise* about C. If numerical evaluation results in a *negative* moment, it is *clockwise* about C. The signs of these moments depend upon hinge location and the direction of drum rotation. Keeping the sign convention in mind, and referring to Figure 16.5(b), a moment summation about C may be written as

$$M_N + M_f - F_a a = 0 \tag{16-46}$$

whence the required actuation force F_a becomes[33]

$$F_a = \frac{M_N + M_f}{a} \tag{16-47}$$

Clearly, if M_N and M_f have the *same* sign, self-energization is not possible. If the friction moment *assists* the actuating force in applying the brake[34], the brake will be *self-energizing* (and may be self-locking), as already discussed in 16.5 and 16.6. From (16-47) then, for counterclockwise drum rotation, the long-shoe brake becomes self-locking if[35]

$$M_N + M_f \leq 0 \tag{16-48}$$

Reversing the direction of drum rotation reverses the sign of M_f, and the actuating force equation for reversed drum rotation becomes

$$F_a = \frac{M_N - M_f}{a} \tag{16-49}$$

Because of the instability inherent in a self-locking brake, it is often recommended that self-energizing long-shoe brakes should be configured so that

$$\left|\frac{M_f}{M_N}\right| \leq 0.7 \tag{16-50}$$

It is well to remember that either reversing the direction of drum rotation or moving the hinge pin location may profoundly affect moment directions and self-energizing capability.

The friction braking torque capacity, T_f, for a long-shoe brake such as the one de-

[33]Compare with (16-13) for a *short-shoe* brake.

[34]That is, M_f has the same sense as $F_a a$.

[35]Compare with (16-14) for a *short-shoe* brake.

picted in Figure 16.5 may be written by summing moments on the drum about its axis of rotation to obtain

$$T_f = \int_{\varphi_1}^{\varphi_2} RF_f = \int_{\varphi_1}^{\varphi_2} R(\mu p_P w_c R d\varphi) \tag{16-51}$$

substituting from (16-36),

$$T_f = \frac{\mu w_c R^2 p_{max}}{(\sin\varphi)_{max}} \int_{\varphi_1}^{\varphi_2} \sin\varphi d\varphi = \frac{\mu w_c R^2 p_{max}}{(\sin\varphi)_{max}}(\cos\varphi_1 - \cos\varphi_2) \tag{16-52}$$

Just as for short-shoe brakes, the design value of friction torque capacity, T_f, must be large enough to safely bring the moving mass of the operating system to zero velocity within the specified response time t_r, and the energy stored in the moving mass must be safely dissipated by the brake package.

For long-shoe brakes, as for short-shoe brakes,[36] the horizontal and vertical hinge reaction forces, for pin location C, may be found from horizontal and vertical force summations on the lever and shoe (taken together as a free body), as illustrated in Figure 16.5(b). However, the expressions are more complicated because of the nonuniform pressure distribution. Referring to Figure 16.5(b), a vertical force summation on the free body shown gives

$$R_v + \int_{\varphi_1}^{\varphi_2} (dN)_v + \int_{\varphi_1}^{\varphi_2} (dF_f)_v - F_a = 0 \tag{16-53}$$

in which the vertical component of the *normal forces* may be found as

$$\begin{aligned} N_v &= \int_{\varphi_1}^{\varphi_2} (dN)_v = \int_{\varphi_1}^{\varphi_2} p_P w_c R \sin(\beta + \varphi)\, d\varphi \\ &= \int_{\varphi_1}^{\varphi_2} \frac{\sin\varphi}{(\sin\varphi)_{max}} p_{max} w_c R \sin(\beta + \varphi)\, d\varphi \\ &= \frac{w_c R p_{max}}{(\sin\varphi)_{max}} \int_{\varphi_1}^{\varphi_2} \sin\varphi \sin(\beta + \varphi)\, d\varphi \end{aligned} \tag{16-54}$$

This may be integrated to find

$$\begin{aligned} N_v = \frac{w_c R p_{max}}{4(\sin\varphi)_{max}} &[2 \sin\beta(\cos 2\varphi_1 - \cos 2\varphi_2) \\ &+ \cos\beta(2\alpha - \sin 2\varphi_2 + \sin 2\varphi_1)] \end{aligned} \tag{16-55}$$

Similarly, the vertical component of the *friction forces* may be written as

$$(F_f)_v = \int_{\varphi_1}^{\varphi_2} (dF_f)_v = -\int_{\varphi_1}^{\varphi_2} \frac{\sin\varphi}{(\sin\varphi)_{max}} p_{max} \mu w_c R \cos(\beta + \varphi)\, d\varphi \tag{16-56}$$

which may be integrated to obtain

$$\begin{aligned} (F_f)_v = \frac{\mu w_c R p_{max}}{4(\sin\varphi)_{max}} &[-2 \cos\beta(\cos 2\varphi_1 - \cos 2\varphi_2) \\ &+ \sin\beta(2\alpha - \sin 2\varphi_2 + \sin 2\varphi_1)] \end{aligned} \tag{16-57}$$

Combining (16-55) with (16-57),

$$R_v = F_a - \frac{w_c R p_{max}}{4(\sin\varphi)_{max}}[-(2\sin\beta - 2\mu\cos\beta)(\cos 2\varphi_1 - \cos 2\varphi_2)$$

[36]See (16-17) and (16-18).

$$+ (\cos\beta + \mu \sin\beta(2\alpha - \sin 2\varphi_2 + \sin 2\varphi_1))] \tag{16-58}$$

Following a similar logic, a horizontal force summation on the free body shown in Figure 16.5(b) gives

$$R_h = \frac{w_c R p_{max}}{4(\sin\varphi)_{max}}[(2\mu \sin\beta - 2\cos\beta)(\cos 2\varphi_1 - \cos 2\varphi_2) + (\mu\cos\beta + \sin\beta)(2\alpha - \sin 2\varphi_2 + \sin 2\varphi_1)] \tag{16-59}$$

As a practical matter, the pressure distribution assumed in (16-36) is often significantly altered during the early "wear-in" period because the region of higher pressure wears more rapidly,[37] thereby redistributing the higher pressure more uniformly. All of the long-shoe equations from (16-41) through (16-59) are affected by pressure redistribution during the early wear-in period. For example, the *torque capacity*, T_f, calculated from (16-52), tends to be smaller than the *actual torque capacity* following initial wear-in, so (16-52) gives a *conservative* result in the design sense. In the final analysis, experimental testing should always be conducted to validate any new brake design proposal.

A variation on the long-shoe brake[38] is the pivoted symmetrical-shoe brake sketched in Figure 16.7. The shoe pivot point, Q, is located on the vertical centerline of the symmetrical shoe at the radius r_f, selected so that the frictional moment about Q due to *normal* forces is also zero. Thus there is no tendency of the shoe to rotate about Q. This is basically a desirable condition since it tends to equalize wear over the whole arc of contact. In fact, however, because the shoe moves progressively closer to the drum as wear takes place, in turn reducing r_f, progressively larger *friction* moments are produced about Q, and accelerated wear is generated at either the *toe* or *heel* of the shoe, depending upon the direction of drum rotation. For these reasons the pivoted symmetrical-shoe brake is not often used. The *single-shoe* external block brake shown in Figure 16.5 is less common than the *two-shoe* design crudely sketched in Figure 16.2(a). The governing equations are the same for the two-shoe design as for the single-shoe design, except that the friction torque on the drum is the *sum* of the torques from the two shoes. Since the direction of drum rotation and hinge pin location directly influence brake performance, these details must be carefully considered in the design of a two-shoe brake.[39] Depending upon these details, both shoes may be self-energizing, neither shoe may be self-energizing, or one shoe may be self-energizing while the other is not self-energizing, for a given direction of drum rotation.

The analysis of *internal expanding* shoe brakes, as crudely sketched in Figure 16.2(b), gives the *same equations* for pressures, torques, forces, and moments that have been de-

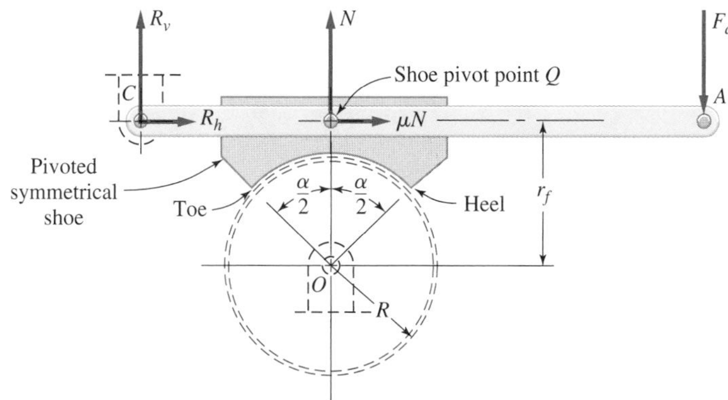

Figure 16.7
External long-shoe brake with pivoted symmetrical shoe.

[37]See (16-27). [38]See Figure 16.5. [39]See (16-13), (16-16), (16-47), and (16-50).

Figure 16.8
Sketch of internal expanding two-shoe brake arrangement typical of rear-wheel automotive applications.

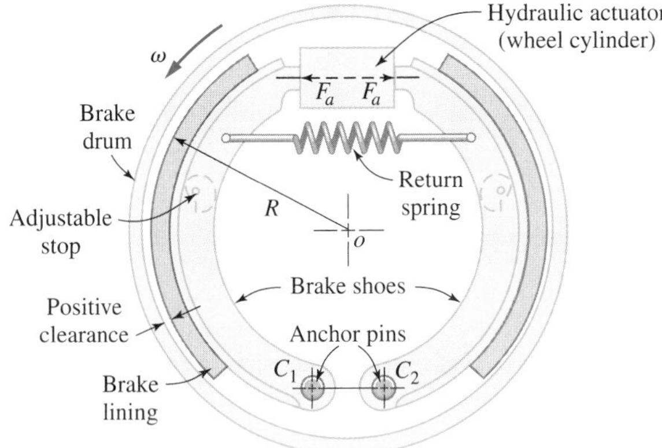

veloped for *external contracting* shoes. Internal expanding two-shoe brake systems are widely used in automotive applications. As sketched in Figure 16.8, each shoe typically pivots at one end about an anchor pin, and is actuated by a hydraulic cylinder at the other end. The typical arrangement, as shown, places the hydraulic actuator between the non-pivoted ends of the two shoes, resulting in one self-energizing shoe in the forward direction of drum rotation and the other shoe self-energizing in the reverse direction. A light return spring is used to retract the shoes against the adjustable stops when the brake is not actuated. The adjustable stops are used to maintain a small positive clearance between the shoes and the drum when the brake is not actuated.

16.8 Band Brakes

A band brake (or clutch) employs a flexible band, usually a thin metal strip lined with a flexible friction material, wrapped around a rigid drum, as shown schematically in Figure 16.2(c). This simple device is sketched in more detail in Figure 16.9, showing a band brake actuated by a lever arrangement. For the clockwise drum rotation shown, the clockwise

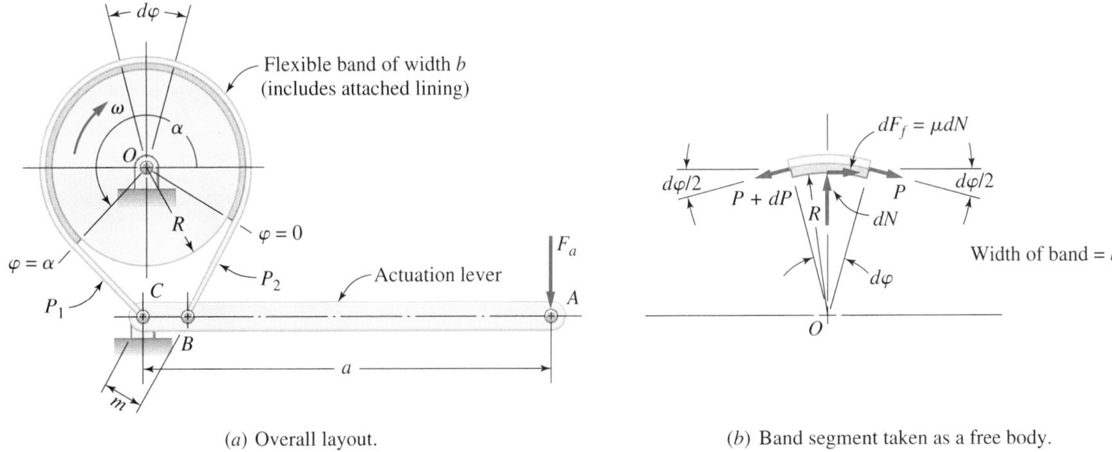

(*a*) Overall layout. (*b*) Band segment taken as a free body.

Figure 16.9
Simple band brake.

friction forces on the band tend to increase tensile force P_1 and decrease tensile force P_2. The angle of wrap α is usually in the range of 270° to 330°. Taking a short elemental segment of the band as a free body, as shown in Figure 16.9(b), it may be noted that the tension force P normal to the right end of the band segment is less than the tension force $P + dP$ normal to the left end, because of the friction force dF_f. If the segment is defined by two radial cutting planes that are $d\varphi$ apart, and chosen to be symmetrical about a vertical centerline, a horizontal force summation on the segment may be written as

$$(P + dP)\cos\frac{d\varphi}{2} - P\cos\frac{d\varphi}{2} - dF_f = 0 \tag{16-60}$$

For the very small angle $d\varphi$,

$$\cos\frac{d\varphi}{2} \approx 1 \tag{16-61}$$

so (16-60) becomes

$$dP - \mu dN = 0 \tag{16-62}$$

A vertical force summation gives

$$dN - (P + dP)\sin\frac{d\varphi}{2} - P\sin\frac{d\varphi}{2} = 0 \tag{16-63}$$

For small angles,

$$\sin\frac{d\varphi}{2} \approx \frac{d\varphi}{2} \tag{16-64}$$

Since $dPd\varphi$ is a higher order differential, (16-63) may therefore be writttten as

$$dN - 2P\left(\frac{d\varphi}{2}\right) = 0 \tag{16-65}$$

or

$$dN = Pd\varphi \tag{16-66}$$

Combining (16-62) and (16-66)

$$dP - \mu P d\varphi = 0 \tag{16-67}$$

or

$$\frac{dP}{P} = \mu d\varphi \tag{16-68}$$

Integrating both sides over wrap angle α

$$\int_{P_2}^{P_1} \frac{dP}{P} = \int_0^{\alpha} \mu d\varphi \tag{16-69}$$

which gives

$$\ln\frac{P_1}{P_2} = \mu\alpha \tag{16-70}$$

or

$$\frac{P_1}{P_2} = e^{\mu\alpha} \tag{16-71}$$

Referring again to Figure 16.9, the braking torque T_f may be written by taking a moment summation about drum center, O, to give

$$T_f = (P_1 - P_2)R = P_2R(e^{\mu\alpha} - 1) \tag{16-72}$$

Also, from (16-66)

$$Pd\varphi = dN = pbRd\varphi \tag{16-73}$$

or

$$p = \frac{P}{bR} \tag{16-74}$$

where p = pressure at any point in the arc of contact
P = tensile force in the band at the same point
R = drum radius
b = width of band

Since pressure is proportional to P, and P is largest at the tight side where the band leaves the drum tangentially,[40] the maximum contact pressure p_{max} is

$$p_{max} = \frac{P_1}{bR} = \frac{P_2 e^{\mu\alpha}}{bR} \tag{16-75}$$

Rearranging (16-75),

$$P_2 = \frac{bRp_{max}}{e^{\mu\alpha}} \tag{16-76}$$

and the braking torque equation (16-72) may be rewritten as

$$T_f = bR^2 p_{max}(1 - e^{-\mu\alpha}) \tag{16-77}$$

As in any brake or clutch application, the maximum pressure p_{max} is limited by the lining material selected.[41]

The actuating force F_a on the lever, as depicted in Figure 16.9, depends on the lever geometry and where the flexible band attachments are made. If the tight side of the band is attached at lever pivot point C, summing moments about C gives

$$F_a a = mP_2 \tag{16-78}$$

whence

$$F_a = \left(\frac{m}{a}\right)\left(\frac{bRp_{max}}{e^{\mu\alpha}}\right) \tag{16-79}$$

This arrangement is usually called a *simple* band brake.

If the lever shown in Figure 16.9 is redesigned so that the tight side of the band is attached to the lever at a point *away* from the lever pivot point, a moment on the lever is produced about C that assists F_a in actuating the brake. This arrangement is usually called a *differential* band brake. Figure 16.10 depicts a differential band brake. For this arrangement, the actuating force F_a may again be found from the lever moment summation about C as[42]

$$F_a = \frac{m_2 P_2 - m_1 P_1}{a} \tag{16-80}$$

For the direction of rotation and lever arrangement shown in Figure 16.10, (16-80) indicates that the differential band brake is self-energizing, and if

$$m_1 P_1 \geq m_2 P_2 \tag{16-81}$$

it is self-locking.

[40] $P = P_1$ at this location. [41] See Table 16.1. [42] Compare with (16-79) for a *simple* band brake.

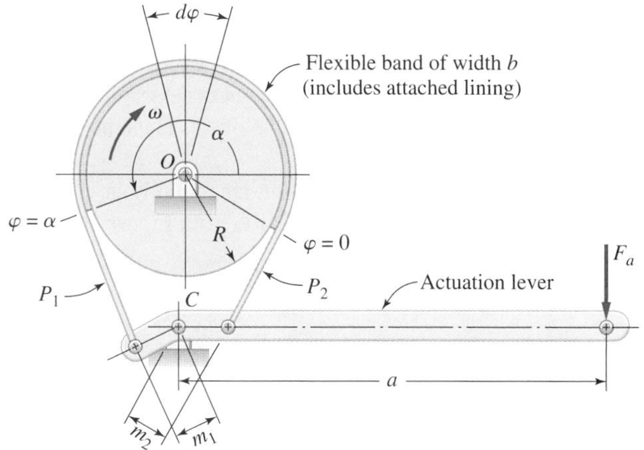

Figure 16.10
Differential band brake.

Example 16.2 Band Brake Design

A band brake is being considered for use in an application that requires a braking torque capacity of 3000 in-lb on a 10-inch-radius cast-iron drum rotating clockwise, as sketched in Figure E16.2. It is desired to design a foot-operated actuation lever to be located below the drum that will require an actuation force no greater than 30 lb. For ergonomic reasons, the lever should be no longer than $a = 20$ inches. Propose a lever design, lining material, and band width that would fulfill the design specifications.

Solution

a. A *simple* band brake will be investigated first. If necessary, a differential band brake will be examined later. The design arrangement shown in Figure E16.2 will be selected as an initial design proposal for the actuation lever.

b. From Table 16.1, tentatively select a woven cotton lining material[43] bonded to a thin steel band attached at B and C, as sketched in Figure E16.2. The coefficient of friction, from Table 16.1, is

$$\mu = 0.47 \tag{1}$$

and the allowable maximum pressure for the woven cotton lining is

$$(p_{max})_{allow} = 100 \text{ psi} \tag{2}$$

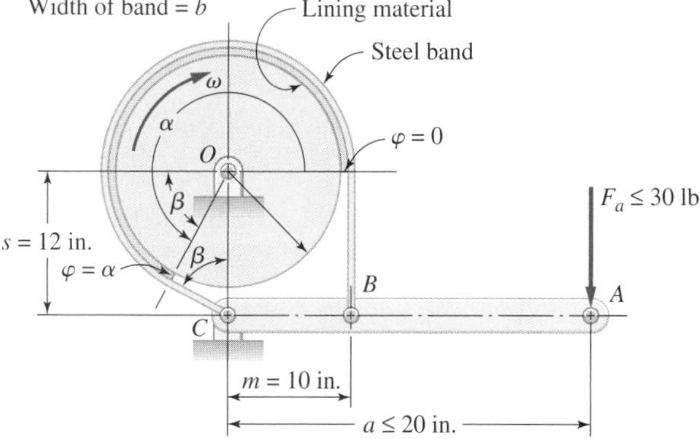

Figure E16.2
Sketch of proposed lever arrangement for simple band brake application.

[43]Because flexibility is an important criterion.

Example 16.2 Continues

c. The initial proposal will be to place lever pivot point C on the vertical centerline of the drum a distance $s = 12$ inches below the center of the drum, O. The band attachment point B will be placed 10 inches to the right of lever pivot point C, making the band at that location vertical and tangent to the drum. Thus the angle of wrap α starts at the horizontal drum centerline ($\varphi = 0$).

The band wraps around the drum, leaving tangentially at $\varphi = \alpha$, and attaches to the frame at lever pivot point C. Geometrically, from Figure E16.2,

$$\alpha = 180 + \beta = 180 + \sin^{-1}\frac{10}{12} = 180 + 56.4 \approx 236° \tag{3}$$

Summarizing the proposal,

$$\begin{aligned} a &= 20 \text{ inches} \\ m &= 10 \text{ inches} \\ R &= 10 \text{ inches} \\ \alpha &= 236° \quad (4.12 \text{ rad}) \end{aligned} \tag{4}$$

d. Utilizing (16-77),

$$3000 = b(10)^2(100)(1 - e^{-0.47(4.12)}) = b(8557.8) \tag{5}$$

or

$$b_{min} = \frac{3000}{8557.8} = 0.35 \text{ inch} \tag{6}$$

If a larger width b were chosen for the band, the 3000 in-lb torque capacity could be produced at a p_{max} value less than the allowable 100 psi.

e. From (16-79) then, for a band of width 0.35 inch, the actuating force would be

$$F_a = \left(\frac{10}{20}\right)\left(\frac{0.35(10)(100)}{e^{0.47(4.12)}}\right) = 25.3 \text{ lb} \tag{7}$$

Since

$$F_a = 25.3 < (F_a)_{allow} = 30 \text{ lb} \tag{8}$$

the design specifications are satisfied.

A width of $b = 0.35$ inch would work, therefore, but as a matter of design judgment, it seems disproportionately small. One might arbitrarily choose $b = 1.0$ inch, with the knowledge from (5) that the design torque of 3000 in-lb could be produced with a p_{max} value lower than[44] the allowable 100 psi, and from (7) that the same (acceptable) actuating force would produce the required braking torque. The proposed design recommendation, therefore, is to adopt the configuration sketched in Figure E16.2, with the following specifications:

1. Use a woven cotton lining material bonded to a 1-inch-wide steel band.
2. Use a lever of length $a = 20$ inches.
3. Place the lever pivot point on the vertical centerline of the drum, 12 inches below the center of the drum.
4. Position the actuation lever horizontally, with the band attachment point B located 10 inches to the right of lever pivot point C.

Of course, many other satisfactory dimensional combinations could be found. Experimental validation would be suggested prior to production.

[44]The value would be $p_{max} = 0.35/1.0(100) = 35$ psi.

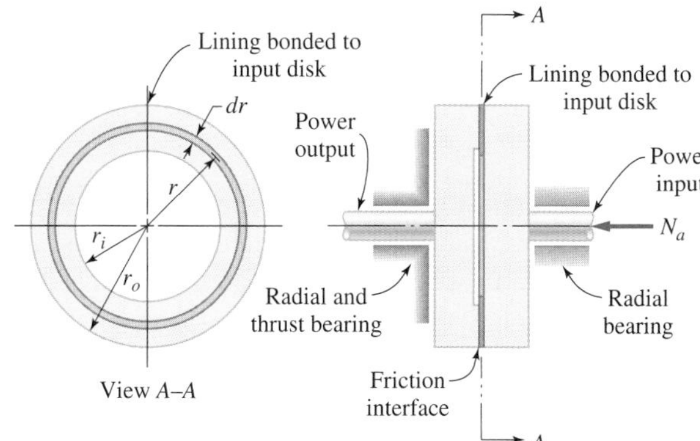

Figure 16.11
Single-disk brake or clutch.

16.9 Disk Brakes and Clutches

A schematic diagram of a disk (plate) brake or clutch is shown in Figure 16.2(d). The unit shown embodies *two* friction interfaces. The unit is actuated by clamping a central rotating disk between two straddling coaxial disks *splined* to a *separate shaft* (clutch), or clamping a central rotating disk between two straddling coaxial disks splined to a *fixed housing* (brake). The device is actuated by applying opposing axial forces to the splined outer disks, causing them to slide and clamp the central disk between them, as shown, to bring all disks to a common velocity.[45]

Other versions of a disk brake or clutch may embody only a single friction interface, as shown in Figure 16.11, or multiple friction interfaces, as shown in Figure 16.12. Disk brakes are generally considered to provide the best performance and best fade-resistance of any type of brake. Since the disk configuration provides freedom from centrifugal force effects,[46] it is widely used for clutch applications. Using multiple-disk versions, a large frictional area may be installed in a small space, and favorable pressure distributions are usually generated by an initial wear-in period.

Due to the concentrated heat generation within a multiple-disk brake package, and the difficulty in adequately dissipating the friction-generated heat, such compact devices (with full annular interfacial contact between disks) are rarely used for brakes in high-power applications because they would usually exceed allowable temperatures. For some installations, effective heat dissipation *is* possible.[47]

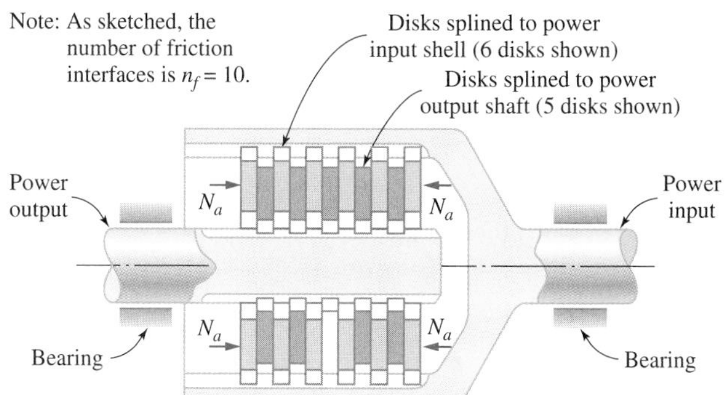

Figure 16.12
Multiple-disk brake or clutch.

[45]For a brake, the common velocity is zero.

[46]If shoe-type devices are used as clutches (both shafts rotating), the shoes are subjected to a centrifugal force field that tends either to increase the pressure between shoes and rim (internal shoes) or to decrease the pressure between shoes and rim (external shoes).

[47]True for caliper brakes, for example, as sketched in Figure 16.13.

Disk brakes do not offer self-energization, and typically require high axial actuation forces; they are therefore often actuated by using power-assist devices. A common clutch configuration is to use a preloaded axial spring arrangement to provide the necessary clamping force for proper engagement, then utilize a lever arrangement or hydraulic actuator to *disengage* the device.

Just as for other types of brakes and clutches, a designer is interested in estimating friction torque capacity; actuating forces required; limitations imposed by pressure, temperature, or wear; strength limitations; and energy dissipation (heat dissipation) capability.

Depending upon construction details and materials chosen, an assumption of either *uniform wear* or *uniform pressure* is usually made in order to estimate pressure distribution, braking torque, and actuation force for a disk brake or clutch.

If the disks tend to be *rigid*, the greatest wear will at first occur over the outer circumferential portion of the disk because tangential velocity is greater there.[48] After a certain amount of initial wear, the pressure distribution will change so as to ultimately result in nearly *uniform wear*.[49]

If the disks tend to be *flexible*, and are held in intimate contact,[50] nominally *uniform pressure* exists between the disk interfaces.[51]

Uniform Wear Assumption

For a design based on the uniform wear assumption, (16-27) may be utilized to give

$$\delta_n = k_w(pV) = k_w(2\pi rn)p = (2\pi nk_w)pr \tag{16-82}$$

where δ_n = normal wear rate
k_w = material pair constant[52]
p = contact pressure at radius r
V = tangential relative sliding velocity between contacting disks, at radius r
r = radius to any selected elemental annular ring of radial thickness dr that lies between the inner radius r_i and outer radius r_o, as illustrated in Figure 16.11
n = relative rotational speed between disks, at the interface, rpm

Adopting the *uniform wear* assumption.

$$\delta_n = \text{constant} \tag{16-83}$$

Combining (16-82) and (16-83) gives

$$pr = \text{constant} \tag{16-84}$$

so it may be observed that the maximum pressure p_{max} will occur at the inner radius, giving, from (16-82),

$$\delta_n = (2\pi nk_w)p_{max}r_i \tag{16-85}$$

Equating (16-82) to (16-85) then yields

$$p = \frac{r_i}{r}p_{max} \tag{16-86}$$

[48] See (16-27).

[49] This occurs when the (pV) product is nominally constant over the whole interface.

[50] Often by suitably distributed axial spring loading.

[51] The uniform wear assumption is more conservative in the design sense.

[52] The material wear constant may be available from the friction material manufacturer.

For the uniform wear assumption, referring to Figure 16.11, axial normal force N_a may be written as

$$(N_a)_{uw} = \int_{r_i}^{r_o} p dA = \int_{r_i}^{r_o} \left(\frac{r_i}{r} p_{max} \right) (2\pi r dr) = 2\pi p_{max} r_i (r_o - r_i) \qquad (16\text{-}87)$$

The friction torque (braking torque) may be found by integrating the product of differential tangential friction force, dF_f, times radius r, for an elemental annular area dA,[53] from inner radius r_i to outer radius r_o. This gives

$$(T_f)_{uw} = \int_{r_i}^{r_o} r dF_f = \int_{r_i}^{r_o} r\mu p dA = \int_{r_i}^{r_o} r\mu \left(\frac{r_i}{r} p_{max} \right) 2\pi r dr \qquad (16\text{-}88)$$

Performing the integration, and utilizing (16-87),

$$(T_f)_{uw} = \mu \pi p_{max} r_i (r_o^2 - r_i^2) = \mu \left(\frac{r_o + r_i}{2} \right) N_a \qquad (16\text{-}89)$$

If a multiple-disk arrangement is used as shown in Figure 16.12, the torque for *each friction interface* is given by (16-88). Therefore, if there are n_f friction interfaces in a multiple-disk unit, the braking torque capacity is given by

$$(T_f)_{uw} = n_f \mu \pi p_{max} r_i (r_o^2 - r_i^2) = n_f \mu \left(\frac{r_o + r_i}{2} \right) N_a \qquad (16\text{-}90)$$

Uniform Pressure Assumption

For a design based on the *uniform pressure* assumption,

$$p = p_{max} \qquad (16\text{-}91)$$

$$(N_a)_{up} = \int_{r_i}^{r_o} p dA = p_{max} \int_{r_i}^{r_o} 2\pi r dr = p_{max} \pi (r_o^2 - r_i^2) \qquad (16\text{-}92)$$

and the friction torque for *one* friction interface is

$$\begin{aligned} (T_f)_{up} &= \int_{r_i}^{r_o} r dF_f = \int_{r_i}^{r_o} r\mu p dA = \mu p_{max} \int_{r_i}^{r_o} r(2\pi r dr) \\ &= 2\pi \mu p_{max} \left(\frac{r_o^3 - r_i^3}{3} \right) = \frac{2\mu (r_o^3 - r_i^3)}{3(r_o^2 - r_i^2)} N_a \end{aligned} \qquad (16\text{-}93)$$

Finally, if a multiple-disk unit with n_f friction interfaces is used,

$$(T_f)_{up} = 2\pi n_f \mu p_{max} \left(\frac{r_o^3 - r_i^3}{3} \right) = \frac{2\mu n_f (r_o^3 - r_i^3)}{3(r_o^2 - r_i^2)} N_a \qquad (16\text{-}94)$$

Caliper disk brakes, illustrated in Figure 16.13, are often used in automotive applications. They utilize friction lining "pads" that extend circumferentially over only a *small sector* of the disk to give crescent-shaped pads[54] squeezed against both sides of the rotating disk. Since most of the rotating disk is directly exposed to the ambient atmosphere, convective heat dissipation is enhanced. In addition, a hollow disk with integral vanes is sometimes used to "pump" air through interior passages of the disk, providing substantial additional cooling. The actuating force F_a, shown in Figure 16.13, is usually provided by a hydraulic actuator. The equations for axial normal force N_a and braking torque T_f may

[53]See Figure 16.11. [54]Other shapes are sometimes used for the pad.

Figure 16.13
Caliper-type disk brake.

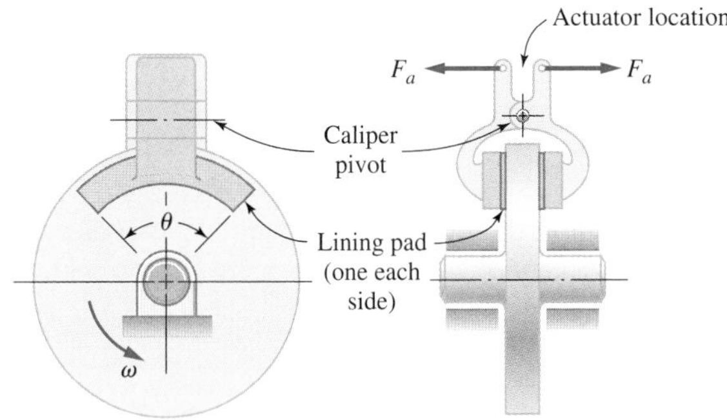

be readily developed from (16-87), (16-89), (16-92), or (16-94), by simply multiplying the selected equation by the ratio $\theta/2\pi$, where θ is the angle subtending the brake pad sector. The angle θ usually lies in the range from $\pi/4$ to $\pi/2$. The inner radius, r_i, for disk brakes, including caliper brakes, usually lies in the range from 0.60 r_o to 0.80 r_o.

Example 16.3 Multiple-Disk Brake Design

A half-ton low-cost drum-type hoist has been designed to lift or lower a payload attached to the end of a flexible cable which is wrapped around a 16-inch-diameter drum, as shown in Figure E16.3. A constant-speed centrifugal governor limits the lowering speed to a maximum velocity of 10 ft/sec. The cast-iron drum weighs 300 lb and has a radius of gyration of 7.25 inches.

Propose a multiple-disk brake of the type shown in Figure 16.12, mounted coaxially with the drum, that has the capacity to completely stop the dropping payload in 0.2 sec. An additional restriction is that the required normal actuating force N_a (see Figure 16.12) must not exceed 2000 lb.

Solution

a. Since this is a low-speed application, and cost is a factor, the tentative material pair selection will be cast iron against cast iron (dry). From Table 16.1, the properties of cast iron[55] on cast iron are

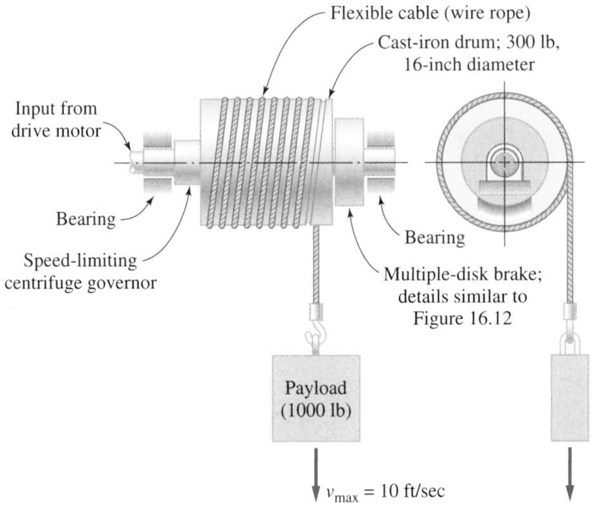

Figure E16.3
Sketch of drum hoist.

Flexible cable (wire rope)
Cast-iron drum; 300 lb, 16-inch diameter
Input from drive motor
Bearing
Speed-limiting centrifuge governor
Bearing
Multiple-disk brake; details similar to Figure 16.12
Payload (1000 lb)
v_{max} = 10 ft/sec

[55]See also Appendix Table A.1.

$$\begin{aligned}\mu &= 0.15 \text{ to } 0.2\\ (p_{max})_{allow} &= 150 \text{ to } 250 \text{ psi}\\ T_{max} &= 600°\text{F}\\ V_{max} &= 3600 \text{ ft/min}\end{aligned} \tag{1}$$

b. Since the disks of cast iron are rigid, the *uniform wear* assumption will be used to estimate the actuating force N_a and friction torque T_f required to meet design specifications. Hence, from (16-87) and (16-90), respectively,

$$(N_a)_{uw} = 2\pi p_{max} r_i (r_o - r_i) \tag{2}$$

and

$$(T_f)_{uw} = n_f \mu \pi p_{max} r_i (r_o^2 - r_i^2) = n_f \mu \left(\frac{r_o + r_i}{2}\right) N_a \tag{3}$$

c. Energy dissipation requirements may be estimated as

$$\begin{aligned}KE &= (KE)_{\substack{rotating\\drum}} + (KE)_{\substack{translating\\payload}} = \frac{1}{2} J_{dr}\omega_{dr}^2 + \frac{1}{2} m_{pl} v_{pl}^2\\ &= \frac{1}{2}\left(\frac{W_{dr}}{g} k_{dr}^2\right)\left(\frac{v_{pl}}{R}\right)^2 + \frac{1}{2}\left(\frac{W_{pl}}{g}\right) v_{pl}^2\end{aligned} \tag{4}$$

Evaluating this expression numerically,

$$\begin{aligned}KE &= \frac{1}{2}\left[\frac{300}{386}(7.25)^2\left(\frac{10 \times 12}{8.0}\right)^2 + \frac{1000}{386}(10 \times 12)^2\right]\\ &= \frac{1}{2}[9192 + 37{,}306] = 23{,}249 \text{ in-lb}\end{aligned} \tag{5}$$

d. To stop the drum rotation in the specified time of $t = 0.2$ sec, assuming uniform deceleration, the angular displacement θ of the drum during the braking period may be estimated as

$$\theta = \omega_{ave} t = \left(\frac{\omega_i + \omega_f}{2}\right) t = \left(\frac{15 + 0}{2}\right) 0.2 = 1.5 \text{ rad} \tag{6}$$

The work that must be done by the braking torque to meet specifications may be estimated utilizing (5) and (6) as

$$T_f = \frac{KE}{\theta} = \frac{23{,}249}{1.5} = 15{,}500 \text{ in-lb} \tag{7}$$

e. To be compatible with the drum radius of 8.0 inches, a somewhat smaller outer radius for the disks will be chosen, say, $r_o = 6.0$ inches. Recalling the thumb rule[56] relationship between r_i and r_o, the inner disk radius will be selected as

$$r_i \approx 0.6 r_o = 0.6(6.0) = 3.6 \text{ inches} \tag{8}$$

f. To fulfill the specification that requires

$$N_a \leq 2000 \text{ lb} \tag{9}$$

and solving (2) for p_{max}, using the results of (e),

$$p_{max} = \frac{2000}{2\pi(3.6)(6.0 - 3.6)} = 37 \text{ psi} \tag{10}$$

[56] See last paragraph of 16.9.

Example 16.3 Continues

This value of P_{max} = 37 psi is much less than the material pair allowable value of 150 to 250 psi, as given in (1), so it is acceptable.

g. Solving (3) for n_f, using the maximum permissible pressure of 37 psi from (f), a conservative value of μ = 0.15 from (1), and the result of (7)

$$n_f = \frac{15{,}500}{(0.15)\,\pi(37)(3.6)(6.0^2 - 3.6^2)} = 10.7 \approx 11.0 \text{ friction surfaces required} \quad (11)$$

h. To estimate the temperature rise, it will be assumed that the friction-generated heat will be transferred to only about 10 percent of the total volume of the 6 cast-iron disks (11 disk surfaces), or the weight of heat-absorbing cast-iron mass would be approximately[57]

$$W = (6)(0.10)\,\pi(6.0^2 - 3.6^2)(0.25)(0.270) \approx 3 \text{ lb} \quad (12)$$

Using (16-23),[58]

$$\Delta\Theta = \frac{H_f}{CW} = \frac{\left(\dfrac{23{,}249}{9336}\right)}{0.12(3)} \approx 7°\text{F} \quad (13)$$

This temperature rise is more than acceptable, since (1) allows a maximum temperature of 600°F. Also, the peripheral tangential velocity of the drum (v = 10 ft/sec = 600 ft/min) assures that the V_{max} = 3600 ft/min criterion of (1) is met.

i. Summarizing, the following design configuration is suggested:

1. Use a multiple-disk configuration, similar to that sketched in Figure 16.12, with n_f = 11 friction surfaces.
2. Make the disks of cast iron, 0.25 inch thick, with outer radius 6.0 inches and inner radius 3.6 inches.
3. Use a hydraulic actuator to provide an axial normal force of 2000 lb to squeeze the cast-iron friction disks together.

Of course, many other equally acceptable design configurations would work just as well. Experimental evaluation of the brake package should be made before production.

16.10 Cone Clutches and Brakes

By comparing the cone clutch (brake) sketched in Figure 16.14 with the disk clutch (brake) shown in Figure 16.11, it may be deduced that the disk clutch is simply a special case of a cone clutch with a *cone angle* α of 90°. In practice, the cone angle, α, usually is selected to lie within the range of about 8° to 15° with 12° commonly chosen. Cone angles less than about 8° tend to produce a self-locking wedge condition, making engagement "grabby" and disengagement difficult. Large cone angles require the use of larger actuating forces to produce a specified friction torque capacity, diminishing one of the principal advantages of selecting a cone clutch or brake in the first place.[59] It should also be noted that construction of cone clutches or brakes with more than one friction interface is usually impractical.

[57]It is assumed that the disks are each about $\frac{1}{4}$ inch thick. Weight density for cast iron is 0.270 lb/in (see Table 3.4).

[58]Also see (16-24), and (5) and (12) above.

[59]The wedging action of a properly designed clutch or brake permits a reduction of normal actuation force to only about 20 percent of the normal actuation force required for an equivalent disk brake or clutch with a single friction interface (n_f = 1).

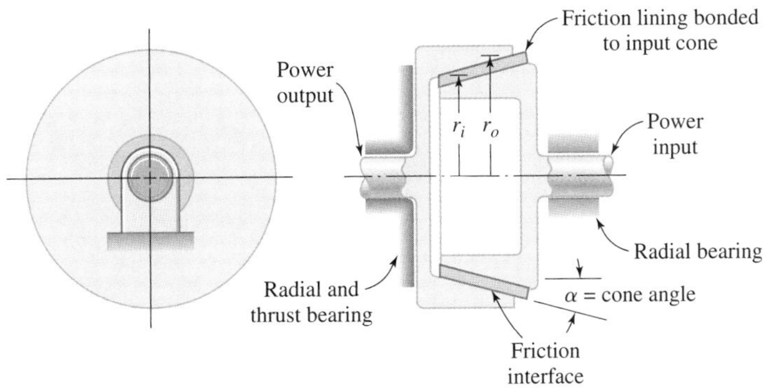

(*a*) Sketch of cone clutch arrangement.

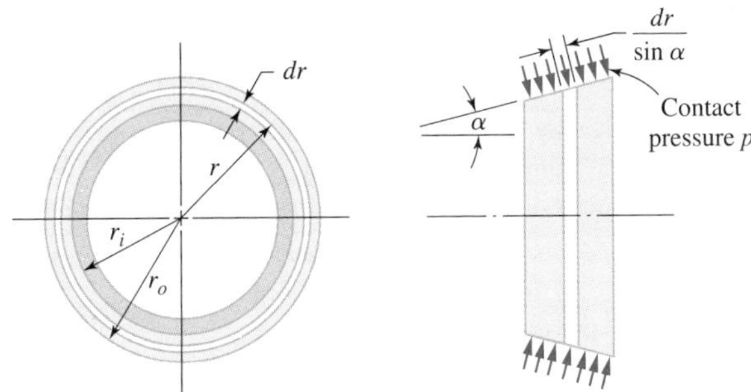

(*b*) Detail of friction lining contact interface.

Figure 16.14
Cone clutch or brake.

From Figure 16.14(b), the surface contact area dA may be written as[60]

$$dA = \frac{2\pi r dr}{\sin\alpha} \tag{16-95}$$

Equations for normal actuating force and friction torque capacity of a cone clutch with cone angle α may be developed by inserting (16-95) into (16-87) and (16-88) if uniform wear conditions are assumed, or into (16-92) and (16-93) if uniform pressure conditions are assumed. Thus, for the *uniform wear* assumption,

$$(N_a)_{uw} = \frac{2\pi p_{max} r_i (r_o - r_i)}{\sin\alpha} \tag{16-96}$$

and

$$(T_f)_{uw} = \frac{\mu\pi p_{max} r_i (r_o^2 - r_i^2)}{\sin\alpha} = \mu\left(\frac{r_o + r_i}{2}\right)\left(\frac{N_a}{\sin\alpha}\right) \tag{16-97}$$

For the *uniform pressure* assumption,

$$(N_a)_{up} = \frac{\pi p_{max}(r_o^2 - r_i^2)}{\sin\alpha} \tag{16-98}$$

and

$$(T_f)_{up} = \frac{2\pi\mu p_{max}\left(\frac{r_o^3 - r_i^3}{3}\right)}{\sin\alpha} = \frac{2\mu(r_o^3 - r_i^3)}{3(r_o^2 - r_i^2)}\left(\frac{N_a}{\sin\alpha}\right) \tag{16-99}$$

[60]Compare with $dA = 2\pi\, r dr$ used for disk clutches and brakes.

Problems

16-1. A Short-shoe block brake is to have the configuration shown in Figure P16.1, with the drum rotating clockwise at 500 rpm, as shown. The shoe is molded fiberglass, the drum is aluminum-bronze, and the entire assembly is continuously water-sprayed. Maximum-allowable contact pressure is 200 psi and the coefficient of friction of wet molded fiberglass on aluminum-bronze is 0.15.

a. Using *symbols only*, derive an expression for actuating force F_a, expressed as a function of p_{max}.

b. If the actuating force must not exceed 30 lb, what minimum lever length d should be used?

c. Using *symbols only*, write an expression for braking torque T_f.

d. Calculate a numerical value for the maximum-allowable braking torque that may be expected from this design.

16-2. Repeat problem 16-1, except that the drum rotates clockwise at 600 rpm, the shoe lining is woven cotton, the drum is cast iron, and the environment is dry. In addition, referring to Figure P16.1, e is 3.0 inches, R is 8.0 inches, the contact area A is 8.0 in^2, and $F_a = 60$ lb is the maximum-allowable value of applied force, vertically downward.

16-3. Classify the short-shoe block brake shown in Figure P16.1 as either "self-energizing" or "non-self-energizing."

16-4. Repeat problem 16-1 for the case where the drum rotates counterclockwise at 800 rpm.

16-5. Repeat problem 16-1, except that the shoe lining is *cermet* and the drum is 1020 steel.

16-6. Repeat problem 16-1, except that the coefficient of friction is 0.2, the maximum-allowable contact pressure is 80 psi, the contact area A is 10.0 in^2, e is 30.0 inches, R is 9.0 inches, c is 12.0 inches, and the maximum-allowable value of F_a is 280 lb.

16-7. A short-shoe block brake is to have the configuration shown in Figure P16.7, with the drum rotating clockwise at 600 rev/min, as shown. The shoe is molded fiberglass, the drum is stainless steel, and the entire assembly is submerged in salt water. Maximum-allowable contact pressure is 125 psi and the coefficient of friction of wet molded fiberglass on stainless steel is 0.18.

a. Using *symbols only*, derive an expression for actuating force F_a as a function of P_{max}.

b. What is the maximum actuating force that should be used for proper operation and acceptable design life?

c. Using *symbols only*, write an expression for braking torque.

d. Calculate a numerical value for maximum braking torque that may be expected from this design.

e. Would you classify this device as "self-energizing" or "non-self-energizing"? Why?

16-8. For the shoe brake shown in Figure P16.8, it is difficult to determine by inspection whether the short-shoe assumption will produce a sufficiently accurate estimate of braking torque upon application of the actuating force F_a.

a. Determine the percent error in calculated braking torque that you would expect in this case if the short-shoe assumption is used for calculation of the braking torque.

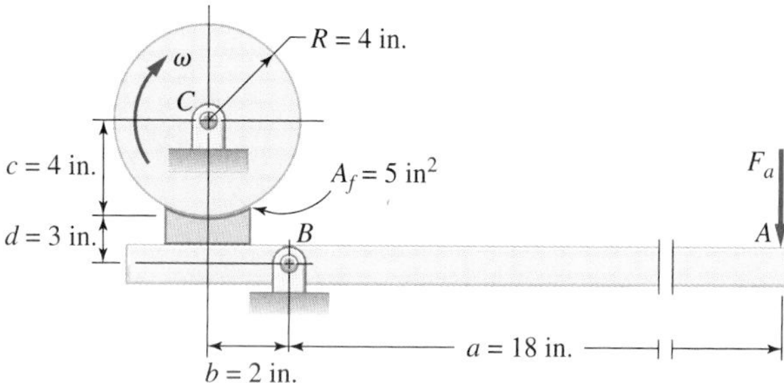

Figure P16.7
Short-shoe block brake submerged in salt water.

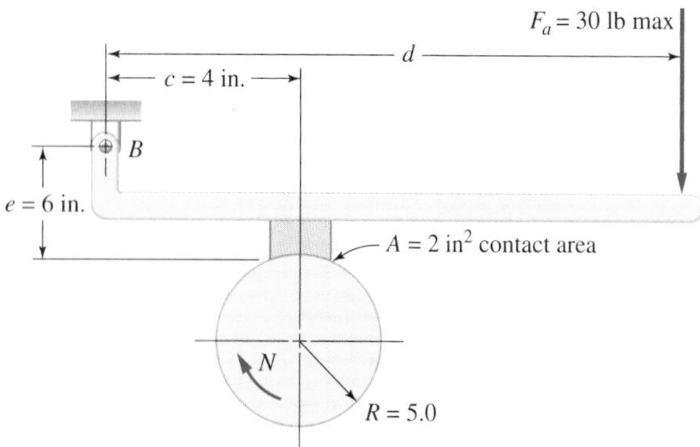

Figure P16.1
Short-shoe block brake.

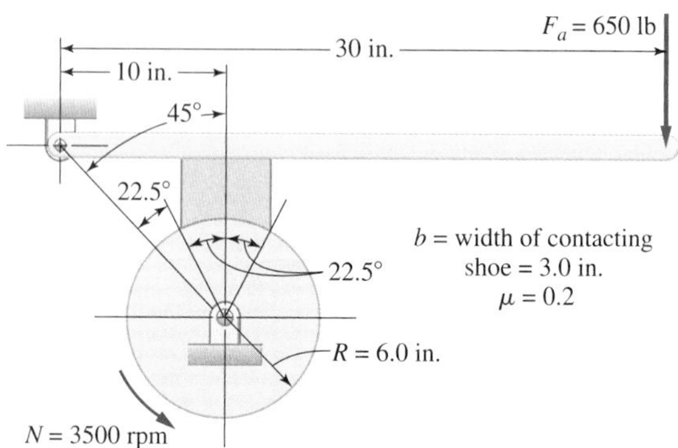

Figure P16.8
Shoe brake assembly.

Base your determination on the premise that the long-shoe equations are completely accurate.

b. Would the error made by using the short-shoe assumption be on the "conservative" side (braking torque calculated by short-shoe assumption is less than the true value of braking torque) or on the "nonconservative" side?

c. Do you consider the error made by using the short-shoe assumption to be significant or negligible for this particular case?

16-9. Repeat problem 16-8, except that a = 36 inches, F_a = 300 lb, α = 22.5°, β = 22.5°, R = 7.0 inches, b = 2.0 inches, μ = 0.25, and the drum rotates counterclockwise at a speed of 2500 rpm. An accurate estimate of 150 psi for the actual value of p_{max} has already been made.

16-10. A short-shoe block brake has the configuration shown in Figure P16.10, with the drum rotating clockwise at 63 rad/sec, as shown The shoe is wood and the drum is cast iron. The weight of the drum is 322 lb, and its radius of gyration is 7.5 inches. The maximum-allowable contact pressure is 80 psi, and the coefficient of friction is μ = 0.2. Other dimensions are shown in Figure P16.10.

a. Derive an expression for the actuating force F_a, and calculate its maximum-allowable numerical value.

b. Derive an expression for braking torque, and calculate its numerical value when the maximum-allowable actuating force is applied.

c. Estimate the time required to bring the rotating drum to a stop when the maximum-allowable actuating force is applied.

d. Would you expect frictional heating to be a problem in this application?

16-11. Repeat problem 16-10 for the case where the drum is rotating *counterclockwise* at 63 radians per second.

16-12. Figure P16.12 shows a 1000-kg mass being lowered at a uniform velocity of 3 m/s by a flexible cable wrapped around a drum of 60-cm diameter. The drum weight is 2 kN, and it has a radius of gyration of 35 cm.

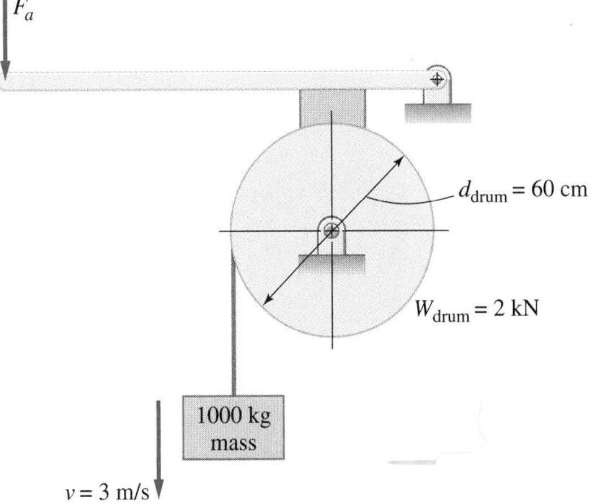

Figure P16.12
Brake-controlled drum hoist.

a. Calculate the kinetic energy in the moving system.

b. The brake shown maintains the rate of descent of the 1000-kg mass by applying the required steady torque of 300 kg-m. What additional braking torque would be required to bring the entire system to rest in 0.5 sec?

16-13. A short-shoe block brake is sketched in Figure P16.13. Four seconds after the 1-kN actuating force is applied, the CW rotating drum comes to a full stop. During this time, the drum makes 100 revolutions. The estimated coefficient of friction between drum and shoe is 0.5. Do the following:

a. Sketch the brake-shoe-and-arm assembly as a free-body diagram.

b. Is the brake self-energizing or self-locking for the direction of drum rotation shown?

c. Calculate the braking torque capacity of the system shown.

d. Calculate the horizontal and vertical reaction forces on the free body at pin location D.

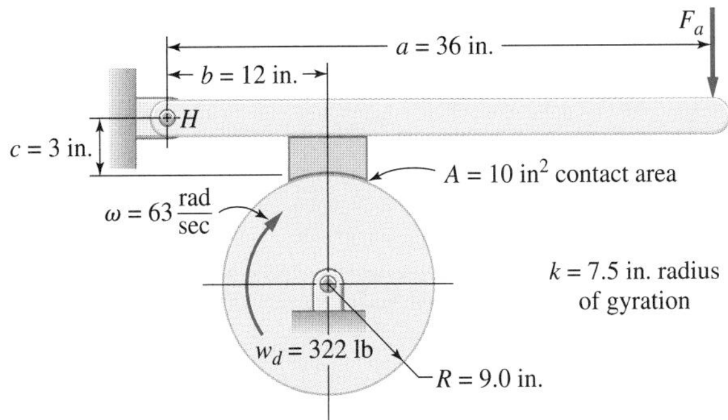

Figure P16.10
Short-shoe block brake with wood shoe.

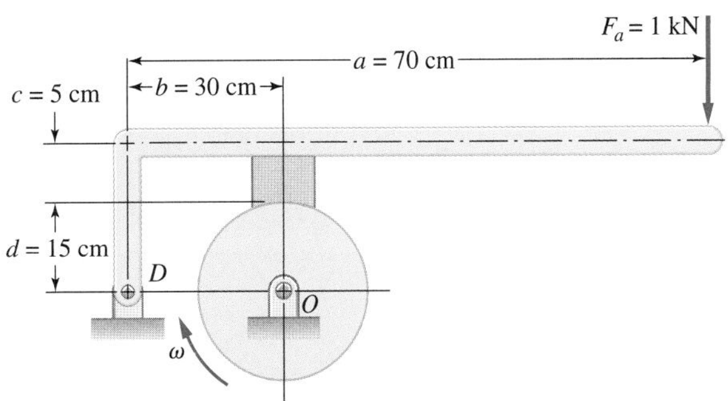

Figure P16.13
Short-shoe block brake.

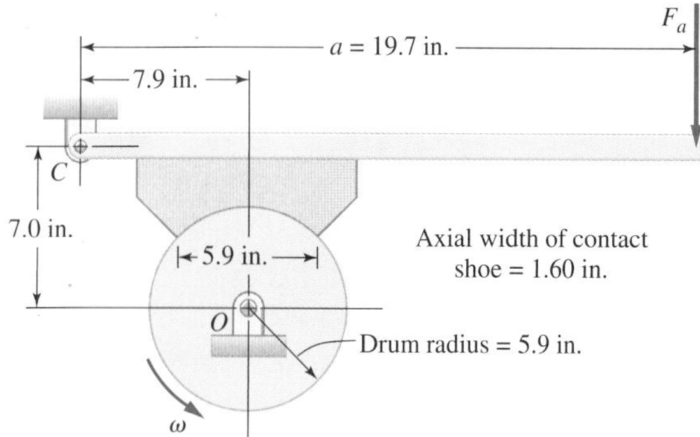

Figure P16.14
Long-shoe block brake.

e. Calculate the energy dissipated (work done by the brake) in bringing the drum to a stop.

f. If it were desired to make the brake self-locking, to what value would dimension d have to be increased?

16-14. A long-shoe brake assembly is sketched in Figure P16.14. The estimated coefficient of friction between shoe and drum is 0.3, and the maximum-allowable pressure for the lining material is 75 psi. Noting the CCW direction of rotation, determine the following:

a. The maximum actuating force F_a that can be applied without exceeding the maximum-allowable contact pressure.

b. The friction braking torque capacity corresponding to F_a calculated in (a).

c. The vertical and horizontal components of the reaction force at pin location C.

d. Is the brake self-energizing or self-locking?

16-15. Repeat problem 16-14 for the case where the drum rotates in the *clockwise* direction.

16-16. The shoe brake shown in Figure P16.16 is to be fabricated using an impregnated asbestos lining material at the contact surface. The lining material should not be operated at maximum pressures higher than 100 psi.

a. What is the largest actuating force, F_a, that should be used with this braking system as now designed?

b. If this largest-allowable actuating force is applied, what braking torque is produced on the rotating drum?

16-17. Repeat problem 16-16 for the case where the drum rotates in the *clockwise* direction.

16-18. A simple band brake of the type shown in Figure 16.9 is to be constructed using a lining material that has a maximum-allowable contact pressure of 600 kPa. The diameter of the rotating drum is to be 350 mm, and the proposed width of the band is 100 mm. The lever length is 900 mm and dimension m is 45 mm. The angle of wrap is to be 270°. Tests of the lining material indicate that a good estimate for the coefficient of friction is 0.25. Do the following:

a. Calculate the tight-side band tension at maximum-allowable pressure.

b. Calculate the slack-side band tension at maximum-allowable pressure.

c. Calculate the maximum torque capacity.

d. Calculate the actuating force corresponding to maximum torque capacity.

16-19. Repeat problem 16-18 for the case where the angle of wrap is 180°.

16-20. A simple band brake is constructed using a 0.050-inch-thick by 2-inch-wide steel band to support the tensile forces. Carbon-graphite material is bonded to the inside of the steel band to provide the friction surface for braking, and the rotating drum is to be a solid-steel cylinder, 16 inches in diameter and 2 inches in axial thickness. The brake band is wrapped around the rotating drum so that it is in contact over 270° of the drum surface. It is desired to bring the drum to a complete stop in one revolution from its operating speed of 1200 rpm. What would be the maximum tensile stress induced in the 0.050-inch-thick steel band during the braking period if the drum were brought to a stop in exactly one revolution? Assume that the rotating drum is the only significant mass in the system, and that the brake is kept dry.

16-21. A differential band brake is sketched in Figure P16.21 The maximum-allowable pressure for the band lining material is 60 psi, and the coefficient of friction between the lining and the drum is 0.25. The band and lining are 4 inches in width. Do the following:

a. If the drum is rotating in the *clockwise* direction, calculate the tight-side tension and slack-side tension at maximum-allowable pressure.

b. For *clockwise* drum rotation, calculate the maximum torque capacity.

c. For *clockwise* drum rotation, calculate the actuating force corresponding to maximum torque capacity.

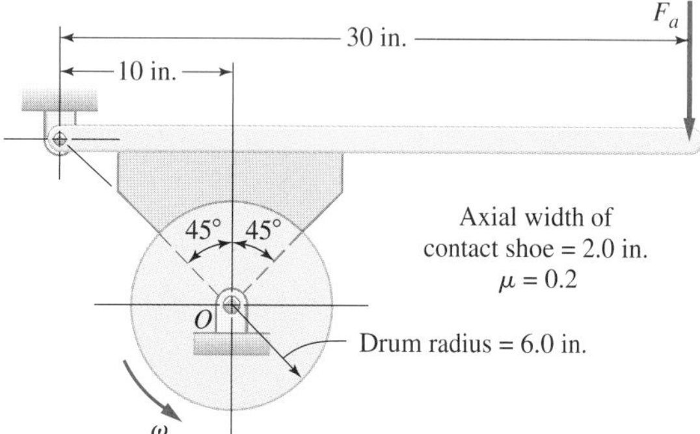

Figure P16.16
Long-shoe block brake.

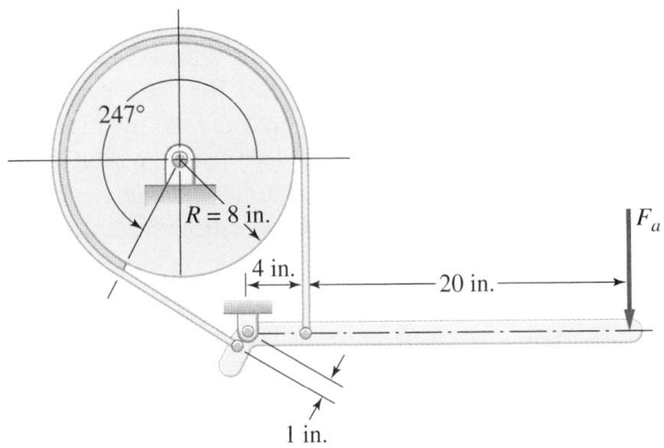

Figure P16.21
Differential band brake.

d. If the direction of drum rotation is changed to *counterclockwise*, calculate the actuating force corresponding to maximum torque capacity.

16-22. In the analysis and design of disk brakes and clutches, it is usual to hypothesize either "uniform wear" or "uniform pressure" as a basis for making calculations.

a. What important information can be derived on the basis of making a proper choice between these two hypotheses?

b. How would you decide whether to choose the uniform wear hypothesis or the uniform pressure hypothesis in any given situation?

16-23. It is desired to replace the long-shoe brake shown in Figure P16.16 with a simple band brake of the same width b. If the materials used are the same at the friction surface (i.e., $\mu = 0.2$ and $p_{max} = 100$ psi are unchanged) and the drum size must remain unchanged, what wrap angle should be used for the simple band brake to produce the same braking torque capacity as the long-shoe brake shown in Figure P16.16?

16-24. A disk clutch is being proposed for an industrial application in which the power-input shaft supplies 15 horsepower steadily at 650 rpm. The patented friction lining material, which is to be bonded to one or more annular disk surfaces, is to be brought into contact with stiff steel disks to actuate the clutch. The outside diameter of the clutch disks must be no larger than 5 inches, and it is desired to configure the annular friction surfaces so that the inner diameter is about $^2/_3$ of the outer diameter. The coefficient of friction between the patented friction lining material and steel is $\mu = 0.32$ and the maximum-allowable contact pressure is 150 psi. What is the minimum number of friction surfaces required for the clutch to function properly?

16-25. A disk brake is to be constructed for use on a high-speed rotor balancing machine. It has been decided that a carbon-graphite friction material will be used against a steel-disk mating surface to provide the braking action. The environment is dry. For clearance reasons the *inner diameter* of the steel brake disk must be 10.0 inches and its thickness is 0.375 inch. Further, the brake must be able to absorb 2.5×10^6 in-lb of kinetic energy in one-half revolution of the disk brake as it brings the high-speed rotor to a full stop. Only one braking surface can be used.

a. What should be the *outside diameter* of this disk brake?

b. What axial normal actuating force N_a will be required for the brake to function properly?

c. Due to the short stopping time, it is estimated that only about 10 percent of the volume of the steel disk constitutes the entire "effective" heat sink for the brake. About how large a temperature rise would you expect in this brake during the stop? Is this acceptable?

16-26. For use in a specialized underwater hoisting application, it is being proposed to design a disk clutch with a 20-inch outside diameter. Hard-drawn phosphor bronze is to be used in contact with chrome-plated hard steel to form the friction interfaces ($\mu = 0.03$, $p_{max} = 150$ psi). The clutch must transmit 150 horsepower continuously at a rotational speed of 1200 revolutions per minute. Following a rule of thumb that says that for good design practice the inner diameter of a disk clutch should be about $^2/_3$ of the outer diameter, determine the proper number of friction interfaces to use for this proposed clutch. Since the device operates under water, neglect temperature limitations.

16-27. It is desired to replace a single-contact-surface *disk brake* used on the end of a rotating drum by a *long-shoe block brake*, as shown in Figure P16.27, without changing the drum. The materials used at the friction interface are the same for both cases. It may reasonably be assumed, therefore, that both brakes will operate at the same limiting pressure p_{max} during actuation. The original disk brake contact surface had an outer radius equal to the drum radius, and an inner radius of two-thirds the outer radius. What width b is required for the new long-shoe brake shown in Figure P16.27 to produce the same braking torque capacity as the old disk brake?

16-28. A disk clutch has a single set of mating annular friction surfaces having an outer diameter of 300 mm and an inner di-

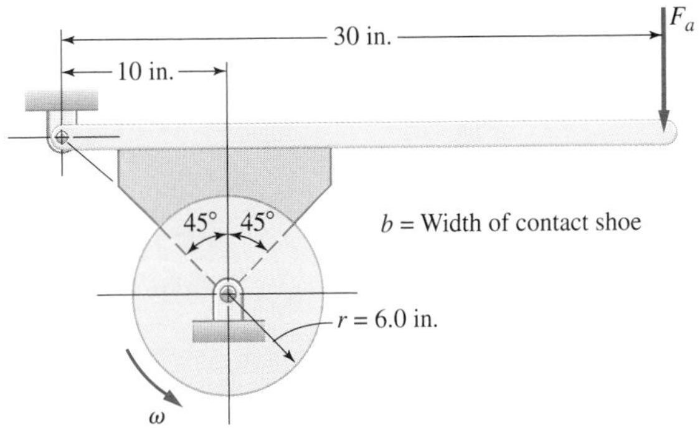

Figure P16.27
Long-shoe block brake replacement for a disk brake.

ameter of 225 mm. The estimated coefficient of friction between the two contacting surfaces is 0.25, and the maximum-allowable pressure for the lining material is 825 kPa. Calculate the following:

a. Torque capacity under conditions that make the uniform wear assumption more nearly valid.

b. Torque capacity under conditions that make the uniform pressure assumption more nearly valid.

16-29. A multiple-disk clutch is to be designed to transmit a torque of 750 in-lb while fully submersed in oil. Space restrictions limit the outside diameter of the disks to 4.0 inches. The tentatively selected materials for the interposed disks are rigid molded-asbestos against steel. Determine appropriate values for the following:

a. Inside diameter of the disks

b. Total numbers of disks

c. Axial normal actuating force

16-30. The wheels of a standard adult bicycle have a rolling radius of approximately 13.5 inches and a radius to the center of the hand-actuated caliper disk brake pads (see Figure 16.13) of 12.5 inches. The combined weight of the bike plus the rider is 200 pounds, equally distributed between the two wheels. If the coefficient of friction between the tires and the road surface is twice the coefficient of friction between the caliper brake pads and the metallic wheel rim, calculate the clamping force that must be applied at the caliper to slide the wheel upon hand brake application.

16-31. A cone clutch having a cone angle α of 10° is to transmit 40 horsepower continuously at a rotational speed of 600 rpm. The contact width of the lining along an element of the cone is 2.0 inches. The lining material is wound asbestos yarn and wire, operating against steel. Assuming that the uniform wear assumption holds, do the following:

a. Calculate the required torque capacity.

b. Calculate the change in radius of the contact cone (i.e., $r_o - r_i$) across the contact width of the lining.

c. Calculate an acceptable value for r_i so that required torque capacity can be satisfied.

d. Calculate the corresponding value of r_o.

Chapter 17

Belts, Chains, Wire Rope, and Flexible Shafts

17.1 Uses and Characteristics of Flexible Power Transmission Elements

In the introductory discussion of gear drives in 15.1, it was noted that the transmission of power or motion from one rotating shaft to another may be accomplished in many different ways. The choices include flat belts, V-belts, toothed timing belts, chain drives, flexible shafts, friction wheel drives, and gears. Wire rope is also used in power transmission, but is usually limited to hoisting or haulage applications, where a motor-driven input shaft rotates a drum so as to lift or lower a payload by winding or unwinding the wire rope. Friction wheel drives and gear drives have already been discussed in Chapter 15. The other power transmission elements just listed are discussed in this chapter.

Belt drives are well suited to applications in which the center distance between rotating shafts is large, and are usually simpler and more economical than other power transmission alternatives. A belt drive often eliminates the need for a more complicated arrangement of gears, bearings, and shafts. With proper design insight, belts are usually quiet, easily replaced, and in many cases, because of their flexibility and damping capacity, they reduce the transmission of unwanted shock and vibration between shafts. Simplicity of installation, minimum maintenance requirements, high reliability, and adaptability to a variety of applications are also characteristics of belt drives. However, because of slip and/or creep,[1] the angular velocity ratio between the two rotating shafts may be inexact, and the power and torque capacities are limited by the coefficient of friction and interfacial pressure between belt and pulley.[2] Belts are commercially available in many different cross sections, as illustrated in Figure 17.1.[3] Typical pulley (sheave) configurations used with various types of belts are shown in Figure 17.2.

Chain and sprocket drives, like belt drives, can span long center distances, and like gears, can provide positive transmission of speed, torque, and power. For a given (average) angular velocity ratio and power capacity, chain drives are usually more compact than belt drives but less compact than gear drives. Mounting and alignment requirements for chain drives are typically more precise than for belt drives, but may be less precise than for gear drives. Properly lubricated chain drives may be expected to have long service lives. The use of chain drives permits many shafts to be simultaneously driven from a single powered-input shaft, as long as the angle of wrap on any given sprocket is at least

[1] *Slip* occurs *uniformly* along the entire surface of contact between belt and pulley (or sheave), while *creep* occurs *differentially* along the surface of contact due to *local* differences in elastic deformation of the belt.

[2] Except for toothed belts (see 17.6 and Figure 17.1).

[3] Also see ref. 1 or ref. 2, for example.

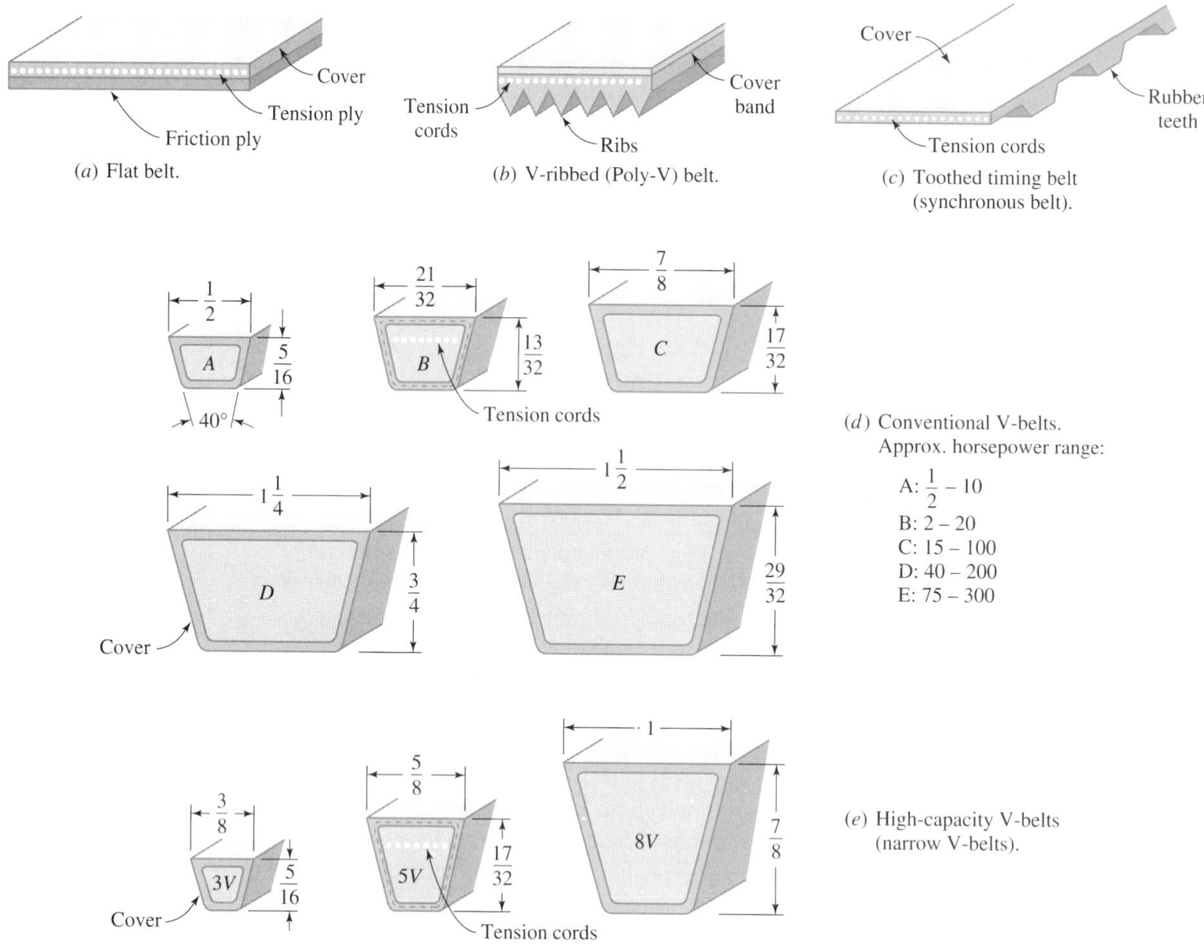

Cover
Tension ply
Friction ply

(*a*) Flat belt.

Tension cords
Cover band
Ribs

(*b*) V-ribbed (Poly-V) belt.

Cover
Rubber teeth
Tension cords

(*c*) Toothed timing belt (synchronous belt).

A: $\frac{1}{2}$, $\frac{5}{16}$, 40°
B: $\frac{21}{32}$, $\frac{13}{32}$
C: $\frac{7}{8}$, $\frac{17}{32}$
Tension cords
D: $1\frac{1}{4}$, $\frac{3}{4}$
E: $1\frac{1}{2}$, $\frac{29}{32}$
Cover

(*d*) Conventional V-belts.
Approx. horsepower range:

A: $\frac{1}{2}$ – 10
B: 2 – 20
C: 15 – 100
D: 40 – 200
E: 75 – 300

3V: $\frac{3}{8}$, $\frac{5}{16}$
5V: $\frac{5}{8}$, $\frac{17}{32}$
8V: 1, $\frac{7}{8}$
Cover
Tension cords

(*e*) High-capacity V-belts (narrow V-belts).

Figure 17.1
Sketches of various types of belts in common usage.

Some examples of belts and chains.

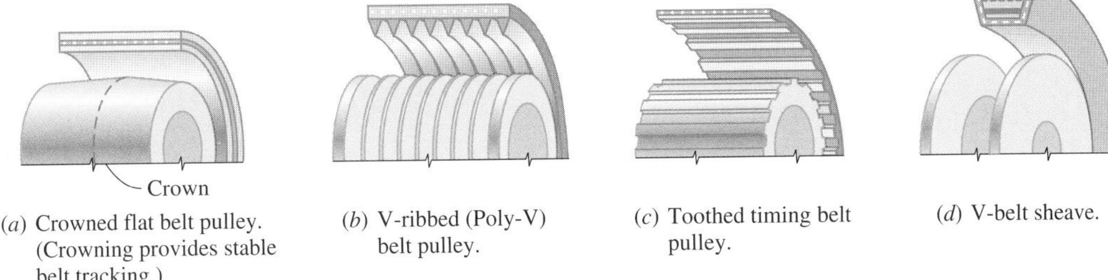

(*a*) Crowned flat belt pulley. (Crowning provides stable belt tracking.)
(*b*) V-ribbed (Poly-V) belt pulley.
(*c*) Toothed timing belt pulley.
(*d*) V-belt sheave.

Figure 17.2
Typical pulley (sheave) configurations used with belts sketched in Figure 17.1.

about 120°. Usually, the cost of a chain drive lies between that of a gear drive (higher cost) and a belt drive (somewhat lower cost) for equivalent power transmission capacity. Many different power transmission chain configurations are available, some of which are shown in Figure 17.3.[4] Typical sprocket configurations are shown in Figure 17.4.

Wire rope is often used for hoisting, haulage, and conveyor applications, in which the wire rope supports tensile loading along its length. The flexibility of wire rope is achieved by using a large number of small-diameter wires (Figure 17.5) twisted around a central core of fiber, independent wire rope, or a single wire strand.[5] Typically, several small wires (e.g., 7, 19, or 37) are first twisted into a strand. Then a number of multiwire strands, usually 6 or 8, are twisted about the core to form the bending-flexible wire rope.[6] The central core, which supports the strands radially, is usually saturated with a lubricant that seeps among the wires so they can more easily slide with respect to each other to minimize fretting and wear. Many standard wire rope sizes are commercially available.[7]

Flexible shafts are used to transmit rotary power or motion along a *curved path* between two machine shafts that are not colinear, or that may have relative motion with respect to each other, making direct connection of driving and driven shafts impractical (see

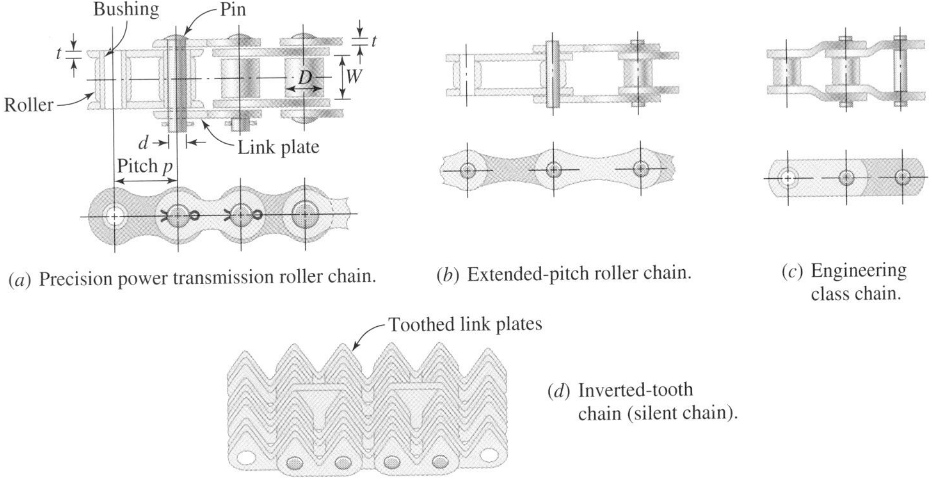

(*a*) Precision power transmission roller chain.
(*b*) Extended-pitch roller chain.
(*c*) Engineering class chain.
(*d*) Inverted-tooth chain (silent chain).

Figure 17.3
Common types of power- and motion-transfer chain configurations.

[4]Also see ref. 3, for example. [5]Designated (FC), (IWRC), and (WSC), respectively.

[6]It is well to note that *nominal* rope classifications may or may not reflect actual construction. For example, the 6×19 class includes constructions such as 6×21 filler wire, 6×25 filler wire, and 6×26 Warrington Seale.

[7]See, for example, Table 17.9 or ref. 4.

Roller chain
(see Figure 17.3)
Roller chain sprocket

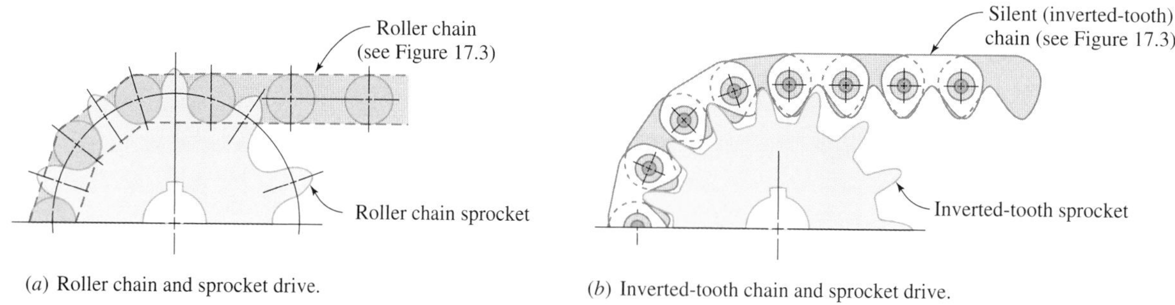

(a) Roller chain and sprocket drive.

Silent (inverted-tooth)
chain (see Figure 17.3)
Inverted-tooth sprocket

(b) Inverted-tooth chain and sprocket drive.

Figure 17.4
Sprocket configurations used with power transmission chain (see Figure 17.3).

Figure 17.6). Flexible shafts may also be used for remote control of machine elements that must be manipulated or adjusted manually or mechanically while operating.

Typically, flexible shafts are built up "nearly solid" by tightly winding one layer of wire over another about a single "mandrel wire" in the center, as shown in Figure 17.7(a). In most applications, flexible shafts are encased in a metal- or rubber-covered, flexible *sheath*, as shown in Figure 17.7(b), which acts as a supporting guide, protects the shaft from dirt or damage, and retains the lubricant. Examples of flexible shaft applications include portable power tools, weed whackers, speedometer drives, upholstery shampoo heads, tractor power-takeoff drives, event-counter drives, and position controls for automotive exterior rear-vision mirrors. Many standard flexible shafts are commercially available.[8]

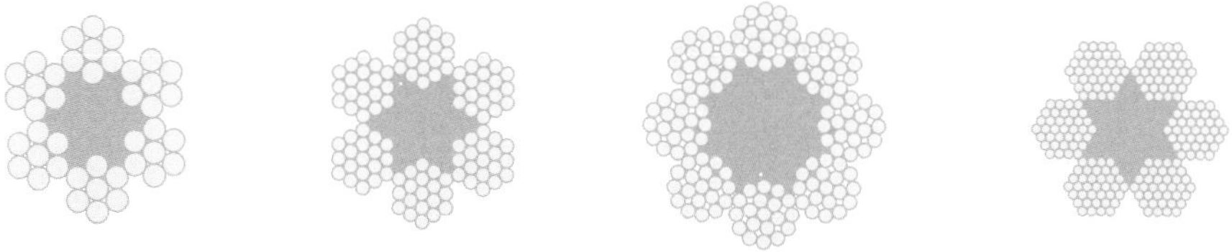

6 × 7
(haulage)
6 × 19
(standard hoisting)
8 × 19
(extra flexible)
6 × 37
(special flexible)

(a) Commonly selected wire rope cross sections.

7 wire
(See above)
19 Warrington
(W)
19 Seale
(S)
25 filler wire
(FW)

(b) Some of the available multiwire strand patterns. Combinations of these are also available.

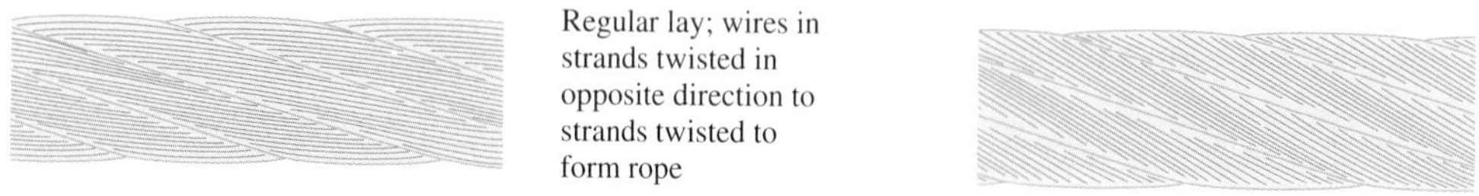

Regular lay; wires in strands twisted in opposite direction to strands twisted to form rope
Lang lay; wires in strands twisted in same direction as strands twisted to form rope

(c) Wire rope winding practice.

Figure 17.5
Various common wire rope configurations.

[8]See, for example, ref. 5 or ref. 6.

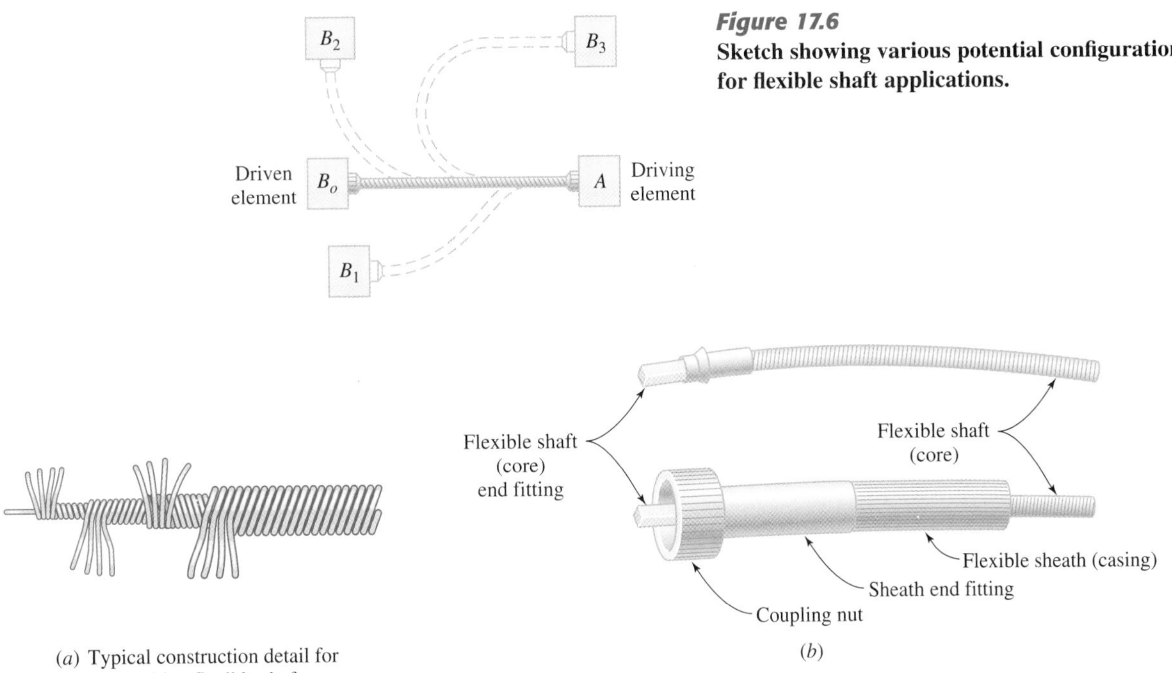

Figure 17.6
Sketch showing various potential configurations for flexible shaft applications.

Figure 17.7
Construction details and support configuration for flexible shafting.

17.2 Belt Drives; Potential Failure Modes

As illustrated in Figure 17.8, all types of belts loop around at least two pulleys that usually have different diameters. Except for toothed timing belts, it is necessary to *pretension* the belt by forcing the pulleys apart, inducing an initial static tensile force of T_o in the tension cords.[9] In turn, the initial tension produces a normal pressure between the belt and each pulley contact surface. This allows power transmission by virtue of the friction force available at each belt/pulley interface. When power is applied to the driver pulley, tension in one side of the belt is increased to a value above the pretension level because of belt stretching, while in the other side of the belt tension decreases to a value below the pretension level. The belt span with increased tension is called the *tight side* or *taut side* (tension T_t) and the span with decreased tension is called the *slack side* (tension T_s). As the moving belt repeatedly passes around the pulleys, at any given cross section the tension cords are subjected to fluctuating loads ranging from T_s to T_t and back with each belt pass, in addition to a constant centrifugally induced tensile force T_c. Fatigue resulting from nonzero-mean[10] *cyclic tension* therefore becomes a potential failure mode in belt drives.

In addition to the fluctuations in tensile loading, a belt is also subjected to *cyclic bending* as it passes about each of the pulleys (sheaves). Hence *bending* fatigue contributes to potential belt failure in addition to *tensile* fatigue. In some cases, adhesive/abrasive wear may be a potential failure mode,[11] and degradation in belt material properties (corrosion),

[9]See Figure 17.1. [10]See 2.6.

[11]For example, if enough wear occurs between a V-belt and the V-pulleys, the belt may bottom in the V-groove to prevent wedging action and proper tensioning of load-carrying cords. Another example of failure by wear is found in high-efficiency flat-belt drives that use an inner high-friction ply [see Figure 17.1(a)] to transmit torque more effectively. If the high-friction ply wears through, excessive slippage may occur and the belt may burn (see ref. 11).

Figure 17.8
Basic configuration and tensile force cycle typical of all belts. (See ref. 9.)

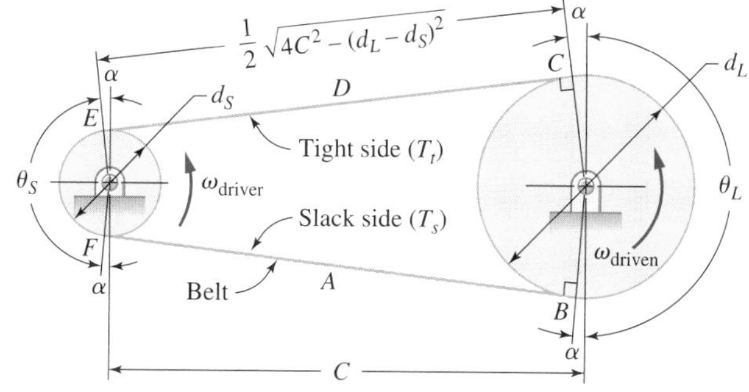

θ_S = angle of wrap (smaller pulley) = $\pi - 2\alpha$
θ_L = angle of wrap (larger pulley) = $\pi + 2\alpha$
d_S = diameter of smaller pulley
d_L = diameter of larger pulley

$$\alpha = \sin^{-1}\frac{d_L - d_S}{2C}$$

$$L = \text{belt length} = \sqrt{4C^2 - (d_L - d_S)^2} + \frac{d_L\theta_L - d_S\theta_S}{2}$$

(For V-belt datum length, use sheave datum diameters.)

(*a*) Belt terminology and geometry.

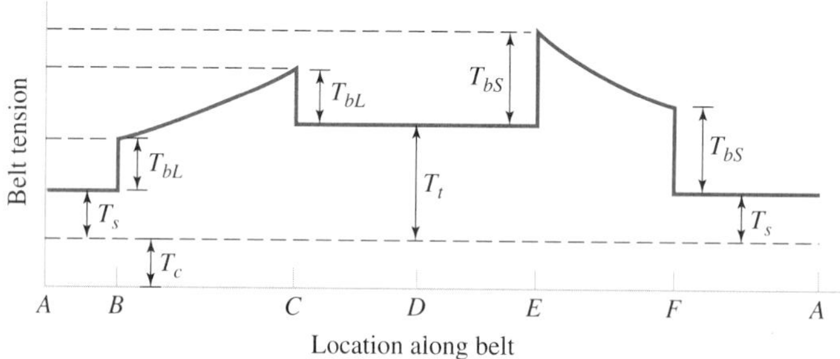

T_c = belt tension induced by centrifugal force
T_s = slack side tension
T_t = taut side tension
T_{bS} = belt tension induced by bending about smaller pulley
T_{bL} = belt tension induced by bending about larger pulley

(*b*) Cyclic tension during one belt pass.

usually because of elevated temperature or adverse environments,[12] is a potential failure mode as well. Since belts may be characterized as *composite* structures, the failure modes just listed may manifest themselves within the tension cords, within the elastomeric matrix, or at the interface between the cords and the matrix. Macroscopically, belt failures may express themselves as cord breakage, cord-rubber separation, radial cracking (due to continued curing of the rubber[13]), or circumferential cover cracking.

[12]For example, oily conditions.

[13]Cyclic strains induced by belt operation may produce hysteresis losses of as much as 10–18 percent of the transmitted power. This, combined with the poor thermal conductivity of belt materials, may lead to significant internal temperature increases, "overcuring" and degradation of belt material properties, and cracking. Maximum acceptable belt temperature is around 180°F. One rule of thumb is that for every 30°F increase above the 180°F level, the belt life is reduced about 50 percent.

17.3 Belts; Materials

Originally, belts were manufactured using prime-quality cotton cords as tension members. The cords were embedded in a matrix of natural rubber to provide flexible constraint for the cords and a higher friction coefficient at the belt surface to enhance the transfer of torque and power. As belts of higher capacity and higher reliability were developed during and after World War II, new tension cord materials of higher strength and higher stiffness were utilized, and synthetic rubber compounds replaced natural rubber in belt construction. Although cotton fibers embedded in natural rubber may still be used in low-performance applications, tension cords of modern high-performance belts (see Figure 17.1) are usually made of polyamide strips or polyester cords for flat belts; polyester cords or aramid cords for V-ribbed (Poly-V) belts; fiberglass or steel strands for toothed timing belts; and polyester, fiberglass, or aramid fibers for conventional or high-capacity V-belts. The matrix for all belt types is typically a synthetic rubber, often Neoprene, to enhance resistance to oil, heat, and ozone. Polyurethane is sometimes used for the friction ply of modern flat belts [see Figure 17.1(a)]. Belt cover material is usually cotton or nylon cloth impregnated with synthetic rubber and molded in place to protect the core and resist belt wear.

17.4 Belt Drives; Flat Belts

Flat belts[14] are the simplest and usually the least expensive type of belt. They are most effective when operating speeds are relatively high and power transmission requirements are relatively low. Operational linear belt velocities typically range from about 2500 ft/min to about 7500 ft/min, with a speed around 4000 ft/min usually regarded as "ideal." At lower speeds, for a given power specification, required belt tension tends to become large, often making required belt width awkwardly large.[15] As a result, shaft size, bearing capacity, and space envelope requirements may become unacceptable in some applications.

At higher speeds, dynamic forces due to belt whip and/or vibration may reduce drive stability and shorten belt life. V-ribbed (Poly-V) belts may be regarded as flat belts with an enhanced friction coefficient due to the wedging action of the ribs as they are forced by belt tension into the pulley grooves until the tension cords essentially "ride" on the outer diameter of the pulley (see Figures 17.1(b) and 17.2(b)).

The basic equation for the limiting torque that can be transmitted by a low-speed flat belt is the same as for a band brake (see Figure 16.9). Adapting the band brake equations (16-71) and (16-72) to the belting notation defined in Figure 17.8, and noting that *slip will occur first at the smaller pulley* because of its smaller angle of wrap,[16] the *slip equation* for a low-speed flat belt drive may be written as

$$\frac{T_t}{T_s} = e^{\mu\theta_s} \tag{17-1}$$

and the friction torque transmitted at the smaller pulley becomes

$$T_f = (T_t - T_s)\frac{d_s}{2} = \frac{T_s d_s}{2}(e^{\mu\theta_s} - 1) \tag{17-2}$$

[14] See Figure 17.1(a).

[15] As required belt width increases, shaft-and-pulley alignment become more critical because belt tension must be uniform over the full width.

[16] This may be an especially critical design issue for belt drives in which pulley diameters are greatly different or pulleys are closely spaced. A wrap angle of at least $\theta_s = 150°$ is recommended for flat belts.

For higher linear velocities, significant centripetal acceleration of the belt mass passing around the pulleys causes a centrifugal-force-induced belt tension (inertia force), T_c, that must be included in the free-body segment of the belt used to derive the slip equation.[17] This inertia force component may be expressed as

$$T_c = m_1 v^2 = \frac{w_1 v^2}{g} \tag{17-3}$$

where w_1 = unit weight of belt, lb/ft
v = belt velocity, ft/sec
g = constant = 32.2 ft/sec^2

Therefore, instead of the static slip equation of (17-1), the slip equation for a moving belt is, based on equilibrium requirements,

$$\frac{T_t - T_c}{T_s - T_c} = e^{\mu\theta_s} \tag{17-4}$$

where θ_s is angle of wrap[18] on the smaller pulley.

Utilizing the horsepower equation (4-32) to express transmissible belt horsepower,

$$hp = \frac{(T_t - T_s)V}{33{,}000} \tag{17-5}$$

where T_t = tight side tension, lb
T_s = slack side tension, lb
$V = 60v$ = linear belt velocity, ft/min

Also, when a flat belt is initially pretensioned to T_o, a free-body analysis gives the relationship

$$T_t + T_s = 2T_o \tag{17-6}$$

Finally, the design-allowable tight-side belt tension may be expressed as[19]

$$(T_t)_d = \frac{T_a b_b}{K_a} \tag{17-7}$$

where T_a is design-allowable tensile force per unit belt width (lb/in), b_b is width of the belt (in), and K_a is a factor that depends upon the application. Table 17.1 provides experience-based limiting design values for T_a for a few materials.[20] The tabled values for T_a are usually divided by an application factor, K_a, to account for shock or impact loading characteristics of the prime mover and the driven machine. Typical application factors are shown in Table 17.2.

[17]That is, referring to Figure 16.9(b), for a *moving* belt, an inertia force $T_c d\varphi$ must be added to the free body, in the radial direction (vertically upward), and included in the vertical force summation [expressed in (16-63) for the static case].

[18]See Figure 17.8(a).

[19]An alternate (and more fundamental) expression might be formulated as $(T_t)_d = (S_N)_c n_c A_c / n_d$, where $(S_N)_c$ is "fatigue strength" of an individual cord, n_c is number of cords, A_c is net area of a cord, and n_d is the design safety factor (see Chapter 5). A cord can be one or more twisted strands, each having hundreds of monofilament fibers twisted together. The problem with this alternative calculation is that few data have been published to support this design approach.

[20]These empirical values, which embody direct tension, bending, and centrifugal force effects, provide little insight into how failure mode interactions take place, what safety factors have been included, or which parameters might be changed to improve belt life. Further, these allowable values are based on *smooth* system operation.

TABLE 17.1 Empirical Experience-Based Design Data[1] for High-Performance Flat Belts[2]

Material	Thickness, t_b, in	Allowable Tension per Unit Width, T_a, lb/in	Minimum Pulley Diameter, d_{min}, in	Specific Weight, lb/in^3	Coefficient of Friction
Polyamide	0.03	10	0.6	0.035	0.5
	0.05	35	1.0	0.035	0.5
	0.07	60	2.4	0.051	0.5
	0.11	60	2.4	0.037	0.8
	0.13	100	4.3	0.042	0.8
	0.20	175	9.5	0.039	0.8
	0.25	275	13.5	0.039	0.8
Urethane	0.06	5	0.38–0.50	0.038–0.045	0.7
	0.08	10	0.50–0.75	0.038–0.045	0.7
	0.09	19	0.50–0.75	0.038–0.045	0.7
Polyester cord	0.04	57–225	1.5		

[1]Data from refs. 2 and 7.
[2]These empirical values, which embody direct tension, bending, and centrifugal force effects, provide little insight into how failure mode interactions take place, what safety factors have been included, or which parameters might be changed to improve belt life. Further, these allowable values are based on the assumption that the system operates smoothly, without impacts or jerks.

TABLE 17.2 Application Factor, K_a

Prime Mover Characteristic	Driven Machine Characteristic: Uniform	Moderate Shock	Heavy Shock
Uniform (e.g., electric motor, turbine)	1.00	1.25	1.75 or higher
Light shock (e.g., multicylinder engine)	1.25	1.50	2.00 or higher
Medium shock (e.g., single-cylinder engine)	1.50	1.75	2.25 or higher

Example 17.1 Flat Belt Selection

A proposed flat belt drive system is being considered for an application in which the input motor shaft speed (*driving* pulley) is 3600 rpm, the approximate *driven* shaft speed requirement is 1440 rpm, the power to be transmitted has been estimated as 0.5 horsepower, and the driven device is known to operate smoothly. The desired center distance between driving and driven pulleys is to be approximately 10 inches. Engineering management has suggested using a belt fabricated by using polyamide tension strips. Select a flat belt for this application.

Solution

From (17-5),

$$hp = \frac{(T_t - T_s)V}{33{,}000} \tag{1}$$

and using (15-23), pitch-line belt velocity V may be written as

Example 17.1 Continues

$$V = \frac{2\pi r_s n_s}{12} \tag{2}$$

where r_s is in inches, n_s is in rpm, and V is in ft/min.

From Table 17.1, for polyamide belts, an initial belt thickness of 0.05 inch will be tried; the minimum recommended pulley diameter for this belt thickness is $d_s = 1.0$ inch. Hence (2) becomes

$$V = \frac{2\pi(0.5)(3600)}{12} = 940 \text{ ft/min} \tag{3}$$

and (1) becomes

$$(0.5) = \frac{(T_t - T_s)(940)}{33{,}000} \tag{4}$$

which gives

$$(T_t - T_s) = \frac{33{,}000(0.5)}{940} = 17.6 \text{ lb} \tag{5}$$

Evaluating (2) as

$$940 = \frac{2\pi(d_L/2)(1440)}{12} \tag{6}$$

the diameter of the larger pulley, d_L, may be calculated as[21]

$$d_L = \frac{(940)(12)(2)}{2\pi(1440)} = 2.5 \text{ inches} \tag{7}$$

Also from Figure 17.8(a), the angle of wrap for the small pulley, θ_s, may be calculated as

$$\begin{aligned} \theta_s &= \pi - 2\alpha = \pi - 2\sin^{-1}\left(\frac{d_L - d_s}{2C}\right) \\ &= \pi - 2\sin^{-1}\left(\frac{2.5 - 1.0}{2(10)}\right) = \pi - 0.15 = 2.99 \text{ rad} \end{aligned} \tag{8}$$

The specific belt weight is given in Table 17.1 for the 0.05-inch belt thickness as 0.035 lb/in^3. If the belt width is taken as b_b, the weight of the belt per unit length, w_1 (lb/ft), may be calculated as

$$w_1 = (b_b)(0.05)(0.035)(12) = (0.021)b_b \text{ lb/ft} \tag{9}$$

Utilizing (17-3), noting that belt velocity v in this expression is in ft/sec, and finding the coefficient of friction from Table 17.1 as $\mu = 0.5$, (17-3) becomes

$$T_c = \frac{w_1 v^2}{g} = \frac{0.021 b_b (V/60)^2}{g} = \frac{0.021 b_b (940/60)^2}{32.2} = 0.16 b_b \tag{10}$$

Then (17-4) may be written as

$$\frac{T_t - 0.16 b_b}{T_s - 0.16 b_b} = e^{0.5(2.99)} = 4.46 \tag{11}$$

Note from Table 17.2 that the application factor for this case is $K_a = 1.0$, because both the electric motor drive and the driven machine are smooth and uniform in operation.[22] Thus (from Table 17.1) the limiting design-allowable tension per unit width for the initially selected belt is

$$T_a = 35 \text{ lb/in} \tag{12}$$

[21]Pitch-line belt velocity must be the same at all points along the belt. [22]See specifications.

Thus (17-7) may be written as

$$(T_t)_d = \frac{35b_b}{K_a} = 35b_b \tag{13}$$

Now, solving (5), (11), and (13) together,

$$\frac{35b_b - 0.16b_b}{(35b_b - 17.6) - 0.16b_b} = 4.46 \tag{14}$$

or

$$\frac{34.84b_b}{34.84b_b - 17.6} = 4.46 \tag{15}$$

which gives a minimum required belt width of

$$b_b = 0.70 \text{ inch} \tag{16}$$

Other information of interest might include belt length L_b and initial tension T_o required for proper operation. From Figure 17.8(a), belt length may be estimated as

$$\begin{aligned} L_b &= \sqrt{4(10)^2 - (2.5 - 1.0)^2} + \frac{2.5(3.29) + 1.0(2.99)}{2} \\ &= 19.94 + 5.61 = 25.5 \text{ inches} \end{aligned} \tag{17}$$

From (17-6) the required initial belt tension T_o is

$$T_o = \frac{T_t + T_s}{2} = \frac{[34.84(0.70)] + [34.84(0.70) - 17.6]}{2} = 15.6 \text{ lb} \tag{18}$$

Based on the initial trial selection of a 0.05-inch-thick polyamide belt, the results all seem reasonable. This belt will be recommended, therefore, but it is clear that many other choices could be found that would also satisfy the design criteria.

To summarize:

1. A polyamide belt of 0.05-inch thickness and 0.70-inch width is recommended.
2. The approximate belt length required is 25.5 inches, based on a small driver pulley diameter of 1.0 inch, a larger driven pulley diameter of 2.5 inches, and a center distance of 10 inches.
3. An initial belt tension of 15.6 lb is recommended to achieve optimum performance.

17.5 Belt Drives; V-Belts

As with flat belts, for each pass of an operating *V-belt* the tension cords are subjected to fluctuating tensile loads ranging from T_s to T_t, cyclic bending that is a function of sheave[23] diameter, and a constant centrifugal force component.[24] These nonzero-mean cyclic forces, just as for flat belts, suggest that fatigue failure is a likely failure mode for V-belts. Further, a variation in tension occurs among the tension cords *across the width* of a V-belt because of the *wedging action* into the narrower groove of a mating sheave, as illustrated in Figure 17.9.[25] Because of this nonuniform distribution of cord tension, the edge cords have

[23]Often pronounced "*shiv*," rhyming with "give." [24]See Figure 17.8.

[25]Typically, the included angle of a standard V-belt section is about 40°, and the smaller included angle of the V-groove in a sheave is usually between about 30° and 38°. For *smaller* sheave angles the tendency is for the belt to self-lock in the V-groove, causing a stick-slip-induced "jerky" behavior. For *larger* sheave angles, the increase in *effective* coefficient of friction in the V-groove (due to wedging) is reduced too much.

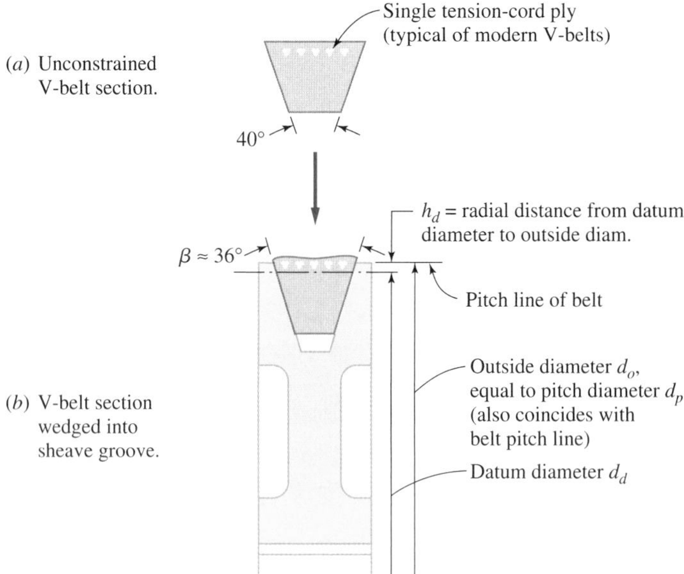

Figure 17.9
Section of a modern V-belt before and after it is wedged into a mating sheave by belt tension. The sketch also illustrates the new standard *datum diameter system* recently adopted by the V-belt industry.

been found to carry higher fluctuating loads per cord than the intermediate cords; hence *the peak fluctuating cord stresses occur in the edge cords*. Often, an edge-cord factor, conceptually similar to a stress concentration factor,[26] is used to estimate *edge cord* stresses from *average* cord stresses. Average cord stresses are readily calculable for any given belt section (see Table 17.3).

Like the fatigue of metallic parts,[27] V-belt fatigue is a function of both the maximum and the minimum cyclic stress experienced by the belt during the nonzero-mean loading of the tension cords.[28]

The slip equation given in (17-4) for flat belts may be modified for use with V-belts by substituting an *effective* coefficient of friction, μ', in place of μ, where[29]

$$\mu' = \frac{\mu}{\sin\left(\dfrac{\beta}{2}\right)} \tag{17-8}$$

For V-belts then, the slip equation becomes

$$\frac{T_t - T_c}{T_s - T_c} = e^{\mu'\theta_s} = e^{[\mu\theta_s/\sin(\beta/2)]} \tag{17-9}$$

The horsepower equation given in (17-5) requires no *basic* modification for use with V-belts, but multiplication of the nominal power requirement by an application factor, K_a, to obtain a "design value" for power required, *is* recommended for V-belt applications (see Table 17.2). The belt pretension equation, (17-6), remains unchanged as well.

V-belt configurations have become well standardized,[30] and widely tested for reliability and life. Because of evolving construction details associated with modern V-belt cross

[26]See 4.8. [27]See 2.6.

[28]Nonzero-mean loading of the tension cords involves contributions from fluctuating direct tension, cyclic bending, and steady centrifugal force.

[29]See Figure 17.9 for the definition of the sheave *included angle* β.

[30]See, for example, the standard sections shown in Figure 17.1(d), (e), and (f). Since most V-belts are manufactured as continuous loops, V-belt lengths have also been standardized. A sampling of standard lengths is shown in Table 17.5. Metric belts have also been standardized, but are not included in this textbook.

TABLE 17.3 Data-based Empirical Constants[1] Developed for Use with (17-10) for Selected[2] V-Belt Cross Sections (All data are for polyester tension cords)

Belt Section	d_d for[3] A,B,C,D; d_o for[4] 3V, 5V, in	K_i	K_o, ksi	K_m, ksi	k	C_1	C_2	C_3	C_4	A_c, in^2	n_c	w_1, lb/ft
A	3	6.13×10^{-8}	19.8	26.4	−1	5.0	111	0.101×10^{-6}	0.20	1.73×10^{-3}	7.0	0.065
	4	6.13×10^{-8}	19.8	26.4	−1	5.0	111	0.101×10^{-6}	0.18	1.73×10^{-3}	7.0	0.065
	5	6.13×10^{-8}	19.8	26.4	−1	5.0	111	0.101×10^{-6}	0.17	1.73×10^{-3}	7.0	0.065
	6	6.13×10^{-8}	19.8	26.4	−1	5.0	111	0.101×10^{-6}	0.16	1.73×10^{-3}	7.0	0.065
	7	6.13×10^{-8}	19.8	26.4	−1	5.0	111	0.101×10^{-6}	0.16	1.73×10^{-3}	7.0	0.065
B	4	1.78×10^{-7}	17.3	26.0	−1	5.2	123	0.133×10^{-6}	0.14	1.73×10^{-3}	7.8	0.112
	5	1.78×10^{-7}	17.3	26.0	−1	5.2	123	0.133×10^{-6}	0.15	1.73×10^{-3}	7.8	0.112
	6	1.78×10^{-7}	17.3	26.0	−1	5.2	123	0.133×10^{-6}	0.15	1.73×10^{-3}	7.8	0.112
	7	1.78×10^{-7}	17.3	26.0	−1	5.2	123	0.133×10^{-6}	0.16	1.73×10^{-3}	7.8	0.112
	8	1.78×10^{-7}	17.3	26.0	−1	5.2	123	0.133×10^{-6}	0.17	1.73×10^{-3}	7.8	0.112
C	6	9.39×10^{-11}	14.1	20.8	0	7.5	200	0.213×10^{-6}	0.15	2.88×10^{-3}	9.0	0.199
	7	9.39×10^{-11}	14.1	20.8	0	7.5	200	0.213×10^{-6}	0.14	2.88×10^{-3}	9.0	0.199
	8	9.39×10^{-11}	14.1	20.8	0	7.5	200	0.213×10^{-6}	0.14	2.88×10^{-3}	9.0	0.199
	10	9.39×10^{-11}	14.1	20.8	0	7.5	200	0.213×10^{-6}	0.13	2.88×10^{-3}	9.0	0.199
	12	9.39×10^{-11}	14.1	20.8	0	7.5	200	0.213×10^{-6}	0.13	2.88×10^{-3}	9.0	0.199
D	7	6.76×10^{-10}	10.8	14.6	0	26.5	256	0.291×10^{-6}	0.12	5.15×10^{-3}	11.0	0.406
	8	6.76×10^{-10}	10.8	14.6	0	26.5	256	0.291×10^{-6}	0.11	5.15×10^{-3}	11.0	0.406
	9	6.76×10^{-10}	10.8	14.6	0	26.5	256	0.291×10^{-6}	0.11	5.15×10^{-3}	11.0	0.406
	10	6.76×10^{-10}	10.8	14.6	0	26.5	256	0.291×10^{-6}	0.11	5.15×10^{-3}	11.0	0.406
	11	6.76×10^{-10}	10.8	14.6	0	26.5	256	0.291×10^{-6}	0.10	5.15×10^{-3}	11.0	0.406
3V	3	1.58×10^{-7}	16.9	28.3	−1	5.0	101	0.094×10^{-6}	0.23	1.73×10^{-3}	5.0	0.049
	4	1.58×10^{-7}	16.9	28.3	−1	5.0	101	0.094×10^{-6}	0.22	1.73×10^{-3}	5.0	0.049
	5	1.58×10^{-7}	16.9	28.3	−1	5.0	101	0.094×10^{-6}	0.21	1.73×10^{-3}	5.0	0.049
	6	1.58×10^{-7}	16.9	28.3	−1	5.0	101	0.094×10^{-6}	0.21	1.73×10^{-3}	5.0	0.049
	7	1.58×10^{-7}	16.9	28.3	−1	5.0	101	0.094×10^{-6}	0.21	1.73×10^{-3}	5.0	0.049
5V	6	9.99×10^{-8}	16.0	29.2	−1	6.0	200	0.202×10^{-6}	0.19	2.88×10^{-3}	6.3	0.141
	7	9.99×10^{-8}	16.0	29.2	−1	6.0	200	0.202×10^{-6}	0.18	2.88×10^{-3}	6.3	0.141
	8	9.99×10^{-8}	16.0	29.2	−1	6.0	200	0.202×10^{-6}	0.18	2.88×10^{-3}	6.3	0.141
	10	9.99×10^{-8}	16.0	29.2	−1	6.0	200	0.202×10^{-6}	0.17	2.88×10^{-3}	6.3	0.141
	12	9.99×10^{-8}	16.0	29.2	−1	6.0	200	0.202×10^{-6}	0.17	2.88×10^{-3}	6.3	0.141

[1] *Correlation coefficients* for all cross sections ranged from 0.81 to 0.99. A typical correlation coefficient was 0.9.

[2] Data for other cross sections available in ref. 10. Note also that data for C_4 are approximate values. C_4 is actually a weak function of tension ratio, as given in ref. 10.

[3] Datum diameter.

[4] Outside diameter.

TABLE 17.4 Relationships Among Datum System Pitch Diameter, Outside Diameter, and Datum Diameter for Classical V-Belts and Sheaves[1]

Belt Cross Section	Datum Diameter Range (previously pitch diameter range)	Difference Between Datum and Outside Diameters,[2] $2h_d$, in	Difference Between Pitch and Outside Diameters,[3] $2\ a_p$, in	Minimum Recommended Datum Diameter, $(d_d)_{min}$, in
A	All	0.250	0.00	3.0
B	All	0.350	0.00	4.6
C	All	0.400	0.00	6.0
D	All	0.600	0.00	12.0

[1]Adopted by International Standards Organization (ISO) and Rubber Manufacturers' Association, *Engineering Standard for Classical V-Belts and Sheaves.* Data extracted from ref. 10.
[2]See Figure 17.9.
[3]For other constructions, a_p may not be zero.

sections, in particular the trend toward moving the tension cords to a more ideal location very near the outside diameter of the sheave, industry standards have recently been changed to adopt the *datum system* instead of the *pitch system* traditionally used for V-belt and sheave specifications.[31] The datum system reduces or eliminates inaccuracies associated with specifications framed in the older pitch system. Table 17.4 gives datum system relationships among datum diameter d_d, pitch diameter d_p, and outside diameter d_o for selected conventional V-belt cross sections.

The following guidelines[32] apply to the use of the datum system for conventional V-belt drives:

1. Datum belt length is calculated using datum diameters of the sheaves.
2. Center distance is calculated using datum diameters and datum length.
3. Linear belt speed is calculated using pitch diameter.
4. Horsepower is calculated using pitch diameter.
5. Speed ratio is calculated using pitch diameter.

Several semiempirical life prediction equations have been successfully formulated for V-belt cross sections.[33] One such formulation, based on the assumptions that edge cord stresses govern V-belt fatigue life and that nonzero-mean stress effects are significant, is given by the expression[34]

$$N_f = K_i[K_o - \sigma_a]^2[K_m - \sigma_m]^2 L_d^{1.75} V^k \tag{17-10}$$

where N_f = belt failure life in cycles attributed to a single pulley[35]
K_i, K_o, K_m, k = empirical constants developed from statistical analysis of experimental data
L_d = belt *datum* length, inches (see Table 17.5)
V = linear belt speed, ft/min
σ_m = actual *mean* edge cord stress, psi, as calculated from (17-11)
σ_a = actual *amplitude* of edge cord stress, psi, as calculated from (17-12)

[31]Adopted by International Standards Organization (ISO) and Rubber Manufacturers' Association as *Engineering Standard for Classical V-Belts and Sheaves* (see ref. 10).
[32]See ref. 10. [33]See refs. 8, 9, 10, and 11. [34]See ref. 10.
[35]Since every belt drive embodies at least two sheaves, some type of *cumulative damage* expression becomes necessary to account for multiple sheaves of different diameters (see 2.6).

TABLE 17.5 Selected Nominal Standard Datum Length L_d, Outside Length L_o, and Inside Length L_i for Several Standard V-Belt Sections[1]

Belt Section	Datum Length, L_d, in	Outside Length[2], L_o, in	Inside Length L_i, in
A	22.3	23.3	21.0
	32.3	33.3	31.0
	42.3	43.3	41.0
	52.3	53.3	51.0
	62.3	63.3	61.0
	72.3	73.3	71.0
	82.3	83.3	81.0
	92.3	93.3	91.0
	101.3	102.3	100.0
	137.3	138.3	136.0
	181.3	183.3	180.0
B	29.8	30.8	28.0
	39.8	40.8	38.0
	49.8	50.8	48.0
	59.8	60.8	59.0
	69.8	70.8	69.0
	89.8	90.8	79.0
	109.8	110.8	89.0
	149.8	150.8	149.0
	211.8	212.8	211.0
C	53.9	55.2	51.0
	62.9	64.2	60.0
	73.9	75.2	71.0
	83.9	85.2	81.0
	99.9	101.2	97.0
	117.9	119.2	115.0
	138.9	140.2	136.0
	160.9	162.2	158.0
	182.9	184.2	180.0
	212.9	214.2	210.0
D	123.3	125.2	120.0
	147.3	149.2	144.0
	176.3	178.2	173.0
	198.3	200.2	195.0
	213.3	215.2	210.0
3V		25.0	
		40.0	
		56.0	
		67.0	
		80.0	
		95.0	
		112.0	
		125.0	
		140.0	
5V		50.0	
		60.0	
		71.0	
		80.0	
		90.0	

TABLE 17.5 (*Continued*)

Belt Section	Datum Length, L_d, in	Outside Length[2], L_o, in	Inside Length L_i, in
5V (*Continued*)		100.0	
		112.0	
		125.0	
		132.0	
8V		100.0	
		150.0	
		200.0	
		250.0	
		300.0	
		400.0	
		500.0	
		600.0	

[1]Many additional standard length choices are available from manufacturers' catalogs. For example, see ref. 10.
[2]For 3V, 5V, and 8V belt sections, the outside length may be assumed equal to the pitch length.

The defining equations for mean and alternating edge cord stresses, based on equilibrium concepts, are given by

$$\sigma_m = \frac{T_{te} + T_{be} + 2T_{ce} + T_{se}}{2A_c} \tag{17-11}$$

and

$$\sigma_a = \frac{T_{te} + T_{be} - T_{se}}{2A_c} \tag{17-12}$$

where T_{te} = actual[36] edge cord component of tight side belt tension T_t
T_{se} = actual edge cord component of slack side belt tension T_s
T_{be} = edge cord tension component due to bending
T_{ce} = edge cord tension component due to centrifugal force
A_c = nominal area of each cord[37]

To calculate the bending-induced edge cord tension component, T_{be}, a semiempirical expression has been developed for various standard belt sections that embody polyester tension cords,[38] as

$$T_{be} = \frac{C_1 + C_2}{d_d} \tag{17-13}$$

where C_1 and C_2 are constants for a given belt section and cord material (see Table 17.3), and d_d is the sheave datum diameter.

To calculate the centrifugal-force-induced edge cord component, T_{ce}, the semiempirical expression is

$$T_{ce} = C_3 V^2 \tag{17-14}$$

where C_3 is a constant for a given belt section (see Table 17.3).

To calculate the tight side edge cord tension component, T_{te}, the empirical relationship is

$$T_{te} = C_4 T_t \tag{17-15}$$

[36]Corrected for nonuniform distribution of fiber loading across the belt width. [37]See Table 17.3.
[38]Data for cords made of other materials are not readily available.

where C_4 is a constant for a given belt section, cord material, and sheave diameter[39] (see Table 17.3).

For purposes of making a reasonable first choice for belt cross section in any given application, Figure 17.10 presents experience-based recommendations for an initial belt-section choice based on horsepower and speed requirements of the application.

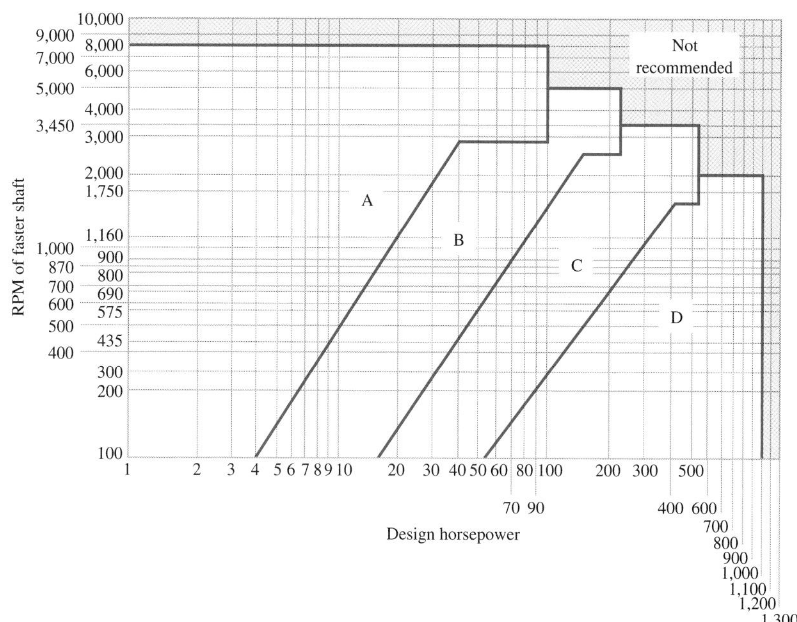

(*a*) Standard section V-belts.

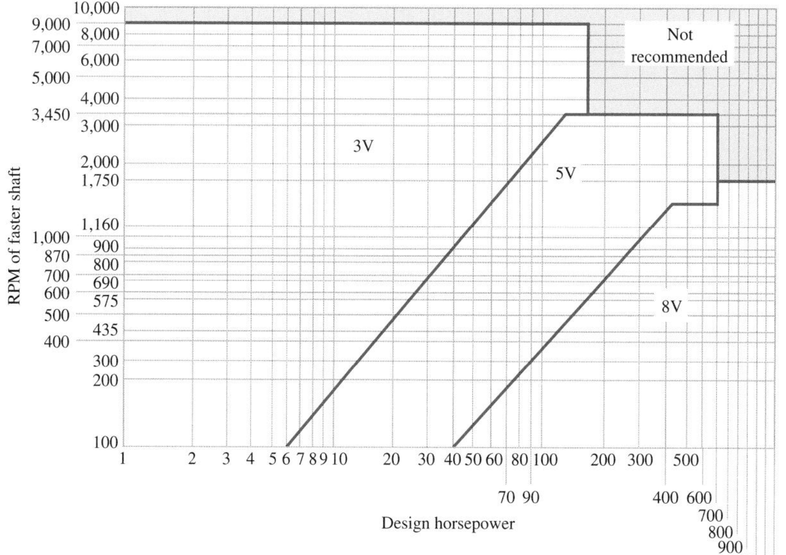

(*b*) Narrow section V-belts.

Figure 17.10
Recommended nominal belt selection as a function of application, horsepower requirement, and speed requirement. *Source:* Dayco Products, Inc. (Adapted from ref. 1.)

[39] C_4 is also a weak function of tension ratio, as shown in ref. 10.

If the first few iterations in the belt selection process seem to converge toward an *unacceptable* solution, consideration should be given to using an arrangement of several parallel belts operating side-by-side on multiply grooved sheaves. If such a multiple-belt drive configuration is adopted, it usually becomes necessary to install *matched sets of belts*[40] to assure that tolerances on the datum lengths of all belts in the set are close enough for uniform load sharing among all the belts. Multiply grooved sheaves are commercially available with 1, 2, 3, 4, 5, 6, 8, or 10 grooves, but it is usually recommended that no more than 5 parallel belts be used. If any belt fails in a multiple-belt drive, all belts must be replaced with a new matched set. In some applications involving pulsating loads, the multiple V-belts may become unstable, whipping or slapping together. For such applications, a *banded V-belt* configuration may solve the problem. Banded V-belts consist of up to five V-belt sections connected by a tie band[41] to provide lateral stability. If the design application requires the use of a *serpentine* belt configuration in which *both sides* of the belt must drive, double-V (double-angle) belt cross sections are commercially available.

In addition to the design procedures just discussed, experience-based recommendations for achieving optimum performance from a proposed V-belt drive configuration include:

1. The center distance[42] C should usually be no less than the datum diameter of the larger sheave, and no more than three times the sum of the sheave datum diameters. Thus

$$3(d_d + D_d) \geq C \geq D_d \tag{17-16}$$

2. For best results a V-belt should run at a linear belt velocity in the range from 1500–6500 ft/min, with 4000 ft/min a usual target. Special applications, such as snowmobile drives, may allow linear belt velocities up to 15,000 ft/min.
3. Within space envelope constraints, it is recommended that selected sheaves be larger than the minimum recommended diameter. This results in longer belt life and reduced maintenance costs.

It should be noted that most V-belt manufacturers publish extensive catalog data and give simple belt selection procedures.[43] While these catalog lookup procedures are recommended for the final phase of belt selection, they provide little insight into how failure mode interactions take place, what safety factors have been included, or which parameters might be changed to improve belt life. In accordance with the objectives of this text, it is recommended that the more fundamental design procedures just presented be implemented to gain insight, and then manufacturers' catalogs be consulted to confirm the selection.

Finally, it is important to recognize that the belt life calculated using (17-10) is the number of cycles attributed to a *single* sheave. All V-belt drives employ at least two sheaves, usually of different diameters. Thus, from (17-13), each different sheave induces its own bending contribution, and its own mean and alternating stress components, as illustrated by (17-11) and (17-12). Consequently, the belt failure life associated with each of the different sheave diameters will be different for each sheave.[44] To determine the resultant belt life when two or more sheaves are utilized in a belt drive,[45] a cumulative damage relationship such as the Palmgren–Miner hypothesis[46] may be utilized.

If n_i is defined as the number of operating cycles at a stress level σ_i, the Palmgren–Miner hypothesis of (2-47) may be written as

[40]See, for example, ref. 1.

[41]A tie band has the configuration of a thin integrally attached outer flat belt connecting the parallel V-belt sections.

[42]See Figure 17.8. [43]See, for example, ref. 1. [44]Assuming that no other sheaves contribute to failure.

[45]All practical belt drives employ at least two sheaves. [46]See (2-47) and related discussion.

Failure is predicted to occur if (FIPTOI)

$$\sum_{i=1}^{m} \frac{n_i}{(N_f)_i} \geq 1 \tag{17-17}$$

where n_i is determined by the application and $(N_f)_i$ may be calculated from (17-10). For example, for a two-sheave drive system, (17-17) gives

FIPTOI

$$\frac{n_1}{(N_f)_1} + \frac{n_2}{(N_f)_2} \geq 1 \tag{17-18}$$

Since each belt pass N_p induces one cycle of σ_1 and one cycle of σ_2,

$$n_1 = n_2 = n_i = N_p \tag{17-19}$$

and from (17-19), the number of belt passes (belt cycles) to failure, $(N_p)_f$, is

$$\frac{1}{(N_p)_f} = \frac{1}{(N_f)_1} + \frac{1}{(N_f)_2} \tag{17-20}$$

Example 17.2 V-Belt Selection

A newly constructed metropolitan transit authority "bus barn" will require the installation of 12 exhaust fans to properly ventilate the facility. Preliminary engineering estimates indicate that belt-driven constant-speed fans running at approximately 800 rpm should provide satisfactory ventilation. It is desired to drive the fans with 1750-rpm electric motors. The center distance between each motor and its corresponding fan shaft would ideally lie in the range 38–42 inches. Estimates of fan power requirements indicate that each fan will nominally require about 5 horsepower on a steady-state basis, and that flutter and flow instabilities will probably produce *moderate shock* loading. The fans are to operate 365 days a year, 24 hours a day.

Using a design safety factor of $n_d = 1.2$ on horsepower, propose a conventional V-belt drive arrangement that will provide a mean life of about 2 years between belt replacements.

Solution

For fan operation 24 hours/day for 2 years at a speed of approximately 800 rpm, the corresponding number of belt passes, N_p, is

$$\begin{aligned} N_p &= \left(800\frac{\text{passes}}{\text{min}}\right)\left(60\frac{\text{min}}{\text{hr}}\right)\left(24\frac{\text{hr}}{\text{day}}\right)\left(365\frac{\text{days}}{\text{yr}}\right)(2\text{ yr}) \\ &= 84 \times 10^8 \text{ belt passes} \end{aligned} \tag{1}$$

For this two-sheave belt drive, each belt pass induces one stress cycle $(\sigma)_{mot}$ at the motor sheave, and one stress cycle $(\sigma)_{fan}$ at the fan sheave.

The design horsepower requirement for each fan system, $(hp)_d$, may be estimated as

$$(hp)_d = (hp)_{nom}K_a n_d = (5)(1.25)(1.2) = 7.5 \text{ horsepower} \tag{2}$$

Based on specifications, the speed ratio R must be

$$R = \frac{n_{mot}}{n_{fan}} = \frac{(d_d)_{fan}}{(d_d)_{mot}} = \frac{1750}{800} = 2.19 \tag{3}$$

Thus

$$(d_d)_{fan} = 2.19(d_d)_{mot} \tag{4}$$

From (17-5)

$$hp = \frac{(T_t - T_s)V}{33{,}000} \tag{5}$$

Example 17.2 Continues

where linear belt velocity is

$$V = \frac{\pi (d_p)_{fan} n_{fan}}{12} = \frac{\pi (d_p)_{fan}(800)}{12} = 209.4(d_p)_{fan} \tag{6}$$

or, combining (4) and (6)

$$V = (209.4)(2.19)(d_d)_{mot} = 458.7(d_d)_{mot} \tag{7}$$

A trial value for motor sheave datum diameter may be based on either Figure 17.1(d) or Figure 17.10, from which a conventional A-section V-belt is suggested. Then from Table 17.4, the minimum recommended datum diameter for an A-sheave is 3.0 inches. Experience dictates that sheave datum diameters larger than the minimum should be used if practical. Hence a trial value for motor sheave datum diameter will be chosen to be

$$(d_d)_{mot} = 4.50 \text{ inches} \tag{8}$$

whence, from (4),

$$(d_d)_{fan} = (2.19)(4.50) = 9.86 \text{ inches} \tag{9}$$

Then from (7)

$$V = 458.7(4.50) = 2064 \text{ ft/min} \tag{10}$$

By rule of thumb[47] this lies in the acceptable range but is toward the low-speed end. Next, from (5),

$$7.5 = \frac{(T_t - T_s)(2064)}{33{,}000} \tag{11}$$

or

$$(T_t - T_s) = \frac{(7.5)(33{,}000)}{2064} = 120 \text{ lb} \tag{12}$$

From Figure 17.8(a), the angle of wrap for the smaller motor sheave, for a nominal 40-inch center distance, is

$$\theta_s = \theta_{mot} = \pi - 2\alpha = \pi - 2\sin^{-1}\left(\frac{9.86 - 4.50}{2(40)}\right) = \pi - 0.13 = 3.01 \text{ rad} \tag{13}$$

and for the larger fan sheave

$$\theta_{fan} = \pi + 0.13 = 3.27 \text{ rad} \tag{14}$$

The nominal belt datum length then is

$$(L_d)_{nom} = \sqrt{4(40)^2 - (9.86 - 4.50)^2} + \frac{9.86(3.27) + 4.50(3.01)}{2} = 102.7 \text{ inches} \tag{15}$$

Checking A-section standard datum length belts usually in stock, shown in Table 17.5, the closest standard datum length shown in 101.3 inches, and this standard length belt will be adopted here.[48] Therefore

$$L_d = 101.3 \text{ inches} \tag{16}$$

Reading from Table 17.3, the specific weight of an A-section belt is

$$w_1 = 0.065 \frac{\text{lb}}{\text{ft}} \tag{17}$$

[47]See text discussion in *guideline 2* following (17-16).

[48]Standard V-belts are usually specified by the section-size letter followed by the nominal inside length of the belt in inches. For the case at hand, this would be an A100 belt.

Next, using (17-3),

$$T_c = \frac{(0.065)\left(\frac{2064}{60}\right)^2}{32.2} = 2.39 \text{ lb} \tag{18}$$

and (17-9) may be evaluated as[49]

$$\frac{T_t - 2.39}{T_s - 2.39} = e^{[0.3(3.01)]/[\sin(36/2)]} = e^{2.92} = 18.58 \tag{19}$$

Solving (12) for T_s,

$$T_s = T_t - 120 \text{ lb} \tag{20}$$

and substitution of this into (19)

$$\frac{T_t - 2.39}{(T_t - 120) - 2.39} = 18.58 \tag{21}$$

or

$$T_t = 129 \text{ lb} \tag{22}$$

and

$$T_s = 9 \text{ lb} \tag{23}$$

Also, it may be noted from (17-6) that the approximate belt pretension required is

$$T_o = \frac{129 + 9}{2} = 69 \text{ lb} \tag{24}$$

Critical point *edge cord* tensions may next be calculated using constants from Table 17.3. For the A-section belt in bending, using (17-13), the edge cord tensions attributable to bending are

$$(T_{be})_{mot} = \frac{5.0 + 111}{4.5} = 25.8 \text{ lb} \tag{25}$$

and

$$(T_{be})_{fan} = \frac{5.0 + 111}{9.86} = 11.8 \text{ lb} \tag{26}$$

For edge cord tension attributable to centrifugal force, (17-14) gives

$$T_{ce} = 0.101 \times 10^{-6}(2064)^2 = 0.4 \text{ lb} \tag{27}$$

For tight-side and slack-side edge cord tensions attributable to power transmission, (17-15) gives

$$T_{te} = 0.175(129) = 22.6 \text{ lb} \tag{28}$$

and

$$T_{se} = 0.175(9) = 1.6 \text{ lb} \tag{29}$$

Placing these values into (17-11), the edge cord *mean stresses* may be calculated as

$$(\sigma_m)_{mot} = \frac{22.6 + 25.8 + 2(0.4) + 1.6}{2(1.73 \times 10^{-3})} = 14{,}680 \text{ psi} \tag{30}$$

and

$$(\sigma_m)_{fan} = \frac{22.6 + 11.8 + 2(0.4) + 1.6}{3.46 \times 10^{-3}} = 10{,}635 \text{ psi} \tag{31}$$

[49]Using $\mu = 0.3$ for dry rubber on steel, from Appendix Table A.1, and assuming an included sheave angle of $\beta = 36°$.

Example 17.2 Continues

Next, inserting these values into (17-12),

$$(\sigma_a)_{mot} = \frac{22.6 + 25.8 - 1.6}{3.46 \times 10^{-3}} = 13{,}525 \text{ psi} \tag{32}$$

and

$$(\sigma_a)_{fan} = \frac{22.6 + 11.8 - 1.6}{3.46 \times 10^{-3}} = 9480 \text{ psi} \tag{33}$$

From (17-10), using constants for an A-section belt from Table 17.3, the number of motor sheave cycles to produce failure would be[50]

$$\begin{aligned}(N_f)_{mot} &= 6.13 \times 10^{-8}(19{,}800 - 13{,}525)^2(26{,}400 - 14{,}680)^2(101.3)^{175}(2064)^{-1} \\ &= 5.19 \times 10^8 \text{ cycles}\end{aligned} \tag{34}$$

Similarly, the number of fan sheave cycles to produce failure would be[51]

$$\begin{aligned}(N_f)_{fan} &= 6.13 \times 10^{-8}(19{,}800 - 9480)^2(26{,}400 - 10{,}635)^2(101.3)^{1.75}(2064)^{-1} \\ &= 2.55 \times 10^9 \text{ cycles}\end{aligned} \tag{35}$$

Finally, using (17-19)

$$\frac{1}{(N_p)_f} = \frac{1}{(N_f)_{mot}} + \frac{1}{(N_f)_{fan}} \tag{36}$$

whence the number of belt passes to failure becomes

$$(N_p)_f = \frac{1}{\dfrac{1}{5.19 \times 10^8} + \dfrac{1}{2.55 \times 10^9}} = 4.31 \times 10^8 \text{ passes} \tag{37}$$

Comparing the result of (37) with the belt-pass requirement for a two-year maintenance-free period given in (1), it is clear that the chosen A-section belt falls short, and would fail early. One simple solution would be to use parallel belts on multigroove sheaves. As can be seen from (28), (30), and (32),[52] a two-belt drive would cut the mean and alternating stress levels by nearly a factor of two, which from (34) would produce a significant increase in belt life. Another solution might be to use larger sheaves. For a larger motor sheave, from (25), the edge cord tension due to bending would be reduced and, consequently, from (30) and (32), the mean and alternating stress levels would be lowered, probably giving a significant increase in belt life.

17.6 Belt Drives; Synchronous Belts

As illustrated in Figures 17.1(c) and 17.2(c), toothed timing belts (synchronous belts) do not rely upon friction for transmission of torque and power; they transmit torque and power by virtue of positive engagement of a toothed belt meshing with toothed sprocket grooves. Thus a toothed belt drive provides a constant angular velocity ratio (no slip or creep), requires minimal belt pretension (only enough to prevent "tooth skipping" when starting or braking), can be operated at high speeds (up to 16,000 ft/min), and can transmit high torques and high power.

[50]Assuming the *motor* sheave is the only source of cyclic loading.

[51]Assuming the *fan* sheave is the only source of cyclic loading.

[52]Noting from (37) that the fan sheave contribution is small compared to the smaller motor sheave.

Because high-modulus tension cords[53] are used in toothed belt construction, little change in belt length or tooth pitch is experienced when the belt is loaded. Hence each belt tooth can correctly mesh with the corresponding pulley groove as they engage and reside until leaving the mesh site. Belt tooth profile and spacing (pitch), as well as pulley tooth shape and pitch, are precisely controlled during manufacture to enhance smooth, uniform operation. The teeth of standard synchronous belts are trapezoidal in profile. For heavy-duty applications, the tooth profile is sometimes modified to provide an enlarged shear cross section with a corresponding reduction of transverse shearing stress in the belt teeth. Neither the tooth shape nor the pitch has been standardized for these specialty belts, and both may vary with the product manufacturer. At normal operating speeds toothed timing belts tend to operate smoothly and quietly, and there is no *chordal-action* speed-variation as there is for chain drives.[54] If noise generation does become a problem, some manufacturers provide proprietary tooth shapes (e.g., parabolic profiles[55]) to reduce noise generation or increase drive capacity. Helical tooth arrangements are also used for high-performance synchronous drive applications for smoother, quieter operation and stronger teeth, just as for helical *gear* teeth.[56] Two-sided toothed belts are also available for serpentine-drive applications in which the belt must drive from both sides.

The design and selection process for synchronous belts is very similar to the V-belt selection process given in 17.5, and will not be repeated here. It is worth noting, however, that most timing belt manufacturers, like V-belt manufacturers, publish extensive catalog data and give simple belt selection procedures.[57]

17.7 Chain Drives; Potential Failure Modes

Figure 17.3 illustrates some of the more commonly used power transmission chain configurations.[58] All precision power transmission chains are manufactured by pin-connecting a continuous series of links which sequentially mesh with sprocket teeth as the chain loops around two or more sprockets. Each pin-and-bushing joint articulates as the chain passes around the sprockets; hence each joint acts as a journal and sleeve bearing. Just as for any other journal-and-sleeve bearing application,[59] proper lubrication of the sliding pin-and-bushing interface is crucial to achieving the *potential* wear life of the chain. Ironically, periodic maintenance schedules, which are intended to clean and relubricate chains to *enhance* wear life, may sometimes actually *reduce* wear life because of local lubricant starvation unless a conscientious effort is made to properly reintroduce lubricant to the relatively inaccessible pin-and-bushing interfaces.

Three different options for lubrication are: Type I—manual or drip lubrication, Type II—oil bath or slinger-disk lubrication, and Type III—directed oil stream or pressure spray lubrication.[60] Chain manufacturers often recommend Type I for applications in which linear chain velocity is between about 170 and 650 ft/min, Type II for applications in which linear chain velocity is between about 650 and 1500 ft/min, and Type III if linear chain velocity exceeds about 1500 ft/min. Oil flow rates required for effective lubrication range from about $^1/_4$ gal/min for a chain drive transmitting 50 horsepower to about 10 gal/min for drives transmitting 2000 horsepower.

The tight side chain tension forces are transferred to the sprocket teeth through rollers or through toothed link plates.[61] As the moving chain passes around the sprockets, the tensile force in the chain fluctuates from the *tight side tension* to the *slack side tension*,[62] and

[53]See 17.3. [54]See 17.7. [55]See ref. 13. [56]See 15.12 and ref. 13. [57]See, for example, ref. 12.

[58]Many other types of chain are commercially available. See, for example, ref. 14.

[59]See Chapter 10. [60]Also see 10.5, 10.6, and 15.10. [61]See Figures 17.3 and 17.4.

[62]Slack side tension is usually near zero.

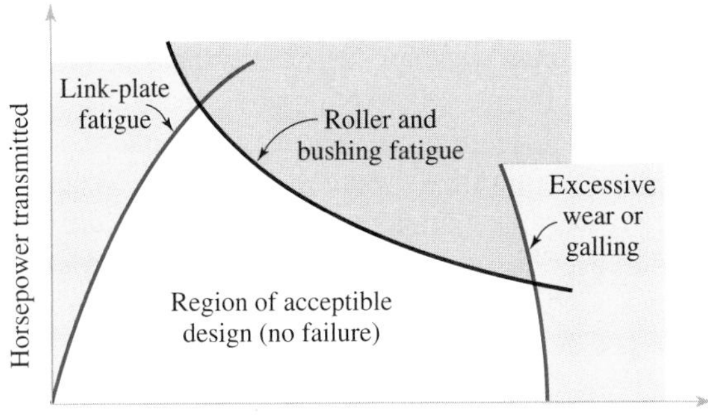

Figure 17.11
Tent-shaped design-acceptable region bounded by limiting failure curves for precision roller chain. See equations (17-20), (17-21), and (17-22).

back, for each chain pass. At chain pass speeds above about 3000 ft/min, centrifugal forces may also add significantly to the tensile forces in the chain. In addition, *superposed* higher-frequency fluctuations in chain tension may be caused by a kinematic consequence of engagement between the chain links and the sprocket teeth. This kinematic behavior is known as *chordal action*.[63] Fatigue, therefore, becomes a primary failure mode candidate for power transmission chains. Fatigue failures may be generated in the roller-bushing link plates, toothed link plates, roller bushings, or tooth surfaces (surface fatigue). In addition, abrasive wear, adhesive wear, fretting wear, or galling between the bushing and pin may be potential failure modes in some circumstances. When wear between pins and bushings or rollers and teeth produces enough dimensional change to cause rollers to climb too high on the sprocket teeth, the chain may skip from one tooth to the next. For wear-based chain elongations of more than about 3 percent, chain replacement is usually necessary.

A sketch showing the probable failure regions as a function of power and speed is shown in Figure 17.11. Empirical relationships defining the boundaries of each failure region are included in 17.9.

17.8 Chain Drives; Materials

Based on material selection guidelines presented in Chapter 3 and the failure mode discussion of 17.7, candidate materials for power transmission chains should have good strength (especially good fatigue strength), high stiffness, good wear resistance, good resistance to surface fatigue, good resilience, and in some applications good corrosion resistance, as well as reasonable cost. Materials meeting these criteria[64] include steels and steel alloys, cast iron, malleable iron, stainless steel alloys, and, for special applications, alloys of brass, bronze and certain polymeric materials. Chain components may be pressed, cast, forged, machined, or welded to obtain the desired geometry. Chain parts may be heat treated or not heat treated, as required for strength. Carburizing and case hardening, through hardening, induction hardening, or plating may be used to enhance wear resistance or provide corrosion protection.

In a standard commercially available roller chain[65] the *link plates* are typically carbon steel, heat treated if the application is demanding, and edges may be induction hardened to improve sliding wear resistance. *Bushings* usually are carbon steel or alloy steel, case

[63]As discussed in 17.9, *chordal action (polygonal action)* results in a rise and fall of the centerline of each link as the sprocket rotates and the links articulate. This causes speed variations and fluctuations in chain tension that range in magnitude from about 1 to 10 percent of the tight side chain tension, depending upon the number of sprocket teeth.

[64]See 3.2. [65]See Figure 17.3.

hardened or through hardened. In certain applications, however, bushings may be made of heat-treated stainless steel, bronze, graphite, or other materials. *Pins* are usually made of carbon steel or alloy steel carburized and case hardened, or through hardened. *Rollers* are generally made of carbon or alloy steels, carburized and surface hardened or through hardened, as the application requires. Sprocket teeth, usually steel, are often surface hardened to about Rockwell C 59–63.

17.9 Chain Drives; Precision Roller Chain

The basic configuration of a *single-strand* standard roller chain is illustrated in Figure 17.3(a). *Multiple-strand* roller chain, consisting of two or more parallel strands of chain assembled on common pins, has also been standardized, as have *double-pitch* chains.[66] In addition, may nonstandard chains are commercially available, including chains with sealed joints or with sintered metal bushings, chains with extra clearance, and chains made of corrosion- or heat-resistant materials. Dimensions and nominal ultimate strengths for single-strand roller chain are shown in Table 17.6.

Multiple-strand roller chain may be utilized to transmit higher horsepower, but because of nonuniform loading among the strands, a *strand factor*, K_{st} (always less than the number of strands), must be introduced to account for the uneven load sharing among the parallel strands (see Table 17.7).

TABLE 17.6 Dimensions and Nominal Ultimate Tensile Strengths for Single-Strand Standard Precision Roller Chain[1]

ANSI Chain No.	Pitch, p, in	Roller Diameter, D, in	Roller Width, W, in	Pin Diameter, d, in	Link-Plate Thickness, t, in	Recommended Min. Center Distance, in	Nominal Ultimate Tensile Strength, lb	Specific Weight, w_1, lb/ft
25	$\frac{1}{4}$	0.130[2]	$\frac{1}{8}$	0.091	0.030		1,050	0.09
35	$\frac{3}{8}$	0.200[2]	$\frac{3}{16}$	0.141	0.050	6	2,400	0.23
41[3]	$\frac{1}{2}$	0.306	$\frac{1}{4}$	0.141	0.050	9	2,600	0.27
40	$\frac{1}{2}$	0.312	$\frac{5}{16}$	0.156	0.060	9	4,300	0.40
50	$\frac{5}{8}$	0.400	$\frac{3}{8}$	0.200	0.080	12	7,200	0.66
60	$\frac{3}{4}$	0.469	$\frac{1}{2}$	0.234	0.094	15	9,800	0.98
80	1	0.625	$\frac{5}{8}$	0.312	0.125	21	17,600	1.69
100	$1\frac{1}{4}$	0.750	$\frac{3}{4}$	0.375	0.156	27	26,400	2.63
120	$1\frac{1}{2}$	0.875	1	0.437	0.187	33	39,000	3.87
140	$1\frac{3}{4}$	1.000	1	0.500	0.219	39	50,900	4.98
160	2	1.125	$1\frac{1}{4}$	0.562	0.250	45	63,200	6.58
180	$2\frac{1}{4}$	1.406	$1\frac{13}{32}$	0.687	0.281		81,500	9.00
200	$2\frac{1}{2}$	1.562	$1\frac{1}{2}$	0.781	0.312	57	105,500	11.38
240	3	1.875	$1\frac{7}{8}$	0.937	0.375	66	152,000	15.89

[1]See Figure 17.3(a); For multiple-strand factors, see Table 17.7.
[2]Bushing diameter; chain is rollerless.
[3]Lightweight chain.

[66]See refs. 15 and 16.

TABLE 17.7 Multiple-Strand Factors

Number of Strands	Strand Factor, K_{st}
1	1.0
2	1.7
3	2.5
4	3.3
5	3.9
6	4.6

Referring again to Figure 17.11, link-plate fatigue governs the failure envelope at lower speeds while roller and bushing fatigue tend to govern at higher speeds. Practical upper limits on linear chain speed are imposed by the onset of excessive wear or galling. About 9000 ft/min is the maximum chain speed that can be successfully used; 2500 ft/min is much more usual.

Empirical expressions have been developed for each of the failure boundaries sketched in Figure 17.11. For *link-plate fatigue* the limiting horsepower is given by

$$(hp_{lim})_{lp} = K_{lp} N_s^{1.08} n_s^{0.9} p^{(3.0-0.07p)} \tag{17-21}$$

where K_{lp} = 0.0022 for no. 41 chain (lightweight chain)
= 0.004 for all other chain numbers
N_s = number of teeth in smaller sprocket
n_s = rotational speed of smaller sprocket, rpm
p = chain pitch, inches

For *roller and bushing fatigue*, the limiting horsepower is

$$(hp_{lim})_{rb} = \frac{1000 K_{rb} N_s^{1.5} p^{0.8}}{n_s^{1.5}} \tag{17-22}$$

where K_{rb} = 29 for chains no. 25 and 35
= 3.4 for chain no. 41
= 17 for chains no. 40 through 240

For excessive *wear or galling*, the limiting horsepower is

$$(hp_{lim})_g = \left(\frac{n_s p N_s}{110.84}\right)(4.413 - 2.073p - 0.0274N_L) - \left(\ln\frac{n_L}{1000}\right)(1.59 \log p + 1.873) \tag{17-23}$$

where N_L = number of teeth in larger sprocket
n_L = rotational speed of larger sprocket

For any given application, the design horsepower must not exceed any of the *horsepower limits* calculated using (17-21), (17-22), and (17-23).

Chordal action (polygonal action), which causes a fluctuation in chain speed and chain tension each time a chain link engages a sprocket tooth, may be a serious limiting design factor in chain performance, especially in high-speed applications. As illustrated in Figure 17.12 for roller chain, chordal action is a kinematic consequence of the fact that the *line of approach* of the chain is *not tangent* to the pitch circle of the sprocket; it is *colinear with a chord* of the pitch circle. Therefore, as the sprocket rotates, the link makes first contact with the sprocket when the link centerline is below the (parallel) tangent to the pitch circle, which has the radius r_p. As a consequence, the link centerline is caused to rise from r_{ch} to r_p, then fall back to r_{ch}, a behavior known as chordal action.

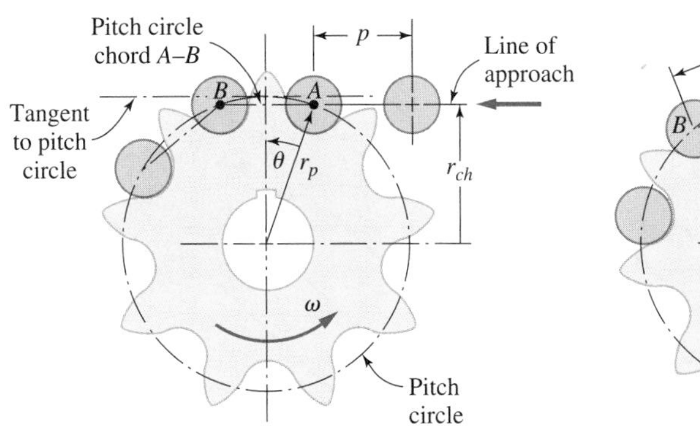

(*a*) First contact of roller *A* with sprocket.

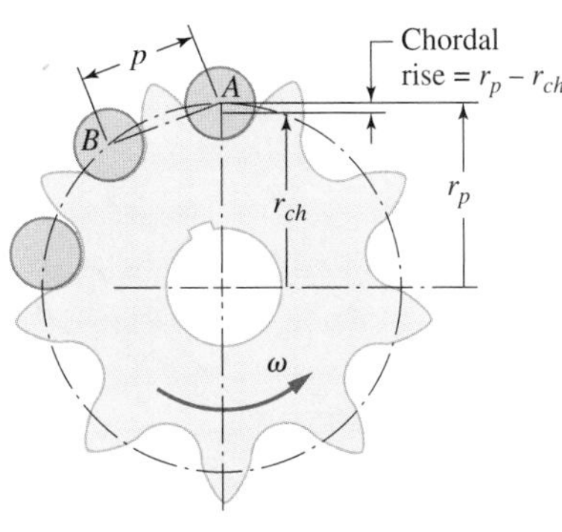

(*b*) Chordal rise produced by sprocket rotation θ.

Figure 17.12
Depiction of *chordal action* in a roller chain.

In effect, chordal action causes the sprocket pitch radius to cyclically fluctuate, resulting in a cyclic fluctuation in chain speed. Thus even if the drive sprocket rotates at constant angular velocity, the driven sprocket experiences a speed fluctuation. Figure 17.13 shows the estimated speed fluctuation as a function of number of sprocket teeth. It may be noted that the larger the number of teeth in a sprocket for a given chain velocity, the smoother the action, the more uniform the chain speed, and the lower the impact loading between chain and sprocket. Normally, however, sprockets should have no more than about 60 teeth. This is because of the difficulty in maintaining proper fit (as wear progresses) for larger numbers of teeth, and because of increased manufacturing costs for sprockets with larger numbers of teeth.

17.10 Roller Chain Drives; Suggested Selection Procedure

In selecting an appropriate roller chain and sprockets for a particular application, design decisions must be made about chain type, size, space allocation, lubricant delivery, mount-

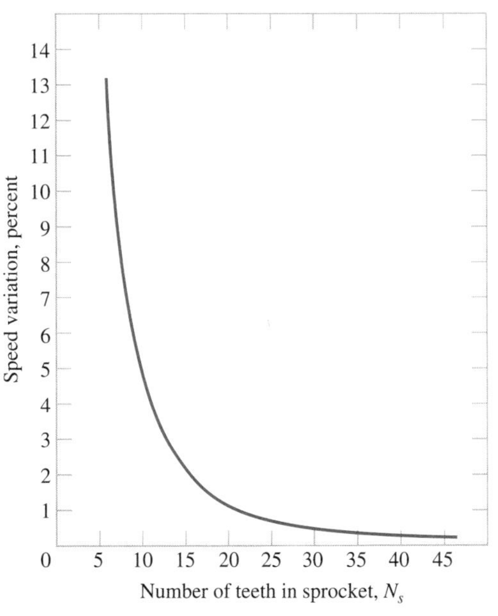

Figure 17.13
Roller chain velocity fluctuation as a function of number of sprocket teeth.

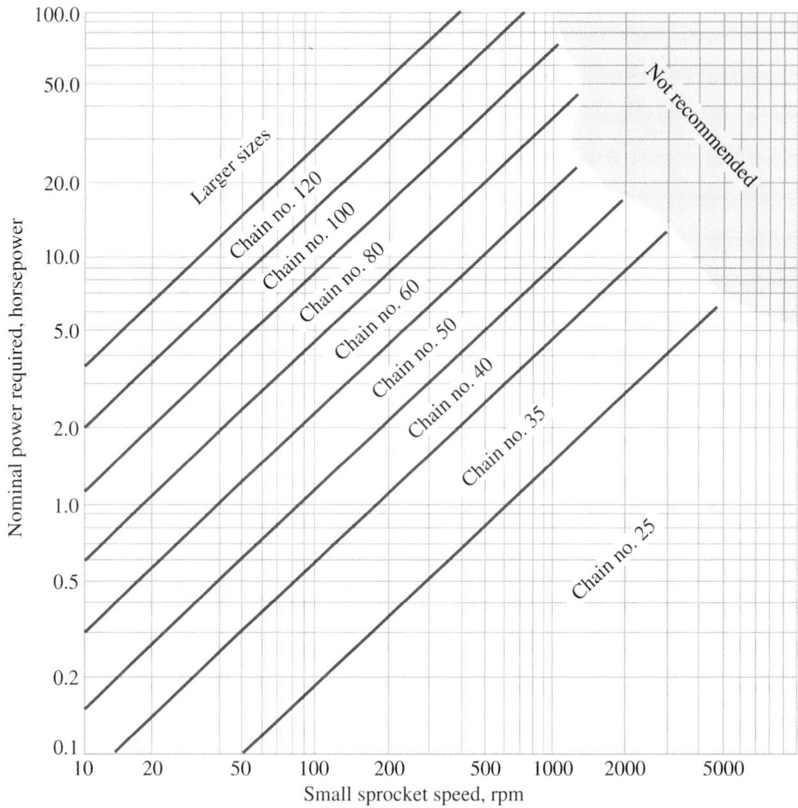

Figure 17.14
First iteration selection of roller-chain pitch as a function of power and speed requirements.

ing methods, and other details. Usually the chain selection process is iterative. To help make an initial choice, the experience-based recommendations of Figure 17.14 often prove effective. A procedure for establishing the elements of a good roller chain drive configuration is listed next.

1. Establish design specifications for the application, including power to be transmitted, input and output shaft speed requirements, allowable speed fluctuation, shaft center distance limitations, space envelope constraints, design life requirements, safety factors, and any other special design criteria.
2. Determine the design horsepower requirement $(hp)_d$ by multiplying an application factor[67] K_a times the nominal design horsepower required, and dividing by a multiple-strand factor[68] K_{st}, giving

[67]See Table 17.2. More precise application factors for chain selection may be found in chain manufacturers' catalogs.

[68]See Table 17.7.

$$(hp)_d = \frac{K_a(hp)_{nom}}{K_{st}} \tag{17-24}$$

3. Tentatively select an appropriate chain *pitch* to use for a first iteration, using Figure 17.14 as a guideline. Check Table 17.6 to make sure that minimum center distance requirements are met. The optimum range for center distance is between about 30 and 50 chain pitches. Center distances greater than about 80 pitches are not recommended.
4. Tentatively select the number of teeth for the small sprocket (usually the driver) using Figure 17.13 as a guide. Even for low-speed applications the small sprocket usually should have at least 12 teeth.
5. Tentatively determine the number of teeth, N_L, for the larger sprocket (usually the driven sprocket). Speed ratios should not exceed about 7:1 (10:1 maximum). If a higher speed ratio is required, a double-reduction drive should be proposed. The number of teeth on the larger sprocket may be calculated from

$$N_L = \frac{N_s n_s}{n_L} \tag{17-25}$$

where N_s = number of teeth in smaller sprocket
n_s = rotational speed of smaller sprocket, rpm
n_L = rotational speed of larger sprocket, rpm

6. Using (17-21), (17-22), and (17-23), calculate the limiting horsepowers associated with link-plate fatigue, roller and bushing fatigue, and wear or galling, respectively. The design horsepower calculated from (17-24) should not exceed any of these three limiting values. On the other hand, design horsepower should be as close to *all* of the limiting values as is practical.
7. If necessary, iterate again until these requirements are satisfied.
8. Calculate linear chain velocity V from

$$V = \frac{pNn}{12} \text{ ft/min} \tag{17-26}$$

where p = pitch, in
N = number of sprocket teeth
n = rotational speed of sprocket, rpm

Chain velocity V should not exceed about 9000 ft/min, with velocities in the range 2000–3000 ft/min being much more common.

9. Calculate chain length L in *pitches* using

$$L = \left(\frac{N_L + N_s}{2}\right) + 2C + \frac{(N_L - N_s)^2}{4\pi^2 C} \tag{17-27}$$

The chain length must be an integral multiple of the pitch, and an *even* number of pitches is recommended to avoid the need for half-links. Note that center distance C in this equation must also be expressed in *pitches*.

10. Select an appropriate type of lubrication based on linear chain velocity calculated from (17-26). Lubrication options as a function of chain velocity are discussed in 17.7.
11. Summarize the recommendations.

Example 17.3 Precision Roller Chain Selection

A precision roller chain drive is being proposed to power a helicopter transmission test stand at 600 rpm, using a 1200-rpm electric motor. It is desired to have a center distance between the driving motor sprocket and driven test-stand sprocket of about 40 inches and the drive should be as compact as is practical. Moderate shock loading is expected. The nominal power to be transmitted is estimated as 20 horsepower. A single-strand chain is preferred. Further, speed fluctuations of no more than 2 percent of the linear chain velocity can be tolerated. Select a suitable roller chain and associated sprockets for this application.

Solution

a. Following the procedure recommended in 17.10, the design horsepower may be calculated from (17-24) as[69]

$$(hp)_d = \frac{K_a(hp)_{nom}}{K_{st}} = \frac{(1.25)(20)}{1.0} = 25.0 \text{ horsepower} \tag{1}$$

b. For 25 design horsepower, a single-strand chain, and a motor sprocket speed (small sprocket speed) of 1200 rpm, Figure 17.14 suggests a no. 60 chain for the first iteration. The pitch of a no. 60 chain, from Table 17.6, is

$$p = 0.75 \text{ inch} \tag{2}$$

c. The minimum center distance for a no. 60 chain is, From Table 17.6,

$$C_{min} = 15.0 \text{ inches} \tag{3}$$

The maximum recommended center distance[70] of 80 pitches is, for a no. 60 chain,

$$C_{max} = 80(0.75) = 60 \text{ inches} \tag{4}$$

Since the specifications prescribe a *desired* center distance value of

$$C = 40 \text{ inches} \tag{5}$$

and this lies in the acceptable range between 15 and 60 inches, a nominal center distance of approximately 40 inches will be adopted. For a no. 60 chain this corresponds to a center distance of about 53 *pitches*.

d. Specifications require that speed fluctuations be no more than 2 percent of linear chain velocity. For Figure 17.13 then, the number of teeth in the motor sprocket must be

$$N_s \geq 16 \text{ teeth} \tag{6}$$

Since compact design is a specified criterion, a driving sprocket with 16 teeth will be adopted for the first iteration.

e. Next, the number of teeth on the large sprocket may be calculated from (17-25) as

$$N_L = \frac{N_s n_s}{n_L} = \frac{(16)(1200)}{(600)} = 32 \text{ teeth} \tag{7}$$

f. Using (17-21), the limiting horsepower based on link-plate fatigue for a *no. 60 chain* is

$$(hp_{lim})_{lp} = 0.004(16)^{1.08}(1200)^{0.9}(0.75)^{[3.0-0.07(0.75)]} = 20.2 \text{ horsepower} \tag{8}$$

Using (17-22), the limiting horsepower based on roller and bushing fatigue for a *no. 60 chain* is

$$(hp_{lim})_{rb} = \frac{1000(17)(16)^{1.5}(0.75)^{0.8}}{(1200)^{1.5}} = 20.8 \text{ horsepower} \tag{9}$$

[69]Values for K_a and K_{st} are from Tables 17.2 and 17.7, respectively. [70]See step 3 of 17.10.

Finally, using (17-23), the limiting horsepower based on excessive wear or galling is

$$(hp_{lim})_g = \left[\frac{(1200)(0.75)(16)}{110.84}\right][4.413 - 2.073(0.75) - 0.0274(32)] \\ - \left[\ln\left(\frac{600}{1000}\right)\right][1.59 \log(0.75) + 1.873] = 258 \text{ horsepower} \tag{10}$$

Clearly, the horsepower limit based on galling is much larger than the required 25 design horsepower, but the limiting power for both link-plate fatigue and roller and bushing fatigue are *exceeded* by the required design horsepower. Therefore, a larger chain must be selected and power limits recalculated.

Repeating (8) and (9) for a *no. 80 chain*,[71]

$$(hp_{lim})_{lp} = 0.004(16)^{1.08}(1200)^{0.9}(1.0)^{[3.0-0.07(1)]} = 47.2 \text{ horsepower} \tag{11}$$

and

$$(hp_{lim})_{rb} = \frac{1000(17)(16)^{1.5}(1.0)^{0.8}}{(1200)^{1.5}} = 26.2 \text{ horsepower} \tag{12}$$

The galling horsepower limit in (10) for a no. 60 chain is already 10 times the required design horsepower, it need not be recalculated for the no. 80 chain.

From (11) and (12) it is clear that for a no. 80 chain, in this application, roller and bushing fatigue would be the governing failure mode. Since the design horsepower (25 hp) does not exceed the roller and bushing fatigue horsepower limit (26.2 hp), a no. 80 chain appears to be a good choice.

g. Repeating step (c) for a no. 80 chain,

$$C_{min} = 21.0 \text{ inches} \tag{13}$$

and

$$C_{max} = 80(1.0) = 80 \text{ inches} \tag{14}$$

The specification center distance goal of

$$C = 40 \text{ inches} \tag{15}$$

therefore remains acceptable, and for a no. 80 chain this corresponds to a center distance of 40 *pitches*.

h. Next, the linear chain velocity may be calculated from (17-24) as

$$V = \frac{(1)(16)(1200)}{12} = 1600 \text{ ft/min} \tag{16}$$

This is an acceptable chain velocity.

i. The chain length in pitches may be calculated from (17-27), to the nearest integral pitch, as

$$L = \left(\frac{32 + 16}{2}\right) + 2(40) + \frac{(32 - 16)^2}{4\pi^2(40)} = 104 \text{ pitches} \tag{17}$$

Since this is an *even* number of pitches, no half-links are required; this chain is therefore adopted.

j. Using the lubrication guidelines of 17.7, for the chain velocity of 1600 ft/min calculated in (16), Type III (directed oil stream or pressure spray) would be recommended but Type II (oil bath or slinger-disk) would also be acceptable.

k. Summarizing the recommendations:

[71]See Table 17.6 for pertinent data.

Example 17.3 Continues

1. Use a single-strand standard no. 80 precision roller chain of length equal to 104 pitches ($p = 1.0$ inch for no. 80 chain).
2. Use a center distance of approximately 40 inches, making provision for installation and tightening.
3. Use a 16-tooth motor drive sprocket and a 32-tooth driven test-stand sprocket.
4. Specify directed oil stream lubrication but note that oil bath or slinger-disk lubrication would probably be acceptable as well.
5. Use chain casings to contain the oil and provide a lubricant sump.
6. Provide adequate safety guards and shields. Chain drives can be dangerous if not shielded.

17.11 Chain Drives; Inverted-Tooth Chain

Illustrated in Figure 17.3(d) is an inverted-tooth chain (silent chain[72]) of a side-by-side series of alternately interlaced, toothed, flat metal link plates extending across the width of the chain. The assemblage of flat link plates is pin-connected to permit articulation. The "tooth" profiles of the links are usually straight sided, but for special applications may have an involute profile. Power is transmitted through positive engagement of the chain teeth with meshing sprocket teeth, typically giving smooth, quiet operation similar to belt drives but with compactness and strength similar to gear drives. Various pin-joint configurations have been developed by silent chain manufacturers, ranging from round pins in round bushings to special rocker-joints[73] designed to minimize sliding friction, compensate for chordal action, and enhance wear life by substituting rolling friction for sliding friction in the joints. Inverted-tooth chain usually embodies guide links either on the sides or in the center[74] to keep the chain from sliding sidewise off of the sprockets.

Inverted-tooth chain (and sprockets), like precision roller chain, has been standardized by the industry. Standard pitch lengths ranging from $^3\!/_8$ inch to 2 inches are commercially available, and standard widths from 0.5 to 6 inches for 0.375-inch pitch chain up to widths from 4 to 30 inches for 2.0-inch pitch chain can be supplied. Just as for precision roller chain, proper lubrication is crucial for long life and minimum wear. The selection procedure for inverted-tooth chain parallels the precision roller chain selection procedure discussed in 17.10.

17.12 Wire Rope; Potential Failure Modes

Wire rope may be vulnerable to failure by any of several possible modes,[75] depending on load, speed, and environment, as well as type, size, construction, and materials selected for the rope. As illustrated in Figure 17.5, wire rope is manufactured by first helically twisting several small wires together to form a multiwire strand, then helically twisting several strands together to form a rope. When tensile loading is applied to the helically twisted wires and strands, the wires tend to stretch and the helixes tend to "tighten." Both of these loading consequences generate Hertz contact stresses and relative sliding motions between and among wires. As loads cycle, and as the wire rope is repeatedly bent around drums or

[72]So named because of its relatively quiet operation.

[73]HY-VO® is the tradename for a unique chain design utilizing rocker-joints, manufactured by Morse Chain Division of Borg-Warner Corp.

[74]See Figure 17.3(d). [75]See 2.3 and ref. 17, pp. 59–62.

sheaves, the conditions just described may induce failure by tensile fatigue, bending fatigue, fretting fatigue, surface fatigue wear, abrasive wear, yielding, or ultimate rupture. Corrosion may also be a factor.

Wire rope winding practice, as shown in Figure 17.5(e), is a significant factor in resistance to fatigue and wear. As sketched, the wires in *regular lay* rope appear to be *nominally aligned* with the axis of the rope; the wires in *lang lay* rope appear to make an angle with respect to the axis of the rope.[76] Either *lay* may be formed by winding strands or rope into a *right-hand* helix or a *left-hand* helix. Lang lay rope displays as much as a 15–20 percent superiority over regular lay rope in terms of fatigue and wear resistance. This superiority results from[77] (1) the fact that *smaller* geometry-dictated bending strains are induced in the exposed outer wires of lang lay rope as the rope passes around a drum or sheave, resulting in lower cyclic bending stresses and longer fatigue life, and (2) the fact that geometry-dictated contact areas between individual wires of the rope are *larger* for lang lay rope; hence the contact pressure is lower, and longer wear life results.

On the other hand, lang lay rope tends to rotate, sometimes severely, when axial loads are applied, unless the rope is secured against rotation at both ends. Also, it is less able to resist crushing action against a drum or a sheave.

Because of the complicated geometry of a wire rope, neither the kinematics nor the stress levels at potential critical points have been well formulated.[78] Standards of the wire rope industry suggest that striking a proper *balance* between resistance to bending fatigue[79] and abrasion resistance is essential to a wise selection of wire rope. The *X-chart* shown in Figure 17.15 illustrates the comparison between bending fatigue resistance and abrasion resistance for several different rope and strand constructions. Experience-based

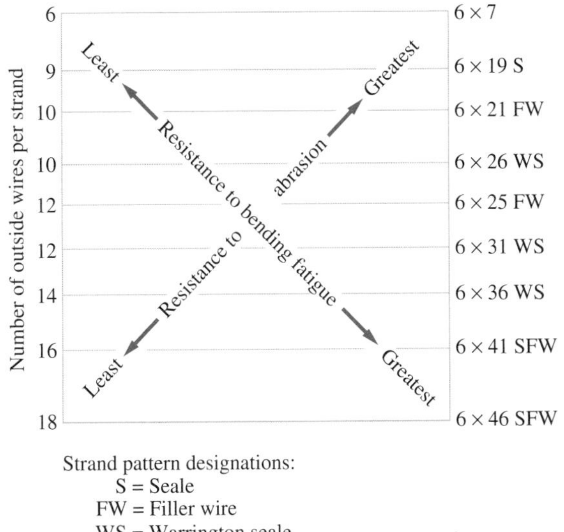

Figure 17.15
X-chart depicting the comparison between bending fatigue resistance and abrasion resistance for several widely used wire rope constructions. See Figure 17.5 for sketches of basic strand patterns. (Adapted from ref. 4, with permission from Wire Rope Technical Board.)

[76]For special applications, an *alternate lay* rope, in which alternating regular and lang lay strands are wound to form the rope, may sometimes be used.

[77]For a more detailed explanation, see ref. 17.

[78]In principle, it is possible to model the cyclic Hertz contact stresses and relative small-amplitude sliding motions within an "ideal" wire rope (e.g., see ref. 17). Relating these quantities to actual failure by fretting fatigue, bending fatigue, tensile fatigue, or wear, however, is very uncertain. Therefore, empirical means of evaluating potential failure, and the use of very large safety factors, are employed to select appropriate wire rope configurations for any given application.

[79]Fretting fatigue consequences are embedded in this quantity.

TABLE 17.8 Symptoms and Causes of Wire Rope Failures[1]

Type of Failure	Symptoms	Possible Causes
Fatigue	Transverse wire breaks—either straight across or Z-shape. Broken ends appear grainy.	Rope may be bent around too small a radius; possible vibration or whipping; wobbly sheaves; reverse bends; bent shafts; tight grooves; small drums or sheaves; incorrect rope construction; improper installation; poor end terminations.
Yielding	Wire breaks exhibit a mixture of cup-and-cone ruptures and diagonal shear ruptures.	Unanticipated overloads; sticky or grabby clutches; jerky operation; loose drum bearings; fast starts or stops; broken sheave flange; incorrect rope size; incorrect material grade; improper end terminations.
Abrasive wear	Outer wires worn smooth, sometimes to a "knife edge," followed by rupture.	Improper dimensions of drum or sheave grooves; frozen or stuck sheaves; material or surface hardness of sheaves or drums too low; misalignment of sheaves or drum; kinks in rope; improperly attached fittings; grit or sand environment; objects imbedded in rope.
Combined fatigue and wear	Reduced cross section of outer wires; wires broken off squarely.	Normal long-term consequence produced by typical operating conditions.
Combined yielding and wear	Reduced cross section of outer wires; wires necked down in a cup-and-cone configuration.	Normal long-term consequence produced by typical operating conditions.
Cut, gouged, or rough wire	Wires pinched down, mashed, and/or cut; rough diagonal shear-like wire rupture.	Mechanical abuse; abnormal or accidental forces during installation.
Twisting	Wire ends show evidence of plastic twisting and/or corkscrew appearance.	Mechanical abuse; abnormal or accidental forces during installation.
Mashing	Wires flattened and spread; followed by rupture.	Mechanical abuse; abnormal or accidental forces during installation. (Usually occurs on the drum.)

[1]Adapted from ref. 4, with permission from Wire Rope Technical Board.

symptoms observed for the various failure modes just discussed are detailed in Table 17.8,[80] together with possible causes.

In the transmission of power by a wire rope, a typical arrangement is to wind the wire rope on a drum, which may either be *grooved* to guide and support the rope as it winds or unwinds, or on a plain (smooth) drum.[81] Rope attachment geometry, drum dimensions and tolerances, and multiple-layer winding practice are all details that require attention in selecting a wire rope system.[82] Limited data for selecting wire rope and for sizing drums and sheaves are given in 17.13 and 17.14. Regular inspection and maintenance of wire rope drives are also important to the efficiency and life expectancy of these systems.[83]

[80]Courtesy of Wire Rope Technical Board; see ref. 4.

[81]For smooth drums, the first layer of rope is typically wound smoothly and uniformly to provide a helical groove that will guide and support successive layers. The first layer should never be unwound.

[82]Wire rope manufacturers are equipped to provide detailed guidelines for rope selection and optimum geometries for drums and sheaves.

[83]Detailed inspection techniques are presented in ref. 4.

17.13 Wire Rope; Materials

Using the material selection guidelines of Chapter 3, and considering the potential failure modes just discussed, it may be deduced that candidate materials for wire rope applications should be resistant to failure by fatigue, fretting fatigue, surface fatigue wear, abrasive wear, yielding, ultimate rupture, and in some cases, corrosion. By far, the most widely used material for wire rope is high-carbon steel. Other materials that may be selected to meet special requirements include iron, stainless steel, Monel, and bronze.

For *steel* wire rope material, industry practice is to specify material strength characteristics by *grade*.[84] Standard grades include *traction steel (TS)*, *mild plow steel (MPS)*, *plow steel (P)*, *improved plow steel (IPS)*, *extra improved plow steel (EIPS)*, and *extra, extra improved plow steel (EEIPS)*. *Plow steel (PS)* strength properties are the basis for calculating the strengths of all steel rope wires. Table 17.9 shows nominal strength data for selected steel wire rope applications.

Stainless steel wire rope is usually made of AISI 302, 304, 316, or 305 stainless steel alloy. For *Monel* wire rope, Type 400 Monel is usually selected. When operating environments suggest *bronze* wire rope, Type A phosphor bronze is a usual alloy selection. Small-diameter galvanized iron and stainless steel wire ropes are commonly *plastic-coated* to protect against corrosion, and, in some cases, to reduce wear. *Plastic filled* wire rope, in which the internal spaces among the wires are filled with a plastic matrix, are also used in some applications to reduce both internal and external wear.

17.14 Wire Rope; Stresses and Strains

Stresses that may play a role in wire rope selection include:

1. Direct tensile stress in the wires of the rope
2. Bending stress in the wires, induced by bending of the rope around drums or sheaves
3. Compressive stress (pressure) between the rope and the drum or sheave

Direct tensile stress, σ_t, may be estimated from

$$\sigma_t = \frac{T}{A_r} \tag{17-28}$$

where T = resultant tensile force on the rope, lb
A_r = approximate metallic cross-sectional area of the rope as a function of rope diameter d_r, in^2 (see Table 17.9)

Force components that may contribute to resultant tensile force T include:

1. Load to be lifted
2. Weight of the rope
3. Inertial effects arising from accelerating the load from a resting velocity of zero to the specified operational lifting velocity
4. Impact loading
5. Frictional resistance

Bending stresses in the wires, induced whenever the rope bends around the drum or a sheave, may be estimated by utilizing the classical equation from elementary strength of

[84]These steel grade names were originated during the early stages of wire rope development, and continue to be used to specify the strength of a particular size and grade of rope.

TABLE 17.9 Material and Construction Data for Selected Wire Rope Classes

Nominal classification		6 × 7	6 × 19	6 × 37	8 × 19
Number of outer strands		6	6	6	8
Number of wires per strand[1]		3–14	15–26	27–49	15–26
Maximum number of outer wires[1]		9	12	18	12
Approx. diameter of outer wires[1], d_w, in		$d_r/9$	$d_r/13$–$d_r/16$	$d_r/22$	$d_r/15$–$d_r/19$
Materials typically available[2,3] (approx. ultimate strength, ksi)	Core: (FC)	IPS (200)	I (80) T (130) IPS (200)	IPS (200)	I (80) T (130) IPS (200)
	Core: (IWRC)	IPS (190)	IPS (190) EIPS (220) EEIPS (255)	EIPS (220) EEIPS (255)	IPS (190) EIPS (220)
Approx. metallic cross section of rope, A_r, in^2	Core: (FC)	$0.384\ d_r^2$	$0.404\ d_r^2$ (S)[4]	$0.427\ d_r^2$ (FW)[4]	$0.366\ d_r^2$ (W)[4]
	Core: (IWRC)	$0.451\ d_r^2$	$0.470\ d_r^2$ (S)[4]	$0.493\ d_r^2$ (FW)[4]	$0.497\ d_r^2$ (W)[4]
Standard nominal rope diameters available, d_r, in		$^1/_4$–$^5/_8$ by $^1/_{16}$ th's; $^3/_4$–$1^1/_2$ by $^1/_8$ th's	$^1/_4$–$^5/_8$ by $^1/_{16}$ th's; $^3/_4$–$2^3/_4$ by $^1/_8$ th's	$^1/_4$–$^5/_8$ by $^1/_{16}$ th's; $^3/_4$–$3^1/_4$ by $^1/_8$ th's	$^1/_4$–$^5/_8$ by $^1/_{16}$ th's; $^3/_4$–$1^1/_2$ by $^1/_8$ th's
Unit weight of rope, lb/ft		$1.50\ d_r^2$	$1.60\ d_r^2$	$1.55\ d_r^2$	$1.45\ d_r^2$
Approx. modulus of elasticity for the rope[3,5], E_r, psi	0–20% of S_u	11.7×10^6 (FC)	10.8×10^6 (FC); 13.5×10^6 (IWRC)	9.9×10^6 (FC); 12.6×10^6 (IWRC)	8.1×10^6 (FC)
	21–65% of S_u	13.0×10^6 (FC)	12.0×10^6 (FC) 15.0×10^6 (IWRC)	11.6×10^6 (FC) 14.0×10^6 (IWRC)	9.0×10^6
Recommended min. sheave or drum diameter, $(d_s)_{min}$, in		$42\ d_r$	$34\ d_r$	$18\ d_r$	$26\ d_r$

[1]While the interior wires of a strand are of *some* significance, a strand's important characteristics relate to the number and size of the *outer* wires.
[2]Typical materials are designated as I (iron), T (traction steel), IPS (improved plow steel), EIPS (extra, improved plow steel), and EEIPS (extra, extra improved plow steel). In wire ropes, the *rope* ultimate strength is a function of rope size, wire size, and construction details, as well as material properties.
[3]Typical core constructions are *fiber core* (FC) and *independent wire rope core* (IWRC).
[4]See Figure 17.5(b) for construction details of *Seale* (S), *Filler Wire* (FW), and *Warrington* (W) strand configurations.
[5]Carefully note that the *rope* modulus E_r is not the same as Young's modulus of elasticity for the material.

materials[85]

$$\frac{M}{E_r I_w} = \frac{1}{\rho} = \frac{2}{d_s} \tag{17-29}$$

where M = applied moment
E_r = modulus of elasticity of the *rope*[86]
I_w = area moment of inertia of the *wire*, about its neutral axis of bending
ρ = radius of curvature of bent rope (also approximately equal to radius of sheave or drum)
d_s = diameter of sheave or drum

Solving (17-29) for I_w

$$I_w = \frac{Md_s}{2E_r} \tag{17-30}$$

Substituting (17-30) into (4-5)

$$\sigma_b = \frac{Mc_w}{I_w} = \frac{M\left(\dfrac{d_w}{2}\right)}{\left(\dfrac{Md_s}{2E_r}\right)} = \frac{d_w}{d_s}E_r \tag{17-31}$$

where d_w = wire diameter.

It is of importance to note from (17-31) that, if necessary, the bending stresses may be reduced by using smaller wires or larger sheaves.

Compressive stress, or *unit radial pressure*, between the rope and the sheave may be estimated in the same way as for plain bearings,[87] by utilizing projected contact area to calculate the nominal pressure p. For the wire rope system configuration shown in Figure 17.16, by equilibrium

$$pA_{proj} = p(d_r d_s) = 2T \tag{17-32}$$

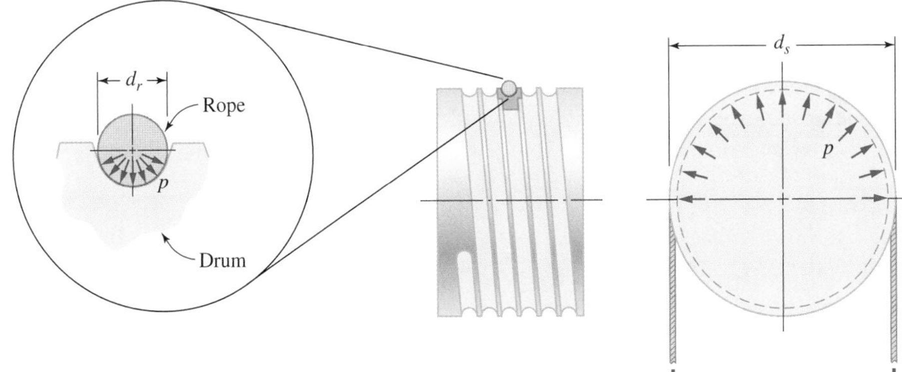

Figure 17.16
Typical wire rope system configuration in which the rope, nestled in an accurately machined groove, passes around a drum.

[85]See, for example, ref. 19, p. 138.

[86]Because each wire in the rope forms a helical spiral around an axis in space, its *axial spring rate* is less than it would be for the same wire if it were straight. Referring to Figures 2.2 and 2.3, spring rate and modulus of elasticity are seen to be related by a constant. The traditional approach used to account for this spring rate difference has been to define a *pseudo-modulus of elasticity* for the rope, E_r, smaller than Young's modulus, which relates stress in the *wire* to strain in the *rope*. Experimentally determined values for E_r are shown in Table 17.9.

[87]See (10-2).

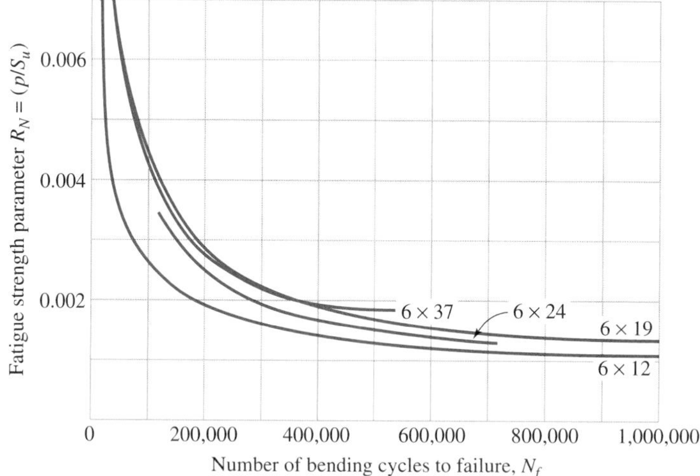

Figure 17.17
Fatigue lives for several wire rope constructions, as a function of fatigue strength parameter R_N.

or

$$p = \frac{2T}{d_r d_s} \tag{17-33}$$

where p = unit radial pressure
A_{proj} = projected contact area = $d_r d_s$
d_r = nominal rope diameter
d_s = diameter of sheave or drum (sometimes called "tread" diameter)

Experimental correlations between a *fatigue strength parameter*, R_N, and the number of bending cycles to failure, N_f, have been established for several rope classes,[88] as illustrated in Figure 17.17. One bending cycle consists of flexing and unflexing the rope once as it passes around the sheave or drum. The fatigue strength parameter R_N may be calculated from

$$R_N = \frac{p}{\sigma_u} \tag{17-34}$$

Also, since *wear* is a function of contact pressure,[89] experience-based limiting values have been established for maximum-allowable bearing pressure, $(p_{max})_{wear}$, as a function of *rope class* and *sheave material.* Some of these wear-based guidelines[90] have been included in Table 17.10.

17.15 Wire Rope; Suggested Selection Procedure

In selecting an appropriate wire rope system, design decisions must be made about material, size, and construction of the rope; geometry of the sheaves and drum; and other details. Usually, the wire rope selection procedure will be iterative. To help make an initial choice, the experience-based recommendations of Figure 17.15 often prove effective. A procedure for selecting a good wire rope system is presented next.

1. Establish design specifications for the application, and prioritize design goals with respect to failure mode, life, safety, cost, and any other special requirements.
2. Based on design priorities established in step 1, tentatively select a rope construction

[88]See ref. 20. [89]See, for example, (2-117) or (2-121). [90]See ref. 4 for additional data.

TABLE 17.10 Experience-based Wear-Related Allowable Maximum Bearing Pressure Between Rope and Drums or Sheaves of Various Materials[1] (psi)

Drum or Sheave Material	Regular Lay Rope				Lang Lay Rope				Comments
	6 × 7	6 × 19	6 × 37	8 × 19	6 × 7	6 × 19	6 × 37	8 × 19	
Wood	150	250	300	350	165	275	330	400	Against end grain of beech, hickory, or gum.
Cast iron	300	480	585	680	350	550	660	800	Minimum hardness of BHN 125.
Cast carbon steel	550	900	1075	1260	600	1000	1180	1450	30–40 points of carbon; minimum hardness of BHN 160.
Manganese steel, induction or flame hardened	1470	2400	3000	3500	1650	2750	3300	4000	

[1]Abridged from ref. 4, with permission from Wire Rope Technical Board.

by interpreting Figure 17.15. Also select a rope material, using methods of Chapter 3, and select a safety factor based on methods of Chapter 5.[91]

3. Using the tentatively selected rope material and class, initially size the rope using (17-28). Be sure to include all potential loading sources. Calculate a tentative rope diameter $(d_r)_{static}$ based on static loading.
4. Using the tentatively selected rope diameter $(d_r)_{static}$, find the recommended minimum sheave diameter, d_s, from Table 17.9.
5. Estimate bending stress in the outer wires using (17-31) and information about wire diameter, d_w, from Table 17.9. It should be noted that (17-31) provides only an approximate value for bending stress in the wire; it usually is not used directly in design calculations.
6. Using the specified design life requirement, N_d, enter Figure 17.17, focus on the curve for the tentatively selected rope class, and read the value of R_N corresponding to N_d. Next, combine (17-33) and (17-34) assuming that the sheave diameter remains unchanged, incorporate the safety factor $n_{fatigue}$ from step 3, and calculate the required rope diameter, $(d_r)_{fatigue}$, based on fatigue.
7. From Table 17.10, find the wear-based limiting pressure for the chosen rope class and sheave or drum material. Utilizing (17-33), calculate the required rope diameter, $(d_r)_{wear}$, based on wear.
8. From the results of steps 4, 6, and 7, identify the *largest* required rope diameter among $(d_r)_{static}$, $(d_r)_{fatigue}$, and $(d_r)_{wear}$, and select the *nominal standard* wire rope diameter that equals or just exceeds this value.[92]
9. Recheck all calculations using the standard wire rope selected. If necessary, modify the selection.

[91]Typical *industry practice* is to select safety factors based only on *static ultimate breaking load* of the proposed rope. Since the influences of fatigue, wear, corrosion, and other factors are not specifically addressed when this *industry practice* is adopted, the chosen safety factor must be *large* to account for the unknown influences. Values of such "static-ultimate-strength based" safety factors are often chosen to be as large as 5–8, or even higher. By contrast, in this text the selection of a safety factor is closely tied to the prediction of the *governing failure mode* in each individual application, as discussed in Chapter 5. In any case, a designer is always responsible for meeting safety factor requirements specified by any and all applicable codes and standards.

[92]See Table 17.9.

10. Summarize results, including:
    a. Standard rope size required
    b. Rope construction (core, number of strands, number of wires per strand, strand configuration, nominal rope diameter, and the lay of the strands and the rope)
    c. Material for rope, sheaves, and drum
    d. Sheave and drum diameters
    e. Other special requirements

Example 17.4 Wire Rope Selection

A small electric one-ton hoist is to be designed to operate as a low-speed overhead lift for a small machine shop. Two lines are to be used to support the load, which is attached to a vertically moving sheave with a swivel hook, as shown schematically in Figure E17.4. If the desired design life of the wire rope is 2 years, and approximately 15 lifts per hour are to be made, 8 hours a day, 250 days a year, select an appropriate wire rope for the application. Occasional suddenly applied loads are possible. Also, specify the diameter and material for the vertically moving sheave. Local safety codes require a safety factor of 5 based on static ultimate strength.

Solution

Following the suggested procedure of 17.15, the following steps may be taken:

1. Based on the specifications given, a balanced design is appropriate, in which the likelihood of failure by fatigue and wear should be about the same. The desired design life of the rope is specified as 2 years. For this application, safety is an important issue, and cost is also important.
2. Figure 17.15 indicates that for a balance between failure by fatigue and failure by wear, a 6×25 FW construction or a 6×31 WS construction would be an appropriate

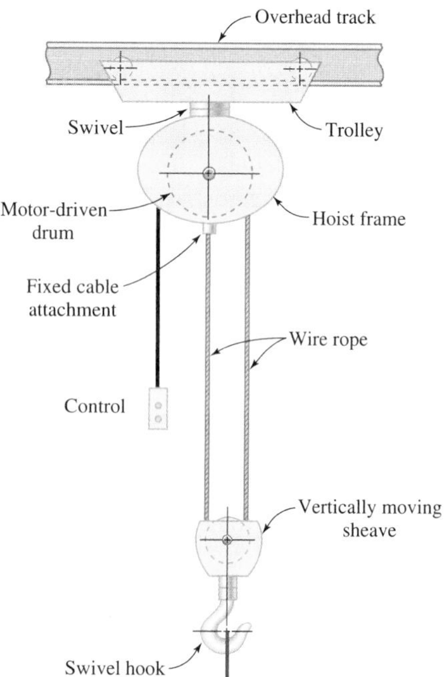

Figure E17.4
Sketch of a proposed one-ton hoist.

choice. From Table 17.9 it may be noted that the 6 × 25 FW rope is classed as a 6 × 19 rope, and the 6 × 31 WS rope is classed as a 6 × 37 rope. For a first iteration, the tentative class selection will be 6 × 37, specifically a 6 × 37 WS rope construction.

To keep the rope small, an *improved plow steel (IPS)* material will be tried first, and, to improve flexibility, a *fiber core (FC)* will be investigated.

Using the methods of Chapter 5,[93] the rating factors selected for this electric-hoist wire rope application are chosen as follows:

Rating Factor	Selected RN
1. Accuracy of loads knowledge	−3
2. Accuracy of stress calculation	+3
3. Accuracy of strength knowledge	0
4. Need to conserve	−2
5. Seriousness of failure consequences	+2
6. Quality of manufacture	−1
7. Conditions of operation	−1
8. Quality of inspection/maintenance	−1

Using (5-3),

$$t = -3 + 3 + 0 - 2 + 2 - 1 - 1 - 1 = -3 \tag{1}$$

then from (5-4)

$$n_d = 1 + \frac{(10 - 3)^2}{100} = 1.5 \tag{2}$$

Summarizing safety factor choices:

$$\begin{aligned} n_{ult} &= 5.0 \text{ (code requirement; based on static ultimate strength)} \\ n_{fatigue} &= 1.5 \text{ [based on result of equation (2)]} \\ n_{wear} &= 1.0 \text{ [since Table 17.10 gives values for } \textit{allowable} \text{ stresses} \\ &\quad (\textit{allowable} \text{ pressures) that already have a safety factor built in]} \end{aligned} \tag{3}$$

3. From (17-28), the static direct tensile stress in the rope is

$$\sigma_t = \frac{T}{A_r} \tag{4}$$

and from (2-99), for suddenly applied loads, and utilizing (4),

$$(\sigma_{max})_{\substack{suddenly\\applied}} = 2\sigma_t = 2\frac{T}{A_r} \tag{5}$$

For this first iteration, the tensile force, T, in the rope, will be based on the load to be lifted and the suddenly applied impact factor of 2 as shown in (5). The rope weight will be initially neglected (small rope), inertial effects will be neglected (low-speed lift), and frictional resistance will be neglected.

From Table 17.9, the approximate metallic cross section for a fiber-core 6 × 37 rope is

$$A_r = 0.427 d_r^2 \tag{6}$$

Since two lines carry the nominal 2000-lb load, each line carries

$$T = \frac{2000}{2} = 1000 \text{ lb/line} \tag{7}$$

[93]Review Example 5.1 to refresh details.

Example 17.4 Continues

and (5) gives

$$(\sigma_{max})_{\substack{suddenly\\applied}} = 2\frac{1000}{0.427d_r^2} \tag{8}$$

From (3)

$$n_{ult} = 5 \tag{9}$$

and from Table 17.9, the static ultimate strength of improved plow steel (IPS) is

$$\sigma_u = 200{,}000 \text{ psi} \tag{10}$$

Hence the design stress σ_d for static loading may be calculated as

$$(\sigma_d)_{static} = \frac{200{,}000}{5} = 40{,}000 \text{ psi} \tag{11}$$

Equating

$$(\sigma_{max})_{\substack{suddenly\\applied}} = \sigma_d \tag{12}$$

and solving (8) for required rope diameter d_r,

$$d_r = \sqrt{\frac{2000}{(40{,}000)(0.427)}} = 0.34 \text{ inch} \tag{13}$$

From Table 17.9, the next larger standard rope diameter is 0.38 inch. Hence, based on static ultimate strength requirements,

$$(d_r)_{static} = 0.38 \text{ inch} \tag{14}$$

4. From Table 17.9, the recommended minimum sheave diameter for this rope is

$$d_s = 18d_r = 18(0.38) = 6.75 \text{ inches} \tag{15}$$

5. As a point of reference, the bending stress in the outer wires may be estimated by using (17-31) and data from Table 17.5 to find[94]

$$\sigma_b = \frac{d_w}{d_s}E_r = \frac{\left(\frac{0.38}{22}\right)}{6.75}(11.0 \times 10^6) = 28{,}150 \text{ psi} \tag{16}$$

This appears to be an acceptable bending stress.

6. The desired design life N_d may be calculated as

$$N_d = \left(15\frac{\text{lifts}}{\text{hr}}\right)\left(8\frac{\text{hr}}{\text{day}}\right)\left(250\frac{\text{days}}{\text{yr}}\right)(2 \text{ yr}) = 6 \times 10^4 \text{ cycles (bends)} \tag{17}$$

Reading into Figure 17.17 with this life, using the curve for 6×37 rope, the value of R_N corresponding to failure in 6×10^4 cycles may be read as

$$R_{N_f} = 0.0057 \tag{18}$$

Next, (17-34) may be used to calculate the value of pressure p corresponding to failure in 6×10^4 cycles as

$$(p)_{N_f} = R_{N_f}\sigma_u = 0.0057(200{,}000) = 1140 \text{ psi} \tag{19}$$

From (3), the fatigue safety factor is $n_{fatigue} = 1.5$, so the design-allowable pressure may be calculated as

[94]Note that Table 17.9 gives two values of E_r, depending on whether the rope is loaded to less than 20 percent of ultimate strength, or more than 20 percent. The stiffer value has been chosen here to give a more conservative (higher) estimate of bending stress in the wires.

$$(p_d)_{fatigue} = \frac{(p)_{N_f}}{n_{fat}} = \frac{1140}{1.5} = 760 \text{ psi} \qquad (20)$$

Inserting this fatigue-based design pressure into (17-33), and assuming the sheave diameter remains unchanged, the rope diameter requirement based on fatigue may be calculated as

$$(d_r)_{fatigue} = \frac{2T}{(p_d)_{fat} d_s} = \frac{2(1000)}{(760)(6.75)} = 0.39 \text{ inch} \qquad (21)$$

Hence, the chosen 0.38-inch-diameter standard rope is (barely) acceptable.

7. From Table 17.10, for a 6 × 37 rope on a cast carbon-steel sheave (BHN 160),[95] the allowable bearing pressure based on wear is

$$(p_d)_{wear} = 1180 \text{ psi} \qquad (22)$$

Inserting this wear-based design pressure into (17-33), the rope diameter requirement based on wear may be calculated as

$$(d_r)_{wear} = \frac{2T}{(p_d)_{wear} d_s} = \frac{2(1000)}{(1180)(6.75)} = 0.25 \text{ inch} \qquad (23)$$

8. The largest diameter requirement, based on $(d_r)_{static} = 0.38$ inch, $(d_r)_{fatigue} = 0.39$ inch, and $(d_R)_{wear} = 0.25$ inch, is dictated by *fatigue* life requirements. It is interesting to note that for this particular project the diameter requirement based on ultimate strength using a code-required safety factor is essentially the same as the diameter requirement based on fatigue strength using a logic-based safety factor.[96] Such good agreement will not always be the case.

Summarizing, the following recommendations are presented:

a. Choose $^3/_8$-inch 6 × 31 WS improved plow steel (IPS) fiber core (FC) rope.[97]

b. Choose a cast carbon-steel sheave material (BHN = 160) with a diameter of 6.75 inches.

17.16 Flexible Shafts

As discussed in 17.1, when misalignment problems or large offsets between two rotating shafts are dictated by operational design requirements, a flexible shaft[98] may provide a good solution for connecting the shafts and transmitting the torque, power, or rotary motion between them. Flexible shafts (cores), as shown, for example, in Figure 17.7, are constructed of 1 to 12 tightly wound helical layers of wire wrapped around a mandrel wire, alternating the hand of each successive layer. The *hand (lay)* of a flexible shaft is determined by the hand of the *outer* helically wound layer of wire. For power transmission in which the direction of rotation is always the same, the direction of the power source should tend to wrap the outer helical layer more tightly; hence, as viewed from the power source, a left-hand lay should be selected if the power source rotates clockwise and vice versa. Flexible shafts are also commercially available for bidirectional operation, but torque and power capacity are significantly lower than for unidirectional operation.[99]

Since the helically wrapped wire construction of flexible shafting is somewhat similar to that of wire rope, it may be deduced that when power is transmitted by the flexible

[95]Brinnel hardness number; see Table 3.13. [96]See Chapter 5.

[97]If unspecified, the rope is assumed to be right regular lay. [98]See Figure 17.7.

[99]Compare data from Tables 17.11 and 17.12.

shaft, Hertz contact stresses are generated between and among wires, and small-amplitude relative sliding motion is induced between wires as well. Hence, the potential failure modes for flexible shafting, as for wire rope, include fatigue, fretting fatigue, wear, and yielding, depending upon the characteristics of the operating loads. Corrosion may also be a factor.

As for any other machine part, materials for flexible shafts may be selected by using the guidelines of Chapter 3. Typically, the wire used for power transmission applications is high-carbon steel heat treated and stress relieved, but any material available in wire form may be specified if needed to satisfy special design requirements. To accommodate such special needs, flexible shafts are usually available in stainless steel, phosphor-bronze, Monel, Inconel, titanium, or even Hastelloy-X.

Characteristics of a flexible shaft of given diameter may be changed by varying the number of wires per layer, number of helically wrapped layers, wire diameter, spacing between helically wrapped wires, and wire material. Just as for any other rotating shaft, the operating torque to be transmitted by a flexible shaft is determined by the power and speed requirements of the application. The basic relationship for power as a function of torque and speed, given in (4-32), may be solved for torque to obtain

$$T = \frac{63{,}025(hp)}{n} \tag{17-35}$$

where T = torque, in-lb
n = shaft speed, rpm
hp = horsepower

Tables 17.11 and 17.12 provide maximum-allowable torque requirements, $(T_{\max})_{allow}$, as a function of bend radius, for several commercially available flexible shaft sizes. These *allowable* torque values have an embedded safety factor of about 4, so if a particular de-

TABLE 17.11 Recommended Maximum Operating Torque[1] for Selected[2] Standard Flexible Shafts of High-Carbon Steel as a Function of Bend Radius R_b, for *Unidirectional* Operation

Shaft (Core) Diam., in	Max.-Allow. Speed, rpm	Min.-Allow. Bend Radius, in	Torsional Deflection, deg/ft/in-lb	Ultimate Torsional Failure[3] Moment, in-lb	Recommended Maximum Torque $(T_{max})_{allow}$ Corresponding to Various Bend Radii,[4] in-lb							
					3	4	6	8	10	12	15	20
0.127	30,000	2.7	21.48	12	0.2	0.7	1.2	1.5	1.6	1.7	1.8	1.9
0.147	20,000	3.2	10.11	30		1.2	2.6	3.3	3.8	4.1	4.4	4.7
0.183	15,000	3.2	7.39	32		1.2	2.8	3.5	4.0	4.3	4.6	4.9
0.245	10,000	3.2	0.97	195			12.8	16.0	18.0	20.0	21.0	23.0
0.304	7,500	3.6	0.44	338			19.0	26.0	30.0	33.0	35.0	38.0
0.370	5,500	6.3	0.17	690				20.0	35.0	45.0	55.0	65.0
0.495	4,500	5.9	0.06	1230				45.0	70.0	86.0	103.0	120.0
0.620	4,000	6.7	0.019	2420				53.0	109.0	147.0	184.0	221.0
0.740	3,000	6.7	0.009	4370				96.0	198.0	265.0	332.0	400.0
0.990	2,500	8.4	0.003	9344					206.0	386.0	567.0	747.0

[1]From ref. 5, courtesy: S. S. White Technologies, Inc. A safety factor of approximately 4 has been embedded in these recommended allowable torque values. The methods of Chapter 5 may be utilized to adjust these allowable torque values if necessary. These values assume that the applied torque acts in the direction that tends to tighten the helical outer layer of wire.
[2]Many other shafts and variations are commercially available. See ref. 5, for example.
[3]Torque at which a flexible shaft will deform permanently or break.
[4]Listed bend radii are in inches.

TABLE 17.12 Recommended Maximum Operating Torque[1] for Selected[2] Standard Flexible Shafts of High-Carbon Steel as a Function of Bend Radius R_b, for *Bidirectional* Operation

Shaft (Core) Diam., in	Max.-Allow. Speed, rpm	Min.-Allow. Bend Radius, in	Torsional Deflection, deg/ft/in-lb		Ultimate Torsional Failure Moment[5], in-lb		Recommended Maximum Torque $(T_{max})_{allow}$ Corresponding to Various Bend Radii,[6] in-lb									
			TOL[3]	LOL[4]	TOL[3]	LOL[4]	3	4	6	8	10	12	15	20	25	Straight
0.127	30,000	3.0	11.65	23.71	16	12	0.4	1.2	2.1	2.6	2.8	3.0	3.2	3.4	3.5	3.9
0.147	20,000	4.0	6.55	12.50	21	26		1.8	4.1	5.2	5.9	6.3	6.8	7.2	7.5	8.6
0.183	15,000	4.0	3.07	6.94	44	41		2.8	6.3	8.0	9.1	9.8	10.5	11.1	11.6	13.2
0.245	10,000	4.0	0.74	1.23	141	121		8.4	19.0	23.0	27.0	29.0	31.0	33.0	34.0	39.0
0.304	7,500	4.5	0.28	0.55	281	207			20.0	32.0	39.0	44.0	48.0	53.0	55.0	67.0
0.370	5,500	6.0	0.11	0.21	515	384			29.0	53.0	67.0	77.0	86.0	96.0	102	125
0.495	4,500	7.0	0.044	0.081	1214	869				75.0	117	145	172	200	217	284
0.620	4,000	8.0	0.015	0.024	2135	1760					188	250	317	381	420	574
0.740	3,000	10.0	0.009	0.018	3533	2441						351	440	529	582	797
0.990	2,500	12.0	0.002	0.003	8513	6763							972	1281	1466	2209

[1]From ref. 5, courtesy: S.S. White Technologies, Inc. A safety factor of approximately 4 has been embedded in these recommended allowable torque values. The methods of Chapter 5 may be utilized to adjust these allowable torque values if necessary.
[2]Many other shafts and variations are commercially available. See ref. 5, for example.
[3]Direction of operating torque tends to *tighten the outer layer* of helically wound wire.
[4]Direction of operating torque tends to *loosen the outer layer* of helically wound wire.
[5]Torque at which a flexible shaft will deform permanently or break.
[6]Listed bend radii are in inches.

sign application requires a different safety factor specification, the tabled allowable torque values should be modified. Also, the allowable torque values for continuous operation are based on an allowable temperature rise in the shaft of about 55° above room temperature[100] when the tangential surface velocity of the rotating shaft is about 500 surface feet per minute. The maximum-allowable shaft speeds shown in Tables 17.11 and 17.12 are based on this criterion. It is also important to note that bend radii smaller than recommended values[101] cause additional temperature increases because of the increased internal friction developed within the flexible shaft as the wires are forced to rub against each other more vigorously in tighter bends. The bend radius R_b may be geometrically estimated for any particular design configuration. For example, if an application requires a large offset between two parallel rotating shafts, as illustrated in Figure 17.18, the bend radius, R_b, may be calculated as

$$R_b = \frac{x^2 + y^2}{4x} \tag{17-36}$$

where x = offset between shaft centerlines
y = distance between components

If recommended limits on torque, speed, and bend radius are adhered to, an operational mean life of approximately 10^8 revolutions may be expected.[102]

The suggested procedure for selecting an appropriate flexible shaft for a particular design scenario is relatively simple. The steps are:

[100]Larger temperature rises significantly reduce shaft fatigue life. [101]See Tables 17.11 and 17.12.
[102]For longer or shorter design lives, contact a flexible shaft manufacturer.

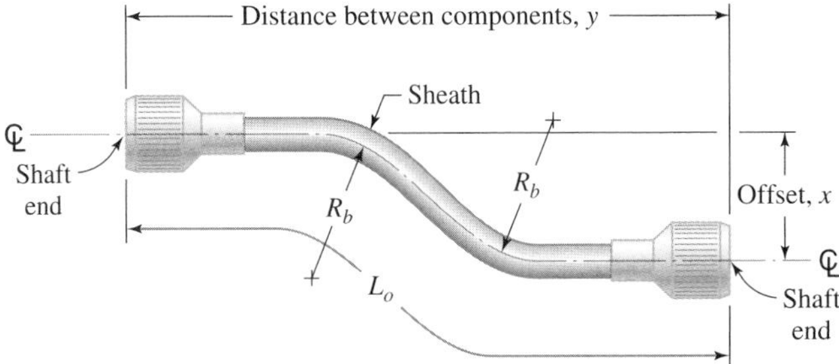

Figure 17.18
Flexible shaft connection between offset parallel shafts. For internal details, see Figure 17.7.

1. Establish whether the application is unidirectional or bidirectional. For unidirectional operation, select data from Table 17.11; for bidirectional operation, use data from Table 17.12.
2. Using (17-35), calculate the torque requirement for the flexible shaft.
3. Using (17-36), or other relationship pertinent to the geometric configuration, determine the bend radius R_b.
4. If torsional stiffness is a design issue, calculate the required torsional stiffness.
5. Utilizing either Table 17.11 or 17.12, depending upon the result of step 1, enter the table with the required torque from step 2 and bend radius from step 3 to tentatively read out an acceptable shaft diameter.[103]
6. With the tentatively selected shaft diameter, check to make sure that torsional stiffness requirements of step 4 are met. Check to make sure that maximum-allowable speed restrictions are not violated. Based on (17-35), it is clear that for a given horsepower requirement, higher speeds correspond to smaller shafts. In other words, flexible shafts like to be run fast. Thus, if the application requires that the driven device be operated at a different speed than the power source, the speed changer should be so located in the power train sequence so that the flexible shaft operates at the higher speed.
7. Determine shaft length, L_s, accurately by measuring along the shaft centerline in a scale layout, or by using appropriate calculations. Shaft lengths of 30 feet or more are commercially available and may be used successfully in many cases as long as the flexible shaft (core) is properly supported in a companion sheath.[104] *Shorter* shaft lengths are usually more critical because of the potentially tight bend radii that might be produced in the installation.
8. Iterate to satisfy all design criteria, and specify the shaft of choice.

Example 17.5 Flexible Shaft Selection

Engineering management of the manufacturer of the rotary lawn mower alluded to in Example 16.1 now believes that a significant increase in engine efficiency could be achieved by operating the engine with the crankshaft axis *horizontal* instead of vertical as shown in Figure E16.1A of example 16.1. Of the various means available to connect the horizontal crankshaft to the vertical rotary bladeshaft, a flexible shaft appears to be most economical. As a consultant, you are asked to investigate the feasibility of using a flexible shaft connection in this situation and propose an acceptable shaft, if possible. It has been determined geometrically that the shaft would have to accommodate a bend radius of about 8 inches, and in normal operation the engine must develop 3.5 horsepower at 2400 rpm. Select an appropriate flexible shaft for this application.

[103] Usually there are many acceptable combinations. [104] See Figure 17.7(b).

Solution

Following the steps just outlined,

1. This application is unidirectional; hence Table 17.11 will be used.
2. From (17-35)

$$T_{req'd} = \frac{63{,}025(3.5)}{2400} = 92 \text{ in-lb} \quad (1)$$

3. By specification,

$$R_b = 8 \text{ inches} \quad (2)$$

4. No stiffness specification has been imposed.
5. Entering Table 17.11 with the required torque of 92 in-lb and bend radius of 8 inches, the required standard flexible shaft diameter is

$$d_{shaft} = 0.74 \text{ inch} \quad (3)$$

6. No stiffness criterion needs to be satisfied.
7. The operating speed of the mower unit is 2400 rpm, safely below the 3000-rpm limit specified for the shaft selected.
8. Shaft length is to be determined by a scale layout of the mower assembly.
9. It appears that an appropriate flexible shaft selection can be made for this application. The recommended shaft is a *0.740-inch-diameter shaft made of high-carbon steel wire.*

Problems

17-1. A flat belt drive system is to be designed for an application in which the input motor shaft speed (*driving* pulley) is 1725 rpm, the *driven* shaft speed is to be approximately 960 rpm, and the power to be transmitted has been estimated as 3.0 horsepower. The driven machine has been evaluated and found to have the characteristics of moderate shock loading during operation. The desired center distance between driving and driven pulleys is approximately 18 inches.

a. If $^1/_8$-inch-thick polyamide belt material were chosen for this application, what belt width would be required?

b. What initial tension would be required for proper operation?

17-2. A flat belt drive consists of two cast-iron pulleys, each 4 feet in diameter, spaced 15 feet apart, center-to-center. The belt is to be made of two-ply oak-tanned leather, each ply being $^5/_{32}$-inch thick, and the specific weight of the leather material is 0.040 lb/in^3. The application involves a water-spray environment in which the belt is constantly subjected to the water spray (see Appendix Table A.1 for coefficients of friction). It has been experimentally determined that the tensile stress in the belt should not exceed 300 psi for safe operation. If 50 horsepower is to be transmitted at a pulley speed of 320 rpm, what belt width should be specified?

17-3. A 1725-rpm 5-horsepower high-torque electric motor is to be used to drive a woodworking table saw in a storm-window manufacturing plant. The saw is to operate 16 hours per day, 5 days per week, at full-rated motor horsepower. A V-belt drive is to be used between the motor and the saw input pulley. Ideally, the center distance between the motor drive sheave and the driven saw sheave should be about 30 inches, and the driven sheave should rotate at approximately 1100 rpm. Saw operation will probably produce moderate shock loading. Propose a V-belt drive arrangement that will provide a mean life of about 1 year between belt replacements.

17-4. A D-section V-belt is to be used to drive the main power shaft of an agricultural combine (an agricultural combine may be regarded as a combination of conveyors, elevators, beaters, and blowers). The power source is a 6-cylinder, 30-hp internal combustion engine which delivers full-rated power to a 12-inch-diameter drive sheave at 1800 rpm. The driven sheave is 26 inches in diameter, and the center distance between sheaves is 33.0 inches. During the harvest season, combines typically operate continuously 24 hours per day.

a. If a D-section V-belt were specified for use in this application, how often would you predict that the belt would require replacement?

b. Based on the knowledge that it takes about five hours to change out the main drive belt, would you consider the replacement interval estimate in (a) to be acceptable?

17-5. a portable bucket elevator for conveying sand is to be driven by a single-cylinder internal-combustion engine operating at a speed of 1400 rpm, using a B-section V-belt. The driving pulley and driven pulley each have a 5.00-inch pitch diameter. If the bucket elevator is to lift two tons per minute (4000 lb/min) of sand to a height of 15 feet, continuously for 10 hours per working day, and if friction losses in the elevator are about 15 percent of operating power, how many working days until failure would you estimate for the B-section belt if it has a datum length of 59.8 inches (B 59 belt)?

17-6. In a set of 5V high-capacity V-belts, each belt has a pitch length (outside length) of 132.0 inches, and operates on a pair of 12-inch diameter multiply grooved sheaves. The rotational speed of the sheaves is 960 rpm. To achieve a mean life expectancy of 20,000 hours, find the number of belts that should be used in parallel to transmit 200 horsepower.

17-7. It is desired to use a compact roller chain drive to transmit power from a dynamometer to a test stand for evaluation of aircraft auxiliary gear boxes. The chain drive must transmit 90 horsepower at a small-sprocket speed of 1000 rpm.

a. Select the most appropriate roller chain size.

b. Determine the minimum sprocket size to be used.

c. Specify appropriate lubrication.

17-8. It is desired to use a roller chain drive for the spindle of a new rotating shaft fatigue testing machine. The drive motor operates at 1750 rpm and the fatigue machine spindle must operate at 2170 rpm. It is estimated that the chain must transmit 11.5 hp. Spindle-speed variation of no more than 1 percent can be tolerated.

a. Select the most appropriate roller chain.

b. What minimum sprocket size should be used?

c. Specify appropriate lubrication.

d. Would it be feasible to use no. 41 lightweight chain for this application?

17-9. A five-strand no. 40 precision roller chain is being proposed to transmit power from a 21-tooth driving sprocket that rotates at 1200 rpm. The driven sprocket is to rotate at one-quarter the speed of the driving sprocket. Do the following:

a. Determine the limiting horsepower that can be transmitted by this drive, and state the governing failure mode.

b. Find the tension in the chain.

c. What chain length should be used if a center distance of approximately 20 inches is desired?

d. Does the 20-inch center distance lie within the recommended range for this application?

e. What type of lubrication should be used for this application?

17-10. It is desired to market a small air-driven hoist in which the load is supported on a single line of wire rope and the rated design load is $^3/_8$ ton (750 lb). The wire rope is to be wrapped on a drum of 7.0 inches diameter. The hoist should be able to lift and lower the full-rated load 16 times a day, 365 days a year, for 20 years before failure of the rope occurs.

a. If a special flexible 6 × 37 improved plow steel (IPS) rope is to be used, what rope size should be specified?

b. With the rope size determined in (a), it is desired to estimate the "additional stretch" that would occur in the rope if a 750-lb load were being lowered at the rate of 2 ft/sec, and when the load reaches a point 10 feet below the 7.0-inch drum, a brake is suddenly applied. Make such an estimate.

17-11. An electric hoist, in which the load is supported on two lines, is fitted with a $^1/_4$-inch 6 × 19 improved plow steel (IPS) wire rope that wraps on an 8-inch-diameter drum and carries an 8-inch sheave with an attached hook for load lifting. The hoist is rated at 1500-lb capacity.

a. If full-rated load were lifted each time, about how many "lifts" would you predict could safely be made with this hoist? Use a fatigue safety factor of 1.25. Note that there are 2 "bends" of the rope for each lift of the load.

b. If the hoist were used in such a way that one-half the time it is lifting full-rated load but the rest of the time it lifts only one-third of rated load, what hypothesis or theory would you utilize for estimating the number of lifts that could safely be made under these circumstances?

c. Numerically estimate the number of lifts that could be safely made under the mixed loading described in (b). Again, use a fatigue safety factor of 1.25 in your estimating procedure.

17-12. It is desired to select a wire rope for use in an automotive tow truck application. A single line is to be used and, considering dynamic loading involved in pulling cars back onto the roadway, the typical load on the rope is estimated to be 7000 lb. It is estimated that approximately 20 cars per day will be pulled back onto the highway (i.e., the rope experiences 20 "bends" per day) under full load. If the truck is used 360 days per year, and a design life of 7 years is desired for the rope:

a. What size IPS wire rope would you specify if the rope is to be of 6 × 19 regular lay construction?

b. What minimum sheave diameter would you recommend?

17-13. A deep-mine hoist utilizes a single line of 2-inch 6 × 19 extra improved plow steel (EIPS) wire rope wrapped on a cast carbon-steel drum that has a diameter of 6 feet. The rope is used to vertically lift loads of ore weighing about 4 tons from a shaft that is 500 feet deep. The maximum hoisting speed is 1200 ft/min and the maximum acceleration is 2 ft/sec^2.

a. Estimate the maximum direct stress in the "straight" portion of the 2-inch single-line wire rope.

b. Estimate the maximum bending stress in the "outer" wires of the 2-inch wire rope as it wraps onto the 6-foot-diameter drum.

c. Estimate the maximum unit radial pressure (compressive stress) between the rope and the sheave. *Hint:* Model the 2-inch single-line rope wrapped around the 6-foot

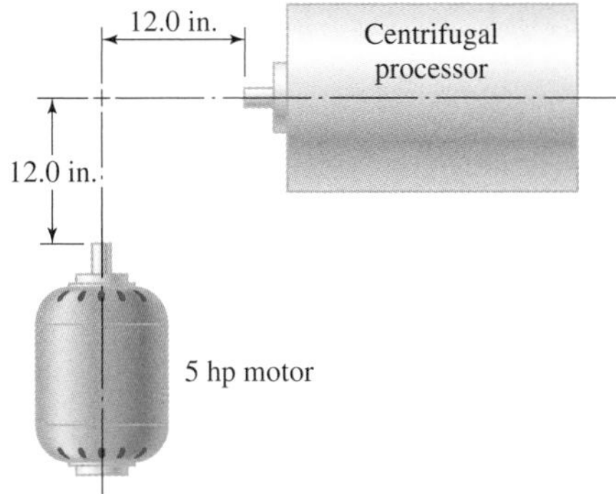

Figure P17.14
Sketch of motor and centrifugal processor to be connected at right angles by a flexible shaft.

sheave as a "band brake," utilizing equation (16-76) with $\alpha = 2\pi$ and $\mu = 0.3$, to find p_{max}.

d. Estimate the fatigue life of the 2-inch wire rope as used in this application.

17-14. It is necessary to mount a 7.5-horsepower 3450-rpm electric motor at right angles to a centrifugal processor as shown in Figure P17.14 (also see arrangement $A - B_2$ of Figure 17.6). It had been planned to use a 1 : 1 bevel gearset to connect the motor to the processor, but a young engineer suggested that a flexible shaft connection might be quieter and less expensive. Determine whether a flexible shaft is a viable alternative, and, if so, specify a flexible shaft that should work.

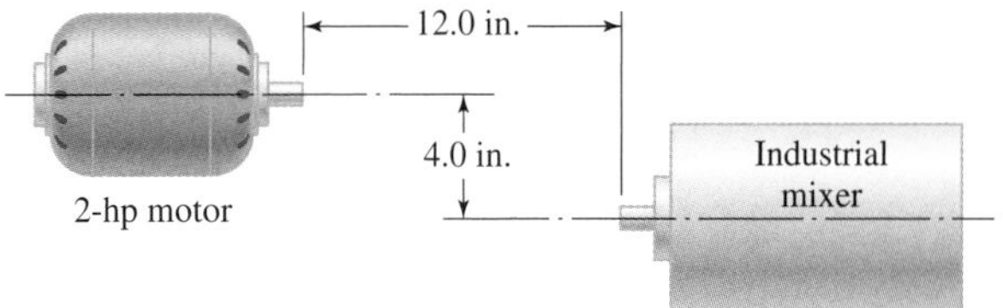

Figure P17.15
Sketch of motor and offset industrial mixer that it must drive.

17-15. To avoid other equipment mounted on the same frame, the centerline of a 2-horsepower 1725-rpm electric motor must be offset from the parallel centerline of the industrial mixer that it must drive.

a. For the offset shown in Figure P17.15, select a suitable flexible shaft (also see arrangement $A - B_1$ of Figure 17.6).

b. To improve mixing efficiency, it is being proposed to replace the standard 2-horsepower motor with a "reversible" motor having the same specifications. With the "reversible" motor, would it be necessary to replace the flexible shaft chosen in (a)? If so, specify the size of the replacement shaft.

c. Comparing results of (a) with (b), can you think of any potential operational problems associated with the flexible shaft when operating in the "reverse" direction?

Chapter 18

Flywheels and High-Speed Rotors

18.1 Uses and Characteristics of Flywheels

Flywheels are rotating masses installed in rotating systems of machine elements to act as a storage reservoir for kinetic energy, as depicted in Figure 18.1. Usually,[1] the primary task of a flywheel is to control, within an acceptable band, the angular velocity and torque *fluctuations* inherent in the power source, the load, or both. Figure 18.2 illustrates the superposed *torque versus angular displacement* curves for a fluctuating *driver* and a fluctuating *load*.

By definition, the *driver* torque, T_d, is considered to be positive when its sense is in the direction of shaft rotation and the *driver is supplying energy* to the shaft-flywheel system. The *load* torque, T_l, is considered to be positive when its sense is in the direction of rotation and the *shaft-flywheel system is supplying energy* to the load. It may be noted that

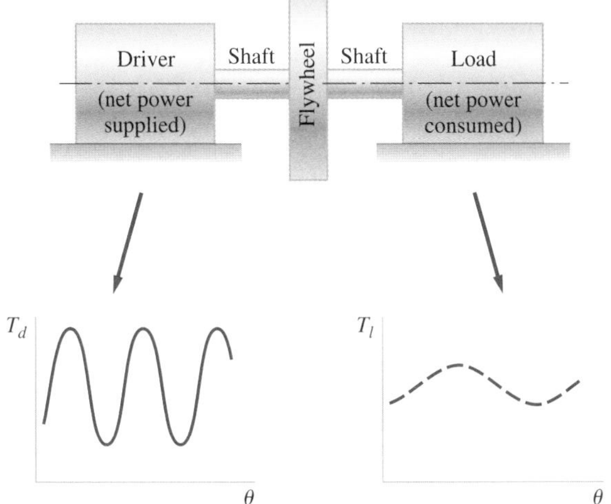

Figure 18.1
Schematic depiction of flywheel system in which either the driver, the load, or both may fluctuate.

[1]Theoretically, however, a flywheel could be used as the sole power source for an operating system if enough kinetic energy could be safely prestored. This might be done by initially "spinning up" its mass to a rotational speed high enough to supply the energy required by the driven machine during an entire mission cycle. As a practical matter, this would be rarely feasible. One alternative currently under intense scrutiny is to couple a flywheel with a compact prime mover such as a turbogenerator to form a *hybrid* power source. Energy from the gas-turbine-driven coupled electric generator would be used to keep the flywheel spinning at the proper speed. Associated with the development of such hybrid systems are, necessarily, the development of gimbal-mounts for the flywheel to prevent transmission of gyroscopic forces to the host vehicle, containment chambers to safety contain any shrapnel in case the flywheel should fail (explode), and an evacuation system for the containment chamber to minimize hydrodynamic drag and improve system efficiency.

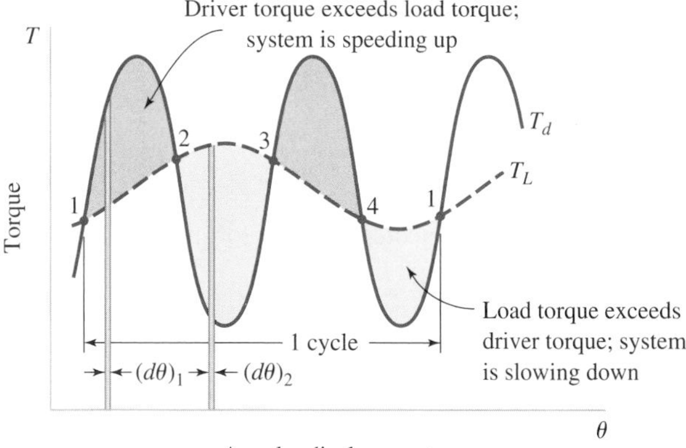

Figure 18.2
Superposed curves of torque versus angular displacement for a machine in which both the drive torque, T_d, and the load torque, T_l, fluctuate, as shown in Figure 18.1.

during increments of time when the *supplied* driver torque *exceeds* the *required* load torque, the flywheel mass is *accelerated* and additional kinetic energy is *stored* in the flywheel. During increments of time when the *required* load torque *exceeds* the *supplied* driver torque, the flywheel mass is *decelerated* and some of the kinetic energy in the flywheel is *depleted.*

When the operational mission of a machine system is dependent upon restricting the speed or torque fluctuations to lie within a defined *control band*, it is possible to estimate the mass moment of inertia and rotational flywheel speed required to accomplish the task. Since all real machine elements have mass, the "flywheel effect" from each significant rotating mass in the system must also be incorporated in the pertinent calculations.

By utilizing a properly designed flywheel, one or more of the following potential advantages may be realized:

1. *Reduced amplitude* of speed fluctuation
2. *Reduced peak torque* required of the driver
3. *Reduced stresses* in shafts, couplings, and possibly other components in the system
4. *Energy automatically stored and released* as required during the cycle

18.2 Fluctuating Duty Cycles, Energy Management, and Flywheel Inertia

If the average angular velocity, ω_{ave}, of a rotating system is to remain unchanged over time, the first law of thermodynamics requires that the *added* kinetic energy stored in the flywheel during one operational cycle must be equal to the kinetic energy *depleted* from the flywheel during the same cycle.[2] Isolating the shaft-flywheel system as a thermodynamic free body,[3] as shown in Figure 18.3, therefore requires that the kinetic energy *supplied to* the flywheel system during each cycle must equal the work *done by* the flywheel system during each cycle,[4] or

$$d(KE) = d(W) \tag{18-1}$$

where KE = kinetic energy supplied to the rotating flywheel system
W = net work done (energy depleted) by the rotating flywheel system

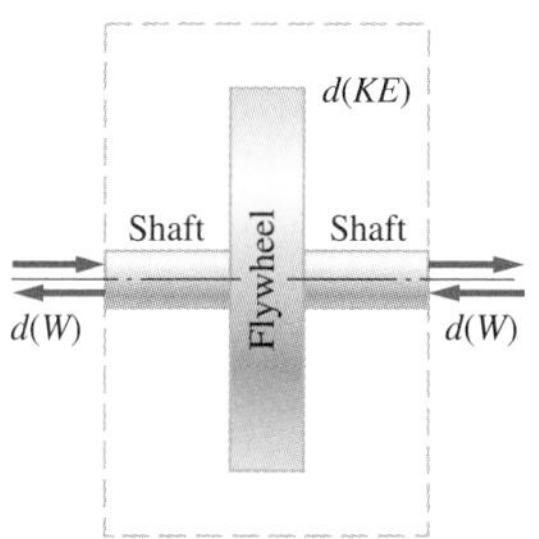

Figure 18.3
Isolated shaft-flywheel system. (See also Figure 18.1.)

[2] Assuming no losses due to friction, windage, etc. [3] Sometimes called *control volume*.
[4] Otherwise, the average rotational velocity of the system would either increase or decrease over time.

For a rotating mass having a mass moment of inertia J and angular velocity ω, it is well known[5] that

$$KE = \tfrac{1}{2}J\omega^2 \tag{18-2}$$

hence

$$d(KE) = \tfrac{1}{2}Jd(\omega^2) \tag{18-3}$$

Also, presuming a rigid shaft subjected to a net torque equal to $(T_l - T_d)$,

$$dW = (T_l - T_d)d\theta \tag{18-4}$$

Combining (18-1), (18-3), and (18-4),

$$\tfrac{1}{2}Jd(\omega^2) = (T_l - T_d)d\theta \tag{18-5}$$

Upon integration, (18-5) becomes

$$\tfrac{1}{2}J\int_{\omega_{min}^2}^{\omega_{max}^2} d(\omega^2) = \int_{\theta_{\omega_{min}}}^{\theta_{\omega_{max}}} (T_l - T_d)d\theta \equiv U_{max} \tag{18-6}$$

where U_{max} = by definition, the maximum change in kinetic energy of the shaft-flywheel system due to speed fluctuation between minimum and maximum values

From (18-6) then,

$$\tfrac{1}{2}J(\omega_{max}^2 - \omega_{min}^2) = U_{max} \tag{18-7}$$

Factoring the left side

$$J\left(\frac{\omega_{max} + \omega_{min}}{2}\right)(\omega_{max} - \omega_{min}) = U_{max} \tag{18-8}$$

but

$$\left(\frac{\omega_{max} + \omega_{min}}{2}\right) = \omega_{ave} \tag{18-9}$$

so (18-8) may be rewritten as

$$J\omega_{ave}(\omega_{max} - \omega_{min}) = U_{max} \tag{18-10}$$

where $(\omega_{max} - \omega_{min})$ is the peak-to-peak fluctuation in angular velocity during each operational cycle.

It is common practice to define a coefficient of speed fluctuation, C_f, as

$$C_f \equiv \frac{\omega_{max} - \omega_{min}}{\omega_{ave}} \tag{18-11}$$

Using (18-11), (18-10) may be rewritten as

$$J\omega_{ave}^2 C_f = U_{max} \tag{18-12}$$

Since ω_{ave} is known, and U_{max} can be determined for any particular system, the required mass moment of inertia, $J_{req'd}$, may be calculated if the coefficient of speed fluctuation for proper system operation is known.[6] A few acceptable experience-based values for C_f are

[5]See any good undergraduate textbook on dynamics (e.g., ref. 1.)

[6]It is well to keep in mind that in actual systems there may be several rotating masses operating at different speeds but connected through gears or other speed changers. The contribution of an *i*th connected rotating mass, rotating at angular velocity ω_i, is proportional to the square of the angular velocity ratio with respect to the operational velocity. Thus an *equivalent* mass moment of inertia, referred to ω_{op}, may be calculated for *any* rotating mass in the system as $J_{ei} = J_i(\omega_i/\omega_{op})^2$.

TABLE 18.1 Experience-Based Coefficients of Speed Fluctuation, C_f, for Successful Equipment Performance[1]

Required Level of Speed Uniformity	C_f
Very uniform	
Gyroscopic control systems	≤ 0.003
Hard disk drives	
Uniform	
AC electrical generators	0.003–0.012
Spinning machinery	
Some fluctuation acceptable	
Machine tools	0.012–0.05
Compressors, pumps	
Moderate fluctuation acceptable	
Excavators	0.05–0.2
Concrete mixers	
Larger fluctuations acceptable	
Crushers	> 0.2
Punch presses	

[1]See refs. 2 and 3.

given in Table 18.1. Expressions for mass moments of inertia as functions of size and shape are included in Appendix Table A.2 for several solid bodies.

To determine U_{max} for any given system, it is first necessary to construct and superpose *driver torque* and *load torque* curves as a function of angular displacement θ over one full operational cycle (see Figure 18.2). Next, the values of θ corresponding to maximum and minimum values of angular velocity during the cycle, $\theta_{\omega_{max}}$ and $\theta_{\omega_{min}}$, respectively, must be identified. In some cases, this may require the preparation of a plot of angular velocity ω versus angular displacement θ. U_{max} is then found as the *net area* enclosed within the torque-displacement diagram[7] between $\theta_{\omega_{max}}$ and $\theta_{\omega_{min}}$. Depending upon the circumstances, the net area, U_{max}, may be found using either analytical, numerical, or graphical methods.

To summarize, an appropriate flywheel configuration may be determined for any given application by completing the following steps:

1. Plot T_l versus θ for one full operational cycle of the machine.
2. Superpose a plot of T_d versus θ for the same operational cycle.
3. Plot ω versus θ.
4. Identify the locations of $\theta_{\omega_{max}}$ and $\theta_{\omega_{min}}$.
5. Find U_{max}.
6. Find $J_{req'd}$ to give the desired C_f.
7. For the selected material and desired geometrical configuration, find the required flywheel dimensions.

Example 18.1 Flywheel Design for Speed Control

A punch press is to be driven by a constant-torque electric motor that operates at 1200 rpm. It is desired to install a flywheel in the rotating system to control speed fluctuations within acceptable bounds. It is being proposed to cut a flywheel disk from a 2-inch-thick steel plate.

[7]T vresus θ diagram.

Example 18.1 Continues

To good approximation, the punch press torque steps from 0 to 10,000 ft-lb, remaining constant for 45°, drops back to zero for the next 45°, steps up to 6000 ft-lb for the next 45°, then drops back to zero for the remainder of the cycle.

You are asked to estimate the flywheel diameter required for satisfactory punch press operation. For this initial estimate, assume there are no other significant rotating masses in the system.[8]

Solution

Following the steps outlined in 18.2, the flywheel diameter requirement may be estimated as follows:

1. Plot punch press torque, T_{pp}, as specified, versus angular displacement, θ, as shown in Figure E18.1A.
2. Plot the constant drive motor torque, T_{mot}, in Figure E18.1A. To find the drive motor torque required for steady-state operation of the system, it may be noted that the total energy supplied by the motor, per cycle, must equal the total energy consumed by the punch press per cycle. Thus, for equilibrium,

$$\int_{1cycle} T_{pp} d\theta = \int_{1cycle} T_{mot} d\theta \tag{1}$$

From the specified punch press torque-displacement cycle, and the knowledge that the motor torque curve has the constant value T_{mot}, (1) may be rewritten as[9]

$$(10{,}000)\frac{\pi}{4} + (6000)\frac{\pi}{4} = T_{mot}(2\pi) \tag{2}$$

so the required motor torque is

$$T_{mot} = \frac{16{,}000\pi}{(4)(2\pi)} = 2000 \text{ ft-lb} \tag{3}$$

3. Without actually knowing the magnitudes of maximum and minimum angular velocities, ω_{max} and ω_{min}, their location in the cycle as a function of angular displacement θ may be deduced from Figure E18.1A, and plotted as shown in Figure E18.1B. Log-

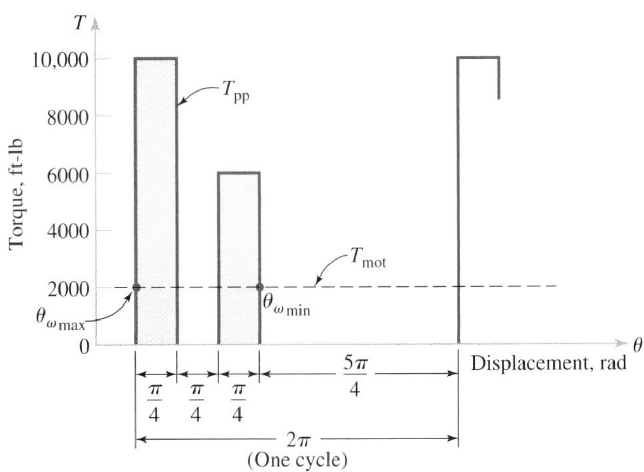

Figure E18.1A
Punch press torque and drive motor torque plotted versus angular displacement over one complete cycle of operation of the rotating system.

[8] This is probably not a very good assumption, but it yields a "conservative" initial value for $J_{req'd}$.

[9] For this case, the integrals may be readily evaluated from Figure E18.1A by summing well-defined rectangular areas.

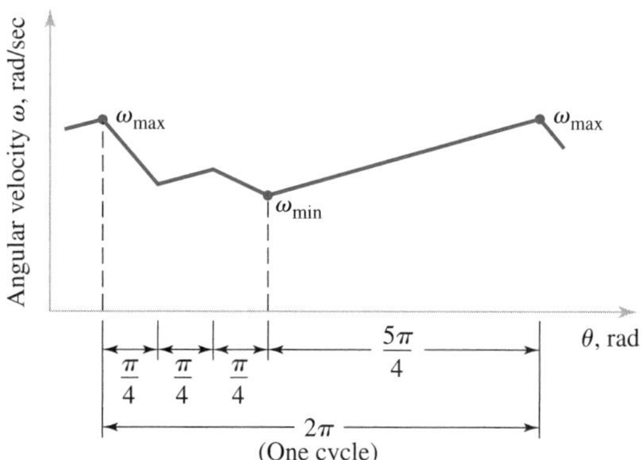

Figure E18.1B
Angular velocitty of the rotating system plotted as a function of angular displacement over one complete cycle of operation. Also see Figure E18.1A.

ically, the maximum angular velocity occurs after the constant motor torque has acted continuously over the long displacement segment of $5\pi/4$ radians. When the punch press torque, T_{pp}, then jumps to 10,000 ft-lb, the system immediately starts to slow, and continues to slow until T_{pp} drops back to zero. The angular velocity then builds again, but only for a displacement segment of $\pi/4$ radians. The second punch press torque-spike then slows the system again, this time to its minimum value in the cycle.

4. For the cycle defined in Figure E18.1B, the maximum angular velocity, ω_{max}, occurs at $\theta = 0$ radians, and the minimum angular velocity, ω_{min}, at $\theta = 3\pi/4$ radians, as indicated.

5. The maximum *change* in kinetic energy stored in the flywheel, U_{max}, may be calculated from (18-6), based on Figure E18.1A, as

$$\begin{aligned} U_{max} &= (10{,}000 - 2000)\frac{\pi}{4} + (0 - 2000)\frac{\pi}{4} + (6000 - 2000)\frac{\pi}{4} \\ &= 7854 \text{ ft-lb } (94{,}248 \text{ in-lb}) \end{aligned} \tag{4}$$

6. From Table 18.1, an acceptable coefficient of speed fluctuation for punch press applications is about

$$C_f = 0.2 \tag{5}$$

Using (18-13),

$$J_{req'd} = \frac{94{,}248}{(0.2)\left[\dfrac{2\pi(1200)}{60}\right]^2} = 29.8 \text{ in-lb-sec}^2 \tag{6}$$

7. From Appendix Table A.2, case 2,

$$J = \frac{1}{2}mr_o^2 = \frac{Wr_o^2}{2g} = \frac{[w(\pi r_o^2 b)]r_o^2}{2g} = \frac{w\pi b r_o^4}{2g} \tag{7}$$

Since the material is specified to be steel ($w = 0.283$ lb/in^3) and the specified thickness is $b = 2$ inches, from (6) and (7) the required outer radius of the solid-disk flywheel is

$$r_o = \sqrt[4]{\frac{2gJ}{w\pi b}} = \sqrt[4]{\frac{2(386)(29.8)}{(0.283)\pi(2.0)}} = 10.67 \text{ inches} \tag{8}$$

The recommendation then is to use a solid-disk flywheel made of steel, with a diameter of 21.3 inches and thickness of 2 inches, to obtain satisfactory operation. However, another iteration to include the "flywheel effect" of other rotating masses in the system should probably be conducted before final specifications are written.

18.3 Types of Flywheels

Since the kinetic energy stored in a rotating mass is given by[10]

$$KE = \frac{1}{2}J\omega^2 \tag{18-13}$$

it is clear that if a large capacity for storing kinetic energy is required, the angular velocity, ω, should be as large as possible, and the mass moment of inertia of the rotating mass should also be as large as possible.[11]

Noting that J is a function of the magnitudes of the mass elements comprising the flywheel and their distances from its axis of rotation, the way to achieve large values of J is to configure the rotating flywheel so that as much mass as possible lies as far as possible from the axis of rotation.[12] This suggests a flywheel configuration in which a relatively heavy circumferential rim is connected by a lightweight structure to a hub secured to the rotating power transmission shaft of the flywheel system. Historically, in the design and construction of flywheels for *low-performance* mechanical systems, these configurational guidelines were translated into the well-known *spoke-and-rim* flywheel configuration, as shown in Figure 18.4(a). For *high-performance* rotating systems in current practice (often rotating at very high angular velocities), it is more common to use uniform-strength flywheel configurations[13] and very-high-strength materials.

Furthermore, the rotating flywheel mass must have an axisymmetric geometry to avoid unbalanced dynamic forces induced by eccentric mass centers. Rotating unbalanced forces typically induce excessive shaft deflections, large-amplitude vibrations, and large forces on supporting bearings and structure.

Axisymmetric geometries that have been successfully implemented for flywheel applications, in addition to the spoke-and-rim configuration, include rotating solid cylindrical disks of constant thickness, "uniformly stressed" solid disks (thickness varies with radius), and "uniformly stressed" solid disks with rims.[14] Some of these configurations are sketched in Figure 18.4.

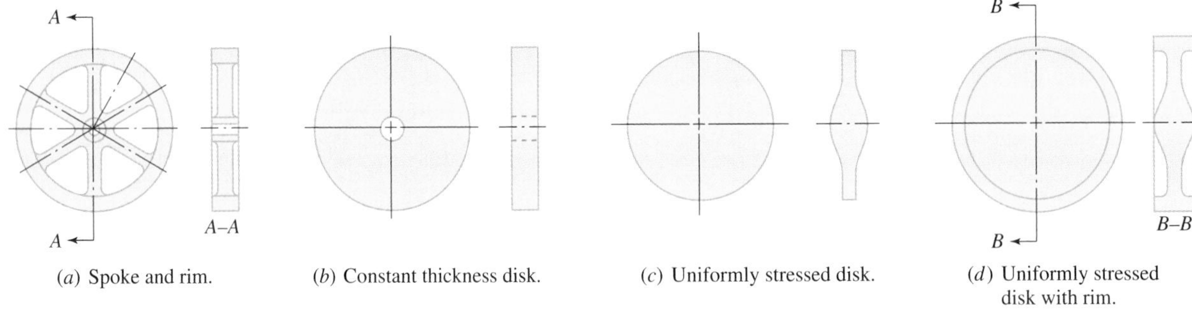

(*a*) Spoke and rim. (*b*) Constant thickness disk. (*c*) Uniformly stressed disk. (*d*) Uniformly stressed disk with rim.

Figure 18.4
Sketches of various types of flywheels.

[10]See (16-20), for example.

[11]However, note the *adverse* effect of large ω and large J on stress levels, as discussed in 18.6, 18.7, and 18.8.

[12]See various expressions for J in Table A.2.

[13]See Figures 18.4(c) and (d).

[14]"Uniformly stressed" configurations cannot be fully achieved in practice because of attachment constraints where the disk meets the rim and the hub, inducing nonuniform local stresses.

18.4 Potential Failure Modes

By far, the most probable potential failure modes for flywheels and (especially) high-speed rotors are brittle fracture or ductile rupture. Because efficient design of high-speed rotors is linked to placement of mass elements at large radii from the axis of rotation in these axisymmetric structures, very large centrifugal forces may be induced, especially at high rotational speeds. In turn, the radial and tangential stresses may become very large, giving rise to the potential for bursting or "exploding" of the rotor into pieces.[15] If a burst occurs, the fragments are typically hurled outward with high velocities because of their high energy content. No matter what material is used, any flywheel will break up if it is rotated fast enough. As a practical safety issue, all high-speed rotating systems muse be surrounded by a *containment enclosure* so that flying shrapnel cannot escape. Spin-testing of high-speed rotors within their containment enclosures is nearly always performed to determine flywheel burst speeds and define containment effectiveness.

In addition to fracture or rupture of a high-speed rotor, other possible failure modes include fretting wear or fretting fatigue at the mounting interface between the flywheel and its shaft, or fatigue failure of the shaft because of cyclic torsion or imbalance-induced cyclic bending.

18.5 Flywheel Materials

Low-speed, low-performance flywheels can be made of virtually any material. Spoke-and-rim flywheels rotating at low speeds have traditionally been made of cast iron to keep the cost as low as possible. Cast steel may also be used for improved strength properties to allow operation at somewhat higher speeds. For higher-performance applications, wrought or forged steels or alloy steels permit much higher operating speeds, especially if careful consideration is given to geometric shape,[16] stress concentration,[17] and residual stress.[18] Steel weldments may be used in some flywheel applications. Recent advances in the use of geometrically tailored filamentary composite materials have shown high promise for filament-wound flywheels. Glass-epoxy, steel-epoxy, and graphite-epoxy composites all provide the potential for safe high-speed operation and high energy-storage capacity, if properly configured. Other materials, such as aluminum, magnesium, and titanium, may be utilized in special applications.

18.6 Spoke-and-Rim Flywheels

A spoke-and-rim flywheel, such as the one illustrated in Figure 18.4(a), is an indeterminate elastic structure with elastic coupling between rim, spokes, and hub. If an accurate analysis is required, a finite element model should probably be developed. However, initial approximations may be made by employing appropriate *simplifying assumptions*.[19] These approximations permit the estimation of operational stresses in the rim, the spokes, and the hub.

Stresses *in the rim*, which include tangential (hoop) stresses and bending stresses, depend greatly upon whether the spokes are small in cross section and radially flexible (soft springs) or large in cross section and radially inflexible (stiff springs). Radially *flexible* spokes provide *little constraint* to the rotating rim, so tangential stresses dominate,[20] and only *small* bending stresses are induced by the spoke constraints. Radially *stiff* spokes pro-

[15]For example, disk or disk-and-rim flywheels usually burst into 3–6 pieces when they fail. [16]See 6.2.
[17]See 4.8. [18]See 4.13. [19]See 6.5. [20]See Figure 9.1.

vide *significant radial constraint* to the rim, so centrifugally induced beam-bending stresses dominate in the rim, and only *small* tangential stresses are induced by rotation of the rim mass. The spoke "springs" and the rim "hoop spring" are operationally in parallel.[21] Therefore, the magnitude of rim bending stress relative to hoop tensile stress depends directly upon the relative stiffnesses of the spokes and the rim.

Stresses *in the spokes* include spoke-axial tensile stresses induced by centrifugal forces on the rim mass, and spoke bending stresses caused by speed and torque fluctuations. The relative magnitudes of these stresses depend greatly upon whether the spokes are bending-flexible or bending-stiff compared to the rim bending stiffness.

Hub stresses are usually not calculated directly, but experience suggests that the hub diameter should be about 2 to 2.5 times the shaft diameter, and the hub length should be about 1.25 times the shaft diameter to meet typical keyway and key strength requirements.

Based on the observations just made, various simplifying assumptions may be used to "bracket" the rim stresses and spoke stresses. For example, it might be assumed that the spokes are so flexible radially that they exert *no* constraint; the rim hoop stresses could then be estimated as hoop stresses in a free ring rotating about its center. Or, it might be assumed that the spokes are so rigid that the rim *cannot expand* radially at the spoke-to-rim attachment sites; the rim could then be modeled as a fixed-fixed beam[22] in bending, uniformly loaded by centrifugal forces acting on the rim mass between spokes. In the case of spoke-axial stresses, it might be assumed that each spoke is centrifugally loaded along its radial axis by its prorated share of rim mass, with no support from rim hoop stiffness. Spoke bending stresses might be estimated by using cantilever bending or double-cantilever bending models, depending upon the bending stiffness of the spokes relative to the rim.[23] Other simplifying assumptions might be made to accommodate special circumstances.

Stresses in a Rotating Free Ring

For the case of a flywheel rim with little or no spoke constraint, the rim can be modeled as a radially thin rotating free ring. For a thin ring the rim *radial* stress is approximately zero, and the rim *hoop* stress is the only one that needs to be calculated. Figure 18.5(a) il-

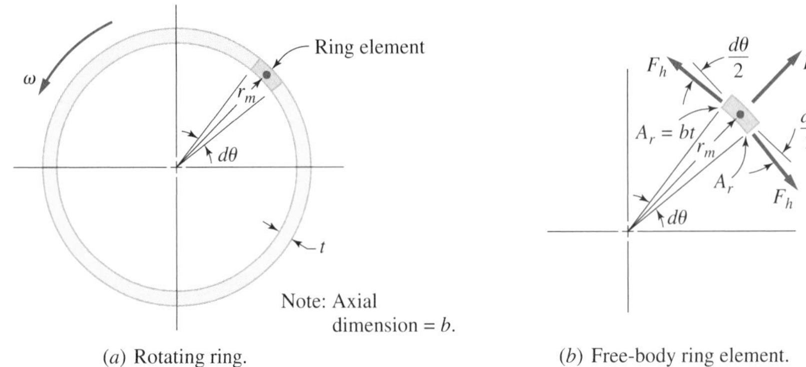

Figure 18.5
Free ring rotating about its center.

[21]See Figure 4.35.

[22]By symmetry, the rim's radial cross section cannot rotate at spoke-attachment sites; thus the rim acts as a fixed-fixed beam between spokes.

[23]If the rim is very stiff to bending, and a torque is induced by speed change, a relative angular displacement between hub and rim will bend each spoke as a combination of two contiguous cantilever beams, one fixed at the hub end and one fixed at the rim end, with the two joined at their "free" ends. This ogee-shaped continuous deflection curve is here termed *double-cantilever* bending. On the other hand, if the rim provides little or no bending resistance, the angular displacement between hub and rim will be resisted by the spoke acting as a *single* cantilever beam fixed at the hub and loaded at its free end by the rim.

lustrates such a ring, and Figure 18.5(b) shows a differential mass element of the ring selected as a free body. The radial centrifugal force, F_r, acting outward on the mass center of the differential element, is

$$F_r = mr_m\omega^2 \tag{18-14}$$

or

$$F_r = \left(\frac{r_m d\theta A_r w}{g}\right)(r_m)\left(\frac{2\pi n}{60}\right)^2 \tag{18-15}$$

and the tangential (hoop) force, F_h, is

$$F_h = \sigma_h A_r \tag{18-16}$$

where m = mass of differential ring element
r_m = mean radius of ring
ω = angular velocity of rotating ring
A_r = bt = cross-sectional area of ring
w = weight density of ring material
n = rotational speed, rpm

Summing radial forces on the selected differential element,

$$F_r - 2F_h \sin\left(\frac{d\theta}{2}\right) = 0 \tag{18-17}$$

Substituting (18-15) and (18-16) into (18-17),

$$\left(\frac{r_m^2 d\theta A_r w}{g}\right)\left(\frac{2\pi n}{60}\right)^2 - 2A_r\sigma_h \sin\left(\frac{d\theta}{2}\right) = 0 \tag{18-18}$$

Imposing the "small angle" assumption that

$$2\sin\left(\frac{d\theta}{2}\right) = 2\left(\frac{d\theta}{2}\right) = d\theta \tag{18-19}$$

and solving (18-18) for the hoop stress σ_h,

$$\sigma_h = \frac{wr_m^2 n^2}{35{,}200} \tag{18-20}$$

where σ_h = hoop stress, psi
r_m = mean rim radius, in
n = rotational speed, rpm
w = weight density of rim material, lb/in^3

An alternative form of (18-20) may be written as

$$\sigma_h = \frac{W_{ur} r_m^2 n^2}{35{,}200 A_r} \tag{18-21}$$

where W_{ur}, the weight per unit circumferential length of the rim, is defined as

$$W_{ur} = wA_r \tag{18-22}$$

Bending Stresses in Flywheel Rim

Centrifugally induced beam-bending rim stresses may be estimated by modeling a section of the rim betwen two adjacent spokes as a straight beam, fixed at both ends, loaded by a uniformly distributed centrifugal force, f_c, as illustrated in Figure 18.6.

Applying case 7 of Table 4.1, the maximum bending moment, M_{max}, that occurs at the fixed supports (spoke centerlines) is

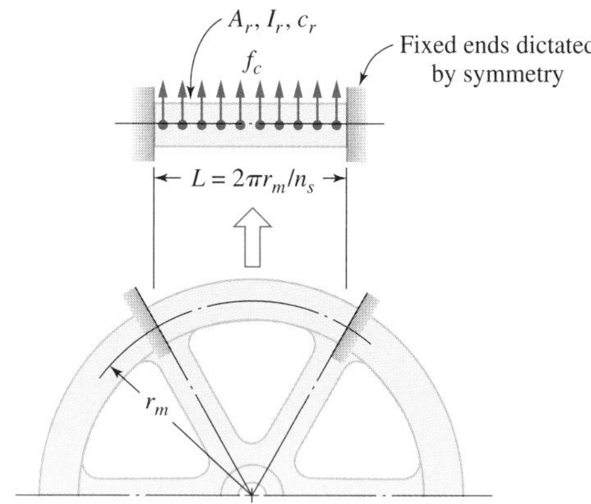

Figure 18.6
Simplifying assumptions illustrated for calculating approximate rim bending stresses.

$$M_{max} = \frac{f_c L^2}{12} \tag{18-23}$$

where the distributed centrifugal force per unit circumferential length may be estimated as

$$f_c = \frac{m r_m \omega^2}{L} = \left(\frac{w A_r L r_m}{gL}\right)\left(\frac{2\pi n}{60}\right)^2 \text{ lb/in} \tag{18-24}$$

whence (18-23) becomes

$$M_{max} = \frac{w r_m n^2 L^2}{(35{,}200)(12)} \text{ in-lb} \tag{18-25}$$

and using (4-5), the maximum bending stress at the outer fibers at the spoke centerlines may be calculated as

$$(\sigma_b)_{max} = \frac{M_{max} c_r}{I_r} = \frac{w A_r r_m n^2 L^2 c_r}{(35{,}200)(12) I_r} \text{ psi} \tag{18-26}$$

Spoke-Axial Tensile Stresses

To estimate spoke-axial tensile stress for stiff spokes, a common simplifying assumption is to ignore hoop-spring support from the rim and consider each spoke to be radially loaded by its prorated share of the rim mass, as illustrated in Figure 18.7. A radial force resolution gives

$$F_r - F_s = 0 \tag{18-27}$$

where F_r = radial centrifugal force component due to rotation of the spoke's assigned rim mass about the flywheel center of rotation
F_s = spoke-axial tensile force on each spoke

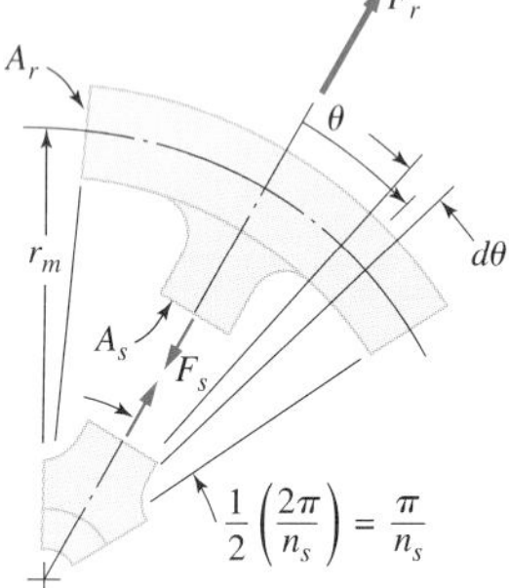

Figure 18.7
Simplifying assumptions illustrated for calculating approximate spoke-axial tensile stresses.

Based on Figure 18.7, (18-27) may be rewritten as

$$2\int_0^{\pi/n_s} \frac{w r_m d\theta A_r}{g}(r_m)\left(\frac{2\pi n}{60}\right)^2 \cos\theta - \sigma_s A_s = 0 \tag{18-28}$$

where σ_s = spoke-axial tensile stress, psi
A_s = spoke cross-sectional area, in^2
w = weight density of rim material, lb/in^3
r_m = mean rim radius, in
$d\theta$ = angle subtending differential element
θ = angle from spoke axis to differential element
A_r = cross-sectional area of rim, in^2
n = rotational speed, rpm

Solving (18-28) for σ_s gives the spoke-axial tensile stress as

$$\sigma_s = \frac{w r_m^2 n^2 \left(\frac{A_r}{A_s}\right) \sin\left(\frac{\pi}{n_s}\right)}{17{,}600} \text{ psi} \tag{18-29}$$

18.7 Disk Flywheels of Constant Thickness

Constant-thickness disk flywheels are often chosen by a designer because they can be easily cut from plate stock, making them inexpensive. The radial and tangential stresses generated in a rotating constant-thickness disk are both axisymmetric, making them very similar in character to the radial and tangential stresses in a thick-walled cylinder subjected to internal and/or external pressure.[24] The primary difference is that stresses in a rotating flywheel are induced by centrifugal *body* forces acting outward on the mass elements of the disk, while the stresses in a thick-walled cylinder are induced by *surface* pressure forces acting on the inner and/or outer cylindrical surfaces.

Just as for thick-walled cylinders, the development of equations for radial and tangential stress distributions in a rotating disk flywheel requires application of the principles of elasticity theory. These include force-equilibrium concepts, force-displacement relationships, geometrical compatibility restrictions, and boundary condition specifications.[25]

A constant-thickness disk flywheel is sketched in Figure 18.8, showing a small differential free-body element within the flywheel mass. All forces acting on the free-body element are shown. Writing a radial *force equilibrium* expression for the differential free-body element

$$(\sigma_r + d\sigma_r)(t)(r + dr)d\varphi - \sigma_r(t)(rd\varphi) - 2\sigma_t(t)dr\sin\frac{d\varphi}{2} + \frac{w(t)(rd\varphi)dr(r)\omega^2}{g} = 0 \tag{18-30}$$

which may be reduced to

$$(\sigma_r - \sigma_t)dr + rdr + \frac{wr^2\omega^2}{g}dr = 0 \tag{18-31}$$

where σ_r = radial stress
σ_t = tangential stress
r = radial distance to element inner surface
ω = angular velocity of flywheel
w = weight density of flywheel material

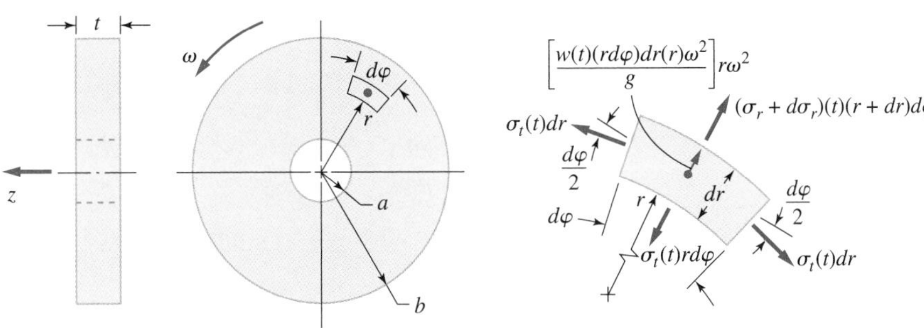

Figure 18.8
Constant-thickness disk flywheel rotating at angular velocity ω.

[24] See 9.7. [25] See 9.5.

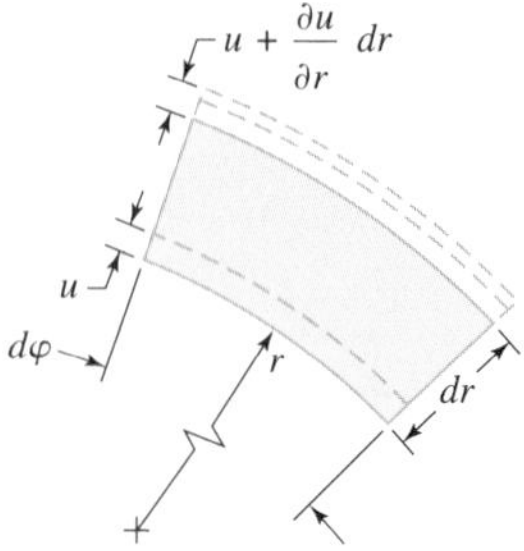

Figure 18.9
Displacements and strains illustrated for a small differential element in a constant-thickness disk flywheel.

The *force-displacement* relationships for the differential element shown in Figure 18.9 are[26]

$$\varepsilon_r = \frac{1}{E}[\sigma_r - \nu\sigma_t] \tag{18-32}$$

$$\varepsilon_t = \frac{1}{E}[\sigma_t - \nu\sigma_r] \tag{18-33}$$

$$\varepsilon_z = \frac{1}{E}[-\nu(\sigma_r + \sigma_t)] \tag{18-34}$$

where $\varepsilon_r, \varepsilon_t, \varepsilon_z$ = normal strains in the radial, tangential, and axial (thickness) directions, respectively
E = Young's modulus of elasticity
ν = Poisson's ratio
σ_r, σ_t = normal stresses in radial and tangential directions, respectively

If u is defined to be *displacement* in the radial direction, as sketched in Figure 18.9, *geometrical compatibility* may be established for the differential volume element through the definition of *engineering strain* as

$$\varepsilon_r = \frac{\Delta L_r}{L_r} = \frac{\left[\left(u + \frac{\partial u}{\partial r}dr\right) - u\right]}{dr} = \frac{\partial u}{\partial r} = \frac{du}{dr} \tag{18-35}$$

and

$$\varepsilon_t = \frac{\Delta L_t}{L_t} = \frac{(r + u)d\varphi - rd\varphi}{rd\varphi} = \frac{u}{r} \tag{18-36}$$

Finally, the pertinent *boundary conditions* may be established as

$$(\sigma_r)_{r=a} = 0 \tag{18-37}$$

and

$$(\sigma_r)_{r=b} = 0 \tag{18-38}$$

Solving Hooke's Law expression (18-32) and (18-33) simultaneously for stresses σ_r and σ_t gives

$$\sigma_r = \frac{E}{1 - \nu^2}[\varepsilon_r + \nu\varepsilon_t] \tag{18-39}$$

$$\sigma_t = \frac{E}{1 - \nu^2}[\nu\varepsilon_r + \varepsilon_t] \tag{18-40}$$

Next, inserting geometrical compatibility expressions (18-35) and (18-36) into (18-39) and (18-40) gives

$$\sigma_r = \frac{E}{1 - \nu^2}\left[\frac{du}{dr} + \nu\left(\frac{u}{r}\right)\right] \tag{18-41}$$

and

$$\sigma_t = \frac{E}{1 - \nu^2}\left[\nu\left(\frac{du}{dr}\right) + \frac{u}{r}\right] \tag{18-42}$$

inserting (18-41) and (18-42) into equilibrium equation (18-31), the result becomes

$$\frac{d}{dr}\left[\frac{1}{r}\frac{d}{dr}(ur)\right] = -Ar \tag{18-43}$$

[26]These are the Hooke's Law relationships given by (4-73), (4-74) and (4-75), with the axial stress σ_z equal to zero since there are no forces in the axial direction.

where

$$A \equiv \frac{w\omega^2(1 - v^2)}{Eg} \tag{18-44}$$

Integrating (18-43) gives

$$\frac{du}{dr} = -\frac{3Ar^2}{8} + C_1 - \frac{C_2}{r^2} \tag{18-45}$$

where C_1 and C_2 are constants of integration.

Substituting (18-45) into (18-41) and (18-42)

$$\sigma_r = \frac{E}{1 - v^2}\left[-\frac{3Ar^2}{8} + C_1 - \frac{C_2}{r^2} + \frac{v}{r}\left(-\frac{Ar^3}{8} + C_1 r + \frac{C_2}{r}\right)\right] \tag{18-46}$$

and

$$\sigma_t = \frac{E}{1 - v^2}\left[v\left(-\frac{3Ar^2}{8} + C_1 - \frac{C_2}{r^2}\right) + \frac{1}{r}\left(-\frac{Ar^3}{8} + C_1 r + \frac{C_2}{r}\right)\right] \tag{18-47}$$

Next, invoking boundary conditions (18-37) and (18-38), the constants C_1 and C_2 may be found by solving (18-46) and (18-47) simultaneously to give

$$C_1 = \frac{(3 + v)}{8(1 + v)}(A)(a^2 + b^2) \tag{18-48}$$

and

$$C_2 = \frac{(3 + v)}{8(1 + v)}(A)(a^2 b^2) \tag{18-49}$$

Finally, putting these expressions back into (18-46) and (18-47) gives

$$\sigma_r = \left[\frac{(3 + v)w\omega^2}{8g}\right]\left[a^2 + b^2 - r^2 - \frac{a^2b^2}{r^2}\right] \tag{18-50}$$

and

$$\sigma_t = \left[\frac{(3 + v)w\omega^2}{8g}\right]\left[a^2 + b^2 - \left(\frac{1 + 3v}{3 + v}\right)r^2 + \frac{a^2b^2}{r^2}\right] \tag{18-51}$$

Noting in these expressions that

$$\left(\frac{1 + 3v}{3 + v}\right) < 1 \tag{18-52}$$

and

$$\frac{a^2 + b^2}{r^2} > 0 \tag{18-53}$$

it may be deduced that for all operating conditions

$$\sigma_t > \sigma_r \tag{18-54}$$

Further, σ_t reaches its maximum value where r is a minimum, so

$$(\sigma_t)_{max} \text{ occurs at } r = a \tag{18-55}$$

It may also be noted that

$$\sigma_r = 0 \quad \text{at} \quad r = a \tag{18-56}$$

To calculate $(\sigma_t)_{max}$, $r = a$ may be substituted into (18-51) to give

$$(\sigma_t)_{max} = \frac{w\omega^2}{4g}[(3 + v)b^2 + (1 - v)a^2] \tag{18-57}$$

and, because $\sigma_r = 0$, the state of stress is uniaxial.

Example 18.2 Constant-Thickness Disk Flywheel Design

Exploratory marketing has led to a *concept proposal* for a new machine to crush oyster shells, which are fed to chickens as a dietary supplement for improving egg-shell strength to minimize egg-shell breakage. The proposed crushing machine has an estimated torque versus angular displacement curve as shown in Figure E18.2, and is to be driven by a constant-torque electric motor at a rotational speed of 3450 rpm. It is being proposed to use a constant-thickness disk flywheel to limit speed fluctuation to an acceptable range. As an initial configuration, it is being proposed to cut the flywheel from a 1.50-inch-thick steel plate of annealed low-carbon steel (S_{yp} = 43,000 psi).

a. Assuming that there are no other significant rotating masses in the system,[27] you are asked to estimate the flywheel diameter for satisfactory crusher operation. A mounting hole at the center of the disk is estimated to be 2.0 inches in diameter.

b. A minimum safety factor of 4 based on the governing failure mode is being proposed by engineering management. Would the constant-thickness disk flywheel resulting from (a) meet this proposed safety factor criterion?

Solution

a. Since, for each cycle, the energy supplied by the drive motor must equal the total energy consumed by the crusher, for steady-state operation

$$\int_{1cycle} T_{mot} d\theta = \int_{1cycle} T_{crush} d\theta \tag{1}$$

Using Figure E18.2, (1) may be evaluated for motor torque as

$$2\pi T_{mot} = 2\left[\frac{8000\left(\frac{\pi}{2}\right)}{2}\right] + (24{,}000)\frac{\pi}{2} = 16{,}000\pi \text{ ft-lb} \tag{2}$$

or

$$T_{mot} = \frac{16{,}000\pi}{2\pi} = 8000 \text{ ft-lb (constant)} \tag{3}$$

Using methods illustrated in Example 18.1, the maximum and minimum angular velocities during each operating cycle occur at

$$\theta_{\omega_{max}} = \frac{\pi}{2} \text{ rad} \tag{4}$$

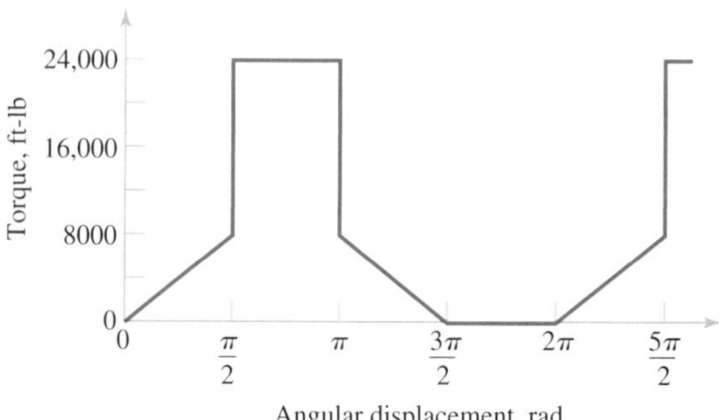

Figure E18.2
Crusher torque versus angular displacement of the rotating system.

[27]This is probably not a very good assumption, but it provides a "conservatrive" initial value for $J_{req'd}$.

and

$$\theta_{\omega_{min}} = \pi \text{ rad} \tag{5}$$

The maximum change in kinetic energy between $\pi/2$ and π, U_{max}, may next be calculated by plotting (3) in Figure E18.2, from which

$$U_{max} = \int_{\pi/2}^{\pi} (T_{crush} - T_{mot})d\theta = (24{,}000 - 8000)\frac{\pi}{2} = 25{,}130 \text{ ft-lb} \tag{6}$$

Selecting a value of $C_f = 0.2$ from Table 18.1 (18-13) gives

$$J_{req'd} = \frac{U_{max}}{C_f \omega_{ave}^2} = \frac{25{,}130}{(0.2)\left[\dfrac{2\pi(3450)}{60}\right]^2} = 1.0 \text{ ft-lb-sec}^2 \ (12.0 \text{ in-lb-sec}^2) \tag{7}$$

b. From Appendix Table A.2, case 2,

$$J = \frac{1}{2}mr_o^2 = \frac{w\pi t r_o^4}{2g} \tag{8}$$

whence

$$(r_o)_{req'd} = \sqrt[4]{\frac{2gJ}{w\pi t}} = \sqrt[4]{\frac{2(386)(12)}{(0.283)\pi(1.5)}} = 9.13 \text{ inches} \tag{9}$$

For this proposed flywheel disk based on (18-54), (18-55), and (18-56), the state of stress is uniaxial (tangential component only) and the critical point is at the edge of the inner hole where $r_i = 2.0/2 = 1.0$ inch. At this location the only stress is,[28] from (18-57),

$$(\sigma_t)_{max} = \frac{0.283(361)^2}{4(386)}[(3 + 0.3)(9.13)^2 + (1 - 0.3)(1.0)^2] = 6588 \text{ psi} \tag{10}$$

Since yielding is the probable failure mode, the safety factor would be

$$n_{yp} = \frac{\sigma_{yp}}{(\sigma_t)_{max}} = \frac{43{,}000}{6588} = 6.5 \tag{11}$$

This is more than adequate to meet the specified minimum safety factor value of 4.

18.8 Disk Flywheels of Uniform Strength

The radial and tangential stresses in a constant-thickness disk flywheel vary as a function of radius, as indicated by (18-50) and (18-51). As a result, some of the material in the flywheel operates efficiently at the specified design stress, but much of the flywheel material operates at lower stresses, resulting in inefficient use of the lightly stressed material.[29] If it is desired to configure a flywheel that utilizes material more efficiently, so that stress everywhere is nearly the same, the condition to be met is

$$\sigma_r = \sigma_t \equiv \sigma \quad \text{(constant)} \tag{18-58}$$

When the condition of (18-58) is met, the flywheel is usually called a flywheel of *constant strength* or *uniform strength*. To satisfy (18-58) it is necessary to specify disk thickness z as a function of radius r, so that the result is uniform stress throughout, then invoke

[28] Assuming no stress concentration effects from mounting constraints.

[29] See Chapter 6.

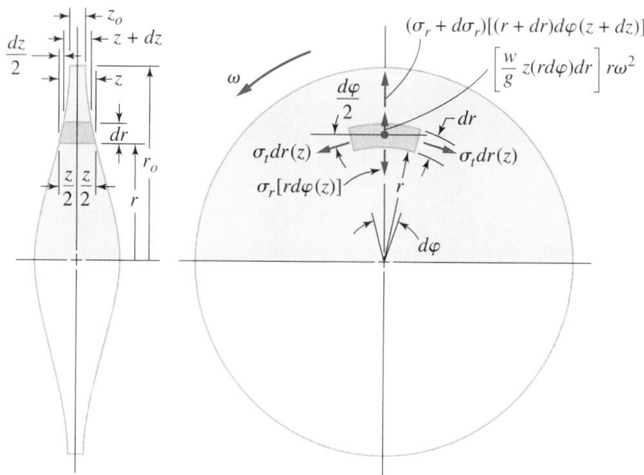

Figure 18.10
Constant-strength disk flywheel (high-speed rotor) operating at angular velocity ω.

the principles of elasticity theory[30] to determine the resulting flywheel thickness profile. Because of increased complexity and cost, constant-strength profiles are usually restricted to *high-performance, high-speed* rotors. Figure 18.10 illustrates a differential free-body volume element from a constant-strength flywheel.[31] A radial force resolution on the free-body element in Figure 18.10 gives

$$(\sigma_r + d\sigma_r)[(r + dr)d\varphi(z + dz)] - \sigma_r[rd\varphi(z)] + \left[\frac{w}{g}z(rd\varphi)dr\right]r\omega^2 - 2\sigma_t dr(z)\sin\frac{d\varphi}{2} = 0 \tag{18-59}$$

Imposing condition (18-58), and using the small angle approximation, (18-59) reduces to

$$\sigma dz + \left(\frac{wz\omega^2}{g}\right)rdr = 0 \tag{18-60}$$

or

$$\frac{dz}{z} = -\left(\frac{w\omega^2}{\sigma g}\right)rdr \tag{18-61}$$

If the disk thickness $z(r)$ is defined to be z_o at the outer radius, r_o, (18-61) may be integrated to give

$$z = z_o e \exp\left[\left(\frac{w\omega^2}{2\sigma g}\right)(r_o^2 - r^2)\right] \tag{18-62}$$

Thus for a constant-strength high-speed rotor, the thickness z is an exponential function of radius between the limits

$$z_{r=r_o} = z_o \tag{18-63}$$

and disk thickness at the center of rotation, z_{max}, given by

$$z_{max} = z_{r=0} = z_o e \exp\left[\frac{w\omega^2 r_o^2}{2\sigma g}\right] \tag{18-64}$$

It should also be noted that no center hole can be used for mounting a constant-strength disk flywheel without destroying the uniform stress pattern.[32]

[30]See 18.7. [31]Compare with Figure 18.8 for a constant-thickness flywheel.

[32]For configuring variable-thickness flywheel profiles that *do* include a central hole, approximate methods are usually employed as discussed in ref. 4, or finite element programs may be utilized.

18.9 Uniform-Strength Disk Flywheel with a Rim

In certain applications it may be desired to increase the polar mass moment of inertia of a constant-strength flywheel by adding a *rim* attached at outer disk radius r_o, as illustrated in Figure 18.11. Assuming the rim to be radially thin ($r_o \gg t$), and noting that at operational angular velocity ω the rim exerts a distributed force of q lb/in outwardly around the flywheel web[33] at r_o, the uniform-strength condition requires that

$$\sigma = \sigma_o = \frac{q}{z_o} \quad \text{at} \quad r = r_o \tag{18-65}$$

A radial force resolution on the free-body element of the *rim*, shown in Figure 18.11, gives

$$-qr_o d\varphi - 2\sigma_{h-rim} A_{rim} \sin\left(\frac{d\varphi}{2}\right) + \left(\frac{wA_{rim} r_o d\varphi}{g}\right) r_o \omega^2 = 0 \tag{18-66}$$

Utilizing (18-65), this may be rewritten as

$$\frac{wr_o^2 A_{rim} \omega^2}{g} - \sigma_{h-rim} A_{rim} - \sigma_o z_o r_o = 0 \tag{18-67}$$

Next, it may be noted that geometrical compatibility requires that at radius r_o the change in rim radius, $(\Delta r)_{rim}$, must be equal to the change in disk radius, $(\Delta r)_{disk}$. Thus

$$(\Delta r)_{rim} = (\Delta r)_{disk} \tag{18-68}$$

Defining C as circumference,

$$r = \frac{C}{2\pi} \tag{18-69}$$

whence

$$\Delta r = \frac{\Delta C}{2\pi} \tag{18-70}$$

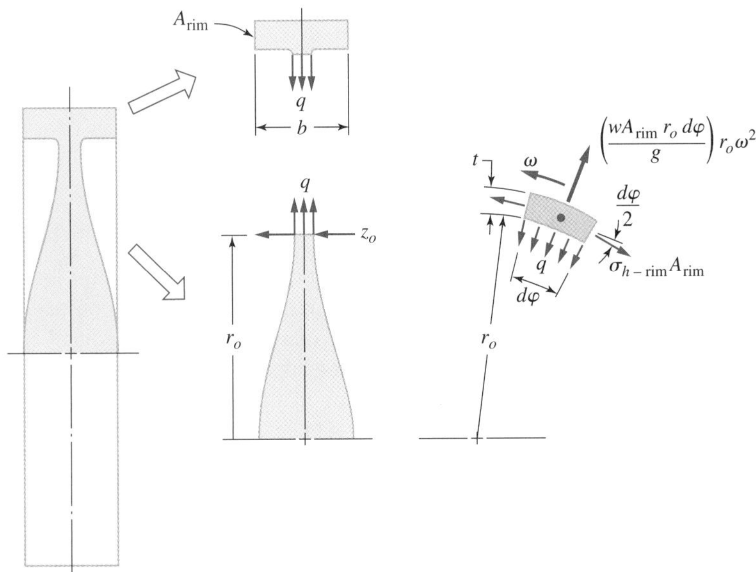

Figure 18.11
Constant-strength disk flywheel with a rim.

[33]Likewise, the flywheel exerts an equal and opposite distributed force q inwardly on the rim.

For the *rim,*[34] utilizing Hooke's Law, the circumferential strain is

$$\frac{\Delta C}{C} \equiv \varepsilon_C = \frac{\sigma_{h-rim}}{E} \tag{18-71}$$

so, at radius r_o,

$$(\Delta r)_{rim} = \frac{\Delta C}{2\pi} = \frac{C\sigma_{h-rim}}{2\pi E} = \frac{\sigma_{h-rim}}{E} r_o \tag{18-72}$$

For the *disk,*[35] utilizing Hooke's Law, the circumferential strain is

$$\frac{\Delta C}{C} \equiv \varepsilon_C = \frac{1}{E}(\sigma - \nu\sigma) \tag{18-73}$$

so

$$(\Delta r)_{disk} = \frac{\Delta C}{2\pi} = \frac{C(1-\nu)\sigma}{2\pi E} = \frac{(1-\nu)\sigma}{E} r_o \tag{18-74}$$

Combining (18-65), (18-72), and (18-74) with (18-68) gives, at $r = r_o$,

$$\left(\frac{\sigma_{h-rim}}{E}\right) r_o = \frac{(1-\nu)\sigma}{E} r_o = \frac{(1-\nu)q}{Ez_o} r_o \tag{18-75}$$

so

$$\sigma_{h-rim} = \frac{(1-\nu)q}{z_o} = (1-\nu)\sigma_o \tag{18-76}$$

and (18-67) may be rewritten as

$$\frac{wr_o^2 A_{rim}\omega^2}{g} - (1-\nu)\sigma_o A_{rim} - \sigma_o z_o r_o = 0 \tag{18-77}$$

Solving for web thickness z_o at the rim-web juncture radius, r_o,

$$z_o = \left[\frac{wr_o^2\omega^2}{g\sigma_o} - (1-\nu)\right] \frac{A_{rim}}{r_o} \tag{18-78}$$

The flywheel profile therefore may be defined, for a constant-strength rotor with a rim, by combining (18-78) and (18-62) to give

$$z = \left[\frac{wr_o^2\omega^2}{g\sigma_o} - (1-\nu)\right] \frac{A_{rim}}{r_o} e \exp\left[\frac{w\omega^2}{2\sigma_o g}\right](r_o^2 - r^2) \tag{18-79}$$

Example 18.3 Constant-Strength Flywheel with a Rim

In a high-speed rotor application, the nominal rotational speed of the rotor is to be n = 12,500 rpm. It has been decided to investigate the feasibility of using a constant-strength flywheel with a rim to provide the required polar mass moment of inertia, already estimated to be about $J_{req'd}$ = 11.5 in-lb-sec^2. As a first approximation, engineering management has authorized the (conservative) assumption that the rim provides all of $J_{req'd}$ and the assumption that the radial rim thickness, t, should be about 10 percent of the mean rim radius r_m. The material being proposed is ultra-high-strength AISI 4340 steel, heat treated to have S_u = 287,000 psi and S_{yp} = 270,000 psi.[36] A safety factor of 2 is desired, based on yielding as the probable failure mode. For this first approximation, it is suggested that

[34] Since the rim is assumed to be thin, the tangential hoop stress, σ_{h-rim}, is uniaxial.

[35] The disk is subjected to a biaxial state of stress in which $\sigma_r = \sigma_t = \sigma$.

[36] See Table 3.3.

stress concentration effects be ignored. Because of running-clearance requirements, surrounding components, and the need for a safety shield around the rotor in case of explosive failure, it is desired that the outer rim radius be limited to about 11.0 inches. Determine a uniform-strength geometry that should satisfy these requirements, and sketch the proposed rotor profile.

Solution

For the selected material, with $S_{yp} = 270{,}000$ psi and a prescribed design safety factor of $n_d = 2$, the design stress becomes

$$\sigma_d = \frac{\sigma_{yp}}{n_d} = \frac{270{,}000}{2} = 135{,}000 \text{ psi} \tag{1}$$

Since the available space envelope limits the outer rim radius, $r_{rim-outer}$, to 11.0 inches, and assuming

$$t = 0.1 r_o \tag{2}$$

the maximum value of r_o may be estimated from

$$r_o + t = 1.1 r_o = 11.0 \text{ inches} \tag{3}$$

as

$$(r_o)_{max} = \frac{11.0}{1.1} = 10.0 \text{ inches} \tag{4}$$

Using case 4 of Appendix Table A.2, for $r_o = 10.0$ inches,

$$J_{r_o=10} = mr_o^2 = \frac{2\pi t b w r_o^3}{g} = \frac{2\pi(1.0)b(0.283)(10.0)^3}{386} = 4.61b \text{ in-lb-sec}^2 \tag{5}$$

Setting $J_{r_o=10}$ equal to $J_{req'd}$,

$$J_{r_o=10} = 4.61b = J_{req'd} = 11.5 \tag{6}$$

and solving for rim width b,

$$b = \frac{11.5}{4.61} = 2.5 \text{ inches} \tag{7}$$

Summarizing, the proposed rim would have an inner radius[37] of $r_o = 10.0$ inches, a radial thickness $t = 1.0$ inch, and an axial width $b = 2.5$ inches.

Next, using (18-78), the web thickness z_o at the rim-web juncture radius r_o becomes

$$z_o = \left[\frac{(0.283)(10.0)^2\left(\dfrac{2\pi(12{,}500)}{60}\right)^2}{386(135{,}000)} - (1 - 0.3)\right]\frac{(1.0)(2.5)}{10} = 0.06 \text{ inch} \tag{8}$$

From a practical standpoint, such a thin web at the point of attachment with the rim may pose serious manufacturing problems. For this first iteration, therefore, a web thickness at r_o will arbitrarily be specified to be

$$z_o = 0.188 \text{ inch} \tag{9}$$

and a generous blend radius between web and rim will also be suggested.

Using (18-79) and (9), the web profile may be defined by

$$z = 0.188e \exp\left[\left(\frac{(0.283)\left(\dfrac{2\pi(12{,}500)}{60}\right)^2}{386(2)(135{,}000)}\right)(10^2 - r^2)\right] = 0.188e^{0.001(100 - r^2)} \tag{10}$$

[37]Because the rim is radially "thin," the mean radius and inner radius may be assumed to be the same.

Example 18.3 Continues

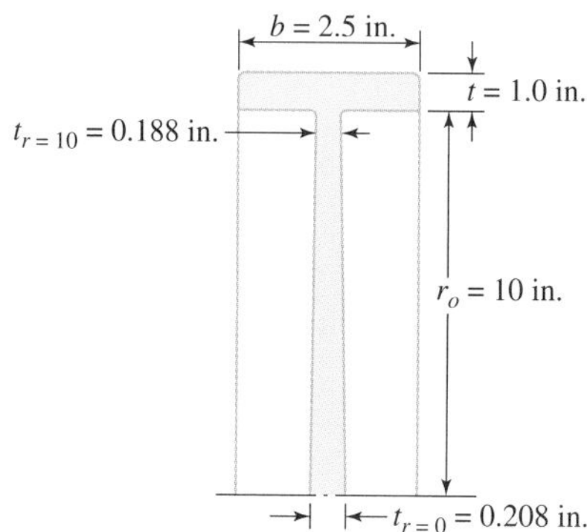

Figure E18.3
Sketch of proposed profile for constant-strength rotor with a rim.

As illustrated in the sketch of Figure E18.3, for this particular case the web thickness variation is small. Thus a straight-sided tapered web should probably be investigated, or perhaps a web of constant thickness should be considered.

18.10 Flywheel-to-Shaft Connections

Like gears, pulleys, sprockets, or any other rotating component, flywheels must be mounted on rotating power transmission shafts accurately and securely, and with a minimum of stress concentration at geometrical discontinuities.[38] In fact, flywheel-to-shaft connections are typically more critical than for other shaft-mounted components because rotational speeds are often very high for flywheels. Also, constant-strength flywheel profiles are based on the assumption that no center hole exists.[39] Mounting is made very difficult if no center hole is allowed. Finite element modeling is often necessary for optimization of the hub and hub-to-web transition geometry.

Several configurations for mounting a flywheel to a shaft are sketched in Figure 18.12. The simplest configuration, shown in Figure 18.12(a), utilizes an interference fit between the shaft and center hole of the constant-thickness flywheel. This configuration, however, induces significant stress concentration[40] at the critical point (c.p.), and may make assembly difficult if the flywheel must be pressed onto the shaft over a long distance. The use of a raised mounting pad with a generous blend radius, as shown in Figure 18.2(b), both reduces stress concentration and eliminates the need to press the flywheel over long distances. A similar concept involves the addition of a locating shoulder at the end of the mounting pad for proper axial placement of the flywheel.

In some cases a tapered connection may be used, such as the one sketched in Figure 18.12(c). Tightening the retaining nut wedges the conical surfaces together to produce highly accurate concentricity of the flywheel and shaft, and also provides high transfer torque capacity between them.[41] An alternative tapered connection is illustrated in Figure 18.12(d), where two matching oppositely tapered[42] concentric rings are nested as shown.

[38]See Chapter 8. [39]See 18.8. [40]See Figure 4.25. [41]Also see Figure 8.10.

[42]One ring is internaly tapered and the other ring is externally tapered. This provides a *cylindrical* interface with both shaft and flywheel.

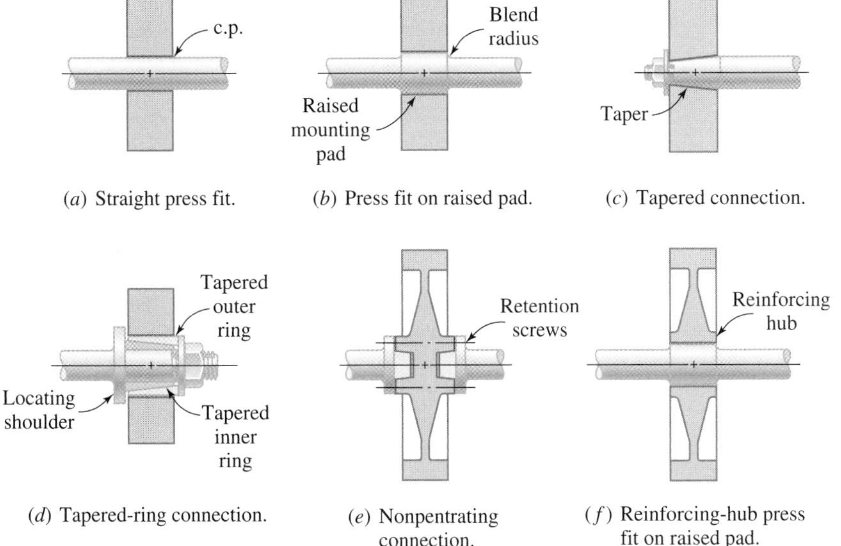

Figure 18.12
Various ways of securing flywheels to supporting shafts.

Because of the taper, the rings deform radially when compressed axially, locking everything in place.

To avoid the penetration of a *constant-strength* flywheel by a center hole, the alternative mounting configuration shown in Figure 18.12(e) has sometimes been used. Careful attention must be directed toward potential stress concentration sites if the configuration is adopted. In other cases the uniform stress pattern of a constant-strength flywheel may be adequately approximated by the use of an appropriately designed hub surrounding the center hole as sketched in Figure 18.12(f).

Problems

18.1. A deep-drawing press is estimated to have the load torque versus angular-displacement characteristics shown in Figure P18.1. The machine is to be driven by a constant-torque electric motor at 3600 rpm. The total change in angular velocity from its maximum value to its minimum value must be controlled to within $\pm$ 3 percent of the average angular velocity of the drive.

a. Compute and sketch the motor input torque versus angular displacement curve.

b. Sketch a curve of angular velocity (qualitative) versus angular displacement (quantitative).

c. On the torque versus angular displacement curve, carefully locate angular displacement values corresponding to maximum and minimum angular velocity.

d. Calculate U_{max}.

e. Calculate the required mass moment of inertia for a flywheel that would properly control speed fluctuation to within $\pm$ 3 percent of average angular velocity, as specified.

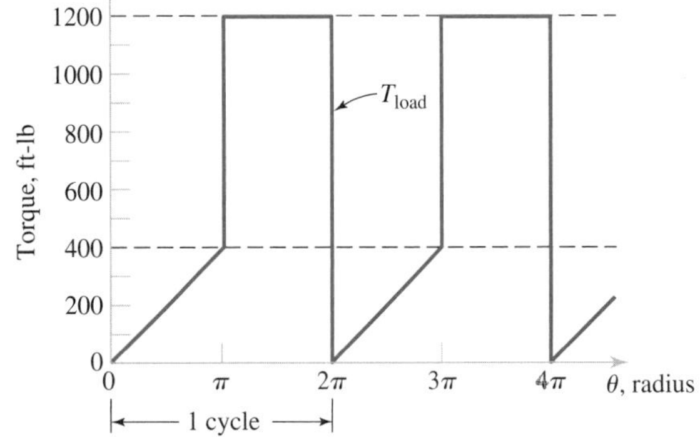

Figure P18.1
Torque versus angular displacement curve for a deep-drawing press.

18.2. A hammermill has the load torque versus angular displacement curve shown in Figure P18.2, and is to be driven by a constant-torque electric motor at 3450 rpm. A flywheel is to be used to provide proper control of the speed fluctuation.

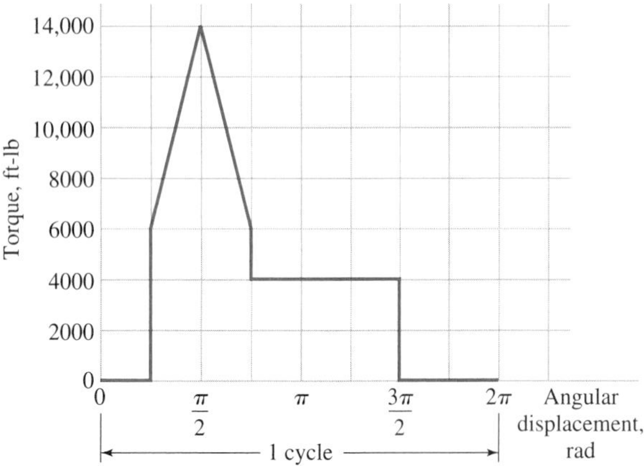

Figure P18.2
Torque versus angular displacement curve for a hammer-mill.

a. Compute and plot the motor input torque versus angular displacement curve.

b. Sketch angular velocity (qualitative) of the shaft-flywheel system as a function of angular displacement (quantitative). Specifically note locations of maximum and minimum angular velocity on the torque versus angular displacement curve.

c. Calculate U_{max}.

d. Calculate the mass moment of inertia required for the flywheel to properly control the speed fluctuation.

18.3. A natural gas engine is to be used to drive an irrigation pump that must be operated within $\pm$ 2 percent of its nominal operating speed of 1000 rpm. The engine torque versus angular displacement curve is the sawtooth T_{engine} curve shown in Figure P18.3. The pump torque verse angular displacement curve is the stepped T_{pump} curve shown. It is desired to use a solid-steel flywheel of 10-inch radius to obtain the desired speed control.

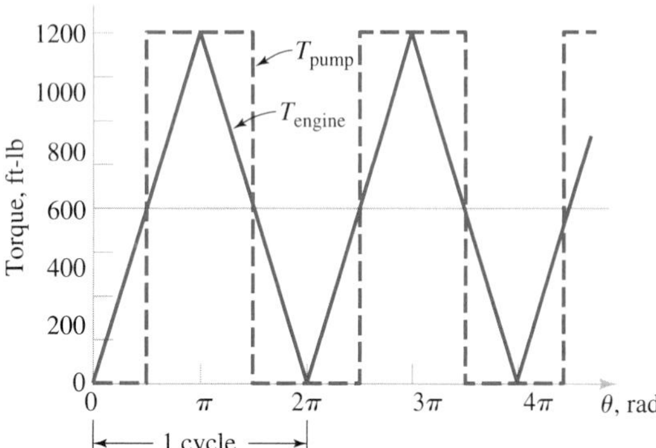

Figure P18.3
Torque versus angular displacement curve for an irrigation pump.

a. Sketch the angular velocity (qualitative) of the flywheel system as a function of angular displacement (quantitative), and identify points of maximum and minimum angular velocity on the torque versus angular displacement curve.

b. Calculate U_{max}.

c. Calculate the mass moment of inertia of the flywheel that would be required to properly control the speed.

d. Of what thickness should the flywheel be made?

18.4. A spoke-and-rim flywheel of the type shown in Figure 18.4(a) is made of steel, and each of the six spokes may be regarded as very stiff members. The mean diameter of the flywheel rim is 38.2 inches. The rim cross section is rectangular, 4 inches in the radial dimension and 2 inches in the axial dimension. The flywheel rotates counterclockwise at a speed of 2800 rpm.

a. Calculate your best estimate of the maximum bending stress generated in the rim.

b. At what critical sections in the rim would this maximum bending stress occur?

18.5. A spoke-and-rim flywheel of the type shown in Figure 18.4(a) has a mean rim diameter of 5 ft and rotates at 300 rpm. The rim cross section is 8 inches by 8 inches. During the duty cycle, the flywheel delivers energy at a constant rate over $^1/_4$ revolution, during which time the flywheel speed drops 10 percent. There are six spokes of elliptical cross section, with major axis twice the minor axis. The major axes of the elliptical spokes are parallel to the circumferential direction. The cast-iron material weighs 280 lb/ft^3 and has a design-allowable stress of 3000 psi.

a. Determine the required dimensions of the spoke cross section.

b. Estimate the hoop stress in the rim.

c. Estimate the bending stress in the rim.

d. Dimension the hub if a 1020 steel shaft is used and is sized only on the basis of transmitted torque.

18.6. A disk flywheel has a 600-mm outside diameter, 75-mm axial thickness, and is mounted on a 60-mm-diameter shaft. The flywheel is made of ultra-high-strength 4340 steel (see Tables 3.3, 3.4, and 3.5). The flywheel rotates at a speed of 10,000 rpm in a high-temperature chamber operating at a constant temperature of 425°C. Calculate the existing safety factor for this flywheel, based on yielding as the governing failure mode.

18.7. A disk-type flywheel, to be used in a punch press application with $C_f = 0.04$, is to be cut from a 1.50-inch-thick steel plate. The flywheel disk must have a central hole of 1.0-inch radius, and its mass moment of inertial must be 50 in-lb-sec^2.

a. What would be the maximum stress in the disk at 3600 rpm?

b. Where would this stress occur in the disk?

c. Would the state of stress at this critical point be multiaxial or uniaxial? Why?

d. Would low-carbon steel with a yield strength of 40,000 psi be an acceptable material for this application?

$$Hint\text{: } J = \frac{\pi t(r_o^4 - r_i^4)}{2g}$$

18.8. A steel disk-type flywheel, to be used in a V-belt test stand operating at 3000 rpm, must have a coefficient of speed fluctuation of $C_f = 0.06$. The flywheel has been analyzed in a preliminary way and it is proposed to use a constant-thickness disk, 3 inches thick, with a central hole of 2-inch radius and an outer radius of 10 inches. Further, it is desired to drill one small hole through the disk at a radius of 8 inches, as shown in Figure P18.8.

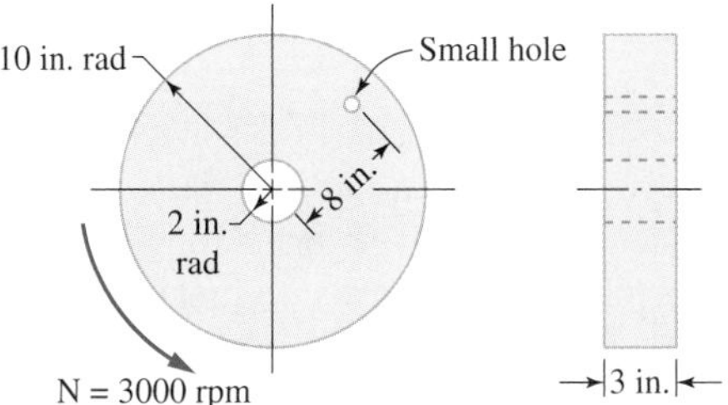

Figure P18.8
Disk flywheel with a small axial hole drilled through the disk at a radius of 8 inches.

a. At the location of the small hole, neglecting stress concentration, what would be the magnitude of the tangential stress?

b. At the location of the small hole, neglecting stress concentration, what would be the magnitude of the radial stress?

c. Would the state of stress at this location ($r = 8$ inches) be uniaxial or multiaxial? Why?

d. At the location of the small hole, *not* neglecting stress concentration, what would be a rough approximation of the magnitude of tangential and radial stress components at the edge of the small hole? (*Hint:* See Figure 4.22.)

18.9. A proposed constant-strength flywheel is to be 1.00 inch in axial thickness at the center of rotation, and 0.10 inch thick at its outer radius, which is 15.00 inches. If the material is AM 350 stainless steel, and the flywheel is operating in a 400°F ambient air environment, estimate the rotational speed in rpm at which yielding should initiate.

18.10. A constant-strength steel flywheel with a rim is being considered for an application in which the allowable stress of the flywheel material is 20,000 psi, the outer radius of the disk is 12 inches, the rim loading is 5000 lb per inch of circumference, and the flywheel rotates at 6000 rpm. Calculate the thickness of the flywheel web at radii of 0, 3, 6, 9, and 12 inches, and sketch the profile of the web cross section.

Chapter 19

Cranks and Crankshafts

19.1 Uses and Characteristics of Crankshafts

In applications where the functional design objective is to conceive a mechanism that can transform rotary motion into rectilinear motion, or the reverse, the *slider-crank mechanism*,[1] sketched in Figure 19.1(a), is a frequent choice. When considering the *kinematics*[2] of a slider-crank mechanism, or any other mechanism, it is usual to first establish *theoretical full-cycle motion capability*. This is done by proposing the types and locations of the joints, and the lengths of all the links. After establishing *theoretical* full-cycle motion capability for a proposed design configuration, it is very important to follow up by examining the potential for *topological interference* between and among the links in an operating mechanism. Topological interference, which is related to the ability of a rotating link to complete a full cycle of motion without interference, is a fundamental property of a linkage configuration, and cannot be avoided by simply reshaping the links. It is based on the physical fact that links cannot pass through each other. It becomes important, therefore, for a designer to investigate the three-dimensional consequences of topological interference, even for planar motion mechanisms. Figure 19.1(b) illustrates a three-dimensional em-

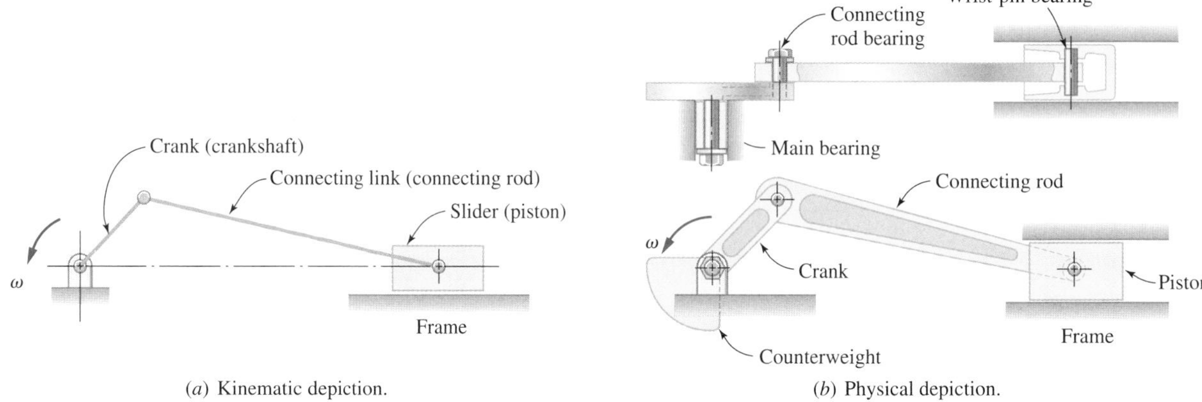

(*a*) Kinematic depiction.

(*b*) Physical depiction.

Figure 19.1
Kinematic and physical representations of a slider-crank mechanism.

[1]See, for example, ref. 1 or ref. 2.

[2]Kinematics is the study of positions, displacements, velocities, and accelerations of points, and the angular positions, angular displacements, angular velocities, and angular accelerations of solid bodies.

bodiment of the planar slider-crank device sketched in Figure 19.1(a) that avoids topological interference.

Concentrating on the link called the *crank*, or *crankshaft*, as depicted in Figure 19.1, the following discussion highlights a useful design procedure for configuring a successful crankshaft. Although textbooks rarely examine the detailed design of a crankshaft, this topic represents an excellent opportunity to integrate virtually all of the basic concepts presented in this text, including failure prevention principles, material selection, safety factor selection, geometry determination, manufacturing aspects, and assembly requirements. A discussion of crankshaft design is therefore included here.

19.2 Types of Crankshafts

Basically, crankshafts may be divided into two types: *side cranks* (overhung cranks) and *center cranks* (straddle-mounted cranks). A side crank variation called a *disk crank* is sometimes used, especially when the required crank radius, or "throw," is small. These variations are all sketched in Figure 19.2.

The advantages of a side crank configuration include geometric simplicity, relative ease of manufacture, relative ease of assembly, ability to use simple slide-on bearings, and relatively low cost. The advantages of a center crank configuration include good stability, balanced forces, and lower stresses, but cost is higher and a split connecting rod bearing is required for assembly. For applications that require properly phased multiple sliders (pistons), a multithrow crankshaft may be developed by placing several center cranks side-by-side, in sequence, along a common centerline of rotation. The throws are rotationally indexed to provide the desired phasing. Multicylinder internal combustion engines utilize such an arrangement.

All types of crankshafts are subjected to dynamic forces generated by the rotating eccentric mass center at each crank pin. Connecting rods and sliders also contribute *their* inertial effects to the resultant dynamic forces on the crank pins. It is usually necessary to utilize counterweights[3] and dynamic balancing to minimize shaking forces and rocking couples generated by these inertia forces. Details of dynamic balancing are beyond the scope of this brief discussion, but have been extensively presented in the literature.[4]

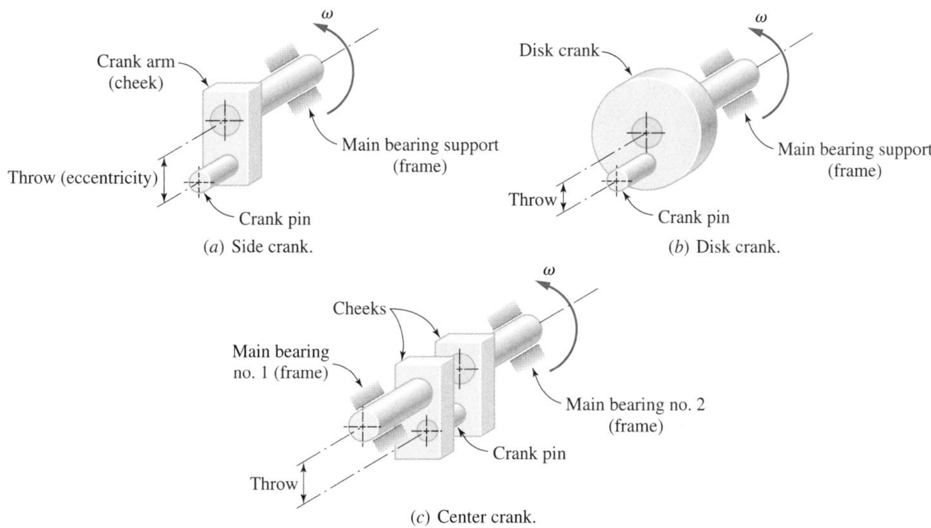

Figure 19.2
Types of crankshafts.

[3]See Figure 19.1(b). [4]See, for example, ref. 1 or ref. 2.

19.3 Potential Failure Modes

As shown in Figure 19.2, typical crankshafts are made up of structural elements (shafts, cheeks) and bearing elements (main bearings, crank pin bearings). Reviewing the list of potential failure modes given in 2.3, and in view of the stress concentration sites generated by a complex geometry and the cyclic nature of the loads acting on a rotating crankshaft, the structural elements are most susceptible to failure by fatigue, brittle fracture, or yielding. Although some crankshaft applications utilize rolling element bearings,[5] the use of plain bearings[6] is more common. Plain bearing elements are most susceptible to failure by adhesive wear, abrasive wear, corrosive wear, surface fatigue wear, galling, seizure, or, in some applications, yielding or brittle fracture. The most probable of these potential failure modes depends upon the specific requirements of each application.

19.4 Crankshaft Materials

Crankshaft materials must be selected to meet both the structural strength requirements and the bearing-site wear requirements. Plain bearing material candidates have been discussed in 10.4, where it was noted that most successful bearing applications embody a bearing pair consisting of a hard material sliding on a replaceable element of soft ductile material. In the typical crankshaft application, soft, ductile sleeves are attached to the connecting rod or the frame, so the crankshaft material must have the ability to provide a hard surface at the bearing sites. Structural strength requirements may be met by many materials, but providing wear resistance at the bearing sites narrows the list of acceptable candidates. Because of the asymmetric geometry, many crankshafts have been manufactured by casting or forging a "blank," to be finish-machined later. Built-up weldments are used in some applications. Traditionally, cast iron, cast steel, and wrought steel have been used for crankshafts. The use of selectively carburized and hardened bearing surfaces is also common. The methods of Chapter 3 should be utilized to select appropriate material candidates for any given crankshaft application.

19.5 Summary of Suggested Crankshaft Design Procedure

Based on the discussions of 19.1 through 19.4, the procedural steps in developing a successful crankshaft design may be summarized as follows:

1. Based on functional specifications and contemplated system configuration, generate a first conceptual sketch for the crankshaft.
2. For the initially proposed configuration, determine displacements, velocities, and accelerations of importance for each (critical) kinematic *phase*[7] of the overall mechanism, throughout a full cycle.
3. Conceive a tentative basic shape for the crankshaft in view of its function and constraints.
4. Perform a global force analysis including all surface and body forces on the mechanism at all (critical) kinematic phases.

[5]See Chapter 11. [6]See Chapter 10.

[7]When the parts of a mechanism have passed through all the possible positions that they can assume after starting from a specified set of relative positions, and have returned to their original relative positions, they have completed a *cycle* of motion. The simultaneous relative positions of a cycle constitute a *phase*.

5. Taking the *crankshaft* as a *free body*, calculate and display all forces acting during each (critical) phase.
6. Select potential critical points in the proposed crankshaft configuration (see 6.3), and perform local force analyses to determine forces and moments acting at each critical section.
7. Identify potential failure modes for the proposed crankshaft (see 19.3).
8. Select a tentative crankshaft material, and identify probable manufacturing processes (see 19.4).
9. Select an appropriate design safety factor (see Chapter 5).
10. Calculate design stresses appropriate to each potential failure mode identified in step 7.
11. Employ appropriate stress and failure prevention calculations for all critical phases and all critical points, iterating until critical dimensions are found that will meet functional design specifications and provide acceptable operating life.
12. Sketch the updated crankshaft configuration so that further refinements may be considered.

As for any design project, certain simplifying assumptions may be required to expedite the calculations. Initial assumptions are often made with respect to forces acting, critical phases, critical sections, critical points, stresses at the critical points, governing failure modes, material properties, or others. Also, see 8.7 for additional useful guidelines.

Example 19.1 Crankshaft Design

It is desired to design a crankshaft for a newly proposed single-piston belt-driven compressor. A preliminary design study has been reported to engineering management suggesting the following design specifications and guidelines:

a. Piston force P due to gas pressure, and including dynamic effects, will range from 3000 lb down on the crankshaft at the connecting rod bearing site to 1000 lb up at the same location.
b. Piston "stroke" should be 5.0 inches (i.e., *crank throw* should be 2.5 inches).
c. Because of stability advantages and life requirements, a straddle-mounted crankshaft is suggested.
d. A preliminary bearing study indicates that an allowable bearing pressure of $\sigma_{w-allow}$ = 500 psi should provide an acceptable wear life for this application.
e. For reasons of standardization, inventory control, and cost reduction, it is desired to make all bearings alike.
f. The proposed V-belt drive pulley is to have a pitch diameter of 8.0 inches.
g. It is known from experience[8] that the ratio of T_t to T_s should be about 10.
h. To accommodate the space envelopes of adjacent components, preliminary centerline spacings have been established as shown in the sketch of Figure E19.1A.

Propose a first-iteration crankshaft configuration, complete with dimensions, that will satisfy these design requirements and guidelines.

Solution

Following the steps suggested in 19.5, a design proposal for the crankshaft may be developed as follows:

[8] Also see Chapter 17.

Example 19.1 Continues

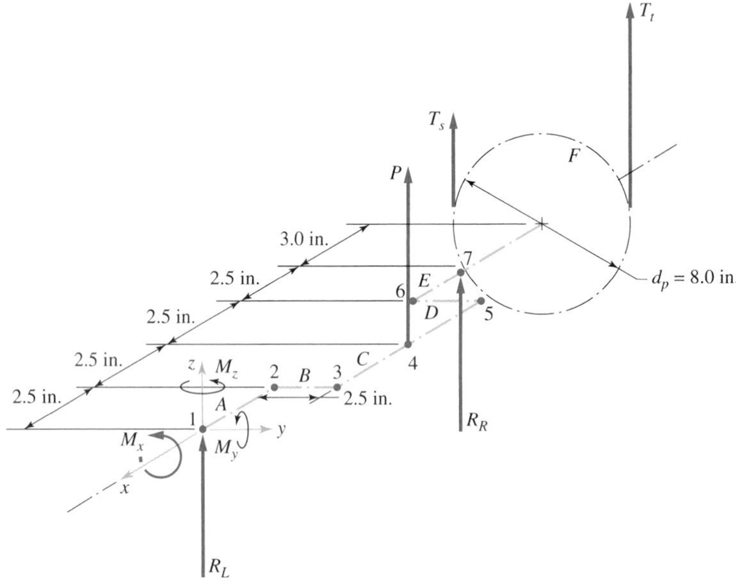

Figure E19.1A
Preliminary sketch of proposed compressor crankshaft configuration, including centerline location dimensions and externally applied forces.

1. Based on the design specifications and contemplated system configuration, a preliminary conceptual sketch may be prepared, as shown in Figure E19.1A.
2. To temporarily avoid the need for a complete kinematic analysis for the proposed compressor, it will be *assumed* (conservatively) that the maximum operating force on the piston acts vertically on the crankshaft when the crankshaft is in its most vulnerable phase. The most vulnerable phase is *assumed* to be the phase for which the axial centerlines of main bearings and crank pin all lie in a common horizontal plane. This phase has been sketched in Figure E19.1A. Later iterations should incorporate a more accurate kinematic analysis.
3. A tentative basic shape for the crankshaft, based on the preliminary design data embodied in the sketch in Figure E19.1A, is shown in Figure E19.1B.

Figure E19.1B
Tentative basic crankshaft shape.

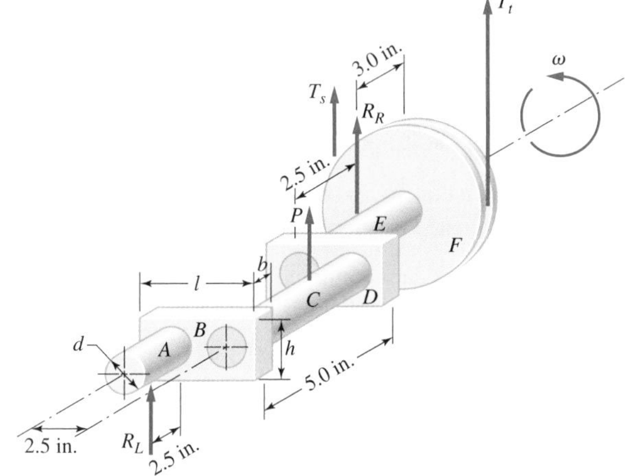

4. A global force analysis is not required for this particular case since sufficient data are provided by the specifications and the assumptions of step 2.
5. A local force analysis for the crankshaft taken as a free body, utilizing Figures E19.1A and E19.1B, may next be conducted. Using the coordinate system defined in Figure E19.1A, the static equilibrium equations of (4-1) may be adapted to this particular crankshaft design scenario to give, based on force equilibrium,

$$\sum F_x \equiv 0 \qquad \text{(identically satisfied)} \tag{1}$$

$$\sum F_y \equiv 0 \qquad \text{(identically satisfied)} \tag{2}$$

and by summing forces in the z-direction,

$$R_L + P + R_R + T_t + T_s = 0 \tag{3}$$

Next, examining moment equilibrium,

$$\sum M_z \equiv 0 \qquad \text{(identically satisfied)} \tag{4}$$

and by summing moments about the x- and y-axes, respectively,

$$2.5P + 4T_t - 4T_s = 0 \tag{5}$$

and

$$10R_R + 5P + 13(T_t + T_s) = 0 \tag{6}$$

Also, by specification,

$$T_t = 10T_s \tag{7}$$

Since P is a known force (maximum value of 3000 lb by specification and vertically downward on crank pin by assumption), the *four* independent equations (3), (5), (6), and (7) may be utilized to solve for the *four unknown forces* R_L, R_R, T_s, and T_t. Substituting $P = -3000$ into (3), (5), (6), and (7) gives

$$R_L + (-3000) + R_R + T_t + T_s = 0 \tag{8}$$

$$2.5(-3000) + 4T_t - 4T_s = 0 \tag{9}$$

$$10R_R + 5(-3000) + 13(T_t + T_s) = 0 \tag{10}$$

$$T_t = 10T_s \tag{11}$$

Substituting (11) into (8), (9), and (10), and solving them simultaneously gives

$$\begin{aligned} T_s &= 208 \text{ lb} \\ T_t &= 2080 \text{ lb} \\ R_R &= -1474 \text{ lb (hence, direction is down)} \\ R_L &= 2186 \text{ lb} \end{aligned} \tag{12}$$

To clarify the results of this force analysis, Figure E19.1C may be prepared to show the magnitudes, directions, and locations of all forces on the crankshaft at its most vulnerable phase.
6. The initial selection of potential critical points in the crankshaft is a function of designer judgment and experience.[9] Tentatively, seven potentially critical sections have been selected for this crankshaft configuration. These potential critical sections have been designated 1 through 7, as shown in Figure E19.1C. Referring to the tentative

[9]Critical points are points within a machine part that have high potential for failure because of high stresses or strains, low strength, or a critical combination of these (see 6.3).

Example 19.1 Continues

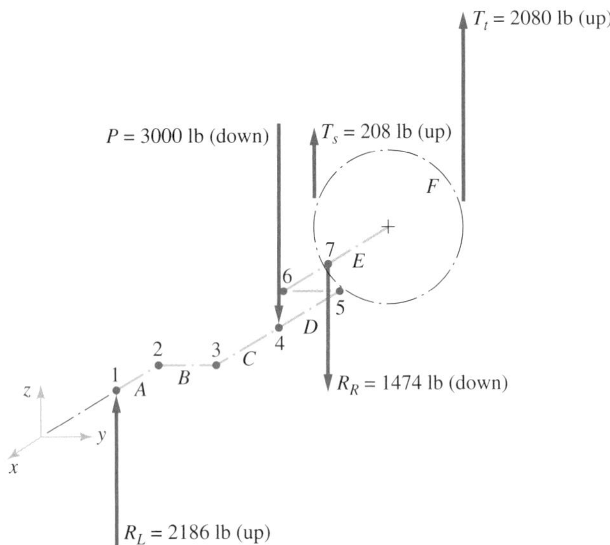

Figure E19.1C
Graphic summary of results from the overall force analysis.

geometry proposed in Figure E19.1B, it is clear that at many of the selected critical sections two different geometric shapes must be examined. For example, at critical section 2, the circular cross section of element *A* and rectangular cross section of element *B* must both be investigated to determine appropriate dimensions. It is also important to notice that the external forces may induce transverse shear, torsion, and/or bending in any of the crankshaft elements *A* through *E*. The basis for critical-section selection in this example has been to assume that a potential critical section exists at each location where either a load is applied or a geometric transition takes place.

Noting from Figure E19.1C that there are no forces in the x- or y-directions, and no moments about the z-axis,[10] the local forces and moments at each of the seven selected critical sections may be calculated on the basis of static equilibrium. Examining Figure E19.1B and E19.1C together, the local force analyses produce the results summarized in Table E19.1A. To determine tentative dimensions for each element at each critical section, calculations may be made for each of the 11 potential critical sections listed in Table E19.1A. However, calculational detail may be reduced by the following logic. By specification, all bearings are to be made the same. Therefore, by reviewing the relative magnitudes of the forces and moments on each circular element tabulated in Table E19.1A, it may be deduced that sections 4C, 6E, and 7E are more critical than any of the other *circular* critical sections. If the rectangular cheeks are also made alike, 3B and 6D are more critical than any of the other *rectangular* sections.

Summarizing, the 11 potential critical sections of Table E19.1A may be reduced to 5: circular elements 4C, 6E, and 7E, and rectangular elements 3B and 6D.

7. Based on the discussion of 19.3, the failure modes to be investigated here should include *wear*, *fatigue*, and *yielding*.
8. Selection of material candidates for this crankshaft application may be expedited by implementing the material selection methods discussed in Chapter 3. As a matter of fact, the detailed material selection process for a crankshaft similar to the one under consideration here has been presented in Example 3.1 of Chapter 3. For the crankshaft under consideration here, based on the methods of Example 3.1, the tentative mate-

[10]See also equations (1), (2), and (4).

TABLE E19.1A Local Forces and Moments[1] at the Seven Selected Critical Sections

Critical Section	Element	Section Shape	$F_{z\text{-}ts}$, lb	M_t, in-lb	M_b, in-lb
1	A	⊕	2186	0	0
2	A	⊕	2186	0	5465
2	B	⊞	2186	5465	0
3	B	⊞	2186	5465	5465
3	C	⊕	2186	5465	5465
4	C	⊕	2186	5465	10,930
5	C	⊕	−814	5465	8895
5	D	⊞	−814	8895	5465
6	D	⊞	−814	8895	7500
6	E	⊕	−814	7500	8895
7	E	⊕	−2214	7500	6860

[1]Note that $F_{z\text{-}ts}$ is transverse shear force on the element in the z-direction, M_t is torsional moment on the element, and M_b is bending moment on the element. The sign convention is as depicted in Figure E19.1A.

rial selection will be forged 1020 steel, case hardened at the bearing sites.[11] The basic core properties for 1020 steel are, from Tables 3.3 and 3.10,

$$S_u = 61{,}000 \text{ psi}$$
$$S_{yp} = 51{,}000 \text{ psi}$$
$$e(2 \text{ in}) = 15 \text{ percent}$$

9. Determination of a design safety factor for this crankshaft application may be based on Chapter 5, and in particular, Example 5.1. For the crankshaft application under consideration here, the selected values for the rating factors[12] are given in Table E19.1B.

From (5-3)

$$t = -1 + 0 + 0 - 2 + 2 + 0 - 2 + 0 = -3 \tag{13}$$

and, since $t \geq -6$, from (5-4) the desired design safety factor may be calculated as

$$n_d = 1 + \frac{(10 - 3)^2}{100} = 1.5 \tag{14}$$

10. To calculate design stresses related to each of the three failure modes listed in step 7, the crankshaft material properties from step 8 and the safety factor from step 9 may

[11]Consultation with a materials engineer is suggested for help in choosing a suitable case-hardening method and specifying appropriate temperatures and times required to develop the desired properties of both case and core.

[12]See 5.2.

Example 19.1 Continues

TABLE E19.1B Rating Factors for Crankshaft Design Application

Rating Factor	Selected Rating Number (RN)
1. Accuracy of loads knowledge	−1
2. Accuracy of stress calculation	0
3. Accuracy of strength knowledge	0
4. Need to conserve	−2
5. Seriousness of failure consequences	+2
6. Quality of manufacture	0
7. Conditions of operation	−2
8. Quality of inspection and maintenance	0

be combined to calculate the wear-based design stress $(\sigma_w)_d$, the fatigue-based design stress $(\sigma_{max-N})_d$, and the yielding-based design stress $(\sigma_{yp})_d$.

The *wear-based* design stress is, by specification,

$$(\sigma_w)_d = 500 \text{ psi} \tag{15}$$

The *fatigue-based* design stress will be predicated upon infinite-life properties[13] so, utilizing the procedure for estimating infinite-life fatigue strength,[14]

$$S_{N=\infty} = S_e = 0.5S_u = 0.5(61{,}000) = 30{,}500 \text{ psi} \tag{16}$$

the fatigue-based design stress may be estimated, using (2-40), as[15]

$$(\sigma_{max-N})_d = \left(\frac{1}{n_d}\right)\left(\frac{(S_e/K_f)}{1 - m_t R_t}\right) \tag{17}$$

where

$$m_t = \frac{S_u(S_e/K_f)}{S_u} \tag{18}$$

and

$$R_t = \frac{\sigma_m}{\sigma_{max}} \tag{19}$$

By specification, $P_{max} = 3000$ lb and $P_{min} = -1000$ lb, so

$$R_t = \frac{\sigma_m}{\sigma_{max}} = \frac{P_m}{P_{max}} = \frac{\left[\dfrac{3000 + (-1000)}{2}\right]}{3000} = 0.33 \tag{20}$$

and arbitrarily selecting a trial value of $K_f = 2$,

$$m_t = \frac{61{,}000 - (30{,}500/2)}{61{,}000} = 0.75 \tag{21}$$

so (17) may be evaluated as

[13] A reasonable first assumption, although not stated explicitly in the specifications.

[14] See "Estimating *S-N* Curves" in 2.6.

[15] Note that an *estimated* strength-reduction factor [see discussion following (4-111)], K_f, may be arbitrarily included to make this first iteration more nearly reflect stress concentration in the final geometry. After initial dimensions are established, Figure 4.24 may be used to improve the estimate of K_f.

$$(\sigma_{max-N})_d = \left(\frac{1}{1.5}\right)\left(\frac{(30{,}500/2)}{1-(0.33)(0.75)}\right) = 13{,}510 \text{ psi} \tag{22}$$

Finally, the *yielding-based* design stress is

$$(\sigma_{yp})_d = \left(\frac{1}{1.5}\right)(51{,}000) = 34{,}000 \text{ psi} \tag{23}$$

11. Next, tentative dimensions will be determined by estimating critical stresses and establishing dimensions so that none of the design stresses calculated in step 10 are exceeded.

a. Based on *wear*, where A_p is projected bearing area,[16] the maximum operating bearing pressure may be calculated as

$$\sigma_{w-max} = \frac{P_{max}}{A_p} \tag{24}$$

Selecting P_{max} to be the force on the most heavily loaded bearing (crank pin bearing), setting σ_{w-max} equal to the wear-based design stress, and solving for A_p,

$$A_p = \frac{P_{max}}{(\sigma_w)_d} = \frac{3000}{500} = 6 \text{ in}^2 \tag{25}$$

It is common to specify a "square" bearing,[17] so

$$A_p = dl = d^2 = l^2 \tag{26}$$

Combining (25) and (26), and solving for diameter,

$$l = d = \sqrt{6} = 2.45 \text{ inches} \tag{27}$$

Tentatively, therefore, it will be proposed that both of the main bearings and the crank pin bearing have a diameter of 2.5 inches and a bearing length of approximately 2.5 inches.

b. Comparing (22) with (23), the fatigue-based design stress is *more* critical (lower) than the yielding-based design stress. Dimensions based on fatigue will therefore automatically satisfy yielding[18] requirements. Equation (22) will be used as the strength basis for this design.

Using the fatigue-based design stress from (22), dimensions may be established at each of the five more-probable critical sections summarized in step 6, namely, circular elements 4C, 6E, and 7E, and rectangular elements 3B and 6D.

At circular section 4C, using data from Table E19.1A, the maximum values of transverse shear, torsion, and bending stresses may be calculated for a 2.5-inch-diameter circular section[19] as

$$\tau_{ts} = \frac{4}{3}\left(\frac{F_{z-ts}}{A}\right) = \frac{4}{3}\left(\frac{2186}{\frac{\pi(2.5)^2}{4}}\right) = 595 \text{ psi} \tag{28}$$

$$\tau_{tor} = \frac{M_t a}{J} = \frac{(5465)(1.25)}{\left(\frac{\pi(2.5)^4}{32}\right)} = 1780 \text{ psi} \tag{29}$$

[16]Also see, for example, (9-4) and (10-2). [17]Terminology for a plain bearing with length equal to diameter.

[18]Such might not be the case if occasional high loads occurred in the loading spectrum.

[19]The value of 2.5 inches diameter is used as a starting point since the wear-based dimensions must be 2.5 inches or larger. See 4.4 for appropriate stress equations.

Example 19.1 Continues

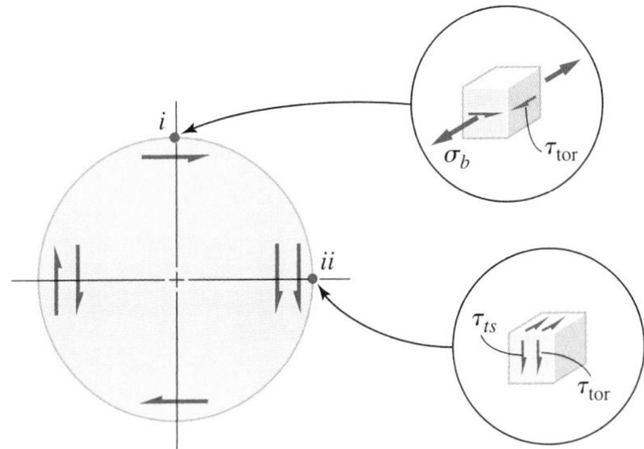

Figure E19.1D
Sketch showing states of stress at critical points in circular critical sections 4C, 6E, and 7E.

and

$$\sigma_b = \frac{M_b c}{I} = \frac{(10{,}930)(1.25)}{\left(\dfrac{\pi(2.5)^4}{64}\right)} = 7125 \text{ psi} \tag{30}$$

Based on the discussions of 4.3, the locations of each of these maximum stresses may be identified as shown in Figure E19.1D. Thus at critical point i of section 4C, bending and torsion combine, and at critical point ii torsion and transverse shear combine. Using (4-96) the *equivalent stress*, or *von Mises stress*, may be calculated at point i as

$$(\sigma_{eq})_i = \sqrt{\frac{1}{2}\left[(\sigma_1 - \sigma_2)^2 + (\sigma_2 - \sigma_3)^2 + (\sigma_3 - \sigma_1)^2\right]} \tag{31}$$

Using stress cubic equation (4-59), the principal stresses for critical point i of section 4C may be calculated using (29) and (30), giving

$$\sigma^3 - \sigma^2(\sigma_b) + \sigma(-\tau_{tor}^2) = 0 \tag{32}$$

or

$$\sigma_1 = 7540 \text{ psi} \tag{33}$$

$$\sigma_2 = 0 \tag{34}$$

$$\sigma_3 = -420 \text{ psi} \tag{35}$$

and (31) becomes

$$(\sigma_{eq})_i = \sqrt{\frac{1}{2}[(7540)^2 + (420)^2 + (7960)^2]} = 7760 \text{ pri} \tag{36}$$

Comparing (36) with (22), it is clear that critical point i is well below the design stress, and the wear-based diameter of 2.5 inches is therefore also acceptable for fatigue and yielding as potential failure modes.

Likewise, for critical point ii of section 4C, (4-59) gives

$$\sigma^3 - \sigma^2(0) + \sigma[-(\tau_{tor} + \tau_{ts})^2] = 0 \tag{37}$$

or

$$\sigma_1 = 2375 \text{ psi} \tag{38}$$

$$\sigma_2 = 0 \tag{39}$$

$$\sigma_3 = -2375 \text{ psi} \tag{40}$$

and (31) becomes, for the critical point *ii*,

$$(\sigma_{eq})_{ii} = \sqrt{\frac{1}{2}[(2375)^2 + (2375)^2 + (-4750)^2]} = 4115 \text{ psi} \tag{41}$$

Comparing (41) with (36), it is clear that critical point *i* governs for section 4C, so, as already observed, the wear-based diameter of 2.5 inches is very acceptable.

Following a similar procedure for critical section 6E, using data from Table E19.1A,

$$\tau_{ts} = \frac{4}{3}\left(\frac{814}{\frac{\pi(2.5)^2}{4}}\right) = 220 \text{ psi} \tag{42}$$

$$\tau_{tor} = \frac{(7500)(1.25)}{\left(\frac{\pi(2.5)^4}{32}\right)} = 2445 \text{ psi} \tag{43}$$

$$\sigma_b = \frac{(8895)(1.25)}{\left(\frac{\pi(2.5)^4}{64}\right)} = 5800 \text{ psi} \tag{44}$$

The reduced version of the stress cubic equation given in (32) is also valid for critical point *i* of section 6E, giving solutions

$$\sigma_1 = 6695 \text{ psi} \tag{45}$$

$$\sigma_2 = 0 \tag{46}$$

$$\sigma_3 = -895 \text{ psi} \tag{47}$$

whence, again using (31) for section 6E at critical point *i*,

$$(\sigma_{eq})_i = \sqrt{\frac{1}{2}[(6695)^2 + (895)^2 + (-7590)^2]} = 7185 \text{ psi} \tag{48}$$

Comparing (48) with (22), it may be observed that critical point *i* of section 6E is well below the design stress, and again the wear-based diameter of 2.5 inches governs the design.

Using (37) to determine principal stresses at critical point *ii* of section 6E,

$$\sigma_1 = 2665 \text{ psi} \tag{49}$$

$$\sigma_2 = 0 \tag{50}$$

$$\sigma_3 = -2665 \text{ psi} \tag{51}$$

and using (31) for critical point *ii* of section 6E,

$$(\sigma_{eq})_{ii} = \sqrt{\frac{1}{2}[(2665)^2 + (2665)^2 + (-5330)^2]} = 4615 \text{ psi} \tag{52}$$

As before, the 2.5-inch diameter dictated by wear requirements results in very acceptable states of stress at section 6E.

By similar reasoning for section 7E, using data from Table E19.1A,

$$\tau_{ts} = \frac{4}{3}\left(\frac{2214}{\left(\frac{\pi(2.5)^2}{4}\right)}\right) = 600 \text{ psi} \tag{53}$$

$$\tau_{tor} = \frac{(7500)(1.25)}{\left(\frac{\pi(2.5)^4}{32}\right)} = 2445 \text{ psi} \tag{54}$$

Example 19.1 Continues

$$\sigma_b = \frac{(6860)(1.25)}{\left(\frac{\pi(2.5)^4}{64}\right)} = 4470 \text{ psi} \tag{55}$$

giving principal stresses calculated by solving (32) at critical point i,

$$\sigma_1 = 5550 \text{ psi} \tag{56}$$

$$\sigma_2 = 0 \tag{57}$$

$$\sigma_3 = -1080 \text{ psi} \tag{58}$$

and again using (31)

$$(\sigma_{eq})_i = \sqrt{\frac{1}{2}[(5550)^2 + (1080)^2 + (-6630)^2]} = 6160 \text{ psi} \tag{59}$$

This too is acceptable

At critical point ii of section 7E the principal stresses are, from (37),

$$\sigma_1 = 3045 \text{ psi} \tag{60}$$

$$\sigma_2 = 2 \tag{61}$$

$$\sigma_3 = -3045 \text{ psi} \tag{62}$$

and (31) gives

$$(\sigma_{eq})_{ii} = \sqrt{\frac{1}{2}[(3045)^2 + (3045)^2 + (-6090)^2]} = 5275 \text{ psi} \tag{59}$$

This being acceptable, the preliminary conclusion is that all circular section critical points are safe from the standpoint of potential failure by either fatigue or yielding if the wear-based diameter of 2.5 inches is used for all circular sections. It may be noted in passing that if weight-saving were important, these bearing cross sections could be "hollowed out" without much loss of strength.[20]

For the critical points in rectangular sections 3B and 6D, identified in step 6 as the more critical rectangular sections, wear is not a factor; the *fatigue design stress* from (22) may be used as the basis for determining rectangular section dimensions.

At rectangular section 3B, illustrated in Figure E19.1E, and using data from Table E19.1A, the maximum values of transverse shear, torsion, and bending stresses may be calculated for the rectangular cross section from[21]

$$\tau_{ts} = \frac{3}{2}\left(\frac{F_{z\text{-}ts}}{A}\right) = \frac{3}{2}\left(\frac{2186}{bh}\right) \tag{64}$$

$$\tau_{tor} = \frac{M_t}{Q} = \frac{5465}{\left(\frac{0.5h^2b^2}{1.5h + 0.9b}\right)} \tag{65}$$

and

$$\sigma_b = \frac{M_b}{\left(\frac{I}{c}\right)} = \frac{5465}{\left(\frac{bh^2}{6}\right)} \tag{66}$$

[20]See 6.2.

[21]See 4.4 for appropriate stress equations. Also note that the expression for Q is derived from Table 4.4, case 3, by setting $a = h/2$ and $b = b/2$, where b and h are defined in figure E19.1E.

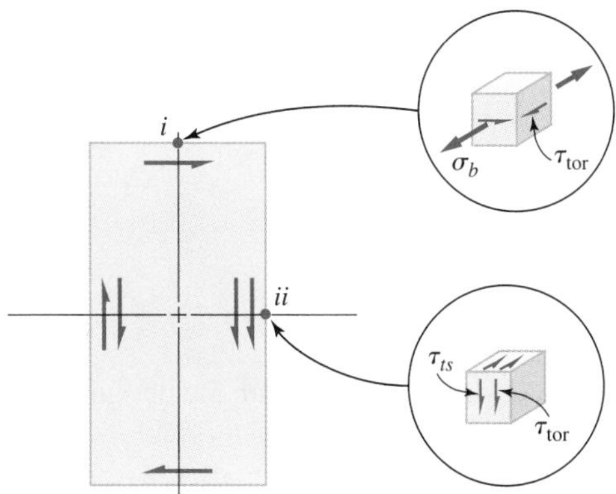

Figure E19.1E
Sketch showing states of stress at critical points in rectangular critical sections 3B and 6D.

Before proceeding further, minor design decisions must be made relative to the *aspect ratio*, *b/h*, of the rectangular cheek section. Referring to Figure E19.1B, it is reasonable to make cheek dimension h a little larger than the bearing diameter d so that a generous "blend" radius may be incorporated at the bearing-to-cheek transition. Since $d = 2.5$ inches, cheek dimension h will be arbitrarily chosen as

$$h = 3.0 \text{ inches} \tag{67}$$

Experience suggests that aspect ratios less than about 0.5 are often vulnerable to failure by buckling. As a first iteration on cheek dimensions, the aspect ratio will be arbitrarily chosen as

$$\frac{b}{h} = 0.5 \tag{68}$$

Using (67) and (68), the cheek width b may then be calculated as

$$b = 0.5(3.0) = 1.5 \text{ inches} \tag{69}$$

The stresses of (64), (65), and (66) may then be explicitly calculated as

$$\tau_{ts} = \frac{3}{2}\left(\frac{2186}{(1.5)(3.0)}\right) = 730 \text{ psi} \tag{70}$$

$$\tau_{tor} = \frac{5465}{\left(\dfrac{0.5(3.0)^2(1.5)^2}{1.5(3.0) + 0.9(1.5)}\right)} = 3160 \text{ psi} \tag{71}$$

and

$$\sigma_b = \frac{5465}{\left(\dfrac{1.5(3.0)^2}{6}\right)} = 2430 \text{ psi} \tag{72}$$

The locations of each of these maximum stresses may be identified as shown in Figure E19.1E.

Thus, at critical point *i* of section 3B, bending and torsion combine, and at critical point *ii* torsion and transverse shear combine.[22] Using the reduced stress cubic equation of (32), the principal stresses for critical point *i* of section 3B may be calculated as

[22]Actually $\tau_{tor} = 3160$ psi, as calculated in (71), is valid only for critical point *ii*. The torsional shear stress at critical point *i* will be smaller (see Figure 4.10). It will be (conservatively) assumed for this calculation that $\tau_{tor} = 3165$ psi is valid for both critical points *i* and *ii*.

Example 19.1 Continues

$$\sigma_1 = 4605 \text{ psi} \tag{73}$$

$$\sigma_2 = 0 \tag{74}$$

$$\sigma_3 = -2175 \text{ psi} \tag{75}$$

and again utilizing (31)

$$(\sigma_{eq})_i = \sqrt{\frac{1}{2}[(4605)^2 + (2175)^2 + (-6780)^2]} = 6000 \text{ psi} \tag{76}$$

This is acceptable when compared with the design stress $(\sigma_{max-N})_d = 13{,}510$ psi given in (22).

Likewise, for critical point *ii* of section 3B, (32) may be used to determine the principal stresses as

$$\sigma_1 = 3890 \text{ psi} \tag{77}$$

$$\sigma_2 = 0 \tag{78}$$

$$\sigma_3 = -3890 \text{ psi} \tag{79}$$

and (31) gives

$$(\sigma_{eq})_{ii} = \sqrt{\frac{1}{2}[(3890)^2 + (3890)^2 + (-7780)^2]} = 6740 \text{ psi} \tag{80}$$

again acceptable when compared to (22).

Finally for rectangular section 6D, using data from Table E19.1A,

$$\tau_{ts} = \frac{3}{2}\left(\frac{814}{(1.5)(3.0)}\right) = 270 \text{ psi} \tag{81}$$

and

$$\tau_{tor} = \frac{8895}{\left(\dfrac{0.5(3.0)^2(1.5)^2}{1.5(3.0) + 0.9(1.5)}\right)} = 5135 \text{ psi} \tag{82}$$

and

$$\sigma_b = \frac{7500}{\left(\dfrac{1.5(3.0)^2}{6}\right)} = 3335 \text{ psi} \tag{83}$$

The principal stresses at critical point i of section 6D, using (32), are

$$\sigma_1 = 7070 \text{ psi} \tag{84}$$

$$\sigma_2 = 0 \tag{85}$$

$$\sigma_3 = -3730 \text{ psi} \tag{86}$$

which gives

$$(\sigma_{eq})_i = \sqrt{\frac{1}{2}[(7070)^2 + (3730)^2 + (-10{,}800)^2]} = 9500 \text{ psi} \tag{87}$$

For critical point *ii* of section 6D,

$$\sigma_1 = 5405 \text{ psi} \tag{88}$$

$$\sigma_2 = 0 \tag{89}$$

$$\sigma_3 = -5405 \text{ psi} \tag{90}$$

TABLE E19.1C Resultant Stresses at Selected Crankshaft Critical Points

Critical Section[1]	Section Shape	Critical Point[2]	Resultant *Equivalent* Stress (von Mises stress), psi
4C		*i*	7760
4C		*ii*	4115
6E		*i*	7185
6E		*ii*	4615
7E		*i*	6160
7E		*ii*	5275
3B	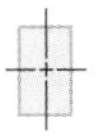	*i*	6000
3B	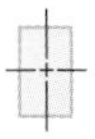	*ii*	6740
6D	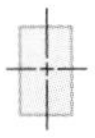	*i*	9500
6D	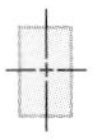	*ii*	9360

[1]See Figure E19.1C.
[2]See Figures E19.1D and E19.1E.

which gives

$$(\sigma_{eq})_{ii} = \sqrt{\frac{1}{2}[(5405)^2 + (5405)^2 + (-10{,}810)^2]} = 9360 \text{ psi} \tag{91}$$

From the above calculations, a summary table of equivalent (von Mises) stresses at the critical points selected in step 6 may be prepared as shown in Table E19.1C.

It may be noted from Table E19.1C that all critical point stresses are safely below the fatigue-based design stress of 13,510 psi, and from step 11, bearing wear specifications are also met.

12. Based on the specifications and tentative dimensions just determined, the crude sketch of Figure E19.1B may now be transformed into a more refined first-iteration sketch, as shown in Figure E19.1F.

Having completed this first-iteration proposal, the next step would be to submit the proposal to a *design review*, then incorporate appropriate changes arising from the review. Next, it would be necessary to go through the entire design calculation procedure again, using more accurate stress analyses, stress concentration factors, and materials properties; incorporating manufacturing specifications; and paying attention to production costs. The process should typically be repeated until management is satisfied. Finally, prototype testing should be conducted.

Example 19.1 Continues

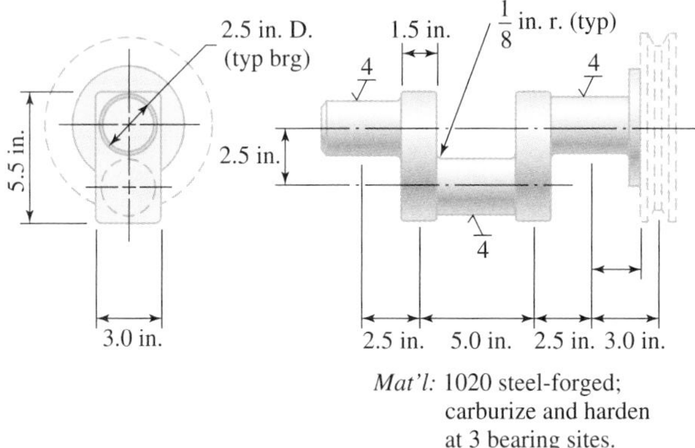

Figure E19.1F
Refined first-iteration sketch of proposed crankshaft configuration. Compare with Figure E19.1B.

Problems

19-1. The cylindrical bearing journal of an overhung crankshaft has been sized for wear and found to require a diameter of 1.50 inches based on a wear analysis. A force analysis of the journal has shown there to be a transverse shear force of 10,000 lb, torsional moment of 9000 in-lb, and bending moment of 8000 in-lb at the critical cross section of the cylindrical journal bearing. If the fatigue-based design stress governs, and has been found to be 39,000 psi, calculate whether the 1.5-inch-diameter journal is safely designed to withstand the fatigue loading.

19-2. The overhung crankshaft shown in Figure P19.2A is supported by bearings at R_1 and R_2, and loaded by P on the crankpin, vertically, as shown. The crank position shown may be regarded as the most critical position. In this critical position, load P ranges from 900 lb up to 1800 lb down. The material properties are given in Figure P19.2B. Based on wear estimates, all cylindrical bearing diameters should be 0.875 inch. Neglecting stress concentration effects, and using a safety factor of 1.5, determine whether the diameter of 0.875 inch at R_1 is adequate if infinite life is desired.

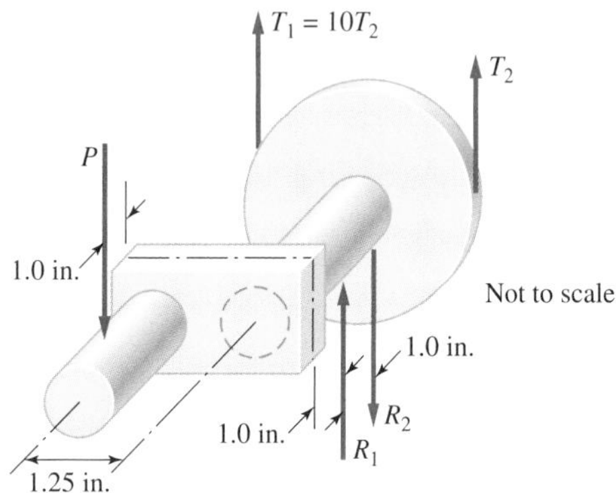

Figure P19.2A
Overhung crankshaft.

19-3. The overhung crankshaft shown in Figure P19.3 is supported on bearings R_1 and R_2 and loaded by force P on the crank pin. This is taken to be the most critical crank position. The load P ranges from 1500 lb up to 1500 lb down. It has been decided to use forged and carburized AISI 4620 steel for the crankshaft material (see Tables 3.3, 3.10). The allowable bearing design pressure, based on wear, is 750 psi. It has been determined that wear is the governing failure mode for bearing journal A. A fillet radius of 0.17-inch is desired where cylindrical journal A blends into rectangular cheek B. The ratio w/h is to be 0.5 (see Figure P19.3). A safety factor of 3 is to be used for all failure modes except wear. For wear, the safety factor is already included in the bearing design pressure as given.

a. For cheek member B, determine the governing design stress.

b. Assuming a "square" bearing, determine diameter and length of journal A.

c. Based on results of (b) and other pertinent data given, find rectangular cross-sectional dimensions, w and h, for cheek B.

d. Identify the most critical section of cheek B, and the critical point location(s) within the critical cross section.

e. For each critical point identified in (d), specify the types of stress acting.

f. Calculate all pertinent forces and moments for these critical points.

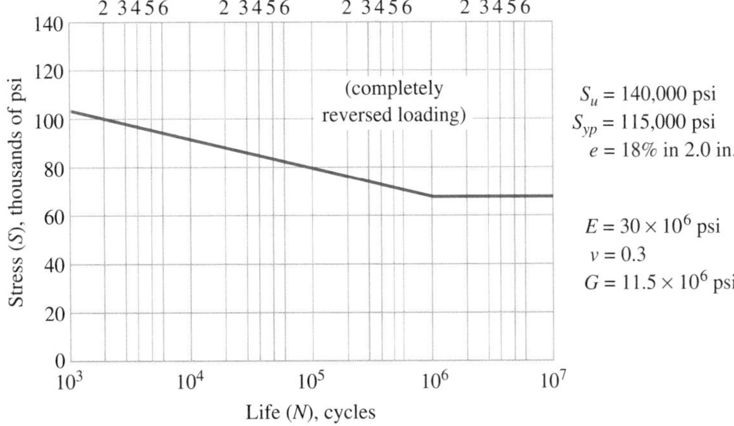

Figure P19.2B
Material properties for overhung crankshaft.

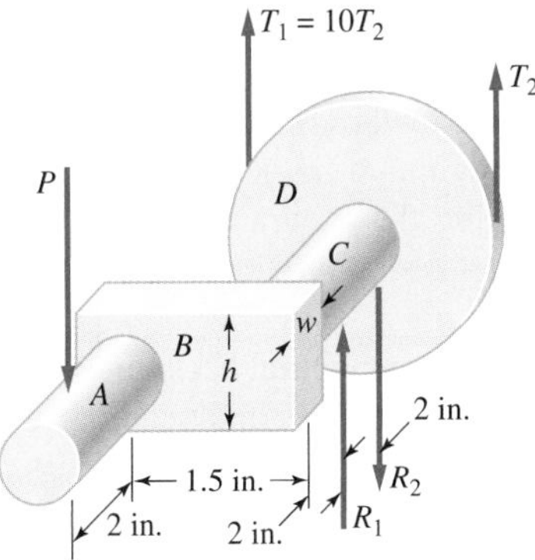

Figure P19.3
Overhung crankshaft.

g. Calculate all pertinent stresses for these critical points.

h. Calculate equivalent combined stresses for these critical points.

i. Is the design of cheek *B* acceptable?

j. Is the design of cheek *B* controlled by yielding, fatigue, journal wear, or something else?

19-4. The cylindrical bearing journal of a straddle-mounted crankshaft has been tentatively sized based on wear requirements, and it has been found that a diameter of 1.38 inches is required for this purpose. A force analysis of the journal at the critical cross section has shown there to be a transverse shear force of 8500 lb, torsional moment of 7500 in-lb, and bending moment of 6500 in-lb. If the governing failure mode is fatigue, and the design stress based on fatigue has been found to be 40,000 psi, calculate whether the 1.38-inch-diameter journal is properly designed to safely withstand the fatigue loading.

19-5. A straddle-mounted crankshaft for a belt-driven single-cylinder refrigeration compressor is to be designed to meet the following specifications:

a. The force on the connecting rod bearing ranges from 3500 lb down to 1300 lb up.

b. The belt tension ratio is $T_1 = 8T_2$.

c. The crank throw is 2.2 inches.

d. Allowable maximum bearing pressure is 700 psi.

e. The pulley pitch diameter = 8.0 inches.

f. All bearings are to be made alike.

g. Main bearings are 12 inches apart center-to-center, with the connecting rod bearing halfway between.

h. The pulley overhangs the main bearing 4 inches.

Design a suitable crankshaft and construct good engineering sketches of the final design.

Chapter 20

Completing the Machine

20.1 Integrating the Components; Bases, Frames, and Housings

Just as for any of the components discussed in Chapters 8 through 19, a machine *base* or *frame* may be designed by giving attention to forces and moments placed on these supporting structures by the mounted components as the machine operates. In some cases the weight of structural frame members themselves must be considered. The tasks of evaluating potential failure modes,[1] selecting appropriate materials,[2] and establishing an effective structural geometry[3] are all important when designing a frame. Necessary attributes of an effective structural geometry include maintaining the accuracy of required spatial relationships between and among mounted components, and providing failure-free component support during the design life of the machine.

The function of a base or a frame is to provide a structural backbone for mounting the various operating components in their prescribed positions and securing them together. Types of frames, as illustrated in Figure 20.1, include baseplates, open-truss frames, shell structures, C-frames, O-frames, housings, and special-purpose combinations of these.

A *baseplate* is the simplest type of machine frame. Baseplates are typically flat plates with mounting holes or T-slots, and have a thickness that is small relative to length and width. Reinforcing ribs on the bottom side are sometimes used to increase rigidity or decrease weight. In some applications, bottom-side ribs may run in both the length-direction and the width-direction to form a waffle-grid rib pattern.

Open-truss spatial structures are widely used as machine frames because they combine good strength and rigidity with minimum weight and efficient use of material.[4] Such structures are often fabricated from rectangular or round tubes, rods, or other rolled shapes, pinned or welded at the joints. The truss configuration can be tailored to any specific application so that mounting pads have the positions and orientations needed for easily attaching the various components. The *configurational guidelines* listed in 6.2 are especially important when configuring an efficient open-truss frame.

When operational forces and moments are small, or when structural deflections are not critical, a simple *table*, with a mounting plate supported by sturdy legs, may sometimes be used. For higher-performance design requirements, however, the cross-braced truss structure is usually more effective.

Shell structures are also used as frames in some instances. Examples of *thin-walled* shell structures, sometimes called *stressed-skin* structures, include airframes, spacecraft structures, and some ground-based transportation vehicle frames. Such structures typically embody bulkheads, stringers, and local stiffeners to prevent the thin skin from buckling.

[1]See Chapter 2. [2]See Chapter 3. [3]See Chapter 6. [4]See 6.2 and Figure 6.3.

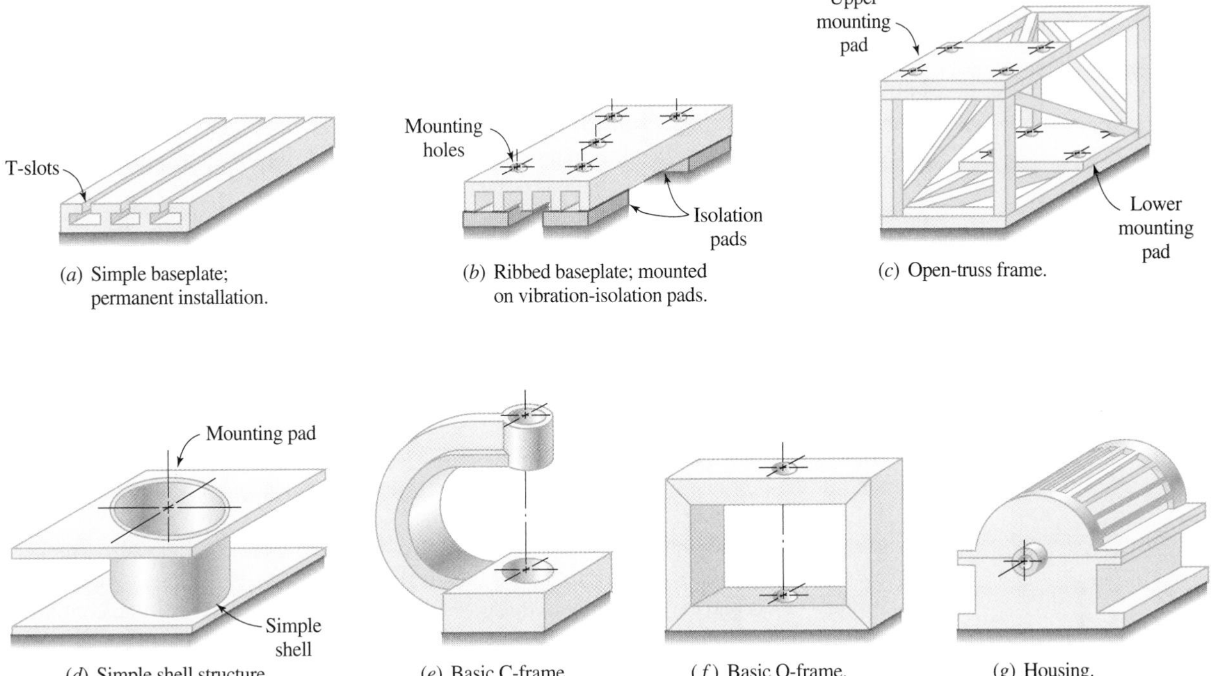

Figure 20.1
Illustration of several common types of machine frames.

The term "stressed-skin" derives from the fact that the thin wall of the shell, efficient in resisting high-tensile membrane stresses, provides direct structural support when properly integrated with bulkheads and stiffeners. Heavier-walled shells may be used as supporting structure for electric motors, gearboxes, and other similar subassemblies where bearings, shafts, and gears must be accurately held in alignment. When shell structures are configured to *protect* the components mounted within the shell from dust, dirt, or other foreign material, they are usually called *housings*. Housings may also be designed to provide a lubricant sump if needed.

Other types of frames, such as *C-frames* or *O-frames*, are sometimes used. For stationary machinery such as drill presses, milling machines, stamping machines, and punch presses, the C-frame (see Figure 20.1) is probably the most widely used. If more stability is required than a C-frame can provide, or if worksite deflections must be held to very small values, an O-frame may be the best choice. Accessibility for C-frame workspace is usually better than for O-frames.

As for any machine element, the potential *failure modes* for machine frames are strongly dependent upon loading characteristics, operating environment, and material properties corresponding to the governing failure mode. Reviewing the potential failure modes listed in 2.3, the more likely failure modes for frames include force-induced elastic deformation, yielding, brittle fracture, fatigue, and corrosion.

Common materials used for baseplates, frames, and housings include cast iron, cast steel, and wrought steel. For high-performance applications, or in corrosive environments, other materials such as titanium, aluminum, or magnesium may sometimes be used. Because of its inherent damping capacity, relatively low cost, and relative ease of obtaining the desired shape, cast iron is frequently chosen. This is especially true if mass production quantities are anticipated. Typical applications include baseplates, electric motor frames, pumps, compressors, and machine tools. If large or medium-size frames having intricate shapes are required, welded construction may be used to build up a frame from wrought steel or cast steel parts. Welded construction, however, must be undertaken with the knowledge that built-in stresses (residual stresses) produced by the welding process may cause significant warpage, sometimes a serious drawback.

In the final analysis, the design procedure for frames follows the same guidelines as for all other components. That is, the size, shape, and selected material must satisfy the functional requirements of supporting the loads without failure and maintaining the necessary dimensional accuracy throughout the prescribed design life, and do it at an acceptable cost.

Example 20.1 Frame Design

To successfully compete in a specialized "sheet goods" production application, it is necessary to continuously control the sheet goods *thickness* to within ± 10 percent all across a production *width* of 20 feet. The nominal *target thickness* of $t = 0.010$ inch is to be controlled in the production process by adjusting a thin rectangular *aperture* through which the slurry material issues before it is dried, trimmed, and rolled. It is being proposed to use a special on-line sheet-thickness measurement and control system to continuously scan the sheet product and, when necessary, adjust the aperture to keep sheet thickness within the specified ± 10 percent of the target thickness. Sheet goods production is to be continuous over long periods of time.

A proposed measurement and control system has already been developed. It integrates a scanning sensor package tied to a computer-controlled aperture-adjustment device. As presently conceived, the scanning sensor package consists of an upper carriage unit and a lower carriage unit, with a clearance space of approximately 0.040 inch between their sensor faces, to allow continuous noncontacting passage of the sheet goods. The fully instrumented upper carriage unit and lower carriage unit each weigh about 50 pounds. Figure E20.1A illustrates the concept. The details of this scanning sensor system have been satisfactorily worked out to synchronously drive both upper and lower scanner carriages together, back and forth across the 20-foot width of the sheet goods, at a scanning rate of 5 feet per second. The sheet goods also *feed* at a rate of 5 feet per second. The required vertical clearance between the frame's upper and lower support beams, to accommodate the two scanning sensor packages, is estimated to be about 2 feet. Special high-precision support rails and rollers have been designed and tested, and a laser-based alignment procedure has been developed to assure vertical (thickness direction) position accuracy of the moving sensors to within ± 0.0005 inch, if the support rails are solidly mounted to the frame.

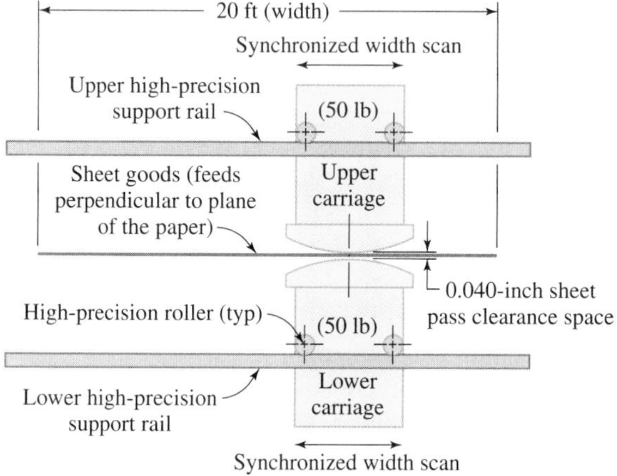

Figure E20.1A
Sketch showing the proposed thickness measurement system for continuous production of 20-foot-wide special sheet goods.

For the system just described, conceive a frame to support the measurement system as it scans back and forth to continuously monitor and adjust production sheet thickness.

Solution

Reviewing the skeleton sketch of the proposed component layout, shown in Figure E20.1A, important design requirements and observations include:

1. The continuously fed sheet has a width of 20 feet. Therefore the selected frame style must accommodate an unobstructed clearance space at least 20 feet wide.
2. Tolerances on sheet thickness are tight ($\pm$ 10 percent of the 0.010-inch target thickness is $\pm$ 0.001 inch). Hence, midspan transverse (vertical) frame deflections must be kept small enough to accommodate moving sensor package accuracy of $\pm$ 0.0005 inch, and still maintain control of sheet thickness to within tolerances of $\pm$ 0.001 inch.

Reviewing these requirements in light of the various types of frames sketched in Figure 20.1, it appears that an *O-frame* should be a good configurational choice because it would be naturally compatible with the passage of wide sheet goods, and it has the potential for supporting the required *small* midspan deflections if transverse beam stiffnesses can be made high enough. For these reasons, a first-iteration choice for frame style will be an O-frame.

Because the tolerances are tight, the span is large, and the sensor package weight is moderate (50 pounds), the weight of the structure itself might, at first, be thought a factor in determining midspan frame deflections as the scanning packages move transversely across the sheet. However, by specification, a laser beam high-precision alignment procedure has been developed for measuring and adjusting the thickness-direction sensor locations to within $\pm$ 0.0005 inch vertically, all across the width. Since the sensor system, together with its precision rails and rollers, can be mounted on the frame (already deflected due to self-weight), and then aligned using the laser beam device, any frame deflection due to self-weight can be "calibrated out." Thus frame deflections due to self-weight need not be considered here.

Combining these observations with the proposed layout of components, as sketched in Figure E20.1A, a first-iteration proposal for an O-frame structure, built up from hollow box-section beams and columns,[5] seems appropriate. Such an arrangement is sketched in Figure E20.1B. As discussed earlier in this section, and in view of the operational requirements for the O-frame proposed in Figure E20.1B, the most likely failure mode

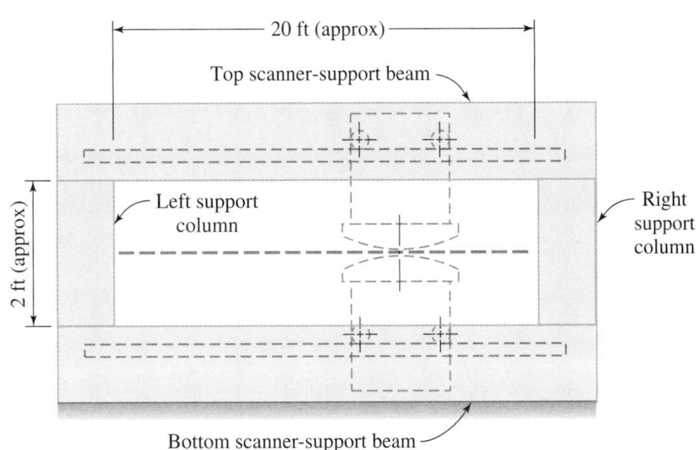

Figure E20.1B
Proposed configuration for O-frame structure built up from hollow box-section beams and columns.

[5]See configurational guidelines of 6.2.

Example 20.1 Continues

would appear to be *force-induced elastic deformation.*[6] Less likely, but worth checking, would be *yielding*.

Since *stiffness* is so important, and large beams and columns of hollow cross section are to be joined to build up the O-frame, an initial material choice of 1020 wrought steel is proposed.[7] Ductility, cost, strength, weldability, and availability, in addition to stiffness, are favorable attributes supporting the choice of low-carbon steel.

Accepting the proposed configuration shown in Figure E20.1B, the next step is to estimate beam section sizes required to maintain the specified sheet goods tolerances. The procedure will be to find minimum values of I (area moment of inertia of hollow beam section) based on scanner package weight, and associated changes in midspan beam deflection as the 50-pound scanners move back and forth along each support beam.

Considering the *top* scanner support beam, shown in Figure E20.1B, the midspan deflection, when the 50-pound scanner is at midspan, may be calculated from

$$y_{midspan} = \left(\frac{1}{K_{ends}}\right)\frac{PL^3}{EI} \tag{1}$$

where P is a concentrated midspan load of 50 lb, L =beam length of 240 inches, E = modulus of elasticity of 30×10^6 psi, I = area moment of inertia, in^4, and K_{ends} is an end condition constant. In this redundant structure, the end supports for the beam are neither fixed nor simple. For this first approximation, it will be *assumed* that the end constraints lie *half-way between fixed and simple*. Using deflection equations from Table 4.1, cases 1 and 6, the value of K_{ends} that lies half-way between fixed and simply supported ends would be

$$K_{ends} = \frac{192 - 48}{2} = 72 \tag{2}$$

so (1) becomes

$$y_{midspan} = \left(\frac{1}{72}\right)\left(\frac{50(240)^3}{(30 \times 10^6)I}\right) \tag{3}$$

Since the allowable sheet thickness t = 0.010 inch must be maintained to within ± 0.0005 inch, and sensor package accuracy is known to be ± 0.0005 inch, the *maximum-allowable* midspan beam deflection is

$$(y_{midspan})_{max} = 0.0010 - 0.0005 = 0.0005 \text{ inch} \tag{4}$$

Using this maximum-allowable deflection in (3) and solving for the minimum required value for area moment of inertia,

$$(I_{req'd})_{min} = \frac{(50)(240)^3}{72(30 \times 10^6)(0.0005)} = 640 \text{ in}^4 \tag{5}$$

Arbitrarily proposing a hollow box beam cross section, as sketched in Figure E20.1C, with aspect ratio of 2 (depth d_o is twice the width b_o), and arbitrarily defining the wall thickness t to be 5 percent of width b_o,[8] the area moment of inertia may be expressed as[9]

$$I_{box} = \frac{b_o d_o^3}{12} - \frac{b_i d_i^3}{12} \tag{6}$$

By initial assumption,

[6]See 2.3. [7]See Chapter 3.

[8]The quality of these assumed initial values usually improves with designer experience, but virtually all sets of assumed cross-sectional dimensions will, through repeated iteration, converge to similar final numerical values.

[9]See Table 4.2 and Example 4.2.

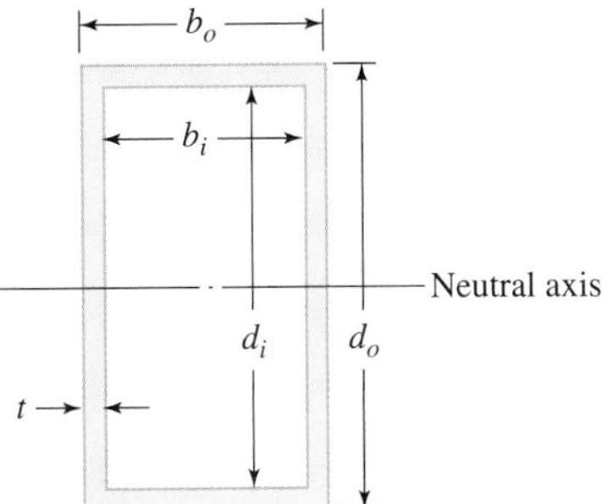

Figure E20.1C
Nominal cross-sectional configuration of proposed scanner top support beam.

$$\begin{aligned} t &= 0.05b_o \\ d_o &= 2b_o \end{aligned} \tag{7}$$

so, based on the geometric details shown in Figure E20.1C, it may be determined that

$$b_i = 0.90b_o \tag{8}$$

and

$$d_i = 0.95d_o \tag{9}$$

The required *minimum* box beam area moment of inertia, based on the initial assumptions just discussed, may next be calculated from (6) as

$$\begin{aligned} I_{box} &= \frac{b_o d_o^3}{12} - \frac{b_i d_i^3}{12} = \frac{1}{12}[b_o d_o^3 - (0.90b_o)(0.95d_o)^3] \\ &= \left(\frac{1 - 0.77}{12}\right) b_o d_o^3 = 0.019 b_o d_o^3 \end{aligned} \tag{10}$$

Starting the iteration process to determine the cross-sectional dimensions of the box beam, a tentative selection of $b_o = 8$ inches would, from (10), give

$$I_{box} = (0.019)(8)[2(8)]^3 = 623 \text{ in}^4 \tag{11}$$

Comparing (11) with (5) to judge the suitability of the proposed box beam dimensions, the question that may be posed is

$$I_{box} \overset{?}{>} (I_{req'd})_{min} \tag{12}$$

Or, is

$$623 \overset{?}{>} 640 \tag{13}$$

The answer is *no*, but I_{box} is very close to being large enough. To satisfy (13), it will be proposed to increase thickness t slightly. From (7) the first-iteration thickness is

$$t_1 = 0.05(8) = 0.40 \text{ inch} \tag{14}$$

For the second iteration, it will be proposed to increase the thickness to a nominal $^7/_{16}$ inch, or

$$t_2 = 0.44 \text{ inch} \tag{15}$$

Thus, the second-iteration proposal for the box beam dimensions will be

$$\begin{aligned} b_o &= 8.0 \text{ inches} \\ d_o &= 16.0 \text{ inches} \\ t_2 &= 0.44 \text{ inch} \end{aligned} \tag{16}$$

Example 20.1 Continues

From (16)

$$b_i = 8.0 - 2(0.44) = 7.12 \text{ inches}$$
$$d_i = 16.0 - 2(0.44) = 15.12 \text{ inches} \qquad (17)$$

Substituting (16) and (17) into (10),

$$I_{box} = \frac{(8)(16)^3}{12} - \frac{(7.12)(15.12)^3}{12} = 680 \text{ in}^4 \qquad (18)$$

For this set of box beam dimensions, the question posed is

$$I_{box} = 680 \overset{?}{>} (I_{req'd})_{min} = 640 \qquad (19)$$

The answer to this question is *yes*, so the proposed top beam cross-sectional dimensions will be

$$b_o = 8.0 \text{ inches}$$
$$d_o = 16.0 \text{ inches} \qquad (20)$$
$$t = 0.44 \text{ inch}$$

Since the bottom beam is no more critical than the top beam, it is suggested that its dimensions be made the same. As a matter of uniformity, it is further suggested that the short columns also be given the same cross-sectional dimensions, but they should be checked for *yielding* potential, as should the top and bottom beams.

To check the columns for yielding, the direct compressive stress may be calculated from the top beam weight,[10] the sensor package weight, and the column cross-sectional area. The top beam weight may be approximated as

$$W_{beam} = WA_{net}L = (0.283)[2(8 + 16)(0.44)](20 \times 12) = 1435 \text{ lb} \qquad (21)$$

so the nominal compressive stress, σ_c, in the column walls would be[11]

$$\sigma_c = \frac{1}{A_{net}}\left[\frac{W_{beam}}{2} + W_{sensor}\right] = \frac{1}{21.1}\left[\frac{1435}{2} + 50\right] = 36 \text{ psi} \qquad (22)$$

Compressive stress in the column walls is, therefore, negligibly small.

Checking next for possible midspan yielding of the top beam, and assuming the top beam to be simply supported (the most vulnerable possibility), Table 4.2, case 3, gives the maximum midspan moment as

$$M_{max} = \frac{wL^2}{8} = \frac{[0.283(8 + 16)(2)(0.44)](240)^3}{8} = 43{,}060 \text{ in-lb} \qquad (23)$$

Using (4-5), with numerical values from (19) and (23),

$$(\sigma_b)_{max} = \frac{M_{max}c}{I} = \frac{43{,}060\left(\frac{16}{2}\right)}{680} = 505 \text{ psi} \qquad (24)$$

This bending stress is also negligibly small compared to the yield strength of 1020 steel (see Table 3.3).

The governing failure mode, therefore, is force-induced elastic deformation of the beam (midspan deflection).

To summarize, an O-frame style is recommended for this application, as sketched in Figure E20.1B. It is proposed that each of the four frame members have a hollow box cross

[10]Using 1020 steel.

[11]Assuming that the scanner unit's weight adds directly to column load when the scanner is positioned at the adjacent edge of the sheet product.

section with an outer width of 8.0 inches, outer depth of 16.0 inches, and wall thickness of $^{7}/_{16}$ inch.

Before proceeding further, this proposal should be reviewed from the standpoints of aesthetics, ergonomics, and cost, as well as functionality, reliability, inspectability, safety, and environmental factors.

Finally, since the design is so sensitive to midspan beam deflection, an astute designer might think to check the possibility of temperature-induced elastic deformation as an additional potential failure mode in this application. Logically, if the sheet, which passes within about a foot of the inner surface of the top beam, were above ambient temperature, an additional thermally induced elastic midspan deflection might render the proposed frame unacceptable.

20.2 Safety Issues; Guards, Devices, and Warnings

A good rule of thumb for designers is that *any machine, machine part, machine function, or machine process that may cause injury must be safeguarded.* Good safeguarding methods[12] should prevent all operator contact with dangerous moving parts, make safeguard removal or tampering difficult, protect against falling objects, create no new hazards, create no interference, and allow safe lubrication. Because each design (or design modification) is unique, the designer needs to give full attention to the safety aspects of the product.

Safety, often defined as "freedom from danger, injury, or damage," is actually a *relative* attribute that may change from time to time, and may be judged differently in different circumstances.[13] Further, two distinctly different activities are required for determining how "safe" a product is: (1) *estimating* or *measuring* risk, and (2) *judging the acceptability* of the estimated risk. The estimation or measurement of risk may be quantitatively expressed as an objective, data-based, probabilistic attribute of the product under consideration. Judging the acceptability of the estimated risk is a much more subjective matter of personal or societal value judgment.[14]

A primary obligation of a designer (and manufacturer) is to make the product "safe," that is, to reduce the risks associated with the product to an acceptable level. Stated in other words, the designer's obligation in this context is to eliminate danger from using the product. *Danger* may be defined as an unreasonable or unacceptable combination of hazard and risk. A *hazard* is defined as a condition or changing set of circumstances that presents a *potential for injury. Risk* is the probability and severity of an adverse outcome. On this basis, a designer would typically proceed by taking the following steps:[15]

1. As far as possible, design all hazards out of the product.
2. If it is not possible to design out all hazards, provide guards or devices to eliminate the danger.
3. If it is not possible to provide proper and complete protection through the use of guards and safeguarding devices, provide appropriate directions and post "clear warnings."

A brief summary of common types of *guards* is given in Table 20.1. Types of *safeguarding devices* are described in Table 20.2. Standards[16] have been developed for accident-prevention signs and warning labels to establish suggestions for effective wording, colors, and placement.

[12] See, for example, ref. 1 or ref. 2.

[13] For example, a machine tool may be judged "safe" in the hands of an adult, but "unsafe" in the hands of a child.

[14] See also 5.4. [15] Also see 1.3. [16] See, for example, refs. 3–7.

TABLE 20.1 Types of Guards[1]

Method	Safeguarding Action	Advantages	Limitations
Fixed	Provides a barrier	Can be constructed to suit many specific applications. In-plant construction is often possible. Can provide maximum protection. Usually requires minimum maintenance. Can be suitable to high-production, repetitive operations.	May interfere with visibility. Can be limited to specific operations. Machine adjustment and repair often require its removal, thereby necessitating other means of protection for maintenance personnel.
Interlocked	Shuts off or disengages power and prevents starting of machine when guard is open; should require the machine to be stopped before the worker can reach into the danger area.	Can provide maximum protection. Allows access to machine for removing jams without time-consuming removal of fixed guards.	Requires careful adjustment and maintenance. May be easy to disengage.
Adjustable	Provides a barrier that may be adjusted to facilitate a variety of production operations.	Can be constructed to suit many specific applications Can be adjusted to admit varying sizes of stock.	Hands may enter danger area—protection may not be complete at all times. May require frequent maintenance and/or adjustment. The guard may be made ineffective by the operator. May interfere with visibility.
Self-adjusting	Provides a barrier that moves according to the size of the stock entering the danger area.	Off-the-shelf guards are often commercially available.	Does not always provide maximum protection. May interfere with visibility. May require frequent maintenance and adjustment.

[1]Reprinted from ref. 2, with permission of John Wiley & Sons, Inc.

TABLE 20.2 Types of Devices[1]

Method	Safeguarding Action	Advantages	Limitations
Photoelectric	Machine will not start cycling when the light field is interrupted. When the light field is broken by any part of the operator's body during the cycling process, immediate machine braking is activated.	Can allow freer movement for operator.	Does not protect against mechanical failure. May require frequent alignment and calibration. Excessive vibration may cause lamp filament damage and premature burnout. Limited to machines that can be stopped.
Radio frequency (capacitance)	Machine cycling will not start when the capacitance field is interrupted. When the capacitance field is disturbed by any part of the operator's body during the cycling process, immediate machine braking is activated.	Can alow freer movement for operator.	Does not protect against mechanical failure. Antennae sensitivity must be properly adjusted. Limited to machines that can be stopped.
Electromechanical	Contact bar or probe travels a predetermined distance between the operator and the danger area. Interruption of this movement prevents the starting of machine cycle.	Can allow access at the point of operation.	Contact bar or probe must be properly adjusted for each application; this adjustment must be properly maintained.
Pullback	As the machine begins to cycle, the operator's hands are pulled out of the danger area.	Eliminates the need for auxiliary barriers or other interference at the danger area.	Limits movement of operator. May obstruct workspace around operator. Adjustments must be made for specific operations and for each individual.

TABLE 20.2 (*Continued*)

Method	Safeguarding Action	Advantages	Limitations
Restraint (holdback)	Prevents the operator from reaching into the danger area.	Little risk of mechanical failure.	Limits movement of operator. May obstruct workspace. Adjustments must be made for specific operations and for each individual. Requires close supervision of the operator's use of the equipment.
Safety trip controls Pressure-sensitive body bar. Safety tripod. Safety tripwire	Stops machine when tripped.	Simplicity of use.	All controls must be manually activated. May be difficult to activate controls because of their location. Only protects the operator. May require special fixtures to hold work. May require a machine brake.
Two-hand control	Concurrent use of both hands is required, preventing the operator from entering the danger area.	Operator's hands are at a predetermined location. Operator's hands are free to pick up a new part after first half of cycle is completed.	Requires a partial-cycle machine with a brake. Some two-hand ontrols can be rendered unsafe by holding with arm or blocking, thereby permitting one-hand operation. Protects only the operator.
Two-hand trip	Concurrent use of two hands on separate controls prevents hands from being in danger area when machine cycle starts.	Operators hands are away from danger area. Can be adapted to multiple operations.	Operator may try to reach into danger area after tripping machine. Some trips can be rendered unsafe by

TABLE 20.2 (*Continued*)

Method	Safeguarding Action	Advantages	Limitations
		No obstruction to hand feeding.	holding with arm or blocking, thereby permitting one-hand operation.
		Does not require adjustment for each operation.	Protects only the operator.
			May require special fixtures.
Gate	Provides a barrier between danger area and operator or other personnel.	Can prevent reaching into or walking into the danger area.	May require frequent inspection and regular maintenance.
			May interfere with operator's ability to see the work.
Automatic feed	Stock is fed from rolls, indexed by machine mechanism, etc.	Eliminates the need for operator involvement in the danger area.	Other guards are also required for operator protection—usually fixed barrier guards.
			Requires frequent maintenance.
			May not be adaptable to stock variation.
Semiautomatic feed.	Stock is fed by chutes, movable dies, dial feed, plungers, or sliding bolster.	(Same as automatic feed—see above)	(Same as automatic feed—see above)
Automatic ejection	Workpieces are ejected by air or mechanical means.	(Same as automatic feed—see above)	May create a hazard of blowing chips or debris.
			Size of stock limits the use of this method.
			Air ejection may present a noise hazard.
Semiautomatic ejection	Workpieces are ejected by mechanical means, which are initiated by operator.	Operator does not have to enter danger area to remove finished work.	Other guards are required for operator protection.
			May not be adaptable to stock variation.

TABLE 20.2 (*Continued*)

Method	Safeguarding Action	Advantages	Limitations
Robots	They perform work usually done by operator.	Operator does not have to enter danger area.	Can create hazards themselves.
		Are suitable for operations where high stress factors are present, such as heat and noise.	Require maximum maintenance.
			Are suitable only to specific operations.

[1]Reprinted from ref. 2, with permission of John Wiley & Sons, Inc.

20.3 Design Reviews; Releasing the Final Design

Reducing the *time-to-market* and completing a product design *within budget* are very important goals for design success. Using cross-functional product design teams[17] and concurrent engineering strategies[18] greatly enhances the likelihood of achieving these goals. In addition, it is prudent to schedule periodic *design reviews* to assess progress, to monitor design and development activities, to ensure that all requirements and specifications are being met, and to provide feedback of information to all concerned. Usually, at least three design reviews will be necessary to meet design goals: one at the *preliminary design* stage, one during the *intermediate design* stage, and one before releasing the *final design* to production.[19] More sophisticated products may require several additional design reviews during the design process.

The preliminary design review (concept review) is probably the most important because it usually has a major impact on the design configuration. Intermediate design reviews and the final design review (before releasing to production) usually have less impact.

A design review is typically conducted by an ad-hoc design review board made up of materials engineers, mechanical design engineers, reliability engineers, electrical engineers, safety engineers, a management representative, a marketing representative, an insurance consultant, a "products-liability" attorney, and "outside experts," as needed. Members of the design review board usually should not be drawn from those involved in the day-to-day design and development of the product under review.

Releasing the final design on-time and on-budget is often crucial to a manufacturer's good reputation, and is usually demanded by the marketplace. It is important, therefore, for a designer to stay focused on meeting specifications effectively without incorporating "nice but unnecessary" features into the product.[20]

However vast the darkness, we must supply our own light.

—Stanley Kubrick

[17]See 1.1. [18]See 7.1. [19]See 1.6.

[20]Unneeded features, no matter how desirable, virtually always increase complexity, cost, and time-to-market. Inexperienced designers often fall prey to this trap.

Problems

20-1. Splined connections are widely used throughout industry, but little research has been done to provide the designer with either analytical tools or good experimental data for spline strength or compliance estimates. In question are such matters as basic spline-tooth strength, shaft strength, notch effect, and spline geometry effects, as well as spline-compliance effects on the torsional spring rate of a system containing one or more splined connections.

It is desired to construct a splined-joint testing setup versatile enough to facilitate both strength and life testing of various splined connections, as well as to perform torsional compliance testing on such joints. The testing setup is to accommodate in-line splined connections, offset parallel shafts connected by double universal joints, and angular shaft connections. Parallel shaft offsets up to 10 inches must be accommodated, and angular shaft centerline displacements up to 45° may be required. Splines up to 3 inches in diameter may need to be tested in the device, and shafting samples, including splined connections, may be up to 40 inches in length. Rotating speeds up to 3600 rpm may be required.

The basic setup, sketched in Figure P20.1, is to utilize a variable-speed drive motor to supply power to the input shaft of the testing setup, and a dynamometer (device for measuring mechanical power) used to dissipate power from the output shaft of the testing setup.

a. Select an appropriate type of frame or supporting structure for integrating the drive motor, testing setup, and dynamometer into a laboratory test stand for investigating splined-connection behavior, as just discussed.

b. Sketch the frame, as you envision it.

c. Identify potential safety issues that should, in your opinion, be addressed before putting the test stand into use.

20-2. You have been given the task of designing a special hydraulic press for removing and replacing bearings in small to medium-size electric motors. You are to utilize a commercially available $1\frac{1}{2}$-ton (3000 lb) capacity hydraulic actuator, mounted vertically, as sketched in Figure P20.2. As shown, the actuator body incorporates a 2-inch-diameter mounting boss. A lower platen for supporting the motor bearing packages is to be incorporated, as sketched. A minimum vertical clearance of 3 inches is required, as indicated, and a minimum unobstructed horizontal clearance of 5 inches between the vertical centerline of the hydraulic press and the closest structural member is also required. The operator's intended position is indicated in Figure P20.2, as well.

a. Select an appropriate type of frame or supporting structure for integrating the hydraulic actuator and support platen into a compact, stand-alone assembly, giving reasons for your selection.

b. Make a neat sketch of the frame, as you envision it.

c. Design the frame that you have sketched in (b) so that it may be expected to operate for 20 years in an industrial setting, without failure. It is estimated that the press will be operated, on average, once a minute, 8 hours each day, 250 days a year.

20-3. Cutting firewood is popular with "do-it-yourselfers" in many parts of the world, but hand-splitting the logs is a less-popular task. You are being asked to design a compact, "portable," moderately priced, firewood splitting machine for "home" use. The device should be capable of handling logs up to 16 inches in diameter and 24 inches long, splitting them into fireplace-size pieces. A cord of wood (a stack of wood 4 ft × 4 ft × 8 ft) should take no longer than an hour to split. Management has decided that a *power-screw-driven* splitting wedge should be investigated as a first choice. The concept is sketched crudely in Figure P20.3. Safety is to be considered, as well as compactness and portability.

a. Select an appropriate type of frame or supporting structure for integrating the power-screw-driven wedge and ad-

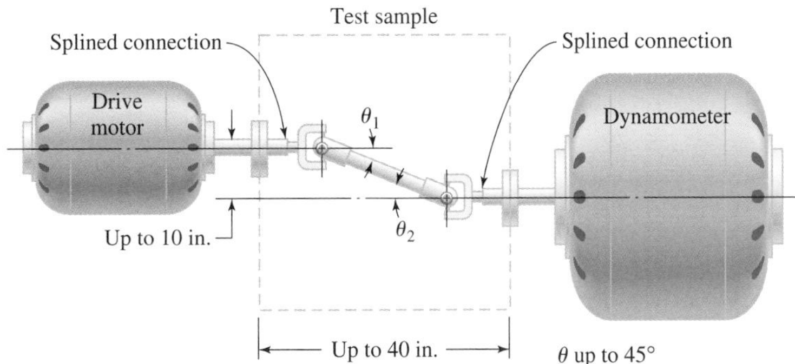

Figure P20.1
Plan view of proposed testing setup for investigating splined connections.

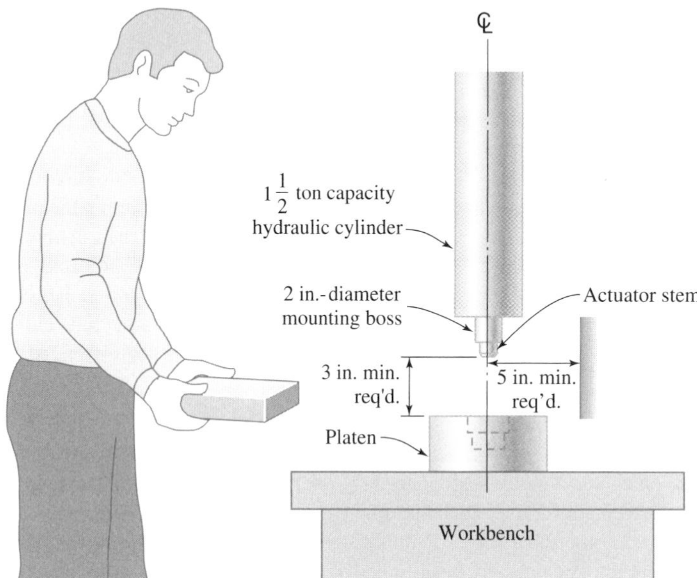

Figure P20.2
Schematic arrangement for proposed hydraulic press.

justable log-support unit into a compact, portable, stand-alone assembly, giving reasons for your selection.

b. Make a neat sketch of the frame, as you envision it.

c. From your sketch, identify each coherent subassembly, and give each subassembly a descriptive name.

d. Make a neat sketch of each coherent subassembly, and, treating each subassembly as a *free body*, qualitatively indicate all significant forces on each subassembly.

e. Design the power-screw subassembly. Preliminary estimates indicate that with a properly shaped splitting wedge, the wedge travel need not exceed half the length of the log, and the "splitting force" required of the power screw need not exceed about 3500 lb in the direction of the screw axis.

f. Design the adjustable log-support unit.

g. Design any other subassembly that you have named in (c).

h. Design the frame that you have sketched in (b).

i. Discuss any potential safety issues that you envision to be important.

20-4. The input shaft of a rotary coal-grinding mill is to be driven by a gear reducer through a flexible shaft-coupling, as shown in Figure P20.4. The output shaft of the gear reducer is to be supported on two bearings mounted 10 inches apart at *A* and *C*, as shown. A 1:3 spur gear mesh is being proposed to drive the gear-reducer output shaft. A spur gear is mounted on the output shaft at midspan between the bearings, as shown, and

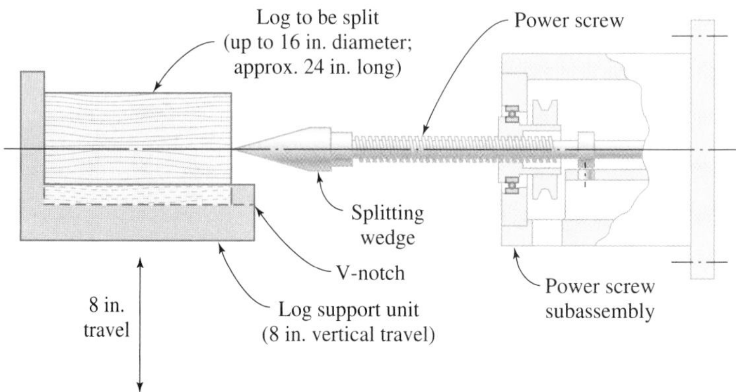

Figure P20.3
Home-use firewood splitting machine.

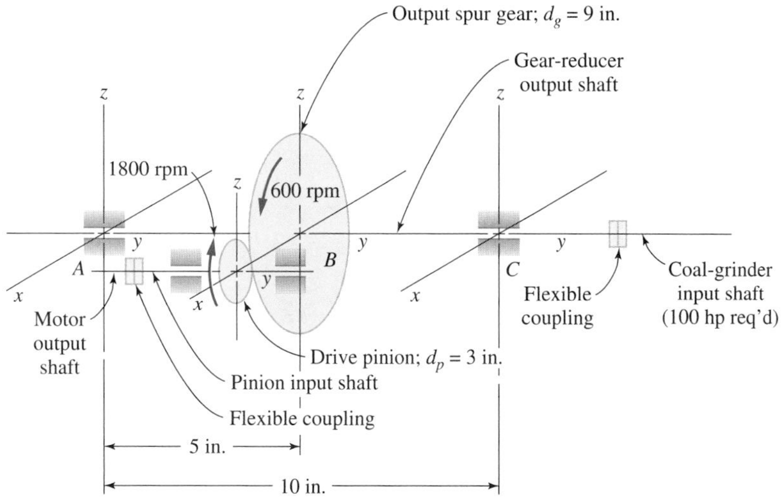

Figure P20.4
Schematic arrangement of components for a proposed gear-reducer subassembly.

is to have a pitch diameter of 9 inches. The pitch diameter of the drive pinion is to be 3 inches. The coal-grinding mill is to be operated at 600 rpm, and requires 100 horsepower continuously at its input shaft.

An 1800-rpm electric motor is to supply power to the pinion input shaft. Concentrating attention on the *spur gear speed reducer* sketched in Figure P20.4, do the following:

a. Select an appropriate type of frame or supporting structure for integrating the gears, shafts, and bearings into a compact stand-alone subassembly, giving reasons for your selection.

b. Make a neat sketch of the frame, as you envision it.

c. Design or select the spur gear set.

d. Design the gear-reducer output shaft.

e. Design the pinion input shaft.

f. Select appropriate bearings for the gear-reducer output shaft.

g. Select appropriate bearings for the pinion input shaft.

h. Specify appropriate lubrication for the gears and bearings.

i. Design the frame that you have sketched in (b).

20-5. a. In the context of mechanical design, define the terms *safety, danger, hazard*, and *risk*.

b. List the actions a designer might take to provide proper *safeguards* before releasing a machine to customers in the marketplace.

c. Make a list of *safeguarding devices* that have been developed to help reduce to an acceptable level the risks associated with engineered products.

Appendix

NSPE Code of Ethics for Engineers[1]

Preamble

Engineering is an important and learned profession. As members of this profession, engineers are expected to exhibit the highest standards of honesty and integrity. Engineering has a direct and vital impact on the quality of life for all people. Accordingly, the services provided by engineers require honesty, impartiality, fairness and equity, and must be dedicated to the protection of the public health, safety, and welfare. Engineers must perform under a standard of professional behavior that requires adherence to the highest principles of ethical conduct.

I. Fundamental Canons

Engineers, in the fulfillment of their professional duties, shall:

1. Hold paramount the safety, health and welfare of the public.
2. Perform services only in areas of their competence.
3. Issue public statements only in an objective and truthful manner.
4. Act for each employer or client as faithful agents or trustees.
5. Avoid deceptive acts.
6. Conduct themselves honorably, responsibly, ethically, and lawfully so as to enhance the honor, reputation, and usefulness of the profession.

II. Rules of Practice

1. Engineers shall hold paramount the safety, health, and welfare of the public.
   a. If engineers' judgment is overruled under circumstances that endanger life or property, they shall notify their employer or client and such other authority as may be appropriate.
   b. Engineers shall approve only those engineering documents that are in conformity with applicable standards.
   c. Engineers shall not reveal facts, data or information without the prior consent of the client or employer except as authorized or required by law or this Code.

[1]*Source:* Reprinted by permission of the National Society of Professional Engineers.

   d. Engineers shall not permit the use of their name or associate in business ventures with any person or firm that they believe are engaged in fraudulent or dishonest enterprise.

   e. Engineers having knowledge of any alleged violation of this Code shall report thereon to appropriate professional bodies and, when relevant, also to public authorities, and cooperate with the proper authorities in furnishing such information or assistance as may be required.

2. Engineers shall perform services only in the areas of their competence.

   a. Engineers shall undertake assignments only when qualified by education or experience in the specific technical fields involved.

   b. Engineers shall not affix their signatures to any plans or documents dealing with subject matter in which they lack competence, nor to any plan or document not prepared under their direction and control.

   c. Engineers may accept assignments and assume responsibility for coordination of an entire project and sign and seal the engineering documents for the entire project, provided that each technical segment is signed and sealed only by the qualified engineers who prepared the segment.

3. Engineers shall issue public statements only in an objective and truthful manner.

   a. Engineers shall be objective and truthful in professional reports, statements, or testimony. They shall include all relevant and pertinent information in such reports, statements, or testimony, which should bear the date indicating when it was current.

   b. Engineers may express publicly technical opinions that are founded upon knowledge of the facts and competence in the subject matter.

   c. Engineers shall issue no statements, criticisms, or arguments on technical matters that are inspired or paid for by interested parties, unless they have prefaced their comments by explicitly identifying the interested parties on whose behalf they are speaking, and by revealing the existence of any interest the engineers may have in the matters.

4. Engineers shall act for each employer or client as faithful agents or trustees.

   a. Engineers shall disclose all known or potential conflicts of interest that could influence or appear to influence their judgment or the quality of their services.

   b. Engineers shall not accept compensation, financial or otherwise, from more than one party for services on the same project, or for services pertaining to the same project, unless the circumstances are fully disclosed and agreed to by all interested parties.

   c. Engineers shall not solicit or accept financial or other valuable consideration, directly or indirectly, from outside agents in connection with the work for which they are responsible.

   d. Engineers in public service as members, advisors, or employees of a governmental or quasi-governmental body or department shall not participate in decisions with respect to services solicited or provided by them or their organizations in private or public engineering practice.

   e. Engineers shall not solicit or accept a contract from a governmental body on which a principal or officer of their organization serves as a member.

5. Engineers shall avoid deceptive acts.

   a. Engineers shall not falsify their qualifications or permit misrepresentation of their or their associates' qualifications. They shall not misrepresent or exaggerate their responsibility in or for the subject matter of prior assignments. Brochures or other presentations incident to the solicitation of employment shall not misrepresent pertinent facts concerning employers, associates, joint venturers, or past accomplishments.

b. Engineers shall not offer, give, solicit or receive, either directly or indirectly, any contribution to influence the award of a contract by public authority, or which may be reasonably construed by the public as having the effect of intent of influencing the awarding of a contract. They shall not offer any gift or other valuable consideration in order to secure work. They shall not pay a commission, percentage, or brokerage fee in order to secure work, except to a bona fide employee or bona fide established commercial or marketing agencies retained by them.

III. Professional Obligations

1. Engineers shall be guided in all their relations by the highest standards of honesty and integrity.
   a. Engineers shall acknowledge their errors and shall not distort or alter the facts.
   b. Engineers shall advise their clients or employers when they believe a project will not be successful.
   c. Engineers shall not accept outside employment to the detriment of their regular work or interest. Before accepting any outside engineering employment they will notify their employers.
   d. Engineers shall not attempt to attract an engineer from another employer by false or misleading pretenses.
   e. Engineers shall not promote their own interest at the expense of the dignity and integrity of the profession.
2. Engineers shall at all times strive to serve the public interest.
   a. Engineers shall seek opportunities to participate in civic affairs; career guidance for youths; and work for the advancement of the safety, health and well-being of their community.
   b. Engineers shall not complete, sign, or seal plans and/or specifications that are not in conformity with applicable engineering standards. If the client or employer insists on such unprofessional conduct, they shall notify the proper authorities and withdraw from further service on the project.
   c. Engineers shall endeavor to extend public knowledge and appreciation of engineering and its achievements.
3. Engineers shall avoid all conduct or practice that deceives the public.
   a. Engineers shall avoid the use of statements containing a material misrepresentation of fact or omitting a material fact.
   b. Consistent with the foregoing, Engineers may advertise for recruitment of personnel.
   c. Consistent with the foregoing, Engineers may prepare articles for the lay or technical press, but such articles shall not imply credit to the author for work performed by others.
4. Engineers shall not disclose, without consent, confidential information concerning the business affairs or technical processes of any present or former client or employer, or public body on which they serve.

a. Engineers shall not, without the consent of all interested parties, promote or arrange for new employment or practice in connection with a specific project for which the Engineer has gained particular and specialized knowledge.

b. Engineers shall not, without the consent of all interested parties, participate in or represent an adversary interest in connection with a specific project or proceeding in which the Engineer has gained particular specialized knowledge on behalf of a former client or employer.

5. Engineers shall not be influenced in their professional duties by conflicting interests.

   a. Engineers shall not accept financial or other considerations, including free engineering designs, from material or equipment suppliers for specifying their product.

   b. Engineers shall not accept commissions or allowances, directly or indirectly, from contractors or other parties dealing with clients or employers of the Engineer in connection with work for which the Engineer is responsible.

6. Engineers shall not attempt to obtain employment or advancement or professional engagements by untruthfully criticizing other engineers, or by other improper or questionable methods.

   a. Engineers shall not request, propose, or accept a commission on a contingent basis under circumstances in which their judgment may be compromised.

   b. Engineers in salaried positions shall accept part-time engineering work only to the extent consistent with policies of the employer and in accordance with ethical considerations.

   c. Engineers shall not, without consent, use equipment, supplies, laboratory, or office facilities of an employer to carry on outside private practice.

7. Engineers shall not attempt to injure, maliciously or falsely, directly or indirectly, the professional reputation, prospects, practice, or employment of other engineers. Engineers who believe others are guilty of unethical or illegal practice shall present such information to the proper authority for action.

   a. Engineers in private practice shall not review the work of another engineer for the same client, except with the knowledge of such engineer, or unless the connection of such engineer with the work has been terminated.

   b. Engineers in governmental, industrial, or educational employ are entitled to review and evaluate the work of other engineers when so required by their employment duties.

   c. Engineers in sales or industrial employ are entitled to make engineering comparisons of represented products with products of other suppliers.

8. Engineers shall accept personal responsibility for their professional activities, provided, however, that Engineers may seek indemnification for services arising out of their practice for other than gross negligence, where the Engineer's interests cannot otherwise be protected.

   a. Engineers shall conform with state registration laws in the practice of engineering.

   b. Engineers shall not use association with a nonengineer, a corporation, or partnership as a "cloak" for unethical acts.

9. Engineers shall give credit for engineering work to those to whom credit is due, and will recognize the proprietary interests of others.

   a. Engineers shall, whenever possible, name the person or persons who may be individually responsible for designs, inventions, writings, or other accomplishments.

b. Engineers using designs supplied by a client recognize that the designs remain the property of the client and may not be duplicated by the Engineer for others without express permission.

c. Engineers, before undertaking work for others in connection with which the Engineer may make improvements, plans, design, inventions, or other records that may justify copyrights or patents, should enter into a positive agreement regarding ownership.

d. Engineers' designs, data, records, and notes referring exclusively to an employer's work are the employer's property. Employer should indemnify the Engineer for use of the information for any purpose other than the original purpose.

As Revised February 2001

"By order of the United States District Court for the District of Columbia, former Section 11(c) of the NSPE Code of Ethics prohibiting competitive bidding, and all policy statements, opinions, rulings or other guidelines interpreting its scope, have been rescinded as unlawfully interfering with the legal right of engineers, protected under the antitrust laws, to provide price information to prospective clients; accordingly, nothing contained in the NSPE Code of Ethics, policy statements, opinions, rulings or other guidelines prohibits the submission of price quotations or competitive bids for engineering services at any time or in any amount."

Statement by NSPE Executive Committee

In order to correct misunderstandings which have been indicated in some instances since the issuance of the Supreme Court decision and the entry of the Final Judgment, it is noted that in its decision of April 25, 1978, the Supreme Court of the United States declared: "The Sherman Act does not require competitive bidding."

It is further noted that as made clear in the Supreme Court decision:

1. Engineers and firms may individually refuse to bid for engineering services.
2. Clients are not required to seek bids for engineering services.
3. Federal, state, and local laws governing procedures to procure engineering services are not affected, and remain in full force and effect.
4. State societies and local chapters are free to actively and aggressively seek legislation for professional selection and negotiation procedures by public agencies.
5. State registration board rules of professional conduct, including rules prohibiting competitive bidding for engineering services, are not affected and remain in full force and effect. State registration boards with authority to adopt rules of professional conduct may adopt rules governing procedures to obtain engineering services.
6. As noted by the Supreme Court, "nothing in the judgment prevents NSPE and its members from attempting to influence governmental action . . ."

NOTE: In regard to the question of application of the Code to corporations vis-à-vis real persons, business form or type should not negate nor influence conformance of individuals to the Code. The Code deals with professional services, which services must be performed by real persons. Real persons in turn establish and implement policies within business structures. The Code is clearly written to apply to the Engineer and it is incumbent on members of NSPE to endeavor to live up to its provisions. This applies to all pertinent sections of the Code.

TABLE A.1 Coefficients of Friction[1]

Material Pair	Application	Static, Sliding, or Rolling	Surface Conditions	Friction Coefficient: Approximate Range	Friction Coefficient: Typical Value[2]
Hard steel on hard steel	general	static	dry	0.15–0.78	0.45
" "	"	sliding	"	0.10–0.42	0.25
" "	"	static	lubricated	0.005–0.23	0.11
" "	"	sliding	"	0.03–0.12	0.08
Mild steel on mild steel	general	static	dry	0.11–0.74	0.50
" "	"	sliding	"	0.11–0.57	0.35
" "	"	static	lubricated	0.09–0.19	0.11
" "	"	sliding	"	0.01–0.09	0.08
Steel on bronze	general	static	dry	0.08–0.10	0.09
" "	"	sliding	"	0.06–0.15	0.08
" "	"	static	lubricated	0.0004–0.06	0.06
" "	"	sliding	"	0.0004–0.03	0.03
Steel on cast iron	general	static	dry	0.15–0.29	0.20
" "	"	sliding	"	0.09–0.39	0.12
" "	"	static	lubricated	0.05–0.18	0.10
" "	"	sliding	"	0.0035–0.13	0.06
Steel on aluminum	general	static	dry	0.61	0.61
" "	"	sliding	"	0.47	0.47
Steel on brass	general	static	dry	0.51	0.51
" "	"	sliding	"	0.44	0.44
Steel on tungsten carbide	general	static	dry	0.50	0.50
" "	"	sliding	"	0.08	0.08
Cast iron on cast iron	general	static	dry	0.16–1.10	0.60
" "	"	sliding	"	0.12–0.25	0.18
" "	"	static	lubricated	0.05–0.15	0.10
" "	"	sliding	"	0.06–0.16	0.10
Cast iron on bronze	general	sliding	dry	0.13–0.22	0.17
" "	"	"	lubricated	0.05–0.08	0.07
Cast iron on brass	general	sliding	dry	0.30	0.30
Steel on steel	conical surfaces	static	dry	0.22	0.22
" "	"	"	shrink fitted	0.13	0.13
Steel on cast iron	conical surfaces	static	dry	0.16	0.16
" "	"	"	shrink fitted	0.33	0.33
Any metal on another metal	conical surfaces	static	lubricated	0.15	0.15
Hard steel on babbitt metal	plain bearings	static	dry	0.42	0.42
" "	"	sliding	"	0.08–0.25	0.17
" "	"	static	lubricated	0.35	0.35
" "	"	sliding	"	0.055–0.14	0.10
Mild steel on cadmium-silver	plain bearings	sliding	lubricated	0.097	0.10

TABLE A.1 (*Continued*)

Material Pair	Application	Static, Sliding, or Rolling	Surface Conditions	Friction Coefficient	
				Approximate Range	Typical Value[2]
Mild steel on phosphor bronze	plain bearings	sliding	dry	0.34	034
" "	"	"	lubricated	0.17	0.17
Steel on Teflon	plain bearings	static	dry	0.04–0.18	0.10
" "	"	sliding	lubricated	0.04	0..04
Typical rolling element bearing materials	deep groove ball bearings	rolling[3]		0.001–0.004	0.002
" "	angular contact ball bearings	rolling[3]		0.002–0.003	0.002
" "	thrust ball bearings	rolling[3]		0.001–0.006	0.002
" "	roller bearings	rolling[3]		0.001–0.003	0.002
" "	tapered roller bearings	rolling[3]		0.002–0.020	0.003
" "	needle bearings	rolling[3]		0.003–0.010	0.003
Typical gearing materials	gearing	sliding	boundary lubrication	0.06–0.20	0.15
" "	"	"	mixed lubrication	0.03–0.07	0.05
" "	"	"	full film lubricaton	0.01–0.04	0.03
Cast iron on cast iron	clutches	sliding	dry	0.15–0.20	0.15
" "	"	"	wet	0.05	0.05
Cast iron on bronze	clutches	sliding	dry	0.13–0.22	0.15
" "	"	"	wet	0.05	0.05
Hard steel on hard steel	clutches	sliding	dry	0.10–0.42	0.15
" "	"	"	wet	0.05	0.05
Woven asbestos on cast iron or steel	clutches	sliding	dry	0.3–0.6	0.30
" "	"	"	wet	0.1–0.2	0.10
Carbon graphite on cast iron	clutches	sliding	dry	0.25	0.25
" "	"	"	wet	0.1–0.15	0.10
Cast iron on cast iron	brakes	sliding	dry	0.20	0.20
" "	"	"	oily	0.07	0.07
Asbestos fabric on metal	brakes	sliding	dry	0.35–0.40	0.35
" "	"	"	oily	0.25	0.25
Steel band on cast iron	belts	sliding	dry	0.18	0.18
Oak–tanned leather on cast iron or steel	belts	sliding	dry	0.25	0.25
" "	"	"	wet	0.20	0.20
" "	"	"	greasy	0.15	0.15

TABLE A.1 (*Continued*)

Material Pair	Application	Static, Sliding, or Rolling	Surface Conditions	Friction Coefficient: Approximate Range	Friction Coefficient: Typical Value[2]
Canvas on cast iron or steel	belts	sliding	dry	0.20	0.20
" "	"	"	wet	0.15	0.15
" "	"	"	greasy	0.12	0.12
Rubber on cast iron or steel	belts	sliding	dry	0.30	0.30
" "	"	"	wet	0.18	0.18

[1]Sources include the following:

1. Baumeister, T., ed., *Mark's Mechanical Engineers Handbood*, 6th ed., McGraw-Hill, New York, 1958.
2. Carmichael, Colin, ed., *Kent's Mechanical Engineer's Handbook*, 12th ed., Wiley, New York, 1950.
3. Lingaiah, K., *Machine Design Data Handbook*, McGraw-Hill, New York, 1994.
4. Peterson, M. B., and Winer, W. O., eds., *Wear Control Handbook*, American Society of Mechanical Engineers, New York, 1980.
5. Hindhede, U., et al., *Machine Design Fundamentals: A Practical Approach*, Wiley, New York, 1983.
6. Szeri, A. Z., *Tribology: Friction, Lubrication, and Wear*, Hemisphere Publishing, McGraw-Hill, 1980.
7. Orlov, P., *Fundamentals of Machine Design* (4 volumes), Mir Publishers, Moscow, 1977.

[2]"Typical" values shown are estimates made by the author of this textbook, and are thought to be appropriate unless more specific information is available. Coefficient of friction (μ) data are remarkably variable and often contradictory. Designers must always accept responsibility for the value chosen, so if specific data are not available, the axiom "friction is always against you" leads to the conservative choice of a high-end-of-range value for μ when low friction forces are critical, and a low-end-of-range value when high friction forces are critical. It is sometimes necessary to conduct well-controlled experiments to obtain a specific value of μ for a given critical application.

[3]Friction torques calculated using this value are to be based on bearing *bore*.

TABLE A.2 Mass Moments of Inertia *J* and Radii of Gyration *k* for Selected Homogeneous Solid Bodies Rotating About Selected Axes, as Sketched[1]

Description of Body		Mass Moment of Inertia, J	Radius of Gyration, k
Case 1. Sphere	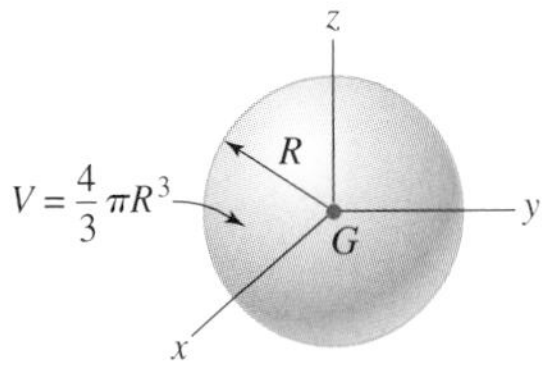 $V = \frac{4}{3}\pi R^3$	$J_x = J_y = J_z = \frac{2}{5}mR^2$	$k_x = k_y = k_z = \sqrt{\frac{2}{5}}R$
Case 2. Cylinder	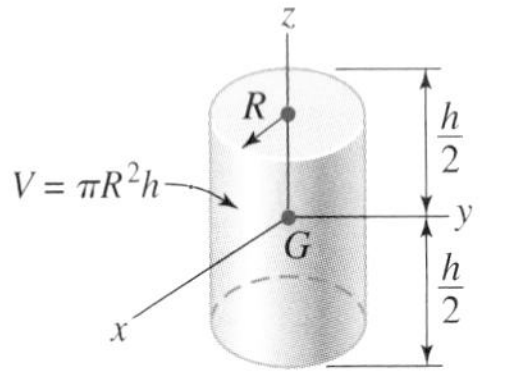 $V = \pi R^2 h$	$J_z = \frac{1}{2}mR^2$	$k_z = \sqrt{\frac{1}{2}}R$
		$J_x = J_y = \frac{1}{12}m(3R^2 + h^2)$	$k_x = k_y = \sqrt{\frac{1}{12}(3R^2 + h^2)}$
Case 3. Thin circular disk	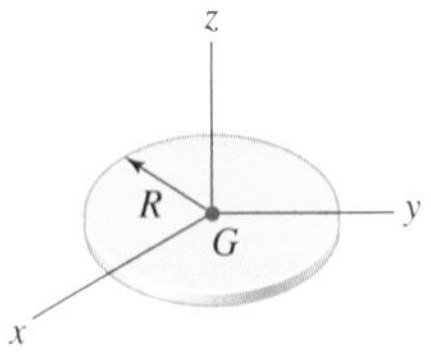	$J_z = \frac{1}{2}mR^2$	$k_z = \sqrt{\frac{1}{2}}R$
		$J_x = J_y = \frac{1}{4}mR^2$	$k_x = k_y = \sqrt{\frac{1}{4}}R$
Case 4. Thin ring	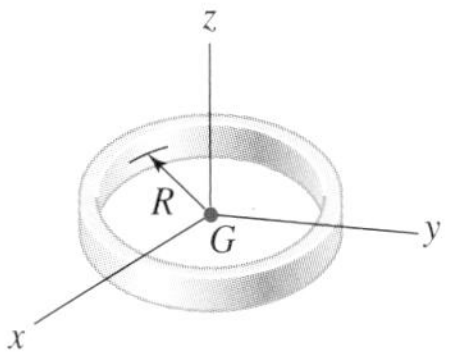	$J_z = mR^2$	$k_z = R$
		$J_x = J_y = \frac{1}{2}mR^2$	$k_x = k_y = \sqrt{\frac{1}{2}}R$
Case 5. Cone	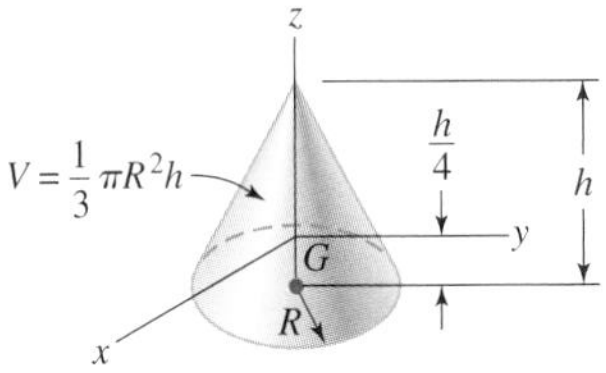 $V = \frac{1}{3}\pi R^2 h$	$J_z = \frac{3}{10}mR^2$	$k_z = \sqrt{\frac{3}{10}}R$
		$J_x = J_y = \frac{3}{80}m(4R^2 + h^2)$	$k_x = k_y = \sqrt{\frac{3}{80}(4R^2 + h^2)}$

[1]Extracted from Hibbler, R. C., *Engineering Mechanics: Dynamics*, 2nd ed., Macmillan, New York, 1978. Also note that m is the mass of the rotating body.

TABLE A.3 Section Properties of Selected W (Wide Flange) Shapes[1]

W
Shapes

Designation	A, in^2	d, in	t_w, in	b_f, in	t_f, in	T, in	k, in	k_1, in	I_{xx}, in^4	r_{xx}, in	I_{yy}, in^4	r_{yy}, in	W, lb/ft
W36 × 300	88.3	36.7	0.95	16.7	1.68	30.9	2.63	1.69	20,300	15.2	1300	3.83	300
W36 × 135	39.7	35.6	0.60	12.0	0.79	32.1	1.54	1.13	7,800	14.0	225	2.38	135
W24 × 162	47.7	25.0	0.71	13.0	1.22	20.8	1.72	1.19	5,170	10.4	443	3.05	162
W24 × 55	16.3	23.6	0.40	7.0	0.51	20.8	1.11	1.00	1,360	9.1	29	1.34	55
W18 × 119	35.1	19.0	0.66	11.3	1.06	15.1	1.46	1.19	2,190	7.9	253	2.69	119
W18 × 35	10.3	17.7	0.30	6.0	0.43	15.5	0.83	0.75	510	7.0	15	1.22	35
W16 × 100	29.7	17.0	0.59	10.4	0.99	13.3	1.69	1.13	1,500	7.1	186	2.50	100
W16 × 57	16.8	16.4	0.43	7.1	0.72	13.6	1.12	0.88	758	6.7	43	1.60	57
W16 × 50	14.7	16.3	0.38	7.1	0.63	13.6	1.03	0.81	659	6.7	37	1.59	50
W16 × 45	13.3	16.1	0.35	7.0	0.57	13.6	0.97	0.81	586	6.7	33	1.57	45
W16 × 40	11.8	16.0	0.31	7.0	0.51	13.6	0.91	0.81	518	6.6	29	1.57	40
W16 × 36	10.6	15.9	0.30	7.0	0.43	13.6	0.83	0.75	448	6.5	25	1.52	36
W16 × 26	7.7	15.7	0.25	5.5	0.35	13.6	0.75	0.75	301	6.3	9.6	1.12	26
W14 × 730	215.0	22.4	3.07	17.9	4.91	10.0	5.51	2.75	14,300	8.2	4,720	4.69	730
W14 × 500	147.0	19.6	2.19	17.0	3.50	10.0	4.10	2.31	8,210	7.5	2,880	4.43	500
W14 × 233	68.5	16.0	1.07	15.9	1.72	10.0	2.32	1.75	3,010	6.6	1,150	4.10	233
W14 × 145	42.7	14.8	0.68	15.5	1.09	10.0	1.69	1.56	1,710	6.3	677	3.98	133
W14 × 90	26.5	14.0	0.44	14.5	0.71	10.0	1.31	1.44	999	6.1	362	3.70	90

TABLE A.3 (*Continued*)

W Shapes

Designation	A, in^2	d, in	t_w, in	b_f, in	t_f, in	T, in	k, in	k_1, in	I_{xx}, in^4	r_{xx}, in	I_{yy}, in^4	r_{yy}, in	W, lb/ft
W14 × 48	14.1	13.8	0.34	8.0	0.60	10.9	1.19	1.00	484	5.9	51	1.91	48
W14 × 22	6.5	13.7	0.23	5.0	0.34	11.6	0.74	0.75	199	5.5	7.0	1.04	22
W12 × 336	98.8	16.8	1.78	13.4	2.96	9.1	3.55	1.69	4,060	6.4	1,190	3.47	336
W12 × 96	28.2	12.7	0.55	12.2	0.90	9.1	1.50	1.13	833	5.4	270	3.09	96
W12 × 50	14.6	12.2	0.37	8.1	0.64	9.3	1.14	0.94	391	5.2	56	1.96	50
W12 × 14	4.2	11.9	0.20	4.0	0.23	10.4	0.53	0.56	89	4.6	2.4	0.75	14
W10 × 112	32.9	11.4	0.76	10.4	1.25	7.5	1.75	1.00	716	4.7	236	2.68	112
W10 × 45	13.3	10.1	0.35	8.0	0.62	7.5	1.12	0.81	248	4.3	53	2.01	45
W10 × 12	3.5	9.9	0.19	4.0	0.21	8.4	0.51	0.56	54	3.9	2.2	0.79	12
W8 × 67	19.7	9.0	0.57	8.3	0.94	5.8	1.33	0.94	272	3.7	89	2.12	67
W8 × 28	8.2	8.1	0.29	6.5	0.47	6.1	0.86	0.63	98	3.5	22	1.62	28
W8 × 10	3.0	7.9	0.17	3.9	0.21	6.5	0.51	0.50	31	3.2	2.1	0.84	10
W6 × 25	7.4	6.4	0.32	6.1	0.46	4.5	0.75	0.56	54	2.7	17	1.52	25
W6 × 9	2.7	5.9	0.17	3.9	0.22	4.5	0.47	0.50	16	2.5	2.2	0.91	9
W5 × 19	5.6	5.2	0.27	5.0	0.43	3.5	0.73	0.44	26	2.2	9.1	1.28	19
W4 × 13	3.8	4.2	0.28	4.1	0.35	2.6	0.60	0.50	11	1.7	3.9	1.00	13

[1]Condensed from *Manual of Steel Construction, Load and Resistance Factor Design,* 3rd ed., Chicago, 2001, with permission from American Institute of Steel Construction, Inc. Many additional commercially available sections are tabulated. Some dimensions have been rounded.

TABLE A.4 Section Properties of Selected S (Standard I) Shapes[1]

S Shapes

Designation	A, in^2	d, in	t_w, in	b_f, in	t_f, in	T, in	k, in	I_{xx}, in^4	r_{xx}, in	I_{yy}, in^4	r_{yy}, in	W, lb/ft
S24 × 121	35.5	24.5	0.80	8.1	1.09	20.5	2.00	3,160	9.4	83	1.53	121
S20 × 75	22.0	20.0	0.64	6.4	0.80	16.8	1.63	1,280	7.6	30	1.16	75
S18 × 70	20.5	18.0	0.71	6.3	0.69	15.0	1.50	923	6.7	24	1.08	70
S15 × 50	14.7	15.0	0.55	5.6	0.62	12.3	1.38	485	5.8	16	1.03	50
S12 × 50	14.6	12.0	0.69	5.5	0.66	9.1	1.44	303	4.6	16	1.03	50
S10 × 35	10.3	10.0	0.59	4.9	0.49	7.8	1.13	147	3.8	8	0.90	35
S8 × 23	6.8	8.0	0.44	4.2	0.43	6.0	1.00	65	3.1	4.3	0.80	23
S6 × 17.25	5.1	6.0	0.47	3.6	0.36	4.4	0.81	26	2.3	2.3	0.68	20
S6 × 12.5	3.7	6.0	0.23	3.3	0.36	4.4	0.81	22	2.5	1.8	0.71	12.5
S5 × 10	2.9	5.0	0.21	3.0	0.33	3.5	0.75	12	2.1	1.2	0.64	10
S4 × 7.7	2.3	4.0	0.19	2.7	0.29	2.5	0.75	6.1	1.6	0.7	0.58	7.7
S3 × 5.7	1.7	3.0	0.17	2.3	0.26	1.8	0.63	2.5	1.2	0.5	0.52	5.7

[1]Condensed from *Manual of Steel Construction, Load and Resistance Factor Design,* 3rd ed., Chicago, 2001, with permission from American Institute of Steel Construction, Inc. Many additional commercially available sections are tabulated. Some dimensions have been rounded.

TABLE A.5 Section Properties of Selected C (Channel) Shapes[1]

C Shapes

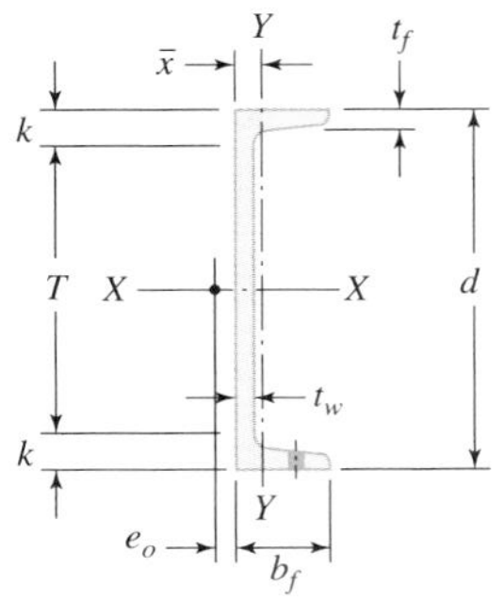

Designation	A, in^2	d, in	t_w, in	b_f, in	t_f, in	T, in	k, in	$\bar{x}$, in	e_o, in	I_{xx}, in^4	r_{xx}, in	I_{yy}, in^4	r_{yy}, in	W, lb/ft
C15 × 50	14.7	15.0	0.72	3.7	0.65	12.1	1.44	0.80	0.58	404	5.2	11	0.87	50
C12 × 30	8.8	12.0	0.51	3.2	0.50	9.8	1.13	0.67	0.62	162	4.3	5.1	0.76	30
C10 × 20	5.9	10.0	0.38	2.7	0.44	8.0	1.00	0.61	0.64	79	3.7	2.8	0.69	20
C9 × 15	4.4	9.0	0.29	2.5	0.41	7.0	1.00	0.59	0.68	51	3.4	1.9	0.66	15
C8 × 11.5	3.4	8.0	0.22	2.3	0.39	6.1	0.94	0.57	0.70	33	3.1	1.3	0.62	11.5
C7 × 9.8	2.9	7.0	0.21	2.1	0.37	5.3	0.88	0.54	0.65	21	2.7	1.0	0.58	9.8
C6 × 10.5	3.1	6.0	0.31	2.0	0.34	4.4	0.81	0.50	0.49	15	2.2	0.9	0.53	10.5
C5 × 9	2.6	5.0	0.33	1.9	0.32	3.5	0.75	0.48	0.43	8.9	1.8	0.6	0.49	9
C4 × 5.4	1.6	4.0	0.18	1.6	0.30	2.5	0.75	0.46	0.50	3.9	1.6	0.3	0.44	5.4
C3 × 4.1	1.2	3.0	0.17	1.4	0.27	1.6	0.69	0.44	0.46	1.7	1.2	0.2	0.40	4.1

[1]Condensed from *Manual of Steel Construction, Load and Resistance Factor Design,* 3rd ed., Chicago, 2001, with permission from American Institute of Steel Construction, Inc. Many additional commercially available sections are tabulated. Some dimensions have been rounded.

TABLE A.6 Section Properties of Selected Equal-Leg L (Angle) Shapes[1]

L Shapes

Designation	t, in	A, in^2	$x = y$, in	$I_{xx} = I_{yy}$, in^4	$r_{xx} = r_{yy}$, in	$r_{zz} = r_{min}$, in	W, lb/ft
L8 × 8 × $1\frac{1}{8}$	1.13	16.8	2.40	98	2.4	1.6	57.2
L8 × 8 × $\frac{1}{2}$	0.50	7.8	2.17	49	2.5	1.6	26.7
L6 × 6 × 1	1.00	11.0	1.86	35	1.9	1.2	37.5
L6 × 6 × $\frac{3}{8}$	0.38	4.4	1.62	15	1.9	1.2	14.9
L5 × 5 × $\frac{7}{8}$	0.88	8.0	1.56	18	1.5	1.0	27.3
L5 × 5 × $\frac{5}{16}$	0.31	3.1	1.35	7.4	1.6	1.0	10.4
L4 × 4 × $\frac{3}{4}$	0.75	5.4	1.27	7.6	1.2	0.8	18.5
L4 × 4 × $\frac{1}{4}$	0.25	1.9	1.08	3.0	1.3	0.8	6.6
L3 × 3 × $\frac{1}{2}$	0.50	2.8	0.93	2.2	0.90	0.6	9.4
L3 × 3 × $\frac{3}{16}$	0.19	1.1	0.81	0.95	0.93	0.6	3.7
L$2\frac{1}{2}$ × $2\frac{1}{2}$ × $\frac{3}{8}$	0.38	1.7	0.76	0.97	0.75	0.5	5.9
L2 × 2 × $\frac{1}{8}$	0.13	0.5	0.53	0.19	0.62	0.4	1.7

[1]Condensed from *Manual of Steel Construction, Load and Resistance Factor Design*, 3rd ed., Chicago, 2001, with permission from American Institute of Steel Construction, Inc. Many additional commercially available sections are tabulated. Some dimensions have been rounded.

References

Chapter 1

1. Dieter, G. E., Volume Chair, *ASM Handbook, vol. 20, Material Selection and Design*, ASM International, Materials Park, OH, 44073, 1997.

2. Kelley, R. E., "In Praise of Followers," *Harvard Business Review*, v. 66, no. 6, Nov–Dec 1988, pp. 142–148.

3. Hauser, J. R., and Clausing, D., "The House of Quality," *Harvard Business Review*, v. 66, no. 3, May–June 1998, pp. 63–73.

4. "3-D Static Strength Prediction Program," version 4.2, Center for Ergonomics, Univ. of Michigan, Ann Arbor, May 1999.

5. Ricks, Thomas E., "Lesson Learned," *Wall Street Journal*, May 23, 1997.

6. American National Standards Institute, 11 West 42nd Street, New York, 10036.

7. International Organization for Standardization, Geneva, Switzerland.

8. Humphreys, K. K., *What Every Engineer Should Know About Ethics*, Marcel Decker, New York, 1999.

9. American Association of Engineering Societies, *Model Guide for Professional Conduct*, Washington, DC, Dec. 13, 1984.

10. National Society of Professional Engineers, *Code of Ethics for Engineers*, Alexandria, VA, July, 2001.

11. Hoversten, P., "Bad Math Added Up to Doomed Mars Craft," *USA Today*, McLean, VA, Oct 1, 1999, p. 4A.

12. *Standard for Use of the International System of Units (SI): The Modern Metric System*, IEEE/ASTM SI ID-1997, ASTM Committee E43 on SI Practice, American Society for Testing and Materials, West Conshohocken, PA., 1997.

Chapter 2

1. Collins, J. A., *Failure of Materials in Mechanical Design: Analysis, Prediction, Prevention*, 2nd ed., Wiley, New York, 1993.

2. Hayden, H. W., Moffatt, W. G., and Wulff, J., *The Structure and Properties of Materials, vol. III, Mechanical Behavior*, Wiley, New York, 1965.

3. Tada, H., Paris, P., and Irwin G., *The Stress Anlysis of Cracks Handbook*, 2nd ed., Paris Productions, St. Louis, MO, 1985.

4. "Standard Test Method for Plane Strain Fracture Toughness of Metallic Materials," Designation: E 399-90, *Annual Book of ASTM Standards*, v. 3.01, American Society for Testing and Materials, Philadelphia, 1992.

5. Hertzberg, R. W., *Deformation and Fracture Mechanics of Engineering Materials*, Wiley, New York, 1976.

6. Gallagher, J., *Damage Tolerant Design Handbook*, MCIC-HB-01R, December 1983.

7. Matthews, W. T. "Plane Strain Fracture Toughness (KIC) Data Handbook for Metals," Report No. AMMRC M573-6, U.S. Army Materiel Command, NTIS, Springfield, VA, 1973.

8. Kanninen, M. F., and Popelar, C. H., *Advanced Fracture Mechanics*, Oxford University Press, New York, 1985.

9. Irwin, G. R., "Fracture Mode Transition for a Crack Traversing a Plate," *Journal of Basic Engineering, Trans. ASME*, v. 82, 1960, pp. 417–425.

10. Rice, R. C., ed., *Fatigue Design Handbook AE-10*, 2nd ed., Society of Automotive Engineers, Warrendale, PA, 1988.

11. Madayag, A. F., *Metal Fatigue Theory and Design*, Wiley, New York, 1969.

12. Grover, H. J., Gordon, S. A., and Jackson, C. R., *Fatigue of Metals and Structures*, Government Printing Office, Washington, DC, 1954.

13. Higdon, A., Ohlsen, E. H., Stiles, W. R., and Weese, J. A., *Mechanics of Materials*, New York, 1967.

14. Fuchs, H. O., and Stephens, R. I., *Metal Fatigue in Engineering*, Wiley, New York, 1980.

15. Juvinall, R. C., *Engineering Considerations of Stress, Strain, and Strength*, McGraw-Hill, New York, 1967.

16. Shigley, J. E., and Mischke, C. R., *Mechanical Engineering Design*, 5th ed., McGraw-Hill, New York, 1989.

17. Bannatine, J. A., Comer, J. J., and Handrock, J. L., *Fundamentals of Metal Fatigue Analysis*, Prentice-Hall, Englewood Cliffs, NJ, 1990.

18. Burr, A. H., and Cheatham, J. B., *Mechanical Analysis and Design*, 2nd ed., Prentice-Hall, Englewood Cliffs, NJ, 1995.

19. *Proceedings of the Conference on Welded Structures*, v. I and II, Welding Institute, Cambridge, England, 1971.

20. *Proceedings of International Conference on Fatigue*, American Society of Mechanical Engineers (jointly with Institution of Mechanical Engineers), New York, 1956.

21. Stulen, F. B., Cummings, H. N., and Schulte, W. C., "Preventing Fatigue Failures—Part 5," *Machine Design*, v. 33, no. 13, 1961.

22. Grover, H. J., *Fatigue of Aircraft Structures*, Government Printing Office, Washington, DC, 1966.

23. Dowling, N. E., "Fatigue Failure Predictions for Complicated Stress-Strain Histories," *Journal of Materials*, v. 7, no. 1, March 1972, pp. 71–87.

24. Shigley, J. E., and Mitchell, L. D., *Mechanical Engineering Design*, McGraw-Hill, New York, 1983.

25. Morrow, J. D., Martin, J. F., and Dowling, N. E., "Local Stress-Strain Approach to Cumulative Fatigue Damage Analysis," Final Report, T. & A.M. Report No. 379, Dept. of Theoretical and Applied Mechanics, University of Illinois, Urbana, January 1974.

26. Wundt, B. M., *Effects of Notches on Low Cycle Fatigue, STP-490*, American Society for Testing and Materials, Philadelphia, 1972.

27. Rolfe, S. T., and Barsom, J. M., *Fracture and Fatigue Control in Structures*, Prentice-Hall, Englewood Cliffs, NJ, 1977.

28. Hoeppner, D. W., and Krupp, W. E., "Prediction of Component Life by Application of Fatigue Crack Growth Knowledge," *Engineering Fracture Mechanics*, v. 6, 1974, pp. 47–70.

29. Paris, P. C., and Erdogan, F., "A Critical Analysis of Crack Propagation Laws," *Journal of Basic Engineering, ASME Transactions*, Series D, v. 85, no. 4, 1963, pp. 528–534.

30. Popov, E. P., *Introduction to Mechanics of Solids*, Prentice-Hall, Englewood Cliffs, NJ 1958.

31. Crandall, S. H., and Dahl, N. C., *An Introduction to the Mechanics of Solids*, McGraw-Hill, New York, 1957.

32. Timoshenko, S. P., and Gere, J. M., *Theory of Elastic Stability*, McGraw-Hill, New York, 1961.

33. Shanley, F. R., *Mechanics of Materials*, McGraw-Hill, New York, 1967.

34. Shanley, F. R., *Strength of Materials*, McGraw-Hill, New York, 1957.

35. *Military Standardization Handbook, Metallic Materials and Elements for Aerospace Vehicle Structures*, MIL-HDBK-5B, Superintendent of Documents, Washington, DC, September 1971.

36. Horton, W. H., Bailey, S. C., and McQuilkin, B. H., *An Introduction to Instability*, Stanford University, Paper No. 219, ASTM Annual Meeting, June 1966.

37. Timoshenko, S. P., *Theory of Elastic Stability*, McGraw-Hill, New York, 1936.

38. Larson, F. R., and Miller, J., "Time-Temperature Relationships for Rupture and Creep Stresses," *ASME Transactions*, v. 74, 1952, pp. 765 ff.

39. Sturm, R. G., and Howell, F. M., "A Method of Analyzing Creep Data," *ASME Transactions*, v. 58, 1936, p. A62.

40. Robinson, E. L., "The Effect of Temperature Variation on the Long-Time Rupture Strength of Steels," *ASME Transactions*, v. 74, 1952, pp. 777–781.

41. Burwell, J. T., Jr., "Survey of Possible Wear Mechanisms," *Wear*, v. 1, 1957, pp. 119–141.

42. Peterson, M. B., Gabel, M. K., and Derine, M. J., "Understanding Wear"; Ludema, K. C., "A Perspective on Wear Models"; Rabinowicz, E., "The Physics and Chemistry of Surfaces"; McGrew, J., "Design for Wear of Sliding Bearings"; Bayer, R. G., "Design for Wear of Lighly Loaded Surfaces"; *ASME Standardization News*, v. 2, no. 9, September 1974, pp. 5–32.

43. Rabinowicz, E., *Friction and Wear of Materials*, Wiley, New York, 1966.

44. Lipson, C., *Wear Considerations in Design*, Prentice-Hall, Englewood Cliffs, NJ, 1967.

45. Shigley, J. E., and Mischke, C. R., *Mechanical Engineering Design*, 5th ed., McGraw-Hill, New York, 1989.

46. Fontana, M. G., *Corrosion Engineering*, 3rd ed., McGraw-Hill, New York, 1986.

47. Fontana, M. G., and Green, N. D., *Corrosion Engineering*, McGraw-Hill, New York, 1967.

48. Collins, J. A., "A Study of the Phenomenon of Fretting-Fatigue with Emphasis on Stress Field Effects," Ph.D. dissertation, Ohio State University, Columbus, 1963.

49. Marin, J., *Mechanical Properties of Materials and Design*, McGraw-Hill, New York, 1942.

50. Hoeppner, D. W., Chandrasekaran, V., and Elliot, C. B., eds., *Fretting Fatigue: Current Technology and Practice, ASTM STP 1367*, American Society for Testing and Materials, West Conshohocken, PA, 2000.

51. Parker, E. R., "Modern Concepts of Flow and Fracture," *Transactions of American Society for Metals*, 1959, p. 511.

52. Lyons, H., "An Investigation of the Phenomenon of Fretting-Wear and Attendant Parametric Effects Towards Development of Failure Prediction Criteria," Ph.D. dissertation, Ohio State University, Columbus, 1978.

Chapter 3

1. *Materials Engineering, Materials Selector*, Penton Publishing, Cleveland, December 1991.

2. *Damage Tolerant Design Handbook*, Metals and Ceramics Information Center, Battelle, Columbus, 1983.

3. Ashby, M. F., *Materials Selection in Mechanical Design*, Butterworth Heinemann, Oxford, UK, 1999.

4. *Machine Design, Materials Reference Issue*, Penton Publishing, Cleveland, 1991.

5. *Marks' Standard Handbook for Mechanical Engineers*, 9th ed., Avallone, E. A., and Baumeister III, T., McGraw-Hill, New York, 1996.

6. *Metallic Materials and Elements for Aerospace Vehicle Structures, MIL-HDBK-5*, Department of Defense, Washington, DC, 1971.

7. *ASM Handbook, vol. 1, Properties and Selection: Irons, Steels, and High Performance Alloys*, ASM International, Materials Park, OH, 1990.

8. *ASM Handbook, vol. 2, Properties and Selection: Nonferrous Alloys and Special Purpose Materials*, ASM International, Materials Park, OH, 1991.

9. *Metals Handbook, Desk Edition*, ASM International, Materials Park, OH, 1985.

10. *ASM Materials Handbook, Desk Edition,* ASM International, Materials Park, OH, 1995.

11. Westbrook, J. H., and Rumble, J. R., eds., *Computerized Materials Data Systems,* Workshop Proceedings, Fairfield Glade, TN, National Bureau of Standards, 1983.

12. Fontana, M. G., and Green, N. D., *Corrosion Engineering,* McGraw-Hill, New York, 1967.

13. Gough, H. J., and Sopwith, D. G., "The Resistance of Some Special Bronzes to Fatigue and Corrosion Fatigue," *J. Inst. Metals*, v. 60, no. 1, 1937, pp. 143–153.

14. Smith, C. O., *ORSORT,* Oak Ridge, Tennessee.

15. Johnson, R. L., *Optimal Design of Mechanical Elements*, Wiley, New York, 1961.

16. "MSC/MVISION Materials Information System," MacNeal-Schwendler Corporation, 815 Colorado Blvd., Los Angeles, CA 90041.

17. "CMS Cambridge Materials Selector," Granta Design Ltd., 20 Trumpington St., Cambridge CB @ 1QA, UK.

18. "MAPP," EMS Software, 2234 Wade Ct., Hamilton, OH 45013.

19. Westbrook, J. H., "Sources of Material Property Data and Information," in *ASM Handbook, Vol. 20, Material Selection and Design,* ASM International, Materials Park, OH 44073, 1997.

20. CMS: *Cambridge Material Selector*, Cambridge University Engineering Department, Cambridge, UK, 1992.

21. PERITUS, Matsel Systems, Ltd., Liverpool, UK.

Chapter 4

1. Timoshenko, S., *Strength of Materials, Part I*, Van Nostrand, New York, 1955.

2. *Manual of Steel Construction*, 8th ed., American Institute of Steel Construction, Inc., 400 North Michigan Ave., Chicago, IL 60611, 1988.

3. Timoshenko, S., and Goodier, J. N., *Theory of Elasticity*, McGraw-Hill, New York, 1989.

4. Young, W. C., *Roark's Formulas for Stress and Strain*, 6th ed., McGraw-Hill, New York, 1989.

5. Lingaiah, K., *Machine Design Data Handbook*, McGraw-Hill, New York, 1994.

6. Collins, J. A., *Failure of Materials in Mechanical Design: Analysis Prediction, Prevention*, 2nd ed., Wiley, New York, 1993.

7. Mohr, O., *Civilingenieur*, 1882, p. 113.

8. Nadai, A., *Theory of Flow and Fracture of Solids,* Vol. I, 2nd ed., McGraw-Hill, New York, 1950.

9. Pilkey, W. D., *Peterson's Stress Concentration Factors*, 2nd ed., Wiley, New York, 1997.

10. Juvinall, R. C., *Stress, Strain and Strength*, McGraw-Hill, New York, 1967.

11. Murray, W. M. ed., *Fatigue and Fracture of Metals*, Wiley, New York, 1952.

12. Juvinall, R. L., and Marshek, K. M., *Fundamentals of Machine Component Design*, 2nd ed., Wiley, New York, 1991.

13. Faupel, J. H., *Engineering Design*, Wiley, New York, 1964.

14. Boresi, A. P., Sidebottom, O. M., Seely, F. B., and Smith, J. O., *Advanced Mechanics of Materials*, 3rd ed., Wiley, New York, 1978.

15. Lyst, J. O., "The Effect of Residual Strains upon Rotating Beam Fatigue Properties of Some Aluminum Alloys," Technical Report No. 9-60-34, Alcoa, Pittsburgh, 1960.

16. Spotts, M. F., *Design of Machine Elements*, 6th ed., Prentice-Hall, Englewood Cliffs, NJ, 1985.

Chapter 5

1. Juvinall, R. C., and Marshek, K. M., *Fundamentals of Machine Component Design*, 2nd ed., Wiley, New York, 1991.

2. Norton, R. L., *Machine Design*, Prentice-Hall, Upper Saddle River, NJ, 1996.

3. Burr, A. H., and Cheatham, J. B., *Mechanical Analysis and Design*, 2nd ed., Prentice-Hall, Englewood Cliffs, NJ, 1991.

4. Spotts, M. F., *Design of Machine Elements*, 6th ed., Prentice-Hall, Englewood Cliffs, NJ, 1985.

5. Shigley, J. E., and Mischke, C. R., *Mechanical Engineering Design*, 5th ed., McGraw-Hill, New York, 1989.

6. Kapur, K. C., and Lamberson, L. R., *Reliability Engineering Design*, Wiley, New York, 1977.

7. Collins, J. A., *Failure of Materials in Mechanical Design: Analysis, Prediction, Prevention*, 2nd ed., Wiley, New York, 1993.

8. Mischke, C. R., "Stochastic Methods in Mechanical Design: Part 1: Property Data and Weibull Parameters" (pp. 1–10); "Part 2: Fitting the Weibull Distribution to the Data" (pp. 11–16); "Part 3: A Methodology" (pp. 17–20); "Part 4: Applications" (pp. 21–28), *Failure Prevention and Reliability*, S. Sheppard, ed., American Society of Mechanical Engineers, New York, 1989.

9. Billinton, R. and Allan, R. N., *Reliability Evaluation of Engineering Systems*, 2nd ed., Plenum Press, New York, 1992.

10. Jenkins, Jr., H. W., "Business World," *Wall Street Journal*, Dow Jones & Co., New York, Dec. 15, 1999.

11. Kopala, D., and Raabe, A., "Redundancy Management: Reducing the Probability of Failure," *Motion*, Applied Motion Products, Inc., Watsonville, CA, Fall 1997.

12. Denson, W., "A Tutorial: PRISM," *J. Reliability Analysis Center (RAC)*, IIT Research Institute, Rome, NY, 3rd qtr., 1997.

Chapter 6

1. Marshek, K. M., *Design of Machine and Structural Parts*, Wiley, New York, 1987.

2. Young, W. C., *Roark's Formulas for Stress and Strain*, 6th ed., McGraw-Hill, New York, 1989.

3. *Preferred Limits and Fits for Cylindrical Parts*, ANSI B4.1-1967, reaffirmed 1999; *Preferred Metric Limits and Fits*, ANSI B4.2-1978, reaffirmed 1999, American Society of Mechanical Engineers, New York.

4. *Dimensioning and Tolerancing for Engineering Drawings*, ANSI Y14.5M-1994, American Society of Mechanical Engineers, New York.

5. Spotts, M. F., *Design of Machine Elements*, 6th ed., Prentice-Hall, Englewood Cliffs, NJ, 1985.

6. Foster, L. W., *Geometric Dimensioning and Tolerancing*, Addison-Wesley, Reading, MA, 1970.

7. Baumeister, T., *Mark's Standard Handbook for Mechanical Engineers*, 7th ed., McGraw-Hill, New York, 1967.

8. DeDoncker, D., and Spencer, A., "Assembly Tolerance Analysis with Simulation and Optimization Techniques," SAE Technical Paper 870263, Society of Automotive Engineers, Warrendale, PA, 1987.

9. "Improve Quality and Reduce Cost Through Controlled Variation and Robust Design," Brochure SW308, Engineering Animation, Inc. (EAI), Ames, IA.

Chapter 7

1. Dixon, J. R., and Poli, C., *Engineering Design and Design for Manufacturing*, Field Stone Publishers, Conway, MA, 1995.

2. Kalpakjian, S., *Manufacturing Processes for Engineering Materials*, Addison-Wesley, Reading, MA, 1984.

3. DeGarmo, E. P., *Materials and Processes in Manufacturing*, 5th ed., MacMillan, New York, 1979.

4. Redford, A., and Chal, J., *Design for Assembly*, McGraw-Hill, London, 1994.

5. Boothroyd, G., and Dewhurst, P., *Product Design for Assembly Handbook*, Boothroyd Dewhurst, Inc., Wakefield, RI, 1987.

6. Boothroyd, G., and Dewhurst, P., "Design for Assembly: Selecting the Right Method," *Machine Design*, Penton Publishing, Cleveland, Nov. 10, 1983, pp. 94–98.

7. Boothroyd, G., and Dewhurst, P., "Design for Assembly: Manual Assembly," *Machine Design*, Penton Publishing, Cleveland, Dec. 8, 1983, pp. 140–145.

8. Dewhurst, P., and Boothroyd, G., "Design for Assembly: Automatic Assembly," *Machine Design*, Penton Publishing, Cleveland, Jan. 26, 1984, pp. 87–92.

9. Dewhurst, P., and Boothroyd, G., "Design for Assembly: Robots," *Machine Design*, Penton Publishing, Cleveland, Feb. 23, 1984, pp. 72–76.

10. Boothroyd, G., and Dewhurst, P., *Design for Assembly Handbook*, University of Mass., Amherst, MA, 1983.

11. McMaster, R. C. (ed.), *Nondestructive Testing Handbook*, 2nd ed., American Society for Nondestructive Testing, Columbus, OH, 1987, vols. 1 and 2.

12. Pere, E., Gomez, D., Langrana, N., and Burdea, G., "Virtual Mechanical Assembly on a PC-Based System," *Proceedings of the 1996 ASME Design Engineering Technical Conferences*, Irvine, CA, August 18–22, 1996.

13. Shyamsundar, N., Ashai, Z., and Gadh, R., "Design for Disassembly Methodoly for Virtual Prototypes," *Proceedings of the 1996 ASME Design Engineering Technical Conferences*, Irvine, CA, August 18–22, 1996.

14. Bulkeley, W. M., "Parametric Resets Its Design Software to Match Engineering's Needs," *Wall Street Journal*, Dow Jones & Co., New York, Dec. 27, 1999.

Chapter 8

1. Den Hartog, J. P., *Mechanical Vibrations*, McGraw-Hill, New York, 1947.

2. Inman, D. F., *Engineeing Vibrations*, Prentice-Hall, Englewood Cliffs, NJ, 1996.

3. Norton, R. L., *Machine Design—An Integrated Approach*, Prentice-Hall, Upper Saddle River, NJ, 1996.

4. Spotts, M. F., *Design of Machine Elements*, 6th ed., Prentice-Hall, Englewood Cliffs, NJ, 1985.

5. Shigley, J. E., and Mischke, C. R., *Mechanical Engineering Design*, 5th ed., McGraw-Hill, New York, 1989.

6. Faires, V. M., *Design of Machine Elements*, 4th ed., Macmillan, New York, 1965.

7. Shigley, J. E., and Mischke, C. R. eds., *Standard Handbook of Machine Design*, McGraw-Hill, New York, 1986.

8. Burr, A. H., and Cheatham, J. B., *Mechanical Analysis and Design*, 2nd ed., Prentice-Hall, Englewood Cliffs, NJ, 1995.

9. Deutschman, A. D., Michels, W. J., and Wilson, C. E., *Machine Design—Theory and Practice*, Macmillan, New York, 1975.

10. Peterson, R. E., *Stress Concentration Factors*, Wiley, New York, 1974.

11. Lingaiah, K., *Machine Design Data Handbook*, McGraw-Hill, New York, 1994.

12. Walsh, R. A., *McGraw-Hill Machining and Metalworking Handbook*, McGraw-Hill, New York, 1994.

13. Pestel, E. C., and Leckie, F. A., *Matrix Methods in Elastomechanics*, McGraw-Hill, New York, 1963.

14. Rieder, W. G., and Busby, H. R., *Introductory Engineering Modeling Emphasizing Differential Models and Computer Simulations*, Wiley, New York, 1986.

Chapter 9

1. *The ASME Boiler and Pressure Vessel Code, Sections I–XI*, American Society of Mechanical Engineers, New York, 1995.

2. Timoshenko, S., and Goodier, J. N., *Theory of Elasticity*, McGraw-Hill, New York, 1951.

3. Pilkey, W. D., *Peterson's Stress Concentration Factors*, 2nd ed., Wiley, 1997.

Chapter 10

1. *Machine Design: Mechanical Drives Reference Issue*, Penton Publishing, Cleveland, Oct. 13, 1988.

2. Hersey, M. D., *Theory and Research in Lubrication*, Wiley, New York, 1966.

3. Burr, A. H., and Cheatham, J. B., *Mechanical Analysis and Design*, 2nd ed., Prentice-Hall, Englewood Cliffs, NJ, 1995.

4. Hamrock, B. J., *Fundamentals of Fluid Film Lubrication*, McGraw-Hill, New York, 1993.

5. Juvinall, R. C., and Marshek, K. M., *Fundamentals of Machine Component Design*, 2nd ed., Wiley, New York, 1991.

6. Raymondi, A. A., and Boyd, J., "A Solution for the Finite Journal Bearing and Its Application to Analysis and Design," Parts I, II, III, *Trans. American Society of Lubrication Engineers*, v. 1, no. 1, pp. 159–209, 1958.

7. Tower, B., "Reports on Friction Experiments," *Proc. Inst. Mech. Engr.*, First Report Nov. 1883, Second Report 1885, Third Report 1888, Fourth Report, 1891.

8. Reynolds, O., "On the Theory of Lubrication and Its Application to Mr. Beauchamp Tower's Experiments," *Phil. Trans. Roy. Soc.* (London), v. 177, p. 157 ff., 1886.

9. Spotts, M. F., *Design of Machine Elements*, 2nd ed., Prentice-Hall, New York, 1953.

10. Shigley, J. E., and Mischke, C. R., *Mechanical Engineering Design*, 5th ed., McGraw-Hill, New York, 1989.

11. Currie, I. G., *Fundamental Mechanics of Fluids*, 2nd ed., McGraw-Hill, New York, 1989.

12. Sommerfeld, A., "Zur Hydrodynamischen Theorie der Schmiermittel-Reibung" ("On the Hydrodynamic Theory of Lubrication"), *Z. Math. Physik*, v. 50, p. 97 ff., 1904.

13. Fuller, D. D., *Theory and Practice of Lubrication for Engineers*, Wiley, New York, 1956.

Chapter 11

1. "Power and Motion Control," *Machine Design*, Penton Publishing, Cleveland, June 1989.

2. *Load Ratings and Fatigue Life for Ball Bearings*, ANSI/AFBMA Standard 9-1990, American National Standards Institute, New York, 1990.

3. *Load Ratings and Fatigue Life for Roller Bearings*, ANSI/AFBMA Standard 11-1990, American National Standards Institute, New York, 1990.

4. Shaft and Housing Fits for Metric Radial Ball and Roller Bearings (Except Tapered Roller Bearings) Conforming to Basic Boundary Plans, ANSI/ABMA Standard 7-1995, American National Standards Institute, New York, 1995.

5. SKF,® "General Catalog 4000 US," SKFUSA, King of Prussia, PA, 1991.

6. Timken,® "The Tapered Roller Bearing Guide," Timken Company, Canton, OH, 1994.

7. Tallian, T. E., "On Competing Failure Modes in Rolling Contact," *Trans. ASLE*, v. 10, pp. 418–439, 1967.

8. Skurka, J. C., "Elastohydrodynamic Lubrication of Roller Bearings," *J. Lubr. Technology*, v. 92, pp. 281–291, 1970.

9. Hamrock, B. J., *Fundamentals of Fluid Film Lubrication*, McGraw-Hill, New York, 1993.

Chapter 12

1. *Acme Screw Threads*, ASME B1.5-1988, American Society of Mechanical Engineers, New York, 1988.

2. *Stub Acme Screw Threads*, ASME B1.8-1988, American Society of Mechanical Engineers, New York, 1988.

3. *Buttress Inch Screw Threads, 7deg/45deg Form with 0.6 Pitch Basic Height of Thread Engagement*, ASME B1.9-1973 (R1992), American Society of Mechanical Engineers, New York, 1992.

4. Lingaiah, K., *Machine Design Data Handbook*, McGraw-Hill, New York, 1994.

5. Parker, L., and Levin, A., " '97 Check Found Stabilizer Piece Worn," *USA Today*, v. 18, no. 107, p. 73A, Arlington, VA, Feb. 14, 2000.

Chapter 13

1. Parmley, R. O., ed., *Standard Handbook of Fastening and Joining*, 2nd ed., McGraw-Hill, New York, 1989.

2. DeGarmo, E. P., *Materials and Processes in Manufacturing*, Macmillan, New York, 1979.

3. *Unified Inch Screw Threads (UN and UNR Thread Form)*, ASME B1.1-1989, American Society of Mechanical Engineers, New York, 1989.

4. *Metric Screw Threads—M Profile*, ASME B1.13M-1995, American Society of Mechanical Engineers, New York, 1995.

5. *Metric Screw Threads—MS Profile*, ASME B1.21M-1978, American Society of Mechanical Engineers, New York, 1978.

6. Ito, Y., Toyoda, J., and Nagata, S., "Interface Pressure Distribution in a Bolt-Flange Assembly," *ASME Paper No. 77-WA/DE-11*, 1977.

7. Little, R. E., "Bolted Joints; How Much Give?," *Machine Design*, Nov. 9, 1967.

8. Bruhn, E. F., *Analysis and Design of Flight Vehicle Structures*, Tristate Offset Company, 817 Main Street, Cincinnati, OH 45202, 1965.

9. Connor, L. P., ed., *Welding Handbook*, 8th ed., v. 1, American Welding Society, Miami, FL, 1987.

10. Norris, C. H., "Photoelastic Investigation of Stress Distribution in Transverse Fillet Welds," *Welding Journal*, v. 24, p. 557, 1945.

11. Deutschman, A. D., Michels, W. J., and Wilson, C. E., *Machine Design, Theory and Practice*, Macmillan, New York, 1975.

12. Wileman, J., Choudhury, M., and Green, I., "Computation of Member Stiffness in Bolted Connections," *Journal of Mechanical Design, Transactions of the American Society of Mechanical Engineers*, v. 113, Dec., 1991, New York.

13. Jenny, C. L., and O'Brien, A., eds., *Welding Handbook*, 9th ed., v. 1, American Welding Society, Miami, FL, 2001.

Chapter 14

1. *Spring Design Manual, AE-21*, 2nd ed., Society of Automotive Engineers, Warrendale, PA, 1996.

2. *Design Handbook: Engineering Guide to Spring Design*, Associated Spring, Barnes Group, Bristol, CT, 1987

3. Wahl, A. M., *Mechanical Springs*, McGraw-Hill, New York, 1963

4. Timoshenko, S., and Goodier, J. N., *Theory of Elasticity*, McGraw-Hill, New York, 1951.

5. Collins, J. A., *Failure of Materials in Mechanical Design*, 2nd ed., John Wiley & Sons, New York, 1993.

6. Juvinall, R. C., and Marshek, K. M., *Fundamentals of Machine Component Design*, 2nd ed., Wiley, New York, 1991.

7. Maier, K. W., "Springs That Store Energy Best," *Product Engineering*, v. 29, no. 45, Nov. 10, 1958.

Chapter 15

1. Dudley, D. W., *Handbook of Practical Gear Design*, McGraw-Hill, New York, 1984.

2. Phelan, R. M., *Fundamentals of Mechanical Design*, 3rd ed., McGraw-Hill, New York, 1970.

3. Wilson, C. E., Sadler, J. P., and Michels, W. J., *Kinematics and Dynamics of Machinery*, Harper & Row Publishers, New York, 1983.

4. Mabie, H. H., and Ocvirk, F. W., *Mechanisms and Dynamics of Machinery*, Wiley, New York, 1975.

5. Juvinall, R. L., and Marshek, K., *Fundamentals of Machine Component Design*, 2nd ed., Wiley, New York, 1991.

6. Norton, R. L., *Machine Design*, Prentice-Hall, Upper Saddle River, NJ, 1996.

7. Houser, D. R., "Gear Noise Sources and Their Prediction Using Mathematical Models," *Gear Design, AE 15*, SAE International, Warrendale, PA, 1990.

8. ANSI/AGMA 1010-E95 (Revision of AGMA 10.04), *American National Standard, Appearance of Gear Teeth—Terminology of Wear and Failure*, American Gear Manufacturers Association, Alexandria, VA, Dec. 13, 1995.

9. Breen, D. H., "Fundamentals of Gear/Strength Relationships; Materials," *Gear Design, AE15*, SAE International, Warrendale, PA, 1990.

10. ANSI/AGMA 2001-C95, *American National Standard, Fundamental Rating Factors and Calculation Methods for Involute and Helical Gear Teeth*, American Gear Manufacturers Association, Alexandria, VA, Jan. 12, 1995.

11. USAS B6.1-1968, *USA Standard System—Tooth Proportions for Coarse Pitch Involute Spur Gears*, American Gear Manufacturers Association, Alexandria, VA, Jan. 27, 1968.

12. AGMA 370.01, *AGMA Design Manual for Fine-Pitch Gearing*, American Gear Manufacturers Association, Alexandria, VA, April 1973.

13. Szczepanski, G. S., Savoy, J. P., Jr., and Youngdale, R. A., "Chapter 14, The Application of Graphics Engineering to Gear Design," *Gear Design, AE 15*, SAE International, Warrendale, PA, 1990.

14. ANSI/AGMA 2000-A88, *American National Standard, Gear Classification and Inspection Handbook*, American Gear Manufacturers Association, Alexandria, VA.

15. DIN, Toleranzen für Stirnrädverzahnungen, DIN 3962 and DIN 3963, Aug. 1978 (German).

16. Lewis, Wilfred, "Investigation of the Strength of Gear Teeth," *Proceedings of Engineers Club*, Philadelphia, 1893.

17. Buckingham, Earle, *Analytical Mechanics of Gears*, McGraw-Hill, New York, 1949.

18. Lipson, C., and Juvinall, R. L., *Handbook of Stress and Strength*, Macmillan, New York, 1963.

19. Juvinall, R. L., *Engineering Considerations of Stress, Strain and Strength*, McGraw-Hill, New York, 1967.

20. AGMA 908-B89, *Geometry Factors for Determining the Pitting Resistance and Bending Strength of Spur, Helical, and Herringbone Gear Teeth*, American Gear Manufacturers Association, Alexandria, VA, April 1989.

21. Drago, R. J., "How to Design Quieter Transmissions," *Machine Design*, Penton Media, Inc., Cleveland, Dec. 11, 1980.

22. ANSI/AGMA 6021-G89, *For Shaft-Mounted and Screw Conveyor Drives Using Spur, Helical and Herringbone Gears*, American Gear Manufacturers Association, Alexandria, VA, November 1989.

23. ANSI/AGMA 2005-C96, *Design Manual for Bevel Gears*, American Gear Manufacturers Association, Alexandria, VA, 1996.

24. Coleman, W., "Guide to Bevel Gears," *Product Engineering*, McGraw-Hill, New York, June 10, 1963.

25. Coleman, W., "Design of Bevel Gears," *Product Engineering*, McGraw-Hill, New York, July 8, 1963.

26. *Straight Bevel Gear Design*, Gleason Works, Machine Division, Rochester, NY, 1980.

27. ANSI/AGMA 2003-B97, *Rating the Pitting Resistance and Bending Strength of Generated Straight Bevel, Zerol Bevel, and Spiral Bevel Gear Teeth*, American Gear Manufacturers Association, Alexandria, VA, 1997.

28. ANSI/AGMA 6022-C93, *Design Manual for Cylindrical Worm Gearing*, American Gear Manufacturers Association, Alexandria, VA, Dec. 16, 1993.

29. ANSI/AGMA 6034-B92, *Practice for Enclosed Cylindrical Worm Gear Speed Reducers and Gearmotors*, American Gear Manufacturers Association, Alexandria, VA, 1992.

30. Buckingham, E., and Ryffel, *Design of Worm and Spiral Gears*, Buckingham Associates, Springfield, VT, 1973. Reprinted 1984 by Hurd's Offset Printing Corp., Springfield, VT.

Chapter 16

1. Shigley, J. E., and Mischke, C. R., *Standard Handbook of Machine Design*, 2nd ed., McGraw-Hill Book Co., New York, 1996.

2. Hibbeler, R. C., *Engineering Mechanics: Dynamics*, 2nd ed., Macmillan, New York, 1978.

Chapter 17

1. *Industrial V-Belt Drives–Design Guide, Publication 102161*, Dayco Products Inc., Dayton, OH, 45401, 1998.

2. Shigley, J. E., and Mischke, C. R, *Standard Handbook of Machine Design*, 2nd ed., McGraw-Hill, New York, 1996.

3. *Whitney Chain Catalog WC97/CAT.R1*, Jeffrey Chain Corp., Morristown, TN, 37813, 1997.

4. *Wire Rope Users Manual*, 3rd ed., Wire Rope Technical Board, (888)289-9782.

5. *Ready-Flex Standard Flexible Shafts and Ratio Drives*, S. S. White Technologies, Inc., Piscataway, NJ, 08854, 1994.

6. *Flexible Shaft Engineering Handbook*, Stow Mfg. Co., Binghamton, NY, 13702, 1965.

7. Shigley, J. E., and Mischke, C. R., *Mechanical Engineering Design*, 5th ed., McGraw-Hill, New York, 1989.

8. Marco, S. M., Starkey, W. L., and Hornung, K. G., "Factors Which Influence the Fatigue Life of a V-belt," *Engineering for Industry, Transactions of ASME, Series, 3*, vol. 82, no. 1, Feb. 1960, pp. 47–59.

9. Worley, W. S., "Design of V-Belt Drives for Mass Produced Machines," *Product Engineering*, vol. 24, 1953, pp. 154–160.

10. Oliver, L. R., Johnson, C. O., and Breig, W. F., "V-Belt Life Prediction and Power Rating," Paper No. 75-WA/DE 26, ASME, 1975.

11. Gerbert, Goran, *Traction Belt Mechanics*, Machine and Vehicle Design, Chalmers University of Technology, 412 96 Goteborg, Sweden, 1999.

12. *Dayco Synchro-Cog® Drive Design, Publication 105180*, Dayco Products, Inc., Dayton, OH, 45401, 1998.

13. *Eagle Pd® Synchronous Belts and Sprockets, Engineering Manual*, Goodyear Tire and Rubber Company, Akron, OH, 1999.

14. *Engineering Class Chain Publication 2M-3/86*, Jeffrey Chain Corp., Morristown, TN, 37813, 1986.

15. ANSI B29.1M-1993, "Precision Power Transmission Roller Chains, Attachments, and Sprockets," ASME, New York, 1993.

16. ANSI B29.3M-1994, "Double-Pitch Power Transmission Roller Chains and Sprockets," ASME, New York, 1994.

17. Starkey, W. L., and Cress, H. A., "An Analysis of Critical Stresses and Mode of Failure of a Wire Rope," *Engineering for Industry, Trans. ASME*, v. 81, 1959, p. 307 ff.

18. Timoshenko, S., *Strength of Materials, Part I*, D. Van Nostrand, New York, 1955.

19. Drucker, D. L., and Tachau, H., "A New Design Criterion for Wire Rope," *Trans. ASME*, v. 67, 1945. p. A-33.

20. Spotts, M. F., *Design of Machine Elements*, 6th ed., Prentice-Hall, Englewood Cliffs, NJ, 1985.

Chapter 18

1. Hibbeler, R. C., *Engineering Mechanics: Dynamics*, 2nd ed., Macmillan, New York, 1978.

2. Lingaiah, K., *Machine Design Data Handbook*, McGraw-Hill, New York, 1994.

3. Shigley, J. H., and Mishke, C. R., *Standard Handbook of Machine Design*, 2nd ed., McGraw-Hill, New York, 1996.

4. Faupel, J. H., and Fisher, F. E., *Engineering Design*, 2nd ed., Wiley, New York, 1981.

Chapter 19

1. Waldron, K. J., and Kinzel, G. L., *Kinematics, Dynamics and Design of Machinery*, Wiley, New York, 1999.

2. Mabie, H. H., and Ocvirk, F. W., *Mechanisms and Dynamics of Machinery*, Wiley, New York, 1975.

Chapter 20

1. Dieter, G. E., Volume Chair, *Volume 20—Material Selection and Design, ASM Handbook*, ASM International, Material Park, OH, 1997.

2. Kutz, M., ed., *Mechanical Engineers' Handbook*, Wiley, New York, 1986.

3. ANSI Z535.4, "American National Standard for Product Safety Signs and Labels," American National Standards Institute, 1991.

4. ANSI Z535.1 "American National Standard Safety Color Code," American National Standards Institute, 1991.

5. ANSI Z535.2, "American National Standard for Environmental and Facility Safety Signs," American National Standards Institute, 1991.

6. ANSI 2535.3, "Criteria for Safety Symbols," American National Standards Institute, 1991.

7. ANSI Z535.5, "Specifications for Accident Prevention Tags," American National Standards Institute, 1991.

Photo Credits

All Chapter Openers

© CORBIS

Chapter 11

Page 411: Courtesy RBC Bearings.

Chapter 12

Page 441: Courtesy RBC Bearings.

Chapter 14

Page 515: Courtesy Associated Spring.

Chapter 15

Page 559: Courtesy Quality Transmission Components.

Chapter 16

Page 660: Photo by George Achorn. Courtesy Swedespeed.

Chapter 17

Page 698: Courtesy Rexnord Corporation.

Index